SLOSHING

This book presents sloshing with marine- and land-based applications, with a focus on ship tanks. It also includes the nonlinear multimodal method developed by the authors and an introduction to computational fluid dynamics. Emphasis is also placed on rational and simplified methods, including several experimental results. Topics of special interest include antirolling tanks, linear sloshing, viscous wave loads, damping, and slamming. The book contains numerous illustrations, examples, and exercises.

Odd M. Faltinsen received his Ph.D. in naval architecture and marine engineering from the University of Michigan in 1971 and has been a Professor of Marine Hydrodynamics at the Norwegian University of Science and Technology since 1976. Dr. Faltinsen has experience with a broad spectrum of hydrodynamically related problems for ships and sea structures, including hydroelastic problems. He has published approximately 300 scientific publications and is the author of the textbooks *Sea Loads on Ships and Offshore Structures* and *Hydrodynamics of High-Speed Marine Vehicles*, published by Cambridge University Press in 1990 and 2005, respectively. Faltinsen is a Foreign Associate of the National Academy of Engineering, USA, and a Foreign Member of the Chinese Academy of Engineering.

Alexander N. Timokha obtained his Ph.D. in fluid dynamics from Kiev University in 1988 and, later, a full doctorate in physics and a mathematics degree (habilitation) in 1993 at the Institute of Mathematics of the National Academy of Sciences of Ukraine. He is now Leading Researcher and Professor of Applied Mathematics at the Institute of Mathematics. Since 2004, he has been a Visiting Professor at CeSOS, Norwegian University of Science and Technology, Trondheim, Norway. In the 1980s, he was involved as a consultant of hydrodynamic aspects of spacecraft applications for the famous design offices of Yuzhnoye and Salut. Dr. Timokha's current research interests lie in mathematical aspects of hydromechanics with emphasis on free-surface problems in general and on sloshing in particular. He has authored more than 120 publications and 2 books.

Sloshing

ODD M. FALTINSEN
Norwegian University of Science and Technology

ALEXANDER N. TIMOKHA
Norwegian University of Science and Technology and
National Academy of Sciences of Ukraine

CAMBRIDGE
UNIVERSITY PRESS

CAMBRIDGE
UNIVERSITY PRESS

32 Avenue of the Americas, New York NY 10013-2473, USA

Cambridge University Press is part of the University of Cambridge.

It furthers the University's mission by disseminating knowledge in the pursuit of education, learning and research at the highest international levels of excellence.

www.cambridge.org
Information on this title: www.cambridge.org/9781107646735

First published 2009
First paperback edition 2014

A catalogue record for this publication is available from the British Library

Library of Congress Cataloguing in Publication data

Faltinsen, O. M. (Odd Magnus), 1944–
Sloshing / Odd M. Faltinsen, Alexander N. Timokha. – 1st ed.
p. cm.
Includes bibliographical references and index.
ISBN 978-0-521-88111-1 (hardback)
1. Sloshing (Hydrodynamics) I. Timokha, A. N. II. Title.
TA357.5.S57F35 2009
620.1′064–dc22 2009006711

ISBN 978-0-521-88111-1 Hardback
ISBN 978-1-107-64673-5 Paperback

Contents

Nomenclature

pair A and $\overline{A}$, or A_i	dominant wave amplitudes in the steady-state analysis of nonlinear three-dimensional sloshing, or wave amplitudes in ocean wave problems
A_{ij}^{Name}	added mass coefficients for three-dimensional statement; *Name* specifies subject [$i, j = 1, \ldots, 6$ and Name = frozen, filled, slosh, etc.]
a_{ij}^{Name}	the same as A_{ij}^{Name}, but for a two-dimensional statement
B	beam (breadth) of a ship or catamaran
pair B and $\overline{B}$	dominant wave amplitudes in the steady-state analysis of nonlinear three-dimensional sloshing
$B_t = L_2$	breadth of tank for three-dimensional sloshing
Bo	Bond number
B_{ij}	elements of the damping matrix [$i, j = 1, \ldots, 6$]
b_{ij}	the same as B_{ij}, but for two-dimensional statement [$i, j = 1, \ldots, 6$]
b_s	effective sloshing breadth
c_0	speed of sound
Ca	Cauchy number
C_E	modified Euler number
C_D	drag coefficient
C_M	mass coefficient
C_v	modified cavitation number
C_{ij}^{Name}	restoring coefficients [$i, j = 1, \ldots, 6$; Name = frozen, filled, slosh, etc.]
D or d	diameter, draft of a ship
$D_0 = 2R_0$	diameter of spherical tank
d^*/dt	$*$-time derivative of a vector function in the body-fixed (noninertial) coordinate system; the superscript asterisk indicates that one should not time-differentiate the unit vectors (see eq. (2.50))
$\boldsymbol{e}_i$ or $\boldsymbol{e}_x, \boldsymbol{e}_y, \boldsymbol{e}_z$	unit vectors of the body (tank)-fixed coordinate system [$i = 1, 2, 3$]
$\boldsymbol{e}_i'$	unit vectors of the Earth-fixed coordinate system [$i = 1, 2, 3$]

$\mathtt{E}$	Young's modulus
$E(t)$, $\langle E \rangle$	energy, time-averaged energy
E_g	work done by gravitational force; bulk modulus of gas
E_k	kinetic energy
E_p	potential energy
E_l	bulk modulus of liquid
E_v	bulk modulus of elasticity
E_{ext}	work done by external forces
E_{in}	internal strain energy of deforming the object
E_{mem}	membrane elasticity
Eu	Euler number
$\boldsymbol{F}^{\mathrm{Name}}(t)$	hydrodynamic force, where Name declares specific conditions on the considered fluid (e.g., filled, frozen) if needed $[= (F_1, F_2, F_3)]$
F_i^{Name}	for $i = 1, 2, 3$, components of $\boldsymbol{F}^{\mathrm{Name}}(t)$; for $i = 4, 5, 6$, components of the hydrodynamic moment $\boldsymbol{M}_O(t)$ in the $Oxyz$-coordinate system
Fn	Froude number
$f_M(x, y)$	wave patterns defined by the natural sloshing modes, $f_M = \varphi_M(x, y, 0)$ [M is integer or a set of integers; e.g., i, j]
$\boldsymbol{g} = \mathsf{g}$	gravitational acceleration vector $[= g_1\boldsymbol{e}_1 + g_2\boldsymbol{e}_2 + g_3\boldsymbol{e}_3]$
g	gravitational acceleration [$= 9.81$ m s^{-2}]
g_i	components of $\boldsymbol{g}$ in the $Oxyz$-coordinate system ($i = 1, 2, 3$)
$\boldsymbol{G}_O(t)$	angular fluid momentum relative to the origin O
h	liquid depth
$\overline{h}$	nondimensional liquid depth scaled by tank breadth or length
H	wave height
H_t	tank height
$H_{1/3}$	significant wave height
$\boldsymbol{I}^0$	inertia tensor for a frozen liquid $[= \{I_{ij}^0\}]$
$\mathtt{I}$	second moment of area with respect to the neutral axis for the beam problem
$\boldsymbol{J}^1(t)$	inertia tensor for sloshing $[= \{J_{ij}^1(t)\}]$
$\boldsymbol{J}_0^1$	linearized inertia tensor (time-independent) for sloshing $[\{J_{0ij}^1\}]$
$J_\alpha(\cdot)$	the Bessel function of the first kind [α is a real nonnegative number]
k or k_M	wave number; if M (integer or several integer indices, e.g., i, j, or a symbol) is present, the wave number for natural sloshing modes
KC	Keulegan–Carpenter number

l	characteristic linear dimension in two-dimensional statement; tank breadth for two-dimensional sloshing problem
l_b	length of a baffle
l_s	effective sloshing length
L	characteristic linear dimension in three-dimensional statement; the length of a ship; a typical dimension in some illustrative examples and exercises
L	Lagrangian
$L_t = L_1$	length of a tank in three-dimensional analysis
L_m	length in model scale
L_p	length in prototype scale
M	mass of an object in a three-dimensional statement
M_l	mass of a contained liquid in three-dimensional statement
$\boldsymbol{M}(t)$	fluid momentum
$\boldsymbol{M}_O^{\text{Name}}(t)$	hydrodynamic moment relative to the origin O in the $Oxyz$-coordinate system; Name declares specific conditions on the considered fluid (e.g., filled, frozen) if needed $[= (M_{O1}, M_{O2}, M_{O3}) = (F_4^{\text{Name}}, F_5^{\text{Name}}, F_6^{\text{Name}})]$
m	mass of an object in a two-dimensional statement, mass per unit length
m_k	spectral moments $[k = 0, 1, 2, \dots]$
m_l	mass of a contained liquid in two-dimensional statement
Ma	Mach number
$\boldsymbol{n} = (n_1, n_2, n_3)$	outer normal vector of a fluid volume
$\boldsymbol{n}^+$	normal vector with positive direction into a fluid volume $[= -\boldsymbol{n}]$
O	origin of the body-fixed coordinate system $Oxyz$
$O(\varepsilon)$	expresses the same order as a small parameter $\varepsilon \ll 1$
O'	the origin of the Earth-fixed (inertial) coordinate system $O'x'y'z'$
$Oxyz$	the body[tank]-fixed coordinate system
$O'x'y'z'$	the Earth-fixed [inertial] coordinate system
$o(\varepsilon)$	expresses higher order than a small parameter $\varepsilon \ll 1$
P	pressure impulse
$p(x, y, z, t)$	pressure
p_0	ullage pressure [= const]
p_a	atmospheric pressure
p_v	liquid vapor pressure
p_D	dynamic pressure
$Q(t)$	the liquid domain (in most cases, the tank liquid)
Q_0	the tank liquid domain in hydrostatic state

r	component of the cylindrical polar coordinate system (r, θ, z)
$\boldsymbol{r} = (x, y, z)$	radius vector of a point in the body-fixed coordinate system
$\boldsymbol{r}'$	radius vector of a point in the Earth-fixed coordinate system $[= \boldsymbol{r}'_O + \boldsymbol{r}]$
$\boldsymbol{r}_{lC}(t)$	radius vector of the mobile mass center of a contained liquid in the $Oxyz$-coordinate system $[= (x_{lC}(t), y_{lC}(t), z_{lC}(t))]$
$\boldsymbol{r}_{lC_0}$	radius vector of a contained liquid in the hydrostatic state in the $Oxyz$-coordinate system $[= (x_{lC_0}, y_{lC_0}, z_{lC_0})]$
$R_0[= \frac{1}{2}D_0]$	radius of a circular cylindrical tank or a circular spherical tank
r_0	radius of internal structures (e.g., poles) inserted into the liquid
$r_{jj}, j = 4, 5, 6$	radii of gyration
Ra	arithmetical mean roughness on the body surface
Rn and RE	Reynolds number, different definitions
Rn_{tr}	transition Reynolds number
$S(t)$	wetted tank surface
S_0	tank surface below the mean free surface
St	Strouhal number
S_Q	boundary enclosing the liquid volume Q [e.g., $\Sigma(t) + S(t)$]
t	time (s)
$\boldsymbol{t}$	tangential vector
T	period
T_0, T_1, and T_2	modal period and mean wave periods
T_M	for sloshing, natural sloshing periods [M is integer or a set of integers, e.g., i, j]
T_s	surface tension
T_d	duration of an external loading
T_{sc}	scantling draft
T_{mem}	membrane tension
T_{st}	tension of a string
u	the Ox-component of $\boldsymbol{v}$
u_1, u_2, u_3	see $\boldsymbol{v}$
u_r	see $\boldsymbol{v}_r$
U	characteristic velocity
U_g	gravity potential $[= -\boldsymbol{g} \cdot \boldsymbol{r} = -gz']$
$U_{sn} = U_n$	normal velocity component of a fluid surface; see $\boldsymbol{n}$
u_n	normal component of the fluid velocity on a fluid surface; see $\boldsymbol{n}$
$\boldsymbol{v}$	absolute fluid velocity $[= u\boldsymbol{e}_1 + v\boldsymbol{e}_2 + w\boldsymbol{e}_3 = (u, v, w) = (u_1, u_2, u_3)]$
$\boldsymbol{v}_r$	relative (with respect to the $Oxyz$-system) fluid velocity $[= u_r\boldsymbol{e}_1 + v_r\boldsymbol{e}_2 + w_r\boldsymbol{e}_3]$

v	the Oy-component of $\boldsymbol{v}$
v_r	see $\boldsymbol{v}_r$
v_O	velocity of the origin O [$= v_{O1}\boldsymbol{e}_1 + v_{O2}\boldsymbol{e}_2 + v_{O3}\boldsymbol{e}_3 = (v_{O1}, v_{O2}, v_{O3}) = (\dot{\eta}_1, \dot{\eta}_2, \dot{\eta}_3)$]
V	entry (vertical) velocity in slamming problems
Vol	fluid volume (area for two-dimensional case)
w	the Oz-component of $\boldsymbol{v}$
$\mathrm{w}(x, t)$	beam deflection
Wn	Weber number
w_r	see $\boldsymbol{v}_r$
W	the action; see eq. (2.80) [$= \int_{t_1}^{t_2} \mathrm{L}dt$]
(x_1, x_2, x_3)	(x, y, z)
$Y_\alpha(\cdot)$	Bessel function of the second kind [α is a real nonnegative number]

Greek symbols

α or α_i	used for definitions of different angles including the phase angle; auxiliary parameters
β	generalized coordinate in Lagrange variational formulation, deadrise angle
β_M	generalized coordinates in Lagrange variational formulation for multidimensional mechanical system, amplitudes of the natural sloshing modes in the modal representation of the free surface [M is integer or a set of integers, e.g., i, j]
χ	void fraction
δ	denotes variation of a functional value or generalized coordinate, e.g., $\delta\beta$, in variational formulations; boundary-layer thickness; a small distance when analyzing proximity effect of structures in Section 4.7.2.2
δ_{ij}	Kronecker delta
ε	formal small parameter in asymptotic analysis; the dimensionless forcing amplitude in multimodal method
$\Phi(x, y, z, t)$	velocity potential of the absolute velocity field $\boldsymbol{v}$ defined in the body-fixed coordinate system $Oxyz$
$\varphi_M(x, y, z)$	natural sloshing modes [M is integer or a set of integers, e.g., i, j]

γ	vortex density
$\eta_i(t)$	translatory ($i = 1, 2, 3$) and angular ($i = 4, 5, 6$) components of motions of the tank [body]-fixed coordinate system $Oxyz$ relative to an inertial coordinate system; also used for global ship motions [$i = 1, \ldots, 6$]
$\iota_{m,i}$	roots of the equation $J'_m(\iota_{m,i}) = 0$
$\kappa_M = \sigma_M^2/g$	spectral parameter of the problem on natural sloshing modes [M is integer or a set of integers, e.g., i, j]
κ	ratio of the specific heat
λ	wavelength
μ	dynamic viscosity coefficient
ν	kinematic viscosity coefficient
θ	component of the cylindrical polar coordinate system (r, θ, z)
Θ	angle measuring the wave propagating direction of elementary wave components in the sea relative to a main wave propagation direction
ρ	fluid density
ρ_l	liquid density
ρ_i	inner and exterior liquid density
ρ_o	ρ_o ullage gas density
ρ_g	gas density
ρ_c	gas density in the cushion
σ	circular forcing frequency or a frequency of an external wave
σ_M	wave frequencies; for sloshing, natural sloshing frequencies [M is integer or a set of integers, e.g., i, j]
σ_e	frequency of encounter
$\Sigma(t)$	free surface of a liquid during sloshing
Σ_0	mean free surface = hydrostatic liquid surface = unperturbed free surface
τ_l	laminar shear stress
τ_τ	turbulent shear stress
$\boldsymbol{\tau} = \{\tau_{ij}\}$	viscous stress components along the $(x_i - x_j)$-components ($i, j = 1, 2, 3$)
$\boldsymbol{\omega}(t)$	instant angular velocity of the tank (the $Oxyz$-coordinate system) with respect to an inertial coordinate system [$= (\omega_1(t), \omega_2(t), \omega_3(t))$]

$\omega_i(t)$	projections of the angular velocity $\boldsymbol{\omega}(t)$-vector in the $Oxyz$-coordinate system; equal to $\dot{\eta}_{i+3}(t)$, $i = 1, 2, 3$, for linear dynamics of the tank
$\boldsymbol{\Omega}(x, y, z, t)$	Stokes–Joukowski potential $[= (\Omega_1(x, y, z, t), \Omega_2(x, y, z, t), \Omega_3(x, y, z, t))]$
$\boldsymbol{\Omega}_0(x, y, z)$	Stokes–Joukowski potential for linear sloshing theory $[= (\Omega_{01}(x, y, z), \Omega_{02}(x, y, z), \Omega_{03}(x, y, z))]$
$\Omega(t)$	gas cushion volume
$\boldsymbol{\varpi}$	vorticity vector
ξ or ξ_M	(M is set of integers) damping ratio(s)
ζ	coefficient of bulk viscosity
ζ_a	amplitude of linear sea waves
$z = \zeta(x, y, t)$	normal representation of the free surface
$Z(x, y, z, t) = 0$	implicitly defined free surface

Preface and Acknowledgment

Our initial motivation for writing this book was to provide background on the analytically based *nonlinear* multimodal method for sloshing developed by the authors. We soon realized that we had to give a broader scope on sloshing and also present material on computational fluid dynamics (CFD), viscous flow, the effect of internal structures, and slamming. Furthermore, experimental results are to a large degree presented to validate the theoretical results and give physical insight.

A broad variety of CFD methods exist, and other textbooks provide details on different numerical methods. Our focus has been on giving an introduction to the many CFD methods that exist. An important aspect has also been to link the material to practical aspects. Our main application is for ship tanks, where sloshing can be very violent and slamming and coupling between sloshing and ship motions are important aspects. However, we have also emphasized links to other engineering fields with applications such as tuned liquid dampers for tall buildings, rollover of tanker vehicles, oil–gas separators used on floating ocean platforms, onshore tanks, and seiching in harbors and lakes; space applications are not addressed. Whenever possible we have tried to provide examples and have emphasized exercises where we provide hints and solutions. This fact has led to the development of simple analytical methods for analysis of, for instance, transient sloshing in spherical and horizontal circular cylindrical tanks, two-phase liquid flow, the effect of tank deformations, wave-induced hydroelastic analysis of a monotower with sloshing of water inside the shaft, flow through screens and swash bulkheads, and hydrodynamic analysis for automatic control of U-tanks.

Sloshing is a fascinating topic, and the first author was deeply involved in theoretical aspects of sloshing in liquefied natural gas tanks from the beginning of the 1970s, when he worked at Det Norske Veritas. Following that period was an approximately 20-year break in his activities with sloshing until he started again at the end of the past century. The second author has worked on spacecraft applications with particular emphasis on sloshing in fuel tanks, and since the beginning of the 1990s he has been involved with mathematical aspects of sloshing at the Institute of Mathematics, National Academy of Sciences of Ukraine, Kiev. It was their common interest in nonlinear multimodal methods for sloshing that brought them together at the Center for Ships and Ocean Structures (CeSOS), Norwegian University of Science and Technology (NTNU), Trondheim.

Mathematics is a necessity in reading the book, but we have tried to also emphasize physical explanations. Knowledge of calculus, including vector analysis and differential equations, is necessary to read the book in detail. The reader should also be familiar with dynamics and basic hydrodynamics of potential and viscous flow of an incompressible fluid. This book is more advanced from a theoretical point of view than the previous books *Sea Loads on Ships and Offshore Structures* and *Hydrodynamics of High-Speed Marine Vehicles* by the first author. Part of the book has been taught to graduate students at the Department of Marine Technology, NTNU. The book should be of interest for both engineers and applied mathematicians working with advanced aspects of sloshing. A pure mathematical language is avoided to better facilitate communication with readers with engineering backgrounds.

Quality control is an important aspect of writing a book, and we received help from both experts in different fields and graduate students. Dr. Svein Skjørdal of the Grenland Group, Sandefjord, and Dr. Martin Greenhow of Brunel University have been critical reviewers of all three books written by the first author. Dr. Skjørdal was helpful in seeing the topics from a practical point of view. The contributions by Dr. Olav Rognebakke, DNV, to several topics in the book are greatly appreciated.

Yanlin Shao read fastidiously through the text and asked many important questions that enabled us to clarify the text. In addition he has controlled calculations and provided solutions to all exercises. The detailed control of Dr. Hui Sun and Xiangjun Kong is also appreciated.

Professor Dag Myrhaug of NTNU and Professor J. M. R. Graham of Imperial College, London, critically reviewed Chapter 6 on viscous wave loads and damping.

Professor Marilena Greco of CeSOS and INSEAN and Professor G. X. Wu of University College provided important contributions to Chapter 10 on CFD and Chapter 11 on slamming. The expert help from Dr. Ould El Moctar of Germanischer Lloyds in reviewing Chapter 10 and by Professor Alexander Korobkin of University of East Anglia in reviewing Chapter 11 is also greatly appreciated.

Many other people should be thanked for their critical reviews and contributions, including Bjørn Abrahamsen, CeSOS; Professor Jørgen Amdahl, NTNU; Dr. Petter Andreas Berthelsen, CeSOS; Dr. Henrik Bredmose, University of Bristol; Dr. Claus M. Brinchmann, Rolls-Royce Intering Products; Dr. Chunhua Ge, Lloyd's Register of Shipping; Mateusz Graczyk, CeSOS; Xiaoyu Guo, CeSOS; Dr. Kjell Herfjord, StatoilHydro; Professor Changhong Hu, Kyushu University; David Kristiansen, CeSOS; Professor Carl Martin Larsen, CeSOS; Dr. Claudio Lugni, INSEAN; Professor J. N. Newman; Jan Arne Opedal, NTNU and DNV; Dr. Csaba Pakozdi, CeSOS; Professor Bjørnar Pettersen, NTNU; Dr. Hang Sub Urm, DNV; Tone Vestbøstad, CeSOS; and Dr. Zhu Wei, CeSOS.

The tedious work of obtaining permissions to use published material was done by Karelle Gilbert, CeSOS.

Acronyms and Abbreviations

AFRA	average freight rate assessment
AP	after perpendicular
BEM	boundary element method
CFD	computational fluid dynamics
CL	centerline
COG	center of gravity
DLWL	designer's load waterline
DWT	deadweight
FDM	finite difference method
FEM	finite element method
FLS	fatigue limit state
FP	forward perpendicular
FPSO	floating production storage and offloading
FVM	finite volume method
IMO	International Maritime Organization
ISSC	International Ship and Offshore Structures Congress
ITTC	International Towing Tank Conference
JONSWAP	Joint North Sea Wave Project
LNG	liquefied natural gas
LPG	liquefied petroleum gas
O/O	ore/oil
OBO	oil/bulk/ore
RANS	Reynolds-averaged Navier–Stokes
RAO	response amplitude operator
RV	regasification vessel
SOLAS	Safety of Life at Sea
SPH	smoothed particle hydrodynamics
TLCD	tuned liquid column damper
TLD	tuned liquid damper
TLP	tension leg platform
TSD	tuned sloshing damper
ULCC	ultralarge crude carrier
ULS	ultimate limit state
VIV	vortex-induced vibration
VLCC	very large crude carrier

1 Sloshing in Marine- and Land-Based Applications

1.1 Introduction

Sloshing must be considered for almost any moving vehicle or structure containing a liquid with a free surface and can be the result of resonant excitation of the tank liquid. Excitation with frequencies in the vicinity of the lowest natural frequencies of the liquid motion is of primary practical interest. Sloshing can also be the result of transient motion, for instance, when we spill coffee from a coffee cup.

Resonant free-surface flows in tanks in aircraft, missiles, and rockets have been the focus of extensive research. For these vehicles, sloshing has a strong influence on their dynamic stability. The fact that sloshing may strongly interact with the dynamics of the carrying "body" is evident when we carry a bucket of water. If sloshing starts in the bucket, it is difficult to stop it.

The important free-surface correction of the metacentric height used in the assessment of the floating stability of ships can be regarded as a special quasi-static tank sloshing phenomenon. It corresponds to tank motion frequencies that are much lower than natural sloshing frequencies. Using the word *sloshing* in this connection may contradict common use of the word. The liquid motion relative to the tank is governed by the fact that the free surface remains horizontal.

On floating oil and gas production platforms, sloshing affects the efficiency of oil–gas separators. Structures fixed onshore may be exposed to sloshing if an earthquake occurs (Fischer & Rammerstorfer, 1999). Sloshing effects are included in design loads for liquefied natural gas (LNG) storage tanks. Large-scale sloshing in a lake with steep sides may be the result of a landslide or earthquake. During dam construction such circumstances should be investigated. Large-scale sloshing may also occur in harbors and lakes and even on an oceanic scale. Sloshing in a container can be used to dampen wind-induced motions of tall buildings. A tuned liquid damper (TLD) is a well-known concept in the civil engineering world. Its physical basis is similar to an antirolling tank onboard a ship.

A partially filled ship tank can experience violent liquid motion when the ship motions contain energy in the vicinity of the highest natural period for the liquid motion inside the tank. Impact between the liquid and the tank structure is then likely to occur. One scenario is tank roof impact (see Figure 1.1). The consequence is wave breaking, spray, and mixing of air (or gas) and liquid. Extreme cases, with air bubbles everywhere in the liquid, have been experimentally observed.

Ingress of water to a damaged ship and the resulting dynamics of the water on large deck areas can result in sloshing and affect the capsizing process.

The hydrodynamics of sloshing is complicated. Its understanding requires a combination of theory, computational fluid dynamics (CFD), and experiments. Tank design is often based on experiments. However, scaling impact slosh pressures from model to full scale is not completely understood. Furthermore, the combination of hydrodynamics and structural mechanics is an important aspect of tank design.

Our focus is primarily on ship applications. However, we broaden the discussion by considering offshore platforms, TLDs, harbor resonance, land transportations, onshore tanks, and space applications.

1.2 Resonant free-surface motions

Potential flow theory of an incompressible liquid can in many practical cases adequately describe sloshing. An infinite number of eigenfrequencies and eigenmodes exist: nontrivial solutions satisfying the field equation (Laplace equation), liquid mass conservation, and boundary conditions with no tank excitation. These eigensolutions satisfy the linear free-surface conditions, but they also play an important role in modeling nonlinear forced sloshing and are the basis for the analytically based multimodal method described in Chapters 7, 8, and 9.

The highest natural periods are of prime importance in assessing the severity of sloshing. The natural frequencies and corresponding modes for

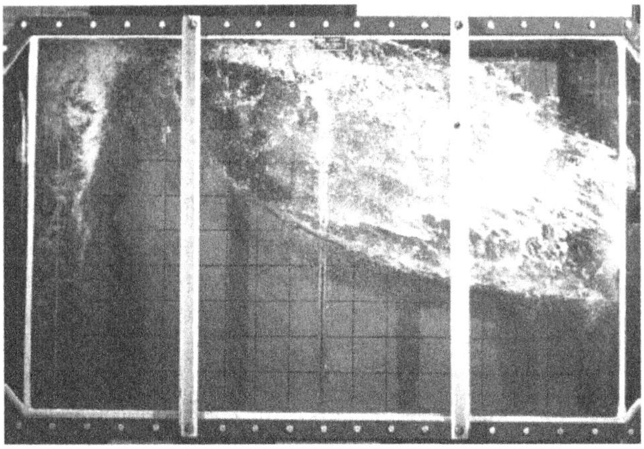

Figure 1.1. Tank roof impact during horizontal excitation of a rectangular tank with excitation frequency close to the lowest natural sloshing frequency.

liquid motion depend on the tank shape and the liquid-depth-to-tank-breadth (length) ratio, which are exemplified in Figure 1.2 for two-dimensional flow in a rectangular tank with liquid depth h and tank breadth (length) l. The highest natural period $T_1 = 2\pi/\sqrt{g\pi \tanh(\pi h/l)/l}$ (see eq. (4.11)), based on linear potential flow theory, is presented for different values of liquid depth as a function of the tank breadth in Figure 1.2. The shallow and deep liquid asymptotic forms of T_1 are $T_1 = 2l/\sqrt{gh}$, $h/l \to 0$ and $T_1 = 2l\sqrt{gh}$, $h/l \to \infty$. The natural period increases with the length for a given liquid depth. The results in Figure 1.2 are relevant for ship tanks. The highest natural period for sloshing (seiching) in lakes is much larger, that is, 14 hours in Lake Erie. The liquid depth can have a significant influence on the natural period for a given tank breadth (length). However, the influence is small when $h/l \gtrsim 1.0$, which are defined here as deep liquid conditions when the liquid motions do not "feel" the tank bottom. Shallow liquid conditions are defined as $h/l \lesssim 0.1$. Intermediate and finite liquid depths correspond to $0.1 \lesssim h/l \lesssim 0.2$–$0.25$ and 0.2–$0.25 \lesssim h/l \lesssim 1.0$, respectively. We have used approximate values for the intervals of h/l to indicate that there are no strict boundaries among shallow, intermediate, finite, and infinite liquid-depth conditions. Strong changes and amplifications in the liquid behavior occur for excitation frequencies close to the lowest natural frequency for liquid depths in the vicinity of the critical depth, $h/l = 0.3368\ldots$ (see Fulzt, 1962; and Sections 8.3.1.1 and 8.5).

A standing wave with a wavelength twice the tank breadth and a node in the middle of the tank is dominant according to the linear theory for a two-dimensional rectangular tank flow in

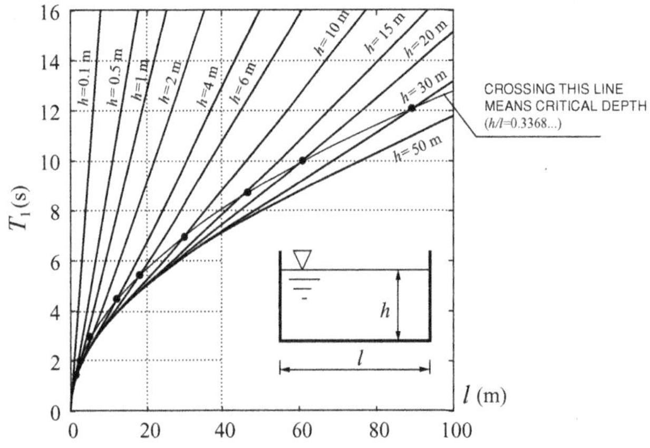

Figure 1.2. First-mode natural sloshing period for a two-dimensional rectangular tank versus the tank breadth. The critical depth-to-breadth ratio is also indicated.

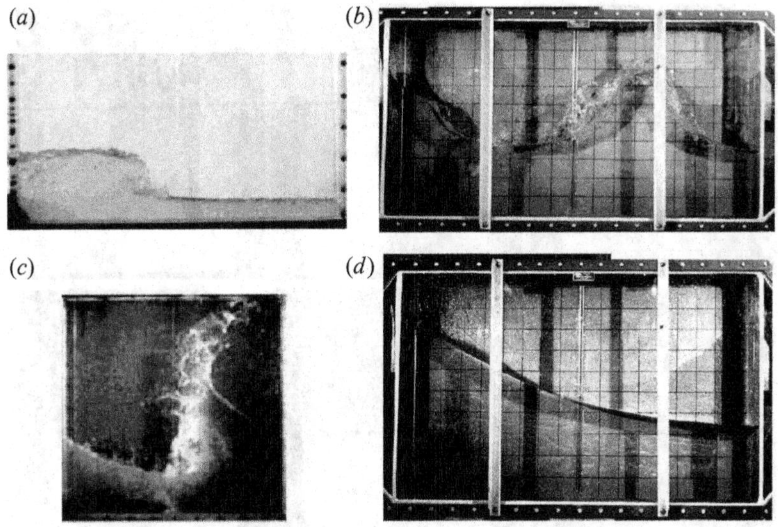

Figure 1.3. Examples of typical periodic free-surface motions for (a) shallow, (b) intermediate, (c) critical, and (d) finite liquid-depth conditions for forced horizontal oscillations with period T in the vicinity of the highest natural sloshing period T_1 of two-dimensional flow in a rectangular tank. Shallow liquid conditions (panel a) are $h/l = 0.125$, $T/T_1 = 1$. The forcing-amplitude-to-tank-breadth ratio is 0.1. Nearly critical depth conditions (panel c) are for $h/l = 0.35$, $T/T_1 = 0.787$.

resonant conditions at the highest natural period. When viscous damping effects are neglected, a linear theory based on the potential flow of an incompressible liquid predicts infinite steady-state response for a forcing frequency equal to a natural frequency of the liquid motion. The reason is zero damping, because the only possible damping source is due to radiated waves which, of course, are impossible for liquid in a container. However, the response is limited at resonant conditions due to the nonlinear transfer of energy between the different modes of liquid motion. The flow at resonant conditions in a tank with no interior structures (i.e., negligible viscous damping) is clearly different for shallow, intermediate, and finite liquid-depth conditions. The infinite-depth case resembles the finite-depth case. These facts are illustrated by photos from experiments in Figure 1.3. Hydraulic jumps traveling back and forth in the tank may occur in shallow liquid. When $h/l \gtrsim 0.4$, the wave motion resembles a standing wave with the largest free-surface elevations at the tank walls. Even though the free-surface elevation is smallest around the middle of the tank, an exact node (i.e., a position with zero free-surface elevation for all time instants as predicted by linear theory) does not exist. When $h/l \gtrsim 0.4$, the damping is small and mainly associated with viscous boundary-layer effects along the tank boundary for clean tanks of practical interest. *Clean tank* means that the tank has no internal structures such as baffles, transverse webs, and horizontal stringers obstructing the flow. The consequence of small damping is that a very long time (e.g., 50 oscillation periods) is needed to reach steady-state (periodic) conditions. Breaking waves occur more easily in intermediate and shallow depths and at the critical depth than when $h/l \gtrsim 0.4$. The consequence of breaking waves is greatly increased damping, so that steady-state conditions are reached sooner. Large breaking waves may occur in the middle of the tank as illustrated in Figure 1.3 for the shallow-, intermediate-, and critical-depth cases, which is a consequence of the fact that several natural modes interact. Many natural modes have significant roles, particularly in the shallow-depth case.

The critical-depth case illustrated in Figure 1.3(c) needs special attention because a steady-state condition with the flow oscillating at the forcing period was not achieved. Subharmonic behavior arose for this particular forcing amplitude and period (Colagrossi *et al.*, 2006).

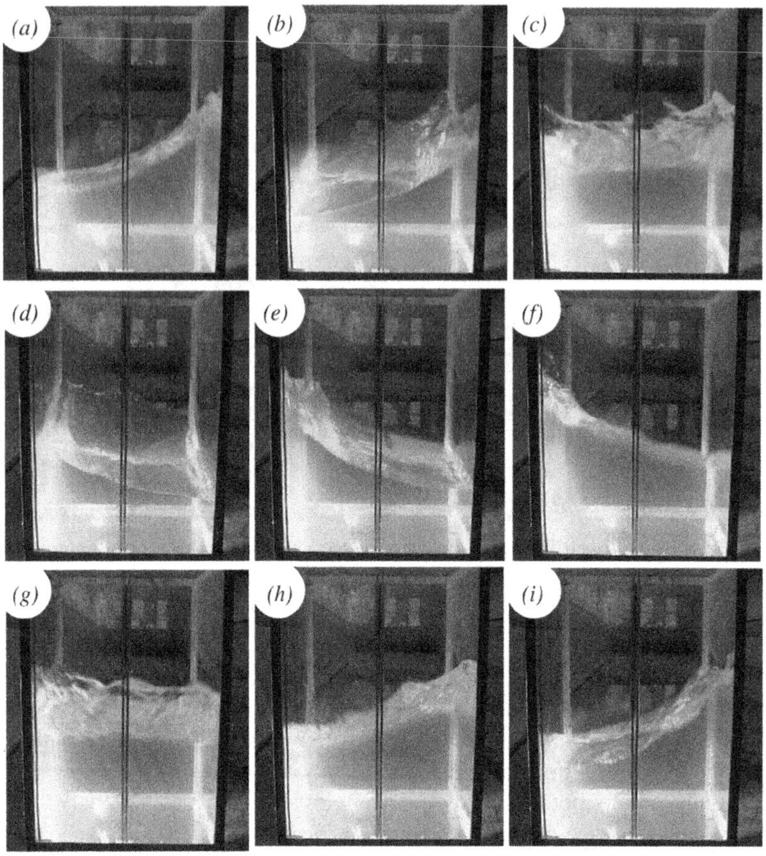

Figure 1.4. Photo time series from (a) to (i) demonstrating counterclockwise "swirling" wave motions in a square-base tank for a water-depth-to-tank-breadth ratio of 0.508. The tank is excited longitudinally along a wall (back-and-forth motions relative to the camera) and the forcing-amplitude-to-tank-breadth ratio is 0.0078. The excitation frequency is equal to the lowest natural frequency.

Swirling (rotary) wave motions may occur during harmonic horizontal excitation of liquid motion in vertically axisymmetric or square-base tanks when the forcing frequency is in the vicinity of the lowest natural frequency. Examples of vertically axisymmetric tanks are spherical tanks and vertical circular cylinders. A consequence of swirling is a lateral hydrodynamic force component that is perpendicular to the forced-oscillation direction. We can easily observe swirling by doing experiments with a cup of coffee or a glass of water. Figure 1.4 demonstrates steady-state swirling for a square-base tank. The wave elevations are largest at the tank corners. The wave motion is counterclockwise in Figure 1.4. However, the rotation direction depends on transient conditions, and the steady-state swirling motion could be clockwise or counterclockwise. Swirling may change its rotation direction during transient conditions. Swirling is a nonlinear phenomenon. If linear theory were used in the case presented in Figure 1.4 with forcing parallel to a wall, there could only be two-dimensional (planar) waves. Nonlinear effects may also cause diagonal waves (i.e., waves where the maximum wave elevations occur at diagonally opposite corners of the tank with small wave elevations at the other two corners). Again, the transient conditions determine the diagonal wave direction. There are conditions depending on the liquid depth and the forcing amplitude and frequency when no steady-state wave motions are possible in a square-base tank. We denote this as irregular motions, or "chaos," in Chapter 9. We use

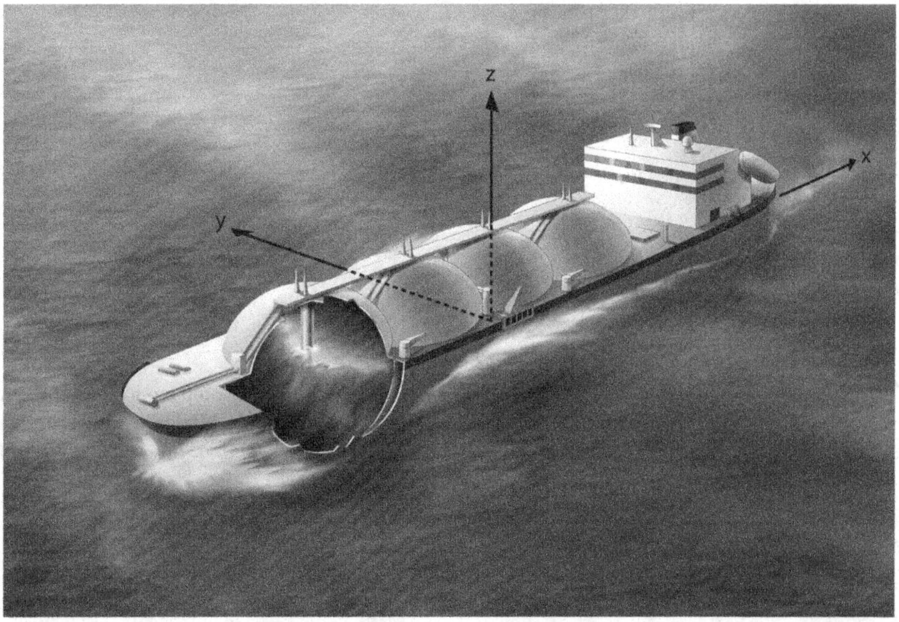

Figure 1.5. Definition of rigid-body motion modes relative to an inertial coordinate system (x, y, z) that moves with steady ship speed and has its origin in the mean free surface. When the ship does not oscillate, the z-axis is often chosen to go through the center of gravity of the ship. The translatory dynamic ship motions along the x-, y-, and z-axis are denoted η_1, η_2, and η_3, respectively, so that η_1 is the surge, η_2 is the sway, and η_3 is the heave displacement. Furthermore, let the angular displacements of the rotational motions about the x-, y-, and z-axis be η_4, η_5, and η_6, respectively, so that η_4 is the roll, η_5 is the pitch, and η_6 is the yaw angle (Artist: Bjarne Stenberg).

quotation marks to indicate that it is not chaos according to a mathematical definition (Smith, 1998).

1.3 Ship tanks

A variety of ship tank shapes exist, comprising rectangular, prismatic, tapered, and spherical as well as horizontal cylindrical tanks. The liquid may be oil, liquefied gas, water, or high-density cargoes like molasses or caustic soda. If potential flow theory of an incompressible liquid can be used to describe the liquid dynamics and the effect of gas pockets can be neglected, the hydrodynamic loads on a rigid tank are proportional to the liquid density. Even though a tank is completely filled with liquid, the angular ship motions cause liquid motion relative to the tank motion (see Chapter 5).

The free surface of the liquid in a tank has a statically destabilizing effect on the roll (heel) moment about the ship's center of gravity for a given heel angle η_4, which must be considered. This consideration may lead to restrictions on the tank breadth or increased width of the ship. The magnitude of the static destabilizing heel moment due to a rectangularly shaped free surface of length L_t and breadth B_t can be expressed as $\frac{1}{12}\rho_l g B_t^3 L_t \eta_4$, where ρ_l is the density of the tank liquid (see Section 3.6.1).

Sloshing in a partially filled ship tank may occur, for instance, as a consequence of a collision between two ships, grounding, or the collision of a ship with ice; however, wave-induced ship motions are the primary excitation mechanism. The linear (small-amplitude) rigid-body motion modes – surge, sway, heave, roll, pitch, and yaw – of a ship are defined in Figure 1.5 relative to an inertial coordinate system translating with the steady speed of a ship on a straight course.

When assessing the severity of ship tank sloshing in sea waves, the first step is to decide on realistic operational areas and use available weather statistics expressed as joint probability

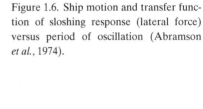

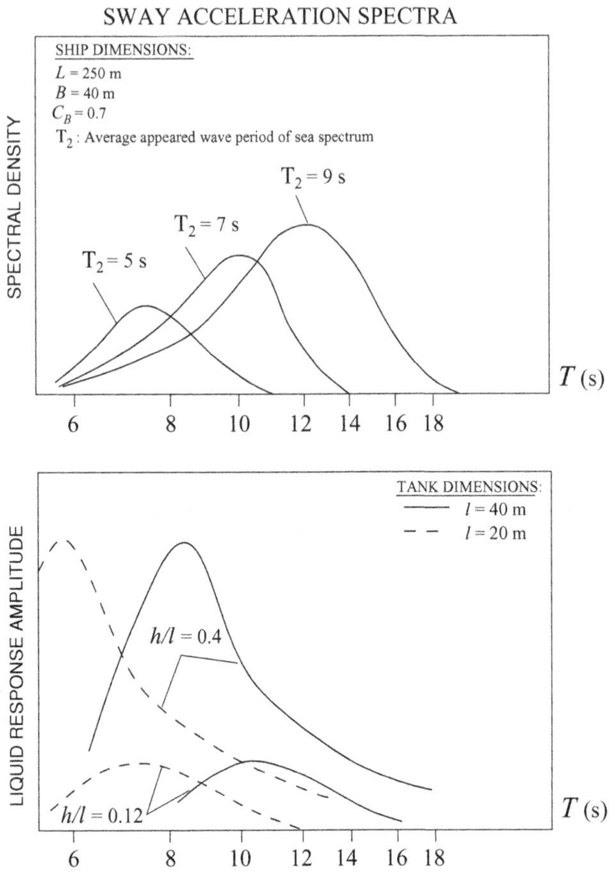

Figure 1.6. Ship motion and transfer function of sloshing response (lateral force) versus period of oscillation (Abramson *et al.*, 1974).

distributions of significant wave height $H_{1/3}$ and mean wave period T_2. Significant wave height may be defined as the mean height of one-third of the highest waves or by using spectral moments. Different definitions of mean wave period exist, as we will see in Section 3.3. The ship motions and resulting sloshing can be determined for a given $H_{1/3}$ and T_2, mean wave direction and ship speed, when a wave spectrum is specified. Standard wave spectra such as the Joint North Sea Wave Project (JONSWAP) and Pierson–Moskowitz spectra are typically used (see Section 3.3). An important parameter for ship motions is the ship length, whereas the resulting sloshing depends on the tank shape, tank breadth, tank length, and liquid depth. Ship rolling is a typical resonance phenomenon. The roll resonance period may be significantly influenced by the loading condition primarily in terms of the transverse metacentric height $\overline{GM_T}$; $\overline{GM_T}\eta_4$ is the uprighting moment arm of the static heel (list) moment about the ship center of gravity for small heel angles η_4 (in radians).

Because sloshing is a typical resonance phenomenon, it is not necessarily the most extreme ship motions or external wave loads that cause the most severe sloshing. Therefore, external wave-induced loads can in many practical cases be described by linear theory. However, nonlinearities must be accounted for when describing the tank liquid motions. Furthermore, an interaction exists between sloshing and the ship motions in a seaway. Because it is the highest sloshing period (natural period) that is of prime interest, vertical tank excitation is of secondary importance. Furthermore, vertical tank motions cannot excite sloshing according to linear theory; lateral and angular tank motions cause the largest liquid response in the frequency range of interest.

Figure 1.6 presents a family of sway acceleration spectra corresponding to different sea-state conditions. No values are given for the

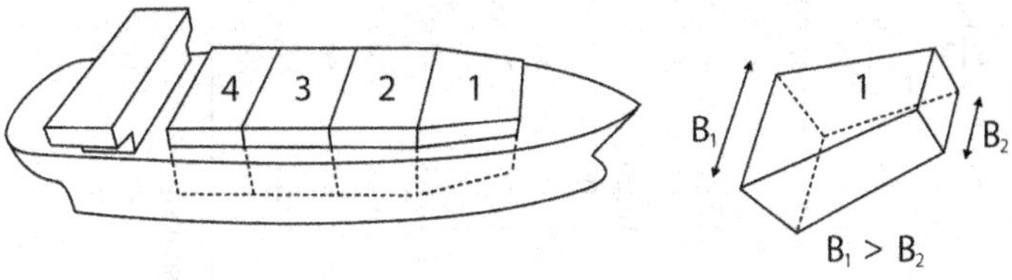

Figure 1.7. Tapered tank in the bow of a ship.

spectral density, which depends on the significant wave height $H_{1/3}$. The spectral values are proportional to $(H_{1/3})^2$ according to linear theory. A representative shape of the transfer function for liquid response is also shown without specifying any values. By transfer function we mean the ratio between the steady-state response amplitude and the tank forcing amplitude as a function of the wave period in regular harmonic waves. If the system is linear, the transfer function does not depend on the forcing amplitude. However, the important point here is when the liquid response becomes large. Figure 1.6 shows that different-sized tanks with different filling levels yield peak response amplitudes within the range of sway periods to be expected. If the liquid-depth-to-tank-breadth (length) ratio h/l is fixed, an increased tank length increases the highest natural period of the liquid flow. We see this from the fact that the highest natural period is $T_1 = 2\pi/\sqrt{g\pi \tanh(\pi h/l)/l}$. As a consequence, higher sea states and larger ship motions excite sloshing around resonance. The fewer internal structures that are present to obstruct the flow in the tank, the more severe is the sloshing.

The tank length, L_t, is an important parameter for sloshing excited by surge and pitch. To assess this properly, we have to go through the previously outlined procedure. However, we can get an indication in the case of rectangular tanks by using Figure 1.2 and replacing tank length l in the figure with L_t. The figure shows that the liquid depth can have a significant influence on the highest natural sloshing period T_1. T_1 must be related to wave periods with significant ship motions for representative operational areas. If the vessel is an oceangoing vessel, roughly speaking we may say that sloshing matters if T_1 is between 5 and 20 s. An important parameter for sloshing to be excited by sway, roll, and yaw motions of the ship is the tank breadth, B_t. We can then make a qualitative assessment similar to that for surge and pitch excitation by replacing B_t with the tank length l in Figure 1.2 in the case of rectangular tanks.

Internal structures have a damping effect on sloshing. Oil tankers, shuttle tankers, liquefied petroleum gas (LPG) carriers, IHI self-supporting prismatic-shape International Maritime Organization type B (SPB) LNG carriers, and floating production storage and offloading (FPSO) vessels have internal structures such as transverse webs and horizontal stringers. The different types of vessels and internal structures are described in detail later in this section, but it requires the more detailed analysis of Chapter 6 to quantify the damping effect. For tapered tanks, it is usual to have clean tanks. A drawing of a tapered tank in the bow of a ship is shown in Figure 1.7. The figure also illustrates the numbering of ship tanks: we start counting from the bow.

Table 1.1 shows typical cargo tank dimensions and tank shapes. Excluded are bulk carriers as well as combination oil/bulk/ore (OBO) carriers. The beam, draft, and length between perpendiculars, L_{PP} of the ships are given. The forward perpendicular is a vertical line through the intersection of the designer's load waterline (DLWL) and the foreside of the stem. The after perpendicular (AP) is a vertical line that passes through the rudder post or the transom profile, implying an ambiguity in the definition of AP. When the vessel has a transom stern and no rudder, AP is at the transom according to the preceding description, where L_{PP} is the same as the length of DLWL. The scantling draft indicated in the table refers to the maximum draft for which the vessel is designed, from a structural strength point of view. The ratio of the cargo tank length to the ship length is high for membrane LNG carriers, LPG carriers, very large crude carriers (VLCCs), and FPSOs compared to other ship types. The fact that membrane LNG carriers also have wide and clean cargo tanks means that sloshing may be a problem due to rolling and transverse ship motions. The same is true for bulk

Table 1.1. *Cargo tank dimensions and capacity for different ship types*

No.	Type	Cargo capacity	Tank Shape	L_{PP}	B	D	T_s	L_t	B_t	H_t	$\frac{L_t}{L_{PP}}\%$	$\frac{B_t}{B}\%$
1–1	Chemical tanker	37,000 t	Rectangular tank	168.5	31.0	15.6	11.8	19.8	8.66	13.60	11.75	27.94
1–2			On-deck cylinder					19.0	5.00	5.00	11.27	16.12
2–1	LPG carrier	84,000 t	Rectangular tank	219.7	36.0	21.9	11.6	43.12	16.88	19.11	19.63	46.89
2–2			Tapered tank					40.81	10.63	19.11	18.57	29.53
3–1	Shuttle tanker	140,000 t	Rectangular	256.5	42.5	22.0	15.5	32.0	18.69	20.12	12.48	43.98
3–2			Tapered tank					28.0	15.00	20.12	10.92	43.98
4–1	VLCC	300,000 t	Rectangular tank	320.0	58.0	31.0	22.0	51.2	21.84	29.20	16.00	37.66
4–2			Partly tapered tank					51.2	21.84	29.20	16.00	37.66
5–1	Suezmax tanker	158,000 t	Rectangular	264.0	48.0	23.2	17.0	33.6	21.70	21.75	12.73	45.21
5–2			Tapered tank					33.6	14.80	21.75	12.73	30.83
6–1	Shuttle tanker	850,000 BBL	Rectangular tank	256.0	46.0	22.4	16.0	30.8	13.36	19.80	12.03	29.04
6–2			Tapered tank					30.8	9.80	19.80	12.03	21.30
7–1	FPSO	850,000 BBL	Rectangular tank	277.0	45.5	28.2	20.0	39.0	18.0	26.4	14.08	39.57
8	LNGc[a]	140,000 m^3	Rectangular tank	268.0	43.0	26.0	12.4	42.840	37.400	27.200	15.99	86.98
9	LNGc[a]	216,000 m^3	Rectangular tank	303.0	50.0	27.0	13.0	40.425	44.164	29.205	13.34	88.33
10	LNGc[b]	256,000 m^3	Rectangular tank	333.0	55.0	27.0	13.7	45.140	48.84	28.59	13.56	88.80

Note: t = tonnes and BBL is the barrel unit for volume (1 BBL = 0.1589873 m^3); L_{PP} = length between perpendiculars, B = beam, D = design ship draft, T_s = scantling draft, L_t = tank length, B_t = tank breadth (B_t for a tapered tank is the largest tank breadth), and H_t = tank height. Length dimensions are in meters. (H.S. Urm, personal communication, 2008).

[a] Cargo containment system thickness of 300 mm is deducted (GTT MKIII system).

[b] Cargo containment system thickness of 530 mm is deducted (GTT NO96 system).

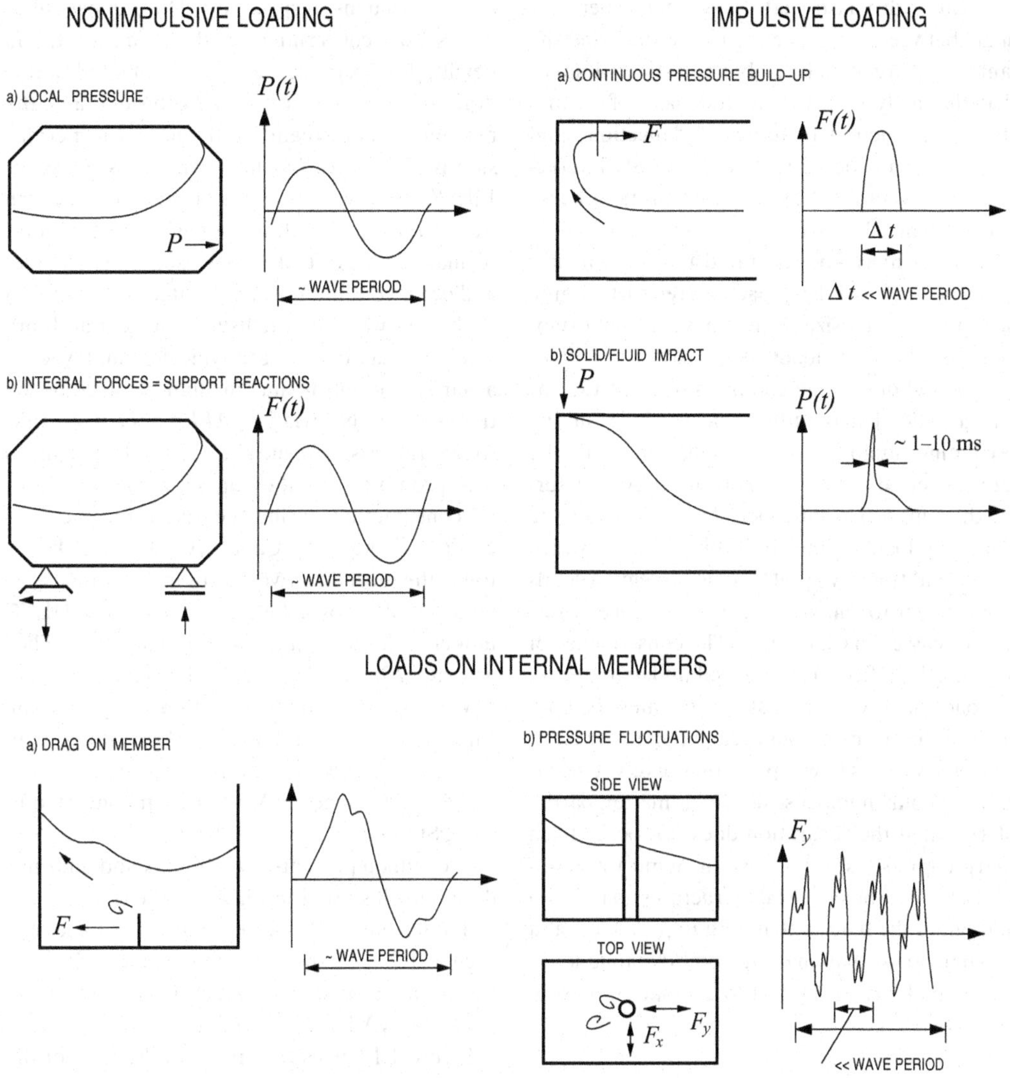

Figure 1.8. Examples of sloshing loads (Olsen, 1976).

(i.e., ore/oil [O/O] and OBO) carriers. Strictly speaking, the membrane LNG tank is not a clean tank due to the presence of the pump tower. However, its damping effect on sloshing is minor. It is common to use centerline longitudinal bulkheads for LPG carriers to reduce the severity of sloshing. Because of the narrow width and internal stiffening of narrow wing ballast tanks on oil tankers, FPSOs, and gas carriers, sloshing is not a problem for their tanks. For ships that have varied loading conditions in a seaway (e.g., FPSOs and shuttle tankers), sloshing considerations are important for proper functioning of the tank level measurement system.

Hydrodynamic loading inside a tank can be classified as either impact loads or "dynamic," nonimpulsive loads. In this context dynamic loads are loads that have dominant time variations on the time scale of the sloshing period, whereas impact loads may last only from 10^{-3} to 10^{-2} s. Sloshing loads must be considered in the context of structural stress response and are of significance for both ultimate limit state (ULS) and fatigue limit state (FLS) assessments. ULS is a limit state (collapse, tearing off, etc.) that is due to a one-time extreme load, whereas FLS is a limit state caused by repetitive loading that results in fatigue. Figure 1.8 lists different load categories.

Local hydroelastic effects matter when the angle between the impacting free surface and the tank structure is small. Hydroelasticity implies that the analysis of the hydrodynamic flow and structural reaction in terms of deflections and stresses cannot be separated. A mutual interaction exists whereby the structural vibrations cause hydrodynamic loads and vice versa.

A gas cavity can be generated as a consequence of the geometry of the impacting free surface and, depending on its size, it may have an important effect on slamming loads. A gas cavity has a natural period due to the compressibility of the gas and an added mass effect due to the resulting oscillating liquid. Pressure oscillations with the natural period of the gas cavity are triggered during the impact. The influence of a gas cavity on the slamming loads cannot be Froude scaled, which is the traditional way of scaling slamming loads from model to full scale. This procedure is further discussed in Chapter 11. The consequence of a boiling liquefied gas is not well understood.

Loads on any internal structures must be considered. In this case high-cycle fatigue may be of concern. Viscous flow separation is likely to occur and Reynolds-number scale effects matter, particularly when the separation does not occur from sharp corners, as in the case of pump towers in LNG tanks. Some internal structures may be in and out of the fluid so that both impact loads and dynamic loads may matter. Total dynamic loads on the tank are of interest to estimate tank support reactions.

1.3.1 Oil tankers

Examples of oil tanker types with their deadweight (DWT) range in parentheses are Panamax (from 55,000 to 75,000 DWT), Aframax (from 75,000 to 120,000 DWT), Suezmax (from 120,000 to 200,000 DWT), VLCC (from 200,000 to 320,000 DWT), and ultralarge crude carrier (ULCC; from 320,000 DWT and up). Deadweight refers to the difference between the total weight of a ship in fully loaded condition and the lightship weight (weight of the ship with no fuel, passengers, cargo, etc.). Furthermore, DWT is an abbreviation for deadweight measured in metric tonnes. One metric tonne is equal to 9.81 kN. The size of Panamax and Suezmax tankers is dictated by the need to traverse the Panama and Suez canals, respectively. Different requirements exist concerning length, beam, draft, and height, for example. The maximum allowable draft is 12.04 m in tropical freshwater and the maximum ship breadth without special permission is 32.26 m for Panama Canal passage. As of July 26, 2001, vessels with a beam of 164 feet can transit the canal with a draft of 62 feet. Details of maximum draft of vessel exceeding 164 feet and other conditions are presented by the official Web site (http://www.lethsuez.com/tg_draft.htm). The Aframax tankers are typically employed on a variety of short- and medium-haul crude oil trades. The abbreviation *AFRA* means average freight rate assessment. The use of VLCC tankers was prompted by the rapid growth in global oil consumption during the 1960s and the 1967 closing of the Suez Canal. A VLCC tanker is today the most effective way of transporting large volumes of oil over long distances. Most ULCC tankers were built in the mid- to late 1970s. They are rather inflexible because they can enter very few ports. Fewer than 40 of these ships remain. The world's largest ULCC, *Jahre Viking*, with length 458 m, beam 69 m, full load draft 24.5 m, and 565,000 tonnes DWT has been converted to an FPSO unit.

A common present-day Aframax and Suezmax design is illustrated in Figure 1.9. The vessel has a double hull and a centerline (CL) bulkhead. Figure 1.9 also shows a double hull with two longitudinal bulkheads, which is common for a present-day VLCC design.

Figure 1.10 gives the structural buildup of the transverse bulkhead of an oil tanker. For ships with horizontal stringers as in Figure 1.10, the stringers are most often located aft in the cargo tanks. It is common to have two to three stringers, depending on the ship's size. For small tankers, two stringers are usually adopted. For VLCCs, three stringers are usually adopted. Rarely, four stringers are arranged. It is unusual for oil tankers designed in European yards to have horizontal stringers located on both sides of transverse bulkheads. Because such a design is not economical, it is also difficult to find such a design built in far eastern shipyards.

Sloshing has always been an important design criterion for oil tankers. Problem areas are transverse bulkheads, stringers, deck plating, webs of transverse deck girders (beams), crossties

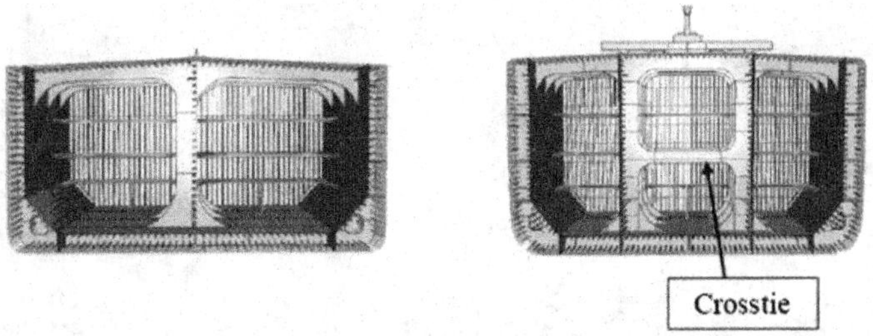

Figure 1.9. Common Aframax and Suezmax oil tanker designs of today with double hull and CL bulkhead (left); common VLCC oil tanker design of today with double hull and two longitudinal bulkheads (right) (Det Norske Veritas AS).

(see Figure 1.9), and piping support. It is often an owner's requirement to have no limitation on partial tank filling.

Environmental concerns have led to requirements for double-hull tankers. Ship owners try to avoid having internal structures in cargo tanks, for cleaning reasons; however, the resulting wide and clean oil tanks increase the probability of severe sloshing. It is usual to have swash (wash) bulkheads in either center or wing cargo tanks. A swash bulkhead is a bulkhead with openings. The swash bulkhead is typically placed in the middle of the tank perpendicular to the main flow direction. If the ratio between the area of the holes and the area of the bulkhead is small, as in Figure 1.18, an important effect is the change in the highest natural sloshing period to a level where sloshing is less severe. The flow through the holes causes flow separation and thereby damping of resonant sloshing. It is desirable that the holes do not change the highest natural period relative to what it would be without holes. This effect has been studied experimentally by Garza (1964) and Abramson and Garza (1965) for forced horizontal excitation of a vertical circular tank that is compartmented into sectors by means of radial walls; 45°-, 60°-, and 90°-sector tanks were investigated. Both the excitation amplitude and perforation of sector walls affect the damping and the highest natural period. Dodge (2000) gives a rule of thumb that if the total area of the perforations exceed 10% of the area, the liquid tends to slosh between the compartments and the slosh natural frequency tends to approach the value of an uncompartmented tank. Having a swash bulkhead instead of a bulkhead has the advantage

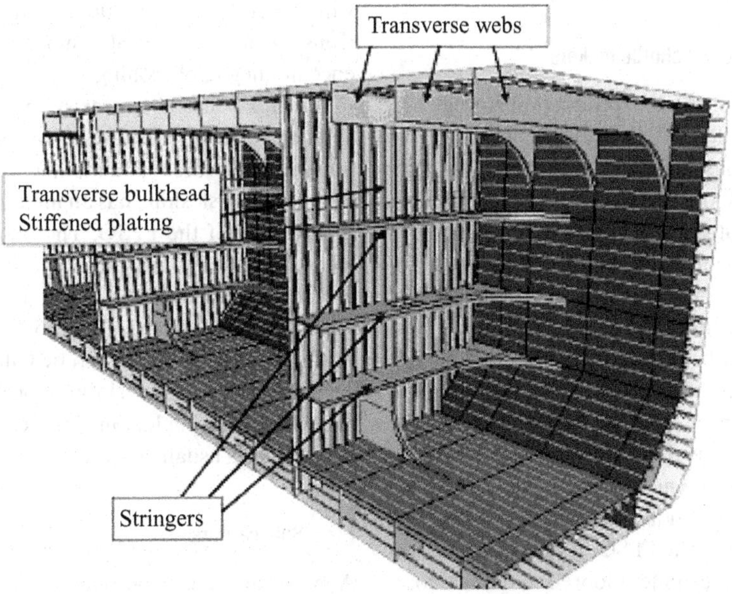

Figure 1.10. Structural buildup of transverse bulkhead of oil tanker (Det Norske Veritas AS).

Figure 1.11. FPSO ship and shuttle tanker in tandem arrangement (Photo: Illustration, StatoilHydro).

of having one pump instead of two pumps for unloading the oil for a given cargo volume. A theoretical analysis of sloshing with swash bulkheads is provided in Sections 6.7.3 and 6.8. Stringers (see Figure 1.10) also provide a damping effect on sloshing, which is discussed in Section 6.5.2.

Sloshing may not be a problem in actual operation of VLCCs and Suezmax tankers, because full-load or empty-tank conditions are very usual.

1.3.2 FPSO ships and shuttle tankers

FPSO stands for floating production storage and offloading (Ridley, 2004). Well stream fluid from subsea reservoirs is first brought up to the production facility onboard a vessel like an FPSO. The FPSO unit does not need to be a ship, as illustrated in Figure 1.11. (Figure 1.30 shows a different concept.) After separation of oil, gas, and water onboard the FPSO, the oil may be transported to an onshore refinery by a shuttle tanker. During transfer of crude oil to the tanker in a harsh environment such as in the North Sea, the FPSO ship and the tanker have to be in a tandem mooring arrangement (Figure 1.11). The FPSO ship is kept in position by a turret mooring system, which allows the FPSO to turn like a weathervane around a cylinder moored to the sea floor. The heading of the FPSO and the position of the tanker is assisted by an automatic control system linked to propellers.

The tanks of an FPSO ship are similar to those of oil tankers; no limitations exist on partial filling, and sloshing is anticipated for all load conditions. Because the FFSO ship is weathervaning so that head or nearly head sea are dominant, pitch motion is important for sloshing excitation. However, some offshore fields may be exposed to swells coming from distant storms and with general wave directions relative to the ship. Consequently, the FPSO ship may also experience nonnegligible rolling.

A mutual interaction exists between sloshing and ship motions. Because mean and slowly varying wave drift forces and moments depend on the ship motions, sloshing has an influence on the stationkeeping of the FPSO. The manner in which mean and slowly varying wave drift forces and moments, in combination with current and wind loads, influence mooring line loads and dynamic position systems is discussed by Faltinsen (1990).

For shuttle tankers, sloshing problems are expected only during loading. No restriction on filling heights is usually required.

1.3.3 Bulk carriers

A bulk carrier can be defined in various ways. SOLAS (Safety of Life at Sea) defines a bulk

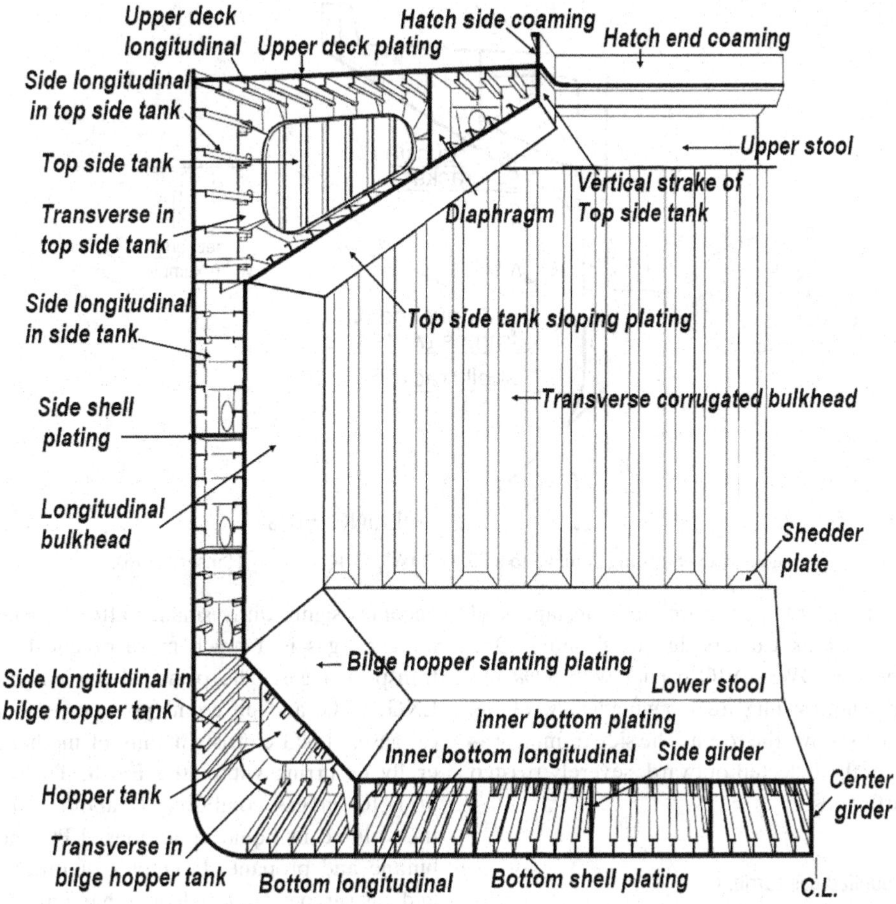

Figure 1.12. Structures and terminology of a bulk carrier.

carrier as "[a] ship constructed with a single deck, top side tanks and hopper side tanks in cargo spaces and intended primarily to carry dry cargo in bulk; an ore carrier; or a combination carrier." Figure 1.12 shows a typical cross-section of a bulk carrier with topside and hopper side tanks indicated. A hatch cover is on the top of the cargo space. Combination carriers are

- OBO (ore/bulk/oil) carriers (i.e., a ship that can carry oil or dry bulk cargoes) and
- O/O (ore/oil) carriers (i.e., a ship that can carry oil or iron ore).

The main types of bulk carriers are small (less than 10,000 DWT), Handysize (from 10,000 to 35,000 DWT), Handymax (from 35,000 to 65,000 DWT), Panamax (from 65,000 to 80,000 DWT), Capesize (from 80,000 to 200,000 DWT), and very large bulk carriers (larger than 200,000 DWT). Capesize refers to vessels incapable of using the Panama or Suez canals, not necessarily because of the tonnage but because of their size. A description of bulk carriers is given by Urm and Shin (2004).

The fact that the OBO and O/O carriers have wide and clean cargo tanks means that sloshing may be a problem due to rolling and transverse ship motions. Limitation of the transverse metacentric height is normally used to control roll natural period relative to the highest natural sloshing period.

Bulk carriers typically have an odd number of holds (e.g., 7 or 9). Ballast water is usually stored in hold number 4 or 5. Because ballast exchange is required outside the port for a bulk carrier, possibilities exist for sloshing damage. The hatch cover is particularly vulnerable. The topside tank bottom is also a problem area (H.S. Urm, personal communication, 2008).

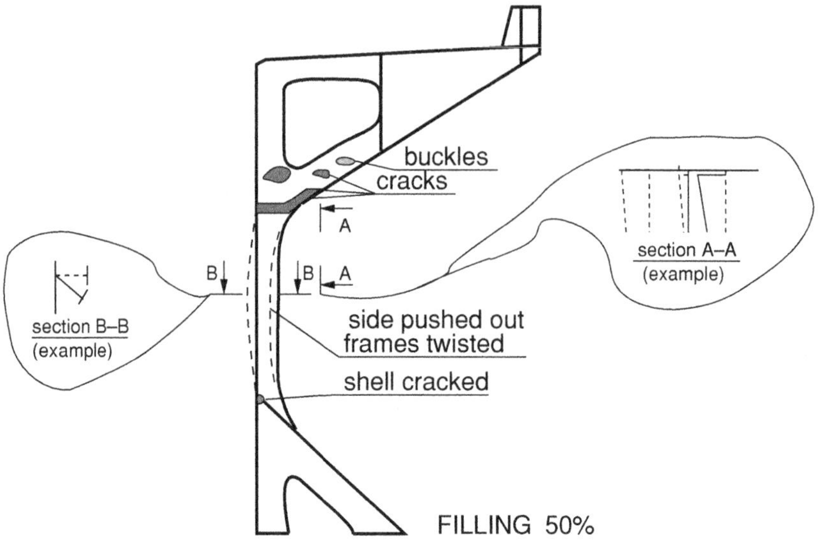

Figure 1.13. Sloshing damage to a 90,000 DWT OBO carrier (Hansen, 1976).

Hansen (1976) reports on three instances of damage on bulk carriers due to sloshing. One was a 90,000 DWT OBO carrier with 50% filling in one hold sailing in seas giving heavy rolling. Figure 1.13 shows the result. The side framing was permanently deflected outward, severely twisted and partly loose.

1.3.4 Liquefied gas carriers

A gas can be liquefied by pressurization and refrigeration. Because the volume of a gas becomes significantly smaller after liquefaction, liquefied gas is a much more practical way to transport a gas. The principal gas cargoes are LNG, LPG, and different types of petrochemical gases. LNG consists mainly of methane naturally occurring within the Earth. The volume of LNG in gas conditions is about 600 times the volume in liquid conditions. LPG includes butane and propane. Examples of other liquefied gas cargoes are ethylene, ethane, propylene, ammonia, and vinyl chloride. Figure 1.14 presents the temperature–vapor pressure relationships of

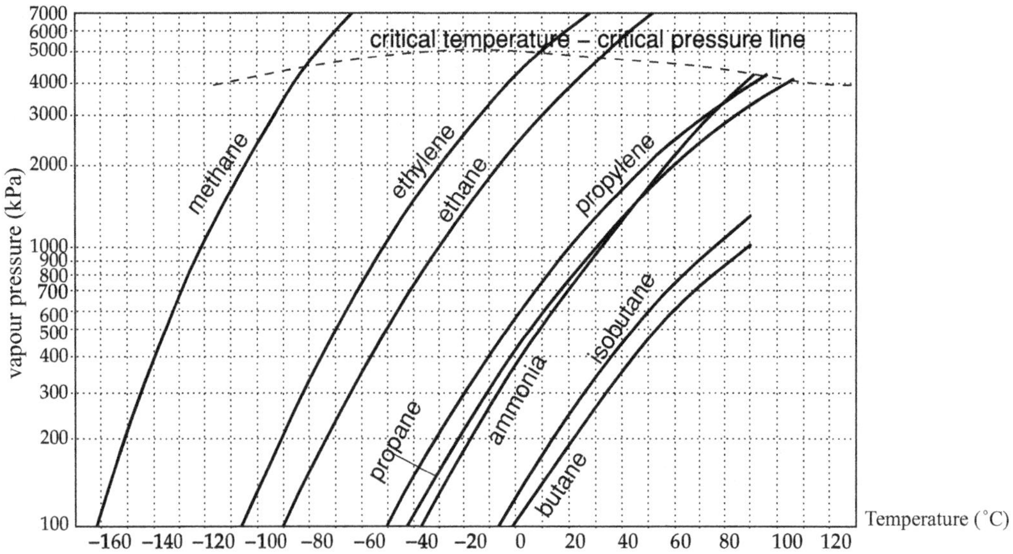

Figure 1.14. Temperature–vapor pressure relationships of various liquefied gases. (Data are based on NIST Chemistry WebBook at http://webbook.nist.gov/chemistry.)

Table 1.2. *Temperatures and densities for some liquefied gas and chemical gas cargoes at atmospheric pressure*

Cargo	Design temperature (°C)	Design density (10^3 kg m^{-3})
Methane	−164	0.42
Ethylene	−104	0.57
Propane	−42	0.58
Propylene	−47	0.61
Butane	−10	0.60
Ammonia, anhydrous (NH_3)	−33	0.68
Acetaldehyde	28.8	0.78
Propylene oxide	33.9	0.86

Note: Data are based on the NIST Chemistry WebBook at http://webbook.nist.gov/chemistry.

various liquefied gases. The figure gives the boiling temperature at different ambient pressures. LNG is not pressurized during transportation and it is carried at its boiling temperature of −162°C. When fully refrigerated (i.e., not pressurized), butane and propane are carried at about −5° and −42°C, respectively (see Figure 1.14). Boiling temperatures and mass densities for some liquefied gas and chemical gas cargoes at atmospheric pressure are presented in Table 1.2. Further details can be found in Lide (2008).

Tank types are divided into independent and membrane tanks. Three categories of independent tanks exist: types A, B, and C have a full secondary barrier, reduced secondary barrier, and no secondary barrier, respectively. Membrane tanks have a full secondary barrier. Figure 1.15 shows an example of an independent C-type fully pressurized tanks. Liquefied gas carriers are described by Emi *et al.* (2004).

1.3.5 LPG carriers

Regional and coastal LPG cargoes are often carried fully pressurized at ambient temperature. The design pressure is about 20 bars. Semipressurized (semirefrigerated) LPG carriers are typically larger than the fully pressurized type and can have cargo capacities up to ~20,000 m^3. The pressure vessels are often bilobe in cross-section (see Figure 1.16) and designed for operating pressures up to 7 bars. Fully refrigerated LPG

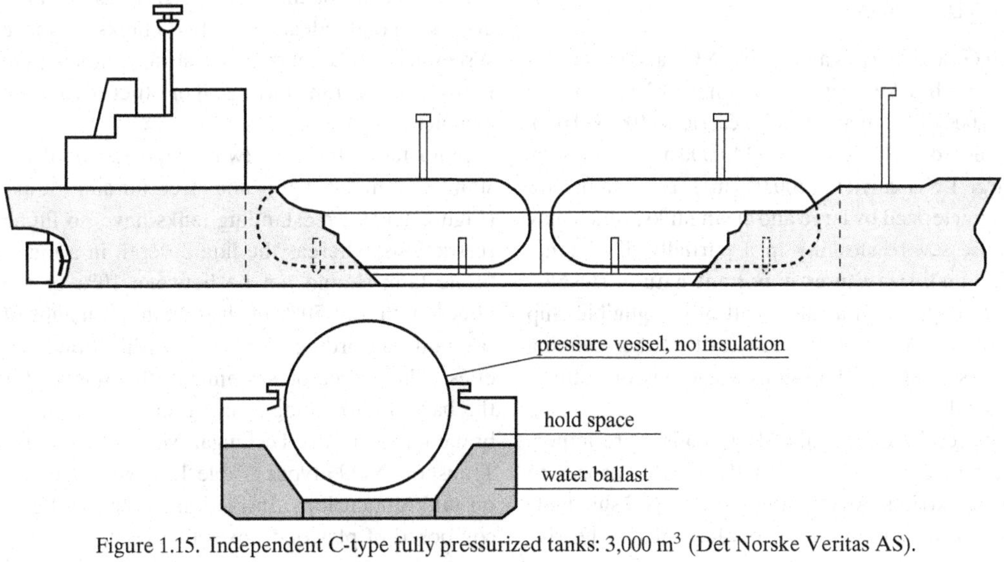

Figure 1.15. Independent C-type fully pressurized tanks: 3,000 m^3 (Det Norske Veritas AS).

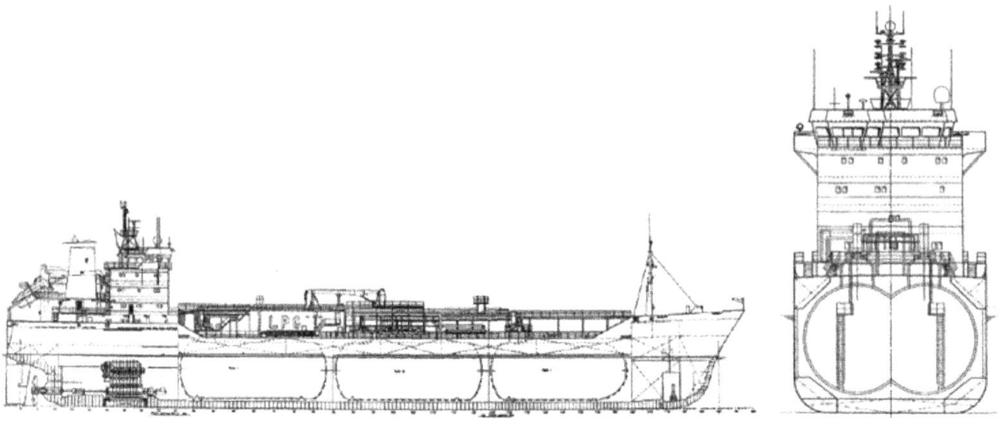

Figure 1.16. LPG carrier with C-type semirefrigerated tanks with bilobe cross-section. Design vapor pressure is 4.5 bar according to the International Maritime Organization. Total capacity is 8,500 m^3; tank length is 25.4 m; and tank breadth is 15.8 m (Senjanovic *et al.*, 2004).

carriers are generally large ships with a cargo capacity up to 100,000 m^3. The temperature may be as low as $-48°$C.

When a CL bulkhead is fitted in an independent tank type A, of the prismatic type, partial filling is allowed. No limitations exist on partial filling when swash bulkheads are installed. Partial filling is possible for bilobe or cylindrical tank type of independent tank type C. No limitations exist on partial filling with swash bulkheads or cylindrical shape. Transverse bulkheads, deck plating, and webs of transverse deck girders (beams) are problem areas due to sloshing (H.S. Urm, personal communication, 2008).

1.3.6 LNG carriers

LNG carrier types and their LNG carrying capacity in brackets are very small LNG carriers (18,000 m^3), Panamax LNG carriers (100,000 m^3), standard LNG carriers (140,000 m^3), and very large LNG carriers (200,000 m^3). LNG tanks are characterized by large and clean tanks, which may cause severe sloshing in a partially filled tank. The tank size causes important natural sloshing periods to be in a range with nonnegligible ship motions. A clean tank implies that the damping of resonant liquid motions due to viscous effects is small.

Several factors make slosh loads more important in LNG ship design than with respect to other carriers. A tank failure in an LNG ship merits special consideration because of (1) the risk of brittle fracture of the primary structure (low-temperature shock), (2) the expensive repair cost of the complicated tank designs, (3) the high out-of-service costs, and (4) a potentially explosive cargo. Partially filled conditions occur because (1) chilldown liquid is needed to maintain cold tanks on return trips, (2) partial unloading is desirable when multiple port stops are made, (3) offshore loading or unloading at sea creates significant time periods at undesirable fill depths, and (4) liquid boil-off occurs. The boil-off rate is 0.15% of the liquid volume per day. However, reliquefaction of the boil-off gas is possible. The liquid depth in prismatic tanks can be as much as 10% of the tank length on return trips. Also the complexity of the tank design in LNG carriers is such that at least some LNG tanks are more susceptible to damage from slosh loading than tanks built for transporting oil or other petroleum products.

LNG tanks fall into two categories: nonfreestanding (membrane) and freestanding tanks (Figure 1.17). Freestanding tanks have no filling restrictions, whereas the liquid depth in a membrane tank should not be between 10% of the tank length and 70% of the maximum height of the tank, according to major classification societies. The effect of sloshing is the reason for the barred tank filling range. Examples of membrane tanks are the Technigaz Mark III and Gaz Transport NO96 types. Table 1.1 gives examples on tank dimensions. Insulation for the NO96 is composed of plywood boxes that contain perlite

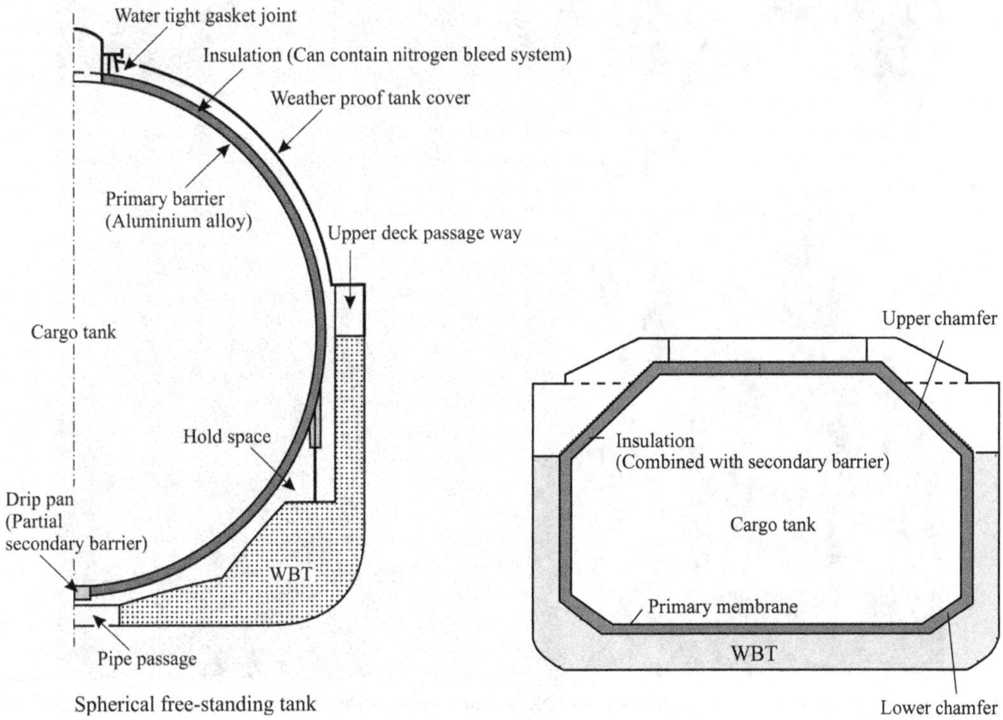

Figure 1.17. LNG carrier with independent spherical tanks of type B (left). LNG Carrier with membrane tanks (right) (Emi *et al.*, 2004). More details about the membrane tank Technigaz Mark III is given in connection with Figure 11.60.

powder. Flat 0.7-mm-thick invar (36% Ni steel) membranes are used as primary and secondary barriers. In Section 11.9.6 we describe the Technigaz Mark III containment system in more detail in connection with slamming analysis. Examples of freestanding tanks of type B are the Moss Kværner spherical tank (see Figure 1.17) and the self-supported prismatic tank IHI SPB (see Figure 1.18). The Moss Kværner tank is supported at its equator by a semiflexible skirt.

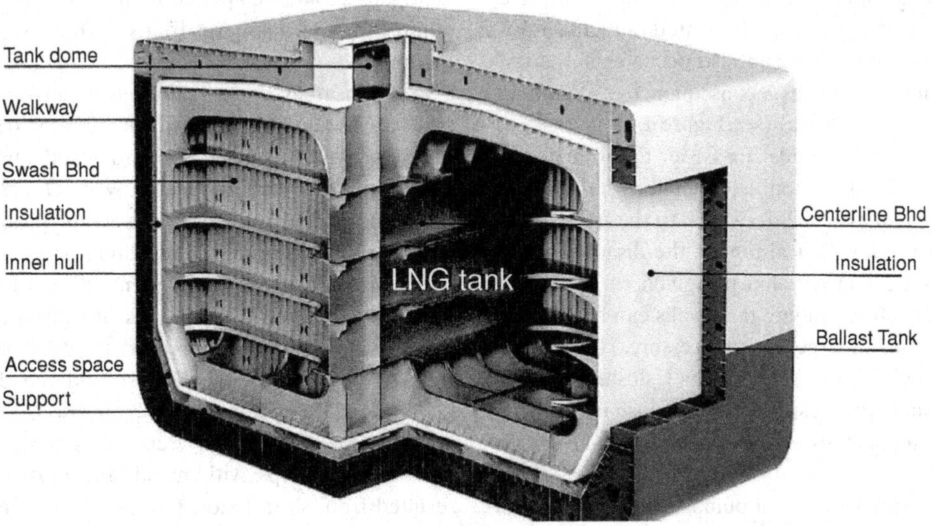

Figure 1.18. The IHI SPB self-supported prismatic type B tank used for LNG cargo.

Figure 1.19. Pump tower inside a prismatic LNG membrane tank (Photo: Gaztransport &Technigaz).

The primary problem associated with sloshing in membrane tanks is the potential damage to tank walls from peak impact slosh pressures. Because this type of tank cannot be analyzed to determine its failure strength, special load tests are performed on representative segments of the structure to determine its load-bearing strength. Severe slosh loads in the membrane tank can occur at small fill depth resulting from steep propagating waves, as illustrated in Figure 1.3(a). Severe slosh loads can also occur near the tank top, for instance, as a consequence of "standing" types of slosh waves (see Figure 1.1).

Freestanding tanks are easier to fabricate, and the insulation system is easier to install than on other systems. One drawback to the freestanding design of a spherical tank is the disadvantage of requiring a larger ship per given cargo volume. Because freestanding tank walls can be designed to withstand large impact pressures, the primary problem associated with LNG sloshing in freestanding tanks results from the slosh loads on the tank support structure and on internal components.

Figure 1.19 shows a pump tower inside a prismatic LNG membrane tank. More structural details are shown in Figure 1.20. The emergency pipes and two discharge pipes are the main strength members of the pump tower trusswork. These columns are connected by intermediate braces and are fitted to the baseplate at the bottom of the tower. Sloshing loads should be applied to the columns and braces located below the liquid dome. Gravity, inertial loads, and thermal loads should be applied to all elements. The pump tower of a spherical tank is illustrated in Figure 1.21.

Problem areas due to sloshing in membrane tanks are the upper and lower chamfer, the transverse bulkhead, the inner deck plating at the transverse bulkhead, and the pump tower. The cargo tank plating and the pump tower are problem areas for spherical freestanding tanks. High-cycle fatigue is of concern for a pump tower. Problem areas for the IHI SPB tank are cargo tank stringers, the transverse bulkhead, the upper deck web at the transverse bulkhead, and the deck plate (H.S. Urm, personal communication, 2008).

Abramson *et al.* (1974) reported on damage to two LNG ships with membrane tanks that resulted from slosh loads. On the *Polar Alaska*, supports of the electric cables supplying the cargo

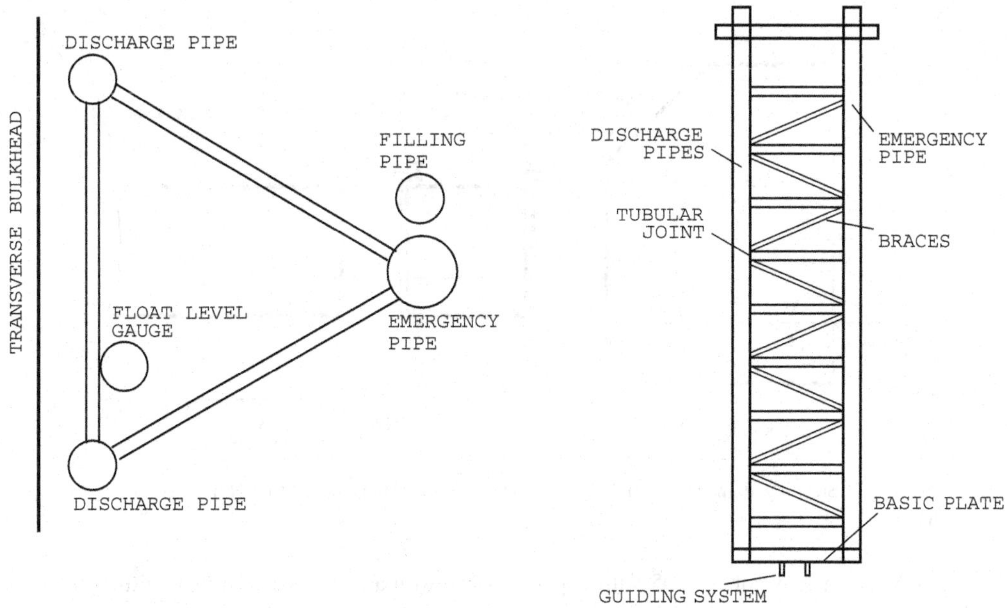

Figure 1.20. Illustration of pump tower layout and main structure in LNG membrane tanks (DNV Classification, 2006).

pumps were broken by liquid sloshing loads, which occurred when the tank was approximately 15–20% full. The broken cable support resulted in damage to the bottom of the membrane tank.

On the *Arctic Tokyo*, a leak in tank 1 was caused by liquid sloshing when the tank was about 20% full. Inspection revealed that the leak was located, along with four deformed points in the membranes, in the aft corners of the transverse and longitudinal bulkheads at about the liquid surface level. More details are presented in Figure 1.22.

Zalar *et al.* (2005) describe damage on the *Larbi Ben M'Hidi* in the late 1970s. After this, the upper chamfer part of NO96 was strengthened using reinforced boxes.

Damage to the cargo containment system occurred for the membrane-type LNG carrier

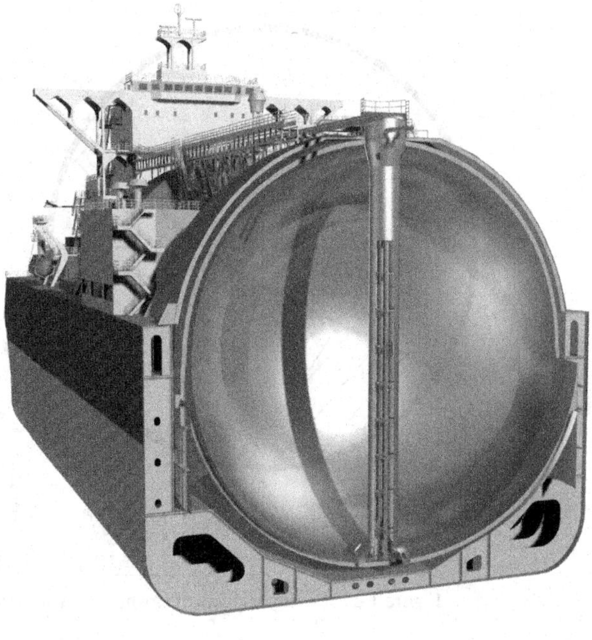

Figure 1.21. Transverse section through LNG carrier with MOSS® spherical containment system. The center column contains piping, stairs, and power cables for the submerged pumps that are installed at the bottom of the tank (Picture courtesy of Moss Maritime, Norway).

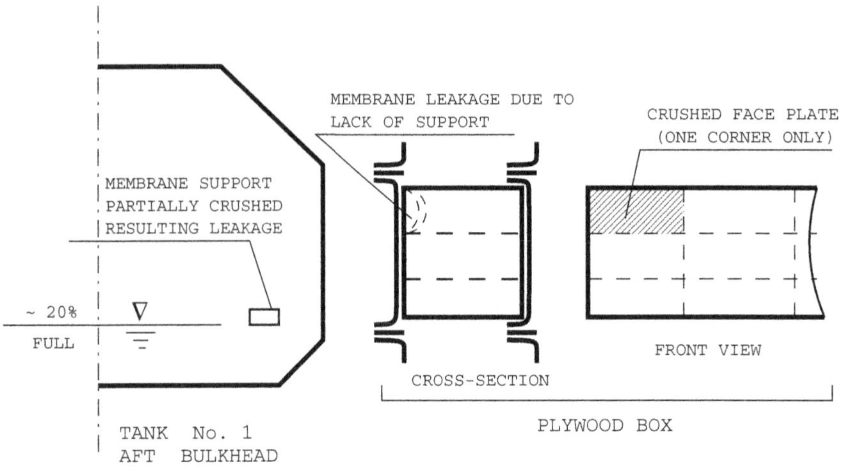

Figure 1.22. Sloshing damage to an LNG membrane tank (Hansen, 1976). (Note: Test pressure ~6 bars.)

Catalunya Spirit in November 2005 with a partial filling around 10% of the tank length in several tanks. The most severe damage occurred in the second tank from the bow, but other tanks were also damaged; however, no leakage occurred. The damaged areas were in the vertical sides of the prismatic tank, close to the lower chamfer. A probable cause is a hydraulic jump-type flow (see Figure 1.3(a)) in the transverse tank cross-section, which results in a breaking wave with a nearly vertical front hitting the tank wall and causing severe impact loads.

The spherical type of LNG carrier has not been subject to any major sloshing damage even though violent free-surface motion is likely due to the tank dimensions and the clean internal surface. Knowing the total hydrodynamic force in the horizontal direction in a partially filled tank is important to avoid buckling (see Figure 1.23).

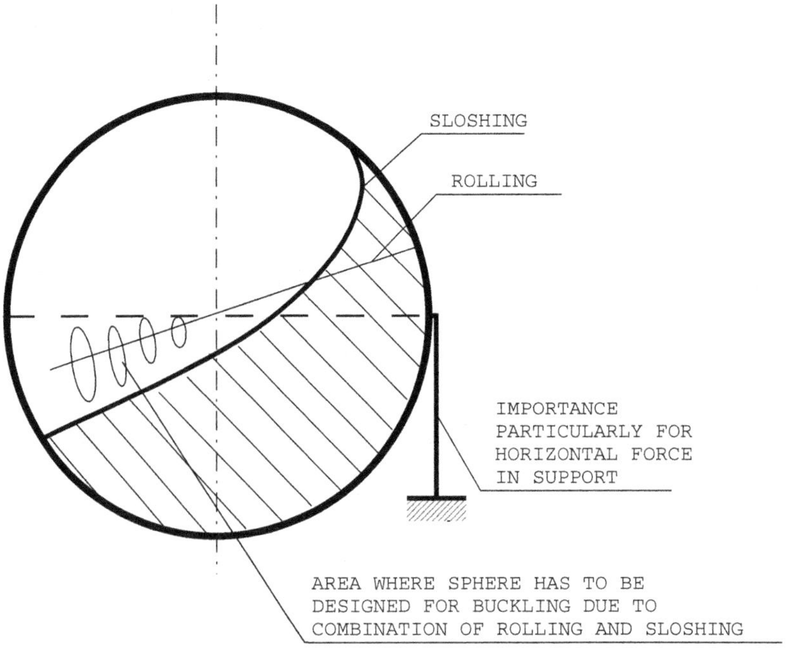

Figure 1.23. Design aspects of sloshing in a spherical LNG tank (Hansen, 1976).

On- and offloading from LNG carriers at offshore sites exposed to high seas require a sloshing assessment of the structural integrity of the tanks and supporting structures. LNG regasification vessels (RVs; Janssens, 2006) operate with partially filled tanks a large fraction of the time. These vessels incorporate equipment for the vaporization of LNG and discharge of the high-pressure gas through a subsea pipeline directly into the consumer grid system. They dock at specially designed offshore locations. These vessels have no restrictions on filling heights; however, operational limits in terms of maximum significant wave height may be imposed, where the ship's capability to keep heading into the waves is taken into account. In some sites, nonparallel wind-generated sea and swell may make partial filling operations difficult. Ship-to-ship transfer of LNG can be done in tandem or side-by-side configuration of the two vessels. This partial filling operation is integral to an operating scenario where standard LNG carriers transfer their cargo to RVs that feed natural gas directly into pipelines for onshore use.

1.3.7 Chemical tankers

The most sophisticated chemical tankers are called chemical parcel tankers and are designed to carry a wide range of liquid cargo. Great variations in cargo density exist (e.g., phosphoric acid (85%) and methanol have a density of $1.69 \cdot 10^3\,\mathrm{kg\,m^{-3}}$ and $0.79 \cdot 10^3\,\mathrm{kg\,m^{-3}}$, respectively. The chemical cargoes can be divided into the following groups (Werner, 2004):

1. petrochemical products,
2. coal-tar products,
3. carbohydrate derivatives (molasses, alcohols, etc.),
4. animal fats and vegetable oils, and
5. heavy chemicals (sulfuric acid, phosphoric acid, hydrochloric acid, caustic soda, etc.).

A chemical product tanker is less sophisticated than the chemical parcel tankers and is designed to carry refined petroleum products such as jet fuel, naphtha, diesel and petrol (gasoline), and chemicals easy to handle such as octane and xylene. Smaller product tankers with capacities around $40 \cdot 10^3\,\mathrm{m^3}$ capacity carry caustic soda.

Oceangoing chemical tankers generally range from 10,000 to 50,000 DWT in size. They often have a heating system to avoid having some cargoes solidify or become so viscous that they cannot easily be moved (e.g., molasses and waxes). Because chemical tankers operate with various cargoes and multiport loadings, partial fillings are usual.

Wide-breadth tanks are common for product tankers. Important sloshing could be excited by the transverse ship motions and rolling. A CL longitudinal bulkhead is common for chemical parcel tankers; it reduces the possibility of sloshing excitation by sway, roll, and yaw. A chemical parcel tanker typically has many small tanks with no sloshing problems. The problem area due to sloshing in on-deck cylinders on chemical parcel tankers is at the end of a cylinder. Sloshing problems are not anticipated for rectangular tank shapes. Sloshing should, in particular, be checked for high-density cargoes (H.S. Urm, personal communication, 2008).

1.3.8 Fish transportation

Fish may be transported in tanks. For instance, salmon are transported in ship tanks from fish farms at sea to a factory for slaughtering and further processing. Stories about seasick salmon have been told. The salmon may occupy up to 50% of the tank volume. The tank is usually filled with water so that there is no sloshing. However, when the ship pitches and/or rolls, the water has a motion relative to the tank even for a completely filled tank (see Sections 5.4.2.2 and 5.3.2.2). An analysis of water motion in a tank with fish is beyond the scope of this book.

1.3.9 Cruise vessels

The operational time of swimming pools on cruise vessels is limited by sloshing. A possible layout of swimming pools on the top deck of a cruise vessel is schematically presented in Figure 1.24. There is typically a wading region with shallow depth along the rim of the pool.

Operational criteria relevant for swimming pools are not established in the same way as for rolling, accelerations, bottom slamming, and green water on the deck of ships (Faltinsen, 2005). A suggested operational criterion for a swimming

Figure 1.24. Sloshing in swimming pools on cruise vessels is of concern (Artist: Bjarne Stenberg).

pool is that the expected largest wave elevation in 3 hours should not exceed the rim of the pool, which typically is 0.25 m above mean water level. The percentage of operational time on a yearly basis could, for instance, be set equal to 60%. A procedure on how to estimate the operational time must be based on available weather statistics for realistic operational areas such as the Gulf of Mexico, the Caribbean Sea, and a route between Haiti and Miami. Ocean wave statistics can be found in Hogben and Lumb (1967) and Det Norske Veritas (DNV, 2007). As before we consider individual sea states with a probability of occurrence. By using the operational criterion of the expected largest wave elevation being at the rim of the pool, one can determine if the swimming pool can operate in a given sea state. The yearly operational time can then be established using ocean wave statistics. If the sloshing in the swimming pool is a linear process, the proposed procedure follows the usual procedure in seakeeping analysis of ships. However, sloshing is a nonlinear process and certain simplifications such as equivalent linearization are necessary to establish the yearly operational time.

1.3.10 Antirolling tanks

Ship motions excite sloshing, which in turn affects the ship motions. Ships equipped with antirolling tanks utilize this effect. Figure 1.25 shows examples of a U-tube tank and a free-surface tank used as antirolling tanks. The sloshing-induced roll moment on the vessel causes roll damping if the highest natural sloshing period is tuned to be close to the roll natural period.

1.4 Tuned liquid dampers

Tall buildings may experience excessive oscillations due to wind and earthquakes. The main problem during wind action is discomfort for people, especially on the highest floors. It is the highest natural periods that are of concern for wind- and earthquake-excited oscillations. Because the structural damping of the lowest modes is small, large amplification of resonant oscillations occurs when no motion dampers are installed. Kareem *et al.* (1999) provided a list of different types of motion dampers used for tall buildings around the

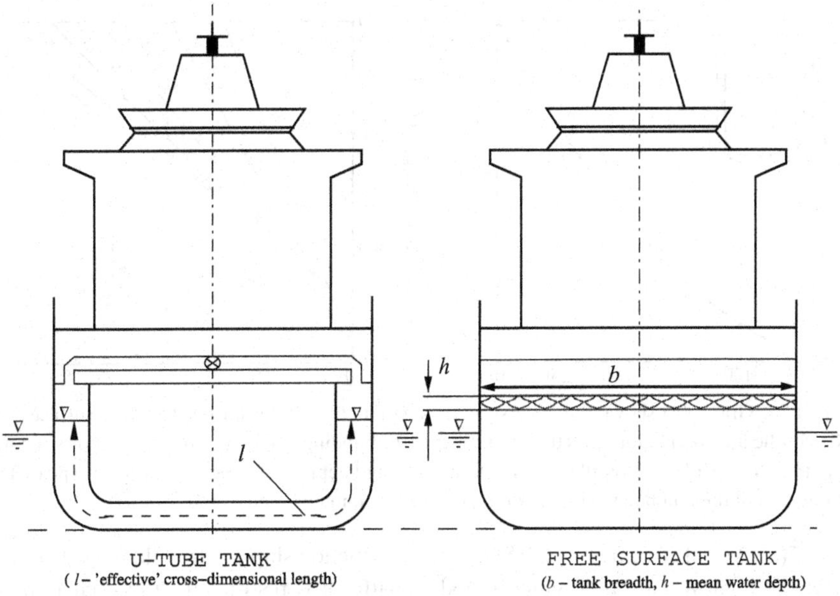

Figure 1.25. Examples on antirolling tanks.

world. Data are given by Kareem *et al.* (1999) for the highest natural periods in two perpendicular transverse directions of a building. The highest torsional natural period is also listed for some of the buildings. Some of their data on the natural periods are presented in Figure 1.26 as a function of the building height H, where the highest natural period T_1 for each building has been selected. A least-squares fit line, T_1 (s) = 0.02116 (s m^{-1}) · H (m) is drawn through the data.

A TLD is one possible damping device for wind- and earthquake-excited oscillations of tall buildings. TLDs are divided into two main types: tuned sloshing dampers (TSDs) and tuned liquid column dampers (TLCDs). A TSD is a free-surface tank and a TLCD is a U-tube tank (i.e., similar to the two types of antirolling tanks shown in Figure 1.25). The principle for a TLD to provide damping is similar to antirolling tanks that damp roll motions. The first step is to select the natural sloshing period equal to the natural period for the oscillation mode that should be damped.

The formula presented in Figure 1.26 can be used to exemplify the size of a liquid damper as a function of the building height. We choose a U-tube with constant cross-sectional area and define an average length L between the two free surfaces of the U-tube in the flow oscillation direction (see Figure 1.25). The natural period of the U-tube can then be expressed as $T_{nu} = 2\pi\sqrt{L/(2g)}$. By choosing $T_{nu}(\mathrm{s}) = T_1(\mathrm{s}) \cdot 0.02116\,(\mathrm{s\,m^{-1}}) \cdot H(m)$, we get $L = 0.00022253 \cdot H^2$ with all lengths in meters. This result is plotted

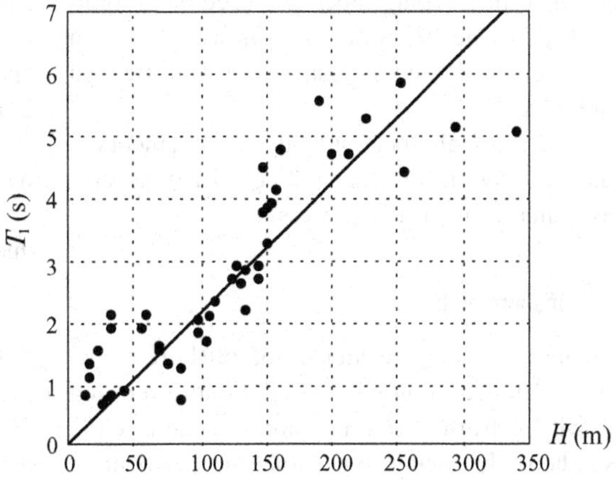

Figure 1.26. Highest natural periods T_1 for the tall buildings listed by Kareem *et al.* (1999) as a function of the building height H. A least-squares fit line is drawn through the data (Kong, personal communication, 2008). • = sample data, solid line = the least-squares fitting, T_1 (s) = 0.02116 (s m^{-1}) · H (m).

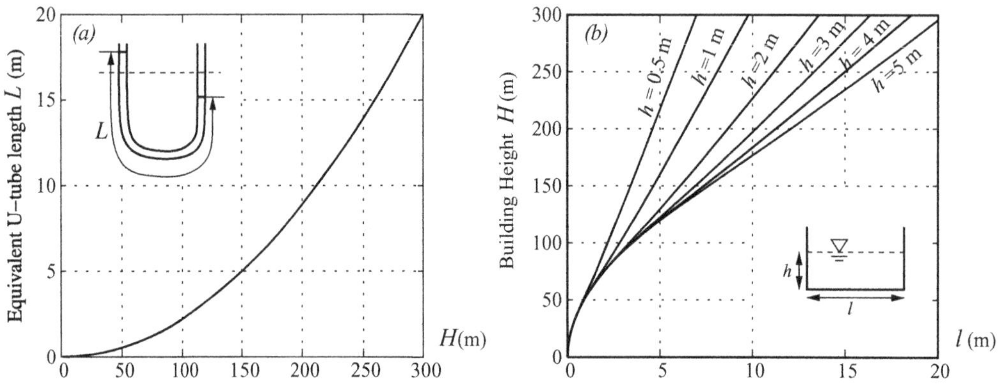

Figure 1.27. Graph (a) shows the necessary average length L between the two free surfaces of a TLCD in the flow oscillation direction as a function of building height H. Graph (b) shows the necessary tank breadth l in the oscillation direction and liquid depth h of a rectangular tank type of TSD as a function of the building height H (Kong, personal communication, 2008).

in Figure 1.27(a). When it comes to a TSD, we would select the highest natural sloshing period to be equal to the natural period of the oscillation mode to be damped. Figure 1.2 illustrates that both the water depth and the tank breadth determine the natural sloshing period for a rectangular tank, which is further illustrated in Figure 1.27(b) by relating the building height to the tank breadth and liquid depth.

The length of the TLD perpendicular to the flow direction determines the magnitude of the oscillation damping caused by the TLD. This topic is discussed for antirolling tanks in Chapter 3. A constraint for a ship is that an oversized free surface of the tank negatively affects the static heel stability. Several TLDs may be used. Damping devices are introduced in a TLD to broaden the period range with satisfactory oscillation damping. Screens perpendicular to the flow directions can be effective sloshing damping devices when properly placed in the tank. A TLCD with damping vanes proposed by Motioneering for the 975-foot-tall Comcast Center in Philadelphia (Pennsylvania, USA), is illustrated in Figure 1.28.

TLDs have also been used by civil engineers to damp vibrations of other tall constructions such as chimneys and bridge towers.

1.5 Offshore platforms

Figure 1.29 shows examples of offshore structures. The jacket, gravity-based monotower, and gravity platform with four columns penetrate the sea floor. The semisubmersible and the floating production ship are free floating. The tension leg platform is restrained from oscillating vertically by tethers, which are vertical anchor lines that are tensioned by the platform buoyancy being larger than the platform weight. Both the ship and the semisubmersible are kept in position by a spread mooring system. An alternative would be to use thrusters and a dynamic positioning system. Pipes (risers) are used as connections between equipment on the sea floor and the platform.

Figure 1.30 shows a novel floating offshore platform by Sevan Marine. The submerged hull is a truncated vertical circular cylinder with a damping device similar to a bilge keel on a ship. One application of the platform is as an FPSO unit. The storage tanks are annular and sectored upright circular tanks (see Figure 4.48 for a definition) obtained by imagining the platform as a cake that is divided into six equal parts. The sector under the living quarters is prohibited from use as an oil storage tank. Figure 1.30 also illustrates the layout of the tanks during the construction phase. As an example, the radii of the inner and outer tank surfaces, R_1 and R_2, are 6 and 27 m, respectively. The corresponding tank height is 24.5 m. The tanks are clean except for stringers on one side of the radial bulkheads (see Figure 1.10). Any liquid depth is allowed under production. When sloshing occurs, the largest free-surface elevations occur at the inner cylinder, where the narrowest mean free surface is located.

The Draugen platform is a gravity-based monotower (see Figure 1.29) located at a water depth of 252.5 m in the Norwegian Sea. The water is flooded to a depth just below mean sea level

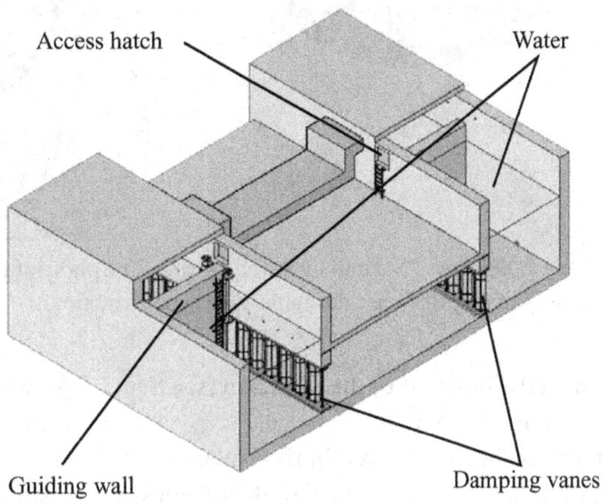

Figure 1.28. TLCD proposed by Motioneering for the 975-ft-tall Comcast Center in Philadelphia, Pennsylvania, USA (photograph of the Comcast Center by Peter Aaron/Esto).

inside the shaft. Several pipes are located inside the shaft. A sloshing period of about 4.3 s has been observed due to sloshing of the water inside the shaft (Drake, 1999). In addition, a resonance period of about 3.9 s was observed, which is associated with global structural elastic response. These two periods are the highest natural periods for sloshing and the platform, respectively. The two periods are not sufficiently close for the sloshing inside the tank to act as a TLD for the

Figure 1.29. Six types of offshore structures. From left to right we have a monotower, jacket, gravity platform, semisubmersible, floating production ship, and tension leg platform (Artist: Bjarne Stenberg).

Figure 1.30. The FPSO unit by Sevan Marine. The photo at the right shows the top view of the annular and sectored upright circular tanks during construction.

wave-induced vibrations of the platform (see Sections 5.4.5.4 and 5.4.5.5).

Sloshing has been observed in the ballast tanks of a tension leg platform. The tank includes a circular cylindrical part with radius 14.2 m in the lower part of the columns of the platform. A dingy was intended to be used during inspection of the ballast tanks; however, the waves in the ballast tanks were so large that nobody dared to use the dingy. A ladder made of composite material was partly torn apart, probably due to wave forces. The water depth in the cylindrical part was estimated to be about 5 m. The highest natural sloshing period for a circular cylindrical tank with radius 14.2 m and water depth 5 m is $T_{1,1} =$ 7.38 s (see Figure 4.10 and eq. (4.41)). Sea states with nonnegligible wave energy at a wave period of 7.38 s occur relatively frequently. Furthermore, the wave-induced horizontal motion of the platform in waves with period 7.38 s is nonnegligible, which is the reason sloshing was observed.

Sloshing can occur in an oil–gas separator on a floating offshore platform. Oil–gas separators are used to separate the oil, gas, and water during oil production. A picture of a separator is shown schematically in Figure 1.31. Typical lengths and radii are about 15 and 2 m, respectively. Separation occurs as the liquids and gas flow through the separator. Because the density of water is greater than that of oil, there exists a lower layer of water with oil above. Waves are generated both on the free surface between oil and gas and on the interface between oil and water. In Section 4.9.2.1, we show, by selecting realistic main dimensions of a separator, that sloshing can be excited by wave-induced motions of the

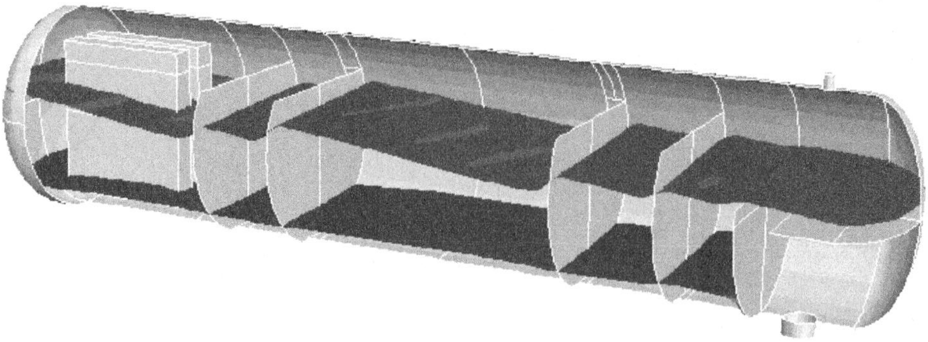

Figure 1.31. Oil–gas separator on a floating platform at sea. The lower liquid is water and the upper liquid is oil. Perforated plates are introduced perpendicular to the flow direction to minimize sloshing (courtesy of National Tank Company, NATCO).

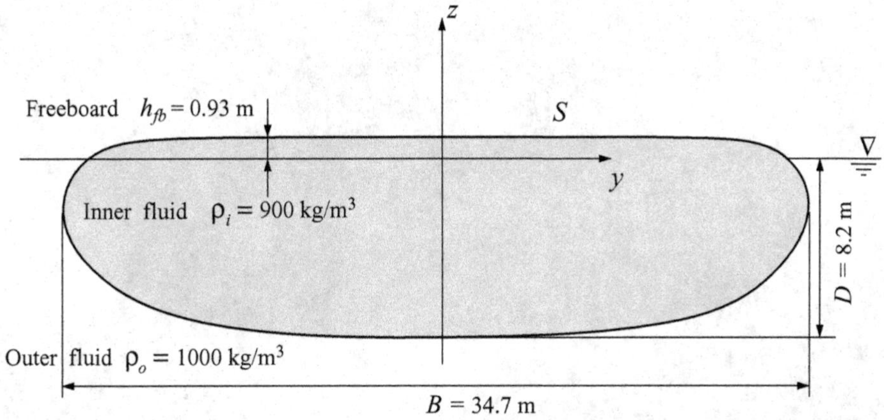

Figure 1.32. Cross-sectional shape of floating fabric container with initial membrane length 76.3 m, membrane elasticity $E_{mem} = 1{,}000$ k Nm^{-1}, outside water density $\rho_o = 1{,}000$ kg m^{-3} and inside fluid density $\rho_i = 900$ kg m^{-3}. Numerical calculations give membrane tension $T_{mem} = 19.04$ kN m^{-1}, freeboard $h_{fb} = 0.93$ m, and internal overpressure $p_o = 0.22$ N m^{-2}. The container has a fractional filling of 60% and the cross-sectional area is 278 m^2, the draft is 8.2 m, and the width is $B = 34.7$ m (Løland and Aarsnes, 1994).

platform. The consequence of sloshing is a mixing effect on the oil, water, and gas that delays oil production. Sloshing damping devices are therefore commonly used to increase operational time. The use of perforated plates placed strategically perpendicular to the flow direction is one way of doing this.

1.6 Completely filled fabric structure

If the tank is completely filled with an incompressible liquid, resonances occur if the tank structure is elastic. We illustrate this effect for a completely filled fabric structure. For this purpose, let us consider long slender containers made of rubber and used for transportation of freshwater and oil as described by Hawthorne (1961). (Nowadays one can use coated fabric instead of rubber.) The word *fabric* means a sandwich structure coated with a polymer on both sides. It is a very flexible structure with negligible bending stiffness. The benefit is that it requires a small storage place when empty. Floating fabric containers of, for instance, 10,000 m^3 have been built for freshwater transportation.

Different parts of the container are sewed together. Because high stress concentrations occur at the ends of the overlaps, fatigue is an important issue. The main dimensions are determined by the required enclosed volume, the towing resistance, and the seakeeping performance. Figure 1.32 gives an example of calculated cross-sectional shape. The shape and the fabric tension depend strongly on the fractional filling, which is defined as the ratio between the actual and the maximum possible filling. When fractional filling increases, the shape approaches a circular cylinder.

Figure 1.33 is from model tests of a fabric structure in waves. We see from the photo that the liquid inside the completely filled fabric structure moves. To find the natural frequencies we must analyze the coupling of the internal and external flow with the elastic behavior of the fabric structure. The analysis can be performed by using the linear frequency domain solution of Zhao and Triantafyllou (1994) and Zhao (1995). Because the freeboard of a fabric structure is very small, nonlinear wave effects become important for relatively small incident wave amplitudes.

1.7 External sloshing for ships and marine structures

Sloshing can also occur in confined liquid spaces that have a free surface and external flow. Examples are resonant oscillations between the hulls of a catamaran or in a moonpool (see Figure 1.34). A moonpool is an opening in the middle of the ship used for marine operations. The

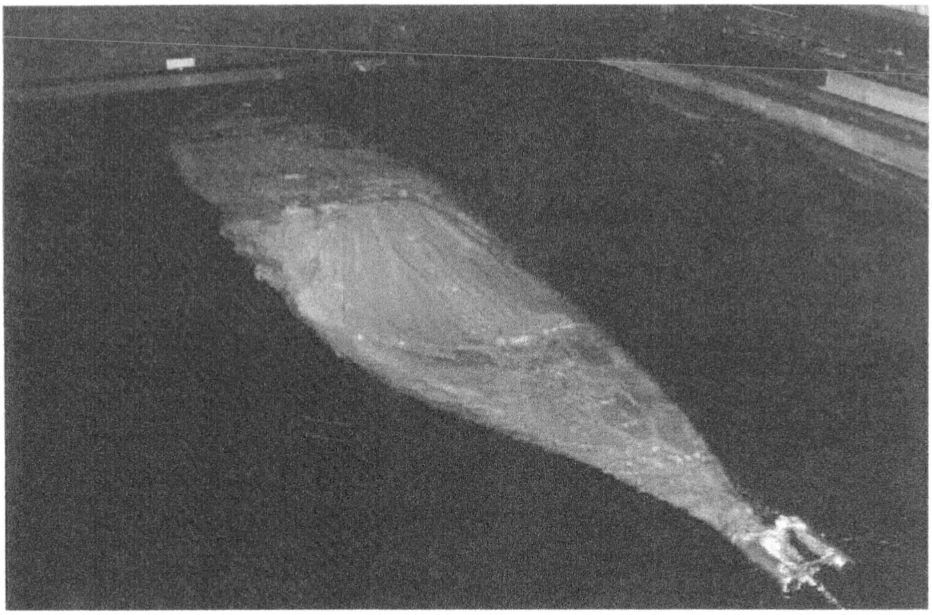

Figure 1.33. Model tests of a completely filled fabric structure in waves.

most important resonance is called piston-mode oscillation. The word *piston* refers to the liquid moving in nearly one dimension as a rigid body in the moonpool. The piston-mode resonance frequency occurs in a frequency range with relatively large vertical ship motions that act as an excitation. A consequence of the fact that the confined space with resonance oscillations is part of the external water domain of the ship is the generation of far-field waves, which causes a damping of the resonance oscillations. However, the damping is not sufficient to prevent a large amplification of free-surface flow in the moonpool relative to the vertical ship motions. A general tendency is that the narrower the horizontal cross-section of the moonpool, the larger the motion of the free surface in the moonpool for given vertical ship motions. Flow separation at sharp corners, for instance at plates in the moonpool opening, can cause nonnegligible damping of the piston-mode resonance oscillations. Another possibility for damping is to use perforated walls in the moonpool as indicated in Figure 1.34. Free-surface nonlinearities are less important for

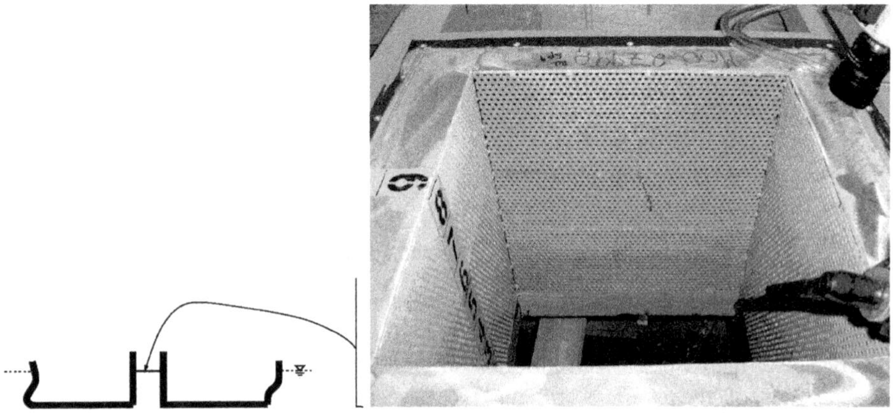

Figure 1.34. Schematic drawing of the centerplane of a ship with a moonpool together with a bird's-eye view of a model of a moonpool with perforated walls.

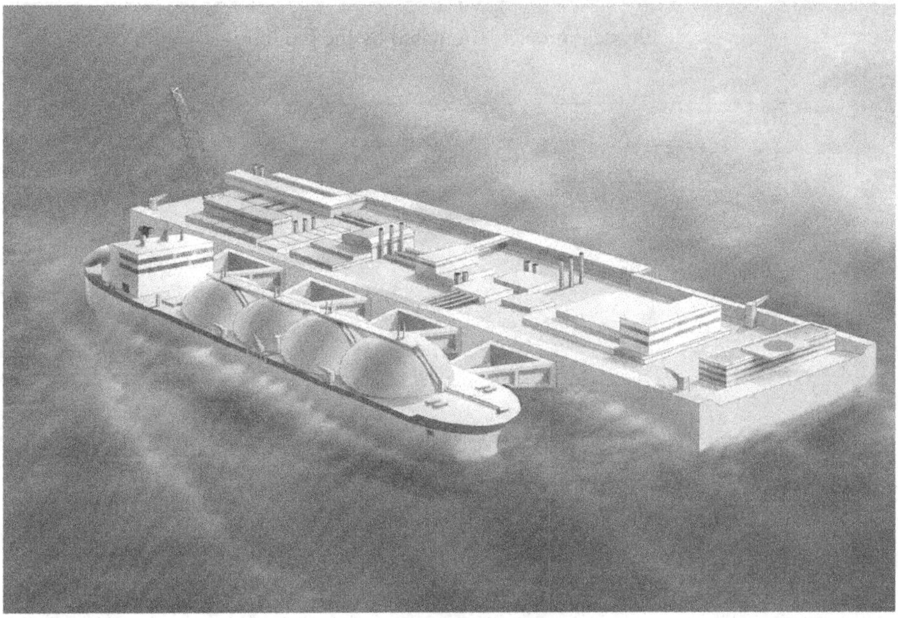

Figure 1.35. LNG ship at a terminal with submerged vertical walls causing resonant water motion in the gap between the ship and the terminal (Artist: Bjarne Stenberg).

piston-mode resonance than for sloshing in a tank.

Moonpool oscillations can also be excited during the forward speed of a ship in calm water. A hypothesis is that the excitation mechanism is flow separation at the edges of the moonpool opening.

Piston-mode resonance can also occur in the water gap between a ship and a terminal with submerged vertical walls (see Figure 1.35). This scenario is relevant for LNG terminals offshore. The LNG carriers offload their cargo to the terminals where the LNG is regasified and transported to land via pipelines. Of particular importance in the design of the fender system and the mooring lines are the wave-induced horizontal ship motions, which are affected by the flow in the water gap.

A very special case where resonant oscillations amplify the problem is shown in Figure 1.36. The bow part of an FPSO vessel was equipped with a submerged "lip" to act as damping device for vertical wave-induced motions of the ship. The idea was that vortex shedding from the lip edges should cause important viscous damping. The upper part of the lip became dry part of the time in a seaway during transit as a consequence of relative vertical motions and local resonant water motion on top of the lip. This subsequently caused high flow velocities around the sharp corners of the lip and damaging slamming impact

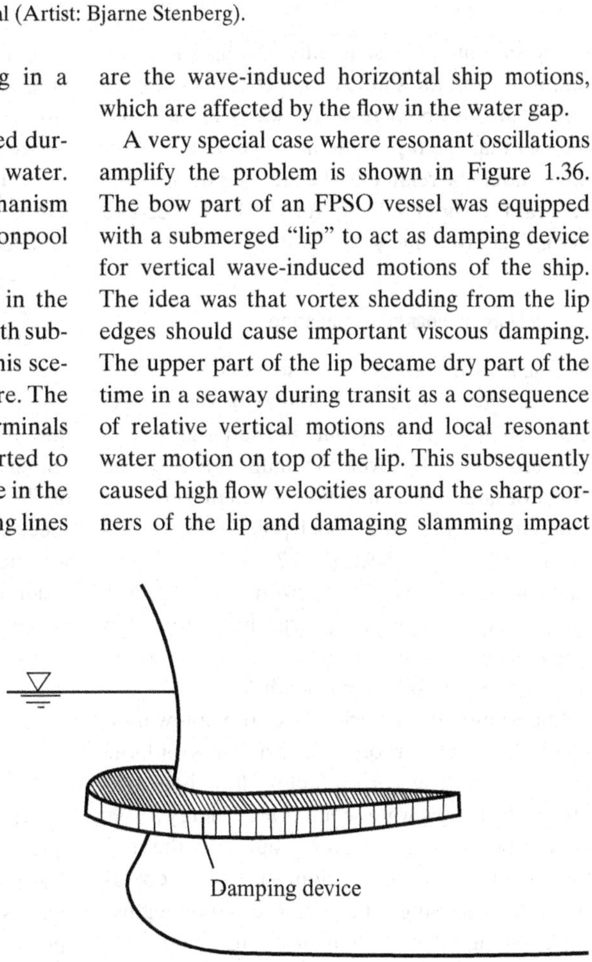

Figure 1.36. Damping device for vertical wave-induced motions of an FPSO.

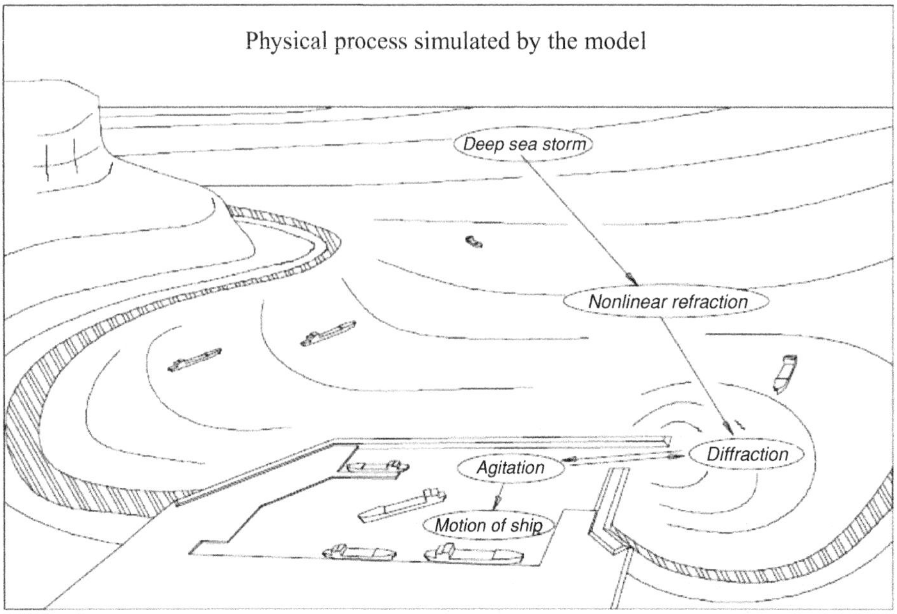

Figure 1.37. A physical process causing harbor resonance (Drimer *et al.*, 2000).

to the ship hull. Consequently, the lip had to be removed from the ship. A conflict exists between our argument about resonant water motion and the fact that the top of the lip got dry. The argument should be related to the fact that the relative vertical threshold velocity for the lip to become dry is decreased due to the resonance.

1.8 Sloshing in coastal engineering

Resonances in harbors are also of concern. The vertical water surface motion may be one or several meters at certain points in the basin at harbor resonance. However, the main concern is the effect of large horizontal flow accelerations on a moored ship near the nodal points of the harbor resonance (see Sections 3.7, 4.4.3 and 4.4.4). The consequence is large horizontal wave forces on the ship with resulting large horizontal ship motions and possible failure of the mooring system, causing the ship to come adrift.

Important natural periods are from a few minutes for artificial harbors to several hours for large natural bays. The corresponding flow is associated with shallow water conditions. Energy losses come from waves propagating out from the harbor opening, bottom friction, and flow separation at the opening. The presence of nonnegligible damping of the dominant natural modes for harbor resonance implies that resonance oscillations in harbors are associated with forced oscillations caused by external disturbances.

Because storm waves may have periods up to 20 s, they cannot directly excite harbor resonance. However, nonlinear wave–wave interaction taking place in the shoaling zone matters (see Figure 1.37). For instance, second-order wave–wave interaction causes difference frequency effects with energy at the important harbor resonance frequencies, where *first order* refers to linear waves. We can split an irregular sea state into linear regular wave components of different circular frequencies σ_i. The wave amplitudes of the wave components are given by the wave spectrum. Second-order wave theory means that we include terms involving products of first-order quantities (e.g., terms with time dependence), $\cos(\sigma_i t + \varepsilon_i)\cos(\sigma_j t + \varepsilon_j)$, which can be reexpressed as

$$\tfrac{1}{2}\{\cos[(\sigma_i - \sigma_j)t + (\varepsilon_i - \varepsilon_j)] + \cos[(\sigma_i + \sigma_j)t + (\varepsilon_i + \varepsilon_j)]\}.$$

We then see that difference frequencies $\sigma_i - \sigma_j$ appear.

Another possible excitation mechanism of harbor resonance is tsunami waves caused by earthquakes that may be thousands of kilometers

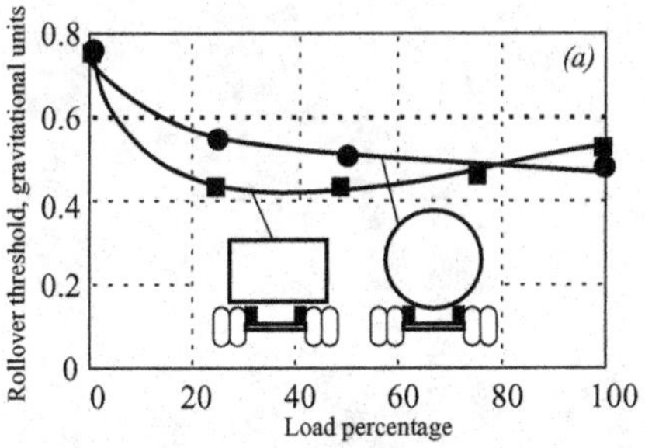

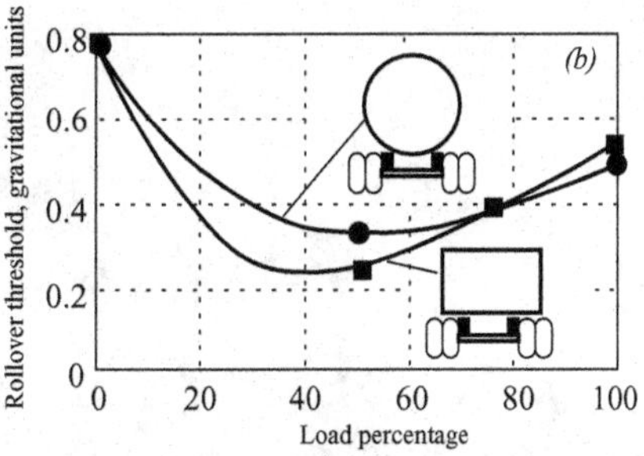

Figure 1.38. Rollover thresholds in (a) steady and (b) transient (0.5 Hz) turn as a function of load of unrestrained liquid and tank shape (adapted from Strandberg, 1978, by Winkler, 2000).

away. Svendsen and Jonsson (1976) also mention the peculiar example of harbor resonance at Jakobshavn on Greenland. The excitation is caused by waves radiating from the floating front of the Jakobshavn glacier when an iceberg is released.

1.9 Land transportation

Romero *et al.* (2005) reported that sloshing of liquid cargo within tank trucks affects the lateral stability of the carrying vehicle and has been directly associated with 4% of heavy-truck road accidents.

A tank vehicle with partially filled tank(s) may roll over due to sloshing as a consequence of a fast transverse maneuver (cornering), such as rapidly changing lanes on a highway. When transverse sloshing has been excited, the sloshing mainly occurs with a period equal to the highest natural sloshing period and continues for many sloshing periods. The sloshing causes an oscillating transverse force and roll moment on the tanker vehicle that may cause the vehicle to roll over. The highest natural period depends on the tank geometry and the filling ratio. For instance, the highest natural period for transverse sloshing in a half-filled tank can be 2 s. If the time scales associated with the transverse maneuver are larger than the order of ten times the highest natural period, sloshing does not occur.

Winkler (2000) has given an overview of rollover of commercial vehicles, where he states: "The basic measure of roll stability is the static rollover threshold, expressed as lateral acceleration in gravitational units (g). The American Association of State Highway and Transportation Officials (AASHTO) guidelines for highway curve design result in lateral accelerations as high as $0.17g$ at the advised speed" (AASHTO Green Book, 1990). Figure 1.38 exemplifies the rollover threshold for tanker vehicles with circular and rectangular tanks. Both steady and transient turns are examined. The free surface is perpendicular to the combined forces of gravity and lateral acceleration during steady turn. The figure illustrates the reduced rollover threshold that occurs during transient conditions relative to the steady conditions.

A sudden stop of the vehicle causes longitudinal sloshing in the tank, which is particularly critical if the tank is unbaffled or is not divided into compartments. The latter is the case for transportation of food (e.g., milk) because of sanitary cleaning reasons. Shallow liquid types of waves develop for such tanks at a sudden stop of the vehicle. When the waves hit the end of the tank, they push the vehicle in the wave direction. The consequence is, for instance, that the waves can shove a stopped truck out into an intersection. Commercial driver's manuals contain warnings about the sloshing phenomena just described.

Sloshing in partially filled containers must also be considered in the dynamics of freight trains and can contribute to derailment and rollover when passing through tight turns. Examples of how to theoretically model a freight train with partially filled containers are presented by Bogomaz *et al.* (1998) and Vera *et al.* (2005). Bogomaz (2004) has written a book in Russian that is devoted entirely to this topic.

1.10 Onshore tanks

Sloshing-induced loads in onshore tanks due to earthquake motions must be considered in the tanks' design. Examples are oil storage and

Figure 1.39. Open-top fire in a naphtha tank caused by earthquake-induced sloshing (photograph courtesy of Sapporo Fire Bureau).

elevated water tanks. Acceleration spectra for vertical and horizontal accelerations define tank excitation, and these spectra are given in standards used by civil engineers (e.g., NS-EN 1998 and NS 3491–12). These spectra vary with geographical location. The main part of the motion energy is in the high-frequency sloshing, but significant energy may be present for frequencies in vicinity of the lowest sloshing frequency.

Housner (1957, 1963) developed a simplified method where an effective fluid mass accelerates with the container while an additional effective fluid mass undergoes resonant motions at the lowest sloshing frequency. He thus splits the forces into two components: an "impulsive" force with a short period and a longer "convective" force that gives the sloshing effect.

Hatayama (2008) described severe earthquake-induced sloshing damage to seven large oil storage tanks with floating roof structures as a consequence of the 2003 Tokachi-oki earthquake in northern Japan, which generated large-amplitude long-period (4–8-s) ground motions. The highest natural sloshing period in the tanks suffering severe damage ranged from 5 to 12 s. The damage occurred in terms of roof sinking and fire. Figure 1.39 illustrates a case resulting in an open-top fire.

1.11 Space applications

The sloshing technology developed for space applications is not directly applicable to ship cargo tanks because emphasis was placed on frequencies and total forces as they relate to control system requirements and, therefore, the effect of impact slosh pressure on structural requirements has not been studied to any extent. Furthermore, the excitation amplitudes considered in space applications are too small for ship-motion simulation (Abramson *et al.*, 1974).

A large percentage of the initial weight of boost or launch vehicles is fuel. A concern is that the dominant fuel-slosh frequencies are close to the control system frequencies and/or the elastic body bending frequencies. The consequence can be dynamic instabilities and large-amplitude response of the launch vehicles (Abramson, 1966). Vertical excitation of propellant tanks is discussed by Abramson (1966) and Dodge (2000).

The problem can be linked to Mathieu-type instabilities.

The fact that spacecraft orbiting the Earth are at close to weightless conditions significantly affects sloshing in fuel tanks. Surface tension plays a dominant role instead of gravity. Spacecraft and rockets can spin to increase stabilization and to moderate the effects of solar heating. Liquid motions in a spinning tank as well as the effects of weightless conditions are extensively discussed by Abramson (1966) and Dodge (2000) and are not discussed here.

1.12 Summary of chapters

Part of the presentation is theoretically oriented. Emphasis is placed on examples and exercises that explain the theory. Furthermore, experimental results are presented and partly used to validate theoretical and numerical results.

Chapter 2 presents the governing equations for sloshing inside a tank that are used in connection with CFD and analytically oriented methods. This approach includes Navier–Stokes equations and the Laplace equation for the velocity potential of irrotational motion of an incompressible liquid. Boundary conditions are presented. A tank-fixed coordinate system is used in the formulation. Needed background on the Lagrange variational formalism for the sloshing problem is also introduced.

Chapter 3 studies the coupling between sloshing and ship motions in a seaway. The first step is to analyze regular and irregular waves. Then a summary is given of how linear exterior hydrodynamic loads can be analyzed in the frequency and time domain and how statistical estimates can be made of response variables in short- and long-term predictions. It is demonstrated that sloshing can significantly affect ship motions in a seaway. The effects of antirolling tanks are examined in detail.

Sloshing in confined spaces with a free-surface exterior to a ship is discussed, including the case of piston-mode resonance in moonpools and between the hulls of a multihull ship.

Chapter 4 is about linear natural sloshing frequencies and modes. The very few analytical solutions (e.g., for two- and three-dimensional rectangular tanks and vertical circular cylinders) are derived. Variational formulations are used to express upper and lower bounds of natural frequencies for a general tank shape. It is shown how to evaluate the effect of interior structures and how to use variational formulation and extension of the solution outside the liquid domain to express natural frequencies in terms of analytical results, for instance for a rectangular tank. Other presented techniques are domain decomposition and shallow liquid approximations. Two-phase resonant liquid flow is also examined.

Chapter 5 presents linear modal theory for forced tank oscillations. Two- and three-dimensional rectangular tanks, upright circular cylinders, and spherical tanks are considered. The basis is irrotational flow of an incompressible liquid. Linear modal theory is of particular practical relevance for transient flow problems and has a special advantage when analytical solutions of the natural modes exist. Lukovsky's formulas are derived for linear hydrodynamic forces and moments on the tank.

A completely filled tank is examined and it is shown that the liquid motions do not behave as if "frozen" for angular tank motions.

Transient sloshing during the collision of two ships, the effect of elastic tank structure deformations on sloshing, and coupling between sloshing and wave-induced vibrations of a monotower are discussed. A simplified modal theory for transient sloshing in a two-dimensional circular tank and a spherical tank is introduced. The two-dimensional formulation is applied to rollover analysis of a tank vehicle.

Chapter 6 is about viscous wave loads and damping. When a tank is clean, the viscous effects are confined to a thin boundary layer along the tank surface. Both laminar and turbulent boundary-layer flows are discussed. It is shown that the associated damping of sloshing is very small for liquids of primary interest. The effect of the boundary layer on the potential flow is in terms of an inflow/outflow from the boundary layer.

The viscous loads on interior structures are formulated in terms of a generalized Morison equation with empirical mass and drag coefficients that depend on the structural shape and flow parameters such as the Reynolds number and the Keulegan–Carpenter number. Viscous damping due to baffles, screens, and vertical circular cylinders and plates are discussed. The damping due to wire-mesh screens is shown to be particularly significant. TLDs consisting of a rectangular tank with different interior structures are discussed. It is demonstrated that linear potential flow theory with viscous damping due to a wire-mesh screen in a rectangular tank can predict resonant wave amplitudes, force amplitudes, and phases for forced longitudinal tank motions very well. However, it is emphasized that the damping is relatively large and the forcing amplitudes are relatively small.

The effect of swash bulkheads and screens with high solidity ratio is discussed based on a theoretical model.

Chapter 7 gives the general theoretical background for the analytically based nonlinear multimodal methods to be described in Chapters 8 and 9. The Bateman–Luke variational theory is used to formulate a general, infinite nonlinear modal theory. Corresponding Lukovsky's formulas for hydrodynamic forces and moments are derived.

Chapter 8 considers two-dimensional flow in a rectangular tank with finite liquid depth. Moiseev's method is presented. How to order terms in an analytically based nonlinear multimodal method is discussed. An important consideration is the presence of secondary resonance. Finite, critical, intermediate, and shallow-liquid conditions are examined. It is shown that steady-state subharmonic behavior may occur at the critical depth. The steady-state hydraulic jump theory by Verhagen and van Wijngaarden (1965) is presented. The Mathieu instability for vertical tank excitation is discussed. The loads on interior structures are also dealt with, and experimental results are provided.

Chapter 9 applies the analytically based nonlinear multimodal method to study sloshing in vertical circular cylinders and three-dimensional rectangular tanks. Swirling, diagonal waves, and "chaos" are discussed. Secondary resonance and its corresponding adaptive modal system are presented. Experimental results for hydrodynamic lateral forces acting on spherical tanks with different filling levels and as a function of excitation amplitude and frequency are presented. Experimental results for forces on a tower inside a spherical tank are also discussed.

Chapter 10 is about CFD and sloshing. The broad variety of CFD methods is discussed, including details about the boundary element, finite difference, finite volume, finite element, and smoothed particle hydrodynamics methods. Special attention is given to interface-tracking and interface-capturing methods. Advantages and disadvantages of CFD methods are discussed.

Chapter 11 deals with slamming loads and their effect on structural stresses. Sidewall impact with wave breaking and roof impact are considered. Gas cushions and flip-through may occur. The effects of viscosity and gas cushion and liquid compressibility are discussed. Model tests are commonly used to assess slamming loads. When gas cavities influence the slamming loads, it is emphasized that both Euler and Froude numbers must be the same at model and full scales. The importance of hydroelasticity is emphasized.

2 Governing Equations of Liquid Sloshing

2.1 Introduction

This chapter contains the governing equations and boundary conditions that will be used in connection with computational fluid dynamics (CFD) and analytically oriented methods for sloshing inside a tank. A broad variety of CFD methods exists that may be based on Navier–Stokes equations, Euler equations, or the Laplace equation for velocity potential (see Chapter 10). The Euler equations follow from the Navier–Stokes equations by neglecting viscosity. A liquid can for most practical cases be considered incompressible, whereas compressibility is important in describing the behavior of a gas. For instance, CFD methods based on either Navier–Stokes or Euler equations may use the finite difference method, the finite volume method, the finite element method, or the smoothed particle hydrodynamics method to solve the field equations. In this chapter we present the governing equations for these methods in a *tank-fixed, noninertial coordinate system.*

The Laplace equation for velocity potential is also the basis for the analytically oriented methods considered in this book. These methods can explain many nonlinear sloshing phenomena in a very efficient way; however, they (e.g., multimodal methods; see Chapters 7, 8, and 9) cannot be applied in cases where wave breaking and flow separation have an important effect on sloshing. Multimodal methods can be applied directly to the free-boundary sloshing problem, but, in major cases, these methods employ variational formulations based on the Lagrange or Bateman–Luke principles. This chapter presents the needed background on these principles.

2.2 Navier–Stokes equations

Navier–Stokes equations are presented in many textbooks on fluid mechanics (e.g., Newman, 1977; Schlichting, 1979; White, 1974). We limit the derivation to two-dimensional flow of an incompressible fluid and refer to the aforementioned textbooks for a detailed and general derivation of Navier–Stokes equations. For our applications, liquids can in most cases be considered incompressible (i.e., sound waves do not matter).

We begin by deriving the Navier–Stokes equations for an incompressible fluid in an inertial system. The Navier–Stokes equations are the basis for our discussions on viscous flow effects in Chapter 6 and on the presentation of the different CFD methods in Chapter 10.

2.2.1 Two-dimensional Navier–Stokes formulation for incompressible liquid

We introduce a Cartesian and inertial coordinate system Oxy. The two-dimensional Navier–Stokes equations for an incompressible fluid without gravity can be written as

$$\frac{\partial u}{\partial t} + u\frac{\partial u}{\partial x} + v\frac{\partial u}{\partial y} = -\frac{1}{\rho}\frac{\partial p}{\partial x} + \nu\left(\frac{\partial^2 u}{\partial x^2} + \frac{\partial^2 u}{\partial y^2}\right), \tag{2.1}$$

$$\frac{\partial v}{\partial t} + u\frac{\partial v}{\partial x} + v\frac{\partial v}{\partial y} = -\frac{1}{\rho}\frac{\partial p}{\partial y} + \nu\left(\frac{\partial^2 v}{\partial x^2} + \frac{\partial^2 v}{\partial y^2}\right). \tag{2.2}$$

The continuity equation is

$$\frac{\partial u}{\partial x} + \frac{\partial v}{\partial y} = 0, \tag{2.3}$$

where u and v are the x- and y-components of the fluid velocity vector $\boldsymbol{v}$, t is the time variable, and p is the pressure, here referred to in an inertial system. Furthermore, ρ is the density of the fluid and ν is the kinematic viscosity coefficient. The relationship between ν and the dynamic viscosity coefficient μ is

$$\nu = \mu/\rho. \tag{2.4}$$

Values of ρ and ν for different liquids used in model tests or as cargo in ships are presented in Tables 11.1 and 11.2, respectively. We have three equations and three unknowns, u, v, and p. We need a set of initial and boundary conditions to solve equations (2.1)–(2.3) (see, e.g., Section 2.2.2).

2.2.1.1 Continuity equation

The continuity equation can be derived by considering a small fluid domain ΔQ that does not change with time. The time rate of change of the mass in the fluid domain is zero for an incompressible fluid, which is the same as saying that the total mass flux into the fluid domain is zero. Because the density is constant for an incompressible fluid, it follows that

$$\int_{S_{\Delta Q}} \boldsymbol{v} \cdot \boldsymbol{n} \, \mathrm{d}S = 0, \qquad (2.5)$$

where $\boldsymbol{n}$ is the outward normal vector of the surface $S_{\Delta Q}$ enclosing ΔQ. Surface integral (2.5) can be rewritten as a volume integral by means of the divergence theorem (see eq. (A.1)). This transforms eq. (2.5) into the following equation:

$$\int_{\Delta Q} \nabla \cdot \boldsymbol{v} \, \mathrm{d}Q = 0.$$

Because the preceding equation should be satisfied for arbitrary ΔQ, it follows that

$$\nabla \cdot \boldsymbol{v} = 0, \qquad (2.6)$$

which is the same as eq. (2.3) for two-dimensional flow.

2.2.1.2 Viscous stresses and derivation of the Navier–Stokes equations

Eqs. (2.1) and (2.2) follow by analyzing the motion inside an arbitrary fluid domain and enforcing that the time rate of change of momentum inside the fluid domain is equal to the sum of forces acting on the fluid domain (i.e., Newton's second law). These are forces due to hydrodynamic pressure and viscous stresses. Concerning the hydrodynamic pressure contribution, the force per unit area due to pressure p acts perpendicularly to a surface element as $-p\boldsymbol{n}$, where $\boldsymbol{n}$ is the outward normal vector.

To introduce the viscous stresses we consider a two-dimensional rectangular fluid domain with sides parallel to the x- and y-axes (see Figure 2.1). On the top side of the domain, AB, we have the viscous stress components τ_{xy} and τ_{yy} along the x- and y-axis, respectively. They can be expressed as

$$\tau_{xy} = \mu\left(\frac{\partial u}{\partial y} + \frac{\partial v}{\partial x}\right) \quad \text{and} \quad \tau_{yy} = 2\mu\frac{\partial v}{\partial y}. \qquad (2.7)$$

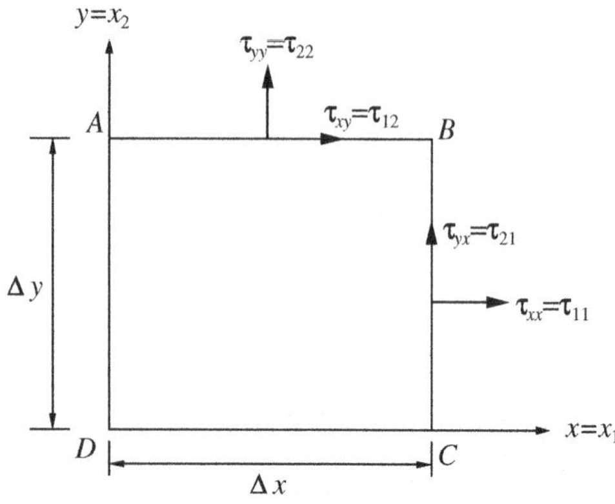

Figure 2.1. Rectangular fluid domain $ABCD$ and viscous stresses acting on sides AB and BC.

The viscous stress components on the vertical side BC are

$$\tau_{yx} = \mu\left(\frac{\partial v}{\partial x} + \frac{\partial u}{\partial y}\right) \quad \text{and} \quad \tau_{xx} = 2\mu\frac{\partial u}{\partial x}, \qquad (2.8)$$

where τ_{yx} and τ_{xx} are stress components directed along the y- and x-axis, respectively.

To express the stresses in a more abbreviated and general way in three dimensions, we change notation so that $x = x_1$, $y = x_2$, $z = x_3$, $u = u_1$, $v = u_2$, and $w = u_3$ and introduce τ_{ij}, where i or j is equal to 1, 2, or 3 when referring to x, y, and z, respectively. This notation is depicted in Figure 2.1 for the two-dimensional case. Here we have also included a third dimension by introducing the z-coordinate of the Cartesian coordinate system (x, y, z) and the velocity component $w = u_3$ along the z-axis. We define a surface element with an outward unit normal vector $\boldsymbol{n} = (n_1, n_2, n_3)$. If this surface element belongs to a side of a fluid domain as in Figure 2.1, then $\boldsymbol{n}$ is pointing outward from the fluid domain. If we consider a surface element on a body surface, then the normal direction is into the fluid domain. The viscous stress (force per unit area) in the ith direction is then

$$\tau_{i1}n_1 + \tau_{i2}n_2 + \tau_{i3}n_3, \qquad (2.9)$$

where

$$\tau_{ij} = \mu\left(\frac{\partial u_i}{\partial x_j} + \frac{\partial u_j}{\partial x_i}\right). \qquad (2.10)$$

We note the symmetry $\tau_{ij} = \tau_{ji}$. The justification of this linear relationship between viscous stresses and derivatives of velocity components is discussed by Newman (1977). Most common fluids, like water and air, satisfy the Newtonian stress relations given by eq. (2.10). Examples of non-Newtonian fluids are molasses (syrup), honey, ketchup, and mud (Chhabra & Richardson, 1999; Chhabra, 2006).

We now demonstrate that eqs. (2.9) and (2.10) are consistent with eq. (2.7). For example, let us consider the top side AB in Figure 2.1, where $n_1 = 0$, $n_2 = 1$, and $n_3 = 0$. This means that we have the viscous stress components $\mu\,(\partial u_1/\partial x_2 + \partial u_2/\partial x_1)$ and $2\mu \partial u_2/\partial x_2$ (i.e., the same as eq. (2.7)). With the same procedure at the bottom side, CD, where $n_1 = 0$, $n_2 = -1$, and $n_3 = 0$, we find the viscous stress components $-\mu\,(\partial u_1/\partial x_2 + \partial u_2/\partial x_1)$ and $-2\mu \partial u_2/\partial x_2$ directed along the x- and y-axis, respectively. These expressions are similar to eq. (2.7) but have opposite signs. This has to be kept in mind for our next derivation of eqs. (2.1) and (2.2).

As mentioned, eqs. (2.1) and (2.2) follow from Newton's second law. We shall focus on eq. (2.1) and have in mind the fluid domain in Figure 2.1. The sides Δx and Δy are assumed small so that all quantities can be approximated by the lowest-order terms in a Taylor expansion about the center of the domain. We first evaluate the forces acting on the domain. The resultant viscous force components in the x-direction acting on AD and BC and along AB and DC can then be approximated as

$$\Delta x \frac{\partial}{\partial x}\left(2\mu \frac{\partial u}{\partial x}\right)\Delta y \quad \text{and} \quad \Delta y \frac{\partial}{\partial y}\mu\left(\frac{\partial u}{\partial y} + \frac{\partial v}{\partial x}\right)\Delta x, \tag{2.11}$$

respectively. The sum of the components can finally be rewritten as

$$\mu\left(\frac{\partial^2 u}{\partial x^2} + \frac{\partial^2 u}{\partial y^2}\right)\Delta x \Delta y \tag{2.12}$$

by means of continuity equation (2.3). By a similar Taylor expansion the pressure force on the surface of the fluid can be approximated as $-\Delta x \Delta y \partial p/\partial x$.

Then we consider the time rate of change of fluid momentum in the x-direction. Part of this is due to momentum flux through AB, BC, CD, and DA. The momentum flux through a surface element that is not moving is

$$\rho \boldsymbol{v}(\boldsymbol{v} \cdot \boldsymbol{n})\, \mathrm{d}S, \tag{2.13}$$

where $\boldsymbol{v} = (u, v, w)$ and $\mathrm{d}S$ is the area of the surface element. Again using a Taylor expansion, we find that the momentum flux in the x-direction through AD and BC and through AB and CD can be approximated as

$$\rho \Delta x \frac{\partial}{\partial x}(u^2)\Delta y \quad \text{and} \quad \rho \Delta y \frac{\partial}{\partial y}(uv)\,\Delta x, \tag{2.14}$$

respectively. The sum can, by means of continuity equation (2.3), be rewritten as

$$\rho\left(u\frac{\partial u}{\partial x} + v\frac{\partial u}{\partial y}\right)\Delta x \Delta y. \tag{2.15}$$

Then we have to add the term $\rho \Delta x \Delta y \partial u/\partial t$ to get all the contributions to the time rate of change of the fluid momentum in the x-direction inside the domain. These contributions must be balanced by the forces acting on the fluid domain. By doing this we find the following equation:

$$\rho\left(\frac{\partial u}{\partial t} + u\frac{\partial u}{\partial x} + v\frac{\partial u}{\partial y}\right)\Delta x \Delta y = -\frac{\partial p}{\partial x}\Delta x \Delta y + \mu\left(\frac{\partial^2 u}{\partial x^2} + \frac{\partial^2 u}{\partial y^2}\right)\Delta x \Delta y,$$

which is a first-order equation valid for small Δx and Δy. By dividing with $\rho \Delta x \Delta y$ on both sides and then letting Δx and Δy go to zero, we see that this leads to eq. (2.1).

2.2.2 Three-dimensional Navier–Stokes equations

Let us now introduce the third dimension and introduce the gravity. The Navier–Stokes equations in vector form can be expressed as

$$\rho \frac{D\boldsymbol{v}}{Dt} = -\nabla p + \rho \boldsymbol{g} + \mu \nabla^2 \boldsymbol{v}, \tag{2.16}$$

where $\boldsymbol{g}$ is the gravitational acceleration vector, $\boldsymbol{v}$ is the fluid velocity, and D/Dt is the material (substantial) derivative, which expresses the rate of change with time when we follow a fluid particle; that is,

$$\frac{D}{Dt} = \frac{\partial}{\partial t} + \boldsymbol{v} \cdot \nabla. \tag{2.17}$$

Equation (2.16) can also be expressed as

$$\rho \frac{D\boldsymbol{v}}{Dt} = -\nabla p + \rho \boldsymbol{g} + \nabla \cdot \left(\tau_{ij} \boldsymbol{e}_i \boldsymbol{e}_j\right), \tag{2.18}$$

where τ_{ij} is defined by eq. (2.10) and $\boldsymbol{e}_i$ and $\boldsymbol{e}_j$ with i and j from 1 to 3 are unit vectors along the coordinate axis. The interpretation of $\tau_{ij}\boldsymbol{e}_i\boldsymbol{e}_j$ is that i and j should be independently summed from 1 to

3; $e_i e_j$ is a dyad (pair of vectors) and not a vector. However, $\nabla \cdot \tau_{ij} e_i e_j$ is a vector. Equation (2.18) is useful when we later derive the equations for conservation of fluid momentum.

To solve the Navier–Stokes equations together with continuity equation (2.6), we need to specify initial and boundary conditions. The boundary condition on a body specifies that there is no slip between the body and the fluid. Furthermore, we must require that the fluid does not penetrate the body. Continuity in the stress is required on the interface between a gas and a liquid. Furthermore, a fluid particle on the free surface will remain on the interface. We will be more specific about initial and boundary conditions when we later consider special cases. Analytical solutions of the Navier–Stokes equations generally do not exist, and we must to some extent rely on numerical CFD techniques. However, CFD has its own limitations due to existing computational speed and the fact that we do not have a complete understanding of all physical phenomena in fluid dynamics such as turbulence, ventilation, and cavitation. It is difficult to solve Navier–Stokes equations because of the nonlinearities associated with the convective acceleration term $\boldsymbol{v} \cdot \nabla \boldsymbol{v}$. Further complications are due to nonlinear free-surface conditions in our problem.

If the dynamic viscosity coefficient μ is set equal to zero, eqs. (2.16) and (2.18) are referred to as *Euler equations*.

2.2.2.1 Vorticity and potential flow

Vorticity is associated with viscous flow. The vorticity is generated by viscous shear effects in the boundary layer, in spilling breakers and when overturning waves hit the underlying liquid. The vorticity $\boldsymbol{\varpi}$ is a vector defined by

$$\boldsymbol{\varpi} = \nabla \times \boldsymbol{v} = \boldsymbol{e}_1 \left(\frac{\partial w}{\partial y} - \frac{\partial v}{\partial z} \right) + \boldsymbol{e}_2 \left(\frac{\partial u}{\partial z} - \frac{\partial w}{\partial x} \right) + \boldsymbol{e}_3 \left(\frac{\partial v}{\partial x} - \frac{\partial u}{\partial y} \right), \quad (2.19)$$

where, as usual, $\boldsymbol{v} = (u, v, w)$ is the fluid velocity and $\boldsymbol{e}_1$, $\boldsymbol{e}_2$, and $\boldsymbol{e}_3$ are unit vectors along the x-, y-, and z-axis, respectively. By using the Navier–Stokes equations, we can derive partial differential equations (White, 1974) that describe how vorticity is advected (convected) with the flow velocity and diffused due to viscosity.

If the flow has no vorticity, a velocity potential Φ can be introduced so that

$$\boldsymbol{v} = \nabla \Phi = \frac{\partial \Phi}{\partial x} \boldsymbol{e}_1 + \frac{\partial \Phi}{\partial y} \boldsymbol{e}_2 + \frac{\partial \Phi}{\partial z} \boldsymbol{e}_3. \quad (2.20)$$

This is a simple consequence of the fact that $\nabla \times \nabla f = 0$ for any analytical function f. We can now take into account expression (2.20) in the continuity equation for incompressible fluid, eq. (2.6):

$$\frac{\partial u}{\partial x} + \frac{\partial v}{\partial y} + \frac{\partial w}{\partial z} = 0, \quad (2.21)$$

which gives

$$\nabla^2 \Phi = \frac{\partial^2 \Phi}{\partial x^2} + \frac{\partial^2 \Phi}{\partial y^2} + \frac{\partial^2 \Phi}{\partial z^2} = 0. \quad (2.22)$$

Equation (2.22) expresses the fact that the velocity potential for an incompressible fluid satisfies the three-dimensional Laplace equation. If we substitute presentation (2.20) into Euler's equations and integrate these equations in space, we get Bernoulli's equation for the pressure p, that is,

$$p + \rho \left[\frac{\partial \Phi}{\partial t} + \tfrac{1}{2} |\nabla \Phi|^2 + U_g \right] = C(t), \quad (2.23)$$

where $C(t)$ is a time-dependent constant to be determined by initial and boundary conditions. Furthermore, U_g is the gravity potential. Let us express U_g by allowing for a general orientation of the gravitational acceleration vector $\boldsymbol{g}$ relative to a coordinate system $Oxyz$. This is important because we later operate with a tank-fixed coordinate system. We then start by expressing the gravitational acceleration vector in the Euler equations as

$$\boldsymbol{g} = g_1 \boldsymbol{e}_1 + g_2 \boldsymbol{e}_2 + g_3 \boldsymbol{e}_3 = \nabla (g_1 x + g_2 y + g_3 z). \quad (2.24)$$

The gravity potential U_g appears after integrating the Euler equations in space to get Bernoulli's equation. It is defined within to a time-varying function and may be expressed as

$$U_g = -g_1 x - g_2 y - g_3 z = -\boldsymbol{g} \cdot \boldsymbol{r}, \quad (2.25)$$

where $\boldsymbol{r} = x\boldsymbol{e}_1 + y\boldsymbol{e}_2 + z\boldsymbol{e}_3$.

The corresponding free-boundary-value problem is formulated in Section 2.4.2 and is the basis for the multimodal method that is elaborated later in the book.

2.2.2.2 Compressibility

Compressibility matters for gas flow. However, we can, for most practical cases of sloshing in ship tanks, consider the liquid incompressible. An exception is the case of a mixture between liquid and gas (see Sections 11.6 and 11.8). A short description of the effect of compressibility on the equations of fluid flow is given, assuming an adiabatic process. The consequence is that pressure p is only a function of the density ρ of the fluid; that is, mathematically, $p = p(\rho)$. The inverse relationship is also true; that is, $\rho = \rho(p)$. The density is, in general, a function of x, y, z, and t.

For a compressible fluid, the continuity equation is modified. As in Section 2.2.1, it can be derived by considering a small fluid control domain ΔQ that does not change with time. The time rate of change of the mass in the fluid domain can be related to the mass flux in and out of the fluid domain and can mathematically be expressed as

$$\frac{d}{dt}\int_{\Delta Q} \rho \, dQ = -\int_{S_{\Delta Q}} \rho \boldsymbol{v} \cdot \boldsymbol{n} \, dS, \tag{2.26}$$

where $\boldsymbol{n}$ is the outward normal vector. Because the fluid control domain does not change with time, we have

$$\frac{d}{dt}\int_{\Delta Q} \rho \, dQ = \int_{\Delta Q} \frac{\partial \rho}{\partial t} dQ$$

and, using the divergence theorem (see eq. (A.1)) in eq. (2.26), we get

$$\int_{\Delta Q} \left(\frac{\partial \rho}{\partial t} + \nabla \cdot (\rho \boldsymbol{v}) \right) dQ = 0$$

for arbitrary domain ΔQ. This means that

$$\frac{\partial \rho}{\partial t} + \nabla \cdot (\rho \boldsymbol{v}) = 0. \tag{2.27}$$

The Navier–Stokes equation is also modified to include a bulk viscosity coefficient. The viscous stress tensor is then expressed as (White, 1974)

$$\tau_{ij} = \mu \left(\frac{\partial u_i}{\partial x_j} + \frac{\partial u_j}{\partial x_i} \right) + \delta_{ij} \zeta \nabla \cdot \boldsymbol{v}, \tag{2.28}$$

where δ_{ij} is the Kronecker delta and ζ is the coefficient of bulk viscosity. Some CFD methods assume numerically that the interface between the liquid and the gas has a finite thickness. The consequence of this assumption is that the viscosity coefficient between liquid and gas varies continuously through the interface between the liquid and the gas. The spatial derivatives of μ and ζ will then matter when eq. (2.28) is inserted into the Navier–Stokes equations.

The equation of state expressing the relationship between pressure and density can be rewritten by the continuity equation. We begin by expressing eq. (2.27) as

$$\frac{\partial \rho}{\partial t} + \boldsymbol{v} \cdot \nabla \rho + \rho \nabla \cdot \boldsymbol{v} = 0.$$

We now use the fact that $\frac{\partial \rho}{\partial t} = \frac{\partial \rho}{\partial p}\frac{\partial p}{\partial t}$ and $\frac{\partial \rho}{\partial x_i} = \frac{\partial \rho}{\partial p}\frac{\partial p}{\partial x_i}$. The continuity equation can then be rewritten as

$$\frac{\partial p}{\partial t}\frac{d\rho}{dp} + \frac{d\rho}{dp}\boldsymbol{v} \cdot \nabla p + \rho \nabla \cdot \boldsymbol{v} = 0,$$

which gives

$$\frac{\partial p}{\partial t} + u_i \frac{\partial p}{\partial x_i} = -\rho c^2 \frac{\partial u_i}{\partial x_i}, \tag{2.29}$$

where

$$c = \sqrt{dp/d\rho}. \tag{2.30}$$

In this book, we will study linear acoustic applications. This assumes linearization of the governing equations relative to the liquid being at rest. To avoid nonlinearities, the quantity $c^2 = dp/d\rho$ in eq. (2.29) must then be considered at $\rho = \rho_0$. The value c then becomes a constant independent of ρ. This constant value c_0 is normally denoted as the speed of sound. It is common to express the speed of sound as

$$c_0 = \sqrt{E_v/\rho}, \tag{2.31}$$

where E_v is the bulk modulus for elasticity. Values of the speed of sound for different liquids used in model tests or as cargo in ships are presented in Tables 11.1 and 11.2, respectively. A mixture of gas and liquid can significantly influence the speed of sound.

As a special case we assume irrotational flow of a compressible fluid. The fact that the flow is irrotational means that a velocity potential Φ exists so that the fluid velocity is given by $\boldsymbol{v} = \nabla \Phi$. The equations are linearized by assuming that p and $\boldsymbol{v}$ are small quantities. Gravity is neglected and the pressure is approximated as $p = -\rho \partial \Phi / \partial t + p_0$, where p_0 is constant ambient pressure. Then eq. (2.29) can be approximated as

$$\frac{\partial^2 \Phi}{\partial t^2} - c_0^2 \nabla^2 \Phi = 0. \tag{2.32}$$

An equation of the preceding form is called a *wave equation*. We refer to eq. (2.32) in

Section 10.5.3 when we study forced acoustic waves in a chamber and in Section 11.8 when we account for acoustic liquid effects on slamming.

2.2.3 Turbulent flow

The Reynolds number, $Rn = UL/\nu$, is an important parameter for determining when turbulent flow occurs; here L is a characteristic length and U is a characteristic velocity. For instance, when we consider steady ambient flow past a circular cylinder, U and L are typically chosen as the ambient flow velocity and the cylinder diameter, respectively. If a nonseparated oscillating boundary-layer flow is examined, then U and L are respectively the flow velocity amplitude and the fluid particle amplitude at the outskirts of the boundary layer (see Section 6.2). When Rn is larger than a transition Reynolds number Rn_{tr} for a specific flow problem, the flow is turbulent.

The flow is turbulent in practical full-scale conditions. When it comes to model test conditions, the flow is laminar in the boundary layer along the tank walls. However, viscous flow away from the boundary layer can easily be turbulent. An example is separated flow from internal objects, for instance, due to cross-flow past a cooling tower in a liquefied natural gas tank or internal baffles. In principle we can directly use Navier–Stokes equations and solve them numerically to study the turbulent flow. However, a very fine time and spatial discretization is needed in the numerical solution. Currently available computer technology limits the possibilities. Instead Reynolds-averaged Navier–Stokes (RANS) formulations are commonly used.

The RANS formulation means that we decompose the flow velocity and pressure into one part, which varies on the time scale of turbulence, and into another part, which is time averaged over the time scale of turbulence. Then we for incompressible fluid insert this result into eqs. (2.6) and (2.16) and time-average the equations over the time scale of turbulence. The so-called Reynolds stress terms then appear from time-averaging the convective acceleration terms. These terms are new unknowns that must be related to the turbulent averaged flow velocities. Then we need new equations. In practice these are empirical (i.e., we need guidance from experiments; see Schlichting (1979) for further details). In the following text we do not include the RANS formulation; we keep with the original version of the Navier–Stokes equations, which are suitable for laminar flow.

2.2.4 Global conservation laws

2.2.4.1 Conservation of fluid momentum

We begin by expressing the equation for conservation of momentum in a fluid, $\boldsymbol{M}(t) = (M_1, M_2, M_3)$, in a general way. The fluid is assumed incompressible. An inertial coordinate system is considered. Let $S_Q(t)$ be a closed surface that encloses a fluid domain $Q(t)$. The momentum inside $Q(t)$ can then be written as

$$\boldsymbol{M}(t) = \int_{Q(t)} \rho \boldsymbol{v} \, \mathrm{d}Q, \tag{2.33}$$

where $\boldsymbol{v} = (u_1, u_2, u_3)$ is the fluid velocity. The enclosing surface S_Q does not need to follow the fluid motion.

By using the definition of a derivative and noting that both the volume and the velocity may change with time, we can apply Reynolds transport theorem (A.2) to $\boldsymbol{M}(t)$:

$$\dot{\boldsymbol{M}} = \rho \int_{Q(t)} \frac{\partial \boldsymbol{v}}{\partial t} \mathrm{d}Q + \rho \int_{S_Q(t)} \boldsymbol{v} U_{sn} \, \mathrm{d}S, \tag{2.34}$$

where U_{sn} is the normal component of the velocity of the surface S_Q. The overdot indicates a time derivative. The volume integral in eq. (2.34) can be rewritten by expressing $\partial \boldsymbol{v}/\partial t$ by the Navier–Stokes equations

$$\rho \left(\frac{\partial \boldsymbol{v}}{\partial t} + \boldsymbol{v} \cdot \nabla \boldsymbol{v} \right) = -\nabla (p + \rho g z) + \nabla \cdot \tau_{ij} \boldsymbol{e}_i \boldsymbol{e}_j, \tag{2.35}$$

where z is vertical coordinate and the z-axis is positive upward with $z = 0$ on the mean free surface.

By using vector algebra, we can show that

$$\nabla \cdot (\boldsymbol{v}\boldsymbol{v}) = \boldsymbol{v} \cdot \nabla \boldsymbol{v} \tag{2.36}$$

for an incompressible fluid. This result follows from using the summation convention, a Cartesian coordinate system $Ox_1x_2x_3$ with unit vectors $\boldsymbol{e}_i$ $(i = 1, 3)$, and expressing $\nabla \cdot (\boldsymbol{v}\boldsymbol{v})$ as $\boldsymbol{e}_i \frac{\partial}{\partial x_i} \cdot (u_j \boldsymbol{e}_j u_k \boldsymbol{e}_k)$, where i, j, and k imply independent summation from 1 to 3. The spatial derivatives of the unit vectors are zero in a Cartesian coordinate system. We perform first the dot multiplication between $\boldsymbol{e}_i$ and $\boldsymbol{e}_j$ and then differentiate the expression, giving $\frac{\partial u_j}{\partial x_j} u_k \boldsymbol{e}_k + u_j \frac{\partial u_k}{\partial x_j} \boldsymbol{e}_k$, which is equal to $(\nabla \cdot \boldsymbol{v})\, \boldsymbol{v} + \boldsymbol{v} \cdot \nabla \boldsymbol{v}$. Because $\nabla \cdot \boldsymbol{v} = 0$ for an incompressible fluid, eq. (2.36) follows.

The volume integral in eq. (2.34) can be reduced to a surface integral by using generalized Gauss theorem (A.1), which gives

$$\begin{aligned}\dot{\boldsymbol{M}} &= -\int_{S_Q} p\boldsymbol{n}\,\mathrm{d}S - \rho g\int_{S_Q} z\boldsymbol{n}\,\mathrm{d}S \\ &\quad -\rho\int_{S_Q} \boldsymbol{v}\,(u_n - U_{sn})\,\mathrm{d}S \\ &\quad +\int_{S_Q} \boldsymbol{n}\cdot\tau_{ij}\boldsymbol{e}_i\boldsymbol{e}_j\,\mathrm{d}S, \end{aligned} \tag{2.37}$$

where u_n is the normal component of the fluid velocity on the surface S_Q.

Let us apply the formula to liquid motion inside a tank. The control domain $Q(t)$ coincides in this case with the liquid domain inside a tank; U_{sn} and u_n are then equal both on the free surface and on the wetted body surface. The air (gas) is assumed stagnant relative to the tank. The free surface condition in the presence of no surface tension is that the tangential hydrodynamic stresses are zero and the normal hydrodynamic stresses at the free surface are equal to the ambient gas pressure p_0. Because a constant pressure p_0 gives zero contribution in the pressure term in eq. (2.37), we can extract p_0 from p in eq. (2.37). The consequence is that the sum of the pressure and viscous stress terms in eq. (2.37) are the same as the negative of the force $\boldsymbol{F}$ due to the liquid acting on the tank boundaries. We can further write $-\rho g\int_{S_Q} z\boldsymbol{n}\,\mathrm{d}S = -M_l g\boldsymbol{e}_3$, where M_l is the liquid mass. As a consequence, the resulting force is

$$\boldsymbol{F} = -\dot{\boldsymbol{M}} - M_l g\boldsymbol{e}_3 = -\dot{\boldsymbol{M}} + M_l\boldsymbol{g}. \tag{2.38}$$

If steady-state conditions with a period T are assumed, the time average of $\dot{\boldsymbol{M}}$ over one period is

$$\begin{aligned}\langle\dot{\boldsymbol{M}}\rangle &= \frac{1}{T}\int_{t_0}^{t_0+T}\dot{\boldsymbol{M}}\,\mathrm{d}t \\ &= \frac{1}{T}\left[\boldsymbol{M}(t_0+T) - \boldsymbol{M}(t_0)\right] = 0.\end{aligned}$$

The consequence is that the time average of force $\boldsymbol{F}$ over one period is the weight of the liquid (i.e., no mean hydrodynamic force acts on the tank). This result is different from exterior fluid problems, where a mean hydrodynamic force exists due to far-field wave radiation by the body (Faltinsen, 1990). Viscous loads may also cause mean hydrodynamic loads in the exterior flow problem.

The FVM (see Section 10.4) expresses the Navier–Stokes equations in an equation that follows from conservation of fluid momentum. We consider a small fluid domain ΔQ and let the boundary surface S_Q be fixed so that $U_{sn} = 0$, which gives

$$\begin{aligned}\frac{d}{dt}\int_{\Delta Q}\rho\boldsymbol{v}\,\mathrm{d}Q &= -\int_{S_Q} p\boldsymbol{n}\,\mathrm{d}S - \rho g\int_{S_Q} z\boldsymbol{n}\,\mathrm{d}S \\ &\quad -\rho\int_{S_Q}\boldsymbol{v}u_n\,\mathrm{d}S + \int_{S_Q}\boldsymbol{n}\cdot\tau_{ij}\boldsymbol{e}_i\boldsymbol{e}_j\,\mathrm{d}S.\end{aligned}$$

The continuity equation for an incompressible fluid is expressed in integral form (2.5), where $\boldsymbol{v}\cdot\boldsymbol{n} = u_n$. Here the normal vector direction is out of the liquid domain.

2.2.4.2 Conservation of kinetic and potential fluid energy

We begin by expressing the conservation of kinetic and potential fluid energy in a general way as was done for the conservation of fluid momentum. The fluid is assumed *incompressible.* The kinetic and potential fluid energy inside the closed surface can be expressed as

$$\begin{aligned}E(t) &= \rho\int_{Q(t)}\left[\tfrac{1}{2}\left(u_1^2 + u_2^2 + u_3^2\right) + gz\right]\mathrm{d}Q \\ &= \rho\int_{Q(t)}\left[\tfrac{1}{2}\boldsymbol{v}\cdot\boldsymbol{v} + gz\right]\mathrm{d}Q.\end{aligned} \tag{2.39}$$

We have chosen the z-axis to be vertical upward. We can show, in a similar way as in deriving eq. (2.37), that

$$\begin{aligned}\dot{E}(t) &= \rho\int_{Q(t)}\boldsymbol{v}\cdot\frac{\partial\boldsymbol{v}}{\partial t}\mathrm{d}Q \\ &\quad +\rho\int_{S_Q}\left[\tfrac{1}{2}\boldsymbol{v}\cdot\boldsymbol{v} + gz\right]U_{sn}\,\mathrm{d}S.\end{aligned} \tag{2.40}$$

The term $\partial\boldsymbol{v}/\partial t$ can be rewritten by means of the Navier–Stokes equations for an incompressible fluid and can be expressed in component form as

$$\frac{\partial u_i}{\partial t} = -u_j\frac{\partial u_i}{\partial x_j} - \frac{1}{\rho}\frac{\partial}{\partial x_i}(p + \rho g x_3) + \frac{1}{\rho}\frac{\partial\tau_{ij}}{\partial x_j}, \tag{2.41}$$

where the viscous stress components τ_{ij} are given by eq. (2.10) and $x_1 = x$, $x_2 = y$, $x_3 = z$. Furthermore, a conventional summation is used, meaning that

$$\frac{\partial\tau_{ij}}{\partial x_j} = \frac{\partial\tau_{i1}}{\partial x_1} + \frac{\partial\tau_{i2}}{\partial x_2} + \frac{\partial\tau_{i3}}{\partial x_3}. \tag{2.42}$$

Based on Landau and Lifschitz (1959) we can write

$$\rho \boldsymbol{v} \cdot \frac{\partial \boldsymbol{v}}{\partial t} = -\rho \boldsymbol{v} \cdot (\boldsymbol{v} \cdot \nabla) \boldsymbol{v} - \boldsymbol{v} \cdot \nabla(p + \rho g z) + u_i \frac{\partial \tau_{ij}}{\partial x_j}$$
$$= -\rho (\boldsymbol{v} \cdot \nabla) \left(\tfrac{1}{2} \boldsymbol{v} \cdot \boldsymbol{v} + \frac{p}{\rho} + g z \right)$$
$$+ \nabla \cdot (\boldsymbol{v} \cdot \boldsymbol{\tau}) - \tau_{ij} \frac{\partial u_i}{\partial x_j}, \tag{2.43}$$

where $\boldsymbol{v} \cdot \boldsymbol{\tau}$ is a vector with components $u_1 \tau_{1j} + u_2 \tau_{2j} + u_3 \tau_{3j}$, since $\boldsymbol{\tau}$ is a dyad (pair of vectors). We can also express eq. (2.43) as

$$\rho \boldsymbol{v} \cdot \frac{\partial \boldsymbol{v}}{\partial t} = -\nabla \cdot \left[\rho \boldsymbol{v} \left(\tfrac{1}{2} \boldsymbol{v} \cdot \boldsymbol{v} + \rho^{-1} p + g z \right) - \boldsymbol{v} \cdot \boldsymbol{\tau} \right]$$
$$- \tau_{ij} \frac{\partial u_i}{\partial x_j}. \tag{2.44}$$

It follows by using eqs. (2.40) and (2.44) and the generalized Gauss theorem that

$$\dot{E} = -\rho \int_{S_Q(t)} \left[(u_n - U_{sn}) \left(\tfrac{1}{2} \boldsymbol{v} \cdot \boldsymbol{v} + g z \right) \right] \mathrm{d}S$$
$$- \int_{S_Q(t)} u_n p \, \mathrm{d}S + \int_{S_Q(t)} \boldsymbol{n} \cdot (\boldsymbol{v} \cdot \boldsymbol{\tau}) \, \mathrm{d}S$$
$$- \int_{Q(t)} \tau_{ij} \frac{\partial u_i}{\partial x_j} \mathrm{d}Q. \tag{2.45}$$

The last term in eq. (2.45) represents viscous dissipation in the fluid.

Let us express the conservation of energy for liquid motion inside a tank. The first step is to rewrite the viscous stress term acting on the enclosing surface, S_Q. The derivation is done for two-dimensional flow. However, the procedure can be generalized to three-dimensional flow. We operate with normal (vector $\boldsymbol{n}$) and tangential (vector $\boldsymbol{t}$) directions on the enclosing surface S_Q:

$$\boldsymbol{v} \cdot \boldsymbol{\tau} = (u_t \boldsymbol{t} + u_n \boldsymbol{n}) \cdot (\tau_{tt} \boldsymbol{tt} + \tau_{tn} \boldsymbol{tn} + \tau_{nt} \boldsymbol{nt} + \tau_{nn} \boldsymbol{nn})$$
$$= u_t \tau_{tt} \boldsymbol{t} + u_t \tau_{tn} \boldsymbol{n} + u_n \tau_{nt} \boldsymbol{t} + u_n \tau_{nn} \boldsymbol{n}.$$

The consequence is $\boldsymbol{n} \cdot (\boldsymbol{v} \cdot \boldsymbol{\tau}) = u_t \tau_{tn} + u_n \tau_{nn}$. A boundary condition on the wetted tank surface is a no-slip condition between the liquid and the tank, which means that the tangential fluid velocity is equal to the tangential velocity of the tank boundary, U_t. Boundary conditions at the free surface in the case of zero surface tension and no gas flow are

$$\tau_{tn} = 0 \quad \text{and} \quad -p + \tau_{nn} = -p_0, \tag{2.46}$$

where p_0 is the constant gas pressure. We redefine the liquid pressure p to mean the excess pressure relative to p_0. Equation (2.45) is then still valid for an incompressible liquid. Using the fact that $u_n = U_{sn} \equiv U_n$ on the enclosing surface S_Q gives

$$\dot{E} = -\int_{S(t)} (p - \tau_{nn}) U_n \, \mathrm{d}S + \int_{S(t)} \tau_{tn} U_t \, \mathrm{d}S$$
$$- \int_{Q(t)} \tau_{ij} \frac{\partial u_i}{\partial x_j} \mathrm{d}Q, \tag{2.47}$$

where S is the wetted tank surface.

2.2.4.3 Examples: two special cases

Case 1. The tank does not move (i.e., U_n and U_t are zero), implying that

$$\dot{E} = -\int_{Q(t)} \tau_{ij} \frac{\partial u_i}{\partial x_j} \mathrm{d}Q, \tag{2.48}$$

which expresses how the time rate of change of the kinetic and potential energy in the liquid is related to the viscous dissipation in the liquid volume. This fact is used in Chapter 6 to express how the wave amplitude decays with time due to viscous dissipation.

Case 2. Steady-state ($T = 2\pi/\sigma$)-periodic oscillations are assumed, implying that time-averaged $\langle \dot{E} \rangle = 0$ or that

$$-\left\langle \int_{S(t)} (p - \tau_{nn}) U_n \mathrm{d}S \right\rangle + \left\langle \int_{S(t)} \tau_{tn} U_t \mathrm{d}S \right\rangle$$
$$- \left\langle \int_{Q(t)} \tau_{ij} \frac{\partial u_i}{\partial x_j} \mathrm{d}Q \right\rangle = 0.$$

If viscosity is not present, the consequence is $\langle \int_{S(t)} p U_n \mathrm{d}S \rangle = 0$. If the forcing velocity of the tank has a time dependence $\cos(\sigma t)$, it tells that the pressure cannot have the $\cos(\sigma t)$ Fourier component. We previously showed under the section of conservation of fluid momentum that hydrodynamic pressure cannot have a time-independent component during steady-state oscillations. The conclusion is that the T-periodic pressure is governed by $p = a_1 \sin(\sigma t) + \sum_{j=2}^{\infty} a_j \cos(j \sigma t + \varepsilon_j)$.

When it comes to subharmonic steady-state solutions caused by the $\cos(\sigma t)$-proportional forcing (see Sections 8.5 and 8.10), we should consider ($2T, 3T, \ldots$)-periodic regimes; therefore, the averaging procedure suggests $\langle \cdot \rangle = (2T)^{-1} \int_0^{2T} \cdot \, \mathrm{d}t$, $\langle \cdot \rangle = (3T)^{-1} \int_0^{3T} \cdot \, \mathrm{d}t$, and so forth.

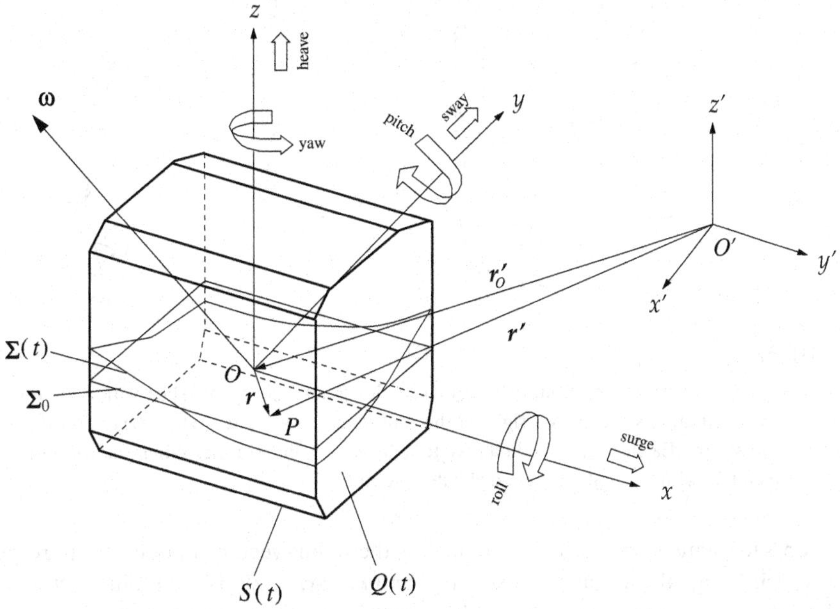

Figure 2.2. Moving rigid tank partially filled with liquid. The $Oxyz$ system is rigidly fixed with the tank. The $O'x'y'z'$ system is an inertial coordinate system. Relative motions of the $Oxyz$ system with respect to the $O'x'y'z'$ system are determined at instant t by the translatory velocity $\boldsymbol{v}_O(t) = d\boldsymbol{r}'_O/dt = \dot{\boldsymbol{r}}'_O$ and angular velocity $\boldsymbol{\omega}(t)$.

For the $2T$ subharmonic regimes, $\langle \int_{S(t)} p\, U_n \mathrm{d}S \rangle = 0$ deduces, for example, that the pressure is governed by $p = a_1 \cos(\frac{1}{2}\sigma t + \varepsilon_1) + a_2 \sin(\sigma t) + \sum_{j=3}^{\infty} a_j \cos(\frac{1}{2} j \sigma t + \varepsilon_j)$.

2.3 Tank-fixed coordinate system

A *rigid* tank moving with an unsteady velocity and partially filled with liquid is considered. Actually, the tank cannot always be considered rigid; the flexibility of the tank may matter in connection with liquid impact (see Chapter 11). Furthermore, collision between ships causes deformations of the tank structure (see Sections 5.4.2.3 and 5.4.2.4) and elastic vibrations have to be considered in connection with earthquake excitation of land-based tanks.

The time-dependent liquid domain $Q(t)$ is bounded by the free surface $\Sigma(t)$ and the wetted tank surface $S(t)$ as shown in Figure 2.2. An inertial (nonaccelerated and nonrotating) coordinate system $O'x'y'z'$ is introduced. The z'-axis is assumed vertical and positive upward. Because Earth's rotation is very slow relative to the time scale in sloshing problems, *we can for all practical problems consider Earth coordinates as an inertial system.* A more appropriate inertial system to use as a reference for a ship tank is the coordinate system defined in Section 3.5 in connection with the analysis of linear wave-induced ship motions, which is a translating coordinate system moving with steady ship speed U.

The tank-fixed coordinate system $Oxyz$ presented in Figure 2.2 is an accelerating and rotating coordinate system. The instantaneous angular velocity vector of the $Oxyz$ system is denoted $\boldsymbol{\omega}(t)$ with components ω_1, ω_2, and ω_3 along the x-, y-, and z-axes, respectively. The origin of the coordinate system has a translatory velocity $\boldsymbol{v}_O(t)$ relative to the inertial system. Furthermore, yaw angle Ψ, pitch angle Θ, and roll angle Φ of the tank, which are called *Euler angles*, must also be introduced. When the angles are finite, it matters in which order they are executed. The usual order is yaw, pitch, and roll, as illustrated in Figure 2.3. We introduce a coordinate system (x'_1, y'_1, z'_1) with the same origin as the tank-fixed coordinate system and imagine that the tank is first oriented so that the x'_1-, y'_1-, and z'_1-axes are parallel to the x'-, y'-, and z'-axes, respectively, of the inertial system.

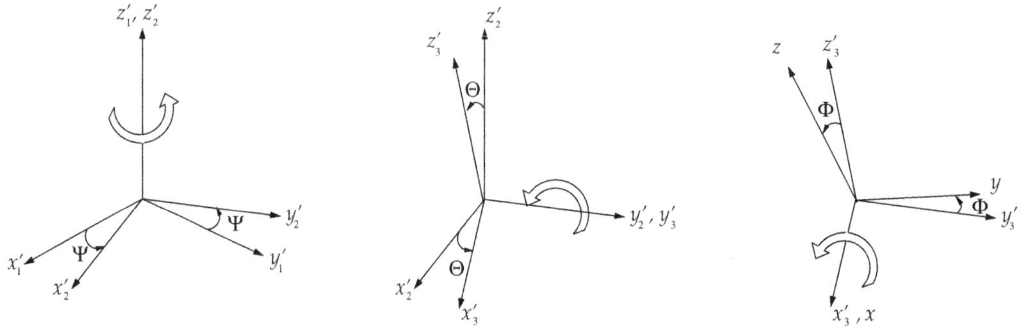

Figure 2.3. The order of the rotation of the Euler angles Φ, Θ, and Ψ. We start with a coordinate system (x_1, y_1, z_1) that has the same origin as the tank-fixed coordinate system $Oxyz$ and that has axes parallel with the Earth-fixed coordinate system $O'x'y'z'$ at initial time. When the rotations are finished, we end up with the body-fixed coordinate system $Oxyz$.

The first step is to rotate the vessel in yaw about the z_1'-axis, which brings the x_1'- and y_1'-axes to the x_2'- and y_2'-axes. The z_2'-axis is the same as the z_1'-axis. The second step is to rotate the vessel in pitch about the y_2'-axis, which brings the x_2'- and z_2'-axes to the x_3'- and z_3'-axes. The y_3'-axis is the same as the y_2'-axis. The final step is to rotate the vessel in roll about the x_3'-axis, which brings the coordinate axis to the x-, y-, and z-axes.

We need additional relationships to proceed with an analysis of finite Euler angles (Etkin, 1959; Faltinsen, 2005). For instance, the time rate of change of Ψ, Θ, and Φ can be expressed in terms of Ψ, Θ, Φ, ω_1, ω_2, and ω_3. Furthermore, we can express the velocity components of the vessel in the Earth-fixed coordinate system in terms of Ψ, Θ, Φ, and the three components of the translatory velocity $\boldsymbol{v}_O(t)$ of the origin of the tank-fixed coordinate system. The procedure is simplified considerably when the oscillatory rigid-body motions are small and the expressions can be linearized. In that case we use η_4, η_5, and η_6 as symbols for roll, pitch, and yaw, respectively, and we can write $\omega_1 = \dot{\eta}_4, \omega_2 = \dot{\eta}_5$, and $\omega_3 = \dot{\eta}_6$, where the overdot indicates a time derivative.

In later applications of the tank-fixed coordinate system we define the origin to coincide with the geometrical center of the free surface of the tank and the z-axis to point upward when the tank and the liquid are at rest.

We now show how the rigid-body velocity of the tank can be expressed in terms of $\boldsymbol{v}_O(t)$ and $\boldsymbol{\omega}(t)$. We define $\boldsymbol{r}'$ as the radius vector of a point P on the rigid tank with respect to O'; $\boldsymbol{r}'_O = \overrightarrow{O'O}$ is the radius vector of point O with respect to O' and $\boldsymbol{r} = (x, y, z)^T$ is the radius vector of point P with respect to O, meaning that $\boldsymbol{r}' = \boldsymbol{r}'_O + \boldsymbol{r}$ or

$$\boldsymbol{r}'(t) = \boldsymbol{r}'_O(t) + x\boldsymbol{e}_1(t) + y\boldsymbol{e}_2(t) + z\boldsymbol{e}_3(t), \quad (2.49)$$

where $\boldsymbol{e}_i$ $(i = 1, 2, 3)$ are unit vectors of the $Oxyz$ system. The velocity relative to the inertial system is obtained by time-differentiating eq. (2.49). We must then recognize the fact that the unit vectors $\boldsymbol{e}_i$ $(i = 1, 2, 3)$ rotate relative to the inertial system (i.e., they are time-dependent). Therefore, we must also account for the time derivative of the unit vectors. Let us present a general differentiation formula that we also apply in Section 2.4.1, where we consider fluid particle velocities and accelerations. We then define a vector $\boldsymbol{B}(t) = b_1(t)\,\boldsymbol{e}_1(t) + b_2(t)\,\boldsymbol{e}_2(t) + b_3(t)\,\boldsymbol{e}_3(t)$. It is possible to show (see, e.g., Faltinsen, 2005) that

$$\begin{aligned}\frac{d\boldsymbol{B}(t)}{dt} &= \dot{\boldsymbol{B}}(t)\\ &= \dot{b}_1(t)\,\boldsymbol{e}_1(t) + b_1(t)\,\dot{\boldsymbol{e}}_1(t) + \dot{b}_2(t)\,\boldsymbol{e}_2(t)\\ &\quad + b_2(t)\,\dot{\boldsymbol{e}}_2(t) + \dot{b}_3(t)\,\boldsymbol{e}_3(t) + b_3(t)\,\dot{\boldsymbol{e}}_3(t)\\ &= \dot{b}_1(t)\,\boldsymbol{e}_1(t) + \dot{b}_2(t)\,\boldsymbol{e}_2(t) + \dot{b}_3(t)\,\boldsymbol{e}_3(t)\\ &\quad + \boldsymbol{\omega} \times \boldsymbol{B}(t)\\ &= \frac{d^*\boldsymbol{B}(t)}{dt} + \boldsymbol{\omega} \times \boldsymbol{B}(t). \end{aligned} \quad (2.50)$$

Equation (2.50) shows that the time derivative $\dot{\boldsymbol{B}}(t)$ of a vector defined in a noninertial coordinate system includes the two components $d^*\boldsymbol{B}/dt = \dot{b}_1(t)\,\boldsymbol{e}_1(t) + \dot{b}_2(t)\,\boldsymbol{e}_2(t) + \dot{b}_3(t)\,\boldsymbol{e}_3(t)$ and $\boldsymbol{\omega} \times \boldsymbol{B}(t)$, where we have used the superscript * in connection with the time derivative to indicate that we should not time-differentiate the unit vectors.

If we now apply eq. (2.50) to eq. (2.49) and assume that the coordinates x, y, and z are a fixed point on the tank, these coordinates are time-independent and the velocity of the tank can be expressed as

$$\boldsymbol{v}_b = \boldsymbol{v}_O + \boldsymbol{\omega} \times \boldsymbol{r}, \quad \boldsymbol{v}_O(t) = d\boldsymbol{r}'_O/dt = \dot{\boldsymbol{r}}'_O. \quad (2.51)$$

It is most convenient in sloshing analysis to refer to the tank-fixed coordinate system instead of an inertial system. However, the governing equations are derived in an inertial system; therefore, special care must be shown when the equations are transferred to an accelerated coordinate system.

2.4 Governing equations in a noninertial, tank-fixed coordinate system

2.4.1 Navier–Stokes equations

A basis for deriving the Navier–Stokes equations given by eq. (2.16) was Newton's second law, which is valid only in an inertial system. We cannot therefore directly apply eq. (2.16) to the accelerating and rotating tank-fixed coordinate system. What we must do is transform each individual term to the tank-fixed coordinate system. The difficulty involved is to express the time derivative. The spatial derivatives are invariant. Indeed, let us introduce the tank-fixed Cartesian coordinate system (x, y, z) with the unit vectors $\boldsymbol{e}_1, \boldsymbol{e}_2, \boldsymbol{e}_3$ along the x-, y-, and z-axes, respectively. Furthermore, an inertial coordinate system $O'x'y'z'$ is superposed with $Oxyz$ at time instant t, implying that all spatial derivatives of the vector and scalar fields are the same and, therefore, the corresponding terms in transformed Navier–Stokes equations (2.16) (from inertial to body-fixed coordinate system $Oxyz$) save their structure. However, the time derivatives of the scalar and vector fields change (see Section 2.3).

The time derivative D/Dt in eq. (2.16) is the material derivative, which expresses the time rate of change as we follow a fluid particle in space and time. The fact that individual fluid particles are followed in time is called a Lagrangian description. To express the acceleration $D\boldsymbol{v}/Dt$ of a fluid particle, we start out by expressing the position $\boldsymbol{r}'(t) = \boldsymbol{r}'_O(t) + \boldsymbol{r}(t)$ of a fluid particle relative to the inertial coordinate system (see Figure 2.2). A Lagrangian description implies that

$$\frac{D\boldsymbol{v}}{Dt} = \frac{D^2\boldsymbol{r}'}{Dt^2}. \quad (2.52)$$

We can, according to the differential formula given by eq. (2.50), write

$$\frac{D\boldsymbol{r}'}{Dt} = \dot{\boldsymbol{r}}'_O + \dot{\boldsymbol{r}} = \boldsymbol{v}_O + \dot{x}\boldsymbol{e}_1 + \dot{y}\boldsymbol{e}_2 + \dot{z}\boldsymbol{e}_3 + \boldsymbol{\omega} \times \boldsymbol{r}$$
$$= \boldsymbol{v}_O + \boldsymbol{v}_r + \boldsymbol{\omega} \times \boldsymbol{r}, \quad (2.53)$$

where $\boldsymbol{v}_r = \dot{x}\boldsymbol{e}_1 + \dot{y}\boldsymbol{e}_2 + \dot{z}\boldsymbol{e}_3 = u_r\boldsymbol{e}_1 + v_r\boldsymbol{e}_2 + w_r\boldsymbol{e}_3$ is the velocity of a fluid particle relative to the tank-fixed coordinate system. Computing the second time derivative using the same differentiation rules and accounting for the equality $\dot{\boldsymbol{\omega}} = d^*\boldsymbol{\omega}/dt + \boldsymbol{\omega} \times \boldsymbol{\omega} = d^*\boldsymbol{\omega}/dt$ give the actual acceleration of a fluid particle in terms of the apparent acceleration in the body-fixed coordinate system:

$$\frac{D^2\boldsymbol{r}'}{Dt^2} = \frac{d^*\boldsymbol{v}_O}{dt} + \boldsymbol{\omega} \times \boldsymbol{v}_O + \dot{\boldsymbol{\omega}} \times \boldsymbol{r}$$
$$+ \boldsymbol{\omega} \times (\boldsymbol{v}_r + \boldsymbol{\omega} \times \boldsymbol{r}) + \frac{d^*\boldsymbol{v}_r}{dt} + \boldsymbol{\omega} \times \boldsymbol{v}_r.$$

The preceding equation contains the translatory acceleration $\boldsymbol{a}_O = d^*\boldsymbol{v}_O/dt$ of the tank coordinate system and the translatory relative acceleration $\boldsymbol{a}_r = d^*\boldsymbol{v}_r/dt$, which could also be expressed as

$$\boldsymbol{a}_O = \frac{d^*\boldsymbol{v}_O}{dt} = \dot{v}_{O1}\boldsymbol{e}_1 + \dot{v}_{O2}\boldsymbol{e}_2 + \dot{v}_{O3}\boldsymbol{e}_3,$$
$$\boldsymbol{a}_r = \frac{d^*\boldsymbol{v}_r}{dt} = \dot{u}_r\boldsymbol{e}_1 + \dot{v}_r\boldsymbol{e}_2 + \dot{w}_r\boldsymbol{e}_3.$$

Finally, we obtain

$$\frac{D^2\boldsymbol{r}'}{Dt^2} = \boldsymbol{a}_O + \boldsymbol{\omega} \times \boldsymbol{v}_O + \dot{\boldsymbol{\omega}} \times \boldsymbol{r} + 2\boldsymbol{\omega} \times \boldsymbol{v}_r$$
$$+ \boldsymbol{\omega} \times (\boldsymbol{\omega} \times \boldsymbol{r}) + \boldsymbol{a}_r, \quad (2.54)$$

where $2\boldsymbol{\omega} \times \boldsymbol{v}_r$ is the Coriolis acceleration and $\boldsymbol{\omega} \times (\boldsymbol{\omega} \times \boldsymbol{r})$ is the centripetal acceleration.

Now we are able to express the Navier–Stokes equation in the tank-fixed coordinate system. It is convenient to use an Eulerian description of $\boldsymbol{a}_r$. An Eulerian description examines the rate of change in time and space at fixed points. We write $\boldsymbol{a}_r$ as

$$\boldsymbol{a}_r = \frac{D\boldsymbol{v}_r}{Dt} = \frac{\partial^*\boldsymbol{v}_r}{\partial t} + \boldsymbol{v}_r \cdot \nabla\boldsymbol{v}_r, \quad (2.55)$$

where $\partial^*\boldsymbol{v}_r/\partial t$ means that we time-differentiate $\boldsymbol{v}_r$ for a fixed point in the tank-fixed coordinate system and that we must not time-differentiate the unit vectors. Those unit vectors do not vary with time relative to the tank-fixed coordinate system. Furthermore, because of eq. (2.51) the absolute velocity is $\boldsymbol{v} = \boldsymbol{v}_O + \boldsymbol{v}_r + \boldsymbol{\omega} \times \boldsymbol{r}$ and $\nabla \cdot (\boldsymbol{v}_O + \boldsymbol{\omega} \times \boldsymbol{r}) = 0$; we can for an incompressible liquid express the Navier–Stokes equations in a tank-fixed coordinate system as

$$\frac{\partial^* \boldsymbol{v}_r}{\partial t} + \boldsymbol{v}_r \cdot \nabla \boldsymbol{v}_r = -\frac{1}{\rho}\nabla p + \boldsymbol{g} + \nu \nabla^2 \boldsymbol{v}_r - \boldsymbol{a}_O - (\boldsymbol{\omega} \times \boldsymbol{v}_O) - \dot{\boldsymbol{\omega}} \times \boldsymbol{r} - 2(\boldsymbol{\omega} \times \boldsymbol{v}_r) - \boldsymbol{\omega} \times (\boldsymbol{\omega} \times \boldsymbol{r}). \tag{2.56}$$

The components of the gravitational vector $\boldsymbol{g} = (g_1, g_2, g_3)$ along the axes of the tank-fixed coordinate system $Oxyz$ in Figure 2.2 can be expressed in terms of the Euler angles defined in Figure 2.3 as

$$g_1 = g \sin \Theta; \quad g_2 = -g \cos \Theta \sin \Phi; \quad g_3 = -g \cos \Theta \cos \Phi. \tag{2.57}$$

We can show this fact as follows. The gravitational acceleration acts along the z_1'-axis. The second drawing in Figure 2.3 shows that the acceleration of gravity has components $g \sin \Theta$ and $-g \cos \Theta$ along the x_3'- and z_3'-axis, respectively; no component exists along the y_3'-axis. We then use the last drawing in Figure 2.3. Because the x-axis coincides with the x_3'-axis, the x-component of $\boldsymbol{g}$ is $g \sin \Theta$, which is consistent with the first of eqs. (2.57). Decomposing the z_3'-component $-g \cos \Theta$ along the y- and z-axes then gives the terms in the second and third of eqs. (2.57).

The derivation of the acceleration in an accelerated coordinate system applies to any particle. When formulating the rigid-body dynamics in an accelerated coordinate system, we can apply the derived acceleration formulas in combination with Newton's second law. A difference from the fluid dynamics case is that the rigid-body case has only external forces. Centripetal acceleration, or centrifugal acceleration which is minus the centripetal acceleration, is what everyone experiences during the turning of a car.

To get a feeling for the order of magnitude of the centrifugal acceleration on board a ship due to rigid-body ship dynamics in waves, we limit ourselves to roll and express the magnitude of the centrifugal acceleration as $\dot{\eta}_{4a}^2 r$, where $\dot{\eta}_{4a}$ is the angular roll velocity amplitude of the ship. Because $\dot{\eta}_{4a}$ is seldom larger than 0.15 radians per second, we choose this value as a test case. The question is then what is r? If the center of roll is a point where no horizontal motions exist, r would be a representative radial distance from the center of the roll axis. However, no center of roll generally exists (Faltinsen, 2005), so we consider a quasi center of roll where there is minimum horizontal motion. We use $r = 30$ m as our case, which gives a centrifugal acceleration of $0.07g$. Actually what we should have done in a more careful analysis would be to consider the coupled vessel motions in six degrees of freedom; then the ambiguity of introducing a center of roll would not be a problem.

Let us also illustrate the Coriolis acceleration with a practical example. Consider a merry-go-round consisting of a rigid disk that rotates with a constant angular velocity. Children are sitting at the outer edge of the disk. Let us imagine that a child is trying to send a ball to another child that is seated opposite him or her. He or she sends the ball in an initially straight line; however, the ball does not continue going in a straight line relative to the disk. The Coriolis acceleration, which acts perpendicular to the ball's velocity relative to the disk, causes the ball to have a curved path when seen from the disk.

2.4.1.1 Illustrative example: application to the Earth as an accelerated coordinate system

We now consider the Earth as an accelerated coordinate system. The rotational vector $\boldsymbol{\omega}$ is assumed constant and goes through the North and South Poles. We set $\boldsymbol{v}_O$ equal to zero, so we can write

$$\frac{D^2 \boldsymbol{r}'}{Dt^2} = 2(\boldsymbol{\omega} \times \boldsymbol{v}_r) + \boldsymbol{\omega} \times (\boldsymbol{\omega} \times \boldsymbol{r}) + \boldsymbol{a}_r,$$

where $\boldsymbol{v}_r$ and $\boldsymbol{a}_r$ are not the true velocity and acceleration, but the velocity and acceleration as observed in the Earth-fixed coordinate system. We see that $\boldsymbol{\omega} \times (\boldsymbol{\omega} \times \boldsymbol{r})$ is a centripetal acceleration directed toward the axis of rotation of the Earth. Let us estimate its maximum magnitude. We set $\omega = 2\pi/T$, where T is 24 hours or 86,400 s. We approximate the Earth diameter as 12,760 km, which gives a maximum value of $0.03\,\mathrm{m\,s^{-2}}$ or about 0.3% of the gravitational acceleration at the equator. This means that the effect of centripetal acceleration is negligible relative to that of gravitational acceleration.

The Coriolis acceleration is perpendicular to the velocity of an object. The component of the angular velocity ω of the Earth that is vertical to the Earth at latitude ϕ can be expressed as $|\omega| \sin\phi$, where $\phi = 0^\circ$ and $\phi = 90^\circ$ are the equator and the North Pole, respectively. The magnitude of the Coriolis acceleration for an object that moves horizontally at a latitude ϕ is therefore $0.00015\,|\boldsymbol{v}_r| \sin\phi$.

Objects deflect to the right in the Northern Hemisphere and to the left in the Southern Hemisphere due to the Coriolis force $-2M(\boldsymbol{\omega} \times \boldsymbol{v}_r)$, where M is the mass of the object. We cannot neglect the effect of the Coriolis force associated with Earth's rotation on large-scale ocean flows or airplane flights. However, the Coriolis acceleration due to Earth's rotation can be disregarded in our application to ship tanks.

2.4.2 Potential flow formulation

2.4.2.1 Governing equations

We now express the governing equations for an irrotational flow and incompressible fluid in the tank-fixed coordinate system. The problem was addressed in an inertial coordinate system in Section 2.3. The flow variables can be expressed in terms of the velocity potential Φ, which satisfies Laplace equation (2.22). The pressure in the inertial coordinate system can be expressed by Bernoulli's equation (2.23). Since the Laplace equation does not include time derivatives, an analysis similar to that of Section 2.4.1, by superposing two coordinate systems at instant t, shows that the Laplace equation is invariant, meaning that

$$\Delta\Phi \equiv \frac{\partial^2\Phi}{\partial x'^2} + \frac{\partial^2\Phi}{\partial y'^2} + \frac{\partial^2\Phi}{\partial z'^2} = \frac{\partial^2\Phi}{\partial x^2} + \frac{\partial^2\Phi}{\partial y^2} + \frac{\partial^2\Phi}{\partial z^2} = 0 \quad \text{in} \quad Q(t). \tag{2.58}$$

Bernoulli's equation is a result of integrating the Euler equations, which are only valid in an inertial system, so we cannot directly apply it to an accelerated coordinate system. We need to study $\partial\Phi/\partial t$ to transform the Bernoulli equation for the pressure to the tank-fixed coordinate system. It is important to stress that $\partial\Phi/\partial t$ in the inertial system means the time derivative of Φ for a fixed point with coordinates (x', y', z'). However, in the tank-fixed coordinate system we want to operate with the time derivative of Φ for a fixed point (x, y, z), which can be expressed as a matter of definition:

$$\left.\frac{\partial\Phi}{\partial t}\right|_{\text{in noninertial } Oxyz} = \lim_{\Delta t\to 0} \frac{\Phi(x, y, z, t+\Delta t) - \Phi(x, y, z, t)}{\Delta t}.$$

We let the point (x, y, z) coincide with a point (x', y', z') at time t. The difference between point (x, y, z) and point (x', y', z') at time $t + \Delta t$ is then approximately equal to $\boldsymbol{v}_b\Delta t$, where $\boldsymbol{v}_b = \boldsymbol{v}_O + \boldsymbol{\omega} \times \boldsymbol{r}$ (see eq. (2.51)) is the velocity of the tank relative to the inertial system. Taylor expansion of $\Phi(x, y, z, t+\Delta t)$ about the point (x', y', z') gives

$$\left.\frac{\partial\Phi}{\partial t}\right|_{\text{in noninertial } Oxyz} = \lim_{\Delta t\to 0} \frac{\Phi(x', y', z', t+\Delta t) + \boldsymbol{v}_b \cdot \nabla\Phi\Delta t - \Phi(x, y, z, t)}{\Delta t} = \left.\frac{\partial\Phi}{\partial t}\right|_{\text{in inertial } O'x'y'z'} + \boldsymbol{v}_b \cdot \nabla\Phi. \tag{2.59}$$

The consequence is that Bernoulli equation (2.23) for pressure p can be expressed as

$$p + \rho\left[\left.\frac{\partial\Phi}{\partial t}\right|_{\text{noninertial}} - \boldsymbol{v}_b \cdot \nabla\Phi + \tfrac{1}{2}(\nabla\Phi)^2 + U_g\right] = C(t), \tag{2.60}$$

where $C(t)$ has to be determined. The gravity potential U_g can, according to eq. (2.25), be expressed as $U_g = -\boldsymbol{g} \cdot \boldsymbol{r}'$, or, remembering that $\boldsymbol{r}' = \boldsymbol{r}'_O + \boldsymbol{r}$ and the gravity potential is defined within to a time-dependent function, as $U_g = -\boldsymbol{g} \cdot \boldsymbol{r}$. We choose $z = 0$ to correspond to the free surface in the tank when there is no sloshing and the tank is in an upright position (i.e., the free surface is horizontal). We assume that the ambient pressure in the gas outside possible gas cavities is constant and equal to p_0. We select $C(t)$ as the pressure p_0 and confirm that this choice is consistent with the pressure distribution in the liquid with no sloshing. Equation (2.60) then gives $p = p_0 - \rho g z$ (i.e., the pressure has a hydrostatic behavior in the liquid with $p = p_0$ on the free surface). This result confirms that our choice of $C(t)$ gives consistent results. Therefore, the pressure can be expressed in the tank-fixed coordinate system as

$$p - p_0 = -\rho\left[\frac{\partial\Phi}{\partial t} + \tfrac{1}{2}(\nabla\Phi)^2 - \nabla\Phi\cdot(\boldsymbol{v}_O + \boldsymbol{\omega}\times\boldsymbol{r}) + U_g\right]. \quad (2.61)$$

We now study the body boundary and free surface conditions.

2.4.2.2 Body boundary conditions

The boundary condition on the wetted tank surface $S(t)$ requires that there is no flow through the tank surface. No requirement exists for a no-slip boundary condition on the tank surface as when solving the Navier–Stokes equations. We can express this mathematically as

$$\boldsymbol{v}\cdot\boldsymbol{n} = \boldsymbol{v}_O\cdot\boldsymbol{n} + [\boldsymbol{\omega}\times\boldsymbol{r}]\cdot n \quad \text{on} \quad S(t), \quad (2.62)$$

where $\boldsymbol{n}$ is the normal vector to $S(t)$ and $\boldsymbol{v}$ is the liquid velocity relative to our inertial system. We can also express $\boldsymbol{v}$ as $\nabla\Phi$. The right-hand side of eq. (2.62) expresses the normal component of the tank velocity, $\boldsymbol{v}_b = \boldsymbol{v}_O + \boldsymbol{\omega}\times\boldsymbol{r}$. We can rewrite eq. (2.62) as

$$\frac{\partial\Phi}{\partial n} = \boldsymbol{v}\cdot\boldsymbol{n} = \boldsymbol{v}_O\cdot\boldsymbol{n} + \boldsymbol{\omega}\cdot[\boldsymbol{r}\times\boldsymbol{n}] \quad \text{on} \quad S(t) \quad (2.63)$$

by using vector algebra. We may also reformulate the body boundary condition as $\boldsymbol{v}_r\cdot\boldsymbol{n} = 0$ on S, where $\boldsymbol{v}_r$ is the liquid velocity relative to the tank.

2.4.2.3 Free-surface conditions

In the following text we assume no gas cavities. This condition is considered in Sections 10.2.1 and 11.6 and requires that we also set up additional equations governing the dynamic behavior of the gas cavity. Surface tension is neglected. On the free surface (i.e., the interface between the gas and the liquid), we must then require that the pressure in the liquid at the free surface is equal to the pressure p_0 in the gas at the free surface. This is the *dynamic free-surface condition*:

$$\frac{\partial\Phi}{\partial t} + \tfrac{1}{2}(\nabla\Phi)^2 - (\boldsymbol{v}_O + \boldsymbol{\omega}\times\boldsymbol{r})\cdot\nabla\Phi + U_g = 0 \quad \text{on} \quad \Sigma(t), \quad U_g = -\boldsymbol{g}\cdot\boldsymbol{r}. \quad (2.64)$$

In addition comes the *kinematic free-surface condition*, which ensures that a particle on the free surface will remain on the free surface. The kinematic free-surface condition can be formulated mathematically as follows. If the tank-fixed coordinate system $Oxyz$ is used, the most general, "implicit" definition of the free surface is associated with the equation

$$Z(x, y, z, t) = 0. \quad (2.65)$$

The fluid particles remain on the free surface for the entire time, meaning that the material derivative of Z in the inertial coordinate system $O'x'y'z'$ should be zero:

$$0 = \frac{D'Z}{Dt} = \left.\frac{\partial Z}{\partial t}\right|_{O'x'y'z'} + \boldsymbol{v}\cdot\nabla Z = \left.\frac{\partial Z}{\partial t}\right|_{Oxyz} - \boldsymbol{v}_b\cdot\nabla Z + \nabla\Phi\cdot\nabla Z \quad \text{on} \quad \Sigma(t), \quad (2.66)$$

where we have used a similar derivation as for eq. (2.59). Furthermore, we should remember eqs. (2.51) and (2.20) and the fact that the normal vector to the free surface $\Sigma(t)$ can be expressed as

$$\boldsymbol{n} = \frac{\nabla Z}{|\nabla Z|} \quad \text{on} \quad \Sigma(t), \quad (2.67)$$

transforming eq. (2.66) to the so-called kinematic boundary condition

$$\frac{\partial\Phi}{\partial n} = \boldsymbol{v}_O\cdot\boldsymbol{n} + \boldsymbol{\omega}\cdot[\boldsymbol{r}\times\boldsymbol{n}] - \frac{\partial Z/\partial t}{|\nabla Z|} \quad \text{on} \quad \Sigma(t), \quad (2.68)$$

where the time derivative is evaluated in the body-fixed coordinate system $Oxyz$.

In practically important cases, the tank wall may be parallel to the z-axis in the free surface, and one can define the z-coordinate to be vertical and positive upward when the tank and the liquid are at rest. This admits the so-called normal form of the free surface. The notation $\zeta(x, y, t)$ is introduced as the z-coordinate of the free surface, where $\zeta(x, y, t)$ is a single-valued function of x and y (i.e., we do not consider overturning waves). The equation

$$Z(x, y, z, t) = z - \zeta(x, y, t) = 0 \quad (2.69)$$

expresses formally the equation of the free surface. Kinematic boundary condition (2.68) can be further rewritten as

$$\frac{\partial\Phi}{\partial z} - \frac{\partial\Phi}{\partial x}\frac{\partial\zeta}{\partial x} - \frac{\partial\Phi}{\partial y}\frac{\partial\zeta}{\partial y} = (\boldsymbol{v}_O + \boldsymbol{\omega}\times\boldsymbol{r})\cdot \left(-\frac{\partial\zeta}{\partial x}, -\frac{\partial\zeta}{\partial y}, 1\right) + \frac{\partial\zeta}{\partial t} \quad \text{on} \quad \Sigma(t). \tag{2.70}$$

Equations (2.64) and (2.70) are the dynamic and kinematic free-surface conditions used in connection with the nonlinear multimodal method to be described in Chapters 7, 8, and 9. Different ways exist to mathematically formulate the free-surface conditions. A common way to satisfy nonlinear free-surface conditions with a boundary element method (see Section 10.2) is to follow liquid particles on the free surface $\Sigma(t)$ and calculate how the velocity potential on the free surface changes with time to satisfy the fact that the liquid pressure is equal to the gas pressure on the free surface (dynamic free-surface condition). The motion of the free-surface particles follows by time-integrating the liquid velocity $\nabla\Phi$ on $\Sigma(t)$. The fact that we follow liquid particles in time on the free surface means that the kinematic free-surface condition is satisfied. More details are provided in Section 10.2.

In Chapters 4 and 5, we consider the linearized problem of sloshing for nonmoving and moving tanks, respectively. The linearization procedure is applied to the boundary conditions on the free surface $\Sigma(t)$. Other governing equations and boundary conditions are linear in terms of the two unknowns ζ and Φ. The linearization can formally be done for forced tank motion by introducing a small parameter ε that expresses the order of magnitudes of angular tank-motion amplitudes and the ratio between translatory tank-motion amplitudes and a characteristic cross-dimension of the tank, L. The free-surface elevation, its slope, and the velocity field are assumed to be $O(\varepsilon)$ and the $o(\varepsilon)$ terms are neglected in boundary conditions (2.64) and (2.70); here $O(\cdot)$ means "the order of magnitude of," whereas $o(\cdot)$ means "of smaller order than what is written inside the brackets." Consider free oscillations (i.e., the case where $\boldsymbol{v}_O = \boldsymbol{\omega} = 0$). We also assume for this case that ζ and Φ are of $O(\varepsilon)$. Because the free-surface elevation is of $O(\varepsilon)$, we may substitute the values of Φ on $\Sigma(t)$ with its values at the mean free surface $z = 0$ at corresponding x- and y-coordinates. This procedure transforms the two boundary conditions to the form

$$\zeta(x, y, t) = -\frac{1}{g}\frac{\partial\Phi}{\partial t} \quad \text{on} \quad z = 0, \tag{2.71}$$

$$\frac{\partial\Phi}{\partial z} = \frac{\partial\zeta}{\partial t} \quad \text{on} \quad z = 0. \tag{2.72}$$

We find by differentiating eq. (2.71) with respect to time and substituting the expression into eq. (2.72) that

$$\frac{\partial^2\Phi}{\partial t^2} + g\frac{\partial\Phi}{\partial z} = 0 \quad \text{on} \quad z = 0. \tag{2.73}$$

These linearized free-surface conditions are also used in Chapter 3 when ocean waves are discussed.

2.4.2.4 Mass (volume) conservation condition

When no liquid inflow or outflow occurs for the sloshing problem, the mass of the incompressible liquid should remain constant; that is,

$$M_l = \int_{Q(t)} \rho \mathrm{d}Q = \rho Vol = \text{const}, \tag{2.74}$$

where M_l and Vol are the liquid mass and volume, respectively. Another way to express this fact is by exchanging the integration surface $S_{\Delta Q}$ in eq. (2.5) with $S(t) + \Sigma(t)$. Equation (2.74) is not satisfied automatically. In fact it restricts the admissible class of functions Z (or ζ) to define the free surface.

2.4.2.5 Free boundary problem of sloshing and initial/periodicity conditions

Equations (2.58), (2.63), (2.64), (2.68), and (2.74) constitute the time-dependent free-boundary-value problem of liquid sloshing in a moving tank. It is called a *free*-boundary-value problem because part of the boundary (the free surface) is free to move. To be mathematically complete, this problem requires either *initial* or, in the case of the periodic vector functions $\boldsymbol{v}_O(t)$ and $\boldsymbol{\omega}(t)$, *periodicity* conditions. Physically, the initial conditions determine an initial scenario; the solution of the initial-value problem implies a transient wave, which is influenced by combined effects of both $(\boldsymbol{v}_O(t), \boldsymbol{\omega}(t))$ and the initial perturbations. The free-boundary problem with a periodicity condition states periodic (steady-state) solutions. Stable steady-state waves are possible after the transients associated with the initial conditions have died out. However, when potential flow in a limited domain is assumed, no mechanisms exist that cancel out transient effects. Because steady-state solutions exist in reality, we must introduce damping due to nonpotential flow effects into the system, which is done in Chapters 6, 8, and 9 by introducing viscous damping terms. However, there exist conditions when no steady-state solutions are possible, which is discussed in Section 9.1.1 in connection with sloshing in square-base, nearly square-base, and vertical circular cylindrical tanks. Certain frequency domains exist that are dependent on the excitation amplitude and the liquid depth, where the flow will remain transient. We call it "chaos."

The typical initial conditions are

$$Z(x, y, z, t_0) = Z_0(x, y, z); \quad \left.\frac{\partial \Phi}{\partial n}\right|_{\Sigma(t_0)} = V_n(x, y, z)|_{\Sigma(t_0)}, \tag{2.75}$$

where the known function Z_0 defines the initial position of the free surface $\Sigma(t_0): Z_0(x, y, z) = 0$, and $V_n(x, y, z)$ implies the initial absolute normal velocities on the already defined $\Sigma(t_0)$. For prescribed motions of the tank, the initial conditions of eq. (2.75) are equivalent to

$$Z(x, y, z, t_0) = Z_0(x, y, z); \quad \left.\frac{\partial Z}{\partial t}\right|_{t=t_0} = Z_1(x, y, z)|_{\Sigma(t_0)}, \tag{2.76}$$

where Z_0 and Z_1 are given functions.

The literature also contains a series of specific initial conditions that correspond to initial scenarios of a special nature. An example is given by Faltinsen *et al.* (2000), who considered a flow that starts from rest with sufficiently small tank oscillations. Fluid accelerations are initially assumed to dominate over gravity. The initial phase can then be described by a linear sloshing theory, where the excess pressure relative to the pressure in the air is expressed as $-\rho\partial\Phi/\partial t$. The initial conditions are zero pressure (i.e., $\Phi = 0$) on an initially horizontal free surface, which can mathematically be formulated as

$$\Sigma(t_0): \quad Z(x, y, z, t_0) = z; \quad \Phi|_{\Sigma(t_0)} = 0. \tag{2.77}$$

When $\boldsymbol{\omega}(t)$ and $\boldsymbol{v}_O(t)$ are periodic vector functions of the forcing period T, the *periodicity conditions*

$$Z(x, y, t+T) = Z(x, y, t); \quad \nabla\Phi(x, y, z, t+T) = \nabla\Phi(x, y, z, t) \tag{2.78}$$

can be used to identify steady-state wave motions. The second condition requires that fluid domain $Q(t+T) = Q(t)$. This requirement is justified by the first of conditions (2.78), which establishes the equivalence of instantaneous free-surface shapes at t and $t+T$.

If we linearize eqs. (2.58), (2.63), (2.64), (2.68), (2.74), and (2.76) in the original nonlinear free-boundary-value problem, it is possible to show that a unique solution exists (Feschenko *et al.*, 1969; Morand & Ohayon, 1995). The mathematical proof of existence and uniqueness of the initial nonlinear boundary-value problem eqs. (2.58), (2.63), (2.64), and (2.68) and eqs. (2.74) and (2.76) in its general form is still an open question (even in the two-dimensional formulation). Being familiar with both former Soviet and Western literature, the present authors were able to find only a limited set of mathematical publications that report local existence theorems for the initial-boundary-value problems. Almost all of these results are documented by Shinbrot (1976), Reeder and Shinbrot (1976, 1979), Ovsyannikov *et al.* (1985), Lukovsky (1990), and Lukovsky and Timokha (1995). As shown later by considering asymptotic solutions, the boundary-value problem of eqs. (2.58), (2.63), (2.64), and (2.68) and eqs. (2.74) and (2.78) for periodic solutions is not uniquely solvable.

Using the solution of the free-boundary-value problem eqs. (2.58), (2.63), (2.64), (2.68), and (2.74) inserted into Bernoulli equation (2.61) makes it possible to compute the pressure $p(x, y, z, t)$ in $Q(t)$. Thereby, one can by defining positive surface normal vector out of the liquid get the resulting hydrodynamic force $\boldsymbol{F}(t)$ and moment $\boldsymbol{M}_O(t)$ (relative to the origin O) acting on the tank:

$$\boldsymbol{F}(t) = \int_{S(t)} (p - p_0)\boldsymbol{n}\,dS = -\rho \int_{S(t)+\Sigma(t)} \left[\frac{\partial \Phi}{\partial t} + \tfrac{1}{2}(\nabla\Phi)^2 - \nabla\Phi\cdot(\boldsymbol{v}_O + \boldsymbol{\omega}\times\boldsymbol{r}) + U_g\right]\boldsymbol{n}\,dS,$$
$$\boldsymbol{M}_O(t) = \int_{S(t)} \boldsymbol{r}\times((p - p_0)\boldsymbol{n})\,dS = \int_{S(t)+\Sigma(t)} \boldsymbol{r}\times((p - p_0)\boldsymbol{n})\,dS = \int_{Q(t)} \boldsymbol{r}\times\nabla(p - p_0)\,dQ, \tag{2.79}$$

where we used $p - p_0 = 0$ on $\Sigma(t)$ and Gauss theorem (A.1) to transform from surface to volume integration for the moment.

2.5 Lagrange variational formalism for the sloshing problem

The free-boundary-value problem given by eqs. (2.58), (2.63), (2.64), (2.68), and (2.74) is the basis for analytically oriented methods that are elaborated in Chapters 7, 8, and 9. These methods employ the so-called modal approach, which assumes a Fourier representation of the free surface and the velocity potential with time-dependent coefficients. General details of the approach are presented in Chapter 7. The derivation of the multimodal systems needs a variational formulation of the free-boundary-value problem. The formulation uses the Lagrange-type formalism and the corresponding Eulerian calculus of variations.

2.5.1 Eulerian calculus of variations

Lagrange, Euler, Maupertus, and Hamilton, among others, imagined the observed perfection in the universe was a consequence of a certain economy in nature rather than a needless expenditure of energy. Thus, the natural motion of objects is governed by minimization (finding the extrema) of a scalar quantity. This quantity has come to be known as the action of an object. The action W has units of [energy] × [time]. If $\beta(t)$ determines a path (trajectory) of the mechanical object, the action should deliver an extremal point of W; that is, we minimize

$$W(\beta) = \int_{t_1}^{t_2} L(\beta, \dot{\beta})\,\mathrm{d}t, \quad \text{where}$$
$$L(\beta, \dot{\beta}) = E_k(\beta, \dot{\beta}) - E_p(\beta). \tag{2.80}$$

The function L, known as the Lagrangian, is defined in terms of the difference between the kinetic and potential energies, denoted by E_k and E_p, respectively, which means that dissipation is neglected. The minimization of the action W should be subject to the two boundary conditions

$$\beta(t_1) = a, \quad \beta(t_2) = b \tag{2.81}$$

at the initial and final positions of the mechanical object in the time interval $[t_1, t_2]$.

The minimization of the action (the energy integral) W provides a generalization that gives the motion of all classical mechanical problems. The path β can be understood as a time-dependent scalar function (generalized coordinate) of a single-dimensional mechanical system, a vector function of t for multidimensional systems, or a function of both spatial coordinates and the time for solid bodies and in fluid dynamics. We demonstrate this fact by a series of examples. In sloshing problems, the path β is associated with instantaneous positions of the free surface $\Sigma(t)$.

Bearing in mind that β is an abstractly defined path of a mechanical system, we should state what we mean as kinetic and potential energies. The *kinetic energy* of an object with mass M is defined as an integral of the inertial force times the distance through which it moves – also known as the inertial work done by an object. This energy is dependent on the square of the velocity that formally may be associated with the time derivative of the displacement (i.e., $\dot{\beta}^2$). The kinetic energy is independent of the sign convention for $\dot{\beta}$. If the inertial force is presented as $F_{\text{inertia}} = M\ddot{\beta}$, the kinetic energy is expressed as

$$E_k = \int F_{\text{inertia}}\mathrm{d}\beta = \int M\ddot{\beta}\,\mathrm{d}\beta = \int M\dot{\beta}\,\mathrm{d}\dot{\beta}$$
$$= \tfrac{1}{2}M\dot{\beta}^2 + \text{const}, \tag{2.82}$$

where eq. (2.82) is a scalar function for a single-dimensional mechanical system and $\dot{\beta}$ depends only on time. For a continuous medium (solid bodies and fluids), the velocity $\dot{\beta}$ is associated with the velocity field $\boldsymbol{v}$, which is also a function of spatial coordinates. Expression (2.82) should include the spatial integral; that is,

$$E_k = \tfrac{1}{2}\int_{Vol} \rho \boldsymbol{v}^2 \mathrm{d}Q + \text{const}, \tag{2.83}$$

where ρ is the mass per unit of Vol. In the three-dimensional case, ρ is the density, but for two-dimensional problems, ρ is the mass per area.

The total potential energy may be defined similarly as the work done by various forces dependent on position rather than its derivative. The potential energy usually is composed of three terms:

$$E_p = E_{in} + E_{ext} + E_g, \tag{2.84}$$

where E_{in} is the internal strain energy of deforming the object (the work done by internal forces

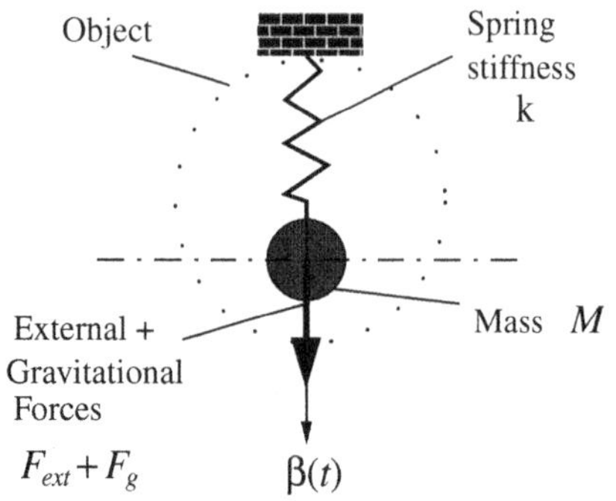

Figure 2.4. Schematic sketch of forces associated with potential energy acting on an object.

F_{in} causing the object to deform), E_{ext} is the work done by external forces, and E_g is the work done by gravitational forces. It is also possible to include the potential energy from fluid compressibility, thermal and magnetic fields, as well as that due to interaction with specific structures. Analytical expressions for these energies change from case to case.

For example, a linear spring with stiffness k is characterized by $F_{in} = \mathrm{k}\beta$:

$$E_{in} = \int F_{in} \mathrm{d}\beta = \int \mathrm{k}\beta \, \mathrm{d}\beta = \tfrac{1}{2}\mathrm{k}\beta^2 + \text{const.} \tag{2.85}$$

As for the energy of external forces (F_{ext}) on the object, this potential energy must have a sign opposite that of the internal potential energy E_{in} for equilibrium to be maintained; that is,

$$E_{ext} = -\int F_{ext} \mathrm{d}\beta. \tag{2.86}$$

The gravitational energy should be included if the path β is not perpendicular to the gravity acceleration g. In the opposite case, when the path β is along the direction of the gravitational field, the gravitational potential energy is

$$E_g = -\int F_g \, \mathrm{d}\beta = -\int Mg \, \mathrm{d}\beta = -Mg\beta + \text{const.} \tag{2.87}$$

When considering a single-dimensional mechanical system, the components of the potential energy can be illustrated by the potential forces as shown in Figure 2.4.

The optimal $\beta(t)$ will be "better" than every other choice $\beta(t) + \delta\beta(t)$ that satisfies the boundary constraints. The variational symbol δ is used here to express an infinitesimal increment and $\delta\beta\,(t)$ is a virtual displacement. Due to conditions (2.81), the variations of the paths satisfy the condition

$$\delta\beta(t_1) = \delta\beta(t_2) = 0. \tag{2.88}$$

Schematically, the perturbed motions $\beta(t) + \delta\beta(t)$ that satisfy eq. (2.88) as well as the corresponding L-values are presented in Figure 2.5. These variations are often called isochronic: the points on the trajectories in the (t, β, L)-space are varied only by β, whereas variations along the time axis are not considered. The isochronic variations introduce the main (extremum for W) $\beta(t)$ and admissible $\beta(t) + \delta\beta(t)$ trajectories, whose points can be compared at any instant t. The test functions $\delta\beta(t)$ in fact determine small fluctuations of the main trajectory.

For a small test function $\delta\beta(t)$ and corresponding $\delta\dot{\beta}$, a Taylor-series expansion of the Lagrangian L with β and $\dot{\beta}$ as two independent

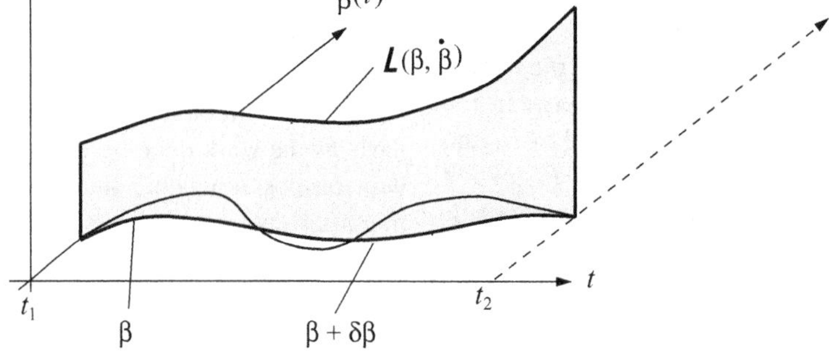

Figure 2.5. Variations of the trajectories (paths) are presented in the (t, β, L)-space, where L is the Lagrangian. Because of conditions (2.81), positions at t_1 and t_2 are the same.

variables can be expressed as

$$L(\beta+\delta\beta,\dot\beta+\delta\dot\beta)=L(\beta,\dot\beta)+\left(\delta\beta\frac{\partial L}{\partial\beta}+\delta\dot\beta\frac{\partial L}{\partial\dot\beta}\right)+O(\delta\beta^2,\delta\dot\beta^2).$$

If higher-order terms are neglected, the integration of this expression gives

$$\delta W=W(\beta+\delta\beta)-W(\beta)=\int_{t_1}^{t_2}\left(\delta\beta\frac{\partial L}{\partial\beta}+\delta\dot\beta\frac{\partial L}{\partial\dot\beta}\right)dt,$$

which is like a Taylor-series expansion of the action W. For an optimal path β, $\delta W=0$ for any small test function $\delta\beta$, which is the same as saying that $dW/d\beta=0$ at the main path β:

$$\int_{t_1}^{t_2}\left(\delta\beta\frac{\partial L}{\partial\beta}+\delta\dot\beta\frac{\partial L}{\partial\dot\beta}\right)dt=0.$$

By using integration by parts, we arrive at

$$\left[\delta\beta\frac{\partial L}{\partial\dot\beta}\right]_{t_1}^{t_2}-\int_{t_1}^{t_2}\delta\beta\left(\frac{d}{dt}\left(\frac{\partial L}{\partial\dot\beta}\right)-\frac{\partial L}{\partial\beta}\right)dt=0.$$

The first term is zero due to boundary conditions (2.88). Because the second term should be zero for any test function $\delta\beta$, the only possibility is that

$$\frac{d}{dt}\left(\frac{\partial L}{\partial\dot\beta}\right)-\frac{\partial L}{\partial\beta}=0. \tag{2.89}$$

Hence, the solution of differential equation (2.89) represents the necessary (but not always sufficient) condition to provide a solution to the original minimization of the action integral. The *Euler–Lagrange equation* (2.89) represents an approach to derive the *dynamic equations* of motion of an object by considering action (energy integrals) rather than forces as in Newton's laws.

The original formulation of equation (2.80) is extendable to a *multidimensional variable* problem (i.e., a mechanical system with multiple degrees of freedom). The statement of equations (2.80) and (2.81) suggests a finite set of generalized coordinates; that is, the path β implies a vector with time-dependent generalized coordinates β_i $(i=1,\ldots,N)$. The action is therefore a function of these coordinates:

$$W(\beta_1,\ldots,\beta_N)=\int_{t_1}^{t_2}L(\beta_1,\dot\beta_1,\ldots,\beta_N,\dot\beta_N)dt \tag{2.90}$$

subject to

$$\beta_i(t_1)=a_i,\quad \beta_i(t_2)=b_i,\quad i=1,\ldots,N. \tag{2.91}$$

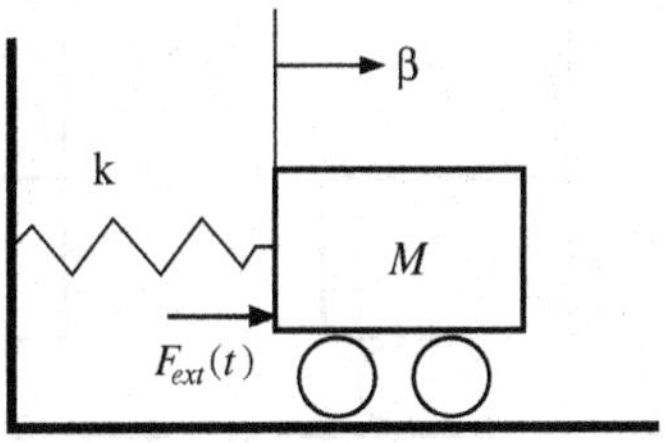

Figure 2.6. One-dimensional horizontal spring–mass system excited by the force $F_{ext}(t)$.

One can see that the zero variation of the action, $\delta W=0$, leads to the Euler–Lagrange equations appearing as a system of ordinary differential equations:

$$\frac{d}{dt}\left(\frac{\partial L}{\partial\dot\beta_i}\right)-\frac{\partial L}{\partial\beta_i}=0,\quad i=1,\ldots,N. \tag{2.92}$$

Rayleigh damping (Rayleigh & Lindsay, 1945) leads to a modification of the Euler–Lagrange equations to include the effect of a *velocity proportional* damping. The procedure assumes the following "dissipation" function:

$$D=\tfrac{1}{2}\sum_{i,j}c_{ij}\dot\beta_i\dot\beta_j, \tag{2.93}$$

which is a quadratic form of the velocity. The modified Euler–Lagrange equations take then the form

$$\frac{d}{dt}\left(\frac{\partial L}{\partial\dot\beta_i}\right)-\frac{\partial L}{\partial\beta_i}+\frac{\partial D}{\partial\dot\beta_i}=0,\quad i=1,2,\ldots. \tag{2.94}$$

Rayleigh damping is widely used in the field of structural analysis. It is a member of the Caughey (1960) family of damping models. Viscous damping associated with flow separation from structures cannot be handled by the Rayleigh damping model.

2.5.2 Illustrative examples

2.5.2.1 Spring–mass systems

Let us consider a *horizontal linear spring–mass system* excited by the force $F_{ext}(t)$, as illustrated in Figure 2.6. The kinetic energy of this system is $E_k=\frac{1}{2}M\dot\beta^2$. The potential energy includes the work done by the force $F_{ext}(t)$ as well as the internal strain energy of the spring. The latter is due to eq. (2.85) and is given by $E_{in}=\frac{1}{2}\mathrm{k}\beta^2$. Hence, the Lagrangian is given by $L=\frac{1}{2}M\dot\beta^2-\left(\frac{1}{2}\mathrm{k}\beta^2-F_{ext}\beta\right)$. Euler-Lagrange equation (2.89)

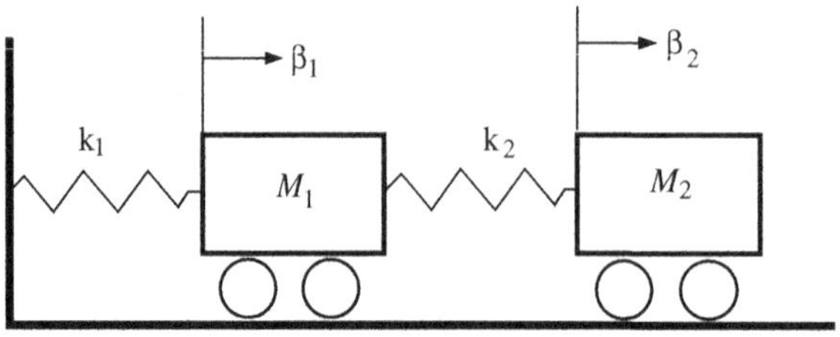

Figure 2.7. A two-mass horizontal spring–mass system.

then deduces the Newtonian equilibrium:

$$\frac{d}{dt}\left(\frac{\partial L}{\partial \dot{\beta}}\right) - \frac{\partial L}{\partial \beta} = M\ddot{\beta} + \mathrm{k}\beta - F_{ext} = 0,$$

where $-\mathrm{k}\beta$ is the restoring force.

Note that the internal strain energy may include additional, nonlinear terms (i.e., $E_{in} = \frac{1}{2}\mathrm{k}\beta^2 + \frac{1}{4}\mathrm{K}\beta^4$). In this case, the Euler–Lagrange equation leads to the equation of the so-called nonlinear Duffing oscillator:

$$\frac{d}{dt}\left(\frac{\partial L}{\partial \dot{\beta}}\right) - \frac{\partial L}{\partial \beta} = M\ddot{\beta} + \mathrm{k}\beta + \mathrm{K}\beta^3 - F_{ext}(t) = 0. \tag{2.95}$$

When the external force is proportional to $\cos(\sigma t + \theta)$ (i.e., a harmonic forcing exists and the forcing frequency σ is close to $\sqrt{\mathrm{k}/M}$, the eigenfrequency of the spring–mass system), nonlinear equation (2.95) is characterized by new features relative to the linear oscillator analyzed in Section 5.3.3. The steady-state resonant solutions of eq. (2.95) are studied in Section 8.2. We study asymptotic ordering of these solutions to be extended to resonant sloshing in a rectangular tank.

Another example is a system of *two coupled horizontal linear springs* depicted in Figure 2.7. The Lagrangian of this system is given by

$$L = \left(\tfrac{1}{2}M_1\dot{\beta}_1^2 + \tfrac{1}{2}M_2\dot{\beta}_2^2\right) - \left(\tfrac{1}{2}\mathrm{k}_1\beta_1^2 + \tfrac{1}{2}\mathrm{k}_2(\beta_2 - \beta_1)^2\right).$$

Applying Euler–Lagrange equations (2.92) leads to the system of ordinary differential equations

$$M_1\ddot{\beta}_1 + \mathrm{k}_1\beta_1 - \mathrm{k}_2(\beta_2 - \beta_1) = 0;$$
$$M_2\ddot{\beta}_2 + \mathrm{k}_2(\beta_2 - \beta_1) = 0,$$

which can be expressed in matrix form as

$$\begin{bmatrix} M_1 & 0 \\ 0 & M_2 \end{bmatrix}\begin{pmatrix} \ddot{\beta}_1 \\ \ddot{\beta}_2 \end{pmatrix} + \begin{bmatrix} \mathrm{k}_1 + \mathrm{k}_2 & -\mathrm{k}_2 \\ -\mathrm{k}_2 & \mathrm{k}_2 \end{bmatrix}\begin{pmatrix} \beta_1 \\ \beta_2 \end{pmatrix} = 0.$$

2.5.2.2 Euler–Bernoulli beam equation

Mechanical systems involving elastic bodies or fluids are often treated as having an infinite number of degrees of freedom. A simple example is the dynamics of a beam, illustrated in Figure 2.8. The beam dynamics is described by an initial boundary-value problem in terms of the horizontal coordinate x and time t. The problem allows for the Lagrange formulation, calculus of variations, and even infinite-dimensional Euler–Lagrange *modal* equations, which are exemplified in Section 7.2. The beam equation is applied in Sections 5.4.5 and 11.9.2.

Let us consider a *beam* of length l as shown in Figure 2.8, where $\mathrm{w}(x,t)$ is the instantaneous transverse deflection of the beam, x is a coordinate along the beam (the beam is modeled as a one-dimensional object), and f is a distributed load (Clough & Penzien, 1993). The Euler–Bernoulli beam model is assumed, which means that the effect of axial forces on bending stresses is neglected. The kinetic energy of the beam is

$$E_k = \frac{1}{2}\int_0^l m\left(\frac{\partial \mathrm{w}}{\partial t}\right)^2 \mathrm{d}x, \tag{2.96}$$

where $m = m(x)$ is the mass per unit length. The potential energy consists of the sum

$$E_p = \underbrace{\frac{1}{2}\int_0^l \mathtt{EI}\left(\frac{\partial^2 \mathrm{w}}{\partial x^2}\right)^2 \mathrm{d}x}_{E_{in}} - \underbrace{\int_0^l f\mathrm{w}\,\mathrm{d}x}_{E_{ext}} + \underbrace{mg\int_0^l \mathrm{w}\,\mathrm{d}x}_{E_g}, \tag{2.97}$$

where $\mathtt{E}$ is the Young modulus of elasticity, $\mathtt{I}$ is the second moment of area with respect to the neutral axis perpendicular to the applied loading (the gravity term disappears when we consider a

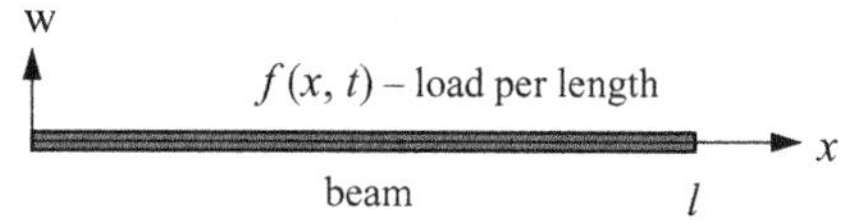

Figure 2.8. Sketch of a beam with forcing $f(x,t)$.

vertical beam), $\partial \mathrm{w}/\partial x$ is the slope of the beam, $\mathrm{EI}(\partial^2 \mathrm{w}/\partial x^2)$ is the bending moment of the beam, and $-\partial\left(\mathrm{EI}\partial^2 \mathrm{w}/\partial x^2\right)/\partial x$ is the shear force in the beam. Details on the calculation of the internal potential energy can be found in books on structural dynamics (e.g., Clough & Penzien, 1993; Ziegler, 1995).

Owing to eqs. (2.96) and (2.97), the Lagrangian is formally presented by the following integral:

$$L = \int_0^l \left[\tfrac{1}{2} m \left(\frac{\partial \mathrm{w}}{\partial t}\right)^2 - \tfrac{1}{2}\mathrm{EI}\left(\frac{\partial^2 \mathrm{w}}{\partial x^2}\right)^2 + f\mathrm{w} - mg\mathrm{w}\right] \mathrm{d}x, \tag{2.98}$$

which includes $\mathrm{w}(x, t)$ and its spatial derivative. It is therefore impossible to consider w as a generalization of the path β from variational formulation (2.80). The procedure must distinguish between variations of w and its spatial derivatives, which are definitely not the same.

The beam equation and some of the boundary conditions can therefore be derived from minimization of (2.80) by direct variations of w. For this purpose, we consider the extrema ($\delta W = 0$) of the action

$$W\left(\mathrm{w}, \frac{\partial \mathrm{w}}{\partial t}\right) = \int_{t_1}^{t_2} L\left(\mathrm{w}, \frac{\partial^2 \mathrm{w}}{\partial x^2}, \frac{\partial \mathrm{w}}{\partial t}\right) \mathrm{d}t \tag{2.99}$$

for variations that satisfy the conditions

$$\delta \mathrm{w}(x, t_1) = \delta \mathrm{w}(x, t_2) = 0 \quad \text{for all } x \in [0, l]. \tag{2.100}$$

The variation gives the variational equation

$$\int_{t_1}^{t_2}\int_0^l \left[m\frac{\partial \mathrm{w}}{\partial t}\frac{\partial \delta \mathrm{w}}{\partial t} - \mathrm{EI}\frac{\partial^2 \mathrm{w}}{\partial x^2}\frac{\partial^2(\delta \mathrm{w})}{\partial x^2} + f\delta \mathrm{w} - mg\delta \mathrm{w}\right] \mathrm{d}x\mathrm{d}t = 0. \tag{2.101}$$

Integration by parts of the first quantity with respect to time (subject to conditions (2.100)) as well as the spatial integration by parts twice of the second quantity reexpresses eq. (2.101) in the form

$$\int_{t_1}^{t_2}\int_0^l \left[-m\frac{\partial^2 \mathrm{w}}{\partial t^2} - \frac{\partial^2}{\partial x^2}\left(\mathrm{EI}\frac{\partial^2 \mathrm{w}}{\partial x^2}\right) + f - mg\right]\delta \mathrm{w}\, \mathrm{d}x\, \mathrm{d}t$$
$$- \int_{t_1}^{t_2}\left[\mathrm{EI}\frac{\partial^2 \mathrm{w}}{\partial x^2}\delta\left(\frac{\partial \mathrm{w}}{\partial x}\right)\right]\Bigg|_0^l \mathrm{d}t$$
$$+ \int_{t_1}^{t_2}\left[\frac{\partial}{\partial x}\left(\mathrm{EI}\frac{\partial^2 \mathrm{w}}{\partial x^2}\right)\delta \mathrm{w}\right]\Bigg|_0^l \mathrm{d}t = 0. \tag{2.102}$$

Because eq. (2.102) must be fulfilled for arbitrary test functions $\delta \mathrm{w}$ including those satisfying (2.105), the first integral reduces to the beam equation:

$$-m\frac{\partial^2 \mathrm{w}}{\partial t^2} - \frac{\partial^2}{\partial x^2}\left(\mathrm{EI}\frac{\partial^2 \mathrm{w}}{\partial x^2}\right) + f - mg = 0. \tag{2.103}$$

Furthermore, the beam equation contains a fourth-order derivative in x. Hence, it is necessary to have four conditions, which are normally boundary conditions. If we do not restrict variations $\partial(\delta \mathrm{w})/\partial x$ and $\delta \mathrm{w}$ at the ends (i.e., $\partial(\delta \mathrm{w})/\partial x$ and $\delta \mathrm{w}$ are arbitrary time-dependent functions at the ends), the last two integrals of eq. (2.102) become zero only when w satisfies the zero bending moment and shear force conditions at the ends:

$$\frac{\partial^2 \mathrm{w}}{\partial x^2} = \frac{\partial}{\partial x}\left(\mathrm{EI}\frac{\partial^2 \mathrm{w}}{\partial x^2}\right) = 0 \quad \text{at} \quad x = 0 \quad \text{and } x = l. \tag{2.104}$$

These conditions are the so-called *natural* boundary conditions of the problem – those that naturally follow from the variational statement.

If the ends (or one of the ends) are clamped, that is,

$$\mathrm{w} = \frac{\partial \mathrm{w}}{\partial x} = 0 \tag{2.105}$$

at the ends (one of the ends), the test functions $\delta \mathrm{w}$ and $\partial \delta \mathrm{w}/\partial x$ do not vary at the corresponding ends of eq. (2.102) and, therefore, variational equation (2.102) does not imply any boundary conditions at these ends. This result means that the clamped-end *boundary conditions are not natural* for the variational statement. These conditions *should therefore be fulfilled prior to considering action* (2.99). In contrast, we must not incorporate free-end conditions (2.104) in the variational methods based on eq. (2.101). These conditions are automatically satisfied as a result of the extrema condition $\delta W = 0$. An example is a cantilever beam that is clamped at one end and free at the other (see Figure 2.9). In this case, the test functions w in functional (2.99) should satisfy the clamped-end conditions at $x = 0$, but the free boundary condition at $x = l$ follows from the necessary extrema condition $\delta W = 0$. The boundary conditions may also include loads caused by other structures, usually model *supports* (i.e., model point loads, moments, or other

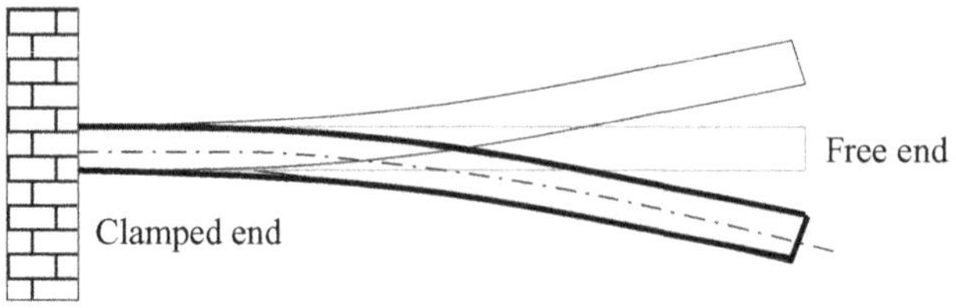

Figure 2.9. Schematic vibrations of a cantilever beam with one clamped end and one free end.

effects). The Lagrange formulation may include many of those boundary conditions.

2.5.2.3 Linear sloshing in an upright nonmoving tank

When it comes to the sloshing problem, we must recall, as it was discussed in the context of Euler–Lagrange equation (2.89), that the Lagrange formulation leads only to the dynamic equations for the considered mechanical system. The sloshing problem also involves the kinematic subproblem consisting of the Laplace equation and the Neumann boundary conditions on the wall and the free surface. This kinematic subproblem determines the liquid velocity field (the velocity potential) from a given wave elevation. The kinematic part of the sloshing problem does not follow from the Lagrange formulation and must be satisfied a priori.

As an example, let us show how the Lagrange variational formulation may be applied to a linearized sloshing problem in a stationary tank with liquid depth h. For that case, $\boldsymbol{\omega} = \boldsymbol{v}_O = \boldsymbol{0}$ and the dynamic and kinematic boundary conditions are given by eqs. (2.71) and (2.72).

As we have already mentioned, the path β of statements (2.80) and (2.81) can then be associated with a function $\zeta(x, y, t)$ describing the instantaneous free-surface positions. The incompressible liquid has zero internal energy. The minimization of the action W means

$$\delta W(\zeta, \partial\zeta/\partial t) = \delta \int_{t_1}^{t_2} L\,\mathrm{d}t = 0, \qquad (2.106)$$

where the Lagrangian $L = E_k - E_p$ and where E_k and E_p are the kinetic and potential energies, respectively. Because we consider a linear problem, it is sufficient to express the kinetic energy in the domain Q_0 below the mean free surface Σ_0, that is, $E_k = \frac{1}{2}\rho\int_{Q_0} (\nabla\Phi)^2\,\mathrm{d}Q$, where Φ is the velocity potential. This expression correctly gives the kinetic energy to $O(\varepsilon^2)$, where the parameter ε characterizes the free-surface elevation ζ and the velocity potential. The difference between the kinetic energy in the instantaneous liquid domain $Q(t)$ and in Q_0 is of $O(\varepsilon^3)$. Introduction of these higher-order terms leads to nonlinear terms in the formulation of the sloshing problem. The potential energy due to sloshing can be expressed as

$$E_p = \rho g \int_{Q(t)} z\,\mathrm{d}Q - \tfrac{1}{2}\rho g h^2 = \tfrac{1}{2}\rho g \int_{\Sigma_0} \zeta^2\,\mathrm{d}S.$$

Formally, we get

$$L(\zeta, \partial\zeta/\partial t) = \underbrace{\tfrac{1}{2}\rho \int_{Q_0} (\nabla\Phi)^2\,\mathrm{d}Q}_{E_k} - \underbrace{\tfrac{1}{2}\rho g \int_{\Sigma_0} \zeta^2\,\mathrm{d}S}_{E_p} + O(\varepsilon^3). \qquad (2.107)$$

To explain why we write that $L = L(\zeta, \partial\zeta/\partial t)$ but not that $L = L(\zeta, \Phi)$, one must recall that the velocity field $\nabla\Phi$ of a limited ideal liquid volume is not an independent parameter but a function of its free-surface motions (see Birkhoff, 1960). As we pointed out in the beginning of the section, the flow field is a function of $\partial\zeta/\partial t$ (i.e., $\Phi = \Phi[\partial\zeta/\partial t]$), which follows from the kinematic subproblem

$$\nabla^2\Phi = 0 \quad \text{in } Q_0; \quad \frac{\partial\Phi}{\partial n} = 0 \quad \text{on } S_0;$$
$$\frac{\partial\Phi}{\partial z} = \frac{\partial\zeta}{\partial t} \quad \text{on } \Sigma_0, \qquad (2.108)$$

meaning that we should solve problem (2.108) and substitute it into Lagrangian (2.107) prior to calculating the variation of the action. As a matter of fact, the Laplace equation and kinematic boundary conditions from eq. (2.108) are not *natural* for the Lagrange formulation, as we discussed in Section 2.5.2.2 for the beam problem.

Variations of the free surface due to general formulation (2.81) should satisfy

$$\delta\zeta(x, y, t_1) = \delta\zeta(x, y, t_2) = 0. \qquad (2.109)$$

Equations (2.106)–(2.109) express the Lagrange formulation for the studied linear sloshing problem.

To show that eq. (2.106) leads to dynamic boundary condition (2.71), let us assume that both the velocity potential and the function ζ are parametrically dependent on a small time-independent number α (i.e., $\Phi = \Phi(x, y, z, t, \alpha)$ and $\zeta = \zeta(x, y, t, \alpha)$) so that $\alpha = 0$ corresponds to the extreme value of the action W. Substitution of the functions $\Phi(x, y, z, t, \alpha)$ and $\zeta(x, y, t, \alpha)$ into eqs. (2.106) and (2.107) causes the action to be a single-variable function; that is, $W(\alpha) = W(\Phi(x, y, z, \alpha), \zeta(x, y, \alpha))$. Formal calculation of its variation, δW, implies the differential of W by α at $\alpha = 0$:

$$dW\Big|_{\alpha=0} = \delta W = \frac{dW}{d\alpha}\Big|_{\alpha=0} \alpha$$

$$= \frac{\partial W}{\partial \Phi}\Big|_{\alpha=0} \underbrace{\frac{\partial \Phi}{\partial \alpha}\alpha}_{\delta\Phi} + \frac{\partial W}{\partial \zeta}\Big|_{\alpha=0} \underbrace{\frac{\partial \zeta}{\partial \alpha}\alpha}_{\delta\zeta},$$

where $\partial\Phi/\partial\alpha$ and $\partial\zeta/\partial\alpha$ are in fact arbitrary functions that yield small arbitrary variations $\delta\Phi$ and $\delta\zeta$ for small nonzero values of α. Consequently, we obtain the following expressions, which are based on Gauss' theorem (A.1) and the fact that both $(\partial\zeta/\partial t, \Phi)$ and $(\delta[\partial\zeta/\partial t], \delta\Phi)$ satisfy problem (2.108):

$$\delta W = \rho \int_{t_1}^{t_2} \left[\int_{Q_0} \nabla\Phi\nabla(\delta\Phi)\, \mathrm{d}Q - g \int_{\Sigma_0} \zeta(\delta\zeta)\mathrm{d}S \right] \mathrm{d}t$$
$$+ [\text{nonlinear in } (\zeta, \Phi)]$$
$$= \rho \int_{t_1}^{t_2} \underbrace{\int_{\Sigma_0} \frac{\partial(\delta\Phi)}{\partial n} \Phi \mathrm{d}S}_{\int_{\Sigma_0} \frac{\partial(\delta\zeta)}{\partial t}\Phi \mathrm{d}S = \frac{d}{dt}\int_{\Sigma_0} \delta\zeta\Phi \mathrm{d}S - \int_{\Sigma_0} \frac{\partial\Phi}{\partial t}\delta\zeta \mathrm{d}S} \mathrm{d}t$$
$$- g\rho \int_{t_1}^{t_2} \int_{\Sigma_0} \zeta(\delta\zeta)\mathrm{d}S\, \mathrm{d}t + [\text{nonlinear in } (\zeta, \Phi)]$$
$$= -\rho \int_{t_1}^{t_2} \int_{\Sigma_0} \left[\frac{\partial\Phi}{\partial t} + g\zeta \right] \delta\zeta\, \mathrm{d}S\, \mathrm{d}t$$
$$+ \rho \underbrace{\int_{\Sigma_0} \delta\zeta(x, y, t_2)\Phi\, \mathrm{d}S - \int_{\Sigma_0} \delta\zeta(x, y, t_1)\Phi\, \mathrm{d}S}_{=0, \text{because of the boundary conditions at } t_1 \text{ and } t_2}$$
$$+ [\text{nonlinear in}(\zeta, \Phi)] = 0. \tag{2.110}$$

Because eq. (2.110) has to be satisfied for arbitrary $\delta\zeta$, it indeed leads to linear dynamic condition (2.71) on the mean liquid plane Σ_0.

2.5.3 Lagrange and Bateman–Luke variational formulations for nonlinear sloshing

The Lagrange variational principle also derives nonlinear dynamic condition (2.64) of the sloshing problem, assuming that eqs. (2.58), (2.63), (2.68), and (2.74) are fulfilled. A mathematical proof of this fact is a nontrivial task. Most probably, Petrov (1964) was the first to give it. However, his derivation contained arithmetic errors that were discussed and corrected by Komarenko (1989). Alternative mathematical proofs were presented by Berdichevsky (1983) and Limarchenko (1978a). Although the Lagrange formulation is not as convenient as the Bateman–Luke variational principle to derive nonlinear modal systems (Chapter 7), a set of publications exists in which the Lagrange formulation was used in multimodal analysis (see Limarchenko, 1978b, 1983, and references therein). We refer interested readers to the original publications for more details.

When formulating the variational principles for nonlinear sloshing, we must use the implicit parameterization of the free surface $\Sigma(t)$ (i.e., $Z(x, y, z, t) = 0$) because of the possibility of nonvertical walls at the free-surface zone and overturning waves.

2.5.3.1 The Lagrange variational formulation

Smooth solutions of the free-boundary-value sloshing problem are associated with the extrema of the action $W(Z)$; that is,

$$W(Z) = \int_{t_1}^{t_2} L\, \mathrm{d}t = 0,$$
$$L = \int_{Q(t)} \left[\tfrac{1}{2}\rho(\nabla\Phi)^2 - U_g \right] \mathrm{d}Q, \tag{2.111}$$

where the gravity potential $U_g = -\boldsymbol{g} \cdot \boldsymbol{r}$ and the velocity potential is the solution of the Neumann boundary-value (kinematic) problem (2.58), (2.63), (2.68), and (2.74). The variations satisfy the boundary conditions

$$\delta Z(x, y, z, t_1) = \delta Z(x, y, z, t_2) = 0; \tag{2.112}$$

that is, the free-surface shape is the same at instants t_1 and t_2.

An alternative to the Lagrange variational formulation is the so-called Bateman–Luke variational principle. In contrast to formulation

(2.111), in which the Lagrangian is the difference between kinetic and potential energy, the Bateman–Luke variational principle associates L with a "pressure integral." We start with the *formal* expression $L = \int_{Q(t)}(p - p_0)\,dQ$ with $p - p_0$, the same expression as Bernoulli equation (2.61). The Bateman–Luke principle *requires* that one considers $\Phi(x, y, z, t)$ and $Z(x, y, z, t)$ as any two independent functional variables ($W = W(Z, \Phi)$). This consideration means that $\Phi(x, y, z, t)$ and $Z(x, y, z, t)$ are not necessarily the velocity potential and the function defining the free-surface elevation, respectively. Moreover, eq. (2.61) based on the test functions $\Phi = \Phi(x, y, z, t)$ and $Z(x, y, z, t)$ is not necessarily the actual sloshing pressure. However, one can show that the necessary extrema condition of the Bateman–Luke action $W(Z, \Phi)$ mathematically derives all the equations and boundary conditions of the nonlinear sloshing problem, meaning that originally independent functions $\Phi(x, y, z, t)$ and $Z(x, y, z, t)$ become the actual velocity potential and the free-surface elevation *only* when the extrema of the Bateman–Luke action $W = W(Z, \Phi)$ is satisfied. The Laplace equation and all the boundary conditions are natural for the Bateman–Luke principle.

Historical aspects on the "pressure-integral" Lagrangian are outlined by Lukovsky and Timokha (1995) and Faltinsen *et al.* (2000). We present the formulation of the Bateman–Luke principle in the form given by Lukovsky (1990).

2.5.3.2 The Bateman–Luke principle

Smooth solutions of the free-boundary-value problem (2.58), (2.63), (2.64), (2.68), and (2.74) coincide with the extrema of the action

$$\begin{aligned} W(Z,\Phi) &= \int_{t_1}^{t_2} L\,dt,\\ L &= \int_{Q(t)} (p - p_0)\,dQ\\ &= -\rho\int_{Q(t)}\left[\frac{\partial\Phi}{\partial t} + \tfrac{1}{2}(\nabla\Phi)^2 - \nabla\Phi\cdot(\boldsymbol{v}_O + \boldsymbol{\omega}\times\boldsymbol{r}) + U_g\right] dQ \end{aligned} \tag{2.113}$$

subject to the test functions satisfying

$$\begin{aligned} &\delta\Phi(x,y,z,t_1) = 0, \quad \delta\Phi(x,y,z,t_2) = 0;\\ &\delta Z(x,y,z,t_1) = 0, \quad \delta Z(x,y,z,t_2) = 0. \end{aligned} \tag{2.114}$$

A compact proof of this variational formulation was presented by Lukovsky and Timokha (1995). Because the formulation will be used for derivations of modal systems, let us describe some details of this proof.

First we assume $\Phi = \Phi(x, y, z, t, \alpha_1)$ and $Z = Z(x, y, z, t, \alpha_2)$, where the two small parameters α_1 and α_2 are independent and their zeros correspond to the minimum of the action W. Substitution of Φ and Z into action (2.113) makes it a function of two real variables α_1 and α_2 (i.e., $W = W(\alpha_1, \alpha_2)$). Calculus of variations at the point $\alpha_1 = \alpha_2 = 0$ implies

$$\begin{aligned} dW(0,0) &= \left.\frac{\partial W}{\partial\alpha_1}\right|_{(0,0)}\alpha_1 + \left.\frac{\partial W}{\partial\alpha_2}\right|_{(0,0)}\alpha_2\\ &= \delta W = \underbrace{\frac{\partial W}{\partial Z}\frac{\partial Z}{\partial\alpha_2}\alpha_2}_{\delta Z} + \underbrace{\frac{\partial W}{\partial\Phi}\frac{\partial\Phi}{\partial\alpha_1}\alpha_1}_{\delta\Phi}\\ &= -\rho\int_{t_1}^{t_2}\left\{-\int_{\Sigma(t)}\left[\frac{\partial\Phi}{\partial t} + \tfrac{1}{2}(\nabla\Phi)^2 - \nabla\Phi\cdot(\boldsymbol{v}_O + \boldsymbol{\omega}\times\boldsymbol{r}) + U_g\right]\frac{\delta Z}{|\nabla Z|}dS\right.\\ &\quad + \int_{Q(t)}\left[\nabla\Phi\cdot\nabla(\delta\Phi) + \frac{\partial(\delta\Phi)}{\partial t}\right.\\ &\quad \left.\left. -\nabla(\delta\Phi)\cdot(\boldsymbol{v}_O + \boldsymbol{\omega}\times\boldsymbol{r})\right]dQ\right\}dt = 0. \end{aligned} \tag{2.115}$$

To simplify integral expressions in eq. (2.115) we use Reynolds transport theorem (A.2), where the differentiation parameters are formally α_1 and α_2. The gravity potential U_g and $(\boldsymbol{v}_O + \boldsymbol{\omega}\times\boldsymbol{r})$ do not depend on α_1 and α_2; therefore, variations of the corresponding terms are zero. Because δZ and $\delta\Phi$ are independent, when posing $\delta\Phi = 0(\alpha_1 = 0)$ and considering arbitrary variations δZ, equality (2.115) implies the dynamic boundary condition on $\Sigma(t)$. This result follows from the integral over $\Sigma(t)$ in eq. (2.115). Furthermore, when $\delta Z = 0$, the remaining integral of eq. (2.115) over $Q(t)$ can be modified by using Green's first identity (A.5) and Gauss's theorem (A.1), which imply the following formulas:

$$\int_{Q(t)}\nabla\Phi\cdot\nabla(\delta\Phi)\,dQ = \int_{S(t)+\Sigma(t)}\delta\Phi\frac{\partial\Phi}{\partial n}\,dS - \int_{Q(t)}\nabla^2\Phi\,\delta\Phi\,dQ,$$

$$\int_{Q(t)} \boldsymbol{v}_O \cdot \nabla(\delta\Phi)\mathrm{d}Q = \int_{Q(t)} \nabla(\boldsymbol{v}_O \cdot \boldsymbol{r}) \cdot \nabla(\delta\Phi)\,\mathrm{d}Q = \int_{S(t)+\Sigma(t)} \delta\Phi(\boldsymbol{v}_O \cdot \boldsymbol{n})\,\mathrm{d}S,$$

$$\int_{Q(t)} (\boldsymbol{\omega} \times \boldsymbol{r}) \cdot \nabla(\delta\Phi)\mathrm{d}Q = \int_{S(t)+\Sigma(t)} \delta\Phi((\boldsymbol{\omega} \times \boldsymbol{r}) \cdot \boldsymbol{n})\,\mathrm{d}S,$$

$$\int_{Q(t)} \frac{\partial(\delta\Phi)}{\partial t}\mathrm{d}Q = \frac{d}{dt}\int_{Q(t)} \delta\Phi\,\mathrm{d}Q + \int_{\Sigma(t)} (\delta\Phi)|_{\Sigma(t)} \frac{\partial Z/\partial t}{|\nabla Z|}\mathrm{d}S.$$

This transforms the last integral quantity of eq. (2.115) to the variational equality

$$\delta W|_{\delta Z=0} = -\rho\int_{t_1}^{t_2}\left\{\int_{S(t)}\left[\frac{\partial\Phi}{\partial n} - (\boldsymbol{v}_O + \boldsymbol{\omega}\times\boldsymbol{r})\cdot\boldsymbol{n}\right]\delta\Phi\,\mathrm{d}S - \int_{Q(t)} \nabla^2\Phi\delta\Phi\,\mathrm{d}Q + \int_{\Sigma(t)}\left[\frac{\partial\Phi}{\partial n} - (\boldsymbol{v}_O + \boldsymbol{\omega}\times\boldsymbol{r})\cdot\boldsymbol{n} + \frac{\partial Z}{\partial t}\Big/|\nabla Z|\right]\delta\Phi\,\mathrm{d}S\right\}\mathrm{d}t - \rho\int_{Q(t)}\delta\Phi\,\mathrm{d}Q\Big|_{t=t_1}^{t=t_2} = 0, \quad (2.116)$$

where the last quantity, $\rho\int_{Q(t)}\delta\Phi\mathrm{d}Q|_{t=t_1}^{t=t_2}$, gives zero contribution because $\delta\Phi = 0$ as $t = t_1, t_2$ due to conditions (2.114). As long as eq. (2.116) is fulfilled for arbitrary $\delta\Phi$, it is equivalent to the original free-boundary problem.

2.6 Summary

Liquid sloshing in a moving tank is analyzed in a body-fixed coordinate system. The chapter starts with background on the Navier–Stokes equations and conservation laws in an inertial coordinate system. We then introduce the tank-fixed coordinate system and gives the main rules on how to recalculate the scalar and vector fields from inertial (Earth-fixed) to noninertial systems (rigidly fixed with the tank).

The explicit form of the Navier–Stokes equations in the tank-fixed coordinate system is presented in Section 2.4. Furthermore, because the majority of analytically based methods use the potential flow formulation, we present details on the boundary conditions of the sloshing problem written in a noninertial coordinate system. Prescribed motions of the tank are assumed. The velocity potential of the absolute velocity is a function of the coordinates of the movable system and time. It satisfies the Laplace equation in the time-dependent domain. The so-called kinematic and dynamic free-boundary conditions are introduced. The first condition represents the normal absolute velocity on the free surface, but the dynamic condition implies that the hydrodynamic pressure is equal to the constant pressure outside of the liquid (e.g., atmospheric pressure). We also discuss periodic and initial conditions.

In addition, the chapter gives background on the Lagrange formalism that is commonly used in analytically oriented solutions of liquid sloshing problems. In particular, Chapter 7 introduces multimodal methods, which are based on the Bateman–Luke variational principle, a variant of the Lagrangian formulation of the nonlinear sloshing problem. This variational principle is extensively discussed in the chapter.

2.7 Exercises

2.7.1 Flow parameters

(a) Assume that a tank is completely filled with liquid. Start with the Navier–Stokes equations expressed by eqs. (2.1) and (2.2), and initial and boundary conditions. Introduce a characteristic velocity U and a characteristic length L. Make the equations nondimensional in terms of U, L, and ρ. Explain the fact that nondimensional stresses and velocities are functions of the *Reynolds number,* $Rn = UL/\nu$. Express the nondimensional stresses and velocities.

(b) Study a potential flow problem for an incompressible liquid with a free surface inside a tank. Make the equations nondimensional by introducing a characteristic velocity U and a characteristic length L. Show that the *Froude number,* $Fn = U/\sqrt{Lg}$, is a flow parameter. How should the nondimensional flow variables be

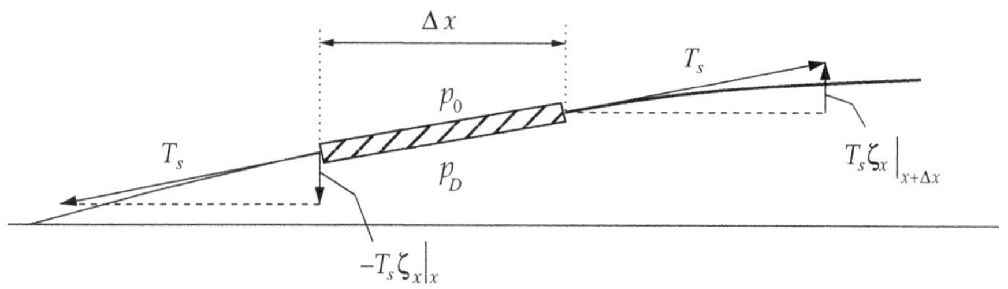

Figure 2.10. Vertical force component of the surface tension T_s acting on a free-surface element of length Δx and infinitesimally small thickness. A small wave slope $\zeta_x = \partial\zeta/\partial x$ and two-dimensional flow are assumed; p_D = liquid pressure, and p_0 = ambient air pressure.

expressed? If the flow is associated with a forcing frequency σ, how should σ be made nondimensional according to Froude scaling?

(c) Use the equations in Subsection 2.2.2.2 to show that the *Mach number*, $Ma = U/c_0$, is a flow parameter for a compressible fluid. Express the *Cauchy number*, $Ca = \rho U^2/E_v$, in terms of the Mach number, where E_v is the bulk modulus for elasticity.

2.7.2 Surface tension

(a) We shall assume a two-dimensional flow situation in the x–z-plane and study the effect of surface tension T_s per unit length (see Figure 2.10). A representative value of T_s for the water–air interface is 0.073 N m^{-1}. The wave slope $\partial\zeta/\partial x$ is assumed small so that we can linearize the free-surface conditions. Figure 2.10 illustrates the vertical force due to surface tension acting on a free-surface element of length Δx and infinitesimal thickness. The ambient air and the hydrodynamic pressures are denoted p_0 and p_D, respectively. Assume potential flow of an incompressible liquid with velocity potential Φ and show that the linear dynamic free-surface condition can be expressed as

$$\frac{\partial\Phi}{\partial t} + g\zeta - \frac{T_s}{\rho}\frac{\partial^2\zeta}{\partial x^2} = 0 \quad \text{on} \quad z = 0.$$

Show by combining the dynamic and kinematic free-surface conditions that

$$\frac{\partial^2\Phi}{\partial t^2} + g\frac{\partial\Phi}{\partial z} - \frac{T_s}{\rho}\frac{\partial^2}{\partial x^2}\frac{\partial\Phi}{\partial z} = 0 \quad \text{on} \quad z = 0.$$

(b) Introduce a characteristic velocity U and a characteristic length L. Make the equations nondimensional in terms of U, L, and ρ. Make the equations nondimensional and show that the *Weber number*, $Wn = U^2 L\rho/T_S$, is a flow parameter.

2.7.3 Kinematic boundary condition

Show the details of the derivation of the kinematic free-surface condition expressed by eq. (2.68).

2.7.4 Added mass force for a nonlifting body in infinite fluid

We study a moving body in an infinite incompressible fluid with no ambient flow. A body-fixed Cartesian coordinate system $Oxyz$ is introduced. The velocity vector of an arbitrary point P on the body is expressed as $\boldsymbol{v}_b = \boldsymbol{v}_O + \boldsymbol{\omega} \times \boldsymbol{r}$, where $\boldsymbol{v}_O(t)$ is the velocity vector of the point O, $\boldsymbol{r}$ is the radius vector from O to P, and $\boldsymbol{\omega}(t)$ is the vector of the angular velocity of rotation of the body. We express $\boldsymbol{v}_O(t) = (\dot\eta_1, \dot\eta_2, \dot\eta_3)$ and $\boldsymbol{\omega}(t) = (\dot\eta_4, \dot\eta_5, \dot\eta_6)$, where $\dot\eta_1$, $\dot\eta_2$, and $\dot\eta_3$ are the components of $\boldsymbol{v}_O(t)$ along the body-fixed coordinate axis; $\dot\eta_4$, $\dot\eta_5$, and $\dot\eta_6$ have similar meaning for $\boldsymbol{\omega}(t)$. The normal vector $\boldsymbol{n} = (n_1, n_2, n_3)$ to the body surface $S(t)$ is positive out of the fluid domain. Furthermore, we define $\boldsymbol{r} \times \boldsymbol{n} = (n_4, n_5, n_6)$. We assume that the flow can be described by a velocity potential Φ and that there is zero circulation around the body (i.e., nonlifting body).

(a) Show that the velocity potential can be expressed as $\Phi = \sum_{j=1}^{6} \varphi_j \dot\eta_j$, where φ_j satisfies $\frac{\partial\varphi_j}{\partial n} = n_j$, $j = 1, \ldots, 6$ on $S(t)$.

(b) Show by using the equations for conservation of fluid momentum (see Section 2.2.4.1) that the hydrodynamic force

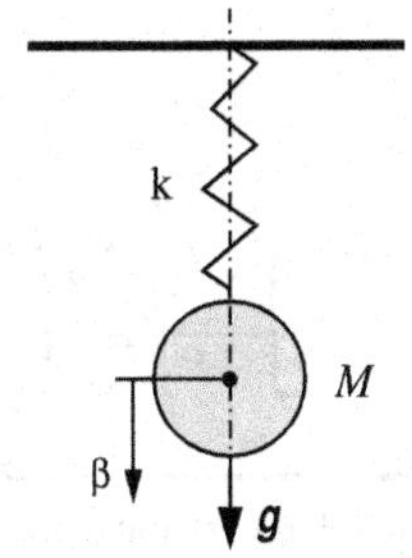

Figure 2.11. A vertical spring–mass system.

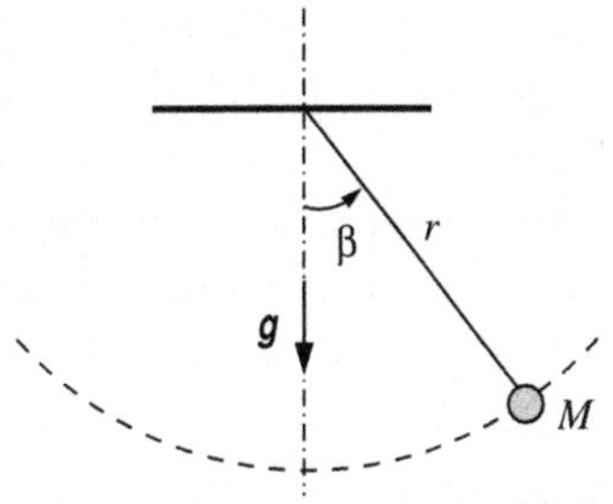

Figure 2.12. A pendulum with a rigid mass M.

on the body can be expressed as $\boldsymbol{F} = -\dot{\boldsymbol{M}}$, where $\boldsymbol{M} = \int\int_S \rho\Phi\boldsymbol{n}\,\mathrm{d}S$. (*Hint*: Use the generalized Gauss theorem to rewrite the expression for the fluid momentum.)

(c) Show that the hydrodynamic force acting on the body can be expressed as $\boldsymbol{F} = -d^*\boldsymbol{B}/dt - \boldsymbol{\omega}\times\boldsymbol{B}$, where $\boldsymbol{B} = (B_1, B_2, B_3)$, $B_j = \sum_{k=1}^{6} A_{jk}\dot{\eta}_k$, $A_{jk} = \rho\int\int_S \varphi_k n_j\,\mathrm{d}S$, and $d^*\boldsymbol{B}/dt$ is defined by eq. (2.50).

(d) Why are A_{jk} called added mass coefficients?

(e) We now set $\dot{\eta}_3 = \dot{\eta}_4 = \dot{\eta}_5 = 0$ and assume that the x–z- and x–y-planes are symmetry planes for the body. Show that the hydrodynamic force $\boldsymbol{F} = (F_1, F_2, F_3)$ can be expressed as

$$\begin{aligned} F_1 &= -A_{11}\ddot{\eta}_1 + (A_{22}\dot{\eta}_2 + A_{26}\dot{\eta}_6)\,\dot{\eta}_6, \\ F_2 &= -A_{22}\ddot{\eta}_2 - A_{26}\ddot{\eta}_6 - A_{11}\dot{\eta}_1\dot{\eta}_6, \\ F_3 &= 0. \end{aligned} \qquad (2.117)$$

(*Hint*: Use symmetry and antisymmetry properties of the flow and the body surface.)

(f) Consider the centrifugal force on the body by choosing O as the center of gravity. Decompose the centrifugal force along the coordinate axes and show which of the terms in eq. (2.117) have similar dependence on $\dot{\eta}_j$ as the centrifugal force components.

2.7.5 Euler–Lagrange equations for finite-dimensional mechanical systems

(a) Consider the vertical spring–mass system in Figure 2.11. In this case the potential energy must include both internal strain and gravitation potential energy. The kinetic energy is the same. Derive the Euler–Lagrange equation for this case.

Answer: The Lagrangian is $L = \frac{1}{2}M\dot{\beta}^2 - \left(\frac{1}{2}k\beta^2 - Mg\beta\right)$, which leads to the Euler–Lagrange equation

$$M\ddot{\beta} + k\beta - Mg = 0. \qquad (2.118)$$

(b) Consider a two-dimensional pendulum with a mass point M as shown in Figure 2.12 and use the angular position function $\beta(t)$ as the generalized coordinate for the admissible trajectories (paths). This approach makes it possible to use Euler–Lagrange equation (2.89) and derive the equation for the pendulum dynamics.

Answer: The Lagrangian is $L = \frac{1}{2}M(r\dot{\beta})^2 + Mgr\cos\beta$, which gives the Euler–Lagrange equation

$$Mr^2\ddot{\beta} + Mgr\sin\beta = 0. \qquad (2.119)$$

(c) This example is a typical case in seismic dynamics. Normally support reactions do not work because the supports do not move. The earthquake ground motion moves the support by $\beta_0(t)$; hence, support reaction $R(t)$ must be included in the total energy as the work done by an external force (see Figure 2.13). Derive the dynamic

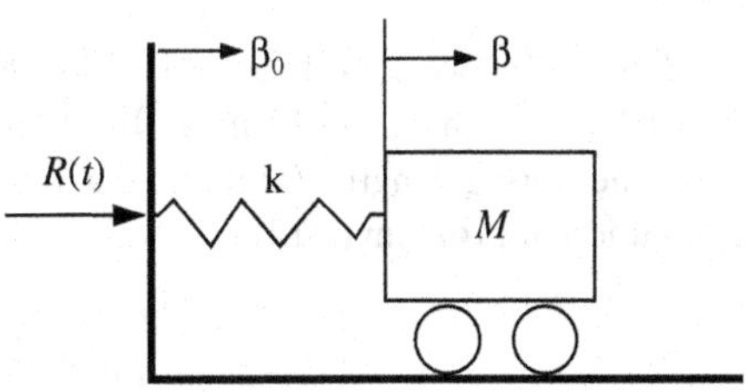

Figure 2.13. A horizontal spring–mass system with support excitation.

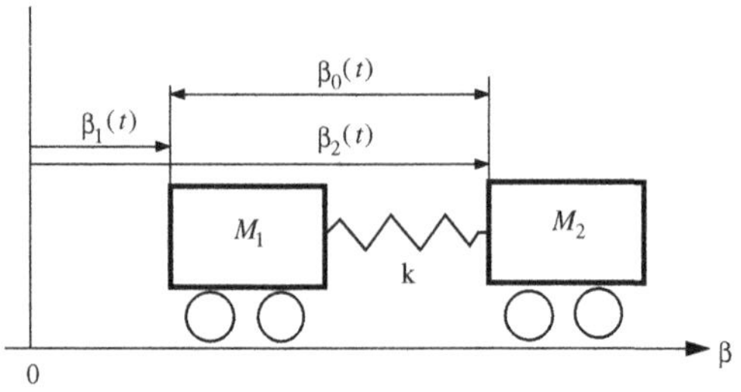

Figure 2.14. Two horizontal spring–mass systems moving in the positive direction of the β-axis.

equations using the Euler–Lagrange equation leading to eq. (2.92).

Answer: The Lagrangian is $L = (\frac{1}{2}M\dot{\beta}^2) - (\frac{1}{2}\mathrm{k}(\beta-\beta_0)^2 - R\beta_0)$. Applying the Euler–Lagrange equations to each degree of freedom gives

$$-\mathrm{k}(\beta-\beta_0) - R = 0;\quad M\ddot{\beta} + \mathrm{k}(\beta-\beta_0) = 0. \tag{2.120}$$

(d) Consider the Euler–Lagrange equation for two masses coupled by a spring as shown in Figure 2.14. The mechanical system moves in the right direction because of an initial disturbance which means that, at the initial time, when the spring is not stressed, the velocity of the left mass is positive and equal to v_{init}, but the right mass has zero velocity.

Derive the governing equations by using the Lagrange formulation and formulate the initial condition.
Answer:

$$\begin{aligned} &M_1\ddot{\beta}_1 - \mathrm{k}(\beta_2-\beta_1-b_0) = 0,\\ &M_2\ddot{\beta}_2 + \mathrm{k}(\beta_2-\beta_1-b_0) = 0,\\ &\beta_1(0) = 0,\quad \dot{\beta}_1(0) = v_{init},\\ &\beta_2(0) = b_0,\quad \dot{\beta}_2(0) = 0, \end{aligned} \tag{2.121}$$

where $b_0 = \beta_0(0)$ is the initial distance between the masses, which can be treated as the spring length. Find the analytical solution of problem (2.121).

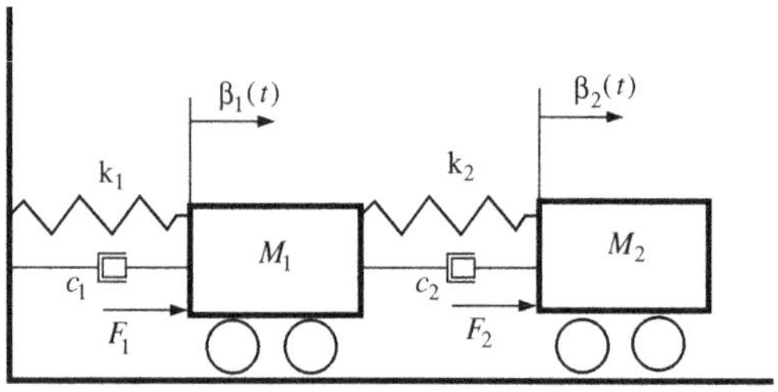

Figure 2.15. A forced horizontal two-spring-damper-mass system.

Answer:

$$\beta_1 = v_d\left[t + \frac{M_2}{M_1\sigma_0}\sin\sigma_0 t\right],$$

$$\beta_2 = b_0 + v_d\left[t - \frac{1}{\sigma_0}\sin\sigma_0 t\right], \tag{2.122}$$

where

$$\sigma_0^2 = \frac{\mathrm{k}(M_1+M_2)}{M_1M_2},\qquad v_d = \frac{v_{init.}M_1}{M_1+M_2}. \tag{2.123}$$

Show that the time $t = T_d = \pi/(2\sigma_0)$ implies the instant when two masses have the same velocity v_d and this spring has minimum length (corresponding to the minimum of $\beta_0(t)$ in Figure 2.14).

(e) Use definitions in Figure 2.15 to derive the Euler–Lagrange equation for the forced horizontal two-spring-damper-mass system in Figure 2.15. Damping is associated with Rayleigh dissipation function (2.93), which takes the following form:

$$D = \tfrac{1}{2}c_1\dot{\beta}_1^2 + \tfrac{1}{2}c_2\left(\dot{\beta}_2 - \dot{\beta}_1\right)^2, \tag{2.124}$$

where c_1 and c_2 are the damping coefficients of the dashpots.

Answer: The Lagrangian is $L = (\frac{1}{2}M_1\dot{\beta}_1^2 + \frac{1}{2}M_2\dot{\beta}_2^2) - (\frac{1}{2}\mathrm{k}_1\beta_1^2 + \frac{1}{2}\mathrm{k}_2(\beta_2-\beta_1)^2 - F_1\beta_1 - F_2\beta_2)$, which gives the system of differential equations

$$\begin{aligned} &M_1\ddot{\beta}_1 + c_1\dot{\beta}_1 - c_2\left(\dot{\beta}_2-\dot{\beta}_1\right) + \mathrm{k}_1\beta_1\\ &\qquad - \mathrm{k}_2(\beta_2-\beta_1) - F_1 = 0,\\ &M_2\ddot{\beta}_2 + c_2(\dot{\beta}_2-\dot{\beta}_1) + \mathrm{k}_2(\beta_2-\beta_1) - F_2 = 0. \end{aligned} \tag{2.125}$$

3 Wave-Induced Ship Motions

3.1 Introduction

Sloshing in a ship tank is the consequence of a mutual interaction between wave-induced ship motions and the flow inside the tank. The first step in describing this framework is to introduce wave theory for ocean waves. We first present linear wave theory in regular harmonic waves in deep and finite water depth. An irregular sea state can be represented as a sum of regular waves of different frequencies, amplitudes, and wave propagation directions. We then give recommended wave spectra that describe the frequency content in irregular waves.

Linear wave theory assumes that the wave slope is asymptotically small. Not all waves occurring in reality can be described by linear wave theory; extreme examples are breaking waves. We see breaking waves on a beach, but they can also occur in the open sea in deep water. A strong current in the opposite direction of the wave propagation direction steepens the waves, which is a phenomenon known in connection with the Agulhas Current outside the east coast of Africa. A typical feature of a nonlinear wave is that the vertical distance between the wave crest and the mean water level is larger than the distance between the mean water level and the wave trough.

Scatter diagrams of significant wave heights and mean wave periods are needed in operational and design studies. These diagrams describe the probability of occurrence of different sea states for a given operational area.

Linear theory describing the external hydrodynamic forces acting on a ship can, to a large extent, describe the wave-induced motions of a ship in operational conditions. However, nonlinear effects matter in severe sea states. Because sloshing is a typical resonance phenomenon, it is not necessarily the most extreme ship motions or external wave-induced loads that cause the most severe sloshing. This finding implies that external wave-induced loads can in many practical cases be described using linear theory. We limit ourselves to linear external hydrodynamic loads. It is then common to use a frequency-domain theory: steady-state solutions with prescribed frequency equal to the frequency of encounter between regular waves and the ship are studied. Because sloshing in a tank is a nonlinear process, we cannot assume that the flow inside the tank oscillates with a given frequency. The consequence is that the external hydrodynamic loads must be transferred to the time domain, which can be done by using the frequency-domain solution. The resulting external hydrodynamic loads are then expressed in terms of convolution integrals that contain memory effects of the ship's behavior.

An important practical matter is to recognize that external hydrodynamic loads are described in an inertial system that moves with the steady forward speed of the ship, which is the *key* coordinate system of this chapter. This point differs from other chapters where the primary focus is on an internal liquid flow in a tank and sloshing and, therefore, the key coordinate system is fixed with the ship (tank) coordinate system. We need to show how to transfer information between these two coordinate systems.

In this book, the *rule* is to use the notation $Oxyz$ for the key coordinate system. Therefore, when it comes to sloshing phenomena in other chapters, the $Oxyz$-coordinate system denotes the body-fixed frame. However, it follows from the rule that the present chapter should adopt $Oxyz$ as the notation for a coordinate system rigidly fixed with respect to the mean oscillatory position of the ship. Furthermore, if we focus exclusively on external flow with no ship, $Oxyz$ is the Earth-fixed coordinate system. When the analysis of this chapter requires consideration of internal liquid motions, the $\overline{Oxyz}$ system is introduced as the notation for the coordinate system rigidly fixed with the tank (or ship). The coordinate system $OXYZ$ is an auxiliary coordinate system whose meaning may change from case to case.

3.2 Long-crested propagating waves

We consider linear long-crested waves propagating in the x-direction in water with infinite

Table 3.1. *Velocity potential, dispersion relationship, wave profile, velocity, and acceleration for regular sinusoidal propagating waves in finite and infinite water depth according to linear theory*

	Finite water depth	Infinite water depth
Velocity potential	$\varphi = \frac{g\zeta_a}{\sigma}\frac{\cosh k(z+h)}{\cosh kh}\cos(\sigma t - kx)$	$\varphi = \frac{g\zeta_a}{\sigma}e^{kz}\cos(\sigma t - kx)$
Connection between wave number k and circular frequency (dispersion relationship)	$\frac{\sigma^2}{g} = k\tanh kh$	$\frac{\sigma^2}{g} = k$
Connection between wavelength λ and wave period T	$\lambda = \frac{g}{2\pi}T^2\tanh\frac{2\pi}{\lambda}h$	$\lambda = \frac{g}{2\pi}T^2$
Wave profile	$\zeta = \zeta_a\sin(\sigma t - kx)$	$\zeta = \zeta_a\sin(\sigma t - kx)$
Hydrodynamic pressure	$p_D = \rho g\zeta_a\frac{\cosh k(z+h)}{\cosh kh}\sin(\sigma t - kx)$	$p_D = \rho g\zeta_a e^{kz}\sin(\sigma t - kx)$
x-component of velocity	$u = \sigma\zeta_a\frac{\cosh k(z+h)}{\sinh kh}\sin(\sigma t - kx)$	$u = \sigma\zeta_a e^{kz}\sin(\sigma t - kx)$
z-component of velocity	$w = \sigma\zeta_a\frac{\sinh k(z+h)}{\sinh kh}\cos(\sigma t - kx)$	$w = \sigma\zeta_a e^{kz}\cos(\sigma t - kx)$
x-component of acceleration	$a_1 = \sigma^2\zeta_a\frac{\cosh k(z+h)}{\sinh kh}\cos(\sigma t - kx)$	$a_1 = \sigma^2\zeta_a e^{kz}\cos(\sigma t - kx)$
z-component of acceleration	$a_3 = -\sigma^2\zeta_a\frac{\sinh k(z+h)}{\sinh kh}\sin(\sigma t - kx)$	$a_3 = -\sigma^2\zeta_a e^{kz}\sin(\sigma t - kx)$

Note: $\sigma = 2\pi/T$, $k = 2\pi/\lambda$, T is the wave period, λ is the wavelength, ζ_a is the wave amplitude, g is the acceleration of gravity, t is the time variable, x is the direction of wave propagation, z is positive upward, $z = 0$ mean water level, h is the average water depth, ρ is the density of the water. Total pressure in the water: $p_D - \rho g z + p_a$ (p_a is the atmospheric pressure).
Source: Faltinsen (1990).

horizontal extent and no obstacles present. The Earth-fixed coordinate system $Oxyz$ is used. The density and temperature are assumed constant in the liquid domain (i.e., no stratification). Assuming harmonic oscillations with circular frequency σ(rad/s), no mean flow, atmospheric pressure on the free surface, linearity in terms of a small wave slope, and expressing the velocity potential as φ with a time dependence $\exp(i\sigma t)$ leads by using eq. (2.73) to the following combined dynamic and kinematic free-surface condition:

$$-\sigma^2\varphi + g\frac{\partial\varphi}{\partial z} = 0 \quad \text{on} \quad z = 0, \tag{3.1}$$

where we have used that $\partial^2\varphi/\partial t^2 = -\sigma^2\varphi$. The boundary condition on the sea floor, $z = -h$, expresses no flow through the horizontal sea bottom; that is,

$$\frac{\partial\varphi}{\partial z} = 0 \quad \text{on} \quad z = -h. \tag{3.2}$$

Furthermore, φ satisfies the two-dimensional Laplace equation in x and z. The solution of this boundary-value problem can be found in many textbooks dealing with water waves and is not derived here. We use the results of Faltinsen (1990), which are presented in Table 3.1. We note that the wave profile, the dynamic pressure $p_D = -\rho\partial\varphi/\partial t$, the water velocity, and the acceleration are linearly dependent on the wave amplitude ζ_a.

A practical fact is that the water motion for deep-water waves is negligible from half a wavelength λ down into the liquid, which results from the exponential factor $\exp(kz) = \exp(2\pi z/\lambda)$ in the deep-water results in Table 3.1. For instance, if $z = -\frac{1}{2}\lambda$, $\exp(kz) = 0.043$ or if $z = -\lambda$, $\exp(kz) = 0.002$.

According to linear theory, a water particle moves in a circle for deep water and in an ellipse for finite water depth. The circle radius is equal to the wave amplitude for a water particle on the

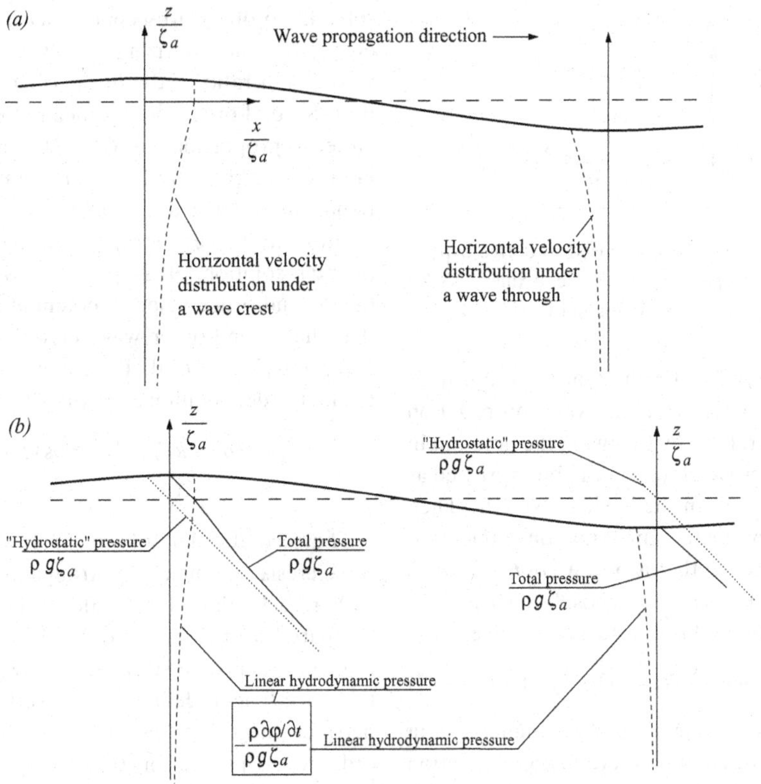

Figure 3.1. (a) Horizontal velocity distribution and (b) pressure variation under a wave crest and a wave trough according to linear wave theory. The x- and z-axes have different scales.

free surface. The two semi-axes of the elliptical motion for finite depth are horizontal and vertical; the horizontal semi-axis is larger. The vertical semi-axis is equal to the wave amplitude for a water particle on the free surface. When the water depth is shallow (i.e., $h/\lambda \lesssim 1/20$), the horizontal water velocity is much larger than the vertical water velocity. This result can be seen by Taylor expansion of the finite-depth expressions of water velocity at $z = -h$ and assuming kh to be small. Furthermore, we can show that the total pressure is hydrostatic relative to the instantaneous free-surface elevation when $kh \to 0$.

It should be noted that the linear theory assumes the velocity potential and water velocity are constant from the mean to the actual free-surface level. This assumption is made when the free-surface conditions are formulated. The horizontal velocity distribution shown in Figure 3.1(a) for the flow under a wave crest is consistent with linear theory. Figure 3.1(a) also shows the velocity under a wave trough, where we have used the analytical velocity distribution up to the free-surface level. It is then implicitly assumed that the difference between the horizontal velocity at the wave trough and the velocity at $z = 0$ is small compared with the velocity itself.

Figure 3.1(b) shows how the pressure varies with depth both under a wave crest and a wave trough. The "hydrostatic" pressure $-\rho g z$ cancels the dynamic pressure $-\rho \partial\varphi/\partial t|_{z=0}$ at the free surface (see eq. (2.71)). This condition is the linear dynamic free-surface condition, which is exactly satisfied at the wave crest in Figure 3.1(b), whereas a higher-order error exists under the wave trough. By "higher-order error" we mean that the error is approximately proportional to $(\zeta_a/\lambda)^n$, where the order is $n \geq 2$. Linear theory is therefore correct to $O(\zeta_a/\lambda)$, where $O(\cdot)$ means order of magnitude.

The expressions in Table 3.1 can be generalized to any wave propagation direction. For this purpose, we consider an auxiliary Earth-fixed coordinate system (X, Y, z), whose vertical axis coincides with Oz, but the OXY frame has an angle β relative to the original Oxy, as

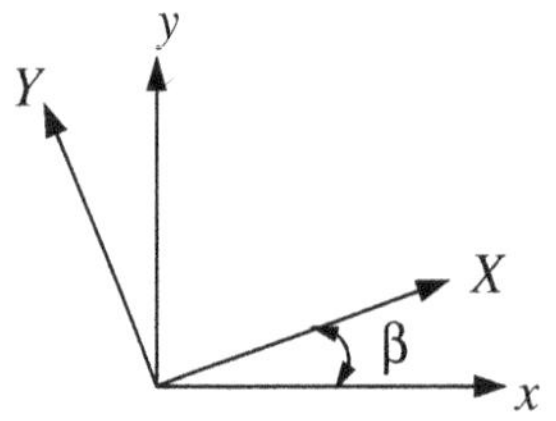

Figure 3.2. Coordinate systems used to derive expressions for waves propagating along the auxiliary X-axis with an angle β relative to the original x-axis.

shown in Figure 3.2. Furthermore, the X-axis is chosen to coincide with the wave propagation direction. Therefore, by following the notation in Table 3.1, the wave elevation can be expressed as $\zeta = \zeta_a \sin(\sigma t - kX)$ in the auxiliary system. Then we make a coordinate transformation to the original (x, y, z) system; that is, $x = X\cos\beta - Y\sin\beta$, $y = X\sin\beta + Y\cos\beta$, $X = x\cos\beta + y\sin\beta$, $Y = -x\sin\beta + y\cos\beta$. This transformation gives

$$\zeta = \zeta_a \sin(\sigma t - kx\cos\beta - ky\sin\beta). \quad (3.3)$$

The results of linear theory summarized in Table 3.1 represent a first-order approximation in satisfying the free-surface conditions. This approximation can be improved by introducing higher-order terms in a consistent manner – a Stokes expansion. The next approximation would solve the problem to second order in the parameter ζ_a/λ, characterizing the wave amplitude/wavelength ratio of the linear (first-order) solution. Second-order theory means that we keep all terms proportional to $O((\zeta_a/\lambda)^2)$ and $O(\zeta_a/\lambda)$ in a consistent way. For sinusoidal unidirectional progressive deep-water waves where the solution in Table 3.1 represents the first-order (linear) solution, it is possible to show that the second-order velocity potential is zero and that the second-order wave elevation ζ_2 is $\zeta_2 = -\frac{1}{2}\zeta_a^2 k\cos[2(\sigma t - kx)]$. By combining this with the first-order solution $\zeta_a \sin(\sigma t - kx)$, we get

$$\zeta = \zeta_a \sin(\sigma t - kx) - \tfrac{1}{2}\zeta_a^2 k\cos[2(\sigma t - kx)] + o((\zeta_a/\lambda)^2). \quad (3.4)$$

The second-order solution (3.4) sharpens the wave crests and makes the trough more shallow. In Figure 3.3 the second-order wave profiles are compared with "infinite"-order wave profiles for four different wave steepnesses H/λ, where H is the wave height, defined as the vertical distance between the trough and the crest. The infinite-order wave profile for $H/\lambda = 0.10$ is given by Schwartz (1974), whereas the theory presented by Bryant (1983) is used to determine the wave profile for the other values of H/λ. The wave elevation is symmetric about $x = 0$ in

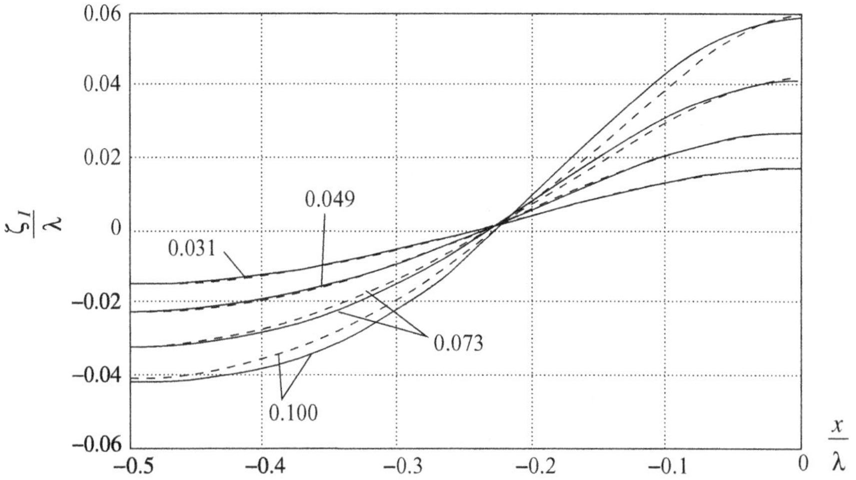

Figure 3.3. Second-order and "infinite"-order wave profiles for deep-water Stokes waves for a given time instant (Greco, personal communication, 2008), where $\zeta_I(x)$ is the wave elevation for a given time instant, H is the wave height, and λ is the wavelength. Solid lines denote a second-order Stokes wave and dashed lines correspond to infinite-order Stokes waves. The lines are labeled by the H/λ values.

Figure 3.3, representing steady-state theories that cannot describe transient plunging breakers. The wave profiles computed by second-order theory and infinite-order theory compare very well for $H/\lambda = 0.031$ and $H/\lambda = 0.049$. For $H/\lambda = 0.073$ the deviation is larger, but still the relative error for the maximum wave elevation is less than 0.8%. When the wave steepness is increased to 0.1, the relative difference between the two different wave profiles becomes more significant. The exact wave profile is more peaked at the crest and flatter at the troughs than in the second-order profile, indicating that linear and second-order wave theories are not sufficient to describe the wave properly for steep waves. However, in the following chapters linear theory is used to a large extent to describe the incident wave elevation and kinematics.

3.3 Statistical description of waves in a sea state

In practice, linear theory is used to simulate irregular seas and to obtain statistical estimates. The wave elevation of a long-crested irregular sea propagating along the positive x-axis can be written as the sum of a large number of wave components:

$$\zeta = \sum_{j=1}^{N} A_j \sin(\sigma_j t - k_j x + \varepsilon_j), \tag{3.5}$$

where each wave component j corresponds to the linear solution from Table 3.1 with different wave amplitude $A_j (= \zeta_a$ for single-component linear wave), angular frequency σ_j, wave number k_j, and random phase angle ε_j. The random phase angles ε_j are uniformly distributed between 0 and 2π and constant with time. For deep and finite-depth water waves, σ_j and k_j are related by the dispersion relationship (see Table 3.1). The wave amplitude A_j can be expressed by a wave spectrum $S(\sigma)$ as

$$\tfrac{1}{2}A_j^2 = S(\sigma_j)\Delta\sigma, \tag{3.6}$$

where $\Delta\sigma$ is a constant difference between successive frequencies. Because the wave energy of regular waves is proportional to the square of the wave amplitude, the wave spectrum describes the energy distribution. The instantaneous wave elevation is Gaussian distributed with zero mean and variance σ_ζ^2 equal to $\int_0^\infty S(\sigma)\,d\sigma$, which can be shown by using the definition of mean value and variance applied to the "signal" represented by eq. (3.5). We find, for instance, that $\sigma_\zeta^2 = \frac{1}{2}\sum_{j=1}^{N} A_j^2$. By using eq. (3.5) and letting $N \to \infty$ and $\Delta\sigma \to 0$, we get $\sigma_\zeta^2 = \int_0^\infty S(\sigma)d\sigma$. The relationship between a time-domain solution of the waves (i.e., eq. (3.5)) and the frequency-domain representation of the waves by means of a wave spectrum $S(\sigma)$ is illustrated in Figure 3.4. Simulation of waves, (e.g., eq. (3.5)) is in practice usually done using the fast Fourier technique (Newland, 1984).

The wave spectrum can be estimated from wave measurements (Kinsman, 1965). It assumes that we can describe the sea as a stationary random process. In practice, this assumption means that we are talking about a limited time period in the range from half an hour to maybe 10 hours. In the literature this is often referred to as a short-term description of the sea.

Recommended sea spectra from the International Ship and Offshore Structures Congress (ISSC) and the International Towing Tank Conference (ITTC) are often used to calculate $S(\sigma)$. For instance, for open-sea conditions, the 15th ITTC recommended the use of ISSC spectral formulation for a fully developed sea:

$$\frac{S(\sigma)}{H_{1/3}^2 \mathrm{T}_1} = \frac{0.11}{2\pi}\left(\frac{\sigma \mathrm{T}_1}{2\pi}\right)^{-5} \exp\left[-0.44\left(\frac{\sigma \mathrm{T}_1}{2\pi}\right)^{-4}\right], \tag{3.7}$$

where $H_{1/3}$ is the significant wave height defined as the mean of the one-third-highest waves and T_1 is a mean wave period defined as $\mathrm{T}_1 = 2\pi m_0/m_1$, where the spectrum moments, m_k, are given by $m_k = \int_0^\infty \sigma^k S(\sigma)\,d\sigma$ $(k = 0, 1, \ldots)$. $H_{1/3}$ is often redefined as

$$H_{1/3} = 4\sqrt{m_0}, \tag{3.8}$$

giving a value that is usually close to the $H_{1/3}$ just defined. Equation (3.7) satisfies eq. (3.8). Strictly speaking this relation is only true for a narrow-band spectrum and when the instantaneous value of the wave elevation is Gaussian distributed.

The spectrum given by eq. (3.7) is the same as the modified Pierson–Moskowitz spectrum,

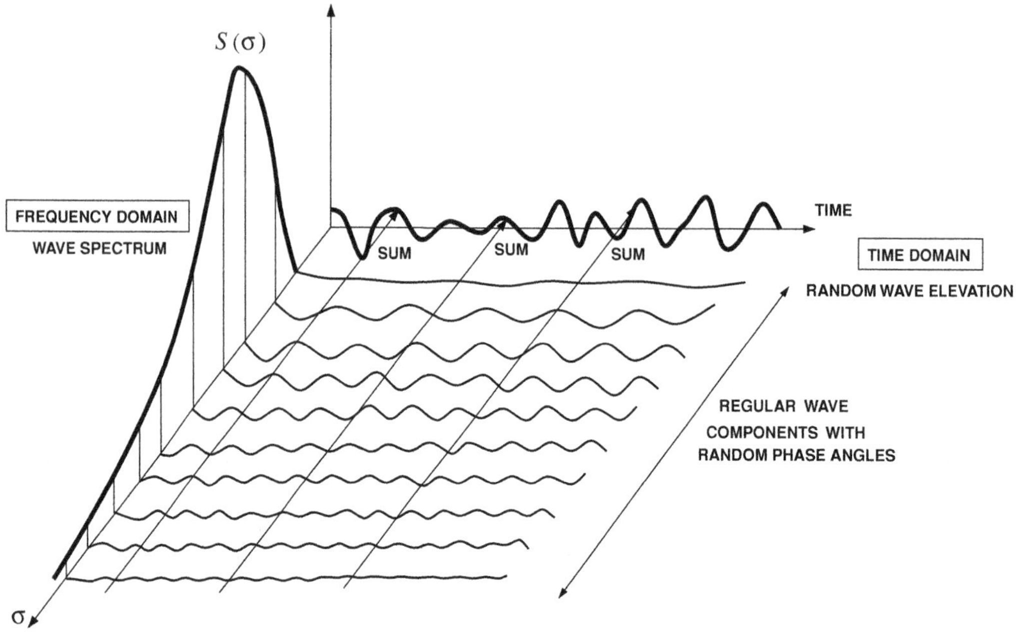

Figure 3.4. Illustration of the connection between frequency-domain and time-domain representations of waves in a long-crested short-term sea state. There are three "sum" lines drawn, illustrating how regular wave components with random phase angles at a given time add up to give the irregular wave elevation at that time instant.

where it is more usual to use the mean wave period T_2 defined as

$$T_2 = 2\pi\,(m_0/m_2)^{1/2}. \tag{3.9}$$

The following relation exists between T_1 and T_2 for the spectrum given by eq. (3.7):

$$T_1 = 1.086T_2. \tag{3.10}$$

The period T_0 corresponding to the peak frequency of the spectrum can be written as

$$T_0 = 1.408T_2. \tag{3.11}$$

The peak period T_0 is also referred to as the modal period.

The spectrum formulation given by eq. (3.7) is shown in Figure 3.5. We note that little energy density exists when $\sigma T_2/(2\pi)$ is less than 0.5 and larger than ~1.5. For large frequencies the wave spectrum decays like σ^{-5}. The 17th ITTC recommended the following Joint North Sea Wave Project (JONSWAP)-type spectrum for limited fetch (e.g., in the North Sea):

$$S(\sigma) = 155\frac{H_{1/3}^2}{T_1^4\sigma^5}\exp\left(\frac{-944}{T_1^4\sigma^4}\right)(3.3)^Y\ (\mathrm{m^2 s}), \tag{3.12}$$

where

$$Y = \exp\left(-\left(\frac{0.191\sigma T_1 - 1}{2^{1/2}\vartheta}\right)^2\right) \quad \text{and}$$

$$\vartheta = \begin{cases} 0.07 & \text{for} \quad \sigma \le 5.24/T_1, \\ 0.09 & \text{for} \quad \sigma > 5.24/T_1. \end{cases}$$

The JONSWAP spectrum formulation can be used with other characteristic periods by the substitution of

$$T_1 = 0.834,\ T_0 = 1.073T_2, \quad T_0 = 1.287T_2. \tag{3.13}$$

We note that eq. (3.13) differs from the relationship between T_1, T_0, and T_2 given by eqs. (3.10) and (3.11) for the Pierson–Moskowitz spectrum.

The JONSWAP spectrum is shown in Figure 3.5. The peak value of the modified Pierson–Moskowitz (ISSC) spectrum occurs at a different ($\sigma T_2/2\pi$) value than the JONSWAP spectrum, which can be seen from eqs. (3.11) and (3.13).

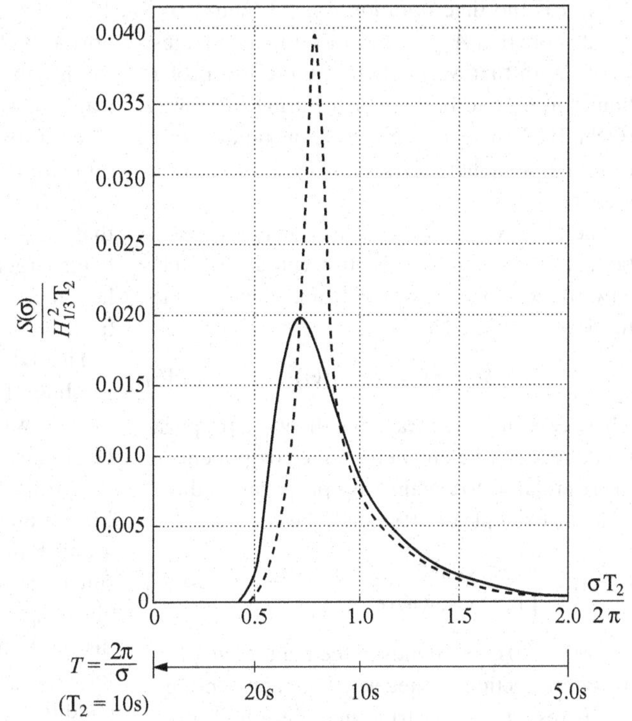

Figure 3.5. Examples of wave spectra ($H_{1/3}$ is the significant wave height, T_2 is the mean wave period). Modified Pierson–Moskowitz spectrum (solid line, eq. (3.7)); JONSWAP spectrum (dashed line, eq. (3.12)) (Faltinsen, 1990).

A good approximation to the probability density function for the wave amplitude maxima (peak values) of the wave elevation (denoted by A) can be obtained from the Rayleigh density distribution given by

$$p(A) = A/m_0 \exp(-A^2/(2m_0)), \qquad (3.14)$$

where m_0 is related to $H_{1/3}$ by eq. (3.8). Strictly speaking, the Rayleigh distribution depends on the wave spectrum being narrowbanded, which is an approximation for the spectra we have discussed. In deriving the Rayleigh distribution, it is also assumed that the instantaneous value of the wave elevation is Gaussian distributed.

We can simulate a seaway by using eq. (3.5), but this expression repeats itself after a time $2\pi/\Delta\sigma$. Hence, a large number of wave components, N, is needed to avoid this problem. A practical way to avoid this is to choose a random frequency in each frequency interval $(\sigma_j - \frac{1}{2}\Delta\sigma,\ \sigma_j + \frac{1}{2}\Delta\sigma)$ and calculate the wave spectrum with those frequencies. The number of wave components ought to be about 1,000. This number depends partly on the selection of the minimum and maximum frequency components. The minimum frequency component σ_{min} is easier to select than the maximum frequency component, σ_{max}. For instance if a Pierson–Moskowitz spectrum is used, $\sigma_{min} \approx \pi/T_2$. The wave energy drops off more slowly for larger frequencies than for small frequencies. We should therefore investigate the results for different values of σ_{max} to ensure that the results do not depend on the selection of σ_{max}. We have only shown how we can simulate the wave elevation ζ. Similarly, for instance, for the horizontal water velocity u and acceleration a_1, one can superpose the results for regular waves given in Table 3.1. The random phase angles ε_j are the same for ζ, u, and a_1.

When simulations are done using the same sea spectrum and the same duration, the results will not be the same because of the random selection of frequencies and phase angles. The largest amplitude in each simulation (realization) is different. By selecting a large number of realizations we find that the extreme values have their own probability distribution. This result was discussed, for instance, by Ochi (1982). In practice the *most probable largest* value A_{max} of A is often used, which can be approximated as

$$A_{max} = (2m_0 \ln(t/T_2))^{1/2}, \qquad (3.15)$$

where t is the time duration (e.g., 3 hours). We should note that A_{max} is the most probable largest value. With that we imply that there is a probability for A_{max} to be exceeded during the time t (Ochi, 1982). The most probable maximum crest-to-trough wave height H_{max} during the same time is simply $2A_{max}$.

The effect of short-crestedness may be important. A short-crested sea is often characterized by a two-dimensional wave spectrum, which in practice is often written as

$$S(\sigma, \Theta) = S(\sigma) f(\Theta), \tag{3.16}$$

where Θ is an angle measuring the wave propagation direction of elementary wave components in the sea relative to a main wave propagation direction. An example of $f(\Theta)$ might be

$$f(\Theta) = \begin{cases} 2\pi^{-1}\cos^2\Theta, & -\frac{1}{2}\pi \leq \Theta \leq \frac{1}{2}\pi \\ 0, & \text{elsewhere,} \end{cases} \tag{3.17}$$

where $\Theta = 0$ corresponds to the main wave propagation direction. Other ways of representing a short-crested sea spectrum may be found in the report of the 10th ISSC. Let us assume as an example that the main wave propagation direction is along the x-axis. Due to eq. (3.3) with $\Theta_k = \beta$, for short-crested sea, eq. (3.5) can be generalized to

$$\zeta = \sum_{j=1}^{N}\sum_{k=1}^{K} \left(2S(\sigma_j, \Theta_k)\,\Delta\sigma_j\Delta\Theta_k\right)^{1/2} \times \sin\left(\sigma_j t - k_j x\cos\Theta_k - k_j y \sin\Theta_k + \varepsilon_{jk}\right). \tag{3.18}$$

Hindcast services providing directional wave spectrum for a given time and ship position are available.

3.4 Long-term predictions of sea states

So far we have discussed a "short-term" description of the sea, which means the significant wave height and the mean wave period are assumed constant during the time period considered. The significant wave height and mean wave period will vary in a "long-term" description of the sea. To construct a long-term prediction of the sea, we need to know the joint frequency of the significant wave height and the mean wave period. This discrete joint frequency distribution is referred to as scatter diagram. An example is given in Table 3.2. These data are representative for the northern North Sea. The frequency table shows, for instance, that the probability of the significant wave height being between 3 and 4 m (the fourth significant wave height interval in the table, fourth row, $j = 4$) and the spectral period being 10 s is $2960/100001 = 0.0296$. It also shows that the probability of the significant wave height being larger than 2 m (the second significant wave height, $j = 2$) is $1 - (8636 + 32155)/100001 = 0.59$.

The table can be used in many ways (e.g., to obtain long-term statistics of the wave amplitude or wave height). For each significant wave-height interval (enumerated with index j) we find the probability of occurrence, p_j, from the table. For instance, the probability that $H_{1/3}$ is between 4 and 5 m is 9118/100001. Since the probability function for the maxima of the wave elevation for given significant wave height follows a Rayleigh distribution (see eq. (3.14)), we can obtain the long-term probability as a simple summation of conditional probabilities over all the significant wave-height intervals from 1 to M; that is,

$$P(H) = 1 - \sum_{j=1}^{M} \exp\left\{-2H^2/\left[H_{1/3}^{(j)}\right]^2\right\} p_j, \tag{3.19}$$

where $P(H)$ is the long-term probability that the wave height does not exceed H. When using eq. (3.14), we have set $H = 2A$. If we use Table 3.2 we see that $M = 15$, $H_{1/3}^{(1)} = 0.5$ m, $H_{1/3}^{(2)} = 1.5$ m, and so forth. The exceedance probability level $Q = 1 - P(H)$ and the number of response cycles, N_0, are related by $Q = 1/N_0$. For instance, during 100 years and assuming an average period of 7 s, we find $N_0 = 100 \cdot 365 \cdot 24 \cdot 3600/7 = 4.5 \cdot 10^8$ (i.e., $Q = 10^{-8.65}$). By using eq. (3.19) we can find the value H for which $Q = 10^{-8.7}$. We have then found the wave height of what is called the "100-year wave height" in offshore engineering. DNV (2007) has provided information as in Table 3.2 for worldwide nautic zones.

The sea state number is often used to classify the sea and gives a more rough description of the relationship between $H_{1/3}$ and mean wave period. This description is shown in Table 3.3 and also gives information about the wind speed. One should note that sea state number is not the same as Beaufort number.

Table 3.2. *Joint frequency of significant wave height and spectral peak period*

Significant wave height (m) (upper limit of) (interval)	Spectral peak period (s)																		
	3	4	5	6	7	8	9	10	11	12	13	14	15	16	17	18	19	20	>20	Sum
1	59	403	1061	1569	1634	1362	982	643	395	232	132	74	41	22	12	7	4	2	2	8636
2	9	212	1233	3223	5106	5814	5284	4102	2846	1821	1098	634	355	194	105	56	30	16	17	32155
3	0	8	146	831	2295	3896	4707	4456	3531	2452	1543	901	497	263	135	67	33	16	15	25792
4	0	0	6	85	481	1371	2406	2960	2796	2163	1437	849	458	231	110	50	22	10	7	15442
5	0	0	0	4	57	315	898	1564	1879	1696	1228	748	398	191	84	35	13	5	3	9118
6	0	0	0	0	3	39	207	571	950	1069	885	575	309	142	58	21	7	2	1	4839
7	0	0	0	0	0	2	27	136	347	528	533	387	217	98	37	12	4	1	0	2329
8	0	0	0	0	0	0	2	20	88	197	261	226	138	64	23	7	2	0	0	1028
9	0	0	0	0	0	0	0	2	15	54	101	111	78	39	14	4	1	0	0	419
10	0	0	0	0	0	0	0	0	2	11	30	45	39	22	8	2	1	0	0	160
11	0	0	0	0	0	0	0	0	0	2	7	15	16	11	5	1	0	0	0	57
12	0	0	0	0	0	0	0	0	0	0	1	4	6	5	2	1	0	0	0	19
13	0	0	0	0	0	0	0	0	0	0	0	1	2	2	1	0	0	0	0	6
14	0	0	0	0	0	0	0	0	0	0	0	0	0	1	0	0	0	0	0	1
15	0	0	0	0	0	0	0	0	0	0	0	0	0	0	0	0	0	0	0	0
Sum	68	623	2446	5712	9576	12799	14513	14454	12849	10225	7256	4570	2554	1285	594	263	117	52	45	100001

Note: Representative data are for the northern North Sea.

Table 3.3. *Annual sea state occurrence in the open ocean of the North Atlantic and North Pacific*

Sea state	Significant wave height (m)		Sustained wind speed (kn)[a]		North Atlantic			North Pacific		
					Percentage probability of sea state	Modal wave period (s)		Percentage probability of sea state	Modal wave period (s)	
Number	Range	Mean	Range	Mean		Range[b]	Most probable[c]		Range[b]	Most probable[c]
0–1	0–0.1	0.05	0–6	3	0.70	–	–	1.30	–	–
2	0.1–0.5	0.3	7–10	8.5	6.80	3.3–12.8	7.5	6.40	5.1–14.9	6.3
3	0.5–1.25	0.88	11–16	13.5	23.70	5.0–14.8	7.5	15.50	5.3–16.1	7.5
4	1.25–2.5	1.88	17–21	19	27.80	6.1–15.2	8.8	31.60	6.1–17.2	8.8
5	2.5–4	3.25	22–27	24.5	20.64	8.3–15.5	9.7	20.94	7.7–17.8	9.7
6	4–6	5	28–47	37.5	13.15	9.8–16.2	12.4	15.03	10.0–18.7	12.4
7	6–9	7.5	48–55	51.5	6.05	11.8–18.5	15.0	7.00	11.7–19.8	15.0
8	9–14	11.5	56–63	59.5	1.11	14.2–18.6	16.4	1.56	14.5–21.5	16.4
>8	>14	>14	>63	>63	0.05	18.0–23.7	20.0	0.07	16.4–22.5	20.0

[a] Ambient wind sustained at 19.5 m above surface to generate fully developed seas. To convert to another altitude, H_2, apply $V_2 = V_1(H_2/19.5)^{1/7}$.
[b] Minimum is 5th percentile and maximum is 95th percentile for periods given wave height range.
[c] Based on periods associated with central frequencies included in Hindcast Climatology.
Source: Lee and Bales (1985).

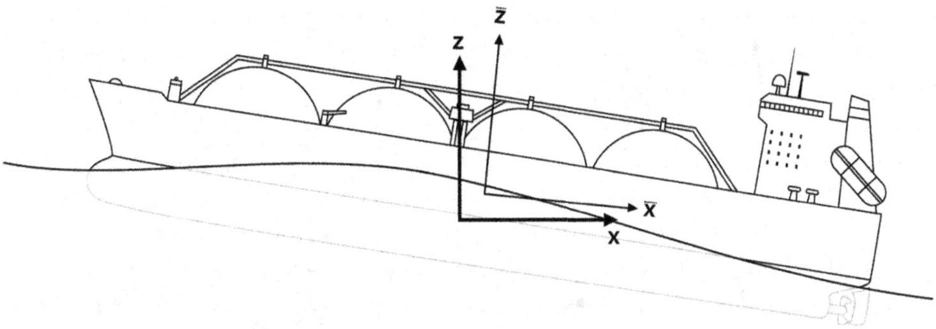

Figure 3.6. Inertial system (x, y, z) moving with the constant ship speed U. Body-fixed coordinate system $(\bar{x}, \bar{y}, \bar{z})$.

3.5 Linear wave-induced motions in regular waves

3.5.1 Definitions

Linear theory can, to a large extent, describe the wave-induced motions of a ship in operational conditions. However, nonlinear effects are more important in severe sea states. Henceforth, we consider a ship in incident regular linear waves of amplitude ζ_a (see Table 3.1). The wave steepness is small (i.e., the waves are far from breaking). Linear theory implies that the wave-induced motion amplitudes are linearly proportional to ζ_a.

A useful consequence of linear theory is that we can obtain results in irregular waves by adding together results from regular waves of different amplitudes, phases, wavelengths, and propagation directions. Therefore, it is sufficient from a hydrodynamic point of view to analyze a ship in incident regular sinusoidal waves of small wave steepness, which is done in the following text. We assume a steady-state condition; no transient effects are present due to initial conditions. This condition implies that the linear dynamic motions and loads on the ship are harmonically oscillating with the same frequency as the wave loads that excite the ship. The hydrodynamic problem in regular waves is normally dealt with as two subproblems:

(a) The forces and moments on the ship when the body is restrained from oscillating and incident regular waves are present. The hydrodynamic loads are called *wave excitation loads* and are composed of the so-called Froude–Kriloff loads and the diffraction forces and moments. Froude–Kriloff loads are due to the pressure field in the incident waves, which are undisturbed by the ship. Newman (1977) refers to what we call the *diffraction problem* as the scattering problem. The diffraction loads in his nomenclature are the sum of the Froude–Kriloff and scattering loads.

(b) The forces and moments on the body when the structure is forced to oscillate in calm water with the wave excitation frequency in any rigid-body motion mode. Incident waves are not present, but the oscillating body causes radiating waves. The hydrodynamic loads are identified as *added mass, damping* and *restoring* forces, and moments. This subproblem is often termed the *radiation problem.*

Due to linearity, the forces obtained in subproblems a and b can be added to give the total hydrodynamic force. One cannot separate the diffraction and radiation problem in a nonlinear theory.

Before we go into detail and describe the different hydrodynamic loads, we should define coordinate systems and rigid-body motion modes. In this section, a right-handed coordinate system (x, y, z) is fixed with respect to the *mean oscillatory position of the ship* with positive z vertically upward through the center of gravity (COG) of the ship; the origin is in the plane of the undisturbed free surface. If the ship moves with a steady forward speed, the coordinate system moves with the same speed. In addition, we define a *body-fixed coordinate system* $(\bar{x}, \bar{y}, \bar{z})$ that coincides with the (x, y, z) system when the ship does not oscillate (see Figure 3.6). We show the connection between these two coordinate systems by

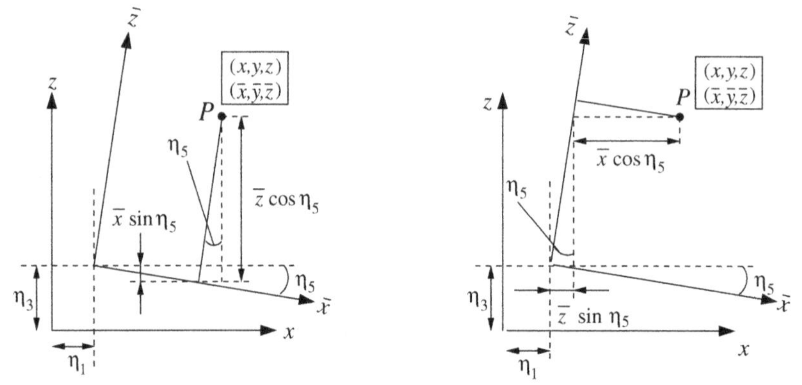

Figure 3.7. Transformation between body-fixed $(\bar{x}, \bar{y}, \bar{z})$ and inertial (x, y, z)-coordinate system.

considering either head or following sea in the x–z-plane. The ship will surge, heave, and pitch.

In a similar way to that described in Section 2.3, we define η_1 (surge) and η_3 (heave) as the translatory motions of the origin of the $(\bar{x}, \bar{y}, \bar{z})$ system along the x- and z-axis, respectively. Positive rotational angle η_5 (pitch) about the y- or $\bar{y}$-axis corresponds to bow up. We consider then a fixed point P on the ship (see Figure 3.7) with coordinates $\bar{x}, \bar{y}, \bar{z}$. The corresponding x- and z-coordinates can be derived as illustrated in Figure 3.7; that is,

$$x = \bar{x}\cos\eta_5 + \eta_1 + \bar{z}\sin\eta_5,$$
$$z = \bar{z}\cos\eta_5 + \eta_3 - \bar{x}\sin\eta_5. \tag{3.20}$$

Because linear theory is considered, we keep only linear terms in η_i, which gives as a first approximation $x = \bar{x}$ and $z = \bar{z}$. Second approximations of eq. (3.20) are

$$x = \bar{x} + \eta_1 + z\eta_5, \quad z = \bar{z} + \eta_3 - x\eta_5. \tag{3.21}$$

The longitudinal and vertical motions of point P on the ship can therefore be expressed in the (x, y, z) system as, respectively, $\eta_1 + z\eta_5$ and $\eta_3 - x\eta_5$. Therefore, we do not need the body-fixed coordinate system in describing the linear motions. Because the (x, y, z)-coordinate system is an inertial system, we can directly apply Newton's second law and Bernoulli's equation in this system. If we had used the body-fixed coordinate system, we would have had to modify these equations. The body-fixed coordinate system is a natural choice to use if the complete nonlinear ship–wave interaction problem were to be solved.

Let us now return to a more general formulation of the linear motions in combination with the (x, y, z)-coordinate system. Based on analysis in Section 2.3 with other notations for inertial and noninertial (body-fixed coordinate system), let the translatory displacements in the x-, y- and z-directions with respect to the origin be η_1, η_2, and η_3, respectively, so that η_1 is the surge, η_2 is the sway, and η_3 is the heave displacement. Furthermore, let the angular displacements of the rotational motions about the x-, y-, and z-axes be η_4, η_5, and η_6, respectively, so that η_4 is the roll, η_5 is the pitch, and η_6 is the yaw angle. The coordinate system and the translatory and angular displacement conventions are shown in Figure 3.8. We should note that the $Oxyz$ system moves on a straight course with the steady forward speed of the ship (i.e., the introduced η_1 does not include this effect).

When adopting the previously introduced notations, it follows from eq. (2.51) for rigid-body velocity and linearization that the motion of any point on the ship can be written as $\boldsymbol{s} = \eta_1\boldsymbol{i} + \eta_2\boldsymbol{j} + \eta_3\boldsymbol{k} + (\eta_4\boldsymbol{i} + \eta_5\boldsymbol{j} + \eta_6\boldsymbol{k}) \times \boldsymbol{r}$, where $\boldsymbol{r} = x\boldsymbol{i} + y\boldsymbol{j} + z\boldsymbol{k}$ and $\boldsymbol{i}$, $\boldsymbol{j}$, $\boldsymbol{k}$ are unit vectors along the x-, y-, and z-axis, respectively, resulting in

$$\boldsymbol{s} = (\eta_1 + z\eta_5 - y\eta_6)\boldsymbol{i} + (\eta_2 - z\eta_4 + x\eta_6)\boldsymbol{j} + (\eta_3 + y\eta_4 - x\eta_5)\boldsymbol{k}. \tag{3.22}$$

We now express η_j in the case of steady-state harmonic oscillations in regular incident waves. We then need an expression to which the incident waves can relate. The wave elevation in the Earth-fixed coordinate system is expressed by eq. (3.3). To avoid confusion with notations, we

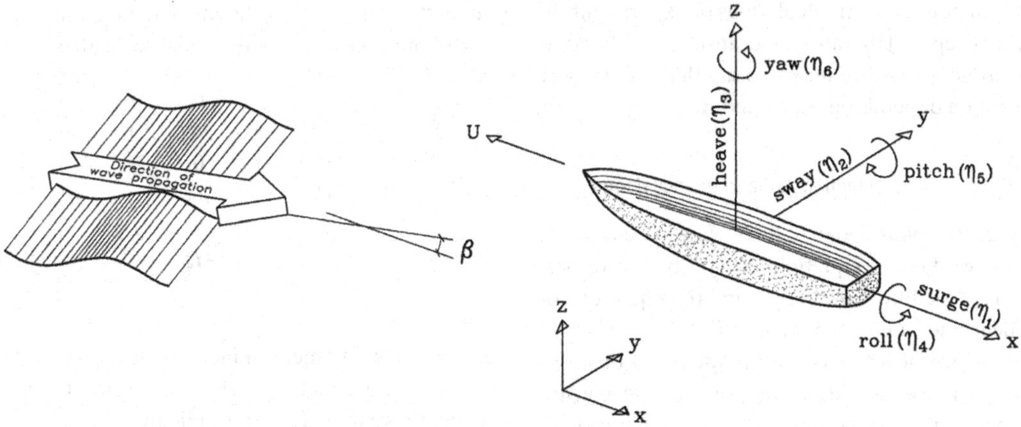

Figure 3.8. Definitions of coordinate system, rigid-body motion modes, and wave propagation direction. The coordinate system moves with the steady forward speed U of the vessel but does not oscillate with the ship. The origin is in the mean free surface. The z-axis passes through the center of gravity of the vessel when the vessel does not oscillate.

termporarily denote the Earth-fixed coordinate system as $OXYZ$ and, thereby, rewrite eq. (3.3) in the form $\zeta = \zeta_a \sin(\sigma_0 t - kX\cos\beta - kY\sin\beta)$, where σ_0 is now the frequency of the waves in the $OXYZ$ system. This frequency is not the same as the frequency of encounter. Indeed, the corresponding coordinate transformation to the xyz system (see Figure 3.9), $X = x - Ut$, $Y = y$, gives

$$\zeta = \zeta_a \sin\left((\sigma_0 + kU\cos\beta)\,t - kx\cos\beta - ky\sin\beta\right), \tag{3.23}$$

whereby the wave elevation oscillates in the xyz system with the frequency $\sigma_0 + kU\,cos\beta$. The frequency of encounter, σ_e, is then

$$\sigma_e = \sigma_0 + kU\cos\beta \quad \text{and} \quad k = \sigma_0^2/g = 2\pi/\lambda. \tag{3.24}$$

Several times later we use σ instead of σ_e as a notation. Here $\beta = 0$, 90°, 180° corresponds to head sea, beam sea, and following sea, respectively. Furthermore, $\beta \approx 45°$ is called bow sea and $\beta \approx 135°$ is quartering sea.

The linear steady-state motion η_j in six degrees of freedom can now be expressed as

$$\eta_j = \eta_{ja}\sin(\sigma_e t + \varepsilon_j), \quad (j = 1, \ldots, 6). \tag{3.25}$$

Positive ε_j means a phase lead relative to the wave elevation at $x = 0$ and $y = 0$. (Note that the literature and computer programs may have different definitions of phases, but as long as we know the definitions we can transform one definition of phase angle into another.) The amplitude η_{ja} is proportional to ζ_a in linear theory. The ratio η_{ja}/ζ_a is called a transfer function (or response amplitude operator [RAO]) for motion mode j. It is a function of σ_e, U, and β and has to be found

Figure 3.9. Inertial coordinate system (x, y, z) translating with steady ship speed U. Earth-fixed coordinate system (X, Y, Z). The z- and Z-axes are vertical upward. The origins of the coordinate systems are in the mean free surface. U is the ship speed, β is the wave propagation angle.

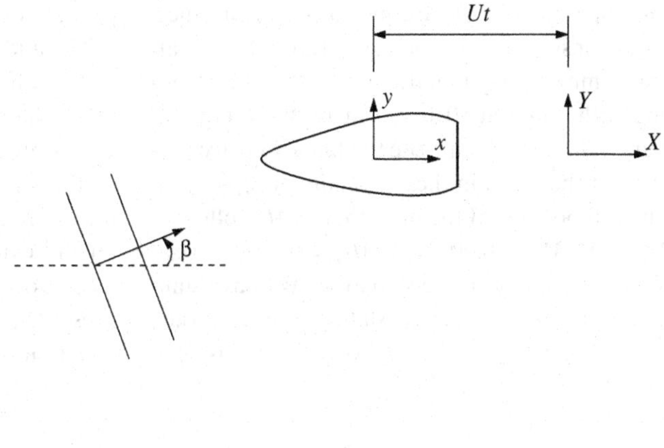

by either numerical calculations or experiments in a seakeeping laboratory. Because σ_e is related to σ_0 by eq. (3.24), we may also say that the transfer function depends on σ_0, U, and β.

3.5.2 Equations of motion in the frequency domain

Once the hydrodynamic forces have been found, we can set up the equations of rigid-body motions. This procedure follows by using the equations of linear and angular momentum. At a later stage we discuss how to incorporate the effect of sloshing in the equations of motion. So the considered hydrodynamic flow is presently exterior to the ship. For steady-state sinusoidal motions we may write

$$\sum_{k=1}^{6} \left[(M_{jk} + A_{jk}) \ddot{\eta}_k + B_{jk}\dot{\eta}_k + C_{jk}\eta_k \right] = F_j e^{i\sigma_e t} \quad (j = 1, \ldots, 6), \quad (3.26)$$

where M_{jk}, A_{jk}, B_{jk}, and C_{jk} are, respectively, the components of the generalized mass, added mass, and damping and restoring matrices of the ship; F_j are the complex amplitudes of the exciting force and moment components. However, obtaining the hydrodynamic forces is by no means trivial, particularly for a ship at forward speed. The subscripts in, for example, $A_{jk}\ddot{\eta}_k$ refer to force (moment) component in the j-direction because of motion in the k-direction. The equations appear as differential equations; however, they are not true differential equations, in general – they apply only for steady-state conditions. The general form of differential equations is discussed in Section 3.8.

The equations for $j = 1, 2, 3$ follows from Newton's second law, which assumes an inertial system like the (x, y, z) system fixed with respect to the ship's steady forward motions. For instance, let us consider $j = 1$. For a structure that has lateral symmetry (symmetric about the x–z-plane) and with COG at $(0, 0, z_G)$ in its static equilibrium position, we can write the linearized acceleration of the COG in the x-direction as $\ddot{\eta}_1 + z_G\ddot{\eta}_5$. The components of the mass matrix M_{jk} follow as $M_{11} = M, M_{12} = 0, M_{13} = 0, M_{14} = 0, M_{15} = Mz_G$, $M_{16} = 0$, where M is the ship mass. We have similar results for the other translatory directions (i.e., $j = 2, 3$). For $j = 4, 5, 6$ we have to use the equations derived from the angular momentum. We can then set up the following mass matrix:

$$M_{jk} = \begin{bmatrix} M & 0 & 0 & 0 & Mz_G & 0 \\ 0 & M & 0 & -Mz_G & 0 & 0 \\ 0 & 0 & M & 0 & 0 & 0 \\ 0 & -Mz_G & 0 & I_{44} & 0 & -I_{46} \\ Mz_G & 0 & 0 & 0 & I_{55} & 0 \\ 0 & 0 & 0 & -I_{46} & 0 & I_{66} \end{bmatrix}, \quad (3.27)$$

where I_{jj} is the moment of inertia in the jth mode and I_{jk} is the product of inertia with respect to the coordinate system (x, y, z). Explicitly,

$$I_{44} = \int (y^2 + z^2)\,\mathrm{d}M; \quad I_{55} = \int (x^2 + z^2)\,\mathrm{d}M,$$
$$I_{66} = \int (x^2 + y^2)\,\mathrm{d}M; \quad I_{46} = \int xz\,\mathrm{d}M, \quad (3.28)$$

where $\mathrm{d}M$ is the mass of an infinitesimally small structural element located at (x, y, z). The integration in eq. (3.28) is over the whole structure and is done in practice by a summation; I_{46} can often be neglected. Furthermore, it is common to express I_{jj} as Mr_{jj}^2, where the radius of gyration, r_{jj} corresponding to pitch and yaw is typically 0.25 times the ship length. The radius of gyration in roll is typically 0.35 to 0.40 times the beam of the ship.

One may wonder why we did not choose the origin of the coordinate system in the COG of the vessel as is typically done for spacecraft systems; that approach would be natural from a ship mass point of view. However, it is more convenient to use the chosen coordinate system with origin in the mean free surface when hydrodynamic problems involving external sea flow are considered. Then we can solve these hydrodynamic problems without considering the location of the vertical position of the COG.

The added mass and damping loads are steady-state hydrodynamic forces and moments due to forced harmonic rigid-body motions. No incident waves are present; however, the forced motion of the structure generates outgoing waves. The forced motion results in oscillating fluid pressure on the exterior wetted hull surface. If the ship has zero forward speed, then it is the dynamic pressure $p_D = -\rho\partial\varphi/\partial t$ that is considered in the equation of added mass and damping loads. The

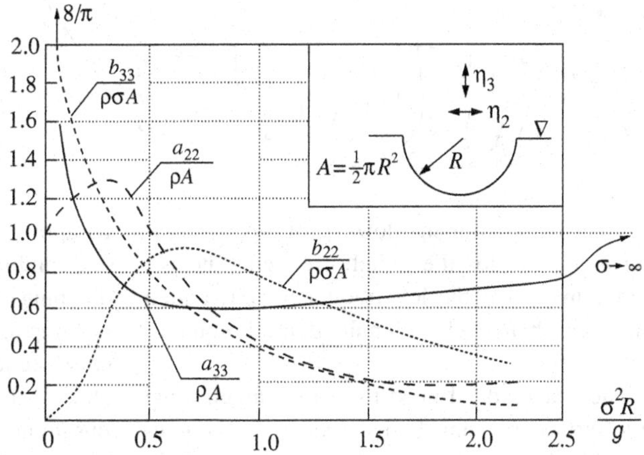

Figure 3.10. Calculated two-dimensional added mass and damping in sway and heave of a hemicircle in deep water: a_{22} is the added mass in sway, a_{33} is the added mass in heave, b_{22} is the damping in sway, b_{33} is the damping in heave, and σ is the circular frequency of oscillation.

velocity potential φ is linearly dependent on the forced motion amplitude and harmonically oscillates with the forcing frequency. Integration of these pressure loads over the mean position of the ship's wetted surface gives the resulting forces and moments on the ship. By defining the force components in the x-, y-, and z-directions as F_1, F_2, and F_3 and the moment components along the same axis as F_4, F_5, and F_6, we can formally write the hydrodynamic added mass and damping loads due to harmonic motion mode η_k as

$$F_j = -A_{jk}\ddot{\eta}_k - B_{jk}\dot{\eta}_k. \tag{3.29}$$

What we have implicitly said is that added mass has nothing to do with a finite mass of the water that is oscillating. The latter is a common misunderstanding. Actually heave-added mass for a catamaran and for various submerged bodies can, for instance, be negative in a certain frequency domain.

In Figure 3.10 we present numerically calculated two-dimensional added mass and damping coefficients in sway and heave of a semisubmerged circular cylinder of radius R with axis in the mean free surface as a function of the nondimensional frequency parameter $\sigma^2 R/g$ for deep-water conditions. Small letters are used as notations for added mass and damping coefficients for two-dimensional conditions. The free-surface condition for the velocity potential φ is given by eq. (3.1); we consider an inertial system but do not account for the effect of the ship's forward speed in the formulation of the linear free-surface conditions. How the ship's forward speed enters into the free-surface conditions is beyond the scope of this book and is discussed elsewhere (Faltinsen, 2005).

When $\sigma \to \infty$, the free-surface condition can be simplified as $\varphi = 0$ on $z = 0$, which is a consequence of the fact that the water flow accelerations dominate over the gravitational acceleration g. Neglecting g in eq. (3.1) gives $\varphi = 0$ on $z = 0$. If as an example we consider heave-added mass a_{33}, we can obtain the desired result by making a mirror image of the submerged body about the mean free surface $z = 0$. The velocity potential associated with forced heave motion of the double body is antisymmetric with respect to $z = 0$ (i.e., $\varphi = 0$ on $z = 0$). The consequence is that the flow around the double body correctly represents the flow in the water domain. Let us follow the procedure of how to determine the added mass in heave. We find the vertical hydrodynamic force in heave by properly integrating the hydrodynamic pressure $-\rho\partial\varphi/\partial t$. We can then understand the fact that the infinite-frequency heave-added mass of a hemicircle in infinite depth is equal to $\frac{1}{2}\rho\pi R^2$ (i.e., half the added mass of a circular cylinder in infinite fluid). Because the free-surface condition $\varphi = 0$ does not allow for free-surface waves to be generated by the oscillating body, the damping coefficients go to zero when $\sigma \to \infty$.

On the other hand, when $\sigma \to 0$, we can approximate the free surface as a rigid wall (i.e., $\partial\varphi/\partial z = 0$ on $z = 0$; see eq. (3.1)). The sway-added mass can then be obtained by making a mirror image of the body about the free surface. The double-body flow, for forced sway, has $z = 0$ as

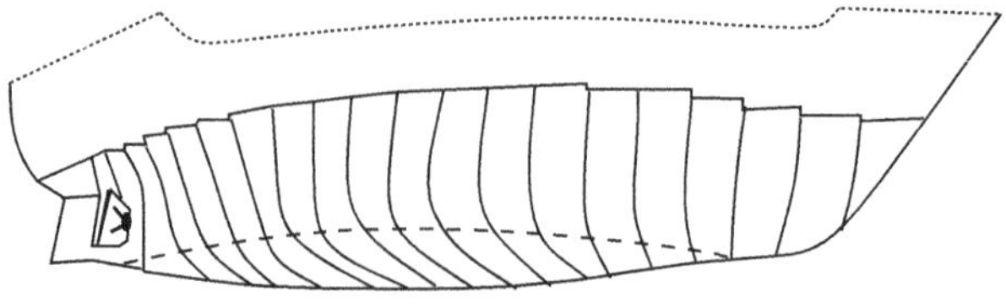

Figure 3.11. Illustration of strip theory for ships.

a symmetry line for the flow. It follows by arguing as we did for the infinite-frequency heave-added mass that the low-frequency sway-added mass for a hemicircle in infinite depth is equal to $\frac{1}{2}\rho\pi R^2$.

The heave-added mass for a two-dimensional free-surface piercing body in deep water goes logarithmically to infinity when $\sigma \to 0$. We cannot obtain that result by simply using the free-surface condition $\partial\varphi/\partial z = 0$ on $z = 0$. We must perform a proper asymptotic analysis by solving the problem for finite σ and then let $\sigma \to 0$. The consequence is a spatially constant term in the velocity potential that depends logarithmically on the frequency. This constant term influences the hydrodynamic force for a free-surface piercing body. However, there is zero contribution to the hydrodynamic force when the body is totally submerged (i.e., the added mass is finite when $\sigma \to 0$). We can in the latter case solve the low-frequency problem directly by using the free-surface condition $\partial\varphi/\partial z = 0$ on $z = 0$. The two-dimensional damping coefficients in heave and sway go to zero when $\sigma \to 0$ for deep-water conditions.

When $\sigma \to 0$, water depth carries increased importance, which is a consequence of the fact that small frequencies are associated with long wavelengths that are affected by the sea bottom. A consequence of the finite depth is that the two-dimensional added mass in heave for a free-surface piercing body becomes finite when $\sigma \to 0$. Furthermore, it can be seen in Exercise 3.11.4 that the two-dimensional heave damping coefficient is finite for a free-surface piercing body in finite depth. This result is once more a consequence of the limiting process $\sigma \to 0$.

Two-dimensional results for added mass and damping can be combined with strip theory to obtain an approximation of the three-dimensional added mass and damping for a ship. The principle is to divide the underwater part of the ship into a number of strips (see Figure 3.11). Two-dimensional added mass and damping coefficients are calculated for each strip and combined according to which added mass and damping coefficients are wanted. Using strip theory implies that the variation of the flow in the cross-sectional plane is much larger than the variation of the flow in the longitudinal direction, but this implication is not true at the ends of the ship. Strip theory is not satisfactory for small frequencies. It is state-of-the-art engineering practice to use three-dimensional numerical methods based on the boundary element method (BEM) to solve the linear frequency-domain problem for ships and offshore structures with zero steady forward speed.

More examples are given in Section 4.7 on two-dimensional frequency-independent added-mass coefficients, including interaction effects between bodies and the effect of a wall. These results are used to estimate the effect of interior tank structures on the natural frequencies of liquid motion in a tank.

Heave, pitch, and roll are response variables where resonance frequencies play an important role. The resonant response amplitude is obviously dependent on damping. Three main sources of damping are caused by the external flow for a displacement ship. They can be categorized as

- wave radiation damping,
- viscous damping, and
- lifting effects due to rudders and fins.

In addition we may mention damping effects due to propeller and thruster race (Gjelsvik, 1983). Wave radiation damping dominates for heave and pitch resonant motions. The lifting effect damping increases with forward speed. It is common to add a viscous roll damping term in eq. (3.26), whereby the term $B_{44}^{visc}\dot{\eta}_4\,|\dot{\eta}_4|$ is added to the left-hand side of equation $j = 4$. This term can be linearized by introducing a linear term $K\dot{\eta}_4$ so that the work done over one period is the same for this term and the nonlinear term. This linearization gives $K = B_{44}^{visc}\eta_{4a}8\sigma_e/(3\pi)$, where η_{4a} is the roll amplitude. The most important effect of viscous damping is associated with vortex shedding from bilge keels and the resulting influence on the pressure distribution on the hull. The effect of skin friction is more important in model scale than in full scale. At model scale it cannot be totally neglected, whereas it is negligible at full scale. For a ship without bilge keels, the bilge radius has an important influence on the eddy-making roll damping for midship sections (Tanaka, 1961). Eddy-making damping can be quite large for rectangular cross-sections, whereas it is of less importance for conventional midship sections. Kato (1966) and Ikeda *et al.* (1977) have given empirical formulas for roll damping due to bilge keels. If no other roll damping devices are used, bilge keel damping can very well amount to 50% or more of the total damping that is dependent on the beam-to-draft ratio.

Different ways exist to calculate the added mass, damping, and wave excitation loads due to potential flow effects which appear in the equations of motions in the frequency domain. These calculations are discussed, for instance, by Faltinsen (1990, 2005).

If we integrate the hydrostatic pressure loads on the ship hull (i.e., the term $-\rho g z$), it results in restoring forces and moments. It is then necessary to integrate over the instantaneous position of the ship. Since the COG of the ship is not chosen as the origin of the coordinate system, one must also consider moments due to the ship's weight, which acts through the COG. We write the linear restoring force and moment components as

$$F_j = -C_{jk}\eta_k. \tag{3.30}$$

The only nonzero restoring coefficients for a ship in an upright condition (i.e., where the x–z-plane is the symmetry plane for the submerged volume) are

$$C_{35} = C_{53} = -\rho g \iint_{A_W} x \, dS,$$

$$C_{44} = \rho g \nabla (z_B - z_G) + \rho g \iint_{A_W} y^2 \, dS = \rho g \nabla \overline{GM_T},$$

$$C_{55} = \rho g \nabla (z_B - z_G) + \rho g \iint_{A_W} x^2 \, dS = \rho g \nabla \overline{GM_L},$$

$$C_{33} = \rho g A_W, \tag{3.31}$$

where A_W is the waterplane area; ∇ is the displaced volume of water; z_G and z_B are the z-coordinates of the COG and center of buoyancy, respectively; $\overline{GM_T}$ is the transverse metacentric height; and $\overline{GM_L}$ is the longitudinal metacentric height. We can, for instance, deduce C_{33} by considering forced heave motion and analyzing the additional buoyancy forces due to hydrostatic pressure $-\rho g z$, which can be linearly approximated as $-\rho g A_W \eta_3$. From this result, C_{33} follows from eq. (3.30). If the vessel is fully submerged like a submarine, A_W is zero and $(z_B - z_G) > 0$ is necessary to have a vessel stable in the upright condition.

If the ship has partially filled tanks, the liquid motion in the tanks influences the dynamics of the vessel. If the behavior of the liquid in the tanks is assumed quasi-steady, reductions occur in the longitudinal and transverse metacentric heights. The effect is always taken into account in floating stability calculations. If the period of oscillation is high relative to the highest natural period of the liquid motion in the tank, this calculation is a good approximation. Because resonant fluid motions (sloshing) in a tank may occur, changing the metacentric heights due to the free surface of tanks would be wrong in general for the equations of motion. A common practice is to include the mass of the tank liquid in determining the total mass of the vessel, its COG, and moments and products of inertia (i.e., the tank liquid is considered "frozen"). When the hydrodynamic forces and moments due to sloshing are analyzed, the "frozen" liquid effects are parts of the expressions (see also Section 5.4.1.3). *Hence, we must be careful not to count the "frozen" liquid effects twice.* An analysis of the coupling between the dynamic liquid motions in the tank and the ship motion is described in Section 3.6 and accounts for proper phasing of the forces and moments caused by sloshing.

For a ship with lateral (port–starboard) symmetry, the six coupled equations of motion reduce to two sets of equations: one set of three coupled equations for surge, heave, and pitch and another set of three coupled equations for sway, roll, and yaw. Thus, for a ship with lateral symmetry, surge, heave, and pitch are not linearly coupled with sway, roll, and yaw.

Equations of motion (3.26) can be solved by substituting $\eta_k = \bar{\eta}_k e^{i\sigma_e t}$ into the left-hand side; $\bar{\eta}_k$

is the complex amplitude of the motion mode k. Dividing by the factor $e^{i\sigma_e t}$, the resulting equations can be separated into real and imaginary parts, leading to six coupled algebraic equations for the real and imaginary parts of the complex amplitudes for surge, heave, and pitch in the frequency domain. A similar algebraic equation system can be set up for sway, roll, and yaw. If a nonlinear roll viscous term is introduced, one solution strategy is to do several calculations with an assumed roll angle in the viscous roll damping term and then for each case find which incident wave amplitude causes the assumed and calculated roll angles to be equal. This procedure has to be modified when an irregular sea is considered (Price & Bishop, 1974). The matrix equations following from the equations of motion in the frequency domain can be solved using standard methods. It is also common to operate with the complex equation system without separating it into real and imaginary parts. When the motions are found, the wave loads can be obtained by using the expressions for hydrodynamic forces, which were discussed previously.

It should once more be stressed that eqs. (3.26) are only generally valid for steady-state sinusoidal motions. For instance, in a transient free-surface problem, the hydrodynamic forces include memory effects and do not depend only on the instantaneous values of body velocity and acceleration (Cummins, 1962; Ogilvie, 1964). This problem is further discussed in Section 3.8.

3.6 Coupled sloshing and ship motions

A partially filled tank may cause important coupling with the ship motions in a frequency range of dominant sloshing. Newman (2005) presented three-dimensional linear potential flow calculations for a stationary ship that illustrated the large mutual interaction that can exist between sloshing and wave-induced ship motions. Furthermore, because ship motions have an important effect on mean wave loads (Faltinsen, 1990), sloshing also matters in a stationkeeping analysis. Newman's calculations demonstrated this fact.

The effect of sloshing in a linear frequency-domain solution based on potential flow is that the added mass matrix A_{jk} due to the external flow is modified (i.e., an added mass term A_{jk}^{tank} associated with sloshing appears). The damping matrix B_{jk} has no modifications because B_{jk} is associated with external wave radiation. We may also have represented the effect of sloshing as a modification of the restoring matrix, which follows from the fact that we may rewrite A_{jk}^{tank} as $-\sigma^2 C_{jk}^{\text{tank}}$ when steady-state oscillations with frequency σ are assumed, where C_{jk}^{tank} denotes restoring coefficients that should be added to the hydrostatic restoring coefficients C_{jk}.

Journee (1983) presented comparisons between experimental and numerical calculations of roll for a liquefied natural gas carrier. The model was free to move in six degrees of freedom. Linear hydrodynamic theory in the frequency domain was used in combination with empirical viscous roll damping. The effect of the filling ratio was investigated for three cargo tanks. Good agreement between theory and experiments was demonstrated. However, the natural roll frequency of the ship was about half the lowest natural frequency of the liquid in the cargo tanks, meaning that nonlinear resonant sloshing is not significant at the roll resonance period of the ship.

In the next subsection we analyze the quasi-steady free-surface effects of a tank. The following detailed discussion of unsteady effects focuses on simplified representations to reveal the important physical effects associated with coupled sloshing and ship motions.

3.6.1 Quasi-steady free-surface effects of a tank

Here we show how the effect of a tank can be included in the linear equations of ship motion when the frequency of encounter is much smaller than the lowest natural frequency of sloshing in the tank. A quasi-steady tank analysis can then be made. The free surface of the tank remains horizontal and the pressure in the tank is hydrostatic relative to the free surface.

We limit ourselves to roll and consider the resulting roll moment on the ship. We define a body-fixed coordinate system $(\bar{x}, \bar{y}, \bar{z})$ and an auxiliary inertial coordinate system (X, Y, Z) where the Z-axis is vertical upward and with $Z = 0$ corresponding to the horizontal free surface Σ of the tank (see Figure 3.12). In practice, the $\bar{x}$–$\bar{z}$-plane is in the ship's centerplane and the tank may be located to the side of the centerplane.

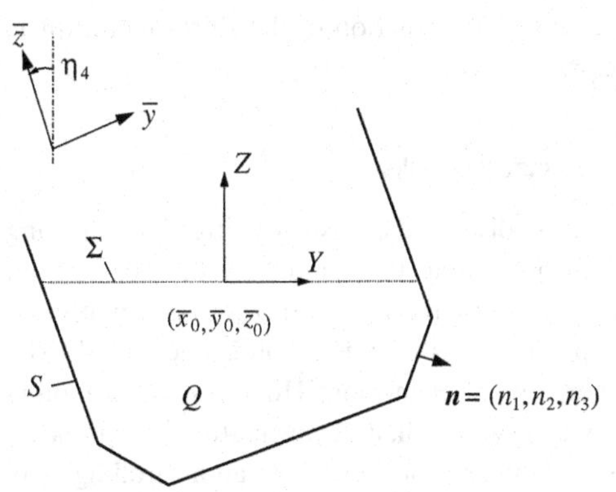

Figure 3.12. Definitions of coordinate systems.

The origin of the (X, Y, Z) system is in $(\bar{x}_0, \bar{y}_0, \bar{x}_0)$. When the roll angle η_4 is zero, the Z-axis and the $\bar{z}$-axis are parallel. Furthermore, when the tank is in an upright position, the X–Z-plane is assumed to be a symmetry plane for the tank. The two coordinate systems can be related to each other by the following relationships:

$$Z = (\bar{y} - \bar{y}_0)\sin\eta_4 + (\bar{z} - \bar{z}_0)\cos\eta_4,$$
$$Y = (\bar{y} - \bar{y}_0)\cos\eta_4 - (\bar{z} - \bar{z}_0)\sin\eta_4,$$

which can for small roll angles be approximated as

$$Z = (\bar{y} - \bar{y}_0)\,\eta_4 + (\bar{z} - \bar{z}_0),$$
$$Y = (\bar{y} - \bar{y}_0) - (\bar{z} - \bar{z}_0)\,\eta_4.$$

The pressure p in the liquid inside the tank can be expressed as $p - p_0 = -\rho_l g Z$, where p_0 is the gas pressure above the liquid free surface and ρ_l is the density of the liquid. The hydrostatic pressure causes a roll moment about the $\bar{x}$-axis that can be expressed as

$$F_4 = -\rho_l g \int_S Z(\bar{y}n_3 - \bar{z}n_2)\,\mathrm{d}S, \qquad (3.32)$$

where S is the wetted tank surface. The normal vector to S, $\boldsymbol{n} = (n_1, n_2, n_3)$, has positive direction out of the liquid (see Figure 3.12), meaning that $p\boldsymbol{n}$ is a force vector per unit area acting *on* the tank boundary. Because $Z = 0$ on Σ, eq. (3.32) can be rewritten as

$$F_4 = -\rho_l g \int_{S+\Sigma} Z(\bar{y}n_3 - \bar{z}n_2)\,\mathrm{d}S. \qquad (3.33)$$

Since the integral in eq. (3.33) is over a closed surface $S + \Sigma$, we can use the Gauss theorem given by eq. (A.1) to reexpress the integral. It follows from eq. (A.1) that

$$\int_{S+\Sigma} \bar{y}n_3\,\mathrm{d}S = 0, \quad \int_{S+\Sigma} \bar{y}^2 n_3\,\mathrm{d}S = 0,$$
$$\int_{S+\Sigma} \bar{z}\bar{y}n_3\,\mathrm{d}S = \int_Q \bar{y}\,\mathrm{d}Q,$$

$$\int_{S+\Sigma} \bar{z}n_2\,\mathrm{d}S = 0, \quad \int_{S+\Sigma} \bar{y}\bar{z}n_2\,\mathrm{d}S = \int_Q \bar{z}\,\mathrm{d}Q,$$
$$\int_{S+\Sigma} \bar{z}^2 n_2\,\mathrm{d}S = 0,$$

which means that $F_4 = -\rho_l g \int_Q \bar{y}\,\mathrm{d}Q + \rho_l g \eta_4 \int_Q \bar{z}\,\mathrm{d}Q$. We reexpress the first integral on the right-hand side as

$$\int_Q \bar{y}\,\mathrm{d}Q = \int_{Q_0} \bar{y}\,\mathrm{d}Q - \eta_4 \int_{L_t}\int_{\bar{y}_1}^{\bar{y}_2} \bar{y}(\bar{y} - \bar{y}_0)\,\mathrm{d}\bar{y}\,\mathrm{d}\bar{x}. \qquad (3.34)$$

The integration of the first integral on the right-hand side is over the "frozen" liquid volume Q_0 (see Figure 3.13). When the roll angle η_4 is small, Q_0 is partly bounded by a plane parallel to the $\bar{x}$–$\bar{y}$-plane that contains the X-axis, which is a consequence of the conservation of liquid mass. The integral over Q_0 in eq. (3.34) is independent of time. The last integral on the right-hand side of eq. (3.34) is a correcting term so that we correctly integrate over Q. We have set $\mathrm{d}Q = (\bar{y} - \bar{y}_0)\,\eta_4\,\mathrm{d}\bar{y}\,\mathrm{d}\bar{x}$ in the last integral. The integration limits $\bar{y}_j$ ($j = 1, 2$) of the last integral are defined in the figure. Furthermore, L_t indicates that we integrate over the tank length in the $\bar{x}$-direction.

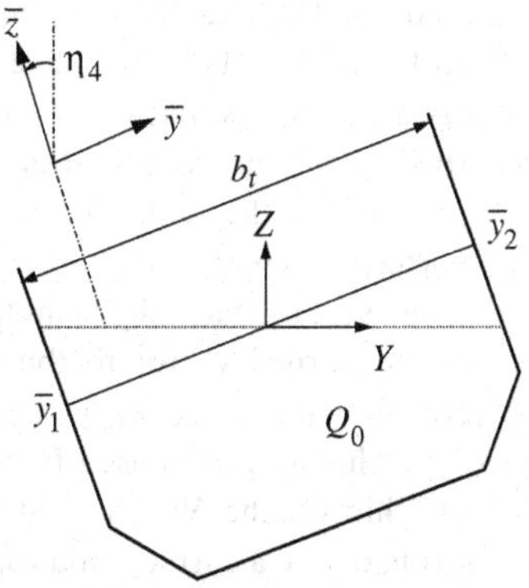

Figure 3.13. Definitions of coordinate systems and the frozen liquid volume Q_0. The tank liquid is below $Z = 0$.

We focus on the time-dependent part of the roll moment that is linearly dependent on η_4, which can be expressed as

$$F_4 = \rho_l g \left(\int_{L_t} \int_{\bar{y}_1}^{\bar{y}_2} (\bar{y} - \bar{y}_0)^2 \, d\bar{y} \, d\bar{x} + \int_{Q_0} \bar{z} \, dQ \right) \eta_4,$$

where we have used the fact that $\int_{\bar{y}_1}^{\bar{y}_2} (\bar{y} - \bar{y}_0) \, d\bar{y} = 0$, which follows from symmetry properties of the tank. The term $\rho_l g \eta_4 \int_{Q_0} \bar{z} \, dQ$ in F_4 represents a "frozen" liquid effect: it is the roll moment caused by the liquid weight. The liquid mass, as previously stated, has been included as part of the vessel's mass and should not be accounted for twice. If we now refer to our ship motion coordinate system defined in Figure 3.8 and denote the tank breadth at the free surface as $b_t(x)$, the correction to C_{44} in eq. (3.31) due to the tank is

$$C_{44}^{\text{tank}} = -\tfrac{1}{12} \rho_l g \int_{L_t} b_t^3(x) \, dx \tag{3.35}$$

and is based on a quasi-steady assumption. This term has a destabilizing effect on the quasi-steady roll moment.

As an example let us consider a rectangular tank with tank breadth $b_t(x)$ and subdivide the tank transversely into N equal compartments with tank breadth $b_t(x)/N$. The correction to C_{44} from all the compartments can then be expressed as $C_{44}^{\text{tank}} = -\frac{1}{12} \rho_l g N^{-2} \int_{L_t} b_t^3(x) dx$. Thereby the destabilizing effect of a tank on the roll moment is reduced by subdividing the tank into compartments.

If we had included the roll moment term due to the liquid weight in C_{44}^{tank}, we would note that the correction term C_{44}^{tank} has a structure similar to that of C_{44} in eq. (3.31). Actually, we could have derived C_{44} in the same way as we have shown for the tank. A difference is that we do not have to be concerned about the conservation of liquid mass as we are for the tank. We note that the signs on corresponding terms in C_{44} and C_{44}^{tank} differ, which comes mathematically from the sign differences of the normal vector to the wetted surface. A term similar to $-\rho g \nabla z_G$ in C_{44} is not present in C_{44}^{tank}. This term is caused by the roll moment of the ship weight. We may, in a similar way as for roll, derive a correction term to C_{55} from the tank. These weight terms are normally taken into the load condition calculations of the ship, and the free-surface term given by eq. (3.35) is given as a correction of the ship's metacentric height.

3.6.2 Antirolling tanks

We now discuss the unsteady effects of sloshing in an antirolling tank and how they can be an effective damping device of roll. As a matter of simplicity we assume that roll is uncoupled from other modes of motion. However, the coupling between sway, roll, and yaw matters. A similarity exists between the physics of an antirolling tank and a tuned liquid damper (TLD) used for suppressing horizontal vibrations of, for instance, tall buildings due to wind and earthquakes.

There are two types of antirolling tanks (see Figure 1.25). They are the U-tube and the free-surface tank with a free surface extending over most of the breadth of the ship. To get the best possible effect, a tank can impose a 15–30% reduction in the transverse metacentric height $\overline{GM_T}$ in quasi-steady conditions. By quasi-steady conditions we mean that the frequency of oscillation is clearly lower than the lowest natural frequency of sloshing (i.e., as described in the previous section).

For the antirolling tank to work properly, it is necessary that the lowest natural frequency for sloshing in the tank be close to the natural roll frequency for the vessel. Some prefer to set these two natural frequencies equal, whereas others choose the lowest sloshing frequency to be 6–10% higher than the natural roll frequency. Then the sloshing motion is out of phase with the roll motion. Typical tanks have a ratio $\overline{\delta GM_T}/\overline{GM_T}$ between 0.15 and 0.3, where $\overline{\delta GM_T}$ is the quasi-steady decrease or free-surface correction in $\overline{GM_T}$ due to the antirolling tank. We can determine the tank dimensions by means of the sloshing frequency and $\overline{\delta GM_T}/\overline{GM_T}$.

The resonance frequency of a free-surface tank can be changed by altering the water level in the tank. This type of antirolling tank is therefore well suited for a ship operating with a wide range of metacentric heights. A change of the natural frequency for sloshing is not so easy to do for a U-tube tank. If the ratio between maximum $\overline{GM_T}$ and minimum $\overline{GM_T}$ is larger than 2 for different loading conditions, it is common to use two antirolling tanks. Besides volume and weight the loss of stability is one of the major drawbacks of

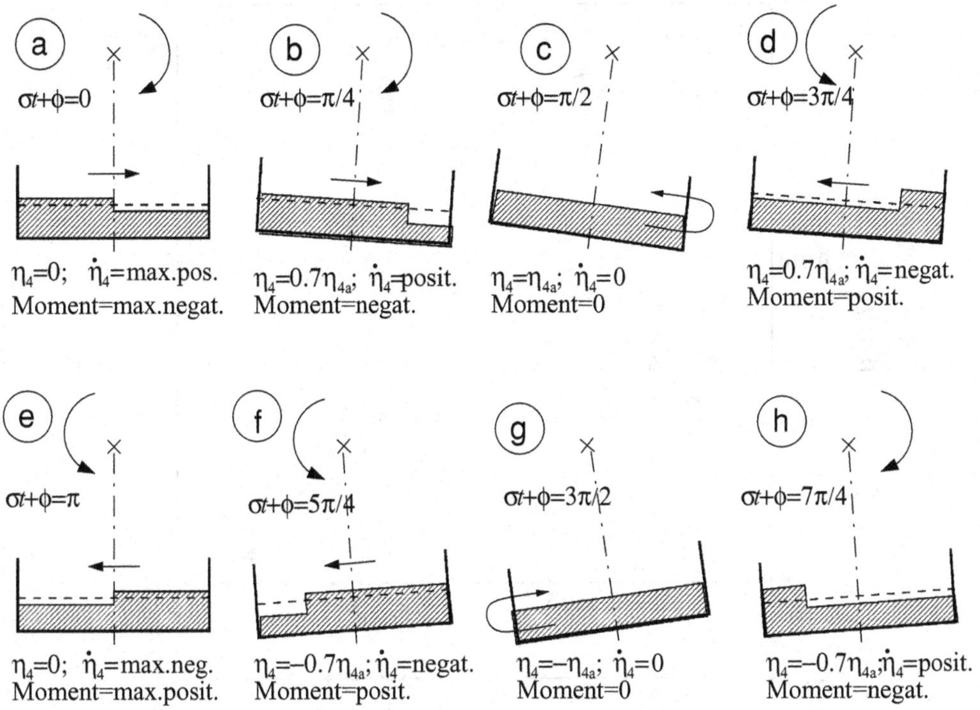

Figure 3.14. Position of hydraulic jump during one period of rolling at tank resonance (van den Bosch & Vugts, 1966).

all antirolling tanks. The loss of stability is permanently present (e.g., when the ship is circling in calm water or when suffering from list due to a strong wind).

We separately discuss free-surface and U-tube tanks in the following text. The flow in a free-surface tank is far more complicated than in a U-tube. Therefore, in this chapter we discuss briefly the free-surface antirolling tank and come back to the hydrodynamic aspects in Chapter 8 after more background has been provided.

3.6.3 Free-surface antirolling tanks

The resonant flow in a free-surface antirolling tank corresponds to strongly nonlinear shallow-water conditions with a hydraulic jump traveling back and forth in the tank. Figure 3.14 illustrates the presence of steady-state hydraulic jumps for a rectangular tank with shallow liquid depth. The tank is forced harmonically in roll with frequency equal to the lowest natural sloshing frequency. The position of the hydraulic jump is in the middle of the tank when the instantaneous roll angle is zero and the roll velocity has either a minimum or a maximum value. The fact that the hydrodynamic pressure in shallow-liquid theory is hydrostatic below the instantaneous free-surface elevation enables us to understand the behavior of the hydrodynamic roll moment indicated in the figure. The roll moment has mimimum and maximum values when the roll velocity is maximum and minimum, respectively. When the tank is used as an antirolling tank on a ship, the consequence is that the sloshing causes a roll damping effect.

If we wish to calculate how large a damping effect an antirolling tank has on roll, we need information about the hydrodynamic roll moment due to the flow in the tank. We assume regular incident waves, and we write the tank roll moment due to forced roll oscillation $\eta_{4a}\sin\sigma t$ as $K_{ta}\sin(\sigma t+\varepsilon_t)$, where K_{ta} and ε_t can be obtained experimentally by doing forced harmonic oscillation tests of the tank. Figure 3.15 presents experimental results for an antirolling tank system consisting of *two* shallow-liquid free-surface tanks. The liquid depth-to-breadth ratio h/l of each tank is 0.035. This gives a nondimensional lowest natural sloshing frequency $\sigma_1\sqrt{l/g}=0.59$ by using

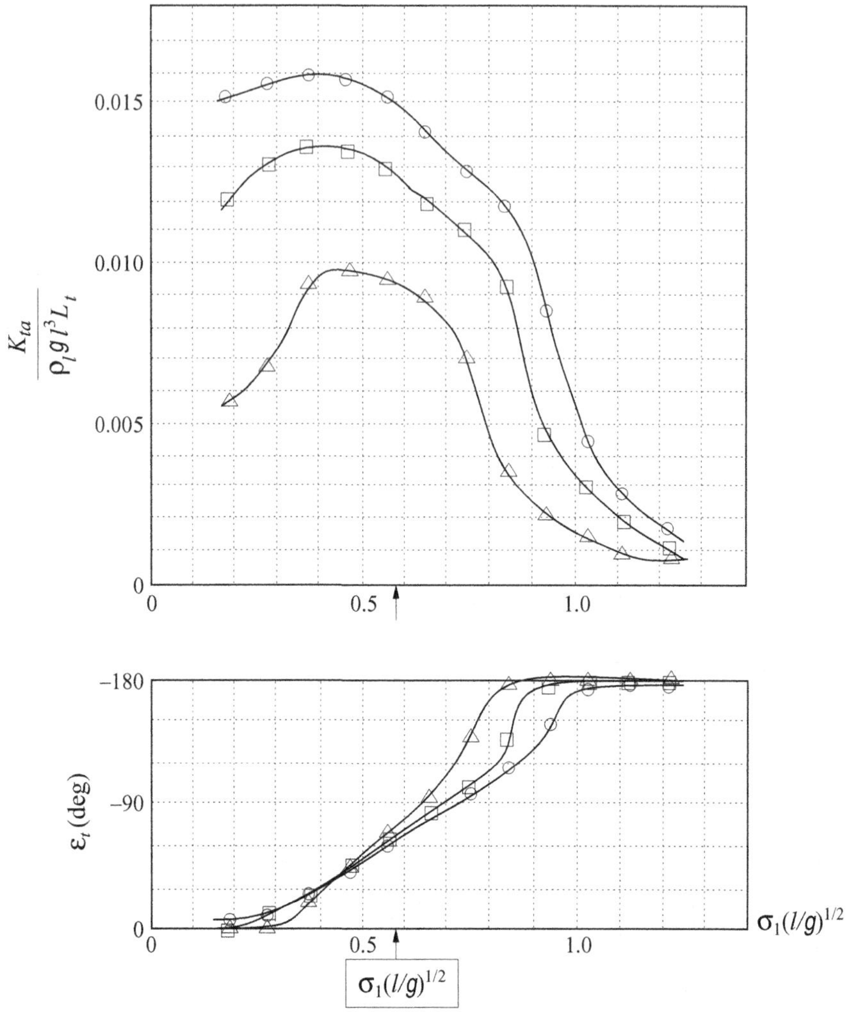

Figure 3.15. Nondimensional amplitude and phase of tank moment for a shallow-liquid-depth free-surface tank system (Vugts, 1968): η_{4a} is the forced roll amplitude in radians; ○, $\eta_{4a} = 0.1$; □, $\eta_{4a} = 0.0667$; Δ, $\eta_{4a} = 0.0333$.

the fact that the shallow depth approximation of the lowest natural frequency can, according to eq. (4.148) (with $m = 1$), be expressed as

$$\sigma_1 = \pi l^{-1}\sqrt{gh}. \quad (3.36)$$

The results for the tank roll-moment amplitude are presented nondimensionally as $K_{ta}/(\rho_l g l^3 L_t)$, where ρ_l is the density of the tank liquid and L_t is the tank length.

The uncoupled roll equation in regular waves can now be expressed as

$$(I_{44} + A_{44})\ddot{\eta}_4 + B_{44}\dot{\eta}_4 + B_v\dot{\eta}_4|\dot{\eta}_4| + C_{44}\eta_4 = F_4^{exc} + F_4^t, \quad (3.37)$$

where I_{44} is the ship's moment of inertia in roll. The effect of the liquid in the tank should not be accounted for in estimating I_{44}; A_{44} is the added moment in roll; B_{44} is the damping in roll due to external wave generation; B_v is a viscous damping coefficient associated with the external flow, which matters in case the vessel is equipped with bilge keels (which is neglected in the following analysis); and C_{44} is the roll restoring coefficient, which is proportional to the metacentric height $\overline{GM_T}$ without any corrections due to an internal free surface in the antirolling tank. We can write $C_{44} = \rho g \nabla \overline{GM_T}$, where the density of the external water is ρ and ∇ is the displaced volume of water of the ship; F_4^{exc} is the roll excitation moment due to the incident waves. Finally,

$$F_4^t = K_{ta}\sin(\sigma t + \varepsilon_t)$$
$$= K_{ta}\sin\sigma t\cos\varepsilon_t + K_{ta}\cos\sigma t\sin\varepsilon_t$$
$$= \frac{K_{ta}}{\eta_{4a}}\cos(\varepsilon_t)\eta_4 + \frac{K_{ta}}{\sigma\eta_{4a}}\sin(\varepsilon_t)\dot{\eta}_4.$$

The only ways that the effect of the liquid in the tank comes into eq. (3.37) is through F_4^t and the fact that the weight of the tank liquid influences the draft of the vessel. We can now write eq. (3.37) as

$$(I_{44}+A_{44})\ddot{\eta}_4 + \left(B_{44} - \frac{K_{ta}}{\sigma\eta_{4a}}\sin\varepsilon_t\right)\dot{\eta}_4 + \left(C_{44} - \frac{K_{ta}}{\eta_{4a}}\cos\varepsilon_t\right)\eta_4 = F_4^{exc}. \quad (3.38)$$

The objective is to reduce ship rolling, which can be obtained by creating a large effective damping $B_{44} - K_{ta}/(\sigma\eta_{4a})\sin\varepsilon_t$ at roll resonance. The term $-K_{ta}\sin\varepsilon_t$ is large if we choose the natural period for the flow inside the tank to coincide with the natural roll period. We see this result in Figure 3.15 by noting that ε_t is close to $-90°$ and K_{ta} is large at the nondimensional natural frequency $\sigma_1\sqrt{l/g} = 0.59$.

We note from eq. (3.38) that the restoring coefficient has also been modified. The term $-K_{ta}\cos\varepsilon_t/\eta_{4a}$ is negative at small oscillation frequency and then represents simply the quasi-steady effect of an internal free surface on the metacentric height. However, $-K_{ta}\cos\varepsilon_t/\eta_{4a}$ is frequency-dependent and changes sign when going from a frequency below the natural frequency for the tank flow to a frequency above. When resonant sloshing occurs, $-K_{ta}\cos\varepsilon_t$ is small because ε_t is close to –90°.

If we set as a condition that the roll natural frequency of the ship $\sigma_n^{roll} = \sqrt{\rho g\nabla\overline{GM_T}/(I_{44}+A_{44})}$ is equal to the lowest natural sloshing frequency given by eq. (3.36), we can determine the water depth in the free-surface tank as $h = (l/\pi)^2\,\rho\nabla\overline{GM_T}/(I_{44}+A_{44})$. It is adequate to use a shallow-water approximation in this context. The tank breadth l should be chosen close to the sectional beam of the vessel because the tank moment is proportional to l^3. Furthermore, the length of the tank is determined by following the previously mentioned recommendations on $\overline{\delta GM_T}/\overline{GM_T}$. Van den Bosch and Vugts (1966) experimentally investigated the influence of the distance s from the tank bottom to the axis of rotation: s is positive if the tank bottom is situated above the roll axis. The measured tank moment amplitude K_{ta} was, for a given absolute value of s, clearly larger for positive than for negative s, suggesting that it is beneficial to place the free-surface tank in the upper parts of the ship. However, a proper analysis implies that we also incorporate the horizontal force and yaw moment due to the tank in the coupled sway–roll–yaw equations.

Because the hydrodynamic roll moment due to sloshing in the tank is a nonlinear function of the roll amplitude (see Figure 3.15), we have to solve eq. (3.38) by successive approximations. However, it is important for accurate estimates to include the coupling effect between sway, roll, and yaw in the analysis.

Figure 3.16 shows examples of free-surface antirolling tanks with screens tested at Marintek's Laboratory in Norway. A broad variety of screen systems are available. We discuss that fact further in Section 6.7.3 in connection with TLDs and discuss how the viscous cross-flow past a screen system affects sloshing behavior. Marintek recommends using three screens: one should be in the middle of the tank and the other two at a quarter-length of the tank from the tank walls. A solidity ratio of about 0.5 is recommended. The *solidity ratio*, *Sn*, is the ratio of the area of the shadow projected by screens on a plane parallel to the screen to the total area contained within the frame of the screen.

We may also use computational fluid dynamics to describe the hydrodynamic effect of an antirolling tank. Another possibility is to use the shallow-liquid theory of Verhagen and van Wijngaarden (1965), which assumes two-dimensional inviscid flow in a rectangular tank without interior structures. We study in Section 8.8.2 the tank roll moments by using the theory of Verhagen and van Wijngaarden (1965) and comparing the results with experiments by van den Bosch and Vugts (1966) for a *single* shallow-liquid tank.

3.6.4 U-tube roll stabilizer

The U-tube is a commonly used antirolling tank but may be more expensive to arrange in a ship compared to a free-surface tank. However, this

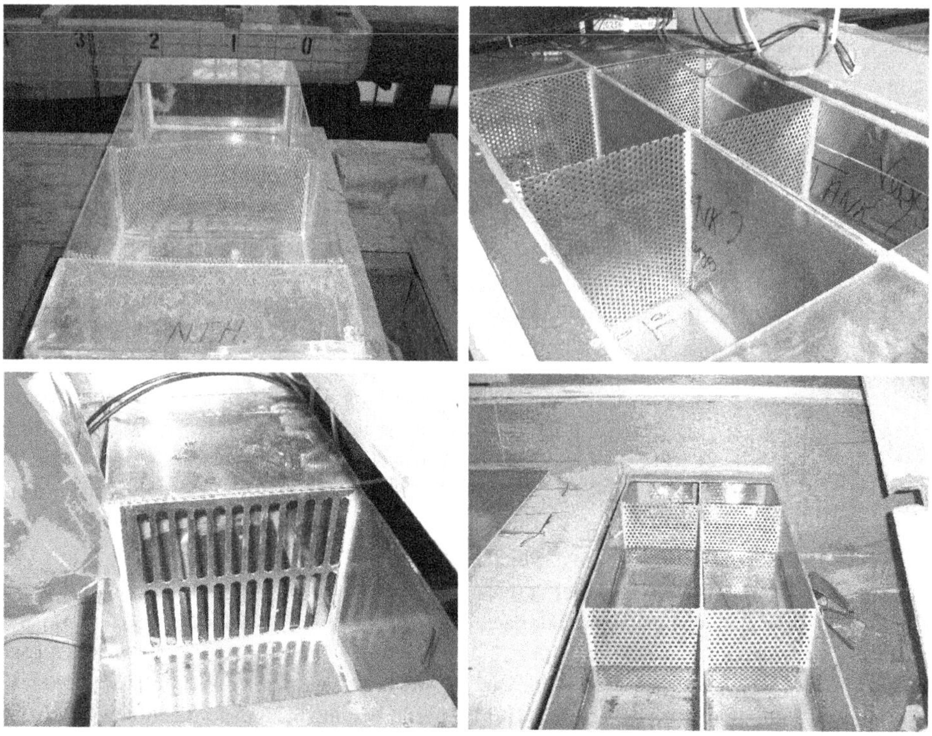

Figure 3.16. Examples of free-surface antirolling tanks with screens tested at Marintek.

arrangement depends very much on the type and design of the ship. Many installations exist where U-shaped tanks are the ideal solution when using existing side tanks. The U-shaped antirolling tank offers antiheeling service for harbor operation as a dual use.

We first describe in detail the hydrodynamic behavior of a U-tube. The analysis uses a tank-fixed coordinate system $O\bar{x}\bar{y}\bar{z}$ with the $\bar{x}$–$\bar{z}$-plane in the longitudinal centerplane of the ship as well as for the tank (see Figure 3.17). Similarly to the notations of Section 2.3, the unit vectors along the $\bar{x}$-, $\bar{y}$-, and $\bar{z}$-axes are denoted $\boldsymbol{e}_x$, $\boldsymbol{e}_y$, and $\boldsymbol{e}_z$, respectively.

The liquid in the tank is assumed incompressible. In addition, we start with the Navier–Stokes equation in a tank-fixed coordinate system (see eq. (2.56)) as a basis to express the behavior of the liquid velocity $\boldsymbol{v}_r$ relative to the tank. However, certain simplifications should be made. The first assumption is that the effect of viscosity is negligible, which means that we consider the Euler equations. This assumption is not true when the boundary-layer flow along the tank walls is analyzed or if flow separation occurs either as a consequence of flow through valves and screens or at high curvature parts of the tank surface. If the tank surface has a sharp corner with an interior angle larger than 180° relative to the liquid domain, the flow will always separate at the corner. The viscous damping due to flow separation is more important than the damping due to viscous energy dissipation in an attached boundary-layer flow. We see in Chapter 6 for rectangular free-surface tanks that viscous boundary-layer damping is very small and that flow through screens may cause a considerable damping.

A next simplification is that the equations should be linearized and only linear terms in the relative liquid velocity $\boldsymbol{v}_r$ and the translatory and angular ship velocities $\boldsymbol{v}_O$ and $\boldsymbol{\omega}$ will be kept. The consequence is, for instance, that the convective acceleration term $\boldsymbol{v}_r \cdot \nabla \boldsymbol{v}_r$, the Coriolis acceleration, and the centripetal acceleration are neglected. The resulting equation is

$$\frac{\partial^* \boldsymbol{v}_r}{\partial t} = -\frac{1}{\rho_l}\nabla p + \boldsymbol{g} - \boldsymbol{a}_O - \dot{\boldsymbol{\omega}} \times \boldsymbol{r}. \quad (3.39)$$

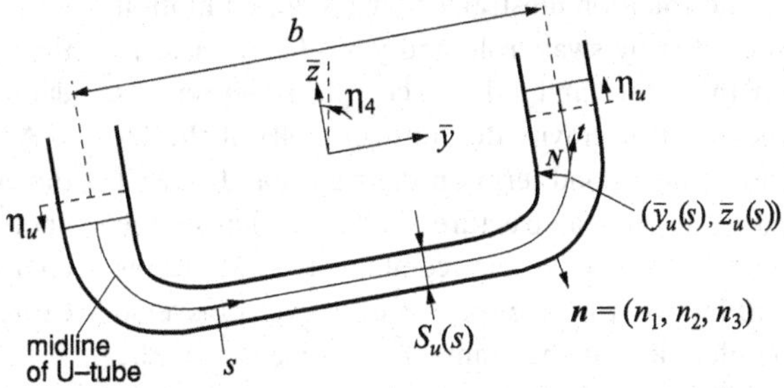

Figure 3.17. Definition of parameters used in general U-tube analysis.

We commented in Section 3.5.2 that surge, heave, and pitch are decoupled from sway, roll, and yaw in a linear formulation of the ship motion problem. The ship acceleration term $\boldsymbol{a}_O + \dot{\boldsymbol{\omega}} \times \boldsymbol{r}$ on the right-hand side of eq. (3.39) includes, therefore, only the effect of sway, roll, and yaw. By using eq. (3.22) and a transformation between the $Oxyz$ coordinate system defined in Figure 3.8 and the $O\bar{x}\bar{y}\bar{z}$ system we can show that it is correct within linear theory to express

$$\boldsymbol{a}_O + \dot{\boldsymbol{\omega}} \times \boldsymbol{r} = \ddot{\eta}_{2t}\boldsymbol{e}_y + \ddot{\eta}_4(-\bar{z}\boldsymbol{e}_y + \bar{y}\boldsymbol{e}_z), \tag{3.40}$$

where a dot above the symbol means time derivative and $\eta_{2t} = \eta_2 - z_O\eta_4 + x_t\eta_6$, where z_O is the z-coordinate of the origin of the $O\bar{x}\bar{y}\bar{z}$-coordinate system and x_t is an average $\bar{x}$-coordinate of the tank, implying that the longitudinal extent of the tank is small relative to the ship's length. Furthermore, we may approximate the acceleration-of-gravity vector in eq. (3.39) as

$$\boldsymbol{g} = -g\boldsymbol{e}_z - g\eta_4\boldsymbol{e}_y. \tag{3.41}$$

The next assumption is that the flow in the U-tube is one-dimensional along a line going through the cross-sectional centers of the U-tube. The curvilinear coordinate along this midline is denoted s and the $\bar{y}$- and $\bar{z}$-coordinates of the midline are $\bar{y}_u(s)$ and $\bar{z}_u(s)$, respectively. We let $s = 0$ and $s = l$ correspond to the instantaneous free surface for negative and positive $\bar{y}$-values, respectively. The vector tangent to the midline of the U-tube, $\boldsymbol{t}$, can be expressed as

$$\boldsymbol{t} = \frac{d\bar{y}_u(s)}{ds}\boldsymbol{e}_y + \frac{d\bar{z}_u(s)}{ds}\boldsymbol{e}_z. \tag{3.42}$$

The instantaneous free-surface elevation relative to the mean free surface is denoted η_u. Positive values of η_u correspond to a higher free surface than the mean free-surface elevation for positive $\bar{y}$, whereas the free surface is lower than the mean free surface for negative $\bar{y}$ (see Figure 3.17). The relative liquid velocity along the midline, v_s, can be expressed in terms of η_u by using continuity of liquid mass. We define $S_u(s)$ as the cross-sectional area of the U-tube at a given value of s. The value of $S_u(s)$ at the free surfaces is denoted S_{u0}. It follows by continuity of liquid mass and one-dimensional flow that $v_s = S_{u0}\dot{\eta}_u/S_u(s)$. We now take the dot product between eq. (3.39) and the tangent vector $\boldsymbol{t}$ and integrate the resulting equation from $s = 0$ to l:

$$\int_0^l \frac{\partial v_s}{\partial t} \mathrm{d}s = \ddot{\eta}_u S_{u0} \int_0^l \frac{\mathrm{d}s}{S_u(s)} \quad \text{and}$$
$$\int_0^l \frac{\partial p}{\partial s} \mathrm{d}s = p(l) - p(0) = 0. \tag{3.43}$$

We have in the last expression used the fact that the pressure is atmospheric on the free surfaces. Furthermore, by using eqs. (3.41) and (3.42), it follows that

$$\int_0^l \boldsymbol{g} \cdot \boldsymbol{t}\, \mathrm{d}s = -g\left[\bar{z}_u(l) - \bar{z}_u(0)\right] - g\eta_4[\bar{y}_u(l) - \bar{y}_u(0)]$$
$$= -2g\eta_u - gb\eta_4,$$

where b is the transverse distance between the centers of the two free surfaces (see Figure 3.17). Finally, eqs. (3.40) and (3.42) give

$$\int_0^l (\boldsymbol{a}_O + \dot{\boldsymbol{\omega}} \times \boldsymbol{r}) \cdot \boldsymbol{t}\, \mathrm{d}s$$
$$= \ddot{\eta}_{2t} b + \ddot{\eta}_4 \int_0^l [-\bar{z}_u(s)\, \mathrm{d}\bar{y}_u(s) + \bar{y}_u(s)\, \mathrm{d}\bar{z}_u(s)].$$

We now collect all terms and get the following second-order differential equation for η_u:

$$\left[S_{u0} \int_0^l \frac{\mathrm{d}s}{S_u(s)}\right] \ddot{\eta}_u + 2g\eta_u$$
$$= -gb\eta_4 - b\ddot{\eta}_{2t}$$
$$-\ddot{\eta}_4 \int_0^l [-\bar{z}_u(s)\mathrm{d}\bar{y}_u(s) + \bar{y}_u(s)\mathrm{d}\bar{z}_u(s)]. \tag{3.44}$$

The equation illustrates how the liquid motion is excited by sway, roll, and yaw. The first term on the right-hand side of eq. (3.44) represents the effect of gravity due to heel (roll) of the U-tube. The second term on the right-hand side of eq. (3.44) has a structure similar to that of the gravity term. One may combine the two terms and interpret $\ddot{\eta}_{2t} + g\eta_4$ as an effective transverse acceleration of the transverse projection of the liquid in the tube. One may of course also include the last term on the right-hand side of eq. (3.44) by including an average transverse acceleration due to roll acceleration $\ddot{\eta}_4$.

The liquid motion η_u expressed by eq. (3.44) is described similarly as a simple mass–spring system with excitation. The mass and spring terms in our case refer to the coefficients ahead of the liquid acceleration $\ddot{\eta}_u$ and displacement η_u, respectively. It is a well-known fact that a mass–spring system has a natural frequency, which we find by setting the excitation equal to zero and assuming a solution of the form $\eta_u = C\exp(\mathrm{i}\sigma_{nu}t)$, where $\mathrm{i} = \sqrt{-1}$ is the complex unit. It follows from eq. (3.44) that the natural frequency is

$$\sigma_{nu} = \sqrt{2g/S_{u0}\int_0^l S_u^{-1}(s)\,\mathrm{d}s}. \tag{3.45}$$

If the cross-sectional area $S_u(s)$ is constant then eq. (3.45) gives $\sigma_{nu} = \sqrt{2g/l}$. Because our mass–spring system has no damping, the response is infinite if the forcing frequency σ is equal to the natural frequency σ_{nu}. So we need a damping term, which physically may be associated with viscous boundary-layer effects and flow separation. The effect of damping of the liquid motion is desirable from a performance point of view of the U-tube as a roll damping device. The effect of boundary layers and flow through screens may be considered in similar way as in Chapter 6. A more complex analysis based on the Navier–Stokes equations is required to include in a rational way the effect of flow separation from the tube surface, especially at bends. It is also possible to control the flow so that effective damping of the liquid motion can be achieved. We do not pursue this point further but just assume that the damping effect can be linearized and included as a term

$$2\left[S_{u0}\int_0^l S_u^{-1}(s)\,\mathrm{d}s\right]\sigma_{nu}\xi_u\dot{\eta}_u \tag{3.46}$$

on the left-hand side of eq. (3.44), where ξ_u is the ratio between the damping and the critical damping.

A different way to find the expression for the natural frequency is to use conservation of energy. We assume no tank motion. The kinetic energy can, by using the expression for v_s, be expressed as

$$E_k = \frac{1}{2}\rho_l S_{u0}^2\dot{\eta}_u^2\int_0^l \frac{\mathrm{d}s}{S_u(s)}.$$

The potential energy is $E_p = \rho_l g S_{u0}\eta_u^2$. It follows by conservation of energy that $d(E_k + E_p)/dt = 0$; that is,

$$\rho_l S_{u0}^2\dot{\eta}_u\ddot{\eta}_u\int_0^l \frac{\mathrm{d}s}{S_u(s)} + 2\rho_l g S_{u0}\eta_u\dot{\eta}_u = 0 \quad \text{or}$$

$$\left[S_{u0}\int_0^l \frac{\mathrm{d}s}{S_u(s)}\right]\ddot{\eta}_u + 2g\eta_u = 0,$$

which is the same as the left-hand side of eq. (3.44).

3.6.4.1 Nonlinear liquid motion

We generalize eq. (3.44) by including nonlinear terms. However, we limit the derivation to only forced sway and roll motions of the U-tube. Furthermore, one-dimensional liquid motion is assumed. We start with the effect of gravitational acceleration. Equation (3.41) can be expressed as $\boldsymbol{g} = -g\cos(\eta_4)\boldsymbol{e}_z - g\sin(\eta_4)\boldsymbol{e}_y$, which leads to the restoring term $2g\eta_u$ in eq. (3.44) being replaced by $2g\cos(\eta_4)\,\eta_u$ and the excitation term $-gb\eta_4$ in eq. (3.44) being replaced by $-gb\sin\eta_4$. The effects of $\partial v_s/\partial t$ and $(\boldsymbol{a}_O + \dot{\boldsymbol{\omega}}\times\boldsymbol{r})$ do not lead to nonlinear terms.

Then we consider the remaining terms of the Euler equations (see eq. (2.56) with $\nu = 0$) that were left out as a part of the linearization. We start with the convective acceleration term $\boldsymbol{v}_r\cdot\nabla\boldsymbol{v}_r$, which for one-dimensional flow can be written as $v_s(\partial v_s/\partial s)\boldsymbol{t}$. We take the dot product of this term with the tangent vector $\boldsymbol{t}$ and then integrate from $s = 0$ to l as we did, for instance, in eq. (3.43), giving $\int_0^l v_s\frac{\partial v_s}{\partial s}\,\mathrm{d}s = \frac{1}{2}[v_s^2(l) - v_s^2(0)] = 0$. The dot product between $\boldsymbol{t}$ and the acceleration term $-\boldsymbol{\omega}\times\boldsymbol{v}_O$ is $-\dot{\eta}_4\dot{\eta}_{2t}d\bar{z}_u(s)/ds$. Integration from $s = 0$ to l gives a restoring term $2\dot{\eta}_4\dot{\eta}_{2t}\eta_u$ in the equation for the liquid motion. The Coriolis acceleration $-2(\boldsymbol{\omega}\times\boldsymbol{v}_r)$ acts perpendicularly to $\boldsymbol{t}$ and, hence, gives zero contribution. Finally, the

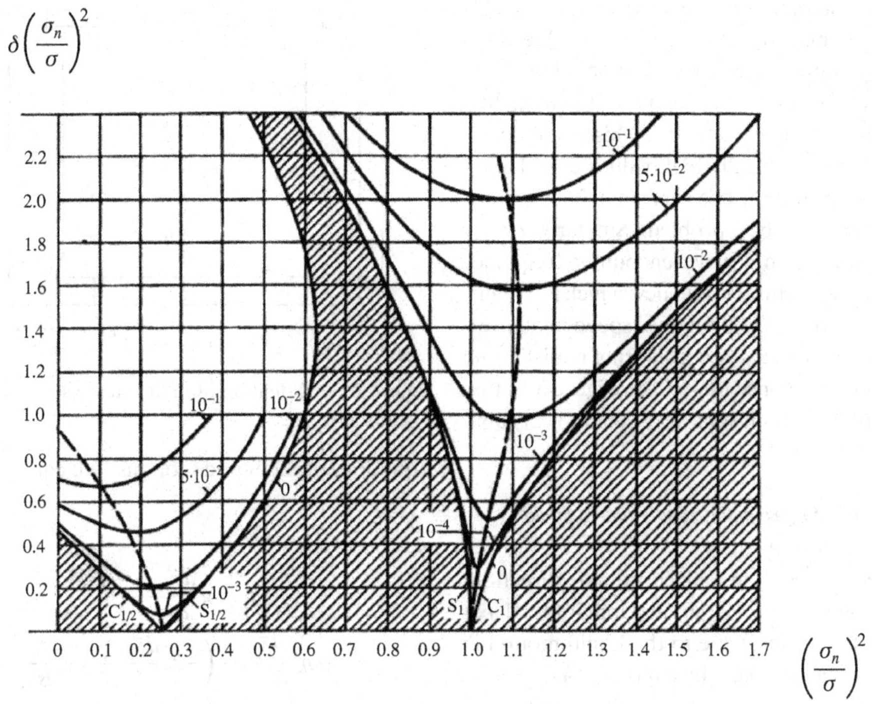

Figure 3.18. Stability diagram for Mathieu equation: σ_n, natural frequency; $\delta\sigma_n^2$, amplitude of harmonically oscillating part of restoring coefficient; σ, frequency of harmonically oscillating part of restoring coefficient. Shaded areas represent stable domains when the ratio ξ between damping and critical damping is zero. Lines are shown with values of $\xi\,(\sigma_n/\sigma)$ equal to 10^{-4}, 10^{-3}, 10^{-2}, $5\cdot10^{-2}$, and 10^{-1}. These lines are boundaries between stable and unstable domains (Klotter, 1978).

dot product between the centripetal acceleration $-\boldsymbol{\omega}\times(\boldsymbol{\omega}\times\boldsymbol{r})$ and $\boldsymbol{t}$ is $\dot{\eta}_4^2\,(\bar{y}_u d\bar{y}_u/ds + \bar{z}_u\, d\bar{z}_u/ds)$. Integration from $s = 0$ to l gives zero contribution. Therefore, eq. (3.44) can be generalized as

$$\left[S_{u0}\int_0^l \frac{ds}{S_u(s)}\right]\ddot{\eta}_u + 2\,(g\cos\eta_4 + \dot{\eta}_4\dot{\eta}_{2t})\,\eta_u$$
$$= -gb\sin\eta_4 - b\ddot{\eta}_{2t}$$
$$-\ddot{\eta}_4\int_0^l[-\bar{z}_u(s)\,d\bar{y}_u(s) + \bar{y}_u(s)\,d\bar{z}_u(s)]. \quad (3.47)$$

We assume harmonic ship motions $\eta_j = \eta_{ja}\cos(\sigma_e t + \varepsilon_j)$ and approximate $\cos\eta_4$ as $1 - \frac{1}{2}\eta_4^2$. The restoring term on the left-hand side of eq. (3.47) can then be expressed as

$$\{[2g - \tfrac{1}{2}g\eta_{4a}^2 + \tfrac{1}{2}\eta_{4a}\eta_{2ta}\sigma_e^2\cos(\varepsilon_4 - \varepsilon_{2t})]$$
$$- g\eta_{4a}^2\cos(2\sigma_e t + 2\varepsilon_4)$$
$$- \eta_{4a}\eta_{2ta}\sigma_e^2\cos(2\sigma_e t + \varepsilon_4 + \varepsilon_{2t})\}\,\eta_u.$$

Let us partly generalize eq. (3.47) with harmonic ship motions by introducing damping. However, we set the excitation equal to zero. The differential equation can then be expressed as

$$\ddot{\eta}_u + 2\xi\sigma_n\dot{\eta}_u + \sigma_n^2[1 + \delta\sin(\sigma t + \beta)]\eta_u = 0, \quad (3.48)$$

where ξ is the ratio between the damping and the critical damping, $\sigma = 2\sigma_e$, and σ_n is the natural frequency for the case of $\delta = 0$. We should note that σ_n is not identical with σ_{nu} given by eq. (3.45). If $\xi = 0$, eq. (3.48) is the classical Mathieu equation. It can be shown that the stability depends on σ_n/σ, δ, and the damping ratio ξ. Klotter (1978) has presented curves that show the instability domains (see Figure 3.18). The unshaded areas represent instability domains when $\xi = 0$. The boundary between the stability and unstability domains are shown for different ξ-values. The domains for dangerous combinations of δ and σ_n/σ are located in the vicinity of $\sigma_n/\sigma = 0.5, 1.0, 1.5, 2.0, \ldots$ when δ is small. Figure 3.18 concentrates on σ_n/σ up to $\sqrt{1.7}$. For instance, $\sigma_n/\sigma = 0.5$ means for our case that $\sigma_e = \sigma_n$, which is a realistic scenario. Damping has a positive influence. The higher the damping, the higher δ has to be for instability to occur. Let us say $\xi = 0.02$, which is a small damping that is possible if no flow separation occurs in the U-tube. Figure 3.18 shows that instabilities can occur for δ slightly larger than 0.8 for $\sigma_n/\sigma = 0.5$. This result corresponds to unrealistically high values of δ. Therefore, the damping must be very small for Mathieu-type instability to be a problem. Similarly $\sigma_n/\sigma = 1.0$ corresponds to a wave encounter frequency that is half the natural frequency, which is a realistic scenario for a ship at forward speed in quartering waves. However, by considering realistic ship motions we note once more from Figure 3.18 that the damping level must be very small for instability to occur.

3.6.4.2 Linear forces and moments due to liquid motion in the U-tube

The force vector $\boldsymbol{F}_u$ in tank coordinate directions and the roll moment F_{4u} about a longitudinal axis through the origin O due to the liquid motion in the U-tube can formally be expressed as

$$\boldsymbol{F}_u = \int_S p\boldsymbol{n}\,dS \quad \text{and} \quad F_{4u} = \int_S (\bar{y}n_3 - \bar{z}n_2)p\,dS, \tag{3.49}$$

where S is the wetted tank surface and $\boldsymbol{n} = (n_1, n_2, n_3)$ is the normal vector to S with positive direction out of the liquid. By p we mean the excess pressure relative to atmospheric pressure. Because p is zero on the free surface, we can rewrite eq. (3.49) as

$$\boldsymbol{F}_u = \int_{S+S_F} p\boldsymbol{n}\,dS, \quad F_{4u} = \int_{S+S_F} (\bar{y}n_3 - \bar{z}n_2)p\,dS, \tag{3.50}$$

where S_F is the free surface. It follows now by Gauss theorem (see eq. (A.1)) that

$$\boldsymbol{F}_u = \int_Q \nabla p\,dQ \quad \text{and}$$
$$F_{4u} = \int_Q \left(\bar{y}\frac{\partial p}{\partial \bar{z}} - \bar{z}\frac{\partial p}{\partial \bar{y}}\right)dQ, \tag{3.51}$$

where Q denotes the liquid domain. We can evaluate eq. (3.51) by using values of $\partial p/\partial \bar{y}$ and $\partial p/\partial \bar{z}$ at the midline and expressing $\partial p/\partial \bar{y}$ and $\partial p/\partial \bar{z}$ in terms of the pressure gradients $\partial p/\partial s$ and $\partial p/\partial N$ along the midline and perpendicular to the midline, respectively. The pressure gradient $\partial p/\partial s$ was already found when we formulated the linear

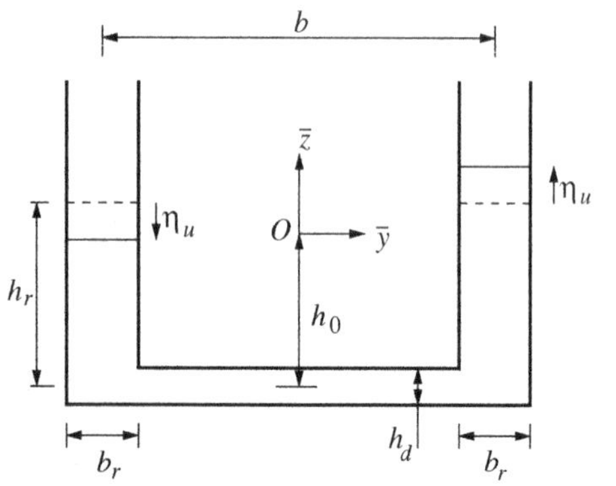

Figure 3.19. Definition of parameters used in detailed U-tube analysis.

equations of the liquid motions. We can write

$$\begin{aligned}\frac{1}{\rho_l}\frac{\partial p}{\partial s} &= \frac{1}{\rho_l}\boldsymbol{t}\cdot\nabla p\\ &= -\frac{S_{u0}}{S_u(s)}\ddot{\eta}_u - g\frac{d\bar{z}_u}{ds} - g\eta_4\frac{d\bar{y}_u}{ds}\\ &\quad - \ddot{\eta}_{2t}\frac{d\bar{y}_u}{ds} - \left(-\bar{z}_u\frac{d\bar{y}_u}{ds} + \bar{y}_u\frac{d\bar{z}_u}{ds}\right)\ddot{\eta}_4.\end{aligned} \tag{3.52}$$

To find $\partial p/\partial N$ we note that the normal vector $\boldsymbol{N}$ (see Figure 3.17) to the midline can be expressed as

$$\boldsymbol{N} = -\frac{d\bar{z}_u(s)}{ds}\boldsymbol{e}_y + \frac{d\bar{y}_u(s)}{ds}\boldsymbol{e}_z. \tag{3.53}$$

It follows now by the fact that $\boldsymbol{v}_r \cdot \boldsymbol{N} = 0$ (i.e., the flow is along the s-direction) and eq. (3.39) that

$$\begin{aligned}\frac{1}{\rho_l}\frac{\partial p}{\partial N} &= \frac{1}{\rho_l}\boldsymbol{N}\cdot\nabla p = -g\frac{d\bar{y}_u}{ds} + g\eta_4\frac{d\bar{z}_u}{ds} + \ddot{\eta}_{2t}\frac{d\bar{z}_u}{ds}\\ &\quad - \ddot{\eta}_4\left(\bar{z}_u\frac{d\bar{z}_u}{ds} + \bar{y}_u\frac{d\bar{y}_u}{ds}\right).\end{aligned} \tag{3.54}$$

3.6.4.3 Lloyd's U-tube model

We consider a U-tube with a constant cross-section along the $\bar{x}$-axis (see Figure 3.19). The longitudinal length of the U-tube is denoted L_t. The U-tube in upright position consists of two reservoirs with constant width b_r and with midline coordinates $\bar{y}_u = \pm\frac{1}{2}b$ and a horizontal duct with constant height h_d. The vertical distance between the midline of the duct and the mean free surface is h_r. The constant $\bar{z}$-coordinate of the horizontal midline is denoted $\bar{z}_u = -h_O$. A conflict exists in assuming one-dimensional flow at the junctions between a reservoir and the duct; however, this effect is considered negligible in the following analysis. We should note the following in the integrations in eqs. (3.44) and (3.51):

- The reservoir at $\bar{y}_u = -\frac{1}{2}b$ has a length of $h_r - \eta_u$. The derivatives of the midline coordinates are $d\bar{z}_u/ds = -1$, $d\bar{y}_u/ds = 0$ and $\partial/\partial\bar{y} = \partial/\partial N$, $\partial/\partial\bar{z} = -\partial/\partial s$.
- The duct at $\bar{z}_u = -h_O$ has a length of b. The derivatives of the midline coordinates are $d\bar{z}_u/ds = 0$, $d\bar{y}_u/ds = 1$ and $\partial/\partial\bar{y} = \partial/\partial s$, $\partial/\partial\bar{z} = \partial/\partial N$.
- The reservoir at $\bar{y}_u = \frac{1}{2}b$ has a length of $h_r + \eta_u$. The derivatives of the midline coordinates are $d\bar{z}_u/ds = 1$, $d\bar{y}_u/ds = 0$ and $\partial/\partial\bar{y} = -\partial/\partial N$, $\partial/\partial\bar{z} = \partial/\partial s$.

Lloyd (1989) analyzed the same U-tube considered in this example by following a slightly different derivation based on Stigter (1966). The resulting linear equations of liquid motion and ship motion are similar to ours. We start with eq. (3.44), which can be expressed as

$$\begin{aligned}(2h_r + bb_r/h_d)\ddot{\eta}_u + B_d\dot{\eta}_u + 2g\eta_u \\ = -gb\eta_4 - b\ddot{\eta}_{2t} - b(h_r + h_O)\ddot{\eta}_4,\end{aligned} \quad (3.55)$$

where a damping term $B_d\dot{\eta}_u$ has been introduced, as in eq. (3.46). We now examine the force given by eq. (3.51). Only linear terms are kept. The force along the $\bar{z}$-axis is simply the weight of the liquid. The force along the $\bar{y}$-axis is

$$\begin{aligned}F_{2u} &= \int_Q \frac{\partial p}{\partial\bar{y}}dQ \\ &= -M_l g\eta_4 - M_l(\ddot{\eta}_{2t} - \bar{z}_{Gl}\ddot{\eta}_4) - \rho_l L_t b_r b\ddot{\eta}_u,\end{aligned} \quad (3.56)$$

where M_l is the liquid mass and $\bar{z}_{Gl}$ is the $\bar{z}$-coordinate of the COG of the liquid mass. The first term on the right-hand side of eq. (3.56) is just a decomposition of the weight of the liquid along the $\bar{y}$-axis of our tank-fixed coordinate system. If this force component is resolved along the y-axis in the inertial coordinate system in Figure 3.8, then we should not include this term. The second term on the right-hand side of eq. (3.56) represents the inertia term due to transverse acceleration of the frozen liquid mass. If the longitudinal extension L_t is small relative to the ship length, then we may express the resulting yaw moment on the ship as $x_t F_{2u}$.

We now consider the roll moment. It follows from eqs. (3.52) and (3.54) and by keeping only linear terms in the liquid motion η_u and the ship's motion that

$$\begin{aligned}F_{4u} = &-\rho_l L_t b_r b(h_r + h_O)\ddot{\eta}_u - \rho_l L_t b_r bg\eta_u \\ &+ M_l\bar{z}_{Gl}g\eta_4 + M_l\bar{z}_{Gl}\ddot{\eta}_{2t} - I_{44l}\ddot{\eta}_4,\end{aligned} \quad (3.57)$$

where I_{44l} is the roll moment of inertia of the frozen liquid with respect to the longitudinal axis through the origin of our considered coordinate system. In the quasi-steady case, eq. (3.57) is an approximation of eq. (3.35), which expresses the reduction of the metacentric height due to liquid in a tank. It follows by using eq. (3.35) that F_{4u} should include the term $\rho_l g L_t[\frac{1}{2}b_r b^2 + \frac{4}{3}(\frac{1}{2}b_r)^3]\eta_4$. Because $\eta_u = -\frac{1}{2}b\eta_4$ in the quasi-steady case, then the term $-\rho_l L_t b_r bg\eta_u$ in eq. (3.57) equals $\frac{1}{2}\rho_l g L_t b_r b^2\eta_4$. This result means that we have neglected a small term of order $(b_r/b)^2$ in eq. (3.57). A reason for this is in our derivation. Equation (3.35) assumes the free surface remains horizontal, whereas the free surface in deriving eq. (3.57) is perpendicular to the U-tube surface.

We now include the effect of the considered U-tube in the equations of ship motion. The inertial coordinate system defined in Figure 3.8 is used. The inertia terms of the frozen liquid given in eqs. (3.56) and (3.57) (i.e., $M_l\bar{z}_{Gl}\ddot{\eta}_{2t}$, $M_l\bar{z}_{Gl}\ddot{\eta}_4$, $M_l\ddot{\eta}_{2t}$, and $I_{44l}\ddot{\eta}_4$) are included as parts of the inertia terms of the ship. We assume the frozen liquid is included in finding the longitudinal and vertical COG of the ship's loading condition. Therefore, the term $M_l\bar{z}_{Gl}g\eta_4$ in eq. (3.57) is not explicitly included in the equations of motion. It follows by generalizing the equations in Section 3.5.2 that the linear sway, roll, and yaw equations without viscous external damping become

$$\begin{aligned}&(M + A_{22})\ddot{\eta}_2 + B_{22}\dot{\eta}_2 + (A_{24} - Mz_G)\ddot{\eta}_4 + B_{24}\dot{\eta}_4 \\ &\quad + A_{26}\ddot{\eta}_6 + B_{26}\dot{\eta}_6 + \rho_l L_t b_r b\ddot{\eta}_u = F_2 e^{i\sigma_e t}, \\ &(A_{42} - Mz_G)\ddot{\eta}_2 + B_{42}\dot{\eta}_2 + (I_{44} + A_{44})\ddot{\eta}_4 + B_{44}\dot{\eta}_4 \\ &\quad + C_{44}\eta_4 + (-I_{46} + A_{46})\ddot{\eta}_6 + B_{46}\dot{\eta}_6 \\ &\quad + \rho_l L_t b_r b(h_r + h_O)\ddot{\eta}_u + \rho_l L_t b_r bg\eta_u = F_4 e^{i\sigma_e t}, \\ &A_{62}\ddot{\eta}_2 + B_{62}\dot{\eta}_2 + (A_{64} - I_{46})\ddot{\eta}_4 + B_{64}\dot{\eta}_4 \\ &\quad + (I_{66} + A_{66})\ddot{\eta}_6 + B_{66}\dot{\eta}_6 + \rho_l L_t b_r bx_t\ddot{\eta}_u = F_6 e^{i\sigma_e t},\end{aligned} \quad (3.58)$$

where $-h_O$ is the z-coordinate of the midline of the horizontal duct in the coordinate system defined in Figure 3.8. It is essential that C_{44} does not include any free-surface corrections due to the tank. The effect of the free surface is accounted for by the terms involving $\ddot{\eta}_u$ and η_u. We note that all three of eqs. (3.58) have four unknowns: η_u, η_2, η_4, and η_6. The necessary fourth equation is eq. (3.55).

We now consider only the coupling between roll and U-tube flow. However, note that it is not recommended in a practical calculation to exclude the coupling with sway and yaw. This simplification reveals how coupling between the roll and the U-tube flow changes the natural frequencies and roll response relative to uncoupled roll. After multiplying eq. (3.55) with $\rho_l L_t b_r$ we can write the following two coupled equations:

$$\begin{aligned}(I_{44}+A_{44})\ddot{\eta}_4 + B_{44}\dot{\eta}_4 + C_{44}\eta_4 + A_{4u}\ddot{\eta}_u + C_{4u}\eta_u \\ = F_4 e^{i\sigma_e t},\\ A_{u4}\ddot{\eta}_4 + C_{u4}\eta_4 + A_{uu}\ddot{\eta}_u + B_{uu}\dot{\eta}_u + C_{uu}\eta_u = 0,\end{aligned} \tag{3.59}$$

where

$$\begin{aligned} A_{uu} &= \rho_l L_t b_r (2h_r + bb_r/h_d), \quad C_{uu} = 2\rho_l L_t b_r g,\\ A_{4u} &= A_{u4} = \rho_l L_t b_r b\,(h_r + h_O),\\ C_{4u} &= C_{u4} = \rho_l L_t b_r b g, \quad B_{uu} = 2\xi_u \sqrt{A_{uu} C_{uu}},\\ B_{44} &= 2\xi_4 \sqrt{(I_{44}+A_{44})\,C_{44}}.\end{aligned}$$

The steady-state solutions of eq. (3.59) are expressed as $\eta_u = \bar{\eta}_u \exp(i\sigma_e t)$, $\eta_4 = \bar{\eta}_4 \exp(i\sigma_e t)$, where $\bar{\eta}_u$ and $\bar{\eta}_4$ are the complex amplitudes of the liquid flow and roll, respectively. Whenever dealing with a such a complex representation, it is the real part, $\mathrm{Re}[\bar{\eta}_u \exp(i\sigma_e t)]$ and $\mathrm{Re}[\bar{\eta}_4 \exp(i\sigma_e t)]$, that has physical meaning. It follows now from eq. (3.59) that

$$\begin{aligned}\bar{\eta}_4 &= \frac{F_4\left(-\sigma_e^2 + 2i\sigma_{nu}\sigma_e\xi_u + \sigma_{nu}^2\right)A_{uu}}{\left(-\sigma_e^2 + 2i\sigma_{n4}\sigma_e\xi_4 + \sigma_{n4}^2\right)(I_{44}+A_{44})A_{uu}\left(-\sigma_e^2 + 2i\sigma_{nu}\sigma_e\xi_u + \sigma_{nu}^2\right) - \left(C_{4u} - \sigma_e^2 A_{4u}\right)^2},\\ \bar{\eta}_u &= -\frac{\left(C_{4u} - \sigma_e^2 A_{4u}\right)}{\left(-\sigma_e^2 + 2i\sigma_{nu}\sigma_e\xi_u + \sigma_{nu}^2\right)A_{uu}}\bar{\eta}_4.\end{aligned} \tag{3.60}$$

We have in eqs. (3.60) denoted the uncoupled natural frequency in roll as $\sigma_{n4} = \sqrt{C_{44}/(I_{44}+A_{44})}$ and the uncoupled natural frequency of the liquid flow in the U-tube as $\sigma_{nu} = \sqrt{C_{uu}/A_{uu}}$. The coupling between the liquid flow and roll causes two natural frequencies to appear. Let us illustrate this fact by neglecting damping. To find the coupled natural frequencies we should set the roll excitation moment $F_4 \exp(i\sigma_e t)$ in eq. (3.59) equal to zero and find possible nontrivial solutions of eq. (3.59). This approach is equivalent to finding the frequencies when the denominator in the expression for $\bar{\eta}_4$ is equal to zero and gives that the two coupled undamped natural frequencies σ_j for the system follow from the equation

$$\left(\sigma_j^2 - \sigma_{n4}^2\right)\left(\sigma_j^2 - \sigma_{nu}^2\right) - \frac{\left(-\sigma_j^2 A_{u4} + C_{u4}\right)^2}{(I_{44}+A_{44})A_{uu}} = 0.$$

For instance, $\sigma_{nu} = \sigma_{n4} \equiv \sigma_n$ gives the following two coupled natural frequencies:

$$\sigma_{1,2}^2 = \frac{\sigma_n^2\sqrt{(I_{44}+A_{44})A_{uu}} \pm C_{u4}}{\sqrt{(I_{44}+A_{44})A_{uu}} \pm A_{u4}}. \tag{3.61}$$

Because the natural roll period can change significantly due to the ship's loading condition, and the natural period of the liquid flow in the U-tube is not sensitive to the reservoir height, the Rolls-Royce/Intering tank stabilizers choose a natural tank period equal to the shortest natural roll period.

Example. We examine a U-tube proposed by Perez (2002) as a roll damping device on a patrol boat. The main data are given in Table 3.4. We have an undamped and uncoupled natural period of 5.93 s for the liquid motion inside the tank.

We now examine the effect of coupling between the U-tube flow and roll. The patrol boat has an overall length $L_{OA} = 52.57$ m, beam $B = 8.6$ m, draft $d = 2.29$ m, and mass $M = 326{,}754$ kg. The vertical distance of the COG

Table 3.4. *Main data for U-tank proposal for a patrol boat*

Name	Value
h_d, duct height	0.45 m
L_t, tank length	3.94 m
b, tank width	6.97 m
h_r, reservoir liquid height	1.215 m
b_r, reservoir width	0.97 m
ρ_l, tank liquid density	1,000 kg m^{-3}
h_O, distance between center duct and COG	2.215 m

Source: Perez (2002).

above the keel is 3.93 m and the roll radius of gyration $r_{44} = 0.31B$ where $I_{44} = Mr_{44}^2$. Furthermore, $C_{44} = Mg\overline{GM_T} = 2.8848 \cdot 10^6$ N m/rad. We can calculate all the hydrodynamic coefficients using a state-of-the-art seakeeping program based on the strip theory of Salvesen *et al.* (1970), for instance, with empirical viscous roll damping coefficients. We do not pursue this line here but just choose some realistic values as an example. We first assume beam seas so that the frequency of encounter, σ_e, is equal to the frequency of the waves, σ_0. We neglect the forward speed and frequency dependence of the hydrodynamic coefficients and simply choose the damping ratio ξ_4 in roll due to wave radiation and viscous effects equal to 0.1. The added moment in roll, A_{44}, is chosen so that the undamped and uncoupled roll natural frequency of the vessel is equal to the undamped and uncoupled natural frequency for the U-tube flow. This choice gives $A_{44} = 0.105 I_{44}$. The roll excitation moment is based on eq. (3.97) (i.e., only the Froude–Kriloff loads with a long-wavelength assumption are considered). However, we multiply the expression with a commonly used factor, $\exp\left(-\frac{1}{2}kd\right)$, to make the Froude–Kriloff approximation better in a broader wavelength range. The latter factor accounts for the exponential decay of the incident wave pressure on the ship hull. Why we consider the pressure at a depth $d/2$ an effective pressure cannot be given a rational explanation. Equation (3.97) can then be modified as $F_4 \exp(i\sigma_e t) = -Mg\overline{GM_T}\exp(-\frac{1}{2}kd)k\zeta_a \cos(\sigma_0 t)$. It is understood that it is the real part of the left-hand side that has physical meaning; F_4 in this particular case is real. If diffraction effects due to the presence of the hull were accounted for, F_4 would generally be complex.

We now use σ as the notation for wave frequency and frequency of encounter. The computed roll amplitude η_{4a} and U-tube motion amplitude η_{ua} of the patrol boat in beam-sea regular waves with amplitude ζ_a, frequency σ, and wave number $k = \sigma^2/g$ are presented in Figure 3.20. A tank damping ratio of $\xi_u = 0.2$ has been used in the calculations. The behavior of the stabilized roll amplitude as a function of frequency has two maxima; we may refer to this as a waterbed behavior, because if we try to push the waterbed down at one place, humps appear nearby. The consequence of minimizing the roll amplitude at the unstabilized roll resonance frequency is that humps appear at the nearby coupled natural frequencies. These frequencies are $\sigma_1 = 0.8$ rad/s and

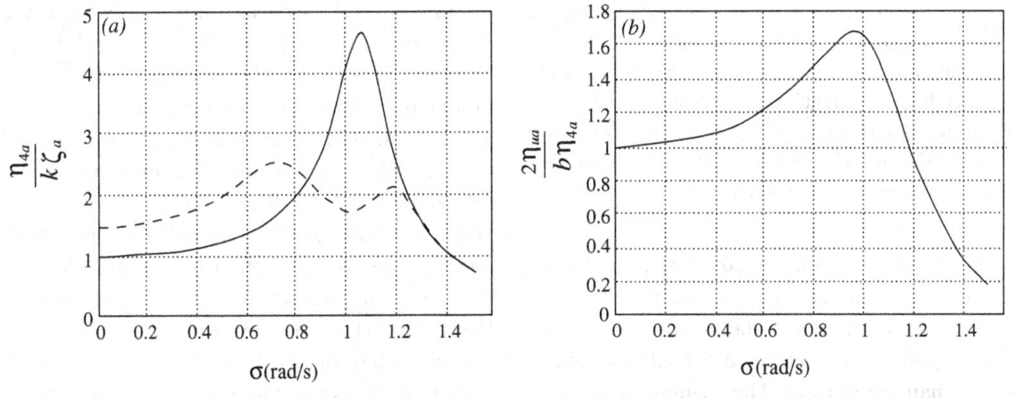

Figure 3.20. Effect of U-tube on (a) the roll amplitude η_{4a} and (b) the U-tube liquid motion amplitude η_{ua} of a 53-m-long patrol boat in beam-sea regular waves with amplitude ζ_a, frequency σ, and wave number $k = \sigma^2/g$. In part (a), the solid line does not account for the roll damping of the U-tube; the effect of the U-tube is represented by a dashed line.

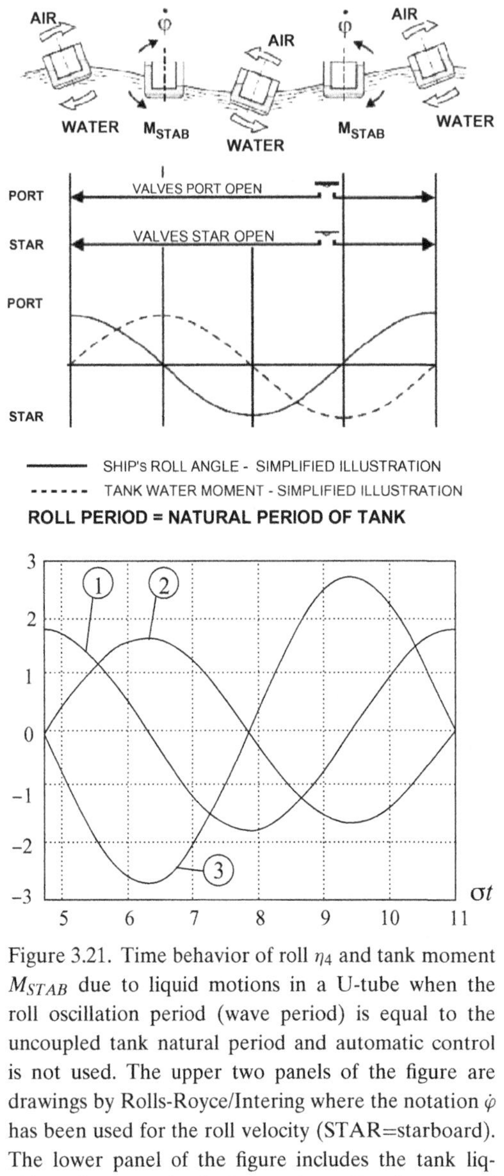

Figure 3.21. Time behavior of roll η_4 and tank moment M_{STAB} due to liquid motions in a U-tube when the roll oscillation period (wave period) is equal to the uncoupled tank natural period and automatic control is not used. The upper two panels of the figure are drawings by Rolls-Royce/Intering where the notation $\dot{\varphi}$ has been used for the roll velocity (STAR=starboard). The lower panel of the figure includes the tank liquid motion η_u and calculations for a 53-m-long patrol boat in beam-sea regular waves with amplitude ζ_a, frequency σ, and wave number $k = \sigma^2/g$. The values $\eta_4/k\zeta_a$, $2M_{STAB}/(\rho_l L_t b_r b^2 g k\zeta_a)$, and $2\eta_u/(bk\zeta_a)$ are presented by graphs 1, 2, and 3, respectively.

$\sigma_2 = 1.2$ rad/s according to eq. (3.61). The humps decrease with increased tank damping. A negative consequence of the stabilization is that the roll amplitude has increased for all frequencies lower than ≈0.85 rad/s. The asymptotic value of $\eta_{4a}/k\zeta_a$ when $\sigma \to 0$ is 1 in the uncoupled analysis. This result is not true for the coupled roll because of the free-surface effect of the U-tube. We can see that result from evaluating $\eta_{4a}/k\zeta_a$ when $\sigma \to 0$ by using eq. (3.60). The result is that

$$\overline{\eta}_4 \to \frac{-Mg\overline{GM}_T k\zeta_a}{Mg\overline{GM}_T - 0.5\rho_l L_t b_r b^2 g} \quad \text{as } \sigma \to 0.$$

The nondimensional U-tube motion amplitude $2\eta_{ua}/(b\eta_{4a})$ does not show a two-hump behavior and the largest value occurs at the natural frequency for the U-tube motion. It follows from eq. (3.60) that $\overline{\eta}_u \to -\frac{1}{2}b\overline{\eta}_4$ when $\sigma \to 0$. This result is consistent with the results in Figure 3.20. The U-tube motion amplitude is an important parameter in assessing whether impact occurs against the tank roof. As a consequence of heavy impact, the dynamic behavior may change to have nonlinear behavior. The vertical distance between the mean free-surface level and the tank roof is typically equal to the reservoir liquid height h_r.

The fact that the considered U-tank increases the response at low frequencies relative to the natural roll frequency implies that it is beneficial to use a controlled U-tank stabilizer as described in the following text.

3.6.4.4 Controlled U-tank stabilizer

The U-tube designs by Rolls-Royce/Intering use control of the tank when the ship rolls with a period larger than the uncoupled natural period for the liquid motion in the tank. We show the principle by first illustrating the physics of when the uncoupled roll period is equal to the uncoupled tank period and the tank is passive. Part of Figure 3.21 is made by Rolls-Royce/Intering and presents the roll angle η_4 together with the roll moment M_{STAB} on the ship due to the liquid motion in the tank as a function of time. We define the term $-\rho_l L_t b_r b(h_r + h_O)\ddot{\eta}_u - \rho_l L_t b_r b g \eta_u$ in eq. (3.57) as M_{STAB}. Our calculations for the 53-m-long patrol boat are presented in the same figure as a function of σt from $\frac{3}{2}\pi$ to $\frac{7}{2}\pi$ when the wave frequency is equal to the uncoupled natural frequencies of roll and the U-tube motions. When $\sigma t = \frac{3}{2}\pi$, our predicted roll angle is a maximum and the liquid motion $\eta_u = 0$ and $M_{STAB} = 0$. This result agrees with the initial time in Rolls-Royce/Intering's drawing. When $\sigma t = 2\pi$, our predicted value of M_{STAB} has a maximum and the tank liquid motion η_u has a minimum, which also agrees with Rolls-Royce/Intering's drawing. Similarly we can also see at later stages the agreement between our calculations and

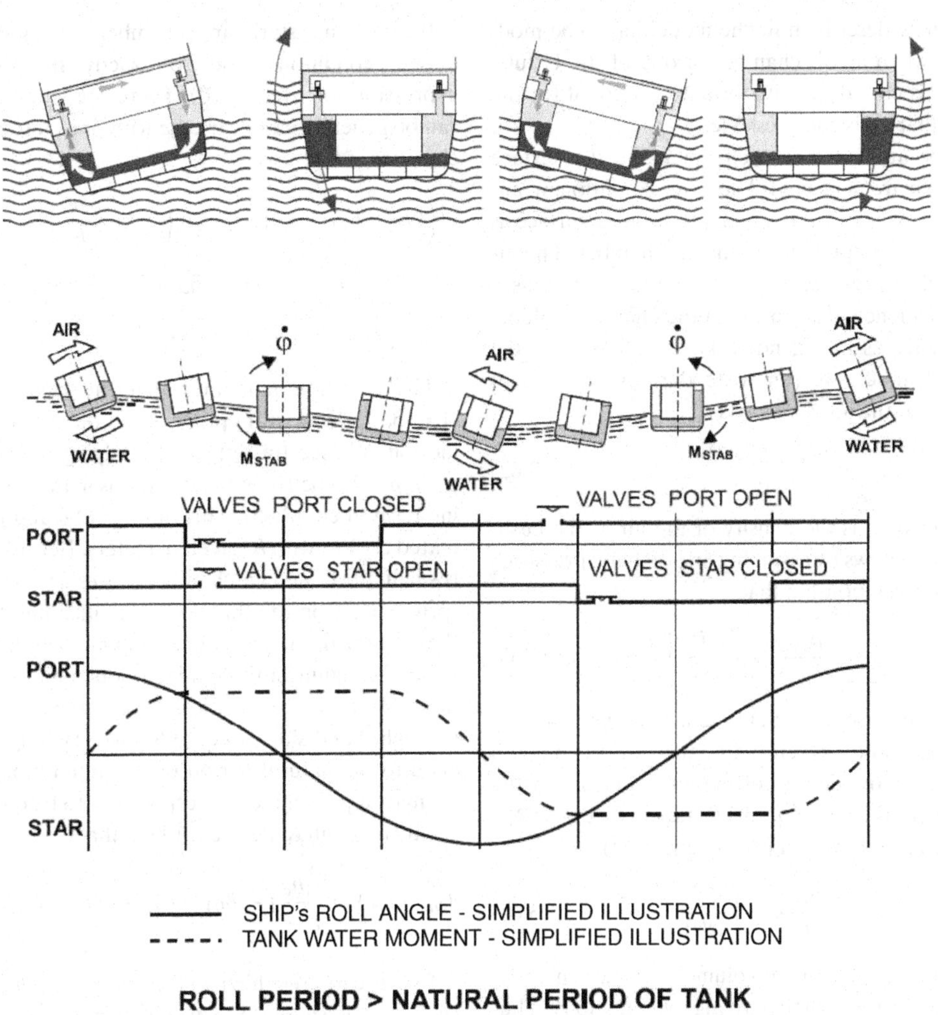

Figure 3.22. Time behavior of roll and the tank moment due to liquid motions in a U-tube when the roll oscillation period is larger than the uncoupled tank natural period and automatic control is used (Rolls-Royce/Intering). The notation $\dot{\varphi}$ has been used for the roll velocity (STAR=starboard).

Rolls-Royce/Intering's drawing. The figure shows that M_{STAB} is 90° out of phase with the roll motion (i.e., it is proportional to minus the roll velocity). When M_{STAB} is moved to the left-hand side of the equation of motion as in the roll equation in eq. (3.58), we see that M_{STAB} causes a damping effect on the roll motion.

When the ship rolls with a period larger than the uncoupled natural period for liquid motion in the tank, valves at the two reservoirs are alternatively opened and closed (see Figure 3.22). When the valves are closed on either the port or the starboard side, it nearly latches the water in the U-tube. Rolls-Royce/Intering denote their system as a passive "controlled U-tank stabilizer." Figure 3.22 illustrates that the control causes a nonsinusoidal time dependence of M_{STAB} even though the roll has sinusoidal time dependence. It follows via Fourier analysis that we can express M_{STAB} as $\sum_{n=1}^{\infty}(a_n \sin n\sigma t + b_n \cos n\sigma t)$. If the time dependence of roll is expressed as $\cos \sigma t$, it is evident due to the phasing of M_{STAB} relative to roll shown in Figure 3.22 that the Fourier term $a_1 \sin \sigma t$ is particularly large.This large term causes important roll damping by the same reason as that for Figure 3.21.

We will try to explain theoretically the hydrodynamic behavior illustrated in Figure 3.22. When both valves are open, the previously described hydrodynamic model for the U-tube can be used.

We now describe how the model has to be modified when an air chamber in one of the U-tube arms is closed. It is important to account for the fact that air is compressible.

The air flow in the chamber and the surface tension are neglected. The latter condition means the pressure in the water at the free surface is equal to the pressure in the air chamber. The air chamber pressure $p_c(t)$ is time-dependent as a consequence of variation in the chamber volume $\Omega(t)$. By assuming no leakage or flow into the air chamber, we can write the conservation of mass for the air chamber as

$$\frac{d\left[\rho_c(t)\Omega(t)\right]}{dt} = 0, \tag{3.62}$$

where $\rho_c(t)$ is the density of the air in the cushion. It follows by assuming an adiabatic pressure–density relationship that

$$\frac{p_c(t)}{p_0} = \left[\frac{\rho_c(t)}{\rho_0}\right]^{\kappa}, \tag{3.63}$$

where the use of $\kappa = 1.4$ for diatomic gases such as nitrogen and oxygen (and hence air) is common and ρ_0 is the density of the air at closure of the cushion. Because eq. (3.62) means that $\rho_c(t)\Omega(t)$ is a constant, we can rewrite eq. (3.63) as

$$\frac{p_c(t)}{p_0} = \left[\frac{\Omega_0}{\Omega(t)}\right]^{\kappa}, \tag{3.64}$$

where Ω_0 is the initial volume of the air chamber at the closure. When deriving the equations of liquid motion for a passive U-tube tank, we assumed that the pressure was the same at the free surfaces of the two U-tube arms so that we could set $p(l) - p(0) = 0$ (see eq. (3.43)). We start by analyzing the case where the port chamber is closed, then $p(l)$ for the starboard chamber is equal to atmospheric pressure p_a, and $p(0)$ changes with time according to the pressure in the closed air chamber. The initial value of $p(0)$ at the closure is assumed atmospheric (i.e., $p_0 = p_a$). The chamber is approximated as rectangular with initial volume $\Omega_0 = h_c L_t b_r$. The instantaneous chamber volume $\Omega(t)$ is $(h_c + \eta_u - \eta_{u0})L_t b_r$, where η_{u0} is the value of η_u at the closure of the air chamber. It follows by using a Taylor expansion that

$$p(l) - p(0) = p_a - p_a\left(\frac{h_c}{h_c + \eta_u - \eta_{u0}}\right)^{\kappa} \approx \frac{p_a\kappa}{h_c}(\eta_u - \eta_{u0}).$$

If only the starboard chamber is assumed closed, conditions lead to exactly the same expression for $p(l) - p(0)$. Therefore, eq. (3.55) can be generalized to the case with one closed air chamber to be

$$\left(2h_r + b\frac{b_r}{h_d}\right)\ddot{\eta}_u + B_d\dot{\eta}_u + \left(2g + \frac{p_a\kappa}{\rho_l h_c}\right)\eta_u = \frac{p_a\kappa}{\rho_l h_c}\eta_{u0} - gb\eta_4 - b\ddot{\eta}_{2t} - b(h_r + h_O)\ddot{\eta}_4 \tag{3.65}$$

The equation illustrates that the closed air chamber adds a stiff spring effect on the liquid motion. We see this effect by studying the ratio $p_a\kappa/(\rho_l h_c 2g)$ between the two terms in the restoring term in eq. (3.65). This ratio can be approximated as $7/h_c$ with h_c given in meters. Because h_c is small, the closed air chamber causes a very stiff spring effect on the liquid motion, meaning that the natural frequency for the liquid motion has increased significantly relative to open-air conditions. Because the duration of the excitation on the right-hand side of eq. (3.65) is very long relative to the natural period of the liquid motion, we may approximate the response η_u to be quasi-steady (Clough & Penzien, 1993); that is,

$$(\eta_u - \eta_{u0}) \approx \frac{\rho_l h_c}{p_a\kappa}\left[-gb\eta_4 - b\ddot{\eta}_{2t} - b(h_r + h_O)\ddot{\eta}_4\right]. \tag{3.66}$$

Because of the high values of $p_a\kappa/(\rho_l h_c 2g)$, $(\eta_u - \eta_{u0})$ is small and the liquid moves with a very small amplitude in the tank (i.e., the liquid is nearly latched) as we also see from the schematic in Figure 3.22.

The hydrodynamic forces and moments due to the liquid motion must also account for the fact that an air chamber is closed. We concentrate on the roll moment and return to the derivation from eq. (3.50). Here we used the fact that the pressure is atmospheric on the free surface so that we could include a free-surface integration in eq. (3.50). What we simply have to do is to subtract the free-surface integration; that is, there is a correction term:

$$F_{4ucorr} = -\int_{S_F}(\bar{y}n_3 - \bar{z}n_2)p\,dS,$$

where $n_3 = 1$ and $n_2 = 0$. We start with the port air chamber being closed and use eq. (3.64) to express the pressure in the air chamber. By using

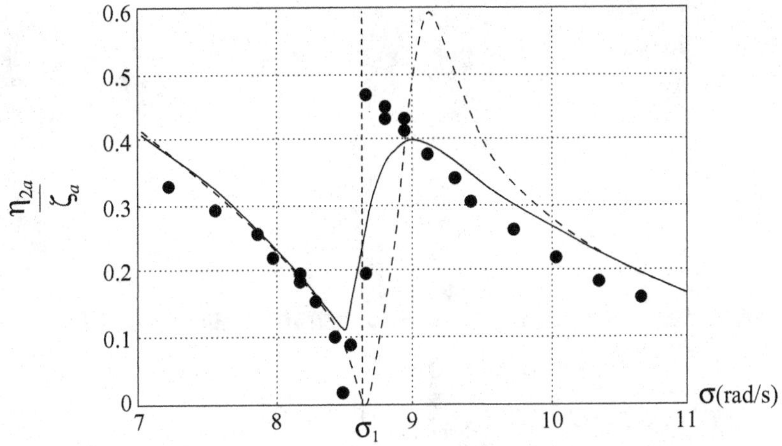

Figure 3.23. Experimental and theoretical results of the effect of sloshing on the steady-state sway amplitude η_{2a} of a two-dimensional model in regular incident waves with frequency σ and amplitude ζ_a. The lowest natural frequency for sloshing is denoted σ_1. One tank with $h = 0.184$ m; • represents experiments, solid line represents nonlinear modal theory, and dashed line represents linear theory.

the approximations that lead to eq. (3.65) we find that

$$F_{4ucorr} = -\tfrac{1}{2} b b_r L_t p_a \kappa \left(\eta_u - \eta_{u0}\right) / h_c. \quad (3.67)$$

The same expression follows if we assume only the starboard chamber is closed. Combining eq. (3.67) with eq. (3.66) gives

$$F_{4ucorr} = \tfrac{1}{2} \rho_l b b_r L_t [g b \eta_4 + b \ddot{\eta}_{2t} + b(h_r + h_O)\ddot{\eta}_4].$$

Because F_{4ucorr} does not depend explicitly on the liquid motion η_u, the expression for M_{STAB} is the same as when the air chambers are open (i.e., $-\rho_l L_t b_r b(h_r + h_O)\ddot{\eta}_u - \rho_l L_t b_r b g \eta_u$). Furthermore, since η_u remains nearly constant when an air chamber is closed, we can understand the nearly constant values of M_{STAB} in Figure 3.22 when the valves on either the port or the starboard side is closed.

3.6.5 Coupled sway motions and sloshing

The mutual interaction between wave-induced ship motions and sloshing in a partly filled tank is further illustrated by the two-dimensional experimental and theoretical studies of Rognebakke (2002) (see also Rognebakke & Faltinsen, 2003). A rectangular cross-section with draft and beam equal to 0.2 and 0.4 m, respectively, was used. Either one or two tanks were applied. The tank breadth ($B_t = l$) and length (L_t) were 0.376 and 0.150 m, respectively. The vessel length was 0.596 m. The model was restrained to move only laterally in sway. Experimental and theoretical results for the sway amplitude η_{2a} divided by the incident wave amplitude ζ_a are presented as a function of frequency σ in Figure 3.23. Results are presented following from using both a linear sloshing theory and the nonlinear multimodal theory of Faltinsen *et al.* (2000). Sloshing has a significant effect on the sway response. To explain the results we introduce the equation of sway motions.

The equations of motion describing the steady-state response in sway can be approximated as

$$[M + A_{22}(\sigma)]\ddot{\eta}_2 + B_{22}(\sigma)\dot{\eta}_2 = F_2^{exc} + F_2^{\text{tank}}, \quad (3.68)$$

where M is the body mass, which does not include the mass of the liquid inside the tank(s); $A_{22}(\sigma)$ and $B_{22}(\sigma)$ are, respectively, the linear sway added mass and damping associated with the external flow; η_2 is the sway motion and a dot expresses the time derivative; σ is the frequency of oscillation; F_2^{exc} is the external linear sway wave excitation force; $A_{22}(\sigma)$, $B_{22}(\sigma)$, and F_2^{exc} have been computed with a state-of-the-art BEM (see Section 10.2); and finally, F_2^{tank} is the sway force caused by the sloshing in the tank and is a function of σ and η_2. We can formally express the sloshing force as

$$F_2^{\text{tank}} = -C_{22}^{\text{tank}}(\sigma, \eta_2)\eta_2. \quad (3.69)$$

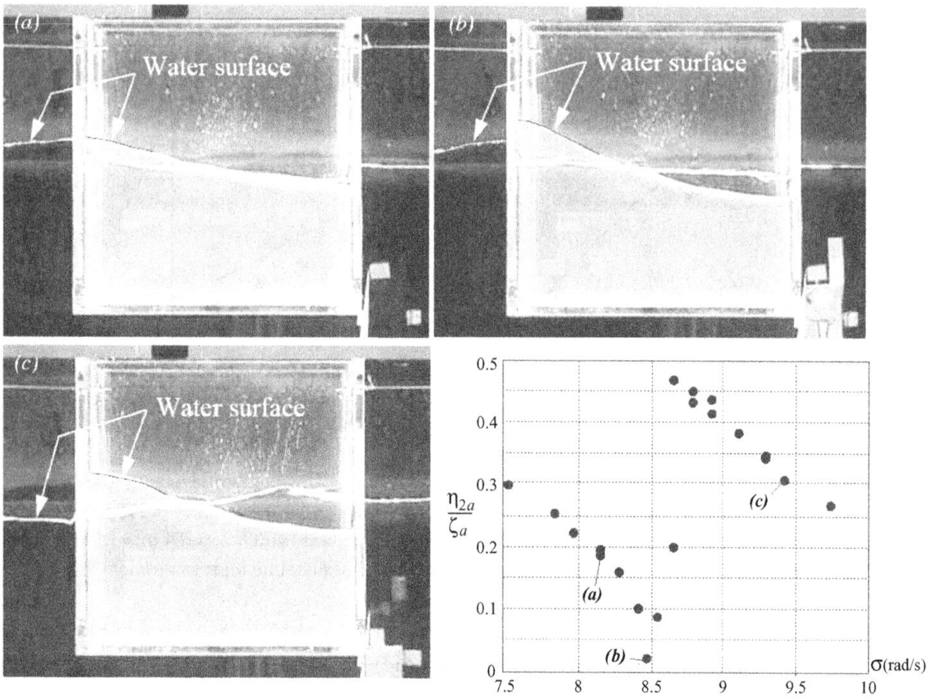

Figure 3.24. Motion of water inside and outside the tank: $h = 0.184$ m, $h/l = 0.489$. One tank is filled.

If a linear theory is used to describe the sloshing forces, then F_2^{tank} is linearly dependent on η_2. It follows from using expressions (5.133) and (5.134) and accounting for two-dimensional flow that

$$F_2^{\text{tank}} = m_l\sigma^2\left[1+\sum_{n=0}^{\infty}\frac{\sigma^2}{\sigma_n^2-\sigma^2}\frac{8l}{h}\frac{\tanh((2n+1)\pi h/l)}{(2n+1)^3\pi^3}\right]\eta_2, \tag{3.70}$$

where σ_n is the natural frequency of the antisymmetric tank modes and can be expressed as $\sigma_n^2 = g((2n+1)/l)\pi\tanh((2n+1)h\pi/l)$ (in notations of eq. (5.134), $\sigma_n = \sigma_{0,2n-1}$). Furthermore, m_l is the two-dimensional mass of the liquid in the tank. We note that eq. (3.70) includes the term $m_l\sigma^2\eta_2 = -m_l\ddot{\eta}_2$ (i.e., the inertia force of the frozen water). Equation (3.70) for the lateral force becomes infinite when $\sigma = \sigma_n$. Furthermore, a term in the sum changes sign by going from a frequency σ below σ_n to above σ_n.

Equations (3.68) and (3.69) show that sloshing acts as a frequency-dependent spring on the sway behavior. Therefore, we can introduce a natural frequency for the sway oscillations. The undamped natural frequency corresponds to the frequency satisfying

$$-\sigma^2[M + A_{22}(\sigma)] + C_{22}^{\text{tank}}(\sigma,\eta_2) = 0. \tag{3.71}$$

We can recognize the presence of a peak (resonance) in the response amplitude for sway in Figure 3.23. This peak is largest for the linear theoretical results when it occurs close to a frequency satisfying eq. (3.71). The linear sloshing model gives zero sway response at the natural frequency for sloshing. We can understand this by combining eqs. (3.68) and (3.69) and noting that C_{22}^{tank} is infinite at a natural tank frequency. We note this fact in Figure 3.23, where the predicted sway response is zero according to the linear sloshing theory at the lowest natural frequency σ_1 for the considered tank.

The change with wave frequency of the phasing between the forces acting on the model is apparent from Figure 3.24. The plot in this figure gives the experimental values for sway motion when one of the tanks is filled with $h = 0.184$ m corresponding to $h/l = 0.489$. Snapshots show the instantaneous position of the free surface, both inside the tank and outside the ship section, for three different wave frequencies. The phasing between the internal and external liquid motion

Figure 3.25. Perspective view of the hemispheroidal hull. The length is 12 m and the midship section is a semicircle of radius $a = 1$ m. Each tank is 2 m long and 62.5 cm deep. The tank breadths are 120, 160, and 120 cm. The free surfaces are at $z = 25$, 12.5, and 25 cm above the exterior waterplane (Newman, 2005).

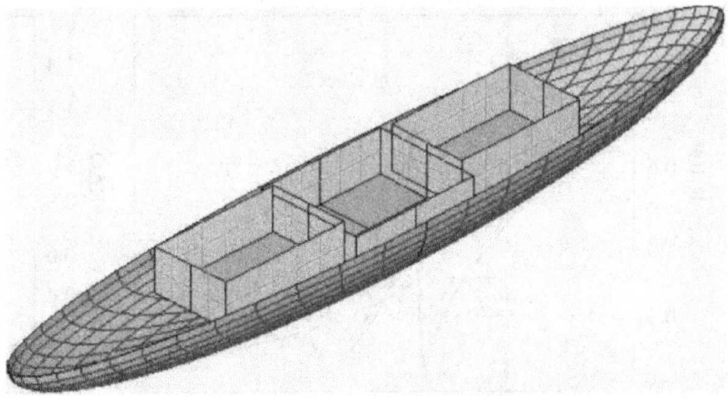

permits us to understand qualitatively why the internal fluid motion can either amplify or reduce the ship motion. Phasing is evident from the relative vertical motion of the free surfaces inside and outside the model.

An interesting phenomenon is observed for wave frequencies close to the resonance for liquid motion in the tanks. When the wave front hits the model, a significant sway motion is initiated, which in turn excites sloshing in the tanks and, thus, a sloshing force starts to counteract the excitation force from the waves. The sway motion decreases until equilibrium is reached. At this stage the sway-induced sloshing force almost balances the excitation force from the waves. However, since $\sigma \approx \sigma_1$, a very small sway motion causes a violent sloshing response.

Even though the results obtained by nonlinear theory are in fair agreement with the experimental results, Rognebakke (2002) reported a sensitivity of the response of a coupled ship motion and sloshing system to the damping of the internal water flow in a frequency range in the vicinity of the natural period for the sway motion. This damping is probably due to viscous effects caused by higher modes and cannot be accurately handled by the nonlinear multimodal method used by Rognebakke (2002).

Our representation of external hydrodynamic loads has been in terms of a frequency-domain solution by assuming that the ship section oscillates in steady-state condition with a single frequency σ. Because the sloshing force is nonlinear with, for instance, force terms oscillating with 2σ, 3σ, and so forth under steady-state conditions, this representation of external hydrodynamic force should instead have been in terms of a convolution integral expressing the time history. This concept is further discussed in Section 3.8.

3.6.6 Coupled three-dimensional ship motions and sloshing in beam waves

Newman (2005) analyzed coupled vessel motion and sloshing of the hemispheroid shown in Figure 3.25. The vessel speed was zero. Beam-sea regular waves in deep water were considered. Both external flow and sloshing were based on linear potential flow theory and steady-state solutions were considered. The hydrodynamic flow was solved numerically using a three-dimensional BEM. The vessel has three internal tanks with the same liquid depths. The tank lengths are the same, but the tank breadths differ. Other details about the tanks are given in the caption of Figure 3.25.

Figure 3.26 shows the sway, heave, and roll amplitudes together with the transverse drift force. Three densities of the tank liquid relative to sea water are considered (i.e., $\rho_{rel} = \rho_l/\rho = 0$, 0.5, 1.0). Here $\rho_l = 0$ corresponds to empty tanks. The total weight of the vessel including the tank liquid is the same for all conditions. When the vessel does not move, the vertical coordinate of the COG is midships in the water plane. The results are presented as functions of the nondimensional wave number $Ka = \sigma^2 a/g$, where a is the maximum radius of the spheroid. The radii of gyration with respect to the Cartesian coordinate system $Oxyz$ with origin in the COG and axes as defined in Figure 3.8 are $r_{44} = 50$ cm and $r_{55} = r_{66} = 3$ m for roll, pitch, and yaw, respectively. The radii of gyration are only associated with the rigid mass distribution of the vessel. The results for sway and heave are presented as RAOs. The RAOs of sway and heave are defined as η_{2a}/ζ_a and η_{3a}/ζ_a, respectively, where ζ_a is the incident wave amplitude and η_{2a} and η_{3a} are the sway and heave amplitudes at the COG. The roll is presented as $\eta_{4a}^{(deg)}/\zeta_a$,

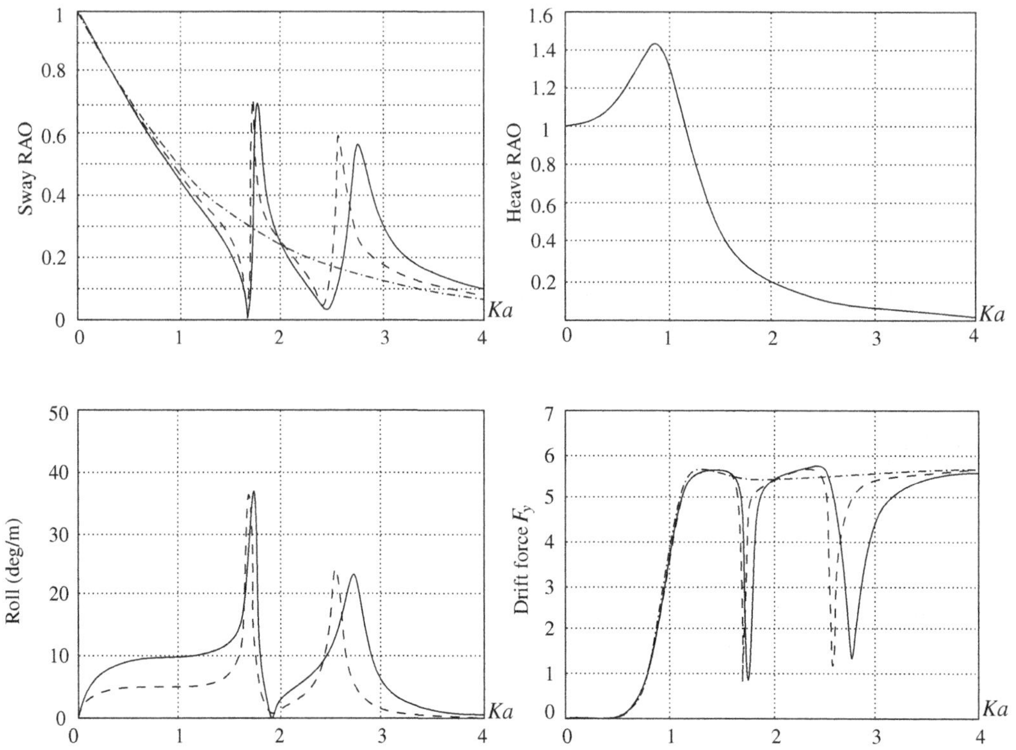

Figure 3.26. Sway, heave, and roll RAOs and nondimensional drift force for the hemispheroidal hull in beam waves (Newman, 2005). Solid line, $\rho_{rel} = 1$; dashed line, $\rho_{rel} = 0.5$; and dashed-dotted line, $\rho_{rel} = 0$.

where $\eta_{4a}^{(deg)}$ is the roll amplitude in degrees. If the roll results are scaled to other similarly sized vessels, we should note that the roll results in Figure 3.26 are the same as $\eta_{4a}^{(deg)} a/\zeta_a$. As an example we consider $Ka = 2.8$ with $\eta_{4a}^{(deg)} a/\zeta_a = 24$, which is close to one of the two maxima of nondimensional roll presented in Figure 3.26. For instance, using $a = 15$ m, wave amplitude-to-wavelength ratio $\zeta_a/\lambda = 0.05$, and the fact that $K = 2\pi/\lambda$ gives $\lambda = 33.7$ m, $\zeta_a = 1.7$ m, and $\eta_{4a}^{(deg)} = 2.7°$. It is common in numerical roll predictions from potential flow models to add empirical viscous roll damping associated with the external flow. This damping was not added in the calculations presented in Figure 3.26. However, because the external flow is not likely to separate for a hemispheroid for the given conditions, the effect of viscous roll damping is small. A more important effect is to account for nonlinearities due to sloshing, as we demonstrated for sway in Section 3.6.5.

The drift force F_y in the figure is nondimensional and defined as $F_y = \overline{F}_2/(\rho g \zeta_a^2 a)$, where $\overline{F}_2$ is the transverse second-order drift force. Second order means that the problem is solved by including hydrodynamic wave–body interaction terms consistent to second order in the incident wave amplitude. However, the second-order velocity potential does not contribute to the drift force. The positive direction of the drift force is in the propagation direction of the incident waves. We note that the predicted drift force is always positive, which agrees with Maruo's (1960) potential flow theory.

Because beam waves are considered and the vessel has fore and aft symmetry, surge, pitch, and yaw are not excited. Furthermore, the hull symmetry about the centerplane and linear theory imply that heave is uncoupled from sway and roll. We note from Figure 3.26 that the heave motion is not affected by the tank liquid. We can understand this by analyzing the vertical force due to vertical tank motion. We then define z_t as the local vertical Earth-fixed coordinate above the mean tank free surface. The velocity potential satisfying the body boundary condition on the mean wetted tank surface and the linear free-surface condition can be expressed as $\varphi = \phi\dot{\eta}_3 = (z_t + 1/K)\dot{\eta}_3$. The resulting linear dynamic pressure follows from the linearized Bernoulli equation, $p = p_0 - \rho_l \partial\varphi/\partial t - \rho_l g\, z_t$, where p_0 is the air pressure in the tank. The pressure term $-\rho_l g z_t$ causes a linear dynamic pressure, $-\rho_l g \eta_3$, on the tank bottom, meaning that the dynamic pressure terms $-\rho_l \partial\varphi/\partial t$ and $-\rho_l g \eta_3$ result in the vertical dynamic force $-\rho_l V_t \ddot{\eta}_3$, where V_t is the liquid volume in the tank. The vertical force is the same as the inertial force resulting from considering a frozen liquid. In Section 5.4.3.2 we present this result as part of a more general analysis for rectangular tanks and show the effect of all modes of rigid-body motions on linear hydrodynamic forces and moments due to the liquid in the tank. Heave cannot excite sloshing according to linear theory so there exists only a "frozen" liquid effect. Because the sum of the weight of the rigid vessel mass and the tank's liquid is kept constant in Newman's calculation, there is no effect of the tank's liquid on the heave motion. In Section 3.5.2 we introduced the equations of motion in the frequency domain and considered the effects of the velocity potential and the hydrostatic pressure. The external hydrodynamic loads came from the normal procedure of associating the dynamic effects of the hydrostatic pressure with restoring coefficients and the velocity potential term in Bernoulli's equation with added mass and damping terms. We could have also done this task for internal tank loads, leading to added mass terms as part of the internal tank loads. This procedure was followed by Newman (2005). However, this procedure gives an added mass in heave due to the tank liquid that had a peculiar behavior that is difficult to understand. In contrast, when we combine the effect of the velocity potential and the hydrostatic pressure we have a plausible physical result (i.e., the tank liquid due to the linear heave excitation behaves as if it is frozen). However, nonlinear effects due to heave excitation exist that lead to sloshing and parametric resonance (Mathieu equation instability), resulting in so-called Faraday waves (see Section 8.10). Generally speaking, we can combine the added mass terms following from the velocity potential and the restoring terms following from the hydrostatic pressure either as combined added mass terms or as combined restoring terms. The latter approach was followed in Section 3.6.5 when coupled sway motions and sloshing were considered. The total dynamic transverse load due to the tank liquid was then represented in terms of a frequency-dependent restoring coefficient.

The hull geometry implies that no normal velocity exists at the submerged hull surface due to roll about the x-axis. Consequently roll motion does not cause any external flow in a potential flow formulation, so the added mass and damping coefficients A_{24}, A_{44}, B_{24}, and B_{44} associated with the external flow are zero. Furthermore, because the pressure loads act normally to the body surface and do not cause any roll moment about the x-axis for the axisymmetric body, the external hydrodynamic coupling terms due to sway, A_{42} and B_{42}, are zero. Another consequence of the fact that the pressure loads cannot cause a nonzero roll moment is that the wave excitation moment in roll is zero. Let us set up the equations of motion in roll when there is no liquid in the tanks. We use the equations of motion in the frequency domain as formulated in Section 3.5.2. Because the vertical z-coordinate, z_G, is zero, we see from eq. (3.27) that the only nonzero component of the mass matrix M_{jk} involved in the roll equation is $I_{44} = Mr_{44}^2$, where M is the mass of the vessel. We can show that the restoring coefficient C_{44} given by eq. (3.31) is zero either by direct calculations or by using the fact that pressure loads (i.e., the hydrostatic pressure component in this case) do not cause a roll moment. Therefore, the roll equation becomes $I_{44}\ddot{\eta}_4 = 0$ (i.e., the roll acceleration is zero when the tanks contain no liquid). Because steady-state conditions are assumed, this equation results in zero roll. This fact is consistent with the results in Figure 3.26. When the tanks contain liquid, the roll motions are caused by sloshing in the tanks and are nonzero.

From Figure 3.26 we note a significant effect of sloshing on sway and roll. For forced sway motions we can express the resulting horizontal force due to the liquid in the tank as in eq. (3.70). Similar results for forced roll and resulting forces and moments are presented in Section 5.3.2. These results can be introduced as either restoring or added mass coefficients on the left-hand sides of the coupled sway and roll equations. The coefficients are infinite at the sloshing frequencies. We showed in the case of a single-degree-of-freedom rigid-body system in sway in Section 3.6.5 that the infinite added mass (restoring coefficient) at a natural sloshing frequency led to zero sway. The lowest natural frequencies in sway and roll of the two tank types of the hemispheroid correspond to $Ka = 1.653$ and 2.427. Figure 3.26 shows that the roll amplitude is not zero at these frequencies. However, it looks from the figure that the sway amplitude is zero at the natural sloshing frequencies. A detailed investigation shows that the sway amplitudes are not zero at the sloshing frequencies, but at a frequency very close to the sloshing frequencies. Newman (personal communication, 2008) has investigated the problem in detail near the sloshing frequencies. We report on some details of his analysis. It follows by starting with eq. (3.26) that the coupled equations of sway and roll can generally be expressed as

$$Z_{22}\eta_2 + Z_{24}\eta_4 = F_2, \; Z_{42}\eta_2 + Z_{44}\eta_4 = F_4, \qquad (3.72)$$

where η_2 and η_4 are the complex amplitudes of sway and roll, and F_2 and F_4 are the complex amplitudes of the excitation force in sway and excitation moment in roll. Furthermore,

$$Z_{jk} = -\sigma^2\left(M_{jk} + A_{jk} + A_{jk}^{\text{tank}}\right) + i\sigma B_{jk} + C_{jk}, \qquad (3.73)$$

where we have used the superscript "tank" to identify the contribution from the liquid in the tank. Here we have included all the dynamic effects of the liquid in the tanks in terms of the added mass coefficients A_{jk}^{tank}. The coefficients in eq. (3.73) are symmetric so that $Z_{24} = Z_{42}$. The solutions of eq. (3.72) are

$$\begin{aligned} \eta_2 &= \left(F_2 Z_{44} - F_4 Z_{24}\right)/D, \\ \eta_4 &= \left(F_4 Z_{22} - F_2 Z_{24}\right)/D, \end{aligned} \qquad (3.74)$$

where $D = Z_{22}Z_{44} - Z_{24}^2$.

The singular part of A_{jk}^{tank} is proportional to $(\sigma - \sigma_n)^{-1}$ at a natural sloshing frequency σ_n ; that is, it follows from eq. (3.74) that sway and roll at $\sigma \approx \sigma_n$ have a solution of the form

$$\eta_k = \frac{a_k + b_k(\sigma - \sigma_n)^{-1}}{c_0 + c_1(\sigma - \sigma_n)^{-1} + c_2(\sigma - \sigma_n)^{-2}},$$

where $a_k(k = 1, 2)$ and $c_i(i = 1, 2, 3)$ are nonsingular. Newman (personal communication, 2008) shows that c_2 is zero for a rectangular tank by using analytical added mass results. We show in Section 5.4.1.3 that this finding is true in the general case. The consequence is finite values of η_k for $\sigma = \sigma_n$.

It follows, as previously discussed for Newman's hemispheroid with three tanks, that

$$\begin{aligned} Z_{24} &= -\sigma^2 A_{24}^{\text{tank}}, \\ Z_{44} &= -\sigma^2\left(M_{44} + A_{44}^{\text{tank}}\right), \quad F_4 = 0, \end{aligned}$$

which implies that

$$\begin{aligned} \eta_2 &= -\sigma^2 F_2\left(M_{44} + A_{44}^{\text{tank}}\right)/D; \\ \eta_4 &= -\sigma^2 F_2 A_{24}^{\text{tank}}/D. \end{aligned}$$

The reason for the zero of η_{2a} in the vicinity of the sloshing frequencies corresponding to $Ka = 1.653$ and 2.427 in Figure 3.26 is that $M_{44} + A_{44}^{\text{tank}} = 0$. Furthermore, we note from Figure 3.26 that $\eta_{4a} = 0$ at $Ka \approx 1.9$. This fact is a consequence of $A_{24}^{\text{tank}} = 0$. The results in Figure 3.26 illustrate that sloshing has caused two resonance frequencies of coupled roll and sway. The explanation is similar to that given for sway in Section 3.6.5.

The transverse drift force is also significantly affected by sloshing. The second-order transverse drift force in potential flow can be expressed in terms of the linear far-field waves generated by the ship (Maruo, 1960). In our case, this value is the sum of the waves caused by diffraction from a fixed ship and radiation due to sway and heave. Because rolling does not cause any external flow in our case, roll contributes only by its influence on sway. The sharp reductions in the sway drift force coincide with the peak sway response. If strip theory is used, the drift force is $\frac{1}{2}\rho g \zeta_a^2 L$ when $Ka \to \infty$, according to Maruo's theory. This result follows from the fact that the ship does not move and the incident waves are reflected by the ship with reflected wave amplitude equal to the incident wave amplitude. This two-dimensional asymptotic formula corresponds to nondimensional drift force $F_y = \overline{F}_2/(\rho g \zeta_a^2 a) = 6$, which is in reasonable agreement with the three-dimensional

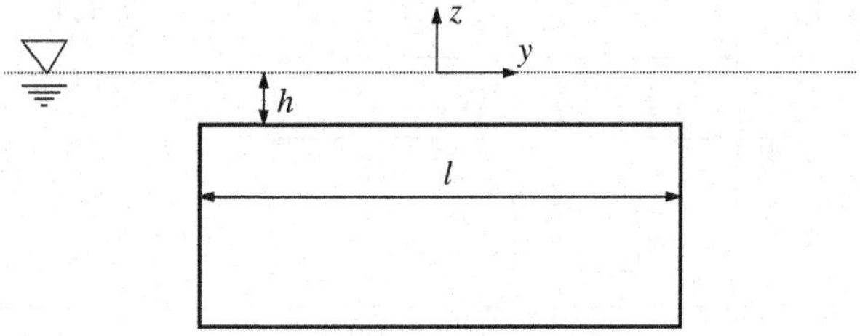

Figure 3.27. Submerged rectangular section with shallow-water region with depth h on the top face.

results in Figure 3.26 for large Ka. The fact that the drift force becomes small due to resonant sloshing in the tanks implies that the wave generation by the ship becomes small.

3.7 Sloshing in external flow

Sloshing can also occur in confined liquid spaces with a free surface in external flow of ships. Examples are resonance oscillations between the hulls of a catamaran or in a moonpool (see Figure 1.34). The fact that the confined space with resonance oscillations is part of the external water domain of the ship causes far-field waves to be generated. The consequence is wave radiation damping of the resonance oscillations. A very special problem involving resonant oscillations was discussed in connection with Figure 1.36. The bow part of a floating production storage and offloading vessel was equipped with a submerged "lip." The upper part of the lip became dry in a seaway during transit as a consequence of relative vertical motions and local resonant water motion on top of the lip.

Another external flow problem with local sloshing is illustrated in Figure 3.27. It involves a submerged rectangular section with a shallow-water region on the top of the rectangle. This problem was analytically investigated by Newman *et al.* (1984). When resonant oscillations occurred in the shallow region, the forces on the rectangle due to forced heave motions were dominated by the pressures on the top face of the rectangle. Negative added mass and sharp peaks in added mass and damping coefficients were found. We will explain why resonance oscillations occur by referring to the coordinate system defined in Figure 3.27. We use shallow-water eq. (4.45) and express the solution of the velocity potential as $\varphi = \overline{\varphi}\exp(i\sigma t)$. If energy is trapped on the top face of the rectangle, no waves occur beyond the top face, so the wave elevation should be zero at $y = \pm\frac{1}{2}l$. We now use the linear dynamic free-surface condition $g\zeta + \partial\varphi/\partial t|_{z=0} = 0$, which relates the wave elevation with the velocity potential at the mean free surface, together with a shallow-water approximation at the top face of the rectangle (i.e., $\overline{\varphi}$ does not vary with the vertical coordinate z). A similar argument is made for the boundary conditions at the harbor entrance in Section 4.4.4. It follows that the boundary conditions at the edges $y = \pm\frac{1}{2}l$ should be nodal points, which means mathematically that $\overline{\varphi} = 0$ at $y = \pm\frac{1}{2}l$. This condition is satisfied by $\overline{\varphi} = C\sin[n\pi(y + \frac{1}{2}l)/l]$, $n \geq 1$. The corresponding natural frequencies σ_n are expressed by $\sigma_n^2 = gh\,(n\pi/l)^2$, where h is the water depth of the shallow region. When we consider the forced heave problem, the velocity potential of the natural mode has to be symmetric in y. The lowest symmetric mode corresponding to $n = 1$ is of particular concern for wave-induced motions. Newman *et al.* (1984) solved the problem by using matched asymptotic expansions. What we discussed earlier was called the internal "outer" region. Additional inner regions exist at the edges $y = \pm\frac{1}{2}l$ as well as an external "outer region" with far-field waves influenced by the flow in the shallow region. We can understand why there is an influence on the outer flow from the flow on the top face of the rectangle by considering the previous eigenvalue problem. The consequence of the fact that the edges $y = \pm\frac{1}{2}l$ are nodal points is a mass flux at the edges. What we are implicitly saying is that the energy cannot be completely trapped on the top face of the rectangle. A coupling exists between the top face flow and the exterior flow.

3.7.1 Piston-mode resonance in a two-dimensional moonpool

Figure 1.34 illustrates a moonpool, which is a hole in a ship's hull that allows direct passing from the working deck to the sea. It is used,

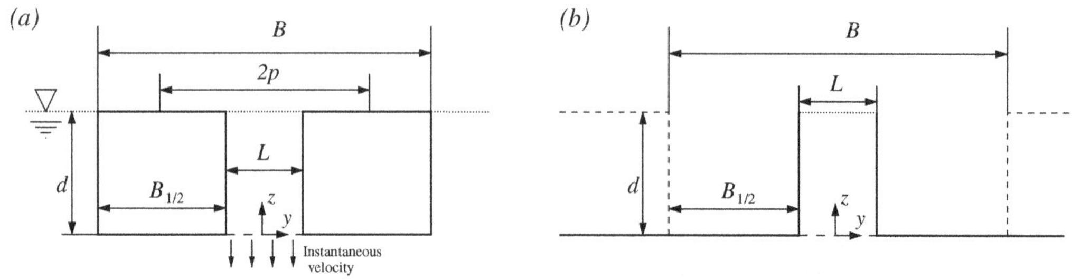

Figure 3.28. Piston-mode resonance between the two hulls, illustrated by instantaneous water velocity vectors indicating large water mass flux in part (a). The velocity is harmonically oscillating with the natural frequency of the resonance. As a first approximation the velocity is constant across the gap between the two hulls. Part (b) introduces definitions of Molin's (2001) model for piston-mode resonance.

for instance, for drilling and marine riser operations. In the case of Figure 1.34, perforated walls are used to damp resonant oscillations in a frequency range of importance for wave-induced vertical ship motions. Of particular concern for resonance oscillations in moonpools and between the hulls of multihull vessels is what Molin (2001) calls piston-mode resonance. His concept is illustrated in Figure 3.28(a) for two-dimensional flow. There is a one-dimensional resonant water motion between the two hulls, which causes an oscillating mass flux with large amplitude at the lower end of the gap between the two hulls. We present Molin's simplified analysis, which does not account for far-field wave generation. A two-dimensional analysis is performed first.

In the analysis we use a coordinate system Oyz, where the y-axis is the plane of the lower horizontal parts of the two rectangular hulls (see Figure 3.28(b)). The origin is in the centerplane of the two hulls and the mean free surface is at $z = d$, where d is the draught. In an eigenvalue analysis we are looking for the nontrivial solutions when there is no forcing, which means the body is restrained from moving.

We assume a one-dimensional water motion in the gap between the two hulls; that is, the velocity potential is approximated as

$$\varphi = A_0 + B_0 z d^{-1}, \tag{3.75}$$

where A_0 and B_0 are independent of y and z. A later analysis determines a relationship between A_0 and B_0. Because it is an eigenvalue problem, we cannot determine the values of both A_0 and B_0. Molin represents the flow for negative z-value by a source distribution along the y-axis. The source density is expressed by the vertical velocity along the y-axis (see, e.g., Faltinsen, 2005). A two-dimensional source in infinite fluid is $(Q/2\pi)\ln r$, where r is the radial distance between the source point and the field point. This representation of the flow implies a source density in the gap expressed by the vertical fluid velocity $\partial\varphi/\partial z$ at $z = 0$ (i.e., for y between $-\frac{1}{2}L$ and $\frac{1}{2}L$), where L is the breadth of the gap. Because $\partial\varphi/\partial z$ is zero on the bottom of each hull, the source density there is also zero.

Thus far no approximations have been made in solving the problem. The difficulty is in representing the vertical velocity at $z = 0$ outside the two hulls; we do not know this value before we solve the complete problem for fluid motion in the whole fluid. It is not sufficient to use only a source distribution for y between $-\frac{1}{2}L$ and $\frac{1}{2}L$, because this approach causes infinite pressure at infinity. It is at this stage that Molin makes a big simplification to ensure that the flow at infinity is not source-like – by placing two sinks at $y = \pm\frac{1}{2}B$, where B is the beam of the catamaran. When using the terms source and sink, we have not been completely precise. The reason is that $\partial\varphi/\partial z$ at the gap is harmonically oscillating and changing between causing a source or a sink effect. The flow seen from infinity must look like a flow with a nonzero source (sink) strength. Molin's procedure ensures that. We can then represent the velocity potential φ at $z = 0$ in the gap as

$$\varphi(y,0,t) = -\frac{1}{\pi}\int_{-\frac{1}{2}L}^{\frac{1}{2}L} \frac{\partial\varphi}{\partial z}(\eta,0,t)\Big[\ln|y-\eta| - \tfrac{1}{2}\ln\left(\tfrac{1}{2}B - y\right) - \tfrac{1}{2}\ln\left(\tfrac{1}{2}B + y\right)\Big]d\eta, \tag{3.76}$$

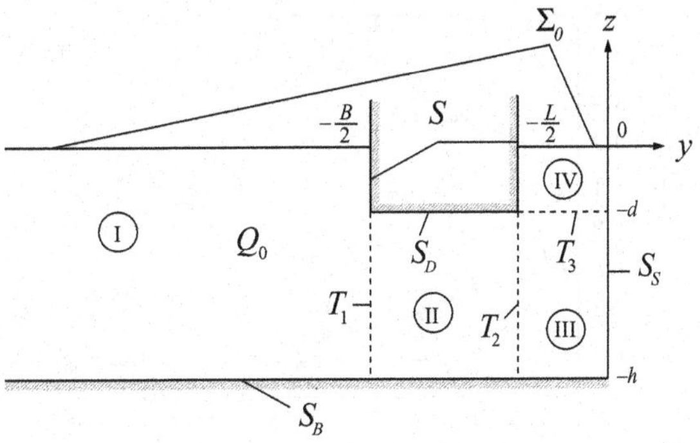

Figure 3.29. Domain decomposition. The water domain for negative y is divided into four domains. Domain I goes to $y = -\infty$.

where the field point and the source (sink) point coordinates are at $(y, 0)$ and $(\eta, 0)$, respectively. The integration is in the η-direction, which is the same as the y-direction in Figure 3.28(b). By using the fact that $|y| \leq 0.5L \ll 0.5B$, eq. (3.76) can be further approximated as

$$\varphi(y, 0, t) = -\frac{1}{\pi} \int_{-\frac{1}{2}L}^{\frac{1}{2}L} \frac{\partial \varphi}{\partial z}(\eta, 0, t) \ln(2|y - \eta|/B)\, d\eta. \tag{3.77}$$

Now we equate eqs. (3.75) and (3.77). We cannot satisfy this procedure exactly, but we can do it in an average way. We first integrate the right-hand side of eq. (3.77) by assuming $\partial\varphi/\partial z = B_0/d$. This expression depends on y. However, it is inconsistent with the left-hand side of eq. (3.76), which, by using eq. (3.75), says that $\varphi(y, 0, t) = A_0$. Therefore, we average by integrating both the left- and the right-hand side from $y = -\frac{1}{2}L$ to $\frac{1}{2}L$ and divide by L, giving

$$A_0 = \frac{1}{\pi}\frac{L}{d} B_0 \left[\tfrac{3}{2} + \ln\left(\tfrac{1}{2}B/L\right)\right]. \tag{3.78}$$

We have found the relationship between A_0 and B_0 but not the natural frequency. We assume A_0 and B_0 to be harmonically oscillating as $\exp(i\sigma_* t)$, where σ_* is the natural frequency, and we use the free-surface condition $-\sigma_*^2\varphi + g\partial\varphi/\partial z = 0$ for $z = d$ and y between $-\frac{1}{2}L$ and $\frac{1}{2}L$. Equation (3.75) then gives $-\sigma_*^2(A_0 + B_0) + g\,B_0/d = 0$ or $\sigma_*^2 = g/d(1 + A_0/B_0)$. Using eq. (3.78) gives Molin's formula for piston-mode resonance frequency:

$$\sigma_* \sqrt{\frac{d}{g}} = \sqrt{\frac{1}{1 + L(1.5 + \ln(B/2L))/\pi d}}. \tag{3.79}$$

The formula gives reasonable estimates for catamaran sections (Faltinsen, 2005). It should be noted that the right-hand side approaches the value 1 for small L and is then in accordance with the solution of Exercise 3.11.7.

Faltinsen *et al.* (2007) conducted experiments with a catamaran section that was forced to oscillate harmonically in heave. Each hull has a rectangular shape. The gap between the two hulls is referred to as a moonpool in the following text. A theoretical method based on domain decomposition was developed. The effects of finite water depth and far-field wave generation were accounted for. Because of the Oz-symmetry, only water motions to the left of the Oz-axis in Figure 3.29 are considered, and the liquid domain is divided into four subdomains (I, II, III, and IV) by auxiliary interfaces T_1, T_2, and T_3.

Figure 3.30 shows experimental and theoretical results that are both based on the domain decomposition method, Molin's formula, and a simplified formula obtained by assuming a small clearance between the hull bottom and the sea floor as well as a small gap between the two hulls. One-dimensional water motion was assumed. Molin's formula overpredicts the natural frequency, in particular for small d/L ratios. Because Molin's formula assumes infinite water depth, we should not expect good results when the draft d is close to the water depth h. The asymptotically small-gap approximation then gives reasonable estimates.

We show the derivation of the simplified formula for a small clearance between the hull bottom and the sea floor and use Figure 3.31 as a reference. The velocity in the vertical column is $\dot{\eta}$. The flow at the corners is two-dimensional. We assume one-dimensional flow in our derivation, which introduces an ambiguity in how to handle the effect of the corners. It follows by continuity of liquid mass that the velocity in the

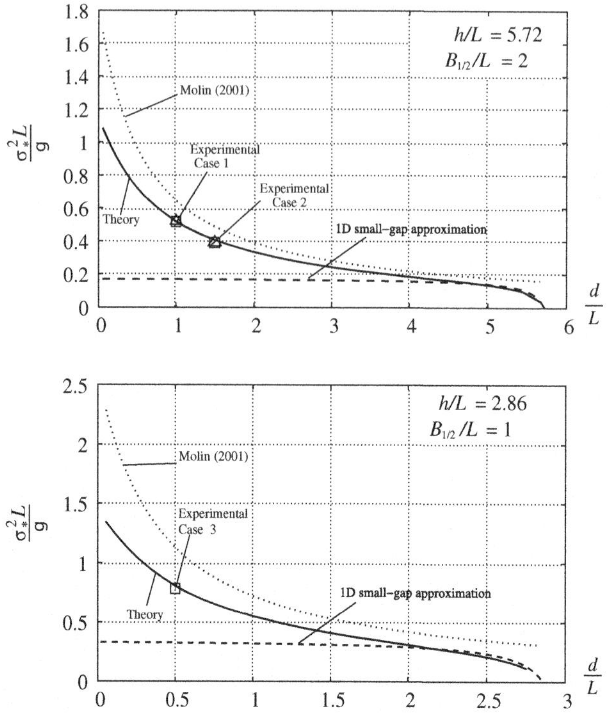

Figure 3.30. Theoretical and experimental predictions of the piston-mode resonance frequency σ_* in a moonpool as a function of the ratio between the draft d and the moonpool width L. The upper graph considers the case $B_{1/2}/L = 2$ and $h/L = 5.72$. The breadth $B_{1/2}$ is defined in Figure 3.28 and h is the water depth. The lower graph is for $B_{1/2}/L = 1$ and $h/L = 2.86$.

horizontal part can be expressed as $\dot{\eta}L/[2(h-d)]$. The kinetic energy of the flow is

$$E_k = \tfrac{1}{2}\rho B_{1/2}(h-d)\left[\tfrac{1}{2}L/(h-d)\right]^2 \dot{\eta}^2 + \tfrac{1}{2}\rho\tfrac{1}{2}Lh\dot{\eta}^2 + \tfrac{1}{2}\rho\tfrac{1}{2}L\eta\dot{\eta}^2.$$

The potential energy is $E_p = \frac{1}{4}\rho g L\eta^2$. We assume in our derivation that $d(E_k + E_p)/dt = 0$. Strictly speaking, this is not true because a communication occurs with the external flow. Our assumption gives

$$\left[\tfrac{1}{4}B_{1/2}L^2/(h-d) + \tfrac{1}{2}hL\right]\ddot{\eta} + \tfrac{1}{2}gL\eta + \tfrac{1}{4}L\dot{\eta}^2 + \tfrac{1}{2}L\eta\ddot{\eta} = 0. \quad (3.80)$$

It follows by linearizing eq. (3.80) and assuming time dependence $\exp(i\sigma_* t)$ that the natural frequency for the water motion is

$$\sigma_* = \sqrt{g(h-d)/\left(\tfrac{1}{2}B_{1/2}L + (h-d)h\right)}.$$

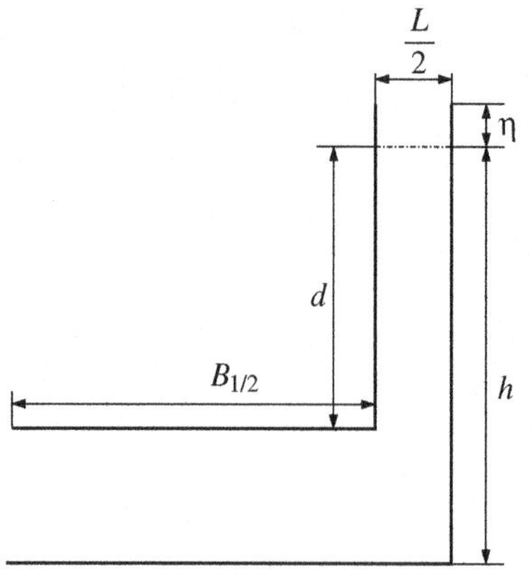

Figure 3.31. Illustration showing piston-mode flow in the moonpool when a narrow gap exists between the bottom of the ship and the sea floor.

It has been shown by an unpublished parametric study with the theory by Faltinsen *et al.* (2007) that the ratio between the maximum mass flux into the moonpool and the maximum mass flux caused by the heave motion during resonant condition was nearly constant. The consequence is that the resonant moonpool amplitude increases with either increasing side-hull beam or decreasing moonpool width. The draft and the water depth are less important parameters.

Faltinsen *et al.* (2007) were not able to get satisfactory agreement between experimental linear theoretical resonant free-surface amplitudes in the moonpool for forced heave motions. Kristiansen and Faltinsen (2008) improved the correlation between theory and experiments by

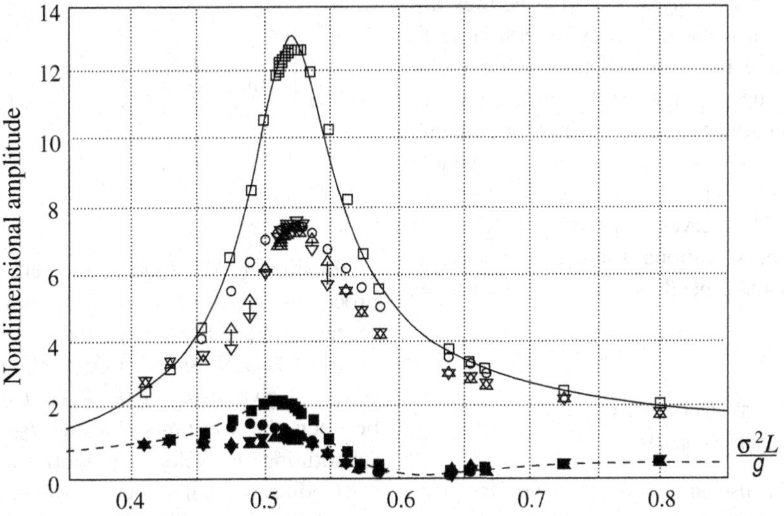

Figure 3.32. Steady-state nondimensional far-field wave amplitude (scaled by η_{3a}) and space-averaged moonpool (piston) amplitude (scaled by η_{3a}) as a function of the forcing frequency σ for case 1 in Figure 3.30 (i.e., $B_{1/2}/L = 2, h/L = 5.72, d/L = 1$). The nondimensional forced heave amplitude is $\eta_{3a}/d = 0.0278$. Piston amplitude: solid line, linear theory; □, nonlinear simulations without flow separation; ○, nonlinear simulations with flow separation, Δ and ∇, two different experiments. Far-field wave amplitude: dashed line, linear theory; ■, nonlinear theory without flow separation; •, nonlinear simulations with flow separation, ▲ and ▼, two different experiments.

accounting for vortex shedding at the sharp edges of the hull section, inclusion of free-surface nonlinearities, and estimates of experimental bias errors due to wave reflection from the wave beach and the wave generator. Free-surface nonlinearities were less important than vortex shedding. Details of the experimental and theoretical predictions of the maximum average amplitude of the free-surface elevation across the moonpool are shown in Figure 3.32 with the far-field wave amplitude. The results are for steady-state conditions and presented as a function of nondimensional squared frequency, $\sigma^2 L/g$. The conditions correspond to case 1 in Figure 3.31 (i.e., $B_{1/2}/L = 2$, $h/L = 5.72$, and $d/L = 1$). The ratio between the forced heave amplitude η_{3a} and the draft is $\eta_{3a}/d = 0.0278$. We note that wave radiation occurs at moonpool resonance. The linear predictions were based on that of Faltinsen *et al.* (2007). The nonlinear simulations were based on potential flow theory, where a BEM was used. The effect of flow separation was accounted for by using a thin-free-shear-layer model. An instantaneous picture of the thin free shear layer is shown in Figure 3.33 at a time instant where the total circulation of the vorticity in the thin free shear layer is close to zero. Flow separation is assumed to occur at the sharp body corners adjacent to the moonpool. The thin free shear layers shown in Figure 3.33 are approximated by two discrete vortices with nearly opposite circulations. Basically, in steady-state flow, four single vortices, or two pairs of vortices, are shed each period. The two single vortices shed every

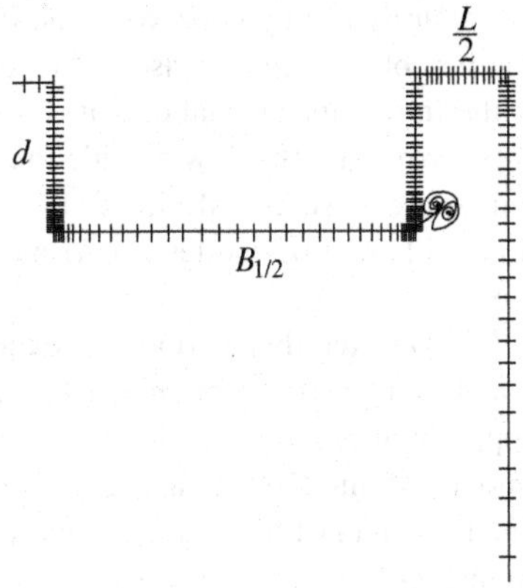

Figure 3.33. Steady-state instantaneous picture of a thin free shear layer obtained by BEM. End points of part of the panels on the free surface, body surface, and the vertical symmetry line of the moonpool are indicated.

half-period form a vortex pair; that is, they are formed in such a way that they remain close to each other and once created they move under mutual influence. A new single vortex starts to develop soon after the vortex pair is formed, and in the numerical model the strategy is to simply remove the pair once this new single vortex is established. This serves in practice as a restart of the system. No rational way exists to account for diffusion and cancellation of vorticity with this type of model.

3.7.2 Piston and sloshing modes in three-dimensional moonpools

Molin (2001) also analyzed resonance frequencies in a three-dimensional moonpool of length L_l, width L, and draft d. The length and width of the barge are assumed infinite in the analysis. The velocity potential in the gap is approximated by the following source distribution:

$$\varphi(x, y, 0, t) = \frac{1}{2\pi}\int_0^{L_l}\int_0^{L}\frac{\partial\varphi(\xi, \eta, 0, t)/\partial z}{\sqrt{(x-\xi)^2+(y-\eta)^2}}\mathrm{d}\eta\,\mathrm{d}\xi. \quad (3.81)$$

The coordinate system $Oxyz$ is centered at one of the lower corners. Equation (3.81) follows by representing the velocity potential below the moonpool as a source distribution along an infinite horizontal plane containing the bottom of the moonpool. Because the vertical velocity is assumed zero on the plane outside the moonpool, the source density is only nonzero at the bottom of the moonpool. Unlike the case in two dimensions, in the three-dimensional case it is not necessary to approximate the flow outside the vessel with point sources (sinks). Molin (2001) considered all possible natural modes occurring in the moonpool.

Molin (2001) stated that a good approximation of natural frequencies is obtained by a single-mode approximation. We start by discussing the piston mode. Molin (2001) gave the following approximate formula for the piston-mode resonance frequency:

$$\sigma_{00} = \sqrt{\frac{g}{d + Lf_3(L/L_l)}}, \quad (3.82)$$

where

$$f_3 = \frac{1}{\pi}\left[\sinh^{-1}\left(\frac{L_l}{L}\right) + \frac{L_l}{L}\sinh^{-1}\left(\frac{L}{L_l}\right) + \frac{1}{3}\left(\frac{L}{L_l} + \frac{L_l^2}{L^2}\right) - \frac{1}{3}\left(1 + \frac{L_l^2}{L^2}\right)\sqrt{\frac{L^2}{L_l^2} + 1}\right].$$

Equation (3.82) follows by generalizing the procedure for a two-dimensional moonpool. We represent the velocity potential in the moonpool by $A_{00} + B_{00}z/d$. Satisfaction of the free-surface condition gives $\sigma_{00}^2 d = g/(1 + A_{00}/B_{00})$. The ratio between the constants A_{00} and B_{00} follows by substituting the velocity potential and the corresponding vertical velocity into eq. (3.81). This equation is only satisfied in an averaged way. So we integrate eq. (3.81) over the cross-section of the moonpool, which is the same as following a Galerkin procedure:

$$\frac{A_{00}}{B_{00}} = \frac{1}{2\pi L_l L d}\int_0^{L_l}\int_0^{L}\left[\int_0^{L_l}\int_0^{L}\frac{\mathrm{d}\eta\,\mathrm{d}\xi}{\sqrt{(x-\xi)^2+(y-\eta)^2}}\right]\mathrm{d}y\,\mathrm{d}x.$$

The preceding integral was analytically derived by Molin (2001), which leads to $A_{00}/B_{00} = Lf_3/d$ with f_3 as given earlier. We have then shown how to obtain eq. (3.82).

Let us now consider a complete analysis that determines all natural frequencies. The velocity potential in the moonpool is expressed as

$$\varphi = \sum_{n=0}^{\infty}\sum_{q=0}^{\infty}\cos\lambda_n x\cos\mu_q y \times (A_{nq}\cosh\nu_{nq}z + B_{nq}\sinh\nu_{nq}z)\exp(i\sigma t), \quad (3.83)$$

where $\lambda_n = n\pi/L_l$, $\mu_q = q\pi/L$, and $\nu_{nq}^2 = \lambda_n^2 + \mu_q^2$. When $n = q = 0$, the hyperbolic functions are replaced with $A_{00} + B_{00}z/d$. We leave it to the reader to control that eq. (3.83) satisfies the Laplace equation and no flow through the vertical walls. In addition comes the free-surface condition $-\sigma^2\varphi + g\partial\varphi/\partial z = 0$ on $z = d$ and the bottom condition expressed by eq. (3.81). Insertion of eq. (3.83) into eq. (3.81) gives

$$A_{00} + \sum\sum \cos \lambda_n x \cos \mu_q y A_{nq}$$
$$= \frac{1}{2\pi} \int_0^{L_l} \int_0^{L}$$
$$\times \frac{B_{00} d^{-1} + \sum\sum \nu_{nq} B_{nq} \cos \lambda_n \xi \cos \mu_q \eta}{\sqrt{(x-\xi)^2 + (y-\eta)^2}} \mathrm{d}\eta\, \mathrm{d}\xi, \tag{3.84}$$

where the double sum is over all values of n and q except $n = q = 0$.

The Galerkin procedure is used to satisfy the free-surface condition and eq. (3.84). Therefore, we multiply the free-surface condition and the moonpool bottom condition with $\cos \lambda_m x \cos \mu_p y$ and integrate in x from 0 to l and in y from 0 to L. We start with the free-surface condition and notice that

$$\int_0^{L_l} \cos \lambda_m x \mathrm{d}x = 0$$
$$\int_0^{L_l} \cos \lambda_n x \cos \lambda_m x \,\mathrm{d}x = 0, n \neq m;$$
$$\int_0^{L} \cos \mu_q y \cos \mu_p y \mathrm{d}y = 0, q \neq p.$$

When $n = q = 0$, this gives

$$-\sigma^2 (A_{00} + B_{00}) + g d^{-1} B_{00} = 0.$$

Otherwise,

$$-\sigma^2 (A_{nq} + B_{nq} \tanh \nu_{nq} d)$$
$$+ g \nu_{nq} (A_{nq} \tanh \nu_{nq} d + B_{nq}) = 0.$$

Application of the Galerkin procedure to eq. (3.84) leads to the evaluation of the integrals

$$I_{mnpq} = \int_0^{L_l} \int_0^{L_l} \int_0^{L} \int_0^{L}$$
$$\frac{\cos \lambda_m x \cos \lambda_n \xi \cos \mu_p y \cos \mu_q \eta}{\sqrt{(x-\xi)^2 + (y-\eta)^2}} \mathrm{d}\eta\, \mathrm{d}y\, \mathrm{d}\xi\, \mathrm{d}x,$$

where I_{mnpq} is nonzero only when both $m + n$ and $p + q$ are even. Molin (2001) describes four types of resonant modes that can appear:

(i) m all even: the modes are symmetric both in x and y (piston modes).
(ii) m, n, p, q, and n odd, p and q even: the modes are antisymmetric in x and symmetric in y (longitudinal sloshing modes).
(iii) m and n even, p and q odd: the modes are symmetric in x and antisymmetric in y (transverse sloshing modes).
(iv) m, n, p, and q all odd: the modes are antisymmetric in x and y.

Symmetry and antisymmetry refer to a coordinate system $OXYz$ with origin in the center of the moonpool bottom (i.e., $X = x - \frac{1}{2}L_l$, $Y = y - \frac{1}{2}L$).

We have now described how to establish linear relationships between A_{nq} and B_{nq}. Formally this gives the linear equation system $\mathbf{C}\boldsymbol{x} = 0$, where $\mathbf{C}$ is a matrix with values depending on the moonpool dimension and $\boldsymbol{x}$ contains the unknowns A_{nq} and B_{nq}. For the equation system to have a solution, it is necessary that the determinant of $\mathbf{C}$ is zero. The latter fact determines the natural frequencies. Each natural frequency has corresponding values of A_{nq} and B_{nq} that are determined, except for a multiplicative constant.

Let us now return to a single-mode approximation and study longitudinal sloshing modes with $q = 0$ and n different from zero. Therefore, we express the velocity potential in the moonpool as $\varphi = \cos \lambda_n x (A_{n0} \cosh \lambda_n z + B_{n0} \sinh \lambda_n z) \exp(i\sigma t)$. Satisfaction of the free-surface condition gives

$$\sigma_{n0}^2 = g\lambda_n \frac{1 + (A_{n0}/B_{n0}) \tanh \lambda_n d}{(A_{n0}/B_{n0}) + \tanh \lambda_n d}.$$

Using a Galerkin procedure as described earlier to satisfy the moonpool bottom condition gives

$$\frac{A_{n0}}{B_{n0}} = \frac{n}{L L_l^2} \int_0^{L} \int_0^{L} \int_0^{L_l} \int_0^{L_l}$$
$$\frac{\cos \lambda_n x \cos \lambda_n \xi}{\sqrt{(x-\xi)^2 + (y-\eta)^2}} \mathrm{d}\xi\, \mathrm{d}x\, \mathrm{d}\eta\, \mathrm{d}y$$
$$= \frac{n}{L L_l^2} I_{nn00}.$$

Molin (2001) shows how to simplify the integral, giving

$$\frac{A_{n0}}{B_{n0}} = \frac{2}{n\pi^2 r} \left\{ \int_0^1 \frac{r^2}{u^2 \sqrt{u^2 + r^2}} \right.$$
$$\times \left[1 + (u-1)\cos(n\pi u) - \frac{\sin(n\pi u)}{n\pi} \right] \mathrm{d}u$$
$$\left. + \frac{1}{\sin \theta_0} - 1 \right\},$$

where $r = L/L_l$ and $\tan \theta_0 = r^{-1}$.

We consider as an example the wellhead barge mentioned by Molin (2001) with a moonpool that is 80 m long, 20 m wide, and has a draft of 5 m. The barge itself may be about three times as long

Table 3.5. *Calculated natural frequencies σ_{nm} (rad/s) for gap resonance of two side-by-side boxes each of length 280 m, breadth 46 m, and draft 16.5 m by Molin's generalized theory and a BEM*

Mode n	1	3	5	7	9	11	13	15
BEM	0.525	0.647	0.758	0.871	0.978	1.081	1.173	1.261
Theory $m = 0$ modes	0.516	0.651	0.774	0.891	1.001	1.103	1.197	1.285
BEM	1.311	1.321	1.340	1.368				
Theory $m = 1$ modes	1.313	1.323	1.343	1.371	1.407	1.449	1.494	1.542

Note: The width of the gap between the hulls is 18 m. The "$m = 0$" modes do not vary across the gap (Eatock Taylor *et al.*, 2008).

and wide as the moonpool. Using Molin's formulas gives a piston-mode resonance period of $T_{00} = 9.40$ s and the natural periods of the four lowest longitudinal sloshing modes equal to $T_{10} = 7.41$ s, $T_{20} = 6.34$ s, $T_{30} = 5.52$ s, $T_{40} = 4.92$ s.

3.7.3 Resonant wave motion between two hulls

Eatock Taylor *et al.* (2008) studied the gap resonance of two side-by-side boxes each of length 280 m, breadth 46 m, and draft 16.5 m. The width of the gap between the hulls is 18 m. Relevant natural frequencies σ_{nm} (rad/s) for beam seas are presented in Table 3.5, where the index n is the number of half sine waves along the gap length. Because of beam seas and the symmetry properties of the hulls, only odd integer numbers are relevant for n. When $m = 0$, no variations exist across the gap and $m = 1$ corresponds to one half-wave transversely across the gap. Generalized Molin theory and a BEM were used. The generalized Molin theory accounts for open ends in the length direction and implies that $\cos \lambda_n x$ in eq. (3.83) is replaced by $\sin \lambda_n x$. The argument is that the eigenvalue problem should satisfy nodal points for the free-surface elevation at the open ends of the gap (see explanations in connection with Figure 3.27). The results of the BEM were obtained by considering the boxes fixed, and the natural frequencies were obtained by inspecting the frequency locations of the peaks of the predicted wave elevations in the gap. Because the maxima predicted by the BEM based on potential flow were finite, far-field wave radiation occurred in a way similar to that discussed for two-dimensional piston-mode resonance in a moonpool in Section 3.7.1. Molin's theory does not account for wave radiation, nor is the breadth of the boxes a parameter. Still we notice that the predictions of natural frequencies by BEM and Molin's generalized theory are in good agreement.

Ronæss (2002) reported on a special resonant wave motion between two ships at equal forward speed U during model tests in a head sea with deep-water conditions. At $\tau = \sigma_e U/g$ slightly above $\frac{1}{4}$, large-amplitude waves built up between the hulls; σ_e is the frequency of encounter. When $\tau > \frac{1}{4}$, all the waves generated by a ship in deep water are downstream according to linear theory (Faltinsen, 2005). When $\tau < \frac{1}{4}$, deep-water waves may be generated upstream of a ship. The phenomenon reported by Ronæss (2002) was observed for two different staggers between the ships. The waves between the hulls were propagating forward and appeared two-dimensional in the longitudinal vertical plane, as sketched in Figure 3.34. On the outside of the largest hull, no such wave pattern was observed, whereas a weak tendency of this pattern was observed outside of the smallest hull.

The wave front appeared to move farther and farther forward during the run but died out before it reached the bows. The rear waves were shorter and steeper than the first waves and nearly

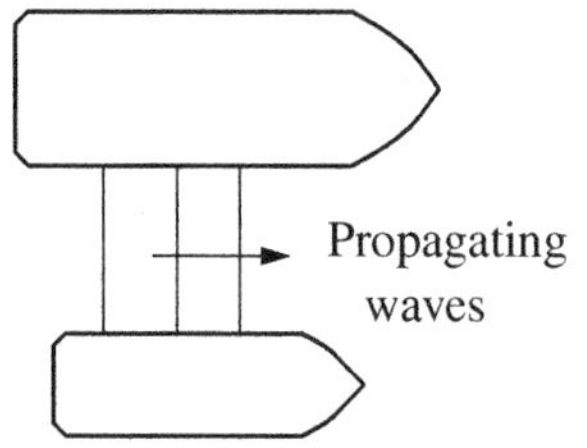

Figure 3.34. Two-dimensional resonant waves between two ships at forward speed in head-sea deep-water waves when $\tau = \sigma_e U/g$ is slightly greater than 1/4 (Ronæss, 2002).

breaking. It took some time to build up these large waves.

3.8 Time-domain response

Sections 3.5 and 3.6 dealt with wave-induced ship response in the frequency domain by analyzing the steady-state solution in regular incident waves. If we consider the behavior of the ship in irregular sea, many excitation frequencies exist. Because added mass and damping are frequency-dependent, we cannot directly use the equation system given by eq. (3.26) in the time domain. However, if we are interested in the steady-state solution, we can circumvent this problem by adding together the responses to regular wave components.

However, scenarios exist where we need the transient response. One example is transient waves generated by a passing ship. Another example is coupling between nonlinear sloshing in a ship tank and ship motion. A third example is wetdeck slamming on a catamaran in regular incident waves. Wetdeck slamming causes a transient vertical force that excites transient response in heave, pitch, and global elastic vibration modes. These examples illustrate when we cannot use eq. (3.26). We must then formulate the equations of motion in a different way. This approach was discussed by Cummins (1962) and Ogilvie (1964). If we consider the case of coupled sloshing and ship motion, we may write the equations of motion as

$$\sum_{k=1}^{6}\left\{[M_{jk}+A_{jk}(\infty)]\ddot{\eta}_k(t)+B_{jk}(\infty)\dot{\eta}_k(t)+\int_0^t h_{jk}(\tau)\dot{\eta}_k(t-\tau)\mathrm{d}\tau+C_{jk}\eta_k\right\}=F_j^{ext}(t)+F_j^{\text{tank}}[\eta_k(t)]\quad(j=1,\ldots,6), \tag{3.85}$$

where F_j^{ext} represents the linear external wave force and moment components. Furthermore, F_j^{tank} are the force and moment components associated with sloshing and are functions of $\eta_k(t)$ and the second time derivative of $\eta_k(t)$. The coupling with nonlinear sloshing is described in Section 8.3.3. We should, as stated earlier, be careful how the frozen liquid effect is incorporated. A common practice is to include the mass of the tank liquid in determining the total mass of the vessel, its COG, and moments of inertia (i.e., the tank liquid is considered "frozen." When the hydrodynamic forces and moments due to sloshing are analyzed, the "frozen" liquid effects are part of the expressions and must not be accounted for twice. The evaluation of F_j^{tank} requires $\eta_k(t)$ to be transformed to the tank-fixed coordinate system. Furthermore, when the hydrodynamic forces and moments on the tank are calculated, they are calculated with respect to the tank-fixed coordinate system and need to be transformed to the ship coordinate system defined in Figure 3.8. It is common to add a viscous roll damping term to eq. (3.85).

The integrals in eq. (3.85) are often referred to as convolution integrals (often used in connection with Laplace and Fourier transforms) or as Duhamel integrals. Here the vessel mass terms M_{jk} and the restoring terms C_{jk} are the same as in eq. (3.26). $A_{jk}(\infty)$ and $B_{jk}(\infty)$ mean infinite-frequency added mass and damping coefficients; $h_{jk}(t)$ are the retardation functions (also referred to as impulse response functions), which can be evaluated by

$$\begin{aligned}h_{jk}(t)&=-\frac{2}{\pi}\int_0^\infty \sigma(A_{jk}(\sigma)-A_{jk}(\infty))\sin\sigma t\,\mathrm{d}\sigma\\&=\frac{2}{\pi}\int_0^\infty (B_{jk}(\sigma)-B_{jk}(\infty))\cos\sigma t\,\mathrm{d}\sigma.\end{aligned} \tag{3.86}$$

Greenhow and White (1997) discussed the convergence of these integrals.

Calculation of $h_{jk}(t)$ requires information on the behavior of either A_{jk} or B_{jk} at all frequencies. It is no problem to calculate A_{jk} and B_{jk} for infinite frequency and for frequencies typical for ship motions. It is more difficult to estimate how A_{jk} and B_{jk} behave asymptotically for high frequencies. Let us illustrate this by referring to a BEM (see Section 10.2). The hull surface is then approximated by panels. The lengths of the panels must be small relative to the wavelength. Different wavelengths are created by an oscillating ship at forward speed. Let us simplify to illustrate our point; we return to Table 3.1. The relationship between wavelength λ and σ is $\lambda = 2\pi g/\sigma^2$. So increasing σ causes small wavelengths and, therefore, small panels. A practical limit exists for how small the panels can be, due to CPU time. Furthermore, the equation system may be ill conditioned at high frequencies due to irregular frequencies (Faltinsen, 1990). Adegeest (1995)

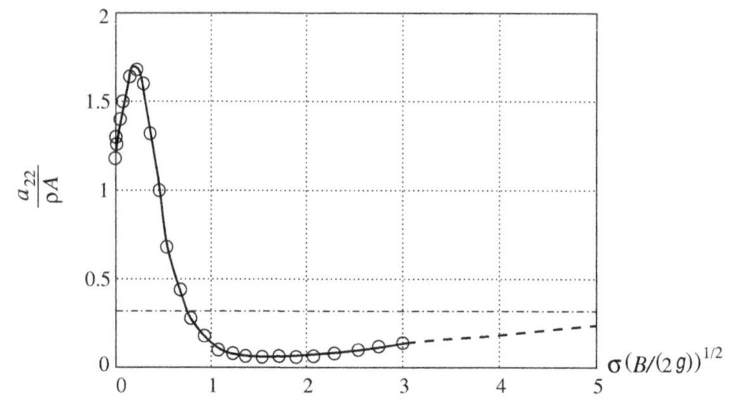

Figure 3.35. Continuous representation of two-dimensional added mass $a_{22}(\sigma)$ of a rectangular cross-section with beam-to-draft ratio 2.0. The spline interpolation of $a_{22}(\sigma)$ is used to draw the solid line through the values with ○. The solid line was matched at $\sigma = \sigma_{high}$ (where $\sigma_{high} = 3$ was assumed) with asymptotic approximation $a_{22}(\sigma) \approx a_{22}(\infty) + a_{22}^{(1)}\sigma^{-2} + a_{22}^{(2)}\sigma^{-4}$ (dashed line). The dashed-dotted line corresponds to the asymptotic limit $a_{22}(\infty) = 0.386\rho A$.

and Kvålsvold (1994) have, for instance, in their studies of nonlinear wave-induced motions and loads on a ship in head sea reported that it is not straightforward to use a convolution integral formulation as in eq. (3.85). However, it has become common to use a formulation like eq (3.85) in combination with simplified nonlinear hydrodynamic loads on the hull to study nonlinear external wave loads on a ship.

We now modify eq. (3.68) to account for a convolution integral formulation of the external hydrodynamic force and include a viscous term for the external flow and spring terms and bearing forces associated with the experimental setup by Rognebakke and Faltinsen (2003) as well. We can write

$$\begin{aligned}&[M + A_{22}(\infty)]\,\ddot{\eta}_2 + B_{22}^{\text{visc}}\dot{\eta}_2|\dot{\eta}_2| \\ &\quad + C_{22}\eta_2 + \int_0^t h_{22}(\tau)\dot{\eta}_2(t-\tau)d\tau \\ &= F_2^{\text{exc}} + F_2^{\text{tank}} + F_2^{\text{bearing}}\frac{\dot{\eta}_2}{|\dot{\eta}_2|},\end{aligned}$$

where we have used the fact that $B_{22}(\infty) = 0$; B_{22}^{visc} represents viscous damping due to external flow, C_{22} is the linear spring coefficient, F_2^{exc} is the horizontal linear wave excitation force, and F_2^{tank} is the horizontal force caused by sloshing. F_2^{bearing} is the constant force acting against the motion from the bearings during the model tests. Once more we should be careful in how the frozen liquid effect is incorporated. When we presented eq. (3.68), we did not include the frozen liquid in M; the effect was part of F_2^{tank}.

Equation (3.86) is used to calculate the retardation function. If, for instance, the added mass is used as a basis for the retardation function calculations, the procedure for finding a continuous representation for the added mass coefficient for all frequencies is crucial and illustrated for sway for the studied rectangular cross-section in Figure 3.35; A is the area of the section, B is the breadth, and ρ is the mass density of water. First the two-dimensional sway added mass $a_{22}(\sigma_k)$ is calculated for N frequencies $0 \le \sigma_k\sqrt{B/(2g)} \le 3$, $(k = 1, \ldots, N)$; $N = 25$ is found to be sufficient. A cubic spline interpolation of $a_{22}(\sigma_k)$ is applied. For $\sigma\sqrt{B/(2g)} > 3$, an asymptotic series formulation in σ is assumed for the added mass. Greenhow (1986) shows that the leading terms for two-dimensional sway added mass are $a_{22}(\sigma) \approx a_{22}(\infty) + a_{22}^{(1)}/\sigma^2 + a_{22}^{(2)}/\sigma^4 + a_{22}^{(3)}\ln\sigma/\sigma^4$. Below $\sigma\sqrt{B/(2g)} = 3$, Romberg integration is used between the zeros of $\sin(\sigma t)$ in the evaluation of eq. (3.86). This method is fast and accurate. The contribution to the integral in eq. (3.86) from $\sigma = 3\sqrt{2g/B}$ to $\sigma = \infty$ is analytically expressed (Greenhow, 1986).

Figure 3.36 shows the calculated retardation functions h_{22} and h_{33} in sway and heave for the rectangular cross-section, where we have used the same notation for the retardation functions as in three dimensions. We note that h_{jj} is practically zero when $t\sqrt{2g/B}$ is larger than 12. A similar

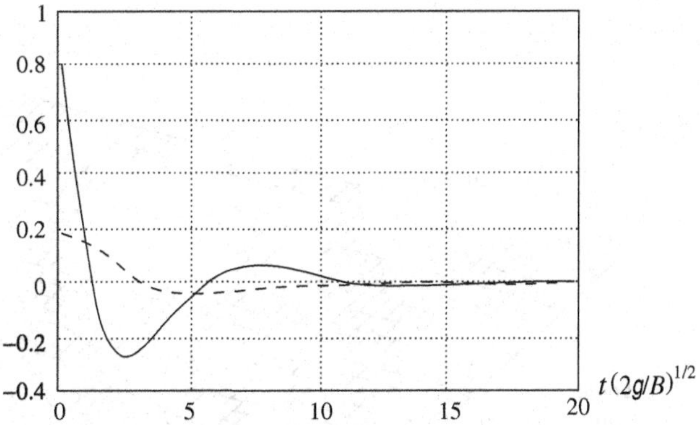

Figure 3.36. Retardation functions in sway and heave for a rectangular cross-section with beam-to-draft ratio 2.0; B is the breadth (beam) of the section and A is the mean submerged area (Rognebakke & Faltinsen, 2003). Solid line, $h_{22}(t)B/(2\rho gA)$, dashed line, $h_{33}(t)B/(2\rho gA)$.

behavior, where h_{jk} is practically zero in a finite time, is also true for h_{jk} in eqs. (3.85). Let us illustrate the consequence of this by examining the terms

$$\int_0^t h_{jk}(\tau)\dot{\eta}_k(t-\tau)\mathrm{d}\tau \tag{3.87}$$

in eq. (3.85). We are going to find $\dot{\eta}_k$ at time t, which corresponds to $\tau = 0$ in eq. (3.87). It is then only necessary to integrate eq. (3.87) over previously found $\dot{\eta}_k$ up to a "cutoff" value $t = t^*$ when h_{jk} is practically zero. The numerical results are not sensitive to t^*. The time-consuming part of the analysis is the calculation of h_{jk}.

The calculation of the retardation function for sway can be verified by noting the relationships of Ogilvie (1964):

$$A_{22}(\sigma) = A_{22}(\infty) - \sigma^{-1}\int_0^\infty h_{22}(t)\sin(\sigma t)\,\mathrm{d}t \quad \text{and}$$

$$B_{22}(\sigma) = \int_0^\infty h_{22}(t)\cos(\sigma t)\,\mathrm{d}t.$$

The same relationships apply in two dimensions, the results of which are shown in the two-dimensional case in Figure 3.37. The original values of $a_{22}(\sigma)$ and $b_{22}(\sigma)$ should be regained, and the discrepancy indicates the level of the error

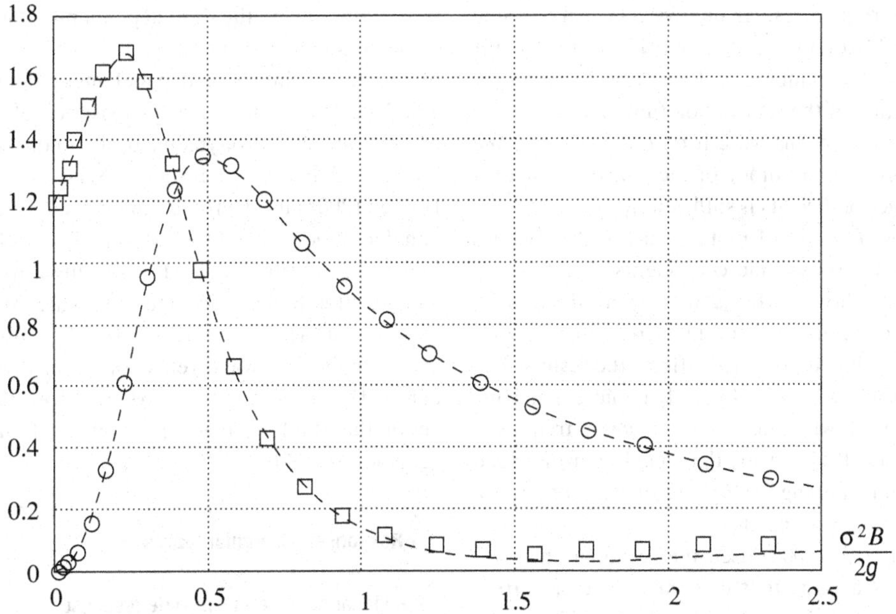

Figure 3.37. Calculation of added mass and damping from the retardation function h_{22}. The damping coefficient $b_{22}(\sigma)$ is used to calculate h_{22}. Rectangular cross-section with beam-to-draft ratio 2.0, where B is the breadth (beam) of the section and A is the mean submerged area. Notations: $\square = a_{22}/(\rho A)$, $\bigcirc = b_{22}/(\rho\sigma A)$; the dashed lines correspond to the values obtained from $h_{22}(t)$.

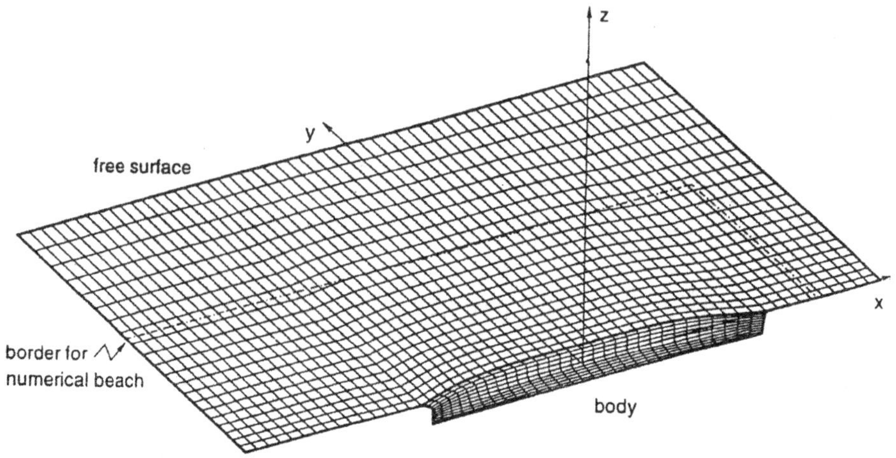

Figure 3.38. Illustration of panels used in the Rankine panel method described by Sclavounos and Borgen (2004).

in the retardation function. Almost exact values are regained for the damping coefficient if the retardation function is calculated based on these values, which is the case for the presented results. Unfortunately, the added mass coefficient has a certain level of error that increases with frequency.

If the retardation function calculations are based on the added mass, the damping coefficient has an error with the largest discrepancy for low frequencies. An unphysical negative value is found around $\sigma = 0$, which is the reason for choosing the damping coefficients as the basis for calculation of the retardation function.

Different means were tried out for improving accuracy. The accuracy of the calculated hydrodynamic coefficients is sufficient for this not to be a factor. The use of more than about 25 values for the hydrodynamic coefficients is not necessary. The choice of frequency, where the numerical calculations and the high-frequency asymptotic solution are patched, affects the results. The asymptotic series for hydrodynamic coefficients converge slowly; thus, a high patching frequency is preferred. However, the calculation of added mass and damping for high frequencies requires a large number of panels.

Nam *et al.* (2006) used the formulation in eq. (3.85) as a basis to study the effect of a free-surface antirolling tank with different filling ratios on the roll motion of a modified S175 hull in beam sea. An equivalent linear viscous roll damping term was added. A soft spring system was included to prevent monotonically increasing surge, sway, and yaw motions. The nonlinear sloshing motion was described by the Euler equation. A finite difference method based on the SOLA-SURF scheme with special treatment of impact loads on the tank ceiling was applied. A similar study was made by Kim (2002). The effect of pillars inside the antirolling tank was investigated. The linear external hydrodynamic loads were based on potential flow and obtained by directly solving the time-dependent boundary-value problem with initial conditions by means of the three-dimensional panel method of Lin and Yue (1990). Other ways exist to solve the exterior linear hydrodynamic problem directly in the time domain. For instance, Sclavounos and Borgen (2004) used a three-dimensional Rankine panel method to study the external flow problem. Figure 3.38 illustrates panels distributed over the mean wetted hull surface and a truncated part of the free surface. A numerical beach is used to avoid unphysical wave reflection from the border of the truncated free surface. More details about the Rankine panel method are found in Sclavounos (1996).

3.9 Response in irregular waves

3.9.1 Linear short-term sea state response

A short-term sea state refers to wave conditions defined by a given and constant significant wave height $H_{1/3}$ and mean wave period T_2. In addition we need to specify a mean wave heading, wave energy spreading, and duration. The duration typically used is 3 or 6 hours.

A useful consequence of linear theory is that we can obtain results in irregular waves by adding together results from regular waves of different amplitudes, wavelengths, and propagation directions.

The so-called frequency-of-encounter wave spectrum is sometimes used in connection with calculations of statistical values in an irregular short-term sea. If we measure the incident wave elevation relative to a coordinate system that is moving with the forward speed of the vessel and then evaluate the spectrum, we will obtain the frequency-of-encounter wave spectrum $S_e(\sigma_e)$. However, if we use a standard wave spectrum like JONSWAP or the Pierson–Moskowitz spectrum, it is more convenient to represent all response variables as a function of the wave frequency, σ_0. We describe how we normally make statistical predictions in a short-term sea. Long-crested seas are assumed. The following procedure is applicable to all linear wave-induced response values, such as the six degrees of motion, accelerations, global wave loads, structure stresses, and so on.

The variance σ_r^2 of a response variable can be expressed as (Faltinsen, 2005)

$$\sigma_r^2 = \int_0^\infty S(\sigma_0) \left|H(\sigma_0, U, \beta)\right|^2 d\sigma_0, \qquad (3.88)$$

where $\left|H(\sigma_0, U, \beta)\right|$ is the transfer function, which is the response amplitude per unit wave amplitude.

By using the Rayleigh distribution we can find the probability of exceeding a given value of, for instance, the heave amplitude. Therefore, we write the probability that the heave amplitude is larger than x:

$$Q(\eta_3 > x) = e^{-x^2/2\sigma_3^2}. \qquad (3.89)$$

We have used the subscript 3 on the variance to indicate heave. By using eq. (3.89) we can calculate the most probable largest value $x_{\max}$ in N oscillations. A good approximation for large N is

$$x_{\max} = \sigma_3\sqrt{2\ln N}, \qquad (3.90)$$

where N can be set equal to t/T_2, where t is the duration of the storm and T_2 the zero upcrossing period.

3.9.2 Linear long-term predictions

By combining the Rayleigh distribution with a joint frequency table (scatter diagram) for $H_{1/3}$ and the modal wave period or mean wave period (see Table 3.2), we can obtain the long-term probability distribution of a response. Summing over both period and wave height gives

$$P(R) = 1 - \sum_{j=1}^{M}\sum_{k=1}^{K} \exp\left(-\tfrac{1}{2}R^2/(\sigma_r^{jk})^2\right) p_{jk}, \qquad (3.91)$$

where $P(R)$ is the long-term probability that the peak value of the response does not exceed R, and σ_r^{jk} is the standard deviation of the response for a mean $H_{1/3}$ and wave period in the significant wave height interval j and the wave period interval k. Furthermore, p_{jk} is the joint probability for a significant wave height and mean wave period to be in interval numbers j and k, respectively. Typically one needs on the order of 100,000 observations to create a reliable scatter diagram. The probability level $Q = 1 - P(R)$ and the number of response cycles N_0 are related by $Q = 1/N_0$. A return period of 100 years corresponds approximately to $Q = 10^{-8.7}$ depending on the response variable and its zero upcrossing period. The corresponding response amplitude R can then be found from eq. (3.91). The procedure is often performed by fitting results to a Weibull distribution (Nordenstrøm, 1973). The Weibull distribution is then used for extreme value predictions.

3.10 Summary

We have discussed the coupled analysis between sloshing and ship motion. Because sloshing is a typical resonance phenomenon, it is not necessarily the most extreme ship motion or external wave-induced loads that cause the most severe sloshing. Linear theory has therefore been used to describe the external hydrodynamic loads on the ship. Linearity gives many advantages in terms of superposition of effects. For instance, an irregular sea state can be represented as a sum of regular waves of different frequencies, amplitudes, and wave propagation directions. If the response of the ship is linear, it means that we may first consider the response in regular waves and afterward we may add the results of regular waves to obtain the response in a sea state. However, it is only the external hydrodynamic loads that may be considered linear. Resonant sloshing in a ship tank has clearly nonlinear behavior.

The fact that we may assume the external hydrodynamic loads are linear enables us to split the external hydrodynamic interaction between the ship and the incident waves into subproblems. We first described the so-called frequency-domain solution; we examined steady-state conditions where the hydrodynamic flow and loads oscillate with the frequency of encounter between regular waves and the ship. We studied two subproblems:

(a) The forces and moments on the ship when the body is restrained from oscillating and incident regular waves are present. The hydrodynamic loads are called *wave excitation loads* and are composed of the so-called Froude–Kriloff loads and diffraction forces and moments. Froude–Kriloff loads are due to the pressure field in the incident waves, which are undisturbed by the ship.

(b) The forces and moments on the body when the structure is forced to oscillate in calm water with the wave excitation frequency in any rigid-body motion mode. No incident waves exist, but the oscillating body causes radiating waves. The hydrodynamic loads are identified as *added mass, damping* and *restoring* forces, and moments. This subproblem is often termed the *radiation problem.*

Because sloshing in a tank is a transient nonlinear process, we cannot assume that the flow inside the tank oscillates sinusoidally with a given frequency. The consequence is that the external hydrodynamic loads must be transferred to the time domain, which can be done by using the frequency-domain solution. The resulting external hydrodynamic loads are then expressed in terms of convolution integrals that contain memory effects of the ship's behavior.

A common practice is to include the mass of the tank liquid in determining the total mass of the vessel, its COG, and moments of inertia (i.e., the tank liquid is considered "frozen." When the hydrodynamic forces and moments due to sloshing are analyzed, the frozen liquid effects are included as parts of the expressions. *We must not count the frozen liquid effects twice.*

We have illustrated in this chapter that there may be strong coupling between the wave-induced ship motions and sloshing. This phenomenon is first analyzed for free-surface and U-tube antirolling tanks. We use the example with the free-surface antirolling tank to show the importance of tuning the lowest natural frequency of the sloshing to the natural frequency of roll. In this way sloshing provides a damping of the roll motion.

It is shown that a closed air chamber significantly influences the liquid behavior in a U-tube. It is almost possible to latch the liquid flow. The flow in the U-tube can thereby be controlled to provide the proper phasing between the roll motion and the roll moment caused by the U-tank flow to provide a damping effect.

We study then coupled sloshing and wave-induced ship motion for a two-dimensional body that is only free to oscillate in the transverse (sway) direction. The presence of sloshing causes a natural frequency of the sway motion in an important wave period range. This case also illustrates that linear sloshing analysis can only qualitatively predict the mutual interaction between wave-induced ship motions and sloshing. The theoretical sway response is zero at the natural sloshing frequency according to linear theory. The reason is the infinite added mass contribution from the sloshing. Even though the added mass coefficients become infinite, it is shown in a linear coupled sway and roll analysis that the sway and roll are nonzero at a natural sloshing frequency.

Sloshing may also occur in confined spaces with a free surface that is exterior to a ship. Examples are resonances between the hulls of catamarans and in moonpools. The piston-mode resonance is of particular concern and is examined in detail by simplified methods that are compared with experiments and exact numerical results. External sloshing generates far-field waves. The consequence is a potential flow damping effect on sloshing. The latter is not true for sloshing inside a tank.

An important practical matter is to recognize that external hydrodynamic loads are described in an inertial system that moves with the steady forward speed of the ship. Because sloshing is described in a tank-fixed coordinate system, we need to show how to transfer information between these two coordinate systems.

Table 3.6. *Scatter diagram showing joint probability of significant wave height $H_{1/3}$ and mean wave period* T_2

	T_2(s)																
$H_{1/3}$ (m)	1	2	3	4	5	6	7	8	9	10	11	12	13	14	15	16	Sum
0.5	0	0	15	70	104	85	50	24	10	4	1	1	0	0	0	0	364
1	0	0	1	17	51	65	49	27	12	5	2	1	0	0	0	0	230
1.5	0	0	0	4	24	44	43	28	13	5	2	1	0	0	0	0	164
2	0	0	0	1	9	24	30	22	12	5	2	1	0	0	0	0	106
2.5	0	0	0	0	3	11	18	16	9	4	1	1	0	0	0	0	63
3	0	0	0	0	1	5	10	10	6	3	1	0	0	0	0	0	36
3.5	0	0	0	0	0	2	5	5	4	2	1	0	0	0	0	0	19
4	0	0	0	0	0	1	2	5	2	1	1	0	0	0	0	0	12
4.5	0	0	0	0	0	0	1	1	1	1	0	0	0	0	0	0	4
5	0	0	0	0	0	0	0	1	1	0	0	0	0	0	0	0	2
5.5	0	0	0	0	0	0	0	0	0	0	0	0	0	0	0	0	0
6	0	0	0	0	0	0	0	0	0	0	0	0	0	0	0	0	0
Sum	0	0	16	92	192	237	208	139	70	30	11	5	0	0	0	0	1000

Note: There is 10% probability for exceedence of $H_{1/3} = 2.5$ m.

3.11 Exercises

3.11.1 Wave energy

Formulas for available wave power in a sea state can be derived by first considering harmonic propagating waves. The incident wave power per unit crest length of regular sinusoidal waves of amplitude ζ_a and circular frequency σ in deep water is

$$P_w = \tfrac{1}{4}\rho g^2 \zeta_a^2/\sigma. \tag{3.92}$$

(a) Generalize eq. (3.92) to a long-crested short-term sea state. The answer is

$$P_w = \frac{1}{2}\rho g^2 \int_0^\infty \sigma^{-1} S(\sigma)\, d\sigma, \tag{3.93}$$

where $S(\sigma)$ is the wave spectrum.

(b) Use the ITTC and ISSC spectrum for fully developed sea given by eq. (3.7) to show that

$$P_w = 0.005535\rho g^2 H_{1/3}^2 T_1. \tag{3.94}$$

(c) Evaluate P_w when $\rho = 1026.9$ kg m^{-3}, $T_1 = 6$ s, and $H_{1/3} = 1$ and 2 m. (*Answer*: P_w is 3.3 and 13.1 kW/m, respectively.)

(d) To determine the average available wave energy for a specific area over a long time (e.g., one year), we need to know the joint probability of $H_{1/3}$ and T_1 for the area. An example that is relevant for a coastal area is given in Table 3.6. However, this is just an example and must not be taken as representative for all coastal areas.

We neglect the fact that the different sea conditions collected in Table 3.6 have different mean wave directions. Show that the average available wave power per unit meter in an operational area can be expressed as

$$\overline{P_w} = \sum_i \sum_j P_w(H_{1/3}, T_2) p_{ij}. \tag{3.95}$$

What does p_{ij} in eq. (3.95) stand for?

(e) Show that $\overline{P_w} = 10.5$ kW/m by using eqs. (3.94) and (3.95) together with the data in Table 3.6 and a mass density of 1,026.9 kg m^{-3}. Put this estimate into the context of energy supply by noting that Norway's electric consumption was 143 TWh in 2000 and that a wave energy device may only have 10–20% efficiency.

3.11.2 Surface tension

We want to study the effect of surface tension on two-dimensional propagating waves in deep water. The free-surface condition is stated in Section 2.7.2. The solution form for the velocity

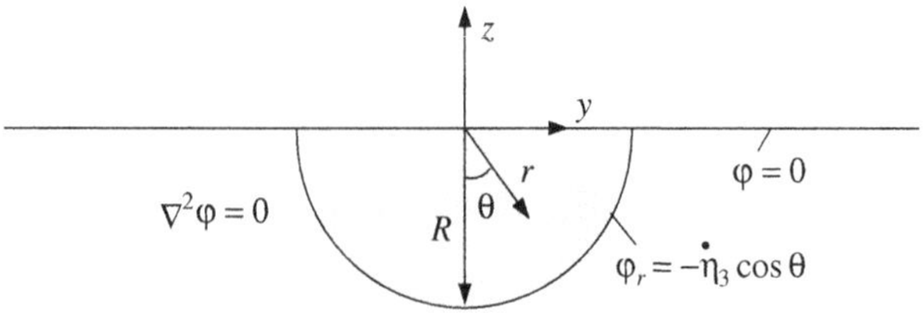

Figure 3.39. The boundary-value problem for high-frequency two-dimensional model added mass in heave for a semisubmerged circular cylinder: $\dot{\eta}_3$, heave velocity, $\varphi_r = \partial\varphi/\partial r$.

potential can be expressed in a similar way as in Table 3.1.

(a) Show that $\sigma^2 = gk + k^3 T_s/\rho$, where σ is the frequency of oscillation and k is the wave number.
(b) Plot the phase speed as a function of wavelength λ for water with $T_s = 0.074\ \mathrm{N\ m^{-1}}$ and $\rho = 1{,}000\ \mathrm{kg\ m^{-3}}$. Confirm from the figure that surface tension does not really matter until λ is less than about 0.05 m.

3.11.3 Added mass and damping

We consider the added mass and damping of a two-dimensional body in deep water.

(a) What are the free-surface conditions when the forcing frequency $\sigma \to 0$ and $\sigma \to \infty$?
(b) Consider a semisubmerged circular cylinder. Derive the sway added mass when $\sigma \to 0$ and the heave added mass when $\sigma \to \infty$. (*Hint:* Consider the boundary-value problem for determining high-frequency two-dimensional heave added mass in heave shown in Figure 3.39.) Show that the velocity potential due to forced heave velocity $\dot{\eta}_3$ can be expressed as $\varphi = \dot{\eta}_3 \left(R^2/r\right)\cos\theta$. Express the vertical force on the cylinder due to the pressure component $-\rho\partial\varphi/\partial t$. Show that this leads to the following two-dimensional heave added mass and damping coefficients $a_{33} = \frac{1}{2}\rho\pi R^2$ and $b_{33} = 0$.
(c) Explain why $a_{22}(0) = a_{33}(\infty)$ as associated with the results presented in Figure 3.35.

3.11.4 Heave damping at small frequencies in finite water depth

We consider a two-dimensional body that can consist of either a submerged single hull or a multihull. The submerged body has the z-axis as a symmetry plane. Furthermore, z is positive upward and $z = 0$ corresponds to the mean water level. The x-axis is in the mean water level. Consider forced harmonic heave motion $\eta_{3a}\cos(\sigma t)$ with small amplitude η_{3a} relative to the cross-dimensional lengths of the body. The ambient flow is zero. The hydrodynamic flow is linearized and proportional to η_{3a}. The mean water depth is constant and equal to h.

(a) Show by conservation of energy that the heave damping coefficient b_{33} can be expressed as

$$b_{33} = 2\rho\frac{g^2}{\sigma^3}\left(\frac{A_3}{\eta_{3a}}\right)^2\frac{kI_1}{\cosh^2(kh)},$$

where $I_1 = \int_{-h}^{0}\cosh^2 k(z+h)\,\mathrm{d}z = (2k)^{-1}\cdot(\sinh(kh)\cosh(kh) + kh)$, k is the wave number, ρ is the mass density of the water, g is the acceleration of gravity, and A_3 is the wave amplitude at infinity. (*Hint:* Start with eq. (2.45) and assume irrotational flow of an incompressible liquid. Show by means of Bernoulli's equation that the time rate of change of kinetic and potential energy is

$$\dot{E}(t) = -\rho\int_{S_Q}\left[\frac{\partial\varphi}{\partial t}u_n - \left(\frac{p-p_0}{\rho} + \frac{\partial\varphi}{\partial t}\right)U_{sn}\right]\mathrm{d}S,$$

where φ is the velocity potential. Choose S_Q to consist of the wetted body surface S_B, two fixed control surfaces S_∞ and $S_{-\infty}$ at $x = \infty$ and $x = -\infty$, respectively, the free surface S_F inside S_∞ and $S_{-\infty}$, and finally a surface S_0 that consists of the sea floor between S_∞ and $S_{-\infty}$. Use the free-surface conditions and the sea-floor condition to simplify the expression. Express the integration of the pressure force on S_B in terms of added mass, damping, and restoring coefficients. Use Table 3.1 to express φ at S_∞ and $S_{-\infty}$. Then time-average $\dot{E}$ over one period of oscillation.)
(b) We now express A_3/η_{3a} in the shallow water limit. This expression is done using

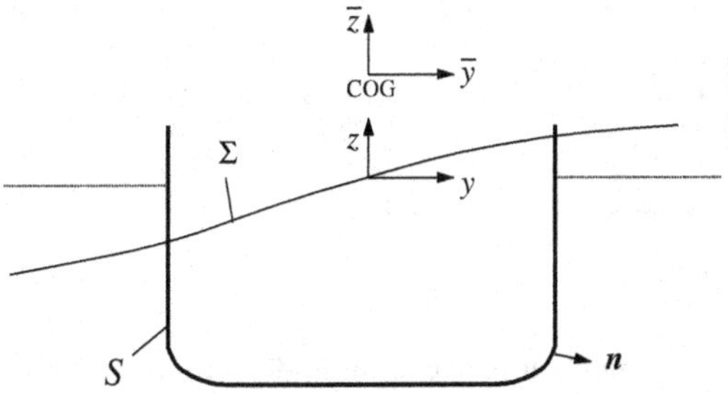

Figure 3.40. Definition of coordinate systems and surfaces Σ and S used in calculating roll excitation moment.

matched asymptotic expansions, where we first define a near field around the body with a length scale smaller that the wavelength. The free-surface condition then becomes the rigid free-surface condition. Due to symmetry properties we limit ourselves to $x > 0$. In the outer expansion of the near-field solution far away from the body, there is a horizontal velocity u that is independent of z.

Show by continuity of water mass that $\frac{1}{2} b_w \sigma \eta_{3a} \sin(\sigma t) = uh$, where b_w is the length of the water plane area of the total body.

(c) We now consider the far-field solution, where we do not see the details of the body. The flow at $x > 0$ can be represented by a wave maker solution. The wave maker is situated at $x = 0$ and its velocity is u. These conditions are consistent with the matching procedure (i.e., the outer expansion of the near-field solution should be equal to the inner expansion of the far-field solution).

The solution of the wave amplitude at infinity can for small kh be expressed as $A_3 = u_a \sigma/(kg)$, where u_a is the velocity amplitude of u.

Show that $b_{33} = \frac{1}{2} \rho b_w^2 \sqrt{g/h}$. (*Hint:* Use the dispersion relationship between σ and k and the expression of I_1 for small kh.)

(d) Discuss the behavior of b_{33}.

3.11.5 Coupled roll and sloshing in an antirolling tank of a barge in beam sea

We consider a barge in regular beam-sea waves. The barge is equipped with a free-surface antirolling tank with a rectangular shape. The tank breadth and length are denoted B_t and L_t, respectively. We study coupled roll motion and sloshing in the tank. The coupling with other modes of motion is neglected. The external dynamic roll moment on the ship is linearized and the tank roll moment accounts for nonlinearities. The roll moments are with respect to an $\bar{x}$-axis through the COG of the barge. The equation of roll motion is formally expressed as

$$(I_{44} + A_{44})\ddot{\eta}_4 + 2\xi\sqrt{(I_{44} + A_{44})\, C_{44}}\dot{\eta}_4 + C_{44}\eta_4 = F_4^{exc} + F_4^t, \qquad (3.96)$$

which is similar to eq. (3.37) except that the damping in eq. (3.96) is expressed in terms of the ratio ξ between the roll damping and the critical damping.

(a) We express the wave excitation moment F_4^{exc} by means of a Froude–Kriloff approximation. Deep water is considered and the incident wavelength λ is assumed long relative to the cross-sectional beam B and draft. Figure 3.40 defines two coordinate systems. The $Oxyz$ system has origin in the mean free surface. The z-axis is vertical upward and the x–z-plane is a symmetry plane for the barge. The $O\bar{x}\bar{y}\bar{z}$ system has its origin in the COG. The incident wave elevation is expressed similarly as in Table 3.1: $\zeta = \zeta_a \sin(\sigma t - ky)$. Show that the liquid pressure on the hull can be approximated as $p = -\rho g(z - \zeta)$ and that we may write the roll excitation moment as

$$F_4^{exc} = \rho g \int_{S+\Sigma} (\bar{z} - \bar{z}_0 - \zeta)(\bar{y} n_3 - \bar{z} n_2)\, \mathrm{d}S,$$

where S is the wetted body surface and Σ is the continuation of the incident wave surface inside the body. The normal vector to S, $\boldsymbol{n} = (n_1, n_2, n_3)$, is positive into the water.

(b) Show by using the Gauss theorem that

$$F_4^{exc} = \rho g \left(\int_Q \bar{y}\, \mathrm{d}Q + \int_Q \bar{z} \frac{\partial \zeta}{\partial y} \mathrm{d}Q \right),$$

where Q is the liquid domain bounded by $S + \Sigma$.

(c) Explain why it is only the free-surface elevation part $\zeta_1 = -\zeta_a \cos\sigma t \sin ky$ that gives a nonzero roll moment.

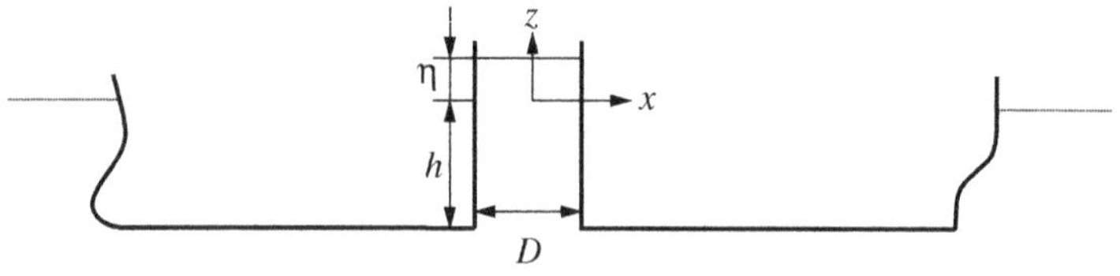

Figure 3.41. Moonpool dimensions.

(d) Show that the linear roll excitation moment can be approximated as

$$F_4^{exc} = \rho g \left(\int_L dx \int_{-\frac{1}{2}B}^{\frac{1}{2}B} y^2 dy + \int_{Q_0} \bar{z} \, dQ \right) \left. \frac{\partial \zeta_1}{\partial y} \right|_{y=0}, \quad (3.97)$$

where B is the cross-sectional beam, L is the barge length, and Q_0 is the volume inside the body below $z = 0$. (*Hint*: Derive eq. (3.97) in a similar way to the analysis of the tank roll moment in Section 3.6.1.)

(e) Consider a barge that has a length $L = 164$ m, beam $B = 32$ m, and a draft of 15 m. The submerged cross-section is rectangular and constant along the ship. Use the fact that the transverse metacentric height $\overline{GM_T}$ is 2.7 m to determine the z-coordinate of COG, z_G. Check the validity of eq. (3.97) as a function of λ/B by using correct linear theory for the Froude–Kriloff roll moment. (*Hint*: Use the expression for the velocity potential given in Table 3.1 by replacing x with y. Calculate the hydrodynamic pressure on the mean submerged hull surface and study the resulting roll moment about COG.)

(f) Introduce a rectangular free-surface antirolling tank on the barge that was defined in part (e). The breadth is $B_t = 28$ m. Make a choice of the vertical tank position in the vicinity of z_G. The undamped and uncoupled natural period in roll without antirolling tank is $T_n^{\text{roll}} = 13.7$ s. Use this fact together with other given information to determine $I_{44} + A_{44}$. Determine the tank length L_t and the water depth h in the tank by following the guidance in Section 3.6.2.

(g) Find the effect of the antirolling tank at resonant conditions as a function of incident wave amplitude within realistic values of wave steepness (see Figure 3.3). The roll damping ratio ξ can be set equal to 0.05 in the example. The roll moment due to sloshing in the tank (relative to the bottom center), $F_4^{\text{bottom}}(t) = F_{4a} \sin(\sigma t - \phi - \psi_{F4})$, should be based on Verhagen and van Wijngaarden's (1965) nonlinear shallow-water theory. An extensive description is given in Section 8.8.2. Use Verhagen and van Wijngaarden's formula (8.137) for the roll moment (corrected by Journee, 2000, and rederived in Section 8.8.2). Pay attention to Ω defined by eq. (8.118) and the limitations of Verhagen and van Wijngaarden's theory, which is only applicable when inequality (8.122) is satisfied.

3.11.6 Operational analysis of patrol boat with U-tube tank

(a) We further analyze the example with the patrol boat from Section 3.6.4.3. Short-term sea states that can be described by the JONSWAP spectrum (see eq. (3.12)) are considered and long-crested beam-sea waves are assumed. The operational limits to be used for the vessel are that the root mean square value (standard deviation) of roll should not be larger than 4° and the probability of impact against the tank roof should be less than or equal to 0.01. Set the vertical distance between the mean free surface level and the tank roof equal to the reservoir liquid height h_r. You should also use the modified long-wave results for the wave excitation moment for shorter wavelengths even though the procedure is questionable. Calculate the maximum operating significant wave height $H_{1/3}$ as a function of the mean wave period T_2.

(b) Use the scatter diagram in Table 3.6 to assess the average operational time in beam sea.

3.11.7 Moonpool and gap resonances

(a) Consider a moonpool as shown in Figure 3.41. The figure represents a longitudinal cross-section of the ship. The moonpool is assumed to have a constant horizontal circular cross-section with diameter D. We assume that the water motion does not vary across the moonpool. Therefore, a constant vertical velocity $d\eta/dt$ exists in the moonpool, where η is the free-surface elevation in the moonpool. Explain the following equation by means of linear potential flow theory and Bernoullis equation:

$$\frac{d^2\eta}{dt^2} + \frac{g}{h}\eta = -\frac{1}{h}\frac{\partial\varphi}{\partial t}\bigg|_{z=-h}.$$

Express the natural frequency for the moonpool motion.

(b) Consider a moonpool with a rectangular cross-section. What are the requirements for the draft-to-moonpool-width ratio and the moonpool length-to-width ratio for a one-dimensional analysis as in part (a) to have an error of 0.01 relative to Molin's formula for piston-mode resonance frequency?

(c) Confirm the theoretical calculations given in Table 3.5 for the natural frequencies in the gap between two side-by-side boxes.

4 Linear Natural Sloshing Modes

4.1 Introduction

This chapter describes how to estimate linear natural sloshing frequencies and modes without using computational fluid dynamics (CFD) methods. The lowest natural frequencies tell us when severe sloshing is expected. Chapter 1 presented a variety of tanks, including rectangular, prismatic (Figures 1.8 and 1.19), tapered (Figures 1.7 and 1.17), spherical (Figures 1.17 and 1.21), and horizontal (Figure 1.31) and vertical cylindrical tanks (Figure 1.39). Horizontal cylindrical tanks may have a circular or a bilobe (Figure 1.16) cross-section. Vertical cylindrical tanks include annular and sectored tanks (Figure 1.30). There exist only a few analytical solutions. Examples are:

- one-dimensional flow in a U-tube,
- two- and three-dimensional rectangular tanks,
- vertical circular cylindrical tanks,
- annular and sectored upright circular tanks, and
- wedge cross-sections with semi-apex angles 45° and 60°.

The U-tube was analyzed in Section 3.6.4. The other solutions are derived in the main text by assuming irrotational flow of an incompressible liquid. Linear free-surface conditions without surface tension are used.

Making a *shallow-liquid* approximation opens new possibilities for deriving analytical solutions. This approach is relevant for seiching in harbors, lakes, and model tank facilities. Dominant longitudinal resonant waves in horizontal cylinders with a high length-to-breadth ratio can also be described by shallow-liquid theory. Applications are for tanks used for land transportation, oil–gas separators (Figure 1.31), and the liquefied petroleum gas tanks illustrated in Figures 1.15 and 1.16. Analytical shallow-liquid solutions are presented for

- two-dimensional flow in parabolic and triangular-shaped basins,
- pumping-mode resonance of a harbor consisting of a basin connected to the open sea by a narrow channel with a constant cross-section, and
- horizontal cylinders with constant cross-section.

The horizontal-cylinder case includes a harbor with an opening to the open sea and two-phase flow involving two liquids (e.g., water and oil) with a gas above.

If the tank domain can be divided into subdomains where an analytical solution form is possible in each subdomain, analytical solutions can be constructed by *domain decomposition*. The solutions in the different subdomains are combined by requiring that both the velocity potential and the normal velocity at the boundary are continuous between the subdomains. The procedure is illustrated in the main text for two-dimensional sloshing in a tank with a shallow-water component. The case is relevant for a swimming pool on a cruise vessel with a wading region along the rim of the pool. Two rectangular subdomains are used in this case.

How can we, in as simple a way as possible, estimate the natural frequencies when no analytical solution exists? First of all, we can use benchmark numerical results, which are presented in this chapter for two-dimensional circular, spherical, and ellipsoidal tanks. Another possibility is to use theoretically derived upper and lower bounds of natural frequencies for a general tank shape following from variational formulations. However, the difference between the upper and lower bounds may be too large in some cases to provide a satisfactory estimate of the natural frequencies. We can also apply variational formulations directly by using test functions that only need to satisfy liquid volume conservation for the lowest eigenmode and, in addition, orthogonality properties for the higher modes. We illustrate this technique for sloshing in two-dimensional circular tanks and spherical tanks. Good agreement with benchmark numerical results is demonstrated. The test functions used are infinite-fluid horizontal dipoles with singularities in the tank's centerplane above the mean free surface. How to choose good test functions is not always obvious.

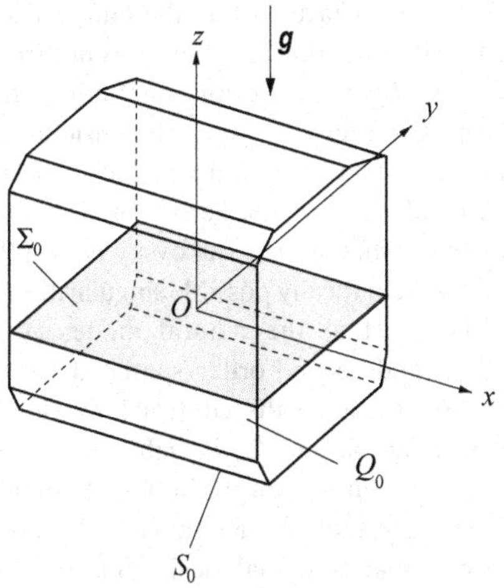

Figure 4.1. Coordinate system, mean liquid domain Q_0, mean free surface Σ_0, and mean wetted tank surface S_0.

The test functions in the aforementioned cases were constructed based on experimental observations of liquid path lines.

One possibility for studying some tank shapes is to use *analytical continuation* of the velocity potential into a domain so that analytical natural modes exist in the sum of the liquid domain and the analytically continued domain. Asymptotic formulas of the natural frequencies in terms of the analytical natural frequencies and modes can then be derived. This procedure is used to estimate the effect of lower chamfers (Figure 1.17) and inclination of the tank bottom in terms of the natural frequencies for a rectangular tank.

Analytical continuation cannot be used to derive asymptotic formulas for natural frequencies in the case of interior structures such as baffles, pump towers, and screens. However, if the interior structures do not have a net source/sink effect on the flow and the cross dimensions of the interior structure are small relative to both the main tank dimensions and the wavelength of the considered sloshing mode, asymptotic formulas can be expressed for the natural frequencies in terms of the added mass coefficients and displaced mass of the interior structure. A similar procedure is described for slender structures with small transverse cross-sections.

4.2 Natural frequencies and modes

The *natural (eigen-)frequencies* and corresponding *natural (eigen-)modes* are nontrivial solutions for zero tank excitation. Liquid mass conservation is a constraint. Figure 4.1 defines the inertial Cartesian coordinate system $Oxyz$ to which we refer. The origin is in the geometrical center of the mean free surface Σ_0. The z-axis is vertical upward; Q_0 and S_0 are the tank volume and the tank surface below Σ_0, respectively. The formulation of the boundary-value problem for the velocity potential of an incompressible liquid in a tank is given in Section 2.4.2. Our focus is on linearized conditions with no tank excitation. The linearized free-surface conditions are given by eqs. (2.71)–(2.73). No tank excitation means that the velocity potential ϕ for the flow in the tank satisfies $\partial\phi/\partial n = 0$ on S_0, where $\boldsymbol{n}$ is the outer normal vector to S_0.

The fact that only the spatial derivatives of ϕ are involved in the boundary condition means we have a *Neumann condition*. (In contrast, a *Dirichlet condition* means that only the value of ϕ is involved in the boundary condition.) We must also require that the liquid volume is conserved, which can be expressed as $\int_{\Sigma_0} \zeta \, dx\, dy = 0$, where $z = \zeta(x, y, t)$ determines the free-surface elevation. Because we are looking for time-periodic solutions with circular frequency σ, we express the solution as

$$\zeta(x, y, t) = \underbrace{f(x, y)}_{\varphi(x,y,0)} \exp(\mathrm{i}\sigma t);$$
$$\phi(x, y, z, t) = \frac{\mathrm{i}g}{\sigma}\varphi(x, y, z)\exp(\mathrm{i}\sigma t), \quad \mathrm{i}^2 = -1. \tag{4.1}$$

We operate with real expressions of f and φ without any loss of generality.

Substituting eq. (4.1) into the linear sloshing problem and combining the kinematic and dynamic boundary conditions as described in Section 2.4.2.3 give the spectral boundary problem

$$\nabla^2\varphi = 0 \quad \text{in} \quad Q_0; \quad \frac{\partial\varphi}{\partial n} = 0 \quad \text{on} \quad S_0;$$
$$\frac{\partial\varphi}{\partial z} = \kappa\varphi \quad \text{on} \quad \Sigma_0 \left(\kappa = \frac{\sigma^2}{g}\right); \quad \int_{\Sigma_0} \varphi \, dx\, dy = 0 \tag{4.2}$$

with the spectral parameter κ in the boundary condition on Σ_0. Our goal is to find nontrivial solutions of homogeneous problem (4.2) with corresponding values of κ (the spectrum)

(Courant & Hilbert, 1953). The last integral condition formally follows from the volume conservation. It guarantees that the spectral problem does not have nontrivial solution $\varphi = f = C \neq 0$, $\kappa = 0$. The integral condition can be omitted when we consider $\kappa > 0$. However, the condition is mathematically needed in forthcoming variational formulations.

Spectral problems with the spectral parameter in parts of the boundary conditions have been studied by a number of mathematicians (see Eastham, 1962; Komarenko *et al.*, 1965; Feschenko *et al.*, 1969; Kopachevski & Krein, 2001; and references in therein). *Spectral theorems* related to eqs. (4.2) are collected in the books by Feschenko *et al.* (1969) and Morand and Ohayon (1995). These theorems show that problem (4.2) has only positive eigenvalues that satisfy

$$0 < \kappa_1 \leq \kappa_2 \leq \cdots \leq \kappa_n \leq \cdots, \quad \kappa_n \to \infty. \tag{4.3}$$

Condition (4.3) may seem trivial. However, it is not straightforward to prove mathematically. The equalities in condition (4.3) are due to the fact that multiple eigenfunctions with the same eigenvalue are possible. Each eigenvalue corresponds to only a finite number of eigenfunctions. If we consider a rectangular tank with a square base, for some natural frequencies there exist two modes corresponding to planar motion in two perpendicular directions parallel to the tank walls.

An infinite number of natural frequencies and modes exist, according to potential flow theory. If viscosity is considered, the viscous solutions have the form $\mathrm{Re}[\exp(\pm \mathrm{i}\sigma_n t - \xi_n \sigma_n t)\varphi_n(x, y, z)]$ $(n = 1, 2, \dots)$, where both the damping factor ξ_n and the "viscous" natural frequency σ_n are real positive numbers. If ξ_n is small, ξ_n is the damping ratio (see Chapter 6). Properties of linear viscous sloshing in nonmoving tanks without interior structures were established by Krein (1964; see also Krein & Langer, 1978a, 1978b). An extended description is found in the book by Kopachevsky and Krein (2003). There exists only a finite (in practice large) number of natural modes with oscillatory behavior (i.e., with nonzero σ_n). The remaining higher modes decay after initial perturbations. The viscous influence on the lower modes and frequencies is negligible for viscosity coefficients of practical interest.

The natural modes φ_n can have singular derivatives at the intersection (contact) curve between the mean free surface Σ_0 and the tank walls due to the fact that the boundary of Q_0 is not smooth at the curve (i.e., an interior angle not equal to 180°; e.g., a corner of a three-dimensional rectangular tank) and the conflict between the free-surface condition and the zero-Neumann condition on the tank surface. Lukovsky *et al.* (1984) studied mathematically possible singularities. The study showed that the natural modes φ_n and their first- and second-order spatial derivatives are continuous along the contact line when the angle between S_0 and Σ_0 is right and the wall has zero curvature at the vicinity of the contact angle. These conditions are satisfied for upright cylindrical tanks (see analytical solutions in Section 4.3). The first-order spatial derivatives of φ_n become singular for the interior angle between the mean free surface and the walls that is larger than 90°. When the interior angle is less than 90°, the first-order spatial derivatives are continuous at the contact curve, but the second-order spatial derivatives of φ_n may not be.

Spectral problem (4.2) does not account for surface tension and associated meniscus at the contact line between the free surface and the tank walls. The surface tension has a negligible effect on the lower natural frequencies when the Bond number, $Bo = gl^2\rho/T_s$, is large (Myskis *et al.*, 1987), where as usual ρ is the density of the liquid and T_s is the surface tension. Furthermore, l is the tank breadth for two-dimensional liquid motion and the larger of the tank length and the tank breadth for three-dimensional flow. A representative value of T_s for the air–water interface is equal to 0.073 N m^{-1}. We saw in Exercise 3.11.2 that the wavelength should be less than ≈0.05 m for surface tension to matter for linear propagating waves. If we transfer this information to a two-dimensional rectangular tank, the lowest mode has a wavelength that is twice the tank breadth l, implying that the effect of surface tension must be considered for the lowest mode when $l < 0.025$ m. This effect corresponds to Bond numbers less than 84 when we use $\rho = 1{,}000$ kg m^{-3} and $T_s = 0.073$ N m^{-1}. However, when considering sloshing in microgravity conditions, Myskis *et al.* (1987) recognized that Bond numbers around 100 still matter for sloshing with contact angles between the free surface and the vertical walls differing from 90°. The reason is the meniscus, which makes the static free surface far from planar at the walls. According to Myskis *et al.* (1987), the size of the meniscus becomes negligible and the related capillary–gravitational sloshing behaves as pure gravitational waves for $Bo \gtrsim 10^4$. The latter conditions are relevant for ship tanks. For instance, if $\rho = 1{,}000\,\mathrm{kg\,m^{-3}}$, $T_s = 0.073\,\mathrm{N\,m^{-1}}$, and $l = 30\,\mathrm{m}$, we get $Bo = 1.2 \cdot 10^8$, indicating that surface tension does not matter for ship tanks. An exception may be in describing high-curvature parts of breaking waves, particularly in model scale. Furthermore, surface tension may matter in describing bubbles in violent sloshing with gas–liquid mixing. Similar discussions showing that surface tension has secondary importance can be made for our other marine applications, tuned liquid dampers, and for tanks used in land transportation. However, surface tension has a dominant effect for fuel tanks on spacecraft in weightless conditions.

The eigenfunctions $\varphi_n(x, y, z)$ are called the linear *natural sloshing modes*. The *natural sloshing periods and frequencies* are given in terms of the eigenvalues $\{\kappa_n\}$ as

$$T_n = 2\pi/\sqrt{g\kappa_n}, \quad \sigma_n = \sqrt{g\kappa_n}. \tag{4.4}$$

The freestanding wave patterns are associated with the natural surface profiles

$$f_n(x, y) = \varphi_n(x, y, 0) = \frac{1}{\kappa_n}\frac{\partial \varphi_n}{\partial z}(x, y, 0). \tag{4.5}$$

These profiles satisfy the volume conservation conditions and are orthogonal to each other; that is,

$$\int_{\Sigma_0} f_i f_j \,\mathrm{d}x\,\mathrm{d}y = \int_{\Sigma_0} \varphi_i(x, y, 0)\,\varphi_j(x, y, 0)\,\mathrm{d}x\,\mathrm{d}y = 0 \quad \text{as} \quad i \neq j. \tag{4.6}$$

Condition (4.6) follows from the mentioned spectral theorems. Moreover, the functions $\{f_i\}$ represent a basis on the mean free surface (i.e., any function $f(x, y)$ that satisfies the volume conservation condition can be represented by a series in $\{f_i\}$). We demonstrate these facts by considering the exact solutions.

4.3 Exact natural frequencies and modes

A few exact solutions exist for natural modes and frequencies. These solutions are, for instance, possible when spectral problem (4.2) can be solved by using separation of spatial variables. The cases include two- and three-dimensional rectangular tanks and upright circular cylindrical tanks.

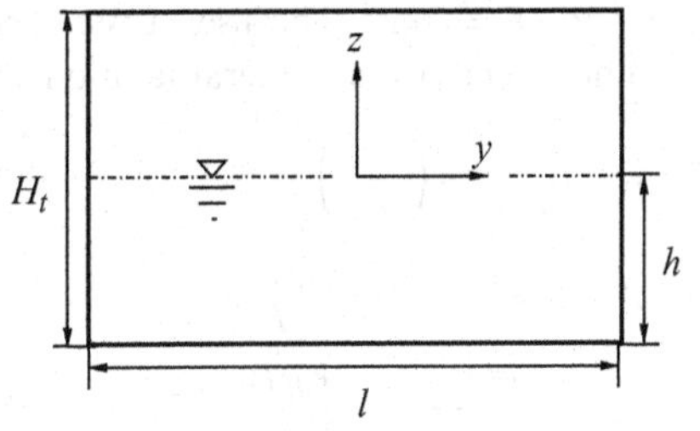

Figure 4.2. Mean liquid shape and notations used for a two-dimensional rectangular tank.

4.3.1 Two-dimensional case

The simplest two-dimensional case with exact analytical natural modes and frequencies is for sloshing in a planar rectangular tank. Review of other possible solutions than those mentioned in the following subsections may be found in Fox and Kuttler (1983) and, in part, Ibrahim (2005).

4.3.1.1 Rectangular planar tank

Problem (4.2) takes the following form for the planar rectangular tank shown in Figure 4.2:

$$\left.\begin{aligned}
&\frac{\partial^2\varphi}{\partial y^2} + \frac{\partial^2\varphi}{\partial z^2} = 0 && \text{in the liquid domain } Q_0,\\
&-\kappa\varphi + \frac{\partial\varphi}{\partial z} = 0 && \text{on } z = 0 \text{ for } |y| \le \tfrac{1}{2}l,\\
&\frac{\partial\varphi}{\partial y} = 0 && \text{on } y = \pm\tfrac{1}{2}l \text{ for } -h \le z \le 0,\\
&\frac{\partial\varphi}{\partial z} = 0 && \text{on } z = -h \text{ for } |y| \le \tfrac{1}{2}l,\\
&\int_{-\frac{1}{2}l}^{\frac{1}{2}l} \varphi(y, 0)dy = 0.
\end{aligned}\right\} \tag{4.7}$$

The z-axis is positive upward and the z-coordinate of the tank bottom is $z = -h$ (where h is the mean liquid depth). The y-coordinates of the vertical tank walls are $y = \pm\frac{1}{2}l$. The mathematical problem does not depend on the density of the liquid, which enters the problem when hydrodynamic pressures, forces, and moments are considered.

The solutions of spectral problem (4.7) can be found by using separation of the two spatial variables y and z; that is, we express $\varphi = Y(y)Z(z)$. The Laplace equation then yields the equality

$Y''/Y = -Z''/Z = \text{const}$. The use of the boundary conditions leads to the general solution

$$\kappa_i = \frac{\pi i}{l}\tanh\left(\frac{\pi i}{l}h\right);$$
$$\varphi_i(y, z) = \cos\left(\frac{\pi i}{l}\left(y + \tfrac{1}{2}l\right)\right) \times \frac{\cosh(\pi i(z+h)/l)}{\cosh(\pi i h/l)}, \quad i \geq 1. \tag{4.8}$$

Eigenfunctions (4.8) are found to within a nonzero multiplier, meaning that $C\varphi_i$ with $C = \text{const} \neq 0$ is also an eigenfunction with the same eigenvalue κ_i. From eq. (4.5), the wave patterns of the freestanding waves are given by

$$f_i(y) = \varphi_i(y, 0) = \cos(\pi i\left(y + \tfrac{1}{2}l\right)/l), \quad i \geq 1. \tag{4.9}$$

Using orthogonality condition (4.6), one can check that eigenfunctions (4.9) represent an orthogonal basis on the waterplane; that is,

$$\int_{-\frac{1}{2}l}^{\frac{1}{2}l} f_i(y) f_j(y)\,\mathrm{d}y = \frac{1}{2}l\delta_{ij}, \quad \delta_{ij} = \begin{cases} 0 & \text{as } i \neq j, \\ 1 & \text{as } i = j, \end{cases} \tag{4.10}$$

where δ_{ij} is the Kronecker delta.

The velocity field associated with freestanding waves following from eq. (4.8) decays from the mean free surface to the tank bottom. Nodal and antinodal vertical lines pass through the liquid volume (see Figure 4.3). A liquid particle moves only horizontally at a nodal line, whereas the motions are only vertical at an antinodal line. The lowest natural mode ($i = 1$) has a node in the middle of the tank and antinodal lines coinciding with the vertical walls. The number of nodal lines is equal to the mode number, i. The nodes divide the interval $[-\frac{1}{2}l, \frac{1}{2}l]$ into $i+1$ subintervals (see Figure 4.3). Antinodes that are not at the vertical walls appear only from the second mode and divide the interval into i subintervals. Figure 4.3 illustrates that the motion of a liquid particle associated with small-amplitude freestanding waves by a natural mode is rectilinear, which differs from linear harmonic long-crested propagating waves. In Section 3.2, we commented that a first approximation of the path of a fluid particle is elliptical in finite liquid depth, circular in infinite depth, and rectilinear in shallow depth.

The natural modes are categorized as either odd (antisymmetric) or even (symmetric). Antisymmetry and symmetry refer to properties of $f_i(y)$ relative to the z-axis. The lowest linear mode $f_1(y)$ has primary importance for ship tank applications. This mode is an antisymmetric standing wave with wavelength twice the tank length and largest wave elevations at the tank walls.

The exact solution given by eqs. (4.8) and (4.9) is consistent with eq. (4.3) that the spectral values κ_i ($i \geq 1$) are positive. Each eigenvalue κ_i corresponds in this case to a single eigenfunction φ_i. The natural sloshing frequencies and periods are

$$\sigma_i = \sqrt{g\frac{\pi i}{l}\tanh\left(\frac{\pi i}{l}h\right)};$$
$$T_i = \frac{2\pi}{\sqrt{g\pi i\tanh(\pi i h/l)/l}}, \quad i = 1, 2, \ldots. \tag{4.11}$$

Figure 1.2 illustrates how the highest natural period of the first sloshing mode for a rectangular tank, T_1, depends on the tank breadth l and the liquid depth h. As shown in later chapters, sloshing behavior, especially in nonlinear descriptions, depends on the depth-to-breadth ratio h/l.

The depth-to-breadth ratio appears in the hyperbolic tangent (tanh) term of expression (4.11) for the natural frequencies. Because $\tanh(\pi i h/l)$ behaves in different ways for small and large values of h/l, one can introduce *shallow-*, *intermediate-*, *finite-*, and *deep-*liquid approximations. As long as

$$\tanh(\pi i h/l) \approx \pi i h/l \tag{4.12}$$

for a reasonable subset of the lower natural modes, we are speaking about shallow-liquid sloshing; therefore, in practice, $h/l \lesssim 0.05$–0.1. Increasing h/l makes eq. (4.12) applicable only for a few of the lowest natural modes (usually 1–5), which corresponds to the so-called intermediate depths, meaning in practice that 0.05–$0.1 \lesssim h/l \lesssim 0.2$–$0.25$. Furthermore, finite depth is a depth larger than the intermediate depth and corresponds to when $\tanh(\pi h/l)$ is not close to 1. The latter means that at least the lowest natural frequency, σ_1, is sensitive to the ratio h/l. Finally, the deep-water approximation implies relatively large h/l so that $\tanh(\pi h/l) \approx 1$.

In the shallow-liquid limit, when h/l goes to zero, eq. (4.11) shows that

$$\sigma_n \approx n\sigma_1, \quad n = 2, 3, \ldots. \tag{4.13}$$

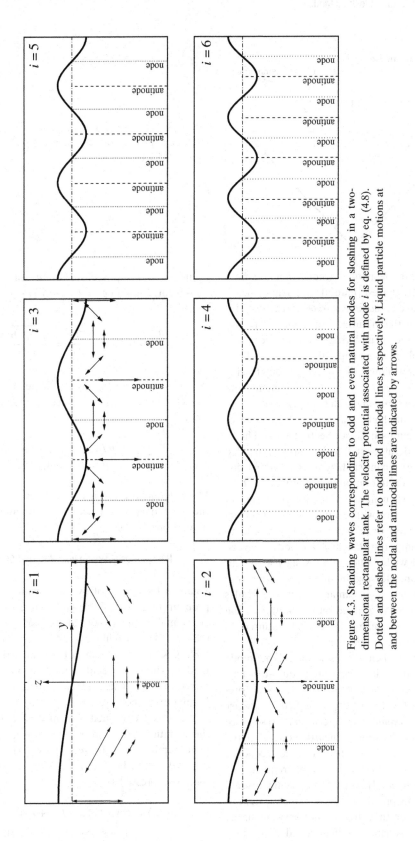

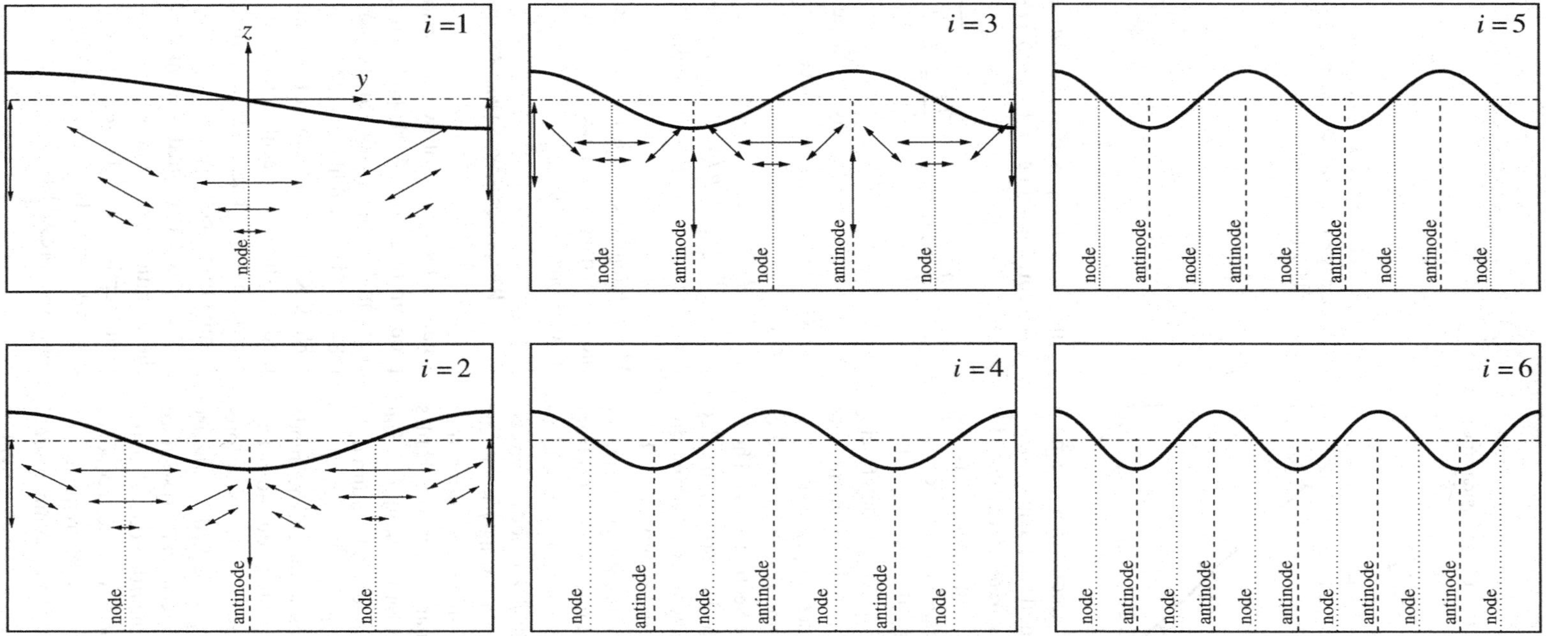

Figure 4.3. Standing waves corresponding to odd and even natural modes for sloshing in a two-dimensional rectangular tank. The velocity potential associated with mode i is defined by eq. (4.8). Dotted and dashed lines refer to nodal and antinodal lines, respectively. Liquid particle motions at and between the nodal and antinodal lines are indicated by arrows.

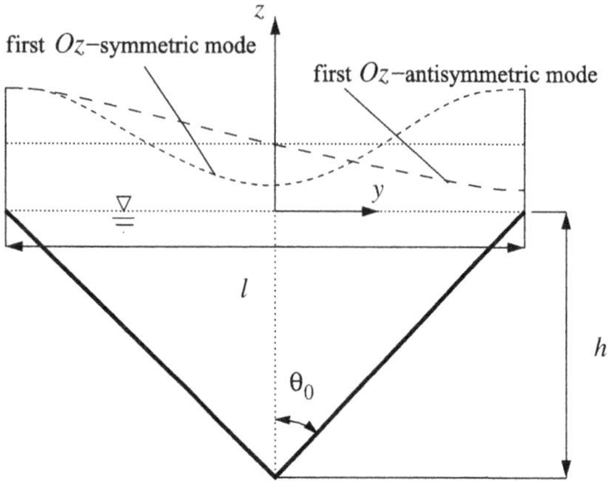

Figure 4.4. Sloshing in a wedge with interior semi-apex angle θ_0. Analytical solutions exist for either $\theta_0 = 45^\circ$ or $\theta_0 = 60^\circ$.

The fact that the natural frequencies are multiples of the lowest natural frequency means that we have a *commensurate spectrum*. This property of the natural frequencies plays an important role in understanding nonlinear surface wave phenomena under the condition of resonant shallow liquid depth. When the forcing frequency $\sigma = 2\pi/T \approx \sigma_1$, the commensurate spectrum and nonlinearities imply that liquid oscillations with higher harmonics $n\sigma$ excite resonance oscillations at σ_n (see Section 8.3.4). Therefore, a considerable set of higher modes is needed to describe resonant shallow-liquid sloshing due to harmonic excitation in the vicinity of the lowest natural frequency. Progressive activation of the higher modes with commensurate spectrum generates steep wave profiles and may lead to hydraulic jumps traveling back and forth between the vertical walls, which is a strongly nonlinear phenomenon. Bearing in mind a Fourier representation of the instantaneous wave profiles for shallow sloshing based on eq. (4.9), one can understand that many modes are required to approximate a hydraulic jump.

The physical phenomenon when a higher mode is *nonlinearly* excited (activated) due to a harmonic excitation of the lowest mode with forcing period $T = 2\pi/\sigma \approx T_1$, is called *secondary (internal) resonance*. The secondary resonance causes an amplification of the nth mode when the nonlinearity causes a superharmonic frequency $n\sigma$ to be close to the nth natural frequency σ_n (i.e., for $n\sigma \approx \sigma_n$). Because of the commensurate spectrum, this process occurs for shallow depths. However, even though h/l is not very small, the secondary resonance of the second mode may occur when T/T_1 is close to the number $i_2 = i_2(h/l)$ corresponding to the ratio $2T_2/T_1$. Therefore,

$$\frac{\sigma_1}{\sigma} = \frac{T}{T_1} \approx i_2(h/l) = \frac{2T_2}{T_1} = \sqrt{\frac{2\tanh(\pi h/l)}{\tanh(2\pi h/l)}}. \tag{4.14}$$

Proceeding in similar way for higher modes, we expect the secondary resonance for the nth mode when

$$\frac{\sigma_1}{\sigma} = \frac{T}{T_1} \approx i_n = \frac{nT_n}{T_1} = \sqrt{\frac{n\tanh(\pi h/l)}{\tanh(n\pi h/l)}}. \tag{4.15}$$

More details are given in Chapters 8 and 9 on the secondary resonance for two- and three-dimensional tanks.

4.3.1.2 Wedge cross-section with 45° and 60° semi-apex angles

Analytical solutions of two-dimensional spectral problem (4.2) for a V-shaped wedge with a 90° apex angle (Figure 4.4) are given by Kelland (1840, 1844), Kirchhoff (1879), and Greenhill (1887). Kelland considered Oz-symmetric eigenmodes, whereas Greenhill focused on antisymmetric modes. These results are discussed by Lamb (1945, Section 258).

We consider inclined straight walls defined by the equation $z = \pm\tan(90^\circ - \theta_0)y - h$, where θ_0 is the semi-apex angle (see Figure 4.4). Our first case is $\theta_0 = 45^\circ$. The *symmetric natural modes* can then be expressed as

$$\varphi(y, z) = \cosh(ky)\cos(k(z+h)) + \cos(ky)\cosh(k(z+h)), \tag{4.16}$$

where k is an unknown constant treated as a wave number. This solution satisfies the Laplace equation and the zero-Neumann boundary conditions on straight inclined walls. The boundary condition at $z = 0$ gives

$$\kappa(\cosh(ky)\cos(kh) + \cos(ky)\cosh(kh)) \\ = k(-\cosh(ky)\sin(kh) + \cos(ky)\sinh(kh)),$$

which has a solution only if

$$\kappa = -\frac{k\sin(kh)}{\cos(kh)} \quad \text{and} \quad \kappa = \frac{k\sinh(kh)}{\cosh(kh)}$$

are simultaneously fulfilled. This solution requires that k is a root of the transcendental equation

$$\tanh(kh) = -\tan(kh), \qquad (4.17)$$

which has an infinite number of roots, including $k = 0$. However, inserting $k = 0$ into eq. (4.16) leads to the solution $\kappa = 0, \varphi = \text{const}$, which does not satisfy the volume conservation condition unless $\varphi = 0$. The latter trivial solution is not an eigenfunction. Furthermore, positive and negative roots of eq. (4.17) have the same modulus and yield the same eigenfunctions (4.16). Therefore, one can concentrate only on positive $k_i > 0 (i \geq 1)$. Sorting in ascending order gives

$$\sigma_i^2/g = \kappa_i = k_i \tanh(k_i h). \qquad (4.18)$$

The first root of eq. (4.17) is $k_1 h = 2.365\ldots$ and, therefore,

$$\sigma_1/\sqrt{g} = 1.5243\ldots/\sqrt{h}. \qquad (4.19)$$

The *antisymmetric modes* may be expressed as

$$\varphi(y, z) = -\sinh(ky)\sin[k(z+h)] \\ - \sin(ky)\sinh[k(z+h)]. \qquad (4.20)$$

Function (4.20) satisfies the Laplace equation and the zero-Neumann condition on the walls. The spectral boundary condition at $z = 0$, as earlier, gives the necessary solvability conditions

$$\kappa = \frac{k\cos(kh)}{\sin(kh)} \quad \text{and} \quad \kappa = \frac{k\cosh(kh)}{\sinh(kh)}, \qquad (4.21)$$

which gives the following transcendental equation for the wave number k:

$$\tanh(kh) = \tan(kh).$$

As for the symmetric modes, positive and negative k have the same modulus and lead to the same eigenfunctions (4.20) (to within the sign).

The case $k = 0$ does not lead to a trivial solution. The limit $k \to 0$ in eq. (4.21) gives

$$\kappa = 1/h. \qquad (4.22)$$

Eigenvalue (4.22) corresponds to the lowest antisymmetric natural mode. The lowest antisymmetric mode can be expressed by a limiting process of eq. (4.20) as

$$\lim_{k \to 0} k^{-2}\varphi(y, z) = -2y(z+h). \qquad (4.23)$$

Greenhill (1887) found eigenvalues and eigenfunctions corresponding to the Oz-symmetric modes of the *triangle making the* $90 - \theta_0 = 30^\circ$ *angle with the mean free surface.* Macdonald (1894, 1896) also studied this problem. An error in Greenhill's results was corrected by Haberman *et al.* (1974) and Packham (1980). A brief discussion also appears in Wehausen and Laitone (1960). The first Oz-symmetric mode is expressed as

$$\varphi_2(y, z) = 1 + (z+h)\left\{[(z+h)/h]^2 \\ - 3(y/h)^2\right\}/(2h), \qquad (4.24)$$

with $\kappa_2 = 2\sqrt{3}/l$ (see Figure 4.4).

The enumeration for the full set of eigenvalues in ascending order associates the first symmetric mode (4.24) with the global enumeration number $m = 2$, and the higher symmetric eigenfunctions are labeled with $m = 2, 4, 6, \ldots$. (Haberman *et al.*, 1974). A similar enumeration was used for the rectangular tank (see Section 4.3.1.1). Haberman *et al.* (1974) distinguished the eigenfunctions that correspond to $m = 6, 10, 14, \ldots$ (φ_{2+4j}, $\kappa_{2+4j}, j = 1, 2, \ldots$) from those with $m = 4, 8, \ldots$ (φ_{4j}, $\kappa_{4j}, j = 1, 2, \ldots$). The first subclass of the symmetric eigenfunctions is

$$\begin{aligned}\varphi(y, z) = {} & \sinh[\alpha_1((z+h)/h - \alpha_2)]\cos[\alpha_1 y/h] \\ & - \sinh\left[\alpha_1((\sqrt{3}y + z + h)/(2h) + \alpha_2)\right] \\ & \times \cos\left[\alpha_1(y - \sqrt{3}(z+h))/(2h)\right] \\ & + \sinh\left[\alpha_1((\sqrt{3}y - z - h)/(2h) - \alpha_2)\right] \\ & \times \cos\left[\alpha_1(y + \sqrt{3}(z+h))/(2h)\right]\end{aligned}$$

with eigenvalues

$$\kappa = 3\alpha_1\left[\cot\left(\sqrt{3}\alpha_1\right) + \sqrt{\tfrac{4}{3} + \cot^2\left(\sqrt{3}\alpha_1\right)}\right], \qquad (4.25)$$

Table 4.1. *Eigenvalues $\kappa = \sigma^2/g$ and other parameters for a triangular channel with $\theta_0 = 60^\circ$*

m	$\sigma^2 l/(2g) = \frac{1}{2} l\kappa$	α_1	α_2
2	1.7321	—	—
4	4.7086	2.7214	−0.3911
6	7.8540	4.5345	−0.4394
8	10.9956	6.3483	−0.4567
$m > 8$	$\approx \dfrac{\pi(m-1)}{2}$	$\approx \dfrac{\pi(m-1)}{2\sqrt{3}}$	$\approx -\frac{1}{2} + \sqrt{3}\ln\left(\dfrac{\sqrt{3}}{(m-1)\pi}\right)$

Source: Haberman *et al.* (1974).

where constants α_1 and α_2 are roots of the transcendental equations

$$\cosh 3\alpha_1 = -\cos(\sqrt{3}\alpha_1) + 2\sec(\sqrt{3}\alpha_1), \quad (4.26)$$

and

$$\sinh \alpha_1(1 + 2\alpha_2) = \sin(\sqrt{3}\alpha_1)/\sqrt{3}. \quad (4.27)$$

The second subclass of symmetric natural modes was found by Habermann *et al.* (1974):

$$\begin{aligned}\varphi(y, z) = {} & \cosh[\alpha_1((z+h)/h - \alpha_2)]\cos[\alpha_1 y/h] \\ & + \cosh\left[\alpha_1((\sqrt{3}y + z + h)/(2h) + \alpha_2)\right] \\ & \times \cos\left[\alpha_1(y - \sqrt{3}(z+h))/(2h)\right] \\ & + \cosh\left[\alpha_1((\sqrt{3}y - z - h)/(2h) - \alpha_2)\right] \\ & \times \cos\left[\alpha_1(y + \sqrt{3}(z+h))/(2h)\right].\end{aligned}$$

The corresponding eigenvalues and constants α_1 are calculated in eqs. (4.25) and (4.26). However, α_2 is a root of the equation $\sinh \alpha_1(1 + 2\alpha_2) = -\sin(\sqrt{3}\alpha_1)/\sqrt{3}$ instead of eq. (4.27). Calculated eigenvalues for symmetric modes are given in Table 4.1.

To the authors' knowledge, no exact solutions are available for the antisymmetric modes in a triangular channel with $\theta_0 = 60^\circ$, only approximate solutions. For instance, the lowest asymmetric mode was computed by Habermann *et al.* (1974) by a variational method to be $\frac{1}{2} l\kappa_1 = 0.710$. This article also gave the approximate values $\frac{1}{2} l\kappa_3 = 3.190$, $\frac{1}{2} l\kappa_5 = 6.278$, and $\frac{1}{2} l\kappa_7 = 9.507$. Finally, analytical approximate solutions were also obtained by Bauer (1982) for small semi-apex angles under the condition that the mean free surface Σ_0 can be replaced by an arc (spherical segment in three-dimensional case). Bauer's expression gives reasonable numerical results for natural frequencies when $\theta_0 < 15^\circ$.

4.3.1.3 *Troesch's analytical solutions*

Troesch (1960) considered a combination of harmonic functions of polynomial type that satisfy $\partial\varphi/\partial z = \kappa\varphi$ on $z = 0$ and then integrated the equation $\partial\varphi/\partial n = 0$ to find S_0. The harmonic functions, by definition, satisfy the Laplace equation. Equation (4.23) is an example of a harmonic function with polynomial structure. Troesch found the first eigenvalue for a family of parabolas in two dimensions and paraboloids of revolution in three dimensions. Levin's (unpublished memorandum; see Troesch, 1960) solution for the cone is a limiting case. Troesch and Weidman (1972), again using the inverse method, found a container shape with a fixed first eigenvalue independent of the filling depth. Solutions using the inverse method (i.e., similarly to that described earlier) were also found by Sen (1927) and Storchi (1949, 1952).

4.3.2 Three-dimensional cases

4.3.2.1 *Rectangular tank*

A three-dimensional rectangular tank is considered. The liquid depth h and the tank length and breadth are defined in Figure 4.5. The origin of the Cartesian coordinate system $Oxyz$ is in the middle point of the mean free surface Σ_0; the z-axis is positive upward.

The analytical solution of spectral problem (4.2) can be found by separating the spatial variables x, y, and z. This approach implies that $\varphi = X(x)Y(y)Z(z)$; the Laplace equation transforms to $X''/X + Y''/Y + Z''/Z = 0$ and,

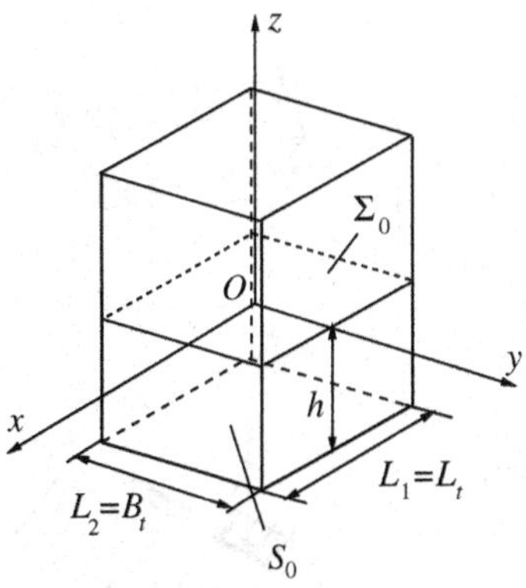

Figure 4.5. Three-dimensional rectangular tank. Geometric notations and coordinate system.

therefore, $X'' + k_x^2 X = 0$, $Y'' + k_y^2 Y = 0$, and $Z'' - (k_k^2 + k_y^2)Z = 0$. The parameters k_x^2 and k_y^2 (treated as wave numbers along the Ox- and Oy-axes, respectively) follow from the boundary conditions on the vertical walls: $X' \equiv dX/dx = 0$ at $x = \pm\frac{1}{2}L_1$ and $Y' \equiv dY/dy = 0$ at $y = \pm\frac{1}{2}L_2$ respectively. These conditions give $k_{xi} = \pi i/L_1$ and $k_{yj} = \pi j/L_2$ with nonnegative integers i and j. The functions $X(x)$ and $Y(y)$ are then associated with

$$f_i^{(1)}(x) = \cos\left[i\pi\left(x + \tfrac{1}{2}L_1\right)/L_1\right] \quad \text{and}$$
$$f_j^{(2)}(y) = \cos\left[j\pi\left(y + \tfrac{1}{2}L_2\right)/L_2\right], \quad i, j \geq 0, \tag{4.28}$$

respectively. The Z-function must satisfy the conditions $Z'(-h) \equiv dZ/dz(-h) = 0$ and $Z'(0) = \kappa_{i,j}Z(0)$, which leads to the following analytical solution:

$$\varphi_{i,j}(x, y, z) = f_i^{(1)}(x) f_j^{(2)}(y) \frac{\cosh\left[k_{i,j}(z+h)\right]}{\cosh(k_{i,j}h)} \tag{4.29}$$

with

$$\sigma_{i,j}^2/g = \kappa_{i,j} = k_{i,j}\tanh(k_{i,j}h);$$
$$k_{i,j} = \pi\sqrt{(i/L_1)^2 + (j/L_2)^2}, \quad i + j \neq 0, \tag{4.30}$$

where $k_{i,j}$ are the wave numbers of the natural modes $\varphi_{i,j}$ for $i, j \geq 1$. When $j = 0$, $k_{i,0} = k_{xi}$; that is, the wave number coincides with that for two-dimensional sloshing along the Ox-axis. Analogously, $i = 0$ gives $k_{0,j} = k_{yj}$. The case $i = j = 0$ is excluded because it corresponds to the solution $\varphi_{0,0} = \text{const}$, which does not satisfy the volume conservation condition.

The wave profiles are determined by the expressions of $\varphi_{i,j}$ on the mean free surface $z = 0$:

$$f_{i,j}(x, y) = \varphi_{i,j}(x, y, 0) = f_i^{(1)}(x) f_j^{(2)}(y). \tag{4.31}$$

Figure 4.6 illustrates the wave profiles $f_{i,j}$ for the nine lowest modes for a nearly square base tank. The wave profiles $f_{i,j}$ fall into several physical subclasses. When $ij = 0$, as we said, natural modes (4.29) correspond to two-dimensional solution (4.8), which is referred to as a *Stokes freestanding wave*. The Stokes waves in the Oxz-plane imply $j = 0$, whereas the Oyz-waves correspond to $i = 0$. The wave patterns are defined as $z = f_i^{(1)}(x)$ and $z = f_j^{(2)}(y)$, respectively. Because Stokes waves occur in either the $Oxz\,(j = 0)$ or the $Oyz\,(i = 0)$ plane, these waves are often called "planar." This terminology is the same as for a pendulum that can perform both planar (in a vertical plane) and rotary motions. An analogy of the pendulum rotary motions is the so-called swirling phenomenon, which occurs due to resonant excitation of the lowest natural frequency in a nearly square base rectangular tank. If $L_1 \approx L_2$ (nearly square base tank), $\sigma_{i,j} \approx \sigma_{j,i}$, which causes complex resonances and three-dimensional waves. Swirling (rotary wave) is the most well-known example of these three-dimensional waves and is discussed extensively in Chapter 9. An example of swirling waves is obtained by superposing the two lowest Stokes waves with a 90° phase difference in time; that is,

$$f_1^{(1)}(x)\cos\sigma_{1,0}t \pm f_1^{(2)}(y)\sin\sigma_{0,1}t. \tag{4.32}$$

The plus/minus sign in the equation corresponds to different rotation directions. Counterclockwise rotating waves in a square-base tank with $L_1 = L_2$ are illustrated in Figure 4.7 by showing the time dependence of the wave elevation at the four corners. The expression used for the wave elevation is

$$\zeta = \cos\left[\pi\left(x + \tfrac{1}{2}L_1\right)/L_1\right]\cos(\sigma_{1,0}t)$$
$$-\cos\left[\pi\left(y + \tfrac{1}{2}L_2\right)/L_2\right]\sin(\sigma_{0,1}t). \tag{4.33}$$

An alternative illustration would be to present the wave elevation along the walls at different time instants. We would then see that the wave form is not conserved as it would be for

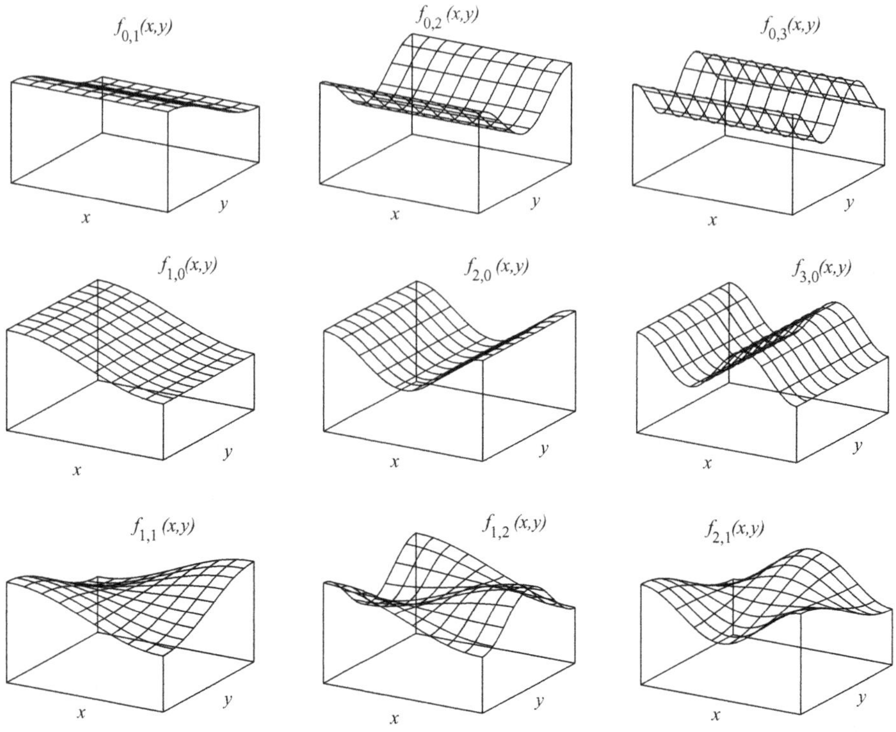

Figure 4.6. The wave patterns associated with the nine lowest natural modes for a nearly square-base three-dimensional rectangular tank.

two-dimensional planar propagating waves. The wave amplitude is largest at the corners.

When analyzing three-dimensional waves, it is also convenient to introduce the wave profiles that involve the pairs of Stokes waves as follows:

$$S_1^i(x,\, y) = f_i^{(1)}(x) - f_i^{(2)}(y);$$
$$S_2^i(x,\, y) = f_i^{(1)}(x) + f_i^{(2)}(y). \tag{4.34}$$

The corresponding three-dimensional patterns are shown in Figure 4.8 for $i = 1$ and a square-base tank. Again, using the pendulum analogy and terminology from the papers by Faltinsen *et al.* (2003) and Miles (1994), we denote these profiles "diagonal" or "squares." Diagonal wave profiles are associated with standing waves oriented in the direction of a diagonal line. The fact

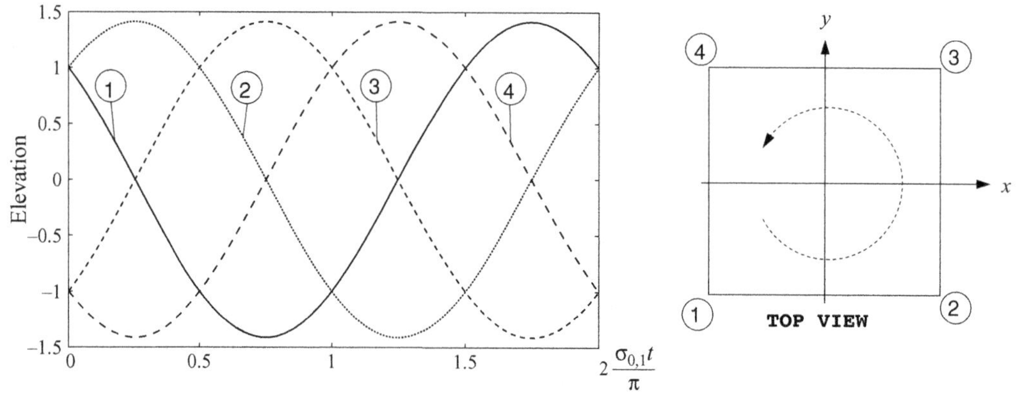

Figure 4.7. Counterclockwise rotating swirling waves in a rectangular tank with a square base. The wave elevations at the four corners given by eq. (4.33) are illustrated as a function of nondimensional time: $\sigma_{0,1}t/\pi = \sigma_{1,0}t/\pi$.

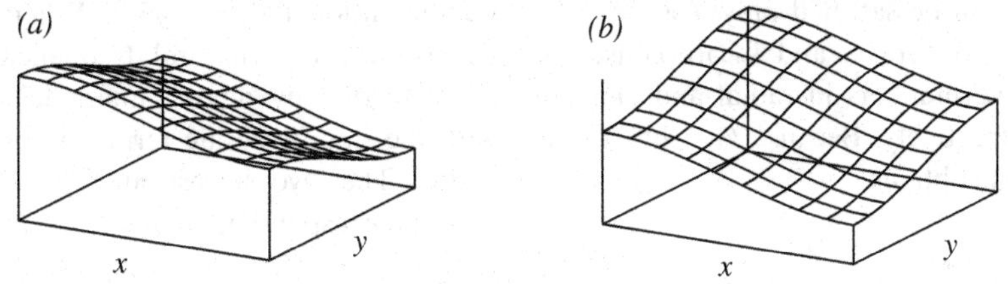

Figure 4.8. Diagonal standing waves defined by eq. (4.34) involving the two lowest Stokes waves for a three-dimensional rectangular tank (parts a and b).

that we have two diagonals is reflected by the plus/minus sign in eq. (4.34). Two opposing corner points of the rectangular surface plane remain fixed, while the two other corner points oscillate in opposite phase for diagonal standing waves. Combining $S_1^i(x, y)$ and $S_2^i(x, y)$ by multiplying each with two different nonzero weight coefficients gives

$$S_i(x, y) = Af_i^{(1)}(x) + Bf_i^{(2)}(y), \quad AB \neq 0, \ |A| \neq |B|, \tag{4.35}$$

which defines the *nearly diagonal* (alternatively, "squares-like" and "diagonal-like") wave patterns.

4.3.2.2 Upright circular cylindrical tank

Natural sloshing frequencies and modes in a vertical circular cylindrical tank (Figure 4.9) are best considered in a cylindrical coordinate system (r, θ, z), which is linked to the Cartesian coordinate system $Oxyz$ by the transformation $x = r\cos\theta$, $y = r\sin\theta$, $z = z$. The spectral problem of natural modes, eq. (4.2), in cylindrical coordinates is

$$\left.\begin{array}{l}
\dfrac{\partial^2\varphi}{\partial z^2} + \dfrac{1}{r}\dfrac{\partial}{\partial r}\left(r\dfrac{\partial\varphi}{\partial r}\right) + \dfrac{1}{r^2}\dfrac{\partial^2\varphi}{\partial\theta^2} = 0, \quad 0 \le r < R_0, \\
\qquad\qquad -h < z < 0, \quad 0 \le \theta < 2\pi, \\
\dfrac{\partial\varphi}{\partial r} = 0, \quad r = R_0, \quad -h < z < 0, \quad 0 \le \theta < 2\pi, \\
\dfrac{\partial\varphi}{\partial z} = 0, \quad z = -h, \quad r < R_0, \quad 0 \le \theta < 2\pi, \\
\dfrac{\partial\varphi}{\partial z} = \kappa\varphi, \quad z = 0, \quad r < R_0, \quad 0 \le \theta < 2\pi, \\
\qquad\qquad \varphi(r, \theta, z) = \varphi(r, \theta + 2\pi, z), \\
\displaystyle\int_0^{R_0} r \int_0^{2\pi} \varphi(r, \theta, 0)\, \mathrm{d}\theta\, \mathrm{d}r = 0.
\end{array}\right\} \tag{4.36}$$

Separation of the spatial variables r, θ, and z is used to express the solution as $\varphi = R(r)\Theta(\theta)Z(z)$. Substitution of this solution in the first of eqs. (4.36) gives

$$\underbrace{\frac{Z''}{Z}}_{k_r^2} + \underbrace{\frac{1}{r}\frac{d}{dr}\left(r\frac{dR}{dr}\right)\frac{1}{R} + \frac{1}{r^2}\underbrace{\frac{\Theta''}{\Theta}}_{-k_\theta^2}}_{-k_r^2} = 0.$$

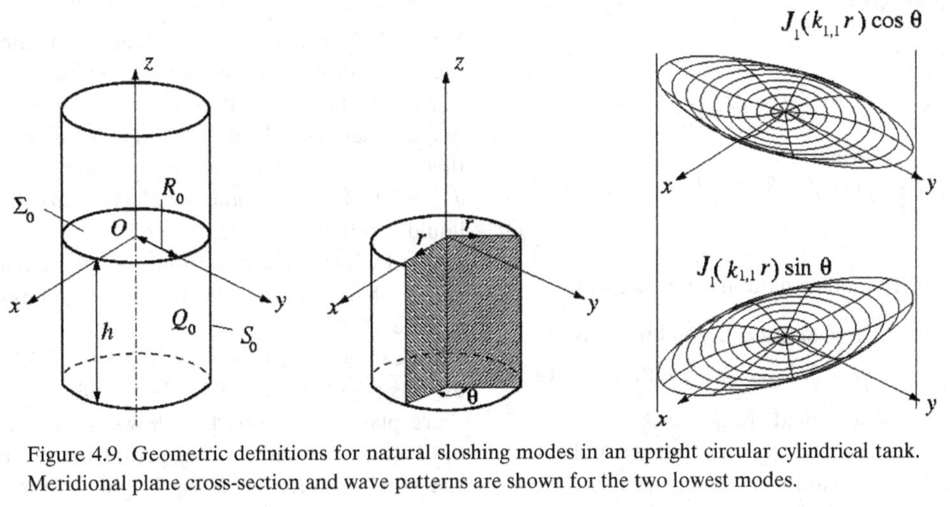

Figure 4.9. Geometric definitions for natural sloshing modes in an upright circular cylindrical tank. Meridional plane cross-section and wave patterns are shown for the two lowest modes.

The equation can be satisfied only if $Z''/Z = k_r^2$ and $\Theta''/\Theta = k_\theta^2$, where k_r and k_θ are constants describing wave numbers in radial and angular directions, respectively. Because $\Theta(\cdot)$ has to be 2π-periodic, one obtains

$$\Theta(\theta) = \begin{cases} \cos(m\theta) \\ \sin(m\theta) \end{cases},$$

where m is a nonnegative integer and $-k_\theta^2 = m^2$. The case $m = 0$ corresponds to pure radial wave profiles (i.e., the wave number in the angular direction is zero).

Bessel functions of the first and second kind satisfy the differential equation for $R(r)$. Because $R(r)$ has to be finite for any r, we can use only Bessel functions of the first kind; that is, $R(r) = J_m(k_r r)$. Since $R'(R_0) = 0$ due to the zero-Neumann condition on the upright walls, the dimensional wave numbers in radial direction $k_r = k_{m,i}$ depend on m. The numbers are related to the nondimensional roots $\iota_{m,i} = k_{m,i}R_0$ of the equation $J'_m(\iota_{m,i}) = 0$. An infinite set of these roots exists for every m. The index i enumerates the roots in ascending order.

Summarizing the results gives the following solutions:

$$\varphi_{m,i}(r,\theta,z) = J_m\left(\iota_{m,i}\frac{r}{R_0}\right)\frac{\cosh(\iota_{m,i}(z+h)/R_0)}{\cosh(\iota_{m,i}h/R_0)} \times \begin{cases} \cos(m\theta) \\ \sin(m\theta) \end{cases}, \quad m = 0, 1, \ldots; \quad i = 1, 2, \ldots. \tag{4.37}$$

The surface wave patterns of natural modes (4.37) are defined as

$$f_{m,i}(r,\theta) = \varphi_{m,i}(r,\theta,0) = J_m\left(\iota_{m,i}\frac{r}{R_0}\right) \times \begin{cases} \cos(m\theta) \\ \sin(m\theta) \end{cases}, \quad m = 0, 1, \ldots; \quad i = 1, 2, \ldots. \tag{4.38}$$

The corresponding natural frequencies are

$$\sigma_{m,i}^2 R_0/g = R_0\kappa_{m,i} = \iota_{m,i}\tanh(\iota_{m,i}h/R_0), \quad m = 0, 1, \ldots; \quad i = 1, 2, \ldots. \tag{4.39}$$

Natural modes and frequencies (4.37)–(4.39) are parameterized by two integer indices, m and i. The natural sloshing frequencies are functions of the liquid-depth-to-radius ratio in nondimensional form (4.39). When $m \neq 0$, all eigenvalues $\kappa_{m,i}$ (natural frequencies $\sigma_{m,i}$) are characterized by double multiplicity associated with the sine or cosine terms in eqs. (4.37) and (4.38). The two lowest antisymmetric surface modes corresponding to $m = i = 1$ (see their patterns in Figure 4.9) can be expressed as

$$z = f_{1,1,1}(r,\theta) = A\frac{J_1(\iota_{1,1}r/R_0)}{J_1(\iota_{1,1})}\cos\theta \quad \text{and}$$
$$z = f_{1,1,2}(r,\theta) = B\frac{J_1(\iota_{1,1}r/R_0)}{J_1(\iota_{1,1})}\sin\theta, \tag{4.40}$$

where A and B are constants and $k_{1,1}R_0 = \iota_{1,1} = 1.841\ldots$. These two modes correspond to the lowest natural frequency $\sigma_{1,1}$, and the relation $\sigma_{1,1} < \sigma_{2,1} < \sigma_{0,1} < \sigma_{1,2} < \cdots$ is true for any liquid depth.

The two lowest surface modes are perpendicular to each other on the mean free surface, which is a condition similar to Stokes waves (associated with $f_1^{(1)}(x)$ and $f_1^{(2)}(y)$) for a rectangular-base tank. In the literature, the amplification of a single lowest natural mode is referred to as *planar* wave motion. However, the flow is not planar (i.e., two-dimensional). A possible reason for the name *planar* may be associated with the fact that pendulum models are used to study sloshing in vertical circular tanks (Ibrahim, 2005; Miles, 1962). The pendulum motions that approximate $f_{1,1,1}(r,\theta)$ and $f_{1,1,2}(r,\theta)$ are planar.

Surge, sway, roll, or pitch excitations with a period T close to the highest natural period are of particular practical concern:

$$T \approx T_{1,1} = \frac{2\pi}{\sqrt{g\iota_{1,1}\tanh(\iota_{1,1}h/R_0)/R_0}}. \tag{4.41}$$

Figure 4.10 presents period $T_{1,1}$ as a function of radius R_0 for different values of the liquid depth h. The effect of the depth is generally large for the cases presented in Figure 4.10. The figure shows that, for a given value of R_0, a lower limit of $T_{1,1}$ exists that corresponds to the case of infinite liquid depth. For instance, when $h/R_0 > 1.0$, $T_{1,1}$ differs by less than 2.5% from the infinite-depth value, whereas $h/R_0 > 1.5$ makes the difference less than 1%.

An example is a tank with radius $R_0 = 14.2$ m used as a ballast tank in a column-stabilized offshore platform. Figure 4.10 shows the range of the highest natural sloshing period for the ballast tank for different filling heights. Nonnegligible sloshing has in practice been experienced

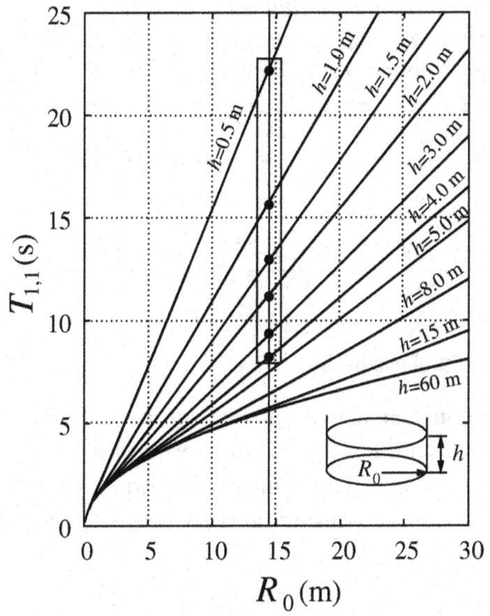

Figure 4.10. The highest natural period $T_{1,1}$ as a function of radius R_0 for an upright circular cylindrical tank. The framed values of $T_{1,1}$ are relevant for the ballast tanks of a column-stabilized offshore platform.

in the ballast tanks. To evaluate if resonant tank oscillations are noticeable, we must know the motion behavior of the platform in different sea states. This behavior can be assessed by first considering the response amplitude operator (RAO) of the horizontal platform motion at the tank position in regular incident waves of different wave directions. The next step is to combine the RAO with wave spectra for realistic sea states as decribed in Chapter 3. This analysis confirms the fact that nonnegligible sloshing in the ballast tanks may occur. However, in practice, nonlinearities matter (see Section 9.3).

Swirling (i.e., a progressive wave motion in the angular direction) is possible in a vertical circular cylinder. Swirling waves follow by superposing two antisymmetric and standing waves with the same mode numbers, m and i, that are orthogonal to each other and are oscillating 90° out of phase with the same natural frequency $\sigma_{m,i}$; that is, the wave motion can be represented as

$$\begin{aligned} &C_{m,i} J_m\left(\iota_{m,i}\frac{r}{R_0}\right)\sin(\pm\sigma_{m,i}t + m\theta + \theta_{m,i}) \\ &= C_{m,i} J_m\left(\iota_{m,i}\frac{r}{R_0}\right)\sin(m(\theta \pm c_{m,i}t) + \theta_{m,i}), \end{aligned} \tag{4.42}$$

where $\theta_{m,i}$ is the phase angle, $C_{m,i}$ determines the amplitude, and $c_{m,i} = \sigma_{m,i}/m$ is the angular phase velocity. The plus/minus sign indicates both counterclockwise and clockwise wave propagation directions.

In Subsection 4.3.1.1, we studied the possibility for natural frequencies in a two-dimensional rectangular tank to form a commensurate spectrum; this occurrence was possible only in the shallow-liquid limit. An infinite number of natural modes is then progressively excited due to the secondary resonance phenomenon and resonant forcing of the lowest natural frequency. In contrast to the rectangular geometry, a finite set of natural frequencies (4.39) may be commensurate at a given finite liquid depth. This fact is best demonstrated by using representation (4.42) of the two lowest natural modes for $m = i = 1$ and two modes with $m > 1$. The corresponding natural frequency, $\sigma_{m,i}$, is by definition commensurate to $\sigma_{1,1}$ and can be amplified due to the secondary resonance when $\sigma_{m,i}/\sigma_{1,1} = m$. Due to eq. (4.39), the latter condition can be written in the form

$$m\sqrt{\iota_{1,1}\tanh(\iota_{1,1}h/R_0)} = \sqrt{\iota_{m,i}\tanh(\iota_{m,i}h/R_0)} \tag{4.43}$$

and can be considered a transcendental equation with respect to h/R_0. Roots of this equation constitute a nonempty set. Solutions up to $m = 4$ are listed by Bryant (1989) and are shown in Table 4.2.

When h/R_0 belongs to this list, we find that the phase velocity in eq. (4.42) is the same as $c_{1,1}$; that is, $c_{m,i} = \sigma_{m,i}/m = \sigma_{1,1} = c_{1,1} = \sqrt{k_{1,1}\tanh(k_{1,1}h)}$, meaning that at least two pairs of freestanding waves may form a synchronized rotary wave.

The previous analysis can be generalized to include a concentric inner cylinder and to an annular and sectored upright circular tank (see Exercise 4.11.3).

4.4 Seiching

The occurrence of resonant waves (sloshing) in lakes and harbors is referred to as seiching. It is characterized by wavelengths λ that are much larger than the water depth h (namely, $\lambda \gg h$). This condition implies that a shallow-water approximation can be applied. An essential assumption in linear shallow-water theory is that the pressure is hydrostatic relative to the instantaneous wave elevation $\zeta(x, y, t)$. If $z = 0$ is the mean free surface and z is, as usual, positive upward, the pressure distribution in the water can

Table 4.2. *Roots of eq (4.43): The roots give the liquid-depth-to-radius ratios h/R_0 of an upright circular cylindrical tank, for which the natural frequencies are partly commensurate ($\sigma_{m,i}/\sigma_{1,1} = m$) and, therefore, the secondary resonance is possible*

$m \backslash i$	2	3	4	5	6	7
2	0.831	–	–	–	–	–
3	0.279	0.455	–	–	–	–
4	0.158	0.249	0.330	0.423	0.542	0.721

Source: Bryant (1989).

be expressed as $-\rho g(z-\zeta)$. If we assume linear potential flow theory and use the dynamic free-surface condition $g\zeta + \partial\varphi/\partial t|_{z=0} = 0$, the pressure distribution can also be expressed as $p = -\rho g z - \rho\, \partial\varphi/\partial t|_{z=0}$. A consequence of the shallow-water approximation is that the velocity potential is, as a first approximation, independent of z. We express the governing equation in an approximate way by using this fact. We assume two-dimensional flow and consider a strip of small length Δx and let it vertically extend from $z = -h(x)$ to the mean free surface $z = 0$ (see Figure 4.11). We express the continuity of liquid mass inside the strip by considering the mass flux into the liquid volume through the vertical surfaces at x and $x + \Delta x$ as well as at $z = 0$ and the sea bottom. Letting $\Delta x \to 0$ results in the following equation:

$$\frac{\partial}{\partial x}\left(h(x)\frac{\partial\varphi}{\partial x}\right) + \left.\frac{\partial\varphi}{\partial z}\right|_{z=0} = 0. \quad (4.44)$$

We now apply the free-surface condition $\partial^2\varphi/\partial t^2 + g\, \partial\varphi/\partial z|_{z=0} = 0$ and generalize eq. (4.44) to also introduce the variation in the y-direction. This change leads to

$$g\left[\frac{\partial}{\partial x}\left(h(x,y)\frac{\partial\varphi}{\partial x}\right) + \frac{\partial}{\partial y}\left(h(x,y)\frac{\partial\varphi}{\partial y}\right)\right] - \frac{\partial^2\varphi}{\partial t^2} = 0. \quad (4.45)$$

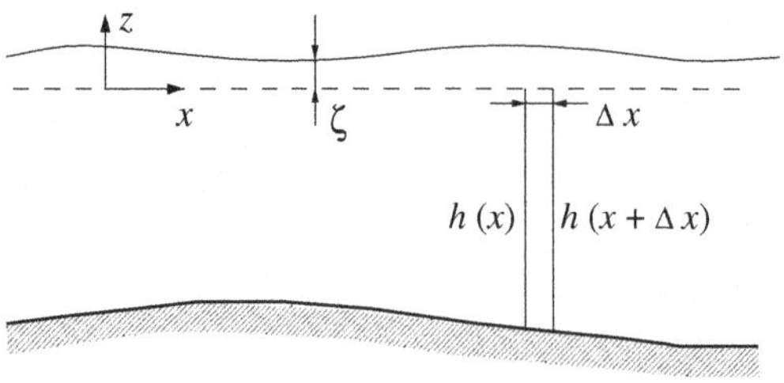

Figure 4.11. Control surface used in formulating the shallow-water equations.

4.4.1 Parabolic basin

We consider a two-dimensional case where the water depth has a parabolic form; that is, $h(x) = h_0(1 - x^2/a^2)$. Harmonic oscillations with frequency σ are considered; that is, $\varphi = \overline{\varphi}\exp(i\sigma t)$, which gives

$$gh_0\frac{\partial}{\partial x}\left(\left(1 - \frac{x^2}{a^2}\right)\frac{\partial\overline{\varphi}}{\partial x}\right) + \sigma^2\overline{\varphi} = 0.$$

The equation has an infinite number of solutions corresponding to the eigensolutions. We start with the lowest natural mode and leave it as an exercise to confirm that $\overline{\varphi}_1 = Cx$ is a natural mode and that the corresponding natural frequency is $\sigma_1 = \sqrt{2gh_0}/a$, where C can be any nonzero constant. It follows from the dynamic free-surface condition that the corresponding free-surface elevation is a straight line. The second lowest natural mode and natural frequency are $\overline{\varphi}_2 = C\left[(x/a)^2 - 1/3\right]$ and $\sigma_2 = \sqrt{6gh_0}/a$, respectively. This mode is a symmetric mode with the maximum value at $x = 0$ and nodal points at $x = \pm a/\sqrt{3}$. The solution for natural mode and natural frequency number n can be expressed as

$$\sigma_n^2 = n(n+1)\frac{gh_0}{a^2}; \quad \overline{\Phi}_n = C \cdot P_n\left(\frac{x}{a}\right),$$

$$P_n(\mu) = \frac{1}{2^n \cdot n!}\frac{d^n}{d\mu^n}\left(\mu^2 - 1\right)^n,$$

where $P_n(\mu)$ is a zonal harmonic (Lamb, 1945). The free-surface elevations for the four lowest modes are shown in Figure 4.12(a).

4.4.2 Triangular basin

Lamb (1945) also presented an analytical solution for the triangular two-dimensional basin shown in Figure 4.12(b). If we take the origin of the coordinate system at one end of the basin as shown in the figure, we can write the natural mode for half of the basin as $\overline{\varphi} = J_0(2\sqrt{\kappa x})$, where J_0 is a Bessel

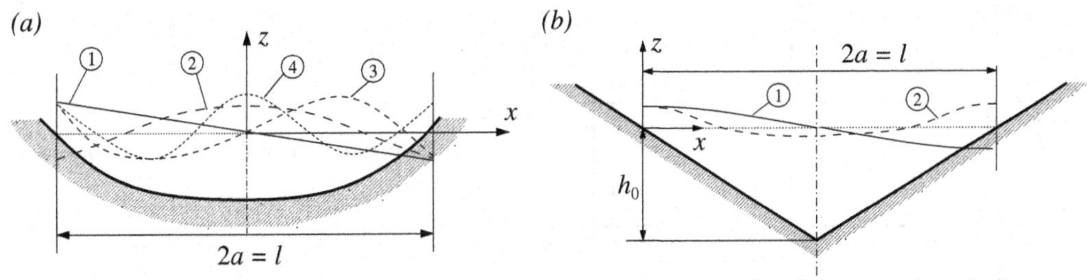

Figure 4.12. Free-surface elevation associated with the four lowest modes for two-dimensional shallow water sloshing (seiching) in a parabolic basin (part a) and the two lowest modes for a triangular basin (part b).

function of the first order, $\kappa = \sigma^2 a/gh_0$, and h_0 is the water depth in the middle of the tank. Let us first consider antisymmetric eigenmodes, which requires $\overline{\varphi} = 0$ at $x = a$ (i.e., $J_0(2\sqrt{\kappa a}) = 0$). This equation has an infinite number of solutions that determine the natural frequencies for the antisymmetric modes. The lowest mode corresponds to a natural period $2\pi/\sigma = 1.30637 \cdot 4a/\sqrt{gh_0}$. The symmetrical modes are determined by requiring that $\partial\overline{\varphi}/\partial x = 0$ at $x = a$, which gives the condition $J_0'(2\sqrt{\kappa a}) = 0$. Figure 4.12(b) shows the free-surface elevation associated with the two lowest modes.

4.4.3 Harbors

The preceding shallow-water analysis can be generalized to a harbor resonance. We must then account for the fact that the harbor basin has a small opening to the sea. Because the wave elevation ζ is zero outside the harbor, $\zeta = -(\partial\varphi/\partial t)/g$ must be zero at the harbor outlet. The consequence is that the harbor outlet is a nodal line with $\varphi = 0$ following from the free-surface condition.

As a special situation, one can consider small cross-sectional dimensions (relative to the harbor length) of the harbor basin. By choosing the x-coordinate in the length dimension, neglecting the velocities in the y-direction, and integrating eq. (4.45) in the y-direction across the harbor, we obtain

$$g\frac{\partial}{\partial x}\left[\mathcal{A}(x)\frac{\partial\varphi}{\partial x}\right] - \mathcal{B}(x)\frac{\partial^2\varphi}{\partial t^2} = 0, \qquad (4.46)$$

where $\mathcal{A}(x)$ and $\mathcal{B}(x)$ are the cross-sectional area and breadth of the harbor basin, respectively.

Equation (4.46) has an analytical solution for the harbor problem with a rectangular basin of length L, which means that $\mathcal{A}$ and $\mathcal{B}$ are constant values and $\mathcal{A} = \mathcal{B}h$. The harbor outlet is situated at $x = L$. The solution with the requirement that $x = L$ corresponds to a nodal line (i.e., $\varphi = 0$), and a vertical wall exists at $x = 0$ (i.e., $\partial\varphi/\partial x = 0$ at $x = 0$) that can be expressed as

$$\varphi = C\cos\left[\tfrac{1}{2}\pi(2n+1)x/L\right]\exp(i\sigma_n t),$$
$$n = 0, 1, 2\ldots,$$

where C is a nonzero constant and the natural frequency σ_n is expressed by $\sigma_n^2 = gh[\tfrac{1}{2}\pi(2n+1)/L]^2$. For the special cases that $\mathcal{A}(x)$ varies either linearly or quadratically with x, we can modify the previous solutions for triangular and parabolic basins by accounting for the outlet condition at $x = L$.

4.4.4 Pumping-mode resonance of a harbor

We consider the harbor configuration in Figure 4.13, which is characterized by a basin with surface area S that is connected to the open sea by a channel of length L and constant cross-sectional area $\mathcal{A}$. A resonant "piston-like" oscillation can occur (i.e., the free surface inside the basin, S,

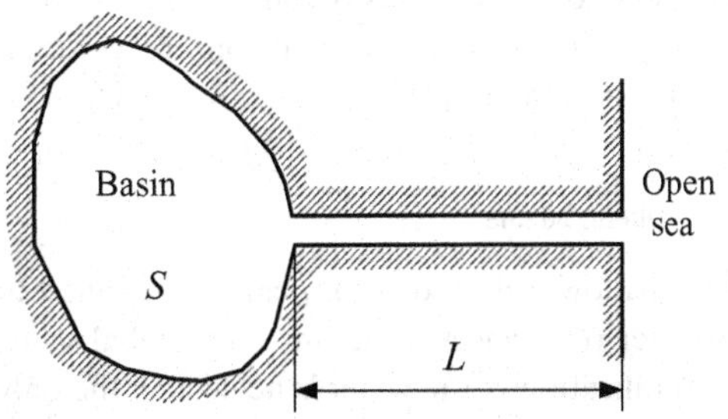

Figure 4.13. Harbor consisting of a basin with surface area S and a channel with length L and constant cross-sectional area $\mathcal{A}$.

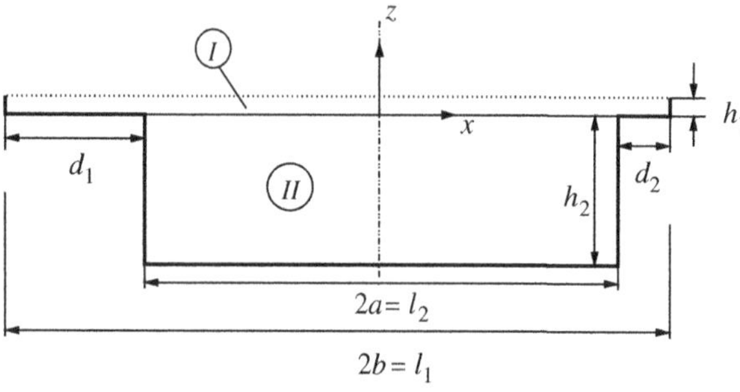

Figure 4.14. Tank with rectangular domains I and II.

oscillates with a spatially constant elevation ζ). We can find this piston resonance frequency by using a shallow-water assumption and beginning with the continuity equation in the basin so that

$$S\dot{\zeta} + U\mathcal{A} = 0, \qquad (4.47)$$

where U is the velocity in the channel.

We need an additional expression of U in terms of ζ. This expression can be found by using the linearized Euler equation $\partial p/\partial x = -\rho \partial U/\partial t$ for the channel flow. Because U is independent of x as a consequence of the continuity equation, it follows from the linearized Euler equation that $\partial p/\partial x$ is a constant along the channel. By setting the free-surface elevation as zero at the outlet of the channel and equal to ζ at the basin and using the fact that the pressure is hydrostatic relative to the instantaneous free surface in a shallow-water approximation gives

$$\begin{aligned}\partial p/\partial x &= (p_{\text{outlet}} - p_{\text{basin}})/L \\ &= (-\rho g z + \rho g(z-\zeta))/L = -\rho g\zeta/L.\end{aligned}$$

It follows now from the linearized Euler equation that $\partial U/\partial t = g\zeta/L$. Combining the latter equation with eq. (4.47) gives

$$S\ddot{\zeta} + L^{-1} g\mathcal{A}\zeta = 0. \qquad (4.48)$$

The solution of eq. (4.48) can be expressed as $\zeta = C\exp(\mathrm{i}\sigma_* t)$, where C is a nonzero constant and the natural frequency σ_* is $\sigma_* = \sqrt{g\mathcal{A}/(SL)}$.

4.4.5 Ocean basins

The shallow-water equations can be modified to consider resonances in oceans on a global scale. If stratification of the water is neglected, the only modification is that the Coriolis acceleration due to Earth's rotation must be accounted for (Platzman, 1972; Rao, 1966; Sorensen, 1993).

4.5 Domain decomposition

A useful technique is to divide a liquid domain into subdomains and then combine the solutions in the different subdomains. This approach is used nowadays in CFD (Quarteroni & Valli, 1999). However, we limit our case to potential flow. Transmission conditions are needed on the boundaries between the subdomains. The conditions are that both the velocity potential and the normal velocity at the boundary are continuous between the adjacent subdomains. The method has a clear advantage when the solutions in different subdomains can be represented analytically. This *domain decomposition method* was quite popular during 1960–1980 to estimate the influence of a conical or a spherical bottom on the natural sloshing frequencies for an upright circular cylindrical tank. The tank shapes were common to spacecraft applications. A review of these results is given by Lukovsky *et al.* (1984).

4.5.1 Two-dimensional sloshing with a shallow-water part

We illustrate the domain decomposition method for a two-dimensional tank with a shallow-water part as illustrated in Figure 4.14. In the figure, domain I has a rectangular shape with height h_1 and length $l_1 = 2a + d_1 + d_2$ and is constructed by extending the shallow part of the tank over the deep part. Domain II also has a rectangular shape with height h_2 and length $2a = l_2$.

A Cartesian coordinate system Oxz is defined with origin at the intersection between the bottom of domain I and the centerplane of domain II. The velocity potential and the free-surface elevation in domain I are denoted φ_1 and ζ_1, respectively. The velocity potential in domain II is φ_2.

We have to formulate boundary conditions that match the solutions in domains I and II. We assume shallow-liquid conditions for domain I, which implies that φ_1 does not depend on z (i.e., it is only a function of x). The continuity equation for domain I for x between $-a$ and a can be expressed as

$$h_1 \frac{\partial^2 \varphi_1}{\partial x^2} + \frac{\partial \zeta_1}{\partial t} - \left.\frac{\partial \varphi_2}{\partial z}\right|_{z=0} = 0, \quad -a \le x \le a, \tag{4.49}$$

which follows by considering a strip of small length Δx and letting it extend vertically from $z = 0$ to the mean free surface. Equation (4.49) can be rewritten as

$$\frac{\partial^2 \varphi_1}{\partial t^2} - g h_1 \frac{\partial^2 \varphi_1}{\partial x^2} + g \left.\frac{\partial \varphi_2}{\partial z}\right|_{z=0} = 0, \quad -a \le x \le a \tag{4.50}$$

by using the dynamic free-surface condition $g\zeta_1 + \partial\varphi_1/\partial t = 0$.

The velocity potential should satisfy the zero-Neumann conditions at the vertical walls:

$$\frac{\partial \varphi_1}{\partial x} = 0 \quad \text{at } x = -a - d_1 \quad \text{and } x = a + d_2. \tag{4.51}$$

Furthermore, φ_2 must satisfy the Laplace equation and the zero-Neumann conditions on the wetted walls and bottom. In addition, continuity of the velocity potential between the domains should also be required; that is,

$$\varphi_1 = \varphi_2, \quad -a \le x \le a, \quad z = 0. \tag{4.52}$$

Let us again assume harmonic time dependence with frequency σ. The solution in domain I can then be expressed as

$$\varphi_1(x, z, t) = \sum_{n=1}^{\infty} A_n \exp(i\sigma t) \times \cos[n\pi(x + a + d_1)/(2b)]. \tag{4.53}$$

This expression satisfies boundary condition (4.51). The general expression for the corresponding velocity potential in domain II that satisfies the Laplace equation and necessary zero-Neumann conditions at $x = \pm a$ and the tank bottom $z = -h_2$ is

$$\varphi_2(x, z, t) = \sum_{n=1}^{\infty} \frac{B_n}{\cosh(n\pi h_2/2a)} \exp(i\sigma t) \times \cos\left[\frac{n\pi}{2a}(x + a)\right] \cosh\left[\frac{n\pi}{2a}(z + h_2)\right]. \tag{4.54}$$

We now substitute eqs. (4.53) and (4.54) into eq. (4.50) and (4.52) and obtain the following linear homogeneous equations with respect to A_n and B_n:

$$\sum_{n=1}^{\infty} A_n[-\sigma^2 + g h_1 (n\pi/(2b))^2] \times \cos[n\pi(x + a + d_1)/(2b)] + g \sum_{n=1}^{\infty} B_n (n\pi/(2a)) \tanh(n\pi h_2/(2a)) \times \cos[n\pi(x + a)/(2a)] = 0, \tag{4.55}$$

$$\sum_{n=1}^{\infty} A_n \cos\left[n\pi(x + a + d_1)/(2b)\right] - \sum_{n=1}^{\infty} B_n \cos\left[n\pi(x + a)/(2a)\right] = 0. \tag{4.56}$$

Equation (4.56) can be used to find B_n as a function of A_n, which is done by multiplying by $\cos(\pi m(x + a)/2a)$ and integrating over the interval $(-a, a)$. The result is

$$B_m = \frac{1}{a} \sum_{k=1}^{\infty} a_{mk} A_k, \quad a_{mk} = \int_{-a}^{a} \cos\left(\frac{\pi m}{2a}(x + a)\right) \times \cos\left(\frac{\pi k}{2b}(x + a + d_1)\right) \mathrm{d}x. \tag{4.57}$$

We now substitute eq. (4.57) into eq. (4.55) and restrict ourselves to a finite number N of unknown A_n. Multiplying the resulting equation by $\cos(\pi m(x + a)/2a)$ and integrating over the interval $(-a, a)$ leads to an $N \times N$ linear equation system for the unknowns A_n. The determinant of the coefficient matrix must be zero for this equation system to have a nontrivial solution. The natural frequencies σ are determined as the solution of

$$\det[A - \sigma^2 B] = 0, \quad B = \{a_{mk}\}_{m,k=1,N}, \quad A = \left\{ g a_{mk} \left[h_1 \left(\frac{\pi k}{2b}\right)^2 + \left(\frac{\pi m}{2a}\right) \tanh\left(\frac{\pi m h_2}{2a}\right) \right] \right\}_{m,k=1,N}. \tag{4.58}$$

If $N = 1$, matrix spectral problem (4.58) leads to the analytical approximation

$$\sigma^2 = g\pi(2a)^{-1} \tanh\left(\tfrac{1}{2}\pi h_2/a\right) + g h_1 \left(\tfrac{1}{2}\pi/b\right)^2. \tag{4.59}$$

The estimate of this lowest natural frequency changes with increasing N. Numerical tests have shown that N must have a value of about 20 to achieve a relative error of the lowest natural frequency that is less than 1%.

4.5.2 Example: swimming pools

Two examples are considered related to rectangular swimming pools of a cruise vessel (see Figure 4.15). The swimming pools do not have horizontal bottoms; however, the effect of the inclined bottom is discussed later. We now focus on two-dimensional sloshing in the longitudinal direction. Relevant parameters for one of the swimming pools are $d_1 = d_2 = 2.0\,\text{m}$, $h_1 = 0.15\,\text{m}$, $h_2 = 1.1\,\text{m}$, and $l_2 = 2a = 8.1\,\text{m}$. The theoretical value based on eq. (4.59) is 4.9 s and calculations with $N = 20$ give 4.74 s. In the other example, $d_1 = 3.0\,\text{m}$, $d_2 = 0.2\,\text{m}$, $h_1 = 0.15\,\text{m}$, $h_2 = 1.5\,\text{m}$, and $2a = 12.5\,\text{m}$. In this case, eq. (4.59) gives 6.48 s and the domain decomposition method gives 6.31 s with $N = 20$. The effect of an inclined bottom on the natural frequencies can be estimated by eq. (4.93) by neglecting the effect of the shallow-water part, which causes approximately 3% higher natural periods. The theoretical results are in reasonable agreement with confidential experimental results. However, damping and three-dimensional waves in the shallow part of the swimming pool were present in the experiments. Possible reasons are nonlinear free-surface effects with wave breaking at the shallow-water parts and vortex shedding at the sharp edges between the shallow- and deep-water parts.

4.6 Variational statement and comparison theorems

Variational (energy-based) formulations of problem (4.2) are powerful tools in finding sloshing eigenfrequencies and eigenmodes. A key element in our presentation of the variational formulations is the *Rayleigh quotient*. The expression for the Rayleigh quotient is derived by using the conservation of kinetic and potential energy. Mathematically the Rayleigh quotient is a functional $K_{Q_0,\Sigma_0}(\varphi)$, which involves integrals of test functions φ (and their spatial derivatives) over the mean free surface Σ_0 and the mean liquid domain Q_0. The test functions must have first-order spatial derivatives that are square-integrable over Q_0. In accordance with standard definitions of the functional (Kolmogorov & Fomin, 1961), K_{Q_0,Σ_0} maps each test function φ into a real number. For the Rayleigh quotient, these real numbers are nonnegative. One can consider the local minima of $K_{Q_0,\Sigma_0}(\varphi)$ on the set of all admissible test functions. Theorems exist (see Feschenko *et al.*, 1969; Morand & Ohayon, 1995) that establish that local minima occur for $\varphi = \varphi_n$ and, moreover, $K_{Q_0,\Sigma_0}(\varphi_n) = \kappa_n$, where φ_n is an eigenmode and κ_n is the corresponding eigenvalue.

We use the fact that $K_{Q_0,\Sigma_0}(\varphi)$ is equal to κ_n when $\varphi = \varphi_n$ to provide estimates of eigenvalues. In particular, we consider a known eigenfunction φ_n (with the corresponding eigenvalue κ_n) as well as small modifications of Q_0 and Σ_0 to Q_0' and Σ_0', respectively. Then the Rayleigh quotient gives the following estimate of the eigenvalue:

$$\kappa_n' = K_{Q_0',\Sigma_0'}(\varphi_n') \approx K_{Q_0',\Sigma_0'}(\varphi_n). \tag{4.60}$$

Other special comparison theorems and asymptotic formulas also exist that establish how κ_n' changes with varying Q_0 and Σ_0.

Furthermore, we use the minima conditions of $K_{Q_0,\Sigma_0}(\varphi)$ to derive *variational equalities* that can be solved by, for instance, the Galerkin method and have solutions that coincide with the natural modes. The *natural* conditions for test functions used in the Rayleigh quotient follow also from the minima conditions of $K_{Q_0,\Sigma_0}(\varphi)$ and from variational formulations. The test functions do not have to satisfy the natural conditions. The natural conditions are the Laplace equation, the zero-Neumann condition on the mean wetted tank surface, and the combined kinematic and dynamic free-surface condition. However, conservation of liquid mass and orthogonality between the modes are not natural conditions and must be satisfied. Furthermore, we stress that no nonphysical restrictions may be imposed on the test functions.

The use of the Rayleigh quotient is not unique for the sloshing problem; it is used, for instance, in structural mechanics. The simplest case is the vibration of a string. We use the latter example to prove the fact that the Rayleigh quotient is a minimum when the eigenmodes are used. The natural conditions for the string problem are also derived.

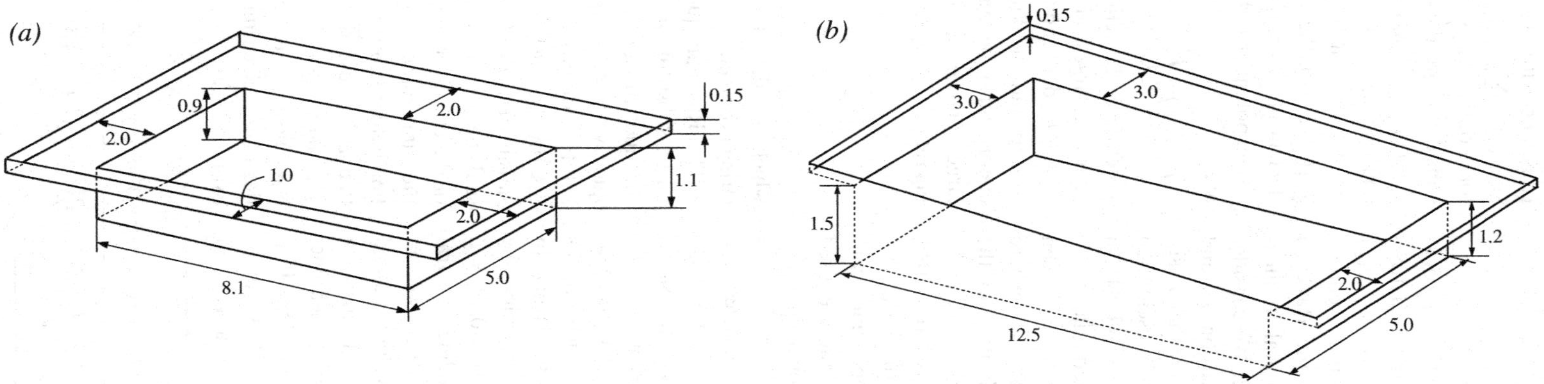

Figure 4.15. Proposed rectangular swimming pools for cruise vessels (dimensions in meters).

4.6.1 Variational formulations

4.6.1.1 Rayleigh's method

We present the Rayleigh method as a background for introducing variational formulations of sloshing. The Rayleigh method is an energy method for estimation of the natural frequencies. A similarity exists between the Rayleigh method and how we estimate natural frequencies for sloshing in Section 4.6.1.2. We consider as an example a string with a tension T_{st} and a constant mass per unit length, m, that is situated between $x = 0$ and l. Our presentation follows closely that of den Hartog (1984) in the case of fixed ends. The differential equation for free vibrations with a frequency σ can be expressed as $T_{st}y''(x) + m\sigma^2 y(x) = 0$, where $y(x,t) = y(x)\sin(\sigma t + \varepsilon)$ is the vibration of the string.

The considered end conditions are either $y(0) = y(l) = 0$ (fixed ends) or $y'(0) = y'(l) = 0$ (zero-slope ends). The zero-slope end conditions are unrealistic in reality but are used to illustrate mathematical properties. For the fixed ends, the eigenfunctions are $y_n(x) = \sin(\pi x n/l)$ and the natural frequencies are $\sigma_n = (\pi n/l)\sqrt{T_{st}/m}$ $(n = 1, 2, \ldots)$. For the zero-slope ends, the eigenfrequencies are the same, but $y_n(x) = \cos(\pi x n/l)$. The string problem with the zero-slope end conditions has the constant solution $y_0 = C \neq 0, \sigma_0 = 0$, which implies rigid-body translatory motions of the string. In both cases, eigenfunctions $y_n(x)$ are mutually orthogonal; that is, $\int_0^l y_n(x)y_k(x)\,\mathrm{d}x = 0,\ k \neq n$. Moreover, due to the corresponding spectral theorem (see Courant & Hilbert, 1953), we can always express $y(x)$, which *satisfies* the fixed-end conditions (i.e., $y(0) = y(l) = 0$), in terms of a Fourier series by the corresponding eigenfunctions $y_n(x)$; that is, $y(x) = \sum_{n=1}^{\infty} a_n y_n(x)$. If a function *does not satisfy* any boundary conditions, the function (but not its derivative) can be presented as a Fourier series by the eigenfunctions obtained with the zero-slope conditions (together with the constant solution y_0); that is, $y(x) = \sum_{n=0}^{\infty} a_n y_n(x)$. As always postulated for Fourier-series theory, the convergence is considered in the mean square metrics (i.e., $\int_0^l (y(x) - \sum_{n=0}^{N} a_n y_n(x))^2\,\mathrm{d}x \to 0$ as $N \to \infty$), and the series may not converge at a fixed point x (e.g., at the end of the interval). When $\int_0^l y(x)\,\mathrm{d}x = 0$, the latter series does not contain the constant solution y_0, namely, $a_0 = 0$.

The potential and kinetic energies of the string are

$$E_p = \tfrac{1}{2}T_{st}\int_0^l (y')^2\,\mathrm{d}x, \quad E_k = \tfrac{1}{2}m\sigma^2\int_0^l y^2\,\mathrm{d}x.$$

When the deflection is largest, the potential energy is largest and the kinetic energy is zero. When the deflection is zero, the potential energy is zero and the kinetic energy is largest. The energy must be equal when the deflection is largest and zero. This condition gives the following ratio:

$$\sigma^2 = K_l(y) = \frac{T_{st}}{m}\frac{\displaystyle\int_0^l (y')^2\mathrm{d}x}{\displaystyle\int_0^l y^2\,\mathrm{d}x}, \tag{4.61}$$

which is called the Rayleigh quotient for the string problem.

The Rayleigh quotient is nonnegative. Obviously, substitution of the eigenfunction y_n into the quotient gives the corresponding eigenfrequency: $\sigma_n^2 = K_l(y_n)$. However, we can also substitute another (test) function, which is not necessarily an eigenfunction. If the test function satisfies the *fixed-end conditions*, it can, as we said, be presented as a Fourier series in terms of the corresponding eigenfunctions. We rewrite the numerator of eq. (4.61) by partial integration before substituting the series into eq. (4.61); that is, we use the fixed-end conditions to get $\int_0^l (y')^2\,\mathrm{d}x = yy'|_0^l - \int_0^l yy''\,\mathrm{d}x = -\int_0^l yy''\,\mathrm{d}x$. It follows now by using the orthogonality properties of the eigenfunctions that eq. (4.61) can be expressed as

$$\left(\frac{\sigma}{\sigma_1}\right)^2 = \frac{\displaystyle\int_0^l \left(a_1^2y_1^2 + (\sigma_2/\sigma_1)^2 a_2^2y_2^2 + (\sigma_3/\sigma_1)^2 a_3^2y_3^2 + \cdots\right)\mathrm{d}x}{\displaystyle\int_0^l \left(a_1^2y_1^2 + a_2^2y_2^2 + a_3^2y_3^2 + \cdots\right)\mathrm{d}x},$$

where the right-hand side is equal to or larger than 1. The minimum value occurs when

$a_m = 0\,(m \geq 2)$; that is, when $y(x)$ coincides with the lowest eigenfunction $y_1(x) = \sin(\pi x/l)$, then $\sigma_1^2 = T_{st}\,(\pi n/l)^2\,/m \leq K_l(y)$ for $y(0) = y(l) = 0$ and the equality is only possible when $y(x) = y_1(x)$. As a matter of fact, the exact eigenfunction always gives a lower value of the quotient than an approximate eigenfunction. We can use the Rayleigh quotient to estimate the lowest natural frequency, σ_1, by assuming a functional form of $y(x)$ satisfying the fixed-end conditions. The result depends on the assumed form. Examples are provided later.

When considering the *zero-slope end* conditions, one must postulate summation from zero in the series $y(x) = \sum_{n=0}^{\infty} a_n y_n(x)$ by the corresponding eigenfunctions; that is, the constant function should be included in the series to provide completeness. As previously stated, *any test function* $y(x)$ can be represented by this series; therefore, we should not give a restriction related to the zero-slope end conditions. However, the constant solution ($y_0 = 1,\ a_0 \neq 0,\ a_i = 0$ for $i \geq 1$) gives a Rayleigh quotient of zero. To exclude the constant solution, we must assume

$$\int_0^l y(x) y_0 \,\mathrm{d}x = 0 \;\Rightarrow\; \int_0^l y(x)\,\mathrm{d}x = 0 \qquad (4.62)$$

for all the tested functions. Repeating the derivation as we did for the case of the fixed-end condition, we obtain that $\sigma_1^2 = T_{st}\,(\pi n/l)^2\,/m \leq K_l(y)$ when eq. (4.62) is fulfilled. As we remarked earlier, this inequality can be used to get a lower bound of the lowest eigenvalue. For the sloshing problem, the volume conservation condition plays the role of eq. (4.62).

As one can see from the two considered cases, we are not free in using the test functions to get the inequality $\sigma_1^2 \leq K_l(y)$. The functions must be restricted by either the fixed-end conditions, or they must satisfy eq. (4.62). However, eq. (4.62) is needed only to exclude the constant solution, which is mathematically possible for the string problem with the zero-slope end conditions, whereas the fixed-end conditions really appear as a specific restriction. To understand why this is so, we now introduce what the so-called natural boundary conditions for a variational formulation. By definition:

> Boundary conditions are called natural for a variational formulation based on a functional, if these follow from the necessary minima (extrema) condition of the functional.

An example and extensive discussion were given on natural boundary conditions in Section 2.5.2.2 for the Euler–Bernoulli beam problem. Another example is in Section 2.5.3, where the Laplace equation and the Neumann boundary conditions did not follow from the Lagrange variational formulation (they are not natural conditions for the formulation) but followed from the Bateman–Luke variational formulation (which become natural conditions for this case).

Let us study the boundary conditions that are natural for Rayleigh quotient (4.61). We consider Rayleigh quotient (4.61) and an eigenfunction $y_n(x)$. Calculus of the extrema (minima) condition for the quotient implies that we consider small deviations of the eigenfunction (i.e., $y_n + \delta y$). Using the fact that $\sigma_n^2 = K_l(y_n)$ and integration by parts gives

$$K_l(y_n + \delta y) - \underbrace{K_l(y_n)}_{\sigma_n^2} = 2\frac{T_{st}\int_0^l y_n'\,(\delta y)'\,\mathrm{d}x - m\sigma_n^2\int_0^l y_n \delta y\,\mathrm{d}x}{m\int_0^l (y_n)^2 \mathrm{d}x} + O[(\delta y)^2].$$

The linear term in δy is the so-called first variation of functional δK_l at point y_n. This variation must be zero for the functional to have an extrema (local minima):

$$T_{st}\int_0^l y_n'\,(\delta y)'\,\mathrm{d}x - m\sigma_n^2\int_0^l y_n\delta y\,\mathrm{d}x \equiv -\int_0^l \left[T_{st}y_n'' + m\sigma_n^2 y_n\right]\delta y\,\mathrm{d}x + y_n'\delta y\big|_0^l = 0 \qquad (4.63)$$

for arbitrary test functions δy. This equation leads to the string equation and the zero-slope boundary conditions $y_n'(0) = y_n'(l) = 0$. These boundary conditions are therefore *natural* conditions for the variational formulation using the Rayleigh quotient. The fixed-end conditions *do not follow* from the local extrema condition (i.e., they are not natural for the studied case). The fixed-end conditions appear, therefore, as a restriction.

Examples. We consider the string problem with fixed-end conditions $y_n(0) = y_n(l) = 0$. Our objective is to find an estimate of the lowest natural frequency by using the fact that eq. (4.61) corresponds to a minimum. Since the end conditions are not natural, we must require that the assumed functional forms satisfy the end conditions. If we as an example assume the parabolic form $y_{PB} = y_0[1 - 4(x - \frac{1}{2}l)^2/l^2]$, eq. (4.61) gives a result that is 0.7% higher than the correct value. Another test function, which satisfies the fixed-end conditions, can be the sawtooth function:

$$y_{ST}(x) = y_0 \begin{cases} x & \text{for } 0 < x < \frac{1}{2}l, \\ \frac{1}{2}l - x & \text{for } \frac{1}{2}l < x < l. \end{cases}$$

Using this function gives the ratio between approximate and exact eigenvalue as $2\sqrt{3}/\pi = 1.103$; that is, the relative error is about 10%.

Let us then consider the string problem with the zero-slope end conditions. These conditions are natural conditions; therefore, it is unnecessary to impose end conditions and using specific end conditions was warned against. However, we need condition (4.62). Earlier test functions $y_{PB}(x)$ and $y_{ST}(x)$ do not satisfy this condition. The simplest possible example is $y_L(y) = y_0(-\frac{1}{2}l + x)$. This test function gives the same ratio between the approximate and the exact eigenvalues as $y_{ST}(x)$ for the fixed-end conditions.

Let us consider the test functions that satisfy both the fixed-end conditions and eq. (4.62). Here the fixed-end condition can be considered an additional restriction. Obviously, these test functions can be presented as Fourier series in terms of $y_n(x) = \sin(\pi x n/l)$. However, simple analysis shows that $y_1(x) = \sin(\pi x/l)$ does not satisfy eq. (4.62); therefore, the variant with $a_1 \neq 0$ and $a_i = 0$ ($i \geq 2$) is not possible. As a result, we will never obtain the actual lowest frequency by varying the Fourier coefficients. Actually, the result will always be clearly larger. A possible solution may be $a_2 \neq 0$ and $a_i = 0\,(i \neq 2)$. However, the estimated eigenvalue becomes twice as large as the correct one. Why we obtain bad approximations of the lowest eigenvalue by imposing unnecessary restrictions can be explained by the different geometric properties of the two lowest eigenfunctions for fixed and zero-slope end conditions. For zero-slope end conditions, y_1 is monotonic in the interval $(0, l)$, while the fixed-end conditions lead to a local extrema at $\frac{1}{2}l$. These properties cannot be fulfilled simultaneously for the test functions.

Natural boundary conditions for the Euler–Bernoulli beam model were discussed in Section 2.5.2.2. We can also derive the natural boundary conditions in a way similar to what we did for the string problem. The first step is to introduce the Rayleigh quotient, which follows from relating the kinetic and potential energies during free vibrations. The kinetic and potential energies are given by eqs. (2.96) and (2.97), respectively. If we express the beam deformations as $w(x, t) = \exp(i\sigma t)W(x)$, neglect gravity, and set the forcing equal to zero, it follows that

$$\sigma^2 = \frac{\int_0^l \mathrm{EI}\,(W'')^2\,dx}{\int_0^l mW^2\,dx},$$

which is the Rayleigh quotient for the beam problem. The next step is to introduce small variations of the Rayleigh quotient together with the necessary minima condition, which leads to the natural boundary conditions for W of zero bending moment and shear force at the beam ends.

4.6.1.2 Rayleigh quotient for natural sloshing

The minimization procedure outlined in the previous section is discussed further for spectral problem (4.2). The quotient with local minima coinciding with eigenvalues κ_n can be constructed by using energy conservation. For this purpose, let us consider the freestanding waves (4.1) and rewrite them as

$$\phi(x, y, z, t) = \varphi_n(x, y, z)\cos(\sigma_n t + \varepsilon), \quad (4.64)$$

where φ_n is the natural sloshing mode. Due to the dynamic boundary condition on Σ_0, the potential energy caused by freestanding wave solution (4.64) is

$$E_p = \tfrac{1}{2}\rho g \int_{\Sigma_0} \zeta^2\,dx\,dy = \frac{\rho\sigma_n^2}{2g}\sin^2(\sigma_n t + \varepsilon)\int_{\Sigma_0} \varphi_n^2(x, y, 0)\,dS.$$

The kinetic energy is

$$E_k = \tfrac{1}{2}\rho\cos^2(\sigma_n t + \varepsilon)\int_{Q_0} (\nabla\varphi_n)^2\,dQ.$$

Considering the energy conservation condition,

$$\frac{d}{dt}(E_k + E_p) \equiv 0$$

$$\Rightarrow \tfrac{1}{2}\rho\sigma_n \sin(2\sigma_n t + 2\varepsilon)\left[\frac{\sigma_n^2}{g}\int_{\Sigma_0} \varphi_n^2(x, y, 0)\,dS - \int_{Q_0} (\nabla\varphi_n)^2\,dQ\right] \equiv 0$$

and, therefore, we get the following formal identity:

$$\kappa_n = \frac{\sigma_n^2}{g} = \frac{\int_{Q_0} (\nabla\varphi_n)^2\,dQ}{\int_{\Sigma_0} \varphi_n^2\,dS}, \quad n = 1, 2, \ldots. \tag{4.65}$$

Based on identity (4.65), one can postulate the required functional for spectral problem (4.2) as follows:

$$K_{Q_0,\Sigma_0}(\varphi) = \frac{\int_{Q_0} (\nabla\varphi)^2\,dQ}{\int_{\Sigma_0} \varphi^2\,dS}. \tag{4.66}$$

Functional (4.66) (i.e., the Rayleigh quotient) is for sloshing proportional to the ratio of the kinetic and potential energies, whereas the Rayleigh quotient for the string and beam problems is proportional to the ratio of the potential and kinetic energies.

By definition, functional (4.66) is defined on admissible test functions φ; that is, on functions that provide finiteness for integrals in both the numerator and the denominator of eq. (4.66). Functional (4.66) is nonnegative and its absolute minimum is achieved with the constant function $\varphi_0 = \text{const} \neq 0$. This condition is the same as that for the string problem in Section 4.6.1.1. Because φ_0 *does not satisfy the liquid mass conservation* condition (see the integral at the end of eq. (4.2)), the constant function does not belong to the eigenfunctions of spectral problem (4.2).

Feschenko *et al.* (1969) gave a rigorous proof of the *variational formulation* based on functional (4.66). We discussed the same at the end of Section 4.6.1.1; it consists of the *two recurrence* items:

Define $\kappa_0 = 0$, $\varphi_0 = \text{const} \neq 0$. *Then*

1. The *lowest eigenvalue* of eq. (4.2) coincides with the absolute minimum of eq. (4.66):

$$\frac{\sigma_1^2}{g} = \kappa_1 = K_{Q_0,\Sigma_0}(\varphi_1) = \min \frac{\int_{Q_0} (\nabla\varphi)^2\,dQ}{\int_{\Sigma_0} \varphi^2\,dS} \tag{4.67}$$

with the requirement that

$$\int_{\Sigma_0} \varphi\varphi_0\,dx\,dy = 0. \tag{4.68}$$

2. Assume that the $n-1$ lower natural frequencies and modes are known (accounting for the multiplicity). Then the nth eigenvalue is

$$\frac{\sigma_n^2}{g} = \kappa_n = K_{Q_0,\Sigma_0}(\varphi_n) = \min \frac{\int_{Q_0} (\nabla\varphi)^2\,dQ}{\int_{\Sigma_0} \varphi^2\,dS}, \tag{4.69}$$

where the test functions must satisfy the orthogonality condition

$$\int_{\Sigma_0} \varphi\varphi_i\,dx\,dy = 0, \quad i = 0, \ldots, n-1. \tag{4.70}$$

A reason for the orthogonality properties given by eq. (4.70) is that each nth mode must be orthogonal to the previous $n-1$ modes due to general orthogonality condition (4.6) following from the spectral theorems. For $i = 0$, the condition expresses conservation of liquid volume.

The *proof* of the variational formulation requires extensive mathematical background in functional analysis and is beyond the scope of this book. An important part of the proof involves the calculus of variation of the Rayleigh quotient, $K_{Q_0,\Sigma_0}(\varphi)$, by the test functions φ. The test functions should not be restricted to functions satisfying the Laplace equation and the boundary conditions such as, for instance, the zero-Neumann boundary condition on the mean wetted tank surface. The Laplace equation and all the boundary conditions are then *natural* (i.e., derivable from the corresponding variational equation). Section 4.6.1.3 derives the variational equation and shows several mathematical details. If we restrict the test functions to satisfy an extra condition (e.g., a Dirichlet condition on a wall), the minimization of the functional can give eigenvalues with too large an error. As we explained in the previous section, this large error occurs because

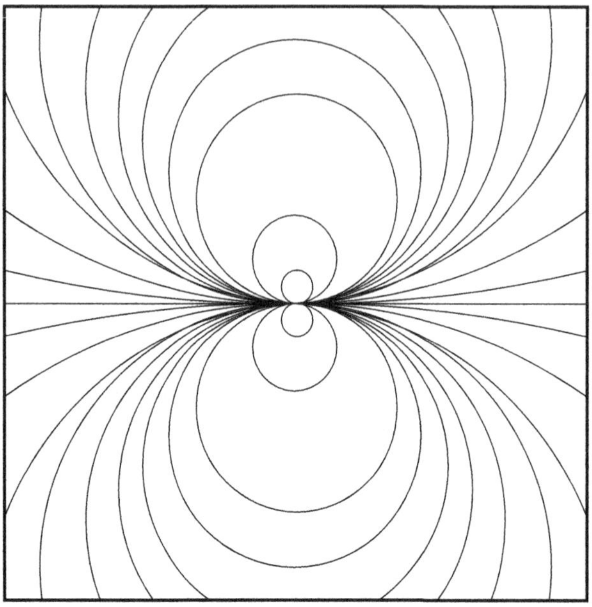

Figure 4.16. Streamlines of a horizontal dipole in infinite fluid.

additional restriction leads to wrong geometric properties of the solution. For example, the zero-Dirichlet condition on the wall for a rectangular tank requires a node on the wall for the natural modes, but we expect an antinode to be present.

The use of functional (4.66) and the fact that the eigenvalues coincide with the local minima of this functional enables us to estimate upper and lower bounds for eigenvalues in the case of complex geometries. Some estimates are quite obvious for the lowest eigenvalue that corresponds to the absolute minimum of the functional on the test functions φ that satisfy the liquid volume conservation (see eqs. (4.67) and (4.68)). For instance, when φ_* is an arbitrary function that fulfills conservation of liquid volume, the absolute minimum obviously does not exceed the value of the functional with φ_* and, therefore,

$$\kappa_1 = \min \frac{\int_{Q_0} (\nabla\varphi)^2 dQ}{\int_{\Sigma_0} \varphi^2 \, dx\, dy} \le \frac{\int_{Q_0} (\nabla\varphi_*)^2 dQ}{\int_{\Sigma_0} \varphi_*^2 \, dx\, dy}. \quad (4.71)$$

The fact that the test functions φ_* in eq. (4.71) do not need to satisfy the boundary conditions and can still provide a good estimate of the natural frequency is demonstrated by the following examples.

Example: approximate natural modes for circular and spherical tanks. Spectral problem (4.2) has no analytical solution for sloshing in a two-dimensional circular tank. Barkowiak *et al.* (1985) conducted a series of sloshing experiments in a circular tank at the lowest natural frequency. An interesting observation was that the path lines of liquid particles resembled the path lines due to an infinite-fluid horizontal dipole with singularity in the tank's centerplane above the mean free surface (i.e., the liquid particles nearly follow streamlines as illustrated in Figure 4.16). We use this observation to derive an estimate of the lowest natural frequency for two-dimensional flow in a circular tank as a function of the ratio between the filling height, h, and the radius, R_0.

We define a Cartesian coordinate system Oyz with origin at the intersection between the mean free surface and the tank's centerplane. The direction of z is positive upward. The velocity potential of an infinite-fluid horizontal dipole with singularity at $(0, a)$ can be expressed as

$$\varphi_* = \frac{y}{y^2 + (z-a)^2}, \quad (4.72)$$

where a is positive. We note that φ_* given by eq. (4.72) satisfies the Laplace equation in the liquid domain and volume conservation condition (4.68). The boundary conditions are generally not satisfied. The body-boundary conditions are only satisfied if the singularity is at the top of the horizontal circular cylinder. We now substitute eq. (4.72) into eq. (4.71) and minimize the expression by varying a. The resulting lowest natural frequency as a function of h/R_0 agrees well with benchmark numerical results and confirms that the estimate cannot be lower than the correct eigenfrequency (see, eq. (4.71)). This result is shown in Table 4.3 (compare the first and the second columns). We may also note that the dipole singularity is, in particular for lower depths, in the vicinity of the top of the horizontal circular cylinder. Furthermore, the satisfactory agreement makes it possible to utilize the dipole-like solution as the first mode in constructing a linear modal theory for sloshing in a circular tank. Section 5.4.6 applies this simplified model to analyze sloshing in a partially filled tank of a tank vehicle during a fast transverse maneuver (cornering) such as a rapid change of lanes on a highway.

Furthermore, the idea of using an infinite-fluid horizontal dipole as an approximate first natural mode is also applicable for a spherical tank. We define a Cartesian coordinate system $Oxyz$ with z upward and as a symmetry axis for the tank. The x- and y-axes are in the mean free surface. We introduce a dipole with singularity at $(0, 0, a)$ and a direction along either the x-axis or the y-axis. Here a is positive. Because the tank is axially symmetric, a cylindrical coordinate system (r, θ, z) is introduced. A general solution of the problem of natural modes in cylindrical coordinates can be represented as eq. (4.144). The lowest natural modes form the sine/cosine pair and are associated with $m = i = 1$. The dipole solutions can be written as

$$\varphi_{*\sin} = \frac{r\sin\theta}{(r^2+(z-a)^2)^{3/2}} \quad \text{and}$$
$$\varphi_{*\cos} = \frac{r\cos\theta}{(r^2+(z-a)^2)^{3/2}}. \tag{4.73}$$

We now substitute eq. (4.73) into functional (4.71) and minimize the expression by varying the parameter a. The resulting spectral parameter κ_{11}, a function of the ratio between the liquid depth and the radius of the sphere, agrees well with the benchmark calculations by McIver (1989) (see Table 4.4). We show in Section 5.4.7.2 how to use this modal representation in a simplified transient analysis of forced lateral sloshing of a spherical tank.

4.6.1.3 Variational equation

As we showed in Section 4.6.1.1, the extrema condition for quotient (4.66) should derive the corresponding variational equation for spectral problem (4.2). Let us consider functional (4.66) defined on smooth functions φ that are not restricted to any special condition. This means in particular that the functions do *not* need to satisfy the Laplace equations and a specific boundary condition. We assume that a function φ_* causes a local minimum of $K_{Q_0,\Sigma_0}(\varphi)$ and introduces small variations of this function (i.e., $\varphi = \varphi_* + \delta\varphi$), where $\delta\varphi$ is an arbitrary small smooth disturbance. The minimum is denoted

$$K_{Q_0,\Sigma_0}(\varphi_*) = \kappa_* = \frac{\int_{Q_0}(\nabla\varphi_*)^2 dQ}{\int_{\Sigma_0}\varphi_*^2\, dS}. \tag{4.74}$$

Substitution of $\varphi = \varphi_* + \delta\varphi$ into functional (4.66) and a Taylor expansion of $1/\int_{\Sigma_0}(\varphi_* + \delta\varphi)^2 dS$ in $\delta\varphi$ gives

$$K_{Q_0,\Sigma_0}(\varphi) = \underbrace{K_{Q_0,\Sigma_0}(\varphi_*)}_{\kappa_*} + \frac{2\int_{Q_0}\nabla\varphi_*\cdot\nabla(\delta\varphi)\,dQ\int_{\Sigma_0}\varphi_*^2\,dS - 2\int_{Q_0}(\nabla\varphi_*)^2 dQ\int_{\Sigma_0}\varphi_*\delta\varphi\,dS}{\left(\int_{\Sigma_0}\varphi_*^2\,dS\right)^2} + O(|\delta\varphi|^2).$$

Because $K_{Q_0,\Sigma_0}(\varphi)$ reaches the local minimum with φ_*, the linear component of the Taylor expansion must disappear. This disappearance occurs if and only if

$$\int_{Q_0}(\nabla\varphi_*\cdot\nabla(\delta\varphi))\,dQ - \kappa_*\int_{\Sigma_0}\varphi_*\delta\varphi\,dS = 0. \tag{4.75}$$

As is usually accepted in the variational analysis, the solution of eq. (4.75) is by definition a nonzero function that satisfies eq. (4.75) for all smooth test functions $\delta\varphi$.

When we derived eq. (4.75), we assumed that $\delta\varphi$ is an arbitrary "small" smooth function. Let us show that, if we do not force the test functions $\delta\varphi$ to satisfy an additional boundary condition, the solution of the variational equation is exactly the set collected from the eigenfunctions of eq. (4.2) and the constant solution $\varphi_0 = \text{const} \neq 0$. Therefore, the Laplace equation and all the boundary condition are *natural* for variational formulation (4.75) as we discussed in Section 4.6.1.1. The parameter κ_* should then be either equal to the eigenvalues or zero for the constant solution. Indeed, one can rewrite eq. (4.75) by first noting the fact that $\nabla\varphi_* \cdot \nabla\delta\varphi = \nabla\cdot(\nabla\varphi_*\delta\varphi) - \delta\varphi\nabla^2\varphi_*$ and then using the divergence theorem (see eq. (A.1)). The result is

$$-\int_{Q_0} \nabla^2\varphi_*\delta\varphi \, \mathrm{d}Q + \int_{S_0} \frac{\partial\varphi_*}{\partial n}\delta\varphi \, \mathrm{d}S + \int_{\Sigma_0}\left(\frac{\partial\varphi_*}{\partial z} - \kappa_*\varphi_*\right)\delta\varphi \, \mathrm{d}S = 0. \quad (4.76)$$

Furthermore, as accepted in the variational analysis, we must first consider all possible smooth test functions $\delta\varphi$. A subset of these functions includes those functions that become zero on the boundaries S_0 and Σ_0 but nonzero in the domain Q_0. This subset annihilates the second and third integrals in eq. (4.76). The first integral is zero for arbitrary smooth test function $\delta\varphi$ if the solution φ_* satisfies the Laplace equation $\nabla^2\varphi_* = 0$, meaning that we can rewrite eq. (4.76) in the form for the solution φ_*:

$$\int_{S_0} \frac{\partial\varphi_*}{\partial n}\delta\varphi \, \mathrm{d}S + \int_{\Sigma_0}\left(\frac{\partial\varphi_*}{\partial z} - \kappa_*\varphi_*\right)\delta\varphi \, \mathrm{d}S = 0.$$

Now we can consider a subset of smooth test functions $\delta\varphi$ that are zero on Σ_0 but nonzero on S_0. The variational equation then gives

$$\int_{S_0} \frac{\partial\varphi_*}{\partial n}\delta\varphi \, \mathrm{d}S = 0$$

for arbitrary smooth traces $\delta\varphi|_{S_0}$ and, therefore, the solution must satisfy the zero-Neumann condition on the mean wetted tank surface (i.e., $\partial\varphi_*/\partial n = 0$ on S_0). Finally, we arrive at

$$\int_{\Sigma_0}\left(\frac{\partial\varphi_*}{\partial z} - \kappa_*\varphi_*\right)\delta\varphi \, \mathrm{d}S = 0$$

to be satisfied by the solution φ_* for smooth test functions with nonzero traces $\delta\varphi|_{\Sigma_0}$.

Summarizing the analysis, the solution φ_* fulfills the Laplace equation and all the boundary condition of eq. (4.2); that is,

$$\nabla^2\varphi_* = 0 \quad \text{in} \quad Q_0; \quad \frac{\partial\varphi_*}{\partial n} = 0 \quad \text{on} \quad S_0;$$
$$\frac{\partial\varphi_*}{\partial z} = \kappa_*\varphi_* \quad \text{on} \quad \Sigma_0. \quad (4.77)$$

The *volume conservation* condition *does not follow* from variational equation (4.75). Therefore, solution φ_* can be either an eigenfunction of eq. (4.2) or a constant function with $\kappa_* = 0$. Obviously, the analysis may fail when admissible solutions φ are restricted to an additional boundary condition (e.g., $\varphi = 0 \Rightarrow \delta\varphi = 0$). We discussed why this is so in the simple example from Section 4.6.1.1.

We can also show eq. (4.75) by starting with multiplying the Laplace equation for φ_* by $\delta\varphi$ and integrating the expression over Q_0 (i.e., $\int_{Q_0} \nabla^2\varphi_*\delta\varphi \, \mathrm{d}Q = 0$). Then we can rewrite the integral as $\int_{Q_0} \nabla\cdot(\nabla\varphi_*\delta\varphi)\,\mathrm{d}Q - \int_{Q_0}\nabla\varphi_* \cdot(\nabla\delta\varphi)\,\mathrm{d}Q = 0$. Application of the divergence theorem and the boundary conditions expressed by eq. (4.77) leads to eq. (4.75). Equation (4.75) is more general, however, because it does not require φ_* to satisfy the Laplace equation or specific boundary conditions.

Variational equation (4.75) is later used to find asymptotic approximations of natural frequencies as a function of small changes of liquid domain Q_0. The equation can also be combined with the Galerkin method (see Section 10.5.1) to find accurate approximations of the natural frequencies and modes (Fox & Kuttler, 1983; Lukovsky *et al.*, 1984). Therefore, we approximate φ_* as

$$\varphi_* = \varphi_i^{(q)} = \sum_{n=1}^{q} b_n\phi_n, \quad (4.78)$$

where $\{\phi_n, n \geq 1\}$ is a complete functional basis and b_n ($n = 1, \ldots, q$) are unknown coefficients. We now substitute eq. (4.78) into equality (4.75) and set $\delta\varphi = A\phi_i$ ($i = 1, \ldots, q$), where A is an arbitrarily small constant, to satisfy the requirement that $\delta\varphi$ should be small. However, as will be shown, the solution does not depend on A. An essential feature of the Galerkin method is that the same ϕ_n used in eq. (4.78) is used in $\delta\varphi$. The result is the spectral matrix problem

$$(\{\alpha_{ij}\} - \kappa^{(q)}\{\beta_{i,j}\})\boldsymbol{b}^{(q)} = \mathbf{0}, \quad (4.79)$$

where the matrices are computed by the functional set $\{\phi_k\}$ as follows:

$$\alpha_{ij} = \int_{Q_0} (\nabla\phi_i \cdot \nabla\phi_j)\,\mathrm{d}Q;$$
$$\beta_{ij} = \int_{\Sigma_0} \phi_i\phi_j \,\mathrm{d}x\,\mathrm{d}y, \quad i, j = 1, \ldots, q.$$

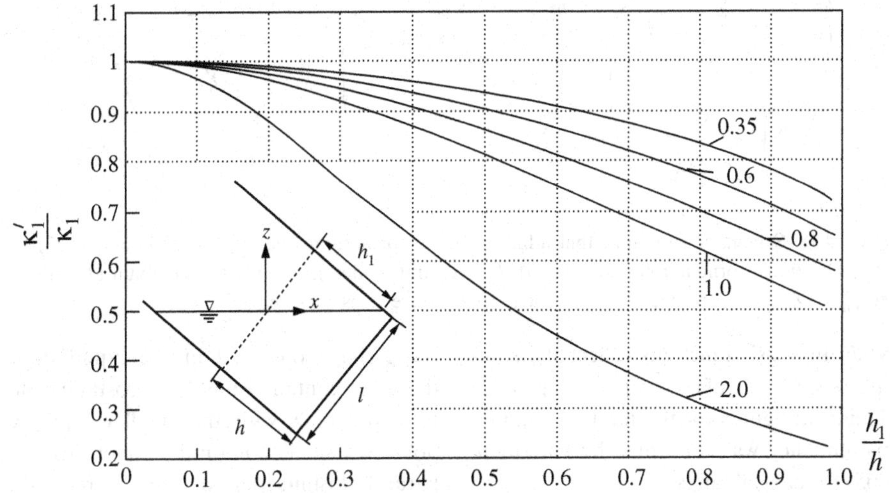

Figure 4.17. The ratio between the first eigenvalue κ_1' for sloshing in an inclined rectangular tank and the eigenvalue κ_1 without any inclination versus the ratio h_1/h. Results are presented for different values of the ratio h/l between the liquid depth and tank breadth without any tank inclination.

The matrices are symmetric. For eq. (4.79) to have a nontrivial solution it is necessary that $\det[\{\alpha_{ij}\} - \kappa^{(q)}\{\beta_{ij}\}] = 0$. This equation determines the real eigenvalues $\kappa_i^{(q)}$ $(i = 1, \ldots, q)$ and eigenvectors $\boldsymbol{b}_k^{(q)} = (b_1^{(k)}, \ldots, b_q^{(k)})$ $(k = 1, \ldots, q)$.

It is important that the Galerkin scheme is equivalent to the minimization in eq. (4.69) and, therefore, the approximate $\kappa_i^{(q)}$ converge monotonically to the value of κ_i from above as $q \to \infty$. The consequence is that the representation of the velocity potential converges toward the correct solution satisfying the Laplace equation and the boundary conditions.

When the functional basis consists of functions satisfying the Laplace equation, this variational approach belongs to the so-called Ritz–Treftz schemes. The matrix elements $\{\alpha_{ij}\}$ may then be reexpressed as

$$\alpha_{ij} = \int_{S_0+\Sigma_0} \frac{\partial \phi_i}{\partial n} \phi_j \, dS.$$

We can show this result by using Green's first identity (see eq. (A.5)). Because the integrals in $\{\alpha_{ij}\}$ have lower dimensions than the original problem (surface integrals for a three-dimensional statement and one-dimensional integrals for a two-dimensional statement), the calculations are more CPU efficient. Furthermore, numerical results of, for instance, Feschenko *et al.* (1969), Lukovsky et al. (1984), and others showed that the Ritz–Treftz scheme provides faster convergence than the original Galerkin method.

As an example of results provided by the Ritz–Treftz scheme, we examine the inclined rectangular tank illustrated in Figure 4.17. The functional basis is described by Timokha (2002). Figure 4.17 shows the calculated ratio between the first eigenvalue in an inclined rectangular tank, κ_1', and the eigenvalue without any inclination, κ_1 (see eq. (4.8)), versus $0 \le h_1/h = \frac{1}{2} l h^{-1} \tan\alpha \le 1$. The inclination of the tank causes a reduction of the lowest natural frequencies. The effect increases with increasing liquid depth. The decrease is 5–20% for intermediate and shallow depths.

Although the Ritz–Treftz schemes for the linear sloshing problem are very CPU-efficient and accurate, they are rarely present in the literature of the past decade because of difficulties in constructing a complete functional basis for many tank shapes, for instance. As long as the functional system is not complete or does not account for analytical properties of the eigenfunctions, the scheme may fail. For instance, it is challenging to account for internal structures properly because of the requirement that flow singularities both interior to the internal structure and at possible sharp corners are considered.

The Ritz–Treftz method is very sensitive to the functional basis and special mathematical studies of the analytical properties of the eigenfunctions are needed to construct the most appropriate

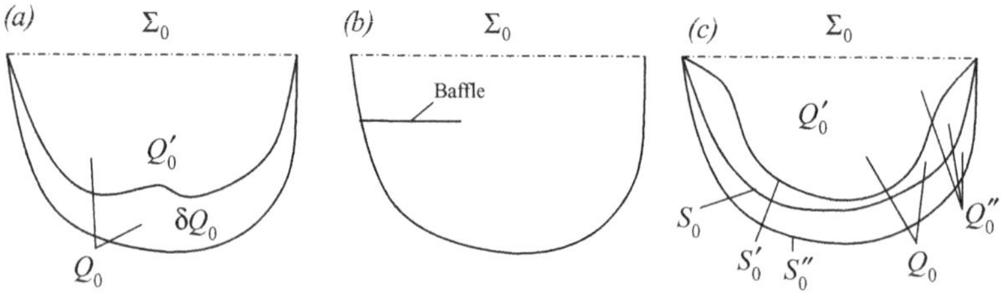

Figure 4.18. Schematic situations that admit estimates of natural frequencies (4.81) and (4.82) by using variational formulation (4.67)–(4.69). The natural modes φ_n are found in domain Q_0, but φ_n' and φ_n'' in Q_0' and Q_0'', respectively. The mean free surface Σ_0 is constant.

basis. When the basis functions reflect the solution properties, 10 to 20 of these functions provide 5 to 6 significant figures for the natural frequencies. This tendency was confirmed by Gavrilyuk *et al.* (2005) for conical tanks.

4.6.2 Natural frequencies versus tank shape: comparison theorems

Let us consider a domain Q_0' that is geometrically close to a shape Q_0 and is contained within Q_0. The corresponding mean free surfaces Σ_0 and Σ_0' are the same. The *lowest natural mode* φ_1 for domain Q_0 is obviously also defined in Q_0'. It follows from eq. (4.71) with integration over Q_0' and with $\varphi_* = \varphi_1$ that

$$\kappa_1' = \frac{\int_{Q_0'} (\nabla\varphi_1')^2\,dQ}{\int_{\Sigma_0} \varphi_1'^2\,dx\,dy} \le \frac{\int_{Q_0'} (\nabla\varphi_1)^2 dQ}{\int_{\Sigma_0} \varphi_1^2\,dx\,dy}. \tag{4.80}$$

Under the same condition, $\Sigma_0 = \Sigma_0'$ (two containers with the same free surface, Figure 4.18(a) and Figure 4.18(b)), the functional formulation of spectral problem (4.2) makes possible a set of more general estimates between *all the eigenvalues* in Q_0 and Q_0'. For these cases, Feschenko *et al.* (1969), Lukovsky *et al.* (1984), and Morand and Ohayon (1995) proved the following:

> Let two mean liquid domains Q_0 and Q_0' be such that $Q_0' \subset Q_0$, $Q_0 = Q_0' \cup \delta Q_0$, and $\Sigma_0' = \Sigma_0$. Then, accounting for the multiplicity (i.e., more than one eigenfunction corresponds to one eigenvalue), the following inequality is true:

$$\kappa_i' \le \kappa_i\,(\sigma_i' \le \sigma_i), \quad i \ge 1. \tag{4.81}$$

Therefore, if the liquid domain of one of two tanks with equal mean free surface can be completely contained in the liquid domain of the other container under hydrostatic conditions, then the tank with the smaller liquid volume possesses smaller natural frequencies or higher natural sloshing period (see Figure 4.18(a) and Figure 4.18(b)).

To find the lower and upper bounds by using inequality (4.81), we should find two appropriate domains that circumscribe/inscribe the considered domain but have the same mean free surface, which is depicted in Figure 4.18(c). Furthermore, if we know the natural frequencies κ_i' and κ_i'' for the two domains Q_0' and Q_0'' ($Q_0' \subset Q_0 \subset Q_0''$, but $\Sigma_0' = \Sigma_0 = \Sigma_0''$), then

$$\kappa_i' \le \kappa_i \le \kappa_i''. \tag{4.82}$$

We demonstrate this fact with the following example.

Example: lowest natural frequency for a half-filled circular tank. Let us illustrate how we can evaluate the first natural frequency in a half-filled two-dimensional circular tank with diameter l as shown in Figure 4.19. No analytical solution exists in this case. To estimate the lowest eigenvalue κ_1 of spectral problem (4.2), we note that the liquid

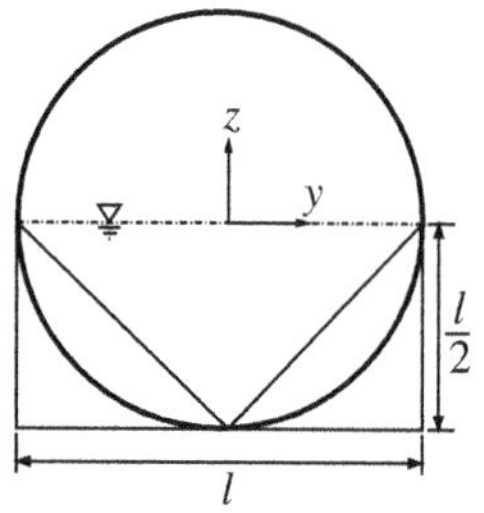

Figure 4.19. Estimate of the lowest natural sloshing frequency for a half-filled circular tank using upper and lower bounds for a rectangle and triangle, respectively.

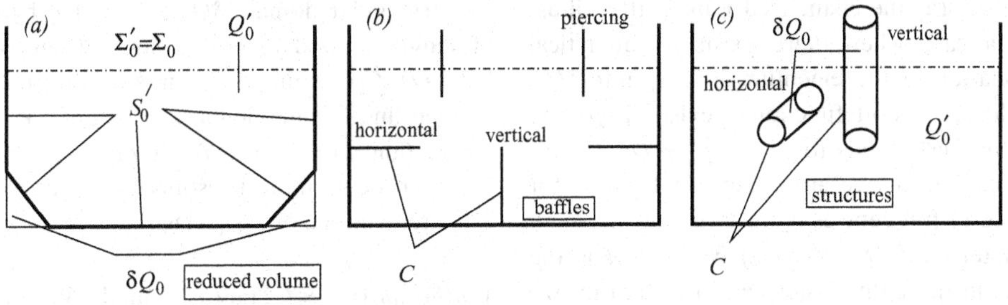

Figure 4.20. Three typical reductions of the contained liquid domain Q_0 of a rectangular tank, $Q_0 = Q_0' \cup \delta Q_0$.

domain is contained within a rectangle of width l and height $\frac{1}{2}l$. Furthermore, a triangle with base l, sides $l/\sqrt{2}$, and semi-apex angle 45° is contained in the liquid domain. Figure 4.19 shows that the mean free surface of the three shapes (triangular, sectorial, and rectangular) is the same. Thus, due to inequality (4.82), $\kappa_1^{\text{triangle}} \leq \kappa_1^{\text{half-circle}} \leq \kappa_1^{\text{rectangle}}$. The lowest eigenvalue for the rectangle was found analytically in Section 4.3.1.1: $\kappa_1^{\text{rectangle}} = \pi l^{-1}\tanh(\frac{1}{2}\pi)$. As shown in Section 4.3.1.2, analytical expressions also exist for eigenvalues in the case of a triangular tank with semi-apex angle 45°. Due to eq. (4.22), the lowest eigenvalue is equal to $\kappa_1^{\text{triangle}} = 2/l$, meaning that

$$2l^{-1} \leq \kappa_1^{\text{half-circle}} = \sigma_1^2 g^{-1} \leq \pi l^{-1}\tanh\left(\tfrac{1}{2}\pi\right). \tag{4.83}$$

For $l = 2\,\text{m}$, inequality (4.83) estimates the lowest circular sloshing frequency between 3.13 rad/s $\leq \sigma_1 \leq$ 3.76 rad/s and the natural period between 1.67 s $\leq T_1 \leq$ 2.01 s. The benchmark numerical results presented in Table 4.3 give $\sigma_1 = 3.65$ rad/s (i.e., closer to the result for the rectangular tank than that for the triangular tank).

4.6.3 Asymptotic formulas for the natural frequencies and the variational statement

4.6.3.1 Small liquid-domain reductions of rectangular tanks

As we mentioned in the beginning of Section 4.6, a rough approximation of the eigenvalues κ_n' in a reduced domain Q_0' may be obtained by using eq. (4.60). When $\Sigma_0 = \Sigma_0'$, a common way is to substitute the eigenfunctions φ_n found for the original, larger domain $Q_0 \supset Q_0'$ into the functional K_{Q_0',Σ_0}. The domains Q_0 and Q_0' should be close to each other. The difference δQ_0 between Q_0 and Q_0' is characterized by a small dimension with respect to the characteristic dimension of the liquid domain Q_0 and the mean free surface Σ_0. The characteristic dimension of Q_0 should be comparable to the considered wavelengths of the natural modes, φ_n and φ_n'. The analysis assumes that the original natural frequencies (eigenvalues κ_n) and modes (eigenfunctions φ_n) are known in an analytical or semianalytical form.

Formula (4.60) does not tell what the error is in terms of the smallness of δQ_0 and due to the replacement of φ_n' by φ_n in the corresponding functional. The formula assumes implicitly that a small reduction of the mean liquid domain causes small changes of the eigenfunctions φ_n' so that it continuously tends to φ_n as $Q_0' \to Q_0$ and $\Sigma_0' \to \Sigma_0$. This assumption is not true in the most general case.

In the next subsections and Section 4.7, we concentrate on the three reductions of the liquid domains of rectangular tanks shown in Figure 4.20. Cases (b) and (c) illustrate modifications of the liquid domain due to fixed interior structures (in this example, baffles and poles). The effect of these structures on the natural sloshing frequencies is analyzed in detail in Sections 4.7.2 and 4.7.3. The flow in reality would separate at the sharp corner of a baffle and for cross-flow past a pole when the Keulegan–Carpenter number is sufficiently large (see Chapter 6). The flow separation is associated with viscous effects, which are neglected in our potential flow analysis. Both the baffles and the poles cause eigenfunctions φ_n' with singularities inside the structure as well as at the sharp edge of a baffle. The singularities do not vanish as the length of a baffle or a pole tends to zero. Therefore, φ_n' does not tend to φ_n as $Q_0' \to Q_0$ and a special technique is needed to derive asymptotic approximations (which is discussed in Section 4.7.1). In contrast, in

Figure 4.20(a) the chamfered tank bottom illustrates a case when there exists an analytical continuation of the eigenfunctions φ_n' into δQ_0. Therefore, a smooth function φ_n' exists in δQ_0 and it continuously tends to φ_n as $Q_0' \to Q_0$. In this case, one can derive an asymptotic formula for the natural frequencies σ_n' in terms of the small parameter $Vol(\delta Q_0)/Vol(Q_0)$ where Vol is the volume in three dimensions and the area in two dimensions. This case is explored in the next subsection.

4.6.3.2 Asymptotic formula for a chamfered tank bottom: examples

We consider cases where the natural modes φ_n' in a domain Q_0' can be analytically continued into a domain Q_0, where analytical solutions φ_n of the natural modes exist. The objective is to express the natural frequencies $\sigma_n' = \sqrt{g\kappa_n'}$ for the domain Q_0' in terms of the natural frequencies $\sigma_n = \sqrt{g\kappa_n}$ for the domain Q_0 with a small correcting factor that accounts for the difference in the liquid domains. The assumptions are as follows:

- The area/volume of the difference δQ_0 between the domains Q_0 and Q_0' is small relative to the area/volume of Q_0.
- The characteristic length dimensions of δQ_0 are small relative to the wavelengths associated with the natural modes φ_n' and φ_n.
- The mean free surfaces Σ_0' and Σ_0 of domains Q_0 and Q_0' are the same.

Our specific cases assume two-dimensional flow. The domain Q_0 corresponds to a rectangular tank. We use the derivation for the chamfered tank bottom shown in Figure 4.21 as an example.

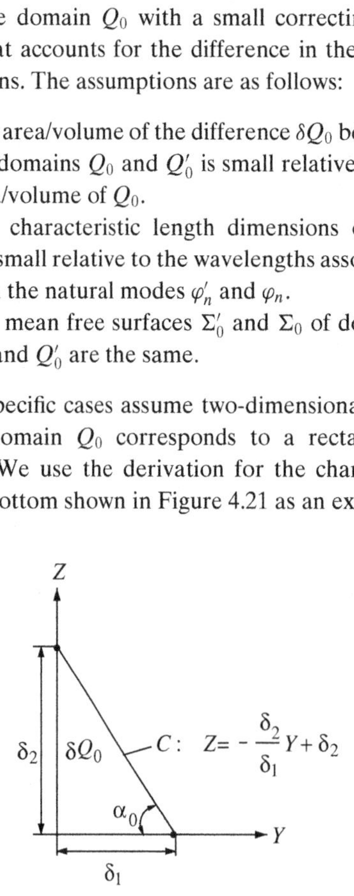

Figure 4.21. Chamfered tank bottom. Local coordinate system and definitions needed for derivation of formula (4.90).

The triangular domain δQ_0 is bounded by sides of length δ_1 and δ_2 and a straight line C. A local OYZ-coordinate system is introduced. The straight line C is defined as $Z = -\tan(\alpha_0)Y + \delta_2$, where $\tan(\alpha_0) = \delta_2/\delta_1$. We *fix the angle* α_0 and consider the asymptotic estimate of σ_n' as $\delta_2 \to 0$. The area/volume of $\delta Q_0[\delta_2]$ is

$$Vol(\delta Q_0[\delta_2]) = \delta_2^2/(2\tan(\alpha_0)) \quad \text{and} \quad Vol(Q_0'[\delta_2]) = Vol(Q_0) - Vol(\delta Q_0[\delta_2]). \tag{4.84}$$

Based on the mathematical arguments from Section 4.6.3.3, let us consider an analytical continuation of $\varphi_n'(y, z)$ through the boundary C into the domain δQ_0, which causes φ_n' to be defined in the domain Q_0. Moreover, from the theorems of Lukovsky *et al.* (1984), integrals of φ_n' and their first derivatives over the domain Q_0 are finite and continuous functions of δ_2. The same holds true for the eigenvalues (i.e., we get the continuous function $\kappa_n'[\delta_2]$). Obviously, $\kappa_n = \kappa_n'[0]$ and $\varphi_n = \varphi_n'[0]$. Our goal is to find an asymptotic expression for $\kappa_n'[\delta_2]$ in the limit $\delta_2 \to 0$.

Let us specify the lowest asymptotic order (in terms of δ_2) in the asymptotic expansion of the eigenvalues $\kappa_n'[\delta_2]$ and eigenfunctions $\varphi_n'[\delta_2]$. For this purpose, we rewrite variational equation (4.75) for Q_0' as follows:

$$\underbrace{\int_{Q_0} (\nabla \varphi_n' \cdot \nabla(\delta\varphi))\,dQ}_{\sim|\delta\varphi|} - \underbrace{\int_{\delta Q_0[\delta_2]} (\nabla \varphi_n' \cdot \nabla(\delta\varphi))\,dQ}_{\sim O(\delta_2^2)|\delta\varphi|} - \kappa_n' \underbrace{\int_{\Sigma_0} \varphi_n' \delta\varphi\,dy}_{\sim|\delta\varphi|} = 0. \tag{4.85}$$

Here and subsequently the integrals over Q_0 exist because there exists an analytical continuation of φ_n' into δQ_0. The order of the two-dimensional integral over δQ_0 follows from eq. (4.84). Comparing the order of the integrals in eq. (4.85) and accounting for eq. (4.75) to be fulfilled for $\varphi_* = \varphi_n$ yields the following asymptotics:

$$\varphi_n'[\delta_2] = \varphi_n + \delta_2^2 F(y, z) + o(\delta_2^2),$$
$$\kappa_n'[\delta_2] = \kappa_n + \delta_2^2 \tilde{\kappa} + o(\delta_2^2). \tag{4.86}$$

Substitution of eqs. (4.86) into equality (4.65) with the domain Q_0' gives

$$\kappa_n'[\delta_2] = \frac{\int_{Q_0} (\nabla\varphi_n + \delta_2^2 \nabla F + \cdots)^2 \mathrm{d}Q}{\int_{\Sigma_0} (\varphi_n + \delta_2^2 F + \cdots)^2 \mathrm{d}S} - \frac{\int_{\delta Q_0[\delta_2]} (\nabla\varphi_n + \delta_2^2 \nabla F + \cdots)^2 \mathrm{d}Q}{\int_{\Sigma_0} (\varphi_n + \delta_2^2 F + \cdots)^2 \mathrm{d}S}$$
$$= I_1 - I_2 + o(\delta_2^2). \quad (4.87)$$

Due to eq. (4.84),

$$I_2 = \frac{\int_{\delta Q_0[\delta_2]} (\nabla\varphi_n)^2 \mathrm{d}Q}{\int_{\Sigma_0} (\varphi_n)^2 \mathrm{d}S} + o(\delta_2^2), \quad (4.88)$$

and I_1 can be approximated by Taylor expansion of $1/\int_{\Sigma_0} (\varphi_n + \delta_2^2 F + \cdots)^2 \mathrm{d}S$ and keeping terms of $O(1)$ and $O(\delta_2^2)$; that is,

$$I_1 = \frac{\int_{Q_0} (\nabla\varphi_n)^2}{\int_{\Sigma_0} (\varphi_n)^2 \mathrm{d}S} + 2\delta_2^2 \frac{\int_{Q_0} \nabla\varphi_n \cdot \nabla F \,\mathrm{d}Q}{\int_{\Sigma_0} (\varphi_n)^2 \mathrm{d}S} - 2\delta_2^2 \frac{\int_{Q_0} (\nabla\varphi_n)^2}{\int_{\Sigma_0} (\varphi_n)^2 \mathrm{d}S} \cdot \frac{\int_{\Sigma_0} \varphi_n F \,\mathrm{d}S}{\int_{\Sigma_0} (\varphi_n)^2 \mathrm{d}S} + o(\delta_2^2).$$

It follows by using eqs. (4.65) and (4.75) with $\varphi_n = \varphi_*$, $\kappa_n = \kappa_*$, and $F = \delta\varphi$ that

$$I_1 = \kappa_n + 2\delta_2^2 \underbrace{\frac{\int_{Q_0} \nabla\varphi_n \cdot \nabla F \,\mathrm{d}Q - \kappa_n \int_{\Sigma_0} \varphi_n F \,\mathrm{d}S}{\int_{\Sigma_0} (\varphi_n)^2 \mathrm{d}S}}_{0} + o(\delta_2^2). \quad (4.89)$$

Summarizing eqs. (4.87)–(4.89), one obtains the required asymptotic formula

$$\boxed{\frac{\sigma_n'^2}{\sigma_n^2} = \frac{\kappa_n'}{\kappa_n} = 1 - \frac{\int_{\delta Q_0} (\nabla\varphi_n)^2 \mathrm{d}Q}{\kappa_n \int_{\Sigma_0} (\varphi_n)^2 \mathrm{d}S}} + \underbrace{o\left(\frac{Vol(\delta Q_0)}{Vol(Q_0)}\right)}_{o(\delta_2^2/l^2)}. \quad (4.90)$$

Even though part of the derivation of eq. (4.90) considered a chamfered tank bottom with two-dimensional flow, the formula is not restricted to that case. We can confirm this fact by using the assumptions stated with bullet points and following a derivation similar to that expressed by eqs. (4.85)–(4.89). One of the essential assumptions is analytical continuation of the velocity potential. A discussion of the possibility of the analytical continuation is presented in Section 4.6.3.3.

All the quantities in eq. (4.90) are scaled by the characteristic wavelength $\lambda \sim l$. For the higher natural modes with short wavelengths, we have the limit $\sim 1/\kappa_n \to 0$ ($\kappa_n = \pi n l^{-1} \tanh(\pi n h l^{-1}) \sim \pi n l^{-1}$, $n \to \infty$ for rectangular tank). Equation (4.90) may, therefore, become invalid with increasing n. The lowest natural frequency can be approximated as

$$\frac{\sigma_1'^2}{g} = \kappa_1' = \min_{\substack{\text{by smooth} \\ \varphi \text{ that satisfy} \\ \int_{\Sigma_0} \varphi \,\mathrm{d}x\,\mathrm{d}y = 0}} \frac{\int_{Q_0'} (\nabla\varphi)^2 \mathrm{d}Q}{\int_{\Sigma_0} (\varphi)^2 \mathrm{d}S}$$
$$\leq \frac{\int_{Q_0'} (\nabla\varphi_1)^2 \mathrm{d}Q}{\int_{\Sigma_0} (\varphi_1)^2 \mathrm{d}S} = \kappa_1 - \frac{\int_{\delta Q_0} (\nabla\varphi_1)^2 \mathrm{d}Q}{\int_{\Sigma_0} (\varphi_1)^2 \mathrm{d}S}. \quad (4.91)$$

Example: chamfered two-dimensional prismatic tank. We use formula (4.90) to evaluate the lowest natural sloshing frequency for the tank illustrated in Figure 4.22. Natural modes (4.8) can be rewritten in the local coordinates of Figure 4.21 as

$$\varphi_n(Y, Z) = \cos\left(\frac{\pi n}{l} Y\right) \frac{\cosh(\pi n Z/l)}{\cosh(\pi n h/l)}.$$

Using the fact that $\nabla \cdot (\varphi_n \nabla\varphi_n) = \nabla\varphi_n \cdot \nabla\varphi_n + \varphi_n \nabla^2\varphi_n$, $\nabla^2\varphi_n = 0$ together with the divergence theorem (see eq. (A.1)) and that $\partial\varphi_n/\partial n$ is only nonzero on the chamfer surface C gives

$$\int_{\delta Q_0} (\nabla\varphi_n)^2 \mathrm{d}Q = \int_C \left.\frac{\partial\varphi_n}{\partial n}\right|_C \varphi_n|_C \,\mathrm{d}S$$
$$= \int_0^{\delta_1} \left[\frac{\delta_2}{\delta_1} \left.\frac{\partial\varphi_n}{\partial Y}\right|_{Z=-\frac{\delta_2}{\delta_1}Y+\delta_2} + \left.\frac{\partial\varphi_n}{\partial Z}\right|_{Z=-\frac{\delta_2}{\delta_1}Y+\delta_2} \right] \varphi_n|_{Z=-\frac{\delta_2}{\delta_1}Y+\delta_2} \,\mathrm{d}Y.$$

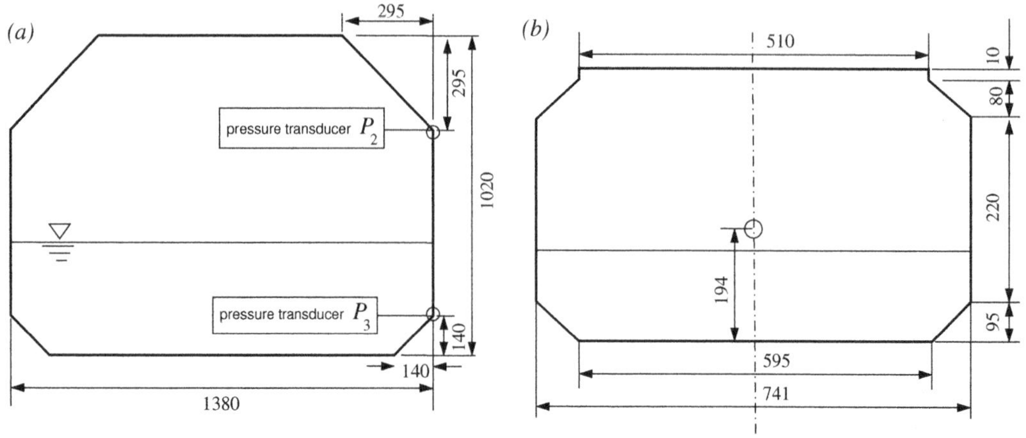

Figure 4.22. Prismatic tanks examined by (a) Abramson *et al.* (1974) and (b) Mikelis *et al.* (1984). Dimensions in millimeters.

Equation (4.90) can then be expressed as

$$\frac{\sigma_n'^2}{\sigma_n^2} = 1 - \frac{\left[\delta_1\delta_2^{-1}\sinh^2(\pi n\delta_2/l) - \delta_2\delta_1^{-1}\sin^2(\pi n\delta_1/l)\right]}{\pi n \sinh(2\pi nh/l)}. \tag{4.92}$$

We note that eq. (4.92) is consistent with the asymptotic assumption in eq. (4.90). Equation (4.92) has the same structure and the lowest-order asymptotic in $\delta_1, \delta_2 \to 0$, but it differs from the expression by Faltinsen and Timokha (2001) due to an arithmetic error in that paper.

Faltinsen and Timokha (2001) also gave a numerical estimate of the lowest eigenfrequency for the tanks in Figure 4.22. The model test with the left tank in Figure 4.22 was performed with $h/l = 0.4$. Equation (4.92) gives $\sigma_1'/\sigma_1 = 0.9996428\ldots$ (i.e., the effect of the lower chamfers is less than 0.1% for the lowest natural frequency). The model tests for the right tank were performed with $h/l = 0.246$. The ratio σ_1'/σ_1 then becomes equal to $0.998462\ldots$; the relative error is less than 0.2%. The relative error due to replacing the tanks in Figure 4.22 by a rectangle therefore gives a negligible error for the lowest natural frequency. This small influence of the chamfered bottom is intuitively expected due to the nearly stagnant flow at the tank corners when the tank bottom is not chamfered.

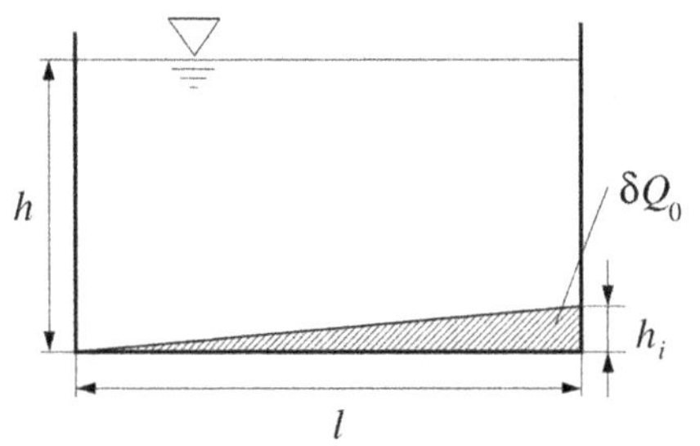

Figure 4.23. Inclined bottom of a swimming pool.

Example: inclined tank bottom. We also apply eq. (4.92) to an inclined bottom as illustrated in Figure 4.23. This scenario is relevant for a swimming pool on a cruise liner (see Figure 4.15). Equation (4.92) can in this special case be expressed as

$$\frac{\sigma_n'^2}{\sigma_n^2} = 1 - \left[\frac{l}{h_i}\sinh^2\left(\pi n\frac{h_i}{l}\right)\right] \Big/ \left[\pi n \sinh\left(\frac{2\pi nh}{l}\right)\right]. \tag{4.93}$$

Example: half-filled circular tank. Analytical continuation from a half-filled circular tank to a rectangular tank with the same free-surface breadth and fill depth is possible, too. It means that eq. (4.91) can be applied with proper definition of δQ_0. The result is $\kappa_1^{\text{half-circle}} = 2.74/l$. Benchmark numerical results by McIver (1989) presented in Table 4.3 give $\kappa_1^{\text{half-circle}} = 2.71/l$.

Example: triangular tank. The asymptotic formula assumes $Vol(\delta Q_0)/Vol(Q_0)$ to be small parameter. What the word "small" means cannot be quantitatively determined and is a question of comparisons with exact or benchmark

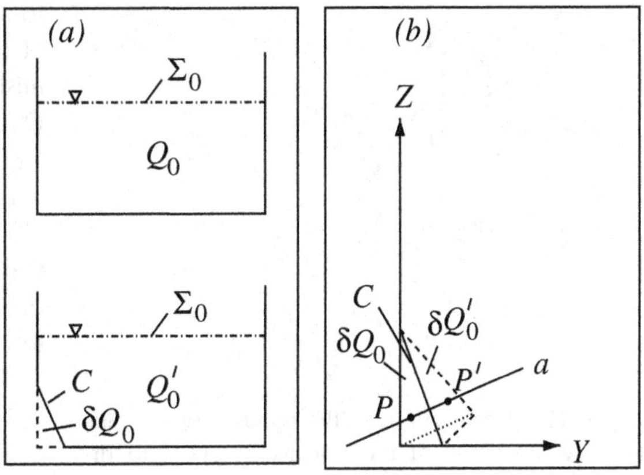

Figure 4.24. Applicability of analytical continuation to a lower chamfer of a prismatic tank.

solutions. The previous case with a half-filled circular tank gave reasonable agreement with $Vol(\delta Q_0)/Vol(Q_0) = 0.215$. Another case with larger $Vol(\delta Q_0)/Vol(Q_0)$ is a triangular tank with semi-apex angle $\theta_0 = 45^0$ (see Figure 4.4). The exact result is $\kappa_1^{\text{triangle}} = 2/l$ according to eq. (4.22), whereas the asymptotic formula with $Vol(\delta Q_0)/Vol(Q_0) = 0.5$ gives $\kappa_1^{\text{triangle}} = 2.20/l$.

4.6.3.3 Discussion on the analytical continuation and the applicability of formula (4.90)

Let us first show that the case of the chamfered tank bottom enables an analytical continuation of natural modes φ_n' into δQ_0 and, as a result, one can use analytical flow in Q_0' to describe the flow in δQ_0, and derived formulas (4.85)–(4.91) are valid. This case is first of all for the chamfered tank bottom in a rectangular tank, which is presented for simplicity in Figure 4.24(a). The following discussion is also relevant for CFD methods using ghost cells (see Section 10.3.9 and Figure 10.10) and ghost particles (see Section 10.6) outside the flow domain to enforce body-boundary conditions.

The mirror symmetry technique. Let us consider a chamfer as shown in Figure 4.24(b). The eigenfunction φ_n' is defined in Q_0'. It satisfies the zero-Neumann boundary condition on the boundary C, which appears due to the chamfer. The analytical continuation into δQ_0 is possible by using a mirror image of φ_n' relative to the straight line C. The mirror-image process uses the fact that the zero-Neumann condition is fulfilled on C by the function φ_n' and, therefore, the continuation of φ_n' must be an even function along a line a that is perpendicular to C (see Figure 4.24(b)). Taking a point P in δQ_0, one may consider the point P', which is symmetric to P relative to C. Furthermore, we have by definition that

$$\varphi_n'(P) = \varphi_n'(P'). \tag{4.94}$$

Equation (4.94) together with φ_n' in Q_0' defines the function φ_n' in the whole Q_0. Obviously, the function φ_n' satisfies the zero-Neumann condition on C and is continuous at C. As a consequence, both the Dirichlet and Neumann transmission conditions are satisfied for C, and the constructed φ_n' is a smooth function satisfying the Laplace equation for the whole Q_0 due to the corresponding transmission theorems (Aubin, 1972).

In the mirror-symmetry technique, definition (4.94) yields the analytical continuation in δQ_0 if and only if the mirror projection of δQ_0 completely belongs to Q_0'. Figure 4.24(b) illustrates this point. The triangle $\delta Q_0'$ is a "donor" of the domain δQ_0. If Q_0' has a smooth boundary and a convex domain, it is also possible to analytically continue the velocity potential in Q_0' to the domain of a rectangular tank, Q_0. Examples are circular and elliptical two-dimensional tanks and spherical tanks with different filling heights.

It is not possible analytically to continue the velocity potential to the domain of an interior tank structure, for instance to the polyhedral domain D illustrated in Figure 4.25. To show that analytical continuation is impossible we pick any point P inside D, as illustrated in the figure. The point defines two points, P^* and P^{**}, outside of D which are symmetric relative to P about the straight portions C' and C'' of the boundary of D. If we apply the mirror-symmetry technique in a similar way as before, then $\varphi(P) = \varphi(P^*)$ and $\varphi(P) = \varphi(P^{**})$. Because $\varphi(P^*)$ is in general different from $\varphi(P^{**})$, we have come to a contradiction:

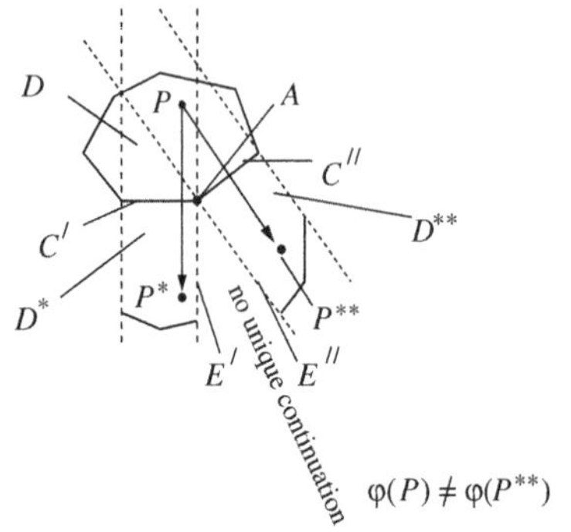

Figure 4.25. Explanations of the impossibility of analytical continuation of the velocity potential to the inside of a polyhedral domain D.

we cannot apply an analytical continuation of the solution outside D to the complete interior of D.

The impossibility of analytical continuation inside an interior body without sharp corners is discussed in Section 4.7.1.

Example: trapezoidal-base (tapered) tank. We consider a trapezoidal-base tank as illustrated in Figure 4.26(a). This type of tank can be situated in the bow of the ship, where it is referred to as a tapered tank (see Figure 1.7). The vertical x–z-plane is a symmetry plane for the tank. Vertical walls are situated at $x = \pm\frac{1}{2}L_1$. The two other vertical tank walls have a small angle α relative to the x–y-plane. Furthermore, the mean free surface and tank bottom are at $z = 0$ and $z = -h$, respectively.

Even though the modifications of the wall (bottom) admit the analytical continuation relative to a three-dimensional rectangular tank, the use of formula (4.90) is questionable and may give misleading conclusions when the deformation of Q_0 changes Σ_0. The reason is that $\Sigma_0 \neq \Sigma_0'$ in this studied case. A direct perturbation technique is needed to find the effect of the deformation on the natural frequencies. The focus is on the eigenvalues $\kappa_{i,0}$ associated with standing waves along the Ox-axis. Due to symmetry of the flow relative to the x–z-plane it is sufficient to analyze half of the tank. The top view of half of the tank is shown in Figure 4.26(b). The first step is the separation of the variable z by presenting the natural modes as $\varphi(x, y, z) = \phi(x, y)\cosh(k_{i,0}(z + h))$. The eigenvalues can be expressed as $\kappa_{i,0} = k_{i,0}\tanh(k_{i,0}h)$ by using the free-surface conditions, where $k_{i,0}$ should be found from a spectral problem in the cross-sectional trapezoidal domain in Figure 4.26(b). This formulation includes the following equation and boundary conditions:

$$\frac{\partial^2\phi}{\partial x^2} + \frac{\partial^2\phi}{\partial y^2} + k_{i,0}^2\phi = 0 \quad \text{in}$$
$$-\tfrac{1}{2}L_1 < x < \tfrac{1}{2}L_1,\ 0 < y < \tfrac{1}{2}L_2 + x\tan\alpha;$$
$$\left.\frac{\partial\phi}{\partial x}\right|_{x=\pm\frac{L_1}{2}} = 0; \quad \left.\frac{\partial\phi}{\partial y}\right|_{y=0} = 0. \tag{4.95}$$

In addition, there is the Neumann boundary condition $\partial\phi/\partial n \equiv n_1\partial\phi/\partial x + n_2\partial\phi/\partial y = 0$ on the inclined line $y = \frac{1}{2}L_2 + x\tan\alpha$, which can be asymptotically simplified by keeping the lowest-order terms in the angle α. In this case $\boldsymbol{n} = (n_1, n_2)$ is the normal vector to the line with

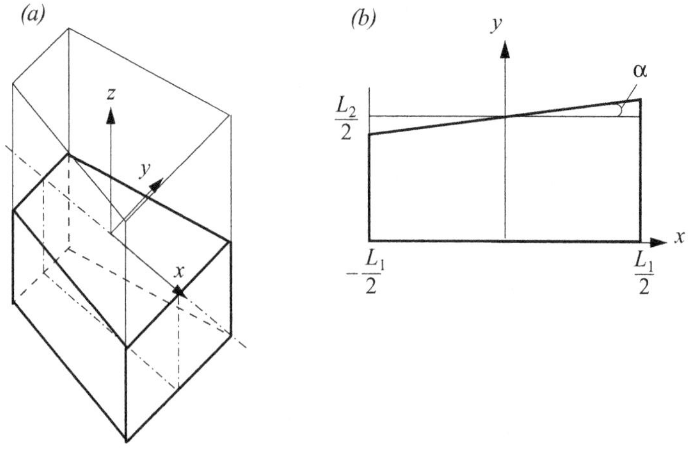

Figure 4.26. (a) Three-dimensional and (b) top view of one-half of a trapezoidal-base tank.

positive direction out of the liquid domain. It follows by using $n_2 \approx 1$, $n_1 \approx -\alpha$ and a Taylor expansion of the boundary condition about $y = \frac{1}{2}L_2$ that

$$\left[\frac{\partial \phi}{\partial y} + \alpha x \frac{\partial^2 \phi}{\partial y^2} - \alpha \frac{\partial \phi}{\partial x}\right]_{y=\frac{1}{2}L_2} = 0. \quad (4.96)$$

Furthermore, as stated earlier, our focus is on almost two-dimensional sloshing in the x–z-plane. A small parameter δ is introduced to express the slow variation of the flow in the y- and z-directions relative to the x-direction; therefore, we write

$$\phi(x, y) = f(x, \delta y) = f(x, y_1), \quad 0 < \delta \ll 1. \quad (4.97)$$

The small parameter δ should be matched with α by using eq. (4.96) and the boundary condition at $y = 0$. We assume first that δ and α are two independent small parameters. Substitution of eq. (4.97) into (4.96) gives

$$\left[\delta \frac{\partial f}{\partial y_1} + \alpha x \delta^2 \frac{\partial^2 f}{\partial y_1^2} - \alpha \frac{\partial f}{\partial x}\right]_{(x,\delta L_2/2)} = 0.$$

The Taylor expansion of this condition about $y_1 = 0$ and the fact that $\partial f/\partial y_1 = 0$ on $y_1 = 0$ gives, by keeping the lowest-order terms in δ and α,

$$\left[\delta^2 \left(\frac{1}{2}L_2 \frac{\partial^2 f}{\partial y_1^2}\right) - \alpha \frac{\partial f}{\partial x}\right]_{(x,0)} = 0. \quad (4.98)$$

The small parameters δ and α are linked by requiring that the two terms in eq. (4.98) have the same order of magnitude (i.e., $\delta^2 \sim \alpha$).

The next step consists of substituting eq. (4.97) into the governing equation of eqs. (4.95), which leads to

$$\frac{\partial^2 f}{\partial x^2} + \delta^2 \frac{\partial^2 f}{\partial y_1^2} + k_{i,0}^2 f = 0. \quad (4.99)$$

Equation (4.99) is now applied at $y_1 = 0$ together with eq. (4.98):

$$\frac{\partial^2 f}{\partial x^2} + \frac{2\alpha}{L_2}\frac{\partial f}{\partial x} + k_{i,0}^2 f = 0 \quad \text{for } y_1 = 0 \quad \text{and} \quad -\tfrac{1}{2}L_1 < x < \tfrac{1}{2}L_1. \quad (4.100)$$

We now look for nontrivial solutions of eq. (4.100) that satisfy $\partial f(x,0)/\partial x = 0$, $x = \pm\frac{1}{2}L_1$, which is the spectral problem with respect to $f = f(x,0)$ and the spectral parameter $k_{i,0}^2$. The problem has the analytical solution

$$k_{i,0}^2 = (\pi i/L_1)^2 + (\alpha/L_2)^2 \quad (4.101)$$

with

$$f(x,0) = \exp\left(-\frac{\alpha}{L_2}x\right)\left[\cos\left(\frac{\pi i}{L_1}\left(x + \tfrac{1}{2}L_1\right)\right) + \frac{\alpha L_1}{\pi i L_2}\sin\left(\frac{\pi i}{L_1}\left(x + \tfrac{1}{2}L_1\right)\right)\right].$$

In summary, eq. (4.101) shows that the sloshing frequencies are, to the lowest order in α, given by

$$\sigma_{i,0}^2/g = k_{i,0}\tanh(k_{i,0}h). \quad (4.102)$$

Formula (4.102) shows that an inclination of the wall leads to increased natural frequencies relative to the natural frequencies for Stokes-like waves along the Ox-axis in a three-dimensional rectangular tank with length L_1 and breadth L_2.

4.7 Asymptotic natural frequencies for tanks with small internal structures

In Sections 4.6.3.1 and 4.6.3.3, we discussed situations that occur when the natural modes φ_n' defined in Q_0' cannot be analytically continued into δQ_0. This type of situation was related to the cases in Figure 4.20(b) and (c), which are discussed in this section. The internal structures are assumed fixed relative to the tank. The internal structure reduces the kinetic energy in the liquid for a given liquid depth. Because the effect on the potential energy is negligible, it follows from eq. (4.65) that the internal structure reduces the natural frequencies.

The following procedure works for any totally submerged small-volume fixed structure. "Small volume" reflects that a characteristic cross-dimensional length of the interior body is small relative to the wavelength of the considered eigenmode. When we denote a surface as *closed* in the following hydrodynamic analysis, it includes thin free-surface piercing bodies and bodies that are thin at the mounting point (line) on the tank boundaries so that neither the mean free-surface area Σ_0 nor the mean wetted tank surface S_0 is affected by the interior structure (see Figure 4.20(b)). Our procedure does not work for the truncated vertical cylinder shown in Figure 4.20(c). However, if the vertical surface-piercing cylinder is standing on the tank bottom and the cross-sectional lengths are small relative to the main tank dimensions, the procedure for a submerged small-volume interior structure can be generalized by using a slender-body theory.

4.7.1 Main theoretical background

We now argue that it is not possible analytically to continue a solution inside the complete interior domain of the body. This fact was discussed for interior closed bodies with sharp corners in Section 4.6.3.3. If the interior body has no sharp corners, we may follow a different argument. The body is assumed to be closed and in an infinite fluid, with no circulation around the body. The flow due to the body can always be represented in terms of a distribution of sources and/or dipoles over the body surface (see Section 10.2). It is also possible to place the flow singularities (sources, dipoles) inside the body. However, we need the flow singularities to represent the flow due to the body. A simple example is steady two-dimensional potential flow of an incompressible fluid past a circular cylinder in an infinite fluid. The total flow can be described as the sum of an ambient flow and a dipole in the direction of the ambient flow, with the singularity at the cylinder center. The singularity prohibits us from analytically continuing the flow from the liquid domain to contain the complete interior of the cylinder. The same is true for any other body shapes.

In the following text, we need to find how the flow caused by the interior structure behaves far away from the structure. Guidance can be provided by following Newman's analysis (1977) for a body in infinite fluid. For sufficiently large radial distance from a three-dimensional body in infinite fluid, r, we can write the velocity potential caused by the body as

$$\varphi = A_0 r^{-1} + \mathbf{A}\cdot\nabla(r^{-1}) + O(r^{-3}), \tag{4.103}$$

where, $4\pi A_0 = -\int_{S_B} \partial\varphi/\partial n \,\mathrm{d}S$ and $\partial/\partial n$ is, as usual, the derivative normal to the body surface S_B. The positive normal direction is into the liquid domain. If the body is rigid and closed, it will later be proven that $A_0 = 0$. The far-field behavior is then dipole-like. The dipole strength $\mathbf{A}$ can be related to the displaced body volume and the added mass coefficients of the body (Newman, 1977). Far-field expressions of the velocity potential of a rigid two-dimensional body can be expressed as

$$\varphi = \mathbf{A}\cdot\nabla(\log r) + O(r^{-2}). \tag{4.104}$$

A natural mode in the original tank without the interior structure with volume δQ_0 is denoted $\varphi_m(x, z)$. A natural mode when a small structure is inserted into the liquid domain (see Figure 4.20(b) and (c)) is called φ'_m. Our analysis starts by applying Green's second identity (see eq. (A.4)) with φ_m and φ'_m over the closed boundary of the mean liquid domain, $S_0 + C + \Sigma_0$, where C is the surface of the interior structure. The result is

$$\int_{\Sigma_0}\left(\varphi'_m\frac{\partial\varphi_m}{\partial n} - \varphi_m\frac{\partial\varphi'_m}{\partial n}\right)\mathrm{d}S + \int_C \varphi'_m\frac{\partial\varphi_m}{\partial n}\mathrm{d}S = 0, \tag{4.105}$$

where the body boundary conditions $\partial\varphi_m/\partial n = 0$ on S_0 and $\partial\varphi'_m/\partial n = 0$ on S_0 and C have been used. Accounting for the boundary conditions $\partial\varphi_m/\partial n = \kappa_m\varphi_m$, $\partial\varphi'_m/\partial n = \kappa'_m\varphi'_m$ on Σ_0 and rearranging, one obtains

$$\kappa'_m = \kappa_m + \frac{\int_C \varphi'_m \frac{\partial\varphi_m}{\partial n}\mathrm{d}S}{\int_{\Sigma_0}\varphi_m\varphi'_m\,\mathrm{d}S} = \kappa_m - \frac{\int_C \varphi'_m \frac{\partial\varphi_m}{\partial n^+}\mathrm{d}S}{\int_{\Sigma_0}\varphi_m\varphi'_m\,\mathrm{d}S}, \tag{4.106}$$

where, as usual, the direction normal to C out of the liquid is n and we introduce n^+ to be positive into the liquid. The positive normal is used in expressing the numerator in eq. (4.106), $I = \int_C \varphi'_m \frac{\partial\varphi_m}{\partial n^+}\mathrm{d}S$, in terms of the added mass coefficients for the interior structure. We write $\varphi'_m = \varphi_m + \varphi^d_m$, where φ^d_m is the velocity potential caused by the presence of the small body. The body boundary condition is

$$\frac{\partial\varphi^d_m}{\partial n^+} = -\frac{\partial\varphi_m}{\partial n^+} \quad \text{on } C. \tag{4.107}$$

The ambient liquid velocity, $\nabla\varphi_m$, varies slowly across the "small" interior structure and can, as a first approximation, be expressed in terms of the ambient velocity at a point O, which we choose as either the geometrical center of the body or the mounting point of a baffle. We can then approximate eq. (4.107) as

$$\frac{\partial\varphi^d_m}{\partial n^+} = -n^+_1\left.\frac{\partial\varphi_m}{\partial x}\right|_O - n^+_2\left.\frac{\partial\varphi_m}{\partial y}\right|_O - n^+_3\left.\frac{\partial\varphi_m}{\partial z}\right|_O + O(r_0), \tag{4.108}$$

where $\boldsymbol{n}^+ = (n^+_1, n^+_2, n^+_3)$. It is now essential that φ^d_m decreases rapidly with distance from the interior body so that the boundary conditions can be approximated. If the interior body is not a closed surface, the far-field behavior of φ^d_m is, in general, source-like. In two dimensions, the flow velocity decays with r^{-1}, where r is the distance from O. When the body surface is closed, the far-field behavior is dipole-like, as already stated, which means the flow velocity in two dimensions decays with r^{-2} (see eq. (4.104)). In practice, therefore, the flow velocity due to φ_m^d is small at a distance on the order of a characteristic length of the body, r_0. If, for instance, a plate is considered, the length r_0 can be chosen as the plate length. We can show mathematically that φ_m^d does not have a far-field source behavior when C is closed by examining the mass flux Q_{mass} per unit liquid density associated with φ_m^d through C, which can be expressed as

$$Q_{\text{mass}} = \int_C \frac{\partial \varphi_m^d}{\partial n^+}\,\mathrm{d}S = -\int_C \frac{\partial \varphi_m}{\partial n^+}\,\mathrm{d}S.$$

It follows by the divergence theorem (see eq. (A.1)) and the fact that φ_m is analytic inside C that $Q_{\text{mass}} = -\int_{\delta Q_0} \nabla^2 \varphi_m\,\mathrm{d}Q = 0$. The fact that Q_{mass} is zero means that φ_m^d cannot have far-field source behavior. If the interior body is thin at the free surface and is free-surface piercing and/or thin at a mounting point (line) on the tank surface, it does not make a difference in the integration if we integrate over the closed surface. In the two latter cases we have to include boundary conditions only on the mean free surface and tank boundary near the interior body, respectively. If the body is not too close to a tank corner or tank-wall-free-surface intersection, both boundary conditions can be approximated by assuming zero normal flow conditions on an infinite plane that contains either the tank surface adjacent to the mounting point (line) of the interior structure at the tank boundary or the adjacent mean free surface of the free-surface piercing body. The reason why we can use a rigid free-surface condition is that the nondimensional natural frequency with characteristic interior body length as a length parameter is small (see Section 3.5.2).

Due to the boundary condition given by eq. (4.108) we can reexpress φ_m^d as

$$\varphi_m^d = -\varphi_m^{d1} \left.\frac{\partial \varphi_m}{\partial x_1}\right|_O - \varphi_m^{d2} \left.\frac{\partial \varphi_m}{\partial x_2}\right|_O - \varphi_m^{d3} \left.\frac{\partial \varphi_m}{\partial x_3}\right|_O + O(r_0), \tag{4.109}$$

where

$$\frac{\partial \varphi_m^{dj}}{\partial n^+} = n_j^+ \quad \text{on } C.$$

In eq. (4.109) we have used x_1, x_2, and x_3 instead of x, y, and z, respectively. It follows that eq. (4.109) is a correct decomposition of φ_m^d by substituting eq. (4.109) into eq. (4.108) and observing that the body boundary conditions are satisfied. Accounting for the fact that $\varphi_m' = \varphi_m + \varphi_m^d$, integral I can be presented as the sum $I_1 + I_2$, where

$$\begin{aligned} I_1 &= \int_C \varphi_m \frac{\partial \varphi_m}{\partial n^+}\,\mathrm{d}S \\ &= \sum_{j=1}^{3} \left[\left.\frac{\partial \varphi_m}{\partial x_j}\right|_O + O(r_0) \right] \int_C \varphi_m n_j^+\,\mathrm{d}S \end{aligned} \tag{4.110}$$

and, due to eq. (4.109),

$$\begin{aligned} I_2 &= \int_C \varphi_m^d \frac{\partial \varphi_m}{\partial n^+}\,\mathrm{d}S \\ &= \sum_{j=1}^{3} \left[\left.\frac{\partial \varphi_m}{\partial x_j}\right|_O + O(r_0) \right] \int_C \varphi_m^d n_j^+\,\mathrm{d}S \\ &= -\sum_{k,j=1}^{3} \left[\left.\frac{\partial \varphi_m}{\partial x_j}\right|_O \left.\frac{\partial \varphi_m}{\partial x_k}\right|_O + O(r_0) \right] \int_C \varphi_m^{dk} n_j^+\,\mathrm{d}S. \end{aligned} \tag{4.111}$$

Integrals in eq. (4.110) may be approximated by using the Gauss theorem (see eq. (A.1)):

$$\begin{aligned} \int_C \varphi_m n_j^+\,\mathrm{d}S &= \int_{\delta Q_0} \frac{\partial \varphi_m}{\partial x_j}\,dQ \\ &= \left[\left.\frac{\partial \varphi_m}{\partial x_j}\right|_O + O(r_0) \right] Vol(\delta Q_0), \\ j &= 1, 2, 3. \end{aligned}$$

Hence,

$$I_1 = Vol(\delta Q_0) \left[\sum_{j=1}^{3} \left(\left.\frac{\partial \varphi_m}{\partial x_j}\right|_O \right)^2 + O(r_0) \right]. \tag{4.112}$$

An essential assumption in using the Gauss theorem to derive eq. (4.112) is that the body surface C is closed and that φ_m is analytic inside C.

We now show that the integrals in eq. (4.111) can be expressed in terms of the added mass coefficients by studying a fictitious forced velocity of the body, $-\left.\partial \varphi_m / \partial x_k\right|_O \cdot \sin \sigma_m t$, in three directions ($k = 1, 2, 3$). This fictitious forced oscillation of the body causes a liquid flow that can be described by the velocity potential $-\varphi_m^{dk}\,\partial\varphi_m/\partial x_k|_O \cdot \sin\sigma_m t$. The corresponding hydrodynamic pressure can be written as a linearized Bernoulli equation:

$$p = \rho\sigma_m\varphi_m^{dk}\left.\frac{\partial\varphi_m}{\partial x_k}\right|_O \cos\sigma_m t.$$

The force in the j-direction due to forcing in the k-direction can then be expressed as

$$F_j = -\int_C p n_j^+\,\mathrm{d}S$$
$$= -\rho\left.\frac{\partial\varphi_m}{\partial x_k}\right|_O \sigma_m\cos(\sigma_m t)\int_C \varphi_m^{dk} n_j^+\,\mathrm{d}S.$$

Now we use our definition of added mass from Section 3.5.2. We can write

$$F_j = A_{jk}\sigma_m\left.\frac{\partial\varphi_m}{\partial x_k}\right|_O \cos(\sigma_m t),$$

where the added mass coefficients are

$$A_{jk} = -\rho\int_C \varphi_m^{dk} n_j^+\,\mathrm{d}S. \tag{4.113}$$

Summarizing eqs. (4.111)–(4.113) and neglecting the higher-order terms gives

$$I = \int_C \varphi'_m\frac{\partial\varphi_m}{\partial n^+}\,\mathrm{d}S$$
$$= \sum_{j,k=1}^{3}\left.\frac{\partial\varphi_m}{\partial x_j}\right|_O\left.\frac{\partial\varphi_m}{\partial x_k}\right|_O\left[\delta_{kj}Vol(\delta Q_0) + \frac{A_{jk}}{\rho}\right], \tag{4.114}$$

where δ_{kj} is the Kronecker delta.

The order of magnitude of nonzero added mass coefficients A_{jk} is $O(r_0^3)$. For the two-dimensional case, we must consider summation in eq. (4.114) up to 2 and introduce the area instead of $Vol(\delta Q_0)$. The two-dimensional added mass coefficients a_{ij} are then of $O(r_0^2)$.

The denominator in eq. (4.106) depends on the ambient and near-field solutions. However, one may show in all the examples that follow that

$$\int_{\Sigma_0}\varphi'_m\varphi_m\,\mathrm{d}S = \underbrace{\int_{\Sigma_0}\varphi_m^2\,\mathrm{d}S}_{O(1)} + o(1) \tag{4.115}$$

that is, the flow due to the presence of the internal structure gives a small contribution to the surface wave profiles. This fact is quite obvious when the structure is small and far from the mean free surface. A justification of eq. (4.115) for a structure close to Σ_0 is that a small-size body influences the wave profiles only in the neighborhood of $O(r_0)$. Mathematically, in all the examples that follow, the local velocity field due to φ_m^d decays rapidly away from O and, in the general case, even though δQ_0 pierces the surface Σ_0 (which then becomes Σ'_0):

$$\int_{\Sigma'_0}\varphi_m\varphi_m^d\,\mathrm{d}S = O(1) \quad \text{and}$$
$$\int_{\Sigma'_0}\varphi_m^2\,\mathrm{d}S = \int_{\Sigma_0}\varphi_m^2\,\mathrm{d}S + o(1). \tag{4.116}$$

Summarizing eqs. (4.106), (4.114), and (4.115) gives the following asymptotic formula for the natural sloshing frequencies:

$$\frac{\sigma_m'^2}{\sigma_m^2} = \frac{\kappa'_m}{\kappa_m}$$
$$= 1 - \frac{\sum_{j,k=1}^{N}\phi_{mj}\phi_{mk}\left[\delta_{kj}Vol(\delta Q_0) + A_{jk}/\rho\right]}{\kappa_m\int_{\Sigma_0}\varphi_m^2\,\mathrm{d}S}, \tag{4.117}$$

where

$$\phi_{mj} = \left.\frac{\partial\varphi_m}{\partial x_j}\right|_O, \quad \phi_{mk} = \left.\frac{\partial\varphi_m}{\partial x_k}\right|_O.$$

The assumptions are that the wavelength of eigenmode m is much larger than the cross-dimensional lengths of the structure and that the structure has a closed submerged surface; $N = 3$ for a finite three-dimensional body, but $N = 2$ and $a_{ij} = A_{ij}$ should be taken for the two-dimensional sloshing. The term $\sum_{j,k=1}^{N}\phi_{mj}\phi_{mk}\delta_{kj}Vol(\delta Q_0)$ in the numerator of eq. (4.117) is caused by a reduction of the kinetic energy of the ambient flow due to the fact that the inserted body has reduced the original liquid domain. The term $\sum_{j,k=1}^{N}\phi_{mj}\phi_{mk}A_{jk}/\rho$ is caused by the fact that the flow due to the inserted body reduces the kinetic energy.

If several substructures are situated far away from each other (i.e., the distance between them is large relative to their cross-dimensions), formula (4.117) can be generalized to include a summation over each inserted structure. The added mass coefficients can be calculated without accounting for the hydrodynamic interaction. However, there are weighting terms $\phi_{mj}\phi_{mk}$ in the numerator of eq. (4.117) associated with the ambient flow velocity at the position of an inserted structure.

One can generalize the derivation to slender bodies with cross-sectional lengths that are small relative to the ambient wavelength of the

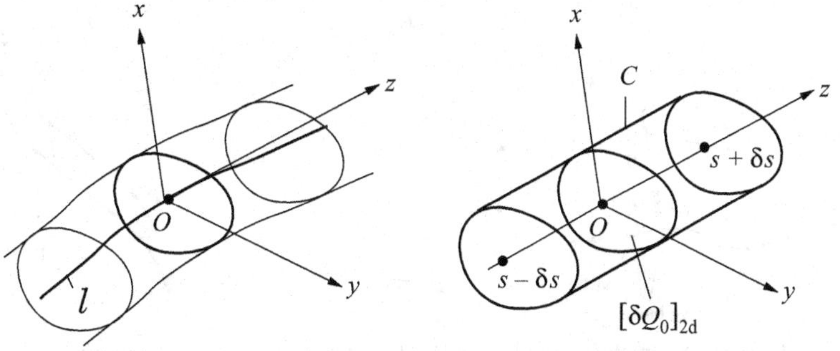

Figure 4.27. Normal parameterization of an element of a slender body.

considered eigenmode. Slender body theory is described by Newman (1977). Near- and far-field solutions are introduced. The near-field solution is governed by the two-dimensional Laplace equation in the cross-sectional plane. If the dominant far-field behavior due to the body is dipole-like with direction in the cross-sectional plane, the hydrodynamic interaction between the flows at different cross-sections is negligible. The latter is the case for our considered examples.

We introduce a curve ℓ that goes through the mean geometrical points O of the cross-sections of the body. The radius of curvature of this line must be large relative to the cross-dimensional lengths. A local Cartesian coordinate system $Oxyz$ with origin O is introduced. The z-axis is chosen tangent to ℓ (see Figure 4.27). Equations (4.108) and (4.109) can be approximated as

$$\frac{\partial \varphi_m^d}{\partial n^+} = -n_1^+ \phi_{m1} - n_2^+ \phi_{m2} + O(r_0) \quad \text{on } C \quad \text{and}$$
$$\varphi_m^d = -\varphi_m^{d1} \phi_{m1} - \varphi_m^{d2} \phi_{m2} + O(r_0) \qquad (4.118)$$

in the vicinity of O. In this case,

$$\left.\frac{\partial \varphi_m}{\partial x}\right|_O = \phi_{m1}(s), \quad \left.\frac{\partial \varphi_m}{\partial y}\right|_O = \phi_{m2}(s), \qquad (4.119)$$

where s is a parameter describing ℓ.

Accounting for eqs. (4.118) and (4.119) in the derivations (4.110)–(4.113) and integrating the results along the curve ℓ by the natural parameter s gives

$$I = \int_\ell \sum_{j,k=1}^{2} \phi_{mj}(s)\phi_{mk}(s) \left[\delta_{kj} Vol([\delta Q_0]_{2d}(s)) + a_{jk}(s)/\rho\right] \mathrm{d}s, \qquad (4.120)$$

where $Vol([\delta Q_0]_{2d}(s))$ is the cross-sectional area and $a_{jk}(s)$ are the two-dimensional added mass coefficients for the cross-section of the slender body. Furthermore, $\phi_{mj}(s)$ is defined by eq. (4.119). Equation (4.120) can, for instance, be applied for a vertical surface-piercing cylinder standing on the tank bottom.

The final formula for the corrected natural frequencies for a slender body is

$$\frac{\sigma_m'^2}{\sigma_m^2} = \frac{\kappa_m'}{\kappa_m} = 1 - \frac{\int_\ell \sum_{j,k=1}^{2} \phi_{mj}\phi_{mk}\left[\delta_{kj} Vol([\delta Q_0]_{2d}) + a_{jk}/\rho\right] \mathrm{d}s}{\kappa_m \int_{\Sigma_0} \varphi_m^2 \, \mathrm{d}S}. \qquad (4.121)$$

The expressions for the effect of an interior structure on the kinetic energy can be related to the force acting on the body, which was made clear in the derivation of eq. (4.121). As special cases (e.g., a vertical cylinder and a baffle), we can recast the formulation in terms of Morison's equation for the force (see Section 6.4). Then we are in a position to use empirically determined mass coefficients that account for viscous flow separation and the effect of flow parameters such as the Keulegan–Carpenter number, Reynolds number, and roughness number. It is difficult to account for the viscous effects a priori. So we need experimental results to validate our potential flow results.

4.7.2 Baffles

4.7.2.1 Small-size (horizontal or vertical) thin baffle

We consider a thin baffle that is mounted on the tank surface and has a right angle to the surface as shown in Figure 4.28(a). A local coordinate system $O\tau\nu$ is introduced with origin O at

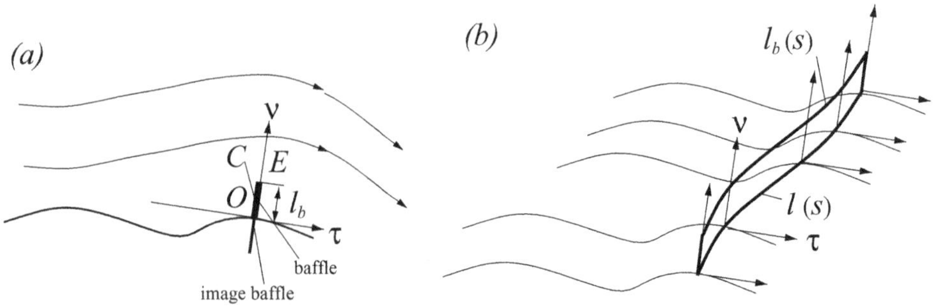

Figure 4.28. Small-size baffles mounted perpendicular to the tank surface.

the mounting point of the baffle. The plate is perpendicular to the $O\tau$-axis. The numbers 2 and 3 are associated with τ and ν, respectively. The baffle width l_b is assumed small relative to the wavelength of the ambient flow, the distances to a tank corner, and the mean free surface.

The ambient velocity components at the point O are denoted

$$\left.\frac{\partial \varphi_m}{\partial \tau}\right|_O = \phi_{m\tau} = \phi_{m2}, \quad \text{but} \quad \left.\frac{\partial \varphi_m}{\partial \nu}\right|_O = \phi_{m3} = 0, \tag{4.122}$$

where we have adopted the notations of Section 4.7.1 for the local coordinate system.

Estimates of the natural frequencies are given by asymptotic formula (4.117). Because the baffle is thin, we approximate the volume of the structure, $Vol(\delta Q_0)$, to be zero. The added mass coefficients, $a_{\nu\nu}$, $a_{\nu\tau}$, and $a_{\tau\nu}$, are also set equal to zero. The added mass coefficient $a_{\tau\tau}$ can be obtained as follows. For a single two-dimensional baffle that is far from other baffles in the tank, the body can be mirror-imaged about the tank boundary to satisfy the boundary condition of no flow through the tank boundary. Therefore, we consider a plate of length $2l_b$ in infinite fluid. The added mass for the baffle is one-half of the added mass for the plate. Figure 4.28(a) gives a graphical illustration of this fact by showing the plate and its mirror image. The result is

$$a_{\tau\tau} = a_{22} = \tfrac{1}{2}\pi\rho l_b^2, \tag{4.123}$$

which gives the following expression:

$$\frac{\sigma_m'^2}{\sigma_m^2} = \frac{\kappa_m'}{\kappa_m} = 1 - \frac{\pi\phi_{m\tau}^2 l_b^2}{2\kappa_m \int_{\Sigma_0} \varphi_m^2 \, dS}. \tag{4.124}$$

The same formula applies for a surface-piercing baffle when the baffle is perpendicular to Σ_0 due to the fact that we can use the rigid-wall condition as the free-surface condition when solving for the velocity potential due to the baffle. The problem is then similar to considering a baffle mounted perpendicular to the tank boundary.

We can generalize formula (4.124) to a slender three-dimensional plate, as illustrated in Figure 4.28(b). In this case, the dimension of the plate, l_b, varies along the mounting line, ℓ. The starting point is the general form of eq. (4.121). When introducing a normal parameterization of the mounting line, eq. (4.121) modifies eq. (4.124) to the form

$$\frac{\sigma_m'^2}{\sigma_m^2} = \frac{\kappa_m'}{\kappa_m} = 1 - \frac{\pi \int_\ell \phi_{m\tau}^2(s) l_b^2(s)\, ds}{2\kappa_m \int_{\Sigma_0} \varphi_m^2 \, dS}, \tag{4.125}$$

where we have accounted for the change of the tangential velocity and cross-dimensional baffle size l_b along mounting line ℓ.

Example: a rigid-ring baffle in an upright circular cylindrical tank. A small rigid-ring baffle installed in an upright circular cylindrical tank is considered (see Figure 4.29). The width of the baffle is l_b and the distance between the baffle and the mean free

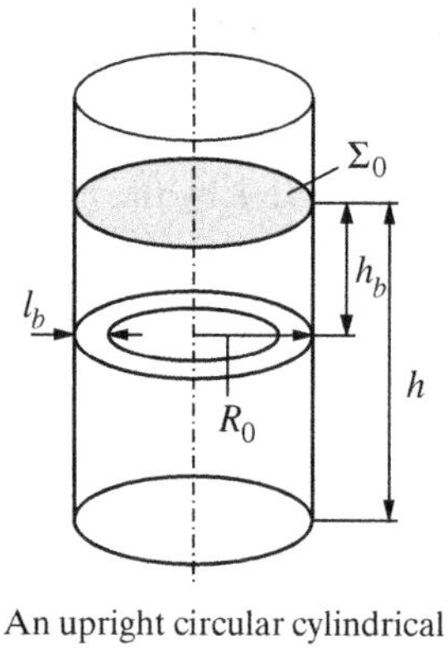

Figure 4.29. An upright circular cylindrical tank with an annular baffle.

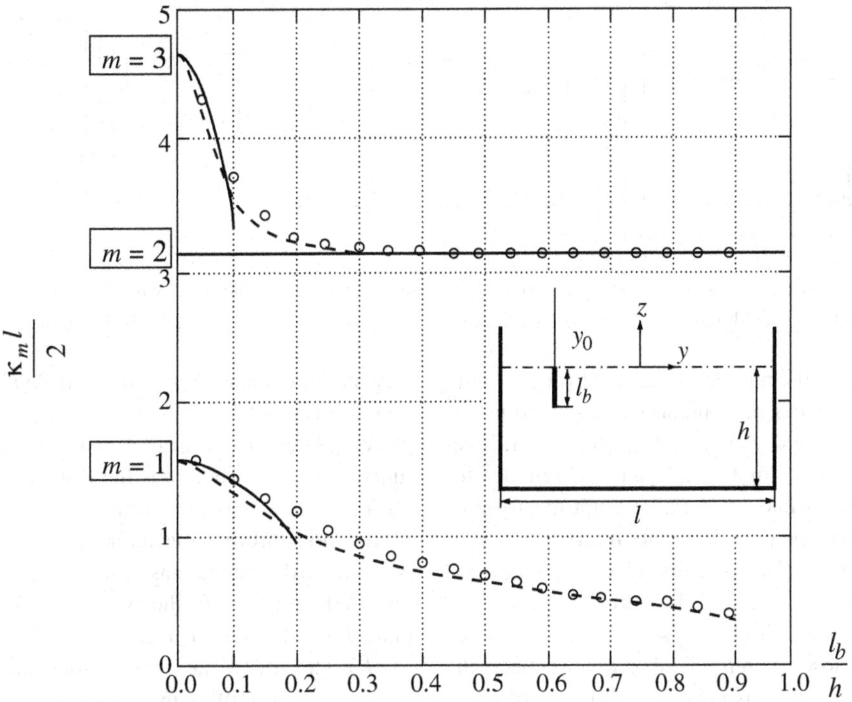

Figure 4.30. The frequency parameter $\frac{1}{2}\kappa_m l$ versus ratio of baffle height to liquid depth, l_b/h, for a surface-piercing baffle at the middle point of the tank ($y_0 = 0$) with ratio of liquid depth to tank breadth $h/l = 1$. The solid line represents asymptotic formula (4.129). The circles are computed from Evans and McIver's (1987) method. The dashed line represents numerical results of a variational method by Jeyakumaran and McIver (1995).

surface is h_b. We assume that

$$\frac{l_b}{R_0} \ll \frac{h_b}{R_0} \quad \text{and} \quad \frac{h}{R_0} = O(1). \tag{4.126}$$

The smallness of h_b/R_0 is discussed in Section 4.7.2.2.

It follows from using eqs. (4.37) and (4.39) that the ambient tangential velocity component is

$$\phi_{m,i\tau} = \left.\frac{\partial \varphi_{m,i}}{\partial z}\right|_{\substack{r=R_0 \\ z=-h_b}} = J_m\left(\frac{\iota_{m,i} r}{R_0}\right)\frac{\iota_{m,i}\sinh(\iota_{m,i}(z+h)/R_0)}{R_0\cosh(\iota_{m,i}h/R_0)} \times \begin{cases}\cos(m\theta) \\ \sin(m\theta)\end{cases}, m = 0, 1, \dots, i = 1, 2, \dots \tag{4.127}$$

Setting $\mathrm{d}s = R_0 \mathrm{d}\theta$ in general formula (4.125) gives

$$\frac{\sigma'^2_{m,i}}{\sigma^2_{m,i}} = \frac{\kappa'_{m,i}}{\kappa_{m,i}} = 1 - \frac{\pi J^2_m(\iota_{m,i})\iota_{m,i}}{\int_0^1 r J^2_m(\iota_{m,i} r)\,\mathrm{d}r} \times \left(\frac{l_b}{R_0}\right)^2 \frac{\sinh^2(\iota_{m,i}(h-h_b)/R_0)}{\sinh(2\iota_{m,i}h/R_0)}. \tag{4.128}$$

Example: vertical free-surface piercing barrier. An asymptotic formula for natural frequencies due to a free-surface piercing barrier (see Figure 4.30) with a small ratio l_b/h between the mean submerged height and the liquid depth follows from eqs. (4.124) and (4.125) by evaluating the ambient tangential velocity $\phi_{m\tau}$ at the mean free surface. The formula for two-dimensional sloshing in a rectangular tank is

$$\frac{\sigma'^2_m}{\sigma^2_m} = 1 - \pi^2 m \frac{\sin^2\left(\pi m\left(y_0 + \frac{1}{2}l\right)/l\right)}{\tanh(\pi m h/l)}\left(\frac{l_b}{l}\right)^2. \tag{4.129}$$

A similar formula was derived by Jeyakumaran and McIver (1995). Comparisons between approximate formula (4.129) and two different numerical solutions based on linear potential flow (valid for any l_b/h) by Evans and McIver (1987) and Jeyakumaran and McIver (1995) are presented in Figure 4.30. Jeyakumaran and McIver report accurate results from the numerical nonasymptotic methods for $l_b/h \le 0.9$. The frequency parameter is plotted against the ratio of baffle height to liquid depth, l_b/h. Because the

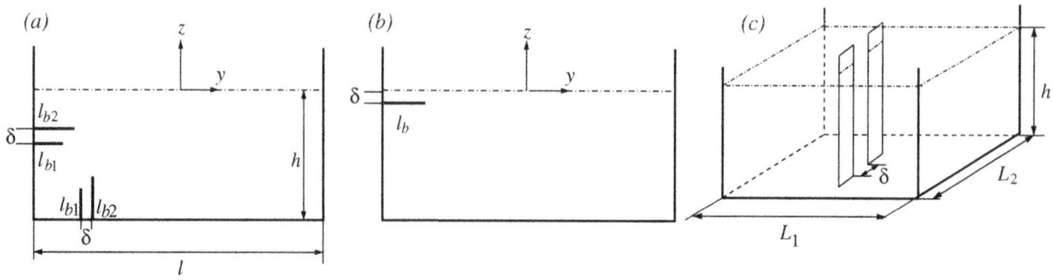

Figure 4.31. Three cases when hydrodynamic interaction between the baffles or with the free surface has to be considered.

barrier is in the middle of the tank, (i.e., $y_0 = 0$), the barrier does not influence the second mode, whereas a significant effect on the natural frequencies exists for the first and third modes for even small values of l_b/h. Asymptotic formula (4.129) agrees well with the numerical results when $l_b/h < 0.2$ for the first eigenfrequency and for $l_b/h < 0.1$ for the third natural frequency. The natural frequencies for the second and third mode become close when $l_b/h \gtrsim 0.3$, which reflects the fact that the flow associated with the third mode is negligible for $|z| \gtrsim 0.3h$ and that the baffle from the third mode's perspective divide the tank into two equal parts. When a gap exists between the lower end of the baffle and the tank bottom ($l_b/h < 1.0$), the first-mode flow shares similarities with the flow in a U-tube.

4.7.2.2 Hydrodynamic interaction between baffles (plates) and free-surface effects

When there are two or more baffles that are close to each other, or when a horizontal baffle becomes close to the mean free surface or when a baffle is close to a tank corner, the added mass coefficients differ from those given by eq. (4.123). Three cases are illustrated in Figure 4.31 for rectangular tanks. Under the given definitions, our focus is on the limit $\delta/l_b \to 0$ (according to previous definitions, $\delta = h_b$ in cases a and b), where δ is defined in the figure for the three cases.

We start by examining the case in Figure 4.31(a) when $l_{b1} = l_{b2}$. The two equally sized baffles are mounted perpendicular to either the tank wall or the tank bottom. The free surface and tank corners are assumed to have negligible influence on the local flow at the baffles. The hydrodynamic problem can then be approximated as potential flow past two baffles mounted perpendicular to an infinite wall. Numerical results by Newman (personal communication, 2007) for the added mass coefficient in the transverse direction of two plates are presented in Figure 4.32 as a function of δ/l_b, where δ is the distance between the plates and l_b is the width of a baffle. When there is no hydrodynamic interaction between the two plates, the added mass coefficient is equal to $\rho\pi l_b^2$. When $\delta/l_b \to 0$, the two plates act as one plate (i.e., the added mass coefficient tends to $\frac{1}{2}\rho\pi l_b^2$). The added mass is monotonically increasing as function of δ in the range between 0 and ∞. A rapid change in the coefficient occurs when $\delta/l_b \lesssim 3$. When $\delta/l_b = 3$, it is equal to $0.96\rho\pi l_b^2$.

A simplified expression for the added mass coefficient and arbitrary δ can be derived if we start out by assuming δ/l_b is large. The considered hydrodynamic problem is equivalent to cross-flow past two plates in a tandem arrangement in an infinite fluid. The problem can be analyzed by first studying one plate and then finding an inflow velocity to the other plate. We introduce a Cartesian coordinate system Oxy (see Figure 4.32). The plate has a forced velocity $V(t)$ along the y-axis. The problem is linearized so that the body boundary condition is imposed on the mean plate position. The mean plate position is situated along the x-axis and has end coordinates $x = \pm l_b$. The same mathematical problem is studied in Section 11.3.1. The y-axis is a symmetry line for the flow (i.e., it behaves similarly to the rigid wall in our original statement of the problem). We consider now the far-field flow (e.g., when y/l_b is large). The far-field flow velocity along the y-axis can be expressed as

$$v = V(t)\left[1 - |y - y_0|\left(l_b^2 + (y - y_0)^2\right)^{-1/2}\right], \tag{4.130}$$

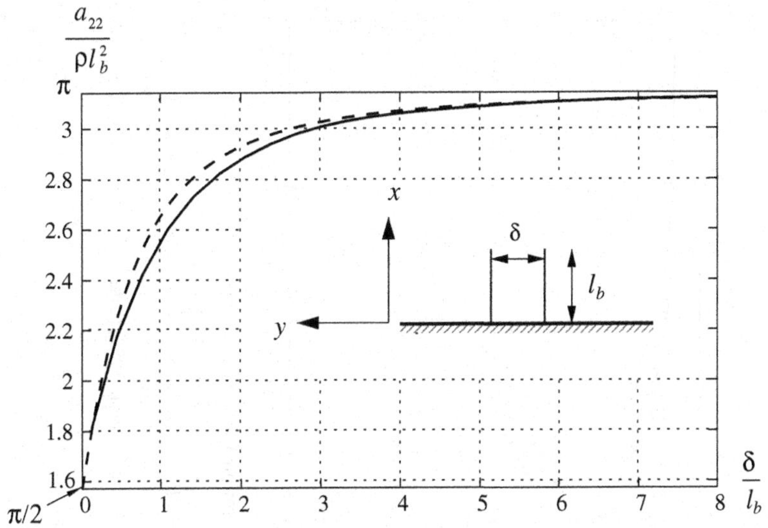

Figure 4.32. The nondimensional added mass coefficient for two plates in tandem arrangement versus the distance between the plates, δ, scaled by the plate width, l_b. Numerical calculations were performed by Newman (personal communication, 2007) using a boundary element method (solid line) and approximate formula (4.131) (dashed line).

where $y_0 = 0$ according to the analysis in Section 11.3.1. When y_0 is of the order of l_b, eq. (4.130) gives the same first-order approximation of the far-field flow velocity as for $y_0 = 0$. The introduction of y_0 is used in further analysis.

The second plate is now considered. This plate is oscillating with forced velocity V and is exposed to an ambient flow due to the other plate that is approximated by eq. (4.130). The flow around the second plate is equivalent to studying the plate with a forced velocity equal to the difference between V and v expressed by eq. (4.130) with $y = \delta$. It follows by the definition of added mass that the added mass for the second plate is

$$a_{22} = \rho\pi l_b^2 \left|\delta - y_0\right| / \sqrt{l_b^2 + (\delta - y_0)^2}, \qquad (4.131)$$

which is also the added mass for the other plate when the interaction effects between the plates are considered. Because the added mass for the two plates mounted on an infinite plane in Figure 4.32 is half the added mass of the two plates in the tandem arrangement, eq. (4.131) is a far-field approximation for the two studied baffles. We now choose y_0 in eq. (4.131) so that the formula is a good approximation in a broad range of δ/l_b. We do that by requiring eq. (4.131) to give the correct value for $\frac{1}{2}\rho\pi l_b^2$ when $\delta = 0$, which gives $y_0 = -0.577 \cdot l_b$. These values are plotted together with the numerical results in Figure 4.32 and show a reasonable agreement.

Based on the preceding results we may assess within which distance the local effect of a single baffle is felt. Equation (4.131), for instance when $\delta = 3l_b$, gives an added mass value that is 0.96 times the added mass with no hydrodynamic interaction. This result is, of course, subjective when we should state that negligible hydrodynamic interaction exists between the local flow due to a baffle and the free surface, tank corners, or other internal structures.

We may generalize the asymptotic procedure for predicting the added mass of a baffle parallel to and near a wall by introducing an image baffle with respect to the wall. When solving the hydrodynamic problem, the image baffle is characterized by velocity opposite that of the original baffle to satisfy the condition of no flow through the wall. We do not pursue this analysis. However, we may expect a small influence from a baffle that is parallel and close to a wall on the sloshing frequency by examining eq. (4.117). This influence is due to the fact that the ambient flow velocity is small.

We now examine the case in Figure 4.31(b), when a horizontal baffle approaches the mean

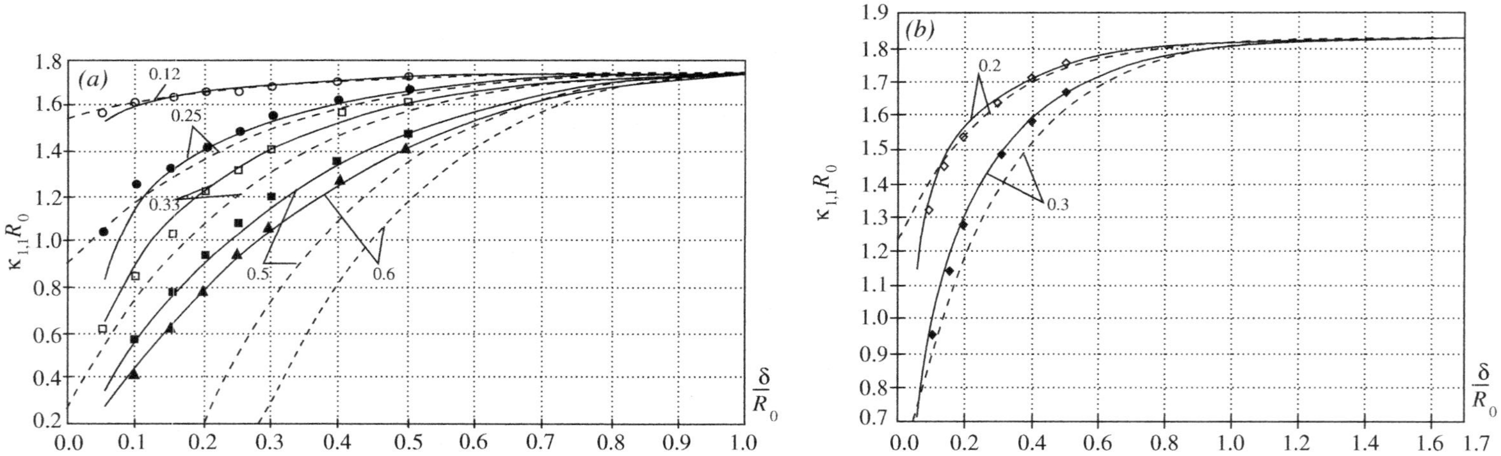

Figure 4.33. The nondimensional lowest eigenvalue versus the vertical location $\delta = h_b$ of the rigid-ring baffle relative to the mean free surface in an upright circular cylindrical tank with radius R_0 (see Figure 4.29). Dashed lines represent asymptotic formula (4.128), assuming $l_b/R_0 \ll 1$ and $\delta/l_b \geq O(1)$. Solid lines represent authors' numerical results. The numbers on the solid and dashed lines are values of l_b/R_0. Points correspond to experimental data by Dorozhkin (personal communication, 1989) (case a, $h/R_0 = 1$) and Miskishev and Churilov (1977) (case b, $h/R_0 = 1.7$).

free surface. To our knowledge, very few mathematical, numerical, and experimental results exist for a baffle in the free-surface zone. Theorems by Gavrilyuk *et al.* (2001) proved for a two-dimensional rectangular tank with two symmetric horizontal baffles that κ'_m decays monotonically with decreasing δ/l (with a fixed finite value of l_b/l), where δ is the plate submergence relative to the mean free surface. Numerical analysis of this limit for both two-dimensional rectangular and circular cylindrical tanks (with a rigid-ring baffle) was done by the authors , who confirmed the monotonic dependence on $\delta/l(\delta/R_0 = h_b/R_0)$ and showed that the theoretical limit should seemingly be zero (i.e., $\kappa'_m \to 0$ as $\delta/R_0 \to 0$). For an annular baffle in an upright circular cylindrical tank (see Figure 4.29 with notation $\delta = h_b$), experimental results are reported by Abramson (1966), Mikishev and Churilov (1977), and Dorozhkin (personal communication, 1989). Comparison of the theoretical predictions from asymptotic formula (4.128), calculations by the authors, and experimental data by Dorozhkin and Mikishev and Churilov are presented in Figure 4.33. They confirmed the applicability of formula (4.128) for $l_b/R_0 \lesssim 0.3$ and finite values of h/R_0. Furthermore, these experiments were done for nonsmall δ/l_b. Mikishev and Dorozhkin (1977) reported experimental difficulties, because the actual sloshing amplitude must be clearly smaller than δ to avoid strong nonlinear shallow-water phenomena on the top of the baffle that may affect the measurements. Abramson (1966) gives experimental results for the nondimensional frequencies for relatively small δ/l_b and different baffle widths. Four ratios of the width to tank radius, l_b/R_0, were tested, ranging between 0.241 and 0.076. The results are consistent with asymptotic formula (4.128) for $\delta/l_b = O(1)$, but the measured frequencies become larger than $\sigma_{1,1}$ (without the baffle) as δ/l_b tends to zero and, as a consequence, $\kappa'_{1,1}$ is not a monotonic function of δ/R_0, which contradicts the results from linear potential flow theory. Possible reasons are local nonlinear free-surface effects and flow separation from the baffle edge.

We can study case c in Figure 4.31 by eq. (4.121) after we derive an analytical expression for the two-dimensional added mass of two side-by-side flat plates of equal length in an infinite fluid. We let the two plates be situated along the x-axis symmetrically with respect to the y-axis. The x-coordinates of the plate ends are $\pm a$, $\pm b$ so that $2b$ is the gap between the two plates and the width of each plate is $a - b$. The arrangement is shown in Figure 4.34 and is similar to that of Figure 11.30. To find the added mass a_{22} we study forced small-amplitude oscillations of the two plates with velocity $V(t)$ along the y-axis. The solution of the resulting flow problem can be found in a similar way as in Section 11.6.3. For $y = 0^-$, $-a < x < -b$, $b < x < a$ we can write that

$$\frac{\partial \varphi}{\partial x} = -\frac{\operatorname{sign}(x)\left[V(t)\,x^2 - C(t)\right]}{\sqrt{(a^2 - x^2)(x^2 - b^2)}},$$

where φ is the velocity potential. The parameter $C(t)$ is determined by the fact that the flow is antisymmetric about the x-axis. The velocity potential must therefore be zero between $-b$ and b and $y = 0$ as well as on $y = 0$ for $|x| \geq a$, which gives the following condition:

$$C(t) = a^2 V(t) E\left[\sqrt{1 - (b/a)^2}\right] \Big/ K\left[\sqrt{1 - (b/a)^2}\right],$$

where K and E are complete elliptic integrals (see Chapter 11). We now follow our definition of added mass; that is, we study the hydrodynamic force, F_2, along the y-axis due to the pressure $-\rho\partial\varphi/\partial t$ and define the added mass a_{22} by $F_2 = -a_{22}\dot{V}$. Noting that φ is antisymmetric with respect to the x-axis and symmetric with respect to the y-axis, we can express the added mass as

$$\begin{aligned} a_{22} &\equiv 4\rho \int_{-a}^{-b} \overline{\varphi}(x, 0^-)\, \mathrm{d}x \\ &= -4\rho \int_{-a}^{-b} x \frac{\partial \overline{\varphi}(x, 0^-)}{\partial x}\, \mathrm{d}x, \qquad (4.132) \end{aligned}$$

where $\overline{\varphi}$ is a normalized velocity potential defined by $\varphi = V(t)\overline{\varphi}$. In eq. (4.132), we used partial integration to facilitate further derivation, which gives the following expression:

$$\begin{aligned} a_{22} = \rho\pi a^2 \Big\{ &1 + [b/a]^2 \\ &- 2E\left[\sqrt{1 - (b/a)^2}\right] \Big/ K\left[\sqrt{1 - (b/a)^2}\right] \Big\}. \end{aligned} \qquad (4.133)$$

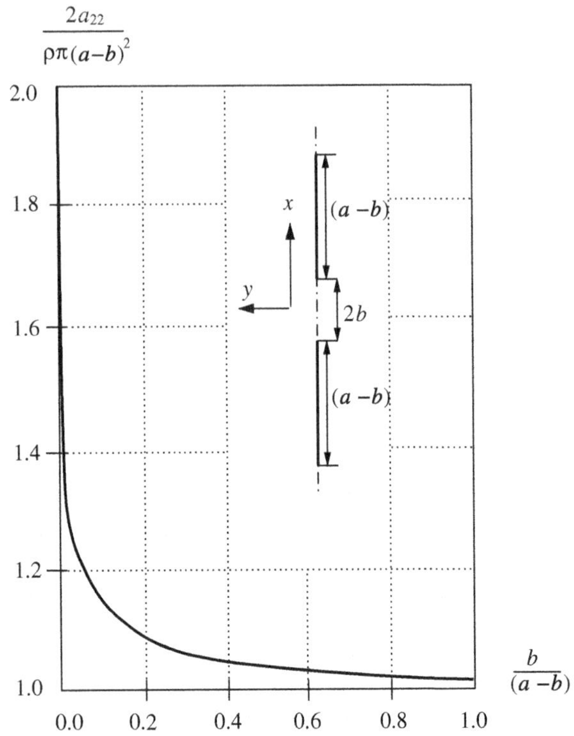

Figure 4.34. Two-dimensional added mass a_{22} for two plates in a side-by-side arrangement in infinite fluid. The gap between the two plates is $2b$. Each plate has width $a-b$.

Numerical results are presented in Figure 4.34, where a_{22} of eq. (4.133) is divided by the added mass for the two plates without plate interference, $\frac{1}{2}\rho\pi(a-b)^2$. These results have been independently confirmed using the boundary element method (Newman, personal communication, 2007). We note from Figure 4.34 a weak interference between the two plates. The interference is much stronger between two circular cylinders. When $b/(a-b) \to 0$, the ratio $E[\sqrt{1-(b/a)^2}]/K[\sqrt{1-(b/a)^2}]$ tends to zero as $1/\ln(b/a)$ with the result that $a_{22} \to \rho\pi a^2$ (i.e., the added mass of a plate with length $2a$). This fact is hardly seen on the scale presented in Figure 4.34, which implies that a rapid change occurs in the added mass when the gap between the plates becomes very small. The asymptotic added mass value when $b/(a-b) \to \infty$ (i.e., $b/a \to 1$) can be analytically confirmed by a series expansion of $E[\sqrt{1-(b/a)^2}]/K[\sqrt{1-(b/a)^2}]$.

4.7.3 Poles

4.7.3.1 Horizontal and vertical poles

Equations (4.117) and (4.121) can be used to study the effect of a horizontal two-dimensional and vertical circular pole with a small radius r_0 relative to the wavelength of the natural mode. The two studied scenarios are shown in Figure 4.35.

Example: submerged small body in a two-dimensional tank. We start by studying a two-dimensional submerged circular body that is *distant from the free surface and the tank surface.* The added mass coefficients are

$$a_{22} = a_{33} = \rho\pi r_0^2, \quad a_{23} = a_{32} = 0, \qquad (4.134)$$

where r_0 is the radius. Using formula (4.117) and the two-dimensional volume $Vol(\delta Q_0) = \pi r_0^2$ gives

$$\frac{\sigma_m'^2}{\sigma_m^2} = 1 - \frac{2\pi r_0^2 \left[\left(\left. \frac{\partial \varphi_m}{\partial y} \right|_{(y_0,z_0)} \right)^2 + \left(\left. \frac{\partial \varphi_m}{\partial z} \right|_{(y_0,z_0)} \right)^2 \right]}{\kappa_m \int_{\Sigma_0} \varphi_m \varphi_m \, dS}, \qquad (4.135)$$

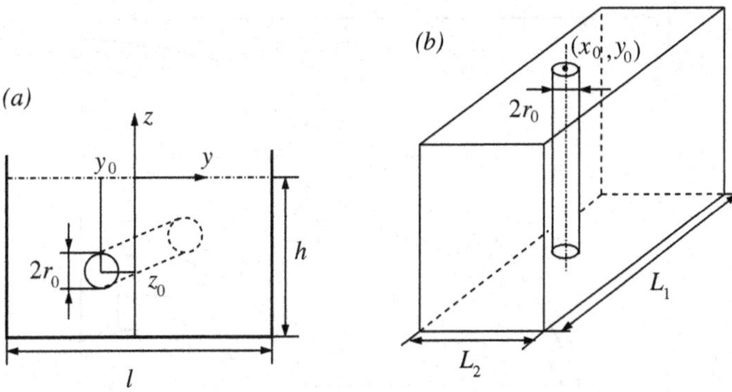

Figure 4.35. Definitions used in two- and three-dimensional analyses of the effect of poles on natural sloshing frequencies.

where (y_0, z_0) is the center of the circle. It follows by using solution (4.8) for the ambient flow in a rectangular tank that formula (4.135) can be expressed as

$$\frac{\sigma_m'^2}{\sigma_m^2} = 1 - 8\pi^2 m \left(\frac{r_0}{l}\right)^2 \times \frac{\sin^2(\pi m(y_0 + \frac{1}{2}l)/l) + \sinh^2(\pi m(z_0 + h)/l)}{\sinh(2\pi m h/l)}, \tag{4.136}$$

where l is the tank length and h is the filling depth.

Formula (4.136) can be generalized to three-dimensional and cross-wave modes, which were derived for three-dimensional rectangular tanks in Section 4.3.2.1. Jeyakumaran and McIver (1995) derived these formulas from a direct calculation of the dipole-type solutions due to the presence of the interior structure. Their results may also be obtained by implementing asymptotic formula (4.121) for a slender body.

Another generalization in two dimensions is for a submerged small-volume body having two symmetry axes such that $a_{23} = a_{32} = 0$. An example is an elliptical cross-section with horizontal and vertical semi-axes. This case leads to

$$\frac{\sigma_m'^2}{\sigma_m^2} = 1 - \frac{Vol.(\delta Q_0)\left[\left(\left.\frac{\partial \varphi_m}{\partial y}\right|_{(y_0,z_0)}\right)^2 + \left(\left.\frac{\partial \varphi_m}{\partial z}\right|_{(y_0,z_0)}\right)^2\right] + \frac{a_{22}}{\rho}\left(\left.\frac{\partial \varphi_m}{\partial y}\right|_{(y_0,z_0)}\right)^2 + \frac{a_{33}}{\rho}\left(\left.\frac{\partial \varphi_m}{\partial z}\right|_{(y_0,z_0)}\right)^2}{\kappa_m \int_{\Sigma_0} \varphi_m \varphi_m \, dy}. \tag{4.137}$$

Example: vertical slender body in a three-dimensional tank. We apply formula (4.121) to study the effect of a vertical pole inserted in a tank as shown in Figure 4.35(b). The pole penetrates the free surface and stands on the tank bottom. The x- and y-coordinates of the pole axis are (x_0, y_0). The ambient natural modes for a vertical cylindrical tank with a constant cross-sectional shape (e.g., rectangular- and circular-base tanks) can be expressed as

$$\varphi_{i,j}(x, y, z) = \frac{\cosh(k_{i,j}(z + h))}{\cosh(k_{i,j}h)} \phi_{i,j}(x, y), \tag{4.138}$$

where $k_{i,j}$ is the wave number.

If the cross-section of the inserted vertical pole has *two perpendicular symmetry axes* such that $a_{12} = a_{21} = 0$, it follows from eq. (4.138) and asymptotic formula (4.121) that

$$\frac{\sigma_{i,j}'^2}{\sigma_{i,j}^2} = 1 - \frac{Vol.(S_{CS})\left.\left(\left(\frac{\partial \phi_{i,j}}{\partial x}\right)^2 + \left(\frac{\partial \phi_{i,j}}{\partial y}\right)^2\right)\right|_{(x_0,y_0)} + \frac{a_{11}}{\rho}\left.\left(\frac{\partial \phi_{i,j}}{\partial x}\right)^2\right|_{(x_0,y_0)} + \frac{a_{22}}{\rho}\left.\left(\frac{\partial \phi_{i,j}}{\partial y}\right)^2\right|_{(x_0,y_0)}}{2k_{i,j}^2 \int_{\Sigma_0} \phi_{i,j}^2 \, dx\, dy} \times \left[1 + \frac{2hk_{i,j}}{\sinh(2k_{i,j}h)}\right], \tag{4.139}$$

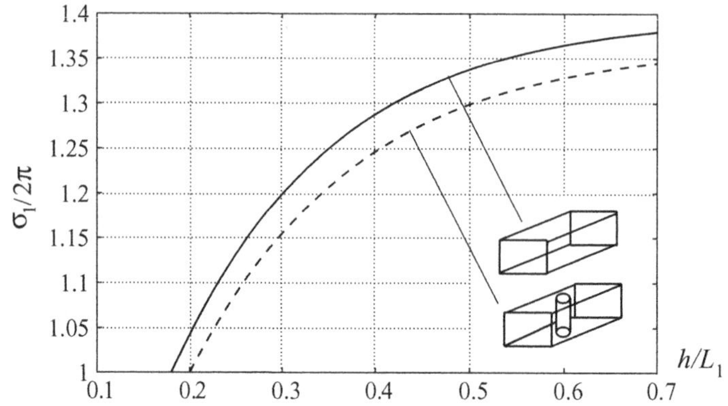

Figure 4.36. Theoretical reduction of the first-mode natural frequency of two-dimensional liquid sloshing due to a single vertical cylindrical pole with radius $r_0 = 0.024$ m in the middle of a rectangular tank with lengths $L_1 = 0.4$ m and $L_2 = 0.2$ m.

where $Vol(S_{CS})$ is the cross-sectional area and $\sigma_{i,j}^2 = gk_{i,j}\tanh(k_{i,j}h)$.

A special case is a rectangular-base tank with a circular vertical pole as shown in Figure 4.35(b). The ambient horizontal velocity components needed in eq. (4.139) can by means of eq. (4.29) be expressed as

$$\frac{\partial\phi_{i,j}}{\partial x} = -\pi i L_1^{-1}\sin\left(\pi i\left(x+\tfrac{1}{2}L_1\right)/L_1\right) \times \cos\left(\pi j\left(y+\tfrac{1}{2}L_2\right)/L_2\right),$$

$$\frac{\partial\phi_{i,j}}{\partial y} = -\pi j L_2^{-1}\cos\left(\pi i\left(x+\tfrac{1}{2}L_1\right)/L_1\right) \times \sin\left(\pi j\left(y+\tfrac{1}{2}L_2\right)/L_2\right). \quad (4.140)$$

The wave number, $k_{i,j}$, is defined by eq. (4.30). For two-dimensional ambient waves along the Ox-axis, the modified natural frequencies due to a single circular pole with radius r_0 and negligible hydrodynamic tank–wall interaction are

$$\frac{\sigma_{i,0}'^2}{\sigma_{i,0}^2} = 1 - 2\pi\frac{L_1}{L_2}\sin^2\left(\frac{\pi i}{L_1}\left(x_0+\tfrac{1}{2}L_1\right)\right) \times \left[1+\frac{2\pi i h/L_1}{\sinh(2\pi i h/L_1)}\right]\left(\frac{r_0}{L_1}\right)^2 \quad (4.141)$$

by using eq. (4.139) and $a_{11}/\rho = \pi r_0^2$. The maximum influence of the vertical pole on a natural frequency is for the lowest mode. When the axis of the pole coincides with antinodes, no influence exists according to the asymptotic formula. The analogous formula for the ratio $\sigma_{0,j}'^2/\sigma_{0,j}^2$ follows from eq. (4.141) by exchanging L_1 and L_2, i and j, and x_0 and y_0, respectively.

One can validate expression (4.141) by using experimental data from Warnitchai and Pinkaew (1998) for the lowest mode and $x_0 = 0$. Warnitchai and Pinkaew (1998) also presented a theoretical formula that agrees with eq. (4.141). Their derivation started out with Morison's equation for the force on the pole to find the reduction in the kinetic energy in the liquid due to the presence of the pole. Figure 4.36 presents the lowest natural frequency with and without a vertical pole versus nondimensional liquid depth h/L_1. The presence of the vertical pole causes a reduction of the lowest natural frequency that is less than ~5% of the natural frequency without the pole for the considered values of h/L_1. The theoretical results are in good agreement with experiments (see Warnitchai & Pinkaew, 1998). However, some differences between experiments and theory are noted even without the presence of the pole.

An asymptotic estimate for the natural frequencies for higher, three-dimensional natural modes with $ij \neq 0$ in a rectangular tank with a vertical circular pole takes, due to eqs. (4.139) and (4.140), the following form:

$$\frac{\sigma_{i,j}'^2}{\sigma_{i,j}^2} = 1 - 4\frac{S_{r_0}}{L_1L_2}\left[1+\frac{2k_{i,j}h}{\sinh(2k_{i,j}h)}\right] \times \left\{-\alpha_i\beta_j + \frac{i^2\alpha_i + (jL_1/L_2)^2\beta_j}{i^2+(jL_1/L_2)^2}\right\},$$

where $\alpha_i = \sin^2(\pi i(x_0+\tfrac{1}{2}L_1)/L_1)$ and $\beta_j = \sin^2(\pi j(y_0+\tfrac{1}{2}L_2)/L_2)$.

4.7.3.2 Proximity of circular poles

The added mass coefficients in formulas (4.117) and (4.121) are influenced by hydrodynamic

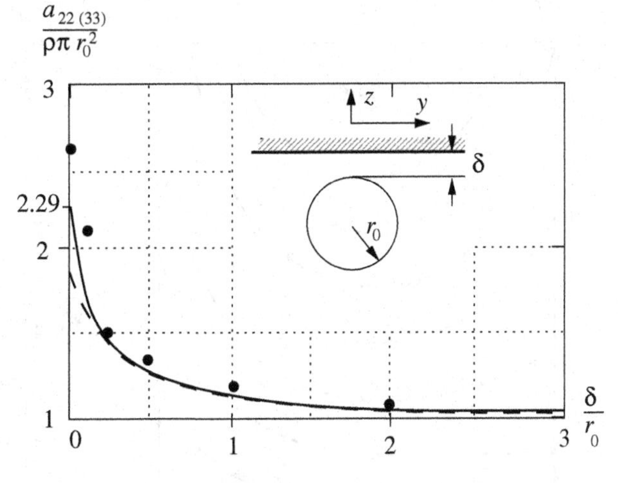

Figure 4.37. Added mass coefficients in heave (a_{33}) or sway (a_{22}) for a circular cross-section at a wall. The experimental data and exact theoretical curves and data are represented by the circles and solid line, respectively (Greenhow & Li, 1987). The dashed line is based on approximate formula (4.143).

interaction effects when a circular cross-section comes close to the free surface, a wall, or another body. Figure 4.37 shows experimentally and theoretically how the added mass in sway (a_{22}) and heave (a_{33}) of a circular cross-section of radius r_0 is influenced by a wall. The distance between the top of the cylinder and the wall is denoted δ. The heave direction is defined perpendicular to the wall. The exact theoretical expressions of a_{22} and a_{33} are according to Venkatesan (1985) and Walton (1986) equal to:

$$\frac{a_{22}}{\rho\pi r_0^2} = \frac{a_{33}}{\rho\pi r_0^2} = \frac{(1-q^2)^2}{q^2}\left\{\tfrac{1}{12} + \tfrac{1}{3}(1-m_1')\frac{K^2(m_1)}{\pi^2} - \frac{E(m_1)K(m_1)}{\pi^2}\right\} - 1, \quad (4.142)$$

where q and $m_1 + m_1' = 1$ are implicit functions of δ and r_0, and $E(\cdot)$ and $K(\cdot)$ are complete elliptic integrals as given by Abramowitz and Stegun (1964, ch. 17). To find q, one should solve the equation $(r_0+\delta)^2 - r_0^2 = r_0^2(1-q^2)^2/4q^2$. The values of m_1 and m_1' are solutions of the equation $q = \exp[-\pi K(m_1')/K(m_1)]$.

The closer the cylinder is to the plane $z = 0$, the greater is the added mass. When the cylinder touches the plane $z = 0$ (i.e., $\delta = 0$), the added mass coefficient is $(\frac{1}{3}\pi^2 - 1)$ or 2.29 times the added mass coefficient for an infinite fluid. The cylinder has to be quite close to $z = 0$ before any influence occurs. For instance, when $\delta/r_0 = 1$, the coefficients increase only by about 20% compared to the case of $\delta/r_0 \to \infty$.

Figure 4.37 also presents an approximate solution derived by Sun (personal communication, 2005) that can be adopted to our problem. The expression is

$$a_{22} = \rho\pi r_0^2(\alpha^4 - 1)/(\alpha^4 - 2\alpha^2), \quad (4.143)$$

where $\alpha = 2(\delta + r_0)/r_0$. An assumption in the derivation is that $(\delta + r_0)/r_0$ is large. However, we see from the figure that the expression remains not too far from the exact solution even though δ/r_0 is small.

The results can be generalized to sway added mass of two side-by-side circular cylinders in an infinite fluid. This follows by making a mirror image of the cylinder in Figure 4.37 with respect to the wall and noting that the wall is a symmetry plane for the flow around the two cylinders. The consequence is, for instance, that the sway added mass is 1.2 times the value with no interaction when the ratio of the distance between the cylinder axes and the cylinder radius is 4.

4.8 Approximate solutions

4.8.1 Two-dimensional circular tanks

Ibrahim (2005) gave a broad review of two-dimensional sloshing in a circular tank. Our presentation has a more limited scope. Budiansky (1960) presented numerical results for the eigenfrequencies of antisymmetric oscillations for a range of fill depths of a circular two-dimensional tank. Kuttler and Sigillito (1984) presented calculations with rigorous error bounds

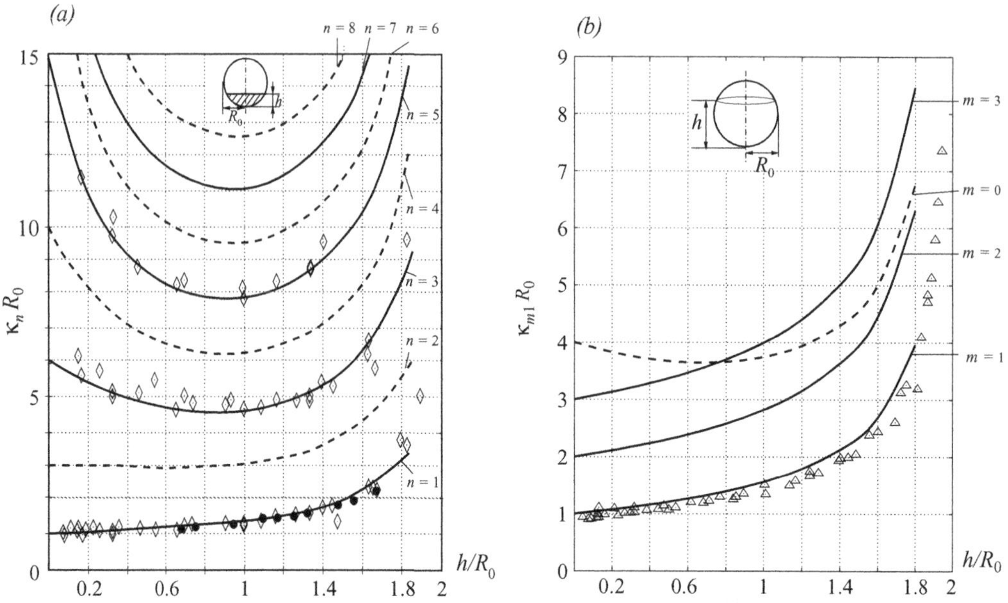

Figure 4.38. (a) Two-dimensional circular and (b) spherical tanks. Nondimensional eigenvalues versus ratio of filling depth to radius. (a) Antisymmetric modes are given by solid lines whereas eigenvalues for symmetric standing waves are presented as dashed lines. (b) We show the four lowest eigenvalues associated with $\kappa_{11}R_0$, $\kappa_{21}R_0$, $\kappa_{01}R_0$, and $\kappa_{31}R_0$, based on theoretical results by McIver (1989). The circles (•) in part (a) represent experimental data by Bogomaz and Sirota (2002) and diamonds (◇) are experimental results by McCarty and Stephens (1960). In part (b), the experimental results by McCarty and Stephens (1960) are denoted by triangles (△).

for antisymmetric and symmetric modes. McIver (1989) also treated this two-dimensional sloshing with results that are in excellent agreement with Kuttler and Sigillito (1984). The solution method was to choose a bipolar coordinate system in which the tank wall and the free surface coincide with coordinate lines and to reformulate eigenvalue problem (4.2) in terms of integral equations that are solved numerically.

McIver (1989) presented nondimensional eigenvalues $\kappa_n R_0$ for the eight lower natural modes for two-dimensional sloshing at different filling heights in a circular container. He distinguished odd (antisymmetric) and even (symmetric) wave profiles, as defined for a rectangular tank in Section 4.3.1.1. The results are presented in Table 4.3 and Figure 4.38(a). Odd values of n correspond to the antisymmetric modes, and even values of n imply symmetric modes. Table 4.3 also includes estimates of the lowest natural frequency by using a horizontal infinite-fluid dipole solution given by eq. (4.72) in combination with a variational formulation. The vertical singularity coordinate a of the dipole above the mean free surface is also given in the table. These results have surprisingly good agreement.

For the limiting case, $h/R_0 = 2$ (completely filled tank), the scaled eigenvalues become infinite because the mean free surface disappears. However, if the eigenfrequencies are scaled by the radius of the mean free surface, $R_s = R_0\sqrt{1-(h/R_0-1)^2}$, the limit $\lim_{h/R_0\to 2}\kappa_n R_s$ is finite and equal 2.00612, 3.45333, 5.12530, 6.62861, 8.25995, 9.78393, 11.3982, and 12.933 for the eight lower modes. The limiting values of $\kappa_n R_0$ as $h/R_0 \to 0$ are given by Lamb (1945) as $\kappa_n R_0 = \frac{1}{2}n(n+1)$, $n = 1, 2, \ldots$. These asymptotic values are used in drawing the graphs in Figure 4.38(a). The experimental data by Bogomaz and Sirota (2002) and McCarty and Stephens (1960) are also presented in the figure and agree well with the numerical results.

4.8.2 Axisymmetric tanks

Conical, spherical, and elliptical tanks are examples of axisymmetric tanks. The cylindrical coordinate system $Or\theta z$ is used to describe the

Table 4.3. *Nondimensional eigenvalues $\kappa_n R_0$ for two-dimensional sloshing in a circular tank as a function of the ratio between filling depth h and radius R_0*

$\frac{h}{R_0}$	$\kappa_1 R_0$	$\kappa_1 R_0\ (a/R_0)$ by eq. (4.72)	$\kappa_2 R_0$	$\kappa_3 R_0$	$\kappa_4 R_0$	$\kappa_5 R_0$	$\kappa_6 R_0$	$\kappa_7 R_0$	$\kappa_8 R_0$
0.2	1.04385	1.044012 (1.836)	2.92908	5.35498	8.03025	10.76724	13.48837	16.1798	18.8477
0.4	1.09698	1.09778 (1.673)	2.89054	4.93704	6.99058	9.00749	11.00134	12.9835	14.9595
0.6	1.16268	1.164845 (1.511)	2.88924	4.69867	6.46064	8.19875	9.92610	11.6490	13.3691
0.8	1.24606	1.25077 (1.348)	2.93246	4.60670	6.23613	7.85373	9.46499	11.07407	12.6813
1.0	1.35573	1.36488 (1.185)	3.03310	4.65105	6.23920	7.81986	9.39668	10.9718	12.5457
1.2	1.50751	1.524338 (1.018)	3.21640	4.85091	6.46747	8.07834	9.68639	11.2932	12.8989
1.4	1.73463	1.765255 (0.842)	3.53751	5.27678	6.99993	8.72206	10.42884	12.1571	13.8722
1.6	2.12374	2.182 (0.651)	4.14328	6.13932	8.10314	10.08074	12.04189	14.0138	15.9749
1.8	3.02140	3.1536 (0.427)	5.62694	8.31388	10.90612	13.55955	16.15857	18.7997	21.4033

Note: Estimate of $\kappa_1 R_0$ based on variational formulation with eq. (4.72) as test functions is also given together with the vertical position a of the dipole singularity above the mean free surface.
Source: (McIver, 1989).

eigenmodes with the Oz-axis along the tank axis. However, we are able to separate only the angular θ-dependence and not the r- and z -dependence as we could do for an upright circular cylindrical tank (see Section 4.3.2.2). The solution of spectral problem (4.2) may be presented in the form

$$\varphi_{m,i}(r, z, \theta) = \phi_{m,i}(r, z) \begin{Bmatrix} \cos(m\theta) \\ \sin(m\theta) \end{Bmatrix}, \quad m = 0, 1, \ldots, \tag{4.144}$$

where m corresponds to the azimuthal wave numbers and $m = 0$ represents axisymmetric modes. The lowest natural mode is commonly associated with $m = i = 1$. For each m there is an infinite sequence of discrete eigenvalues $\kappa_{m,i}$ that follow by solving a spectral problem in the r–z meridional plane. The meridional-plane cross-section is a rectangle for circular cylindrical upright tanks (see Section 4.3.2.2), a triangle for conical tanks (see Figure 4.39), a circle for spherical shapes, and an ellipse for elliptical tanks. Wave profiles corresponding to $i = 1$ and different values of m for a conical tank are shown in Figure 4.39.

4.8.2.1 Spherical tank

Moiseev and Petrov (1965) and Feschenko *et al.* (1969) reported calculations for the two lowest modes for liquid in a spherical tank with $m = 1$ by using variational techniques. Budiansky (1960) and Chu (1964) used integral equation methods to calculate the three lowest natural modes ($i =$ 1, 2, 3) with $m = 1$. An extensive set of numerical data for natural frequencies with $m = 0, 1, 2, 3$ were provided by McIver (1989).

Eigenvalues $\kappa_{mi} R_0$ for a spherical tank are given in Table 4.4 for a range of fill depths and for azimuthal wave numbers $m = 0, 1, 2, 3$; for each value of m, only the lowest four modes are computed; they are based on results by McIver (1989). Moreover, $\kappa_{mi} R_0 \to \infty$ as $h/R_0 \to 2$, but $\kappa_{mi} R_s$ (where R_s is the radius of the mean free surface) is finite. These limits for the four lowest modes (κ_{m1}, $m = 0, 1, 2, 3$) are 4.1213, 2.75475, 4.1213, and 5.40002, respectively. The limiting case when the filling depth of a spherical tank tends to the diameter is equivalent to sloshing in a circular aperture in a rigid plate with zero draft bounding a semi-infinite space filled with liquid; that is,

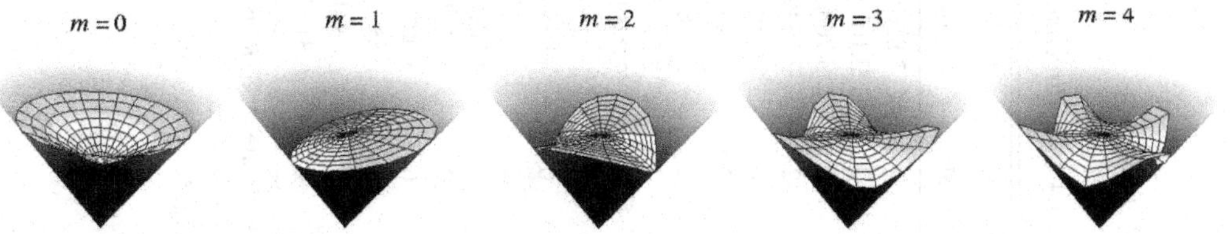

Figure 4.39. Wave profiles for natural modes in a conical tank for different angular wave numbers m (Gavrilyuk *et al.*, 2005).

Table 4.4. *Nondimensional eigenvalues $\kappa_{mi}R_0$ for a spherical tank as a function of the ratio between filling depth h and radius R_0*

$\frac{h}{R_0}$	m	$\kappa_{m1}R_0$	$\kappa_{m2}R_0$	$\kappa_{m3}R_0$	$\kappa_{m4}R_0$	m	$\kappa_{m1}R_0$	$\kappa_{11}R_0$ (a/R_0) by eq. (4.73)	$\kappa_{m2}R_0$	$\kappa_{m3}R_0$	$\kappa_{m4}R_0$
0.2	0	3.82612	9.25613	14.75561	20.1188	1	1.07232	1.07233(2.7328)	6.20081	11.88212	17.3589
	2	2.10792	8.39523	14.29444	19.8090	3	3.12949		10.48832	16.58021	22.1585
0.4	0	3.70804	7.91895	11.94118	15.9077	1	1.15826	1.15833(2.465)	5.67422	9.85513	13.8685
	2	2.23491	7.42178	11.65248	15.6994	3	3.28209		9.05998	13.36738	17.4637
0.6	0	3.65014	7.26596	10.74498	14.1964	1	1.26251	1.2628(2.1953)	5.36832	8.94181	12.4233
	2	2.38767	6.88669	10.50818	14.0217	3	3.46642		8.31214	12.00637	15.5636
0.8	0	3.65836	6.98858	10.23113	13.4553	1	1.39239	1.3933(1.9245)	5.24058	8.55088	11.7995
	2	2.57671	6.65230	10.01555	13.2950	3	3.69610		7.98544	11.41863	14.7386
1.0	0	3.74517	6.97636	10.14748	13.3042	1	1.56016	1.5625(1.6527)	5.27555	8.50444	11.6835
	2	2.81969	6.65941	9.94129	13.1499	3	3.99416		7.97281	11.31996	14.5669
1.2	0	3.93812	7.21881	10.45215	13.6727	1	1.78818	1.794(1.3791)	5.49298	8.77928	12.0208
	2	3.14918	6.91454	10.25082	13.5208	3	4.40308		8.26743	11.66454	14.9711
1.4	0	4.30102	7.80055	11.25589	14.6984	1	2.12320	2.1371(1.1019)	5.97283	9.47622	12.9380
	2	3.63358	7.50871	11.05830	14.5474	3	5.01225		8.97089	12.5789	16.104
1.6	0	5.00753	9.01565	12.9748	16.9191	1	2.68635	2.7218(0.81625)	6.95709	10.95568	14.9158
	2	4.45122	8.73902	12.78174	16.7692	3	6.05468		10.43251	14.5358	18.562
1.8	0	6.76418	12.1139	17.396	22.657	1	3.95930	4.0723(0.5076)	9.45348	14.75484	20.0224
	2	6.31547	11.8582	17.21013	22.5093	3	8.46310		14.13739	19.5653	24.912

Note: Estimate of $\kappa_{11}R_0$ based on variational formulation with eq. (4.73) as test functions is also given together with the vertical position a of the dipole singularity above the mean free surface.

Source: (McIver, 1989).

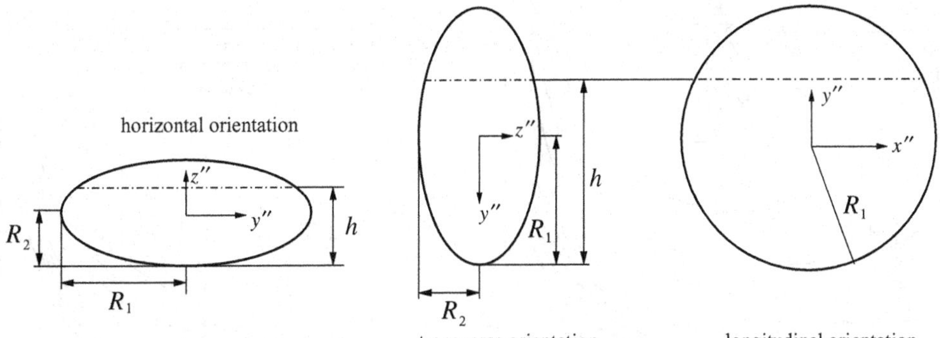

Figure 4.40. Orientations of oblate spheroidal containers.

we can physically imagine the problem as sloshing in a small circular hole of sea ice covering a large area. The surprising result that the infinite sequences of frequencies for azimuthal wave numbers $m = 0$ and $m = 2$ are identical was also confirmed by Henrici *et al.* (1970) and Miles (1972). The dependence on h/R_0 for κ_{m1} ($m = 0, 1, 2, 3$) is graphically displayed in Figure 4.38(b). Table 4.4 also includes estimates of the lowest natural frequency by using a horizontal infinite-fluid dipole solution given by eq. (4.73) in combination with a variational formulation. The vertical singularity coordinate a of the dipole above the mean free surface is also given in the table. The results have surprisingly good agreement.

The analytical solution for the empty limit, $h/R_0 = 0$, can be described by shallow-liquid theory (Lamb, 1945). The result for $m = 0$ is $\kappa_{0i}R_0 = 2i(i-1)$, $i = 2, 3, 4, \ldots$. and for $m = 1$ we can write $\kappa_{1i}R_0 = 2i^2 - 1$, $i = 1, 2, \ldots$. These relations facilitate calculations of eigenvalues for these most important antisymmetric and symmetric modes.

The antisymmetric motions for the lowest mode as $h/R_0 \to 0$ may be found by considering the liquid moving as a rigid body with mass M_l. We denote ϑ the angular motion of the center of gravity of the liquid mass and apply Newton's second law in the angular direction. By assuming small ϑ and approximating the center of gravity of the frozen liquid mass to have a radial coordinate R_0, we get $M_l R_0 \ddot{\vartheta} + M_l g \vartheta = 0$ (i.e., the liquid behaves as a pendulum). Assuming a solution $\vartheta = A\cos(\sigma_{11}t)$ gives

$$\sigma_{11}^2/g = \kappa_{11} = 1/R_0, \quad T_{11} = 2\pi\sqrt{R_0/g}.$$

Abramson (1966) documented a set of results reported by Mikishev and Dorozhkin (1961), Leonard and Walton (1961), Stofan and Armstead (1962), Abramson *et al.* (1963), and Chu (1964). Based on the measured natural frequency of the first mode, Mikishev and Dorozhkin (1961) gave empirical formulas for certain ranges of h/R_0. A review on prolate spheroidal tanks is given by Ibrahim (2005).

4.8.2.2 Ellipsoidal (oblate spheroidal) container

Relevant ellipsoidal containers are flattened spheres. The horizontal cross-sections are circles for a flattened sphere with horizontal orientation, whereas the horizontal cross-sections are ellipses for flattened spheres with transverse and longitudinal orientations. Figure 4.40 shows these three orientations. The origin of the $Ox''y''z''$-coordinate system corresponds to the center of the three-dimensional container. The flattening occurs along the Oz''-axis.

Using an accurate variational method, Feschenko *et al.* (1969) computed eigenvalues for sloshing in ellipsoidal tanks with horizontal orientation. The results from their book are presented in Figure 4.41 for different filling depths and ratios R_2/R_1 between the semi-axes of the ellipse in the meridional plane cross-section. The depth and the lowest eigenfrequency κ_{11} are scaled by R_1. The case $R_2/R_1 = 1$ corresponds to the spherical tank analyzed in Section 4.8.2.1. These results are in good agreement with experiments performed for spacecraft by various Soviet scientists.

Feschenko *et al.* (1969) also presented an asymptotic formula for the lowest eigenvalue in the shallow-liquid limit: $\lim_{h/R_1 \to 0} \kappa_{11} R_1 = (R_1/R_2)^3$. A review of experimental results for

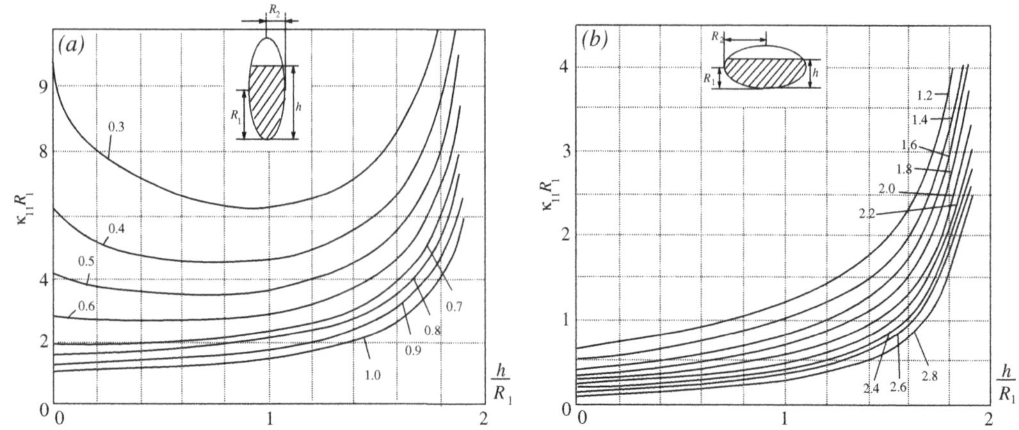

Figure 4.41. Nondimensional eigenvalues $\kappa_{11}R_1$ for the lowest antisymmetric mode in elliptical tanks versus the ratio of filling depth to semi axis. The graphs are labeled with the value R_2/R_1 (based on calculations by Feschenko *et al.*, 1969).

numerical natural sloshing frequencies for an ellipsoidal tank is also given by Ibrahim (2005).

4.8.3 Horizontal cylindrical container

4.8.3.1 Shallow-liquid approximation for arbitrary cross-section

Let us consider a horizontal cylindrical tank, which is illustrated for a circular cross-section in Figure 4.42. The tank is partially filled with a liquid of filling depth h (varying in the interval $0 < h < 2R_0$ for a circular cross-section with radius R_0). The liquid domain has the Oxz- and Oyz-planes as two symmetry planes. The mean free surface Σ_0 is rectangular with length dimensions B_s and L_1 in the transverse and longitudinal directions of the cylinder. A constant cross-section makes it possible to separate the dependence on the x-variable in the velocity potential. The eigenfunctions can then be expressed as

$$\varphi_{mj}(x, y, z) = \phi_{mj}(y, z)\cos\left[\pi m\left(x + \tfrac{1}{2}L_1\right)/L_1\right],$$
$$m = 0, 1, 2, \ldots; \quad j = 1, 2, \ldots. \qquad (4.145)$$

We note that φ_{mj} satisfies the boundary condition at $x = \pm\frac{1}{2}L_1$. Substitution of eq. (4.145) into the original problem yields the m-parametric family of the spectral problems in the y–z cross-section of the cylinder:

$$\frac{\partial^2\phi_{mj}}{\partial y^2} + \frac{\partial^2\phi_{mj}}{\partial z^2} - \left(\frac{\pi m}{L_1}\right)^2\phi_{mj} = 0 \quad \text{in } G,$$
$$\frac{\partial\phi_{mj}}{\partial n} = 0 \quad \text{on } B_1, \quad \frac{\partial\phi_{mj}}{\partial z} = \kappa_{mj}\phi_{mj} \quad \text{on } B_0 \qquad (4.146)$$

(see Figure 4.42 for definitions of G, B_0, and B_1), where j enumerates the eigenvalues κ_{mj} in ascending order for a given m. When $m = 0$, problem (4.146) needs the volume conservation condition $\int_{B_0}\phi_{0j}\,dx = 0$.

The case $m = 0$ corresponds to pure transverse, two-dimensional liquid motions in the circular cross-section. As long as $m \neq 0$, the natural modes following from original spectral problem (4.2) are, generally speaking, three-dimensional. Pure longitudinal wave patterns (i.e., those that occur along the Ox-axis), which do not depend on y and z and vary as $\cos(\pi m(x + \frac{1}{2}L_1)/L_1)$ as was observed for a rectangular cross-section (see Section 4.3.2.1), are not possible in general. The reason is that eigenfunctions following from spectral problem (4.146) depend on y and z for nonrectangular G. However, the dependence on y and z can, as a first approximation, be neglected in a

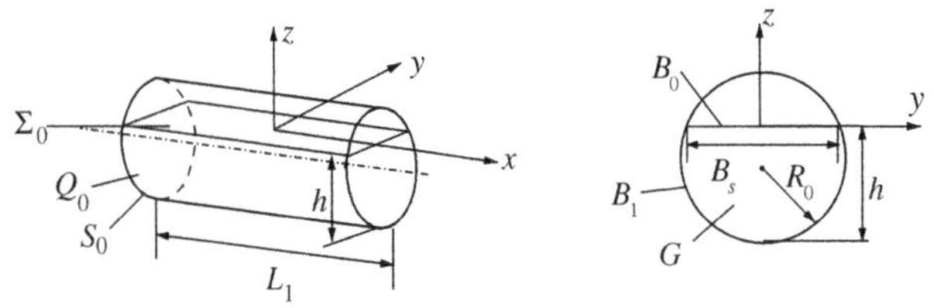

Figure 4.42. Three-dimensional view and cross-section of a horizontal tank.

shallow-liquid approximation for relatively long horizontal tanks (i.e., $B_s \ll L_1$).

A shallow-liquid approximation for the longitudinal flow can be obtained by using eq. (4.46) as a basis, which gives, with constant cross-sectional area of the liquid, $\mathcal{A}$, that

$$g\mathcal{A}\frac{\partial^2 \varphi}{\partial x^2} - B_s\frac{\partial^2 \varphi}{\partial t^2} = 0. \tag{4.147}$$

Substitution of $\varphi = C\cos[\pi m(x+\frac{1}{2}L_1)/L_1]\exp(\mathrm{i}\sigma_{m1}t)$ into eq. (4.147) with C as a nonzero constant gives

$$\sigma_{m1}^2 B_s/g = (\pi m/L_1)^2\,\mathcal{A}. \tag{4.148}$$

If the cross-section is rectangular, we can write

$$\sigma_{m1}^2 L_1/g = (\pi m)^2 h/L_1 \approx \pi m \tanh(\pi m h/L_1). \tag{4.149}$$

By using the hyperbolic tangent function we have an expression that is valid for any liquid depth. We use this "trick" for any cross-section and rewrite eq. (4.148) as

$$\sigma_{m1}^2 L_1/g = (\pi m)^2\mathcal{A}/(L_1B_s) = (\pi m)^2 h\mathcal{A}/(L_1B_sh)$$
$$\approx \pi m\mathcal{A}\tanh(\pi m h/L_1)/(B_sh). \tag{4.150}$$

Furthermore, eq. (4.150) means that the shallow-liquid expression for the natural mode with $\phi_{m1}(y,z) = \text{const}$ in eq. (4.145) should also be approximated with the following natural modes:

$$\varphi_{m1}(x,y,z) = \cos\left[\frac{\pi m}{L_1}\left(x+\tfrac{1}{2}L_1\right)\right] \times \frac{\cosh\left[\pi m\,(z+h)\,/L_1\right]}{\cosh\left(\pi m h/L_1\right)} \tag{4.151}$$

for a rectangular cross-section (see Section 4.3.2.1).

Our objective with eq. (4.150) is to obtain an expression for the natural frequency that gives a good estimate for nonshallow liquid depths. An example is indeed the case for a circular cross-section. Obviously, approximation (4.151) of the longitudinal natural modes for nonrectangular tanks satisfies only the zero-Neumann conditions on the mean wetted tank surface at $x = \pm\frac{1}{2}L_1$. Furthermore, the substitution of eq. (4.151) into the boundary condition on Σ_0 leads to $\sigma_{m1}^2 L_1/g = \pi m\tanh(\pi m h/L_1)$, which is, once again, consistent with eq. (4.150) only for the rectangular tank. The argumentation on why we can use eq. (4.151) as an approximate natural mode is the same as in Sections 4.8.1 and 4.8.2.1 for infinite-fluid dipole approximations of the lowest natural modes for sloshing in circular and spherical tanks. Therefore, we should insert expression (4.151) into Rayleigh quotient (4.66), which gives

$$\frac{\sigma_{m1}^2}{g} = \frac{(\pi m/L_1)^2\frac{1}{2}L_1}{\frac{1}{2}L_1B_s}\int_G \frac{\cosh(2\pi m(z+h)/L_1)}{\cosh^2(\pi m h/L_1)}\,\mathrm{d}S$$
$$= \frac{\pi m}{2B_sL_1}\int_G \frac{\sinh'(2\pi m(z+h)/L_1)}{\cosh^2(\pi m h/L_1)}\,\mathrm{d}S, \tag{4.152}$$

where the cross-sectional area G is shown in Figure 4.42 and the prime denotes the derivative with respect to z. Equation (4.152) is exact for a rectangular cross-section and agrees with formula (4.148) for shallow-liquid sloshing. Approximate formula (4.150) can be obtained by assuming small filling depth relative to the free-surface breadth in eq. (4.152) so that $\sinh'(2\pi m(z+h)/L_1) \approx \sinh(2\pi m h/L_1)/h$ on the interval $[-h, 0]$. We show in the following section that, even though the depth is not small, formula (4.150) gives a reasonable approximation of the natural frequencies for different filling levels in horizontal tanks with a circular cross-section.

Expression (4.151) is used in a simplified analysis of sloshing during acceleration or braking of either a tank vehicle or a freight train in Section 5.6.6.

4.8.3.2 Shallow-liquid approximation for circular cross-section

In formula (4.148), B_s and $\mathcal{A}$ are functions of the ratio h/R_0 for a circular cross-section (see Figure 4.42). Indeed,

$$B_s = 2R_0\sqrt{1-(1-h/R_0)^2} \tag{4.153}$$

and the expression for the area is

$$\mathcal{A} = R_0^2\begin{cases} \arctan\left(\dfrac{\sqrt{1-(1-h/R_0)^2}}{(1-h/R_0)}\right) - \left(1-\dfrac{h}{R_0}\right)\sqrt{1-\left(1-\dfrac{h}{R_0}\right)^2}, & 0<\dfrac{h}{R_0}\le 1, \\ \pi + \arctan\left(\dfrac{\sqrt{1-(1-h/R_0)^2}}{(1-h/R_0)}\right) - \left(1-\dfrac{h}{R_0}\right)\sqrt{1-\left(1-\dfrac{h}{R_0}\right)^2}, & 1<\dfrac{h}{R_0}<2. \end{cases} \tag{4.154}$$

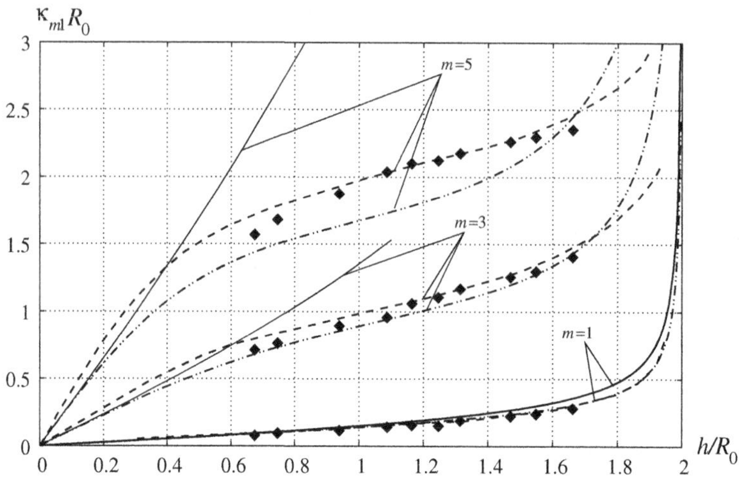

Figure 4.43. Nondimensional eigenvalues $\kappa_{m1}R_0$ ($m = 1, 3, 5$) for a horizontal cylindrical tank with circular cross-section versus the ratio of filling depth to radius, h/R_0; $R_0/L_1 = 0.139$. Asymptotic formula (4.148) (solid lines), its generalization eq. (4.149) (dashed-and-dotted lines), direct numerical calculations by Lukovsky *et al.* (1984) (dashed line), and experimental results by Bogomaz and Sirota (2002) are used. The eigenvalues correspond to the three lowest antisymmetric almost-longitudinal modes.

Let us illustrate the applicability of eqs. (4.148) and (4.150) by comparing with numerical and experimental cases in the books by Lukovsky *et al.* (1984) and Bogomaz and Sirota (2002). The model test results reported by Bogomaz and Sirota (2002) dealt with identification of antisymmetric longitudinal waves ($m = 1, 3, 5$) and their natural frequencies for the ratio of filling depth to tank radius, h/R_0, from 0.6 to 1.7. The model container had dimensions $L_1 = 0.95$ m and $R_0 = 0.1323$ m, which correspond to $R_0/L_1 = 0.139$.... These geometrical proportions are representative for tanks used in railway transportation (Bogomaz, 2004).

The nondimensional measured experimental data of the natural frequencies by Bogomaz and Sirota (2002) are denoted by solid diamonds in Figure 4.43. The dashed lines in Figure 4.43, which represent numerical results by Lukovsky *et al.* (1984), are in good agreement with experiments for nonshallow depths. The numerical results were obtained by the Ritz–Treftz method described in Section 4.6.1.3. Shallow-liquid formula (4.148) shows good agreement with experiments for the lowest longitudinal frequency ($m = 1$), but the results for higher frequencies with $m = 3$ and $m = 5$ are satisfactory only for $h/R_0 \lesssim 0.5$. Generalized shallow-liquid formula (4.149) demonstrates a clear improvement for $m = 3$ and $m = 5$ up to $h/R_0 \approx 1.8$.

Let us further validate eq. (4.148) by using the experimental data of McCarty and Stephens (1960), who studied natural frequencies for the first four antisymmetric modes associated with $m = 1, 3, 5, 7$ ($n = 1, 2, 3, 4$ is the notation used by McCarty and Stephens). These were scaled as follows:

$$\gamma_n = \sigma_{(2n-1)1}\sqrt{\frac{L_1}{g\tanh(\pi nh/L_1)}}, \qquad n = 1, 2, 3, 4. \tag{4.155}$$

Different values of R_0 and L_1 were tested to show that the experimental $\gamma_n = \gamma_n(h/R_0)$ is a weakly dependent function on h/R_0 and almost independent of the ratio R_0/L_1. Equation (4.150) gives

$$\gamma_n = \sqrt{\pi(2n-1)\frac{\tanh[\pi(2n-1)h/L_1]}{\tanh(\pi nh/L_1)}\frac{\mathcal{A}}{B_s h}}, \tag{4.156}$$

where γ_1 is only a function of the ratio h/R_0 and is independent of R_0/L_1.

Equation (4.156) is compared with the experimental data by McCarty and Stephens (1960) in Figure 4.44. The solid line gives our theoretical γ_1 as a function of h/R_0. The theoretical prediction is in quite good agreement with smoothed experimental values for the four tested tanks. Furthermore, satisfactory agreement with these experiments is established for the theoretical value $\gamma_2(h/R_0, R_0/L_1)$. Eq. (4.156) underpredicts the

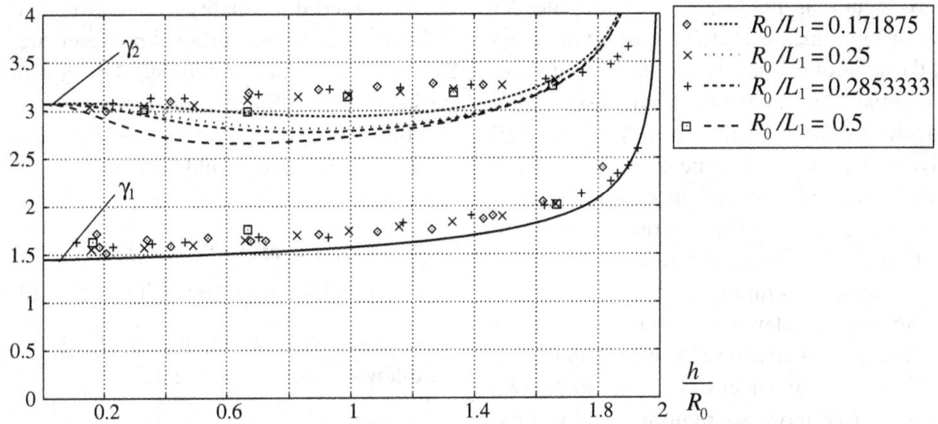

Figure 4.44. Experimental nondimensional frequency parameters γ_1 and γ_2 by eq. (4.155) for the two first antisymmetric longitudinal modes in horizontal circular cylinders with four different radius-to-length ratios R_0/L_1 (McCarty & Stephens, 1960). Comparison with theoretical predictions by eq. (4.156) is shown (single solid line) for the lowest modes. The frequency parameter γ_2 changes with R_0/L_1 (several dashed and dotted lines).

experimental values of McCarty and Stephens (1960). A constant difference of 8% exists between eq. (4.156) and the measurements for γ_1 in $0 < h/R_0 < 1.8$.

An approximation for strongly three-dimensional natural modes associated with the eigenvalues κ_{lj} ($l \geq 1$, $j \geq 2$) needs a numerical method to solve eigenvalue problem (4.146). We have already mentioned the Ritz–Treftz methods developed by Lukovsky *et al.* (1984), which give a semianalytical solution and provide fast convergence with a small number of polynomials satisfying the Laplace equation as a functional basis. Some useful upper bounds and limit formulas for the higher modes can be found in McIver and McIver (1993).

When no free surface exists (i.e., $h/R_0 = 2$), no natural frequency exists for an incompressible liquid inside a rigid tank. This result is evident from Figures 4.43 and 4.44 because the natural frequencies appear to go to infinity when $h/R_0 \to 2$. If liquid compressibility is considered for a completely filled tank, acoustic resonances occur. This fact is elaborated in Exercise 10.8.1. If the tank is completely filled with an incompressible liquid, resonances occur if the tank structure is elastic. This scenario is discussed for a completely filled fabric structure in Section 1.6.

4.9 Two-layer liquid

4.9.1 General statement

The subject of this section is the flow of two incompressible liquids with an interface between them. These liquids may completely fill the tank, as illustrated in Figure 4.45(a), or the upper liquid layer may have a free surface as an interface with a gas (air), as shown in Figure 4.45(b). A free shear layer exists, where the viscosity between

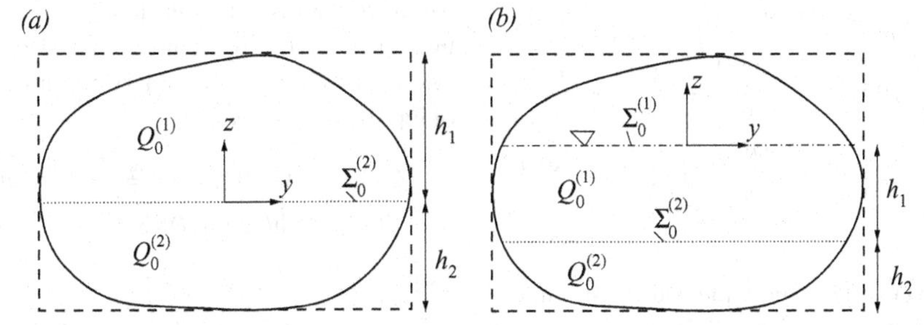

Figure 4.45. (a) Interfacial sloshing and (b) a mixed interfacial-free surface sloshing.

the two liquids is important. The presence of the shear layer means that the tangential velocity at the sides of the interface is discontinuous in the potential flow model. Our further assumption is that the thickness of this shear layer is small relative to typical linear dimensions of the tank and the considered wavelengths, which implies that the pressure is continuous through the free shear layer. Furthermore, the velocity normal to the shear layer is continuous.

The spectral problem on interface flows are of more complicated structure than problem (4.2). It involves different velocity potentials in the two domains. Furthermore, we require equality of the pressures at the interface. The pressures in the two liquid domains can by means of the linearized Bernoulli equation be expressed as

$$p_j = -\rho_j\left[\frac{\partial\phi^{(j)}}{\partial t} + gz\right] + C_j, \quad j = 1, 2. \qquad (4.157)$$

The constants C_j are determined by considering the liquids at rest. If we have a free surface at $z = 0$ between the gas and the liquid and denote $j = 1$ as the upper liquid, then C_1 is equal to the gas pressure, p_0. It follows by requiring continuity of pressure at the interface between the two liquids that $C_2 = (\rho_1 - \rho_2)gh_1 + p_0$, where h_1 is the depth of the upper liquid layer. If there is no free surface but there is contact with the atmosphere, then the constants must be consistent with this fact.

Let us consider the linear theory of freestanding waves on the interface of two liquids with notations defined in Figure 4.45(a). There is *no free surface* in this case. A Cartesian coordinate system $Oxyz$ with origin at the mean interface between the liquids, with densities ρ_1 and ρ_2, is introduced. The linear boundary value problem can be formulated as

$$\begin{gathered}
\nabla^2\phi^{(j)} = 0 \quad \text{in} \quad Q_0^{(j)}, \\
\frac{\partial\phi^{(j)}}{\partial n} = 0 \quad \text{on} \quad S_0^{(j)}, \quad j = 1, 2; \\
\int_{\Sigma_0^{(2)}} \zeta \, dS = 0, \quad \frac{\partial\phi^{(1)}}{\partial z} = \frac{\partial\phi^{(2)}}{\partial z} = \frac{\partial\zeta}{\partial t}, \\
\rho_1\left[\frac{\partial\phi^{(1)}}{\partial t} + g\zeta\right] = \rho_2\left[\frac{\partial\phi^{(2)}}{\partial t} + g\zeta\right] \quad \text{on} \quad \Sigma_0^{(2)},
\end{gathered} \qquad (4.158)$$

where $z = \zeta(x, y, t)$ is the equation for the interface elevation and $S_0^{(j)}$ ($j = 1, 2$) are the mean wetted tank surfaces in the two liquid domains. We have used the equality of the normal velocities and unsteady pressures on the mean interface $\Sigma_0^{(2)}$ as the interface conditions. The equality of the steady part of the pressure is accounted for by the constants C_j.

Furthermore, one should pose the freestanding waves as

$$\begin{gathered}
\zeta(x, y, t) = \exp(i\sigma t) f(x, y), \\
\phi^{(j)}(x, y, z, t) = i\sigma \exp(i\sigma t)\varphi^{(j)}(x, y, z).
\end{gathered} \qquad (4.159)$$

Substitution of eq. (4.159) into eq. (4.158) leads to the following spectral problem:

$$\begin{gathered}
\nabla^2\varphi^{(j)} = 0 \quad \text{in} \quad Q_0^{(j)}, \quad \frac{\partial\varphi^{(j)}}{\partial n} = 0 \quad \text{on} \quad S_0^{(j)}, \\
\int_{\Sigma_0^{(2)}} \varphi^{(j)} \, dS = 0, \quad j = 1, 2, \quad \frac{\partial\varphi^{(1)}}{\partial z} = \frac{\partial\varphi^{(2)}}{\partial z} \\
= \kappa\left[\frac{\rho_2}{\rho_2 - \rho_1}\varphi^{(2)} - \frac{\rho_1}{\rho_2 - \rho_1}\varphi^{(1)}\right] \\
\text{on} \quad \Sigma_0^{(2)},
\end{gathered} \qquad (4.160)$$

where the pair $\varphi^{(1)}$, $\varphi^{(2)}$ plays the role of eigenfunctions and $\kappa = \sigma^2/g$ is an eigenvalue. If $\rho_1 = \rho_2$, we have one fluid (i.e., $\varphi^{(1)} = \varphi^{(2)}$) and the last condition in eq. (4.160) becomes redundant. The resulting boundary-value problem has only the trivial solutions $\varphi^{(1)} = \varphi^{(2)} = 0$.

Eigenvalue problem (4.160) has an analytical solution for some particular tank cases. The typical example is presented by dashed lines in Figure 4.45(a) for a two-dimensional rectangular shape. The solution is obtained by using separation of the spatial variables:

$$\begin{gathered}
\varphi^{(1)}(y, z) = C_m^{(1)} f_m(y)\frac{\cosh(\pi m(z - h_1)/l)}{\cosh(\pi m h_1/l)}, \\
\varphi^{(2)}(y, z) = f_m(y)\frac{\cosh(\pi m(z + h_2)/l)}{\cosh(\pi m h_2/l)}, \\
f_m(y) = \cos\left(\pi m\left(y + \tfrac{1}{2}l\right)/l\right) \quad m = 1, 2, \ldots,
\end{gathered} \qquad (4.161)$$

where h_1 and h_2 are the depths of the two liquids and l is the horizontal tank size. Inserting eqs. (4.161) into the boundary conditions on $\Sigma_0^{(2)}$ of eqs. (4.160) gives

$$\begin{gathered}
C_m^{(1)} = -A_m/B_m, \quad A_m = \tanh(\pi m h_2/l), \\
B_m = \tanh(\pi m h_1/l)
\end{gathered} \qquad (4.162)$$

and

$$\frac{\sigma_m^2}{g} = \kappa_m = \frac{\pi m}{l}\frac{(\rho_2 - \rho_1)A_m B_m}{\rho_2 B_m + \rho_1 A_m}. \qquad (4.163)$$

This formula was also derived by Landau and Lifschitz (1959), who followed a different approach. When $\rho_1/\rho_2 \ll 1$, eq. (4.163) can, as a first approximation, be approximated as eq. (4.8) with $h = h_2$. The assumption $\rho_1/\rho_2 \ll 1$ is relevant if we let the upper layer be a gas instead of a liquid. Second, formula (4.163) simplifies for deep liquids (i.e., $h_1/l, h_2/l \to \infty$), which causes $A_m, B_m \to 1$ and the natural frequencies are expressed as

$$\frac{\sigma_m^2}{g} = \frac{\pi m}{l}\frac{\rho_2 - \rho_1}{\rho_2 + \rho_1}. \qquad (4.164)$$

An analysis of formulas (4.163) and (4.164) shows that the natural interfacial waves are unstable when $\rho_2 < \rho_1$. Indeed, this makes the squares of natural frequencies negative and, in view of representation (4.159), the interfacial waves increase exponentially with time. This result is consistent with the intuitive feeling that the heaviest fluid should be lowest in the tank.

When the two-layer liquid has a free surface, as in Figure 4.45(b), one can also derive a corresponding spectral problem for the natural modes of freestanding waves. However, the case is more complicated, as follows. Using the notation of Figure 4.45(b), the linearized problem relative to mean free surface $\Sigma_0^{(1)}$ and mean interface $\Sigma_0^{(2)}$ requires two functions that define instant positions of the free surface and the liquid interface (i.e., $z = \zeta(x, y, t)$ for the free surface $\Sigma_0^{(1)}$ and $z = -h_1 + \zeta'(x, y, t)$ for the interface $\Sigma_0^{(2)}$). The pressures in the two liquids can, as previously discussed, be expressed as

$$p_1 = -\rho_1\left[\frac{\partial\phi^{(1)}}{\partial t} + gz\right] + p_0;$$
$$p_2 = -\rho_2\left[\frac{\partial\phi^{(2)}}{\partial t} + gz\right] + g(\rho_1 - \rho_2)h_1 + p_0 \qquad (4.165)$$

by using the linearized Bernoulli equation. The unsteady problem couples $\phi^{(1)}$, $\phi^{(2)}$, ζ, and ζ' (henceforth, the prime does not involve differentiation) as follows:

$$\nabla^2\phi^{(j)} = 0 \quad \text{in} \quad Q_0^{(j)};$$
$$\frac{\partial\phi^{(j)}}{\partial n} = 0 \text{ on } S_0^{(j)};$$
$$\int_{\Sigma_0^{(1)}} \zeta\, \mathrm{d}S = \int_{\Sigma_0^{(2)}} \zeta'\, \mathrm{d}S = 0,$$
$$\frac{\partial\phi^{(1)}}{\partial n} = \frac{\partial\zeta}{\partial t}; \quad \frac{\partial\phi^{(1)}}{\partial t} + g\zeta = 0 \quad \text{on} \quad \Sigma_0^{(1)},$$
$$\frac{\partial\phi^{(1)}}{\partial z} = \frac{\partial\phi^{(2)}}{\partial z} = \frac{\partial\zeta'}{\partial t};$$
$$\rho_1\left[\frac{\partial\phi^{(1)}}{\partial t} + g\zeta'\right] = \rho_2\left[\frac{\partial\phi^{(2)}}{\partial t} + g\zeta'\right] \quad \text{on} \quad \Sigma_0^{(2)}. \qquad (4.166)$$

Looking for freestanding waves in the form $\zeta(x, y, t) = \exp(i\sigma t)f(x, y)$, $\zeta'(x, y, t) = \exp(i\sigma t) \times f'(x, y)$, $\phi^{(j)}(x, y, z, t) = i\sigma\exp(i\sigma t)\varphi^{(j)}(x, y, z)$ transforms eqs. (4.166) to the following spectral problem:

$$\nabla^2\varphi^{(j)} = 0 \quad \text{in} \quad Q_0^{(j)}; \quad \frac{\partial\varphi^{(j)}}{\partial n} = 0 \quad \text{on} \quad S_0^{(j)};$$
$$\int_{\Sigma_0^{(1)}} \varphi^{(1)}\, \mathrm{d}S = 0, \quad \frac{\partial\varphi^{(1)}}{\partial z} = \kappa\varphi^{(1)} \quad \text{on} \quad \Sigma_0^{(1)};$$
$$\frac{\partial\varphi^{(1)}}{\partial z} = \frac{\partial\varphi^{(2)}}{\partial z} = \kappa\big[\rho_2\varphi^{(2)} - \rho_1\varphi^{(1)}\big]/(\rho_2 - \rho_1) \quad \text{on} \quad \Sigma_0^{(2)}, \qquad (4.167)$$

where the spectral parameter κ is present in the boundary conditions on $\Sigma_0^{(2)}$ and $\Sigma_0^{(1)}$.

Problem (4.167) is characterized by positive eigenvalues for $\rho_2 > \rho_1$. Otherwise, the hydrostatic state of the two-layer liquid is not stable. Let us demonstrate for a two-dimensional rectangular tank as illustrated with dashed lines in Figure 4.45. The solution of problem (4.167) can be expressed as

$$\varphi_m^{(2)} = f_m(y)\frac{\cosh\left(\frac{\pi m}{l}(z + h_1 + h_2)\right)}{\cosh(\pi m h_2/l)},$$
$$\varphi_m^{(1)} = C_m f_m(y)\frac{\frac{\kappa l}{\pi m}\sinh\left(\frac{\pi m}{l}z\right) + \cosh\left(\frac{\pi m}{l}z\right)}{\cosh(\pi m h_1/l)}, \qquad (4.168)$$

where $f_m(y)$ is from eq. (4.161). Solution (4.168) satisfies all the conditions except those on $\Sigma_0^{(2)}$. Equality of the $\partial/\partial z$-derivatives on $\Sigma_0^{(2)}$ gives $C_m = A_m/(\kappa l/\pi m - B_m)$, where $0 < A_m, B_m < 1$ are given by eq. (4.162). The remaining condition leads to the following quadratic equations with respect to κ:

$$\kappa^2\left[\frac{\rho_2 + \rho_1 A_m B_m}{\rho_2 - \rho_1}\frac{l}{\pi m}\right] - \kappa\left[\frac{\rho_2(A_m + B_m)}{\rho_2 - \rho_1}\right] + \frac{\pi m}{l}A_m B_m = 0, \quad m = 1, 2, \ldots. \qquad (4.169)$$

The discriminant of the quadratic equation (4.169) is

$$\frac{\rho_2^2(A_m+B_m)^2}{(\rho_2-\rho_1)^2}-4\frac{A_mB_m\left(\rho_2+\rho_1A_mB_m\right)}{\rho_2-\rho_1}$$
$$=\frac{\rho_1^2}{(\rho_2-\rho_1)^2}\left[\left(\frac{\rho_2}{\rho_1}\right)^2(A_m-B_m)^2+4A_mB_m\right.$$
$$\left.\times\left(\frac{\rho_2}{\rho_1}(1-A_mB_m)+A_mB_m\right)\right]>0.$$

Because $A_m, B_m < 1$, the discriminant of this equation is always positive, which means that the roots of eq. (4.169) are always real numbers. Analyzing the signs of the coefficients in eq. (4.169) and using Vieta's theorem (Vinberg, 2003), we conclude that both roots $\kappa_m^{1,2}=(\sigma_m^{1,2})^2/g$ are strongly positive for $\rho_2>\rho_1$ and, therefore, the two-layer system is stable and can perform standing-wave motions with modes (4.168). In contrast, the condition $\rho_2<\rho_1$ (heaviest liquid in the upper layer) in eq. (4.169) and Vieta's theorem shows that the real roots κ_m^1 and κ_m^2 have different signs and, therefore, the two-layer system is unstable.

Landau and Lifshitz (1959) analyzed the case of infinite liquid depth for the lower liquid (i.e., $h_2\to\infty$). Equation (4.169) then becomes simplified because $A_m\to 1$ in this limit. Moreover, the equation has the following roots:

$$\frac{\left(\sigma_m^1\right)^2}{g}=\kappa_m^{(1)}=\frac{\pi m}{l},\quad \frac{\left(\sigma_m^2\right)^2}{g}=\kappa_m^{(2)}$$
$$=\frac{\pi m}{l}\frac{(\rho_2-\rho_1)(1-\exp(-2\pi m h_1 l^{-1}))}{\rho_2+\rho_1+(\rho_2-\rho_1)\exp(-2\pi m h_1 l^{-1})},\qquad (4.170)$$

where the first eigenvalue (natural frequency) is the same as for sloshing in a rectangular tank with infinite depth; this frequency is caused by the wave motions of the lower liquid. The second natural frequency is more complex and is caused by coupled motions of the two liquids. However, the limit $h_1\to\infty$ makes the second natural frequency independent of the motions of the lower liquid.

In the preceding text we have examined the number of eigenfrequencies for a given eigenmode. However, in practice it is important to know the number of eigenmodes for a given eigenfrequency. At least two eigenmodes exist for a given eigenfrequency. When we say "at least," we have in mind, for instance, square-base and circular-base tanks, where we showed in Sections 4.3.2.1 and 4.3.2.2 that two eigenmodes exist for a given eigenfrequency when the tank is filled with one liquid.

4.9.2 Two-phase shallow-liquid approximation

We consider a long tank of length L_t and choose the x-axis to be in the longitudinal direction with $x=\pm\frac{1}{2}L_t$ at the tank ends. The total liquid depth and breadth are assumed small relative to L_t. We use subscripts 1 and 2 for the upper and lower liquids, respectively. The mean cross-sectional areas of the upper and lower liquid layers are denoted $\mathcal{A}_i(x)$ $(i=1,2)$. The breadths of the free surface and the liquid interface are $B_{si}(x)(i=1,2)$. It follows from the shallow-liquid approximation that the velocity potential for the flow in two layers is a function of x only (i.e., $\varphi_i(x)$, $i=1,2$). The dynamic free-surface condition gives that the free-surface elevation can be expressed as

$$\zeta_1=-g^{-1}\partial\varphi_1/\partial t.\qquad (4.171)$$

The dynamic condition at the interface gives the following condition:

$$\rho_2\left(\partial\varphi_2/\partial t+g\zeta_2\right)=\rho_1\left(\partial\varphi_1/\partial t+g\zeta_2\right),$$

where ζ_2 is the interface elevation. It follows by using eq. (4.171) that

$$\rho_2\left(\partial\varphi_2/\partial t+g\zeta_2\right)=\rho_1 g\left(\zeta_2-\zeta_1\right).\qquad (4.172)$$

We now use continuity of liquid mass separately for the two layers and consider a volume of length Δx. For the upper layer we get, after division with Δx,

$$\frac{\partial}{\partial x}\left[\mathcal{A}_1(x)\frac{\partial\varphi_1}{\partial x}\right]+\frac{\partial\zeta_1}{\partial t}B_{s1}-\frac{\partial\zeta_2}{\partial t}B_{s2}=0.\qquad (4.173)$$

Similarly for the lower layer we obtain

$$\frac{\partial}{\partial x}\left[\mathcal{A}_2(x)\frac{\partial\varphi_2}{\partial x}\right]+\frac{\partial\zeta_2}{\partial t}B_{s2}=0.\qquad (4.174)$$

The preceding equations can be combined so that only the elevations ζ_i appear. This is done by time-differentiating eqs. (4.173) and (4.174) and then using eqs. (4.171) and (4.172). By means of eq. (4.171), eq. (4.173) can be reexpressed as

$$-g\frac{\partial}{\partial x}\left[\mathcal{A}_1(x)\frac{\partial\zeta_1}{\partial x}\right]+\frac{\partial^2\zeta_1}{\partial t^2}B_{s1}-\frac{\partial^2\zeta_2}{\partial t^2}B_{s2}=0.\qquad (4.175)$$

Furthermore, we can reexpress eq. (4.174) by means of eq. (4.172) as

$$\frac{g}{\rho_2}\frac{\partial}{\partial x}\left\{\mathcal{A}_2(x)\frac{\partial}{\partial x}\left[\rho_1(\zeta_2-\zeta_1)-\rho_2\zeta_2\right]\right\} + \frac{\partial^2\zeta_2}{\partial t^2}B_{s2}=0. \qquad (4.176)$$

We now assume that the cross-sectional areas of the two liquids, $\mathcal{A}_i(x)$ $(i=1,2)$, are constant. Our objective is to find the natural frequencies σ_m. To find them, we substitute into eqs. (4.175) and (4.176)

$$\zeta_i = C_i\exp(\mathrm{i}\sigma_m t)\cos[\pi m(x+\tfrac{1}{2}L_t)/L_t].$$

We see from eqs. (4.171) and (4.172) that the x-dependence of the wave elevation is consistent with the fact that no flow occurs through the tank end walls at $x=\pm\frac{1}{2}L_t$. The result is

$$\left[g\mathcal{A}_1(\pi m/L_t)^2-\sigma_m^2B_{s1}\right]C_1+\sigma_m^2B_{s2}C_2=0,$$
$$g(\rho_1/\rho_2)\mathcal{A}_2(\pi m/L_t)^2C_1 + \left[(1-(\rho_1/\rho_2))g\mathcal{A}_2(\pi m/L_t)^2-\sigma_m^2B_{s2}\right]C_2=0.$$

The condition for this equation system to have nontrivial solutions is that the coefficient determinant is zero, which gives the desired expression for the natural frequencies as the following quadratic algebraic equation in σ_m^2: $a\sigma_m^4+b\sigma_m^2+c=0$ with $a=B_{s1}B_{s2}$, $b=-g(\pi m/L_t)^2[B_{s1}(1-\rho_1/\rho_2)\mathcal{A}_2+B_{s2}(\mathcal{A}_1+\mathcal{A}_2\rho_1/\rho_2)]$, and $c=g^2(\pi m/L_t)^4(1-\rho_1/\rho_2)\mathcal{A}_1\mathcal{A}_2$. Possible solutions are

$$\sigma_m^2=\left(-b\pm\sqrt{b^2-4ac}\right)/(2a), \qquad (4.177)$$

for which σ_m is proportional to $\pi m/L_t$. We are looking for real solutions of σ_m, which requires that σ_m^2 is real and positive. The requirement for real solutions of σ_m^2 is $b^2-4ac\geq 0$; that is,

$$B_{s1}^2(1-\rho_1/\rho_2)^2\mathcal{A}_2^2+B_{s2}^2(\mathcal{A}_1+\mathcal{A}_2\rho_1/\rho_2)^2 + 2B_{s1}B_{s2}(1-\rho_1/\rho_2)(-\mathcal{A}_1+\mathcal{A}_2\rho_1/\rho_2)\mathcal{A}_2\geq 0.$$

We assume that the lower liquid is heavier than the upper liquid (i.e., $\rho_1/\rho_2<1$), which implies that b and c are always negative and positive, respectively. The consequence of real solutions of σ_m^2 is that the solutions of σ_m^2 are positive (i.e., the solutions of σ_m are real).

Two special cases exist for two different liquids.

A. The two liquids have the same density:

$$\sigma_m^2B_{s1}/g=(\pi m/L_t)^2(\mathcal{A}_1+\mathcal{A}_2).$$

B. The two liquids have no free surface:

$$\frac{\sigma_m^2B_{s2}}{g}=\left(\frac{\pi m}{L_t}\right)^2\frac{(\rho_2-\rho_1)\mathcal{A}_1\mathcal{A}_2}{(\rho_1\mathcal{A}_2+\rho_2\mathcal{A}_1)}.$$

4.9.2.1 Example: oil–gas separator

We consider a horizontal circular cylinder with constant cross-section and internal radius R_0. The upper and lower liquid layers are oil and water, respectively. We denote the filling heights of the two liquids as h_i $(i=1,2)$. It then follows from geometry (see eqs. (4.153) and (4.154)) that

$$B_{s1}=2R_0\sqrt{1-[1-(h_1+h_2)/R_0]^2},$$
$$B_{s2}=2R_0\sqrt{1-[1-h_2/R_0]^2}$$

$$\mathcal{A}_2=R_0^2\begin{cases}\arctan\left(\dfrac{\sqrt{1-(1-h_2/R_0)^2}}{(1-h_2/R_0)}\right) -\left(1-\dfrac{h_2}{R_0}\right)\sqrt{1-\left(1-\dfrac{h_2}{R_0}\right)^2}, & 0<\dfrac{h_2}{R_0}\leq 1,\\ \pi+\arctan\left(\dfrac{\sqrt{1-(1-h_2/R_0)^2}}{(1-h_2/R_0)}\right) -\left(1-\dfrac{h_2}{R_0}\right)\sqrt{1-\left(1-\dfrac{h_2}{R_0}\right)^2}, & 1<\dfrac{h_2}{R_0}<2.\end{cases}$$

Furthermore, $\mathcal{A}_1$ follows by replacing h_2 with h_1+h_2 in the expression for $\mathcal{A}_2$ and then deducing the original expression for $\mathcal{A}_2$ with h_2. We use as an example $R_0=2.0$ m, $L_t=14.0$ m, $\rho_1=830\ \mathrm{kg\,m^{-3}}$, $\rho_2=1,000\ \mathrm{kg\,m^{-3}}$, and $h_1+h_2=R_0$.

Figure 4.46 shows the two calculated natural periods for $m=1$ as a function of h_2. The results when the two liquids have the same density, as well as when there are no free-surface wave effects, are plotted. One of the natural periods for the case with two liquid phases becomes very large, and resonant oscillations may be excited due to slow-drift motions of the platform where the oil–gas separator is installed. This period is close to what we predict by neglecting the free-surface effect. The other natural period with two liquid phases is excited by linear wave-frequency motion of the platform. The lowest natural period deviates little from the calculated natural period without accounting for two liquid phases. We may use this fact to estimate this natural period

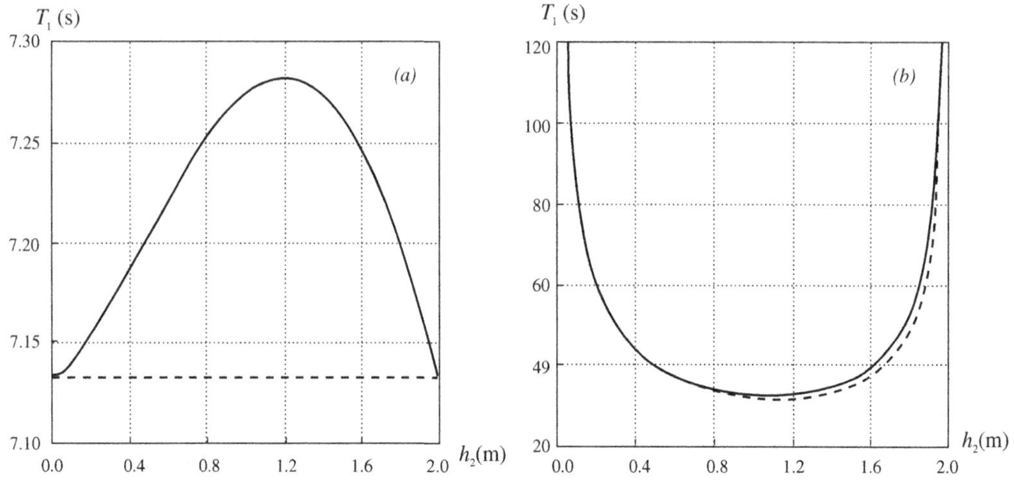

Figure 4.46. Calculated natural periods for the lowest eigenmode of a gas–oil separator half-filled with oil and water as liquids. The tank length L_t is 14.0 m and the cross-sectional radius R_0 is 2.0 m. The results are presented as a function of the filling height of water, h_2. Solid lines are for two-phase liquid flow. The dotted line in graph (a) is based on neglecting the interface between the liquids and the graph considers one liquid. The dotted line in graph (b) is based on neglecting the free-surface waves and using the rigid free-surface condition.

for nonshallow depth via the analysis in Section 4.8.3.1.

Figure 4.47 illustrates the effect of different density ratios ρ_1/ρ_2 between the two liquids as a function of the filling ratio $h_2/(h_1+h_2)$ when $h_1+h_2=R_0$. We keep the original values of R_0 and L_t. The results are not sensitive to h_1+h_2. We note that ρ_1/ρ_2 has to be less than 0.1 for the lowest natural period to differ more than 15% from the natural period obtained by considering the two liquids to have the same density. As expected, the second natural period is strongly influenced by the free surface when $h_2/(h_1+h_2)$ is close to 1. When $h_2/(h_1+h_2)\to 1$, the ratio T_1/T_1^{**} is finite, but far from 1 except when ρ_1/ρ_2 is close to 1. Here T_1^{**} is the natural period based

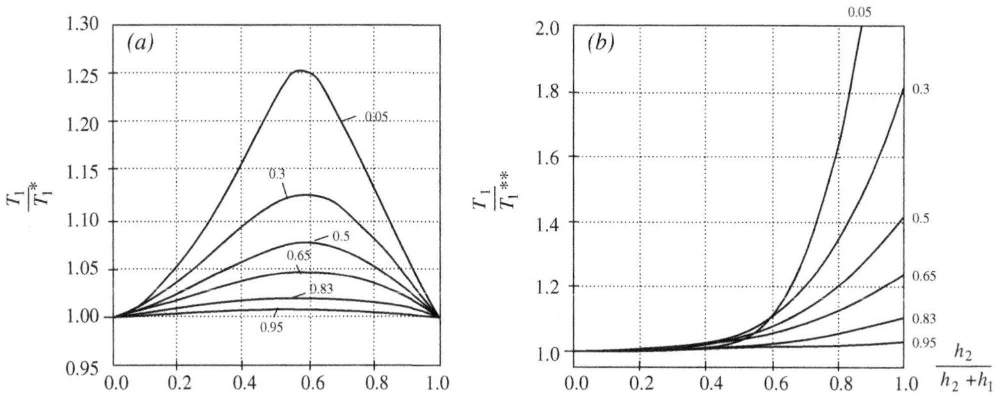

Figure 4.47. The two calculated natural periods T_1 for the lowest eigenmode for two-phase liquid flow in a horizontal circular cylinder of radius $R_0 = 2.0$ m and length $L_t = 14$ m. T_1^* is the natural period based on neglecting the interface between the liquids and considering one liquid; T_1^{**} is based on neglecting free-surface waves and using the rigid free-surface condition. The ratios of T_1 corresponding to T_1^* and T_1^{**} are presented as a function of the filling ratio $h_2/(h_1+h_2)$ for different density ratios ρ_1/ρ_2 between the two liquids. In the calculations, $h_1+h_2=R_0$.

on neglecting the free-surface waves and using the rigid free-surface condition. A small value of ρ_1/ρ_2 implies the limit case of one liquid with air (gas) above it. In this case, the natural frequency for the liquid should not be influenced by the gas. This fact is consistent with eq. (4.177), which gives independent natural periods for fluids with densities ρ_1 and ρ_2 when $\rho_1/\rho_2 \to 0$. Figure 4.47(b) shows that $T_1/T_1^{**} \to \infty$ as $\rho_1/\rho_2 \to 0$ and $h_2/(h_1 + h_2) \to 1$ simultaneously.

The fact that wave-induced platform motions may cause resonance oscillations inside an oil–gas separator and result in important operational limitations means that damping devices are normally introduced inside the separator. An example is the use of perforated vertical plates placed at different positions along the separator (see Figure 1.31). The damping is associated with flow separation and is further discussed in the context of tuned liquid dampers and swash bulkheads in Section 6.7.

4.10 Summary

Linear natural sloshing frequencies and modes were described by the potential flow theory of incompressible liquids without surface-tension effects. Analytical solutions valid for any liquid depth are available when the velocity potential can be expressed as the product of functions that depend on only one spatial variable. This solution is possible for two- and three-dimensional rectangular tanks, vertical cylinders with constant circular cross-section, and annular and sectored upright circular cylindrical tanks. Analytical solutions also exist for more special tank forms such as a two-dimensional wedge with semi-apex angles 45° and 60°.

A shallow-liquid approximation makes it possible to derive additional analytical solutions. In this context a shallow-liquid approximation means that the pressure is hydrostatic relative to the instantaneous free-surface elevation and requires that the considered wavelength is larger than the order of ten times the liquid depth. The considered problem is relevant for seiching in harbors and lakes. Analytical solutions are presented for two-dimensional triangular- and parabolic-shaped basins. If a shallow-liquid tank is horizontal and has a length much larger than the cross-dimensions, longitudinal natural modes and frequencies can be expressed in a simple analytical form for a general cross-section. A formula was proposed to generalize these natural frequencies to any liquid depth. Experimental results for horizontal tanks with a circular constant cross-section were used to validate the formula.

Domain decomposition makes it possible to combine analytical solutions for different subdomains of the liquid. The technique was illustrated for a swimming pool with a shallow-water part.

Variational formulations were used to find natural sloshing frequencies. A Rayleigh quotient that is proportional to the ratio between the kinetic energy and the potential energy was introduced. The natural sloshing frequencies were given by the Rayleigh quotient when eigenmodes were used in the quotient. If any other function (test function) satisfying liquid volume conservation for the lowest mode and the corresponding orthogonality properties for the higher modes are used in the Rayleigh quotient, the corresponding estimates of the natural frequencies are always higher than the natural frequencies. The test functions should not necessarily be restricted to functions satisfying the Laplace equation and the boundary conditions such as, for instance, the zero-Neumann boundary conditions on the mean wetted tank surface. The Laplace equation and all the boundary conditions are then *natural* (i.e., derivable from the corresponding variational equation). If we restrict the test functions to satisfy an extra condition (e.g., they satisfy a Dirichlet condition on a wall), the minimization of the Rayleigh quotient can give eigenvalues with too-large error.

The latter variational formulation was used to estimate the lowest natural sloshing frequency for a two-dimensional circular tank and a spherical tank. A horizontal dipole in infinite fluid with singularity above the mean free surface and in the tank's centerplane was used as a test function. The Rayleigh quotient was minimized by varying the singularity position. The results are in very good agreement with benchmark numerical results.

Variational formulations can also be used to find upper and lower bounds of natural frequencies for a general tank shape. The upper and lower bounds can be estimated by established analytical and numerical benchmark results.

Useful analytical methods exist that show how internal structures and small deviations of the tank geometry affect the natural frequencies. The consequence is that analytical results for rectangular tanks, for instance, give an important basis for a broader range of tank geometries.

An analytical continuation of the solution outside the liquid domain sets severe restrictions on the shape of the hull surface. However, it is possible to apply the method to a chamfered tank bottom and an inclined tank bottom. Examples were given of how to calculate natural frequencies analytically for small modifications of rectangular tanks in terms of the chamfer at the tank corners and for an inclined tank bottom. It was demonstrated that the effect of a chamfer has a very small influence on natural frequencies and modes. If the boundary is smooth and the domain is convex, it is also possible to analytically continue the velocity potential to a rectangular tank domain. Examples are circular and elliptical two-dimensional tanks and spherical tanks with different filling heights.

Analytical continuation cannot be used to estimate the effect of internal structures such as baffles and poles on natural frequencies. Another technique is possible when the internal structure can be considered "small" and the effect of the internal structure on the flow is not source-like. A small structure means that the wavelength of the considered eigenmode is either large relative to the dimensions of the structure or to the cross-dimensions of a slender body. The requirement of a non-source-like effect is satisfied for any closed rigid structure, which is also true when the body is very thin at the intersection of the tank surface and the free surface. Finally, the requirement is satisfied for a vertical cylinder of constant cross-section standing on the tank bottom and penetrating the free surface. Formulas were presented in terms of the added mass coefficients and the volume of the internal structure. The importance of mutual hydrodynamic interactions between internal structures satisfying the aforementioned requirements was discussed in terms of presented formulas for added mass. The hydrodynamic interaction is small when the distance between two bodies is of the order of the cross-dimensional lengths of the bodies.

Two-phase liquid flow was analyzed and an example relevant for a gas–oil separator was studied. The two liquids are oil and water. Two natural periods exist for a given natural mode. The lowest natural mode was examined and it was shown that one of the natural periods is dominated by the free-surface wave effect and the other is dominated by the effect of the interface between the two liquids.

Benchmark numerical results were presented for two-dimensional circular, spherical, and elliptical tanks.

4.11 Exercises

4.11.1 Irregular frequencies

When solving the linear exterior unsteady flow problem for ships and large-volume structures in regular waves, it is common practice to use a boundary element method. We focus on the case when a source distribution method is used, where the source satisfies the linear free-surface conditions. So-called irregular frequencies then occur where the method fails mathematically (John, 1950). The irregular frequencies σ_{irr} are the solutions of a spectral problem similar to problem (4.2) but with a zero Dirichlet condition; that is,

$$\nabla^2\varphi = 0 \quad \text{in } Q_0; \quad \varphi = 0 \quad \text{on } S_0;$$
$$\frac{\partial\varphi}{\partial z} = \eta\varphi \quad \text{on } \Sigma_0, \tag{4.178}$$

where $\eta = \sigma_{\mathrm{irr}}^2/g$, Σ_0 is the water plane of the structure and Q_0 is the liquid domain interior to the structure below Σ_0 and bounded by the hull surface.

(a) Explain why the Dirichlet condition $\varphi = 0$ on S_0 implies that we, in general, cannot satisfy volume conservation.

(b) Derive eigenvalues and eigenfunctions from problem (4.178) for rectangular and upright circular cylindrical tanks. Compare with the solutions in Section 4.3. For a two-dimensional rectangular tank, compare nodal and antinodal lines.

4.11.2 Shallow-liquid approximation for trapezoidal-base tank

Assume shallow-liquid conditions for the trapezoidal-base tank shown in Figure 4.26(a). Derive an equation similar to eq. (4.102) by starting with a shallow-liquid formulation

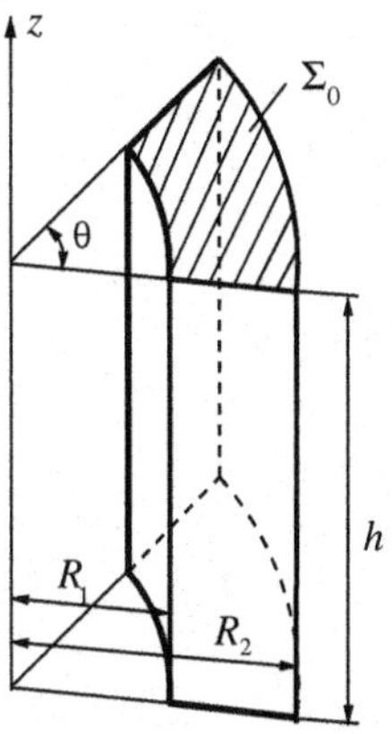

Figure 4.48. Annular and sectored upright circular tank.

as in eq. (4.45). (*Answer*: $\sigma_{i,0}^2 = [(\pi i/L_1)^2 + (\alpha/L_2)^2]gh$.)

4.11.3 Annular and sectored upright circular tank

The natural modes and frequencies can be found in an analytical form for annular and sectored upright circular tanks.

(a) Use Figure 4.48 for variable definitions and explain that the boundary-value problem for natural modes and frequencies can be expressed as

$$\frac{\partial^2\varphi}{\partial z^2} + \frac{1}{r}\frac{\partial}{\partial r}\left(r\frac{\partial\varphi}{\partial r}\right) + \frac{1}{r^2}\frac{\partial^2\varphi}{\partial\theta^2} = 0,$$
$$0 \le R_1 < r < R_2, \quad -h < z < 0, \quad 0 < \theta < \theta_1,$$
$$\frac{\partial\varphi}{\partial r} = 0, \quad r = R_2, \quad \text{and} \quad r = R_1,$$
$$-h < z < 0, \quad 0 < \theta < \theta_1,$$
$$\frac{\partial\varphi}{\partial z} = 0, \quad z = -h, \quad 0 \le R_1 < r < R_2,$$
$$0 < \theta < \theta_1,$$
$$\frac{\partial\varphi}{\partial z} = \kappa\varphi, \quad z = 0, \quad 0 \le R_1 < r < R_2,$$
$$0 < \theta < \theta_1,$$
$$\frac{\partial\varphi}{\partial\theta} = 0, \quad -h < z < 0, \quad 0 \le R_1 < r < R_2,$$
$$\theta = 0 \quad \text{and} \quad \theta = \theta_1,$$
$$\int_{R_1}^{R_2}\int_0^{\theta_1} r\varphi(r,\theta,0)\,d\theta\,dr = 0, \qquad (4.179)$$

where $0 < \theta_1 < 2\pi$ is the angle of the sector and R_1 is the internal radius. The case $R_1 = 2\pi$ corresponds to a nonannular, but possibly sectored, tank. The boundary conditions at $r = R_1$ must then be omitted. The case $\theta_1 = 2\pi$ corresponds to a nonsectored tank, which also requires that the boundary condition at $\theta = 0$ and $\theta = \theta_1$ be rejected by changing them to the periodicity condition $\varphi(r,\theta,z) = \varphi(r,\theta+2\pi,z)$.

(b) Solve problem (4.179) by using separation of variables. The *answer* is

$$\varphi = \cos\left(\frac{m\theta}{2\alpha}\right)\frac{\cosh[\iota_{mn}(z/R_2 + h/R_2)]}{\cosh[\iota_{mn}h/R_2]} \times C_{\frac{m}{2\alpha}}\left(\iota_{mn}\frac{r}{R_2}\right),$$

where the r-dependent function is

$$C_{\frac{m}{2\alpha}}\left(\iota_{mn}\frac{r}{R_2}\right) = \det\begin{vmatrix} J_{\frac{m}{2\alpha}}\left(\iota_{mn}\frac{r}{R_2}\right) & Y_{\frac{m}{2\alpha}}\left(\iota_{mn}\frac{r}{R_2}\right) \\ J'_{\frac{m}{2\alpha}}(\iota_{mn}) & Y'_{\frac{m}{2\alpha}}(\iota_{mn}) \end{vmatrix}$$

and where $J_\mu(\cdot)$ and $Y_\mu(\cdot)$ are Bessel functions of the first and second kind, respectively, and ι_{mn} are roots of the equation $C_{\frac{m}{2\alpha}}(\iota_{mn}R_1/R_2) = 0$, which are ordered in ascending order for a fixed m, $\alpha = \theta_1/2\pi$.

(c) Show that the natural frequencies can be expressed as

$$R_2 g^{-1}\sigma_{mn}^2 = \iota_{mn}\tanh(\iota_{mn}h/R_2),$$
$$m = 0, 1, \ldots; \quad n = 1, 2, \ldots.$$

Use the dimensions for the floating production storage and offloading unit by Sevan Marine in Figure 1.30 and calculate the lowest natural sloshing frequency. Discuss where the associated wave elevation is largest. (*Answer*: The lowest mode is associated with the parameter $\iota_{11} = 4.197282\ldots$ for the sectored tank depicted in Figure 1.30. The lowest natural sloshing frequency is therefore defined by $R_2\sigma_{11}^2/g = 4.197282 \tanh(4.197282 \cdot h/R_2)$.))

4.11.4 Circular swimming pool

Consider the circular swimming pool with a shallow-water part illustrated in Figure 4.49 and find the highest sloshing period based on the procedure in Section 4.5. (*Hint*: The steps in solving the problem are as follows.)

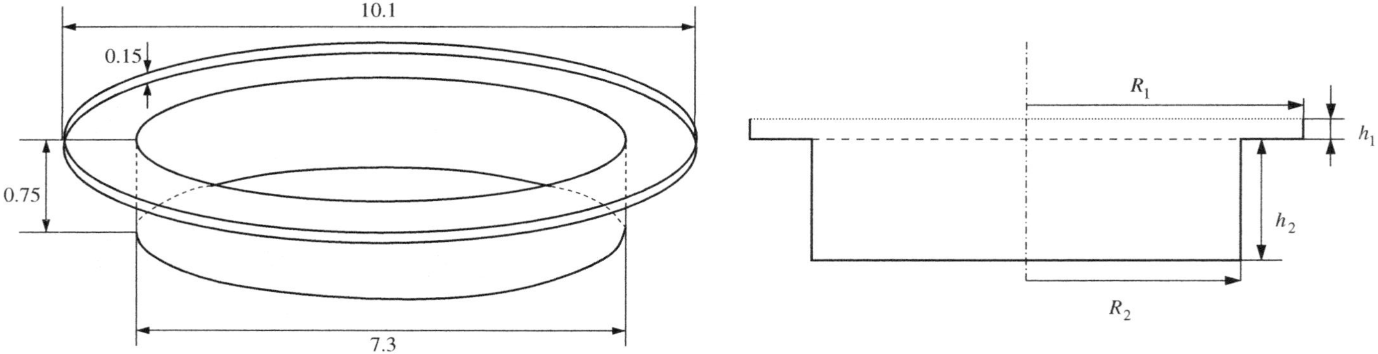

Figure 4.49. Circular swimming pool with a shallow-water part (dimensions in meters).

(a) Set up a procedure based on domain decomposition and find an analogy of continuity equation (4.50). (*Answer*: For an arbitrary upright container with a shallow part of the height h_1, the analogy to two-dimensional eq. (4.50) takes the form

$$\frac{\partial^2\varphi_1}{\partial t^2} - gh_1\left(\frac{\partial^2\varphi_1}{\partial x^2} + \frac{\partial^2\varphi_1}{\partial y^2}\right) + g\left.\frac{\partial\varphi_2}{\partial z}\right|_{z=0} = 0, \quad (x,y)\in\Sigma_2,$$

where Σ_2 is the interface flat surface between upper and lower parts.)

(b) Generalize the representation of the solution similar to eqs. (4.53) and (4.54). Use definitions in Figure 4.49. (*Hint*: For the lower part, use the series by the functions (4.37), but for the upper part, these should not have z-dependence.)

Answer:

$$\varphi_2(r,\theta,z,t) = \exp(i\sigma t)\sum_{m=0}^{\infty}\sum_{i=1}^{\infty} a_{mi}^{(2)} J_m\left(k_{mi}^{(2)} r\right) \times \frac{\cosh\left(k_{mi}^{(2)}(z+h_2)\right)}{\cosh\left(k_{mi}^{(2)}h_2\right)} \begin{Bmatrix}\cos(m\theta)\\ \sin(m\theta)\end{Bmatrix},$$

$$\varphi_1(r,\theta,t) = \exp(i\sigma t)\sum_{m=0}^{\infty}\sum_{i=1}^{\infty} a_{mi}^{(1)} J_m\left(k_{mi}^{(1)} r\right)\begin{Bmatrix}\cos(m\theta)\\ \sin(m\theta)\end{Bmatrix},$$

where $k_{mi}^{(1)} = \iota_{mi}R_1^{-1}$ and $k_{mj}^{(2)} = \iota_{mj}R_2^{-1}$ to satisfy the wall conditions, where ι_{mi} are defined in Section 4.3.2.2, and $a_{mi}^{(l)}$ are unknowns.

(c) Generalize spectral problem (4.58) (*Hint*: Use the orthogonality of the trigonometric functions as well as the following orthogonality of the Bessel functions:

$$\int_0^{R_2} rJ_m\left(k_{mi}^{(2)}r\right)J_m\left(k_{mj}^{(2)}r\right)\mathrm{d}r = 0 \text{ as } i \neq j,$$

which follows from the orthogonality of the natural modes in eq. (4.37). The orthogonality is argued by the general theorem on the natural modes discussed in the beginning of the chapter. Alternatively, the orthogonality of the Bessel functions follows from the formulas (Watson, 1944)

$$J_i'(kr) = k\left(-J_{i+1}(kr) + (kr)^{-1}J_i(kr)\right)$$

$$\times\int_0^R rJ_i(k_1r)J_i(k_2r)\,\mathrm{d}r = -\frac{R}{k_1^2-k_2^2}[-k_2J_i(k_1R)J_{i-1}(k_2R) + k_1J_{i-1}(k_1R)J_i(k_2R)]$$

to be used with corresponding i, k, k_1, k_2, and R.)

Answer:

$$\det\left[A - \sigma^2 B\right] = 0; \quad B = \{a_{ij}\}_{i,j=1,N};$$

$$A = \left\{ga_{ij}\left[h_1\left(k_{mj}^{(1)}\right)^2 + k_{mi}^{(2)}\tanh\left(k_{mi}^{(2)}h_2\right)\right]\right\}_{i,j=1,N}, \quad (4.180)$$

where $a_{ij} = \int_0^{R_2} rJ_m(k_{mj}^{(1)}r)J_m(k_{mi}^{(2)}r)\mathrm{d}r$

(d) Estimate the lowest natural period for sloshing based on the dimensions provided in Figure 4.49. (*Hint*: Remember that the lowest natural frequency in a circular cross-section is associated with $m = 1$.)
Answer: In the one-term approximation ($N = 1$ in eq. (4.180)), the formula for the lowest natural frequency is $\sigma_{1,1}^2 = g[h_1(k_{m1}^{(1)})^2 + k_{m1}^{(2)}\tanh(k_{m1}^{(2)}h_2)]$. It gives the highest natural period for the case equal to 4.46 s. Increasing N to 10 in spectral problem (4.180) corrects the natural period to 4.27 s. Note that a further increase of N in the computational formulas may cause numerical difficulties, because we represent the solution by two Fourier-like series which may weakly converge.

4.11.5 Effect of pipes on sloshing frequencies for a gravity-based platform

Sloshing has been observed inside the Draugen platform, which is a gravity-based platform located at a water depth of 252.5 m in the Norwegian Sea. Figure 4.50(a) gives a schematic view of the submerged part of the platform. The shaft has a circular cross-section with an internal diameter of 15 m at the waterline. The water is flooded to a depth just below mean sea level. Several pipes are located inside the shaft, only the largest of which are illustrated in Figure 4.50(b). A sloshing frequency of about 4.3 s has been observed

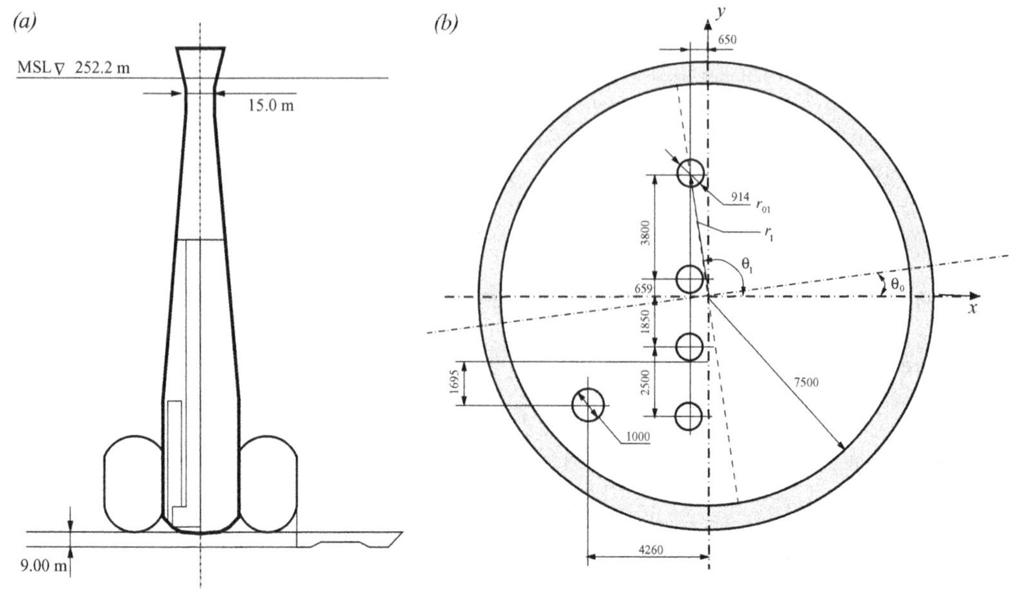

Figure 4.50. (a) View of the Draugen platform and (b) plan of some of the pipes below sea level inside the shaft.

(Drake, 1999). In addition, a resonance period of about 3.9 s was observed, which is associated with global structural elastic response. The following questions concentrate on sloshing.

(a) Argue why you can analyze the sloshing frequency inside the tank by considering a vertical circular tank of constant cross-section with pipes inside.
(b) Does hydrodynamic interaction between the pipes matter according to potential flow theory?
(c) Express the effect of one individual pipe on the natural sloshing frequencies. Assume that the single pipe is situated with the coordinates (x_{0i}, y_{0i}) or (r_i, θ_i) (in a cylindrical coordinate system as shown in Figure 4.50) and has radius r_{0i}. The following solution procedure is suggested.

Comment 1. Note that the natural sloshing modes for an upright circular cylinder given by eq. (4.37) and those for a three-dimensional rectangular tank given by eq. (4.29) have the same hyperbolic cosine dependence along the vertical axis represented by eq. (4.138). Therefore, it is possible to use eq. (4.139) for each of the circular poles with

$$\phi_{m,n} = J_m(k_{m,n}r)\cos(m\theta - \alpha_{1,2}), \quad J'_m(k_{m,n}R_0) = 0 \tag{4.181}$$

and $Vol(S_{CS}) = a_{11}/\rho = a_{22}/\rho = \pi r_{0i}^2$. The two angles $\alpha_{1,2}$ define mutually perpendicular wave patterns; that is,

$$\alpha_1 = \theta_0, \quad \alpha_2 = \theta_0 - \tfrac{1}{2}\pi, \tag{4.182}$$

where θ_0 is an arbitrary angle in the range $0 \le \theta_0 < 2\pi$ (see Figure 4.50).

Comment 2. Use eq. (4.139) to predict frequencies corresponding to almost symmetric modes with $m = 0$.

Answer:

$$\sigma_{0,n}'^2 = gk_{0,n}\tanh(k_{0n}h)\left[1 - \frac{J_0'^2(k_{0,n}r_i)}{2\int_0^1 rJ_0^2(k_{0,n}R_0r)\mathrm{d}r} \times \left\{1 + \frac{2hk_{0,n}}{\sinh(2k_{0,n}h)}\right\}\left(\frac{r_{0i}}{R_0}\right)^2\right].$$

Comment 3. In the derivation of an analogous formula for the case $m \neq 0$, we should take into account that the two modes with the same natural frequency $\sigma_{m,n}$ (associated with eq. (4.181)) split into orthogonal modes with different natural frequencies due to insertion of a single pipe. Therefore, excitations along two perpendicular axes (for brevity, axes Ox and Oy) will trigger different resonance frequencies. Use eq. (4.181) in eq. (4.139) to get a formal asymptotic expression.

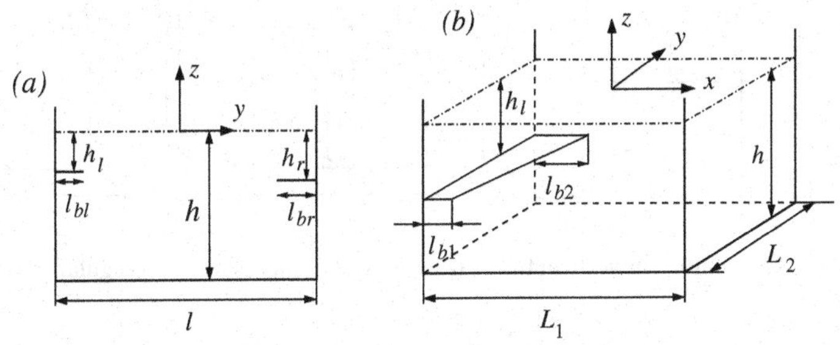

Figure 4.51. Horizontal baffles mounted at the walls of two- and three-dimensional rectangular tanks.

Answer:

$$\sigma'^2_{m,n} = gk_{m,n}\tanh(k_{m,n}h) \times \left[1 - \frac{\alpha_{m,n,i}\cos^2(m\theta_i - \alpha_{1,2}) + \beta_{m,n,i}}{\int_0^1 rJ_m^2(\iota_{m,n}r)\,\mathrm{d}r} \times \left\{1 + \frac{2hk_{m,n}}{\sinh(2k_{m,n}h)}\right\}\left(\frac{r_{0i}}{R_0}\right)^2\right], \quad (4.183)$$

where

$$\alpha_{m,n,i} = J'^2_m\left(\iota_{m,n}\frac{r_i}{R_0}\right) - \beta_{m,n,i};$$

$$\beta_{m,n,i} = \frac{m^2}{\iota^2_{m,n}}\left(\frac{R_0}{r_i}\right)^2 J_m^2\left(\iota_{m,n}\frac{r_i}{R_0}\right) \quad (4.184)$$

and where $\iota_{m,n}$ is defined in Section 4.3.2.2. Note that the answer given by eq. (4.183) varies with θ_0 in eq. (4.182). We now concentrate on the two natural sloshing frequencies associated with $m = n = 1$. Compare your result with that in the following comment.

Comment 4. As long as $m = n = 1$ and $\alpha_1 = \theta_0$, eqs. (4.183) and (4.184) take the form ($\iota_{1,1} = k_{1,1}R_0 = 1.841\ldots$is the lowest root of eq. (4.181))

$$\sigma'^2_{1,1} = \frac{\iota_{1,1}g}{R_0}\tanh\left(\iota_{1,1}\frac{h}{R_0}\right) \times \left[1 - \frac{\alpha_{1,1,i}\cos^2(\theta_i - \theta_0) + \beta_{1,1,i}}{\int_0^1 rJ_1{}^2(\iota_{1,1}r)\,\mathrm{d}r} \times \left\{1 + \frac{2\iota_{1,1}h/R_0}{\sinh(2\iota_{1,1}h/R_0)}\right\}\left(\frac{r_{0i}}{R_0}\right)^2\right], \quad (4.185)$$

$$\alpha_{1,1,i}\left(\frac{r_i}{R_0}\right) = J'^2_1\left(\iota_{1,1}\frac{r_i}{R_0}\right) - \beta_{1,1,i}\left(\frac{r_i}{R_0}\right),$$

$$\beta_{1,1,i}\left(\frac{r_i}{R_0}\right) = \frac{1}{\iota^2_{1,1}}\left(\frac{R_0}{r_i}\right)^2 J_1^2\left(\iota_{1,1}\frac{r_i}{R_0}\right),$$

and $\alpha_{1,1,i} \le 0$ for $0 \le r_i/R_0 < 1$. Which value of θ_0 gives minimum value of $\sigma'_{1,1}(\theta_0)$ defined by eq. (4.185)? Make an assessment of the highest natural sloshing period based on the pipe arrangement shown in Figure 4.50. (*Hint*: Use the summation by i in eq. (4.185) and assume large h/R_0.) The *answers* are 4.111 s and $\theta_0 = (2\pi - 0.326)$ rad.)

4.11.6 Effect of horizontal isolated baffles in a rectangular tank

(a) Consider a two-dimensional *rectangular tank* with two horizontal baffles, where each baffle is mounted on opposite tank walls. Geometrical dimensions are shown in Figure 4.51(a). Assume that l_{bl}/l and l_{br}/l are small so that there are no hydrodynamic interactions between the plates. Furthermore, the baffles are sufficiently far away from the tank bottom to avoid hydrodynamic interaction of the tank bottom on the flow caused by the baffles. Find the effect of the baffles on the natural frequencies.

Answer:

$$\frac{\sigma'^2_m}{\sigma_m{}^2} = 1 - \frac{2\pi^2 m}{\sinh(2\pi mh/l)} \times \left[\left(\frac{l_{bl}}{l}\right)^2\sinh^2(\pi m(h-h_l)/l) + \left(\frac{l_{br}}{l}\right)^2\sinh^2(\pi m(h-h_r)/l)\right].$$

(b) Consider a three-dimensional rectangular tank with a single trapezoidal baffle as shown in Figure 4.51(b) and find the corrected natural frequencies.

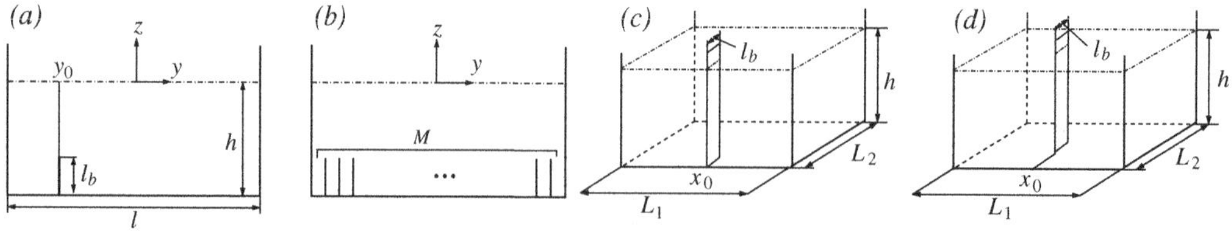

Figure 4.52. Different vertical baffle (plate) arrangements in a rectangular tank.

Answer:

$$\sigma'^2_{i,0} = g\frac{\pi i}{L_1}\left(\tanh\left(\frac{\pi i h}{L_1}\right) - \frac{\pi^2 i}{3}\left(\frac{\sinh(\pi i(h-h_l)/L_1)}{\cosh(\pi i h/L_1)}\right)^2 \times \left[\left(\frac{l_{b1}}{L_1}\right)^2 + \left(\frac{l_{b2}}{L_1}\right)^2 + \frac{l_{b1}l_{b2}}{L_1^2}\right]\right)$$

for the Stokes modes in the Oxz-plane ($j = 0$),

$$\sigma'^2_{0,j} = g\frac{\pi j}{L_2}\left(\tanh\left(\frac{\pi j h}{L_2}\right) - \frac{L_1}{L_2}\frac{1}{12j} \times \left(\frac{\sinh(\pi j(h-h_l)/L_2)}{\cosh(\pi j h/L_2)}\right)^2 \times \left[3(l_{b1}-l_{b2})^2 L_1^{-2} + 2(\pi j)^2 \left((l_{b1}/L_1)^2 + (l_{b2}/L_1)^2 + l_{b1}l_{b2}/L_1^2\right)\right]\right),$$

for the Stokes modes in the Oyz-plane ($i = 0$), and

$$\sigma'^2_{i,j} = g k_{i,j}\left(\tanh(k_{i,j}h) - \frac{k_{i,j}L_1}{6\pi j^2} \times \left(\frac{\sinh(k_{i,j}(h-h_l))}{\cosh(k_{i,j}h)}\right)^2 \times \left[3(l_{b1}-l_{b2})^2 L_1^{-2} + 2(\pi j)^2 \times \left((l_{b1}/L_1)^2 + (l_{b2}/L_1)^2 + l_{b1}l_{b2}/L_1^2\right)\right]\right),$$

for purely three-dimensional modes with $ij \neq 0$.

4.11.7 Isolated vertical baffles in a rectangular tank

(a) Consider the effect of a bottom-mounted vertical baffle on *two-dimensional sloshing* in a rectangular tank (see Figure 4.52(a)). Assume $l_b/l \ll 1$ and finite h/l.

Answer:

$$\frac{\sigma'^2_m}{\sigma^2_m} = 1 - \frac{2\pi^2 m \sin^2(\pi m(y_0 + \frac{1}{2}l)/l)}{\sinh(2\pi m h/l)}\left(\frac{l_b}{l}\right)^2.$$

(b) Generalize the latter result to a set of M equivalent vertical baffles uniformly mounted along the bottom with the mounting points $(il/(M+1) - \frac{1}{2}l,\ -h)$ as shown in Figure 4.52(b). Assume that there is no hydrodynamic interaction between the baffles and with the walls.

Answer:

$$\frac{\sigma'^2_m}{\sigma^2_m} = 1 - \frac{2\pi^2 m}{\sinh(2\pi m h/l)} \times \sum_{i=1}^{M} \sin^2(\pi m i/(M+1))\left(\frac{l_b}{l}\right)^2.$$

(c) Consider the case in Figure 4.52(c) with ambient two-dimensional flow in the x–z-plane.

Answer:

$$\frac{\sigma'^2_{i,0}}{\sigma^2_{i,0}} = 1 - \frac{\pi L_1}{2L_2}\sin^2\left(\frac{\pi i}{L_1}\left(x_0 + \tfrac{1}{2}L_1\right)\right) \times \left[1 + \frac{2\pi i h/L_1}{\sinh(2\pi i h/L_1)}\right]\left(\frac{l_b}{L_1}\right)^2.$$

(d) Modify the latter formula for the case in Figure 4.52(d) when the baffle is far from the walls.

Answer:

$$\frac{\sigma'^2_{i,0}}{\sigma^2_{i,0}} = 1 - \frac{\pi L_1}{4L_2}\sin^2\left(\frac{\pi i}{L_1}\left(x_0 + \tfrac{1}{2}L_1\right)\right) \times \left[1 + \frac{2\pi i h/L_1}{\sinh(2\pi i h/L_1)}\right]\left(\frac{l_b}{L_1}\right)^2. \tag{4.186}$$

Discuss the wall effects.

5 Linear Modal Theory

5.1 Introduction

Modal theories for sloshing are used in Chapters 8 and 9 to study forced oscillations of two- and three-dimensional rectangular tanks and upright circular cylinders. A modal theory transforms the original free boundary-value problem involving eqs. (2.58), (2.63), (2.64), (2.68), and (2.74) to a multidimensional system of ordinary differential equations. The equations are termed *modal* because their unknowns (henceforth, modal functions $\beta_i(t)$, $i \geq 1$) are generalized coordinates of the natural sloshing modes used in describing the free-surface elevation. All physical variables of the liquid can be described in terms of $\beta_i(t)$, $i \geq 1$. The assumption is potential flow of an incompressible liquid. A discussion of how to include viscous effects is given in Chapter 6. Because an infinite set of natural modes exists, the modal systems are theoretically infinite-dimensional. Truncation of the modal system may be done in different ways, such as a naive truncation that accounts for the fact that the higher modes are strongly damped due to viscous effects and, therefore, their contribution is small. Chapters 8 and 9 also show that the truncation of nonlinear modal systems can be based on asymptotic procedures, which identify the leading modes as lower order than that of the forcing amplitude.

The subject of this chapter is a general modal theory for linear sloshing in an oscillating tank. It is based on a linearized version of the original free boundary-value problem involving eqs. (2.58), (2.63), (2.64), (2.68), and (2.74). Both the oscillatory tank motion and the liquid motion are assumed small relative to a characteristic linear dimension of the mean free surface. Linear potential flow theory is relevant for *nonresonant liquid motions of a ship tank in a seaway* and for *transient tank motions* for example due to sea quakes, collision between two ships, and collision between a ship and ice. It is also a transient "tank" motion when we spill coffee from a coffee cup. Resonant steady-state solutions based on linear potential flow theory give infinite response at resonance. The reason is zero damping. Finite response at resonance in a clean tank is generally caused by nonlinear potential flow effects. Cases exist when resonant liquid motion can be predicted satisfactorily by including viscous damping in a linear potential flow method. An example of a tuned liquid damper used to suppress vibrations of tall buildings is presented in Chapter 6. The damping due to internal screens is then relatively high. Moreover, the studied case involves relatively small tank motions.

To demonstrate the idea of multidimensional linear modal modeling and get an example of a modal system, we start by studying forced surge oscillations of a two-dimensional rectangular tank. In this example and elsewhere in this chapter, the modal equations and hydrodynamic coefficients are accompanied by indices that can be associated with either integer numbers (e.g., m) when the modes are naturally counted by a single number (for example, for a two-dimensional rectangular tank), or a set of integer numbers (e.g., (m, k)), when the natural modes are enumerated using two or more integer numbers (three-dimensional rectangular tank, an upright circular cylinder, etc.). For brevity, in the derivation of the general modal system, the modes are counted using a single number. In addition, the text introduces the Stokes–Joukowski potential for linear sloshing, $\boldsymbol{\Omega}_0$. The zero index means that $\boldsymbol{\Omega}_0$ is the zero-order approximation of the "nonlinear" Stokes–Joukowski potential $\boldsymbol{\Omega}$, which is defined in the time-varying liquid domain $Q(t)$ (see Chapter 7 for details). The zero index is also present in all hydrodynamic coefficients that depend on $\boldsymbol{\Omega}_0$, including λ_{0km} and the components of the inertia tensor J^1_{0mk}.

5.2 Illustrative example: surge excitations of a rectangular tank

A two-dimensional rectangular tank with breadth l and mean liquid depth h that is forced with horizontal velocity $v_{O1}(t)$ is considered. Linear potential flow theory of an incompressible liquid is assumed. The boundary-value problem for velocity potential Φ and corresponding free-surface elevation ζ can be expressed as

$$\frac{\partial^2\Phi}{\partial x^2}+\frac{\partial^2\Phi}{\partial z^2}=0\quad\left(-\tfrac{1}{2}l<x<\tfrac{1}{2}l,\,-h<z<0\right),$$

$$\int_{-\frac{1}{2}l}^{\frac{1}{2}l}\zeta\,\mathrm{d}x=0,$$

$$\left.\frac{\partial\Phi}{\partial z}\right|_{z=-h}=0,\quad \boxed{\left.\frac{\partial\Phi}{\partial x}\right|_{x=\pm\frac{l}{2}}=v_{O1}},$$

$$\left.\frac{\partial\Phi}{\partial z}\right|_{z=0}=\frac{\partial\zeta}{\partial t},\quad \left.\frac{\partial\Phi}{\partial t}\right|_{z=0}+g\zeta=0. \tag{5.1}$$

Initial conditions are also needed. The framed nonhomogeneous term in the Neumann conditions on the vertical walls at $x=\pm\frac{1}{2}l$ expresses that the horizontal liquid velocity on the walls is the same as for the solid tank.

The free-surface elevation, ζ, can be expressed in terms of a Fourier series on the interval $-\frac{1}{2}l<x<\frac{1}{2}l$. A complete representation that satisfies liquid volume conservation can be written as

$$z=\zeta(x,t)=\sum_{j=1}^{\infty}\beta_j(t)\cos\left(\pi j\left(x+\tfrac{1}{2}l\right)/l\right)$$
$$=\sum_{j=1}^{\infty}\beta_j(t)f_j(x). \tag{5.2}$$

The velocity potential can be decomposed into two parts:

$$\Phi(x,z,t)=v_{O1}(t)x+\varphi(x,z,t), \tag{5.3}$$

where the term $v_{O1}(t)x$ satisfies the framed nonhomogeneous body-boundary condition in eq. (5.1). As a consequence, the function φ satisfies a zero-Neumann condition on the walls and the bottom as well as the Laplace equation. Because these conditions are also satisfied by the natural sloshing modes, we compose the solution of φ as the following sum of the natural modes $\varphi_n(x,z)$ (Section 4.3.1.1 gives these modes in analytical form):

$$\varphi(x,z,t)=\sum_{n=1}^{\infty}R_n(t)\cos\left(\pi n\left(x+\tfrac{1}{2}l\right)/l\right)\times\frac{\cosh(\pi n(z+h)/l)}{\cosh(\pi nh/l)}$$
$$=\sum_{n=1}^{\infty}R_n(t)\,\varphi_n(x,z). \tag{5.4}$$

Solution (5.2)–(5.4) satisfies all of conditions (5.1) except the kinematic and dynamic free-surface conditions on the mean free surface $z=0$. We start by substituting eqs. (5.2)–(5.4) into the kinematic free-surface (Neumann) condition $\partial\Phi/\partial z|_{z=0}=\partial\zeta/\partial t$ on the mean free surface. The next steps are to multiply the resulting expression by $\cos(\pi j(x+\frac{1}{2}l)/l)$ and integrate from $x=-\frac{1}{2}l$ to $\frac{1}{2}l$. Using orthogonality properties of the natural surface modes, that is, $\int_{-\frac{1}{2}l}^{\frac{1}{2}l}\cos[\pi n(x+\frac{1}{2}l)/l]\cos[\pi j(x+\frac{1}{2}l)/l]\mathrm{d}x=0$ when $j\neq n$, gives the following relation between unknowns β_j and R_j, $j\geq 1$:

$$\dot\beta_j=\kappa_jR_j,\quad j\geq 1, \tag{5.5}$$

where κ_j is given by eq. (4.8).

Substitution of eqs. (5.2)–(5.4) into the dynamic boundary condition of eqs. (5.1) and use of eq. (5.5) to replace R_j with β_j leads to

$$\sum_{j=1}^{\infty}\left[\dot R_j(t)+g\beta_j(t)\right]\cos\left(\pi j\left(x+\tfrac{1}{2}l\right)/l\right)$$
$$=\sum_{j=1}^{\infty}\left[\ddot\beta_j(t)+g\kappa_j\beta_j(t)\right]\kappa_j^{-1}\cos\left(\pi j\left(x+\tfrac{1}{2}l\right)/l\right)$$
$$=-\dot v_{O1}(t)x, \tag{5.6}$$

where $\sigma_j=\sqrt{g\kappa_j}$ $(j\geq 1)$ are the natural sloshing frequencies. Multiplying eq. (5.6) by $\cos(\pi n(x+\frac{1}{2}l)/l)$, integrating the resulting expression from $x=-\frac{1}{2}l$ to $\frac{1}{2}l$, and using the orthogonality of these functions leads to the following *modal system* for modal functions β_n:

$$\ddot\beta_n+\sigma_n^2\beta_n=K_n(t)=-\dot v_{O1}(t)\times\underbrace{\left[\frac{2}{n\pi}\tanh\left(\frac{\pi n}{l}h\right)\left((-1)^n-1\right)\right]}_{P_n},$$
$$n=1,2,\ldots. \tag{5.7}$$

Equation (5.7) describes independent linear oscillators analogous to an uncoupled set of mass-spring systems without damping. The excitation is proportional to tank acceleration $\dot v_{O1}$. Nonzero forcing terms exist only on the right-hand side of eqs. (5.7) for the odd, antisymmetric modes ($n=2i-1$). Activation of the even modes due to horizontal forcing is therefore of a nonlinear nature. This important fact is discussed extensively in the context of nonlinear theories and internal

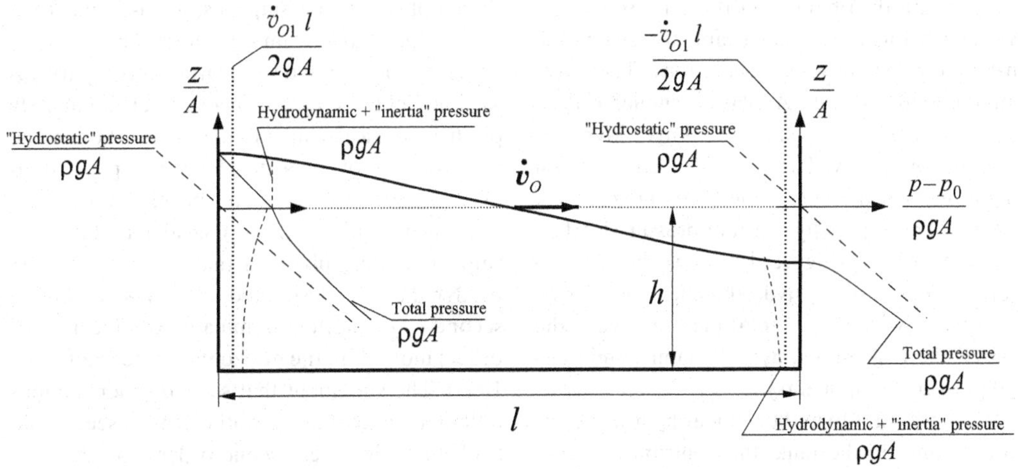

Figure 5.1. Pressure distribution on the tank walls at a given time instant for forced surge oscillations of a two-dimensional rectangular tank. Total pressure is the sum of the hydrostatic, hydrodynamic, and inertial pressure components.

(secondary) resonance phenomena in Chapters 8 and 9.

Equations (5.7) may have analytical solutions, for example, for harmonic excitations with a circular frequency σ (i.e., $\dot{v}_{O1} = -\varepsilon l \sigma^2 \cos(\sigma t)$, where $\varepsilon = \eta_{1a}/l \ll 1$ to ensure validity of the linear theory and where η_{1a} is the surge amplitude). When the considered horizontal excitation is harmonic, the solution does not reach a steady-state (periodic) regime oscillating with the forcing period $T = 2\pi/\sigma$, which is because considered mechanical system (5.7) has no damping terms according to potential flow theory. Solution parts always exist that oscillate with σ and σ_n. If we had considered an exterior potential flow problem, there would have been damping associated with radiated waves. In reality, sloshing-damping occurs due to viscous effects, in particular due to boundary-layer flow along the wetted tank surface when no breaking waves occur. More details on the general solution of eqs. (5.7), steady-state regimes, and transients are presented in the next sections.

When an analytical solution of eqs. (5.7) is not available, a fourth-order Runge–Kutta scheme may be used, for instance, to simulate β_j for generally arbitrary forcing $v_{O1}(t)$. As a result, both transient and steady-state sloshing regimes are handled with modal system (5.7).

Once modal functions β_j have been found, one can calculate all the hydrodynamic characteristics such as liquid velocity, hydrodynamic pressure, force, and moment in terms of the forcing and modal functions. The total pressure in the liquid according to linear theory can be expressed as $p = -\rho\partial\Phi/\partial t - \rho g z + p_0$, where Φ is given by eqs. (5.3) and (5.4), p_0 is the pressure in the air (gas), and $-\rho g z$ is the hydrostatic pressure when there is no liquid motion.

The terms $-\rho g z$ and $-\rho\partial\Phi/\partial t = -\rho\dot{v}_{O1}x - \rho\partial\varphi/\partial t$ must be combined to describe the pressure distribution in the free-surface zone. The last expression is the sum of the hydrodynamic pressure due to sloshing, $-\rho\partial\varphi/\partial t$, and the inertia pressure caused by the translatory liquid acceleration as a solid body. The velocity potential is constant from the mean free surface to the actual free-surface level in a linear theory. This assumption was made when the free-surface conditions were formulated. The latter fact was exemplified for deep-sea propagating two-dimensional waves with amplitude ζ_a in Figure 3.1(b), which shows how the pressure varies with depth under both a wave crest and a wave trough. The "hydrostatic" pressure, $-\rho g z$, cancels the dynamic pressure, $-\rho\partial\varphi/\partial t|_{z=0}$, at the free surface. This condition is the linear dynamic free-surface condition, which is exactly satisfied at the wave crest in Figure 3.1(b), whereas a higher-order error exists under the wave trough. The reason for using quotation marks for the "hydrostatic" pressure is that $-\rho g z$ is only the hydrostatic pressure in a static condition. We must combine $-\rho g z$ with the other pressure components in dynamic flow conditions. Figure 5.1 illustrates in a similar way the pressure distribution on the tank walls of a two-dimensional rectangular tank at a given time instant for forced surge oscillations. The wave amplitude at the wall, A, plays a similar role as ζ_a in Figure 3.1(b). The maximum hydrodynamic pressure on the wall is ρgA. The contribution of the inertial pressure, $-\rho\dot{v}_{O1}x$, is indicated in Figure 5.1 as a spatially constant pressure load on each tank wall. The figure also shows the distribution of the sum of the hydrodynamic and inertial pressure as well as the "total pressure" (i.e., the sum of the hydrostatic, hydrodynamic, and inertial pressure components).

When we want to find the linear hydrodynamic force acting on the tank, the contribution from integrating the pressure over the instantaneous wetted surface contains higher-order terms than those determined when integrating the pressure over the mean wetted surface. By properly integrating the pressure over the mean wetted surface, we get the following two-dimensional horizontal hydrodynamic force:

$$F_1(t) = -m_l\dot{v}_{O1}(t) + m_l\frac{l}{\pi^2 h}\sum_{n=1}^{\infty}\ddot{\beta}_n(t)\frac{1+(-1)^{n+1}}{n^2}, \tag{5.8}$$

where the first term on the right-hand side represents the inertial force of the "frozen" liquid mass ($m_l = \rho lh$ for two dimensions).

5.3 Theory

5.3.1 Linear modal equations

5.3.1.1 Six generalized coordinates for solid-body, linear dynamics

Rigid-body motions in six degrees of freedom for a ship were introduced in Section 3.5, where we operated with two coordinate systems. The first coordinate system is an inertial system that translates with the steady speed of the ship, U, on a straight course. The other coordinate system was ship-fixed and is therefore an accelerated coordinate system. Linear ship motions were assumed and the translatory motions along the axes of the first coordinate system were denoted η_1, η_2, and η_3, respectively. The angular displacements of the rotational motions about these axes were denoted η_4, η_5, and η_6, respectively (see Figure 3.8 and discussion in Section 2.3). The linear-motion assumption implies that the oscillatory ship motion at any ship position is small relative to the ship's beam and draft. The meaning of "small" is in practice a matter of comparisons with model tests. Linear theory often adequately predicts vertical ship motions even though the bow goes out of the water. However, predictions of ship accelerations would be less satisfactory in such a scenario. If we consider a stationary ship or floating platform that is either moored or dynamically positioned in waves, nonlinear second-order motions dominate over linear (first-order) motions in the horizontal plane (Faltinsen, 1990). The statement that second-order motions dominate over linear motions may seem contradictory. However, second-order hydrodynamic loads are smaller than linear loads and it is a resonance phenomenon that amplifies second-order motions. Similar phenomena can happen in waves for heave, roll, and pitch of floating platforms with a small waterplane area, such as semisubmersibles. When sloshing in tanks is considered, it is not the translatory motions but the translatory accelerations and angular velocities and accelerations of the tank that matter. The slowly varying velocities are not necessarily dominant relative to the linear velocities. However, the angular motions excite sloshing. We leave out further considerations of nonlinear ship and platform motion effects in waves.

As we mentioned in Section 2.3, the consequence of linearity is that higher-order terms in η_i, for instance, as products between η_i and η_j, are small and negligible relative to the linear terms. Due to linearity we could also use η_i as definitions of translatory and angular motions in the ship-fixed coordinate system (see Section 3.5); we could do the same for velocity components $\dot{\eta}_i$. When it comes to linear acceleration terms, care must be shown in handling the gravitational acceleration vector, $\boldsymbol{g}$, which in a body-fixed coordinate system has, in general, components along all three body-fixed coordinate axes due to the rotation of the body-fixed coordinate system relative to the Earth.

When sloshing in a tank is studied, it is most convenient to operate with a tank-fixed coordinate system as defined in Section 2.3. The translatory motions of the origin O of the tank-fixed coordinate system can be expressed in terms of the translatory and angular motions of the ship. If the coordinate axes of the tank-fixed coordinate system are parallel to the axes of the global ship-fixed coordinate system, then the rotational angles η_j ($j = 4, 5, 6$) remain the same. *When we study prescribed tank motions, we use Oxyz as the notation for the tank-fixed coordinate system (see Figure 2.2). Furthermore, we use the notation* $\dot{\eta}_j$ ($j = 1, 2, 3$) *for the components of the translatory velocity of origin O*; that is, the translatory velocity vector of O can be expressed as

$$\begin{aligned} \boldsymbol{v}_O(t) &= v_{O1}(t)\boldsymbol{e}_1 + v_{O2}(t)\boldsymbol{e}_2 + v_{O3}(t)\boldsymbol{e}_3 \\ &= \dot{\eta}_1\boldsymbol{e}_1 + \dot{\eta}_2\boldsymbol{e}_2 + \dot{\eta}_3\boldsymbol{e}_3. \end{aligned} \tag{5.9}$$

The angular velocities of the $Oxyz$ system relative to the absolute frame $O'x'y'z'$ is denoted

$$\begin{aligned} \boldsymbol{\omega}(t) &= \omega_1(t)\boldsymbol{e}_1 + \omega_2(t)\boldsymbol{e}_2 + \omega_3(t)\boldsymbol{e}_3 \\ &= \dot{\eta}_4\boldsymbol{e}_1 + \dot{\eta}_5\boldsymbol{e}_2 + \dot{\eta}_6\boldsymbol{e}_3. \end{aligned} \tag{5.10}$$

A basis for the preceding expressions was that the ship motions are small relative to the beam and draft of the ship. However, we must also assume in the following analysis that the ship motions are small relative to a characteristic dimension of the tank. When we consider two-dimensional tank motions, the characteristic dimension L can be chosen as the breadth l of the mean free surface Σ_0. If three-dimensional motions are considered, the characteristic dimension L is related to both the length and the breadth of Σ_0. To express the smallness of the ship motions relative to the tank, we introduce a small dimensionless parameter $\varepsilon \ll 1$ so that

$$\eta_i/L = O(\varepsilon), \quad i = 1, 2, 3 \tag{5.11}$$

$$\eta_i(t) = O(\varepsilon), \quad i = 4, 5, 6. \tag{5.12}$$

As previously stated, special attention should be paid to the *gravitational acceleration vector* in the body-fixed coordinates, namely, $\boldsymbol{g} = g_1(t)\boldsymbol{e}_1(t) + g_2(t)\boldsymbol{e}_2(t) + g_3(t)\boldsymbol{e}_3(t)$. Further dynamic equations need expressions for its projections, g_i. In the linear case, these projections can be expressed as

$$g_1 = g\eta_5, \quad g_2 = -g\eta_4, \quad g_3 = -g. \tag{5.13}$$

5.3.1.2 Generalized coordinates for liquid sloshing and derivation of linear modal equations

In the following derivations it should be recalled that the mean free surface, Σ_0, is fixed relative to the tank-fixed coordinate system $Oxyz$ and defined to be part of the x-y-plane. The free surface relative to Σ_0 is expressed by the equation $z = \zeta(x, y, t)$.

The boundary-value problem for the velocity potential $\Phi(x, y, z, t)$ can be linearized for nonresonant and nonviolent transient conditions. The liquid velocity is then assumed to be of the same order as the tank velocity, which is of the order η_i divided by a time T_1 characterizing the forcing period and the highest natural sloshing period. The liquid velocity, $|\nabla\Phi|$, can be represented nondimensionally as $|\nabla\Phi|/\sqrt{Lg}$ by recognizing the fact that gravity is a parameter and that Froude scaling applies. We can then write

$$O(|\nabla\Phi|/\sqrt{Lg}) = O(\eta_i/L)O(1/(T_1\sqrt{g/L})).$$

Because $T_1\sqrt{g/L}$ is of order 1, $|\nabla\Phi|$ is $O(\varepsilon)$. Since spatial derivatives do not change the ordering, $\Phi(x, y, z, t)$ is also $O(\varepsilon)$. Because the free-surface elevation ζ can be related to $\Phi(x, y, z, t)$ by the free-surface condition, it follows that $\zeta/L = O(\varepsilon)$.

Even though a nondimensional formulation of eqs. (2.58), (2.63), (2.64), (2.68), and (2.74) may be mathematically useful, we work with the original dimensional notations. Neglecting terms that are smaller than $O(\varepsilon)$ is mathematically equivalent to crossing out the nonlinear terms in the original free-boundary-value problem; for instance, velocity-squared terms are neglected. Then we make a Taylor expansion of the free-surface condition about Σ_0. The consequence of the linearization is that the linear free-surface conditions can be formulated on Σ_0 instead of on the instantaneous free-surface position which is unknown a priori.

The linearized boundary-value problem for the velocity potential Φ is

$$\frac{\partial^2\Phi}{\partial x^2} + \frac{\partial^2\Phi}{\partial y^2} + \frac{\partial^2\Phi}{\partial z^2} = 0 \quad \text{in } Q_0, \tag{5.14}$$

$$\begin{aligned} \left.\frac{\partial\Phi}{\partial n}\right|_{S_0} &= (\boldsymbol{v}_O + \boldsymbol{\omega} \times \boldsymbol{r}) \cdot \boldsymbol{n} \\ &= \boldsymbol{v}_O \cdot \boldsymbol{n} + \boldsymbol{\omega} \cdot (\boldsymbol{r} \times \boldsymbol{n}), \end{aligned} \tag{5.15}$$

$$\left.\frac{\partial\Phi}{\partial n}\right|_{\Sigma_0} = \boldsymbol{v}_O \cdot \boldsymbol{n} + \boldsymbol{\omega} \cdot (\boldsymbol{r} \times \boldsymbol{n}) + \frac{\partial\zeta}{\partial t}, \tag{5.16}$$

$$\left.\frac{\partial\Phi}{\partial t}\right|_{\Sigma_0} - g_1 x - g_2 y - g_3\zeta = 0, \qquad (5.17)$$

$$\int_{\Sigma_0} \zeta \, dx \, dy = 0, \qquad (5.18)$$

where Q_0 is the mean liquid domain, S_0 is the mean wetted tank surface, $\boldsymbol{n} = n_1\boldsymbol{e}_1 + n_2\boldsymbol{e}_2 + n_3\boldsymbol{e}_3$ is the outer normal vector, and g_i $(i = 1, 2, 3)$ are projections of the gravitational acceleration given by eq. (5.13). Body-boundary condition (5.15) expresses that the normal component of the total liquid velocity on the mean wetted tank surface, $\partial\Phi/\partial n$, is equal to the normal component of the tank velocity. Kinematic free-surface condition (5.16) states that the normal component of the total liquid velocity, $\partial\Phi/\partial n$, on the mean free surface Σ_0 is equal to the sum of the free-surface velocity $\partial\zeta/\partial t$ relative to the tank and the normal component of the tank velocity on Σ_0. Kinematic and dynamic boundary conditions (5.16) and (5.17) follow by linearizing eqs. (2.68) and (2.64), respectively. Equation (5.18) expresses the fact that the free-surface elevation, ζ, must be consistent with conservation of liquid volume. Problem (5.14)–(5.18) requires initial conditions for the free-surface elevation and the velocity potential on the mean free surface Σ_0.

Again, we direct attention to the fact that $Oxyz$ is the tank-fixed coordinate system and Φ is the absolute velocity potential. These conditions are not the same as those in wavemaker problems solved in an Earth-fixed coordinate system, for instance. As a consequence, dynamic boundary condition (5.17) includes the inhomogeneous term $(-g_1 x - g_2 y)$, which obviously disappears when $Oxyz$ does not perform angular motions or is an Earth-fixed system. Furthermore, ζ describes the wave elevation in the movable, tank-fixed system and, as a result, $\partial\zeta/\partial t$ represents only the *relative velocity component*. Because $\partial\Phi/\partial n|_{\Sigma_0}$ is the normal component of the absolute velocity, the term $\boldsymbol{v}_O \cdot \boldsymbol{n} + \boldsymbol{\omega} \cdot (\boldsymbol{r} \times \boldsymbol{n})$ is required to make kinematic condition (5.16) consistent with our definitions. Excluding the latter term from eq. (5.16), as was typical for older publications on sloshing (Abramson, 1966; Feschenko *et al.*, 1969), does not indicate a wrong formulation, but $\partial\zeta/\partial t$ must then be considered the absolute velocity.

The infinite set of generalized coordinates β_i and $R_i(t)$ is introduced by expanding the free-surface elevation ζ and the velocity potential Φ as a series in terms of the natural sloshing modes; that is,

$$\zeta(x, y, t) = \sum_{i=1}^{\infty} \beta_i(t)\varphi_i(x, y, 0) = \sum_{i=1}^{\infty} \beta_i(t) f_i(x, y), \qquad (5.19)$$

$$\Phi(x, y, z, t) = \boldsymbol{v}_O(t) \cdot \boldsymbol{r} + \boldsymbol{\omega}(t) \cdot \boldsymbol{\Omega}_0(x, y, z) + \underbrace{\sum_{i=1}^{\infty} R_i(t)\,\varphi_i(x, y, z)}_{\varphi(x,y,z,t)}, \qquad (5.20)$$

where φ_i $(i \geq 1)$ are the natural sloshing modes defined by eigenvalue problem (4.2) and $f_i(x, y) = \varphi_i(x, y, 0)$. Representation (5.20) introduces the vector function $\boldsymbol{\Omega}_0(x, y, z) = (\Omega_{01}(x, y, z), \Omega_{02}(x, y, z), \Omega_{03}(x, y, z))$, which is the so-called Stokes–Joukowski potential and appears in the problem that involves liquid motions in a completely filled tank (Joukowski, 1885). In the linear case, this vector function is independent of time and, by definition, it is the solution of the Neumann boundary-value problem

$$\nabla^2\boldsymbol{\Omega}_0 = 0 \quad \text{in } Q_0,$$
$$\frac{\partial\Omega_{01}}{\partial n} = yn_3 - zn_2, \qquad \frac{\partial\Omega_{02}}{\partial n} = zn_1 - xn_3,$$
$$\frac{\partial\Omega_{03}}{\partial n} = xn_2 - yn_1 \quad \text{on } S_0 \cup \Sigma_0. \qquad (5.21)$$

The terms $\boldsymbol{v}_O(t) \cdot \boldsymbol{r}$ and $\boldsymbol{\omega}(t) \cdot \boldsymbol{\Omega}_0(x, y, z)$ in eq. (5.20) take care of the $\boldsymbol{v}_O(t) \cdot \boldsymbol{n}$ and $\boldsymbol{\omega} \cdot (\boldsymbol{r} \times \boldsymbol{n})$ terms in eqs. (5.15) and (5.16), respectively, but φ is responsible *only* for the relative velocity component. The consequence is that $\partial\varphi/\partial n$ is zero on S_0 and equal to $\partial\zeta/\partial t$ on Σ_0.

As we discussed earlier, we can follow Abramson (1966) or Feschenko *et al.* (1969) and other publications of the 1960s and 1970s and omit the term $\boldsymbol{v}_O \cdot \boldsymbol{n} + \boldsymbol{\omega} \cdot (\boldsymbol{r} \times \boldsymbol{n})$ in eq. (5.16). We remarked that, when removing this term, ζ is the wave elevation in an absolute coordinate system and $\partial\zeta/\partial t = \partial\Phi/\partial n$ is associated with the absolute velocity of the free surface. In this case, substitution of eq. (5.20) into the Neumann boundary condition does not make the boundary condition homogeneous on the mean free surface Σ_0. We have got that $\partial\varphi/\partial n$ (i.e., the normal relative velocity component) is equal to $\partial\zeta/\partial t - [\boldsymbol{v}_O \cdot \boldsymbol{n} + \boldsymbol{\omega} \cdot (\boldsymbol{r} \times \boldsymbol{n})]$, where the square brackets denote the normal component of the solid-body velocity on Σ_0.

The natural modes (eigenfunctions from problem (4.2)) describe patterns of freestanding waves that occur in a nonoscillating tank. Each freestanding wave has a natural frequency, σ_i. The squares of the natural frequencies, σ_i^2, are related to the eigenvalues κ_i as $\sigma_i^2 = g\kappa_i$ (eq. (4.2)). The eigenfunctions constitute the basis $\{\varphi_i(x,y,0) = f_i(x,y)\}$ on Σ_0 and satisfy the orthogonality conditions

$$\int_{\Sigma_0} f_i f_j \, dS = 0 \quad \text{when } i \neq j. \tag{5.22}$$

As long as representations (5.19) and (5.20) are postulated, solving the original problem of eqs. (5.14)–(5.18) means finding appropriate $\beta_i(t)$ and $R_i(t)$ $(i \geq 1)$. Laplace equation (5.14), boundary condition (5.15) on the mean wetted tank surface, and volume conservation condition (5.18) are automatically satisfied due to the structure of eqs. (5.19) and (5.20) and the properties of the natural modes. For instance, volume conservation, $\int_{\Sigma_0} \zeta dS = 0$, follows because the integration over the mean free surface of each $\varphi_i(x,y,0)$ in eq. (5.19) is zero due to properties of the normal modes. However, we must still satisfy free-surface conditions (5.16) and (5.17).

Substitution of eqs. (5.20) and (5.19) into eq. (5.16) gives $\sum_{j=1}^{\infty} R_j \kappa_j \varphi_j(x,y,0) = \sum_{j=1}^{\infty} \dot{\beta}_j f_j(x,y)$ and, due to orthogonality condition (5.22),

$$\dot{\beta}_j = \kappa_j R_j, \quad j \geq 1. \tag{5.23}$$

The consequence of eq. (5.23) is that it is sufficient to use time-dependent functions $\{\beta_j\}$ as the *generalized coordinates for liquid motion.* Accounting for relation (5.23) in representations (5.19) and (5.20) as well as definitions (5.9) and (5.10), dynamic free-surface condition (5.17) leads to the following equation:

$$\sum_{j=1}^{\infty} \left(\ddot{\beta}_j + \sigma_j^2 \beta_j\right) \kappa_j^{-1} \varphi_j(x,y,0) + x(\ddot{\eta}_1 - g\eta_5) + y(\ddot{\eta}_2 + g\eta_4) + \sum_{k=4}^{6} \ddot{\eta}_k \Omega_{0(k-3)}(x,y,0) = 0, \quad (x,y) \in \Sigma_0, \tag{5.24}$$

where $\boldsymbol{r}|_{\Sigma_0} = (x,y,0)$ is used. Multiplying equation (5.24) with $\rho\varphi_m(x,y,0) = \rho f_m(x,y)$ (where ρ is the liquid density) and integrating and using orthogonality condition (5.22) gives the following infinite set of uncoupled linear differential equations for the generalized coordinates $\{\beta_i\}$ (*linear modal equations*):

$$\mu_m\left(\ddot{\beta}_m + \sigma_m^2 \beta_m\right) + \lambda_{1m}(\ddot{\eta}_1 - g\eta_5) + \lambda_{2m}(\ddot{\eta}_2 + g\eta_4) + \sum_{k=4}^{6} \ddot{\eta}_k \lambda_{0(k-3)m} = 0, \quad m = 1,2,\ldots. \tag{5.25}$$

The linear modal equations include a set of *hydrodynamic coefficients*, which are independent of time and may be computed prior to studying linear sloshing. Computation of the hydrodynamic coefficients needs only the natural sloshing modes and the Stokes–Joukowski potential:

$$\sigma_m^2 = g\kappa_m, \quad \mu_m = \frac{\rho}{\kappa_m}\int_{\Sigma_0} \varphi_m^2 \, dx\,dy = \frac{\rho}{\kappa_m}\int_{\Sigma_0} f_m^2 \, dx\,dy, \quad \lambda_{1m} = \rho\int_{\Sigma_0} f_m x \, dx\,dy,$$
$$\lambda_{2m} = \rho\int_{\Sigma_0} f_m y \, dx\,dy,$$
$$\lambda_{0km} = \rho\int_{\Sigma_0} f_m \Omega_{0k} \, dx\,dy, \quad k = 1,2,3; \quad m = 1,2,\ldots. \tag{5.26}$$

As we mentioned in the introduction, the index m may be not only a positive integer, as happens for a two-dimensional rectangular tank, but also a *pair of integers* (i.e., $m = (i,j)$), as occurs for three-dimensional tanks.

Because the solutions of eq. (5.25) do not depend on ρ, it appears unnecessary that ρ is used in the definitions of hydrodynamic coefficients. The reason for introducing ρ is that the same hydrodynamic coefficients appear in expressions for the hydrodynamic force and moment, which are linearly dependent on ρ.

5.3.1.3 *Linear modal equations for* prescribed *tank motions*

When rigid-body motions η_i are known, we rearrange eq. (5.25) to have known variables on the right-hand side; that is,

$$\ddot{\beta}_m + \sigma_m^2 \beta_m = K_m(t), \quad m = 1,2,\ldots, \tag{5.27}$$

where the right-hand side is prescribed as

$$K_m(t) = -\frac{\lambda_{1m}}{\mu_m}(\ddot{\eta}_1(t) - g\eta_5(t)) - \frac{\lambda_{2m}}{\mu_m}(\ddot{\eta}_2(t) + g\eta_4(t)) - \sum_{k=4}^{6}\frac{\ddot{\eta}_k(t)\lambda_{0(k-3)m}}{\mu_m}, \quad m = 1,2,\ldots. \tag{5.28}$$

Equations (5.27) and (5.28) indicate that all the results of linear sloshing theory can be deduced from the theory of linear oscillators based on consideration of modal equations (5.27). Because K_m does not involve η_3, heave cannot linearly excite sloshing.

If we want to study the mutual interaction between sloshing and ship motion, we need to introduce the hydrodynamic forces and moments due to sloshing in the equations of ship motion. The following section derives formulas for the integrated hydrodynamic loads in terms of generalized coordinates β_i $(i \geq 1)$ and η_j $(j = 1, 6)$. When setting up the fully coupled equation system, it is important not to count the "frozen" tank liquid effects twice (as also noted in Chapter 3).

Knowing the pressure distribution in addition to the hydrodynamic forces and moments is important for structural stress analysis. We discussed the pressure distribution in Section 5.2. We can use eq. (2.61) to express the linear pressure in the tank-fixed coordinate system:

$$p = p_0 + \rho \boldsymbol{g} \cdot \boldsymbol{r} - \rho\, \partial\Phi/\partial t. \qquad (5.29)$$

The components of the gravitational acceleration vector, $\boldsymbol{g}$, are given by eq. (5.13). Furthermore, $\boldsymbol{r} = x\boldsymbol{e}_1 + y\boldsymbol{e}_2 + z\boldsymbol{e}_3$, and Φ is given by eq. (5.20), leading to

$$\begin{aligned} p &= p_0 - \rho g\,(z + y\eta_4 - x\eta_5) - \rho\,[x\ddot{\eta}_1 + y\ddot{\eta}_2 + z\ddot{\eta}_3 \\ &\quad + \Omega_{01}(x, y, z)\ddot{\eta}_4 + \Omega_{02}(x, y, z)\ddot{\eta}_5 \\ &\quad + \Omega_{03}(x, y, z)\ddot{\eta}_6] - \rho \sum_{i=1}^{\infty} \ddot{\beta}_i \kappa_i^{-1} \varphi_i(x, y, z), \end{aligned} \qquad (5.30)$$

where we have used eq. (5.23) to relate β_i and R_i. When evaluating eq. (5.30) in the free-surface zone, we should recognize the fact that linear theory assumes the velocity potential to be constant in the free-surface zone. Having obtained the pressure, we can obtain expressions for linear hydrodynamic force and moment by properly integrating the pressure over the mean wetted tank surface. A different approach is followed in the next section, where we start with formal expressions of hydrodynamic force and moment and then rewrite these expressions by means of integral theorems. The pressure distribution is further discussed for a completely filled tank in Section 5.4.2.2.

5.3.2 Resulting hydrodynamic force and moment in linear approximation

Most general "modal" formulas for hydrodynamic force and moment caused by liquid sloshing were derived by Lukovsky (1990). He considered a general, nonlinear case. (These formulas are rederived in Section 7.3.) Lukovsky's modal formulas represent the force and moment in terms of $\boldsymbol{v}_O$, $\boldsymbol{\omega}$, and their *-time derivatives (Section 2.3) as well as the position, velocity, and acceleration of the liquid mass center evaluated in the tank-fixed $Oxyz$ system. The liquid mass center can be expressed by modal functions β_i $(i \geq 1)$, because the instantaneous position of the mass center of an incompressible liquid, and its velocity and acceleration, depend uniquely on free-surface motions. Generally speaking, we can use the Lukovsky formulas and linearize them by β_i $(i \geq 1)$ and η_j $(j = 1, 6)$ to get the required linear approximation of hydrodynamic force and moment. However, this section represents an independent derivation for the linear case.

5.3.2.1 Force

The derivation of the hydrodynamic force on the tank starts with eq. (2.38), which expresses the force in terms of the hydrodynamic momentum. The linearized momentum is

$$\begin{aligned} \boldsymbol{M}(t) &= \rho \int_{Q_0} \boldsymbol{v}\, \mathrm{d}Q \\ &= \rho Vol\, \boldsymbol{v}_O + \rho\boldsymbol{\omega} \times \int_{Q_0} \boldsymbol{r}\, \mathrm{d}Q + \rho \int_{Q_0} \boldsymbol{v}_r\, \mathrm{d}Q, \end{aligned} \qquad (5.31)$$

where $\boldsymbol{v}_r$ is the relative velocity of the liquid (see Chapter 2) and Vol is the liquid volume.

The second integral is equal to $\rho\boldsymbol{\omega} \times \boldsymbol{r}_{lC_0}$, where $\boldsymbol{r}_{lC_0}$ is the hydrostatic position of the liquid mass center in the $Oxyz$ system; that is,

$$\boldsymbol{r}_{lC_0} = \frac{\rho}{M_l} \int_{Q_0} \boldsymbol{r}\, \mathrm{d}Q = x_{lC_0}\boldsymbol{e}_1 + y_{lC_0}\boldsymbol{e}_2 + z_{lC_0}\boldsymbol{e}_3,$$

$$x_{lC_0} = \frac{1}{Vol} \int_{Q_0} x\, \mathrm{d}Q, \quad y_{lC_0} = \frac{1}{Vol} \int_{Q_0} y\, \mathrm{d}Q,$$

$$z_{lC_0} = \frac{1}{Vol} \int_{Q_0} z\, \mathrm{d}Q. \qquad (5.32)$$

The last integral in eq. (5.31) can be transformed by using the Gauss theorem (see eq. (A.1)) and the fact that $\boldsymbol{v}_r \cdot \boldsymbol{n} = 0$ on S_0. Before doing that, we note the fact that $\nabla \cdot \boldsymbol{v}_r = 0$ and $\nabla \cdot (\boldsymbol{v}_r \boldsymbol{r}) = (\nabla \cdot \boldsymbol{v}_r)\boldsymbol{r} + \boldsymbol{v}_r = \boldsymbol{v}_r$. The result is

$$\begin{aligned} \rho \int_{Q_0} \boldsymbol{v}_r\, \mathrm{d}Q &= \rho \int_{\Sigma_0} \boldsymbol{r}(\boldsymbol{v}_r \cdot \boldsymbol{n})\, \mathrm{d}S \\ &= \rho \int_{\Sigma_0} \boldsymbol{r} \sum_{n=1}^{\infty} R_n \frac{\partial \varphi_n}{\partial n} \mathrm{d}S, \end{aligned}$$

where the last sum in modal representation (5.20) has been used to express the velocity potential for the relative velocity. Using eq. (5.23) and the spectral boundary condition for the natural modes, $\partial\varphi_n/\partial n = \kappa_n\varphi_n$, transforms the last expression to

$$\rho\int_{\Sigma_0} \boldsymbol{r}\sum_{n=1}^{\infty} R_n\frac{\partial\varphi_n}{\partial n}\mathrm{d}S = \rho\int_{\Sigma_0}\boldsymbol{r}\sum_{n=1}^{\infty}\dot{\beta}_n\varphi_n\,\mathrm{d}S$$
$$= \rho\int_{\Sigma_0}\frac{d}{dt}\left(\sum_{j=1}^{\infty}\beta_j f_j\right)\boldsymbol{r}\,\mathrm{d}S$$
$$= \rho\int_{\Sigma_0}\frac{\partial\zeta}{\partial t}\boldsymbol{r}\,\mathrm{d}S = M_l\frac{d^*\boldsymbol{r}_{lC_1}}{dt},$$

where $\boldsymbol{r}_{lC_1}(t)/L = O(\varepsilon)$ describes the linear displacements of the liquid mass center in the $Oxyz$ system; that is,

$$\boldsymbol{r}_{lC_1}(t) = \boldsymbol{e}_1\frac{\rho}{M_l}\int_{\Sigma_0} x\zeta(x,y,t)\,\mathrm{d}S + \boldsymbol{e}_2\frac{\rho}{M_l}\int_{\Sigma_0} y\zeta(x,y,t)\,\mathrm{d}S. \quad (5.33)$$

The linearized momentum can, together with definitions (5.32) and (5.33), be expressed as

$$\boldsymbol{M} = M_l\left(\boldsymbol{v}_O + \boldsymbol{\omega}\times\boldsymbol{r}_{lC_0} + \frac{d^*\boldsymbol{r}_{lC_1}}{dt}\right). \quad (5.34)$$

Substitution of eq. (5.34) into eq. (2.38) gives the following linear version of the Lukovsky formula:

$$\boldsymbol{F}(t) = -M_l g\boldsymbol{e}_3 + M_l g[\eta_5\boldsymbol{e}_1 - \eta_4\boldsymbol{e}_2] - M_l\left(\frac{d^*\boldsymbol{v}_O}{dt} + \dot{\boldsymbol{\omega}}\times\boldsymbol{r}_{lC_0} + \frac{d^{*2}\boldsymbol{r}_{lC_1}}{dt^2}\right) + [\text{nonlinear}], \quad (5.35)$$

for which we should remember time-differentiation rule (2.50). In this formula, only the term $d^{*2}\boldsymbol{r}_{lC_1}/dt^2$ is caused by sloshing. Other quantities are associated with the liquid weight ($-M_l g\boldsymbol{e}_3$), the dynamic loads of a "frozen" liquid due to inclination of the tank ($M_l g[\eta_5\boldsymbol{e}_1 - \eta_4\boldsymbol{e}_2]$), and translatory inertial ($-M_l d^*\boldsymbol{v}_O/dt$) and rotational inertial ($-M_l\dot{\boldsymbol{\omega}}\times\boldsymbol{r}_{lC_0}$) forces.

Substitution of modal solution (5.19) into eq. (5.35) gives the desired expression for the resulting *linear* hydrodynamic force in terms of the introduced generalized coordinates:

$$\boldsymbol{F}(t) = \underbrace{-M_l g\boldsymbol{e}_3}_{\text{weight}} + \boldsymbol{e}_1\underbrace{\left[M_l(g\eta_5 - \ddot{\eta}_1 - \ddot{\eta}_5 z_{lC_0} + \ddot{\eta}_6 y_{lC_0}) - \sum_{j=1}^{\infty}\lambda_{1j}\ddot{\beta}_j\right]}_{F_1(t)}$$
$$+\boldsymbol{e}_2\underbrace{\left[M_l(-g\eta_4 - \ddot{\eta}_2 - \ddot{\eta}_6 x_{lC_0} + \ddot{\eta}_4 z_{lC_0}) - \sum_{j=1}^{\infty}\lambda_{2j}\ddot{\beta}_j\right]}_{F_2(t)} + \boldsymbol{e}_3\underbrace{[M_l(-\ddot{\eta}_3 - \ddot{\eta}_4 y_{lC_0} + \ddot{\eta}_5 x_{lC_0})]}_{F_3(t)}. \quad (5.36)$$

The j-summation in formula (5.36) changes to summations over two (or more) integer indices for many three-dimensional tanks when the natural modes and, therefore, the modal functions are naturally enumerated, for example, by the pair (j_1, j_2) (i.e., β_j is associated with β_{j_1,j_2}). Formula (5.36) shows that computations of the hydrodynamic force need information on the liquid mass, the hydrostatic mass center, and the hydrodynamic coefficients λ_{1j} and λ_{2j}. These coefficients appeared in linear modal equations (5.27) and (5.28). Thus, as long as we know the hydrodynamic coefficients (5.26) and have computed the solution of eqs. (5.27) and (5.28) for prescribed η_i, we can evaluate the hydrodynamic forces along the $Oxyz$-axes via eq. (5.36). Formula (5.36) becomes simpler for particular tank shapes. Examples are given in Section 5.4.

When the tank is *completely filled*, the mass center does not move; therefore, the β_j-dependent quantities vanish in eq. (5.36). We can then write the *linear* hydrodynamic force for a filled tank as follows:

$$\boldsymbol{F}^{\text{filled}}(t) = \underbrace{-M_l g\boldsymbol{e}_3}_{\text{weight}} + \boldsymbol{e}_1\underbrace{[M_l(g\eta_5 - \ddot{\eta}_1 - \ddot{\eta}_5 z_{lC_0} + \ddot{\eta}_6 y_{lC_0})]}_{F_1^{\text{filled}}(t)} + \boldsymbol{e}_2\underbrace{[M_l(-g\eta_4 - \ddot{\eta}_2 - \ddot{\eta}_6 x_{lC_0} + \ddot{\eta}_4 z_{lC_0})]}_{F_2^{\text{filled}}(t)} + \boldsymbol{e}_3\underbrace{[M_l(-\ddot{\eta}_3 - \ddot{\eta}_4 y_{lC_0} + \ddot{\eta}_5 x_{lC_0})]}_{F_3^{\text{filled}}(t)}. \quad (5.37)$$

Formula (5.37) shows that the linear hydrodynamic force acting on the tank surface of the completely filled tank is the *same as the inertial force due to the "frozen" liquid.*

The term $M_l g[\eta_5 \boldsymbol{e}_1 - \eta_4 \boldsymbol{e}_2]$ in eq. (5.36) is simply a consequence of the frozen liquid weight term $-M_l g \boldsymbol{e}_3'$ in the *tank-fixed* coordinate system. Linear motions, forces, and moments in ship applications are described in an inertial system (i.e., not a tank-fixed coordinate system). Equation (5.36) can be expressed correctly to $O(\varepsilon)$ in the global *inertial* coordinate system $O'x'y'z'$ as

$$\boldsymbol{F}(t) = \underbrace{-M_l g \boldsymbol{e}_3'}_{\text{weight}} + \boldsymbol{e}_1' \underbrace{\left[M_l(-\ddot{\eta}_1 - \ddot{\eta}_5 z_{lC_0} + \ddot{\eta}_6 y_{lC_0}) - \sum_{j=1}^{\infty} \lambda_{1j} \ddot{\beta}_j \right]}_{F_1(t)} + \boldsymbol{e}_2' \underbrace{\left[M_l(-\ddot{\eta}_2 - \ddot{\eta}_6 x_{lC_0} + \ddot{\eta}_4 z_{lC_0}) - \sum_{j=1}^{\infty} \lambda_{2j} \ddot{\beta}_j \right]}_{F_2(t)} + \boldsymbol{e}_3' \underbrace{\left[M_l(-\ddot{\eta}_3 - \ddot{\eta}_4 y_{lC_0} + \ddot{\eta}_5 x_{lC_0}) \right]}_{F_3(t)}. \quad (5.38)$$

The only difference between eqs. (5.36) and (5.38) is the weight term.

5.3.2.2 Moment

The derivation of the linear version of the Lukovsky formula for the moment starts with expression (2.79), where we account for the fact that $p = p_0$ on the free surface. Using Bernoulli equation (2.61) gives the hydrodynamic moment vector relative to the x -, y -, and z-axes as

$$\boldsymbol{M}_O(t) = -\rho \int_{S(t)+\Sigma(t)} \boldsymbol{r} \times \left(\left[\frac{\partial \Phi}{\partial t} + \tfrac{1}{2}(\nabla \Phi)^2 - \nabla \Phi \cdot (\boldsymbol{v}_O + \boldsymbol{\omega} \times \boldsymbol{r}) - \boldsymbol{g} \cdot \boldsymbol{r} \right] \boldsymbol{n} \right) \mathrm{d}S. \quad (5.39)$$

Almost all the quantities on the right-hand side of eq. (5.39) are nonlinear in terms of the introduced generalized coordinates. We linearize term by term in eq. (5.39) and start with

$$-\rho \int_{S(t)+\Sigma(t)} \boldsymbol{r} \times \left(\frac{\partial \Phi}{\partial t} \boldsymbol{n} \right) \mathrm{d}S = -\rho \int_{S_0+\Sigma_0} \boldsymbol{r} \times \left(\frac{\partial \Phi}{\partial t} \boldsymbol{n} \right) \mathrm{d}S + [\text{nonlinear}] = \frac{d^*}{dt} \left[-\rho \int_{Q_0} \boldsymbol{r} \times (\nabla \Phi) \, \mathrm{d}Q \right] + [\text{nonlinear}]. \quad (5.40)$$

We have first set the time derivative outside the surface integral and then used eq. (A.6) to rewrite the surface integral as a volume integral. Furthermore,

$$\rho \int_{S(t)+\Sigma(t)} \boldsymbol{r} \times ((\boldsymbol{g} \cdot \boldsymbol{r}) \boldsymbol{n}) \, \mathrm{d}S = \rho \int_{Q(t)} \boldsymbol{r} \times (\nabla (\boldsymbol{g} \cdot \boldsymbol{r})) \, \mathrm{d}Q = \rho g \int_{Q(t)} \boldsymbol{r} \times (\eta_5 \boldsymbol{e}_1 - \eta_4 \boldsymbol{e}_2 - \boldsymbol{e}_3) \, \mathrm{d}Q + o(\varepsilon) = -M_l g \boldsymbol{r}_{lC_0} \times \boldsymbol{e}_3 + M_l g \boldsymbol{r}_{lC_0} \times (\eta_5 \boldsymbol{e}_1 - \eta_4 \boldsymbol{e}_2) - \rho g \boldsymbol{e}_1 \int_{\Sigma_0} y \zeta(x, y, t) \, \mathrm{d}S + \rho g \boldsymbol{e}_2 \int_{\Sigma_0} x \zeta(x, y, t) \, \mathrm{d}S + [\text{nonlinear}]. \quad (5.41)$$

In the transformation of eq. (5.41), we used eq. (A.6), definition (5.32) of the hydrostatic liquid mass center, and the linear approximation for g_i (i= 1, 2, 3) given by eq. (5.13). Furthermore, the liquid volume $Q(t)$ has been expressed as $Q_0 + \Delta Q$, where the volume integration over ΔQ can, as a consequence of the linearization, be expressed in terms of integrals over the mean free surface Σ_0. The rest of the terms in eq. (5.39) give nonlinear contributions.

The right-hand side of eq. (5.40) needs more simplification to get an expression similar to that for the resulting force. Inserting eq. (5.20) into the square brackets of eq. (5.40) and once more using eq. (A.6) gives

$$-\rho \int_{Q_0} \boldsymbol{r} \times \boldsymbol{v}_O \, \mathrm{d}Q - \rho \int_{Q_0} \boldsymbol{r} \times \nabla (\boldsymbol{\omega} \cdot \boldsymbol{\Omega}_0) \, \mathrm{d}Q - \rho \int_{Q_0} \boldsymbol{r} \times \nabla \left(\sum_{j=1}^{\infty} R_j \varphi_j \right) \mathrm{d}Q = -M_l \boldsymbol{r}_{lC_0} \times \boldsymbol{v}_O - \rho \int_{S_0+\Sigma_0} (\boldsymbol{\omega} \cdot \boldsymbol{\Omega}_0) \underbrace{(\boldsymbol{r} \times \boldsymbol{n})}_{\frac{\partial \boldsymbol{\Omega}_0}{\partial n}} \mathrm{d}S - \rho \int_{S_0+\Sigma_0} \underbrace{(\boldsymbol{r} \times \boldsymbol{n})}_{\frac{\partial \boldsymbol{\Omega}_0}{\partial n}} \left(\sum_{j=1}^{\infty} R_j \varphi_j \right) \mathrm{d}S. \quad (5.42)$$

The last integral in eq. (5.42) can be rewritten by using Green's first identity (A.5) and eq. (5.21) as follows:

$$-\rho \int_{\Sigma_0} \boldsymbol{\Omega}_0 \left(\sum_{j=1}^{\infty} R_j \frac{\partial \varphi_j}{\partial n} \right) \mathrm{d}S = -\sum_{j=1}^{\infty} \ddot{\beta}_j \left(\rho \int_{\Sigma_0} \boldsymbol{\Omega}_0 f_j \,\mathrm{d}S \right). \quad (5.43)$$

The second-to-last integral of eq. (5.42) reduces to a compact form by introducing the inertia tensor, $\boldsymbol{J}_0^1$, in the manner of Joukowski (1885), who studied motions of a solid body with a liquid-filled cavity. Elements of the symmetric inertia tensor are, by definition,

$$J_{0ij}^1 = \rho \int_{S_0+\Sigma_0} \Omega_{0i} \frac{\partial \Omega_{0j}}{\partial n} \mathrm{d}S. \quad (5.44)$$

Adopting eq. (5.44) computes the penultimate integral of eq. (5.42) as $-\dot{\boldsymbol{\omega}} \cdot \boldsymbol{J}_0^1$. Because $J_{0ij} = J_{0ji}$, $\dot{\boldsymbol{\omega}} \cdot \boldsymbol{J}_0^1 = \boldsymbol{J}_0^1 \cdot \dot{\boldsymbol{\omega}}$.

Accounting for eq. (5.40)–(5.44), the resulting hydrodynamic moment takes the form

$$\begin{aligned} \boldsymbol{M}_O(t) &= -M_l g \boldsymbol{r}_{lC_0} \times \boldsymbol{e}_3 + \Big[M_l g \boldsymbol{r}_{lC_0} \times (\eta_5 \boldsymbol{e}_1 - \eta_4 \boldsymbol{e}_2) \\ &\quad - M_l \boldsymbol{r}_{lC_0} \times \frac{d^* \boldsymbol{v}_O}{dt} - J_0^1 \cdot \dot{\boldsymbol{\omega}} - \rho g \boldsymbol{e}_1 \\ &\quad \times \int_{\Sigma_0} y\zeta(x,y,t)\,\mathrm{d}S + \rho g \boldsymbol{e}_2 \int_{\Sigma_0} x\zeta(x,y,t)\,\mathrm{d}S \\ &\quad - \rho \int_{\Sigma_0} \boldsymbol{\Omega}_0 \frac{\partial^2 \zeta}{\partial t^2} \mathrm{d}S \Big] + [\text{nonlinear}]. \quad (5.45) \end{aligned}$$

Equation (5.45) is exactly the same as the linearized Lukovsky formula. The first terms in eq. (5.45) represent the moment of the frozen liquid weight. The three integral terms represent the effect of sloshing. The first three terms in the square brackets are the same as in the work by Joukowski (1885) (linearized in $\boldsymbol{v}_O, \boldsymbol{\omega}$).

If a *quasi-steady approximation* is done as in Section 3.6.1, the free surface remains horizontal in the Earth-fixed coordinate system. We may then neglect $\partial^2 \zeta / \partial t^2$. Because $\int_{\Sigma_0} \zeta \,\mathrm{d}S$ must be zero due to conservation of liquid volume, $\int_{\Sigma_0} x \,\mathrm{d}S = 0$ and $\int_{\Sigma_0} y \,\mathrm{d}S = 0$ (i.e., $\zeta(x, y, t)$ can be approximated as $-y\eta_4 + x\eta_5$). This approximation leads to the following free-surface integral terms in eq. (5.45):

$$\rho g \boldsymbol{e}_1 \left(\eta_4 \int_{\Sigma_0} y^2 \,\mathrm{d}S - \eta_5 \int_{\Sigma_0} yx \,\mathrm{d}S \right) + \rho g \boldsymbol{e}_2 \left(\eta_5 \int_{\Sigma_0} x^2 \,\mathrm{d}S - \eta_4 \int_{\Sigma_0} yx \,\mathrm{d}S \right). \quad (5.46)$$

These results are consistent with the results in Section 3.6.1, where we only studied the effect of roll and pointed out that the free-surface integral in terms of the area moment of inertia of the tank's mean free-surface plane has a quasi-steady destabilizing effect on roll.

Formula (5.45) can be rewritten in terms of the introduced generalized coordinates:

$$\begin{aligned} \boldsymbol{M}_O(t) &= \underbrace{\left[M_l g (x_{lC_0} \boldsymbol{e}_2 - y_{lC_0} \boldsymbol{e}_1) \right]}_{\text{static moment}} \\ &+ \boldsymbol{e}_1 \underbrace{\left[M_l (g z_{lC_0} \eta_4 + z_{lC_0} \ddot{\eta}_2 - y_{lC_0} \ddot{\eta}_3) - \sum_{k=4}^{6} J_{01(k-3)}^1 \ddot{\eta}_k - \sum_{j=1}^{\infty} (g \lambda_{2j} \beta_j + \lambda_{01j} \ddot{\beta}_j) \right]}_{M_{O1}(t) = F_4(t)} \\ &+ \boldsymbol{e}_2 \underbrace{\left[M_l (g z_{lC_0} \eta_5 + x_{lC_0} \ddot{\eta}_3 - z_{lC_0} \ddot{\eta}_1) - \sum_{k=4}^{6} J_{02(k-3)}^1 \ddot{\eta}_k - \sum_{j=1}^{\infty} (-g \lambda_{1j} \beta_j + \lambda_{02j} \ddot{\beta}_j) \right]}_{M_{O2}(t) = F_5(t)} \\ &+ \boldsymbol{e}_3 \underbrace{\left[M_l \left(-g (x_{lC_0} \eta_4 + y_{lC_0} \eta_5) + y_{lC_0} \ddot{\eta}_1 - x_{lC_0} \ddot{\eta}_2 \right) - \sum_{k=4}^{6} J_{03(k-3)}^1 \ddot{\eta}_k - \sum_{j=1}^{\infty} \lambda_{03j} \ddot{\beta}_j \right]}_{M_{O3}(t) = F_6(t)}. \quad (5.47) \end{aligned}$$

We should recall that the j-summation changes to a sum over two (or more) indices for some three-dimensional shapes when the natural modes are naturally parameterized by a set of integer subscripts. One can see that the computations of the hydrodynamic moment require hydrodynamic coefficients (5.26) as well as the symmetric inertia tensor with elements (5.44). In many practical cases, due to special properties of the inertia tensor, formula (5.47) has many zero terms and therefore becomes simpler. Further simplifications of the formulas can be made in choosing the origin O in such a way that $x_{lC_0} = y_{lC_0} = 0$.

The special case of a *completely filled tank* (i.e., $\beta_j = 0$), gives

$$\boldsymbol{M}_O^{\text{filled}}(t) = \underbrace{[M_l g(x_{lC_0}\boldsymbol{e}_2 - y_{lC_0}\boldsymbol{e}_1)]}_{\text{static moment}} + \boldsymbol{e}_1 \underbrace{\left[M_l(g z_{lC_0}\eta_4 + z_{lC_0}\ddot{\eta}_2 - y_{lC_0}\ddot{\eta}_3) - \sum_{k=4}^{6} J^1_{01(k-3)}\ddot{\eta}_k \right]}_{M_{O1}^{\text{filled}}(t)=F_4^{\text{filled}}(t)}$$
$$+ \boldsymbol{e}_2 \underbrace{\left[M_l(g z_{lC_0}\eta_5 + x_{lC_0}\ddot{\eta}_3 - z_{lC_0}\ddot{\eta}_1) - \sum_{k=4}^{6} J^1_{02(k-3)}\ddot{\eta}_k \right]}_{M_{O2}^{\text{filled}}(t)=F_5^{\text{filled}}(t)}$$
$$+ \boldsymbol{e}_3 \underbrace{\left[M_l(-g(x_{lC_0}\eta_4 + y_{lC_0}\eta_5) + y_{lC_0}\ddot{\eta}_1 - x_{lC_0}\ddot{\eta}_2) - \sum_{k=4}^{6} J^1_{03(k-3)}\ddot{\eta}_k \right]}_{M_{O3}^{\text{filled}}(t)=F_6^{\text{filled}}(t)}. \tag{5.48}$$

In contrast to the hydrodynamic force, formula (5.48) shows that the hydrodynamic moment in a completely filled tank *cannot be related to a "frozen" liquid*, which is an important classical result from theoretical hydrodynamics (Moiseev & Rumyantsev, 1968). Corrections are associated with the inertia tensor $\boldsymbol{J}_0^1$ by eq. (5.44), which is not the same as the inertia tensor $\boldsymbol{I}^0$ for the frozen liquid. This inertia tensor for the frozen liquid is, by definition, given by

$$\boldsymbol{I}^0 = \begin{Vmatrix} I^0_{11} & I^0_{12} & I^0_{13} \\ I^0_{21} & I^0_{22} & I^0_{23} \\ I^0_{31} & I^0_{32} & I^0_{33} \end{Vmatrix} = \rho \begin{Vmatrix} \int_{Q_0}(y^2+z^2)\,dQ & -\int_{Q_0} xy\,dQ & -\int_{Q_0} xz\,dQ \\ -\int_{Q_0} xy\,dQ & \int_{Q_0}(x^2+z^2)\,dQ & -\int_{Q_0} yz\,dQ \\ -\int_{Q_0} xz\,dQ & -\int_{Q_0} zy\,dQ & \int_{Q_0}(y^2+x^2)\,dQ \end{Vmatrix}. \tag{5.49}$$

Therefore, to get the moment for the frozen liquid, we should replace J^1_{0ij} by I^0_{ij} in eq. (5.48). We can make an analogy between a hard-boiled egg and an egg that is not boiled to understand that the inertia tensor for the frozen liquid is not the same as for liquid in a completely filled tank. When the hydrodynamic moment is used in a coupled analysis of ship motions and sloshing, we need to evaluate the moment relative to the coordinate system used in the global ship motion analysis. We can then use the following general transformation. If P is a fixed point in the $Oxyz$ system, the recalculations can be done by utilizing the formula

$$\boldsymbol{M}_P = (F_4|_P, F_5|_P, F_6|_P) = \boldsymbol{r}_{PO} \times \boldsymbol{F} + \boldsymbol{M}_O, \tag{5.50}$$

where $\boldsymbol{F}$ and $\boldsymbol{M}_O$ are computed by eqs. (5.36) and (5.47), respectively, and $\boldsymbol{r}_{PO}$ is the radius vector of origin O with respect to P. The symbol P is introduced in the moment components to distinguish them from F_4, F_5, and F_6, which are defined relative to the origin of our coordinate system.

5.3.3 Steady-state and transient motions: initial and periodicity conditions

Linear modal theory is used to describe either steady-state or transient waves due to external prescribed forcing or coupling with global body (ship) motions. *Steady-state solutions* mean that the forcing terms $K_m(t)$ are T-periodic ($K_m(t+T) = K_m(t)$) and the problem consists of identifying the periodic solutions of system (5.27):

$$\beta_m(t+T) = \beta_m(t), \quad m = 1, 2, \ldots. \tag{5.51}$$

Transient analysis requires an initial free-surface position and initial values of the velocity potential on the mean free surface. The corresponding initial conditions for eqs. (5.27) follow from initial conditions (2.76) after substitution of eq. (5.19) and by using orthogonality condition (5.22), which gives

$$\beta_{0m} = \frac{\int_{\Sigma_0} f_m \zeta \, \mathrm{d}S}{\int_{\Sigma_0} f_m^2 \, \mathrm{d}S},$$

$$\beta_{1m} = \frac{\int_{\Sigma_0} f_m \, \partial\zeta/\partial t \, \mathrm{d}S}{\int_{\Sigma_0} f_m^2 \, \mathrm{d}S}, \quad m = 1, 2, \ldots \tag{5.52}$$

in the commonly used initial conditions

$$\beta_m(0) = \beta_{0m}, \quad \dot{\beta}_m(0) = \beta_{1m}, \quad m = 1, 2, \ldots. \tag{5.53}$$

As we discussed in Section 2.4.2.5, a prediction of the initial shape and velocities (associated with ζ and $\partial\zeta/\partial t$, respectively) is in practical cases a complicated task. In the majority of publications, these values are assumed to be zero (i.e., the liquid is assumed to be in a static state at the initial time). Section 2.4.2.5 also discusses so-called impulsive initial conditions (2.77). For these conditions, the initial liquid shape is flat, but the initial velocities are determined from the equality $\Phi(x, y, 0, 0) = 0$. Remembering representation (5.20), we get

$$v_{O1}(0)x + v_{O2}(0)y + \boldsymbol{\omega}(0) \cdot \boldsymbol{\Omega}_0(x, y, 0) + \sum_{m=1}^{\infty} \dot{\beta}_m(0)\kappa_m^{-1} f_m(x, y) = 0, \tag{5.54}$$

which gives

$$\beta_m(0) = 0, \quad \dot{\beta}_m(0) = -\frac{\kappa_m \int_{\Sigma_0} f_m \left[v_{O1}(0)\, x + v_{O2}(0)\, y + \boldsymbol{\omega}(0) \cdot \boldsymbol{\Omega}_0(x, y, 0)\right] \mathrm{d}S}{\int_{\Sigma_0} f_m^2 \, \mathrm{d}S}, \quad m = 1, 2, \ldots. \tag{5.55}$$

The liquid motion will in reality be *damped*. A convenient way to account for this fact is to introduce additional terms, $2\xi_m \sigma_m \dot{\beta}_m$, so that eqs. (5.27) take the form

$$\ddot{\beta}_m + 2\xi_m \sigma_m \dot{\beta}_m + \sigma_m^2 \beta_m = K_m(t), \quad m = 1, 2, \ldots. \tag{5.56}$$

The reason why there is no damping in eqs. (5.27) is the potential flow assumption. In reality, damping exists due to viscous effects, particular because of the boundary-layer flow along the wetted tank surface for a clean tank without interior structures (see Sections 6.2 and 6.3). However, interior structures such as baffles, screens, and poles can introduce a more significant nonlinear damping effect due to flow separation. Furthermore, turbulent energy dissipation in breaking waves can cause considerable damping. Strictly speaking, we cannot justify the way that we introduced damping in eq. (5.56). If it were correct, it would have consequences for the free-surface condition that would not be physically correct. The reason is that eq. (5.56) with $\xi_m = 0$ is a consequence of satisfying the free-surface conditions. If we add terms, we do not satisfy the linear free-surface conditions based on potential flow theory. Therefore, the only way we can argue for eq. (5.56) is that it is simple and can simulate correctly the effect of damping in terms of, for instance, damping rates of natural modes. This fact is demonstrated in Chapter 6.

If damping is neglected, eqs. (5.27) have the following solution:

$$\beta_m(t) = A_m \cos(\sigma_m t) + B_m \sin(\sigma_m t) + \sigma_m^{-1} \int_0^t K_m(\tau) \sin[\sigma_m(t-\tau)] \, \mathrm{d}\tau, \tag{5.57}$$

where constants A_m and B_m follow from imposing initial conditions (5.53).

If transient loading of duration T_d is considered, the ratio T_d/T_m plays an important role (see, e.g., Clough & Penzien, 1993), where $T_m = 2\pi/\sigma_m$ is the natural period of mode m.

If $T_d/T_m \lesssim 0.25$, the force impulse

$$I_m = \int_0^{T_d} K_m(t)\,\mathrm{d}t \tag{5.58}$$

determines the maximum response.

If the forcing is harmonic with a frequency $\sigma \neq \sigma_m$, solution (5.57) will never reach the steady-state (periodic) solution. Indeed, if

$$K_m(t) = \sigma^2 P'_m \cos(\sigma t), \tag{5.59}$$

the mentioned *steady-state solution (regime)* is associated with the $(2\pi/\sigma = T)$-periodic solution

$$\beta_m(t) = \frac{P'_m\sigma^2}{\sigma_m^2 - \sigma^2}\cos(\sigma t), \tag{5.60}$$

while general solution (5.57) is equal to

$$\beta_m(t) = a'_m \cos(\sigma_m t) + b'_m \sin(\sigma_m t) + \frac{P'_m\sigma^2}{\sigma_m^2 - \sigma^2}\cos(\sigma t). \tag{5.61}$$

Obviously, solution (5.61) becomes T-periodic, if and only if $a'_m = b'_m = 0$. This outcome is realized only with specific initial values $\beta_m(0) = P'_m\sigma^2/(\sigma_m^2 - \sigma^2)$ and $\dot\beta_m(0) = 0$. However, we can also directly consider periodic solutions with frequency σ as the steady-state wave motions.

If the initial conditions $\beta_m(0) = \dot\beta_m(0) = 0$ are used, it follows from eq. (5.61) that

$$\beta_m(t) = \frac{P'_m\sigma^2}{\sigma_m^2 - \sigma^2}(\cos(\sigma t) - \cos(\sigma_m t)), \tag{5.62}$$

which means that the transient solution contains two harmonics (i.e., σ and σ_m). Again, the physical reason is that our system has no damping. Solution (5.61) causes so-called beating, demonstrated in Figure 5.2. Beating is characterized by larger maximum magnitudes with respect to those in steady-state regimes and a so-called beating period T_b as shown in Figure 5.2. The maximum magnitude becomes infinite as $\sigma \to \sigma_m$ (resonant case). This limit also causes an infinite beating period. Let us elaborate more from a mathematical point of view on the behavior of the solution. When $\sigma = \sigma_m$, the formal limit in eq. (5.62) gives

$$\beta_m|_{\sigma=\sigma_m} = \lim_{\sigma\to\sigma_m} \frac{P'_m\sigma^2}{\sigma_m^2 - \sigma^2}[\cos(\sigma t) - \cos(\sigma_m t)] = \tfrac{1}{2}P'_m\sigma_m t\sin(\sigma_m t) \tag{5.63}$$

and, therefore, the amplitude of oscillations tends to infinity linearly with time. The beating can mathematically be illustrated by rewriting eq. (5.62) in terms of a slowly varying amplitude. Let us consider $\sigma = \sigma_m + \delta$, where $0 < |\delta| \ll 1$. Then

$$\beta_m(t) = \frac{2P'_m\sigma^2}{\delta(\sigma_m + \sigma)}\sin\left(\tfrac{1}{2}\delta t\right)\sin\left[\left(\sigma_m - \tfrac{1}{2}\delta\right)t\right] \approx \frac{P'_m\sigma^2}{\delta\sigma_m}\sin(\tfrac{1}{2}\delta t)\sin(\sigma_m t). \tag{5.64}$$

Indeed, when $0 < t \ll \delta^{-1}$, we confirm eq. (5.63) because $\delta^{-1}\sin(\frac{1}{2}\delta t) \approx \frac{1}{2}t$. However, focusing on a situation that occurs for larger time (e.g., $t \geq \delta^{-1}$), one can see that the term $\delta^{-1}\sin(\frac{1}{2}\delta t)$ begins oscillating with large period $4\pi/\delta$ and amplitude δ^{-1}. This result is exactly what we see as modulated waves in Figure 5.2.

Results for harmonically forced oscillators can be generalized to the case when the modal equations include damping (i.e., for modal equations (5.56)). The general solution then takes the form

$$\beta_m(t) = \exp(-\xi_m\sigma_m t)\left[a_m\cos(\sigma'_m t) + b_m\sin(\sigma'_m t)\right] + c_m\cos(\sigma t) + d_m\sin(\sigma t), \tag{5.65}$$

where

$$\sigma'_m = \sqrt{1 - \xi_m^2}\,\sigma_m, \quad c_m = \frac{P'_m\sigma^2\left(\sigma_m^2 - \sigma^2\right)}{\left[\sigma_m^2 - \sigma^2\right]^2 + 4\xi_m^2\sigma_m^2\sigma^2},$$

$$d_m = \frac{2\alpha_m\sigma_m\sigma^3 P'_m}{\left[\sigma_m^2 - \sigma^2\right]^2 + 4\xi_m^2\sigma_m^2\sigma^2}.$$

The two coefficients a_m and b_m are found from the initial conditions.

Because the damping is in reality nonzero due to energy dissipation, the first term in solution (5.65) vanishes as $t \to \infty$ and, therefore, solution (5.65) tends to

$$\beta_m(t) = \frac{P'_m\sigma^2}{\left[\sigma_m^2 - \sigma^2\right]^2 + 4\xi_m^2\sigma_m^2\sigma^2} \times \left[\left(\sigma_m^2 - \sigma^2\right)\cos(\sigma t) + 2\xi_m\sigma_m\sigma\sin(\sigma t)\right]. \tag{5.66}$$

If no wave breaking occurs and no interior structures obstruct the flow and cause flow separation, the damping coefficients for the lower modes are small under linear flow conditions. As a consequence, solution (5.66) is close to eq. (5.60) for large time and we may, for brevity, use steady-state expression (5.60) to estimate the wave amplitude response and the hydrodynamic loads. However, the equation demonstrates that

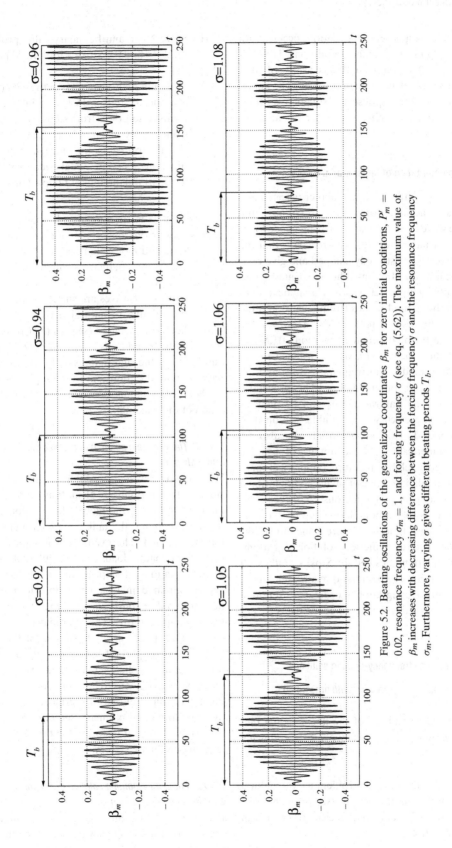

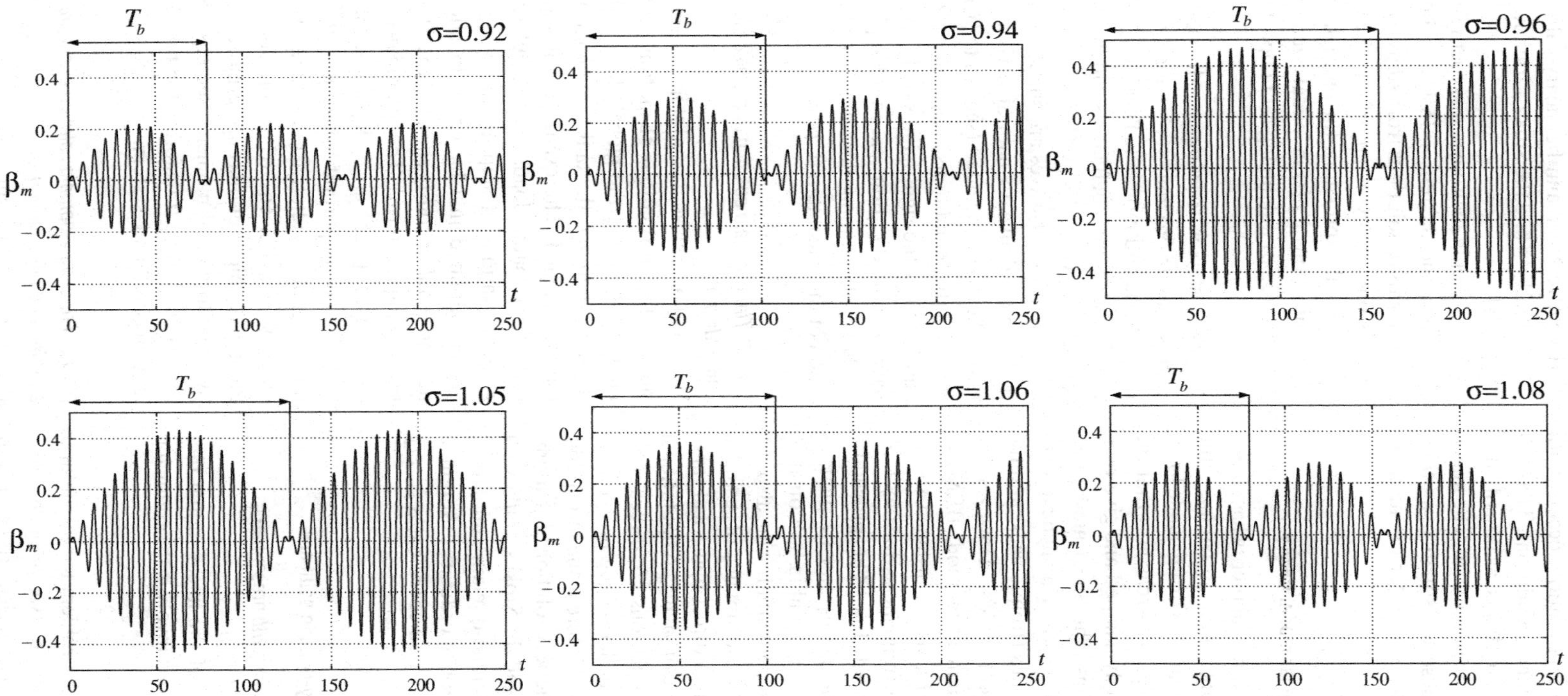

Figure 5.2. Beating oscillations of the generalized coordinates β_m for zero initial conditions, $P'_m = 0.02$, resonance frequency $\sigma_m = 1$, and forcing frequency σ (see eq. (5.62)). The maximum value of β_m increases with decreasing difference between the forcing frequency σ and the resonance frequency σ_m. Furthermore, varying σ gives different beating periods T_b.

very large response amplification occurs with small damping at resonance conditions. The consequence is that nonlinear effects, as described in Chapters 8 and 9, must be introduced. The analysis shows that the response at resonant conditions is limited due to nonlinear transfer of energy to other modes.

5.4 Implementation of linear modal theory

Practical use of the linear modal theory has shown reasonable accuracy when used to compute the surface waves and associated hydrodynamic loads with a few natural modes. The reason is that the higher modes are strongly damped and give relatively small contributions. The linear modal theory is especially efficient when analytical solutions of the natural modes (frequencies) and the Stokes–Joukowski potential exist. However, we may use a boundary element method (BEM) or finite element method (FEM), for instance, to find the natural modes and frequencies for any tank shape based on potential flow. The procedure of how to find natural modes and frequencies with a BEM is discussed in detail in Section 10.2. The Stokes–Joukowski potential can also be found numerically. The advantage of using a modal theory relative to a direct numerical time-domain simulation is one of saving computational time for long time simulations. A direct solution requires that a boundary-value problem be solved for each time step whereas the modal formulation requires numerical solutions for a limited number of modes and the Stokes–Joukowski potential prior to the time-domain simulation. The numerical integration of ordinary differential equations for the generalized coordinates is very fast.

5.4.1 Time- and frequency-domain solutions

5.4.1.1 Time-domain solution with prescribed *tank motion*

The derivation and use of the linear modal theory for *prescribed* motions involves the following steps:

(i) Solve the spectral problem for the natural modes (4.2). Seven to twelve of the lowest modes and frequencies are needed in practice.
(ii) Solve Neumann boundary-value problem (5.21) to find the Stokes–Joukowski potential.
(iii) Compute hydrodynamic coefficients (5.26) and inertia tensor (5.44) as integrals in terms of solutions from steps (i) and (ii).
(iv) Compute by means of eq. (5.25) the analytical expressions for the right-hand-side functions $K_m(t)$ in modal equations (5.27) or (5.56) if damping is of importance.

The final step is to compute all coefficients in expressions (5.36) and (5.47) for the hydrodynamic force and moment. These are functions of the hydrodynamic coefficients and the inertia tensor (see step (iii)).

When transient waves are simulated on a long time scale relative to natural periods, damping terms should be included (see eq. (5.56)). Examples of initial conditions are given by eqs. (5.53) and (5.55). If the right-hand-side functions $K_m(t)$ have a short duration and the damping is relatively small, damping does not matter for calculation of the maximum response.

5.4.1.2 Time-domain solution of coupled sloshing and body motion

When coupled sloshing and ship motions are considered, transformations of motions, forces, and moments between the global ship motion coordinate system and the tank-fixed coordinate system are needed. The evaluation of external linear hydrodynamic loads and formulation of the global equations of ship motion were discussed and exemplified in Chapter 3. In addition, we need to simultaneously solve eq. (5.25) with additional damping terms. The hydrodynamic coefficients in eq. (5.25) are independent of the tank motions. An important aspect is that eq. (5.25) involves motions relative to the tank-fixed coordinate system and needs to be expressed in terms of the global ship motion. Furthermore, the hydrodynamic force and moment (see eqs. (5.36) and (5.47)) need to be transformed to the global ship motion coordinate system.

5.4.1.3 Frequency-domain solution of coupled sloshing and body motion

If coupled sloshing and ship (platform) motions are solved in the frequency domain, the effect of sloshing may be represented in terms of either added mass or restoring coefficients. In fact a combination of added mass and restoring coefficients is also used. If we follow the general procedure described in Section 3.5, we start by studying forced harmonic oscillations of the tank for a given mode and then examine the resulting steady-state linear hydrodynamic force and moment components on the ship due to sloshing. We leave out *two important steps* in the following analysis that we revisit in Section 5.4.3.2: (1) in the equations of ship motion we must refer to the motions in the global *inertial* coordinate systems and (2) the hydrodynamic tank forces and moments must be transformed to the global *inertial* coordinate system.

To be consistent with the previous analysis, we first operate in the *tank-fixed* coordinate system and express motion mode j as

$$\eta_j(t) = \eta_{ja}\cos(\sigma t + \varepsilon_j), \quad j = 1, \dots, 6. \quad (5.67)$$

The resulting flow in the tank follows from the previous analysis. Damping is neglected. The flow is completely defined by modal functions β_j, which are solutions of linear modal equations (5.27), where the forcing term $K_m(t)$ is expressed as

$$K_m(t) = \sigma^2 \underbrace{\tilde{P}_{mj}\eta_{ja}}_{P'_{mj}}\cos(\sigma t + \varepsilon_j). \quad (5.68)$$

The amplitude coefficients P'_{mj} can, by means of eq. (5.28), be expressed as

$$\begin{aligned}
&\tilde{P}_{m1} = \lambda_{1m}/\mu_m, \quad \tilde{P}_{m2} = \lambda_{2m}/\mu_m, \quad \tilde{P}_{m3} = 0,\\
&\tilde{P}_{m4} = (-g\lambda_{2m}\sigma^{-2} + \lambda_{01m})/\mu_m,\\
&\tilde{P}_{m5} = (g\lambda_{1m}\sigma^{-2} + \lambda_{02m})/\mu_m, \quad \tilde{P}_{m6} = \lambda_{03m}/\mu_m.
\end{aligned} \quad (5.69)$$

The coefficients $\tilde{P}_{m4}$ and $\tilde{P}_{m5}$ become formally infinite as the forcing frequency σ vanishes, but the forcing term $K_m(t)$ remains finite, because these amplitude terms are multiplied by σ^2 in eq. (5.68).

Furthermore, linear modal equations (5.27) with the right-hand side of eq. (5.68) have the analytical solution given by eq. (5.60), meaning that we have found the tank flow. The resulting steady-state hydrodynamic force and moment components $F_k(k = 1, \dots, 6)$ acting on the tank follow by inserting the solution into formulas (5.38) and (5.47). The components are henceforth considered in the *inertial* coordinate system and, therefore, $k = 1, 2, 3$ indicates the force components along the axes x', y', and z', and $k = 4, 5, 6$ indicates the moment components around these axes. The added mass coefficients, A_{kj}, are defined by the relationship

$$F_k(t) = -A_{kj}\ddot{\eta}_j(t) = A_{kj}[\sigma^2\eta_{ja}\cos(\sigma t + \varepsilon_j)], \quad k, j = 1, \dots, 6, \quad (5.70)$$

or we could instead have defined restoring coefficients C_{kj} by $F_k = -C_{kj}\eta_j$ (i.e., $-\sigma^2 A_{kj} = C_{kj}$). When it comes to the final effect of the mutual interaction between steady-state sloshing and ship motion, it does not make any difference if we use either A_{kj} or C_{kj}. We present the added mass coefficients for three cases: (1) the liquid is frozen (i.e., we consider an equivalent solid body of the same geometry and density as the unperturbed liquid – an equivalent solid cargo); (2) the tank is completely filled by the liquid; and (3) a free surface exists with linear sloshing.

As we remarked before eq. (5.49), the added mass coefficients for cases 1 and 2 are not the same because of different inertia tensors. For the case of the "frozen" liquid, the use of the formulas (5.37) and (5.48) (with inertia tensor $\boldsymbol{I}^0$ instead of $\boldsymbol{J}_0^1$) gives the added mass coefficients

$$\begin{aligned}
&A_{11}^{\text{frozen}} = M_l, \quad A_{22}^{\text{frozen}} = M_l, \quad A_{33}^{\text{frozen}} = M_l,\\
&A_{44}^{\text{frozen}} = M_l\sigma^{-2}gz_{lC_O} + I_{11}^0,\\
&A_{55}^{\text{frozen}} = \sigma^{-2}M_lgz_{lC_O} + I_{22}^0,\\
&A_{15}^{\text{frozen}} = M_lz_{lC_0}, \quad A_{16}^{\text{frozen}} = -M_ly_{lC_0},\\
&A_{24}^{\text{frozen}} = -M_lz_{lC_0}, \quad A_{26}^{\text{frozen}} = M_lx_{lC_0},\\
&A_{34}^{\text{frozen}} = M_ly_{lC_0}, \quad A_{35}^{\text{frozen}} = -M_lx_{lC_0},\\
&A_{42}^{\text{frozen}} = -M_lz_{lC_O}, \quad A_{43}^{\text{frozen}} = M_ly_{lC_O},\\
&A_{51}^{\text{frozen}} = M_lz_{lC_O}, \quad A_{53}^{\text{frozen}} = -M_lx_{lC_O},\\
&A_{61}^{\text{frozen}} = -M_ly_{lC_0}, \quad A_{62}^{\text{frozen}} = M_lx_{lC_0},\\
&A_{12}^{\text{frozen}} = A_{21}^{\text{frozen}} = A_{13}^{\text{frozen}} = A_{31}^{\text{frozen}} = A_{14}^{\text{frozen}}\\
&\qquad = A_{41}^{\text{frozen}} = A_{23}^{\text{frozen}} = A_{32}^{\text{frozen}} = A_{25}^{\text{frozen}}\\
&\qquad = A_{52}^{\text{frozen}} = A_{36}^{\text{frozen}} = A_{63}^{\text{frozen}} = 0,\\
&A_{64}^{\text{frozen}} = -M_lgx_{lC_0}\sigma^{-2} + I_{31}^0, \quad A_{65}^{\text{frozen}}\\
&\qquad = -M_lgy_{lC_0}\sigma^2 + I_{32}^0, \quad A_{45}^{\text{frozen}} = I_{12}^0,\\
&A_{54}^{\text{frozen}} = I_{21}^0, \quad A_{46}^{\text{frozen}} = I_{13}^0,\\
&A_{56}^{\text{frozen}} = I_{23}^0, \quad A_{66}^{\text{frozen}} = I_{33}^0,
\end{aligned} \quad (5.71)$$

where the inertia tensor $\boldsymbol{I}^0$ is defined by eq. (5.49). We must be aware that the preceding coefficients are *relative to an inertial coordinate system*. If we had operated with a *tank-fixed* coordinate system, the additional terms $M_l g\sigma^{-2}$ in A_{15}^{frozen} and $-M_l g\sigma^{-2}$ in A_{24}^{frozen} would appear as a consequence of the weight components along the x- and y-axes. The $g\sigma^{-2}$ terms appearing for A_{44}^{frozen}, A_{55}^{frozen}, A_{65}^{frozen}, and A_{64}^{frozen} are due to a quasi-steady moment. The added mass for the *completely filled tank* may be obtained from eq. (5.71) by replacing tensor $\boldsymbol{I}^0$ with $\boldsymbol{J}_0^1$. Therefore,

$$A_{(3+k)(3+j)}^{\text{filled}} = A_{(3+k)(3+j)}^{\text{frozen}} + \left(J_{0kj}^1 - I_{kj}^0\right), \quad k, j = 1, 2, 3;$$
$$A_{kj}^{\text{filled}} = A_{kj}^{\text{frozen}} (\text{when } k \text{ or } j \le 3). \tag{5.72}$$

Finally, *sloshing* will also *correct* the added mass coefficients as

$$A_{ij} = A_{ij}^{\text{filled}} + A_{ij}^{\text{slosh}}, \tag{5.73}$$

where

$$A_{1k}^{\text{slosh}} = \sum_{m=1}^{\infty} \lambda_{1m} \frac{\tilde{P}_{mk}\sigma^2}{\sigma_m^2 - \sigma^2},$$
$$A_{2k}^{\text{slosh}} = \sum_{m=1}^{\infty} \lambda_{2m} \frac{\tilde{P}_{mk}\sigma^2}{\sigma_m^2 - \sigma^2},$$
$$A_{3k}^{\text{slosh}} = 0, \quad k = 1, \dots 6,$$
$$A_{43}^{\text{slosh}} = A_{53}^{\text{slosh}} = A_{63}^{\text{slosh}} = 0,$$
$$A_{4(3+j)}^{\text{slosh}} = \sum_{m=1}^{\infty} \left(-\frac{g\lambda_{2m}}{\sigma^2} + \lambda_{01m}\right) \frac{\tilde{P}_{m(3+j)}\sigma^2}{\sigma_m^2 - \sigma^2},$$
$$A_{5(3+j)}^{\text{slosh}} = \sum_{m=1}^{\infty} \left(\frac{g\lambda_{1m}}{\sigma^2} + \lambda_{02m}\right) \frac{\tilde{P}_{m(3+j)}\sigma^2}{\sigma_m^2 - \sigma^2},$$
$$A_{6(3+j)}^{\text{slosh}} = \sum_{m=1}^{\infty} \frac{\lambda_{03m}\tilde{P}_{m(3+j)}\sigma^2}{\sigma_m^2 - \sigma^2}, \quad j = 1, 2, 3,$$
$$A_{41}^{\text{slosh}} = \sum_{m=1}^{\infty} \left(-\frac{g\lambda_{2m}}{\sigma^2} + \lambda_{01m}\right) \frac{\tilde{P}_{m1}\sigma^2}{\sigma_m^2 - \sigma^2},$$
$$A_{42}^{\text{slosh}} = \sum_{m=1}^{\infty} \left(-\frac{g\lambda_{2m}}{\sigma^2} + \lambda_{01m}\right) \frac{\tilde{P}_{m2}\sigma^2}{\sigma_m^2 - \sigma^2},$$
$$A_{51}^{\text{slosh}} = \sum_{m=1}^{\infty} \left(\frac{g\lambda_{1m}}{\sigma^2} + \lambda_{02m}\right) \frac{\tilde{P}_{m1}\sigma^2}{\sigma_m^2 - \sigma^2},$$
$$A_{52}^{\text{slosh}} = \sum_{m=1}^{\infty} \left(\frac{g\lambda_{1m}}{\sigma^2} + \lambda_{02m}\right) \frac{\tilde{P}_{m2}\sigma^2}{\sigma_m^2 - \sigma^2},$$
$$A_{61}^{\text{slosh}} = \sum_{m=1}^{\infty} \frac{\lambda_{03m}\tilde{P}_{m1}\sigma^2}{\sigma_m^2 - \sigma^2}, \quad A_{62}^{\text{slosh}} = \sum_{m=1}^{\infty} \frac{\lambda_{03m}\tilde{P}_{m2}\sigma^2}{\sigma_m^2 - \sigma^2}. \tag{5.74}$$

The coefficients in the expressions for the added mass coefficients, eqs. (5.74), are defined by eqs. (5.69) and (5.26) and can be evaluated when the Stokes–Joukowski potential, natural modes, and frequencies are known. Accounting for eq. (5.69), we can see that $A_{k3}^{\text{slosh}} = 0$ $(k = 1, \dots, 6)$. This fact means that heave cannot excite sloshing in a linear model. When the tank has the x–z-plane as a symmetry plane, the surge, heave, and pitch are uncoupled from sway, roll, and yaw. This means that A_{12}, A_{14}, A_{16}, A_{21}, A_{23}, A_{25}, A_{32}, A_{34}, A_{36}, A_{41}, A_{43}, A_{45}, A_{52}, A_{54}, A_{56}, A_{61}, A_{63}, and A_{65} are zero.

The superscript "filled" used for A_{ij}^{filled} in eq. (5.73) means fictitious lid is placed on the mean free surface. The A_{ij}^{slosh} coefficients contain the quasi-steady hydrostatic effect of the fact that the free surface remains horizontal in the Earth-fixed coordinate system. The latter terms can be expressed by eq. (5.46) as

$$A_{44}^{\text{quasi}} = \rho g\sigma^{-2} \int_{\Sigma_0} y^2\, dS, \quad A_{45}^{\text{quasi}} = A_{54}^{\text{quasi}} = -\rho g\sigma^{-2} \int_{\Sigma_0} yx\, dS,$$
$$A_{55}^{\text{quasi}} = \rho g\sigma^{-2} \int_{\Sigma_0} x^2\, dS. \tag{5.75}$$

The other terms are zero. The coefficients become infinite as $\sigma \to 0$. Accounting for eq. (5.69) in expressions (5.74), we see that values of A_{ij}^{slosh} are finite as σ tends to zero except for A_{44}^{slosh}, A_{45}^{slosh}, and A_{55}^{slosh}, which are proportional to σ^{-2}. The asymptotics is

$$A_{44}^{\text{slosh}} = \sigma^{-2} g^2 \sum_{m=1}^{\infty} \frac{\lambda_{2m}^2}{\sigma_m^2 \mu_m} + O(1);$$
$$A_{55}^{\text{slosh}} = \sigma^{-2} g^2 \sum_{m=1}^{\infty} \frac{\lambda_{1m}^2}{\sigma_m^2 \mu_m} + O(1),$$
$$A_{45}^{\text{slosh}} = A_{54}^{\text{slosh}} = -\sigma^{-2} g^2 \sum_{m=1}^{\infty} \frac{\lambda_{1m}\lambda_{2m}}{\sigma_m^2 \mu_m} + O(1). \tag{5.76}$$

We can show that the main terms in eq. (5.76) are the same as in eq. (5.75) by inserting the definitions for hydrodynamic coefficients (5.26) and using the Fourier series for x and y:

$$x = \sum_{m=1}^{\infty} \frac{\int_{\Sigma_0} x\varphi_m\, dS}{\int_{\Sigma_0} \varphi_m^2\, dS} \varphi_m; \quad y = \sum_{m=1}^{\infty} \frac{\int_{\Sigma_0} y\varphi_m\, dS}{\int_{\Sigma_0} \varphi_m^2\, dS} \varphi_m.$$

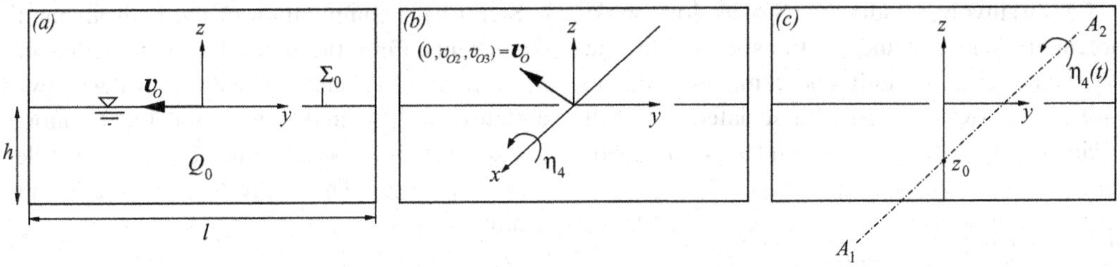

Figure 5.3. Forced motion of a two-dimensional rectangular tank: (a) sway excitation; (b) most general excitation with sway velocity v_{O2}, heave velocity v_{O3}, and roll η_4; (c) angular excitations $\eta_4(t)$ about an axis through the point $(0, z_0)$.

For instance,

$$g^2 \sum_{m=1}^{\infty} \frac{\lambda_{2m}^2}{\sigma_m^2 \mu_m} = g\rho \sum_{m=1}^{\infty} \frac{\int_{\Sigma_0} y\varphi_m \,\mathrm{d}S \int_{\Sigma_0} y\varphi_m \,\mathrm{d}S}{\int_{\Sigma_0} \varphi_m^2 \,\mathrm{d}S}$$

$$= g\rho \int_{\Sigma_0} y \left(\sum_{m=1}^{\infty} \frac{\int_{\Sigma_0} y\varphi_m \,\mathrm{d}S}{\int_{\Sigma_0} \varphi_m^2 \,\mathrm{d}S} \varphi_m \right) \mathrm{d}S.$$

The use of eqs. (5.71)–(5.74) has a clear advantage when analytical expressions can be derived that are, for instance, true for two- and three-dimensional rectangular tanks and circular upright cylinders. In Section 5.4.3 we present explicit expressions for the added mass coefficients of a three-dimensional rectangular tank. It is an unnecessary detour for a frequency-domain solution to numerically determine the Stokes–Joukowski potential, natural modes, and frequencies for tank shapes where analytical solutions are not known. What we should do is directly solve the added mass problem.

The added mass coefficients are frequency dependent with either positive or negative values and are infinite when $\sigma = \sigma_n$ (where n is a fixed integer). The singular behavior is $O[(\sigma - \sigma_n)^{-1}]$ as $\sigma \to \sigma_n$. Even though the product of two added mass coefficients is $O[(\sigma - \sigma_n)^{-2}]$ as $\sigma \to \sigma_n$, combinations of products of the added mass coefficients may lead to a singular behavior $O[(\sigma - \sigma_n)^{-1}]$ as $\sigma \to \sigma_n$. In particular, Newman (personal communication, 2008) showed by simulations (and derivations for a rectangular tank) that the singularity of $A_{44}A_{22} - A_{24}^2$ is $O[(\sigma - \sigma_n)^{-1}]$. This fact was essential when Newman explained why the coupled sway and roll of the hemispheroid presented in Figure 3.25 are finite at the resonance frequency of sloshing.

Using expressions (5.74) with eq. (5.69) offers a straightforward explanation of this point. Indeed, the primary singular terms of the added mass at $\sigma = \sigma_n$ are

$$A_{44} = \frac{1}{\mu_n} \left(-\frac{g\lambda_{2n}}{\sigma^2} + \lambda_{01n} \right)^2 \frac{\sigma^2}{\sigma_n^2 - \sigma^2};$$

$$A_{22} = \frac{\lambda_{2n}^2}{\mu_n} \frac{\sigma^2}{\sigma_n^2 - \sigma^2},$$

$$A_{24} = \frac{\lambda_{2n}}{\mu_n} \left(-\frac{g\lambda_{2n}}{\sigma^2} + \lambda_{01n} \right) \frac{\sigma^2}{\sigma_n^2 - \sigma^2}.$$

These terms cancel out when we consider $A_{44}A_{22} - A_{24}^2$. This fact applies for any two-dimensional and three-dimensional tank.

5.4.2 Forced sloshing in a two-dimensional rectangular tank

5.4.2.1 Hydrodynamic coefficients

A two-dimensional rectangular tank with breadth l and mean liquid depth h is assumed. Forced sway (η_2), heave (η_3), and roll excitation (η_4) are considered, as shown in Figure 5.3. To analyze linear forced sloshing, let us follow the algorithm from Section 5.4.1.1. The obtained results can easily be expanded to linear sloshing in the Oxz-plane for a rectangular tank with length l by changing the forcing parameters $\eta_2 \to \eta_1$ and $\eta_4 \to -\eta_5$.

Step (i) is to find the natural sloshing modes. These natural sloshing frequencies and modes are given by eqs. (4.11) and (4.8)–(4.9):

$$\varphi_m(y, z) = f_m(y) \frac{\cosh(\pi m(z+h)/l)}{\cosh(\pi m h/l)},$$

$$f_m(y) = \cos\left(\frac{\pi m}{l}\left(y + \tfrac{1}{2}l\right)\right), \quad m = 1, \ldots. \qquad (5.77)$$

Step (ii) involves finding the Stokes–Joukowski potential. When liquid motions occur in the Oyz-plane, we need only one component of the Stokes–Joukowski potential associated with roll excitations (i.e., Ω_{01}). The boundary-value problem for $\Omega_{01}(x, z)$ is (see eq. (5.21))

$$\frac{\partial^2 \Omega_{01}}{\partial y^2} + \frac{\partial^2 \Omega_{01}}{\partial z^2} = 0 \quad \text{in } Q_0,$$

$$\left.\frac{\partial \Omega_{01}}{\partial y}\right|_{y=\pm\frac{l}{2}} = -z, \qquad \left.\frac{\partial \Omega_{01}}{\partial z}\right|_{z=0;z=-h} = y. \quad (5.78)$$

The first step in establishing the solution of eq. (5.78) is to find a particular function satisfying the Laplace equation that takes care of the nonhomogenous boundary conditions at $y = \pm\frac{1}{2}l$. This particular solution can be chosen as $-yz$; that is, we can write $\Omega_{01}(y, z) = -yz + F(y, z)$. The function F should then satisfy the Laplace equation and $\partial F/\partial y|_{y=\pm\frac{l}{2}} = 0$, $\partial F/\partial z|_{z=0;z=-h} = 2y$. To find F, we must establish a full set of solutions of the Laplace equation that satisfy the zero-Neumann condition at $y = \pm\frac{1}{2}l$. This set can be found by using separation of spatial variables. The solution takes the form

$$F(y, z) = \sum_{i=1}^{\infty} f_i(y)[C_{1i} \exp(\pi i z/l) + C_{2i} \exp(-\pi i z/l)], \quad (5.79)$$

where $f_i(y)$ is defined by eq. (5.77) and the unknown constants are in the general case computed from the boundary conditions on $z = 0$ and $z = -h$. However, because these conditions are identical, we can simplify eq. (5.79) by remembering that identical conditions are possible only for the z-dependent components in eq. (5.79) that are odd relative to the $z = -\frac{1}{2}h$ line. Therefore,

$$F(y, z) = \sum_{i=1}^{\infty} C_i f_i(y) \frac{\sinh(\pi i(z + \frac{1}{2}h)/l)}{\cosh(\pi i h/2l)}.$$

To find C_i we now need only the condition $\partial F/\partial z|_{z=0} = \sum_{i=1}^{\infty} C_i \pi i l^{-1} f_i(y) = 2y$. Due to orthogonality of functions f_j, coefficients C_i follow from multiplication by $f_k(y)$ and integration over $[-\frac{1}{2}l, \frac{1}{2}l]$. The final result is

$$\Omega_{01}(y, z) = -yz + 4\sum_{j=1}^{\infty} \frac{l^2[(-1)^j - 1]}{(j\pi)^3} \times f_j(y) \frac{\sinh(\pi j(z + \frac{1}{2}h)/l)}{\cosh(\pi j h/2l)}. \quad (5.80)$$

Step (iii) is computation of the hydrodynamic coefficients. Only the hydrodynamic coefficients μ_m, λ_{2m}, and λ_{01m} matter for the considered two-dimensional flows in the Oyz-plane. Furthermore, the inertia tensor has only one nonzero scalar element (i.e., J^1_{011}). The hydrodynamic coefficients can be expressed by eq. (5.26) as

$$\mu_m = \frac{\rho}{\kappa_m} \int_{-\frac{1}{2}l}^{\frac{1}{2}l} f_m^2 \, dy = \frac{\rho l}{2\kappa_m} = \frac{\rho l^2}{2\pi m \tanh(\pi m h/l)},$$

$$\lambda_{2m} = \rho \int_{-\frac{1}{2}l}^{\frac{1}{2}l} y f_m(y) \, dy = \rho \left(\frac{l}{m\pi}\right)^2 [(-1)^m - 1], \quad \sigma_m^2 = g\kappa_m. \quad (5.81)$$

The remaining coefficient needed in eqs. (5.27) is

$$\lambda_{01m} = 2\rho l^3 \frac{(-1)^m - 1}{(m\pi)^3} \tanh\left(\frac{\pi m h}{2l}\right). \quad (5.82)$$

The inertia tensor element can be derived analytically by integration of eq. (5.44) and use of solution (5.80) (Faltinsen & Timokha, 2001). The result is

$$J^1_{011} = \rho l \left\{ \tfrac{1}{3}h\left(h^2 - \tfrac{1}{4}l^2\right) - \frac{16l^3}{\pi^5} \sum_{i=1}^{\infty} \frac{1}{(2i-1)^5} \times \left[\frac{\pi(2i-1)h}{l} - 4\tanh\left(\frac{\pi(2i-1)h}{2l}\right)\right]\right\}$$
$$= \rho l \left\{ \tfrac{1}{3}h^3 - \tfrac{1}{4}hl^2 + \frac{64l^3}{\pi^5} \sum_{i=1}^{\infty} \frac{1}{(2i-1)^5} \times \tanh\left(\frac{\pi(2i-1)h}{2l}\right)\right\}, \quad (5.83)$$

where we used that $\sum_{j=1}^{\infty}(2j-1)^{-4} = \frac{1}{96}\pi^4$. The numerical series in eq. (5.83) converges rapidly; therefore, no numerical problems exist to impede an accurate computation of J^1_{011}.

The next step involves modal equations and formulas for hydrodynamic loads. Accounting for results (5.81) and (5.82), the right-hand-side function $K_m(t)$ is due to eq. (5.28) as follows:

$$K_m(t) = -\frac{\lambda_{01m}}{\mu_m}\ddot{\eta}_4(t) - \frac{\lambda_{2m}}{\mu_m}(\ddot{\eta}_2(t) + g\eta_4(t)) = -P_m\left[\ddot{\eta}_2(t) + S_m\ddot{\eta}_4(t) + g\eta_4(t)\right], \quad (5.84)$$

where

$$P_m = \frac{\lambda_{2m}}{\mu_m} = \frac{2}{m\pi}\tanh\left(\frac{\pi m h}{l}\right)((-1)^m - 1),$$

$$S_m = \frac{\lambda_{01m}}{\lambda_{2m}} = \frac{2l}{\pi m}\tanh\left(\frac{\pi m h}{2l}\right). \quad (5.85)$$

For pure sway excitation, eqs. (5.84) and (5.85) give

$$K_m(t) = \frac{\lambda_{2m}}{\mu_m}\ddot{\eta}_2(t)$$
$$= -\left[\frac{2}{m\pi}\tanh\left(\frac{\pi m}{l}h\right)((-1)^m - 1)\right]\ddot{\eta}_2(t)$$
$$= -P_m\ddot{\eta}_2(t).$$

This expression is the same as that obtained in eq. (5.7) in the example of Section 5.2 by direct integration.

As a special case let us assume the tank rotates with a small roll angle $\eta_4(t)$ about the axis A_1A_2 perpendicular to the Oyz-plane (see Figure 5.3(c)); that is, the translatory motion is zero at the coordinates $y = 0$ and $z = z_0$. We commented in Chapter 3 that such a roll axis does not necessarily exist for a ship in waves due to the phasing between the sway and roll motion. However, introducing a roll axis simplifies the presentation of the flow and pressure in a completely filled tank.

We need to express the translatory motion $\eta_2(t)$ of O to use eq. (5.84). We find it by using the fact that the y-component of the motion of any point can be expressed as $\eta_2 - z\eta_4$, according to linear theory. Requiring this expression to be zero at $z = z_0$ gives

$$\eta_2(t) = z_0\eta_4(t), \qquad (5.86)$$

which means $K_m(t)$ can be expressed as

$$K_m(t) = -P_m\left[(z_0 + S_m)\ddot{\eta}_4(t) + g\eta_4(t)\right]. \quad (5.87)$$

Equation (5.87) is consistent with the formulas of Graham and Rodriguez (1952), who considered the special case $z_0 = -\frac{1}{2}h$, as well as expressions by Faltinsen (1974) and Faltinsen and Timokha (2001), who analyzed harmonic angular forcing of a rectangular tank in the context of nonlinear steady-state sloshing.

The hydrodynamic force follows from eq. (5.36) because we note that $x_{lC_0} = y_{lC_0} = 0$ and $z_{lC_0} = -\frac{1}{2}h$. The horizontal and vertical forces are then expressed as

$$F_2(t) = m_l\left(-g\eta_4(t) - \tfrac{1}{2}h\ddot{\eta}_4(t) - \ddot{\eta}_2(t)\right) - \sum_{j=1}^{N}\lambda_{2j}\ddot{\beta}_j(t), \quad F_3(t) = -m_l\ddot{\eta}_3(t), \qquad (5.88)$$

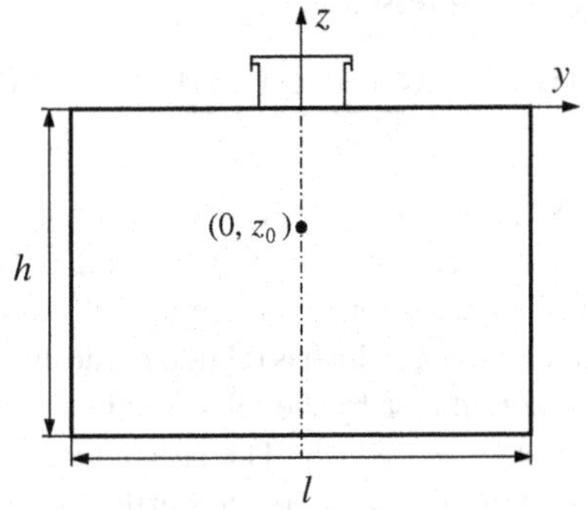

Figure 5.4. Coordinate system and parameter definitions for a completely filled tank. The roll axis goes through the point $(0, z_0)$.

where N is the number of considered natural modes and m_l is the liquid mass for the two-dimensional statement. One should note that η_3 is absent in the modal equations due to the remark in Section 5.3.1.3, but it is important for the vertical force in terms of a frozen liquid effect.

Using general formula (5.47) gives the following formula for the moment about the Ox-axis:

$$F_4(t) = M_{O1}(t)$$
$$= \tfrac{1}{2}hm_l(-g\eta_4(t) - \ddot{\eta}_2(t)) - J^1_{011}\ddot{\eta}_4(t)$$
$$-\sum_{j=1}^{N}\left(g\lambda_{2j}\beta_j(t) + \lambda_{01j}\ddot{\beta}_j(t)\right), \qquad (5.89)$$

where all the coefficients have already been found in analytical form.

5.4.2.2 Completely filled two-dimensional rectangular tank

When the tank is completely filled, the solution of the problem follows by neglecting the R_i terms in eq. (5.20). The velocity potential can be expressed as

$$\Phi(y, z, t) = \dot{\eta}_2(t)y + \dot{\eta}_4(t)\Omega_{01}(y, z), \qquad (5.90)$$

where the Stokes–Joukowski potential Ω_{01} for a two-dimensional rectangular tank is given by eq. (5.80). The coordinate system and parameter definitions are presented in Figure 5.4.

Let us exemplify the streamlines of the velocity field determined by eq. (5.90) when the tank rolls about the point $(0, z_0)$ on the z-axis. In this case, η_2 is defined by eq. (5.86). Equation (5.90) can then be expressed as

$$\Phi(y,z,t) = \underbrace{(z_0 y + \Omega_{01}(y,z))}_{S_{z_0,01}(y,z)} \dot{\eta}_4(t), \qquad (5.91)$$

where $S_{z_0,01}(y,z)$ is the velocity potential of the absolute velocity per unit roll velocity $\dot{\eta}_4(t)$. We can introduce for two-dimensional motion a stream function Ψ, which is related to the absolute velocity potential Φ by the relationships $\partial\Phi/\partial y = \partial\Psi/\partial z$, $\partial\Phi/\partial z = -\partial\Psi/\partial y$. The streamlines of the absolute flow (relative to an Earth-fixed coordinate system) are described by constant values of Ψ. We now introduce the normalized stream function $\Psi_{z_0,01}(y,z)$ associated with the normalized velocity potential $S_{z_0,01}(y,z)$. It follows by using eq. (5.80) and the relationship between the stream function and the velocity potential that

$$\Psi_{z_0,01}(y,z) = \tfrac{1}{2}y^2 - \tfrac{1}{2}z^2 + z_0 z + \frac{8l^2}{\pi^3}\sum_{j=1}^{\infty} \frac{\sin\left(\pi(2j-1)\left(y+\frac{1}{2}l\right)/l\right)}{(2j-1)^3} \times \frac{\cosh\left(\pi(2j-1)\left(z+\frac{1}{2}h\right)/l\right)}{\cosh(\pi(2j-1)h/2l)}. \qquad (5.92)$$

The velocity field and streamlines calculated by eq. (5.92) are shown in Figure 5.5. The physical velocity has been made nondimensional by dividing with $l\dot{\eta}_4(t)$. The nondimensional velocity is indicated by an arrow and its size is indicated for each condition in the figure. The absolute velocity field changes strongly with the ratio of tank height to breadth, h/l, as well as with the position of the roll axis, $z = z_0$. For smaller heights, the vertical velocity component dominates, while the streamlines become almost horizontal with increasing h/l. Very peculiar flow patterns are present when $z_0 = -\frac{1}{2}h$. If the liquid had been "frozen," the streamlines of the absolute liquid flow would be part of circles with center in $z = z_0$; however, the streamlines in Figure 5.5 do not resemble this shape.

We commented earlier that a roll axis generally does not exist, so we must add a translatory velocity field to the flow shown in Figure 5.5. The total linear pressure in a completely filled tank with a roll axis through $y = 0$, $z = z_0$ can be derived based on eq. (5.30). A constant pressure (p_0) is assumed at an opening of the tank at $y = 0$, $z = 0$ that is small relative to the cross-sectional tank areas. This assumption leads to the following expression:

$$p = p_0 - \rho g(z + y\eta_4) - \rho S_{z_0,01}(y,z)\ddot{\eta}_4, \qquad (5.93)$$

where $-\rho g\,(z + y\eta_4)$ is the hydrostatic pressure expressed in the tank-fixed coordinate system. The time-dependent pressure term $-\rho g y\eta_4$ gives spatially constant pressure at the tank walls at $y = \pm\frac{1}{2}l$. If we integrate this pressure term to get the force acting on the tank, the force represents the liquid weight component along the y-axis due to the tank inclination η_4. Figure 5.6 presents the nondimensional hydrodynamic pressure on the tank surface, $-S_{z_0,01}\,(y,z)\,/l^2$, as a function of the roll-axis position and tank height-to-breadth ratios studied in Figure 5.5. The general tendency is that the maximum normalized hydrodynamic pressure increases with increasing depth and decreases when the rotation center is closer to the midpoint of the liquid.

Berstad *et al.* (1997) used a similar analysis in assessing fatigue crack growth in the side longitudinals of a ship in the forepart and midship area close to the waterline. A motivation for their work was the damage observed for oil tankers on the west coast of North America and on the east coast of South Africa. Dynamic loads in the structural analysis included both the internal tank loads and the external hydrodynamic pressure distribution close to the waterline.

Based on the velocity potential expressed by eq. (5.91), we can find the hydrodynamic roll moment of a completely filled rectangular tank. The roll moment relative to origin O at the mean free surface is associated with the roll added mass coefficient A_{44}^{filled} represented by eq. (5.72), which depends on the inertia tensor component J_{011}^{1} defined by eq. (5.83). Using this inertia tensor component, we assume that the roll motions occur relative to origin O. When the roll is relative to the point $(0, z_0)$, it follows from eqs. (5.72) and (5.91) that the roll added mass coefficient is

$$\begin{aligned} A_{44}^{\text{filled}}\Big|_{(0,z_0)} &= m_l g(z_{lC_O} - z_0)\sigma^{-2} + J_{011}^{1}\Big|_{(0,z_0)} \\ &= m_l g(z_{lC_O} - z_0)\sigma^{-2} \\ &\quad + \rho\int_{\Sigma_0+S_0} S_{z_0,01}\frac{\partial S_{z_0,01}}{\partial n}\,dS. \end{aligned}$$

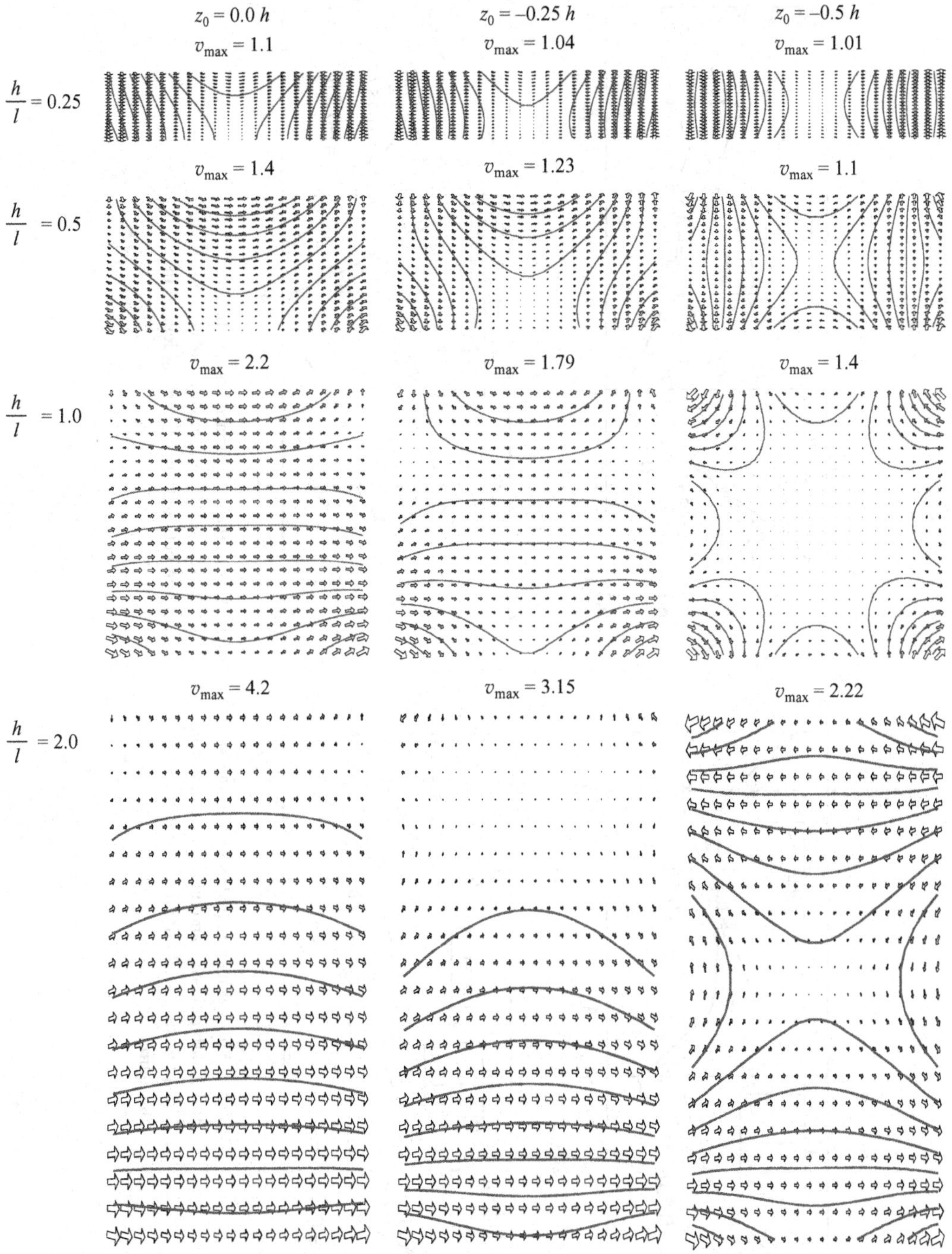

Figure 5.5. Nondimensional absolute velocity fields (arrows) and streamlines for completely filled tank with ratio of tank height to breadth h/l and rolling about an axis at $y = 0$ and $z = z_0$. The coordinate system is defined in Figure 5.4. The scaled values v_{max} of the maximum velocity are presented. The scaling is with respect to $l\dot{\eta}_{4a}$.

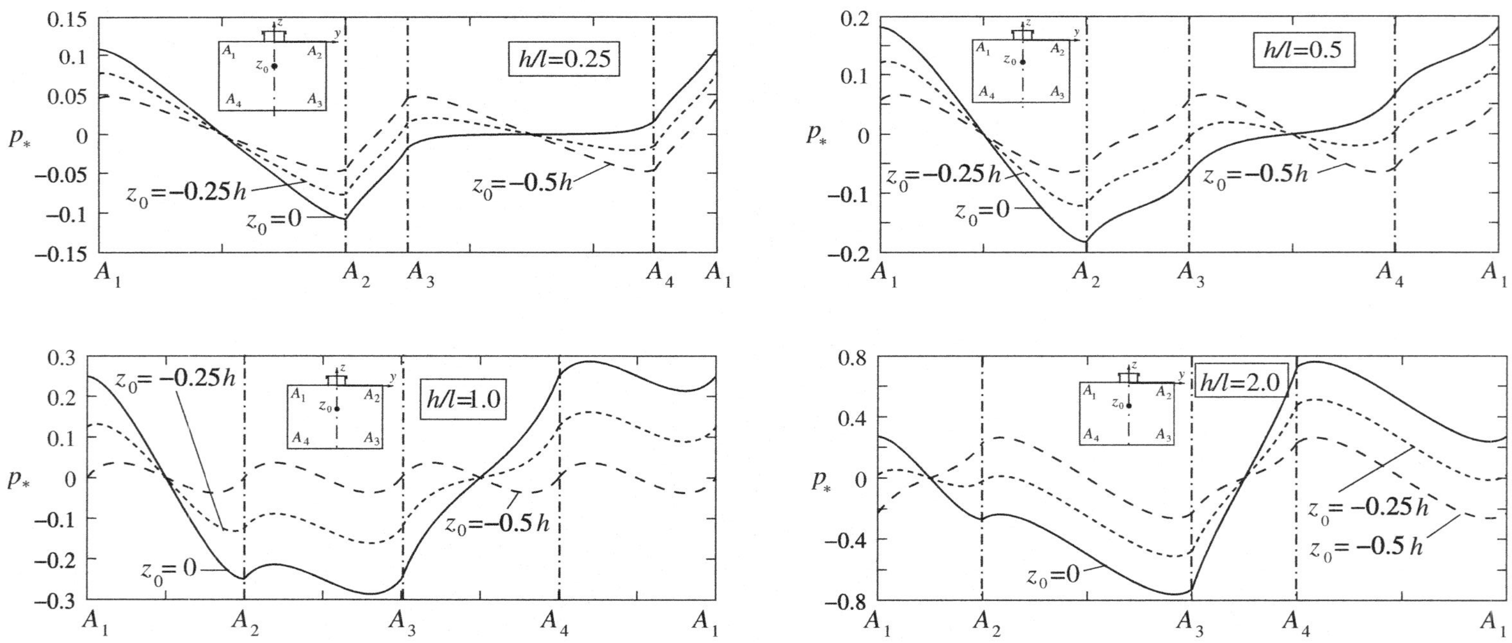

Figure 5.6. Normalized hydrodynamic pressure $p_* = p/(l^2 \rho \ddot{\eta}_{4a})$ on the tank surface for four height-to-breadth ratios h/l and different vertical coordinates z_0 of the roll axis for a completely filled tank. The pressure distribution is presented in the clockwise direction between the corners A_i defined in the figure.

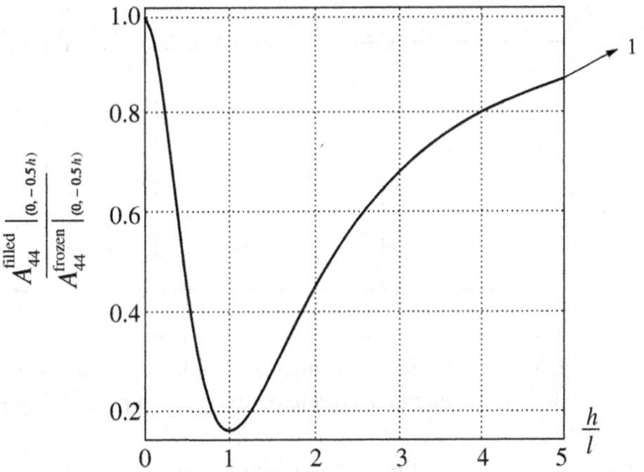

Figure 5.7. The ratio between added moment in roll of a completely filled tank, $A_{44}^{\text{filled}}|_{(0,-\frac{1}{2}h)}$, and the moment of inertia of the frozen liquid, $A_{44}^{\text{frozen}}|_{(0,-\frac{1}{2}h)}$, presented as a function of the ratio of tank height to tank length h/l (see eq. (5.98)). The moment is with respect to the center of the tank liquid.

Using eq. (5.80) gives

$$A_{44}^{\text{filled}}\Big|_{(0,z_0)} = m_l g\left(z_{lC_O} - z_0\right)\sigma^{-2} + \rho l h z_0\left[z_0 + h\right] + J_{011}^1, \quad (5.94)$$

where J_{011}^1 is given by eq. (5.83).

Henceforth, we focus on the case when $z_0 = z_{lC_0} = -\frac{1}{2}h$ (i.e., when the roll motions occur relative to the liquid mass center in an unperturbed state). For this case, eq. (5.94) becomes

$$A_{44}^{\text{filled}}\Big|_{(0,-\frac{1}{2}h)} = A_{44}^{\text{filled}}\Big|_{C_0} = J_{011}^1\Big|_{C_0} = -m_l z_{C_0}^2 + J_{011}^1. \quad (5.95)$$

Equation (5.95) is the same as the *parallel axis theorem* (Steiner's theorem) for a solid body, which expresses the moment of inertia, I_{11}, for an axis parallel to an axis through the center of mass as

$$I_{11} = I_{11}|_C + m_b|OC|^2, \quad (5.96)$$

where $I_{11}|_C$ is the moment of inertia with respect to the axis through the mass center C, m_b is the mass of the body, and $|OC|$ is the distance between the parallel axes. Steiner's theorem can be proven for a completely filled tank by using the formulas presented for the force and moment.

The added mass coefficients of the "frozen" liquid are defined by eq. (5.71) as

$$A_{44}^{\text{frozen}}\Big|_{(0,-\frac{1}{2}h)} = I_{11}^0\Big|_{(0,-\frac{1}{2}h)} = \tfrac{1}{12}\rho h l(l^2 + h^2) \quad (5.97)$$

and the ratio $A_{44}^{\text{filled}}|_{(0,-\frac{1}{2}h)}/A_{44}^{\text{frozen}}|_{(0,-\frac{1}{2}h)}$ becomes the following function of the ratio of tank height to breadth, $\overline{h} = h/l$:

$$\frac{A_{44}^{\text{filled}}\Big|_{\left(0,-\frac{1}{2}h\right)}}{A_{44}^{\text{frozen}}\Big|_{\left(0,-\frac{1}{2}h\right)}} = 1 - \frac{4}{1+\overline{h}^{-2}} + \frac{768}{\pi^5\overline{h}(1+\overline{h}^{-2})} \times \sum_{j=1}^{\infty}\frac{\tanh\left(\frac{1}{2}\pi(2j-1)\overline{h}\right)}{(2j-1)^5}. \quad (5.98)$$

The graph of this function is presented in Figure 5.7. As we remarked in Section 5.4.1.3, $A_{44}^{\text{frozen}}|_{(0,-\frac{1}{2}h)}$ and $A_{44}^{\text{filled}}|_{(0,-\frac{1}{2}h)}$ differ by inertia tensors $\boldsymbol{I}^0$ and $\boldsymbol{J}_0^1$ in the expressions. The graph shows that these tensors are asymptotically equal as $h/l \to 0$ or $h/l \to \infty$. The minimum value of the ratio $A_{44}^{\text{filled}}|_{(0,-\frac{1}{2}h)}/A_{44}^{\text{frozen}}|_{(0,-\frac{1}{2}h)}$, 0.1565, occurs when $h/l = 1$.

Journee (2000) presented systematic experimental results of the roll moment for forced harmonic roll motion with amplitude $\eta_{4a} = 0.10$ rad of the liquefied natural gas (LNG) tank in Figure 5.8(a). The roll axis is indicated in the figure. The reported filling levels were 15, 45, 70, 90, 97.5, and 100% of the tank height. Our focus is on the higher fill levels to check the applicability of formula (5.94). However, eq. (5.94) does not include the quasi-steady hydrostatic pressure effect due to the fact that the free surface remains horizontal in an Earth-fixed coordinate system (see, e.g., Section 3.6.1). Eq. (5.94) assumes that the free surface is fixed relative to the tank. Furthermore, the free surface acts only as a rigid wall when the frequency is small relative to the lowest natural

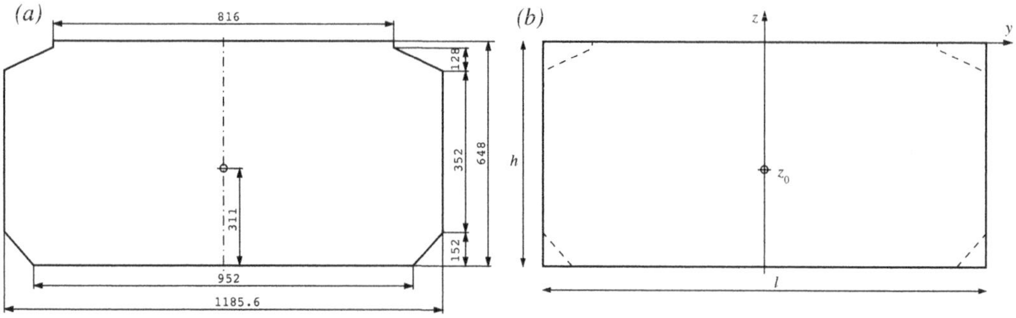

Figure 5.8. Experimental prismatic tank from Journee (1983, 2000) and an equivalent completely filled rectangular tank used in our theoretical predictions. Dimensions are in millimeters.

frequency. If we define a Cartesian coordinate system Oyz with origin at the roof center (see Figure 5.8(b)), the part of the small-frequency nondimensional moment amplitude relative to the point $(0, z_0)$ (scaled by $\rho g l^3 \eta_{4a}$) following from eq. (5.94) is

$$M_{\max,1}^{\text{filled}} = \frac{F_{4a}^{\text{filled}}\big|_{(0,z_0)}}{\rho g l^3 \eta_{4a}} = \frac{A_{44}^{\text{filled}}\big|_{(0,z_0)}\sigma^2}{\rho g l^3 \eta_{4a}} = \underbrace{\frac{h(z_{lC_0} - z_0)}{l^2}}_{c_0}$$
$$+ \left(\sigma\sqrt{\frac{l}{g}}\right)^2 \underbrace{\left[\frac{h z_0(z_0+h)}{l^3} + \frac{1}{3}\left(\frac{h}{l}\right)^3 - \frac{1}{4}\frac{h}{l} + \frac{64}{\pi^5}\sum_{i=1}^{\infty}\frac{\tanh(\pi(2i-1)h/(2l))}{(2i-1)^5}\right]}_{c_2}. \tag{5.99}$$

The first term in eq. (5.99) is the static moment due to the product of the weight and an arm following from the distance between the roll axis and the center of gravity. The nondimensional results of formula (5.99) agree reasonably well with the experiments performed by Journee (1983, 2000) for a completely filled tank (see Figure 5.9). Our estimates for c_0 and c_2 defined by eq. (5.99) are 0.006 and 0.02338, respectively, for $l = 1.1856$ m, $h = 0.648$ m, and $z_0 = -0.337$ m, corresponding to the experimental conditions by Journee (1983). An overprediction is attributable

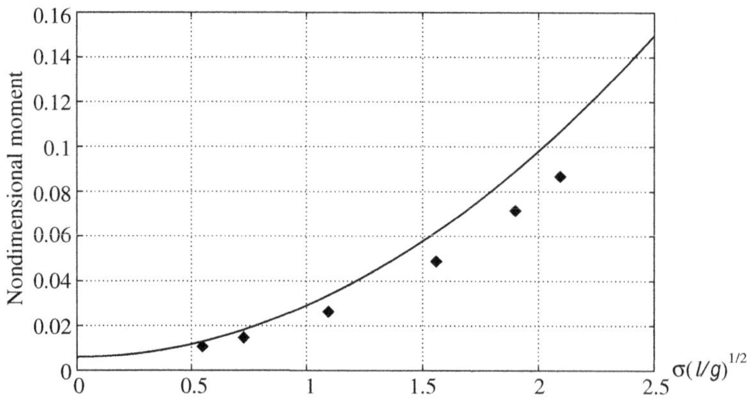

Figure 5.9. Nondimensional maximum roll moment $\max|F_4^{\text{filled}}(t)|_{(0,z_0)}|/(\rho g l^3 \eta_{4a})$ for the completely filled tank in Figure 5.8 versus the nondimensional frequency $\sigma\sqrt{l/g}$. Experimental data by Journee (1983), ♦ for tank in Figure 5.8(a), and the theoretical prediction by formula (5.99) for the tank in Figure 5.8(b) within the framework of linear theory.

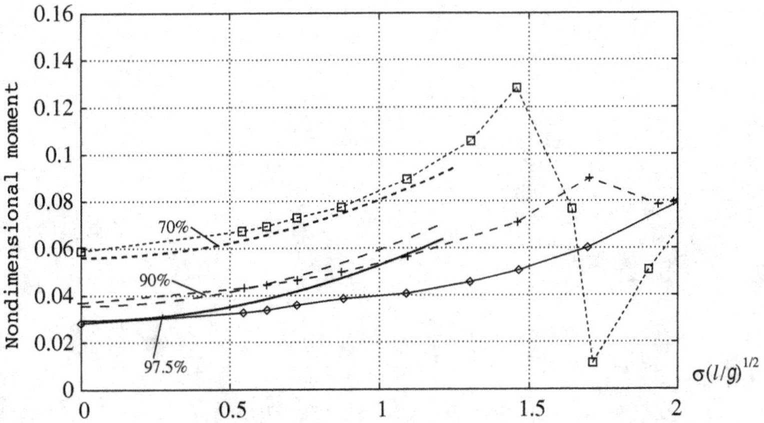

Figure 5.10. Comparison of formula (5.100) with experimental data by Journee (1983) for the nondimensional moment amplitude: ◇ (and corresponding solid lines, $l_1 = 0.816$ m), 97.5% filling height; + (and corresponding dashed lines, $l_1 = 0.885$ m), 90% filling height; and □ (and corresponding dashed lines, $l_1 = 1.1865$ m), 70% filling height. The experimental limits as $\sigma \to 0$ were extrapolated from the experiments by Journee (1983).

to larger c_2. (Journee reported the experimental value $c_2 = 0.01778$.) The main factor that causes a decrease of theoretical c_2 is chamfering. A speculative estimation of the effect can be made by considering the roll moment for the frozen liquid. For that case, the theoretical value for the equivalent rectangular tank in Figure 5.8(b) is $c_2 = 0.0685$, but Journee (1983) gives this value for a chamfered tank in Figure 5.8(a) as $c_2 = 0.05043$, which gives a ratio of 0.7362 between the roll moment for the chamfered and rectangular tanks with frozen liquid. The multiplication of this ratio with our theoretical value of 0.02338 gives 0.01721, which is very close to the experimental value of 0.01778. Journee (2000) was able to show satisfactory agreement between theory and experiment by solving the problem directly with a BEM for the chamfered tank.

Equation (5.99) can be revised by adding a quasi-static roll moment component due to free-surface effects (see Chapter 3 and discussion before eq. (5.47)); that is, we write

$$\frac{F_{4a}^{\text{quasi-static}}\Big|_{(0,z_0)}}{\rho g l^3 \eta_{4a}} = M_{\max,1}^{\text{quasi-static}} = M_{\max,1}^{\text{filled}} + \tfrac{1}{12}\left(l_1/l\right)^3, \tag{5.100}$$

where l_1 is the breadth of the mean free surface; $M_{\max,1}^{\text{filled}}$ is defined by eq. (5.99). No corrections are included for chamfer. This revision corrects the c_0-component to get values close to the experimental results by Journee (1983). The corresponding comparison of the moment amplitude is presented in Figure 5.10 for three different fillings of the tank in Figure 5.8(a). The results are reasonable for small frequencies. For larger frequencies the estimate becomes wrong due to sloshing and roof impact, which are not accounted for by our approximate approach. However, the experimental data for 90 and 97.5% filling height do not show evidence of strong sloshing effects on the moment; that is, the moment amplitude shows no strong change near the lowest linear resonance frequency. However, we should be careful in our conclusion about the free-surface effect on all sloshing loads. We describe in Section 11.5 a case with high filling ratio where flip-through of the free surface at the tank wall occurs. The consequence is high slamming loads.

5.4.2.3 Transient sloshing during collision of two ships

The collision of two ships can lead to strongly transient sloshing in partially filled tanks. Zhang and Suzuki (2007) numerically analyzed the collision between a container vessel and a crude oil cargo tank of a double-hull very large crude carrier (VLCC). The striking ship collides at a right angle with the stationary VLCC amidships as illustrated in Figure 5.11. They accounted for both plastic structural deformations and sloshing.

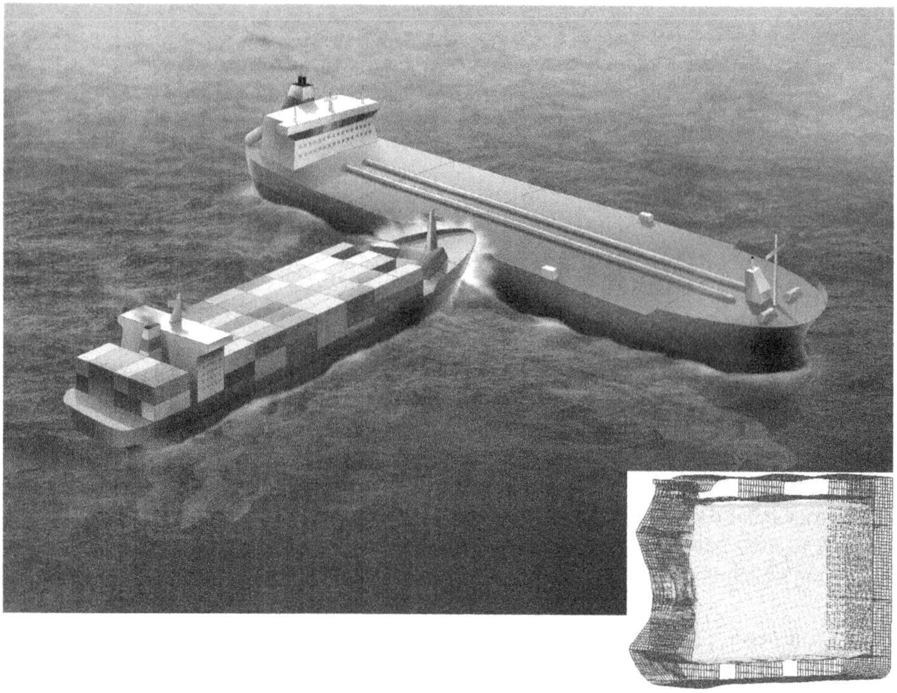

Figure 5.11. Collision between container ship and oil tanker (Artist: Bjarne Stenberg). Detailed drawing of resulting wall deformations in a tank partially filled with crude oil (Zhang & Suzuki, 2007).

An instantaneous picture of the predicted deformations of the tank structure is shown in Figure 5.11. The deformations of the double hull occur during the first 0.3 s of the collision. Because their numerical procedure was very time-consuming, only a limited time was simulated for the collision.

Our following approach for analysis of transient sloshing between two colliding ships uses linear modal theory. We assume a scenario as shown in Figure 5.11. To simplify our calculations we assume that all the deformation energy in the collision is in the bow of the striking container vessel; the oil tank only gets a rigid-body sway velocity. The focus is on the tank flow and resulting free-surface elevation and hydrodynamic tank force as a consequence of the impact.

The global effect of the deformations on the ship motion can be related to a simplified mechanical system with two lump masses linked by a spring. The initial velocity for the "striking" mass is v_{init} and zero for the second mass. The linear spring represents the bow deformations of the striking ship. We do not know the spring coefficient, which in reality is nonlinear and has to be found via nonlinear plastic structural analysis. The analysis introduces the time when the spring reaches its minimum length, T_d (i.e., when it absorbs maximum strength energy); T_d can be related to the spring coefficient. A correct evaluation of external hydrodynamic load is not pursued. The external lateral hydrodynamic loads on the struck ship can be expressed initially in terms of the high-frequency sway added mass A_{22}. The internal lateral hydrodynamic loads due to sloshing can also be expressed in terms of an added mass coefficient. The surge added mass of the striking ship may be 5 to 10% of the mass of the striking ship, M_1, and is neglected.

Using the solution for the second lump mass from Exercise 2.7.5(d) for the time range $0 < t < T_d$ gives the following transverse velocity of the struck ship:

$$\dot{\eta}_2(t) = v_d\left[1 - \cos\left(\tfrac{1}{2}\pi t/T_d\right)\right], \quad 0 \le t \le T_d. \tag{5.101}$$

The velocity v_d at $t = T_d$ can be expressed as

$$v_d = v_{\text{init}} M_1 / \left(M_1 + M_2 + A_{22}\right), \tag{5.102}$$

where M_2 is the mass of the struck ship. Equation (5.102) states that the momentum $v_{\text{init}} M_1$

at the initial impact time $t = 0$ is equal to the momentum of the two ships and the water, $(M_1 + M_2 + A_{22})\,v_d$, at $t = T_d$. Because the bow is deformed plastically, the spring force between the two masses stops at $t = T_d$, which implies a discontinuity in the accelerations of the two vessels at $t = T_d$.

In the following example we use $M_1 = 70{,}000$ tonnes and $M_2 = 340{,}000$ tonnes and simply set $A_{22} \approx 0.3M_2$. A precise estimate of the added mass coefficient requires detailed information about the ship geometry. We assume the striking ship has a speed of 10 m s^{-1}. Equation (5.102) then gives $v_d = 1.367$ m s^{-1}. Amdahl (personal communication, 2008) has estimated that the deformation of the striking ship lasts for $T_d = 2$ s. The considered cargo tank is assumed rectangular with breadth 27 m and length 50 m. The liquid depth h is 23 m and the density of the crude oil, ρ_l, is 860 kg m^{-3}. We assume the height between the tank roof and the mean free surface is *only* 1.2 m. After the deformation of the bow we assume the two ships move as one body without any rotation.

Generally speaking, one can estimate the velocity law after $t > T_d$ by using the differential equation following from Newton's second law for the sway:

$$(M_1 + M_2 + A_{22})\,\ddot{\eta}_2 = -\tfrac{1}{2}\rho_{\text{sea}} C_D D L \dot{\eta}_2^2 \;(t > T_d); \quad \dot{\eta}_2(T_d) = v_d, \tag{5.103}$$

where L and D are the length and draft, respectively, of the struck ship. Furthermore, we need to know the drag coefficient C_D for steady ambient cross-flow past the struck ship and the added mass A_{22}. The influence of the striking vessel on the drag force may be neglected. Strictly speaking, we cannot use the high-frequency added mass when time t becomes clearly larger than T_d.

As long as we obtain the mentioned values, the solution of initial-value problem (5.103) is

$$\dot{\eta}_2(t) = \frac{M_1 + M_2 + A_{22}}{\frac{1}{2}\rho_{\text{sea}} C_D D L t + C_1}, \quad t \geq T_d;$$
$$C_1 = \frac{M_1 + M_2 + A_{22}}{v_d} - \tfrac{1}{2}\rho_{\text{sea}} C_D D L T_d. \tag{5.104}$$

Based on this solution, with realistic values in the preceding formulas, one can find that $\ddot{\eta}_2(t)$ is approximately zero on the time scale of the sloshing periods. Therefore, the acceleration $\ddot{\eta}_2(t)$ undergoes a jump at $t = T_d$, as we have already anticipated. The forthcoming analysis assumes that $\dot{\eta}_2 = v_d$ for $t > T_d$.

Figure 5.12 explains the simplifications of our modeling. A graph that represents the transverse velocity and acceleration of the ship is given in Figure 5.12(a). According to our assumptions, the velocity increases rapidly during the first 2 s, but after that it remains equal to v_d. The transverse ship acceleration $\ddot{\eta}_2$ is a maximum at $t = T_d$ and then jumps to zero. Modal equations (5.56) with damping were used. The damping was estimated by unsteady laminar boundary-layer theory with steady-state periodic conditions (see Section 6.3.1). However, the damping was very small and turns out to be unimportant during the simulated time interval. The crude oil was assumed to be at rest before the collision. Because the sway motion of the struck ship does not start abruptly, we use the initial conditions $\beta_m(0) = \dot{\beta}_m(0) = 0$ $(m \geq 1)$ for the generalized coordinates. The right-hand sides of the modal equations can be expressed as $K_m(t) = -P_m\ddot{\eta}_2(t)$, where P_m is given by eq. (5.85). Because $\ddot{\eta}_2(t)$ is discontinuous at $t = T_d = 2$ s, a special numerical treatment was necessary. As the first step, the solution was found in the interval [0, 2] (in seconds) with zero initial conditions. This solution gives $\beta_m(T_d)$ and $\dot{\beta}_m(T_d)$, which provide initial conditions for the simulations when $t \geq T_d$.

Because $\dot{\eta}_2(t)$ has a sinusoidal time dependence during $0 \leq t \leq T_d$, we may use an analytical solution for the generalized coordinates during that time interval (see eq. (5.65)). Although an analytical solution exists for the time interval [0, 2], the analysis that follows is based on a fourth-order Runge–Kutta integration. Typical wave elevations at the vertical walls are shown in Figure 5.12(b) for liquid depth $h = 23$ m. The linear theory demonstrates that in the initial stage the liquid flows in the left direction (i.e., opposite the impact velocity) so that the maximum elevation is observed at the left wall. After that, the liquid shows almost periodic sloshing with the frequency close to the lowest natural frequency.

The sloshing analysis does not account for the fact that tank-roof slamming occurs. Because the predicted wave amplitudes at the tank walls are about 1.5 m and the height of the tank roof

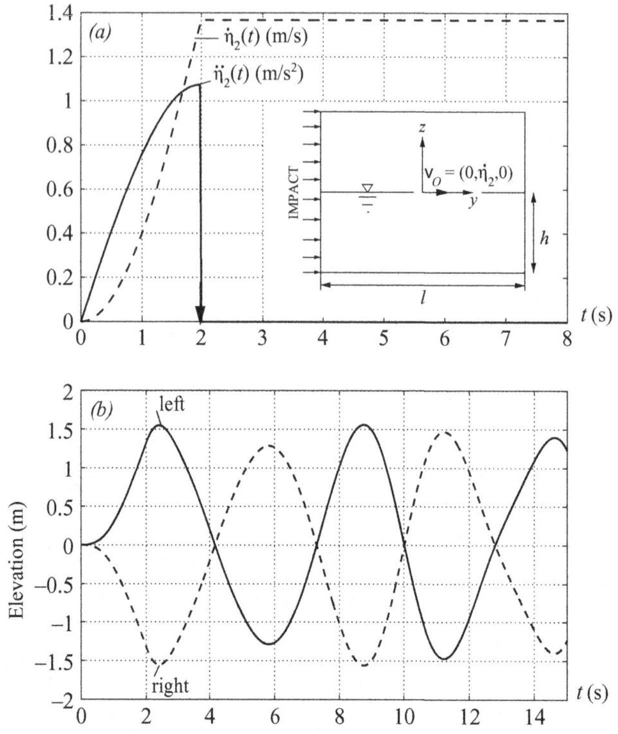

Figure 5.12. (a) Graphs of the impact velocity and related acceleration. The acceleration is discontinuous at $t = 2$ s. (b) The wave elevations at the left and right walls of the oil tank for liquid depth $h = 23$ m.

relative to the mean free surface is 1.2 m, slamming does occur. The slamming impact velocity for the first impact is estimated to be 1.175 m s^{-1}. Slamming load effects depend on the velocity and the shape of the impacting free surface relative to the tank surface. We need to know structural details to assess the severity of slamming, and hydroelasticity may be a factor (see Chapter 11). However, qualitatively speaking we would not expect an impact velocity of about 1.2 m s^{-1} to cause serious slamming problems.

The time history of the resulting hydrodynamic force acting on the tank as calculated by eq. (5.88) is influenced by the actual values of the liquid depth. Figure 5.13 shows the time history of the horizontal hydrodynamic force during the first 15 s after collision as a function of the liquid depth. The horizontal "frozen" liquid force is also presented in the figure (as a dashed line), demonstrating that sloshing matters relative to results obtained by considering the liquid "frozen." The maximum absolute value of the hydrodynamic force during the first 2 s is lower than if the liquid is considered frozen; part of the impact energy transfers to liquid sloshing motions. The influence of sloshing is smaller for larger liquid depths. The direction of the horizontal force is opposite that of the external acceleration and has an almost inertial character (i.e., it is proportional to the tank acceleration). The smaller liquid mass "absorbs" the impact energy and transfers it into liquid sloshing much more rapidly.

The hydrodynamic pressure on a deformed tank wall is a more important measure for hull structure deformations than is the total integrated hydrodynamic force on the tank. To assess the hydrodynamic pressure on the tank wall one should note that the liquid impact on the tank roof causes pressure everywhere in the liquid (Chapter 11). Because we have left out a detailed analysis of slamming in this context, the hydrodynamic pressure is estimated by assuming infinite tank roof height. The predicted wave amplitude A at the tank wall is then a measure of the maximum hydrodynamic pressure $\rho g A$ on the tank wall (see Figure 5.1 and accompanying discussion). Therefore, a wave amplitude of 1.5 m (see Figure 5.12) and a liquid density $\rho_l = 860$ kg m^{-3} causes a maximum hydrodynamic pressure of 12.7 kN m^{-2} (i.e., less than 0.2 bar). This order of magnitude of pressure is not expected to have a significant influence on the deformation of the hull structure

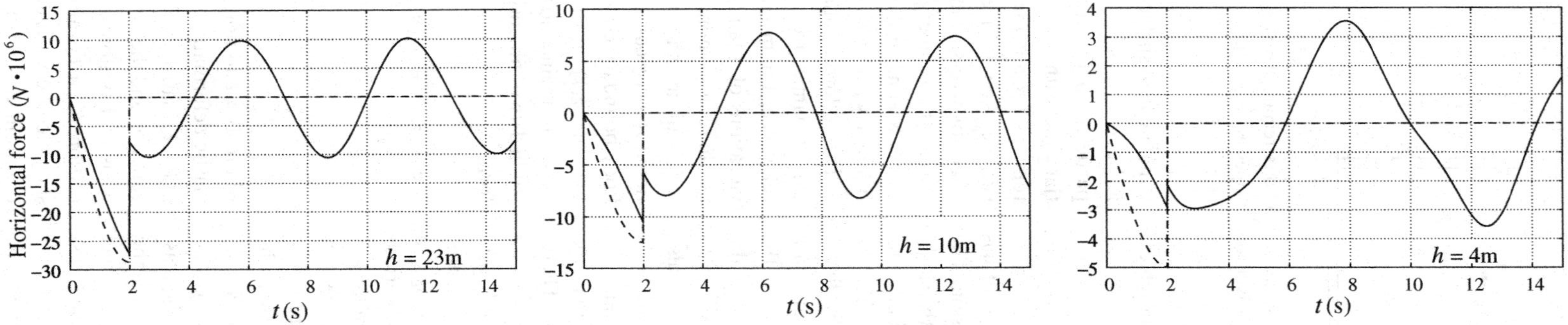

Figure 5.13. Time history of the transient horizontal hydrodynamic force due to ship collision. Dashed line shows the force for a "frozen" liquid, but the solid line accounts for sloshing by linear modal theory.

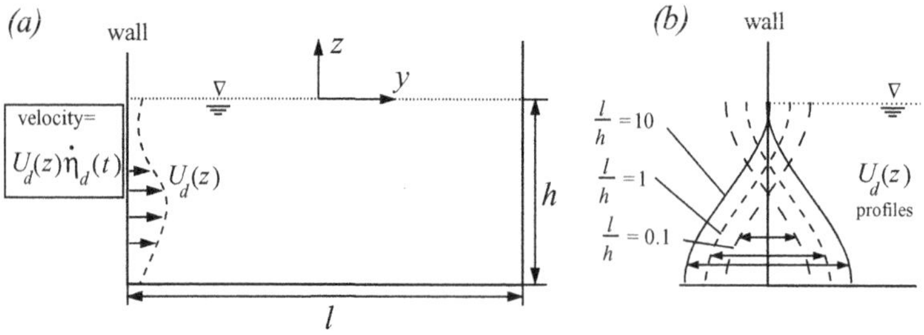

Figure 5.14. (a) Schematic deflection of the left wall of a rectangular two-dimensional tank and the adopted notations. (b) Instantaneous wavemaker shapes at two different time instants that do not excite the first modes due to eq. (5.114).

due to collision (Amdahl, personal communication, 2008).

5.4.2.4 Effect of elastic tank wall deflections on sloshing

We see from Figure 5.11 that the tank wall has deformed as a consequence of the collision. The time history of the deflections influence the sloshing. As long as the deflections are small relative to the main tank dimensions, their effect on the contained liquid can be handled by linear modal theory. For brevity, we concentrate on a two-dimensional rectangular tank and express the prescribed time history of the deflections of the left wall as $U_d(z)\eta_d(t)$. The situation is schematically shown in Figure 5.14. The scenario is also relevant for elastic vibrations of the tank wall. However, typical elastic resonance frequencies for a ship tank are significantly higher than the dominant natural sloshing frequencies. Therefore, the tank can be considered rigid when we analyze wave-induced sloshing in a ship tank, with an important exception when slamming occurs. The reason is that the duration of slamming can be of the order of magnitude of natural periods for elastic vibrations. Hydroelastic slamming is further analyzed in Chapter 11. When the natural frequencies for the elastic vibrations of the tank structure are significantly higher than the dominant natural sloshing frequencies, it is common to use a high-frequency free-surface condition. The velocity potential φ caused by the elastic vibrations is then set equal to zero on the mean free surface. The argument is that the effect of gravity does not matter, so the linear pressure can be approximated as $p_0 - \rho\partial\varphi/\partial t$, where p_0 is the pressure in the air (gas) above the free surface. This approximation implies that the dynamic free-surface condition can be simplified as $\partial\varphi/\partial t = 0$ on $z = 0$. If we now follow liquid particles on the free surface, they start at initial time with $\varphi = 0$. Because $\partial\varphi/\partial t = 0$, $\varphi = 0$ remains for all time as a condition on the free surface. An intuitive argument for neglecting the effect of free-surface waves at high frequencies is that the submergence where the free-surface waves become important goes to zero when the frequency goes to infinity.

The following analysis uses the complete linear dynamic free-surface condition and has relevance, for instance, for earthquake-induced vibrations of a land-based storage tank. As long as there is a nonzero small deflection of the left wall as shown in Figure 5.14, an additional velocity component $U_d(z)\dot{\eta}_d(t)$ exists in the body-boundary condition for the velocity potential $\Phi(y, z, t)$ at the left wall. This condition needs a modification of eq. (5.20) which, for the studied two-dimensional case, has the following form:

$$\Phi(y, z, t) = \Omega_d(y, z)\dot{\eta}_d(t) + (\dot{\eta}_2 y + \dot{\eta}_3 z) + \Omega_{01}(y, z)\dot{\eta}_4 + \sum_{k=1}^{\infty} R_k(t)\varphi_k(y, z), \tag{5.105}$$

where the additional term Ω_d is related to the elastic deformations of the tank wall. We require that Ω_d satisfies the two-dimensional Laplace equation and the boundary conditions on the tank walls and bottom associated with elastic deformations. Furthermore, the flow caused by Ω_d should conserve liquid mass in the tank. The latter requirement leads to a free-surface condition for Ω_d, which becomes more evident after controlling conservation of liquid mass. The resulting Neumann boundary-value problem is

$$\frac{\partial^2 \Omega_d}{\partial y^2} + \frac{\partial^2 \Omega_d}{\partial z^2} = 0 \quad \text{for} -\tfrac{1}{2}l < y < \tfrac{1}{2}l, \quad -h < z < 0;$$

$$\left.\frac{\partial \Omega_d}{\partial y}\right|_{y=\frac{1}{2}l} = 0 \quad -h < z < 0;$$

$$\left.\frac{\partial \Omega_d}{\partial z}\right|_{z=-h} = 0 \quad -\tfrac{1}{2}l < y < \tfrac{1}{2}l,$$

$$\left.\frac{\partial \Omega_d}{\partial y}\right|_{y=-\frac{1}{2}l} = U_d(z) \quad -h < z < 0;$$

$$\left.\frac{\partial \Omega_d}{\partial z}\right|_{z=0} = \frac{1}{l}\int_{-h}^{0} U_d(z)\,\mathrm{d}z = \text{const} \quad -\tfrac{1}{2}l < y < \tfrac{1}{2}l. \qquad (5.106)$$

The total free-surface elevation is

$$\zeta(y,t) = l^{-1}\eta_d(t)\int_{-h}^{0} U_d(z)\mathrm{d}z + \sum_{j=1}^{\infty}\beta_j(t)f_j(y). \qquad (5.107)$$

Equations (5.105) and (5.107) adopt the natural modes in the form of eq. (5.77). We can confirm that eq. (5.107) is consistent with conservation of liquid mass by using the kinematic free-surface condition $\partial\Phi/\partial z = \partial\zeta/\partial t$, evaluating the free-surface flux $\int_{-\frac{1}{2}l}^{\frac{1}{2}l}(\partial\Phi/\partial z)\mathrm{d}y = \dot{\eta}_d(t)\int_{-h}^{0} U_d(z)\,\mathrm{d}z$, and noting the fact that the latter flux is equal to the flux associated with the tank boundary conditions for the total velocity potential Φ. If $\int_{-h}^{0} U_d(z)\,\mathrm{d}z \neq 0$, a constant change of the free surface exists for a given nonzero $\eta_d(t)$.

To get an analytical solution, let us present the velocity profile function $U_d(z)$ via the Fourier series

$$U_d(z) = \alpha_0 + \sum_{k=1}^{\infty}\alpha_k\cos(\pi k(z+h)/h). \qquad (5.108)$$

Because the deflections are prescribed, the coefficients α_k $(k = 0, 1, \ldots)$ are assumed to be known.

The α_0-coefficient of representation (5.108) is responsible for the piston-like in- and outflows caused by the wall deformation and associated fluctuations of the liquid level. Postulating eq. (5.108) in eq. (5.106) gives the following analytical solution:

$$\Omega_d(y,z) = -\frac{\alpha_0}{2l}\left[\left(y-\tfrac{1}{2}l\right)^2 - (z+h)^2\right] - \sum_{k=1}^{\infty}\frac{\alpha_k h}{\pi k}\cos\left(\frac{\pi k}{h}(z+h)\right) \times \frac{\cosh\left(\pi k\left(y-\frac{1}{2}l\right)/h\right)}{\sinh(\pi k l/h)}. \qquad (5.109)$$

We can confirm that eq. (5.109) is the solution Ω_d by first noting that each term in the sum and also the remaining first term satisfy the Laplace equation and the body-boundary conditions on the bottom and the walls. Furthermore, the free-surface condition $\partial\Omega_d/\partial z|_{z=0} = \int_{-h}^{0} U_d(z)\,\mathrm{d}z/l = \alpha_0 h/l$ is satisfied due to the first term in eq. (5.109), while each term of the sum in eq. (5.109) has zero z-derivative on $z = 0$. We must now ensure that the kinematic and dynamic free-surface conditions are satisfied by generalizing the described procedure in Section 5.3.1 and including the effect of Ω_d. The presence of the velocity field associated with eq. (5.109) modifies the right-hand side of modal system (5.27), which according to eq. (5.28) takes the following form for our rectangular tank case:

$$K_m(t) = -\frac{\lambda_{2m}}{\mu_m}(\ddot{\eta}_2 + g\eta_4) - \frac{\lambda_{01m}}{\mu_m}\ddot{\eta}_4 - \frac{\lambda_{dm}}{\mu_m}\ddot{\eta}_d, \qquad (5.110)$$

where μ_m, λ_{2m}, and λ_{01m} are determined by eqs. (5.81) and (5.82). However, the new hydrodynamic coefficients, λ_{dm}, are due to solution (5.109) as follows:

$$\frac{\lambda_{dm}}{\rho} = \int_{-\frac{1}{2}l}^{\frac{1}{2}l}\Omega_d(y,0)\cos\left(\pi m\left(y+\tfrac{1}{2}l\right)/l\right)\mathrm{d}y = -\frac{\alpha_0 l^2}{\pi^2 m^2} - \frac{1}{\pi^2}\sum_{k=1}^{\infty}\frac{(-1)^k\alpha_k l^2 h^2}{(k^2l^2+m^2h^2)}. \qquad (5.111)$$

To find the actual deflections we would need to couple our hydrodynamic analysis with a structural analysis, which would mean a plasticity analysis for the case of Figure 5.11 that is beyond the scope of our book.

Let us turn the problem around. *Will a particular deflection of the tank wall cause no excitation of the lowest sloshing mode?* If we want to cancel the excitation of the lowest sloshing mode, we must require that $K_1(t)$ given by eq. (5.110) is zero. Let us restrict ourselves to two Fourier components of the velocity profile function $U_d(z)$; that is,

$$U_d(z) = \alpha_0 + \alpha_1 \cos(\pi(z+h)/h). \quad (5.112)$$

The parameter λ_{d1} is needed in the calculation of $K_1(t)$. By using eq. (5.111), this calculation can be expressed as $\lambda_{d1}/\rho = -\alpha_0 l^2\pi^{-2} + \alpha_1 l^2h^2\pi^{-2}/(l^2+h^2)$. It follows from eqs. (5.81), (5.82), and (5.110) that the requirement of $K_1(t) = 0$ is equivalent to

$$\left[\tfrac{1}{2}\alpha_0 - \frac{\tfrac{1}{2}\alpha_1 h^2}{(l^2+h^2)}\right]\ddot{\eta}_d = -(\ddot{\eta}_2 + g\eta_4) - \frac{2l}{\pi}\tanh\left(\frac{\pi h}{2l}\right)\ddot{\eta}_4, \quad (5.113)$$

for which it is possible to satisfy for a piston-like excitation (i.e., $\alpha_0 \neq 0$, $\alpha_i = 0$ $(i \geq 1)$). When the ship motions are known and α_0 is given, we can determine $\eta_d(t)$ by eq. (5.113). Hence, theoretically we can think in terms of monitoring the ship motions and then use automatic control to determine $\eta_d(t)$. Of course, considerable uncertainty exists concerning the practical and economic feasibility of doing so. The benefit is that the most important sloshing mode is avoided. However, this outcome could also be achieved either by dividing the tank into two equal parts or by using a swash bulkhead in the middle of the tank (see Sections 6.7 and 6.8).

If the tank is not moving, we can achieve zero excitation of the lowest sloshing mode by setting the coefficient on the left-hand side of eq. (5.113) equal to zero; that is, the profile of the wavemaker defined by eq. (5.112) should satisfy

$$\alpha_0 = \alpha_1 h^2/(l^2+h^2). \quad (5.114)$$

The shapes of the wavemaker at two time instants that satisfy eq. (5.114) for three different ratios of tank breadth to liquid depth, l/h, are presented in Figure 5.14(b). An analysis similar to that just given can be applied to the other tank wall. How to analyze the effect of deflections of the tank bottom is left as an exercise (see Section 5.6.3). Due to linearity we can superimpose the effects of the tank walls and the tank bottom.

5.4.3 Forced sloshing in a three-dimensional rectangular-base tank

5.4.3.1 Hydrodynamic coefficients

We show how to evaluate the terms in the modal equations for a rectangular-base tank by following the steps described in Section 5.4.1.1. The tank length and width along the Ox- and Oy-axes are denoted L_1 and L_2, respectively. As a first step we need to know the natural sloshing modes, which were found in Section 4.3.2.1.

The second step involves finding the Stokes–Joukowski potential $\boldsymbol{\Omega}_0(x,y,z) = (\Omega_{01}, \Omega_{02}, \Omega_{03})$. The boundary-value problem is formulated in eq. (5.21). The components $\Omega_{01}(x,y,z)$ and $\Omega_{02}(x,y,z)$ satisfy the following boundary conditions for a rectangular-base tank:

$$\left.\frac{\partial\Omega_{01}}{\partial x}\right|_{x=\pm\frac{1}{2}L_1} = 0, \quad \left.\frac{\partial\Omega_{01}}{\partial y}\right|_{y=\pm\frac{1}{2}L_2} = -z,$$
$$\left.\frac{\partial\Omega_{01}}{\partial z}\right|_{z=0;z=-h} = y; \quad \left.\frac{\partial\Omega_{02}}{\partial y}\right|_{y=\pm\frac{1}{2}L_2} = 0,$$
$$\left.\frac{\partial\Omega_{02}}{\partial x}\right|_{x=\pm\frac{1}{2}L_1} = z, \quad \left.\frac{\partial\Omega_{02}}{\partial z}\right|_{z=0;z=-h} = -x. \quad (5.115)$$

Because the boundary conditions for Ω_{01} depend only on y and z, it follows that $\Omega_{01} = \Omega_{01}(y,z)$. This solution is the same as that given by eq. (5.80) for a two-dimensional rectangular tank with L_2 instead of l. Similarly we can state that Ω_{02} is only a function of x and z; that is, $\Omega_{02} = \Omega_{02}(x,z)$. The solution follows by replacing y and z in Ω_{01} with x and z, using L_1 instead of L_2 and noting the sign difference in the boundary conditions. We can then write

$$\Omega_{02}(x,z) = xz - 4\sum_{j=1}^{\infty}\frac{L_1^2[(-1)^j-1]}{(j\pi)^3} \times \cos\left(\frac{\pi j}{L_1}\left(x+\tfrac{1}{2}L_1\right)\right)\frac{\sinh\left(\pi j\left(z+\tfrac{1}{2}h\right)/L_1\right)}{\cosh(\pi j h/2L_1)}. \quad (5.116)$$

The boundary conditions for the harmonic function $\Omega_{03}(x,y,z)$ for the rectangular-base tank can, by means of eq. (5.21), be expressed as

$$\left.\frac{\partial\Omega_{03}}{\partial x}\right|_{x=\pm\frac{1}{2}L_1} = -y, \quad \left.\frac{\partial\Omega_{03}}{\partial y}\right|_{y=\pm\frac{1}{2}L_2} = x,$$
$$\left.\frac{\partial\Omega_{03}}{\partial z}\right|_{z=0;z=-h} = 0. \tag{5.117}$$

Consequently Ω_{03} is only a function of x and y. Separation of variables then gives the following solution:

$$\Omega_{03} = \Omega_{03}(x, y) = -xy + 4\sum_{j=1}^{\infty}\frac{L_1^2[(-1)^j - 1]}{(j\pi)^3}$$
$$\times\cos\left(\frac{\pi j}{L_1}(x + \tfrac{1}{2}L_1)\right)\frac{\sinh(\pi j y/L_1)}{\cosh(\pi j L_2/2L_1)}$$
$$= xy - 4\sum_{j=1}^{\infty}\frac{L_2^2[(-1)^j - 1]}{(j\pi)^3}$$
$$\times\cos\left(\frac{\pi j}{L_2}(y + \tfrac{1}{2}L_2)\right)\frac{\sinh(\pi j x/L_2)}{\cosh(\pi j L_1/2L_2)}. \tag{5.118}$$

Step (iii) involves finding the hydrodynamic coefficients. A novelty relative to the two-dimensional case is that the natural modes constitute a two-parameter family and the surface wave profiles are defined by $f_{i,j}(x, y) = f_i^{(1)}(x)f_j^{(2)}(y)$, where the multiplying functions are determined by eq. (4.28). Therefore, the hydrodynamic coefficients must also depend on two indices; that is,

$$\mu_{m,n} = \frac{\rho}{\kappa_{m,n}}\int_{\Sigma_0}\left(f_m^{(1)}(x)f_n^{(2)}(y)\right)^2 \mathrm{d}x\,\mathrm{d}y,$$
$$\lambda_{1(m,n)} = \rho\int_{\Sigma_0} x f_m^{(1)}(x)f_n^{(2)}(y)\,\mathrm{d}x\,\mathrm{d}y,$$
$$\lambda_{2(m,n)} = \rho\int_{\Sigma_0} y f_m^{(1)}(x)f_n^{(2)}(y)\,\mathrm{d}x\,\mathrm{d}y,$$
$$\lambda_{0k(m,n)} = \rho\int_{\Sigma_0} f_m^{(1)}(x)f_n^{(2)}(y)\,\Omega_{0k}\,\mathrm{d}x\,\mathrm{d}y,$$
$$k = 1, 2, 3; \quad m, n = 0, 1, 2, \ldots;$$
$$m^2 + n^2 \neq 0, \tag{5.119}$$

where the spectral parameters $\kappa_{m,n}$ are given in Section 4.3.2.1 and the Stokes–Joukowski potential is defined by eqs. (5.80), (5.116), and (5.118). Integrals (5.119) can be found in analytical form. For example,

$$\mu_{m,n} = \begin{cases} \frac{1}{4}\rho L_1 L_2/\kappa_{m,n}, & \text{if } nm \neq 0, \\ \frac{1}{2}\rho L_1 L_2/\kappa_{m,n}, & \text{if } nm = 0. \end{cases} \tag{5.120}$$

Furthermore, the hydrodynamic coefficients may be zero depending on combinations of m and n. For example, $\lambda_{1(m,n)} = \lambda_{2(m,n)} = 0$ when $nm \neq 0$. For the remaining indices,

$$\lambda_{1(m,0)} = \rho L_2\int_{-\frac{1}{2}L_1}^{\frac{1}{2}L_1} x f_m^{(1)}(x)\,\mathrm{d}x$$
$$= \rho L_2\left(\frac{L_1}{m\pi}\right)^2[(-1)^m - 1],$$
$$\lambda_{2(0,n)} = \rho L_1\int_{-\frac{1}{2}L_2}^{\frac{1}{2}L_2} y f_n^{(2)}(y)\,\mathrm{d}y$$
$$= \rho L_1\left(\frac{L_2}{n\pi}\right)^2[(-1)^n - 1]. \tag{5.121}$$

Since $\Omega_{02} = \Omega_{02}(x, z)$, $\lambda_{02(m,n)} = 0$ for $n \neq 0$. Similarly, $\lambda_{01(m,n)} = 0$ when $m \neq 0$. The remaining nonzero $\lambda_{02(m,0)}$ and $\lambda_{01(0,n)}$ are given by the formulas

$$\lambda_{01(0,n)} = 2\rho L_1 L_2^3\frac{(-1)^n - 1}{(n\pi)^3}\tanh\left(\frac{\pi n h}{2L_2}\right),$$
$$\lambda_{02(m,0)} = -2\rho L_2 L_1^3\frac{(-1)^m - 1}{(m\pi)^3}\tanh\left(\frac{\pi m h}{2L_1}\right). \tag{5.122}$$

Computations of $\lambda_{03(m,n)}$ involve two different expressions for $\Omega_{03} = \Omega_{03}(x, y)$ given by eq. (5.118). When $m = 0$ and $n \neq 0$, substitution of the second expression provides zeros due to $\int_{-\frac{1}{2}L_1}^{\frac{1}{2}} \bullet\,\mathrm{d}x$ in eq. (5.119) for $\lambda_{03(0,n)}$. The same is true for the condition $m \neq 0$, $n = 0$. When $mn \neq 0$, substitution of solution (5.118) into eq. (5.119) gives

$$\lambda_{03(m,n)} = \rho\left(\frac{L_1}{\pi m}\right)^2\left(\frac{L_2}{\pi n}\right)^2[(-1)^m - 1][(-1)^n - 1]$$
$$\times\frac{(nL_1)^2 - (mL_2)^2}{(nL_1)^2 + (mL_2)^2}, \tag{5.123}$$

which confirms the preceding conclusion of zeros when $mn = 0$.

Moreover, $\lambda_{03(m,n)}$ is zero for the case when m or n is even and for some specific combinations of the indices so that

$$\frac{n}{m} = \frac{L_2}{L_1}. \tag{5.124}$$

We can, for instance, see that condition (5.124) is fulfilled for a square-base tank and a saddle mode $f_{1,1}(x, y)$ with $n = m = 1$ (see Figure 4.6). This fact means that $\lambda_{03(1,1)} = 0$ and, therefore, the mode $f_{1,1}(x, y)$ is not excited. However, for a non-square-base tank with $L_2/L_1 \neq 1$, we come to condition $\lambda_{03(1,1)} \neq 0$ and the saddle mode $f_{1,1}(x, y)$ is directly excited by yaw.

The final series of derivations is related to the inertia tensor, for which computations can be done by using eqs. (5.83), (5.116), and (5.118). Taking into account the antisymmetry by the corresponding coordinates for the Stokes–Joukowski potential, we get

$$J^1_{0ij} = 0 \quad \text{for } i \neq j. \tag{5.125}$$

The diagonal elements are

$$\begin{aligned} J^1_{011} &= \rho L_1 L_2 \left\{ \tfrac{1}{3}h^3 - \tfrac{1}{4}hL_2^2 + \frac{64L_2^3}{\pi^5} \right. \\ &\quad \left. \times \sum_{i=1}^{\infty} \frac{1}{(2i-1)^5} \tanh\left(\frac{\pi(2i-1)h}{2L_2}\right) \right\}, \\ J^1_{022} &= \rho L_1 L_2 \left\{ \tfrac{1}{3}h^3 - \tfrac{1}{4}hL_1^2 + \frac{64L_1^3}{\pi^5} \right. \\ &\quad \left. \times \sum_{i=1}^{\infty} \frac{1}{(2i-1)^5} \tanh\left(\frac{\pi(2i-1)h}{2L_1}\right) \right\}, \\ J^1_{033} &= \rho L_1 h \left\{ \tfrac{1}{12}L_2^3 - \tfrac{1}{4}L_2 L_1^2 + \frac{64L_1^3}{\pi^5} \right. \\ &\quad \left. \times \sum_{i=1}^{\infty} \frac{1}{(2i-1)^5} \tanh\left(\frac{\pi(2i-1)L_2}{2L_1}\right) \right\}. \end{aligned} \tag{5.126}$$

When the hydrodynamic coefficients have been determined, we can establish the modal equations (step (iv)). These equations have different structures for the indices satisfying $mn = 0$ and $mn \neq 0$. In the first case, the modal equations couple the Stokes two-dimensional modes along the Ox- or Oy-axis. These modes are directly excited by *surge and pitch or sway and roll.* The equations are the same as if they were derived for a two-dimensional case (i.e., for the Stokes modes along the Ox-axis). Therefore,

$$\begin{aligned} &\ddot\beta_{m,0} + \sigma^2_{m,0}\beta_{m,0} = K_{m,0}(t), \\ &K_{m,0}(t) \\ &\quad = -\frac{\lambda_{02(m,0)}}{\mu_{m,0}}\ddot\eta_5(t) + \frac{\lambda_{1(m,0)}}{\mu_{m,0}}(-\ddot\eta_1(t) + g\eta_5(t)) \\ &\quad = -P_{m,0}\left[\ddot\eta_1(t) - S_{m,0}\ddot\eta_5(t) - g\eta_5(t)\right], \end{aligned} \tag{5.127}$$

where

$$\begin{aligned} P_{m,0} &= \frac{2}{m\pi}\tanh\left(\frac{\pi m h}{L_1}\right)((-1)^m - 1), \\ S_{m,0} &= \frac{2L_1}{\pi m}\tanh\left(\frac{\pi m h}{2L_1}\right). \end{aligned} \tag{5.128}$$

For the Stokes modes along the Oy-axis,

$$\begin{aligned} &\ddot\beta_{0,n} + \sigma^2_{0,n}\beta_{0,n} = K_{0,n}(t), \\ &K_{0,n}(t) \\ &\quad = -\frac{\lambda_{01(0,n)}}{\mu_{0,n}}\ddot\eta_4(t) - \frac{\lambda_{2(0,n)}}{\mu_{0,n}}(\ddot\eta_2(t) + g\eta_4(t)) \\ &\quad = -P_{0,n}\left[\ddot\eta_2(t) + S_{0,n}\ddot\eta_4(t) + g\eta_4(t)\right], \end{aligned} \tag{5.129}$$

where

$$\begin{aligned} P_{0,n} &= \frac{2}{n\pi}\tanh\left(\frac{\pi n h}{L_2}\right)((-1)^n - 1), \\ S_{0,n} &= \frac{2L_2}{\pi n}\tanh\left(\frac{\pi n h}{2L_2}\right). \end{aligned} \tag{5.130}$$

The three-dimensional modes, which are associated with the indices $mn \neq 0$, may be excited only by *yaw*. Surge, sway, roll, and pitch forcing do not linearly excite these modes. The corresponding modal equations take the form

$$\begin{aligned} &\ddot\beta_{m,n} + \sigma^2_{m,n}\beta_{m,n} = K_{m,n}(t), \\ &\qquad K_{m,n}(t) = -\frac{\lambda_{03(m,n)}}{\mu_{m,n}}\ddot\eta_6(t) = -P_{m,n}\ddot\eta_6(t), \end{aligned} \tag{5.131}$$

where

$$\begin{aligned} P_{m,n} &= \frac{4}{\pi^3 m^2 n^2}\left[(-1)^m - 1\right]\left[(-1)^n - 1\right] \\ &\quad \times \frac{(nL_1)^2 - (mL_2)^2}{\sqrt{(nL_1)^2 + (mL_2)^2}}\tanh(k_{m,n}h). \end{aligned} \tag{5.132}$$

Summarizing the results of the linear modal equations makes it possible to draw important conclusions about forced sloshing in a rectangular-base tank. The first conclusion is that surge-and-pitch excitation always leads to the two-dimensional linear waves considered in Section 5.4.2. Similarly, sway-and-roll excitations also give two-dimensional waves, but these waves occur in the Oyz-plane. Moreover, these excitations amplify only the antisymmetric modes shown in Figure 4.3. The combined

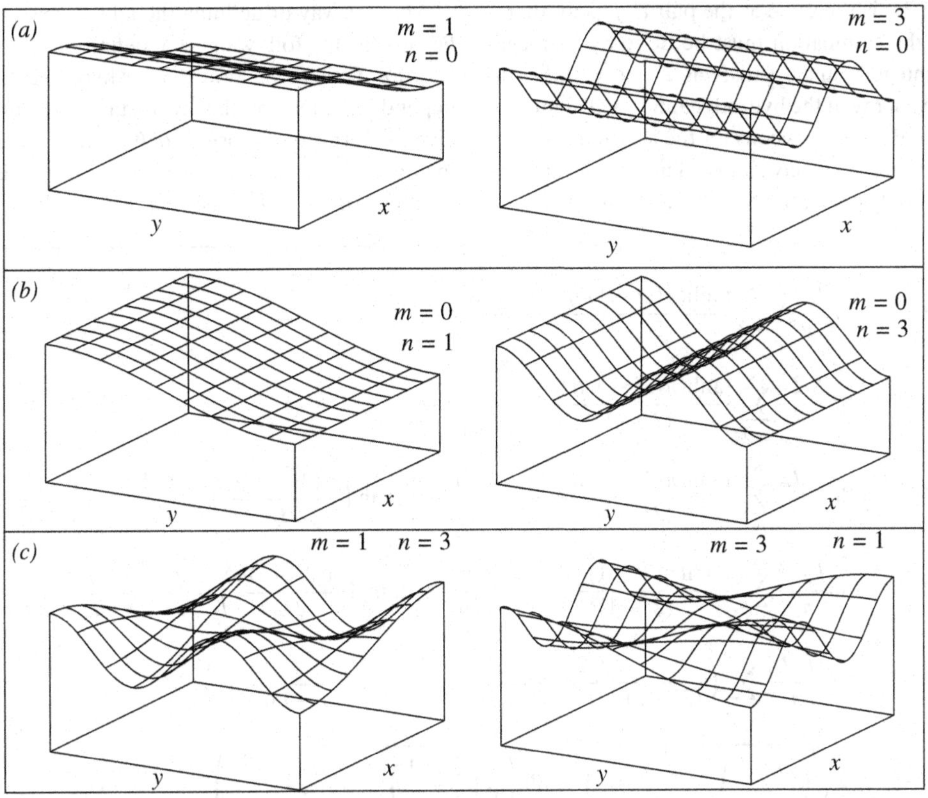

Figure 5.15. Examples of the surface mode shapes that can be excited by (a) surge/pitch, (b) sway/roll, and (c) yaw for a square-base tank.

surge–sway–roll–pitch forcing may cause diagonal modes to recombine the Stokes modes but would never excite three-dimensional modes with $nm \neq 0$. These combinatory modes were extensively discussed in Section 4.3.2.1 and in the context of Figure 4.8. Furthermore, heave excitation does not cause any forced linear waves. When condition (5.124) is not fulfilled (an example is the mode $f_{1,1}(x, y)$ for nonsquare tanks), yaw excitations may excite three-dimensional modes that are antisymmetric relative to both the Oxz- and Oyz-planes. Some of these modes are illustrated in Figure 5.15 for the case of a square-base tank.

5.4.3.2 Added mass coefficients in ship applications

Following eqs. (5.71) and (5.72), we can get explicit expressions for the added mass coefficients of a completely filled tank in an *inertial* coordinate system. The same coordinate system is later used for the case of a free surface. Because of the two symmetry planes (Oxz and Oyz), almost all the coefficients are zero except

$$
\begin{aligned}
A_{11}^{\text{filled}} &= A_{22}^{\text{filled}} = A_{33}^{\text{filled}} = \rho L_1 L_2 h, \quad A_{66}^{\text{filled}} = J_{033}^1,\\
A_{44}^{\text{filled}} &= -\tfrac{1}{2}\rho L_1 L_2 h^2 g\sigma^{-2} + J_{011}^1,\\
A_{55}^{\text{filled}} &= -\tfrac{1}{2}\rho L_1 L_2 h^2 g\sigma^{-2} + J_{022}^1;\\
A_{42}^{\text{filled}} &= -A_{51}^{\text{filled}} = \tfrac{1}{2}\rho L_1 L_2 h^2;\\
A_{15}^{\text{filled}} &= -A_{24}^{\text{filled}} = -\tfrac{1}{2}\rho L_1 L_2 h^2,
\end{aligned}
\tag{5.133}
$$

where the inertia tensor components are defined by eq. (5.126). The term $-\frac{1}{2}\rho L_1 L_2 h^2 g\sigma^{-2}$ in eqs. (5.133) expresses the quasi-steady component of an equivalent "frozen" liquid.

The sloshing-related components of the added mass coefficients are associated with formulas (5.74) with $\tilde{P}_{mk}$ defined by eq. (5.69). Substitution of eq. (5.69) into eq. (5.74) shows that the added mass coefficients are functions of the hydrodynamic coefficients. Explicit expressions for the hydrodynamic coefficients are given by eqs. (5.120)–(5.123). The coefficients depend on the two indices i and j, which means as remarked after eq. (5.26) that the summation index m in eq. (5.74) is the same as the pair (i, j) and, therefore, the summation must be done by nonnegative integer i and j satisfying $i^2 + j^2 \neq 0$. Fortunately, many of the hydrodynamic coefficients are zero and the expressions for the added mass coefficients are relatively simple. The double summation is only presented for A_{66}^{slosh}. The result is

$$A_{11}^{\text{slosh}} = 8\rho \frac{L_2 L_1^2}{\pi^3} \sum_{i=1}^{\infty} \frac{\tanh(\pi(2i-1)h/L_1)}{(2i-1)^3} \frac{\sigma^2}{\sigma_{2i-1,0}^2 - \sigma^2},$$

$$A_{22}^{\text{slosh}} = 8\rho \frac{L_1 L_2^2}{\pi^3} \sum_{j=1}^{\infty} \frac{\tanh(\pi(2j-1)h/L_2)}{(2j-1)^3} \frac{\sigma^2}{\sigma_{0,2j-1}^2 - \sigma^2},$$

$$A_{44}^{\text{slosh}} = 8\rho \frac{L_1 L_2^2}{\pi^3} \sum_{j=1}^{\infty} \frac{\tanh(\pi(2j-1)h/L_2)}{(2j-1)^3} \left[\frac{2L_2}{\pi(2j-1)} \tanh\left(\frac{\pi(2j-1)h}{2L_2} \right) - \frac{g}{\sigma^2} \right]^2 \frac{\sigma^2}{\sigma_{0,2j-1}^2 - \sigma^2},$$

$$A_{55}^{\text{slosh}} = 8\rho \frac{L_2 L_1^2}{\pi^3} \sum_{i=1}^{\infty} \frac{\tanh(\pi(2i-1)h/L_1)}{(2i-1)^3} \left[\frac{2L_1}{\pi(2j-1)} \tanh\left(\frac{\pi(2i-1)h}{2L_1} \right) - \frac{g}{\sigma^2} \right]^2 \frac{\sigma^2}{\sigma_{2i-1,0}^2 - \sigma^2},$$

$$A_{66}^{\text{slosh}} = 64\rho \frac{L_1^3 L_2^3}{\pi^7} \sum_{i,j=1}^{\infty} \left(\frac{((2i-1)L_1)^2 - ((2j-1)L_2)^2}{(2i-1)^2 (2j-1)^2 \left[((2i-1)L_1)^2 + ((2j-1)L_2)^2 \right]} \right)^2$$
$$\times \sqrt{\left(\frac{2i-1}{L_1}\right)^2 + \left(\frac{2j-1}{L_2}\right)^2} \tanh\left(\sqrt{\left(\frac{2i-1}{L_1}\right)^2 + \left(\frac{2j-1}{L_2}\right)^2}\, h \right) \frac{\sigma^2}{\sigma_{2i-1,2j-1}^2 - \sigma^2},$$

$$A_{51}^{\text{slosh}} = A_{15}^{\text{slosh}} = 8\rho \frac{L_2 L_1^2}{\pi^3} \sum_{i=1}^{\infty} \frac{\tanh(\pi(2i-1)h/L_1)}{(2i-1)^3}$$
$$\times \left[-\frac{2L_1}{\pi(2j-1)} \tanh\left(\frac{\pi(2i-1)h}{2L_1} \right) + \frac{g}{\sigma^2} \right] \frac{\sigma^2}{\sigma_{2i-1,0}^2 - \sigma^2},$$

$$A_{42}^{\text{slosh}} = A_{24}^{\text{slosh}} = 8\rho \frac{L_1 L_2^2}{\pi^3} \sum_{j=1}^{\infty} \frac{\tanh(\pi(2j-1)h/L_2)}{(2j-1)^3}$$
$$\times \left[\frac{2L_2}{\pi(2j-1)} \tanh\left(\frac{\pi(2i-1)h}{2L_2} \right) - \frac{g}{\sigma^2} \right] \frac{\sigma^2}{\sigma_{0,2j-1}^2 - \sigma^2}. \tag{5.134}$$

The added mass coefficients in eqs. (5.134) constitute a *symmetric* tensor, A_{ij}^{slosh}. The total added mass coefficients are $A_{ij} = A_{ij}^{\text{filled}} + A_{ij}^{\text{slosh}}$, where $A_{ij} = A_{ji}$ for $i \neq j$.

Equations (5.134) are consistent with quasi-steady approximations of A_{ij}^{slosh} as defined by eqs. (5.46) and (5.75) (see, also Section 3.6.1). Letting $\sigma \to 0$, we can show that

$$A_{44}^{\text{quasi}} = \tfrac{1}{12}\rho L_1 L_2^3 g\sigma^{-2}, \quad A_{55}^{\text{quasi}} = \tfrac{1}{12}\rho L_2 L_1^3 g\sigma^{-2},$$
$$A_{45}^{\text{quasi}} = A_{54}^{\text{quasi}} = 0. \tag{5.135}$$

Another way of defining the added mass coefficients is by following the definition for the exterior flow in Section 3.5.2 where the time-dependent effect of the hydrostatic pressure is given in terms of restoring coefficients by assuming quasi-steady tank motions. Section 3.6.1 is an example of such a quasi-steady analysis where roll is studied. Based on the expression for F_4 from this section, one can derive the resulting roll moment as $-(A_{44}^{\text{quasi}} - \tfrac{1}{2}\rho L_1 L_2 h^2 g\sigma^{-2})\ddot{\eta}_4$, where A_{44}^{quasi} is given by (5.135) (see also (5.75)) and the term $-\tfrac{1}{2}\rho L_1 L_2 h^2 g\sigma^{-2}$ is explained in Section 3.6.1 (see discussion ahead of eq. (3.35)) as "frozen" liquid effect (roll moment caused by the liquid weight). Similar analysis can be done for pitch, that is,

$$A_{44}^{\text{dyn}} = A_{44} - A_{44}^{\text{quasi}} + \tfrac{1}{2}\rho L_1 L_2 h^2 g\sigma^{-2};$$
$$A_{55}^{\text{dyn}} = A_{55} - A_{55}^{\text{quasi}} + \tfrac{1}{2}\rho L_1 L_2 h^2 g\sigma^{-2}. \tag{5.136}$$

There are also modifications in terms involving heave and coupled heave and pitch. We will henceforth introduce the superscript "dyn" in the added mass coefficients that are consistent with definitions from Section 3.5.2.

It is of interest to get limits of the added mass coefficients for large and small excitation frequencies. When $\sigma \to \infty$, series (5.134) has no mathematical limit due to the simple fact that an infinite number of distinct natural frequencies exists. However, the limit exists when we use finite sums in expressions (5.134) (i.e., we neglect the contribution of the very high modes and associated sloshing resonances). A physical argument for neglecting the very high modes is related to viscous effects. When the limit $\sigma \to \infty$ has been used in the expressions, we can afterward let $N \to \infty$. The final result is the following added mass coefficients:

$$A_{11}^{\text{slosh}} = 8\rho\frac{L_2L_1^2}{\pi^3}\sum_{i=1}^{\infty}\frac{\tanh(\pi(2i-1)h/L_1)}{(2i-1)^3}; \quad A_{22}^{\text{slosh}} = 8\rho\frac{L_1L_2^2}{\pi^3}\sum_{j=1}^{\infty}\frac{\tanh(\pi(2j-1)h/L_2)}{(2j-1)^3},$$

$$A_{44}^{\text{slosh}} = 32\rho\frac{L_1L_2^4}{\pi^5}\sum_{j=1}^{\infty}\frac{\tanh(\pi(2j-1)h/L_2)}{(2j-1)^5}\tanh^2\left(\frac{\pi(2j-1)h}{2L_2}\right),$$

$$A_{55}^{\text{slosh}} = 32\rho\frac{L_2L_1^4}{\pi^5}\sum_{i=1}^{\infty}\frac{\tanh(\pi(2i-1)h/L_1)}{(2i-1)^5}\tanh^2\left(\frac{\pi(2i-1)h}{2L_1}\right),$$

$$A_{66}^{\text{slosh}} = 64\rho\frac{L_1^3L_2^3}{\pi^7}\sum_{i,j=1}^{\infty}\left(\frac{((2i-1)L_1)^2-((2j-1)L_2)^2}{(2i-1)^2(2j-1)^2\left[((2i-1)L_1)^2+((2j-1)L_2)^2\right]}\right)^2$$
$$\times\sqrt{\left(\frac{2i-1}{L_1}\right)^2+\left(\frac{2j-1}{L_2}\right)^2}\tanh\left(\sqrt{\left(\frac{2i-1}{L_1}\right)^2+\left(\frac{2j-1}{L_2}\right)^2}\,h\right),$$

$$A_{51}^{\text{slosh}} = A_{15}^{\text{slosh}} = -16\rho\frac{L_2L_1^3}{\pi^4}\sum_{i=1}^{\infty}\frac{\tanh(\pi(2i-1)h/L_1)}{(2i-1)^4}\tanh\left(\frac{\pi(2i-1)h}{2L_1}\right),$$

$$A_{42}^{\text{slosh}} = A_{24}^{\text{slosh}} = 16\rho\frac{L_1L_2^3}{\pi^4}\sum_{j=1}^{\infty}\frac{\tanh(\pi(2j-1)h/L_2)}{(2j-1)^4}\tanh\left(\frac{\pi(2i-1)h}{2L_2}\right). \quad (5.137)$$

The same result can be obtained by starting with the dynamic free-surface condition $\varphi = 0$ on $z = 0$.

When we consider the case $\sigma \to 0$, A_{11}^{slosh} and A_{22}^{slosh} tend to zero due to the fact that the force due to surge and sway is the inertia force of frozen liquid mass in this limit. Roll and pitch yield an infinity due to the $O(\sigma^{-2})$ quasi-steady components in eqs. (5.133) and (5.134). However, the "dynamic" components of these added mass coefficients are finite and

$$A_{44}^{\text{dyn}} = \rho L_1L_2\left\{\tfrac{1}{3}h^3 - \tfrac{1}{4}hL_2^2 + \frac{96L_2^3}{\pi^5}\times\sum_{i=1}^{\infty}\frac{1}{(2i-1)^5}\tanh\left(\frac{\pi(2i-1)h}{2L_2}\right)\right\},$$

$$A_{55}^{\text{dyn}} = \rho L_1L_2\left\{\tfrac{1}{3}h^3 - \tfrac{1}{4}hL_1^2 + \frac{96L_1^3}{\pi^5}\times\sum_{i=1}^{\infty}\frac{1}{(2i-1)^5}\tanh\left(\frac{\pi(2i-1)h}{2L_1}\right)\right\} \quad (5.138)$$

as $\sigma \to 0$. Finally,

$$A_{42} = \tfrac{1}{2}\rho L_1L_2\left(h^2 - \tfrac{1}{6}L_2^2\right);$$
$$A_{51} = -\tfrac{1}{2}\rho L_1L_2\left(h^2 - \tfrac{1}{6}L_1^2\right) \quad (5.139)$$

in the limit.

When the added mass coefficients are used in a ship motion analysis, we have to remember that the ship motions are considered in the ship

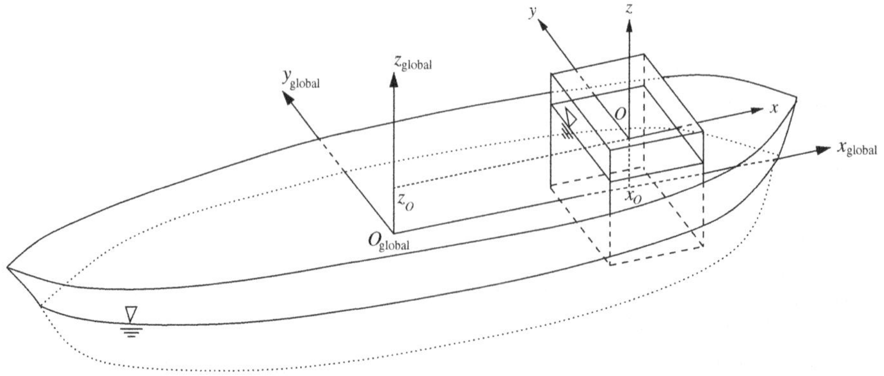

Figure 5.16. Global ship motion coordinate system $O_{global}x_{global}y_{global}z_{global}$ and tank coordinate system $Oxyz$.

motion coordinate system $O_{global}x_{global}y_{global}z_{global}$, which differs from the local tank coordinate system $Oxyz$ that has its origin in the middle of the mean free surface of the tank. Let us assume that the Oxz-plane of the local tank coordinate system coincides with the $O_{global}x_{global}z_{global}$-plane of the global ship motion coordinate system and the origin of the tank coordinate system has coordinates $(x_O, 0, z_O)$ in the ship motion coordinate system as shown in Figure 5.16. With these definitions, the global motion modes $\eta_i^G(t)$ are in a linear approximation linked with the aforementioned $\eta_i(t)$ as follows:

$$\eta_1 = \eta_1^G + z_O\eta_5;\quad \eta_2 = \eta_2^G + x_O\eta_6 - z_O\eta_4;$$
$$\eta_3 = \eta_3^G - x_O\eta_5;\quad \eta_4 = \eta_4^G;\quad \eta_5 = \eta_5^G;\quad \eta_6 = \eta_6^G. \tag{5.140}$$

The hydrodynamic force in the global and tank coordinate systems is the same although the moments differ. The moments in the global ship coordinate system should be recalculated using eq. (5.50). For the studied case, this implies

$$F_1^G = F_1;\quad F_2^G = F_2;\quad F_3^G = F_3;$$
$$F_4^G = F_4 - z_OF_2;\quad F_5^G = F_5 + z_OF_1 - x_OF_3;$$
$$F_6^G = F_6 + x_OF_2. \tag{5.141}$$

By definition, the just-derived added mass coefficients A_{ij} give $F_i = -\sum_{j=1}^{6} A_{ij}\ddot{\eta}_j$. Many of the added mass coefficients are zero. Using eqs. (5.140) and (5.141) gives immediately, for translatory components,

$$F_1^G = -A_{11}\left(\ddot{\eta}_1^G + z_O\ddot{\eta}_5^G\right) - A_{15}\ddot{\eta}_5^G;$$
$$F_2^G = -A_{22}\left(\ddot{\eta}_2^G - z_O\ddot{\eta}_4^G + x_O\ddot{\eta}_6^G\right) - A_{24}\ddot{\eta}_4^G,$$
$$F_3^G = -A_{33}\left(\ddot{\eta}_3^G - x_O\ddot{\eta}_5^G\right), \tag{5.142}$$

which can be used in eq. (5.141) to get the hydrodynamic moment in the global ship coordinates.

Our aim is to express force and moment components due to η_j^G as $F_k^G = -\sum_{j=1}^{6} A_{kj}^G\ddot{\eta}_j^G$. Inserting eqs. (5.142) into eqs. (5.141) and collecting terms at $\ddot{\eta}_j^G$ gives

$$A_{11}^G = A_{11};\quad A_{15}^G = A_{15} + z_OA_{11};\quad A_{22}^G = A_{22};$$
$$A_{24}^G = A_{24} - z_OA_{22};\quad A_{26}^G = x_OA_{22};$$
$$A_{33}^G = A_{33};\quad A_{35}^G = -x_OA_{33};\quad A_{42}^G = A_{42} - z_OA_{22},$$
$$A_{44}^G = A_{44} - 2z_OA_{24} + z_O^2A_{22};$$
$$A_{46}^G = x_OA_{42} - z_Ox_OA_{22},$$
$$A_{51}^G = A_{51} + z_OA_{11};\quad A_{53}^G = -x_OA_{33};$$
$$A_{55}^G = A_{55} + z_O^2A_{11} + x_O^2A_{33} + 2z_OA_{15},$$
$$A_{62}^G = x_OA_{22};\quad A_{64}^G = x_OA_{24} - z_Ox_OA_{22};$$
$$A_{66}^G = A_{66} + x_O^2A_{22}. \tag{5.143}$$

Equations (5.143) use the fact that $A_{24} = A_{42}$, $A_{15} = A_{51}$. We used the notation A_{jk}^{tank} instead of A_{jk}^G in Chapter 3. Furthermore, we warned that the frozen liquid effect must not be accounted for twice when the equations of ship motion were formulated.

Abramson (1966) solved the boundary-value problem for forced surge and pitch and used direct pressure integration over the walls and the bottom to get expressions for forces and pitch moments due to surge and pitch; that is, in our definitions, A_{11}^G, A_{51}^G, and A_{55}^G with $(x_O, 0, z_O) = (0, 0, \frac{1}{2}h)$ (the liquid mass center). He also omitted the quasi-steady terms in A_{ij}^{filled}. A tedious derivation without quasi-steady terms shows that our expressions give the same results in this particular case as those of Abramson (1966).

5.4.3.3 Tank added mass coefficients in a ship motion analysis

In reality we deal with several tanks. When we introduce the global coordinate system, the corresponding added mass coefficients caused by the liquid in these tanks can be computed by the scheme from the previous section. However, the liquid density, length, width, depth, and x_O and z_O are generally different for each tank; therefore, we have to know $\rho^{(k)}$, $L_1^{(k)}$, $L_2^{(k)}$, $h^{(k)}$, $x_O^{(k)}$, and $z_O^{(k)}$ $(k = 1, \ldots, N)$, where N is the number of tanks. Adding the superscript (k) to these parameters in formulas (5.126), (5.133), and (5.134) gives the added mass coefficients $A_{ij}^{(k)}$ (or, if necessary $(A_{ij}^{\mathrm{dyn}})^{(k)}$) for each tank in the local tank coordinate system, but eq. (5.143) gives the total added mass coefficients caused by all the tanks as

$$
\begin{aligned}
&A_{11}^G = \sum_{k=1}^N A_{11}^{(k)}; \quad A_{15}^G = \sum_{k=1}^N \left(A_{15}^{(k)} + z_O^{(k)} A_{11}^{(k)}\right);\\
&A_{22}^G = \sum_{k=1}^N A_{22}^{(k)}, \quad A_{24}^G = \sum_{k=1}^N \left(A_{24}^{(k)} - z_O^{(k)} A_{22}^{(k)}\right);\\
&A_{26}^G = \sum_{k=1}^N x_O^{(k)} A_{22}^{(k)}; \quad A_{33}^G = \sum_{k=1}^N A_{33}^{(k)},\\
&A_{35}^G = -\sum_{k=1}^N x_O^{(k)} A_{33}^{(k)}; \quad A_{42}^G = \sum_{k=1}^N \left(A_{42}^{(k)} - z_O^{(k)} A_{22}^{(k)}\right),\\
&A_{44}^G = \sum_{k=1}^N \left(A_{44}^{(k)} - 2 z_O^{(k)} A_{24}^{(k)} + \left(z_O^{(k)}\right)^2 A_{22}^{(k)}\right);\\
&A_{46}^G = \sum_{k=1}^N x_O^{(k)} \left(A_{42}^{(k)} - z_O^{(k)} A_{22}^{(k)}\right),\\
&A_{51}^G = \sum_{k=1}^N \left(A_{51}^{(k)} + z_O^{(k)} A_{11}^{(k)}\right); \quad A_{53}^G = -\sum_{k=1}^N x_O^{(k)} A_{33}^{(k)},\\
&A_{55}^G = \sum_{k=1}^N \left(A_{55}^{(k)} + \left(z_O^{(k)}\right)^2 A_{11}^{(k)}\right.\\
&\qquad\left. + \left(x_O^{(k)}\right)^2 A_{33}^{(k)} + 2 z_O^{(k)} A_{15}^{(k)}\right);\\
&A_{62}^G = \sum_{k=1}^N x_O^{(k)} A_{22}^{(k)};\\
&A_{64}^G = \sum_{k=1}^N x_O^{(k)} \left(A_{24}^{(k)} - z_O^{(k)} A_{22}^{(k)}\right);\\
&A_{66}^G = \sum_{k=1}^N \left(A_{66}^{(k)} + \left(x_O^{(k)}\right)^2 A_{22}^{(k)}\right).
\end{aligned}
\tag{5.144}
$$

Based on eq. (5.144), we can propose an alternative semianalytical solution to the tank added mass coefficients in Newman (2005) for the case in Figure 3.25. Newman's analysis was based on the WAMIT code with particular analytical control of the added-mass coefficients (Newman, personal communication, 2008) by solving the boundary-value problem for velocity potentials due to sway and roll and using direct pressure integration to obtain hydrodynamic forces and moments. The calculations were made in the global coordinate system and the corresponding figures in Newman (2005) present results for the added mass coefficients that account for both external ship flow and sloshing. Newman's definition of added mass coefficients did not include the effect of integrating hydrostatic pressures at the instantaneous ship position; the quasi-steady components of the added mass coefficients caused by sloshing were neglected. Our focus is only on the added mass coefficients due to sloshing; that is, we exclude from Newman's definitions of the added mass the contribution due to external flow. Translating the numerical results by Newman to the language of our modal method, our interest is in the added mass coefficients, which can be obtained from eq. (5.144) with $(A_{ij}^{\mathrm{dyn}})^{(k)}$ instead of $A_{ij}^{(k)}$ (i.e., by neglecting the quasi-steady components). We use the notation A_{jk}^{Newman} for this purpose. Numerical results were nondimensionalized by Newman (2005) as

$$
\frac{A_{11}^{\mathrm{Newman}}}{\rho\nabla}, \quad \frac{A_{15}^{\mathrm{Newman}}}{a\rho\nabla}, \quad \frac{A_{55}^{\mathrm{Newman}}}{a^2\rho\nabla}, \quad \frac{A_{22}^{\mathrm{Newman}}}{\rho\nabla},
$$
$$
\frac{A_{24}^{\mathrm{Newman}}}{a\rho\nabla}, \quad \frac{A_{44}^{\mathrm{Newman}}}{a^2\rho\nabla}, \quad \frac{A_{66}^{\mathrm{Newman}}}{a^2\rho\nabla}, \tag{5.145}
$$

where ρ is the density of the sea as well as the tank liquid, ∇ is the displaced volume of water of the ship, and a is the radius of the semicircular cross-section at midships.

Furthermore, we should transform these coefficients to the global coordinate system as in eqs. (5.144). The results are presented in Figure 5.17 and show generally good agreement with Newman (2005).

Because the tank lengths are the same, surge and pitch excitation lead to the same natural sloshing frequencies, $\sigma_{2i-1,0}$, for each tank. The lowest resonant sloshing frequency is at $\sigma_{1,0}^2 a/g = 1.184$, which is detected in Figure 5.17(a) for A_{11}^{Newman}, A_{15}^{Newman}, and A_{55}^{Newman}. Furthermore,

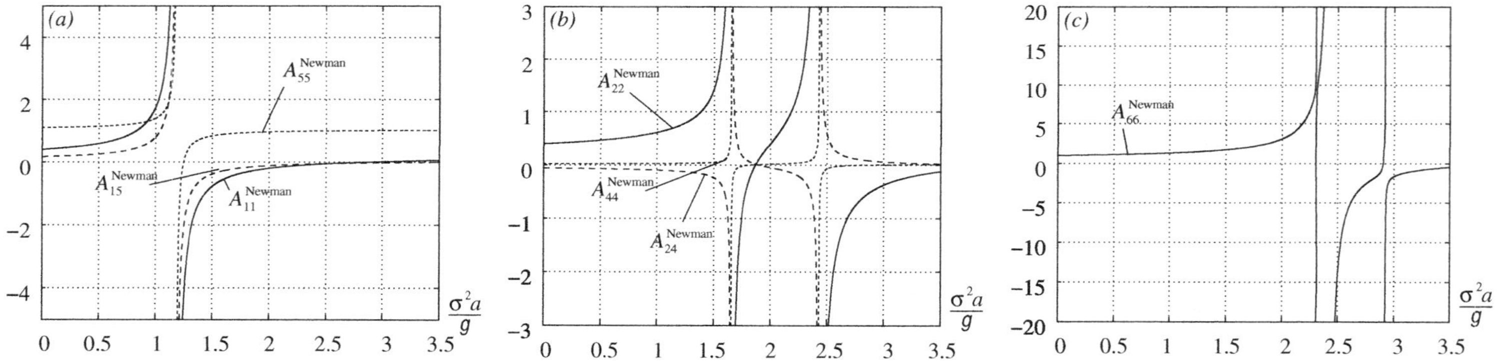

Figure 5.17. Contribution to the nondimensional "dynamic" added mass coefficients A_{ij}^{Newman} from the tanks in the ship shown in Figure 3.25. Quasi-steady components and external flow effects are excluded. The coefficients are made nondimensional as in eq. (5.145).

sway and roll may excite the natural frequencies $\sigma_{0,2i-1}$ of the tanks, which are different for the center and fore/aft tanks. Figure 5.17(b) demonstrates this fact at $\sigma^2 a/g = 1.653$ (center tank) and 2.427 (fore/aft tanks). Finally, the yaw motions cause excitation of the diagonal modes with $\sigma_{2i-1,2j-1}(i, j \geq 1)$ that are different for the tanks. Furthermore, yaw excites the sway mode of the fore/aft tanks. Because of that, Figure 5.17(c) shows three resonances: at $\sigma^2 a/g = 2.31$ (caused by the mode with $\sigma_{1,1}$ for the center tank), 2.427 (first sway mode of the fore/aft tanks), and 2.922 (due to the (1,1) modes of the fore/aft tanks). The last two resonances are clearly present in Figure 3 of Newman (2005), but the first was only indicated by a small irregular feature in his figure. The reason is that the first resonance is very narrow and that Newman's figure was made with a 0.01 increment in $\sigma^2 a/g$, which is too large to clearly detect it. Finally, we must remark that Newman's Figure 3 demonstrates a pronounced hump for the yaw added mass, whereas our Figure 5.17(c) does not show a similar behavior; the hump is associated with the added mass due to the external flow around the ship, whereas we concentrate only on the added mass coefficients caused by sloshing.

5.4.4 Hydrodynamic coefficients for an upright circular cylindrical tank

Based on the natural modes in Section 4.3.2.2 and the Stokes–Joukowski potential, which can be found in an analytical form in the cylindrical coordinate system (Lukovsky *et al.*, 1984), we can compute the hydrodynamic coefficients for a vertical circular cylindrical tank. Traditionally, the procedure implies a normalization of natural modes (4.37) and (4.38) in such a way to provide unit amplitude at the wall for the surface profiles. Because two different natural modes are associated with the cosine and sine terms of the azimuthal angle for each natural frequency $\sigma_{m,i}$ $(m \neq 0)$, we need an additional index, which is denoted by 1 for the cosine term and 2 for the sine term:

$$\varphi_{m,n,1}(r,\theta,z) = R_{m,n}(r)\frac{\cosh(k_{m,n}(z+h))}{\cosh(k_{m,n}h)}\cos(m\theta),$$
$$\varphi_{m,n,2}(r,\theta,z) = R_{m,n}(r)\frac{\cosh(k_{m,n}(z+h))}{\cosh(k_{m,n}h)}\sin(m\theta),$$
$$m = 0, 1, \ldots; \quad n = 1, 2, \ldots \tag{5.146}$$

and, therefore,

$$f_{m,n,1}(r,\theta) = \varphi_{m,n,1}(r,\theta,0) = R_{m,n}(r)\cos(m\theta),$$
$$f_{m,n,2}(r,\theta) = \varphi_{m,n,2}(r,\theta,0) = R_{m,n}(r)\sin(m\theta), \tag{5.147}$$

where

$$R_{m,n}(r) = \frac{J_m(k_{m,n}r)}{J_m(k_{m,n}R_0)}; \quad J'_m\underbrace{(k_{m,n}R_0)}_{\iota_{m,n}} = 0;$$
$$\kappa_{m,n} = \frac{\sigma^2_{m,n}}{g} = k_{m,n}\tanh(k_{m,n}h). \tag{5.148}$$

The next step consists of finding the Stokes–Joukowski potential. After transformation to the cylindrical coordinate system, Neumann boundary-value problem (5.21) yields the following boundary conditions:

$$\frac{\partial \Omega_{01}}{\partial n} = -(zn_r - rn_z)\sin\theta;$$
$$\frac{\partial \Omega_{02}}{\partial n} = (zn_r - rn_z)\cos\theta; \quad \frac{\partial \Omega_{03}}{\partial n} = 0 \tag{5.149}$$

on the surfaces S_0 and Σ_0, where n_z and n_r are the normal vector components along the z-axis and the radial direction, respectively. Due to these conditions, the Stokes–Joukowski potential admits separation of the variables (r, z) and θ and the solutions are as follows:

$$\Omega_{01} = -F(r,z)\sin\theta;$$
$$\Omega_{02} = F(r,z)\cos\theta; \quad \Omega_{03} = 0. \tag{5.150}$$

The boundary problem for $F(r, z)$ follows from the corresponding problem (5.21) after separation of the variables. The problem is stated in the meridional plane cross-section and satisfies

$$\frac{\partial^2 F}{\partial z^2} + \frac{\partial^2 F}{\partial r^2} + \frac{1}{r}\frac{\partial F}{\partial r} - \frac{1}{r^2}F = 0 \quad \text{in}$$
$$0 < r < R_0, \quad -h < z < 0;$$
$$\frac{\partial F}{\partial z} = -r \quad \text{on} \quad 0 < r < R_0 \quad \text{for} \quad z = 0 \quad \text{and} \quad z = -h;$$
$$\frac{\partial F}{\partial r} = z \quad \text{on} \quad -h < z < 0 \quad \text{for} \quad r = R_0;$$
$$|F(0,z)| < \infty. \tag{5.151}$$

The problem has an analytical solution if we follow a similar procedure as for the rectangular tank (see Section 5.4.2.1). The procedure involves the substitution $F(r, z) = rz + F_1(r, z)$ that transforms eqs. (5.151) to the Neumann boundary problem with respect to $F_1(r, z)$. The problem has the same governing equation, but it is characterized by a zero condition of the r-derivative at $r = R_0$. The boundary conditions on the bottom and the mean free surface are $\partial F_1/\partial z|_{z=-h;z=0} = -2r$. This problem is solved by the method of separation of the spatial variables r and z (similar to the procedure in Section 5.4.2.1). The final result is

$$F(r, z) = zr - 4R_0^2 \sum_{j=1}^{\infty} \frac{R_{1,j}(r)}{(\iota_{1,j}^2 - 1)\iota_{1,j}} \times \frac{\sinh\left(k_{1,j}\left(z + \frac{1}{2}h\right)\right)}{\cosh\left(\frac{1}{2}k_{1,j}h\right)}. \quad (5.152)$$

Equation (5.152) is found by direct substitution in problem (5.151) by remembering that the term zr satisfies the Laplace equation and boundary conditions on the vertical wall in eq. (5.151). To compute the coefficients in the sum, we use properties of the Bessel function J_1.

When one knows natural modes (5.146)–(5.148) and the Stokes–Joukowski potential eqs. (5.150) and (5.152), one can analytically find the hydrodynamic coefficients defined by eq. (5.26). Because the Stokes–Joukowski potential has uniquely defined $\sin\theta$ and $\cos\theta$ components and $x = r\cos\theta$, $y = r\sin\theta$ in expressions (5.26), the only nonzero hydrodynamic coefficients are $\mu_{1,j,1}$, $\mu_{1,j,2}$, $\lambda_{1(1,j,1)}$, $\lambda_{2(1,j,2)}$ and $\lambda_{01(1,j,2)}$, $\lambda_{02(1,j,1)}$ ($j = 1, 2, \ldots$). Therefore, the right-hand-side terms in modal equations (5.27) are zero except for the modes with $m = 1$. Only these modes can be excited within the framework of linear theory due to external forcing. The nonzero forcing terms (5.28) in modal equations (5.27) are

$$\mu_{1,j,1} = \mu_{1,j,2} = \frac{\rho\pi}{\kappa_{1,j}} \int_0^{R_0} r R_{1,j}^2(r)\,dr = \frac{\rho\pi R_0^2}{\kappa_{1,j}} \frac{\iota_{1,j}^2 - 1}{2\iota_{1,j}^2} = \frac{\rho\pi R_0^3(\iota_{1,j}^2 - 1)}{2\iota_{1,j}^3 \tanh(\iota_{1,j}h/R_0)},$$

$$\lambda_{1(1,j,1)} = \lambda_{2(1,j,2)} = \rho\pi \int_0^{R_0} r^2 R_{1,j}(r)\,dr = \frac{\rho\pi R_0^3}{\iota_{1,j}^2},$$

$$\lambda_{02(1,j,1)} = -\lambda_{01(1,j,2)} = \rho\pi \int_0^{R_0} r R_{1,j}(r) F(r, 0)\,dr = -\frac{2\pi\rho R_0^4}{\iota_{1,j}^3} \tanh\left(\frac{\iota_{1,j}h}{2R_0}\right). \quad (5.153)$$

Furthermore, analytically we can also find the inertia tensor, which has only two nonzero diagonal elements:

$$J_{011}^1 = J_{022}^1 = \rho\pi R_0^2 \times \left[\frac{1}{3}h^3 - \frac{3}{4}hR_0^2 + 16R_0^2 \sum_{j=1}^{\infty} \frac{\tanh(\iota_{1,j}h/2R_0)}{\iota_{1,j}^3(\iota_{1,j}^2 - 1)}\right]. \quad (5.154)$$

For $m = 1$, the modal equations are formulated relative to $\beta_{1,j,1}$ and $\beta_{1,j,2}$, respectively. The flow associated with excitation in the x–z-plane (i.e., due to surge and pitch) has to be symmetric with respect to the x–z-plane. Because the generalized coordinate $\beta_{1,j,2}$ is associated with a mode varying as $\sin\theta$ (i.e., antisymmetric with respect to the x–z-plane), $\beta_{1,j,2}$ cannot be excited by surge and pitch. Similarly we can argue that sway and roll cannot excite $\beta_{1,j,1}$. Accounting for eq. (5.28) and explicit expressions (5.153), the modal equations for $m = 1$ take the form

$$\ddot\beta_{1,j,1} + \sigma_{1,j}^2 \beta_{1,j,1} = -P_j\left[\ddot\eta_1(t) - g\eta_5(t) - S_j\ddot\eta_5(t)\right],$$
$$\ddot\beta_{1,j,2} + \sigma_{1,j}^2 \beta_{1,j,2} = -P_j\left[\ddot\eta_2(t) + g\eta_4(t) + S_j\ddot\eta_4(t)\right],$$
$$j = 1, 2, \ldots, \quad (5.155)$$

where

$$P_j = \frac{2\iota_{1,j}\tanh(\iota_{1,j}h/R_0)}{\iota_{1,j}^2 - 1},$$
$$S_j = \frac{2R_0\tanh(\iota_{1,j}h/2R_0)}{\iota_{1,j}}. \quad (5.156)$$

Other modal equations are homogeneous (i.e., the right-hand sides of the equations are zero).

The use of modal system (5.155) makes it possible to estimate a deep-water limit in terms of the ratio h/R_0 by evaluating the hydrodynamic coefficients P_j, S_j and the natural frequencies $\sigma_{i,j}$. These values are proportional to a hyperbolic tangent term that becomes close to 1 as $h/R_0 \to \infty$. The limits are

$$\frac{\sigma_{m,n}^2 R_0}{g} \to \iota_{m,n}, \quad P_j \to \frac{2\iota_{1,j}}{\iota_{1,j}^2 - 1} \quad \text{and} \quad S_j \to \frac{2R_0}{\iota_{1,j}}. \quad (5.157)$$

The difference between the finite-liquid-depth values of $\sigma_{m,n}^2$ and P_j is less than 1% for $h/R_0 > 1.5$. As for S_j, it requires $h/R_0 > 3$ to get the same agreement with the infinite depth value. Based on these estimates, we can apply the deep-liquid limit for linear sloshing in a circular cylindrical vertical tank when $h > 3R_0$.

5.4.5 Coupling between sloshing and wave-induced vibrations of a monotower

We considered sloshing inside the shaft of the Draugen platform in Exercise 4.11.5. The focus was on the effect of internal pipes on the lowest natural sloshing frequency. We now consider the coupling between external wave-induced elastic vibrations of the platform and sloshing. The effect of the internal pipes is neglected. The water level inside the shaft is assumed to coincide with the outer mean free surface. Because our attention is on the highest natural period for the elastic vibrations, the structural model is simplified by using a cantilever beam model with a rigid mass at the free end (Clough & Penzien, 1993). The concrete column of the Draugen platform is represented as a beam with varying bending stiffness and mass per unit length. The rigid mass at the top of the beam is an idealization of the deck. The platform and the idealized structural model are illustrated in Figure 5.18. The rotational inertial effect of the rigid mass of the deck is neglected, which means that one of the boundary conditions at the free end of the beam is zero bending moment. Furthermore, the shear force at the free end of the beam is equal to mass times the acceleration of the rigid mass. The fact that we use a cantilever beam with zero deflection and slope at the top of the caisson means that we neglect the influence of the motions of the caisson and its interaction with the soil (e.g., in terms of soil damping). Structural damping is also neglected.

When it comes to both external and internal wave effects, the considered wavelengths are of the order of the diameter of the tower at the waterplane. The wave effects are therefore concentrated near the free surface. Because the ratio between the submerged tower height and the maximum cross-sectional diameter is large, the nonwave part of the flow can be handled by slender body theory without free-surface effects. A first approximation of the flow from a far-field point of view is a distribution of horizontal dipoles along the axis of the submerged part of the tower. The consequence is that strip theory can be used (Newman, 1977). Strip theory implies no hydrodynamic interaction between the cross-sections.

5.4.5.1 Theory

We use the Earth-fixed coordinate system $Oxyz$ and the notations defined in Figure 5.18. The origin is at the intersection between the mean free surface and the shaft axis when the platform is at rest. The z-axis is positive upward. The lower part of the deck and the top of the caisson correspond to $z = h_1$ and $z = -h$, respectively. The Euler–Bernoulli beam equation and the boundary conditions can be expressed as

$$\begin{aligned} m\ddot{\mathtt{w}} + (\mathtt{EIw}'')'' &= f_{\text{ext}} + f_{\text{slosh}} \qquad -h < z < h_1; \\ \mathtt{w}(-h,t) = \mathtt{w}'(-h,t) &= 0; \quad \mathtt{w}''(h_1,t) = 0; \\ (\mathtt{EIw}'')'(h_1,t) &= M_0\ddot{\mathtt{w}}(h_1,t), \end{aligned} \tag{5.158}$$

where the function $\mathtt{w}(z,t)$ describes the beam deflections in the Oyz-plane. Dots are used for time derivatives and primes for derivatives with respect to z. The derivation of the beam equation and appropriate boundary conditions is partly described in Section 2.5.2.2 as an example of using the Lagrange variational principle. Both the structural mass per unit length, m, and the bending stiffness $\mathtt{EI}$ are functions of z; $\mathtt{E}$ is Young's modulus of elasticity and $\mathtt{I}$ is the second moment of the structural cross-sectional area with respect to the x–z-plane. The notation for the deck's mass is M_0. Furthermore, f_{ext} and f_{slosh} are horizontal forces per unit length due to external hydrodynamic loads and sloshing, respectively; therefore, f_{ext} includes the effects of the added mass, wave radiation damping, and the wave excitation loads. The formulation of the boundary conditions at $z = h_1$ uses the facts that the bending moment is $-\mathtt{EIw}''$ and the shear force is the z-derivative of the bending moment. We denote ρ_e and ρ_i as the density of the external and internal liquids, respectively.

For simplicity in the forthcoming formulas, let us define the Heaviside function:

$$\chi_{[a,b]}(z) = \begin{cases} 1, & a \le z \le b, \\ 0, & \text{otherwise.} \end{cases}$$

Our focus is now on the internal liquid (i.e., on what we denote as the tank). The effect of the beam on the internal liquid is associated with small deflections of the vertical walls. The only nonhomogeneous term in the modified problem

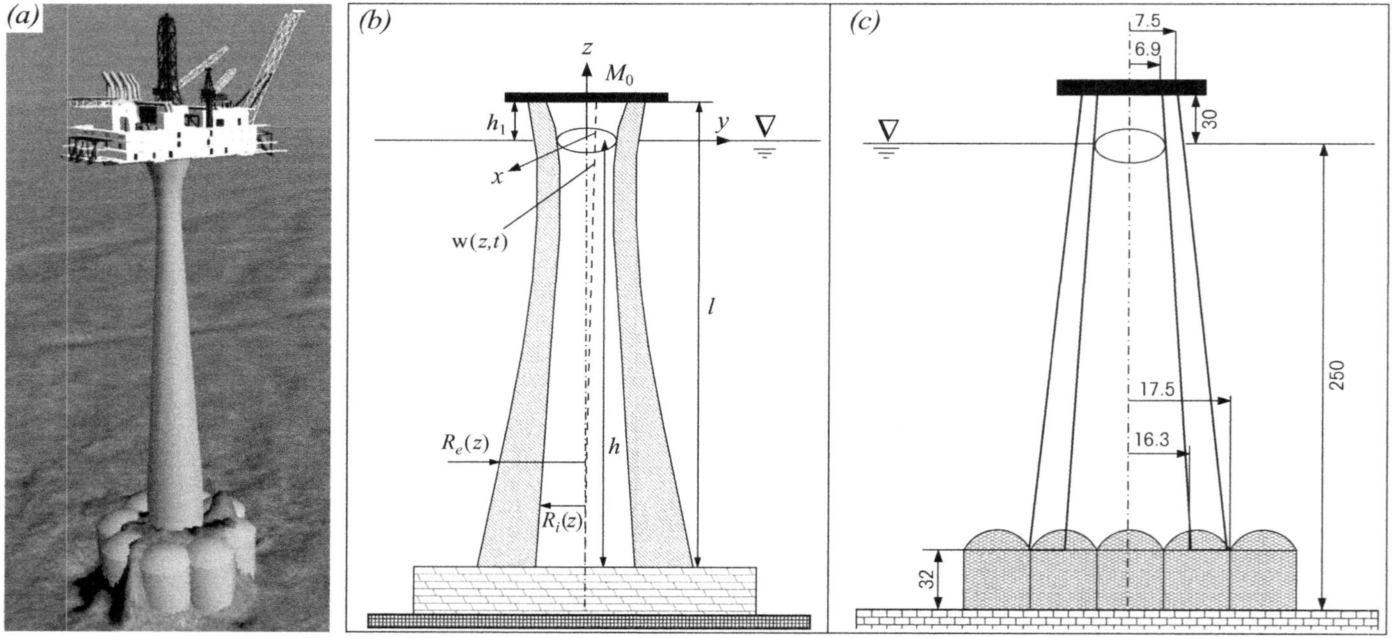

Figure 5.18. (a) The Draugen platform (Artist: Bjarne Stenberg). (b) and (c) Cantilever beam models with a rigid mass at the free end representing structurally the Draugen platform. Dimensions are in meters.

(5.14)–(5.18) (rewritten in the cylindrical coordinate system) is the following Neumann condition on the tank walls:

$$\left.\frac{\partial \Phi}{\partial r}\right|_{r=R_i(z)} = n_y \dot{w}(z,t) = \sin\theta\, \dot{w}(z,t), \qquad (5.159)$$

where $R_i(z)$ is the internal radius (see Figure 5.18) and θ is the angular coordinate in the cylindrical coordinate system $Or\theta z$. In deriving eq. (5.159), we have used the fact that the vertical tower is a slender body, which implies that the z-component of the normal surface vector is negligible. By following the description in Section 5.4.2.4 of small deflections of the tank walls (here associated with the beam deflections), we should represent the velocity potential as follows:

$$\Phi(z,r,\theta,t) = \Omega_d(z,r,\theta,t) + \sum_M R_M(t)\varphi_M(z,r,\theta), \qquad (5.160)$$

where the last sum represents the modal solution (consistent with general scheme (5.20)). The function Ω_d should satisfy condition (5.159), zero-Neumann conditions on the bottom and represents the effect of the liquid domain deforming as a solid body due to the beam deflections. In addition, we assume the modal solution for the free surface in the form of eq. (5.19). Volume conservation is satisfied by the solution, which follows by first using the boundary conditions for Ω_d and requiring that the net mass flux in or out of the liquid domain due to Ω_d is zero. Furthermore, the modal solutions satisfy volume conservation by definition.

Because the considered body is slender, no wave effects exist, and no net source effects are associated with Ω_d, we may find Ω_d by using strip theory:

$$\Omega_d = \underbrace{r\sin\theta}_{y} \cdot \dot{w}(z,t). \qquad (5.161)$$

The modal technique in Section 5.3.1.3 combined with eqs. (5.160) and (5.161) leads to the following modal equations:

$$\begin{aligned}\ddot{\beta}_M + \sigma_M^2 \beta_M &= -\mu_M^{-1}\rho_l \int_{\Sigma_0} \left.\frac{\partial \Omega_d}{\partial t}\right|_{z=0} f_M\, dS \\ &= -\ddot{w}(0,t)\mu_M^{-1}\rho_l \int_{\Sigma_0} y f_M\, dS \\ &= -\ddot{w}(0,t)\lambda_{2M}\mu_M^{-1}. \qquad (5.162)\end{aligned}$$

Comparing eq. (5.162) with eqs. (5.27) and (5.28) shows that sloshing in our case is the same as for a tank that moves horizontally with the acceleration $\ddot{\eta}_2(t) = \dot{v}_{O2}(t) = \ddot{w}(0,t)$ along the Oy-axis, which is a consequence of the fact that the sloshing modes φ_M decay exponentially from the mean free surface toward the bottom. The velocity field component due to sloshing is localized in the interval $[-h_e, 0]$, where h_e is the effective liquid depth. According to Section 4.3.2.2, the latter is about $1.5R_i(0)$ for the studied case and is small relative to the actual tank depth h (i.e., $h_e/h \ll 1$). This fact and a small inclination of the tank walls at $z = 0$ make it possible to use the natural modes for an upright circular cylindrical tank (see Section 5.4.4) as an approximation of $\varphi_M = \varphi_{m,j,i}$, $M = (m,j,i)$. We subsequently use the notation R_{i0} for internal radius $R_i(0)$ at the mean free surface. The modal equations with nonzero forcing terms (see eqs. (5.155) and (5.162)) are as follows:

$$\ddot{\beta}_{1,j,2} + \sigma_{1,j}^2 \beta_{1,j,2} = -P_j \ddot{w}(0,t), \quad j = 1,2,\ldots, \qquad (5.163)$$

where P_j and $\sigma_{1,j}^2$ are expressed by eq. (5.157) in the deep-water limit; that is, $P_j = 2\iota_{1,j}/(\iota_{1,j}^2 - 1)$ and $\sigma_{1,j}^2 = g\iota_{1,j}/R_{i0} = gk_{1,j}$.

From the definitions of Section 5.4.4, we now know the approximate solution of the sloshing problem in terms of the velocity potential:

$$\Phi(z,r,\theta,t) = r\sin\theta \cdot \dot{w}(z,t) + \sin\theta \cdot R_{i0} \times \sum_{j=1}^{\infty} \beta_{1,j,2}(t) \frac{J_1(k_{1,j}r)}{J_1(\iota_{1,j})} \frac{\exp(k_{1,j}z)}{\iota_{1,j}}, \qquad (5.164)$$

where we have used the deep-water limit for the natural sloshing modes; that is, $\cosh(k_{1,j}(z+h))/\sinh(k_{1,j}h) \approx \exp(k_{1,j}z)$. It follows from the linearized Bernoulli equation that the pressure is

$$p = p_0 - \rho_l g z - \rho_l r \sin\theta \ddot{w}(z,t) - \rho_l \sin\theta \cdot R_{i0} \times \sum_{j=1}^{\infty} \ddot{\beta}_{1,j,2}(t) \frac{J_1(k_{1,j}r)}{J_1(\iota_{1,j})} \frac{\exp(k_{1,j}z)}{\iota_{1,j}}. \qquad (5.165)$$

The resulting horizontal force per unit length is

$$\begin{aligned} f_{slosh.}(z,t) &= \int_0^{2\pi} r n_y p(z, R_i(z), \theta, t)\, d\theta \\ &\approx -\rho_l \pi \chi_{[-h,0]}(z)\Big[R_i^2(z)\ddot{w}(z,t) + R_{i0}^2 \times \sum_{j=1}^{\infty} \ddot{\beta}_{1,j,2}(t)\exp(k_{1,j}z)/\iota_{1,j}\Big], \end{aligned} \qquad (5.166)$$

where the first term is the inertia force per unit length due to "frozen" water. The second term, together with modal equations (5.163), describes the sloshing effect.

The substitution of eq. (5.168) into eqs. (5.158), (5.163), (5.166), and (5.167) results in the following spectral boundary problem with respect to $W(z)$:

$$-\sigma^2 \underbrace{\left[m(z) + \pi\chi_{[-h,0]}(z)\left(\rho_e R_e^2(z) + \rho_i R_i^2(z)\right)\right]}_{M_1(z)} W + (\mathrm{EI}(z)W'')''$$
$$= \underbrace{\sigma^2 W(0)\chi_{[-h,0]}(z)\,\rho_i \pi R_{i0}^2 \sum_{j=1}^{\infty} \frac{P_j}{(\sigma_{1,j}/\sigma)^2 - 1} \cdot \frac{\exp(k_{1,j}z)}{\iota_{1,j}}}_{M_2(\sigma^2,z)}, \quad -h < z < h_1;$$
$$W(-h) = W'(-h) = 0; \quad W''(h_1) = 0; \quad (\mathrm{EI}(z)W'')'\big|_{z=h_1} = -\sigma^2 M_0 W(h_1). \tag{5.169}$$

5.4.5.2 Undamped eigenfrequencies of the coupled motions

When analyzing the eigenfrequencies, the external wave excitation loads are set equal to zero, which means that f_{ext} in eq. (5.158) is due to only the added mass and wave radiation damping. Our focus in this section is on undamped natural frequencies. The added mass is frequency dependent due to the fact that far-field waves are generated, which is mathematically a consequence of the Kramers–Kronig relations between added mass and damping (Kotik & Mangulis, 1962) and the fact that damping is associated with wave radiation. However, the wave effect is felt only in the immediate vicinity of the free surface, as we already have pointed out. Added mass effects exist along the total submerged part of the beam. Therefore, we assume that the wave effects on the added mass can be neglected. This assumption enables us to use a strip theory as argued earlier:

$$f_{\mathrm{ext}}(z,t) = -\rho_e \pi R_e^2(z)\ddot{\mathrm{w}}(z,t)\chi_{[-h,0]}(z), \tag{5.167}$$

where $R_e(z)$ is the external cylinder radius.

Let us present the solution of the problem from Section 5.4.5.1 as

$$\beta_{1,j,2} = B_j \exp[\mathrm{i}(\sigma t - \alpha)], \quad j = 1, 2, \ldots;$$
$$\mathrm{w} = W(z)\exp[\mathrm{i}(\sigma t - \alpha)], \tag{5.168}$$

where $W(z)$ is an eigenmode and σ is the corresponding eigenfrequency of the coupled motions. The α-value is the phase shift. An infinite number of eigenfrequencies exists; however, our focus is on the lowest eigenfrequencies. In practice, the higher modes are expected to be strongly structurally damped.

Because of the sloshing effect, the problem depends nonlinearly on the spectral parameter σ^2. Furthermore, when the eigenfrequency is close to the sloshing frequencies $\sigma_{1,j}$, the denominator on the right-hand side of the equation becomes small, making the sloshing influence of primary importance.

5.4.5.3 Variational method

The solution of spectral problem (5.169) is based on a variational formulation. For this purpose, let us consider a solution of the problem and a complete set of smooth test functions $U(z)$. We use an analogy to the FEM described in Section 10.5, where $U(z)$ denotes weighting functions. What we now do in our variational formulation is to multiply both eq. (5.169) and all boundary conditions except $W(-h) = W'(-h) = 0$ by a test function $U(z)$. The next step is to integrate the product of the governing equation in eq. (5.169) and $U(z)$ from $z = -h$ to $z = h_1$ and add terms from the product of boundary conditions and the test function and its derivative. This formulation gives the following variational equality:

$$\int_{-h}^{h_1} \left[\sigma^2 M_1(z)W - (\mathrm{EI}(z)W'')''\right.$$
$$\left. + \sigma^2 M_2(\sigma^2, z)\chi_{[-h,0]}(z)W(0)\right] U(z)\,\mathrm{d}z$$
$$- \left[\mathrm{EI}(z)W'' \cdot U'\right]\Big|_{-h}^{h_1}$$
$$+ \left[\left((\mathrm{EI}(z)W'')' + \sigma^2 M_0 W\right) U\right]\Big|_{-h}^{h_1} = 0. \tag{5.170}$$

We can check that the last two terms involving values at $z = h_1$ and $z = -h$ are zero due to the boundary conditions. The requirement for $W(z)$ to be a solution is that eq. (5.170) must be true for all U and W that satisfy the clamped-end boundary conditions

$$W(-h) = W'(-h) = U(-h) = U'(-h) = 0. \tag{5.171}$$

The variational equality can be transformed to a symmetric form relative to W and U by using integration by parts, the fact that $h_e \ll h$, and the deep-liquid limit. The latter means that

$$\begin{aligned}
&\int_{-h}^{h_1} [\sigma^2 M_2(\sigma^2, z)\chi_{[-h,0]}(z)W(0)]\, U(z)\,\mathrm{d}z \\
&= \sigma^2 W(0) \int_{-h}^{0} M_2(\sigma^2, z)\, U(z)\,\mathrm{d}z \\
&\approx \sigma^2 W(0) \int_{-h_e}^{0} M_2(\sigma^2, z)\, U(z)\,\mathrm{d}z \\
&\approx \sigma^2 W(0)U(0) \int_{-h_e}^{0} M_2(\sigma^2, z)\,\mathrm{d}z \\
&= \sigma^2 \underbrace{\rho_i \pi R_{i0}^3 \sum_{j=1}^{\infty} \frac{2}{(\sigma_{1,j}^2/\sigma^2 - 1)\iota_{1,j}(\iota_{1,j}^2 - 1)}}_{M_3(\sigma^2)} \\
&\quad \times W(0)U(0).
\end{aligned} \tag{5.172}$$

The final variational expression is as follows:

$$\begin{aligned}
&\int_{-h}^{h_1} [\sigma^2 M_1(z)W(z)U(z) - \mathrm{EI}(z)W''(z)U''(z)]\,\mathrm{d}z \\
&\quad + \sigma^2 M_3(\sigma^2)W(0)U(0) + \sigma^2 M_0 W(h_1)U(h_1) = 0.
\end{aligned} \tag{5.173}$$

We now introduce the test functions $\phi_k(z)$, which we require to satisfy $\phi_k(-h) = \phi_k'(-h) = 0$. An approximation of the solution is represented in terms of $\phi_k(z)$:

$$W(z) = \sum_{k=1}^{N} a_k \phi_k(z), \quad \phi_k(-h) = \phi_k'(-h) = 0. \tag{5.174}$$

Substitution of approximate solution (5.174) into variational equality (5.173) and separate use of the test functions ϕ_n $(n = 1, \ldots, N)$ leads to the homogeneous matrix problem $A\boldsymbol{a} = 0$, where $\boldsymbol{a} = (a_1, \ldots, a_N)^T$ denotes the unknowns and A is a symmetric matrix that can be expressed as

$$\begin{aligned}
A = \Bigg\{ &\int_{-h}^{h_1} [\sigma^2 M_1 \phi_k \phi_n - \mathrm{EI}(z)\phi_n''\phi_k'']\,\mathrm{d}z \\
&+ \sigma^2 \phi_n(0)\phi_k(0)M_3(\sigma^2) \\
&+ \sigma^2 M_0 \phi_n(h_1)\phi_k(h_1) \Bigg\}_{n,k=1,\ldots,N}.
\end{aligned} \tag{5.175}$$

The natural modes imply nontrivial solutions of the matrix problem. These solutions are possible for eigenfrequencies σ that correspond to the roots of the equation

$$\det A(\sigma^2) = 0. \tag{5.176}$$

The simplest way to construct a complete basis for the variational procedure is to use polynomials, which make it possible to satisfy the clamped-end conditions at $z = -h$. In addition, to improve convergence we also choose polynomials that satisfy the zero-moment condition at $z = h_1$ (i.e., $\phi_k''(h_1) = 0$). Examples of these polynomials are

$$\begin{aligned}
&\phi_k(z) = (z+h)^{k+1}\,[1 + \alpha_k(z+h)]\,/N_k; \\
&\alpha_k = -\frac{k}{l(k+2)}; \quad N_k = \sqrt{\int_{-h}^{h_1} \phi_k^2\,\mathrm{d}z}, \\
&k = 1, 2, \ldots,
\end{aligned} \tag{5.177}$$

where $l = h_1 + h$ (see Figure 5.18(b)). The variational scheme with basis (5.177) is applied to the pyramid-like beam shown in Figure 5.18(c). This figure is a simplified model of the actual Draugen platform from Figure 5.18(a). We do not have the exact structural data available, but we are guided by the fact that the analysis should lead to realistic eigenfrequencies. In our study we neglect the effects of the motions of the caisson and its interaction with the soil. The following values are used based on Larsen (personal communication, 2008): $\rho_c = 2.4 \cdot 10^3$ kg m^{-3} (density of the concrete), $\rho_i = 1.025 \cdot 10^3$ kg m^{-3} (density of the internal liquid), $\rho_e = 1.025 \cdot 10^3$ kg m^{-3} (density of the external liquid), and $\mathrm{E} = 6 \cdot 10^{10}$ N m^{-2} (E-modulus for reinforced concrete). The mass M_0 of the deck is $2 \cdot 10^7$ kg. The external and internal radii, $R_e(z)$ and $R_i(z)$, are approximated as linear functions of the z-coordinate by using the dimensions shown

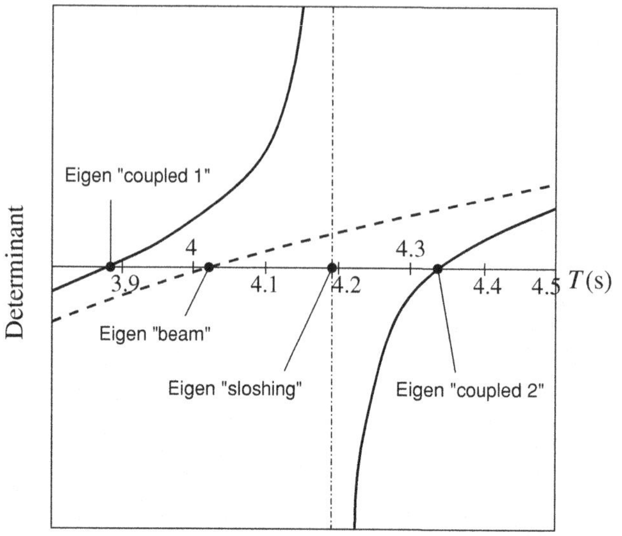

Figure 5.19. Determinant (5.176) as function of $T = 2\pi/\sigma$. Zeros correspond to the highest eigenperiods of the system. The dashed line shows the result when sloshing is ignored. The solid line is for coupled oscillations. Calculations were made with six test functions. The dashed-and-dotted line denotes pure sloshing resonance.

in Figure 5.18(c). The structural mass m per unit length and the second moment $\mathtt{I}$ of the structural cross-sectional area with respect to the x–z-plane can be expressed as $m(z) = \pi\rho_c(R_e^2(z) - R_i^2(z))$ and $\mathtt{I}(z) = \frac{1}{4}\pi(R_e^4(z) - R_i^4(z))$. The integrals in eq. (5.175) can be found analytically because the integrals in eq. (5.175) can be represented as the sum of integrals over the intervals $-h \le z \le 0$ and $0 < z \le h_1$, in which, after substitution of polynomial solution (5.174), (5.177), the integrands are polynomials formed by $\mathtt{I}(z)$, $m(z)$, and $\phi_k(z)$ (and their derivatives).

For the given structure, the variational method is very efficient in computing the lower eigenfrequencies. Calculations show that five to six basis functions (5.177) provide at least five significant figures for the two lowest eigenfrequencies of the coupled system.

The zeros of determinant (5.176) shown in Figure 5.19 correspond to the two highest eigenperiods of the system. The sloshing effect is seen by comparing the dashed and solid lines that correspond to the case when sloshing is ignored and the actual coupled system, respectively. The highest eigenperiod for the structure is 4.01 s when sloshing is neglected. The highest uncoupled sloshing period is 4.189 s. When coupling between sloshing and structural vibrations are considered, the two highest natural periods are 3.88 and 4.336 s, respectively, which are close to the full-scale experimental values 3.9 and 4.3 s.

5.4.5.4 Wave excitation

The ocean wave excitation of the elastic vibrations of the Draugen platform with a period of about 4 s can be approximated by linear theory. The wave excitation loads follow by assuming that the platform is restrained from vibrating. If a frequency-domain solution is studied and we approximate the tower as a circular cylinder with constant cross-section and radius R_{e0}, we can use MacCamy and Fuchs' (1954) theory. We consider incident regular waves with a wave profile $\zeta = \zeta_a \sin(\sigma t - ky)$, where $y = 0$ is at the cylinder axis and $\sigma^2 = gk \tanh kh$. If we were to have been consistent, we would have used the actual water depth and not the depth to the top of the caisson. However, the wave loads do not feel the effect of the water depth and the caisson top for the considered wave periods in 250-m water depth. The horizontal excitation force per unit vertical length in the incident wave propagation direction can be expressed as

$$f_{exc.}(z,t) = \frac{4\rho_e g\zeta_a}{k}\frac{\cosh k(z+h)}{\cosh kh} \times A\,(kR_{e0})\cos(\sigma t - \alpha), \qquad (5.178)$$

where

$$A(kR_{e0}) = \left\{[J_1'(kR_{e0})]^2 + [Y_1'(kR_{e0})]^2\right\}^{-1/2},$$
$$\tan\alpha = J_1'(kR_{e0})/Y_1'(kR_{e0}) \qquad (5.179)$$

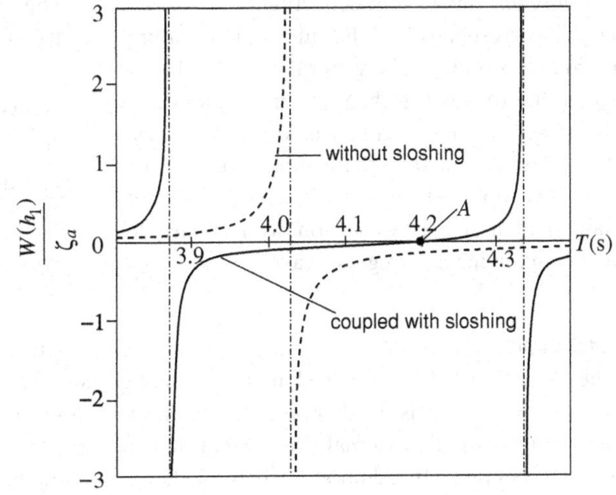

Figure 5.20. Horizontal vibration amplitude $W(h_1)$ of the deck of a monotower divided by the incident wave amplitude ζ_a versus the period T of the incident regular waves. Damping is neglected. The point of zero deflection is denoted A.

and where J_1 is the Bessel function of the first kind and first order and Y_1 is the Bessel function of the second kind and first order. Furthermore, J_1' and Y_1' mean the derivative of J_1 and Y_1 with respect to their argument. Because the wavelengths of interest are not very large relative to the diameter, we cannot make further approximations, such as applying the Morison equation.

When the incident wave frequency is close to the eigenfrequencies of the coupled vibrations, wave excitation loads are localized in a neighborhood of the waterplane. Similar to the considerations of Section 5.4.5.1 for internal sloshing, the water depth in expression (5.178) can be replaced by an effective depth h_e, which is the order of three cylinder radii. Because the actual value of $R_e(z)$ slowly varies in the interval $[-h_e, 0]$, one can apply formula (5.178) with $R_{e0} = R_e(0)$. The external hydrodynamic load term in eq. (5.158) with a deep-water approximation can be expressed as

$$\begin{aligned} f_{\text{ext}}(z,t) &= f_{\text{add}}(z,t) + f_{\text{exc}}(x,t) \\ &= -\chi_{[-h,0]}(z)\rho_e \pi R_e^2(z)\ddot{\mathrm{w}}(z,t) \\ &\quad + \chi_{[-h,0]}(z)\frac{4\rho_e g \zeta_a}{k}\exp(kz)A \\ &\quad \times (kR_{e0})\cos(\sigma t - \alpha). \end{aligned} \tag{5.180}$$

The forced $2\pi/\sigma$-periodic solution of the problem is related to the following functions:

$$\begin{aligned} \beta_{1,j,2} &= B_j \cos(\sigma t - \alpha), \quad j = 1, 2, \ldots; \\ \mathrm{w} &= W(z)\cos(\sigma t - \alpha). \end{aligned} \tag{5.181}$$

Positive and negative values of $W(z)$ correspond to the deflection $w(z)$ being either in phase or in antiphase, respectively, with the wave excitation. Phasing of the response is a consequence of the fact that damping is neglected. In deep water the forcing frequency can be approximated as $\sigma = \sqrt{gk} = \sqrt{2\pi g/\lambda}$, where λ is the wavelength of the incident waves.

Repeating the analysis from Section 5.4.5.3 reduces the approximate solution of the problem to

$$\boldsymbol{A}\boldsymbol{a} = \boldsymbol{b}, \tag{5.182}$$

in which the matrix is defined by eq. (5.175), $\boldsymbol{a} = (a_1, \ldots, a_N)^T$ is composed from the coefficients in eq. (5.174), and the right-hand-side vector $\boldsymbol{b} = (b_1, \ldots, b_N)^T$ consists of the elements

$$b_n = 4\rho_e g \zeta_a k^{-2} A(kR_{e0})\phi_n(0), \quad n = 1, \ldots, N. \tag{5.183}$$

The normalized amplitude parameter for the deck deflection versus the forcing period, $W(h_1)/\zeta_a = \sum_{k=1}^{N} a_k \phi_k(h_1)/\zeta_a$, is presented in Figure 5.20. Calculations are done with and without the effect of sloshing.

Two distinct natural frequencies are associated with the elastic vibrations and sloshing. Sloshing has an obvious influence on the response at the uncoupled natural frequency of vibrations because the coupling affects the natural frequencies. In the next section we include the effect of damping to predict a more realistic response at resonance conditions.

Figure 5.20 shows zero deflection at $T = 4.18$ s, which corresponds to the uncoupled natural period for sloshing. The generalized added mass effect due to sloshing then becomes infinite and the tower, as a consequence, does not move. This effect is similar to that from Section 3.6 for linearly coupled steady-state sloshing and sway motion of a ship cross-section with a partially filled tank in incident regular waves.

5.4.5.5 Damping

The system has three hydrodynamic damping sources, which consist of damping due to sloshing (Chapter 6) and external wave radiation and viscous damping. In addition, structural damping is also present. Our focus is on *wave radiation damping*. The damping due to external viscous effects and due to linear sloshing in a tank without internal structures obstructing the flow is very small. However, resonant sloshing may be affected in reality by nonlinearities. If no internal structures exist, such as pipes, nonlinear free-surface effects transfer energy between liquid motion modes and limit the wave elevation relative to linear theory (Chapters 8 and 9). Furthermore, wave breaking may occur inside the tank with resulting damping. If internal structures are present, cross-flow drag due to pipes, for instance, may cause important damping (Chapter 6).

Modifications of eqs. (5.158) and (5.180) due to wave radiation are approximated by starting out with the following relationship derived by Newman (1962):

$$F_{2a} = \zeta_a \sqrt{4\rho_e g^3 B_{22}/\sigma^3}, \tag{5.184}$$

where F_{2a} is the amplitude of the horizontal wave excitation force on a vertical circular cylinder, and B_{22} is the rigid-body wave radiation damping in sway for the cylinder. The wave force amplitude, F_{2a}, can be obtained by using McCamy and Fuchs' (1954) theory (i.e., by integrating eq. (5.178) from $z = -h$ to $z = 0$). As before, we take into account the fact that the force is localized in the interval $[-h_e, 0]$, which is small relative to the beam length. Integration of eq. (5.178) by z gives

$$\begin{aligned} F_{2a} &= 4\rho_e g \zeta_a k^{-2} A\left(kR_{e0}\right) \\ \Rightarrow B_{22} &= 4\rho_e \sigma k^{-3} A^2\left(kR_{e0}\right). \end{aligned} \tag{5.185}$$

The damping term modifies the external loads as follows:

$$f_{\text{ext}}(z,t) = f_{\text{add}}(z,t) + f_{\text{exc}}(z,t) + f_{\text{rad}}(z,t), \tag{5.186}$$

where, due to eq. (5.185), the wave radiation term can be approximated as

$$f_{\text{rad}}(z,t) = -\chi_{[-h_e,0]}(z) B_{22} h_e^{-1} \dot{w}(z,t). \tag{5.187}$$

Analysis of the frequency-domain solution is simplified when we use complex variables in describing the time dependence. Accounting for eq. (5.178), we represent the sum of the wave damping load and the wave excitation load as

$$\begin{aligned} &f_{\text{rad}}(z,t) + f_{\text{exc}}(z,t) \\ &\quad = -\chi_{[-h_e,0]}(z) B_{22} h_e^{-1} \dot{w}(z,t) + \chi_{[-h,0]}(z) 4\rho_e g \zeta_a k^{-1} \\ &\qquad \times \exp(kz) A(kR_{e0}) \exp(\mathrm{i}\sigma t - \mathrm{i}\alpha) \end{aligned} \tag{5.188}$$

and assume a solution in the form of eq. (5.168). The variational scheme then leads to a matrix problem similar to eq. (5.182) with the following modified complex-variable matrix A:

$$\begin{aligned} A = \Bigg\{ &\int_{-h}^{h_1} [\sigma^2 M_1 \phi_k \phi_n - \mathtt{EI}\phi_n'' \phi_k''] \, dz \\ &+ \sigma^2 \phi_n(0) \phi_k(0) M_3(\sigma^2) - \sigma \mathrm{i} B_{22} \phi_n(0) \phi_k(0) \\ &+ \sigma^2 M_0 \phi_n(h_1) \phi_k(h_1) \Bigg\}_{n,k=1,\ldots,N}. \end{aligned} \tag{5.189}$$

Figure 5.21 presents calculations of the ratio between the wave-induced deflection amplitude of the deck, $|W(h_1)|$, and the incident wave amplitude ζ_a as a function of the incident wave period. The wave radiation damping has a clear influence at both natural periods. We still see that the deflection is zero at the uncoupled natural sloshing period $T = 4.18$ s. If damping associated with sloshing is introduced, the deflection is nonzero at the uncoupled natural sloshing period. Furthermore, the maximum responses become lower when we account for other damping sources.

Results for the shear force at the beam–deck intersection follow by multiplying the results in Figure 5.21 with the deck mass M_0 times $(2\pi/T)^2$. Another important response variable is the bending moment at the clamped beam end, $-\mathtt{EI}w''$,

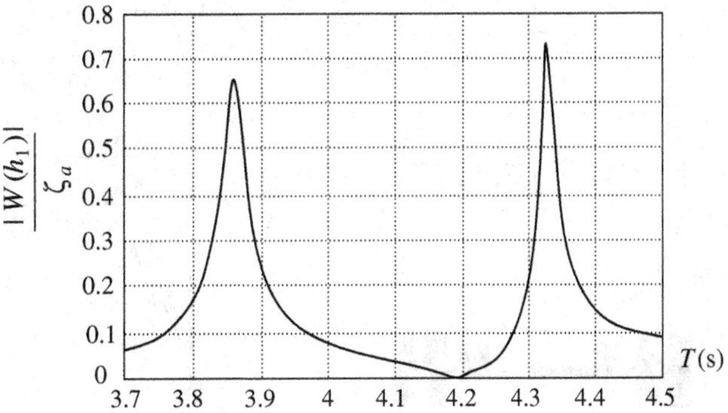

Figure 5.21. Wave-induced horizontal vibration amplitude of the deck, $|W(h_1)|$, divided by the incident wave amplitude ζ_a as a function of the incident wave period $T(\mathrm{s})$. Only damping due to wave radiation is taken into account.

at $z=-h$, which follows by differentiating the expression for w in terms of the polynomials ϕ_k defined by eq. (5.177), but is not further pursued here. Section 3.9.1 describes how to apply results as in Figure 5.21 to short-term sea states described by wave spectra.

5.4.6 Rollover of a tank vehicle

A tank vehicle with partially filled tanks may roll over due to sloshing as a consequence of a fast transverse maneuver (see Section 1.9). The hydrodynamic part of a rollover analysis is exemplified by a linear modal approach using the approximate lowest natural mode in a two-dimensional circular tank given by eq. (4.72). The horizontal acceleration of the tank is expressed as $\ddot{\eta}_2(t)$. A Cartesian coordinate system (y,z) with origin at the intersection between the tank's centerplane and the calm free surface is introduced. Positive z is upward (see Figure 5.22). We define

$$\varphi_1(y,z)=\frac{R_0 y}{y^2+(z-a)^2},\quad f_1(y)=\frac{R_0 y}{y^2+a^2}, \tag{5.190}$$

where φ_1 is the approximate lowest natural mode expressed in terms of a horizontal dipole with singularity at $z=a>0$. The parameter a is a function of the liquid depth h and is found by a variational formulation (see Section 4.6.1.2). The scaled values $\overline{a}=a/R_0$ are presented in Table 4.3. This mode describes the dominant antisymmetric flow of the liquid due to horizontal forcing. The effect of higher modes is neglected in the following analysis.

We use the single-mode approximation of wave elevations and the absolute velocity potential as prescribed by the linear modal theory eqs. (5.19) and (5.20):

$$\zeta(y,t)=\beta_1(t)\frac{R_0 y}{y^2+a^2};$$

$$\Phi(y,z,t)=\dot{\eta}_2(t)y+R_1(t)\frac{R_0 y}{y^2+(z-a)^2},$$

where we used the fact that the tank performs only horizontal motions. The single-dimensional linear modal system (5.27) reduces to the form

$$\ddot{\beta}_1(t)+\sigma_1^2\beta_1(t)=-P_1^0\ddot{\eta}_2(t), \tag{5.191}$$

where the lowest natural frequency $\sigma_1=\sqrt{g\kappa_1}=\sqrt{g(R_0\kappa_1)/R_0}$ and nondimensional values $\kappa=\kappa_1 R_0$ are presented in Table 4.3. The nondimensional hydrodynamic coefficient P_1^0 in eq. (5.191) can be expressed analytically as

$$P_1^0=\frac{\lambda_{21}}{\mu_1}=\frac{\kappa_1\int_{-R_0\sqrt{1-\overline{\varepsilon}^2}}^{R_0\sqrt{1-\overline{\varepsilon}^2}} y f_1\,dy}{\int_{-R_0\sqrt{1-\overline{\varepsilon}^2}}^{R_0\sqrt{1-\overline{\varepsilon}^2}} f_1^2\,dy}$$

$$=\frac{2(\kappa_1 R_0)\left[\sqrt{1-\overline{\varepsilon}^2}-\overline{a}\arctan\left(\overline{a}^{-1}\sqrt{1-\overline{\varepsilon}^2}\right)\right]}{\left[\arctan\left(\overline{a}^{-1}\sqrt{1-\overline{\varepsilon}^2}\right)/\overline{a}-\dfrac{\sqrt{1-\overline{\varepsilon}^2}}{\overline{a}^2+1-\overline{\varepsilon}^2}\right]},$$

$$\overline{\varepsilon}=\frac{h}{R_0}-1. \tag{5.192}$$

It follows from eq. (5.36) that a one-mode approximation of the two-dimensional horizontal hydrodynamic force is given by

$$F_2=-m_l\ddot{\eta}_2(t)-\lambda_{21}\ddot{\beta}_1(t)$$
$$=-\rho R_0^2\left[A_0(\overline{\varepsilon})\ddot{\eta}_2(t)+A_1(\overline{\varepsilon})\ddot{\beta}_1(t)\right], \tag{5.193}$$

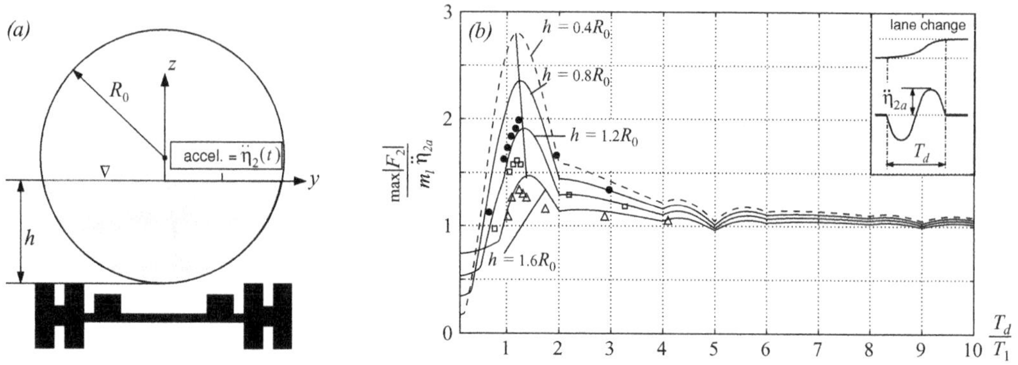

Figure 5.22. Simulation of lane change of a tank vehicle with a circular cylindrical tank. The transverse acceleration of the vehicle is $\ddot{\eta}_2(t)$. The maximum horizontal hydrodynamic force, $\max|F_2|$, divided by the maximum horizontal frozen liquid force, $m_l\ddot{\eta}_{2a}$, is presented as a function of the ratio between the duration of the maneuver, T_d, and the highest natural sloshing period T_1 for different ratios of liquid depth to cylinder radius, h/R_0. The lane-change acceleration is idealized as a sinusoidal function of time (see panel (b)). Along with linear theoretical prediction by the one-mode model (solid and dashed lines), part (b) presents numerical results by Moderassi-Tehrani *et al.* (2006) obtained by using the commercial CFD code FLUENT. FLUENT data correspond to $h/R_0 = 0.8$ (•), 1.2 (□), and 1.6 (Δ).

where $A_0(\bar{\varepsilon})$ is the R_0-scaled area of the liquid, which takes the form

$$A_0(\bar{\varepsilon}) = \begin{cases} -\arctan\left(\sqrt{1-\bar{\varepsilon}^2}/\bar{\varepsilon}\right) \\ \quad +\bar{\varepsilon}\sqrt{1-\bar{\varepsilon}^2}, \quad -1 < \bar{\varepsilon} \le 0, \\ \pi - \arctan\left(\sqrt{1-\bar{\varepsilon}^2}/\bar{\varepsilon}\right) \\ \quad +\bar{\varepsilon}\sqrt{1-\bar{\varepsilon}^2}, \quad 0 < \bar{\varepsilon} < 1, \end{cases} \tag{5.194}$$

and

$$A_1(\bar{\varepsilon}) = 2\left[\sqrt{1-\bar{\varepsilon}^2} - \bar{a}\arctan\left(\sqrt{1-\bar{\varepsilon}^2}/\bar{a}\right)\right]. \tag{5.195}$$

When the transverse maneuver of the tank vehicle, $\ddot{\eta}_2(t)$, is known, the generalized coordinate β_1 of the free-surface elevation follows from solving eq. (5.191) with initial conditions $\beta_1(0)$ and $\dot{\beta}_1(0)$. Substitution of this solution into eq. (5.193) gives the transient hydrodynamic force.

Let us present an example where a tank vehicle changes lanes on a highway. The transverse acceleration of the tank vehicle, $\ddot{\eta}_2(t)$, is given as (see Figure 5.22(b))

$$\ddot{\eta}_2(t) = \begin{cases} 0, & t < 0 \quad \text{or} \quad T_d < t; \\ -\ddot{\eta}_{2a}\sin(2\pi t/T_d), & 0 \le t \le T_d. \end{cases} \tag{5.196}$$

The liquid moves with a constant velocity at $t = 0$ and, therefore, the simulations by eq. (5.191) with the acceleration given by eq. (5.196) should start with zero initial conditions $\beta_1(0) = \dot{\beta}_1(0) = 0$. This condition may be postulated or derived as a consequence of condition (5.55). The solution of eq. (5.191) with eq. (5.196) and zero initial conditions can be expressed as

$$\beta_1(t) = -\frac{P_1^0\ddot{\eta}_{2a}}{\sigma_1^2 - \sigma^2}\left[\frac{\sigma}{\sigma_1}\sin(\sigma_1 t) - \sin(\sigma t)\right], \quad \sigma = \frac{2\pi}{T_d}. \tag{5.197}$$

when t is in the interval $[0, T_d]$. The solution for $t \ge T_d$ represents a free vibration with starting conditions $\beta_1(T_d)$ and $\dot{\beta}_1(T_d)$ following from eq. (5.197). The time-dependent nondimensional horizontal force can, by using eq. (5.193), be expressed as

$$\frac{F_2}{m_l\ddot{\eta}_{2a}} = -\left[\frac{\ddot{\eta}_2(t)}{\ddot{\eta}_{2a}} + \frac{A_1(\bar{\varepsilon})}{\ddot{\eta}_{2a}A_0(\bar{\varepsilon})}\ddot{\beta}_1(t)\right], \tag{5.198}$$

where $m_l\ddot{\eta}_{2a}$ is the maximum horizontal force when the liquid is frozen. The corresponding nondimensional maximum horizontal hydrodynamic force, $\max|F_2|/(m_l\ddot{\eta}_{2a})$, is presented in Figure 5.22(b) as a function of the ratio between the duration of the transverse maneuver, T_d, and the highest natural sloshing period, T_1.

Figure 5.22(b) shows that sloshing matters when $T_d/T_1 \lesssim 5$. The maximum amplification relative to the horizontal frozen liquid force occurs for T_d slightly larger than T_1. Furthermore, increasing the liquid filling decreases T_1 but increases the value of T_d/T_1 corresponding to the maximum amplification. Because of this fact, the actual values of T_d for the maximum are located in a neighborhood of a fixed value for all the fillings. For example, the maximum amplification occurs at $T_d = 2$ s for a tank of radius 1 m.

Figure 5.22(b) shows that the maximum value of $\max|F_2|/(m_l\ddot{\eta}_{2a})$ is close to 3 when $h = 0.4R_0$ and decreases with increasing filling ratio. The values presented in Figure 5.22 are in fair agreement with computational fluid dynamics (CFD) calculations presented by Moderassi-Tehrani *et al.* (2006) based on using Navier–Stokes equations with nonlinear free-surface conditions. Generally speaking, our values predicted by the simplified linear response model are higher than the CFD calculations at the primary resonance zone. However, the relative differences are less than 15%. It is difficult to speculate on the reason for the lower results in the calculations by Moderassi-Tehrani *et al.* (2006). One reason can be due to the fact that we do not in our calculations account for the contribution from all linear modes. Because of computational time, our simplified approach has clear advantages relative to CFD when systematic simulations are made of coupled sloshing and tank vehicle dynamics.

Figure 5.22(b) also illustrates that $\max|F_2|/(m_l\ddot{\eta}_{2a})$ is less than 1 for small values of T_d/T_1. To study this limit, let us first consider how the lateral force behaves as a function of time for different durations T_d. An illustration of the behavior is presented in Figure 5.23 for six different values of T_d/T_1 and $h/R_0 = 1$. When T_d/T_1 is sufficiently small, the maximum force occurs in the time interval $[0, T_d]$. We can show by using eqs. (5.197) and (5.198) that the force behaves approximately as $\sin(t/T_d)$ for $0 < t < T_d$ and is close to zero for $t > T_d$ for small T_d/T_1, because $\beta_1(T_d), \dot{\beta}_1(T_d) \to 0$ as $T_d \to 0$. It also follows in the limit $T_d \to 0$ that $\frac{\max|F_2|}{m_l\ddot{\eta}_{2a}} \to |1 - \frac{P_1^0 A_1}{A_0}|$.

Furthermore, the graphs in Figure 5.23 show that increasing T_d/T_1 to larger than 0.5 causes large-amplitude sloshing for $t > T_d$, so that the lateral force reaches its first maximum somewhere in the range $[T_d, 2T_d]$ (see Figure 5.23(b)). The position of the maximum tends to $t = T_d$ as T_d/T_1 approaches 1 from below.

Figure 5.22(b) demonstrates a piecewise smooth change (discontinuities in the slope) when $T_d = mT_1$, where m is an integer. This result is especially clear for $m = 2$, 4, and 5. These cases are of special interest, because they cause no sloshing after $t > T_d$, as shown in Figure 5.23(f).

When the hydrodynamic results are combined with a rollover analysis, we should note the fact that there is no hydrodynamic roll moment about the axis of a circular tank. This observation follows from the fact that hydrodynamic loads are due to pressure loads acting perpendicular to the tank surface. The consequence is that the center of hydrodynamic pressure is at the tank axis. Furthermore, no vertical hydrodynamic force due to sloshing exists according to linear theory. Even though rollover does not occur, the changing sign of the hydrodynamic force makes it difficult to control the vehicle. Because our model does not include damping, the sloshing force does not decay in the free-vibration phase after forcing has stopped.

5.4.7 Spherical tanks

Angular velocities about an axis system through the center of a spherical tank cannot cause any liquid motions according to potential flow theory. Because vertical forcing cannot cause sloshing according to linear theory, lateral forcing of the tank is most important. To our knowledge no analytical linear theory exists for sloshing in a spherical tank. However, it is possible to find analytical solutions for hydroelastic vibrations of a spherical tank when a high-frequency free-surface condition is used in combination with thin-shell theory (Vinje, 1972, 1973). The shell theory is described, for instance, in the textbook by Kraus (1967).

5.4.7.1 Hydroelastic vibrations of a spherical tank

Let us consider Vinje's case. The spherical shell has mean radius R and constant thickness h_0. The shell is assumed to be a linear elastic material with Young modulus E, Poisson ratio ν_{Poisson}, and density ρ_s. The ratio h_0/R is small; therefore, thin-shell theory can be used. Rotational inertia and shear deformations are neglected and axisymmetric vibrations are assumed. The dynamic free-surface condition for a partially filled shell with high-frequency vibrations can be

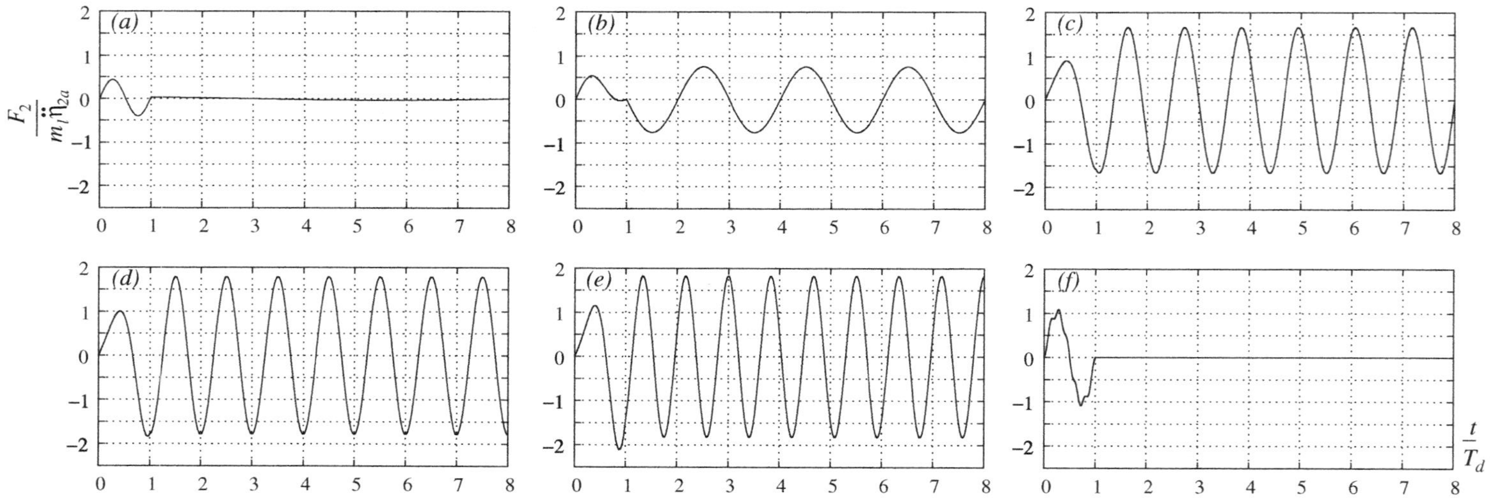

Figure 5.23. Normalized time histories of the lateral hydrodynamic force on a circular tank for $h/R_0 = 1$: (a) $T_d/T_1 = 0.1$; (b) $T_d/T_1 = 0.5$; (c) $T_d/T_1 = 0.9$; (d) $T_d/T_1 = 1.0$; (e) $T_d/T_1 = 1.2$; (f) $T_d/T_1 = 6$. See Figure 5.22 for definitions of the scenario.

approximated as $\varphi = 0$ on the mean free surface, where φ is the velocity potential of the flow. The liquid can be assumed incompressible in marine applications (Vinje, 1972, 1973). The nondimensional natural frequencies, $\hat{\sigma}_n = \sigma_n R\sqrt{\rho/\mathrm{E}}$, are functions of the parameters $M^+ = (\rho_l/\rho_s)(R/h_0)$ and $\chi = 12(R/h_0)^2$, and the filling ratio. When the shell is completely filled, the flow can be represented in terms of Legendre polynomials. The equation system for natural frequencies and modes follows from the dynamic equations of the shell and the requirement of no flow through the shell. The natural frequencies are determined by

$$\det \begin{vmatrix} -\frac{\gamma_n}{c_1}\left(1+\frac{1}{\chi}\right)-\hat{\sigma}_n^2 & -\frac{1}{c_1}\left(c_2-\frac{\gamma_n}{\chi}\right) \\ \eta_n\left(1-\frac{\gamma_n}{\chi c_2}\right) & 2+\frac{\eta_n\gamma_n}{\chi c_2}-c_3\left(1+\frac{M^+}{n}\right)\hat{\sigma}_n^2 \end{vmatrix} = 0, \tag{5.199}$$

where $c_1 = 1-\nu_{\mathrm{Poisson}}^2$, $c_2 = 1+\nu_{\mathrm{Poisson}}$, $c_3 = 1-\nu_{\mathrm{Poisson}}$, $\eta_n = -n(n+1)$, and $\gamma_n = \eta_n + c_3$. Two eigenfrequencies exist for a given n. Equation (5.199) was derived by Vinje (1972). The corresponding equation in Vinje (1973) contains misprints.

A half-liquid-filled tank is also analyzed in detail by Vinje (1972, 1973), who expresses the solution in terms of Legendre polynomials and uses the Galerkin method. An approximate method is followed for the lowest natural frequency of other partially filled shells. The lowest natural frequency is expressed by using conservation of the sum of the total potential energy and the total kinetic energy, and the mode shape is approximated as the mode shape of the completely filled shell.

A spherical tank, such as an LNG tank on a ship, is realistically constrained (i.e., we must modify the previous analysis). Rayleigh and Lindsay (1945) state that the nth natural frequency σ_n^* of a nondissipative system, which is constrained by a spring-like constraint, satisfies the following inequality:

$$\sigma_n \le \sigma_n^* \le \sigma_{n+1}, \tag{5.200}$$

where σ_n is the nth natural frequency of the unconstrained system. Let us exemplify eq. (5.200) for a spherical LNG tank by considering a completely filled tank where σ_n is determined by eq. (5.199). An aluminum shell with $R = 18$ m, $h_0 = 40$ mm, $\nu_{\mathrm{Poisson}} = 0.3$, $\mathrm{E} = 70$ GPa, and $\rho_s = 2{,}700$ kg m^{-3} is considered. Furthermore, the density of the LNG, ρ_l, is set equal to 474 kg m^{-3}, which gives $\chi = 2.43 \cdot 10^6$ and $M^+ = 79.0$. It follows from eq. (5.199) that the two lowest nonzero nondimensional natural frequencies are associated with $n = 2$ and 3. We get $\hat{\sigma}_2 = 0.136$ and $\hat{\sigma}_3 = 0.179$, which gives $f_2 = \sigma_2/2\pi = 6.12$ Hz and $f_3 = \sigma_3/2\pi = 8.06$ Hz. The propeller is an excitation source. For instance, let us consider a 250-m-long LNG ship equipped with a four-bladed propeller running at 1.6 Hz. This situation gives a frequency of excitation of 6.4 Hz.

5.4.7.2 Simplified two-mode modal system for sloshing in a spherical tank

Let us now assume the spherical tank is rigid and consider linear sloshing; we construct a simple linear modal theory by adopting the approximate lowest modes from eqs. (4.73). A similar approach was used in Section 5.4.6 for a two-dimensional circular tank. The infinite-fluid horizontal dipole approximations of the two conjugate lowest modes from eqs. (4.73) are expressed as

$$\varphi_{11,\sin}(r,\theta,z) = \frac{R_0^2 r \sin\theta}{(r^2+(z-a)^2)^{3/2}};$$

$$\varphi_{11,\cos}(r,\theta,z) = \frac{R_0^2 r \cos\theta}{(r^2+(z-a)^2)^{3/2}}, \tag{5.201}$$

where R_0 is the radius of the sphere and a is the vertical position of the dipole relative to the mean free surface. The nondimensional values of a are presented in Table 4.4 for different liquid depths ($0 < h < 2R_0$). The table also gives the corresponding values of the frequency parameter, $\kappa_{11}R_0$, that determine the lowest natural frequency, $\sigma_{11}^2 = g\kappa_{11}$. The analysis concentrates on lateral forcing (i.e., we consider $\eta_i = 0$ ($i = 3$, 4, 5, 6).

According to linear modal theory and the analysis in Section 4.8.2.1, the free-surface elevation is approximated as

$$z = \zeta(r, \theta, t) = \sum_{n=0}^{\infty}\sum_{i=1}^{\infty} R_{ni}(r)\,(\beta_{ni,\cos}(t)\cos n\theta + \beta_{ni,\sin}(t)\sin n\theta), \qquad (5.202)$$

where the natural surface profiles are associated with $R_{ni}(r)\cos n\theta$ and $R_{ni}(r)\sin n\theta$, which are the projections of the corresponding natural modes $\varphi_{ni,\sin}(r,\theta,z)$ and $\varphi_{ni,\cos}(r,\theta,z)$ on the mean free surface $z = 0$. The three-dimensional modes in the spherical tank are enumerated by the pair of indices n and i.

When adopting approximate solution (5.201) for the lowest modes and the two-mode approximation of the free-surface elevation, we get $R_{11}(r) = R_0^2 r/(r^2+a^2)^{3/2}$ and eq. (5.202) reduces to

$$\zeta(r,\theta,t) = \frac{R_0^2 r}{(r^2+a^2)^{3/2}}\,(\beta_{11,\cos}(t)\cos\theta + \beta_{11,\sin}(t)\sin\theta)\,. \qquad (5.203)$$

To derive simplified linear modal equations (5.27) based on expression (5.203), we have to find the hydrodynamic coefficients μ_{11}, $\lambda_{1(11)\cos}$, $\lambda_{1(11)\sin}$, $\lambda_{2(11)\cos}$, and $\lambda_{2(11)\sin}$. Due to azimuthal symmetry, we see that $\lambda_{2(11)\cos} = \lambda_{1(11)\sin} = 0$ and $\lambda = \lambda_{1(11)\cos} = \lambda_{1(11)\sin} \neq 0$. The integration in eq. (5.26) over the mean free surface Σ_0, which is a circle of radius R_s, gives

$$\mu = \mu_{11} = \frac{\rho\pi R_0^3}{4\,(\kappa_{11}R_0)\,\bar{r}_s^2\bar{a}^2(1+\bar{a}^2)^2},$$

$$\lambda = \frac{\rho\pi R_0^3\bar{r}_0}{\sqrt{1+\bar{a}^2}\left(\bar{a}+\sqrt{1+\bar{a}^2}\right)^2}, \qquad (5.204)$$

where $\kappa_{11}R_0$ is the nondimensional frequency parameter (whose numerical values are given in Table 4.4), $\bar{a} = a/r_s$, and

$$\bar{r}_s = \frac{R_s}{R_0} = \begin{cases} \sqrt{1-(1-\bar{h})^2}, & 0<\bar{h}\le 1,\\ \sqrt{1-(2-\bar{h})^2}, & 1<\bar{h}<2,\end{cases}$$

where $\bar{h} = h/R_0$ is the nondimensional depth.

Substituting eqs. (5.204) into eq. (5.28) and accounting for the orthogonality of the two modes from eqs. (5.201) lead to the modal equations

$$\ddot{\beta}_{11,\cos} + \sigma_{11}^2\beta_{11,\cos} = -\frac{\lambda}{\mu_{11}}\ddot{\eta}_1 = -P_{11}\ddot{\eta}_1,$$

$$\ddot{\beta}_{11,\sin} + \sigma_{11}^2\beta_{11,\sin} = -\frac{\lambda}{\mu_{11}}\ddot{\eta}_2 = -P_{11}\ddot{\eta}_2, \qquad (5.205)$$

where $P_{11} = 4(\kappa_{11}R_0)\bar{r}_s^3\bar{a}(1+\bar{a}^2)^{3/2}/(a+\sqrt{1+\bar{a}^2})^2$. Furthermore, we can find the hydrodynamic force components based on the Lukovsky formula. This approach gives

$$F_1 = -m_l\ddot{\eta}_1 - \lambda_{11}\ddot{\beta}_{11,\cos};$$

$$F_2 = -m_l\ddot{\eta}_2 - \lambda_{11}\ddot{\beta}_{11,\sin}. \qquad (5.206)$$

Based on eq. (5.206), or using eqs. (5.69), (5.71), and (5.74), we can find the added mass coefficients A_{11} and A_{22} that are equal in our case due to symmetry. The result is

$$A_{11} = A_{22} = m_l + \frac{\lambda_{11}^2}{\mu_{11}}\frac{\sigma^2}{\sigma_{11}^2-\sigma^2} = m_l + \frac{4\pi\rho R_0^3\,(\kappa_{11}R_0)\,\bar{r}_s^3(1+\bar{a}^2)}{\left(\bar{a}+\sqrt{1+\bar{a}^2}\right)^4}\cdot\frac{\sigma^2}{\sigma_{11}^2-\sigma^2}, \qquad (5.207)$$

where σ is the forcing frequency.

Applications of these formulas could be for an LNG carrier with spherical tanks. One example scenario is collision with ice.

5.4.8 Transient analysis of tanks with asymptotic estimates of natural frequencies

We showed in Chapter 4 how to estimate natural frequencies by asymptotic methods for the following cases: (a) interior structures in rectangular and upright circular cylinders; (b) small deviations of rectangular sections when analytical continuation of the flow to a rectangular section is possible.

A transient analysis with forced motions of these tanks can be performed as follows:

1) Start with the modal equations (5.25) for the generalized free-surface coordinates.
2) Estimate the natural frequencies in the modal equations by using the asymptotic formulas.
3) Calculate the other hydrodynamic coefficients given by eq. (5.26) by means of the natural modes and Stokes-Joukowski potential for either a rectangular tank or upright circular cylinder.
4) Use the Lukovsky formulas for hydrodynamic force and moment.

5.5 Summary

A modal theory assumes potential flow of an incompressible liquid and decomposes the velocity potential in terms of an infinite series of the natural modes and terms that take care of nonhomogeneous terms in the body boundary conditions due to the tank motions. Elastic deflections of the tank can be included. Generalized coordinates $R_n(t)$ are associated with each natural mode. The free-surface elevation can be represented as an infinite sum of the projections of the natural modes on the mean free surface. Generalized coordinates for the contribution of each natural mode to the free-surface elevation are denoted $\beta_n(t)$ and can be related to $R_n(t)$ by the kinematic free-surface condition. The dynamic free-surface condition leads to an infinite set of ordinary differential equations for $\beta_n(t)$.

Hydrodynamic forces and moments on the tank can be explicitly expressed in terms of $\beta_n(t)$ by means of the Lukovsky formulas, which facilitates the analysis of coupling between sloshing and global body (ship) motions.

The principle of a modal theory is shown in this chapter by assuming linear theory. The differential equations for $\beta_n(t)$ are then uncoupled. If the tank motions are prescribed, the differential equation for each generalized coordinate $\beta_n(t)$ is similar to a mass-spring system. The system does not have damping. The consequence is that beating due to terms oscillating with natural frequency and terms oscillating with forcing frequency does not die out over time. Damping due to viscous effects must be added to reach steady-state conditions.

A modal theory has a special advantage when analytical solutions of the natural modes exist. Detailed analytical expressions needed for analysis of two- and three-dimensional rectangular tanks and upright circular cylinders were given.

A completely filled tank was examined. When studying the effect of translatory tank motions, the liquid behaves as it is "frozen." However, this effect is not the case for angular tank motions.

Applications of linear theory were presented. One case involved a collision between two ships and it is suggested that sloshing in a partially filled tank does not have a dominant influence on hull damage. The case was used to analyze the effect of tank wall deflections of a two-dimensional rectangular tank on sloshing. The derived expressions were used to discuss wavemaker shapes at a tank wall that do not excite the lowest sloshing mode.

A hydroelastic analysis of wave-excited vibrations of a monotower installed at 252-m water depth in the North Sea was presented. The external hydrodynamic loads and internal sloshing loads were taken into account. The platform was structurally represented as a cantilever beam with a free-end rigid mass that accounts for the deck structure. Sloshing provides important coupling when the forcing frequency is in the vicinity of the highest natural structural period.

The transverse hydrodynamic force needed in rollover analysis of a tank vehicle with a partially filled horizontal circular cylindrical tank was studied. A simplified analysis was made by considering only one mode, which was approximated by an infinite-fluid horizontal dipole solution from Chapter 4. The same type of simplified transient hydrodynamic analysis in terms of an infinite-fluid horizontal dipole was presented for a spherical tank with different filling ratios.

5.6 Exercises

5.6.1 Moments by direct pressure integration and the Lukovsky formula

We consider a two-dimensional rectangular tank. Derive the pitch moment by direct pressure integration and compare the result with the Lukovsky formula.

5.6.2 Transient sloshing with damping

An artificial way to introduce damping in a potential flow model is to add a Rayleigh damping term to the combined dynamic and kinematic free-surface condition. The procedure becomes simple for horizontal forced motions. In this case, neither the dynamic (5.17) nor the kinematic (5.16) conditions include forcing terms. The modified linear free-surface condition can then be expressed as

$$\frac{\partial^2 \Phi}{\partial t^2} + \mu \frac{\partial \Phi}{\partial t} + g \frac{\partial \Phi}{\partial z} = 0, \quad z = 0.$$

As an example, consider the forced harmonic surge oscillation $\eta_{1a} \sin \sigma t$ of a two-dimensional rectangular tank with liquid depth h and breadth

l. In problem (5.1), you must replace the dynamic and kinematic free-surface conditions with the aforementioned condition.

(a) Use representation (5.3) and find $\varphi(x, z, t)$. (*Hint*: Expand x as a Fourier series and use separation of variables.)

Answer:

$$\varphi = \sum_{n=0}^{\infty} \sin[(2n+1)\pi x/l]\cosh[(2n+1)\pi(z+h)/l]$$
$$\times\left[\exp\left(-\tfrac{1}{2}\mu t\right)\left(A_n \cos\left[t\sqrt{\sigma_n^2 - \tfrac{1}{4}\mu^2\sigma^2}\right]\right.\right.$$
$$\left.\left. + B_n \sin\left[t\sqrt{\sigma_n^2 - \tfrac{1}{4}\mu^2\sigma^2}\right]\right) + C_n \cos\sigma t + D_n \sin\sigma t\right],$$

where σ_n is the natural frequency for mode n with zero damping and

$$C_n = \frac{\left[(\sigma_n^2 - \sigma^2) - \mu^2\right]}{\left(\sigma_n^2 - \sigma^2\right)^2 + \mu^2\sigma^2}\sigma K_n,$$

$$D_n = \frac{\left[(\sigma_n^2 - \sigma^2) + \sigma^2\right]}{\left(\sigma_n^2 - \sigma^2\right)^2 + \mu^2\sigma^2}\mu K_n$$

$$K_n = \frac{\sigma^2 \eta_{1a}(-1)^n}{\cosh[(2n+1)\pi h/l]}\frac{4}{l}\left[\frac{l}{(2n+1)\pi}\right]^2$$

and it is assumed that the tank walls are at $x = \pm\frac{1}{2}l$.

(b) Use the initial conditions $\Phi = 0$ and $\partial\Phi/\partial t = 0$ on the mean free surface $z = 0$ at $t = 0$ to show that

$$A_n = -C_n - \frac{4\sigma\eta_{1a}(-1)^n}{l\cosh[(2n+1)\pi h/l]}\left[\frac{l}{(2n+1)\pi}\right]^2,$$
$$B_n = \left(\sigma_n^2 - \tfrac{1}{4}\mu^2\right)^{-1/2}\left(\tfrac{1}{2}\mu A_n - \sigma D_n\right).$$

5.6.3 Effect of small structural deflections of the tank bottom on sloshing

Consider a two-dimensional rectangular tank and assume the tank walls are rigid while the deflection of the tank bottom in the z-direction can be expressed as $U_d(y)\eta_d(t)$. This problem is relevant for earthquake excitation of a land-based tank. Generalize the derivation in Section 5.4.2.4. When do the deflections not excite the lowest mode?

(*Hint*: Present $U_d(y)$ as a Fourier series:

$$U_d(y) = a_0 + \sum_{j=1}^{\infty} a_j f_j(y).) \qquad (5.208)$$

Answer: The function $\Omega_d(y, z)$ takes the form

$$\Omega_d(y, z) = a_0 z - \sum_{j=1}^{\infty} a_j \frac{l}{\pi j}\cos\left(\frac{\pi j}{l}\left(y + \tfrac{1}{2}l\right)\right) \times \frac{\cosh(\pi j z/l)}{\sinh(\pi j h/l)}, \qquad (5.209)$$

which gives

$$\zeta(y, t) = a_0\,\eta_d(t) + \sum_{j=1}^{\infty}\beta_j(t) f_j(y), \qquad (5.210)$$

where $\beta_j(t)$ are solutions of modal system (5.27) with eq. (5.110) and

$$\frac{\lambda_{dm}}{\rho} = -\frac{a_m l^2}{2\pi m \sinh(\pi m h/l)}. \qquad (5.211)$$

The lowest mode is not excited within the framework of the linear (nonresonant) sloshing theory when $a_1 = 0$.

5.6.4 Effect of elastic deformations of vertical circular tank

(a) Use Section 5.4.2.4 to describe liquid sloshing in a vertical circular cylindrical tank due to axisymmetric deformations of the walls. Define the deformation as $U_d(z)\eta_d(t)$. What is the function Ω_d? Derive the analogy of eqs. (5.106) and (5.109).

Answer: The function takes the form

$$\Omega_d(r, z) = a_0\left[\tfrac{1}{2}r^2 - (z+h)^2\right] + \sum_{j=1}^{\infty} a_j I_0(\pi j r/h)\cos(\pi j z/h),$$
$$a_0 = \frac{1}{R_0 h}\int_{-h}^{0} U_d(z)\,dz,$$
$$a_j = \frac{2}{\pi j I_0'(\pi j R_0/h)}\int_{-h}^{0}\cos(\pi j z/h)\,U_d(z)\,dz. \qquad (5.212)$$

where I_0 is the modified Bessel function of the first kind.

(b) We assume $U_d = 0$ for $z = -h$. Based on eq. (5.212), find a possible analytical form for U_d and Ω_d that satisfies this condition.

Answer: A possible solution is

$$U_d = 1 + \cos(\pi z/h);$$
$$\Omega_d = \frac{1}{R_0}\left[\tfrac{1}{2}r^2 - (z+h)^2\right] + \frac{h I_0(\pi r/h)}{\pi I_0'(\pi R_0/h)}\cos\left(\frac{\pi z}{h}\right).$$

(c) Will the first antisymmetric modes be excited? Derive the modal equations for the axisymmetric modes using Section 5.4.4.

Answer: The modal equations take the form $\ddot{\beta}_{0,j} + \sigma_{0,j}^2 \beta_{0,j} = -P_{dj}\ddot{\eta}_d$, where $P_{dj} = \lambda_{dj}/\mu_{0,j}$ and

$$\mu_{0,j} = \frac{2\rho\pi}{\kappa_{0,j}} \int_0^{R_0} r R_{0,j}^2(r)\,\mathrm{d}r = \frac{\rho\pi R_0^3}{\iota_{0,j}\tanh(\iota_{0,j}h/R_0)},$$

$$\lambda_{dj} = 2\rho\pi \int_0^{R_0} r R_{0,j}\Omega_d(r,0)\,\mathrm{d}r$$

$$= 2\rho\pi\left[a_0\frac{R_0^4}{\iota_{0,j}^2} + R_0 \sum_{n=1}^{\infty} \frac{a_n(\pi n/h)\,I_1(\pi n R_0/h)}{(\pi n/h)^2 + (\iota_{0,j}/R_0)^2}\right].$$

5.6.5 Spilling of coffee

Consider a half-filled coffee cup with a circular cylindrical shape. Its inner diameter and height are 9 and 7 cm, respectively. Assume the time history of the prescribed horizontal velocity of the coffee cup is nonzero during $0 \le t \le T_d$ and can be expressed as $A\sin(\pi t/T_d)$. You may neglect damping in the analysis. Which combinations of A and T_d would cause the coffee to spill over the rim?

5.6.6 Braking of a tank vehicle

Following Section 5.4.6 and the approximate solution in Section 4.8.3.2, use the modal technique to describe linear longitudinal sloshing in a horizontal cylindrical tank that occurs due to longitudinal acceleration (along the horizontal Ox-axis; see Figure 4.42). The problem has relevance for braking of a tank vehicle. Use the expression for approximate natural frequencies (4.150) with eqs. (4.153) and (4.154) and approximate natural modes (4.151).

Show that the modal system takes the form

$$\ddot{\beta}_{m1} + \sigma_{m1}^2 \beta_{m1} = -P_m \ddot{\eta}_1, \qquad (5.213)$$

where

$$\sigma_{m1}^2 = g\frac{\pi m \mathcal{A}}{L_1 B_s h}\tanh\left(\frac{\pi m h}{L_1}\right),$$

$$P_m = \frac{2\mathcal{A}}{\pi m B_s h}\tanh\left(\frac{\pi m h}{L_1}\right)\left[(-1)^m - 1\right],$$

and the cross-sectional area $\mathcal{A}$ and the free surface width B_s are computed by eqs. (4.153) and (4.154), respectively. Furthermore, the longitudinal force is represented by

$$F_1(t) = -M_l\ddot{\eta}_1(t) - \rho B_s L_1^2 \pi^{-2} \sum_{m=1}^{\infty} \frac{(-1)^m - 1}{m^2}\ddot{\beta}_{m1}(t).$$

5.6.7 Free decay of a ship cross-section in roll

Consider a two-dimensional ship cross-section with a rectangular tank partially filled with a liquid. No incident waves are present. The cross-section is given an initial small roll angle $\eta_4(0)$ and is free to roll and sway. Use linear theory in the time domain to formulate the equations for sloshing. A linear frequency-domain solution can be used to express the external added mass and damping coefficients.

(a) What are the initial conditions for the generalized coordinates β_j for the free-surface motion in the tank?

(b) Set up the complete equation system for the coupled ship motions and sloshing. Be careful how the frozen liquid effect in the tank is handled and the fact that a tank-fixed coordinate system is used in combination with an inertial system for the ship.

6 Viscous Wave Loads and Damping

6.1 Introduction

Viscous flow phenomena are important for flow past bluff bodies inside a tank. Examples are cross-flow past stiffeners, girders, baffles, screens, interior pipes, and pump towers as illustrated in Figures 1.9, 1.10, 1.19, 1.20 and 1.21. The resulting loads are needed for both ultimate limit state (ULS) and fatigue limit state (FLS) assessments. Furthermore, the presence of interior structures provides damping of resonant liquid motion. If no interior bluff bodies exist inside a tank in addition to no breaking waves, the main viscous effect is in the boundary layer along the wetted tank surface. We show that this effect can be small; however, the effect is needed in numerical simulations to get rid of transient effects and achieve steady-state conditions.

We start by discussing nonseparated flow and then go on to present Morison's equation, which is a pragmatic way of expressing the wave loads when flow separation occurs. Generalization of the equation is shown for a tank-fixed coordinate system. The equation requires empirical mass and drag coefficients. The original formulation of Morison's equation assumes a vertical circular cylinder, but the equation can be generalized to any structure. A limitation is that the incident flow field has to vary slowly across the cross-dimension of the considered structure. Another way of saying this is that the structure is a poor gravity wave generator (Faltinsen, 1990). We then continue by discussing the damping of the liquid motion due to flow separation and its effect on resonant liquid response. Special attention is given to tuned liquid dampers (TLDs), which are used to suppress horizontal vibrations of structures (e.g., tall buildings). A swash bulkhead is then analyzed to understand how natural sloshing frequencies and the liquid response are influenced by openings in the bulkhead and the forcing amplitude. Finally, vortex-induced vibration (VIV) is addressed. We could in principle have handled viscous flow by applying computational fluid dynamics (CFD) and solving the Navier–Stokes equations. However, this approach is CPU-demanding and it may be numerically challenging. CFD is dealt with in Chapter 10.

6.2 Boundary-layer flow

In this section we discuss how to predict forces on bodies in nonseparated viscous flow. A solid-body surface is assumed with no penetration of fluid, which means our discussions are not relevant for the perforated walls illustrated in Figure 1.34 that are used to damp piston-mode resonance inside a moonpool. The Reynolds number, Rn, is assumed sufficiently high so that the dominant viscous effect is in a boundary layer along the body's surface. Boundary-layer approximations of the Navier–Stokes equations can then be derived. This is first shown for two-dimensional laminar flow based on eqs. (2.1)–(2.3). The x-axis is chosen along the body surface and the y-axis is perpendicular to the body surface. A measure of the boundary-layer thickness, δ, is assumed small relative to the radius of curvature of the body's surface. If the body does not move, the fluid velocity varies rapidly across the boundary layer from zero on the body to the free-stream velocity U_e at $y = \delta$, which implies that $\partial u/\partial y$ is much larger than $\partial u/\partial x$. The consequence is that the term $\partial^2 u/\partial x^2$ in eq. (2.1) can be neglected relative to the term $\partial^2 u/\partial y^2$. The flow in the boundary layer varies, in general, with both x and y. If we neglect one of the terms in continuity equation (2.3), this statement is not true. Because $\partial v/\partial y$ is of the order of v divided by δ and $\partial u/\partial x$ is of the order of u, v is of the order of $u\delta$ (i.e., v is smaller than u). This statement implies that terms $u\partial u/\partial x$ and $v\partial u/\partial y$ in the convective acceleration of eq. (2.1) are of the same order. For the diffusion term on the right-hand side of eq. (2.1) to be of the same order as the convective acceleration term on the left-hand side, it is required that the boundary-layer thickness is $O(1/\sqrt{Rn})$. Furthermore, we assume for the moment that $\partial u/\partial t$ is of the same order as $u\partial u/\partial x$. From eq. (2.2) it follows that $\partial p/\partial y$ is of the order of $u\delta$. Therefore, as a first approximation, in eq. (2.2) we can set $\partial p/\partial y = 0$, whereby p in eq. (2.1) is the same as p at $y = \delta$. Thus, as long as the boundary layer has a small thickness δ, p can, as a first step, be calculated from the flow outside the boundary layer. There the fluid is accurately described by potential flow theory; that is, the fluid can be modeled as inviscid and in irrotational motion. This estimate of p can be done by neglecting the boundary layer and finding the tangential velocity at the body surface, U_e. The version of eq. (2.1) based on potential flow gives

$$\rho\frac{\partial U_e}{\partial t} + \rho U_e\frac{\partial U_e}{\partial x} = -\frac{\partial p}{\partial x}. \tag{6.1}$$

In this way we end up with the following unsteady two-dimensional boundary-layer equations for laminar flow:

$$\frac{\partial u}{\partial t} + u\frac{\partial u}{\partial x} + v\frac{\partial u}{\partial y} = \frac{\partial U_e}{\partial t} + U_e\frac{\partial U_e}{\partial x} + \nu\frac{\partial^2 u}{\partial y^2} \quad \text{and}$$
$$\frac{\partial u}{\partial x} + \frac{\partial v}{\partial y} = 0. \tag{6.2}$$

For turbulent flow we may introduce Reynolds-averaged Navier–Stokes equations. We then insert $u = \overline{u} + u'$, $v = \overline{v} + v'$, $w = \overline{w} + w'$, and $p = \overline{p} + p'$ into the three-dimensional Navier–Stokes equations. The overbar means that the values are time-averaged over the time scale of turbulence and the prime denotes the variations on the time scale of turbulence. The equations are then time-averaged over the time scale of turbulence. The two-dimensional boundary layer version becomes

$$\frac{\partial \overline{u}}{\partial t} + \overline{u}\frac{\partial \overline{u}}{\partial x} + \overline{v}\frac{\partial \overline{u}}{\partial y}$$
$$= \frac{\partial U_e}{\partial t} + U_e\frac{\partial U_e}{\partial x} + \frac{\partial}{\partial y}\left(\nu\frac{\partial \overline{u}}{\partial y} - \overline{u'v'}\right) \quad \text{and}$$
$$\frac{\partial \overline{u}}{\partial x} + \frac{\partial \overline{v}}{\partial y} = 0. \tag{6.3}$$

In this case,

$$\frac{1}{\rho}\frac{\partial}{\partial y}\left(\mu\frac{\partial \overline{u}}{\partial y} - \rho\overline{u'v'}\right) = \frac{1}{\rho}\frac{\partial}{\partial y}(\tau_l + \tau_t), \tag{6.4}$$

where

$$\tau_l = \mu\frac{\partial \overline{u}}{\partial y} \tag{6.5}$$

is the viscous (also called laminar) shear stress and

$$\tau_t = -\rho\overline{u'v'} \tag{6.6}$$

is the turbulent stress.

Measurements show that a domain exists very close to the body surface where $\tau_l \gg \tau_t$. We can understand this by noting that τ_t is zero on the body surface, which is a consequence of the body boundary condition (i.e., $u' = v' = 0$ on the body surface). However, τ_l is not zero on the body surface. The domain where τ_l dominates is called the viscous sublayer. An outer layer can also be defined where turbulent shear dominates and an overlap layer defined where both types of shear are important (White, 1974). To solve the equations we need to relate the turbulent stresses to the mean flow. This process introduces empirical coefficients. The simplest way of doing it is to write

$$\tau_t = \mu_e(y)\frac{\partial \overline{u}}{\partial y}, \tag{6.7}$$

where $\mu_e(y)$ is an empirically determined eddy viscosity coefficient that depends on the distance y from the body surface.

6.2.1 Oscillatory nonseparated laminar flow

To evaluate the velocities and shear forces for *laminar* flow we further simplify the two-dimensional boundary-layer equations. What we describe is referred to as the Stokes second problem and is described in many textbooks on fluid dynamics (see, e.g., Schlichting, 1979, pp. 428–9). We assume that the outer flow velocity outside the boundary layer, U_e, can be written as $U_e(x, t) = U_0(x)\cos\sigma t$. If the oscillatory amplitudes are small, we may neglect quadratic terms in the boundary-layer equations. It is also permissible to neglect the convective acceleration terms if the streamwise gradients of velocity are small, in which case the result can be applied to large-amplitude oscillatory flow, giving

$$\frac{\partial u}{\partial t} - \nu\frac{\partial^2 u}{\partial y^2} = \frac{\partial U_e}{\partial t}. \tag{6.8}$$

A steady-state solution of this equation with the boundary conditions $u = 0$ at $y = 0$ and $u = U_e$ at $y \to \infty$ can be found by substituting the complex solution form $u = U_e(x, t) + A(x)\exp(\beta y)\exp(i\sigma t)$ into eq. (6.8), where A is complex and it is implied to be the real part of the expression for u that has physical meaning. This procedure is convenient when we operate with a linear system. Linearity means that eq. (6.8) and the boundary conditions include only terms that are linearly dependent on u and its

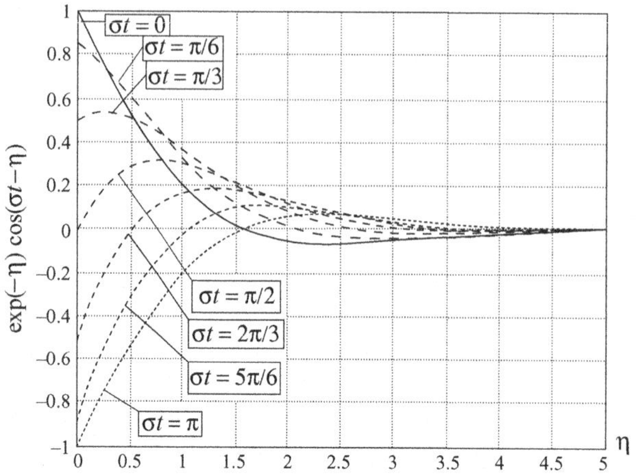

Figure 6.1. Behavior in the boundary layer of the damped-wave part of the boundary-layer solution for horizontal velocity divided by U_0 at the following time instants: $\sigma t = 0,\ \frac{1}{6}\pi,\ \frac{1}{3}\pi,\ \frac{1}{2}\pi,\ \frac{2}{3}\pi,\ \frac{5}{6}\pi,\ \pi$; $\eta = \sqrt{\sigma/(2\nu)}y$ with $y = 0$ at the wall.

derivatives. By disregarding an exponential solution that increases exponentially when $y \to \infty$, we find that the solution must have the form

$$u = U_e(x,t) + A(x) \times \exp\left(-\sqrt{\sigma/2\nu}(1+\mathrm{i})y\right)\exp(i\sigma t). \quad (6.9)$$

This solution gives by using the boundary conditions that

$$u = U_0(x)\left[\cos\sigma t - \exp(-\eta)\cos(\sigma t - \eta)\right], \quad \eta = y\sqrt{\sigma/2\nu}. \quad (6.10)$$

The velocity does not have the same phase at all y-coordinate points in the boundary layer. This fact is illustrated in Figure 6.1, which shows the part of the horizontal velocity that behaves as a damped wave propagating through the boundary layer.

Equation (6.10) can be used to find the boundary-layer thickness, for which many different definitions exist. We use a geometrical measure and consider the point where $U_0(x) \times \exp[-\sqrt{\sigma/(2\nu)}y] = 0.01 \cdot U_0(x)$, which corresponds to $\sqrt{\sigma/(2\nu)}y = 4.6$. For $\nu = 10^{-6}$ m^2 s^{-1} (water) and $T = 10$ s, this means $y = 0.008$ m. Castor oil, with $\nu = 1.03 \cdot 10^{-3}$ m^2 s^{-1}, is considered another example. This value gives a geometrical boundary-layer thickness of 26 cm for laminar flow and $T = 10$ s. This period is a realistic period for full-scale conditions of a tank. The estimated boundary-layer thickness is very small relative to realistic full-scale tank dimensions; for instance, a tank breadth B of 30 m gives $\delta/B \approx 0.009$. However, the possibility exists that the boundary layer is turbulent in this case. Turbulent boundary layers are considered in Section 6.2.3. If we consider model test conditions, the estimated boundary-layer thickness is larger relative to the tank dimensions. If we choose castor oil and $T = 3$ s, the geometrical boundary-layer thickness is 14 cm. The tank breadth in model scale may be of the order of 1 m (i.e., $\delta/B = 0.14$).

Boundary-layer flow causes viscous dissipation. A general expression is given by the last term in eq. (2.45). Using the laminar boundary-layer approximation gives the time rate of change of viscous energy dissipation in the fluid per unit length of the plate:

$$\dot{E}_{vd} = -\mu\int_0^\infty\left(\frac{\partial u}{\partial y}\right)^2 \mathrm{d}y, \quad (6.11)$$

where

$$\frac{\partial u}{\partial y} = U_0(x)\sqrt{\sigma/\nu}\exp\left(-y\sqrt{\sigma/2\nu}\right) \times \cos\left(\sigma t + \tfrac{1}{4}\pi - y\sqrt{\sigma/2\nu}\right). \quad (6.12)$$

We are interested in the average of the time rate of viscous energy dissipation over one period $T = 2\pi/\sigma$. Finding this average involves using the fact that $T^{-1}\int_0^T \cos^2(\sigma t + \varepsilon)\,\mathrm{d}t = \frac{1}{2}$. The final result for the average of the time rate of viscous energy dissipation over one period and per unit length of the plate is

$$\langle\dot{E}_{vd}\rangle = -\frac{\mu}{2}\sqrt{\frac{\sigma}{2\nu}}U_0^2(x). \quad (6.13)$$

This formula is later used for viscous energy dissipation in a rectangular tank.

The wall skin friction force (shear force) per unit area for laminar flow can be written as

$$\tau_w = \mu\left.\frac{\partial u}{\partial y}\right|_{y=0} = \mu U_0\sqrt{\frac{\sigma}{\nu}}\cos\left(\sigma t + \tfrac{1}{4}\pi\right), \quad (6.14)$$

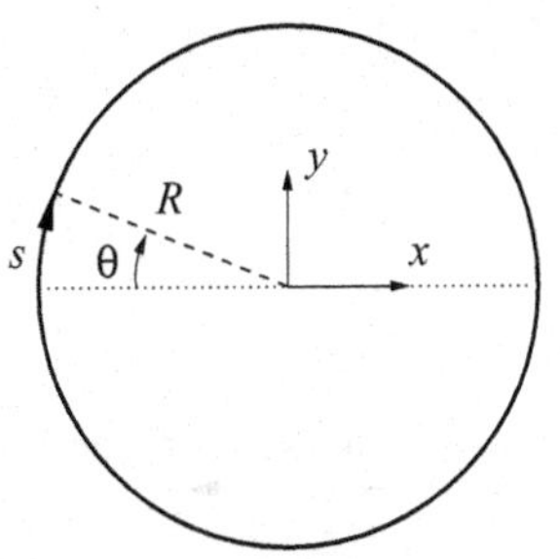

Figure 6.2. Coordinate system for two-dimensional flow past a circular cylinder in infinite fluid.

which oscillates with a phase lead of 45° relative to the external stream velocity U_e.

6.2.2 Oscillatory nonseparated laminar flow past a circular cylinder

Flow around a circular cylinder is generally separated, but regimes exist for small ratios between the ambient oscillation amplitude and the diameter for which the flow stays attached. Our following analysis assumes nonseparated flow. We later consider the effect of flow separation.

Let us apply the solution in the previous section to nonseparated oscillatory laminar flow past a circular cylinder with radius R and diameter D in an infinite fluid. The free-stream velocity far away from the body is written as $U_\infty \cos\sigma t$. The solution outside the boundary layer, U_e, can then be written as $2U_\infty \sin\theta\cos\sigma t$, where θ is the angle of a polar coordinate system (r, θ) and $r = 0$ corresponds to the center of the cylinder. The positive ambient flow direction is along $\theta = \pi$ (see Figure 6.2).

From eq. (6.14), the shear stress acting on the cylinder can be written as

$$\tau_w = \mu 2U_\infty\sqrt{\sigma/\nu}\sin\theta\cos\left(\sigma t + \tfrac{1}{4}\pi\right). \qquad (6.15)$$

The resulting force on the cylinder in the incident flow direction can be expressed by

$$F_1 = \int_0^{2\pi} \tau_w \sin\theta R d\theta$$
$$= \mu U_\infty\sqrt{\sigma/\nu}\cos\left(\sigma t + \tfrac{1}{4}\pi\right) D\pi. \qquad (6.16)$$

By writing $\cos(\sigma t + \frac{1}{4}\pi) = (\sqrt{2}/2)\cos\sigma t - (\sqrt{2}/2)\sin\sigma t$, we see that one force component is in phase with the velocity and another force component is in phase with the acceleration. It should be noted that the force is linear with respect to U_∞. The drag force is normally said to be proportional to the square of the velocity. If we calculate the drag coefficient, we find

$$C_D^F = \frac{\mu U_\infty\sqrt{\sigma/\nu}D\pi\sqrt{2}/2}{\frac{1}{2}\rho U_\infty^2 D} = 2\pi\sqrt{\frac{\pi}{RnKC}}. \qquad (6.17)$$

However, this equation does not give the total viscous drag force. Viscosity also influences the normal stresses (Batchelor, 1970, p. 355). According to Stokes (1851), this influence results in a total drag force on a circular cylinder that is twice as large as that in eq. (6.17). Let us show this. We introduce a coordinate s along the cylinder surface so that $s = R\theta$, which means $U_0(s) = 2U_\infty \sin(s/R)$. The tangential velocity in the boundary layer can be expressed as

$$u = U_0(s)\left[1 - \exp\left(-\sqrt{\sigma/(2\nu)}\,(1+\mathrm{i})\,n\right)\right]\exp(\mathrm{i}\sigma t),$$
$$\mathrm{i} = \sqrt{-1}. \qquad (6.18)$$

It is understood that it is the real part of this expression and the following expressions that have physical meaning. The normal flow velocity v in the boundary layer follows from the continuity equation

$$\frac{\partial v}{\partial n} = -\frac{\partial u}{\partial s}, \qquad (6.19)$$

where n denotes the normal to the body surface with $n = 0$ on the surface. Integrating this equation gives

$$v = -\frac{dU_0}{ds}\left\{n + \sqrt{\frac{2\nu}{\sigma}}\frac{1}{(1+\mathrm{i})}\right.$$
$$\left.\times\left[\exp\left(-\sqrt{\frac{\sigma}{2\nu}}(1+\mathrm{i})\,n\right) - 1\right]\right\}\exp(\mathrm{i}\sigma t). \qquad (6.20)$$

The term $-(dU_0/ds)\,n\exp(\mathrm{i}\sigma t)$ is a consequence of the potential flow part of the solution. We then let $n \to \infty$ in the viscous part of the solution, giving an out-/inflow velocity caused by the boundary layer that is

$$v_0 = \frac{dU_0}{ds}\sqrt{\frac{2\nu}{\sigma}}\frac{\exp(\mathrm{i}\sigma t)}{(1+\mathrm{i})}$$
$$= 2U_\infty\cos\left(\frac{s}{R}\right)\sqrt{\frac{2\nu}{\sigma}}\frac{\exp(\mathrm{i}\sigma t)}{(1+\mathrm{i})R}. \qquad (6.21)$$

This fact is illustrated in Figure 6.3.

So now we have to solve a potential flow problem with normal velocity equal to v_0 on the cylinder surface. The solution of the velocity potential

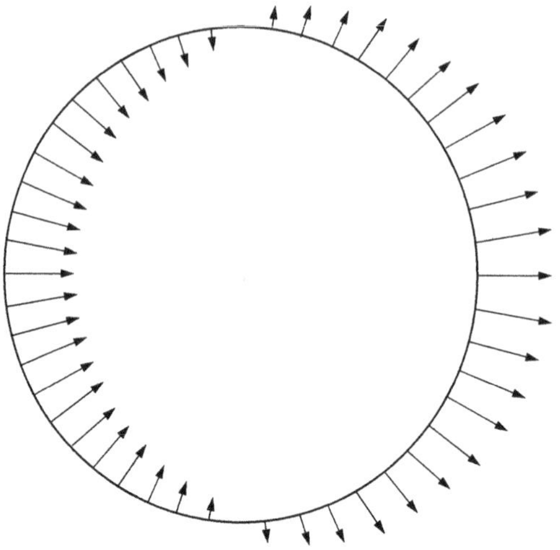

Figure 6.3. Instantaneous inflow/outflow velocity from the laminar boundary layer associated with harmonically oscillating nonseparated flow past a circular cylinder. Illustrated by lines perpendicular to the cylinder surface.

and the hydrodynamic pressure on the body can be expressed as

$$\varphi = -\sqrt{\frac{2\nu}{\sigma}}\frac{2U_\infty R\cos(\theta)\exp(\mathrm{i}\sigma t)}{(1+\mathrm{i})\,r} \quad \text{and}$$
$$p = -\rho\frac{\partial\varphi}{\partial t} = \rho\sqrt{\frac{2\nu}{\sigma}}\frac{2U_\infty \mathrm{i}\sigma\cos(\theta)\exp(\mathrm{i}\sigma t)}{(1+\mathrm{i})}. \tag{6.22}$$

We multiply the pressure by the x-component of the interior surface normal, $n_1 = \cos\theta$, and integrate around the body surface to find the x-component of the pressure force, which gives

$$F_1 = \mathrm{Re}\big[\rho U_\infty\sqrt{2\nu\sigma}(1+\mathrm{i})\,R\pi\exp(\mathrm{i}\sigma t)\big]$$
$$= \rho U_\infty D\pi\sqrt{\nu\sigma}\cos\left(\sigma t + \tfrac{1}{4}\pi\right), \tag{6.23}$$

where D is the cylinder diameter. This pressure force is exactly the same as the contribution from the shear force which is consistent with the previously mentioned well-known fact about the influence of the boundary layer on the pressure distribution. A similar analysis was done by Bearman *et al.* (1985) for a rectangular cross-section.

Similar results may be obtained for a general unsteady laminar boundary-layer flow (see Exercise 6.11.4).

6.2.3 Turbulent nonseparated boundary-layer flow

The boundary-layer flow is likely to be turbulent under full-scale conditions. Jonsson (1978) gave empirical formulas for shear stress that apply to turbulent flow along fixed-plane surfaces. Outside the boundary layer the flow is oscillating harmonically. Jonsson defines the Reynolds number as

$$RE = U_0^2/\sigma\nu, \tag{6.24}$$

where U_0 is the maximum tangential velocity just outside the boundary layer and U_0/σ is the characteristic length. $RE = 10^5$ is proposed as an engineering criterion for transition to turbulence for a smooth surface. For more details on the transition to turbulence one should also consult Fredsøe and Deisgaard (1993, pp. 30–32) and Jensen *et al.* (1989). When the surface is smooth, Jonsson writes the maximum wall shear stress τ_{wm} as

$$2\tau_{wm}/\rho U_0^2 = 0.09\cdot RE^{-0.2}. \tag{6.25}$$

For instance, if $RE = 10^4$, eq. (6.25) gives a maximum wall shear that is 71 times the predictions based on laminar flow using eq. (6.14).

The boundary-layer thickness, δ_w, can according to Smith (1977) be expressed as $\delta_w = (u_*)_m/\sigma$, where $(u_*)_m$ is the maximum wall friction velocity which, by definition, can be written as $\sqrt{\tau_{wm}/\rho}$. It follows by using eq. (6.25) that

$$\delta_w = 0.21\cdot U_0/(\sigma RE^{0.1}). \tag{6.26}$$

This equation gives, for instance, $\delta_w = 0.07\cdot U_0/\sigma$ when $RE = 10^5$. Let us further elaborate what this means by assuming $\nu = 10^{-6}$ m^2 s^{-1} and $\sigma = 0.6$ rad/s. The definition of RE gives

$U_0 = 0.245$ m s^{-1} or $\delta_w = 0.029$ m. If laminar flow had been the case for the same conditions, the geometrical measure of the boundary layer would be 0.008 m. The boundary-layer thickness is larger for turbulent flow than for laminar flow due to the larger exchange of fluid momentum. We note that the boundary-layer thickness depends on U_0 in the case of turbulent flow but it is independent of U_0 in the case of laminar boundary-layer flow.

If the force due to oscillating nonseparated turbulent flow past a circular cylinder is considered, it is not sufficient to account for only the shear stresses. The boundary layer causes inflow/outflow to the potential flow domain, which was relatively easy to derive for a laminar boundary layer (see Section 6.2.2). A turbulent analysis needs an empirical model for the turbulent stress τ_t. One way is to use eq. (6.7). Smith (1977) followed such a model. In addition, he neglected the viscous stresses and assumed the eddy viscosity coefficient depended linearly on y; therefore,

$$\tau_t = \rho K_0 y \, \partial u / \partial y, \quad K_0 = \kappa (u_*)_m, \tag{6.27}$$

where $\kappa = 0.4$ is the von Karman constant. This turbulence model can be justified in the overlap layer.

If the oscillatory amplitudes are small, we get the following unsteady boundary-layer model for tangential velocity u in the boundary layer:

$$\frac{\partial u}{\partial t} = \frac{\partial U_e}{\partial t} + K_0 \frac{\partial}{\partial y}\left(y \frac{\partial u}{\partial y} \right). \tag{6.28}$$

Inserting the steady-state solution $u = U_0(x)\,[1 - Y(y)]\exp(i\sigma t)$ into eq. (6.28) gives the ordinary differential equation

$$\frac{-i\sigma Y}{K_0} + \frac{dY}{dy} + y\frac{d^2Y}{dy^2} = 0. \tag{6.29}$$

The solution can be expressed in terms of the zero-order Kelvin functions ker and kei (Abramowitz & Stegun, 1964):

$$u = U_0(x)\,[\cos\sigma t - F_1(\xi, \xi_0)\cos\sigma t + F_2(\xi, \xi_0)\sin\sigma t], \tag{6.30}$$

where

$$F_1(\xi, \xi_0) = \frac{(\ker\xi)(\ker\xi_0) + (\mathrm{kei}\,\xi)(\mathrm{kei}\,\xi_0)}{\ker^2\xi_0 + \mathrm{kei}^2\,\xi_0},$$

$$F_2(\xi, \xi_0) = -\frac{(\ker\xi)(\mathrm{kei}\,\xi_0) - (\mathrm{kei}\,\xi)(\ker\xi_0)}{\ker^2\xi_0 + \mathrm{kei}^2\,\xi_0},$$

$$\xi = 2\sqrt{\sigma y / K_0}, \quad \xi_0 = 2\sqrt{\sigma y_0 / K_0}. \tag{6.31}$$

and where $y_0 > 0$ is a point at a very short distance from the wall where the boundary condition $u = 0$ is imposed. Because expression (6.30) is infinite at $y = 0$, it is not valid there. Christoffersen and Jonsson (1985) presented the following expression for y_0:

$$y_0 = \tfrac{1}{30}k_{Ni}\,[1 - \exp(-k_{Ni}\,(u_*)_m / (27\nu))] + \nu / (9.025\,(u_*)_m), \tag{6.32}$$

where k_{Ni} is Nikuradse's equivalent sand roughness of the surface (i.e., the characteristic dimension of the physical roughness of the surface) (Schlichting, 1979). However, k_{Ni} may be very different from what the physical roughness of the surface would suggest. The coordinate y_0 for a smooth surface follows simply by setting $k_{Ni} = 0$ in eq. (6.32). To find the shear stress on the wall, τ_w, the local shear stress at y_0 is assumed approximately equal to τ_w. From eq. (6.27) we get

$$\left.\frac{\partial u}{\partial y}\right|_{y_0} = \frac{\tau_w}{\rho K_0 y_0}. \tag{6.33}$$

Using eq. (6.30) to express $\partial u/\partial y$ together with eq. (6.33) gives

$$\tau_w = \tfrac{1}{2}\rho U_0 K_0 \xi_0\,[-F_1'(\xi_0)\cos\sigma t + F_2'(\xi_0)\sin\sigma t], \tag{6.34}$$

where

$$F_1'(\xi_0) = [(\ker_1\xi_0 + \mathrm{kei}_1\,\xi_0)\ker\xi_0 + (\mathrm{kei}_1\,\xi_0 - \ker_1\xi_0)\mathrm{kei}\,\xi_0] / [\sqrt{2}(\ker^2\xi_0 + \mathrm{kei}^2\,\xi_0)],$$

$$F_2'(\xi_0) = [-(\ker_1\xi_0 + \mathrm{kei}_1\,\xi_0)\ker\xi_0 + (\mathrm{kei}_1\,\xi_0 - \ker_1\xi_0)\ker\xi_0] / [\sqrt{2}(\ker^2\xi_0 + \mathrm{kei}^2\,\xi_0)]. \tag{6.35}$$

In these expressions $\ker_1$ and kei_1 are first-order Kelvin functions (Abramowitz & Stegun, 1964). Because y_0 and K_0 depend on $(u_*)_m = \sqrt{\tau_{wm}/\rho}$, ξ_0 is a function of the amplitude of τ_w (see eq. (6.31)). It follows from eq. (6.34) that

$$(u_*)_m = \left\{\tfrac{1}{2}U_0 K_0 \xi_0 \left[[F_1'(\xi_0)]^2 + [F_2'(\xi_0)]^2\right]^{1/2}\right\}^{1/2}. \tag{6.36}$$

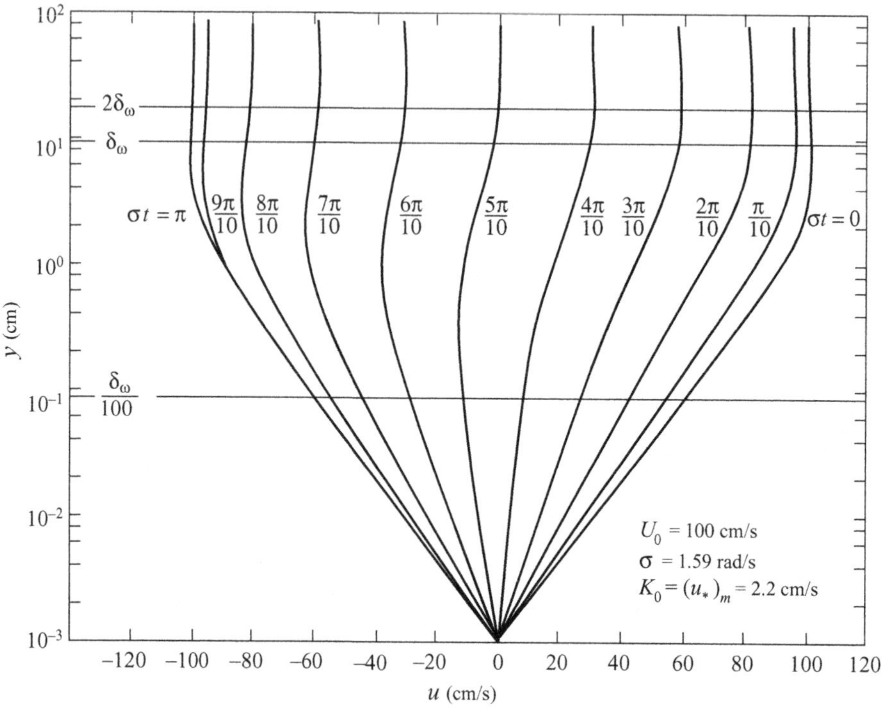

Figure 6.4. An example of velocity field for the oscillatory turbulent boundary layer. The thickness for the boundary layer, given as $\delta_\omega = (u_*)_m/\sigma$, is shown on the left side of the diagram. Note that by $2\delta_\omega$ the velocity field is essentially that of the inviscid flow and that below $\frac{1}{100}\delta_\omega$ the velocity fields are logarithmic; moreover, at the times of maximum flow they are logarithmic to $\frac{1}{10}\delta_\omega$. (*Source:* Smith, 1977; Copyright© 1977 by John Wiley & Sons, Inc., Wiley Interscience, New York. Reprinted with permission of John Wiley & Sons, Inc.)

One way to solve the problem is via an iterative procedure where $(u_*)_m$ in the first step is estimated by using Jonsson's expression for τ_{wm} given by eq. (6.25). Let us illustrate with an example. We consider $\sigma = 1.0$ rad/s, $\nu = 10^{-6}$ m^2 s^{-1}, $U_0 = 1$ m s^{-1}, and $\rho = 10^3$ kg m^{-3}. This gives $RE = 10^6$ and $(u_*)_m = 0.053$ m s^{-1} according to Jonsson's formula. Using this value of $(u_*)_m$ gives, according to eq. (6.32) for a smooth surface, $y_0 = 0.208 \cdot 10^{-5}$ m. Furthermore, we find that $K_0 = 0.021$ m s^{-1} according to eq. (6.27). Equation (6.31) gives $\xi_0 = 0.0198$. Equation (6.36) then gives $(u_*)_m = 0.051$ m s^{-1}, which is close to the first estimate. We can continue with the iteration until satisfactory convergence has been obtained. Then we can calculate u and τ_w by means of eqs. (6.30) and (6.34), respectively.

Figure 6.4 considers an example from Smith (1977). The velocity u in the boundary layer is presented at different time instants σt. A semi-logarithmic presentation has been chosen to emphasize the logarithmic velocity profile near the wall.

6.2.3.1 Turbulent energy dissipation

The average time rate of turbulent energy dissipation over one period and per unit length of the plate can be expressed as

$$\begin{aligned}\langle \dot{E}_d \rangle &= -\frac{1}{T}\int_0^T \tau_w(t) U_0 \cos(\sigma t)\, dt \\ &= \tfrac{1}{4}\rho U_0^2 K_0 \xi_0 F_1'(\xi_0). \qquad (6.37)\end{aligned}$$

Equation (6.37) can be shown by considering a plate that moves tangentially with a velocity $U_t = U_0 \cos \sigma t$ and where the fluid is at rest outside the boundary layer. Application of eq. (2.47) gives eq. (6.37).

When the viscous energy dissipation in a laminar boundary layer was studied, a different formula was used as a basis (see eq. (6.11)). We leave it as an exercise to show that the application of eq. (6.37) to laminar boundary-layer flow gives the same result as eq. (6.13).

6.2.3.2 Oscillatory nonseparated flow past a circular cylinder

The previous analysis for turbulent boundary-layer flow can be applied to oscillatory nonseparated turbulent flow past a circular cylinder, for instance (see Figure 6.2). We saw in the case of laminar flow that it is not sufficient to consider the contribution from shear forces to the integrated force on the body for a circular cylinder. Pressure forces provided an equal contribution to the integrated force. This effect was due to the inflow/outflow from the boundary layer caused by viscous effects.

Let us show how to proceed for a turbulent boundary layer around a circular cylinder. We use the continuity equation given by eq. (6.19). As we did in the laminar case, we neglect the contribution from the potential flow solution. By using eq. (6.30), we find that the inflow/outflow velocity from the boundary layer is

$$\begin{aligned} v_0 &= -\frac{dU_0}{ds}\left[-\int_{y_0}^{\infty} F_1(\xi,\xi_0)dy\cos\sigma t \right. \\ &\quad \left. + \int_{y_0}^{\infty} F_2(\xi,\xi_0)dy\sin\sigma t\right] \\ &= -\frac{K_0}{2\sigma}\frac{dU_0}{ds}\left[-\int_{\xi_0}^{\infty} F_1(\xi,\xi_0)\xi d\xi\cos\sigma t \right. \\ &\quad \left. + \int_{\xi_0}^{\infty} F_2(\xi,\xi_0)\xi d\xi\sin\sigma t\right]. \end{aligned} \tag{6.38}$$

Because the viscous sublayer is very thin and we do not have a description of the flow there, we have neglected its contribution to the inflow/outflow velocity. We should then calculate the potential flow due to this inflow/outflow from the cylinder. Because F_1 and F_2 depend on U_0 through ξ_0, a simple θ-dependence does not exist as it does for laminar flow. So we should solve the potential flow numerically and find the pressure by using Bernoulli's equation.

An alternative is to find the resulting pressure force on the cylinder by Green's second identity in two dimensions (see eq. (A.4)). The surface normal direction is defined as positive out of the liquid domain in eq. (A.4). However, the relationship is the same if the positive direction of the surface normal vector is into the liquid. The latter definition is used in the following text. We denote the velocity potential due to the inflow/outflow from the boundary layer as φ_{io}. The hydrodynamic pressure on the cylinder surface can be approximated as $-\rho\partial\varphi_{io}/\partial t$. The x-component of the force on the cylinder is then

$$F_1 = \rho\int_{S_B}\frac{\partial\varphi_{io}}{\partial t}n_1\, dS, \tag{6.39}$$

where $n_1 = -\cos\theta$ is the x-component of the exterior surface normal. We introduce eq. (A.4), denote $\psi = \partial\varphi_{io}/\partial t$, and let velocity potential φ satisfy $\partial\varphi/\partial n = n_1$ on S_B. Furthermore, S consists of the body surface S_B and a closed surface S_∞ far away from the cylinder. The solution of φ is simply $\varphi = R^2/r\cos\theta$, where (r,θ) are polar coordinates and r has the same direction as n. We can rewrite eq. (6.39) by eq. (A.4) by noting that integration along S_∞ provides zero contribution:

$$\begin{aligned} F_1 &= \rho\int_{S_B}\frac{\partial\varphi_{io}}{\partial t}n_1\, dS = \rho\int_{S_B}\frac{\partial\varphi_{io}}{\partial t}\frac{\partial\varphi}{\partial n}\, dS \\ &= \rho\int_{S_B}\varphi\frac{\partial}{\partial n}\frac{\partial\varphi_{io}}{\partial t}\, dS. \end{aligned} \tag{6.40}$$

We now use the body boundary condition for φ_{io} given by eq. (6.38), which gives

$$F_1 = \rho\int_{S_B}\varphi\frac{\partial v_0}{\partial t}\, dS = \rho R^2\int_0^{2\pi}\cos\theta\frac{\partial v_0}{\partial t}\, d\theta. \tag{6.41}$$

Equation (6.41) can be further simplified; we start with eq. (6.38). We can write

$$\frac{\partial v_0}{\partial t} = -\frac{dU_0}{ds}\int_{y_0}^{\infty}\frac{\partial u_{\text{visc}}}{\partial t}\, dy, \tag{6.42}$$

where $U_0 u_{\text{visc}} = u - U_e$. Using eq. (6.28) then gives

$$\frac{\partial v_0}{\partial t} = \frac{1}{U_0}\frac{dU_0}{ds}K_0 y_0\left.\frac{\partial u}{\partial y}\right|_{y_0} = \frac{1}{U_0}\frac{dU_0}{ds}\frac{\tau_w}{\rho}. \tag{6.43}$$

Inserting eq. (6.43) into eq. (6.41) and using the fact that $U_0 = 2U_\infty\sin\theta$ gives $F_1 = R\int_0^{2\pi}\tau_w\frac{\cos^2\theta}{\sin\theta}\, d\theta$. In addition, we have a contribution from the shear stresses that gives the total viscous force as

$$F_{1v} = R\int_0^{2\pi} \tau_w \frac{(\cos^2\theta + \sin^2\theta)}{\sin\theta}\,d\theta$$
$$= R\int_0^{2\pi} \frac{\tau_w}{\sin\theta}\,d\theta. \tag{6.44}$$

We should note that τ_w depends on $U_0 = 2U_\infty \sin\theta$ and we assume turbulent flow everywhere along the cylinder surface. In practice, a region always exists where U_0 is so small that the criterion for turbulent condition is not locally met. However, turbulence associated with the wake from the previous half-cycle tends to flow back into these low-velocity regions each time the flow direction reverses. Furthermore, a threshold Reynolds number $Rn = U_\infty D/\nu$ must exist where the error due to a small laminar region can be neglected. We should note that τ_w is time-dependent. Equation (6.41) has to be numerically integrated. Jonsson's expression given by eq. (6.25) is used to get a feeling for the magnitude of F_{1v}; that is,

$$|F_{1v}| \approx \rho 2U_\infty^2 R \cdot 0.09 \int_0^{2\pi} \frac{\sin\theta}{RE^{0.2}}\,d\theta$$
$$= \frac{0.63\rho U_\infty^2 R}{\left[U_\infty^2/(\sigma\nu)\right]^{0.2}}, \tag{6.45}$$

where we have assumed that the stress has the same phasing along the cylinder surface. This cannot be assumed a priori and must be investigated based on the previous detailed analysis, which implies that eq. (6.45) is a conservative estimate. The ratio between the value given by eq. (6.45) and that by eq. (6.23) based on laminar flow is $0.63/(2\pi)\cdot(U_\infty^2/(\sigma\nu))^{0.3}$. If $U_\infty^2/(\sigma\nu) = 10^4$, the ratio is equal to 1.59.

6.3 Damping of sloshing in a rectangular tank

6.3.1 Damping due to boundary-layer flow (Keulegan's theory)

Two-dimensional linear flow is assumed outside the boundary layers in the tank in the Oxz-plane defined as in the Oyz-plane in Figure 4.2. The width of the tank in the y-direction is denoted B. The length of the tank in the x-direction is denoted either l or $2a$. Our focus in this chapter is on *antisymmetric* natural modes. We assume no excitation of the tank and concentrate on the velocity potential of a natural mode that is antisymmetric with respect to the z-axis. When using the natural sloshing modes $\varphi_i(x,z)$ and frequencies σ_i from Section 4.3.1.1, the velocity potential and the free-surface motion (caused by an antisymmetric mode) can be expressed as

$$\phi_n(x,z,t) = \frac{gA}{\sigma}\varphi_{2n+1}(x,z)\cos(\sigma_{2n+1}t)$$
$$= \frac{gA}{\sigma}\frac{\cosh(k(z+h))}{\cosh(kh)}\sin(kx)\cos(\sigma t),$$
$$\zeta_n(x,t) = -\frac{1}{g}\frac{\partial\phi_n}{\partial t}\bigg|_{z=0} = A\sin(kx)\sin(\sigma t), \tag{6.46}$$

where, for brevity, we denote $\sigma = \sigma_{2n+1}$ and $k = \pi(2n+1)/l$ with $n = 0, 1, \ldots$. The value of A is the wave amplitude at the tank walls. The liquid velocity components are given by

$$\frac{\partial\phi_n}{\partial x} = \underbrace{\sigma A\frac{\cosh k(z+h)}{\sinh(kh)}\cos(kx)}_{u_0(x,z)}\cos(\sigma t),$$
$$\frac{\partial\phi_n}{\partial z} = \underbrace{\sigma A\frac{\sinh k(z+h)}{\sinh(kh)}\sin(kx)}_{w_0(x,z)}\cos(\sigma t), \tag{6.47}$$

where we used $\sigma^2 = gk\tanh(kh)$ for the natural frequencies.

The potential energy of the wave motion given by eq. (6.46) can be expressed as

$$E_p = \tfrac{1}{2}\rho g B\int_{-\frac{1}{2}l}^{\frac{1}{2}l} \zeta_n^2\,dx = \tfrac{1}{4}\rho g B A^2 l\sin^2(\sigma t). \tag{6.48}$$

The kinetic energy is

$$E_k = \tfrac{1}{2}\rho B\int_{-\frac{1}{2}l}^{\frac{1}{2}l}\int_{-h}^{0}\left[\left(\frac{\partial\phi_n}{\partial x}\right)^2 + \left(\frac{\partial\phi_n}{\partial z}\right)^2\right]dz\,dx$$
$$= \tfrac{1}{4}\rho g B A^2 l\cos^2(\sigma t). \tag{6.49}$$

The sum of the kinetic and potential flow energy is

$$E = E_k + E_p = \tfrac{1}{4}\rho g A^2 l B \tag{6.50}$$

according to linear potential flow theory.

Boundary-layer flow causes viscous dissipation of energy. This viscous energy dissipation is assumed to occur on a much longer time scale than the natural period of oscillation. We then apply a result derived previously for a flat surface to the vertical walls and the bottoms of the tank (see eq. (6.13)). Laminar boundary-layer flow is assumed. The average of the time rate of viscous dissipation over one period and per unit area of the plate can be approximated as

$$\langle \dot{E}_{vd} \rangle = -\tfrac{1}{2}\mu\sqrt{\sigma/2\nu}\, U_0^2(x), \tag{6.51}$$

where we have assumed harmonic oscillations (i.e., we have neglected the effect of the time scale of viscous energy dissipation).

When considering the boundary layer along the vertical walls we should, according to eq. (6.47), use

$$\begin{aligned} U_0^2(z) &= w_0^2\left(\pm\tfrac{1}{2}l, z\right) \\ &= \sigma^2 A^2 \sinh^2(k(z+h))/\sinh^2(kh) \end{aligned} \tag{6.52}$$

in eq. (6.51). Therefore, the two vertical walls at $x = \pm\frac{1}{2}l$ cause a contribution to the viscous energy dissipation over one period; that is,

$$\langle \dot{E}_{wa} \rangle = -\mu\sqrt{\frac{\sigma}{2\nu}}\sigma^2 A^2 B \int_{-h}^{0} \frac{\sinh^2(k(z+h))}{\sinh^2(kh)} dz, \tag{6.53}$$

where $\int_{-h}^{0} \sinh^2(k(z+h))dz = (\frac{1}{4}\sinh(2kh) - \frac{1}{2}kh)/k$.

When studying the boundary layer along the tank bottom, eq. (6.51) should use

$$U_0^2(x) = u_0^2(x, -h) = \sigma^2 A^2 \cos^2(kx)/\sinh^2(kh). \tag{6.54}$$

The tank bottom, therefore, causes a contribution to the viscous energy dissipation over one period; that is,

$$\begin{aligned} \langle \dot{E}_b \rangle &= -\frac{\mu}{2}\sqrt{\frac{\sigma}{2\nu}}\sigma^2 A^2 B \int_{-\frac{1}{2}l}^{\frac{1}{2}l} \frac{\cos^2(kx)}{\sinh^2(kh)} dx \\ &= -\frac{\mu}{4}\sqrt{\frac{\sigma}{2\nu}}\sigma^2 A^2 B \frac{l}{\sinh^2(kh)}. \end{aligned} \tag{6.55}$$

The effect of the two tank end walls lying in the x–z-plane is now considered. The contribution to viscous energy dissipation from the boundary-layer flow at the two tank end walls is

$$\begin{aligned} \langle \dot{E}_{ew} \rangle &= -\mu\sqrt{\frac{\sigma}{2\nu}} \int_{-\frac{1}{2}l}^{\frac{1}{2}l} \int_{-h}^{0} \left[u_0^2(x, z) + w_0^2(x, z)\right] dx\, dz \\ &= -\mu\sqrt{\frac{\sigma}{2\nu}} \frac{\sigma^2 A^2 l \cosh(kh)}{2k\sinh(kh)}. \end{aligned} \tag{6.56}$$

We can now argue as did Keulegan (1959). Using previous expressions (6.53), (6.55), and (6.56), we can write

$$\langle \dot{E} \rangle \equiv \langle \dot{E}_{wa} \rangle + \langle \dot{E}_b \rangle + \langle \dot{E}_{ew} \rangle = -2\alpha T^{-1} E. \tag{6.57}$$

Integration of eq. (6.57) gives $E/E_0 = \exp(-2\alpha t/T)$, where E_0 is the wave energy at $t = 0$. The derivation may sound contradictory because in the derivation we have used harmonic oscillation without any decay over one period. As long as α is small, this derivation can be defended on the time scale of one period. This fact is true in reality. Because E is proportional to the square of the wave amplitude A, we can also write $A/A_0 = \exp(-\alpha t/T)$. The damping ratio is therefore

$$\xi = \xi^{\text{tank'surface}} = \alpha/(2\pi). \tag{6.58}$$

Using expressions (6.53), (6.55), and (6.56) to get $\langle \dot{E} \rangle$ as well as eq. (6.50) for the total energy, we can find α:

$$\alpha = \pi\sqrt{\frac{2\nu}{\sigma l^2}} \frac{\left(\frac{1}{2}\sinh(2kh) - kh\right) + \frac{1}{2}kl}{\sinh(kh)\cosh(kh)} + \pi\sqrt{\frac{2\nu}{\sigma B^2}} \tag{6.59}$$

based on eq. (6.57) and the previous analysis. This finding leads to

$$\alpha = \frac{\sqrt{\pi\nu T}}{B}\left[\left(\frac{B}{l}\right)\left(1 + \frac{\frac{1}{2}kl - kh}{\sinh(kh)\cosh(kh)}\right) + 1\right], \tag{6.60}$$

which introduces the nondimensional small parameter $\sqrt{\nu T}/B$ used by Keulegan (1959).

Even though we considered only antisymmetric modes, expression (6.60) for α and associated damping ratio (6.58) can be adopted for any natural mode φ_i. We should simply substitute the corresponding natural period T_i and $k = \pi i/l$ in the corresponding formulas. This fact and analogies of eq. (6.60) for lower to intermediate depths within the framework of a Boussinesq-type multimodal method were extensively discussed by Faltinsen and Timokha (2002). Faltinsen *et al.* (2005a, 2005b, 2006b) also derived the corresponding formulas for three-dimensional natural sloshing modes in a square-base tank. Derivation of the latter formulas are given as an exercise in this chapter.

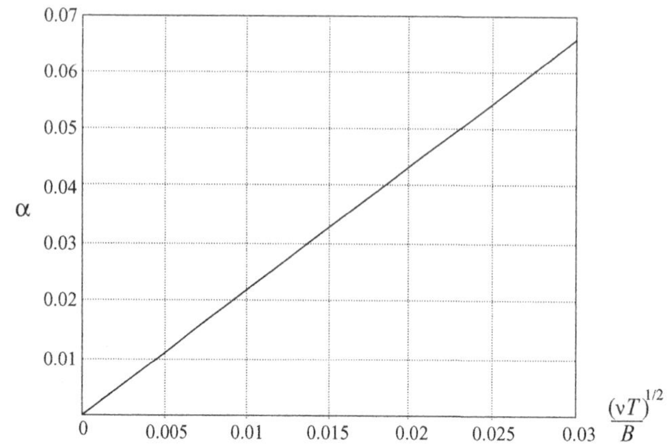

Figure 6.5. Modulus of decay, α, due to laminar boundary layer for the lowest linear sloshing mode of rectangular tanks with $B/l = 0.217$ and $h/l = 0.425$. T is the natural period.

The asymptotic values for deep-liquid conditions ($kh \to \infty$) and shallow-liquid conditions ($kh \to 0$) are

$$\alpha = \frac{\sqrt{\pi \nu T}}{B}\left[\left(\frac{B}{l}\right) + 1\right], \quad kh \to \infty,$$

$$\alpha = \frac{\sqrt{\pi \nu T}}{B}\left[\left(\frac{B}{2h}\right) + 1\right], \quad kh \to 0. \quad (6.61)$$

It then follows from eqs. (6.58) and (6.61) that the ratio between the damping ratios for shallow- and deep-liquid conditions is

$$\frac{\xi_{h\to 0}}{\xi_{h\to\infty}} = \frac{B/2h + 1}{B/l + 1}. \quad (6.62)$$

The preceding equation demonstrates a larger damping ratio for shallow-liquid than for deep-liquid conditions for given values of l and B.

Keulegan (1959) designed experiments on the modulus of decay of the lowest mode ($k = \pi/l$), where B/l and h/l were nearly constant, but l varied from 0.238 to 2.42 m. We set $B/l = 0.217$ and $h/l = 0.425$ as representative values for his studies. Distilled water, glycerol aqueous solutions, ethyl alcohol, xylene, and a mixture of xylene and heavy mineral oil were used. Figure 6.5 shows α as a function of $\sqrt{\nu T}/B$ in the parameter range studied by Keulegan (1959).

For instance, water with $\nu = 10^{-6}$ m^2 s^{-1} and $B = 0.526$ m gives $T = 1.89$ s and $\sqrt{\nu T}/B = 0.0026$, which corresponds to $\alpha = 0.0057$ (i.e., very small damping). A possible error source is the presence of turbulence. The calculation for the Reynolds number, $RE = U_0^2/(\sigma\nu) = 10^5$ is according to Jonsson's (1978) engineering criterion for transition to turbulence for a smooth surface, where $U_0 = \sigma A_0$ is the maximum tangential velocity just outside the boundary layer. Therefore, the amplitude of motion, A_0, just outside the boundary layer should be less than 0.17 m. Keulegan carried out his free-decay experiments for this case by using initial wave amplitude at the wall equal to 0.25 h = 0.26 m. Even though this value does not ensure laminar flow, we should recognize that other parts of the tank boundary below the free surface meet the criterion for laminar boundary-layer flow. Furthermore, the wave amplitude decays with time. Keulegan's (1959) experiments are in reasonable agreement with the theoretical estimate of α for this case. Larger differences occur with viscous laminar boundary-layer theory as the size of the basin decreases. This occurrence is thought to be due to surface tension effects. Lucite and glass basins influenced the results as shown in Figure 6.6. By using an aerosol, Keulegan was able to influence the surface tension to get similar results for Lucite and glass basins.

6.3.2 Incorporation of boundary-layer damping in a potential flow model

We can include the effect of the boundary layer in the potential flow solution in a way similar to what we did for the ambient harmonic oscillatory flow past a circular cylinder. The boundary layer causes an inflow/outflow to the potential flow domain. Let us illustrate this for the case of the two-dimensional rectangular tank with no forcing. We represent the velocity potential of the relative liquid motion (without the boundary-layer effect) as $\Phi(x, z, t) = \sum_{m=1}^{\infty} R_m(t)\varphi_m(x, z)$, where φ_i are the natural modes by eq. (4.8)

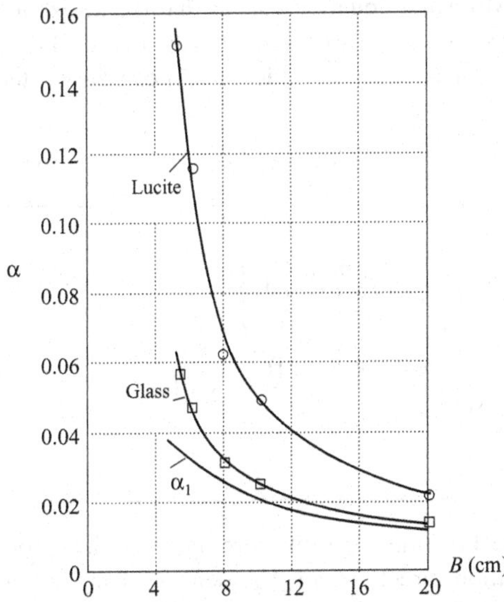

Figure 6.6. Modulus of decay, α, in glass and Lucite basins with distilled water; α_1 is the estimate of modulus of decay due to laminar boundary-layer flow (Keulegan, 1959).

(see eq. (5.20)). Using linear free-surface condition (2.73) or, alternatively, modal equations (5.23)–(5.25) (without forcing terms) gives

$$\ddot{R}_m(t) + \sigma_m^2 R_m(t) = 0, \quad m = 1, 2, \ldots. \tag{6.63}$$

The solution of modal equations (6.63) is $R_m(t) = A_m g \sigma^{-1} \cos(\sigma_m t + \varepsilon)$. This solution is the same as in eq. (6.46), where we studied the damping ratio for antisymmetric modes. In the following we set the phase angle ε equal to zero without loss of generality. We can find the corresponding horizontal and vertical liquid velocities as

$$U = \frac{\partial \Phi}{\partial x} = \sum_{m=0}^{\infty} R_m(t) \frac{\partial \varphi_m}{\partial x},$$
$$W = \frac{\partial \Phi}{\partial z} = \sum_{m=0}^{\infty} R_m(t) \frac{\partial \varphi_m}{\partial z}.$$

An infinite number of modes exists. However, due to linearity, we may consider each mode separately; we can independently study sloshing with velocities

$$U = \frac{\partial \Phi}{\partial x} = R_m(t) \frac{\partial \varphi_m}{\partial x},$$
$$W = \frac{\partial \Phi}{\partial z} = R_m(t) \frac{\partial \varphi_m}{\partial z} \quad \text{for} \quad m = 1, 2, \ldots.$$

Let us start with the vertical mean wetted wall at $x = \frac{1}{2} l$. We consider the following vertical velocity at the outer part of the boundary layer:

$$W_e(z, t) = W_0(z) \cos(\sigma t),$$
$$W_0(z) = \sigma_m A_m \frac{\sinh(\pi m (z + h) / l)}{\sinh(\pi m h / l)}$$

(see analogous formulas in the previous section for antisymmetric modes). When introducing the complex notations, the inflow/outflow velocity from the boundary layer along the positive direction of the x-axis can be expressed as

$$u_0 = -\frac{dW_0}{dz} \sqrt{\frac{2\nu}{\sigma}} \frac{\exp(i\sigma t)}{(1 + i)}.$$

Furthermore, we can do something similar at the other tank boundaries. As a consequence, we get a three-dimensional boundary-value problem due to the fact that there is inflow/outflow at the tank end walls. This problem should then be solved to get the required solution of the general type.

The indicated boundary-value problem is not a convenient way of performing the analysis. We instead construct a solution that ensures the correct decay of the wave amplitude. We let the generalized coordinates $R_m(t)$ satisfy

$$\ddot{R}_m(t) + 2\xi_m \sigma_m \dot{R}_m(t) + \sigma_m^2 R_m(t) = 0,$$
$$m = 1, 2, \ldots, \tag{6.64}$$

where ξ_m is the damping ratio. Alternatively, analogous equations are given by eq. (5.56) for the generalized free-surface coordinates $\beta_m(t)$ with the same damping ratio. Equation (5.56) describes the wave elevation. Using results from Section 5.3.3, we can show that the solution of eq. (6.64) (or eq. (5.56) with zero right-hand side) gives the correct wave amplitude decay. When ξ_m is small, this procedure causes a small error but one that is acceptable in the free-surface condition.

6.3.3 Bulk damping

Viscous energy dissipation in the liquid outside the boundary layers can be estimated by starting with eq. (2.45) and expressing the liquid velocities by potential flow theory. The following derivation assumes two-dimensional potential flow in the Oxz-plane and notations from Section 6.3.1. Because the boundary layer is thin relative to the length dimensions of the liquid domain and we use linear wave theory, we can let the liquid volume in eq. (2.45) consist of x- and z-coordinates satisfying the unperturbed shape $-\frac{1}{2}l \leq x \leq \frac{1}{2}l$, $-h \leq z \leq 0$. For brevity, we assume antisymmetric natural modes with velocity potential ϕ_n given by eq. (6.46). It follows that the time rate of viscous energy dissipation is

$$\dot{E}_{bd} = -\mu B \int_{-\frac{1}{2}l}^{\frac{1}{2}l} \left\{ \int_{-h}^{0} \left[2\left(\frac{\partial^2 \phi_n}{\partial x^2}\right)^2 + 4\left(\frac{\partial^2 \phi_n}{\partial x \partial z}\right)^2 + 2\left(\frac{\partial^2 \phi_n}{\partial z^2}\right)^2 \right] dz \right\} dx$$

$$= -4\mu B \left(\frac{k^2 A g}{\sigma}\right)^2 \frac{l}{2} \int_{-h}^{0} \frac{\sinh^2(k(h+z)) + \cosh^2(k(h+z))}{\cosh^2(kh)} dz \cdot \sin^2(\sigma t)$$

$$= 2\mu B l\, g A^2 k^2 \cdot \sin^2(\sigma t).$$

We used the subscript bd to indicate that the considered viscous energy dissipation is related to bulk damping. We can now follow a similar procedure as in Section 6.3.1 and, based on eq. (6.57), consider $\langle \dot{E}_{bd} \rangle / \langle E \rangle = -2\alpha T^{-1} = 4\nu k^2$. Using definition (6.58) gives the bulk damping ratio:

$$\xi^{\text{bulk}} = \frac{2\nu k^2}{\sigma} = \frac{T \nu k^2}{\pi} \tag{6.65}$$

with notations accepted as in Section 6.3.1.

Again, as in Section 6.3.1, even though we simplified the derivation by assuming only antisymmetric modes, formula (6.65) is true for other modes; the m th mode in eq. (6.65) should use $k = \pi m / l$ and $\sigma = \sigma_m = \sqrt{gk \tanh(kh)}$. We use the subscript m to indicate the mode number and note that ξ_m^{bulk} is proportional to the kinematic viscosity coefficient ν while $\xi_n^{\text{tank'surface}}$ is proportional to $\sqrt{\nu}$. Because ν is a small quantity, it indicates that the viscous bulk damping is smaller than the viscous boundary-layer damping. Bulk damping is in general small relative to boundary-layer damping and negligible for the lower modes. However, it should not be neglected for higher modes. This fact was also discussed by Miles and Henderson (1998) for circular-base tanks.

6.4 Morison's equation

Morison's equation (Morison *et al.*, 1950) is often used to calculate wave loads on vertical circular cylindrical structural members of fixed offshore structures when viscous forces matter. It is necessary that the incident wavelength λ is large relative to the cylinder diameter D. Faltinsen (1990) recommends $\lambda > 5D$.

Morison's equation tells us that the horizontal force dF on a strip of length dz of a vertical rigid circular cylinder (see Figure 6.7) can be written as

$$dF = \tfrac{1}{4}\rho \pi D^2 dz C_M a_1 + \tfrac{1}{2}\rho C_D D dz\, |u|\, u. \tag{6.66}$$

The direction of positive force is in the wave propagation direction; ρ is the density of the water, D is the cylinder diameter, and u and $a_1 = \partial u / \partial t$ are the horizontal undisturbed fluid velocity and acceleration at the midpoint of the strip, respectively. The mass and drag coefficients, C_M and C_D, have to be empirically determined and are dependent on, for instance, the *Keulegan–Carpenter number* (KC), the *Reynolds number* (Rn), and the *roughness number.*

The Keulegan–Carpenter number can be expressed as $KC = U_M T / D$ for ambient oscillatory planar flow with velocity $U_M \sin(\sigma t + \varepsilon)$ past a fixed body with a characteristic length (i.e., diameter D). Because $T = 2\pi/\sigma$ and $A = U_M/\sigma$ is the ambient amplitude of the fluid oscillations, in this case we may write $KC = 2\pi A / D$. We then see that KC expresses the distance moved by a free-stream fluid particle relative to the body diameter. The Reynolds number is $Rn = U_M D / \nu$ and the roughness number is Ra/D, where Ra is the arithmetic mean roughness on the body surface. The Reynolds number is an important parameter

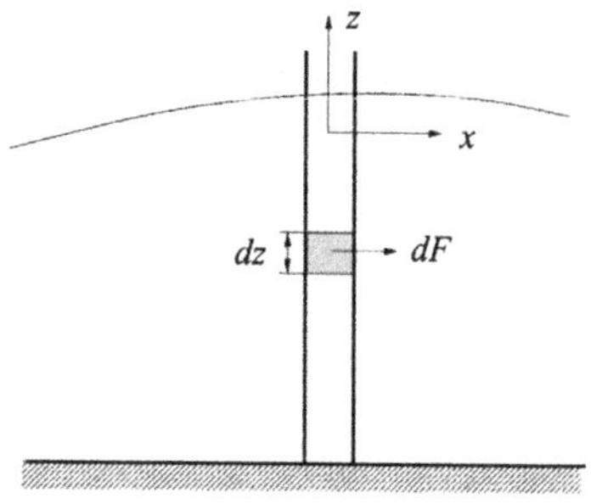

Figure 6.7. Horizontal submerged cross-section of a vertical cylinder.

for flow in the boundary layer. If Rn is below a critical number, the boundary-layer flow is laminar. The boundary-layer flow is turbulent in transcritical flow. Laminar boundary-layer flow along a surface without sharp corners separates more easily than turbulent boundary-layer flow. The critical Reynolds number is the order of 10^5 for steady ambient flow past a circular cylinder (i.e., when $KC \to \infty$). The surface roughness affects the critical Reynolds number (i.e., the transition to turbulence in the boundary layer). Even though the boundary-layer flow is laminar, the separated wakes resulting from the motion are more likely to be turbulent (because wakes are much more unstable than boundary layers with respect to transition to turbulence).

Sometimes $\beta = Rn/KC = D^2/(\nu T)$ is also used to characterize the flow. The presence of a current and nonplanar ambient motion may also matter. Detailed discussions of many of the aforementioned factors can be found in Sarpkaya and Isaacson (1981) and Faltinsen (1990).

If viscous effects are negligible, it is possible to show analytically that Morison's equation is the correct asymptotic solution for large ratios between wavelength and cylinder diameter as long as linear theory applies. The value of C_M should then be 2.0 for a circular cross-section. A derivation is given by Faltinsen (1990). We then start with the undisturbed pressure force, which means we integrate the pressure field that is at the position of the cylinder if the cylinder is not there. This gives the so-called Froude–Kriloff force and half of the contribution to C_M. The fact that the Froude-Kriloff force cannot be the total force follows from noting that the undisturbed pressure field causes a fluid transport through the cylinder wall. The cylinder must therefore set up a pressure field that counteracts the normal component of the undisturbed velocity field at the cylinder wall. If fluid acceleration can be neglected, Morison's equation is a reasonable empirical formulation for the time-averaged force. A nonlinear potential flow theory for wavelengths that are large relative to the cross-dimensions and for sufficiently small wave amplitudes to avoid flow separation is presented by Faltinsen *et al.* (1995).

The application of Morison's equation in the free-surface zone requires accurate estimates of the undisturbed velocity distribution under a wave crest. A straightforward application of Morison's equation implies that the absolute value of the force per unit length is largest at the free surface; however, this scenario is not physically feasible because the pressure is constant on the free surface. Therefore, the force per unit length has to go to zero at the free surface. It should also be noted that the position of the free surface at the cylinder is affected by a wave run up at the upstream side of the cylinder and a wave depression on the downstream side. The vertical position of the maximum absolute value of the local force has to be experimentally determined. The order of magnitude of the position of maximum force per unit length may be 25% of the wave amplitude down into the fluid from the free surface. We should note that Morison's equation cannot predict the oscillatory forces due to vortex shedding in the lift direction (i.e., forces orthogonal to the wave propagation direction and in the cross-sectional plane). The main concern of oscillatory lift forces is in connection with VIV (see Section 6.9). The Keulegan–Carpenter number must be larger than ~7–8 for antisymmetric vortex shedding to occur in purely two-dimensional flow (Bearman, 1985). By antisymmetric we mean that a lift force is present.

6.4.1 Morison's equation in a tank-fixed coordinate system

We modify Morison's equation for a vertical cylinder inside a longitudinally oscillating tank. We consider a strip of height dz (see Figure 6.7). Two-dimensional potential flow in the Oxz-plane is assumed when the cylinder is not there. The x-component of the flow velocity relative to the tank at the mean position of the strip when the cylinder is not there is denoted u_r. Furthermore, an ambient pressure p_{inc} exists at the position of the strip. This incident pressure field causes a force on the strip that acts in the x-direction. We can formally write this Froude–Kriloff force as

$$dF_1^{FK} = \left(- \int_{C(z)} p_{inc} n_1 \mathrm{d}s \right) dz, \qquad (6.67)$$

where $C(z)$ is the cross-sectional surface and n_1 is the x-component of the normal vector to the surface. The positive normal direction is into the liquid. Following from Gauss's theorem (see eq. (A.1)) and the fact that the ambient pressure, p_{inc}, is analytical inside the cylinder,

$$dF_1^{FK} = \left(-\int_{Q(z)} \frac{\partial p_{\text{inc}}}{\partial x} dQ \right) dz,$$

where $Q(z)$ is the cross-sectional area. The pressure gradient $\partial p_{\text{inc}}/\partial x$ is assumed to vary slowly across $Q(z)$ so that

$$dF_1^{FK} \approx \left(-\tfrac{1}{4}\pi D^2 \left. \frac{\partial p_{\text{inc}}}{\partial x} \right|_{x_m} \right) dz, \qquad (6.68)$$

where x_m is the x-coordinate of the cylinder axis. We now apply the Euler equations in a tank-fixed coordinate system (see eq. (2.56)) with $\nu = 0$ and keep linear terms in the velocity and acceleration:

$$dF_1^{FK} = \tfrac{1}{4}\rho\pi D^2 \left(\dot{v}_{O1} + \left. \frac{\partial u_r}{\partial t} \right|_{x_m} \right) dz. \qquad (6.69)$$

If we had introduced nonlinear terms, an additional term, $\frac{1}{4}\rho\pi D^2 (u_r \partial u_r/\partial x + v_r \partial u_r/\partial y)_{x_m} dz$, would be included on the right-hand side of eq. (6.69). One reason for neglecting this term is that the empirical Morison equation does not contain the convective acceleration term $(u_r \partial u_r/\partial x + v_r \partial u_r/\partial y)_{x_m}$.

We assume potential flow theory and consider the consequence of the fact that the undisturbed pressure field causes liquid transport through the cylinder wall. This scenario is not physically possible and the cylinder must therefore set up a pressure field that counteracts the normal component of the undisturbed velocity field at the cylinder wall. This added mass force part can be expressed as

$$dF_1^{AM} = \left(\tfrac{1}{4}\rho\pi D^2 \left. \frac{\partial u_r}{\partial t} \right|_{x_m} \right) dz. \qquad (6.70)$$

We then introduce the effect of the viscosity in the spirit of Morison's equation and express the longitudinal force dF_1 on a strip of length dz of a vertical rigid circular cylinder inside a longitudinally oscillating tank as

$$dF_1 = \left(\tfrac{1}{4}\rho\pi D^2 \dot{v}_{O1} + \tfrac{1}{4}\rho\pi D^2 C_M \left. \frac{\partial u_r}{\partial t} \right|_{x_m} + \tfrac{1}{2}\rho C_D D \left| u_r \right| u_r \right) dz. \qquad (6.71)$$

The values of C_D and C_M in eq. (6.71) are not necessarily the same as in eq. (6.66). We note that eq. (6.66) is a special case of eq. (6.71).

The derived Morison equation in the tank-fixed coordinate system can be generalized to include the effect of other rigid-body modes of the tank by means of eq. (2.56). Let us consider the transverse force when the tank oscillates in sway with a small acceleration $\dot{v}_{O2}$ and with small angular motions η_4 and η_6 in roll and yaw, respectively. Two-dimensional potential flow in the Oyz-plane is assumed when the cylinder is not there. The y-component of the flow velocity relative to the tank at the mean position of the strip when the cylinder is not there is denoted v_r. We follow a similar procedure as before. The Froude–Kriloff force dF_2^{FK} in the y-direction on a strip of height dz can be expressed as

$$\begin{aligned} dF_2^{FK} &\approx \left(-\tfrac{1}{4}\pi D^2 \left. \frac{\partial p_{inc}}{\partial y} \right|_{y_m} \right) dz \\ &= \tfrac{1}{4}\rho\pi D^2 \left(\dot{v}_{O2} + x\ddot{\eta}_6 - z\ddot{\eta}_4 + g\eta_4 + \left. \frac{\partial v_r}{\partial t} \right|_{y_m} \right) dz, \end{aligned}$$

where y_m is the y-coordinate of the cylinder axis. It follows by a generalized Morison equation formulation that the force dF_2 in the y-direction force on a strip of length dz of the considered circular cylinder inside the tank can be expressed as

$$\begin{aligned} dF_2 &= \left(\tfrac{1}{4}\rho\pi D^2 \left(\dot{v}_{O2} + x\ddot{\eta}_6 - z\ddot{\eta}_4 + g\eta_4 \right) \right) dz \\ &\quad + \left(\tfrac{1}{4}\rho\pi D^2 C_M \left. \frac{\partial v_r}{\partial t} \right|_{y_m} \right) dz \\ &\quad + \left(\tfrac{1}{2}\rho C_D D \left| v_r \right| v_r \right) dz. \end{aligned}$$

The Morison equation is discussed further in Section 8.3.2.4, where the single-dominant multimodal method for finite liquid depth is shown to evaluate the ambient liquid velocity. The difficulty in applying Morison's equation in the free-surface zone is stressed once more in this connection. The difficulty becomes particularly evident at lower-intermediate and shallow-liquid depths when steep waves (e.g., hydraulic jumps) occur. The effect of a nearly vertical impacting free surface must be handled by a slamming model and is discussed in Section 8.9. Section 9.4.2 gives experimental results for tower forces in a spherical tank. However, no attempt is made to fit the latter results to a Morison equation formulation.

6.4.2 Generalizations of Morison's equation

Morison's equation and its generalization to tank flow can also be applied to inclined members on truss work, for instance. To demonstrate, let us consider a cylinder inclined in a plane parallel to the incident flow field. The approach would be to decompose the undisturbed velocity and acceleration into components both normal and parallel to the cylinder axis and then to use Morison's equation with normal components of velocity and acceleration. The force direction is normal to the cylinder axis. In the case of potential flow, this expression can be proven to be the correct expression. In the viscous case, we use the "cross-flow" (independence) principle. Actually what we propose to do for an inclined cylinder is not different from the vertical cylinder case. In the latter case, an undisturbed tangential velocity and acceleration component also exist in the fluid.

When the cylinder axis is not in the plane of the wave propagation direction, different possibilities for formulations exist. Let us illustrate this using a submerged horizontal circular cylinder in waves where the wave propagation direction is orthogonal to the cylinder axis. A straightforward generalization of Morison's equation in an Earth-fixed coordinate system would be to write the horizontal and vertical force on a strip of length dy as

$$dF_1 = \left(\tfrac{1}{4}\rho\pi D^2 C_M a_1\right) dy + \left(\tfrac{1}{2}\rho C_D D u (u^2 + w^2)^{1/2}\right) dy, \quad (6.72)$$

$$dF_3 = \left(\tfrac{1}{4}\rho\pi D^2 C_M a_3\right) dy + \left(\tfrac{1}{2}\rho C_D D w (u^2 + w^2)^{1/2}\right) dy, \quad (6.73)$$

where w and a_3 are vertical undisturbed fluid velocity and acceleration components at the midpoint of the strip. Chaplin (1988) has shown that improved correlation with experiments can be obtained if different coefficients C_M and C_D are assigned to the horizontal and vertical components.

When intersection of structural members is considered, care should be shown in the derivation of the Froude–Kriloff loads. Our previous derivation based on Gauss's theorem required a closed body surface. For a closed body surface, typically cancellation effects occur for counteracting large pressure loads.

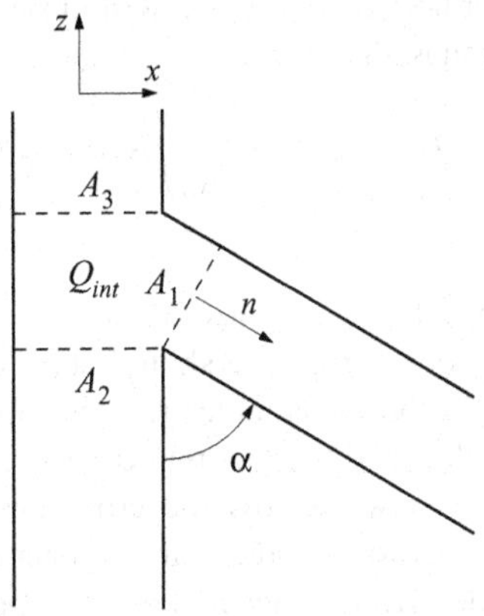

Figure 6.8. Intersection between two structural members.

We illustrate how to calculate the x-component of the Froude–Kriloff force for the intersection between the two structural members in Figure 6.8 that are inside a horizontally oscillating tank. We consider the sum of the vertical member bounded by cross-sectional areas A_2 and A_3 and the adjacent part of the oblique member bounded by A_1. The enclosed volume is denoted Q_{int}. We first pretend that the surface enclosing Q_{int} is wet. We can proceed as we did in Section 6.4.1 by using Gauss's theorem and the Euler equations in an accelerated coordinate system. The generalization of eq. (6.69) becomes

$$F_1^{FK_1} = \rho Q_{\text{int}} \left(\dot{v}_{O1} + \left.\frac{\partial u_r}{\partial t}\right|_{x_m} \right). \quad (6.74)$$

Then we have to correct for the fact that A_1, A_2, and A_3 are not wet. We start with cross-section A_1. The normal vector $\boldsymbol{n}$ to A_1 with positive direction out of Q_{int} has components along the x- and z-axis: $\sin\alpha$ and $-\cos\alpha$, respectively. In this case α is the angle between the two cylinder axes. Therefore, we have to subtract the force component $-p_{inc}A_1 \sin\alpha$ from eq. (6.74). Because the normal vectors of A_2 and A_3 are parallel to the z-axis and we consider the horizontal force, no corrections are needed at A_2 and A_3. The final answer for horizontal Froude–Kriloff force of the intersection is, therefore,

$$F_1^{FK} = \rho Q_{\rm int}\left(\dot{v}_{O1} + \left.\frac{\partial u_r}{\partial t}\right|_{x_m}\right) + p_{\rm inc}A_1 \sin\alpha. \tag{6.75}$$

If we want to generalize Morison's equation to other body forms, we should start with the structure of the loads under the assumptions of potential flow and the fact that the incident flow field varies slowly across the cross-dimensions. We then get one force part associated with the Froude–Kriloff loads and another part from added mass forces. We may even account for structural vibrations of the interior tank structure. The resultant potential flow force may not be in the incident flow direction. Having obtained the potential flow part we adopt empirical mass coefficients and add empirical drag loads. The relative velocity component in the cross-dimensional direction is then used.

It is of course easy to criticize Morison's equation. However, nothing better exists from a practical point of view because of the very complicated flow picture that occurs for separated flow around marine structures.

In the following text we base our discussion on viscous loads written in terms of Morison's equation, and we discuss how C_D and C_M depend on the parameters mentioned earlier. Our discussion starts with Reynolds number (Rn) ranges typical for model test and full-scale conditions in marine structural applications (i.e., $Rn \gtrsim 10^3$). We then focus on very small Reynolds numbers.

6.4.3 Mass and drag coefficients (C_M and C_D)

Bearman (1985) has reported the behavior of flow around a circular cylinder for different KC numbers when the ambient flow is planar and harmonically oscillating. The data by Bearman (1988) show that the flow is symmetric for $KC \lesssim 5$. Symmetric flow implies zero lift force. However, we cannot set up precise limits for the onset of asymmetric flow. Bearman (1985) reports that a change in the vortex shedding appears at $KC = 7$ or 8. The majority of vortex shedding takes place only on one side of the cylinder. The flow has a strong memory in this range. Between KC values of 15 and 25, a vortex is shed before the end of a half-cycle. In addition, a weaker, second vortex forms from the same shear layer. Vortices shed previously combine with new vortices so that vortex pairs travel away in two trails at roughly 45° to the direction of the main flow. The directions can also be switched, which depends on the starting condition. At $KC > 25$, at least three full vortices are shed per half-cycle. The wake resembles a wake in steady flow (i.e., the vortices tend to form a street behind the body).

Graham (1980) has done an interesting analysis for $KC \to 0$, which can be used to explain experimental results for $KC \lesssim 10$. His analysis includes only the effect of separated flow on the pressure distribution around the body. The attached boundary-layer flow with viscous shear forces and its effect on the pressure have to be added. Graham argued that the different behavior of drag coefficients of different sections at small values of KC is due to the relative strengths of the vortex shedding (e.g., a circular cylinder is obviously much weaker than a flat plate). He assumed that the vortex flow for a small Keulegan–Carpenter number depends only on the local flow around each sharp edge. The edge is characterized by its internal angle, δ; for a flat plate $\delta = 0$ and for a square section $\delta = \frac{1}{2}\pi$. The vortex force F_V on a sharp edge acts along the perpendicular to the bisector of the edge angle; F_V can be shown to be proportional to KC^{η}, where $\eta = (2\delta - \pi)/(3\pi - 2\delta)$.

Therefore, the drag coefficient C_D due to vortex shedding for small Keulegan–Carpenter numbers should vary as (1) $C_D \propto KC^{-1/3}$ for a flat plate and (2) $C_D \propto KC$ for a circular cylinder (regarded loosely as a sharp-edged section with $\delta = \pi$). These variations are in quite good agreement with measured data for $KC < 10$ for bodies in planar ambient flow (Graham, 1980). The experimental results for laminar boundary-layer conditions show that

$$\begin{aligned} C_D &= 8.0KC^{-1/3}\ \text{(flat plate)} \quad \text{and} \\ C_D &= 0.2KC\ \text{(circular cylinder)}. \end{aligned} \tag{6.76}$$

The results for a circular cylinder are only of a qualitative nature, which is evident by comparing with Sarpkaya's (1986) experimental results (see Figure 6.9). Sarpkaya examined a circular cylinder in planar oscillatory flow of small amplitude.

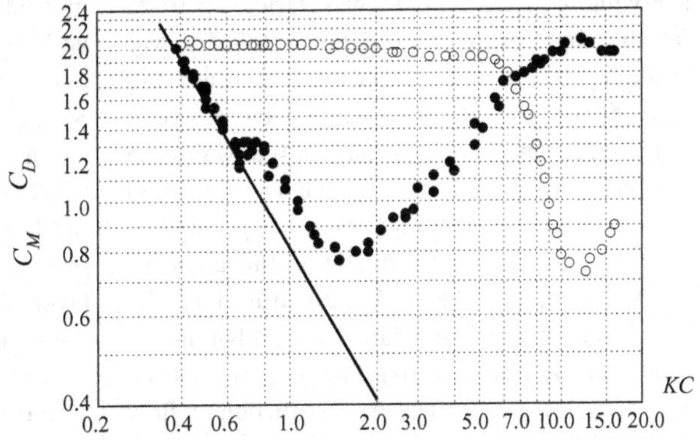

Figure 6.9. Drag and mass coefficients versus Keulegan–Carpenter number for a smooth circular cylinder at $\beta = 1{,}035$. Experiments by Sarpkaya (1986): ○ = C_M, • = C_D; theory (solid line) by Wang (1968).

The results were presented for different values of KC and β ($\beta = Rn/KC$). When boundary-layer flow was laminar and the flow did not separate (i.e., $KC \lesssim 1$) Wang's (1968) formula agreed well with the experimental results. Our previous formula for the drag coefficient (see Section 6.2.2) also agrees well. A difference between the results by Wang and from Section 6.2.2 is that Wang included steady streaming and sum frequency effects. However, the two last effects do not contribute to the drag coefficient. When $KC \gtrsim 2$, values of C_D start to increase with Keulegan–Carpenter number, which is shown by Graham's asymptotic solution. However, one should note that his theory is based on a fixed separation point, which is not the case for separated flow around a circular cylinder, particularly at small Keulegan–Carpenter numbers. Figure 6.9 also shows the values of C_M. We note that C_M remains close to 2 for values of KC up to about 6.

Typical values of C_D and C_M for transcritical flow past a smooth circular cylinder at $KC \gtrsim 40$ are 0.7 and 1.8. A roughness number, Ra/D, of 0.02 can cause more than a 100% increase in C_D relative to the value of C_D for a smooth circular cylinder in ambient oscillatory flow (Sarpkaya, 1985). Therefore, the effect of roughness is more significant in oscillatory ambient flow than in steady incident flow (see Figure 6.10).

The importance of the drag term relative to the mass term increases with the Keulegan–Carpenter number. For instance, Sortland (1986) performed two-dimensional experiments with a smooth circular cylinder in a U-tube where the ambient flow velocity is sinusoidal and one-dimensional. Sortland (1986) showed in his experiments that the drag term becomes larger than the mass term when $KC \gtrsim 8$–10. The fact that the threshold value of KC varies for drag to be dominant is a Reynolds-number effect.

The nature of the ambient flow is important, as demonstrated by results of Chaplin (1984, 1988). Chaplin (1988) presented data for a submerged horizontal circular cylinder in beam waves at small Keulegan–Carpenter numbers. The ambient flow was either nearly circular or elliptical. Chaplin argued that a circulation is set up around the cylinder, which has a large influence on the mass coefficients. In the particular case of a deeply submerged cylinder in nearly circular orbital flow he found that

$$C_M \approx 2 - 0.21KC^2. \tag{6.77}$$

The Keulegan–Carpenter number was less than 2 and the Reynolds number was of the order of 10^4. Equation (6.77) shows a very strong influence of KC on C_M. If the ambient oscillatory flow is planar, the influence of KC on C_M is not strong for small KC (see Figure 6.9).

Contamination effects running in the lengthwise flow direction along the "leading edge" can cause early transition on highly swept cylinders (i.e., small angle between cylinder axis and main flow direction). For example, a cylinder whose leading end is attached to another cylinder or a wall could be affected. Turbulence coming from upstream on the wall or separated flow at the junction is convected along the cylinder, causing transition. Ersdal and Faltinsen (2006) studied experimentally a long, deeply submerged horizontal circular cylinder that was towed axially with velocity U and forced with transverse harmonic oscillations with velocity amplitude U_M.

The value of KC, $U_M T/D$, was varied up to 20 and the angle of attack, $\alpha = \tan^{-1}(U_M/U)$, was varied from 0° up to 20°. If the angle of attack had been 90° (cross-flow), C_D would have a strong dependence on KC, as we have already seen in Figure 6.9. However, KC-dependence was small in the case of Ersdal and Faltinsen's experiments. The most important parameter is the angle of attack. Furthermore, Ersdal and Faltinsen (2006) found that the appropriate Reynolds-number definition was $UD/(\nu \sin\alpha)$. The transition of the boundary layer from laminar to turbulent flow appeared to happen for $UD/(\nu \sin\alpha)$ between $2.0 \cdot 10^5$ and $3.5 \cdot 10^5$.

Bearman *et al.* (1979) presented measurements of a series of two-dimensional bodies in plane oscillatory flow for values of KC between 3 and 70. They concentrated on bluff bodies with sharp-edged separation and measured the in-line force on a flat plate, square, diamond, and circular cylinder. Beyond a KC value of about 10 to 15, the curves for the flat plate and circular and diamond cylinders are remarkably similar; by $KC = 50$, the values for C_D were all only a little higher than their steady flow values, C_D^∞. The square cylinder showed a different trend.

Faltinsen (1990) tried to explain the results by estimating the increased effective incident flow due to returning eddies. The resulting formula is

$$C_D/C_D^\infty = [1 + 0.58\exp(-0.064KC)]^2, \quad KC \geq 10, \qquad (6.78)$$

which is in good agreement with Bearman's results. It even agrees with the experimental values at $KC = 10$. Due to the quasi-steady assumptions in Faltinsen's analysis one should not expect agreement for values of KC below 25. One should be careful in applying eq. (6.78) to new situations since several major assumptions were made. However, Haslum (2000) suggested the following procedure for a circular cylinder. Based on the roughness and Reynolds number, a value for C_D^∞ is chosen from the literature. Equation (6.78) is then followed down to $KC = 10$ (step 1 in the calculations).

For moderate KC ($\approx 1 < KC < 10$), it is assumed that $C_D \propto KC$. The proportionality factor is found using $C_D|_{KC=10}$ (step 2). For very low KC, the flow is attached. If the boundary-layer flow is laminar, we can apply eq. (6.17) by multiplying its value by 2. The moderate and very low KC-number representations are then patched together to give the same value at one KC number (step 3).

An alternative to step 3 is simply to add together the formula for step 2 and the force on the body for attached flow. Attached flow matters for small KC numbers and laminar boundary layer. When the formula from step 2 is important, the attached flow is less important.

We illustrate the behavior of C_D^∞ by classical results for cross-flow past a two-dimensional fixed circular cross-section with diameter D in infinite fluid of density ρ. An ambient incident steady flow with velocity U_∞ is present. The mean force F_1 in the incident flow direction can be expressed as

$$F_1 = \tfrac{1}{2}\rho C_D U_\infty^2 D. \qquad (6.79)$$

Figure 6.10 shows experimental values for C_D over a range of Reynolds numbers ($Rn = U_\infty D/\nu$) from $\approx 2 \cdot 10^4$ to $\approx 4 \cdot 10^6$. The effect of surface roughness is included.

In the figure we note a very distinct drop in C_D in a certain Reynolds-number range. This is referred to as the critical flow regime and is particularly marked for flow around a smooth cylinder. It is common practice to divide this dependence on Reynolds number into different flow regimes. However, many different definitions exist in the literature. We refer to the following four different flow regimes: subcritical, critical, supercritical, and transcritical flow in Figure 6.10. For flow around a smooth circular cylinder the subcritical flow regime is for Reynolds number less than about $2 \cdot 10^5$. The critical flow regime is for $\approx 2 \cdot 10^5 < Rn < 5 \cdot 10^5$. The supercritical flow regime is for $\approx 5 \cdot 10^5 < Rn < 3 \cdot 10^6$ and the transcritical flow is for Reynolds numbers larger than $3 \cdot 10^6$. In subcritical flow the boundary layer is always laminar, whereas in supercritical and transcritical flow, the boundary layer is turbulent upstream of the separation point.

The Reynolds-number range shown in Figure 6.10 is not sufficient for all marine applications. For instance, later in this chapter we discuss wire-mesh screens in TLDs (see Section 6.7.3). The Reynolds number with the wire diameter as a characteristic length can then be very small. Figure 6.11 is therefore included. Very high drag coefficients occur for small Reynolds number. The viscous friction force on the cylinder

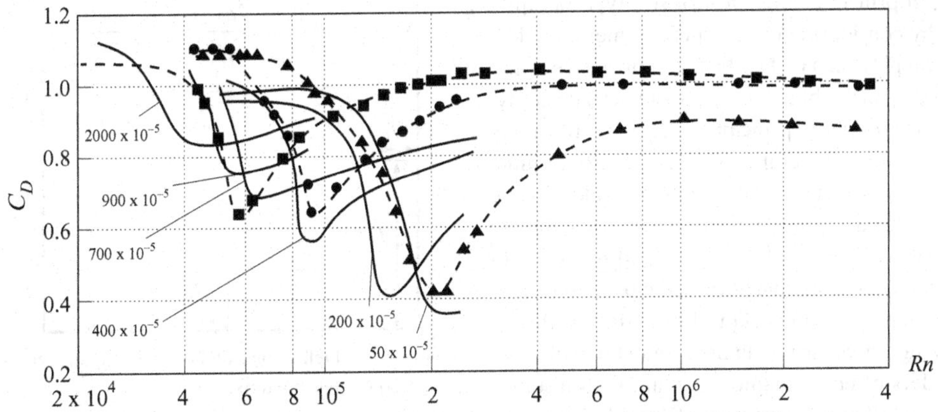

Figure 6.10. Drag coefficient C_D of rough circular cylinders in steady incident flow for different surface roughness coefficients Ra/D (where Ra is average height of surface roughness, D is cylinder diameter, $Rn = U_\infty D/\nu$, and U_∞ is incident flow velocity). Experimental results by Fage and Warsap (1929) (solid lines labeled with Ra/D values) and by Achenbach (1971) (dashed lines with ▲$= Ra/D = 111 \cdot 10^{-5}$, • $= Ra/D = 450 \cdot 10^{-5}$, ■ $= Ra/D = 900 \cdot 10^{-5}$).

is significant for $Rn < 200$ and is small for $Rn >$ 350–400. Because many CFD methods assume laminar flow, the state of the wake at different Reynolds number is important to note. A summary is given by Zdravkovich (1997). The wake is laminar for $Rn < 200$. No separation occurs for $0 < Rn <$ 4–5. A closed wake occurs when 4–5 $< Rn <$ 30–48. Alternate vortex shedding occurs for Reynolds numbers higher than 30–48. Transition to turbulence in the distant wake occurs for 180–200 $< Rn <$ 220–250, whereas the transition happens in the local wake for Reynolds numbers from 220–250 to 350–400. In the $Rn =$ 180–250 and 250–400 regimes, the first process is development of three-dimensional instability (modes A and B) in the two-dimensional von Karman vortex street immediately behind the cylinder. These quite regular structures gradually break down into turbulence as they convect downstream; the transition point moves toward the cylinder as Rn increases. A good description of modes A and B is given by Williamson (1996). The contribution from frictional stress to the drag coefficients in the Reynolds-number range considered in Figure 6.10 is small, although this is not the case for $Rn \lesssim 10^3$. When C_D starts to increase with decreasing Rn for $Rn \lesssim 10^3$, the contribution from pressure loads to C_D remains nearly constant until the wake becomes closed for $Rn \lesssim 50$.

The drag coefficient may be affected by the tank walls and the bottom. This scenario is illustrated experimentally by Sarpkaya and O'Keefe (1996) for cross-flow past a plate attached to a wall and for a free plate in infinite fluid. This study has relevance for estimations of the viscous loads on baffles inside a tank, which are considered in Section 6.5.

A pump tower in an liquefied natural gas membrane tank is more complicated structurally than the vertical cylinder considered in presenting the Morison equation (see Figures 1.19–1.20). We

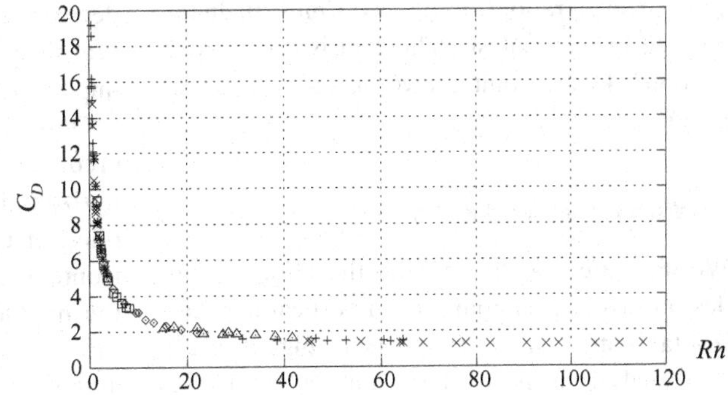

Figure 6.11. Experimental drag coefficient C_D for a circular cylinder in ambient steady flow versus Reynolds number $Rn = U_\infty D/\nu$. The different symbols correspond to different specimens (fibers) (Tritton, 1959).

may, in principle, apply a Morison type of equation by considering each structural member. The vertical pipes can be handled as shown earlier and the loads on the braces can be evaluated by applying the cross-flow principle except for the junctions between structural members. Loads at junctions between structural members could also be handled as discussed in Section 6.4.2. A difficulty is to properly estimate the drag and mass coefficients if hydrodynamic interaction between structural members is important. Let us discuss this by focusing on vertical cylinders and start with two cylinders of equal diameter D and with distance l between the cylinder axes. When l/D is large, (e.g., $l/D \gtrsim 10$), as it is for the two pipes next to the transverse bulkhead in Figure 1.20, an important aspect is to consider if the wake from one of the pipes changes the inflow to another pipe. The motion of the wake at a distance from the pipe is advected with the ambient flow velocity. Let us introduce A as a measure of the amplitude of an ambient liquid particle and express the KC number as $2\pi A/D$; which means hydrodynamic interaction is important for large l/D when $A/l = KC \cdot D/(2\pi l) \gtrsim 1$. However, we must also consider the flow direction of the wake. For instance, considering the two pipes in Figure 1.20 that are next to the transverse bulkhead. The ambient flow velocity is nearly parallel to the bulkhead and, hence, the wake causes an inflow to the other pipe if the preceding equation is satisfied. The two pipes farthest away from the transverse bulkhead are very close and hydrodynamic interaction must always be considered.

The natural frequencies of a pump tower are in a range where structural vibrations of the pump tower may be excited by hull girder and engine vibrations. The consequence is that the pump tower can be considered rigid when sloshing loads are evaluated. However, we cannot rule out that VIV due to sloshing can occur for a pump tower and elastic vibrations of the tower must be incorporated in the analysis. This fact is discussed in Section 9.4.2 for a pump tower in a spherical tank (see Figure 1.21).

6.5 Viscous damping due to baffles

We illustrate how to estimate the effect of baffles on sloshing damping. Both vertical and horizontal baffles are studied (see Figure 6.12). The analysis neglects the effect of the baffles

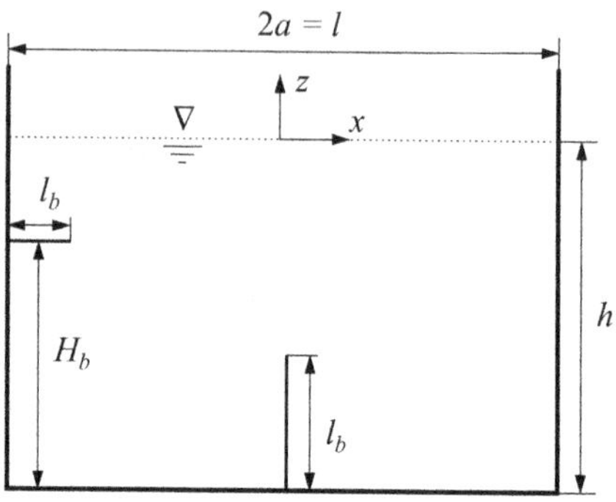

Figure 6.12. Definitions of horizontally and vertically mounted baffles in a tank.

(internal structures) on the natural frequencies and modes, which is studied in Section 4.7.2. For a two-dimensional rectangular tank, Exercise 4.11.6 shows that the effect of horizontal baffles mounted on the vertical walls depends on the vertical position of the baffle and its size. The analysis neglects the effect of vortex shedding. As long as the length of a baffle is less than $0.15l$ and the distance to the waterplane is more than $0.1l$ relative to the horizontal dimension of the tank, l, the influence of the baffle on the lowest natural frequency is less than 6.1% for a rectangular tank and 5% for a vertical circular cylindrical tank. Furthermore, the positions of small-size baffles also give negligible influence, $O(l_b/l)^2$, on the free-surface shape, which suggests that the forthcoming analysis is applicable for small-amplitude sloshing in tanks with baffles of relatively small size. The baffles should not be close to the free surface.

If vertical baffles are mounted to the bottom of a rectangular tank and the liquid depth is finite, their effect on the natural mode is, in general, negligible as long as the distance between their edges and the mean free surface is larger than $0.2l$ of the horizontal dimension of the tank. When the liquid depth becomes smaller or the distance decreases, the effect of the baffling on the natural sloshing modes and frequencies becomes of importance.

Only the *lowest antisymmetric mode* is considered. In the middle of the tank there are only horizontal velocities whereas at the tank walls only vertical velocities exist. The maximum horizontal velocity for a given z-coordinate occurs in the middle of the tank. Its value is larger than the maximum vertical velocity at the tank walls at the same z-coordinate.

6.5.1 Baffle mounted vertically on the tank bottom

A vertical baffle on the bottom of a rectangular tank is first considered. The height of the plate is l_b. The centerplane is selected as the position of the baffle (see Figure 6.12). The force per unit width on the baffle is represented in terms of Morison's equation. It is the drag force part,

$$F_D = \tfrac{1}{2}\rho C_D l_b \,|u_r|\, u_r, \qquad (6.80)$$

that causes viscous dissipation, where u_r is a representative horizontal inflow velocity to the baffle. One way to estimate the drag coefficient, C_D, is to first mirror the baffle about the tank bottom and then consider the baffle and its mirror image in infinite fluid. We use the drag coefficient for cross-flow past a flat plate of length $2l_b$ in infinite fluid. Small Keulegan–Carpenter number is assumed, which means

$$C_D = 8.0KC^{-1/3}, \quad KC = U_{rm}T/(2l_b), \qquad (6.81)$$

where U_{rm} is the maximum value of u_r. What the word "small" means in the context of KC number is a matter of comparison with experiments. Let us say that $KC < O(10)$ is a reasonable assumption. This procedure does not account for the physical restraint that the tank bottom has on the vortex shedding around the plate. Let us illustrate by considering the baffle and its mirror image together with an infinite extension of the tank bottom. This infinitely extended tank bottom forces the flow due to the double baffle to be symmetric about the tank bottom. However, this is not the case for the double baffle in infinite fluid when the value of KC is higher than a certain threshold value. Alternate vortex shedding then occurs from the two edges. The consequence is an asymmetric flow pattern relative to a line coinciding with the infinitely extended tank bottom. The increase of C_D for a wall-mounted plate relative to a free plate can be as much as 50% in the range $1 < KC < 30$. However, an additional complication occurs for the higher KC values due to the influence of the tank walls on the C_D value. Experimental results for a wall-mounted plate are presented by Sarpkaya and O'Keefe (1996).

No excitation of the tank is assumed and the lowest linear mode is examined. The presence of the baffle on the natural mode and frequency is neglected. It is convenient in the derivation of the energy dissipation rate to let the baffle move with the velocity u_r in ambient calm liquid. The rate of energy dissipation during one period of oscillation, T, is the work done during one period of oscillation divided by T. Therefore,

$$D_B = \frac{1}{T}\int_0^T \tfrac{1}{2}\rho C_D l_b \,|u_r|\, u_r u_r \, dt. \qquad (6.82)$$

This is the same as the rate of decay of kinetic and potential energy, $-\dot{E}$.

Because we limit ourselves to the lowest sloshing mode, we can adopt definitions from eqs. (6.46) and (6.47), which include this mode as a particular case. The relative horizontal liquid velocity is then expressed as

$$\frac{\partial \varphi_r}{\partial x} = \frac{\partial \phi_1}{\partial x} = u_0(x, z)\cos(\sigma_1 t), \qquad (6.83)$$

where we used eq. (6.47) with $k = \pi/l$, where A is the maximum amplitude of the free-surface elevation and $\sigma = \sigma_1$. In our case we set $x = 0$ and $z = -h$. This condition gives

$$u_r = \sigma_1 A/[\sinh(\pi h/l)]\cos\sigma_1 t \equiv U_{rm}\cos\sigma t. \qquad (6.84)$$

When evaluating the energy dissipation rate by eq. (6.82), we use $T = T_1$ and the fact that $T^{-1}\int_0^T \cos^2\sigma t\,|\cos\sigma t|\,dt = \frac{4}{3}\pi^{-1}$, which gives

$$D_B = \tfrac{2}{3}\rho C_D l_b \left[\sigma_1 A/\sinh(\pi h/l)\right]^3/\pi, \qquad (6.85)$$

where C_D is estimated by eq. (6.81) and U_{rm} is given by eq. (6.84).

We proceed as we did in the section on damping due to viscous flow in the boundary layers at the tank walls. The damping ratio of the lowest mode, ξ_1, is found from

$$\dot{E} = -D_B = -2\xi_1\sigma_1 E; \quad E = \tfrac{1}{4}\rho g A^2 l, \qquad (6.86)$$

which gives

$$\xi_1 = \tfrac{4}{3}C_D(l_b A/l^2)/[\cosh(\pi h/l)\sinh^2(\pi h/l)],$$
$$C_D = 8\left[A\pi/(l_b \sinh(\pi h/l))\right]^{-1/3}. \qquad (6.87)$$

Formulas (6.87) are demonstrated by a case studied by Isaacson and Premasiri (2001). A vertical baffle is placed on the tank bottom at the centerplane of the tank. Our results are similar to theirs; however, differences occur due to the fact that different estimates of drag coefficients are used. Figure 6.13(a) shows the effect of varying the ratio between maximum wave amplitude and liquid depth. The ratio of baffle height to liquid depth, l_b/h, is 0.05. The damping ratio, ξ_1,

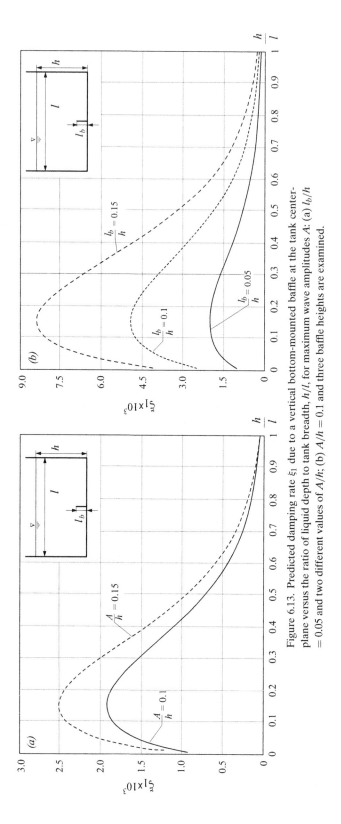

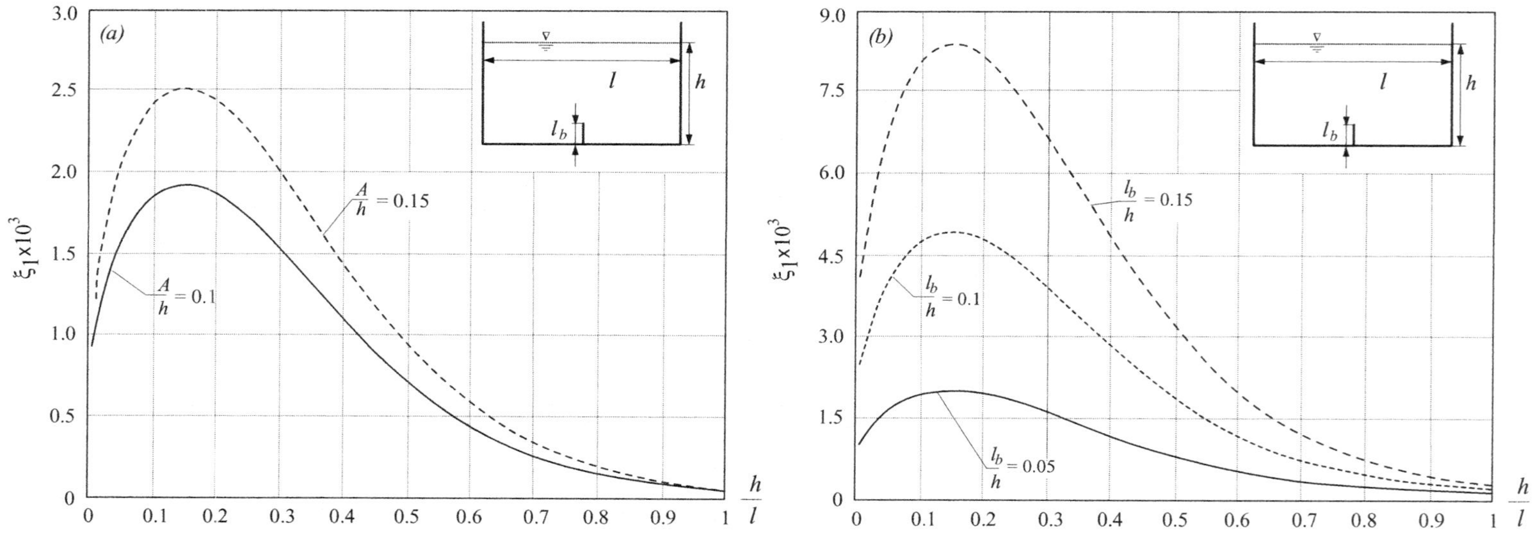

Figure 6.13. Predicted damping rate ξ_1 due to a vertical bottom-mounted baffle at the tank center-plane versus the ratio of liquid depth to tank breadth, h/l, for maximum wave amplitudes A: (a) $l_b/h = 0.05$ and two different values of A/h; (b) $A/h = 0.1$ and three baffle heights are examined.

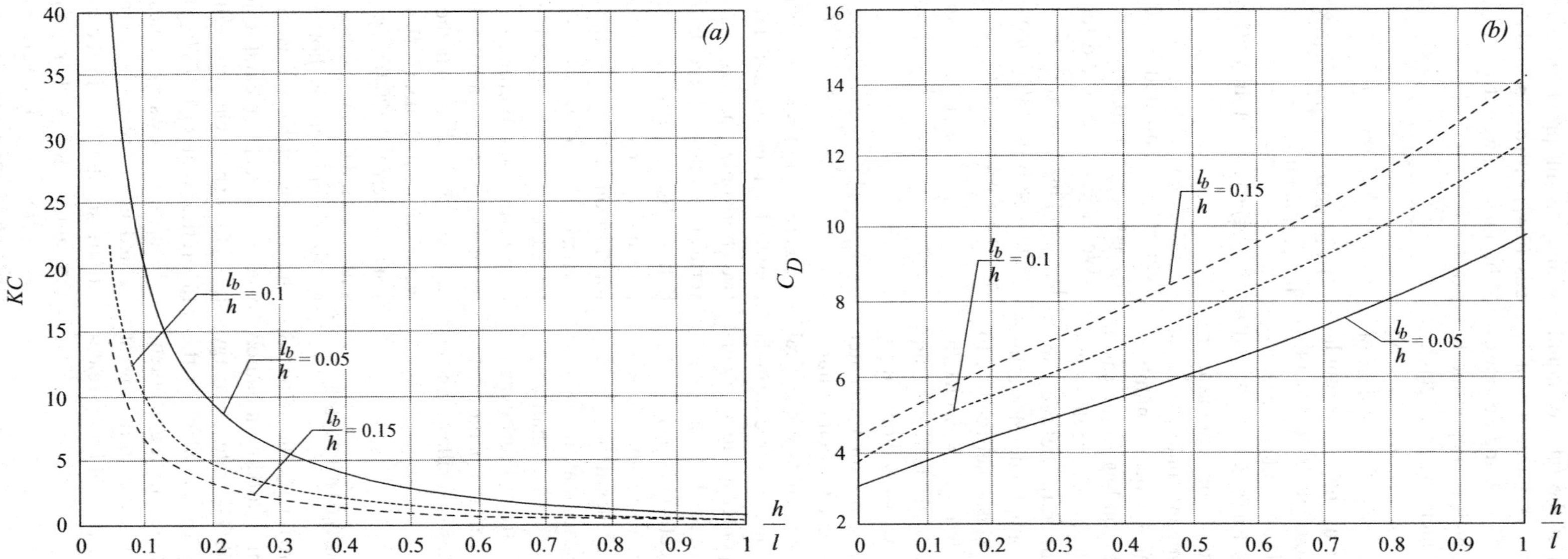

Figure 6.14. (a) Keulegan–Carpenter numbers and (b) drag coefficients associated with the predictions in Figure 6.13(b).

is presented as a function of the ratio of liquid depth to tank breadth, h/l. A maximum of ξ_1 occurs in the vicinity of $h/l = 0.2$ and decreases for higher values of h/l, which is simply due to the fact that the horizontal incident fluid velocity at the baffle decreases. Figure 6.13(b) shows the effect of increasing the baffle height with values of l_b/h from 0.05 to 0.15. The maximum wave amplitude, A, is equal to 0.1 times the liquid depth. We note that the predicted values of ξ_1 are small, with a maximum of 0.008 for $l_b/h = 0.15$ and $h/l \approx 0.2$, which implies a strong amplification of the liquid motion at resonant conditions. Figure 6.14 show the Keulegan–Carpenter numbers and drag coefficients corresponding to the cases in Figure 6.13(b). The Keulegan–Carpenter number also tells us about the ratio between the ambient horizontal-motion amplitude, A_h^i, at the position of the baffles. If A_h^i is small relative to the tank breadth, we can write $A_h^i = l_b \cdot KC/\pi$. The values of KC increase with decreasing liquid depth. Strictly speaking we should only apply our drag formula for small Keulegan–Carpenter numbers and all the Keulegan–Carpenter numbers are not that small. We should be skeptical about applying our empirical drag force formula to the high Keulegan–Carpenter numbers occurring for small h/l. Furthermore, the wall, according to Sarpkaya and O'Keefe (1996), increases the drag coefficient for KC larger than 1. If we want to place more vertical baffles at the tank bottom, we should realize the fact that the ambient horizontal liquid velocity is smaller than at the centerplane. Furthermore, strong hydrodynamic interaction may exist between the baffles if the distance between two baffles is smaller than the order of A_h^i.

6.5.2 Baffles mounted horizontally on a tank wall

We generalize the procedure for a baffle mounted vertically on a tank bottom to a baffle mounted horizontally on a tank wall. The rate of energy dissipation is given by eq. (6.82), which is the same as the rate of decay of the kinetic and potential energy, $-\dot{E}$. We limit ourselves to the lowest sloshing mode. The relative vertical liquid velocity can be expressed as

$$\frac{\partial \varphi_r}{\partial z} = \frac{\partial \phi_1}{\partial z} = w_0(x, z)\cos(\sigma_1 t), \qquad (6.88)$$

where, as in previous derivations, we adopt eqs. (6.47) with $\sigma = \sigma_1$ and $k = \pi/l$.

In our case, we set $x = \frac{1}{2}l$, $z = -h + H_b$, which gives

$$\begin{aligned} w_r &= \sigma_1 A \sinh(\pi H_b/l)/[\sinh(\pi h/l)]\cos(\sigma_1 t) \\ &= W_{rm}\cos(\sigma_1 t). \end{aligned} \qquad (6.89)$$

By using $u_r = w_r$ in eq. (6.82), we get the energy dissipation rate,

$$D_B = \tfrac{2}{3}\rho C_D l_b \left[\sigma_1 A \sinh(\pi H_b/l)/\sinh(\pi h/l)\right]^3/\pi, \qquad (6.90)$$

where C_D is estimated by eq. (6.81) with $U_{rm} = W_{rm}$ given by eq. (6.89).

We proceed as we did for the vertical baffle. The damping ratio ξ_1 of the lowest mode for *two* symmetrically placed horizontal baffles at the two tank walls is then found:

$$\xi_1 = \tfrac{8}{3}C_D\left(\frac{l_b A}{l^2}\right)\frac{\sinh^3(\pi H_b/l)}{\cosh(\pi h/l)\sinh^2(\pi h/l)},$$

$$C_D = 8\left[\left(\frac{A}{l_b}\right)\frac{\pi\sinh(\pi H_b/l)}{\sinh(\pi h/l)}\right]^{-1/3}. \qquad (6.91)$$

Figure 6.15 shows a case similar to that investigated by Isaacson and Premasiri (2001) with two horizontal baffles on a tank wall. The vertical position of the baffles from the tank bottom, H_b, is varied. The theoretical results indicate that it is possible to get higher damping ratios for horizontal baffles at the tank wall than for vertical baffles at the tank bottom if the horizontal baffles are relatively closer to the free surface than to the tank bottom. However, care should be shown in placing a baffle too close to the free surface. The risk is that the baffle goes out of water with subsequent risk of slamming. Isaacson and Premasiri (2001) compared experimental and predicted damping ratios for horizontal wall-mounted and vertical bottom-mounted baffles. The agreement for the vertical baffles is good, whereas the experimental results for the horizontal baffles show clearly smaller values than the theoretical values for $H_b/h \gtrsim 0.7$.

Figure 6.16 shows the values of KC for the cases studied in Figure 6.15. We note that values of KC are smaller than in the cases with a vertical bottom-mounted baffle presented in Figure 6.14 when h/l is small. The values of KC presented in Figure 6.16 are within a range where our empirical drag formula is valid.

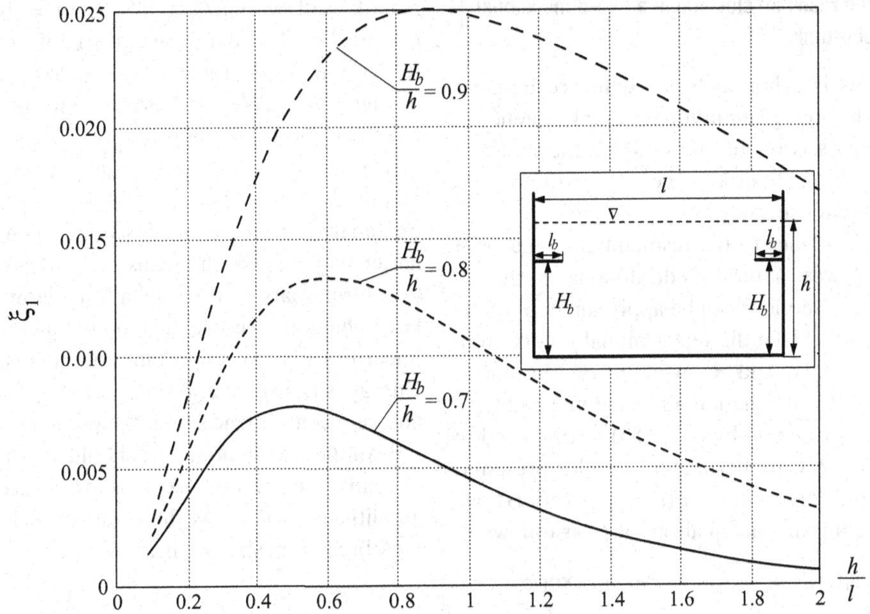

Figure 6.15. Predicted damping ratio ξ_1 due to two horizontal tank-wall-mounted baffles versus ratio of liquid depth to tank breadth, h/l, for maximum wave amplitudes A equal to 0.1 times the liquid depth. In the calculations, $l_b/l = 0.05$, but the vertical position of the baffles from the tank bottom, H_b, is varied.

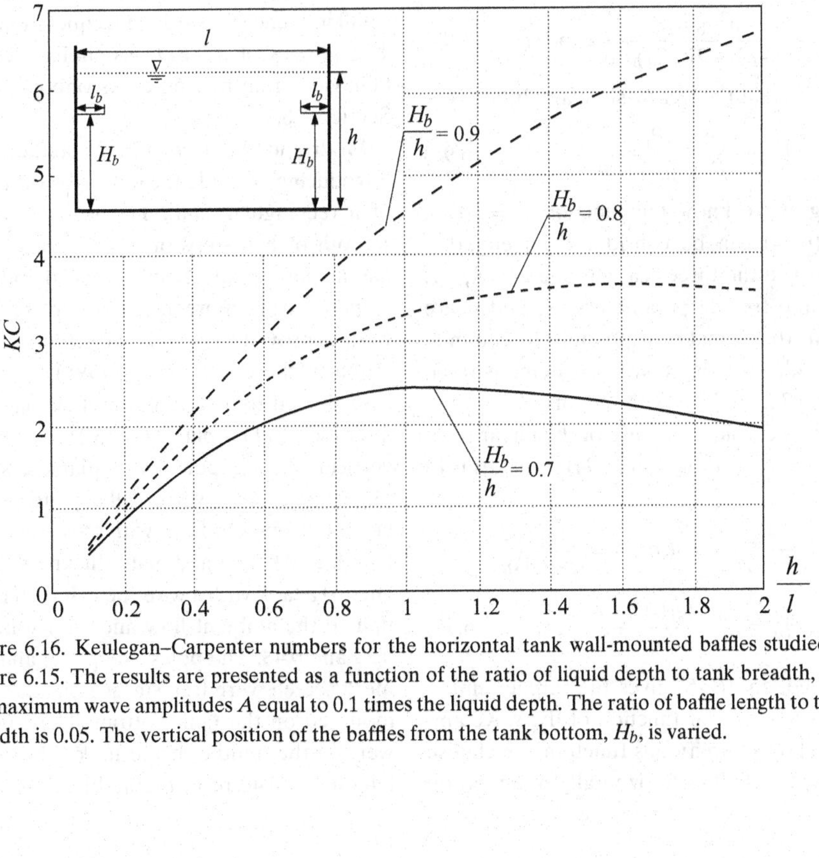

Figure 6.16. Keulegan–Carpenter numbers for the horizontal tank wall-mounted baffles studied in Figure 6.15. The results are presented as a function of the ratio of liquid depth to tank breadth, h/l, for maximum wave amplitudes A equal to 0.1 times the liquid depth. The ratio of baffle length to tank breadth is 0.05. The vertical position of the baffles from the tank bottom, H_b, is varied.

6.6 Forced resonant sloshing in a two-dimensional rectangular tank

Our focus is a horizontally excited rectangular tank with forcing amplitude η_{1a} and frequency σ. The analysis is related to studying the effect of damping due to internal structures on the steady-state resonant sloshing.

When it comes to the resonant case (i.e., $\sigma \approx \sigma_1$), the lowest natural mode dominates; that is, free-surface motions can be approximated by $\zeta = \beta_1(t)\sin(\pi x/l)$ (see the linear modal theory from Chapter 5). To find β_1, we can use the corresponding modal equation with damping and forcing terms presented by eq. (5.56) as discussed in Section 6.3.2. Considering the complex presentation of the forcing (i.e., $\eta_1(t) = \eta_{1a}\exp(i\sigma t)$), we can write this modal equation for β_1 as follows:

$$\ddot{\beta}_1 + 2\xi_1(A)\sigma_1\dot{\beta}_1 + \sigma_1^2\beta_1 = P_1\sigma^2\eta_{1a}\exp(i\sigma t), \tag{6.92}$$

where $P_1 = -4\pi^{-1}\tanh(\pi h/l)$. The damping ratio $\xi_1 = \xi_1(A)$ is assumed to be a known function of the steady-state amplitude of β_1. Examples of those functions are given by eqs. (6.87) and (6.91).

Equation (6.92) has the complex solution

$$\beta_1(t) = \frac{\sigma^2\eta_{1a}P_1}{\sigma_1^2 - \sigma^2 + 2i\xi_1(A)\sigma_1\sigma}\exp(i\sigma t), \tag{6.93}$$

which means that the steady-state amplitude is

$$A = \left|\frac{\sigma^2\eta_{1a}P_1}{\sigma_1^2 - \sigma^2 + 2i\xi_1(A)\sigma_1\sigma}\right|. \tag{6.94}$$

As long as we know the function $\xi_1 = \xi_1(A)$, equality (6.94) can be considered an equation with respect to the three real parameters: A, η_{1a}, and σ. If the forcing parameters η_{1a} and σ are known, eq. (6.94) makes it possible to compute the amplitude A. How it works is demonstrated in Section 6.7.5 for TLD with screens.

Let us now consider the case of the linear resonance with $\sigma = \sigma_1$. Equation (6.94) transforms to the form

$$A = \frac{2\eta_{1a}}{\pi\xi_1(A)}\tanh(\pi h/l) = \frac{2\sigma_1^2\eta_{1a}l}{g\xi_1(A)\pi^2}$$

$$\Rightarrow \eta_{1a} = \frac{\pi A\xi_1(A)}{2\tanh(\pi h/l)}. \tag{6.95}$$

As we see, eq. (6.95) gives the forcing amplitude η_{1a} as a nonlinear function of the wave amplitude A. To exemplify this function, we choose the case of two baffles horizontally mounted on a tank wall with $h/l = 0.5$, $H_b/h = 0.7$, and $l_b/l = 0.05$. The damping ratio is given by eq. (6.91). The results for A versus η_{1a} are presented in Figure 6.17. We note the very strong amplification of the liquid motion. For instance, when η_{1a}/l is about 0.03, the maximum predicted wave amplitude is equal to the liquid depth. Obviously our analysis becomes inconsistent when A/h is larger than 0.3, which means the baffles become dry during part of the oscillations. However, in later chapters, we see that potential flow nonlinearities play an important role at resonance. Energy is transferred to other modes of motions and, as a consequence, limit the oscillation amplitude of the lowest mode. We should also note that in reality we are interested in forced oscillation amplitudes as large as the order of 10% of the tank breadth in ship applications.

6.7 Tuned liquid damper (TLD)

TLDs are used to suppress horizontal vibrations of structures. The TLD consists of a tank partially filled with water. The lowest frequency of sloshing is tuned to a structural natural frequency. High damping of the sloshing with a small influence on sloshing frequency is desirable. The physics of a TLD has similarities with the behavior of antirolling tanks onboard ships (see Section 3.6.2).

Modi and Akinturk (2002) studied TLDs by introducing twin identical wedges on the bottom of a rectangular tank. The wedges were either smooth or had steps or holes. The wedge angle has an important effect on the damper performance. Furthermore, holes or perforations in the wedges proved to be quite effective.

Warnitchai and Pinkaew (1998) experimentally investigated four different rectangular tank configurations: (1) a plain tank; (2) a tank with two vertical circular poles with diameters 0.022 m; (3) a tank with a vertical flat plate perpendicular to the ambient flow, with cross-sectional plate length of 0.050 m and plate thickness of 0.003 m; and (4) a tank with a wire-mesh screen perpendicular to the ambient flow and with solidity ratios 0.29 and 0.48. The poles, the plate, and the wire-mesh screen were free-surface piercing and were mounted on the tank bottom. Their placements were in the middle of the tank. The tank had a length $l = 0.40$ m in the ambient flow direction

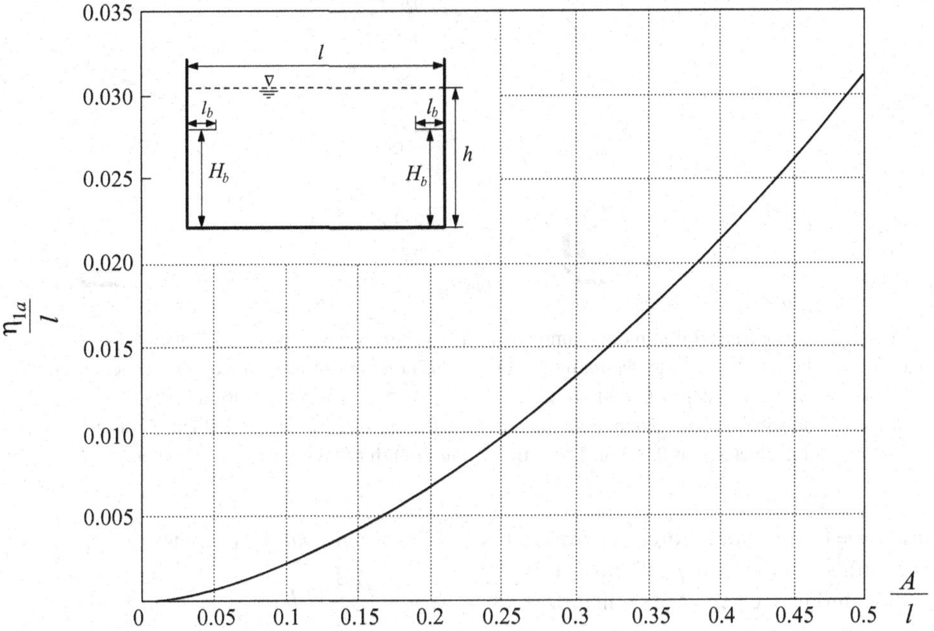

Figure 6.17. Surge amplitude η_{1a} as a function of maximum free-surface amplitude A at forced harmonic surge oscillations under steady-state conditions with frequency equal to the lowest natural frequency for the liquid motion: $h/l = 0.5$, $H_b/h = 0.7$, and $l_b/l = 0.05$.

and width $B = 0.20$ m. The ratio of water depth to tank length was $h/l = 0.3$ in cases 1, 2, and 3. Case 4 was studied with $h/l = 0.2$ and 0.3. Tait *et al.* (2004a, 2004b, 2005) also investigated the influence of a screen on the performance of a TLD. We later see that the plate and the wire-mesh screen in Warnitchai and Pinkaew's (1998) study provide significant damping. However, the circular poles also cause clearly higher damping than a plain tank. We try to estimate the damping ratios and follow a somewhat different procedure than Warnitchai and Pinkaew (1998). However, both procedures are based on Morison's equation. The influence of the interior structures on the natural frequencies is neglected.

Warnitchai and Pinkaew (1998) proposed a formula for the lowest natural frequency that is applicable to the cases with two circular poles and a flat plate as in Figure 6.18. This formula is consistent with our analysis in Exercise 4.11.7 and Section 4.7.3.1 (see also eq. (4.141)). The two vertical poles and the vertical plate at the considered water depth cause, respectively, 1.5 and 3.7% lower first natural frequency than the plain tank. Because the effectiveness of a TLD is sensitive to frequency tuning, relatively small differences in natural frequencies are important. Later we show implicitly by compairing theory and experiments that the wire mesh has a negligible effect on the lowest sloshing frequency.

When it comes to the plain tank, we can use the formulas previously derived based on Keulegan (1959) to find the damping ratio (see Section 6.3.1). Only the viscous damping due to the boundary layers is considered. We get $\xi_1 = \alpha/2\pi = 0.0022$, which is less than the formula used by Warnitchai and Pinkaew (1998). Their formula for viscous damping gives $\xi_1 = 0.0034$ and appears to agree better with the experimental values. However, we do not know about the possible influence from surface tension as noted by Keulegan.

We now present the effect of the interior structures used by Warnitchai and Pinkaew (1998) on the damping ratio; we begin by examining the vertical plate and the poles (i.e., we need to express the drag force on a vertical structure standing on the tank bottom and penetrating the free surface). The position of the center of the structure is at the middle longitudinal position of

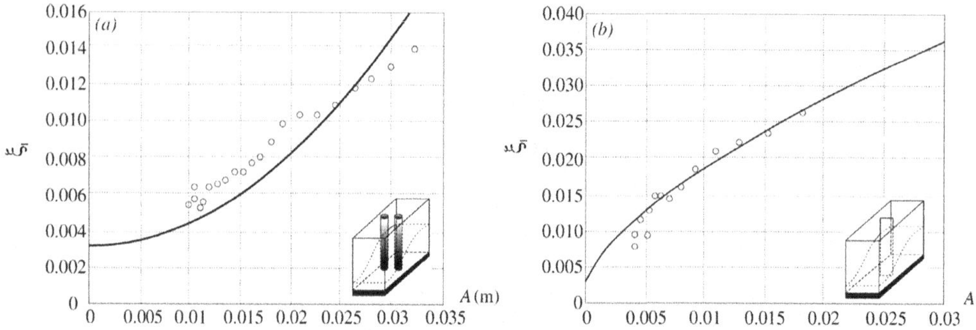

Figure 6.18. Theoretical first-mode damping ratio ξ_1 versus wave amplitude A for two TLD configurations. The effect of energy dissipation at tank walls and bottom is included. Experimental points by Warnitchai and Pinkaew (1998) are marked by ∘. Tank length is $l = 0.40$ m in the ambient flow direction and width is $B = 0.20$ m. Pole diameter is 0.022 m. The cross-sectional length of the plate is 0.050 m and its thickness is 0.003 m. The ratio of water depth to tank length is $h/l = 0.3$.

the tank, $x = 0$. If the drag force is expressed in terms of a drag coefficient C_D, the rate of energy dissipation during one period of oscillation is

$$D_B = \frac{1}{T}\int_0^T\int_{-h}^0 \tfrac{1}{2}\rho C_D(z) \times D(z)\left|u_r(z,t)\right| u_r(z,t)\, u_r(z,t)\, dt\, dz, \quad (6.96)$$

where $D(z)$ is the projected length of the structure at z on a plane perpendicular to the ambient flow. We concentrate on the lowest mode and $u_r(z,t)$ follows from eq. (6.83) by setting $x = 0$:

$$u_r(z,t) = u_0(0,z)\cos\sigma_1 t = \sigma_1 A \frac{\cosh[\pi(z+h)/l]}{\sinh(\pi h/l)}\cos(\sigma_1 t). \quad (6.97)$$

The time integration in eq. (6.96) is similar to that of eq. (6.82). Because C_D may depend on the Keulegan–Carpenter number, for instance, we must allow for a z-dependence of C_D. We can express the Keulegan–Carpenter number for the cross-section at z as

$$KC(z) = \frac{2\pi A\cosh[\pi(z+h)/l]}{D\sinh(\pi h/l)}. \quad (6.98)$$

6.7.1 TLD with vertical poles

If a circular cylinder is considered and $KC(0)$ is less than or equal to 10, we may express C_D due to flow separation as $C_D(z) = C_1 \cdot KC(z)$, where the constant C_1 is 0.2 for subcritical flow with a laminar boundary layer. The energy dissipation rate for *one cylinder* becomes

$$D_B = \frac{\rho}{2}\frac{4}{3\pi}C_1 D\frac{\sigma^4 A^4}{\sinh^4(\pi h/l)}\frac{T}{D}\frac{l}{\pi} \times \left[\frac{\sinh(4\pi h/l)}{32} + \frac{\sinh(2\pi h/l)}{4} + \frac{3}{8}\frac{\pi}{l}h\right]. \quad (6.99)$$

The damping ratio of the lowest mode is then

$$\xi_1 = 2D_B/[\sigma_1 \rho g A^2 lB], \quad (6.100)$$

which means

$$\xi_1 = \frac{8}{3\pi}C_1\frac{\pi\tanh(\pi h/l)}{l\sinh^4(\pi h/l)}\frac{A^2}{B} \times \left[\tfrac{1}{32}\sinh(4\pi h/l) + \tfrac{1}{4}\sinh(2\pi h/l) + \tfrac{3}{8}\pi h/l\right]. \quad (6.101)$$

When flow separation does not occur, we may evaluate the drag force for subcritical flow as linearly proportional to the ambient flow velocity u_r. The corresponding energy dissipation rate for *one cylinder* can be expressed as

$$D_B^V = \frac{1}{T}\int_0^T\int_{-h}^0 \rho\sqrt{2\nu\sigma}\pi D u_r(z,t)u_r(z,t)\, dt\, dz = \rho\sqrt{\frac{\nu\sigma}{2}}\pi D\frac{l}{\pi}\frac{\sigma^2 A^2}{\sinh^2(\pi h/l)} \times \left[\tfrac{1}{4}\sinh(2\pi h/l) + \tfrac{1}{2}\pi h/l\right]. \quad (6.102)$$

The damping ratio of the lowest mode is then

$$\xi_1 = \sqrt{\frac{2\nu}{\sigma_1}}\frac{\pi D}{lB}\frac{\tfrac{1}{4}\sinh(2\pi h/l) + \tfrac{1}{2}\pi h/l}{\cosh(\pi h/l)\sinh(\pi h/l)}. \quad (6.103)$$

The theoretical results for a TLD with *two vertical poles* tested by Warnitchai and Pinkaew

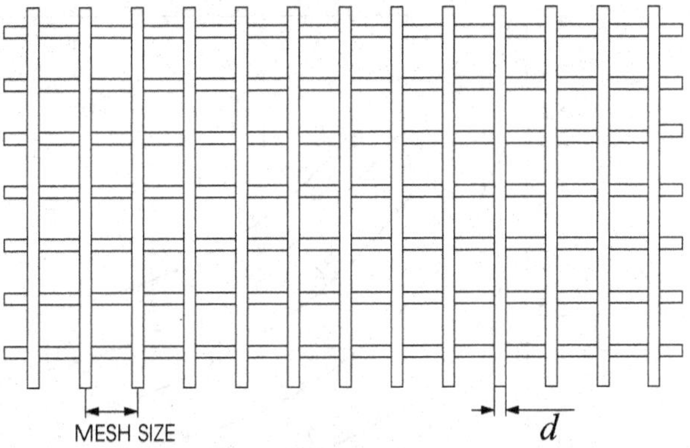

Figure 6.19. Definition of mesh size m for a plane netting panel. Each twine (wire) has projected cross-sectional width d.

(1998) are presented in Figure 6.18(a) with the experimental results. What we have done is to assume no hydrodynamic interaction between the two poles. This approximation is a reasonable one when the distance between the cylinder axes is larger than twice the cylinder diameter and the cylinders are in a side-by-side arrangement (Herfjord, 1996). Furthermore, we have added the expressions for attached flow and separated flow (eq. (6.102)) and the damping due to the plain tank. Because the contribution from separated flow is limited to $KC(0) < 10$, only results up to a wave amplitude of approximately 0.03 m are shown. Equation (6.78) must be included for larger Keulegan-Carpenter numbers. Our results are in reasonable agreement with the experimental values by Warnitchai and Pinkaew (1998). They assumed in their theoretical model that C_D was a constant along the cylinder and found the value of C_D by fitting their theoretical curve to the experiments giving $C_D = 1.5$.

6.7.2 TLD with vertical plate

For the vertical plate we express the drag coefficient as $C_D = 8.0KC^{-1/3}$, where $KC(z)$ is defined by eq. (6.98). The damping ratio of the lowest mode becomes equal to

$$\xi_1 = \frac{32}{3\pi\,(2\pi)^{1/3}}\frac{\tanh(\pi h/l)}{\sinh^{8/3}(\pi h/l)} \times \frac{D^{4/3}A^{2/3}}{lB}\int_0^{\pi h/l}\cosh^{8/3} u\, du, \quad (6.104)$$

where D is the cross-sectional length of the plate. Figure 6.18(b) shows comparisons with the experimental results by Warnitchai and Pinkaew (1998). The agreement is good. Warnitchai and Pinkaew's own theoretical predictions assumed that C_D was a constant along the plate. They found a C_D value of 6.9 by fitting their theoretical curve to the experiments.

6.7.3 TLD with wire-mesh screen

The *solidity ratio*, Sn, is the ratio of the area of the shadow projected by wire meshes on a plane parallel to the screen to the total area contained within the frame of the screen. Figure 6.19 defines the mesh size m and the projected cross-sectional width d of a wire. The solidity ratio of the screen in Figure 6.19 can be expressed as

$$Sn = 2d/m - (d/m)^2. \quad (6.105)$$

Taylor (1944) was the first to investigate flow through and around screens placed in an infinite fluid. He approximated the screen by a number of resistance elements (source elements). Reviews on flow through screens have been provided by Laws and Livesey (1978) and Roach (1987). Figure 6.20 shows the drag coefficient as a function of Reynolds number and the solidity ratio for cross-flow past a finite-plane netting panel in infinite fluid presented by Fridman (1998). The drag force is normalized with respect to the projected area of the netting yarns (wires). The Reynolds number is defined as $U_\infty d/\nu$ with the twine (wire) diameter d as a characteristic length, where U_∞ is the ambient flow velocity without the presence of the netting panel. The results show a clear dependence on the solidity ratio, which means that we cannot consider the drag force on individual wires in ambient flow velocity U_∞ and adding the contributions from each wire to get the total drag force on the netting panel.

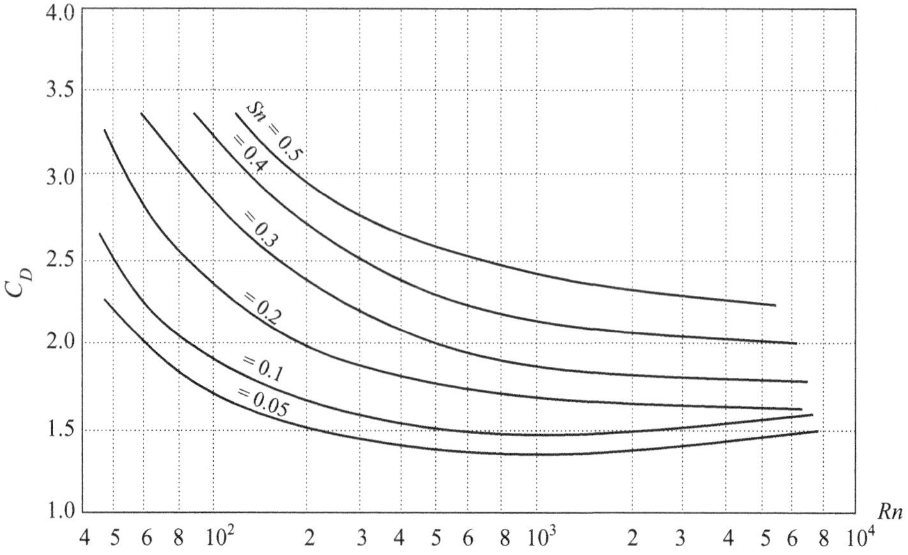

Figure 6.20. Drag coefficient C_D of netting panels in cross-flow at various solidity ratios Sn as a function of Reynolds number, $Rn = U_\infty d/\nu$. Based on Fridman (1998).

One way to estimate the effect of the solidity ratio is to note the fact that the average velocity through a wire mesh of infinite extent is $U_{av} = U_\infty/(1 - Sn)$. This estimate follows from continuity of fluid mass. When considering the drag force on each wire, we use U_{av} as a characteristic velocity instead of U_∞. This approach gives a higher inflow velocity to each wire. The Reynolds number can be redefined as $U_\infty d/(\nu(1 - Sn))$. Let us illustrate this further by using an empirical formula by White (1974) that expresses the drag coefficient for a circular cross-section in infinite fluid. This formula states that

$$C_D \approx 1.0 + 10.0 \cdot Rn^{-2/3}, \quad 1.0 < Rn < 2 \times 10^5, \tag{6.106}$$

where $Rn = U_\infty d/\nu$. The drag force per unit length can then be expressed as $\frac{1}{2}\rho(1.0 + 10.0 \cdot Rn^{-2/3})d \cdot U_{av}^2$. Using the fact that $U_{av} = U_\infty/(1 - Sn)$ and defining the drag coefficient by expressing the drag force per unit length as $0.5\ \rho C_D d \cdot U_\infty^2$ gives the following modified drag coefficient:

$$C_D \approx \left[1.0 + 10.0 \cdot (U_{av} d/\nu)^{-2/3}\right]/(1 - Sn)^2, \quad 1.0 < U_{av} d/\nu < 2 \times 10^5, \tag{6.107}$$

which accounts for the solidity ratio.

This procedure breaks down when $Sn \to 1$. One reason follows by considering two circular cylinders. We commented in Section 6.7.1 that the hydrodynamic interaction had to be considered when the distance between the cylinder axes is smaller than twice the cylinder diameter and the cylinders are in a side-by-side arrangement (Herfjord, 1996). The consequence for this case is that m/d cannot be smaller than 2, which means that our procedure is questionable when $Sn > 0.75$. Another source of error is the three-dimensional flow effects. The local flow at the wire crossing is always three-dimensional. Furthermore, the assumption of strip theory requires that m/d is not too small. What that means in practice is that one must make comparisons with model tests. An additional error source exists due to the fact that we have assumed that all the flow goes through the netting panel. The flow is partly diverted outside the netting panel. For instance, the case of $Sn = 1.0$ corresponds to cross-flow past a plate. The flow separation at the plate edges is then responsible for the drag coefficient of the plate. In Section 6.8 we analyze the case of a wire screen with large solidity ratio by introducing a pressure loss coefficient instead of a drag coefficient.

Equation (6.107) is in fair agreement with the results in Figure 6.20 except for the higher solidity ratios 0.4 and 0.5. However, the formula is more appropriate for a wire mesh that covers the cross-section of a tank. All flow then must pass through the wire mesh. Therefore, later in the section

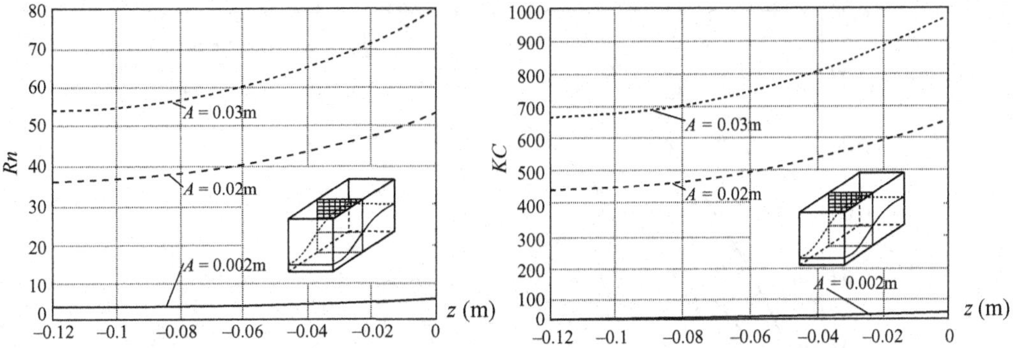

Figure 6.21. Reynolds number and Keulegan–Carpenter number versus depth for a wire with diameter $d = 0.00026$ m in the experiments by Warnitchai and Pinkaew (1998).

we generalize eq. (6.107) to our considered wire mesh in a TLD. Warnitchai and Pinkaew (1998) presented experimental results for damping ratios in three cases with a wire mesh. We show what the Reynolds and Keulegan–Carpenter numbers are for screens with $d = 0.00026$ m. Because the wire diameter is very small, the corresponding Reynolds numbers are very small, which is illustrated in Figure 6.21(a) by not accounting for the solidity ratio. The corresponding Keulegan-Carpenter numbers are shown in Figure 6.21(b). Because the Keulegan–Carpenter numbers are very high, the effect on the drag coefficient can be neglected (see eq. (6.78)).

The damping ratio of the lowest mode, ξ_1, can be derived by following a procedure similar to that in the previous sections and by generalizing eq. (6.107). However, we note a difference if we use eq. (6.96) as a reference. The drag force in this case uses a modified inflow velocity that accounts for the blockage of the wire screen. When the energy dissipation during one period is evaluated, the drag force is multiplied with the ambient cross-flow velocity (i.e., the velocity without the wire screen). The result is

$$\xi_1 = \frac{4\tanh\left(\frac{\pi}{l}h\right)Ad}{(1-Sn)^2\,3l^2B\sinh^3\left(\frac{\pi}{l}h\right)}\left\{\begin{array}{l} N_V\int_{-h}^{0}\left(1+\dfrac{10.0}{Rn_{av}^{2/3}(z)}\right)\cosh^3\left[\dfrac{\pi}{l}(z+h)\right]\mathrm{d}z \\ +B\displaystyle\sum_{j=1}^{N_H}\left(1+\dfrac{10.0}{Rn_{av}^{2/3}\left(z_j^H\right)}\right)\cosh^3\left[\dfrac{\pi}{l}\left(z_j^H+h\right)\right]\end{array}\right\}, \tag{6.108}$$

where

$$Rn_{av}(z) = \frac{\sigma_1 A\cosh[\pi(z+h)/l]d}{(1-Sn)\,\nu\sinh(\pi h/l)}. \tag{6.109}$$

The numbers of vertical and horizontal wires are denoted N_V and N_H, respectively. The z-coordinates of the horizontal wires are z_j^H ($j = 1,\ldots,N_H$). We set $z_j^H = -h - 0.5m + jm$ ($j = 1,\ldots,N_H$). The derivation of eq. (6.109) assumes strip theory with a two-dimensional flow assumption in the local cross-sectional plane for each wire. This implies, as already stated, that m/d cannot be too small.

The theoretically predicted results of the damping ratio for the three wire-mesh cases that were experimentally investigated by Warnitchai and Pinkaew (1998) are shown in Figure 6.22. One case is for the ratio of water depth to tank length, $h/l = 0.3$; wire diameter $d = 0.00026$ m; and solidity ratio, $Sn = 0.29$. This gives a mesh size $m = 6.35d$, $N_H = 72$, and $N_V = 121$. The two other cases use $Sn = 0.48$ and $d = 0.00009$ m, which means $m = 3.59\,d$. The h/l values are 0.2 and 0.3 in the latter two cases with $N_V = 619$ for both cases and $N_H = 247$ and $N_H = 371$ for $h/l = 0.2$ and 0.3, respectively.

Figure 6.22 shows generally good agreement between our theory and the experimental results of Warnitchai and Pinkaew (1998). However, differences occur at very low wave amplitudes. The damping ratios for the cases with a solidity ratio of 0.48 are relatively high, which is beneficial for a

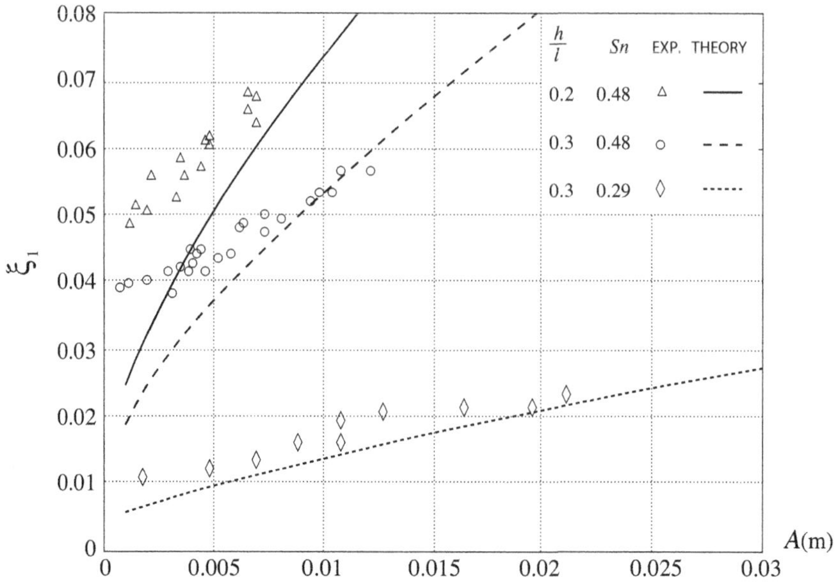

Figure 6.22. Theoretical first-mode damping ratio ξ_1 for the TLD with *wire meshes* with solidity ratios $Sn = 0.29$ (wire diameter 0.00026 m) and $Sn = 0.48$ (wire diameter 0.00009 m) tested by Warnitchai and Pinkaew (1998) as a function of wave amplitude. The ratio of water depth to tank length, h/l, has been varied from 0.2 to 0.3. The tank length is 0.4 m and the tank width is 0.2 m. The plain tank is rectangular. The effect of energy dissipation at tank walls and bottom is included.

TLD as long as the influence on the natural sloshing frequency is small. Furthermore, the results indicate that the ratio between water depth and tank length is an important parameter.

6.7.4 Scaling of model tests of a TLD

The damping ratios due to the vertical poles and wire mesh are clearly Reynolds-number–dependent, which implies that larger tanks with geometrically similar structures should have smaller damping ratios. The effect of the Reynolds number is less significant for vertical plates because the separation lines are fixed (i.e., not Reynolds-number dependent). The energy dissipation at the tank walls and bottom are Reynolds-number–dependent, but this contribution relative to the effect of the internal structure is small.

6.7.5 Forced longitudinal oscillations of a TLD

Forced harmonic horizontal oscillations of a rectangular tank with an interior structure are considered. Our focus is on an excitation frequency σ near the lowest natural frequency, σ_1. In this case, the lowest mode dominates and, based on results from Section 6.6, the steady-state resonant behavior of the lowest mode is given by eq. (6.93) (in complex form). The effect of the internal structures is accounted for by the damping ratio, $\xi_1 = \xi_1(A)$. The wave amplitude A is determined by eq. (6.94). For the studied TLD, the damping ratio $\xi_1(A)$ is presented by eq. (6.108). As we discussed in Section 6.6, substituting eq. (6.108) in eq. (6.94) leads to a nonlinear equation with respect to A, η_{1a}, and σ. When the forcing amplitude η_{1a} is given, eq. (6.94) gives the dependence between the forcing frequency and wave amplitude, A.

The higher modes give lower contributions and are not affected by resonance when σ is close to σ_1. We can therefore set $\xi_i = 0$ $(i \geq 2)$ in the modal equations for the higher modes as long as we are interested in the response near the lowest natural frequency. Based on the modal representation of the free surface and the complex forcing term, we now find that first-mode motions are determined by eq. (6.93), but a steady-state solution for the higher modes is given by $\beta_{2i+1}(t) = P_{2i+1}\sigma^2\eta_{1a}/(\sigma_{2i+1}^2 - \sigma^2)\exp(i\sigma t)$ and $\beta_{2i}(t) = 0$ $(i \geq 1)$, where the coefficients P_k are given by

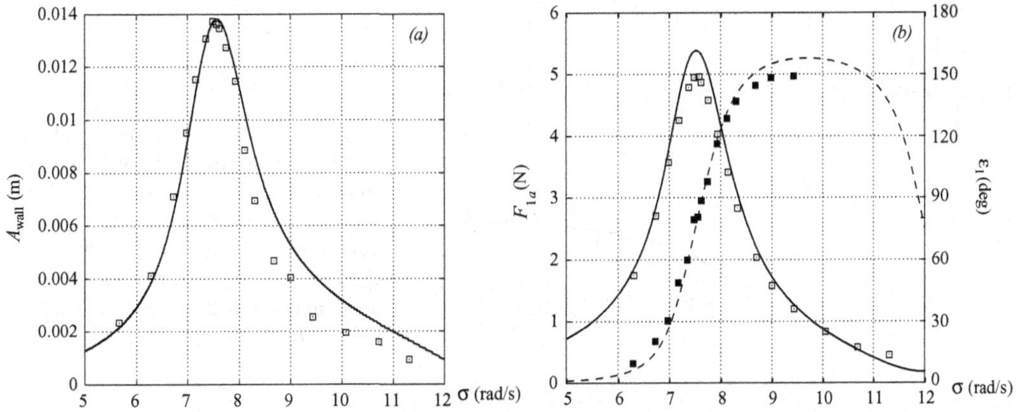

Figure 6.23. Steady-state wave amplitude A_{wall}, longitudinal hydrodynamic force amplitude F_{1a} (solid lines), and phase ε_1 (dashed line) of a rectangular tank with wire mesh ($Sn = 0.48$) in the middle of the tank that is forced longitudinally with amplitude $\eta_{1a} = 0.002$ m and circular frequency σ. The ratio of water depth to tank length is $h/l = 0.3$. The tank breadth B is 0.2 m. The experimental values by Warnitchai and Pinkaew (1998) are denoted by □ (wave and force amplitudes) and ■ (phase).

eq. (5.85). Using this modal solution with eq. (6.93) and modal presentation (5.2) in formula (5.88) for the horizontal force, we can now evaluate the wave elevations at the wall and the horizontal hydrodynamic force acting on the TLD. The maximum wave elevation at the wall is then given as

$$A_{\text{wall}} = \sigma^2 \eta_{1a} \left| \frac{P_1}{\sigma_1^2 - \sigma^2 + 2i\xi_1(A)\sigma_1\sigma} + \sum_{i=1}^{\infty} \frac{P_{2i+1}}{\sigma_{2i+1}^2 - \sigma^2} \right|. \tag{6.110}$$

The horizontal hydrodynamic force can be presented in the following complex form:

$$F_1(t) = B\left[m_l \sigma^2 \eta_{1a} + \frac{\sigma^4 \lambda_1 P_1 \eta_{1a}}{\sigma_1^2 - \sigma^2 + 2i\xi_1(A)\sigma\sigma_1} + \sum_{i=1}^{\infty} \frac{\sigma^4 \lambda_{2i+1} P_{2i+1} \eta_{1a}}{\sigma_{2i+1}^2 - \sigma^2} \right] \exp(i\sigma t), \tag{6.111}$$

where B is the tank width. The hydrodynamic coefficients, λ_{2i+1}, are explicitly derived in Sections 5.4.2.1 and 5.4.3.1 and take the form

$$\lambda_{2i+1} = -\frac{2\rho l^2}{(\pi(2i+1))^2}, \quad i = 0, 1, \ldots. \tag{6.112}$$

The first term in the square brackets of eq. (6.111) represents the inertia force per unit width of the rigid liquid mass m_l. In addition comes the force acting on the interior structure and the viscous force acting on the tank wall. The latter effects are neglected in the following calculations.

Again, when the forcing amplitude η_{1a} is given, we can consider the hydrodynamic forcing amplitude (modulus of the square bracket in eq. (6.111)) as a function of σ and $A = A(\sigma)$.

The force can be split up into one part, which is in phase with the tank acceleration, and another part that is in phase with the tank velocity. If the tank motion is expressed as $\eta_{1a} \sin \sigma t$ and the horizontal sloshing force as $F_1 = F_{1a} \sin(\sigma t - \varepsilon_1)$, then the horizontal sloshing force can be reexpressed in terms of added mass coefficient A_{11} and damping coefficient B_{11} (see definition of A_{11} and B_{11} in Chapter 3) as follows:

$$\begin{aligned} F_1 &= F_{1a} \cos \varepsilon_1 \sin \sigma t - F_{1a} \sin \varepsilon_1 \cos \sigma t \\ &\equiv -A_{11}\ddot{\eta}_1 - B_{11}\dot{\eta}_1 \\ &= A_{11}\sigma^2 \eta_{1a} \sin \sigma t - B_{11}\sigma\eta_{1a} \cos \sigma t, \end{aligned} \tag{6.113}$$

which means

$$\begin{aligned} A_{11} &= F_{1a} \cos \varepsilon_1 / (\sigma^2 \eta_{1a}), \\ B_{11} &= F_{1a} \sin \varepsilon_1 / (\sigma \eta_{1a}). \end{aligned} \tag{6.114}$$

Let us consider as an example a rectangular tank with length $l = 0.40$ m and width $B = 0.20$ m. The ratio of water depth to tank length is 0.3. A wire mesh with solidity ratio $Sn = 0.48$ and wire diameter $d = 0.00009$ m is arranged. This case has been experimentally studied by Warnitchai and Pinkaew (1998) and the damping ratio is presented in Figure 6.22. The tank is forced to oscillate harmonically in the longitudinal direction with amplitude 0.002 m, which corresponds

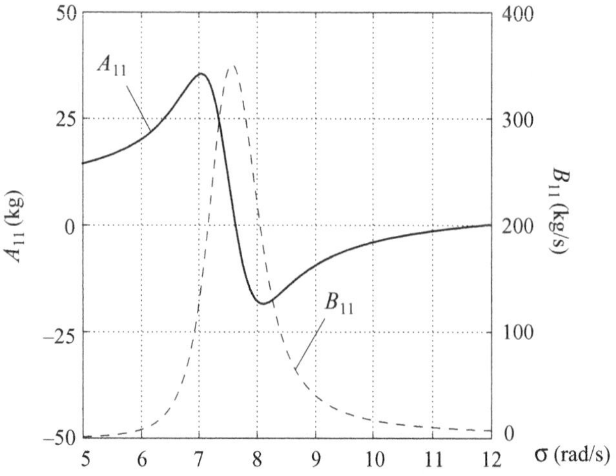

Figure 6.24. Added mass A_{11} (solid line) and damping B_{11} (dashed line) of a rectangular tank with wire mesh ($Sn = 0.48$) in the middle of a tank that is forced longitudinally with amplitude $\eta_{1a} = 0.002$ m and circular frequency σ. The ratio of water depth to tank length is $h/l = 0.3$. The tank breadth B is 0.2 m.

to a ratio of excitation amplitude to tank length of 0.005, which is rather low compared to ship applications. This ratio for ships can be as much as 0.1. Calculated values of the steady-state wave amplitude at the tank wall, A_{wall}, horizontal hydrodynamic force amplitude, and phase as a function of the forcing frequency in the vicinity of the lowest natural frequency are presented in Figure 6.23. Pinkaew (personal communication, 2007) confirmed that there were misprints in the original paper. Therefore, the experimental force amplitudes by Warnitchai and Pinkaew (1998) have been multiplied by a factor of 10. Furthermore, the sign of the experimental force phases in the original paper has been changed.

The experimental frequency of maximum response agrees very well with the potential flow predictions of the lowest natural frequency without accounting for the influence of the wire mesh on the natural frequency. We note that linear potential flow theory with viscous damping is able to predict the wave amplitude, force amplitudes, and phases satisfactorily relative to the model test. One reason is the fact that the damping is relatively high. In later chapters it is shown that nonlinear potential flow effects play an important role for larger excitation amplitudes and for smaller damping.

The corresponding added mass, A_{11}, and damping, B_{11}, coefficients are presented in Figure 6.24. The material mass of the water is 9.6 kg and is inherently accounted for in the added mass. We note that A_{11} comes close to the material water mass for the smaller frequencies of oscillation in Figure 6.24. The added mass is frequency dependent and is negative for σ larger than the lowest sloshing frequency in Figure 6.24. Because added mass must not be interpreted as a physical mass, there is nothing wrong with a negative added mass. A negative added mass only means that the "inertia force" acts in the opposite direction of the tank acceleration, or that we have an apparent reduction in the system mass. Negative added mass is a well-known phenomenon for exterior flow problems around a ship. However, the damping (B_{11}) is positive, as it must be. The largest values of B_{11} occur in a frequency band around the lowest sloshing frequency. If the damping ratio of the sloshing had been larger, this band with large values of B_{11} would have been broader.

When considering the TLD as a damping device for structural vibrations, the forcing amplitude of the tank is a consequence of the equations for the global structural motions. The TLD is then part of the analysis. We showed earlier that the damping ratio for a given excitation amplitude of the tank is nonlinear. We must now do an extra iteration of the damping ratio for different excitation amplitudes to find what the global motions are. An alternative would be to do this in the time domain instead of the frequency domain, as we have done. A time-domain simulation has advantages when the global excitation of the structure is stochastic, as it would be in the case of wind loads due to turbulence or water waves.

A high damping coefficient in as broad as possible a frequency range near the lowest natural sloshing period of the TLD is desirable for the global structural motions that the TLD is supposed to damp. This scenario is illustrated when antirolling tanks are considered in Chapter 3. A difference then is that nonlinear free-surface effects are important for free-surface tanks, whereas here we have considered only linear free-surface effects.

It is not straightforward to generalize the described procedure to stochastic excitation due to wind turbulence or water waves. The reason is the presence of nonlinearities in the damping of the sloshing motions. If nonlinear damping is quadratic, there exists an established equivalent linearization procedure to obtain standard deviations (Price & Bishop, 1974).

6.8 Effect of swash bulkheads and screens with high solidity ratio

We now focus on cases where the solidity ratio can be large. A relevant case for ship applications is the use of a swash (wash) bulkhead in a tank. It is common to have swash bulkheads in either center or wing cargo tanks. A swash bulkhead is a bulkhead with openings. The swash bulkhead is typically placed in the middle of the tank perpendicular to the main flow direction. If the ratio between the area of the holes and the area of the bulkhead is small, as in Figure 1.18, an important effect is the change in the highest natural sloshing period to a level where sloshing is less severe. Flow through the holes causes flow separation and thereby damping of resonant sloshing. The effect of perforated walls has been studied experimentally by Garza (1964) and Abramson and Garza (1965) for forced horizontal excitation of a vertical circular tank that is compartmented into sectors by means of radial walls. Sector tanks of 45°, 60°, and 90° were investigated. Both the excitation amplitude and the perforation of sector walls affect the damping and the highest natural period. The natural period is defined here to correspond to when the liquid response has a maximum as a function of the forced frequency. The fact that the excitation amplitude matters is a nonlinear effect associated with flow through perforated walls. This physical effect will be clearer after the analysis in this section. As a rule of thumb, if the total area of the perforations exceeds 10% of the area, the liquid tends to slosh between the compartments and the slosh natural frequency tends to approach the value of a nondivided tank (Dodge, 2000).

The effect of perforated plates and screens with high solidity ratio on natural sloshing frequencies and associated damping rates can be evaluated by using formulas representing the pressure loss through a screen. The formulas deal with the case when $u_r = u_r(t)$ is a uniform ambient cross-flow velocity relative to the tank. The uniform pressure loss at the screen is empirically expressed as

$$\frac{p^- - p^+}{\frac{1}{2}\rho u_r |u_r|} = K, \tag{6.115}$$

where p^- and p^+ are the pressures at two opposite sides of the screens and $K \geq 0$ is the so-called loss coefficient, which is a function of solidity ratio Sn, Reynolds number, and the screen structure. Equation (6.115) follows from generalizing expressions for steady ambient flow (Blevins, 1992; Roach, 1987) and assuming a small influence of Keulegan–Carpenter number. The latter fact was argued in Section 6.7.3 for a wire-mesh screen. The pressure loss coefficient changes from 0 when there is no screen or perforated wall ($Sn = 0$) to ∞ when the screen or perforated plate becomes a solid wall ($Sn = 1$).

The following discussion concentrates on a wire-mesh screen. As an example we consider a rectangular tank with length l and liquid depth h. The screen is mounted in the middle, as shown in Figure 6.25. The tank is forced with sway motions

$$\eta_2(t) = \eta_{2a}\cos(\sigma t). \tag{6.116}$$

This harmonic forcing excites only "odd" modes in a rectangular tank without screens (see Section 4.3.1.1). In the case of no screens (i.e., $Sn = 0$), the forcing leads to resonance at the frequencies

$$\sigma = \sigma_k^{Sn=0} = \sqrt{g\pi(2k-1)l^{-1}\tanh(\pi(2k-1)h/l)}, \quad k = 1, 2, \ldots. \tag{6.117}$$

When $Sn = 1$ (i.e., the wall divides the tank into two equal tanks with length $a = \frac{1}{2}l$), the resonance occurs at the frequencies

$$\sigma = \sigma_k^{Sn=1} = \sqrt{2g\pi(2k-1)l^{-1}\tanh(2\pi(2k-1)h/l)}, \quad k = 1, 2, \ldots. \tag{6.118}$$

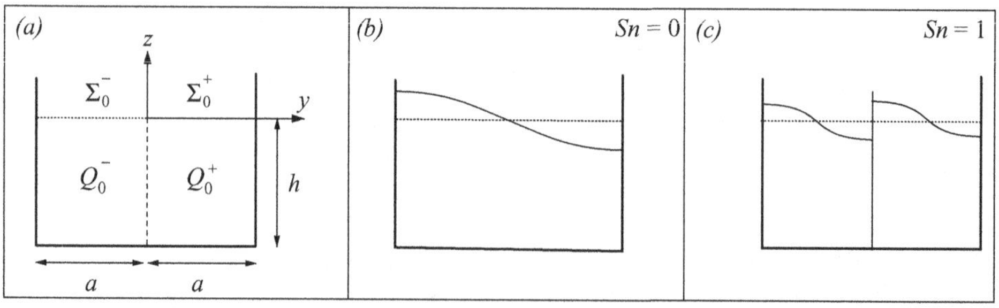

Figure 6.25. A two-dimensional rectangular tank with a screen in the middle. Parts (b) and (c) illustrate the lowest natural modes in the limit cases $Sn = 0$ and $Sn = 1$, respectively.

The lowest natural modes from eqs. (6.117) and (6.118) are associated with the wave profiles in Figure 6.25(b) and (c), respectively.

For a two-dimensional rectangular tank, we can analyze how the lowest natural frequency changes from $\sigma_1^{Sn=0}$ to $\sigma_1^{Sn=1}$ as Sn varies in the interval [0,1]. For this purpose, we use the domain decomposition method with notations in Figure 6.25(a). For each domain Q_0^+ and Q_0^-, we introduce the absolute velocity potentials, Φ^+ and Φ^-, and the surface waves by $z = \zeta^{\pm}(y, t)$ in the tank-fixed coordinate system. Typical boundary conditions of linear sloshing theory should be fulfilled for $\Phi^{\pm}$ and $\zeta^{\pm}$ on the wall, bottom, and mean free surface based on the sway motion η_2 given by eq. (6.116). The horizontal absolute velocities on the screen are equal; that is,

$$U(z,t) = \dot{\eta}_2(t) + u_r(z,t) = \left.\frac{\partial \Phi^+}{\partial y}\right|_{y=0} = \left.\frac{\partial \Phi^-}{\partial y}\right|_{y=0}, \tag{6.119}$$

where $u_r(z, t)$ is the relative velocity. Equation (6.119) gives the *first transmission condition* on the screen for our domain decomposition method. The relative velocity is not uniform but a function of the z-coordinate. In the most general case, it can be represented as the following Fourier series:

$$u_r(z,t) = \dot{\alpha}_0(t) + \sum_{k=1}^{\infty} \dot{\alpha}_k(t) \cos\left(\frac{\pi k}{h}(z+h)\right). \tag{6.120}$$

The z-dependence of u_r is consistent with the fact that the flow is irrotational, that is, $\partial u_r/\partial z = \partial w_r/\partial y$, and that the relative vertical velocity $w_r = 0$ on $z = -h$.

The linear pressure is expressed as $p^{\pm} = -\rho(\partial\Phi^{\pm}/\partial t + gz)$, whereby a pressure loss rule along the screen couples Φ^+ and Φ^- and constitutes the *second transmission condition*. When assuming that U (or u_r) is known and the corresponding sloshing problems in Q_0^- and Q_0^+ accompanied by transmission condition (6.119) are solved, the velocity potential and pressure become functions of u_r. The use of the velocity potential in the second transmission condition reduces the whole problem to a (nonlinear) equation with respect to U (or u_r).

An example of the pressure loss rule is eq. (6.115). However, this equation is formulated for a uniform ambient flow field, while we operate with z-dependent pressures and relative velocities. To employ eq. (6.115), one should assume that the ambient flow field varies slowly along the screen and that an approximate equivalent uniform velocity is associated with the z-averaged cross-flow along the screen; that is,

$$u_r(t) \approx \dot{\alpha}_0(t) = \frac{1}{h}\int_{-h}^{0} u_r(z,t)\,\mathrm{d}z \tag{6.121}$$

(where α_0 is the same as in representation (6.120)). Under this assumption, eq. (6.119) transforms to

$$\begin{aligned} U(z,t) &\approx U(t) + u_r(t) = \dot{\eta}_2(t) + \dot{\alpha}_0(t) \\ &\approx \left.\frac{\partial \Phi^+}{\partial y}\right|_{y=0} = \left.\frac{\partial \Phi^-}{\partial y}\right|_{y=0}. \end{aligned} \tag{6.122}$$

We adopt eq. (6.122) and consider a steady-state flow that oscillates harmonically with forcing frequency σ. Therefore,

$$\begin{aligned} \alpha_0(t) &= a_{01}\cos(\sigma t) + a_{02}\sin(\sigma t), \\ \dot{\alpha}_0(t) &= \sigma\left[-a_{01}\sin(\sigma t) + a_{02}\cos(\sigma t)\right] \\ &= -\sigma\sqrt{a_{01}^2 + a_{02}^2}\,\sin(\sigma t - \theta_0) \\ &= -\rho_0\sigma\sin(\sigma t - \theta_0). \end{aligned} \tag{6.123}$$

With eqs. (6.116) and (6.123), we can find the solutions of linear sloshing problems in the domains Q_0^- and Q_0^+ by using the modal technique from Section 5.4.2.4. In that section, we considered sloshing due to deformation of the tank wall so that the normal velocity component of the deformation was $U_d(z)\dot{\eta}_d(t)$. When studying the sloshing in Q_0^+ with eq. (6.122), we get a situation similar to that in Section 5.4.2.4 with $U_d\dot{\eta}_d(t) = \dot{\alpha}_0(t)$.

Using this solution for Q_0^+ and Q_0^- gives

$$\begin{aligned}\Phi^{\pm}&(y, z, t)\\ &= y\dot{\eta}_2(t) + \Omega_d^{\pm}(y, z)\dot{\alpha}_0(t) \mp \dot{\alpha}_{-1}(t)\\ &\quad + \sum_{k=1}^{\infty} R_k^{\pm}(t)\cos\left(\frac{\pi k y}{a}\right)\frac{\cosh(\pi k(z+h)/a)}{\cosh(\pi k h/a)},\end{aligned} \tag{6.124}$$

where Ω_d^+ is the solution of Neumann problem (5.106) in the rectangular domain $[0, \frac{1}{2}l] \times [-h, 0]$ with $U_d(z) = 1$. The analogous function Ω_d^- is defined in $[-\frac{1}{2}l, 0] \times [-h, 0]$ and is antisymmetric to Ω_d^+. We have introduced a time-dependent function $\alpha_{-1}(t)$ because any solution of Neumann problems is defined within a constant. This time-dependent function did not affect sloshing due to wall deformation in Section 5.4.2.4 and, therefore, was set equal to zero. For the screen problem, the function $\alpha_{-1}(t)$ influences the pressure difference between Q_0^+ and Q_0^-, and the pressure loss at the screen. The result is

$$\Omega_d^{\pm}(y, z)\dot{\alpha}_0(t) = \mp\frac{\dot{\alpha}_0(t)}{2a}\left[(y \mp a)^2 - (z+h)^2\right].$$

The free-surface elevation is expressed separately in the two liquid domains by means of Fourier series. Because of the flux $h\alpha_0(t)$ through the screen and the continuity of liquid mass for each liquid domain, spatially mean free-surface elevations $h\alpha_0(t)/a$ and $-h\alpha_0(t)/a$ exist in the liquid domains denoted with + and −, respectively. Representations of the free-surface elevations $\zeta^{\pm}(y, t)$ in the two liquid domains that are consistent with conservation of liquid mass in each domain and also for the whole tank can be expressed as

$$\zeta^{\pm}(y, t) = \pm\frac{h}{a}\alpha_0(t) + \sum_{j=1}^{\infty}\beta_j^{\pm}(t)\cos\left(\frac{\pi j y}{a}\right). \tag{6.125}$$

The generalized coordinates $R_k^{\pm}(t)$ in the velocity potential given by eq. (6.124) and the generalized coordinates $\beta_j^{\pm}(t)$ of the free-surface elevation can be related by the kinematic free-surface condition in the same way as shown in Section 5.3.1.2; that is,

$$R_j^{\pm}(t) = \frac{a\dot{\beta}_j^{\pm}(t)}{\pi j\tanh(\pi j h/a)}, \quad \sigma_{aj}^2 = \frac{g\pi j}{a}\tanh\left[\frac{\pi j h}{a}\right].$$

The next step, as in Section 5.3.1.2, is to use the linear dynamic free-surface condition, which leads to the following relationship for $\alpha_{-1}(t)$ and differential equations for β_j^+ and β_j^- :

$$\begin{aligned}\ddot{\alpha}_{-1}(t) &= -\tfrac{1}{2}a^{-1}\ddot{\alpha}_0(t)\left[\tfrac{1}{3}a^2 - h^2\right]\\ &\quad + \tfrac{1}{2}a^{-1}\ddot{\eta}_2(t) + a^{-1}gh\alpha_0(t),\\ \ddot{\beta}_j^{\pm} + \sigma_{aj}^2\beta_j^{\pm} &= \pm K_j(t),\end{aligned} \tag{6.126}$$

where

$$K_j(t) = \frac{2}{\pi}\tanh\left(\frac{\pi j h}{a}\right)\left[-\frac{(-1)^j - 1}{j}\ddot{\eta}_2 + \frac{\ddot{\alpha}_0}{j}\right] \tag{6.127}$$

and

$$\sigma_{aj}^2 = \frac{g\pi j}{a}\tanh\left(\frac{\pi j h}{a}\right). \tag{6.128}$$

Because our focus is on steady-state solutions, modal system (6.126) has an analytical solution of the form $\beta_j^{\pm}(t) = b_{1j}^{\pm}\cos(\sigma t) + b_{2j}^{\pm}\sin(\sigma t)$. The coefficients are functions of η_{2a} (the known motion of tank) and a_{01}, a_{02} (two unknowns that characterize the z-averaged relative velocity on the screen). Using the analytical $\beta_j^{\pm}$ in eqs. (6.124) and (6.125) gives pressures $p^+(z, t)$ and $p^-(z, t)$ also as harmonically oscillating functions of t that depend on η_{2a} and a_{01}, a_{02}.

The found pressures are not uniform. We must consider the z-averaged values of the pressures over the screen to implement pressure loss condition (6.115) and to be consistent with assumption (6.122). Remembering the definition of $\dot{\alpha}_0(t)$ given by eq. (6.121), we should transform eq. (6.115) to the form

$$\frac{1}{h}\int_{-h}^{0}\left(\left.\frac{\partial\Phi^+}{\partial t}\right|_{y=0} - \left.\frac{\partial\Phi^-}{\partial t}\right|_{y=0}\right)\mathrm{d}z = \tfrac{1}{2}K\dot{\alpha}_0\,|\dot{\alpha}_0|\,. \tag{6.129}$$

Furthermore, we perform an equivalent linearization of the nonlinear term on the right-hand side of eq. (6.129). Gathering $\cos(\sigma t)$ and $\sin(\sigma t)$

gives the following nonlinear system with respect to a_{01} and a_{02}:

$$\eta_{2a}C + a_{01}C_1 = \tfrac{1}{2}K\frac{h}{\pi\sigma}\int_0^{2\pi/\sigma}\cos(\sigma t)\dot{\alpha}_0|\dot{\alpha}_0|dt = K\frac{4h}{3\pi}a_{02}\sqrt{a_{01}^2 + a_{02}^2},$$

$$a_{02}C_1 = \tfrac{1}{2}K\frac{h}{\pi\sigma}\int_0^{2\pi/\sigma}\sin(\sigma t)\dot{\alpha}_0|\dot{\alpha}_0|dt = -K\frac{4h}{3\pi}a_{01}\sqrt{a_{01}^2 + a_{02}^2}, \quad (6.130)$$

where

$$C = ah + \frac{4a^2}{\pi^3}\sum_{k=1}^{\infty}\frac{(1+(-1)^{k+1})\tanh(\pi kh/a)}{k^3(\sigma_{ak}^2/\sigma^2 - 1)},$$

$$C_1 = \tfrac{2}{3}h\left(a + \frac{h^2}{a}\right) + \frac{4a^2}{\pi^3}\sum_{k=1}^{\infty}\frac{\tanh(\pi kh/a)}{k^3(\sigma_{ak}^2/\sigma^2 - 1)} - \frac{2gh^2}{a\sigma^2}.$$

The solution of system (6.130) can be presented as

$$a_{01} = -\frac{\eta_{2a}CC_1}{C_1^2 + \rho_0^2(4Kh/(3\pi))^2},$$

$$a_{02} = \frac{\eta_{2a}C(4Kh/3\pi)\rho_0}{C_1^2 + \rho_0^2(4Kh/(3\pi))^2}, \quad (6.131)$$

where $\rho_0 = \sqrt{a_{01}^2 + a_{02}^2}$ is a real positive root of the nonlinear equation

$$\eta_{2a}^2C^2\left[\frac{C_1^2}{\rho_0^2} + \left(K\frac{4h}{3\pi}\right)^2\right] = \left(C_1^2 + \rho_0^2\left(K\frac{4h}{3\pi}\right)^2\right)^2. \quad (6.132)$$

Equation (6.132) can be rewritten as the biquadratic equation $(K\frac{4h}{3\pi})^2\rho_0^4 + C_1^2\rho_0^2 - \eta_{2a}^2C^2 = 0$, which makes it possible analytically to find the positive solution of ρ_0.

Based on solutions (6.131) and eq. (6.125) we can, for instance, find the steady-state elevation at the right wall:

$$\zeta^+(a,t) = \frac{h}{a}\alpha_0(t) + \sum_{j=1}^{\infty}\beta_j^+(t)(-1)^j = [\eta_{2a}D + a_{01}D_1]\cos(\sigma t) + a_{02}D_1\sin(\sigma t),$$

$$D = \frac{2}{\pi}\sum_{m=1}^{\infty}\frac{(1+(-1)^{m+1})\tanh(\pi mh/a)}{m(\sigma_{am}^2/\sigma^2 - 1)},$$

$$D_1 = \frac{h}{a} - \frac{2}{\pi}\sum_{m=1}^{\infty}\frac{(-1)^m\tanh(\pi mh/a)}{m(\sigma_{am}^2/\sigma^2 - 1)} \quad (6.133)$$

as well as other hydrodynamic steady-state characteristics.

Blevins (1992) provided pressure loss coefficients for woven screens, rod screens, perforated plates, and grillage. The arrangement for the woven screen is similar to that shown in Figure 6.19 and is chosen as a basis for our studies. The loss coefficient can be approximated (Blevins, 1992; Roach, 1987) by

$$K = \frac{\beta(1-(1-Sn)^2)}{(1-Sn)^2}, \quad (6.134)$$

where β is a function of Reynolds number, $Rn = U_a d/\nu$ (see Section 6.7.3); that is, in our case, it varies with the amplitude U_a of the relative z-averaged cross-flow velocity u_r. The Reynolds number is small for the following case and, generally speaking, causes a variation of β. Based on Blevins (1992), we assume β is constant. The corresponding table in Blevins (1992) gives $\beta \approx 1.3$ for $Rn = 20$; this is the lowest value in the table for Rn, which is adopted later.

The first series of numerical examples is presented in Figure 6.26 for maximum wave elevation at the wall, max $|\zeta^+(a,t)|$. This series includes resonant responses (scaled by the forcing amplitude) for $h/l = h/(2a) = 0.3$ and different solidity ratios. When the solidity ratio is far from 1 (e.g., $Sn = 0.3$), Figure 6.26 clearly demonstrates resonance at $\sigma = \sigma_1^{Sn=0}$ and $\sigma = \sigma_3^{Sn=0}$ associated with antisymmetric modes in a tank without screens. The larger nondimensional forcing amplitude $\varepsilon = \eta_{2a}/l$ gives larger damping (i.e., the peaks in Figure 6.26 become smaller). When Sn increases, the peaks disappear. However, we see an emerging peak between them corresponding to $\sigma = \sigma_1^{Sn=1}$, which occurs earlier for larger forcing.

When $Sn \geq 0.85$ for the presented results, the largest nondimensional response occurs for the largest forcing (i.e., a decreasing flow exists through the screen with increasing forcing). This tendency is the opposite of what happens at lower solidity ratios. If we want to get rid of resonant oscillations in the tank at its lowest natural frequency without a screen, the results imply that the solidity ratio should not be too close to 1 so that sloshing is minimized.

The described method also makes it possible to establish the lowest sloshing frequency versus the solidity ratio. The lowest natural frequency can be found by using the fact that wave elevation at the wall is 90° out of phase with the sway motion at

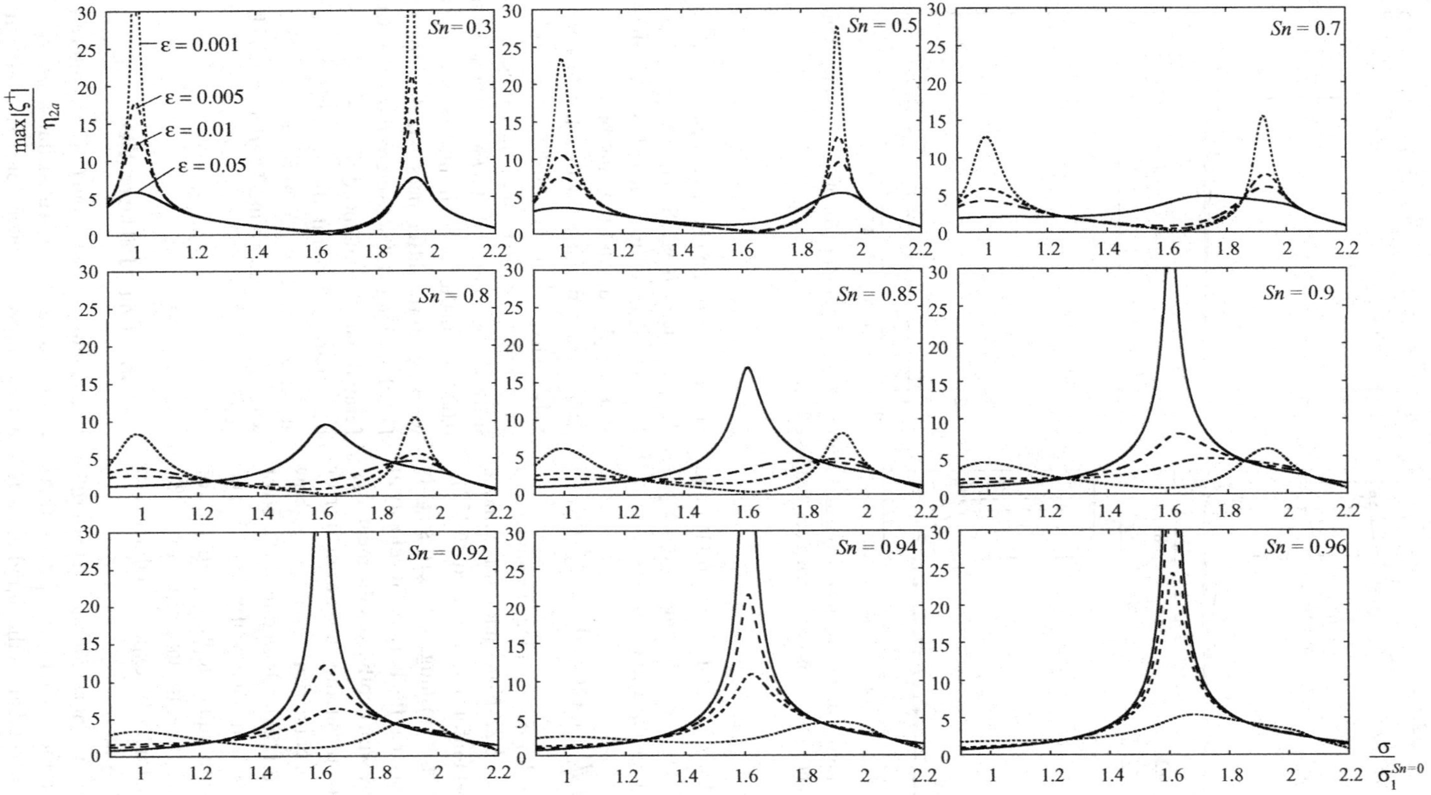

Figure 6.26. Maximum nondimensional wave elevation at the wall versus forcing frequency for $h/l =$ 0.3 and different solidity ratios for a two-dimensional rectangular tank with a screen in the middle. Computational results for different nondimensional lateral forcing amplitudes given by $\varepsilon = \eta_{2a}/l$.

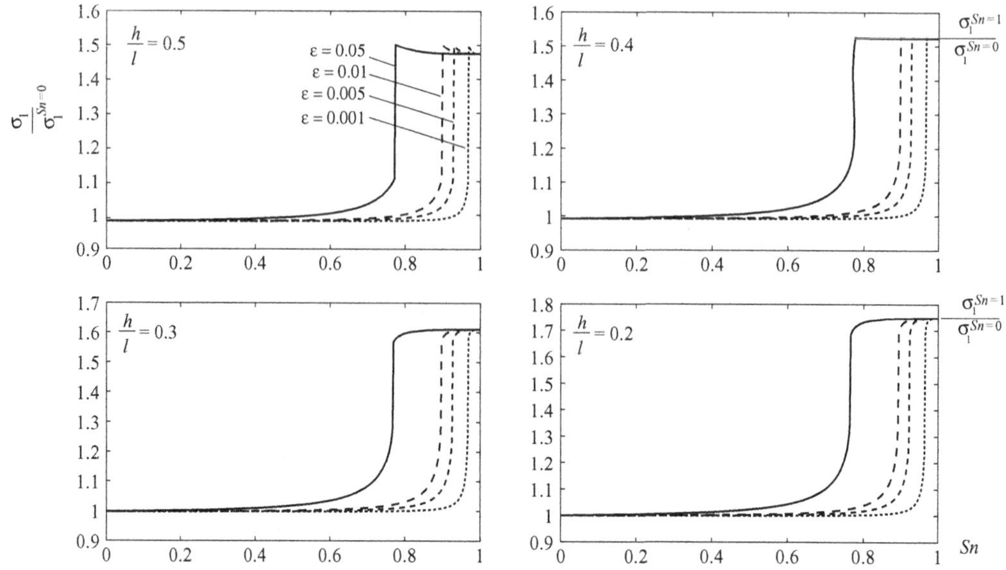

Figure 6.27. Lowest natural frequency versus solidity ratio and different ratios of liquid depth to tank length, h/l, and nondimensional forcing amplitudes $\varepsilon = \eta_{2a}/l$.

resonance. It follows by relating eq. (6.116) and eq. (6.133) that the lowest sloshing frequency is the lowest root of the equation

$$\eta_{2a} D + a_{01} D_1 = 0. \tag{6.135}$$

Results for the lowest sloshing frequency are presented in Figure 6.27. Note that our technique assumes an almost uniform flow (i.e., eq. (6.122)), which is not true for sloshing with zero solidity ratio. The reason is that the horizontal velocity at $y = 0$ should decay with depth from the mean free surface. Therefore, the method gives a small error of $\sigma_1^{Sn=0}$ as $Sn \to 0$. Numerical experiments show that assumption (6.122) leads to only 0.1% error for the lowest natural frequency if $h/l \leq 0.5$. The error increases with liquid depth (up to 1% for fairly deep water). It is also larger, up to 5–7%, for higher modes, for which exponential decay of the velocity matters more.

The results do not confirm Dodge's (2000) rule of thumb that if the area of the perforations exceeds 10% of the total area, the liquid tends to slosh between compartments and the slosh natural frequency tends to approach the value of a nondivided tank. One reason may be that part of the results in Figure 6.27 is for higher forcing amplitudes than those on which Dodge (2000) based his conclusions. For instance, if the total area of the opening in the screen is less than about 20% of the cross-sectional area of the tank (i.e. $Sn >\sim 0.8$) and the ratio ε between the forced sway amplitude and the tank breadth is equal to 0.05, Figure 6.27 shows that the lowest nondivided natural frequency disappears. The highest sway amplitude divided by the cylinder diameter in the experiments that Dodge referred to was 0.00833. Furthermore, it must be investigated to what extent the differences in tank shape, and the fact that Dodge (2000) referred to tests with perforated plates, influenced the results. Loss coefficients for a perforated plate with square edges are clearly different from the loss coefficients of wire screens with round fibers or rods (Blevins, 1992).

6.9 Vortex-induced vibration (VIV)

Vortex-induced vibration (VIV) of interior structures should not be ruled out a priori. VIV is a well-known problem for a structure in ambient steady flow in many fields of engineering. In the marine field, VIV is of concern for risers, pipelines, and Spar buoys and bridges in civil engineering. Vortex shedding is the source of the problem. The fluid–structure interaction associated with VIV is complex and cannot be

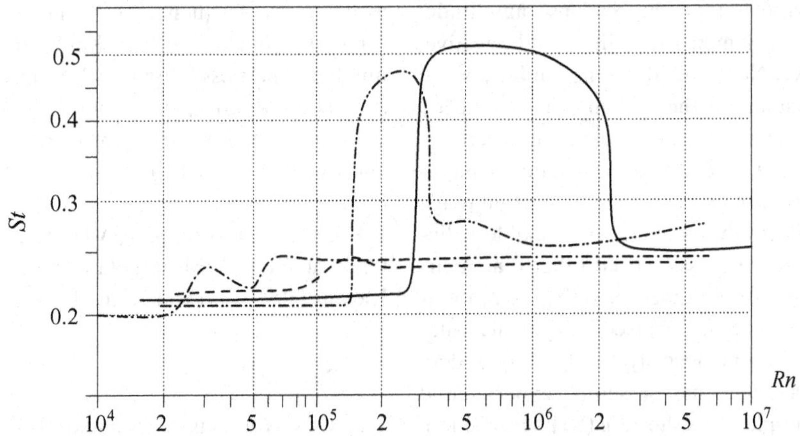

Figure 6.28. Strouhal number of rough circular cylinders in steady incident flow for different surface roughness values Ra/D (where Ra is the average height of surface roughness, D is the cylinder diameter, $St = f_v D/U_\infty$, f_v is the vortex shedding frequency, U_∞ is the incident flow velocity, and $Rn = U_\infty D/\nu$). Solid line, $Ra/D = 7.5 \cdot 10^{-4}$; dashed line, $Ra/D = 3 \cdot 10^{-3}$; dashed-and-dotted line, $Ra/D = 9 \cdot 10^{-3}$; and dashed-and-dot-dotted line, $Ra/D = 3 \cdot 10^{-2}$ (Achenbach & Heinecke, 1981).

described by a linear system. Blevins' (1990) book is a good reference on the subject.

To explain VIV we present classical results for two-dimensional cross-flow past a rigid nonmoving circular cylinder in ambient steady current. We have already presented results for the mean drag coefficient in Figures 6.10 and 6.11. In addition to the mean force expressed by eq. (6.79), unsteady forces also exist that are associated with the vortex shedding frequency, f_v. The nondimensional vortex shedding frequency may be represented by the Strouhal number (St), which is defined by

$$St = f_v D/U_\infty. \tag{6.136}$$

Vortex shedding results in oscillatory forces on the body in both drag and lift directions. If a single vortex shedding frequency exists, the force in the lift direction can be approximated as

$$F_L(t) = |F_L| \cos(2\pi f_v t + \alpha), \tag{6.137}$$

where α is a phase angle. The lift force amplitude, $|F_L|$, is normally expressed in terms of a lift coefficient C_L defined in a similar way as C_D. The force in the drag direction can be approximated as

$$F_D(t) = \overline{F_D} + A_D \cos(4\pi f_v t + \beta), \tag{6.138}$$

where $\overline{F_D}$ is time independent and the same as F_1 in eq. (6.79). Note that the time-dependent part of F_D oscillates at twice the frequency of F_L. Phase angle β in eq. (6.138) is different from α in eq. (6.137).

The first indication of a risk of VIV of a structure is if one of the natural frequencies of the structure, f_n, is in the vicinity of either the vortex shedding frequency, f_v, or twice the vortex shedding frequency. In the former case, resonant oscillations occur in the direction transverse to the incident flow direction (cross-flow VIV), whereas in the latter case oscillations occur in the same direction as the incident flow (i.e., in-line VIV). The reduced velocity, U_R, is an important parameter, which can be expressed as $U_R = U_\infty/(f_n D)$ for an elastically mounted cylinder with diameter D, natural frequency f_n for small-amplitude oscillations with no ambient flow, and ambient steady flow velocity U_∞.

The range of U_R is an important parameter for VIV. The DNV *Guidelines* (1998) for free-spanning pipelines specify that the onset of cross-flow VIV occurs at a value of U_R between 3.0 and 5.0, whereas the maximum vibration levels occur at values between 5 and 7. The upper limit of U_R is set equal to 16. The reason for this broad range of VIV is the fact that VIV affects the added mass and, hence, the natural period of the structure so that resonant conditions occur in a broad range. A simplified explanation of the magnitude of the response is that hydrodynamic damping is affected by VIV and therefore the damping is amplitude dependent. For higher and lower amplitudes than the response amplitude, hydrodynamic damping is positive and negative, respectively. Negative damping indicates an unstable system, so the system finds its steady vibration level when the total damping is zero.

Sarpkaya (1978) analyzed the maximum amplitude of transverse oscillations in the lock-in range for both elastically mounted and flexible cylinders. However, if cross-sections other than a circular cylinder are considered, galloping motions may occur. Galloping is caused by dynamic instabilities and occurs typically for U_R larger than 10. Overvik (1982) has presented experimental data for multiple risers showing the phenomenon. A very important consequence of vortex-induced oscillations is that the drag force increases. Skop *et al.* (1977) proposed an empirical formula for a rigid circular cylinder.

In-line VIV also matters. If we use a similar simplified analysis, as was done previously for the cross-flow VIV, by setting $f_n = 2f_v$ in the in-line case we find that the maximum value occurs at half the value of U_R of the maximum for cross-flow VIV. Because the onset value of U_R is lower for in-line VIV than for cross-flow VIV, in-line VIV occurs more frequently than cross-flow VIV. However, the maximum value of in-line VIV is of the order of 10% of the cross-flow VIV. In reality one would have a combination of in-line and cross-flow VIV.

Another practical matter is the scale effect during model tests, which is most pronounced in the case of a smooth cylinder surface. Model test conditions most often correspond to subcritical flow with laminar boundary-layer flow. The full-scale condition may be in the critical or supercritical flow regimes with turbulence. Because the vortex shedding frequency for a smooth surface is clearly dependent on the flow regimes, scaling VIV from a model to full scale may cause difficulties. However, Reynolds-number effects are less important in the case of moderate surface roughness.

All the cases discussed have steady incident flow. However, vortex-induced oscillations may also occur in ambient oscillatory flow, which is of practical significance in a tank. For large Keulegan–Carpenter numbers, Verley (1982) and Bearman *et al.* (1984) proposed a quasi-steady model to calculate the vortex-induced transverse force on a unit length of a fixed circular cylinder in oscillatory flow. This approach can be used to estimate the frequency components of the lift force as a function of the Keulegan–Carpenter number. The possibility of VIV can be assessed by relating the frequency components of the lift force to the natural frequencies of the structure and can be assessed similarly for unsteady drag force.

It is possible to suppress VIV by using various types of spoilers. More details may be found in Blevins (1990) and Sumer and Fredsøe (1997).

6.10 Summary

The viscous effects of boundary-layer flow and cross-flow past bluff bodies such as stiffeners, girders, baffles, screens, interior pipes, and pump towers were investigated.

The boundary-layer flow is generally laminar in model tests and turbulent in full-scale conditions. The influence of boundary-layer flow on potential flow is in terms of an inflow/outflow from the boundary layer. If a modal theory is used for the potential flow, it is not convenient to solve directly for the boundary-layer effect. Instead, the ordinary differential equations for the principal coordinates of the modal theory are modified by including damping terms.

The damping of the lowest sloshing modes due to boundary-layer flow is small. However, damping is needed in a potential flow model of sloshing to reach steady-state conditions from initial conditions.

The effect of interior bluff bodies was studied by generalizing Morison's equation. The equation needs empirical mass and drag coefficients that depend on Reynolds number, Keulegan–Carpenter number, surface roughness number, and body shape, for instance. An inherent assumption is that the ambient flow varies slowly across the body cross-sections; that is, the body is a poor wavemaker. Care should be shown in applying a Morison equation formulation at the junctions between structural members and when a submerged body with nonzero thickness is mounted to the tank surface. The fact that the tank flow is described in a tank-fixed coordinate system needs proper consideration of what ambient flow velocities and accelerations should be used in a Morison equation–type formulation.

The damping effect of baffles mounted on the walls and bottom of a two-dimensional rectangular tank was investigated. The theoretical results indicate higher damping ratios for horizontal baffles on the tank wall than for vertical baffles at the tank bottom if the horizontal baffles are relatively closer to the free surface than the tank bottom. Clearly larger damping than the boundary-layer damping can generally be achieved with baffles.

The TLD model tests by Warnitchai and Pinkaew (1998) were used to investigate the first-mode damping effect in a three-dimensional rectangular tank due to (a) two vertical circular poles, (b) a vertical flat plate perpendicular to the ambient flow, or (c) a wire-mesh screen. In all cases the interior structures are placed in the middle of the tank. Predictions based on strip theory and Morison's equation with Reynolds-number– and Keulegan-Carpenter-number–dependent empirical coefficients agree reasonably with the experiments. The wire-mesh screen and the vertical flat plate cause more significant damping than the two poles. The studied wire-mesh screen is expected to have larger Reynolds-number scale effects than the vertical plate.

Theoretical predictions of frequency-dependent wave amplitudes, force amplitudes, and phases agree well with experimental results by Warnitchai and Pinkaew (1998) for forced longitudinal excitation of a three-dimensional rectangular tank with wire-mesh screen. The forcing frequencies include the lowest natural sloshing frequency. Theoretical studies use linear potential flow theory with viscous damping effects. The reasons for the good agreement are relatively small forcing amplitudes and large damping.

The TLD studies are relevant for antirolling tanks. A high damping coefficient in as broad a frequency range as possible near the lowest natural sloshing frequency, with a small influence on sloshing frequency, is desirable for good performance of both TLDs and antirolling tanks.

A swash bulkhead was analyzed to understand how natural sloshing frequencies and the liquid response are influenced by the openings in the bulkhead and the forcing amplitude.

6.11 Exercises

6.11.1 Damping ratios in a rectangular tank

(a) Boundary-layer damping of two-dimensional sloshing in a plain rectangular tank was considered in the case of antisymmetric modes. Construct a similar derivation for symmetric modes.

(b) Consider a three-dimensional rectangular tank with lengths L_1 and L_2 along the Ox- and Oy-axis, respectively. Assume three-dimensional linear sloshing associated with $i \neq 0$ and $j \neq 0$ in eq. (4.29) and generalize eq. (6.46) for the velocity potential. Based on this representation, generalize the results for the damping ratios from Sections 6.3.1 and 6.3.3. Show that the bulk damping can be expressed as

$$\xi_{i,j} = \frac{2\nu k_{i,j}^2}{\sigma_{i,j}}, \tag{6.139}$$

where we have used the definition given by eq. (4.30).

Show that viscous boundary-layer damping can be written as

$$\xi_{i,j}^{\text{tank's surface}} = \sqrt{\frac{\nu T_{i,j}}{\pi}} \frac{1}{2L_2} \left[\frac{i^2\left(2 + R_L^{-1}\right) + j^2 R_L \left(2 + R_L\right)}{i^2 + R_L^2 j^2} + 2\pi \frac{i^2 R_L^{-1}\{\frac{1}{2} - \bar{h}\} + j^2 R_L \{\frac{1}{2} - \bar{h} R_L\}}{\sqrt{i^2 + R_L^2 j^2}\,\sinh\left(2\pi\sqrt{i^2 + R_L^2 j^2}\,\bar{h}\right)} \right], \tag{6.140}$$

where $R_L = L_1/L_2, \bar{h} = h/L_1$ and $T_{i,j} = 2\pi/\sigma_{i,j}$.

6.11.2 Morison's equation

Morison's equation for the horizontal force dF on a strip of length dz of a vertical rigid circular cylinder inside a horizontally oscillating tank is expressed by eq. (6.71). Concentrate on the two terms associated with the relative horizontal fluid acceleration, $\dot{u}_r$, and the relative horizontal fluid velocity, u_r. Denote the corresponding part of dF as dF_u. Assume that $\dot{u}_r$ and u_r can be described by the lowest mode for sloshing in a two-dimensional rectangular tank with length l and mean liquid depth h. Place the cylinder in the center of the tank.

(a) Discuss at what time instants the maximum of dF_u occurs as a function of the wave amplitude at the walls, A, as well as the liquid depth, h, tank length, l, and the mass and drag coefficients, C_M and C_D.

(b) Discuss the relative importance of the terms $\dot{u}_r$ and u_r as a function of Keulegan–Carpenter number.

6.11.3 Scaling of TLD with vertical poles

Consider the TLD with vertical poles studied experimentally by Warnitchai and Pinkaew (1998) and handled theoretically in Section 6.7.1. Combine information about the dependence of C_D on Reynolds number and Keulegan–Carpenter number due to separated flow described in the main text together with formulas for attached flow. In the case of a turbulent boundary layer, one can use eq. (6.45) to describe the attached flow effect for small KC.

Discuss the effect of scale on the damping rate.

6.11.4 Effect of unsteady laminar boundary-layer flow on potential flow

We discussed (e.g., in Section 6.2.2) the effect of the inflow/outflow velocity, v_0, on the potential flow arising from a steady-state harmonically oscillating laminar boundary-layer flow. We now study a general unsteady laminar boundary-layer flow that starts from rest at time $t = 0$. We denote the tangential potential flow velocity at the outer part of the boundary layer as $\partial\varphi(s, t)/\partial s$.

(a) Show that the inflow/outflow velocity at time t can be expressed as

$$v_0(s, t) = \sqrt{\frac{\nu}{\pi}} \int_0^t \frac{\partial^2 \varphi(s, \tau)}{\partial s^2} \frac{d\tau}{\sqrt{t - \tau}}. \tag{6.141}$$

(*Hint*: Start by taking the Laplace transform with respect to time of the governing equation and boundary conditions. Express the inverse Laplace transform by using a convolution integral.)

(b) Confirm that the result in (a) is consistent with the steady-state harmonically oscillating case.

(c) Consider a stationary circular cylinder in ambient oscillating flow. Derive an expression for the pressure force due to inflow/outflow from an unsteady boundary layer.

6.11.5 Reduction of natural sloshing frequency due to wire-mesh screen

Investigate the influence of the wire-mesh screen used in Warnitchai and Pinkaew's (1998) TLD studies on the lowest natural sloshing frequency (see Section 6.7.3).

(*Hints*: Use strip theory in the same way as for the drag force in Section 6.7.3 with formulas in Section 4.7.3.)

7 Multimodal Method

7.1 Introduction

The dynamics of a contained liquid with a free surface can be of a strongly nonlinear nature, especially for resonant forcing with frequencies close to the lowest natural frequency. Even though in Chapter 6 we presented a case with a relatively small excitation amplitude and large damping, where linear modal theory involving nonlinear damping agreed with experiments in resonant conditions, nonlinear free-surface effects limit the applicability of the linear sloshing theory. This is true, for instance, for typical ship applications with relatively large forcing amplitudes and resonant conditions.

The derivation of nonlinear modal equations is a tedious and complicated mathematical task. This chapter presents necessary details of the most general, infinite-dimensional nonlinear modal theory by Faltinsen *et al.* (2000), which can be considered a theoretical background for nonlinear modal modeling of potential flow of an incompressible liquid. This modeling is applicable to arbitrary forcing and a broad class of tank shapes. The corresponding modal system is derived by using the Bateman–Luke variational statement (see Section 2.5.3.2).

Historically, the first nonlinear modal theories operated with only a few modes. Second- and higher-order contributions due to other, nonleading modes were neglected. Later on, nonlinear modal analysis adopted an asymptotic intermodal ordering between the leading modes and some of the second-order modes. The first example of such an asymptotic finite-dimensional nonlinear modal system was proposed by Narimanov (1957). He assumed the lowest natural modes to be the dominant ones and reduced the mathematical derivation of the modal system to a series of Neumann boundary-value problems based on Taylor-series expansions of boundary conditions. The original derivations were illustrated for an upright circular cylindrical tank, but the formulas were given for a vertical cylinder of arbitrary cross-section. The scheme has been further used by Stolbetsov (1967a, 1967b) for a rectangular cross-section. Lukovsky (1975) and Narimanov *et al.* (1977) generalized this scheme to tanks of complex, noncylindrical shapes. Narimanov *et al.* (1977) also corrected some minor arithmetic errors from the original paper by Narimanov (1957). Similar arithmetic errors were also present in other low-dimensional nonlinear modal equations (e.g., that by Hutton 1963; see discussion by Miles 1984b). Interested readers may find different versions of the Narimanov-type asymptotic modal systems in Dodge *et al.* (1965), Komarenko and Lukovsky (1974), Dokuchaev (1976), Narimanov *et al.* (1977), and Ganiev (1977); these systems are also reviewed by Lukovsky (1990). Upright circular cylindrical tanks and resonant wave regimes due to lateral excitation are common for such analyses.

A new era of nonlinear modal methods started in the middle of the 1970s with the use of variational procedures. These procedures made it possible to obtain *multidimensional* nonlinear modal equations. Instead of solving a series of Neumann boundary-value problems, derivation of the modal equations needs the determination of a few integrals over the instantaneous liquid domain $Q(t)$. The multidimensional modal theories do not need an asymptotic ordering, but they may be further asymptotically reduced to a finite-dimensional form by adopting asymptotic relations between the modes. The procedure consists of asymptotic expansions of the aforementioned integrals. This process is demonstrated for adaptive modal systems in Chapters 8 and 9. The key ideas of variational multimodal methods were proposed by Miles (1976) and Lukovsky (1976). Miles (1976) demonstrated the applicability of the Lagrange and Bateman–Luke principles to derive infinite-dimensional nonlinear modal systems, whereas Lukovsky (1976) focused on Luke's approach and derived several finite-dimensional examples of such multimodal systems for a vertical circular cylindrical tank. He also compared them with Narimanov's scheme. The purely Lagrange formulation was combined with an asymptotic scheme in the works by Limarchenko (1978b, 1983) and Limarchenko and Yasinskii (1996).

Multimodal methods require an extensive background of Lagrange variational formalism (see introductory details in Section 2.5). The idea of multimodal methods consists of using the *Fourier* (henceforth, *modal*) *approximate solution of the free-surface elevation* with time-dependent coefficients $\beta_i(t)$ $(i = 1, \ldots, N)$. The solution should be substituted into the Lagrangian, which becomes a function of $\beta_i(t)$ $(i = 1, \ldots, N)$. As a matter of fact, the original mechanical problem reduces to a consideration of N generalized coordinates. By calculating the variations of the Lagrangian, we get an *infinite-dimensional system of ordinary differential equations [modal system] for* $N \to \infty$. Again, the system accounts for the full set of nonlinearities and any asymptotic assumptions on the smallness of the wave elevations or liquid depth are not required to derive it.

7.2 Nonlinear modal equations for sloshing

Following Faltinsen *et al.* (2000), we employ the Bateman–Luke principle and the modal approach to derive fully nonlinear modal equations for liquid sloshing. The original works by Miles (1976) and Lukovsky (1976) dealt with translatory motions of upright circular cylindrical tanks. Other analytical techniques that make it possible to account for angular tank motions were reported by Lukovsky (1990), Lukovsky and Timokha (1995), Faltinsen *et al.* (2000), and La Rocca *et al.* (2000). The following content is based on the original work by Faltinsen *et al.* (2000).

7.2.1 Modal representation of the free surface and velocity potential

The multimodal method uses a Fourier representation of the solution with time-dependent unknown coefficients that can be formally considered generalized coordinates. Because the sloshing problem deals with two functions describing the free-surface elevation and the velocity potential (i.e., Z and Φ), two Fourier representations should be postulated.

Following Faltinsen *et al.* (2000), let us present the free surface $\Sigma(t)$ by the equation $z = \zeta(x, y, t)$ (i.e., $Z(x, y, z, t) = z - \zeta(x, y, t) = 0$) in a tank-fixed coordinate system $Oxyz$. This setup is not always possible and limits the applicability of the modal method (see discussion in Section 7.2.3). Furthermore, the function $\zeta(x, y, t)$ is represented as the Fourier series

$$\zeta(x, y, t) = \sum_{i=1}^{\infty} \beta_i(t) f_i(x, y). \tag{7.1}$$

The modal basis $f_i(x, y)$ does *not* necessarily represent the natural sloshing modes, as assumed in Chapter 5 for linear sloshing. However, representation (7.1) must be mathematically correct; that is, it must be possible to express any instant free surface by this Fourier series. The time-dependent functions $\{\beta_i\}$ are interpreted as *generalized coordinates (modal functions)*. A requirement is that liquid mass conservation is satisfied; that is, $\int_{\Sigma_0} f_i(x, y) dx\, dy = 0$, where Σ_0 is the mean free surface.

A Fourier-type representation is also needed for the function $\Phi(x, y, z, t)$. Faltinsen *et al.* (2000) used a representation in the most general case that can be considered a generalization of modal presentation (5.20) for the linear case; that is,

$$\Phi(x, y, z, t) = \boldsymbol{v}_0(t) \cdot \boldsymbol{r} + \boldsymbol{\omega}(t) \cdot \boldsymbol{\Omega}(x, y, z, t) + \underbrace{\sum_{n=1}^{\infty} R_n(t)\, \varphi_n(x, y, z)}_{\varphi(x,y,z,t)}, \tag{7.2}$$

where the basis $\{\varphi_n\}$ is *not* necessarily associated with the natural modes and the Stokes–Joukowski potential, $\boldsymbol{\Omega} = (\Omega_1(x, y, z, t), \Omega_2(x, y, z, t), \Omega_3(x, y, z, t))^T$, is a function of both spatial coordinates and time, satisfying

$$\nabla^2 \boldsymbol{\Omega} = 0 \text{ in } Q(t),$$
$$\frac{\partial \Omega_1}{\partial n} = y n_3 - z n_2; \quad \frac{\partial \Omega_2}{\partial n} = z n_1 - x n_3,$$
$$\frac{\partial \Omega_3}{\partial n} = x n_2 - y n_1 \quad \text{on} \quad S(t) \cup \Sigma(t). \tag{7.3}$$

The functional set $\{\varphi_n(x, y, z)\}$ in representation (7.2) must be *complete* for any admissible liquid shapes $Q(t)$. This additional restriction is discussed in some detail in Section 7.2.3. When it comes to practice, the set $\{\varphi_n(x, y, z)\}$ typically coincides with natural sloshing modes, which means that Φ given by eq. (7.2) automatically satisfies the Laplace equation and the body boundary conditions. However, the free-surface conditions are not satisfied. The use of eq. (7.2) in the Lagrange variational formulation with natural modes $\{\varphi_n(x, y, z)\}$ provides the dynamic free-surface condition, which is natural for the Lagrange formulation, but the kinematic free-surface condition must be a priori satisfied, because it is not natural (see Section 2.5.2.3). Because both the kinematic and the dynamic free-surface conditions are natural in a Bateman–Luke variational formulation (see Section 2.5.3), the latter approach is therefore preferred.

After introducing representations (7.1) and (7.2), the Bateman–Luke variational formulation can be considered a tool to obtain a system of ordinary differential equations with respect to the time-dependent functions $\beta_i(t)$ and $R_n(t)$, which are the so-called nonlinear modal equations.

7.2.2 Modal system based on the Bateman–Luke formulation

By substituting eq. (7.2) in eq. (2.113), the pressure-integral Lagrangian, L, takes the following form:

$$L = -\rho \int_{Q(t)} \left[\frac{d^* \boldsymbol{v}_O}{dt} \cdot \boldsymbol{r} + \frac{\partial}{\partial t}(\boldsymbol{\omega} \cdot \boldsymbol{\Omega}) + \tfrac{1}{2}\nabla(\boldsymbol{\omega} \cdot \boldsymbol{\Omega}) \right. \cdot \nabla(\boldsymbol{\omega} \cdot \boldsymbol{\Omega}) - \boldsymbol{\omega} \cdot (\boldsymbol{r} \times \nabla(\boldsymbol{\omega} \cdot \boldsymbol{\Omega})) - \tfrac{1}{2}\boldsymbol{v}_O^2 - \boldsymbol{\omega} \cdot (\boldsymbol{r} \times \boldsymbol{v}_O) - \boldsymbol{\omega} \cdot (\boldsymbol{r} \times \nabla\varphi) \left. + \nabla(\boldsymbol{\omega} \cdot \boldsymbol{\Omega}) \cdot \nabla\varphi \right] \mathrm{d}Q + L_r, \tag{7.4}$$

where

$$L_r = -\rho \int_{Q(t)} \left[\frac{\partial \varphi}{\partial t} + \tfrac{1}{2}(\nabla\varphi)^2 + U_g \right] \mathrm{d}Q. \tag{7.5}$$

The integral expressions in eq. (7.4) need some simplification. It follows by vector differentiation, the Gauss theorem (see eq. (A.1)), and the Neumann boundary condition of eq. (7.3) that the two last integrand terms in the square brackets of eq. (7.4) cancel each other; that is,

$$\int_{Q(t)} [-(\boldsymbol{\omega} \times \boldsymbol{r}) \cdot \nabla\varphi + \nabla(\boldsymbol{\omega} \cdot \boldsymbol{\Omega}) \cdot \nabla\varphi]\, \mathrm{d}Q = \int_{S(t)+\Sigma(t)} \left(\frac{\partial(\boldsymbol{\omega} \cdot \boldsymbol{\Omega})}{\partial n} - (\boldsymbol{\omega} \times \boldsymbol{r}) \cdot \boldsymbol{n} \right) \varphi\, \mathrm{d}S = 0. \tag{7.6}$$

Other quantities that appear in the integral of eq. (7.4) may be written in a compact form by introducing the inertia tensor $\boldsymbol{J}^1 = \boldsymbol{J}^1(x, y, z, t)$ (see also the definition of the inertia tensor for linear sloshing in Chapter 5). The actual expressions for the symmetric inertia tensor are

$$J_{11}^1 = \rho \int_{Q(t)} \left(y\frac{\partial \Omega_1}{\partial z} - z\frac{\partial \Omega_1}{\partial y} \right) \mathrm{d}Q = \rho \int_{S(t)+\Sigma(t)} \Omega_1 \frac{\partial \Omega_1}{\partial n} \mathrm{d}S,$$

$$J_{22}^1 = \rho \int_{Q(t)} \left(z\frac{\partial \Omega_2}{\partial x} - x\frac{\partial \Omega_2}{\partial z} \right) \mathrm{d}Q = \rho \int_{S(t)+\Sigma(t)} \Omega_2 \frac{\partial \Omega_2}{\partial n} \mathrm{d}S,$$

$$J_{33}^1 = \rho \int_{Q(t)} \left(x\frac{\partial \Omega_3}{\partial y} - y\frac{\partial \Omega_3}{\partial x} \right) \mathrm{d}Q = \rho \int_{S(t)+\Sigma(t)} \Omega_3 \frac{\partial \Omega_3}{\partial n} \mathrm{d}S,$$

$$J_{12}^1 = J_{21}^1 = \rho \int_{Q(t)} \left(z\frac{\partial \Omega_1}{\partial x} - x\frac{\partial \Omega_1}{\partial z} \right) \mathrm{d}Q = \rho \int_{Q(t)} \left(y\frac{\partial \Omega_2}{\partial z} - z\frac{\partial \Omega_2}{\partial y} \right) \mathrm{d}Q = \rho \int_{S(t)+\Sigma(t)} \Omega_1 \frac{\partial \Omega_2}{\partial n} \mathrm{d}S = \rho \int_{S(t)+\Sigma(t)} \Omega_2 \frac{\partial \Omega_1}{\partial n} \mathrm{d}S,$$

$$J_{13}^1 = J_{31}^1 = \rho \int_{Q(t)} \left(x\frac{\partial \Omega_1}{\partial y} - y\frac{\partial \Omega_1}{\partial x} \right) \mathrm{d}Q = \rho \int_{Q(t)} \left(y\frac{\partial \Omega_3}{\partial z} - z\frac{\partial \Omega_3}{\partial y} \right) \mathrm{d}Q = \rho \int_{S(t)+\Sigma(t)} \Omega_1 \frac{\partial \Omega_3}{\partial n} \mathrm{d}S = \rho \int_{S(t)+\Sigma(t)} \Omega_3 \frac{\partial \Omega_1}{\partial n} \mathrm{d}S,$$

$$J_{23}^1 = J_{32}^1 = \rho \int_{Q(t)} \left(x\frac{\partial \Omega_2}{\partial y} - y\frac{\partial \Omega_2}{\partial x} \right) \mathrm{d}Q = \rho \int_{Q(t)} \left(z\frac{\partial \Omega_3}{\partial x} - x\frac{\partial \Omega_3}{\partial z} \right) \mathrm{d}Q = \rho \int_{S(t)+\Sigma(t)} \Omega_2 \frac{\partial \Omega_3}{\partial n} \mathrm{d}S = \rho \int_{S(t)+\Sigma(t)} \Omega_3 \frac{\partial \Omega_2}{\partial n} \mathrm{d}S. \tag{7.7}$$

The expressions are generalizations of linear case (5.44) that follow by replacing Q_0 by $Q(t)$, S_0 by $S(t)$, and Σ_0 by $\Sigma(t)$. Different expressions for J_{ij}^1 in eq. (7.7) are obtained by using the definition of the Stokes–Joukowski potential (eq. (7.3)), the Green first identity (see eq. (A.5)), and the Gauss theorem. The inertia tensor is associated with the following quadratic form:

$$
\begin{aligned}
&-\tfrac{1}{2}\omega_1^2 J_{11}^1 - \tfrac{1}{2}\omega_2^2 J_{22}^1 - \tfrac{1}{2}\omega_3^2 J_{33}^1 - \omega_1\omega_2 J_{12}^1 \\
&- \omega_1\omega_3 J_{13}^1 - \omega_2\omega_3 J_{23}^1 \\
&= -\tfrac{1}{2}\rho \int_{S(t)+\Sigma(t)} (\boldsymbol{\omega}\cdot\boldsymbol{\Omega})\left(\frac{\partial \boldsymbol{\Omega}}{\partial n}\cdot\boldsymbol{\omega}\right) \mathrm{d}S \\
&= \rho \int_{Q(t)} \left(\tfrac{1}{2}\nabla(\boldsymbol{\omega}\cdot\boldsymbol{\Omega})\cdot\nabla(\boldsymbol{\omega}\cdot\boldsymbol{\Omega})\right. \\
&\left. - \boldsymbol{\omega}\cdot(\boldsymbol{r}\times\nabla(\boldsymbol{\omega}\cdot\boldsymbol{\Omega}))\right) \mathrm{d}Q. \qquad (7.8)
\end{aligned}
$$

The equality between the surface and volume integral is a consequence of the Gauss theorem (A.1). The latter volume integral is part of eq. (7.4). After the simplifications have been made, Lagrangian (7.4) takes the form

$$
\begin{aligned}
L = &-\big[\dot v_{O1} l_1 + \dot v_{O2} l_2 + \dot v_{O3} l_3 + \dot\omega_1 l_{1\omega} + \dot\omega_2 l_{2\omega} \\
&+ \dot\omega_3 l_{3\omega} + \omega_1 l_{1\omega t} + \omega_2 l_{2\omega t} + \omega_3 l_{3\omega t} \\
&- \tfrac{1}{2}\left(\omega_1^2 J_{11}^1 + \omega_2^2 J_{22}^1 + \omega_3^2 J_{33}^1\right) - \omega_1\omega_2 J_{12}^1 \\
&- \omega_1\omega_3 J_{13}^1 - \omega_2\omega_3 J_{23}^1 - \tfrac{1}{2} M_l \left(v_{O1}^2 + v_{O2}^2 + v_{O3}^2\right) \\
&+ (\omega_2 v_{O3} - \omega_3 v_{O2}) l_1 + (\omega_3 v_{O1} - \omega_1 v_{O3}) l_2 \\
&+ (\omega_1 v_{O2} - \omega_2 v_{O1}) l_3\big] + L_r, \qquad (7.9)
\end{aligned}
$$

where M_l is the liquid mass and

$$
\begin{aligned}
l_{k\omega} &= \rho \int_{Q(t)} \Omega_k \,\mathrm{d}Q; \quad l_{k\omega t} = \rho \int_{Q(t)} \frac{\partial \Omega_k}{\partial t}\mathrm{d}Q, \\
l_1 &= \rho \int_{Q(t)} x \,\mathrm{d}Q; \quad l_2 = \rho \int_{Q(t)} y \,\mathrm{d}Q, \\
l_3 &= \rho \int_{Q(t)} z \,\mathrm{d}Q. \qquad (7.10)
\end{aligned}
$$

The vectors $\boldsymbol{l}(t) = (l_1, l_2, l_3)$, $\boldsymbol{l}_\omega(t) = (l_{1\omega}, l_{2\omega}, l_{3\omega})$ and $\boldsymbol{l}_{\omega t}(t) = (l_{1\omega t}, l_{2\omega t}, l_{3\omega t})$ are functions of $\{\beta_i\}$ and $\{\dot\beta_i\}$. Furthermore, component (7.5) should account for modal representation (7.2) and, therefore,

$$
\begin{aligned}
L_r &= -\rho \int_{Q(t)} \left[\sum_{n=1}^{\infty} \dot R_n \varphi_n \right. \\
&\left. + \tfrac{1}{2}\sum_{n,k=1}^{\infty} R_n R_k (\nabla\varphi_n\cdot\nabla\varphi_k) + U_g\right] \mathrm{d}Q \\
&= -\left[\sum_{n=1}^{\infty} D_n \dot R_n + \tfrac{1}{2}\sum_{n,k=1}^{\infty} D_{nk} R_n R_k \right. \\
&\left. - g_1 l_1 - g_2 l_2 - g_3 l_3 - m_l \boldsymbol{g}\cdot\boldsymbol{r}'_O\right], \qquad (7.11)
\end{aligned}
$$

where

$$
\begin{aligned}
D_n &= \rho \int_{Q(t)} \varphi_n \mathrm{d}Q, \\
D_{nk} &= D_{kn} = \rho \int_{Q(t)} (\nabla\varphi_n\cdot\nabla\varphi_k)\mathrm{d}Q. \qquad (7.12)
\end{aligned}
$$

The Bateman–Luke variational principle (eqs. (2.113) and (2.114)) considers L a function of two *independent* variables $\zeta(x, y, t)$ and $\Phi(x, y, z, t)$. After substitution of representations (7.1) and (7.2), L is expressed via eqs. (7.9) and (7.11) and becomes a function of the independent generalized coordinates $\{\beta_i\}$ and $\{R_n\}$. Note that the integrals in eqs. (7.7), (7.10), and (7.12) are over $Q(t)$ and $\Sigma(t)$, which are defined by eq. (7.1); that is, the integrals are functions of $\{\beta_i\}$. The independent variations of the action $W = \int_{t_1}^{t_2} L\mathrm{d}t$ by $\delta\zeta$ and $\delta\Phi$ should therefore be related to independent variations by $\delta\beta_i$ and δR_n, respectively, which gives the following variational equality:

$$
\begin{aligned}
\delta W = \int_{t_1}^{t_2} \Bigg[&\sum_n D_n \delta\dot R_n + \sum_{n,k} D_{nk} R_k \delta R_n \\
&+ \Bigg(\sum_n \dot R_n \frac{\partial D_n}{\partial \beta_i} + \omega_1 \frac{\partial l_{1\omega t}}{\partial \beta_i} + \omega_2 \frac{\partial l_{2\omega t}}{\partial \beta_i} + \omega_3 \frac{\partial l_{3\omega t}}{\partial \beta_i} \\
&+ \tfrac{1}{2}\sum_{n,k} R_n R_k \frac{\partial D_{nk}}{\partial \beta_i} + \dot\omega_1 \frac{\partial l_{1\omega}}{\partial \beta_i} + \dot\omega_2 \frac{\partial l_{2\omega}}{\partial \beta_i} \\
&+ \dot\omega_3 \frac{\partial l_{3\omega}}{\partial \beta_i} + (\dot v_{O1} - g_1 + \omega_2 v_{O3} - \omega_3 v_{O2}) \frac{\partial l_1}{\partial \beta_i} \\
&+ (\dot v_{O2} - g_2 + \omega_3 v_{O1} - \omega_1 v_{O3}) \frac{\partial l_2}{\partial \beta_i} \\
&+ (\dot v_{O3} - g_3 + \omega_1 v_{O2} - \omega_2 v_{O1}) \frac{\partial l_3}{\partial \beta_i} \\
&- \tfrac{1}{2}\omega_1^2 \frac{\partial J_{11}^1}{\partial \beta_i} - \tfrac{1}{2}\omega_2^2 \frac{\partial J_{22}^1}{\partial \beta_i} - \tfrac{1}{2}\omega_3^2 \frac{\partial J_{33}^1}{\partial \beta_i} \\
&- \omega_1\omega_2 \frac{\partial J_{12}^1}{\partial \beta_i} - \omega_1\omega_3 \frac{\partial J_{13}^1}{\partial \beta_i} - \omega_2\omega_3 \frac{\partial J_{23}^1}{\partial \beta_i} \Bigg) \delta\beta_i \\
&+ \left(\omega_1 \frac{\partial l_{1\omega t}}{\partial \dot\beta_i} + \omega_2 \frac{\partial l_{2\omega t}}{\partial \dot\beta_i} + \omega_3 \frac{\partial l_{3\omega t}}{\partial \dot\beta_i} \right) \delta\dot\beta_i \Bigg] \mathrm{d}t = 0, \\
i \geq 1. \qquad (7.13)
\end{aligned}
$$

The terms proportional to $\delta\dot R_n$ and $\delta\dot\beta_i$ in expression (7.13) can be integrated by parts, and using the fact that $\delta R_n(t_1) = \delta R_n(t_2) = \delta\beta_i(t_1) = \delta\beta_i(t_2) = 0$ leads to terms involving δR_n and $\delta\beta_i$ instead of $\delta\dot R_n$ and $\delta\dot\beta_i$. We obtain the following infinite system of nonlinear differential equations with respect to the *modal functions* $\{R_n(t)\}$ and

$\{\beta_i(t)\}$ by noting that the multipliers at the independent variations $\delta\beta_i$ and δR_n must be zero to get zero variations in eq. (7.13); that is,

$$\sum_i \frac{\partial D_n}{\partial \beta_i}\dot{\beta}_i - \sum_k R_k D_{nk} = 0, \quad n = 1, 2, \ldots, \tag{7.14}$$

$$\left.\begin{aligned}
&\sum_n \dot{R}_n \frac{\partial D_n}{\partial \beta_i} + \tfrac{1}{2}\sum_n\sum_k \frac{\partial D_{nk}}{\partial \beta_i} R_n R_k + \dot{\omega}_1 \frac{\partial l_{1\omega}}{\partial \beta_i} \\
&+ \dot{\omega}_2 \frac{\partial l_{2\omega}}{\partial \beta_i} + \dot{\omega}_3 \frac{\partial l_{3\omega}}{\partial \beta_i} + \omega_1 \frac{\partial l_{1\omega t}}{\partial \beta_i} + \omega_2 \frac{\partial l_{2\omega t}}{\partial \beta_i} \\
&+ \omega_3 \frac{\partial l_{3\omega t}}{\partial \beta_i} - \frac{d}{dt}\left(\omega_1 \frac{\partial l_{1\omega t}}{\partial \dot{\beta}_i} + \omega_2 \frac{\partial l_{2\omega t}}{\partial \dot{\beta}_i} + \omega_3 \frac{\partial l_{3\omega t}}{\partial \dot{\beta}_i}\right) \\
&+ (\dot{v}_{O1} - g_1 + \omega_2 v_{O3} - \omega_3 v_{O2}) \frac{\partial l_1}{\partial \beta_i} \\
&+ (\dot{v}_{O2} - g_2 + \omega_3 v_{O1} - \omega_1 v_{O3}) \frac{\partial l_2}{\partial \beta_i} \\
&+ (\dot{v}_{O3} - g_3 + \omega_1 v_{O2} - \omega_2 v_{O1}) \frac{\partial l_3}{\partial \beta_i} - \tfrac{1}{2}\omega_1^2 \frac{\partial J_{11}^1}{\partial \beta_i} \\
&- \tfrac{1}{2}\omega_2^2 \frac{\partial J_{22}^1}{\partial \beta_i} - \tfrac{1}{2}\omega_3^2 \frac{\partial J_{33}^1}{\partial \beta_i} - \omega_1\omega_2 \frac{\partial J_{12}^1}{\partial \beta_i} - \omega_1\omega_3 \frac{\partial J_{13}^1}{\partial \beta_i} \\
&- \omega_2\omega_3 \frac{\partial J_{23}^1}{\partial \beta_i} = 0, \quad i = 1, 2, \ldots.
\end{aligned}\right\} \tag{7.15}$$

If the tank has a vertical general cylindrical shape at the mean free surface Σ_0 in its upright position, the values of $\partial l_k/\partial \beta_i$ can be expressed as

$$\frac{\partial l_3}{\partial \beta_i} = \rho \int_{\Sigma_0} f_i^2\, dS\beta_i = \lambda_{3i}\beta_i; \quad \frac{\partial l_2}{\partial \beta_i} = \rho \int_{\Sigma_0} y f_i\, dS = \lambda_{2i},$$
$$\frac{\partial l_1}{\partial \beta_i} = \rho \int_{\Sigma_0} x f_i\, dS = \lambda_{1i}. \tag{7.16}$$

7.2.3 Advantages and limitations of the nonlinear modal method

The infinite-dimensional system (7.14) and (7.15) nonlinearly couples the generalized coordinates $\beta_i(t)$ and $R_n(t)$. It *does not use any assumptions about the smallness* of the surface wave amplitudes and liquid depth and can be used for modeling different liquid–structure interaction problems with irrotational flow of an incompressible liquid. The first subsystem, eq. (7.14), is responsible for the *kinematics*. Its linear analogy is $\dot{\beta}_n = \kappa_n R_n$ (see eq. (5.23)). Nonlinear subsystem (7.15) follows from the *dynamic boundary condition.* Its linear analogy has the form of eq. (5.24). Under certain circumstances the system is equivalent to the original free boundary problem (2.58), (2.63), (2.64), and (2.68) for arbitrary rigid-body motions.

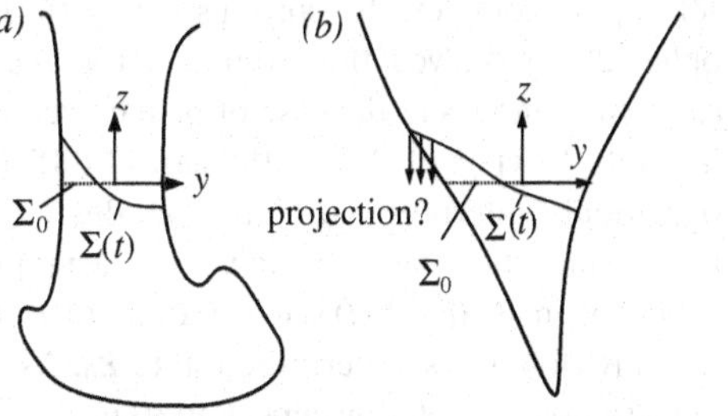

Figure 7.1. Geometric limitations of the modal method.

The following limitations of modal equations (7.14) and (7.15) are mostly associated with those for the modal representations (7.1) and (7.2).

Geometrical: Defining the free surface in the form $z = \zeta(x, y, t)$ means that tank roof impact and overturning waves do not occur. Furthermore, the tank bottom has to be wet. To account for contact of the free surface and the tank roof (roof impact), the modal method needs to be combined with a slamming model (see Chapter 11). Furthermore, representation (7.1) requires that $\{f_i(x, y)\}$ is a Fourier basis, whose domain by definition is time-independent and, generally speaking, coincides with Σ_0. Figure 7.1 illustrates cases when the modal representation is valid as long as the free surface is limited to the vicinity of the free surface where the time-dependent wetted tank surface is parallel to the z-axis (Figure 7.1(a)) and may fail (Figure 7.1(b)). A non-Cartesian parameterization is needed to extend the modal method to tanks with nonvertical walls in Figure 7.1(b). These results were reported by Gavrilyuk *et al.* (2005).

Functional: Modal representations (7.1) and (7.2) need functional sets $\{f_i\}$ and $\{\varphi_n\}$. These representations must admit higher derivatives and be complete on the mean free surface Σ_0 and in the whole tank domain, respectively. It may sometimes be difficult to find the functional basis $\{\varphi_n\}$. Practical choice would consist of linear natural modes; however, the natural modes are theoretically defined only in the unperturbed hydrostatic domain Q_0. Vertical cylindrical tanks with either a circular or a rectangular base have, for instance, natural modes that are defined outside of Q_0. Furthermore, $\{f_i = \varphi_i|_{\Sigma_0}\}$ of these natural modes constitute a complete functional set to represent the free surface. However, it does not mean that the set $\{\varphi_n\}$ is complete for any instant $Q(t)$. Timokha (2002) showed the latter fact for analytical natural modes in the case of planar rectangular tank. In view of this mathematical difficulty, one should interpret the natural modes $\{\varphi_n\}$ as an "asymptotic" basis; namely, the assumption should be made that $\Sigma(t)$ is to some extent asymptotically close to its unperturbed state Σ_0. Assuming asymptotic relationships between $\beta_i(t)$ for nonsmall liquid depths truncates the modal system to a finite-dimensional form. Such an asymptotic truncation may, for instance, be based on the so-called Moiseev asymptotics (see Faltinsen, 1974; Faltinsen *et al.*, 2000; Lukovsky, 1990; Miles, 1984a, 1984b; Moiseev, 1958; Ockendon *et al.*, 1996), whose applicability has been justified in the case of finite liquid depths. Different asymptotic relationships of "shallow-liquid sloshing" as well as corresponding solution methods were reported by Chester (1968), Chester and Bones (1968), Ockendon and Ockendon (1973, 2001), and Ockendon *et al.* (1986). Faltinsen and Timokha (2002) used a Boussinesq-type asymptotic relationship and developed the multimodal method for small liquid depths in a two-dimensional rectangular tank. We return to their results in Chapter 8. An important fact is that asymptotically truncated systems may use natural modes, because the procedure needs only the completeness of $\{\varphi_n\}$ in the unperturbed liquid domain. Furthermore, when the ratio of liquid depth to tank breadth, h/l, becomes small, the small parameter h/l must be linked to the order of magnitude of $\beta_i(t)$ and $R_i(t)$. We see in Section 8.6 that this link keeps the asymptotic modal system infinite-dimensional. A truncation of this system can be based on the assumptions that higher modes are highly damped. The corresponding damping terms should then be included, as we explained in Section 6.3.2.

Numerical: To find natural modes that can be used in the general modal system, a numerical method may be used (Gavrilyuk *et al.*, 2005, 2006). However, the accuracy of the multimodal method is reasonable only when these approximate modes satisfy the zero-Neumann condition on the time-dependent wetted tank surface in the free-surface zone and when the time-dependent wetted tank surface is parallel to the z-axis. Otherwise, a numerical error on the mentioned part of the walls implies an inflow/outflow through $S(t)$. Another numerical problem appears due to a naive truncation of modal system (7.14) and (7.15). First, the truncated system of nonlinear differential equations is very stiff and, therefore, certain numerical methods for solving the equations are numerically unstable unless the time step taken is extremely small (Hairer & Wanner, 1996). It has proved difficult to formulate a precise definition of stiffness, but the main idea is that the equation includes some terms that can lead to rapid variation in the solution. As shown in the papers by Perko (1969), La Rocca *et al.* (2000, 2002), and Shankar and Kidambi (2002), who used truncated modal systems, the stiffness is caused by very high harmonics, which are generated by nonlinearities. Special care is required in this case for the damping terms incorporated into the modal system to provide a robust simulation. This unrealistically stiff system is caused by neglecting a series of physical phenomena and asymptotic intermodal relations between the generalized coordinates. Sections 8.8.1 and 9.2.6.2 report on the numerical stiffness problem for multidimensional modal systems in modeling transient waves.

An important remark is that, in some cases (e.g., for finite liquid depth), system (7.14) may be considered a linear system of algebraic equations with respect to $\{R_k\}$:

$$\sum_k D_{nk}(\beta_i)R_k = \frac{d}{dt}D_n(\beta_i), \quad n = 1, 2, \ldots.$$

By inverting this system, we can find $\{R_n\}$ as a function of $\{\beta_i\}$ and $\{\dot{\beta}_i\}$. After substituting R_n into eq. (7.15), one gets a system of second-order nonlinear differential equations with respect to $\{\beta_i\}$. Therefore, eqs. (7.14) and (7.15) imply a set of coupled nonlinear oscillators, whose displacements are described by $\{\beta_i\}$.

7.3 Modal technique for hydrodynamic forces and moments

In the linear sloshing case (Chapter 5), we showed formulas that give the hydrodynamic force and moment as a function of the forcing and modal functions. For the nonlinear case, those formulas were first derived by Lukovsky (1990). The following text gives new, simpler derivations of these formulas. The importance of these formulas for internal hydrodynamic loads lies in their joint use with solutions of modal system (7.14) and (7.15) to describe both the resulting forces and moment on the tank and the coupling with rigid-body tank dynamics. Furthermore, these formulas can be adopted by other methods that do not use the modal representation, because the formulas involve the tank's translatory and angular velocities responsible for the forcing loads, $\boldsymbol{v}_O$ and $\boldsymbol{\omega}$, together with the center of the mass vector of the liquid, $\boldsymbol{r}_{lC}(t)$, and its derivatives. The mass center is easily reexpressed by $\beta_i(t)$, but it can be redefined in another way for other approximate methods.

7.3.1 Hydrodynamic force

7.3.1.1 General case

To derive Lukovsky's formula for the hydrodynamic force, we use eq. (2.38), which expresses the force in terms of the hydrodynamic momentum of the tank liquid. The momentum is defined as

$$\boldsymbol{M}(t) = \rho \int_{Q(t)} \boldsymbol{v}\, dQ. \tag{7.17}$$

Our derivation is partly similar to that for linear force in Section 5.3.2.1. The first step is to derive an expression for $\boldsymbol{M}$, which accounts for the fact that the absolute velocity of the liquid particles is $\boldsymbol{v} = \boldsymbol{v}_O + \boldsymbol{\omega} \times \boldsymbol{r} + \boldsymbol{v}_r$, where $\boldsymbol{v}_r$ is the relative velocity of the liquid (see Chapter 2); that is,

$$\boldsymbol{M} = \rho Vol\, \boldsymbol{v}_O + \rho \boldsymbol{\omega} \times \int_{Q(t)} \boldsymbol{r}\, dQ + \rho \int_{Q(t)} \boldsymbol{v}_r\, dQ, \tag{7.18}$$

where Vol is the liquid volume. The last integral in eq. (7.18) can be transformed by using the Gauss theorem (A.1) and Reynolds transport theorem (A.2):

$$\rho \int_{Q(t)} \boldsymbol{v}_r\, dQ = \rho \int_{\Sigma(t)} \boldsymbol{r}(\boldsymbol{v}_r \cdot \boldsymbol{n})\, dS + \rho \int_{S(t)} \boldsymbol{r} \underbrace{(\boldsymbol{v}_r \cdot \boldsymbol{n})}_{0} dS$$

$$= \rho \frac{d^*}{dt} \int_{Q(t)} \boldsymbol{r}\, dQ. \tag{7.19}$$

We have used the fact that $\nabla \cdot (\boldsymbol{v}_r \boldsymbol{r}) = \boldsymbol{v}_r$ for an incompressible liquid and that the normal relative velocity on the tank surface is zero ($\boldsymbol{v}_r \cdot \boldsymbol{n} = 0$ on $S(t)$). Furthermore, we have also used the *-time differentiation rule (2.50) and the fact that the integrand $\boldsymbol{r}$ is independent of time. We introduce the vector

$$\boldsymbol{r}_{lC}(t) = \left(\boldsymbol{e}_1 \int_{Q(t)} x\, dQ + \boldsymbol{e}_2 \int_{Q(t)} y\, dQ + \boldsymbol{e}_3 \int_{Q(t)} z\, dQ \right) \Big/ \underbrace{\int_{Q(t)} dQ}_{Vol}, \tag{7.20}$$

which gives the position of the *mass center in the body-fixed coordinate system.* Therefore, eq. (7.18) can be rewritten in the form

$$\boldsymbol{M} = M_l \left(\boldsymbol{v}_O + \boldsymbol{\omega} \times \boldsymbol{r}_{lC} + \frac{d^* \boldsymbol{r}_{lC}}{dt} \right), \tag{7.21}$$

where $M_l = \rho Vol$ is the liquid mass.

Finally, remembering time-differentiation rule (2.50) and the fact that $\dot{\boldsymbol{\omega}} = d^*\boldsymbol{\omega}/dt$, substitution of eq. (7.21) into formula (2.38) and gathering similar terms gives

$$\boldsymbol{F}(t) = M_l \boldsymbol{g} - M_l \left[\frac{d^* \boldsymbol{v}_O}{dt} + \boldsymbol{\omega} \times \boldsymbol{v}_O + \boldsymbol{\omega} \times (\boldsymbol{\omega} \times \boldsymbol{r}_{lC}) + \dot{\boldsymbol{\omega}} \times \boldsymbol{r}_{lC} + 2\boldsymbol{\omega} \times \frac{d^* \boldsymbol{r}_{lC}}{dt} + \frac{d^{*2} \boldsymbol{r}_{lC}}{dt^2} \right], \tag{7.22}$$

where $\boldsymbol{\omega} \times (\boldsymbol{\omega} \times \boldsymbol{r}_{lC})$ is the centripetal acceleration and $2\boldsymbol{\omega} \times d^*\boldsymbol{r}_{lC}/dt$ is the Coriolis acceleration. Equation (7.22) is the Lukovsky (1990) formula for the force. Lukovsky derived it only for potential flow of an incompressible liquid. Because we have not introduced the velocity potential in the derivation, *readers can see that we have shown the validity of this formula for arbitrary incompressible ideal liquid including the case of rotational flows.* Modal representations (7.1) and (7.2) are also not used, which means that Lukovsky's formula (7.22) is valid for any inviscid sloshing in tanks. It makes it possible to compute the resulting force even for the case of overturning waves, and so on within the framework of the inviscid theory. However, eq. (7.22) needs modification if we want to include the effect of surface tension or gas cavities with pressures that differ from p_0.

As will be shown, the presence of *-time derivatives in eq. (7.22) does not make formula (7.22) complicated in practice but rather simplifies computations for multimodal methods. The reason is the existence of very simple formulas for the mass center in a body-fixed coordinate system that leads to sums of Fourier coefficients in the representation of the free surface (7.1).

7.3.1.2 Completely filled closed tank

A particular case occurs when the closed tank is completely filled with a liquid. In this case, $Q(t)$ in eq. (7.20) coincides with the whole tank cavity; therefore,

$$\boldsymbol{r}_{lC} = \boldsymbol{r}_{lC_0} = \text{const}, \quad \frac{d^*\boldsymbol{r}_{lC}}{dt} = 0.$$

That is, the liquid mass center is motionless relative to the tank and coincides with the hydrostatic liquid mass center $\boldsymbol{r}_{lC_0}$ defined by eq. (5.32). Formula (7.22) can then be simplified to the form

$$\boldsymbol{F}^{\text{filled}}(t) = M_l\boldsymbol{g} - M_l\left[\frac{d^*\boldsymbol{v}_O}{dt} + \boldsymbol{\omega}\times\boldsymbol{v}_O + \boldsymbol{\omega}\times(\boldsymbol{\omega}\times\boldsymbol{r}_{lC_0}) + \dot{\boldsymbol{\omega}}\times\boldsymbol{r}_{lC_0}\right]. \quad (7.23)$$

Neglecting the nonlinear term in $\boldsymbol{\omega}$ and $\boldsymbol{v}_O$ transforms eq. (7.23) to formula (5.37) derived for linear dynamics.

7.3.2 Moment

Let us now give a new derivation of Lukovsky's formula for the hydrodynamic moment relative to the origin O. First, we present a proof of the formula that computes the moment by means of the angular momentum and its time derivative. Similar to the analysis in Section 7.3.1, this formula is valid for any ideal incompressible liquid. Furthermore, the use of this formula makes it possible to obtain a new expression in terms of β_i, R_n, and external forcing for the case of potential flow of an incompressible fluid.

7.3.2.1 Hydrodynamic moment as a function of the angular momentum

Consider the angular momentum

$$\boldsymbol{G}_O(t) = \rho\int_{Q(t)} \boldsymbol{r}\times\boldsymbol{v}\,\mathrm{d}Q = \rho\int_{Q'(t)} (\boldsymbol{r}' - \boldsymbol{r}_O)\times\boldsymbol{v}\,\mathrm{d}Q' \quad (7.24)$$

where $\boldsymbol{r} = \boldsymbol{r}' - \boldsymbol{r}_O$ and $Q'(t)$ is the liquid domain in an inertial coordinate system $O'x'y'z'$ ($\mathrm{d}Q' = \mathrm{d}x'\mathrm{d}y'\mathrm{d}z'$). Using Reynolds transport theorem (A.2), the time derivative of $\boldsymbol{G}_O$ may be computed in an inertial coordinate system as

$$\dot{\boldsymbol{G}}_O = \frac{d^*\boldsymbol{G}_O}{dt} + \boldsymbol{\omega}\times\boldsymbol{G}_O = \rho\int_{Q'(t)} (\boldsymbol{r}' - \boldsymbol{r}_O(t)) \times \left.\frac{\partial\boldsymbol{v}}{\partial t}\right|_{\text{in } O'x'y'z'} \mathrm{d}Q' - \underbrace{\rho\dot{\boldsymbol{r}}_O(t)\times\int_{Q'(t)}\boldsymbol{v}\,\mathrm{d}Q'}_{\boldsymbol{v}_O\times\boldsymbol{M}} + \rho\int_{S'(t)+\Sigma'(t)} (\boldsymbol{r}' - \boldsymbol{r}_O(t))\times\boldsymbol{v}\,U_n\,\mathrm{d}S', \quad (7.25)$$

where we can recall $\boldsymbol{r} = \boldsymbol{r}' - \boldsymbol{r}_O(t)$ and definition (7.17) to get alternative expression for the second integral term. Furthermore, we rewrite the first integral on the right-hand side of eq. (7.25) by using the Euler equations for an incompressible liquid (which follows from eq. (2.35) by setting the viscous stress term τ_{ij} equal to zero) and adopting the convective acceleration term as in eq. (2.36). Changing to integration by $Q(t)$ gives

$$\rho\int_{Q(t)} \boldsymbol{r}\times\left.\frac{\partial\boldsymbol{v}}{\partial t}\right|_{\text{in } O'x'y'z'}\mathrm{d}Q = -\rho\int_{Q(t)}\boldsymbol{r}\times\nabla\cdot(\boldsymbol{v}\boldsymbol{v})\mathrm{d}Q - \underbrace{\rho\int_{Q(t)}\boldsymbol{r}\times\nabla(p-p_0)\mathrm{d}Q}_{\boldsymbol{M}_O(t)} + \underbrace{\rho\int_{Q(t)}\boldsymbol{r}\times\boldsymbol{g}\mathrm{d}Q}_{M_l\,\boldsymbol{r}_{lC}\times\boldsymbol{g}}.$$

Here, the integral $-\rho\int_{Q(t)}\boldsymbol{r}\times\nabla\cdot(\boldsymbol{v}\boldsymbol{v})\mathrm{d}Q$ can be shown by using a formula similar to eq. (A.6) to be equal to $-\rho\int_{S(t)+\Sigma(t)}\boldsymbol{r}\times\boldsymbol{v}\,(\boldsymbol{v}\cdot\boldsymbol{n})\,\mathrm{d}S$. Because $\boldsymbol{v}\cdot\boldsymbol{n} = U_n$ on $S(t)+\Sigma(t)$ and $\boldsymbol{r} = \boldsymbol{r}' - \boldsymbol{r}_O(t)$, it follows that the first integral is equal to minus the last integral on the right-hand side of eq. (7.25). In summary, eq. (7.25) gives the expression

$$\boldsymbol{M}_O = M_l\,\boldsymbol{r}_{lC}\times\boldsymbol{g} - \frac{d^*\boldsymbol{G}_O}{dt} - \boldsymbol{\omega}\times\boldsymbol{G}_O - \boldsymbol{v}_O\times\boldsymbol{M}. \quad (7.26)$$

Formula (7.26) is true for any inviscid incompressible liquid, including the case of rotational flows. Using this formula with the previously derived formula for the force (eq. (7.22)), one can compute the hydrodynamic moment relative to another body-fixed point P as

$$\boldsymbol{M}_P(t) = \boldsymbol{r}_{PO}\times\boldsymbol{F}(t) + \boldsymbol{M}_O(t), \quad (7.27)$$

where $\boldsymbol{r}_{PO}$ is the radius-vector of O relative to P.

Using the definition of the inertia tensor (eqs. (7.7) and (7.8)), we can always represent the velocity field as the sum of $\boldsymbol{v}_0 + \nabla(\boldsymbol{\omega}\cdot\boldsymbol{\Omega})$ and $\boldsymbol{v}_1$:

$$\boldsymbol{v} = \boldsymbol{v}_0 + \nabla(\boldsymbol{\omega}\cdot\boldsymbol{\Omega}) + \boldsymbol{v}_1. \quad (7.28)$$

This representation is similar to eq. (7.2). The term $\boldsymbol{v}_0\cdot\boldsymbol{r} + \boldsymbol{\omega}\cdot\boldsymbol{\Omega}$ satisfies the Laplace equation and describes the potential flow of an incompressible liquid in a completely filled tank. Substitution of eq. (7.28) into eq. (7.24) gives

$$\boldsymbol{G}_O = M_l\,\boldsymbol{r}_{lC}\times\boldsymbol{v}_O + \boldsymbol{\omega}\cdot\boldsymbol{J}^1 + \rho\int_{Q(t)}\boldsymbol{r}\times\boldsymbol{v}_1\mathrm{d}Q. \quad (7.29)$$

Henceforth, we consider the case of potential flow (i.e., $\boldsymbol{v} = \nabla\Phi$). This case allows for simplifications in formula (7.29).

7.3.2.2 Potential flow

When inserting eq. (7.2) into eq. (7.25) ($\boldsymbol{v} = \nabla\Phi$), using definitions (7.3), the Gauss theorem and the Green first identity, and eq. (A.6), one gets

$$\begin{aligned}
\boldsymbol{G}_O(t) &= \rho\int_{Q(t)} \boldsymbol{r}\times\boldsymbol{v}_O\,\mathrm{d}Q + \rho\int_{Q(t)} \boldsymbol{r}\times\nabla(\boldsymbol{\omega}\cdot\boldsymbol{\Omega})\mathrm{d}Q \\
&\quad + \rho\int_{Q(t)} \boldsymbol{r}\times\nabla\varphi\,\mathrm{d}Q \\
&= M_l\,\boldsymbol{r}_{lC}\times\boldsymbol{v}_O + \rho\int_{S(t)+\Sigma(t)} (\boldsymbol{\omega}\cdot\boldsymbol{\Omega})\,(\boldsymbol{r}\times\boldsymbol{n})\mathrm{d}S \\
&\quad + \rho\int_{S(t)+\Sigma(t)} (\boldsymbol{r}\times\boldsymbol{n})\varphi\,\mathrm{d}S \\
&= M_l\,\boldsymbol{r}_{lC}\times\boldsymbol{v}_O + \rho\int_{S(t)+\Sigma(t)} (\boldsymbol{\omega}\cdot\boldsymbol{\Omega})\frac{\partial\boldsymbol{\Omega}}{\partial n}\mathrm{d}S \\
&\quad + \rho\int_{S(t)+\Sigma(t)} \frac{\partial\boldsymbol{\Omega}}{\partial n}\varphi\,\mathrm{d}S. \qquad (7.30)
\end{aligned}$$

Remembering the definition of the inertia tensor in eq. (7.8) and modifying the last integral as

$$\begin{aligned}
\rho\int_{S(t)+\Sigma(t)} \frac{\partial\boldsymbol{\Omega}}{\partial n}\varphi\,\mathrm{d}S &= \rho\int_{S(t)+\Sigma(t)} \boldsymbol{\Omega}\frac{\partial\varphi}{\partial n}\,\mathrm{d}S \\
&= -\rho\int_{\Sigma(t)} \boldsymbol{\Omega}\frac{\partial\xi/\partial t}{|\nabla\xi|}\mathrm{d}S \\
&= \frac{d^*}{dt}\rho\int_{Q(t)} \boldsymbol{\Omega}\,\mathrm{d}Q - \rho\int_{Q(t)} \frac{\partial\boldsymbol{\Omega}}{\partial t}\mathrm{d}Q,
\end{aligned}$$

we get

$$\boldsymbol{G}_O = M_l\,\boldsymbol{r}_{lC}\times\boldsymbol{v}_O + \boldsymbol{\omega}\cdot\boldsymbol{J}^1 + \frac{d^*\boldsymbol{l}_\omega}{dt} - \boldsymbol{l}_{\omega t}, \quad (7.31)$$

where vectors $\boldsymbol{l}_\omega = (l_{1\omega}, l_{2\omega}, l_{3\omega})$ and $\boldsymbol{l}_{\omega t} = (l_{1\omega t}, l_{2\omega t}, l_{3\omega t})$ are defined by eq. (7.10) and depend on integrals involving $\boldsymbol{\Omega}(x, y, z, t)$.

Under the previously given definitions, using eqs. (7.31) and (7.21) gives the formula by Lukovsky (1990):

$$\begin{aligned}
\boldsymbol{M}_O &= M_l\boldsymbol{r}_{lC}\times\left(\boldsymbol{g} - \boldsymbol{\omega}\times\boldsymbol{v}_O - \frac{d^*\boldsymbol{v}_O}{dt}\right) \\
&\quad - \boldsymbol{J}^1\cdot\dot{\boldsymbol{\omega}} - \frac{d^*\boldsymbol{J}^1}{dt}\cdot\boldsymbol{\omega} - \boldsymbol{\omega}\times(\boldsymbol{J}^1\cdot\boldsymbol{\omega}) \\
&\quad - \frac{d^{*2}\boldsymbol{l}_\omega}{dt^2} + \frac{d^*\boldsymbol{l}_{\omega t}}{dt} - \boldsymbol{\omega}\times\left(\frac{d^*\boldsymbol{l}_\omega}{dt} - \boldsymbol{l}_{\omega t}\right). \quad (7.32)
\end{aligned}$$

Once again, the formula is not restricted to the modal representation and the structure of the free surface. It is *valid for arbitrary motions of a perfect incompressible liquid with irrotational flows.* The free surface may touch the roof, and breaking or overturning waves are possible as long as the wave does not impact on the underlying liquid.

7.3.2.3 Completely filled closed tank

When the tank is completely filled, the expression for the hydrodynamic moment becomes simpler because $d^*\boldsymbol{l}_\omega/dt = \boldsymbol{l}_{\omega t} = 0$ and the inertial tensor is independent of time:

$$\begin{aligned}
\boldsymbol{M}_O^{\text{filled}}(t) &= M_l\boldsymbol{r}_{lC_0}\times\left(\boldsymbol{g} - \boldsymbol{\omega}\times\boldsymbol{v}_O - \frac{d^*\boldsymbol{v}_O}{dt}\right) \\
&\quad - \boldsymbol{J}^1\cdot\dot{\boldsymbol{\omega}} - \boldsymbol{\omega}\times(\boldsymbol{J}^1\cdot\boldsymbol{\omega}). \qquad (7.33)
\end{aligned}$$

This formula is consistent with the results of eq. (5.48) within the framework of linear dynamics when neglecting nonlinear terms in $\boldsymbol{\omega}$ and $\boldsymbol{v}_O$.

7.4 Limitations of the modal theory and Lukovsky's formulas due to damping

Lukovsky's formulas are used extensively in connection with the multimodal method. The reason is that modal representation (7.1) provides a fast and convenient way to find the instant position of the liquid mass center by means of eq. (7.20). However, a conflict may appear between the facts that the Lukovsky formulas are not valid for viscous liquid flows and that in Chapters 6, 8, and 9 we introduce viscous damping terms in the multimodal method to reach steady-state conditions. This approach requires special attention when using viscous damping terms in the multimodal method with the forthcoming use of the modal solution in Lukovsky's formulas. We can distinguish several variants of using the damping terms. In the first variant (see Sections 8.4 and 8.5), we employ the damping term only to reach the steady-state numerical solution, and final calculations are made with very small damping ratios. Because our goal is to obtain these final calculations and the effect of damping is negligible, such a procedure is consistent with the assumptions of the modal theory and Lukovsky's formulas.

However, three-dimensional simulations by an adaptive modal method in Section 9.2.6 do not assume vanishing damping ratios. For the three-dimensional case, the number of activated modes increases and we cannot reduce the damping ratios to zero, especially for higher modes, which are strongly damped due to boundary-layer flow and breaking waves. Faltinsen *et al.* (2005b) noted that one can assume very small damping for simulations of nearly steady-state conditions, but only for lower dominant modes. The higher modes must be considered highly damped to get an adequate prediction. The argument for using the damping terms is that the adaptive modal theory is of an asymptotic nature and the higher modes contribute less than the lowest dominant modes. Moreover, we compare numerical results with experiments to confirm this asymptotic argument.

Finally, the third variant is for shallow and intermediate liquid depths (Sections 8.6 and 8.7). The corresponding modal theory also needs nonvanishing damping terms both to reach steady-state solutions and to obtain an accurate description of transients. All the modes are of the same order of magnitude in this modal theory and, therefore, we cannot argue the presence of nonvanishing damping terms by employing the asymptotic reasons as we did for simulations of three-dimensional sloshing. A supporting argument for using the modal method and the Lukovsky formulas is comparison with experimental data. Section 8.8.1.2 gives such comparisons for steady-state solutions and boundary-layer damping (Section 6.3). Very good agreement is established for the experimental case by Chester and Bones (1965) for wave elevation. In these experiments, low-viscosity liquid (i.e., "freshwater") was used. For fresh water, Faltinsen and Timokha (2002) also conducted simulations with and without damping terms and compared them with experimental data for transients. Although the presence of damping improves the agreement, this improvement is not very important; this modal theory with the corresponding damping ratios included with related Lukovsky's formulas are applicable for low-viscosity liquids. The situation may change for more viscous liquids. In Figure 8.34(b), we showed that shear forces on the tank bottom and the tank wall can be important to obtain better predictions of hydrodynamic force. When we consider the modal theory and include viscous damping terms due to boundary-layer flow, the latter damping can be related to the inflow/outflow from the boundary layers along the tank surface (see Section 6.3). We also include viscous damping outside the boundary layer; however, this effect is secondary. The Lukovsky formulas ought to consider the effect of a porous tank surface to be consistent with our viscous damping modeling, which would lead to additional components in the expressions for the force and moment by the Lukovsky formulas. Generally speaking, the order of magnitude of these additional force and moment terms are the same as that for the contribution from viscous shear stresses along the tank surface. However, the latter contribution is generally secondary relative to the pressure loads following from the Lukovsky formulas. An exception is a shallow-liquid case with small excitation amplitude, which is discussed in Section 8.8.1.2.

7.5 Summary

The linear modal analysis from Chapter 5 can be extended to the nonlinear case. A convenient way to derive a nonlinear modal system consists of using a Lagrangian-type variational technique. Following Faltinsen *et al.* (2000), this chapter employs the Bateman–Luke variational formulation, which uses the pressure integral as the Lagrangian. For that statement, the free-surface position and the velocity potential are considered two independent variables. Furthermore, we adopt the modal representation of the free surface (7.1) and generalize the modal solution for the velocity potential from eq. (5.20) to eq. (7.2). Substitution into the variational statement makes time-dependent functions β_i and R_n the unknowns and variables of the action. Calculation of the extremum of the action gives the modal system (7.14) and (7.15) (i.e., the infinite-dimensional system of ordinary differential equations that couples β_i and R_n). Discussion of the applicability of this system to describe sloshing was presented in Section 7.2.3.

Bearing in mind the coupled dynamics of the tank and the contained liquid, which requires the resulting hydrodynamic force and moment, in this chapter we presented an independent derivation of the formulas by Lukovsky (1990). These formulas admit simple expressions for the force and moment via the generalized coordinates β_i and R_n.

7.6 Exercises

7.6.1 Modal equations for the beam problem

Part 1. Consider the approximate solution of the beam problem from Section 2.5.2.2:

$$\mathrm{w}(x,t)=\sum_{j=1}^{N}\beta_j(t)\phi_j(x)=\boldsymbol{\beta}^T\boldsymbol{\phi};$$
$$\boldsymbol{\beta}(t)=(\beta_1(t),\ldots,\beta_N(t))^T,$$
$$\boldsymbol{\phi}(x)=(\phi_1(x),\ldots,\phi_N(x))^T, \qquad (7.34)$$

where $\{\phi_j\}$ is a basis system of functions.

(a) What is the Lagrangian in terms of representation (7.34)?

Answer:

$$L=\tfrac{1}{2}\dot{\boldsymbol{\beta}}^T M\dot{\boldsymbol{\beta}}-\tfrac{1}{2}\boldsymbol{\beta}^T K\boldsymbol{\beta}+\boldsymbol{\beta}^T\boldsymbol{F}+\boldsymbol{\beta}^T\boldsymbol{F}_g, \qquad (7.35)$$

where $\boldsymbol{\phi}$ yields the following matrices and vectors:

$$M=\int_0^l m\boldsymbol{\phi}\boldsymbol{\phi}^T\,\mathrm{d}x,\quad K=\int_0^l \mathrm{EI}\left[\frac{\partial^2\boldsymbol{\phi}}{\partial x^2}\right]\left[\frac{\partial^2\boldsymbol{\phi}}{\partial x^2}\right]^T\mathrm{d}x,$$
$$\boldsymbol{F}=\int_0^l f\boldsymbol{\phi}\,\mathrm{d}x,\quad \boldsymbol{F}_g=g\int_0^l m\boldsymbol{\phi}\,\mathrm{d}x. \qquad (7.36)$$

(b) What is the action W due to Lagrangian (7.35) and the Euler–Lagrange equations (2.89) in terms of representation (7.34)?

Answer:

$$\frac{d}{dt}\left(\frac{\partial L}{\partial\dot{\boldsymbol{\beta}}}\right)-\frac{\partial L}{\partial\boldsymbol{\beta}}=M\ddot{\boldsymbol{\beta}}+K\boldsymbol{\beta}-\boldsymbol{F}-\boldsymbol{F}_g=0. \qquad (7.37)$$

(*Hint*: Use the following formulas for differentiation involving the vectors $\boldsymbol{a}$ and $\boldsymbol{b}$ and the matrix A:

$$\frac{\partial}{\partial\boldsymbol{a}}\left(\boldsymbol{a}^T A\boldsymbol{a}\right)=\left(A+A^T\right)\boldsymbol{a},\quad \frac{\partial}{\partial\boldsymbol{a}}\left(\boldsymbol{a}^T\boldsymbol{b}\right)=\boldsymbol{b}.)$$

Euler–Lagrange equations (7.37) with eq. (7.36) are the *modal equations* for the beam problem. Modifications of the Lagrange variational formulations for beam problems are possible for specific systems of basis functions.

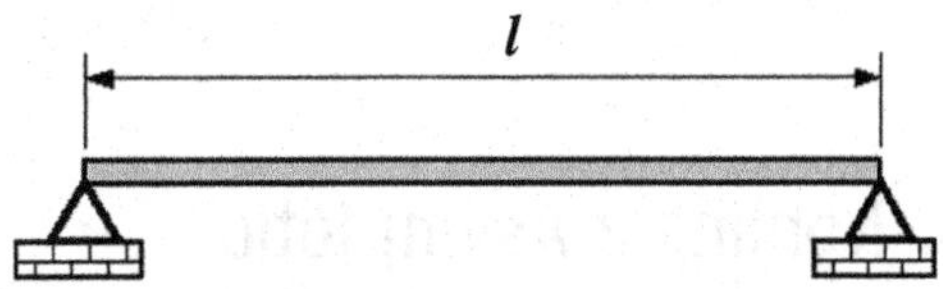

Figure 7.2. Vibration of a simply supported beam.

Part 2. Consider a simply supported beam with constant m, E, and I as in Figure 7.2 and an incomplete modal solution

$$\mathrm{w}(x,t)=\beta_0(t)+\beta_1(t)\sin(\pi x/l)+\beta_2(t)\cos(\pi x/l)+\beta_3(t)\sin(2\pi x/l). \qquad (7.38)$$

Show that the support conditions require $\beta_0(t)=\beta_2(t)=0$ and use results of Part 1 to derive the system of linear modal equations for $\beta_1(t)$ and $\beta_3(t)$. Furthermore, find the eigenvalues from this system that correspond to natural frequencies.

Answer: The modal system takes the form

$$\frac{ml}{2}\begin{bmatrix}1&0\\0&1\end{bmatrix}\begin{pmatrix}\ddot{\beta}_1\\\ddot{\beta}_3\end{pmatrix}+\frac{\mathrm{EI}\pi^4}{2l^3}\begin{bmatrix}1&0\\0&16\end{bmatrix}\begin{pmatrix}\beta_1\\\beta_3\end{pmatrix}=mgl\begin{pmatrix}\frac{2}{\pi}\\0\end{pmatrix}. \qquad (7.39)$$

The eigenfrequencies are

$$\sigma_1=\frac{\pi^2}{l^2}\sqrt{\frac{\mathrm{EI}}{m}},\quad \sigma_2=4\frac{\pi^2}{l^2}\sqrt{\frac{\mathrm{EI}}{m}}. \qquad (7.40)$$

7.6.2 Linear modal equations for sloshing

Part 1. Use the Lagrange formulation from Section 2.5.2.3 and the modal methods developed in Section 7.2 to derive the linear modal equations (5.25).

Part 2. Derive a linear formula for force and moment using nonlinear expresions (7.22) and (7.32). Compare the results with those in Section 5.3.2.

8 Nonlinear Asymptotic Theories and Experiments for a Two-Dimensional Rectangular Tank

8.1 Introduction

This chapter discusses, by means of theory and experiments, how the flow in a two-dimensional rigid rectangular tank behaves as a function of the liquid depth when the forcing frequency is close to the lowest natural sloshing frequency. A main focus is on an analytical nonlinear modal theory, where the free-surface elevation and the velocity potential are represented in terms of a Fourier series and the linear natural modes, respectively. The theoretical background is given in Chapter 7. The modal functions $\beta_i(t)$ and $R_n(t)$ are generalized coordinates for the free surface and the velocity potential, respectively. Linear modal theory assumes $\beta_i(t)$ and $R_n(t)$ to be small and proportional to the forcing (see Chapter 5). However, the latter assumption fails for resonant excitations. Some modal functions $\beta_i(t)$ and $R_n(t)$ then become of lower order relative to the forcing. How such modal functions are ordered in terms of importance and how to construct corresponding asymptotic modal theories is presented in this chapter.

The representation of free-surface elevation in terms of a Fourier series implies that:

- the free surface is a single-valued function of the lateral tank coordinates along the mean free surface (i.e., an overturning wave, as shown at the tank wall in Figure 8.1, cannot be included).
- the free surface intersects a tank wall perpendicularly (i.e., the method cannot predict run-up along a tank wall in the form of a thin liquid layer). The Fourier series representation of the free surface does not converge at the tank walls, but good convergence is obtained away from the walls, except for the case of very steep waves.

We should also recall theoretical assumptions such as irrotational flow of an incompressible inviscid liquid in a rectangular tank with infinite tank roof height. Important questions include the following:

- How restrictive is it to assume infinite tank roof height?
- Why can the liquid be considered incompressible?
- How valid is it to neglect vorticity and viscosity?

We address such questions in the following text.

Obviously, tank roof impact easily occurs for large filling ratios. Our focus, therefore, is on small and moderate filling ratios. However, wave elevation is important in assessing the possibility of slamming against the tank roof. Furthermore, the multimodal method can be included as part of a tank roof impact load analysis as long as we do not analyze impact due to run-up along the tank walls. What is needed from the multimodal method analysis is the free-surface shape and velocity at the impact position.

If no impact occurs, we can argue as follows about the importance of liquid compressibility. A characteristic time scale for acoustic (compressibility) effects is the highest natural period for standing acoustic waves, which in a two-dimensional rectangular tank with finite or lower depths is $T_1^{ac} = 2l/c_0$, where l is the tank breadth and c_0 is the speed of sound (see Exercise 10.8.1). If no mixture occurs between gas and liquid, representative values for c_0 are listed in Tables 11.1 and 11.2. These values range from 975 to 1,774 m s^{-1}. Because T_1^{ac} is clearly smaller than the first-mode natural sloshing period for a two-dimensional rectangular tank presented in Figure 1.2, we may rule out the effect of liquid compressibility.

Furthermore, how valid is it to assume irrotational flow? First of all, a boundary layer exists along the tank surface where viscosity and therefore vorticity matter. Boundary-layer flow causes a damping of sloshing and affects forces on the tank. Both effects can be incorporated into the multimodal method. In Section 6.3.2 we showed how to incorporate boundary-layer damping in the case of laminar flow. The procedure can be generalized to turbulent boundary-layer flow, which is needed in full-scale conditions. The

Figure 8.1. Experimental study of forced horizontal oscillation of a rectangular tank illustrating an overturning wave near a tank wall.

boundary-layer damping ratios are small for the lowest natural modes. However, their presence helps a potential flow model to reach steady-state flow conditions. The contribution of viscous forces is small relative to potential flow pressure forces and is generally neglected. An exception is shown in Section 8.8.1.2 for a shallow-liquid case with small excitation amplitude and liquid response.

Vorticity associated with breaking waves is a more severe limitation. In coastal engineering, breaking waves are divided into different categories (i.e., spilling, plunging, collapsing, and surging waves; Sorensen, 1993). Plunging waves are overturning waves. Spilling waves are described as follows: "turbulence and foam first appear at the wave crest and spread down the front face as the wave propagates forward." Surging waves are associated with the front of the wave as it surges up the beach slope and returns. A collapsing wave "is an intermediate form between the plunging and surging form." Turbulence occurs for a much lower Reynolds number in a free flow with vorticity away from a solid boundary such as for breaking waves than it does in a boundary-layer flow, which is a general consequence of the linear stability analysis of steady laminar parallel flow (Schlichting, 1979). The results for cross-flow past a circular cylinder illustrate this fact about transition to turbulent flow. Free flow where vorticity is associated with the wake becomes turbulent at a significantly lower Reynolds number than the boundary-layer flow (see Section 6.4.3).

We have no means to calculate the damping effect due to turbulent energy dissipation in breaking waves. We can only suggest in an indirect way that wave-breaking damping can be much larger than boundary-layer damping by examining how long it takes in experiments to reach steady-state conditions. If the liquid depth is finite with a ratio of liquid depth to tank breadth larger than ≈ 0.4, wave breaking does not easily occur and it takes a very long time for steady-state conditions to be established. However, this is not the case for wave breaking at smaller depths.

An additional effect of plunging waves is that they disturb the potential flow conditions as they impact on the underlying liquid. If one looks at the problem from a potential flow model, the effect occurs as the potential flow model gets new initial conditions. This effect is believed to be important in some cases with subharmonic oscillations at critical depth, which are described in Section 8.5. The smoothed particle hydrodynamics (SPH) method (see Section 10.6) can describe well cases with wave breaking where steady-state oscillations are interrupted. The considered SPH method solves the Euler equations and does not consider turbulence (i.e., it cannot describe the turbulent energy dissipation). Solving the Euler equations implies that vorticity can be generated. No rational ways exist to introduce new "initial" conditions into the multimodal method due to the plunging–breaking wave's sudden disturbance of the flow.

The error in neglecting viscous effects can be illustrated through the steady-state experimental results presented by Abramson *et al.* (1974) for forced horizontal harmonic oscillations of a liquefied natural gas (LNG) tank with two-dimensional flow. The LNG tank has upper and lower chamfers (see definitions in Figure 1.17). Our focus is first on the horizontal force amplitude per unit length in a shallow-liquid condition with ratio of liquid depth to tank breadth of $h/l = 0.12$. Reginol oil, water, and different mixtures of glycerol and water were used as liquids (properties are given in Table 11.1). Experimental results are presented in Figure 8.2 for ratio of sway amplitude to tank breadth of $\eta_{2a}/l =$ 0.01 and 0.1. Because the force is nondimensional with respect to the liquid density, the nondimensional potential flow results are the same for all liquids. When $\eta_{2a}/l = 0.01$, the forces are very small and the nondimensional measured values depend on viscosity (see Figure 8.2(a)). When $\eta_{2a}/l = 0.1$ (Figure 8.2(b)), viscosity has

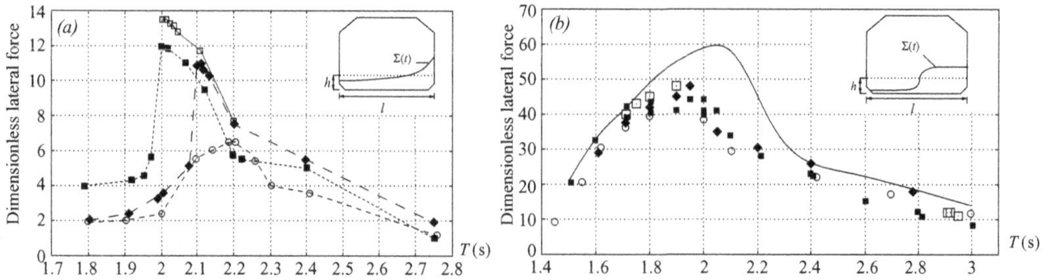

Figure 8.2. Dimensionless lateral hydrodynamic force amplitude $1000F_{2a}/(\rho g l^2 B)$ with small fill depth $h/l = 0.12$, where B is the tank length perpendicular to the forcing direction. The prismatic tank model is illustrated in Figure 4.22(a). Experimental data by Abramson *et al.* (1974) show the influence of viscosity: ■, freshwater; ○, reginol–oil; □, glycerol–water 63%, ◆, glycerol–water 85% (see Table 11.1). Lines are drawn through the measurement points. (a) Viscosity matters for the smaller excitation amplitude $\eta_{2a}/l = 0.01$, whereas the larger forcing amplitude $\eta_{2a}/l = 0.1$ in (b) shows experimental results that are close for different liquids. (b) Theoretical prediction of lateral force by the Boussinesq-type multimodal theory developed by Faltinsen and Timokha (2002) for intermediate and shallow depths indicated by a solid line. The theoretical results account for linear boundary-layer damping in freshwater. Theoretical predictions for part (a) by the same theory with different linear viscous damping rates corresponding to the tested liquids are presented in Figure 8.34.

a small effect on the measured nondimensional forces.

Because no wave breaking is believed to occur for the smaller excitation amplitude, viscosity is a dominant effect only in the boundary layer, which may be assumed to be laminar in the model test conditions. Wave breaking occurs for some forcing frequencies at the higher excitation amplitude $\eta_{2a}/l = 0.1$. The viscous energy dissipation of a free turbulent flow is much smaller than the turbulent energy dissipation. When it comes to converting the results to full scale, the change of the boundary-layer flow from laminar to turbulent presents a clear scale effect. However, the boundary-layer effect is not believed to be a major global effect, as already discussed. When it comes to scaling of free turbulent flow, an analogy can be made to the change of laminar to turbulent flow along a flat plate. If turbulent flow is established, the change with scale relative to the change from laminar to turbulent flow is less severe.

We now examine the finite-liquid-depth case where the liquid depth h is $0.4l$. The forced oscillation amplitude, $\eta_{2a} = 0.01l$, is discussed first. Figure 8.3(a) shows numerical and

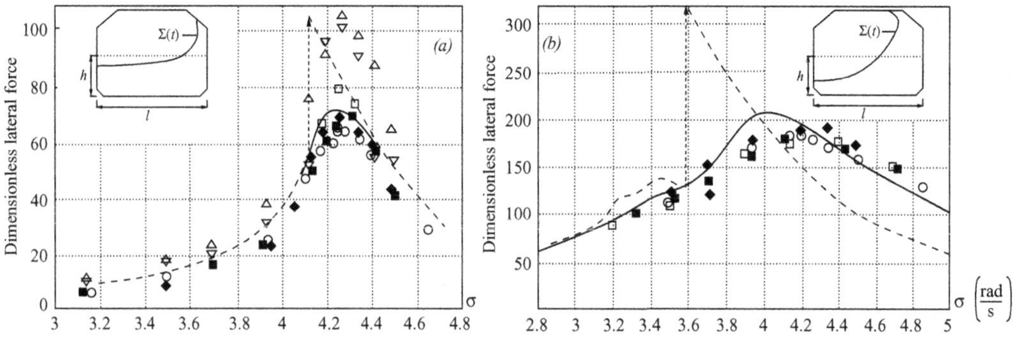

Figure 8.3. The same as Figure 8.2, but for $h = 0.4l$. Dashed lines represent results of the adaptive multidominant multimodal theory by Faltinsen and Timokha (2001) with no roof impact damping. Solid lines are from the same theory extended to account for roof impact as described by Faltinsen and Rognebakke (2000) and Rognebakke and Faltinsen (2000). Numerical results of the viscous CFD code FLOW-3D obtained for freshwater with different internal parameters of the code: Δ for "ALPHA=0.5, EPSDJ=0.01" and ∇ for "ALPHA=1.0, EPSDJ=0.01" (Solaas, 1995).

experimental predictions of steady-state maximum horizontal force amplitude as a function of the forced oscillation frequency, σ. The lowest natural frequency, σ_1, is 4.36 rad/s. The different liquids used in the experiments have a small influence on the nondimensional force. Two different types of theoretical curves are based on the adaptive multimodal method by Faltinsen and Timokha (2001) reported in Section 8.4. Dashed lines assume an infinite tank roof height and solid lines account for tank roof damping in the way described by Faltinsen and Rognebakke (2000) and Rognebakke and Faltinsen (2000). The results show that tank roof impact matters for the global sloshing behavior. The results based on Faltinsen and Timokha (2001) that account for tank roof impact are in good agreement with experiments. The simulations with infinite tank roof height give jumps between different solution branches at certain frequencies. These jumps disappear when tank-roof impact damping is introduced.

The tank-roof damping model is based on energy considerations. The kinetic and potential energy in the jet created by the tank roof impact is assumed to be dissipated after the jet has overturned and impacted on the underlying free surface (see Section 11.3.2). The impact-induced horizontal force is not included. The effect of the two lower chamfers submerged in the liquid is neglected; the error in doing so is small (see Section 4.6.3.2). Results from two simulations by the viscous computational fluid dynamics (CFD) code FLOW-3D published by Solaas (1995) are also presented in Figure 8.3(a). FLOW-3D is a commercial code based on the finite difference method. A volume-of-fluid method is used to capture the free surface. The grid size was $\Delta x = 0.0276$ m and $\Delta z = 0.02782$ m corresponding to 50×37 elements. The effect of the lower chamfers is taken into account. The two simulations correspond to ALPHA = 1.0 and 0.5, respectively. In both cases EPSADJ = 0.01. ALPHA and EPSADJ are parameters associated with the numerical techniques used in FLOW-3D. The presented results show a clear influence of ALPHA. Results obtained with the default value ALPHA = 1.0 are farthest away from the experimental results. The agreement between FLOW-3D and the experiments is fair.

Figure 8.3(b) shows comparisons between theory and experiments for the larger sway excitation amplitude $\eta_{2a} = 0.1l$ and represents a realistic design excitation for ships. The maximum horizontal hydrodynamic force on the tank is about half the weight of the liquid. Only analytically based sloshing models are used. The calculations by Faltinsen and Timokha (2001) for infinite tank roof height showed that the free-surface elevation reached the tank roof and the upper chamfers in the frequency range 3.11 rad/s $< \sigma <$ 6.23 rad/s. When 3.46 rad/s $< \sigma <$ 5.12 rad/s, the calculations by Faltinsen and Rognebakke (2000) and Rognebakke and Faltinsen (2000) showed a significant effect of tank roof damping. Accounting for tank roof impact was not straightforward for $\eta_{2a} = 0.1l$. Very violent motions occurred initially. An artificial damping coefficient was therefore introduced in the transient phase of the numerical simulations. When steady-state oscillations were achieved, the tank roof damping model was switched on. Figure 8.3(b) demonstrates good agreement between theory and experiments.

The results from using one frequency can be generalized to include a large number of harmonic *forcing* components. We should recall that the *forcing* frequency is assumed to be in the vicinity of the lowest natural frequency (i.e., the main attention is paid to a range $\breve{\sigma} < \sigma < \widehat{\sigma}$ with $\sigma/\sigma_1 \approx 1$). When a spectrum of several forcing frequencies $\sigma^{(i)}$ $(i = 1, \ldots, N)$ exists, the generalization requires the same range $\breve{\sigma} < \sigma^{(i)} < \widehat{\sigma}$ for all frequencies in the spectrum, namely, we must not allow any of the harmonic forcing frequency components to be in the vicinity of higher natural sloshing frequencies.

The main advantages of the multimodal method are as follows. Because the coefficients in the equation system are analytically derived and only ordinary differential equations for the generalized coordinates must be numerically time-integrated, numerical errors are generally negligible. The consequence is a very small CPU time relative to that for CFD methods. Coupling between ship motion and sloshing can, therefore, be performed in a stochastic seaway with very long time series to properly evaluate probability density functions for sloshing-related response variables. Furthermore, physical stability analysis can easily be performed and jump phenomena between different steady-state solution branches can be analytically detected. When breaking waves do not occur, the multimodal method can be used as a "verification" of CFD methods. We use quotation marks due to the fact that most CFD methods, except for the boundary element method (BEM), use governing equations other than the multimodal method.

Considering a rectangular tank is somewhat restrictive for practical tank shapes. However, we showed in Section 4.6.3.2 that lower chamfers have a negligible effect on natural modes. This fact is also implicitly confirmed by the comparisons with model tests for the LNG tank in Figures 8.2 and 8.3. Furthermore, we can study the wave loads on interior structures, such as a pump tower, by using calculations with no interior structures as ambient flow in a Morison equation–type of analysis.

The framework of the presented asymptotic theories is not restricted to rectangular tanks. We can consider any two-dimensional tank shape with vertical walls in the free-surface zone and infinite tank roof height. However, we must rely on numerical methods such as BEM to find the parameters (e.g., coefficients in the system of ordinary differential equations for the generalized coordinates of the free-surface elevation).

We start with the infinite-dimensional system (7.14) and (7.15), which includes the most general nonlinear differential equations coupling the time-dependent modal functions β_i and R_n. The equations assume no asymptotic ordering between the modal functions. If the orders of $\beta_i(t)$ and $R_n(t)$ are the same as the nondimensional forcing terms, neglecting terms of higher order than the forcing leads to linear modal equations (5.25). Because the linear sloshing model fails for resonant sloshing, we need to introduce a different ordering for the modal functions. Different ways of doing this depend on the liquid depth and the magnitude of the forcing amplitude. We start by applying the asymptotic technique by Moiseev (1958) and present the steady-state (periodic) resonant solutions of two-dimensional sloshing in a rectangular tank with a finite liquid depth. The theory assumes that the translatory forcing amplitude divided by the tank breadth and the angular forcing amplitude are small and of $O(\varepsilon)$. The primary excited mode (i.e., the lowest mode in our case) has the order $\varepsilon^{1/3}$. This order is the same as that for the Duffing equation, which describes the weakly nonlinear behavior of a spring–mass system (see eq. (2.95)). A brief background on the steady-state solution of the Duffing equation and its stability is given in the beginning of Section 8.2. Furthermore, we apply intermodal asymptotics for the sloshing problem in a rectangular tank and give details of the modal system derivations by Faltinsen *et al.* (2000), which adopt the asymptotics for functions β_i and R_n and reduce general modal equations (7.14) and (7.15) to the corresponding modal equations by neglecting the $o(\varepsilon)$-terms. This approach results in a finite-dimensional asymptotic nonlinear modal subsystem. Although the Duffing (Moiseev) asymptotics characterizes the steady-state (periodic) solutions, comparisons with model tests show that the asymptotic modal system is able to handle time-domain problems with nonlinear transient waves. Because we operate with different asymptotic multimodal methods, we denote the described method a single-dominant multimodal method.

The studied nonlinear steady-state solutions are completely analytical and contain terms oscillating with forcing frequency σ and superharmonics $n\sigma$. Subharmonic steady-state solutions are also possible, which we confirm via experiments and time-domain solutions in the vicinity of the critical depth $h = 0.3368\ldots \cdot l$. The steady-state response changes from hard-spring to soft-spring response at the critical depth, where "hard-spring" and "soft-spring" refer to the behavior of solutions of the Duffing equation. The latter theoretical value of the critical depth is a consequence of the Moiseev-type multimodal method with one dominant mode. Experiments by Fultz (1962) estimated the critical depth at $h = 0.28l$. The discrepancy is actually caused by amplification of higher modes occurring in resonant conditions at the critical depth (see Section 8.5). Mathematically, according to Hermann and Timokha (2008), the critical depth is a monotonically decreasing function of the ratio of forcing amplitude to breadth, ε, and the Moiseev-based theoretical value $h = 0.3368\ldots \cdot l$ is simply the limit value as $\varepsilon \to 0$. Hermann and Timokha (2008) used model 2 from Section 8.4.3 to show that the theoretical critical-depth value becomes consistent with the experimental value $h = 0.28l$ for Fultz's experimental forcing amplitudes.

Modifications of the asymptotic modal system needed for critical, intermediate, and shallow depths and for increasing forcing amplitude are partly based on resonant amplification of higher modes. The amplification occurs due to the secondary (internal) resonance; secondary resonance was discussed in Section 4.3.1.1. Occurrence of the secondary resonance makes the contribution of some of the higher modes comparable with the primary excited lowest mode, which changes the asymptotic ordering. Details on the asymptotic modal systems dealing with the secondary resonance are given in Section 8.3.4. When the liquid depth is in the lower intermediate and shallow depth ranges, the ratio of liquid depth to tank breadth has to be introduced as a small parameter. Because a similarity exists between this method and the Boussinesq-type of methods used to describe water waves in coastal engineering, this particular version of the multimodal method is denoted the *Boussinesq-type multimodal method.* Such multimodal method performs very well for resonant sloshing in intermediate depths with ratios of liquid depth to tank breadth of $\overline{h} = h/l \lesssim 0.2$. The agreement with experiments is also very good when the depth is not too small (i.e., $\overline{h} \gtrsim 0.05$) and the forcing amplitude is not too large. For instance, several jumps between steady-state solution branches in shallow water are well predicted. Even though the theory assumes the free-surface elevation to be of the same order as the liquid depth, the theory shows convergence problems in describing hydraulic jumps due to the many Fourier components needed to adequately describe the free-surface shape. The steady-state hydraulic-jump theory by Verhagen and van Wijngaarden (1965) is therefore presented to discuss these phenomena. The theory provides bounds for when hydraulic jumps occur that agree qualitatively with experiments. Calculated hydrodynamic forces and moment by the hydraulic-jump theory agree reasonably well with experiments.

The major part of the chapter is devoted to lateral and angular forcing. However, parametric resonance and Faraday waves due to vertical excitation are also discussed.

8.2 Steady-state resonant solutions and their stability for a Duffing-like mechanical system

8.2.1 Nonlinear spring-mass system, resonant solution, and its stability

8.2.1.1 Steady-state solution

Let us consider the nonlinear spring–mass system described by differential equation (2.95). We assume an excitation force that oscillates harmonically with the circular frequency $\sigma = 2\pi/T$ (i.e., $F_E(t) = F_{Ea}\cos(\sigma t) = -M\sigma^2\eta_a\cos(\sigma t)$), where M is the point mass and η_a is a parameter.

The following asymptotic analysis requires all the *geometric parameters to be scaled* by a characteristic length of the spring, L. Equation (2.95) can then be written as

$$\ddot{\beta} + \sigma_0^2(\beta + K\beta^3) = -(\eta_a/L)\sigma^2\cos(\sigma t), \quad (8.1)$$

where $\sigma_0 = \sqrt{\kappa/M}$ is the linear eigenfrequency and $K = \kappa/(\sigma_0^2 M)$ is a given coefficient associated with the cubic nonlinear spring term. The position function (generalized coordinate) $\beta(t)$ is also scaled by L (i.e., $\beta := \beta/L$).

The resonant $T = 2\pi/\sigma$-periodic solutions of eq. (8.1) can be found in many textbooks (see, e.g., Bogoljubov & Mitropolski, 1961). Our focus is on an asymptotic solution in terms of the nondimensional forcing amplitude. The resonant condition implies that $\sigma^2 - \sigma_0^2 \to 0$. The asymptotic technique introduces the following two small nondimensional parameters:

$$\varepsilon = \eta_a/L \ll 1, \quad \delta = \left(\sigma_0^2 - \sigma^2\right)/\sigma^2 \ll 1, \quad (8.2)$$

which means that we should find an asymptotic T-periodic solution of eq. (8.1) with two a priori independent small parameters, ε and δ, introduced by condition (8.2). The procedure involves a *matching* of these parameters. Furthermore, a *secularity (necessary solvability) condition* is used to get a finite amplitude when $\sigma = \sigma_0$ (i.e., for the case when the linear analysis predicts infinite response). In the analysis, K is assumed $O(1)$.

First, we should recall (see Section 5.3.3) that the linear approximation of the periodic solution is

$$\beta(t) = -(\varepsilon/\delta)\cos(\sigma t). \quad (8.3)$$

Equation (8.3) introduces the ratio between ε and δ and, bearing in mind the asymptotic limit $\varepsilon \to 0$ for the constructed resonant steady-state solution, we remark that this solution is only valid when $\varepsilon/\delta \ll 1$. Furthermore, solution (8.3) is infinite when $\sigma = \sigma_0$. A matching procedure for ε and δ is based on the following arguments. We do not want the amplitude ε/δ in eq. (8.3) to be $O(1)$, because then we cannot follow a perturbation procedure in the derivation. However, ε/δ must be of lower order than ε to reflect a strong resonant amplification. Therefore, we assume that the amplitude ε/δ in eq. (8.3) is $O(\varepsilon^\alpha)$, where the nondimensional parameter α is between 0 and 1; that is,

$$O(\varepsilon) \ll O(\varepsilon/\delta) = O(\varepsilon^\alpha) \ll 1, \quad 0 < \alpha < 1. \tag{8.4}$$

Equation (8.4) relates δ and α. The solution is then expressed as a power series in ε^α:

$$\beta(t) = \varepsilon^\alpha \beta^{(1)}(t) + \varepsilon^{2\alpha}\beta^{(2)}(t) + \varepsilon^{3\alpha}\beta^{(3)}(t) + \cdots. \tag{8.5}$$

The next steps are to substitute eq. (8.5) into eq. (8.1) and collect terms of equal asymptotic order. The lowest-order equation becomes $\ddot{\beta}^{(1)} + \sigma_0^2\beta^{(1)} = 0$. The solution is $\beta^{(1)} = \mathrm{A}\cos(\sigma_0 t + \theta)$, which does not satisfy the fact that the solution has to be steady-state with a period $2\pi/\sigma$. To achieve this we have to backtrack and rearrange eq. (8.1) by subtracting and adding the term $\sigma^2\beta$ on the left-hand side of the equation. Equation (8.1) can be reorganized as

$$\begin{aligned}
&\varepsilon^\alpha(\ddot{\beta}^{(1)} + \sigma^2\beta^{(1)}) + \varepsilon^{2\alpha}\left(\ddot{\beta}^{(2)} + \sigma^2\beta^{(2)}\right) \\
&\quad + \varepsilon^{3\alpha}\left(\ddot{\beta}^{(3)} + \sigma^2\beta^{(3)} + \sigma_0^2 K\left(\beta^{(1)}\right)^3\right) \\
&\quad + \left(\sigma_0^2 - \sigma^2\right)\varepsilon^\alpha\beta^{(1)} + o(\varepsilon^{3\alpha}) = -\varepsilon\sigma^2\cos\sigma t.
\end{aligned} \tag{8.6}$$

The lowest-order equation becomes $\ddot{\beta}^{(1)} + \sigma^2\beta^{(1)} = 0$. The solution can be expressed as $\mathrm{A}\cos(\sigma t + \theta)$, where the amplitude parameter A and the phase angle θ are unknown at this stage. Our next step is to determine α. The value of α is not arbitrary; our choice must lead to meaningful results. What we do is assume the two last terms on the left-hand side of eq. (8.6) have the same order as the term on the right-hand side. The order of $\sigma^2 - \sigma_0^2$ or δ follows by using eq. (8.4); that is, $\delta = O(\varepsilon^{1-\alpha})$, which means that $\alpha = \frac{1}{3}$, $\delta = O(\varepsilon^{2/3})$ and leads to

$$\begin{aligned}
\ddot{\beta}^{(2)} + \sigma^2\beta^{(2)} &= 0, \\
\ddot{\beta}^{(3)} + \sigma^2\beta^{(3)} &= -\sigma^2\cos\sigma t - \tfrac{3}{4}\sigma_0^2 K\mathrm{A}^3\cos(\sigma t + \theta) \\
&\quad - \Lambda\sigma^2\mathrm{A}\cos(\sigma t + \theta) \\
&\quad - \tfrac{1}{4}\sigma_0^2 K\mathrm{A}^3\cos 3(\sigma t + \theta),
\end{aligned} \tag{8.7}$$

where

$$\Lambda = \left(\sigma_0^2 - \sigma^2\right) / \left(\varepsilon^{2/3}\sigma^2\right) \tag{8.8}$$

is the *detuning factor*. In formulating eq. (8.7) we have used the fact that $\cos^3(\sigma t + \theta) = \frac{1}{4}[\cos 3(\sigma t + \theta) + 3\cos(\sigma t + \theta)]$. Because the differential equation for $\beta^{(2)}$ given by eq. (8.7) is the same as that for $\beta^{(1)}$, it is unnecessary to use a separate solution for $\beta^{(2)}$.

Let us then examine the differential equation for $\beta^{(3)}$. The right-hand side contains terms oscillating with the lowest harmonics σ. If the sum of these terms is nonzero, they would lead to an infinite response of $\beta^{(3)}$ when $\sigma = \sigma_0$. This response is avoided via the secularity condition. We simply require that the sum of all terms oscillating with frequency σ on the right-hand side of eq. (8.7) is zero, which leads to θ being either 0 or π. We change the term involving σ_0^2 on the right-hand side of the differential equation for $\beta^{(3)}$ before applying the secularity condition. Because $\sigma^2 = \sigma_0^2 + O(\varepsilon^{2/3})$, the replacement of σ_0^2 with σ^2 causes a higher-order negligible effect. When we choose $\theta = \pi$, the solvability (secularity) condition is

$$\tfrac{3}{4}K\mathrm{A}^3 + \Lambda\mathrm{A} - 1 = 0. \tag{8.9}$$

If $\theta = 0$, the condition would differ by the sign of the last term on the left-hand-side; therefore, A for $\theta = \pi$ is the opposite of A obtained with $\theta = 0$. The consequence is that $\mathrm{A}\cos(\sigma t + \theta)$ is the same for both $\theta = 0$ and π. In contrast to the linear case, eq. (8.9) causes a finite solution at resonance; that is, when $\sigma = \sigma_0$, we get $\mathrm{A} = (\frac{3}{4}K)^{-1/3}$.

For any fixed forcing frequency σ, or detuning factor Λ, cubic algebraic equation (8.9) has either one or three real roots. These roots with $K = O(1)$ are presented by the two branches in Figure 8.4 for $K < 0$ (soft spring) and $K > 0$ (hard spring), respectively. For each case, the two branches meet at infinity. When Λ coincides with the abscissa of the turning point T of the lower solution branch, two of the three real roots are the same.

Figure 8.4 also shows the so-called backbone (dashed line) that separates the branches. The backbone represents nontrivial solutions of eq. (8.9) without the forcing term:

$$\Lambda\mathrm{A} + \tfrac{3}{4}K\mathrm{A}^3 = 0. \tag{8.10}$$

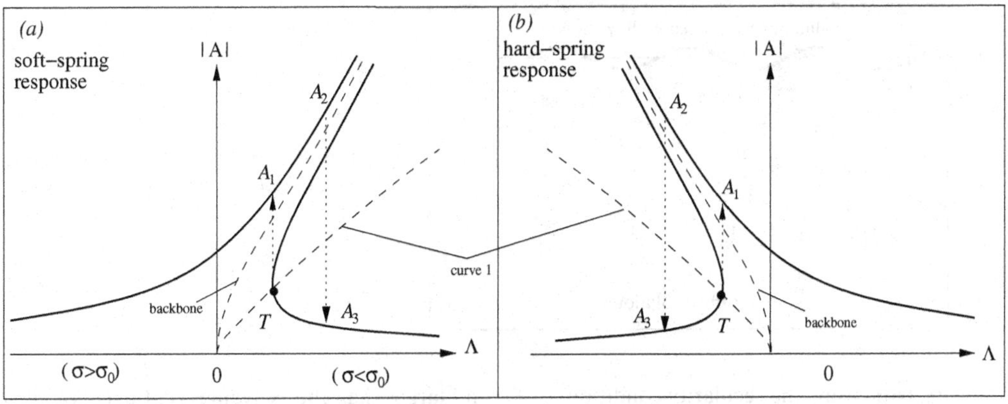

Figure 8.4. The modulus of the amplitude parameter | A | versus the detuning parameter Λ following from secularity condition (8.9) for the Duffing equation (or eq. (8.32) for the sloshing problem): (a) soft-spring behavior ($K < 0$ or for depths larger than the critical depth for a Moiseev-type method; i.e., $m_1^{(0)} < 0$) and (b) hard-spring behavior ($K > 0$ or for depths smaller than the critical depth for a Moiseev-type method; i.e., $m_1^{(0)} > 0$). The subbranch in the area between "backbone" and "curve 1" corresponds to unstable steady-state (periodic) solutions.

Each point on the branches of Figure 8.4 corresponds to a steady-state solution, which may be stable or unstable. The reason why the solution is not infinite at the resonant frequency $\sigma = \sigma_0$ is associated with the presence of the cubic quantity $K\beta^3$ in eq. (8.1). The ordering of the terms is due to the cubic nonlinear spring law and K is not zero or is small.

8.2.1.2 Stability

We now discuss the stability of the steady-state solutions of the Duffing equation. The procedure starts with a perturbation of the steady-state solution and introduces the sum of the steady-state and perturbed solution into governing equation (8.1). A slowly varying time scale is introduced for the perturbed solution and a linear differential equation is derived, again using the secularity condition. If the perturbed solution remains limited as time goes to infinity, the solution is said to be stable. This stability definition indeed implies that the perturbed solution does not need to approach zero as time goes to infinity. We show that a stable solution corresponds to a perturbed solution that is purely oscillatory in the slowly varying time scale.

We start the mathematical derivation by introducing the nondimensional slowly varying time scale $\tau = \sigma\delta_1 t$, where the small parameter $\delta_1 \ll 1$ is later related to ε. A first approximation of the sum of the steady-state and perturbed solution is expressed as

$$\beta = \varepsilon^{1/3}\left[\mathrm{A} + \gamma(\tau)\right]\cos(\sigma t) + \varepsilon^{1/3}\overline{\gamma}(\tau)\sin(\sigma t) + o\left(\varepsilon^{1/3}\right), \quad (8.11)$$

where $\gamma(\tau)$ and $\overline{\gamma}(\tau)$ are assumed to be small relative to A and linear differential equations with respect to $\gamma(\tau)$ and $\overline{\gamma}(\tau)$ are derived as follows. Equation (8.11) is inserted into eq. (8.1). Then the resulting equations at order ε that have a rapidly varying nondimensional time scale σt are examined. We start with the term $\ddot{\beta}$, which can be approximated as

$$\ddot{\beta} = -\varepsilon^{1/3}\sigma^2\left[(\mathrm{A}+\gamma)\cos\sigma t + \overline{\gamma}\sin\sigma t\right] + 2\sigma^2\delta_1\varepsilon^{1/3}\left(-\frac{d\gamma}{d\tau}\sin\sigma t + \frac{d\overline{\gamma}}{d\tau}\cos\sigma t\right) + o\left(\delta_1\varepsilon^{1/3}\right).$$

We want the second term in the expression for $\ddot{\beta}$ to give an $O(\varepsilon)$ contribution to the equations, which means $\delta_1 = O(\varepsilon^{2/3})$.

Let us then study the term involving β^3 in eq. (8.1). The linear terms in γ and $\overline{\gamma}$ in the expression for β^3 are $3\varepsilon\mathrm{A}^2\left[\gamma(\tau)\cos^3\sigma t + \overline{\gamma}(\tau)\cos^2\sigma t\sin\sigma t\right]$. To apply the secularity condition, we extract terms oscillating with σt, giving $\varepsilon\mathrm{A}^2[\frac{9}{4}\gamma(\tau)\cos\sigma t + \frac{3}{4}\overline{\gamma}(\tau)\sin\sigma t]$ as the β^3 contribution in the formulation of the secularity condition. The final result,

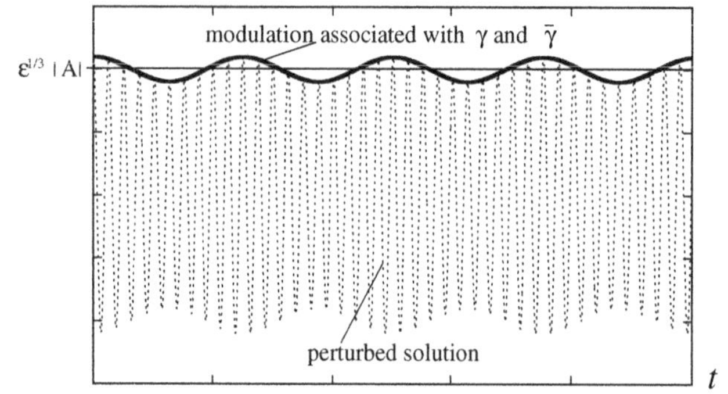

Figure 8.5. Stable modulated oscillations associated with perturbed solution (8.11).

following from using the secularity condition, is

$$\begin{cases} -2\delta_1\varepsilon^{1/3}\dfrac{d\gamma}{d\tau}+\left[\varepsilon^{1/3}\delta+\frac{3}{4}\varepsilon\dfrac{\sigma_0^2}{\sigma^2}K\mathrm{A}^2\right]\bar{\gamma}+o(\varepsilon)=0; \\ 2\delta_1\varepsilon^{1/3}\dfrac{d\bar{\gamma}}{d\tau}+\left[\varepsilon^{1/3}\delta+\frac{9}{4}\varepsilon\dfrac{\sigma_0^2}{\sigma^2}K\mathrm{A}^2\right]\gamma+o(\varepsilon)=0. \end{cases} \tag{8.12}$$

We impose for brevity

$$\delta_1=\tfrac{1}{2}\varepsilon^{2/3}. \tag{8.13}$$

The lowest-order terms of eq. (8.12) then give the following linear system of differential equations with constant coefficients:

$$\begin{bmatrix} \dfrac{d\gamma}{d\tau} \\ \dfrac{d\bar{\gamma}}{d\tau} \end{bmatrix} = \begin{bmatrix} 0 & \Lambda+\frac{3}{4}\dfrac{\sigma_0^2}{\sigma^2}K\mathrm{A}^2 \\ -\Lambda-\frac{9}{4}\dfrac{\sigma_0^2}{\sigma^2}K\mathrm{A}^2 & 0 \end{bmatrix} \begin{bmatrix} \gamma \\ \bar{\gamma} \end{bmatrix}. \tag{8.14}$$

The solution of system (8.14) is $(\gamma,\bar{\gamma})=\exp(\eta\tau)$ with

$$\begin{aligned}\eta&=\pm\sqrt{-\left(\Lambda+\frac{3}{4}\frac{\sigma_0^2}{\sigma^2}K\mathrm{A}^2\right)\left(\Lambda+\frac{9}{4}\frac{\sigma_0^2}{\sigma^2}K\mathrm{A}^2\right)}\\ &=\pm\sqrt{-\left(\Lambda+\tfrac{3}{4}K\mathrm{A}^2\right)\left(\Lambda+\tfrac{9}{4}K\mathrm{A}^2\right)}+o(\varepsilon^{2/3}).\end{aligned} \tag{8.15}$$

The condition shows that the steady-state solutions are unstable when

$$\left(\Lambda+\tfrac{3}{4}K\mathrm{A}^2\right)\left(\Lambda+\tfrac{9}{4}K\mathrm{A}^2\right)<0, \tag{8.16}$$

which means that one of the possible η-values is a positive real number. This condition leads to exponential growth of the perturbed solutions $(\gamma,\bar{\gamma})=\exp(\eta t)$. Otherwise, we have pure imaginary numbers η providing nonvanishing modulated oscillations for the perturbed motions illustrated in Figure 8.5.

Condition (8.16) is satisfied for steady-state solutions when the amplitude parameter A^2 is lower than the A^2-values on the backbone curve and larger than the A^2-values on the curve given by

$$\Lambda+\tfrac{9}{4}K\mathrm{A}^2=0. \tag{8.17}$$

The backbone and the curve given by eq. (8.17) are illustrated in Figure 8.4 together with the steady-state solutions. Both curves exist only in the frequency domain with $\Lambda K<0$, namely, in the domain where three different roots of eq. (8.9) are possible (three steady-state solutions). Intersection of the curve of eq. (8.17) with the lower branch indicates the *turning point* on the branch, T, which separates stable and unstable steady-state solutions.

When two stable steady-state solutions are possible for a given excitation frequency, the transient behavior determines which of the solutions is realized. However, the transient analysis requires that damping is introduced for the effect of the initial conditions to be damped out. This means that a direct numerical integration of eq. (8.1) with initial conditions never reaches a steady-state solution. The presence of a small damping causes a hysteresis, which is formed by jumps between the response curves in Figure 8.4. We illustrate it in part (a) of the figure. We begin to decrease $\Lambda>0$ (slowly increase of the forcing frequency $\sigma<\sigma_0$), starting from a value at the abscissa of A_3 shown in Figure 8.4, and follow the path of the lower subbranch to approach the turning point, T. The subbranch corresponds to stable steady-state solutions. A further decrease of Λ leads to a jump from T to the solution associated with the point A_1 on the upper branch. This solution is also stable. Let us now slowly increase Λ (decrease σ); we then follow the path of steady-state solutions along the upper branch with increasing absolute value of the amplitude

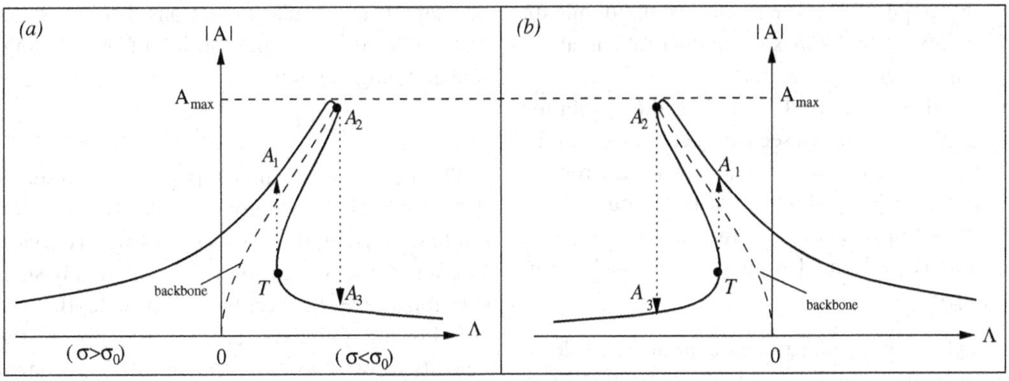

Figure 8.6. The response curves for Duffing equation with damping (see eq. (8.18)) based on secularity condition (8.21); A is the nondimensional amplitude and Λ is the detuning parameter defined by eq. (8.8).

parameter A. However, due to damping, the stable upper branch ends at the point A_2. The consequence is a jump in the solution to the stable solution on the lower subbranch at point A_3.

The presence of the hysteresis implies that the soft-spring nonlinearity causes a maximum response at a frequency lower than σ_0. However, the hard-spring case leads to a maximum for a frequency higher than σ_0. This fact is often adopted as a definition of the *hard- and soft-spring behavior* of the nonlinear mechanical systems, for example, in experimental studies (see Fultz, 1962; and determination of the critical liquid depth, Section 8.3.1.1).

8.2.1.3 Damping

Equation (8.1) can be modified by introducing a linear damping term, which gives

$$\ddot{\beta} + 2\xi\sigma_0\dot{\beta} + \sigma_0^2(\beta + K\beta^3) = -\varepsilon\sigma^2\cos(\sigma t). \quad (8.18)$$

The asymptotic steady-state solution of eq. (8.18) follows by using the same scheme as in the previous section. This solution takes the form of eq. (8.5) with $\alpha = \frac{1}{3}$ and $\beta^{(1)}(t) = \mathrm{A}\cos(\sigma t + \theta)$. Remembering that $\cos(\sigma t) = \cos\theta\cos(\sigma t + \theta) + \sin\theta\sin(\sigma t + \theta)$ on the right-hand side of eq. (8.18), we again follow the asymptotic scheme to derive the secularity condition

$$\tfrac{3}{4}K\mathrm{A}^3 + \Lambda\mathrm{A} = -\cos\theta, \quad \mathrm{A}\gamma_1 = 2\xi\mathrm{A}/\varepsilon^{2/3} = \sin\theta, \quad (8.19)$$

which determines both amplitude parameter A and phase shift θ. Here we have introduced the parameter $\gamma_1 = 2\xi/\varepsilon^{2/3} = O(1)$.

Our focus is on the amplitude parameter, A. Excluding $\cos\theta$ and $\sin\theta$ from eq. (8.19) gives the following secularity equation for A:

$$\mathrm{A}^2\left[\left(\tfrac{3}{4}K\mathrm{A}^2 + \Lambda\right)^2 + \gamma_1^2\right] = 1, \quad (8.20)$$

which can be rearranged as

$$\Lambda = -\tfrac{3}{4}K\mathrm{A}^2 \pm \sqrt{\mathrm{A}^{-2} - \gamma_1^2} \quad (8.21)$$

to simplify the drawing of the amplitude response curves analogous to those in Figure 8.4. Equation (8.21) needs the condition $\mathrm{A}^{-2} - \gamma_1^2 \ge 0$, which limits the amplitude parameter when $\gamma_1 \neq 0$. The maximum value of A is $\mathrm{A}_{max} = \gamma_1^{-2}$. The response curves of the steady-state solution of eq. (8.21) are presented in Figure 8.6. This figure demonstrates hysteresis due to damping.

One can repeat the stability analysis of the steady-state solution by using the technique from Section 8.2.1.2. A theory and procedures on how to identify stability for nonlinear oscillators of the general type are given by Bogoljubov and Mitropolski (1961). Interested readers are referred to this book, which shows that the curve TA_2 corresponds to unstable motions. The other parts of the solution are stable.

8.2.2 Steady-state resonant sloshing due to horizontal excitations

The key points in the asymptotic technique of Section 8.2.1 are as follows:

- The dominant part of the solution is an eigensolution of lower order than the excitation.

- The amplitude and the phase of the dominant solution follow from solving the problem at the same order as the excitation.
- The technique used in finding the amplitude and the phase was a secularity condition requiring the solution to be bounded when the forcing frequency is equal to the natural frequency.
- The stability analysis of the steady-state solutions is performed by a two-time scale technique.

Such a technique can be generalized to sloshing in a two-dimensional rectangular tank with breadth l and liquid depth h. The tank may oscillate horizontally or angularly with circular frequency $\sigma \to \sigma_1$, where σ_1 is the first natural frequency.

For brevity, we focus on the horizontal excitation $\eta_2(t) = \eta_{2a} \cos \sigma t$. We introduce the small parameter ε as the ratio between the forcing amplitude and l (i.e., $\varepsilon = \eta_{2a}/l$). Similarly to relation (8.2) we consider the small parameter

$$\delta = \left(\sigma_1^2 - \sigma^2\right)/\sigma^2 \ll 1, \qquad (8.22)$$

which expresses the closeness of σ to σ_1 (resonant condition). The small parameter δ should be matched with the forcing parameter ε.

One can show that the matching asymptotics (8.8) can be applied to sloshing. If the ratio of liquid depth to tank breadth is finite (i.e., $\overline{h} = h/l = O(1)$), one can use the linear modal system (5.27) and (5.28) and the results in Section 5.3.3 on periodic solutions of the linear modal systems to show that the lowest-order term of the linear solution of the free-surface elevation is

$$\zeta(x, t)/l \sim (\varepsilon/\delta) \cos\left(\pi\left(y + \tfrac{1}{2}l\right)/l\right) \cos \sigma t, \qquad (8.23)$$

where the Cartesian coordinate system Oyz has its origin at the intersection between the mean free surface and the tank's centerplane. The z-axis points upward.

Equation (8.23) implies that the first natural mode dominates and its amplitude is characterized by the lowest-order asymptotic term, $O(\varepsilon/\delta)$. The situation is similar to that for the Duffing equation given by relation (8.3). Using this analogy, Moiseev (1958) carried out a pioneering analytical study on nonlinear resonant sloshing. His results were independently extended by Ockendon and Ockendon (1973) only for horizontal excitation and by Faltinsen (1974) for both horizontal and angular excitations. The investigations confirmed the applicability of the Duffing-like detuning asymptotics:

$$\delta\varepsilon^{-2/3} = \Lambda = O(1). \qquad (8.24)$$

The detuning parameter is called "Moiseev's detuning." The asymptotic secularity condition implies, as usual, the necessary solvability condition and finiteness of the nonlinear steady-state solution as $\sigma = \sigma_1$, except at a critical depth.

Following Faltinsen (1974), we express the velocity potential for steady-state resonant motions as

$$\Phi(y, z, t) = \varphi(y, z, t) - \sigma\varepsilon l \sin(\sigma t) y, \quad -\tfrac{1}{2}l < y < \tfrac{1}{2}l, \quad -h < z < 0. \qquad (8.25)$$

According to Moiseev, the component φ is expressed as $\varphi = \sum_{n=1}^{\infty} \varepsilon^{n/3}\phi_n$ and the free-surface elevation is written as $\zeta = \sum_{n=1}^{\infty} \varepsilon^{n/3}\zeta_n$.

The problem is solved correctly to $O(\varepsilon)$. The dynamic and kinematic free-surface conditions are expressed at the mean free surface by a Taylor expansion and by consistently collecting terms of the same order. We start the boundary-value problem with the lowest-order term, ϕ_1. Because the excitation is $O(\varepsilon)$, a homogeneous boundary-value problem appears without any excitation. For instance, the combined dynamic and kinematic free-surface condition for ϕ_1 is

$$\frac{\partial}{\partial t}\left(\frac{\partial \phi_1}{\partial t} + \frac{\sigma^2}{\sigma_1^2} g \zeta_1\right) = \frac{\partial^2 \phi_1}{\partial t^2} + \frac{\sigma^2}{\sigma_1^2} g \frac{\partial \phi_1}{\partial z} = 0 \quad \text{on} \quad z = 0. \qquad (8.26)$$

The only possible solution is an eigensolution that can be expressed as

$$\phi_1 = \alpha \sin[\pi y/l] \cosh[\pi(z + h)/l] \sin(\sigma t), \qquad (8.27)$$

which is similar to the first nondimensional mode given by eq. (4.8). However, a difference is that the eigensolution oscillates with the lowest natural frequency, σ_1, and not with the forcing frequency as in eq. (8.27). Consistent formulations of the free-surface condition to the order $O(\varepsilon^{1/3})$ other than eq. (8.26) are possible. However, we have constructed the free-surface condition so that eq. (8.27) satisfies eq. (8.26) . This approach is similar to the lowest-order problem for the Duffing equation. The consequence is a correcting term in the free-surface condition for the $O(\varepsilon)$

problem. Furthermore, α in eq. (8.27) is presently unknown and $\alpha\varepsilon^{1/3}$ is the dominant amplitude. The dimensional amplitude parameter, α, should be determined later. Furthermore, the phasing of the asymptotic solution may not be evident at this stage.

The assumption of infinite liquid depth is made to simplify the derivation. We can then write $\phi_1 = \alpha \sin(\pi y/l)\exp(\pi z/l)\sin\sigma t$. The corresponding free-surface elevation,

$$\begin{aligned}\zeta_1 &= -\alpha\frac{\sigma_1^2}{\sigma g}\sin(\pi y/l)\cos\sigma t\\ &= l\mathrm{A}\cos\left(\pi\left(y+\tfrac{1}{2}l\right)/l\right)\cos\sigma t, \quad \mathrm{A} = \frac{\alpha\sigma_1^2}{\sigma g l},\end{aligned} \tag{8.28}$$

follows from the dynamic free-surface condition $\partial\phi_1/\partial t + (\sigma/\sigma_1)^2 g\zeta_1 = 0$ on $z = 0$, where the nondimensional amplitude parameter $\mathrm{A} = O(1)$ is introduced.

The formulation of the boundary-value problem for ϕ_2 also gives a homogeneous Neumann body-boundary condition, whereas the free-surface condition contains a right-hand side with products of terms coming from the first-order approximations ϕ_1 and ζ_1; that is, the dynamic and kinematic free-surface conditions on $z = 0$ are

$$\frac{\partial\phi_2}{\partial t} + \frac{\sigma^2}{\sigma_1^2}g\zeta_2 = -\tfrac{1}{2}\left(\frac{\partial\phi_1}{\partial y}\right)^2 - \tfrac{1}{2}\left(\frac{\partial\phi_1}{\partial z}\right)^2 - \zeta_1\frac{\partial^2\phi_1}{\partial z\partial t},$$

$$\frac{\partial\zeta_2}{\partial t} - \frac{\partial\phi_2}{\partial z} = -\frac{\partial\phi_1}{\partial y}\frac{\partial\zeta_1}{\partial y} + \zeta_1\frac{\partial^2\phi_1}{\partial z^2}.$$

They can be combined as in eq. (8.26) and expressed as

$$\frac{\partial^2\phi_2}{\partial t^2} + \frac{\sigma^2}{\sigma_1^2}g\frac{\partial\phi_2}{\partial z} = -\left(\frac{\pi}{l}\right)^2\alpha^2\sigma\sin(2\sigma t) \quad \text{on } z = 0. \tag{8.29}$$

Because the right-hand side of the free-surface condition oscillates as $\sin(2\sigma t)$ and is proportional to α^2, ϕ_2 also oscillates as $\sin(2\sigma t)$ and is proportional to α^2. Since the right-hand side of eq. (8.29) does not vary with y, from the Laplace equation it follows that the solution ϕ_2 does not vary in space. When ϕ_2 is found, the dynamic free-surface condition determines ζ_2. The result is

$$\phi_2 = \frac{1}{4\sigma}\left(\frac{\pi}{l}\right)^2\alpha^2\sin(2\sigma t),$$

$$\zeta_2 = -\frac{\sigma_1^2}{4\sigma^2 g}\left(\frac{\pi}{l}\right)^2\cos\left(\frac{2\pi y}{l}\right)\alpha^2\left[1+\cos(2\sigma t)\right].$$

The solution of ζ_2 is consistent with conservation of liquid volume (i.e., $\int_{-\frac{1}{2}l}^{\frac{1}{2}l}\zeta_2\,dy = 0$). We note that ϕ_2 causes a pressure component $-\rho\partial\phi_2/\partial t$ that does not decay with depth. The corresponding velocity, $\nabla\phi_2$, is zero. Similar behavior occurs for the second-order solution of two oppositely propagating long-crested linear deep-water wave systems with the same amplitude and frequency (Longuet-Higgins, 1953). We should not be surprised about this similarity. The velocity potential ϕ_1 can be rewritten as two linear long-crested wave systems with the same amplitude and frequency that are propagating in opposite directions in deep water. The solution of ϕ_2 for finite depth is more complicated. A spatially constant velocity potential term $\alpha_0 t$ must be added to the particular solution that satisfies the inhomogeneous free-surface condition to satisfy conservation of liquid volume.

The formulation of the boundary-value problem for ϕ_3 also gives a homogeneous Neumann body-boundary condition, whereas the free-surface conditions contain a right-hand side with the effect of forcing and products of terms coming from ϕ_1, ϕ_2, ζ_1, and ζ_2. The boundary conditions on $z = 0$ due to the dynamic and kinematic free-surface conditions can be expressed as

$$\begin{aligned}\frac{\partial\phi_3}{\partial t} + \frac{\sigma^2}{\sigma_1^2}g\zeta_3 = &-\frac{\Lambda\sigma^2}{\sigma_1^2}g\zeta_1 - \frac{\partial\phi_1}{\partial y}\frac{\partial\phi_2}{\partial y} - \frac{\partial\phi_1}{\partial z}\frac{\partial\phi_2}{\partial z}\\ &-\zeta_1\frac{\partial^2\phi_2}{\partial z\partial t} - \zeta_2\frac{\partial^2\phi_1}{\partial z\partial t} - \zeta_1\frac{\partial\phi_1}{\partial y}\frac{\partial^2\phi_1}{\partial z\partial y}\\ &-\zeta_1\frac{\partial\phi_1}{\partial z}\frac{\partial^2\phi_1}{\partial z^2} - \tfrac{1}{2}\zeta_1^2\frac{\partial^3\phi_1}{\partial z^2\partial t}\\ &+l\sigma^2 y\cos\sigma t,\end{aligned}$$

$$\begin{aligned}\frac{\partial\zeta_3}{\partial t} - \frac{\partial\phi_3}{\partial z} = &-\frac{\partial\phi_1}{\partial y}\frac{\partial\zeta_2}{\partial y} - \frac{\partial\phi_2}{\partial y}\frac{\partial\zeta_1}{\partial y} - \zeta_1\frac{\partial^2\phi_1}{\partial z\partial y}\frac{\partial\zeta_1}{\partial y}\\ &+\zeta_1\frac{\partial^2\phi_2}{\partial z^2} + \zeta_2\frac{\partial^2\phi_1}{\partial z^2} + \tfrac{1}{2}\zeta_1^2\frac{\partial^3\phi_1}{\partial z^3},\end{aligned}$$

where Λ is defined by eq. (8.24). Furthermore, y in the dynamic free-surface condition can be uniquely expanded by the following Fourier series:

$$y = \sum_{n=0}^{\infty}\frac{4l(-1)^n}{(2n+1)^2\pi^2}\sin\left(\frac{2n+1}{l}\pi y\right)$$

and the combined kinematic and dynamic free-surface conditions on $z = 0$ take the form

$$\frac{\partial^2 \phi_3}{\partial t^2} + \frac{\sigma^2}{\sigma_1^2} g \frac{\partial \phi_3}{\partial z}$$
$$= \left[-\Lambda\sigma^2\alpha - \frac{4l^2}{\pi^2}\sigma^3 + \tfrac{1}{4}\left(\frac{\pi}{l}\right)^4 \alpha^3\right] \sin\left(\frac{\pi y}{l}\right) \sin\sigma t$$
$$+ \tfrac{5}{4}\left(\frac{\pi}{l}\right)^4 \alpha^3 \sin\left(\frac{\pi y}{l}\right) \sin 3\sigma t$$
$$- l\sigma^3 \sin\sigma t \sum_{n=1}^{\infty} \frac{4l(-1)^n}{(2n+1)^2\pi^2} \sin\left(\frac{2n+1}{l}\pi y\right). \quad (8.30)$$

The term proportional to $\sin(\pi y/l)\sin(\sigma t)$ in the inhomogeneous part of the combined kinematic and dynamic free-surface condition for ϕ_3 is of particular concern. A particular solution of the Laplace equation that satisfies the free-surface condition with this inhomogeneous term is proportional to $\sin(\pi y/l)\exp(\pi z/l)\sin\sigma t$. The latter solution satisfies the homogeneous free-surface condition $\partial^2\phi_3/\partial t^2 + (\sigma^2/\sigma_1^2) g\, \partial\phi_3/\partial z = 0$ on $z = 0$. As a result, if the coefficient ahead of the term $\sin(\pi y/l)\sin(\sigma t)$ in eq. (8.30) is not zero, it causes infinite ϕ_3 when $\sigma = \sigma_1$. Therefore, a solvability (secularity) condition simply states that its coefficient must be zero; that is,

$$\tfrac{1}{4}\left(\frac{\pi}{l}\right)^4 \alpha^3 - \Lambda\sigma^2\alpha - \frac{4l^2\sigma^3}{\pi^2} = 0. \quad (8.31)$$

The solvability condition becomes more complicated in finite depth. It is characterized by a coefficient at α^3, which depends on the ratio of liquid depth to tank breadth, $\bar{h} = h/l$. Using the modal approach from Section 8.3.1.1, we can write solvability condition (8.31) for the finite depth:

$$P_1 = \Lambda \mathrm{A} + m_1^{(0)}(\bar{h})\mathrm{A}^3, \quad (8.32)$$

where A is the nondimensional amplitude parameter defined by eq. (8.28) and P_1 is defined by eq. (5.85) (see eq. (8.46)). The modal approach from Section 8.3.1.1 derives solvability condition (8.54), which becomes the same as eq. (8.32) when we take into account the lowest order of m_1 by eq. (8.56), the definition of the detuning parameter Λ, and the fact that the amplitude parameter A from the modal analysis is linked with A in eq. (8.32) as $A = \varepsilon^{1/3}\mathrm{A}$.

The coefficient $m_1^{(0)}$ is a monotonically decreasing function of $\bar{h}$ (see Section 8.3.1.1). When $m_1^{(0)}(\bar{h}) = 0$, A is single-valued with infinite value when the forcing frequency is equal to the lowest natural sloshing frequency, which happens at the so-called critical depth. When $\bar{h} \to \infty$, $m_1^{(0)} \to -\frac{1}{4}\pi^2$. This result is consistent with eq. (8.31).

The solution for the third-order potential ϕ_3 and third-order free-surface elevation ζ_3 can be expressed as

$$\phi_3 = -\tfrac{5}{32}\sigma^{-2}(\pi/l)^4 \sin(\pi y/l)\exp(\pi z/l)\,\alpha^3 \sin(3\sigma t)$$
$$- \sin\sigma t \sum_{n=1}^{\infty} \frac{2l^2\sigma(-1)^n}{n(2n+1)^2\pi^2} \sin\left[\frac{(2n+1)\pi y}{l}\right]$$
$$\times \exp\left[\frac{(2n+1)\pi z}{l}\right],$$

$$\zeta_3 = \sigma_1^2\sigma^{-3}g^{-1}(\pi/l)^4\alpha^3\left[\tfrac{1}{32}\sin(\pi y/l)\cos\sigma t\right.$$
$$+ \tfrac{9}{32}\sin(3\pi y/l)\cos\sigma t + \tfrac{1}{16}\sin(\pi y/l)\cos(3\sigma t)$$
$$\left. + \tfrac{3}{32}\sin(3\pi y/l)\cos(3\sigma t)\right] + \frac{l\sigma_1^2}{g}\cos\sigma t$$
$$\times \sum_{n=1}^{\infty} \frac{2l(-1)^n}{n(2n+1)\pi^2} \sin\left[\frac{(2n+1)\pi y}{l}\right].$$

The physical mechanism that causes finite amplitude at the lowest resonance frequency is nonlinear energy transfer between the three lowest modes. Secularity equation (8.32) for amplitude parameter A is similar to that expressed by eq. (8.9) for a Duffing-type spring–mass system. The coefficient $m_1^{(0)}$ is a function of the nondimensional depth, $\bar{h}$, and becomes zero at $\bar{h} = h_* = 0.3368\ldots$. Therefore, the dependence of amplitude parameter A on Λ is schematically the same as in Figure 8.4. Panel (a) corresponds to $\bar{h} > h_*$ and panel (b) corresponds to $\bar{h} < h_*$. The change from hard-spring to soft-spring behavior occurs as $\bar{h}$ increases through h_*. This change has been experimentally observed by Fultz (1962). However, his measurements gave $h_* = 0.28$. A reason for the discrepancy between theoretical and experimental values is amplification of the second mode, whose contributions becomes comparable to that of the first mode. The modeling of liquid sloshing with critical depth needs an adaptive modal technique. We show later that this technique makes it possible to obtain good agreement with experiments; more details are given in Section 8.5. In addition, we refer to Hermann and Timokha (2008), who used the adaptive modal method (described later in the text) to show that the theoretical value of the critical depth is a monotonically decreasing function of the forcing amplitude and the value $\overline{h} = h_* = 0.3368\ldots$ is the asymptotic limit value for infinitesimal forcing. Accounting for the actual experimental forcing amplitudes makes the results by Fultz (1962) consistent with the adaptive multimodal method. (Refer to Exercise 8.12.3 for further details.)

We refer to Faltinsen (1974) for details about the stability analysis, which leads to results similar to that for the Duffing equation.

The important conclusion for $\overline{h} \neq h_*$ is that the second mode is of the order $\varepsilon^{2/3}$ and all other modes are $O(\varepsilon)$; that is,

$$\text{mode}\,1 \sim \varepsilon^{1/3}, \quad \text{mode}\,2 \sim \varepsilon^{2/3}, \quad \text{mode}\,(2i+1) \sim \varepsilon, \quad i \geq 1. \tag{8.33}$$

The remaining even modes are not excited and depend only on the initial conditions. Their contribution is less than $O(\varepsilon)$. This way of ordering terms is fundamentally different from the ordering in weakly nonlinear exterior flow with finite- and infinite-depth conditions, where the lowest-order term is $O(\varepsilon)$, the second-order term is $O(\varepsilon^2)$, and so on.

8.3 Single-dominant asymptotic nonlinear modal theory

8.3.1 Asymptotic modal system

The previously described asymptotic theory for steady-state (periodic) sloshing is generalized to the time domain by using modal equations (7.14) and (7.15). The problem is reduced to a finite-dimensional system of nonlinear ordinary differential equations with respect to the modal functions β_i. Initial conditions are then needed. An essential assumption of this section is that there is only one dominant mode that is associated with the first modal function β_1 as stated in eq. (8.33). This assumption is relaxed in later sections.

The similarities and differences relative to the steady-state analysis based on the Moiseev method are as follows:

- The l-scaled modal functions β_i and R_n are characterized by ordering (8.33).
- The forcing frequency is in a neighborhood of the lowest natural frequency σ_1. However, the difference between the frequencies (i.e., $(\sigma_1^2 - \sigma^2)/\sigma^2 = \delta$), is not restricted to be small and linked with the nondimensional forcing magnitude ε as for the Moiseev method.
- The solution technique does not use a secularity condition, as required by the direct perturbation technique of the dynamic and kinematic free-surface conditions from Section 8.2.2. In contrast, the multimodal technique that follows operates with asymptotic expansions of the integral terms D_n and D_{nk} of system (7.14) and (7.15) in terms of β_i.

The differential equations for the generalized coordinates β_i of the free-surface elevation are based on a nondimensional formulation. The tank breadth, l, is used to make all length dimensions nondimensional. The natural modes presented in Section 4.3.1.1 can be expressed nondimensionally as

$$\overline{\varphi}_n = \overline{f}_n(y)\frac{\cosh\left(\pi n(z+\overline{h})\right)}{\cosh(\pi n\overline{h})},$$
$$\overline{f}_n(y) = \cos\left(\pi n\left(y+\tfrac{1}{2}\right)\right), \tag{8.34}$$

where, as usual, $\overline{h} = h/l$ is the scaled liquid depth. Furthermore, we use modal representation (7.1) and (7.2) with natural modes (8.34) and modal system (7.14) and (7.15), in which all the geometric quantities are *scaled by l.* The small parameter ε is introduced to characterize forcing in a way similar to that for linear sloshing in Chapter 5 (see eqs. (5.11) and (5.12)); that is,

$$\eta_2/l \sim \eta_3/l = O(\varepsilon); \quad \eta_4 = O(\varepsilon). \tag{8.35}$$

Due to eq. (8.33), we also postulate

$$\beta_1 \sim R_1 = O\left(\varepsilon^{1/3}\right), \quad \beta_2 \sim R_2 = O\left(\varepsilon^{2/3}\right),$$
$$\beta_3 \sim R_3 = O(\varepsilon); \quad \beta_n \sim R_n \leq O(\varepsilon), \quad n \geq 4. \tag{8.36}$$

Terms of $o(\varepsilon)$ are neglected.

A derivation of the corresponding modal system is given by Faltinsen *et al.* (2000). Following this derivation, we should first consider subsystem (7.14), which is treated as a linear algebraic system with respect to R_n:

$$\sum_k D_{nk}R_k = \sum_k \frac{\partial D_n}{\partial \beta_k}\dot{\beta}_k, \quad n = 1, 2, \ldots. \tag{8.37}$$

Nondimensional D_{nk} and D_k in eq. (8.37) are the following integrals (see definition by formulas

(7.12) and the explicit expressions for the natural modes φ_k):

$$D_n = \int_{-\frac{1}{2}}^{\frac{1}{2}} \int_{-\bar{h}}^{\sum_{k=1}^{\infty} \beta_k \bar{f}_k} \bar{\varphi}_n dz dy,$$

$$D_{nk} = \int_{-\frac{1}{2}}^{\frac{1}{2}} \int_{-\bar{h}}^{\sum_{i=1}^{\infty} \beta_i \bar{f}_i} \nabla\bar{\varphi}_n \cdot \nabla\bar{\varphi}_k dz dy, \quad (8.38)$$

where the original expressions (7.12) are scaled by the liquid density, ρ. From the integration limits in eq. (8.38) it follows that D_{nk} and D_k are functions of β_i. These functions can therefore be expanded as Taylor series in β_i consistently to $O(\varepsilon)$. Details of this expansion are presented by Faltinsen *et al.* (2000) and Faltinsen and Timokha (2001), and we refer interested readers to these papers.

Furthermore, we express R_n as

$$R_n = \sum_i \gamma_i \dot{\beta}_i + \sum_{ij} \gamma_{ij} \dot{\beta}_j \beta_i + \sum_{ijk} \gamma_{ijk} \dot{\beta}_i \beta_j \beta_k + \cdots$$

and substitute it into eq. (8.37). Explicit values of γ_i, γ_{ij}, and γ_{ijk} are found by gathering similar terms. The result is

$$\begin{aligned}
R_1 &= \frac{\dot{\beta}_1}{2E_1} + \frac{E_0}{E_1^2}\dot{\beta}_1\beta_2 - \frac{E_0}{E_1E_2}\dot{\beta}_2\beta_1 \\
&\quad + \frac{E_0}{E_1}\left(-\tfrac{1}{2} + \frac{4E_0}{E_1E_2}\right)\beta_1^2\dot{\beta}_1, \\
R_2 &= \frac{1}{4E_2}\left(\dot{\beta}_2 - \frac{4E_0}{E_1}\beta_1\dot{\beta}_1\right), \\
R_3 &= \frac{\dot{\beta}_3}{6E_3} - \frac{E_0}{E_1E_3}\dot{\beta}_1\beta_2 - \frac{E_0}{E_2E_3}\dot{\beta}_2\beta_1 \\
&\quad + \dot{\beta}_1\beta_1^2\left(\frac{4E_0^2}{E_1E_2E_3} - \frac{E_0}{2E_3}\right), \\
R_i &= \frac{\dot{\beta}_i}{2iE_i}, \quad i \geq 4
\end{aligned} \quad (8.39)$$

and

$$\begin{aligned}
\dot{R}_1 &= \frac{\ddot{\beta}_1}{2E_1} + \frac{E_0}{E_1^2}\ddot{\beta}_1\beta_2 - \frac{E_0}{E_1E_2}\ddot{\beta}_2\beta_1 \\
&\quad + \dot{\beta}_1\dot{\beta}_2\left(\frac{E_0}{E_1^2} - \frac{E_0}{E_1E_2}\right) \\
&\quad + \frac{E_0}{E_1}\left(-\tfrac{1}{2} + \frac{4E_0}{E_1E_2}\right)\beta_1^2\ddot{\beta}_1 \\
&\quad + 2\frac{E_0}{E_1}\left(-\tfrac{1}{2} + \frac{4E_0}{E_1E_2}\right)\dot{\beta}_1^2\beta_1, \\
\dot{R}_2 &= \frac{1}{4E_2}\left(\ddot{\beta}_2 - \frac{4E_0}{E_1}\left(\beta_1\ddot{\beta}_1 + \dot{\beta}_1^2\right)\right), \\
\dot{R}_3 &= \frac{\ddot{\beta}_3}{6E_3} - \frac{E_0}{E_1E_3}\ddot{\beta}_1\beta_2 - \frac{E_0}{E_2E_3}\ddot{\beta}_2\beta_1 \\
&\quad - \left(\frac{E_0}{E_1E_3} + \frac{E_0}{E_2E_3}\right)\dot{\beta}_1\dot{\beta}_2 \\
&\quad + \left(\ddot{\beta}_1\beta_1^2 + 2\dot{\beta}_1^2\beta_1\right)\left(\frac{4E_0^2}{E_1E_2E_3} - \frac{E_0}{2E_3}\right), \\
\dot{R}_i &= \frac{\ddot{\beta}_i}{2iE_i}, \quad i \geq 4,
\end{aligned} \quad (8.40)$$

where

$$E_0 = \tfrac{1}{8}\pi^2, \quad E_i = \tfrac{1}{2}\pi\tanh(\pi i\bar{h}), \quad i \geq 1. \quad (8.41)$$

The linearized analogy (5.23) of relationships (8.39) between β_i and R_i were obtained in Chapter 5 by using the kinematic free-surface condition.

Expressions (8.39) and (8.40) should be substituted into subsystem (7.15), which takes the following form for two-dimensional liquid motion:

$$\sum_n \dot{R}_n \frac{\partial D_n}{\partial \beta_i} + \tfrac{1}{2}\sum_{n,k} \frac{\partial D_{nk}}{\partial \beta_i} R_n R_k + \dot{\omega}\frac{\partial l_{1\omega}}{\partial \beta_i} + \omega\frac{\partial l_{1\omega t}}{\partial \beta_i}$$
$$- \frac{d}{dt}\left(\omega\frac{\partial l_{1\omega t}}{\partial \dot{\beta}_i}\right) + (\dot{v}_{O2} - g_2)\lambda_{2i} - g_3\beta_i\lambda_{3i} = 0 \quad (8.42)$$

with definitions given in connection with eqs. (7.12) and (7.16). The latter expression includes nonhomogeneous terms associated with the forcing (proportional to $\boldsymbol{\omega} = (\omega_1, 0, 0) = (\dot{\eta}_4, 0, 0)$ and $\boldsymbol{v}_O = (0, v_{O2}, v_{O3}) = (0, \dot{\eta}_2, \dot{\eta}_3)$ for the studied two-dimensional case). The forcing terms are, as earlier, of $O(\varepsilon)$. Furthermore, the two sums should be rearranged by substitution of eqs. (8.39) and (8.40) and the corresponding expressions for D_{nk} and D_k. Remembering the ordering (8.35) and keeping the $O(\varepsilon)$-terms leads to the following three nonlinear modal equations:

$$\begin{aligned}
&(\ddot{\beta}_1 + \sigma_1^2\beta_1) + d_1(\ddot{\beta}_1\beta_2 + \dot{\beta}_1\dot{\beta}_2) + d_2(\ddot{\beta}_1\beta_1^2 + \dot{\beta}_1^2\beta_1) \\
&\quad + d_3\ddot{\beta}_2\beta_1 = \tilde{K}_1(t), \\
&(\ddot{\beta}_2 + \sigma_2^2\beta_2) + d_4\ddot{\beta}_1\beta_1 + d_5\dot{\beta}_1^2 = 0, \\
&(\ddot{\beta}_3 + \sigma_3^2\beta_3) + q_1\ddot{\beta}_1\beta_2 + q_2\ddot{\beta}_1\beta_1^2 + q_3\ddot{\beta}_2\beta_1 \\
&\quad + q_4\dot{\beta}_1^2\beta_1 + q_5\dot{\beta}_1\dot{\beta}_2 = \tilde{K}_3(t).
\end{aligned} \quad (8.43)$$

The first equation contains terms both of $O(\varepsilon^{1/3})$ and $O(\varepsilon)$. The terms in the second and third equations are only $O(\varepsilon^{2/3})$ and $O(\varepsilon)$, respectively. The linear equations that describe higher modes are

$$\ddot{\beta}_i + \sigma_i^2\beta_i = \tilde{K}_i(t), \quad i = 4, \ldots, N, \quad (8.44)$$

where σ_i are the natural sloshing frequencies defined by eq. (4.11) and $\tilde{K}_i(t)$ are the l-scaled forcing terms $K_i(t)$ given by eq. (5.84); that is,

$$\tilde{K}_i(t) = K_i(t)/l = -P_i\left[\frac{\ddot{\eta}_2(t)}{l} + \tilde{S}_i\ddot{\eta}_4(t) + \frac{g}{l}\eta_4(t)\right], \quad (8.45)$$

where

$$P_i = [2/i\pi]\tanh(\pi i\bar{h})\left((-1)^i - 1\right), \quad \tilde{S}_i = S_i/l = [2/i\pi]\tanh(\tfrac{1}{2}\pi i\bar{h}). \quad (8.46)$$

Finally, the coefficients d_k and q_k are expressed as functions of the ratio of liquid depth to tank breadth; that is,

$$d_1 = 2\frac{E_0}{E_1} + E_1; \quad d_2 = 2E_0\left(-1 + \frac{4E_0}{E_1E_2}\right), \quad d_3 = -2\frac{E_0}{E_2} + E_1, \quad d_4 = -4\frac{E_0}{E_1} + 2E_2, \quad d_5 = E_2 - 2\frac{E_0E_2}{E_1^2} - \frac{4E_0}{E_1} \quad (8.47)$$

and

$$q_1 = 3E_3 - \frac{6E_0}{E_1}; \quad q_2 = -3E_0 - 9\frac{E_0E_3}{E_1} + 24\frac{E_0^2}{E_1E_2}, \quad q_3 = 3E_3 - \frac{6E_0}{E_2}, \quad q_4 = -6E_0 - 24\frac{E_0E_3}{E_1} + 48\frac{E_0^2}{E_1E_2} + 24\frac{E_0^2E_3}{E_1^2E_2}, \quad q_5 = 6\left(\tfrac{1}{2}E_3 - \frac{E_0}{E_1} - \frac{E_0E_3}{E_1E_2} - \frac{E_0}{E_2}\right). \quad (8.48)$$

The expressions for q_2 and q_4 correct the misprinted formulas in the original paper by Faltinsen *et al.* (2000). The actual values of these coefficients and P_i and $\tilde{S}_i$ are presented in Table 9.1. One can easily obtain dimensional results for nonlinear sloshing based on studies of system (8.43), in which the values are l-scaled. Therefore, the original generalized dimensional coordinates are simply obtained by multiplying nondimensional β_i with l. When employing these dimensional coordinates, we can get other dimensional characteristics including the velocity field, pressure, hydrodynamic force, and moment as described in Section 8.3.2.

The first two nonlinear equations of eqs. (8.43) couple β_1 with β_2 and do not depend on β_3. Nonlinear energy transfer between the lowest symmetric and antisymmetric modes relative to the tank's centerplane is the mechanism that causes a finite steady-state response at the lowest natural frequency. An exception to this fact is for the critical depth considered in the next section. Equation (8.43) shows that the third mode component, β_3, is excited by the rigid-body motions and the first and second modes.

The nonlinear modal system (8.43) and (8.44) can be numerically integrated by, for instance, a fourth-order Runge–Kutta method. Initial conditions for β_i and its first-order derivative are needed, as was described in Section 5.3.3. The system may also adopt linear damping terms (see Section 6.3.2) to describe energy dissipation.

8.3.1.1 Steady-state resonant waves: frequency-domain solution

The asymptotic periodic (steady-state) solution of the asymptotic nonlinear modal system (8.43) is derived by Faltinsen *et al.* (2000). We start by considering sway forcing with frequency σ so that

$$\eta_2(t)/l = (\eta_{2a}/l)\cos(\sigma t) = \varepsilon\cos(\sigma t), \quad \eta_k(t) = 0, \quad k \neq 2. \quad (8.49)$$

The steady-state waves are associated with periodic solutions of system (8.43) satisfying the periodicity condition:

$$\beta_i(t + 2\pi/\sigma) = \beta_i(t), \quad \dot{\beta}_i(t + 2\pi/\sigma) = \dot{\beta}_i(t), \quad i = 1, 2, 3. \quad (8.50)$$

To construct the asymptotic periodic solution, we express a first approximation of the primary mode as

$$\beta_1(t) = A\cos\sigma t + o(A), \quad A = O\left(\varepsilon^{1/3}\right) \quad (8.51)$$

and substitute it into the second equation of system (8.43). Accounting for periodicity condition (8.50) yields

$$\beta_2(t) = A^2(l_0 + h_0\cos(2\sigma t)) + o(A^2), \quad (8.52)$$

where

$$l_0 = \frac{d_4 - d_5}{2\bar{\sigma}_2^2}; \quad h_0 = \frac{d_5 + d_4}{2\left(\bar{\sigma}_2^2 - 4\right)}, \quad \bar{\sigma}_i = \frac{\sigma_i}{\sigma}, \quad i = 1, 2. \quad (8.53)$$

Furthermore, the dominant amplitude A of the primary mode can be found by substituting eqs. (8.51) and (8.52) into the first equation of system (8.43) and collecting terms that contain the first harmonics. Each harmonic term must be zero. Setting the first harmonic term equal to

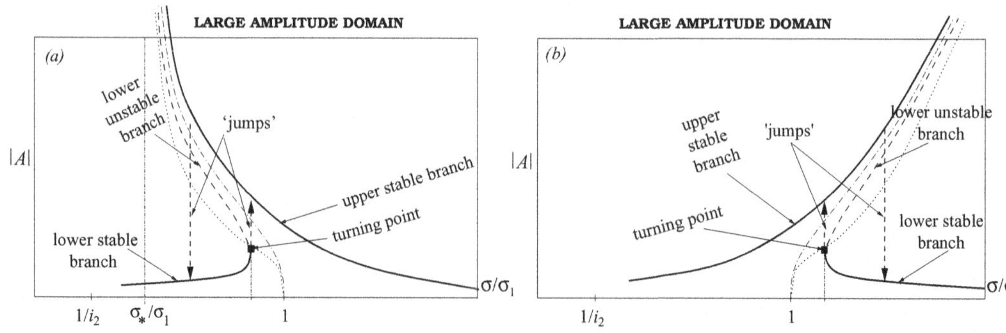

Figure 8.7. Schematic response curves for (a) soft-spring ($\overline{h} > h_* = 0.3368\ldots$) and (b) hard-spring behavior ($\overline{h} < h_* = 0.3368\ldots$) within the framework of the single-dominant multimodal method. The solid line represents stable steady-state regimes. The dashed line identifies an unstable subbranch.

zero gives the following nondimensional equation, coupling the primary mode amplitude A, the forcing frequency σ, the sway amplitude η_{2a}, and the nondimensional liquid depth $\overline{h}$:

$$(\overline{\sigma}_1^2 - 1)\,A + m_1(\overline{h},\overline{\sigma}_1)A^3 - P_1\eta_{2a}/l = 0, \quad (8.54)$$

where

$$m_1(\overline{h},\overline{\sigma}_1) = d_1\left(-l_0 + \tfrac{1}{2}h_0\right) - \tfrac{1}{2}d_2 - 2d_3h_0. \quad (8.55)$$

Equations (8.54) and (8.55) should be compared with the results of the Moiseev method outlined in Section 8.2.2 (for more details see Faltinsen, 1974). A difference is that m_1 is a function of both $\overline{h}$ and (σ_1/σ), whereas $m_1^{(0)}$ following from the Moiseev method is only a function of $\overline{h}$. The reason is that the Moiseev method assumes $\sigma \approx \sigma_1$, whereas our present derivations do not require σ to be close to σ_1 and, as a consequence, l_0 and h_0 in eq. (8.55) are functions of $\overline{\sigma}_1 = \sigma_1/\sigma$. Faltinsen *et al.* (2000) discussed the fact that

$$m_1(\overline{h},\overline{\sigma}_1) = m_1^{(0)} + \left.\frac{\partial m_1}{\partial\overline{\sigma}_1}\right|_{\overline{\sigma}_1=1}(\overline{\sigma}_1 - 1) + o(\overline{\sigma}_1 - 1);$$
$$m_1^{(0)}(\overline{h}) = m_1(\overline{h},1), \quad (8.56)$$

the ability of eq. (8.54) to better predict the response curves when the forcing frequency is slightly away from the primary resonance, and the occurrence of the secondary resonances. Similar to the Moiseev method, eq. (8.56) implies that the response is infinite at the critical depth when the forcing frequency is equal to the lowest natural sloshing frequency. The graph in Figure 9.7 shows the behavior of $m_1^{(0)}$ as function of $\overline{h}$.

When the forced roll is in phase with sway (i.e., $\eta_4(t) = \eta_{4a}\cos(\sigma t)$), eq. (8.54) can be generalized as follows:

$$(\overline{\sigma}_1^2 - 1)\,A + m_1A^3$$
$$-P_1\left[\eta_{2a}/l + \eta_{4a}(\tilde{S}_1 - g/(l\sigma^2))\right] = 0. \quad (8.57)$$

Generally speaking for combined lateral and angular excitations, one should consider the forcing

$$\eta_2(t) = \eta_{2a}\cos(\sigma t + \varepsilon_2) \quad \text{and}$$
$$\eta_4(t) = \eta_{4a}\cos(\sigma t + \varepsilon_4), \quad (8.58)$$

where the phase angles are not equal (i.e., $\varepsilon_2 \neq \varepsilon_4$). Even though the phases differ for sway and roll, considering eq. (8.58) does not change the structure of eq. (8.57). New variables P and ψ are defined by

$$P_1\sigma^2\left[\frac{\eta_{2a}}{l}\cos(\sigma t + \psi_2) + \eta_{4a}\cos(\sigma t + \psi_4)\right.$$
$$\left.\times\left\{\tilde{S}_1 - \frac{g}{l\sigma^2}\right\}\right] = P\sigma^2\cos(\sigma t + \psi).$$

The latter term appears in the expression for $\tilde{K}_1(t)$ in (8.43). Equation (8.57) then takes the form

$$(\overline{\sigma}_1^2 - 1)\,A + m_1(\overline{h},\overline{\sigma}_1)A^3 - P = 0. \quad (8.59)$$

The typical response curves for $\overline{h} > h_*$ and $\overline{h} < h_* = 0.3368\ldots$ are exemplified in Figure 8.7. The response curves change from hard- to soft-spring behavior at the critical depth $h = h_* l = 0.3368\ldots \cdot l$. The response curves are the same for the hard-spring response, but they differ slightly from the branches in Figure 8.6, following from the Moiseev method and eq. (8.32) for the soft-spring case. The reason for the difference is a vertical asymptote in Figure 8.7(a) away from the linear resonance $\overline{h} > h_*$. The appearance of the asymptote, at which two of the three real roots A of eq. (8.59) tend to infinity, is caused by the limit $m_1(\overline{h}, \overline{\sigma}_1) \to 0$ as $\overline{\sigma}_1 \to \sigma_1/\sigma_*$. The limit transforms the left-hand side of expression (8.59) from cubic to linear polynomial form. Two of the roots of the cubic equation, therefore, disappear by tending to plus and minus infinity, respectively, but the third is finite; it becomes equal to $A = P/(\overline{\sigma}_1^2 - 1)$. For the hard-spring response, σ_1/σ_* is in a domain where the single-dominant theory is not applicable. Because of the presence of damping, the asymptotic behavior has no relevance in reality.

When a root A of cubic algebraic equation (8.54) is known, one can construct an asymptotic solution of modal system (8.43). This procedure is straightforward but tedious and consists of gathering the coefficients at the first three Fourier harmonics responsible for the third-order terms of the A-value. We have used Maple to perform it. The result is

$$\begin{aligned}\beta_1(t) &= A\cos(\sigma t) + \tilde{n}_1 A^3 \cos(3\sigma t) + O(A^5),\\ \beta_2(t) &= A^2(l_0 + h_0\cos(2\sigma t)) + O(A^4),\\ \beta_3(t) &= \cos(\sigma t)\left[\tilde{N}_1 A^3 - \varepsilon P_3/\left(1-\overline{\sigma}_3^2\right)\right]\\ &\quad + \tilde{N}_2 A^3\cos(3\sigma t) + O(A^5),\end{aligned} \tag{8.60}$$

where

$$\begin{aligned}\tilde{n}_1 &= -\frac{d_2 + h_0(3d_1 + 4d_3)}{2\left(9 - \overline{\sigma}_1^2\right)},\\ \tilde{N}_1 &= \frac{-3q_2 + q_4 + 2h_0\left(-q_1 - 4q_3 + 2q_5\right) - 4q_1 l_0}{4\left(1 - \overline{\sigma}_3^2\right)},\\ \tilde{N}_2 &= -\frac{q_2 + q_4 + 2h_0(q_1 + 4q_3 + 2q_5)}{4\left(9 - \overline{\sigma}_3^2\right)}.\end{aligned} \tag{8.61}$$

The stability analysis based on the two-time-scale technique from Section 8.2.1.2 can also be combined with the multimodal theory. The solution is expressed as

$$\beta_1 = (A + \gamma(\tau))\cos\sigma t + \overline{\gamma}(\tau)\sin\sigma t + o\left(\varepsilon^{1/3}\right), \tag{8.62}$$

where A is a solution of system (8.57) and the perturbations $\gamma, \overline{\gamma}$ of the steady-state solution depend on the slowly varying time $\tau = \frac{1}{2}\varepsilon^{2/3}\sigma t$. Inserting eq. (8.62) into eq. (8.43), gathering terms of the lowest asymptotic order, and keeping linear terms in γ and $\overline{\gamma}$ lead to the following linear system of ordinary differential equations:

$$\frac{d\mathbf{c}}{d\tau} = \mathbf{C}\mathbf{c}, \tag{8.63}$$

where $\mathbf{c} = (\gamma, \overline{\gamma})^T$ and the quadratic matrix $\mathbf{C}$ consists of the elements

$$\begin{aligned}c_{11} &= c_{22} = 0; \quad c_{12} = \left[\overline{\sigma}_1^2 - 1 + m_1A^2\right],\\ c_{21} &= -\overline{\sigma}_1^2 + 1 - 3m_1A^2.\end{aligned}$$

Because the time dependence of a solution of eq. (8.63) can be expressed as $\exp(\eta\tau)$, a necessary condition for eq. (8.62) to have nontrivial solutions is $\det[-\eta\mathbf{E} + \mathbf{C}] = 0$, where "det" indicates the determinant and $\mathbf{E}$ is the identity matrix. By evaluating the determinant it follows that η must satisfy

$$\eta^2 + \left(\overline{\sigma}_1^2 - 1 + m_1A^2\right)\left(\overline{\sigma}_1^2 - 1 + 3m_1A^2\right) = 0. \tag{8.64}$$

Bearing in mind that instability is associated with real roots η of eq. (8.64), we can now give a new treatment of the stability by introducing two auxiliary branches, which are denoted by dotted and dashed-and-dotted curves in Figure 8.7. Such curves are defined by the equations

$$\begin{aligned}&\left(\overline{\sigma}_1^2 - 1\right) + 3m_1A^2 = 0 \quad \text{and}\\ &\left(\overline{\sigma}_1^2 - 1\right) + m_1A^2 = 0,\end{aligned} \tag{8.65}$$

respectively. These branches delimit the domain in the $(\sigma/\sigma_1, A)$-plane, where the steady-state regimes are unstable.

The second of eqs. (8.65) is an analogy of the backbone discussed in Section 8.2.1 within the framework of the Duffing equation. The backbone corresponds in this case to nonlinear free-standing waves. In a similar way as explained for the Duffing equation, jumps appear between the branches. The first of eqs. (8.65) represents the curve that crosses the turning point on the lower branch for various excitation amplitudes.

8.3.1.2 Time-domain solution and comparisons with experiments

We discuss the applicability and limitations of modal theory with one dominant mode by comparing with experimental results for forced harmonic sway oscillations of a rectangular tank. Even though intermodal relations (8.33) that lead to the modal system (8.43) and (8.44) assume steady-state periodic solutions, the theory can describe transients as well. The experimental tank model was placed on a wagon that could slide back and forth and was controlled by a hydraulic cylinder. The hydraulic system was strong enough to ensure that the motion inside the tank had a

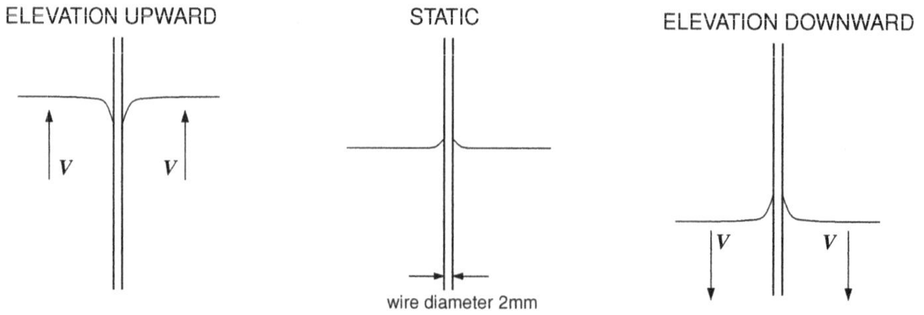

Figure 8.8. Water surface at one of the steel wires of a wave probe as the water surface is rising and falling (Langbein, 2002; Sortland, 1986).

small effect on the tank motion. The tank height, breadth, and length were 1.05, 1.73, and 0.2 m, respectively. The observed free-surface elevation was close to constant in the length direction. A reason why we say "close to constant" is the presence of meniscus effects due to surface tension on the tank walls and wave probes (see, Figure 8.8) that cause three-dimensional capillary waves. The amplitude η_{2a} of sway excitation was between 0.02 and 0.08 m, which corresponds to nondimensional $\varepsilon = \eta_{2a}/l$ from 0.0116 to 0.046. Different forcing frequencies around the lowest resonance frequency σ_1 were tested. The model tests were done with nondimensional water depths $h/l =$ 0.1156, 0.173, 0.289, and 0.3468. Furthermore, the tank was equipped with three wave probes, referred to as FS1, FS2, and FS3 (see Figure 8.9). Probes FS1 and FS2 consist of adhesive copper tape placed directly on the tank wall. Probe FS3 is made of steel wire and stands 0.05 m from the left wall. An error source is associated with the steel wires due to the meniscus effect caused by the surface tension, as illustrated in Figure 8.8.

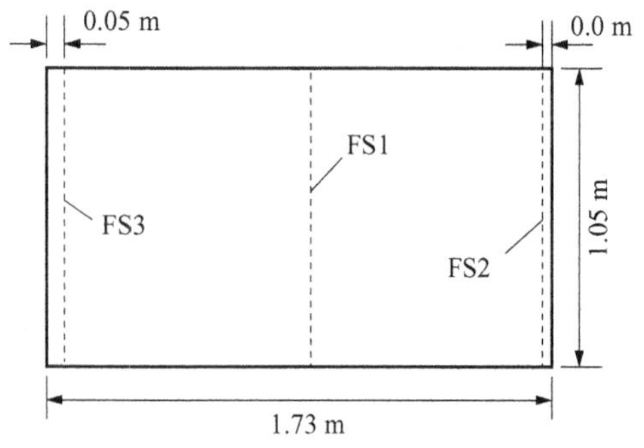

Figure 8.9. Tank dimensions and wave probe positions used in the experiments (Faltinsen *et al.*, 2000).

Roughly speaking we may say that the meniscus effect causes a bias error in the measured wave elevation that is less than the diameter of the steel wire (i.e., 2 mm in our case).

The multimodal method does not predict run-up with a thin liquid layer along a tank wall. Wave probe FS3 is therefore used instead of FS2 for the following comparisons between experimental and theoretical wave elevation predictions near a tank wall.

The sampling frequency was 50 Hz and the measured transient waves were typically 50 s long. Video recordings and visual observation of longer simulations, up to 5 minutes, showed that clear steady-state regimes were not achieved. The video recording for the two largest depths, $h/l = 0.289$ and 0.3468, indicates small damping. No wave breaking was observed, so the small damping can mainly be related to the viscous boundary layer along the wetted tank surface for a clean tank (see Section 6.3).

The most interesting transient stage occurs during the first 50 s. After this time, typical sloshing behavior is repeated with a beating effect.

In the experiments, the free-surface elevation was small and in some cases zero *before the tank's excitation started.* In a time interval lasting up to one-and-a-half forcing periods, the forced tank motion is of transient character. Furthermore, when the latter transient tank motion disappears, the tank moves with nearly constant amplitude η_{2a} and frequency σ. The following numerical simulations do not consider the initial phase when the tank motion has a clearly transient nature. The time history of the sway motion of the tank is prescribed as $\eta_2(t) = \eta_{2a}\sin(\sigma t)$. The initial time of the numerical tank motions, $t = 0$,

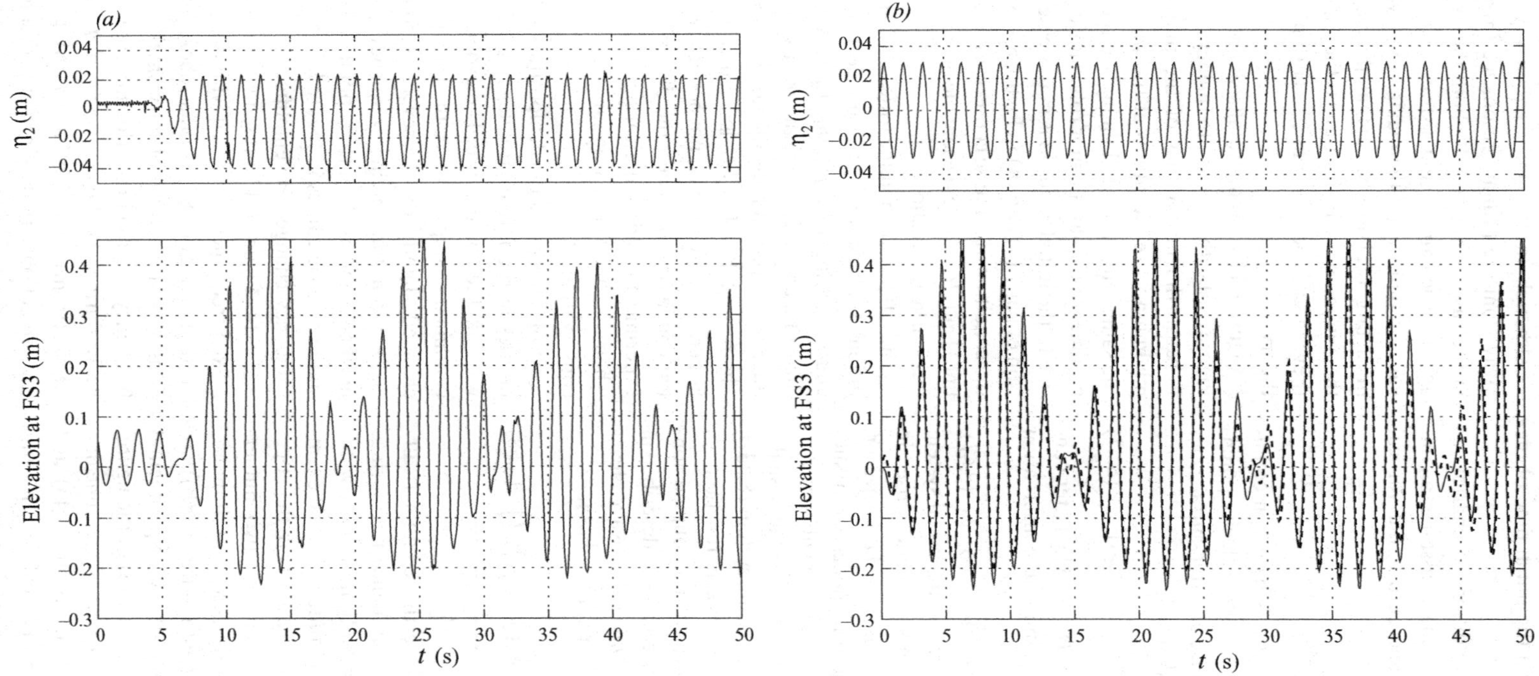

Figure 8.10. (a) Measured tank position and free-surface elevation at wave probe FS3 for $\overline{h} = 0.3468$ and $\sigma/\sigma_1 = T_1/T = 1.11$. The nondimensional sway amplitude is $\eta_{2a}/l = 0.0173$. (b) Calculations by the single-dominant multimodal method for two different initial conditions (Faltinsen *et al.*, 2000). The solid line in (b) corresponds to zero initial conditions, while the dashed line corresponds to impulse conservation condition (8.67).

is therefore different from the initial time of the experiments.

Two different initial conditions (scenarios) at $t = 0$ are tested. The first scenario assumes an initially calm free surface and zero initial vertical velocity:

$$\beta_i(0) = \dot{\beta}_i(0) = 0, \quad i \geq 1. \tag{8.66}$$

The second initial scenario is based on conditions (5.55), which states that the velocity potential on the mean free surface and free-surface elevation are initially zero, that is:

$$\beta_i(0) = 0, \quad \dot{\beta}_i = -\sigma P_i \eta_{2a}/l, \quad i \geq 1, \tag{8.67}$$

where P_i is defined by eqs. (8.46). The latter conditions are referred to as impulse conservation conditions.

Based on initial conditions (8.66) or (8.67), one can numerically time-integrate modal system (8.43) and (8.44). The number of modes used in the following comparisons is $N = 11$. However, the linearly modeled natural modes from subsystem (8.44) give, in all cases, a small contribution so that the graphs do not change visually when using $N = 3$ (i.e., using only system (8.43)). When the nondimensional modal functions β_i have been numerically solved by the coupled nonlinear ordinary differential equations described in the previous section, the *dimensional* wave elevations at $y \in [-\frac{1}{2}l, \frac{1}{2}l]$ can be expressed as

$$\zeta(y, t) = l \sum_{i=1}^{\infty} \beta_i(t) \underbrace{\cos\left(\pi i \left(y + \frac{1}{2}l\right)/l\right)}_{f_i(y)} \tag{8.68}$$

Faltinsen *et al.* (2000) investigated numerically how the initial conditions influence free-surface elevation for different excitation periods, water depths, and excitation amplitudes. The first example in Figure 8.10 refers to a case where the effect of initial conditions is secondary; two numerical series conducted with zero and impulse-conservation conditions (see Figure 8.10(b)) are in satisfactory agreement with the experimental measurements presented in part (a) of the figure. The experimental amplitude decays with time due to damping. Because modal system (8.43) and (8.44) is a conservative mechanical system without dissipation and the numerical damping is negligible, the effect of initial conditions does not decay with time.

Although the forcing frequency was slightly away from the primary resonance ($\sigma/\sigma_1 = 1.11$) in the case of Figure 8.10, both the experiments and the calculations demonstrate strongly nonlinear transient waves: the maximum elevations are much larger (almost twice) than the magnitude of the minimum wave elevation at the wall, whereas they should be the same according to linear theory.

Figure 8.11 presents results similar to those in Figure 8.10 with a forcing frequency $\sigma/\sigma_1 = 1.283$. Because the magnitudes of the maximum and minimum wave elevations at the wall are nearly the same after a short "buildup" of the response, nonlinearities play a minor role. Sloshing is sensitive to the initial conditions in this case. Comparison of parts (a) and (b) in Figure 8.11 shows that the condition of impulse conservation (the dashed line) leads to a more reasonable description of the free-surface elevation.

The effects of nonlinearities and initial conditions become stronger with decreasing depth and increasing forcing amplitude. We illustrate this in Figure 8.12, where the relative forcing frequency is approximately the same as in the case of Figure 8.11 but $\bar{h} = h/l$ is smaller and η_{2a}/l is larger. The agreement is not perfect, with zero initial conditions; however, the difference between experimental and numerical simulation is smaller when the initial conditions are based on impulse conservation.

Faltinsen *et al.* (2000) demonstrated that uncertainties in the forcing frequency affect transient waves in the initial phase. The experimental data contained examples where the forcing frequency changed slightly during several seconds, especially when the forcing frequency was in the neighborhood of the turning point on the steady-state response curves in Figure 8.7. The effect of frequency change is exemplified in Figures 8.13 and 8.14. Figure 8.13 presents simulations with a fixed frequency compared with the experimental results. Figure 8.14 shows theoretical predictions with the effect of varying the excitation frequency; the forcing period increases monotonically during the first 12 s, from $T = 1.76$ to 1.875 s (from $\sigma/\sigma_1 = 0.997$ to 0.9356). The theoretical predictions are better than with a fixed forcing amplitude and frequency. Separate numerical experiments on similar transients showed that variations of the forcing amplitude have less

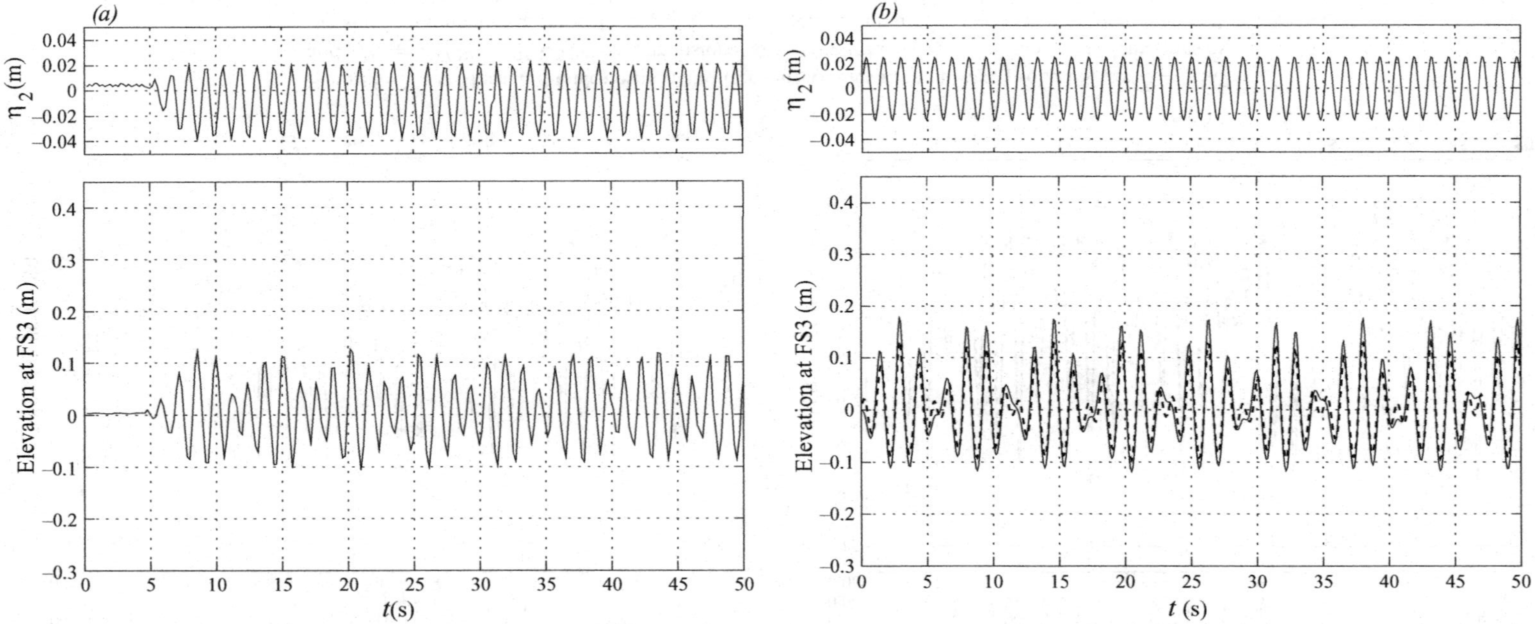

Figure 8.11. The same as in Figure 8.10, but for $\sigma/\sigma_1 = T_1/T = 1.283$ and $\eta_{2a}/l = 0.01445$ (Faltinsen *et al.*, 2000).

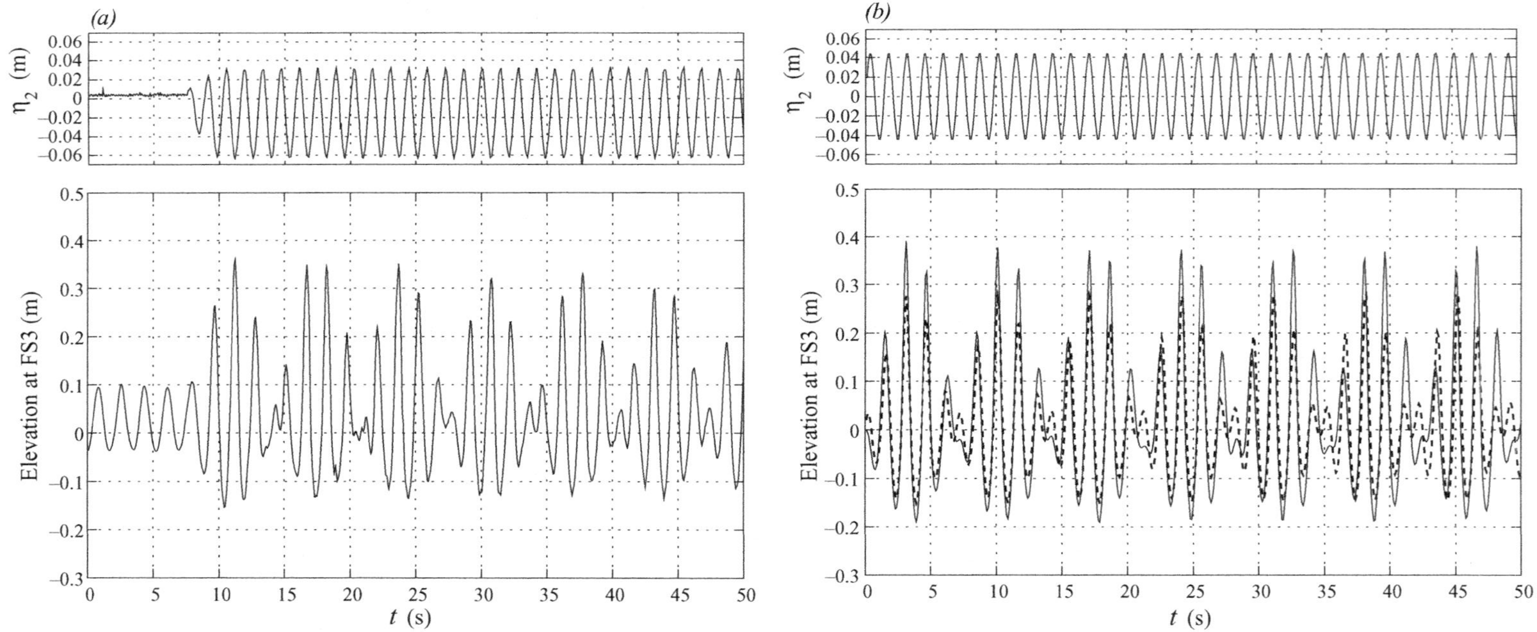

Figure 8.12. (a) Measured tank position and free-surface elevation at wave probe FS3 for $\overline{h} = 0.289$ and $\sigma/\sigma_1 = T_1/T = 1.253$. The nondimensional sway amplitude is $\eta_{2a}/l = 0.026$. (b) Calculations by the single-dominant multimodal method for two different initial conditions (Faltinsen *et al.*, 2000). The solid line in (b) corresponds to zero initial conditions, while the dashed line corresponds to impulse conservation condition (8.67).

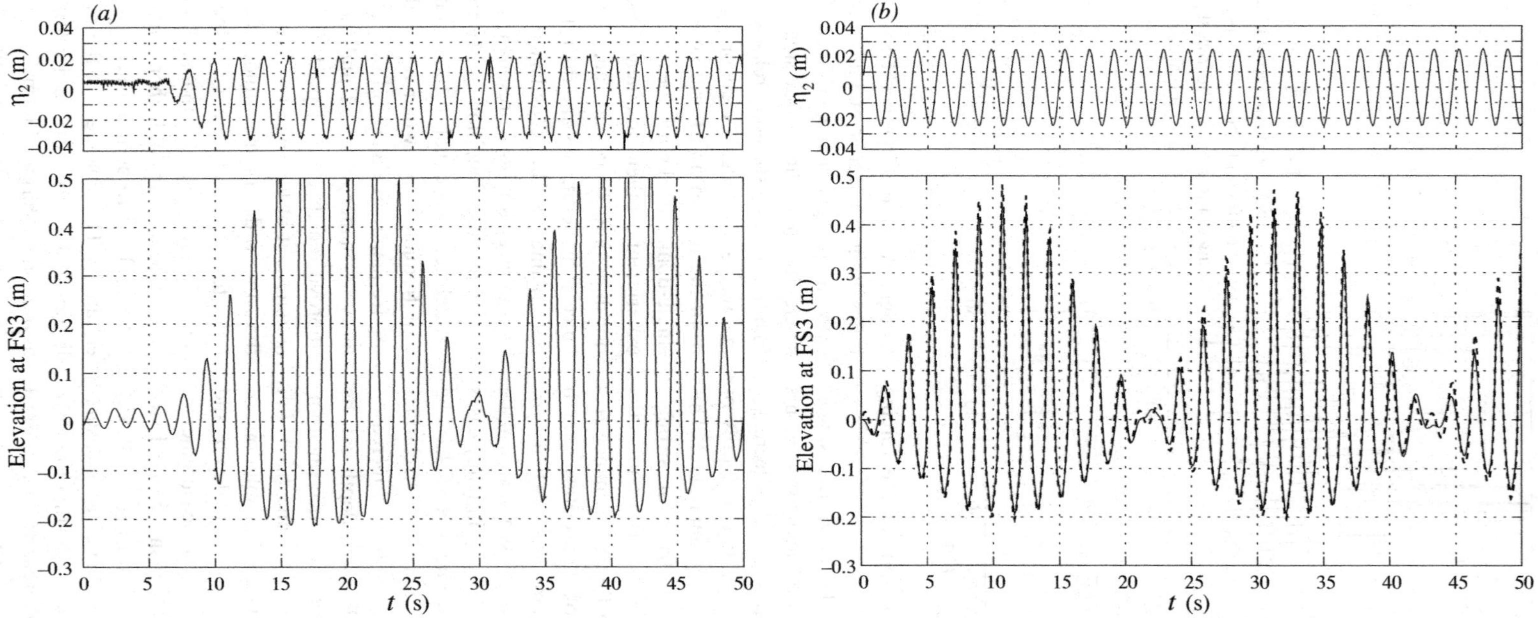

Figure 8.13. Measured tank positions and free-surface elevations at FS3: $\overline{h} = 0.289$, $\eta_{2a}/l = 0.01445$. The calculations in (b) assume that the ratio between the forcing frequency and the first natural frequency is fixed and equal to $\sigma/\sigma_1 = 0.936$ (Faltinsen *et al.*, 2000). The solid line in (b) corresponds to zero initial conditions, while the dashed line corresponds to impulse conservation condition (8.67).

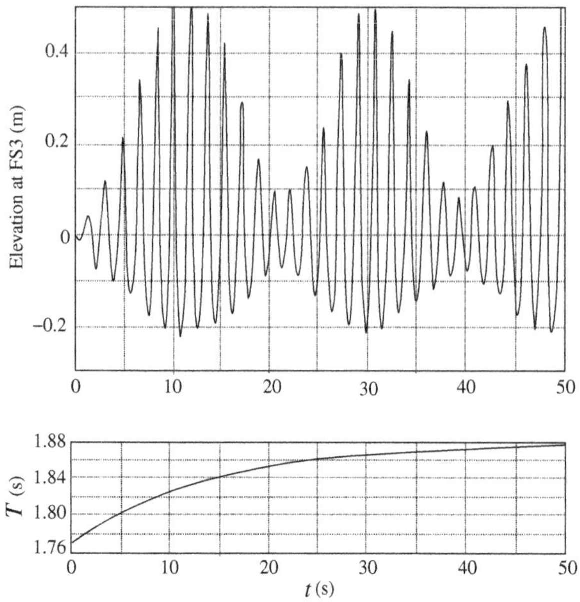

Figure 8.14. The same simulations as in Figure 8.13(b) but accounting for the time variations of the forcing period, T, according to the curve in the lowest graph (Faltinsen *et al.*, 2000). Zero initial conditions are used.

of an effect than the variations of the forcing frequency.

Extensive comparisons between the single-dominant multimodal method and experiments showed increasing disagreement as h/l decreases for intermediate and shallow liquid depths or as η_{2a}/l increases (large forcing amplitude). This disagreement is particularly true for strongly resonant cases with $\sigma/\sigma_1 \approx 1$. Furthermore, the modal system fails at the linear resonance frequency in the vicinity of the critical depth $h/l \approx 0.3368\ldots$ Nonresonant cases, when σ/σ_1 is not too close to the main resonance, can be described even for nonsmall η_{2a}/l, in many cases by the linear modal theory from Section 5.4.2. The asymptotic character of the modal system clarifies why the solution breaks down as η_{2a}/l increases. The situation with a small liquid depth can be explained by the secondary resonance phenomena discussed in Sections 4.3.1.1 and 8.3.4. Let us demonstrate how the theoretical solutions of system (8.43) behave for intermediate depths and compare them with experimental measurements.

The first example is presented in Figure 8.15 for $h/l = 0.1734$, $\sigma/\sigma_1 = 0.96$, and $\eta_{2a}/l = 0.02832$. Comparisons with the experimental measurements of the wave elevations show that the wave crest is well predicted, whereas the theoretical values for the trough are clearly lower than in the experiments. A linear theory (see Chapter 5) predicts equal magnitudes for the crest and the trough. A large difference between the magnitudes of the wave crest and the wave trough indicates a strong nonlinear effect. The nonlinear single-dominant theory (see steady-state solution (8.60) as an illustration) relates the difference in crest and trough behavior to contributions from the second and third modes. Because the single-dominant theory does not agree with the experiments, other nonlinear intermodal relationships must be considered. The equal importance of higher modes and the primary mode can be explained by the secondary resonance caused by nonlinear forcing of resonance oscillations of higher modes. Condition (4.14) gives $i_2^{-1} = 0.9$. Therefore, 2σ is equal to the second-lowest natural frequency, σ_2, when $i_2^{-1} = 0.9$. The forcing frequency in Figure 8.15 is between the primary resonance $\sigma/\sigma_1 = 1$ and the i_2^{-1}-value. To improve the theoretical predictions we have to assume that at least the two lowest modes have the same order of magnitude. This assumption indicates a complete change of the equation system is needed, implying that additional nonlinear quantities, in terms of higher generalized coordinates β_i, have to be introduced in the nonlinear equations. The smallness of the nondimensional liquid depth must also be accounted for in the asymptotic ordering of terms.

The second example in Figure 8.16 shows that the simulation by asymptotic modal system (8.43) may break down for intermediate liquid depths.

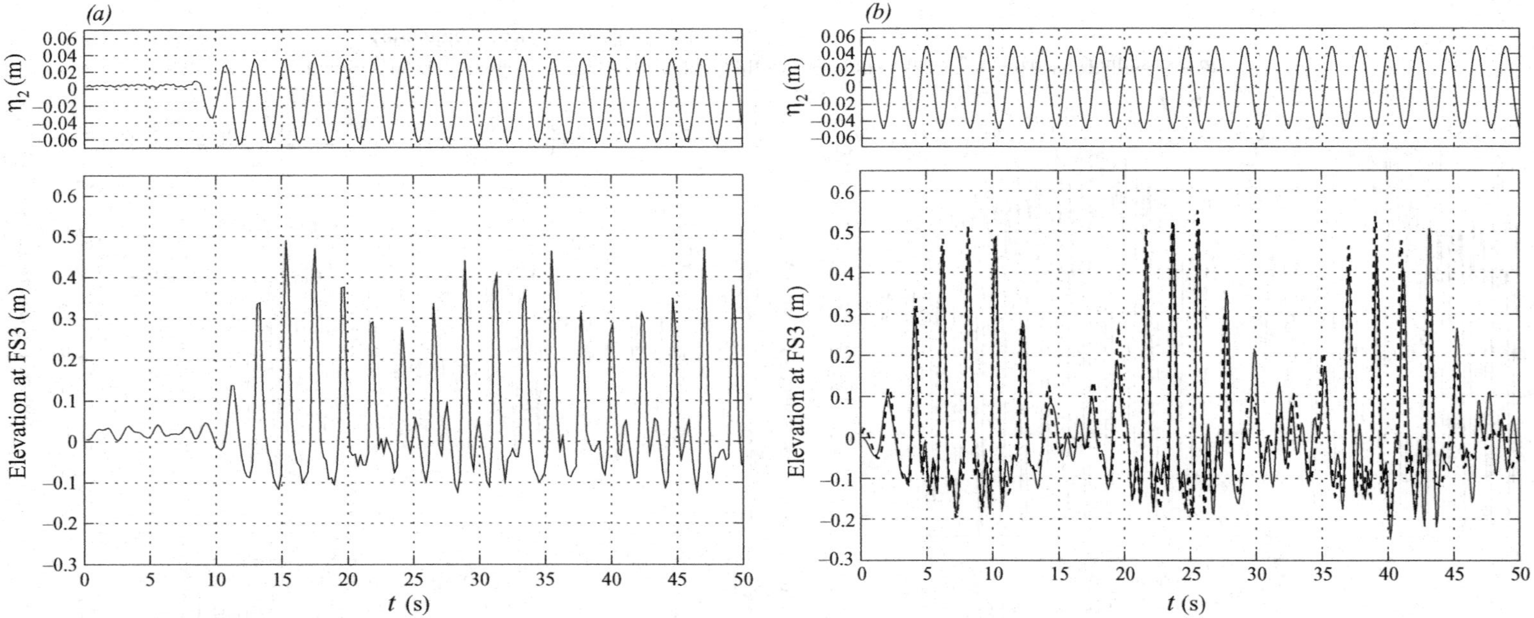

Figure 8.15. (a) Measured and (b) calculated dimensional wave elevation at FS3 for $\overline{h} = 0.1734$, $\sigma/\sigma_1 = 0.96$, and $\eta_{2a}/l = 0.02832$ (Faltinsen *et al.*, 2000). The solid line in (b) corresponds to zero initial conditions, while the dashed line corresponds to impulse conservation condition (8.67).

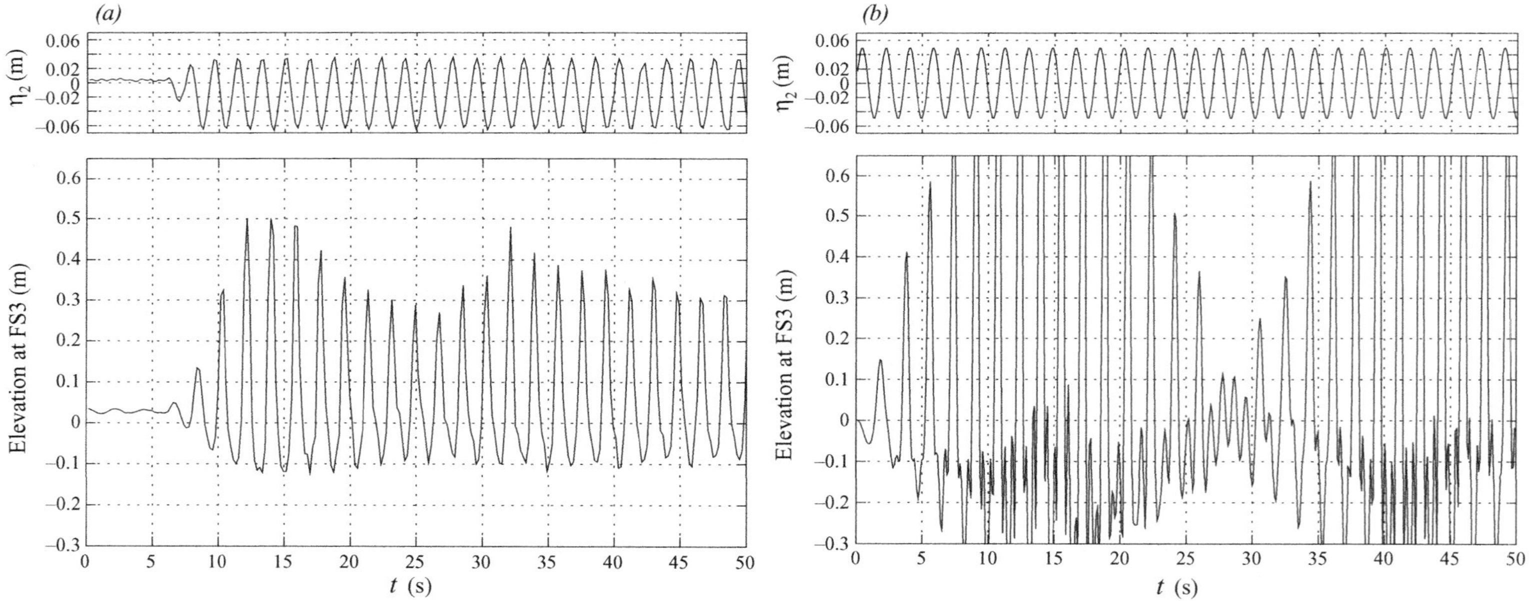

Figure 8.16. The same as in Figure 8.15, but with $\sigma/\sigma_1 = 1.1734$ and $\eta_{2a}/l = 0.0289$ (Faltinsen *et al.*, 2000). Zero initial conditions are used in the numerical simulations in (b).

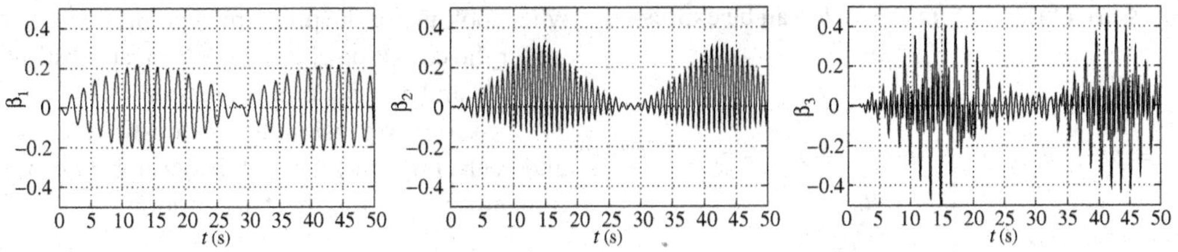

Figure 8.17. Contribution of the lower three modes (nondimensional $\beta_i(t)$) to the calculated free-surface elevation in Figure 8.16 using single-dominant modal theory (Faltinsen *et al.*, 2000).

In this case, $\sigma/\sigma_1 = T_1/T = 1.174$ and $h/l = 0.1734$. The forcing frequency is away from the secondary resonance with the ratio σ/σ_1 somewhere to the left of the abscissa of the turning point in Figure 8.7(b). We expect a steady-state solution of large amplitude associated with a point on the upper branch. This solution increases both the dominant mode and some of the higher modes due to nonlinear energy transfer. The last fact can be shown by presenting the contributions of higher modes from modal system (8.43) and (8.44), as illustrated in Figure 8.17.

As one can see from the previous examples, increasing the forcing amplitude and related response of the primary excited mode or decreasing the liquid depth causes amplification of higher modes and failure of the single-dominant modal system. These ways of amplification, as well as the critical depth case, were considered by Faltinsen and Timokha (2001, 2002) and later on, for three-dimensional sloshing, by Faltinsen *et al.* (2005a, 2005b, 2006b). As long as a higher mode becomes amplified (together with the primary mode), asymptotic ordering (8.33) is not valid. A way to capture the amplification leads to special *adaptive modal systems.* These systems account for different kinds of nonlinear energy transfer from the primary mode to some higher modes. We give more details in Section 8.4.3.

8.3.2 Nonimpulsive hydrodynamic loads

8.3.2.1 Hydrodynamic pressure

Nonimpulsive pressures on the tank surface follow by using Bernoulli's equation expressed in the tank-fixed coordinate system (see eq. (2.61)). Based on modal representation (7.2) and keeping terms of consistent order, the dimensional velocity potential is given by (see expressions (8.39) for R_i)

$$\Phi = \dot\eta_2(t)y + \dot\eta_3(t)z + \dot\eta_4(t)\Omega_{01}(y,z) + l^2\sum_{n=1}^{\infty} R_n(t) \underbrace{\times \cos\left[\frac{\pi n}{l}\left(y+\tfrac{1}{2}l\right)\right]\frac{\cosh(\pi n(z+h)/l)}{\cosh(\pi nh/l)}}_{\varphi_n(y,z)} \tag{8.69}$$

where the Stokes–Joukowski potential Ω_{01} can be expressed by eq. (5.80) as

$$\Omega_{01}(y,z) = -yz + 4\sum_{j=1}^{\infty}\frac{l^2\left[(-1)^j-1\right]}{(j\pi)^3} \times \cos\left[\frac{\pi j}{l}\left(y+\tfrac{1}{2}l\right)\right]\frac{\sinh(\pi j\,(z+h/2)\,/l)}{\cosh(\pi jh/2l)}.$$

By using the Bernoulli equation it follows that the total pressure in the liquid can be approximated as

$$p(y,z,t) = p_0 - \rho g(z+y\eta_4) - \rho[y\ddot\eta_2 + z\ddot\eta_3 + \Omega_{01}(y,z)\,\ddot\eta_4] - \rho l^2\sum_{i=1}^{\infty}\dot R_i\varphi_i(y,z) - \rho l^4\left(\tfrac{1}{2}R_1^2|\nabla\varphi_1|^2 + R_1R_2(\nabla\varphi_1\cdot\nabla\varphi_2)\right), \tag{8.70}$$

where $\dot R_i$ are expressed by eq. (8.40) and the $o(\varepsilon)$-terms in R_i are omitted. Using a solution of modal system (8.43) and (8.44) and formula (8.70), we can find the pressure asymptotically, up to the $O(\varepsilon)$-terms. Care must be shown in using the formula directly in the *free-surface* zone, where the tank surface is not wetted all the time. Below such a zone, we can directly substitute the tank surface coordinate into eq. (8.70). For instance, the pressure at the bottom ($z = -h$) can be expressed as

$$
\begin{aligned}
p(y, -h, t)
&= p_0 + \rho h (g + \ddot{\eta}_3) - \rho y \left[\ddot{\eta}_2 + g\eta_4 + h\ddot{\eta}_4\right] \\
&\quad - \rho l^2 \sum_{n=1}^{\infty} f_n(y) \left[\frac{\dot{R}_n}{\cosh(\pi n h/l)} \right. \\
&\quad \left. - \ddot{\eta}_4 \frac{4((-1)^n - 1)}{(\pi n)^3} \tanh\left(\frac{\pi n h}{2l}\right) \right] \\
&\quad - \rho l^2 \frac{\pi^2 \sin(\pi(y+\frac{1}{2}l)/l)}{4E_1 \cosh(\pi h/l)} \left[\frac{\dot{\beta}_1^2 \sin(\pi(y+\frac{1}{2}l)/l)}{2E_1 \cosh(\pi h/l)} \right. \\
&\quad \left. + \left(\dot{\beta}_1\dot{\beta}_2 - \frac{4E_0}{E_1}\beta_1\dot{\beta}_1^2\right) \frac{\sin(2\pi(y+\frac{1}{2}l)/l)}{E_2 \cosh(2\pi h/l)} \right],
\end{aligned}
\tag{8.71}
$$

where E_i is expressed by formulas (8.41). Furthermore, $\dot{R}_i$ is given by eq. (8.40).

When we want to find pressures in the free-surface zone, the expressions of $\varphi_i(y, z)$ and Ω_{01} in eq. (8.70) must be expanded in a Taylor series about the mean free surface $z = 0$ and consistently kept to $O(\varepsilon)$. Therefore,

$$
\begin{aligned}
\sum_{i=1}^{\infty} \dot{R}_i\varphi_i(y, z) &\sim \sum_{i=1}^{\infty} \dot{R}_i\varphi_i(y, 0) \\
&+ z\dot{R}_1 \left.\frac{\partial\varphi_1}{\partial z}\right|_{z=0} + \tfrac{1}{2}z^2\dot{R}_1 \left.\frac{\partial^2\varphi_1}{\partial z^2}\right|_{z=0} + z\dot{R}_2 \left.\frac{\partial\varphi_2}{\partial z}\right|_{z=0}.
\end{aligned}
$$

If we did not perform a Taylor-series expansion but rather directly substituted z-values into eq. (8.70), φ_n would be exponentially large for positive z and large n. We can see this by noting that

$$
\frac{\cosh[\pi n (z+h)/l]}{\cosh(\pi n h/l)} \sim \exp\left(\frac{\pi n z}{l}\right)
$$

for large n. Let us present consistent expressions for the total pressure in the free-surface zone at the tank walls when the considered point is wetted. We introduce the instantaneous contact points of the free surface, $z_{\mp}(t) = \zeta(\mp\frac{1}{2}l, t)$, and the vertical walls and take into account $z_{\mp}/l \le O(\varepsilon^{1/3})$. Furthermore, we consider points $(\mp\frac{1}{2}l, z_{p\mp})$ on the wall located in the free-surface zone somewhere beneath the instantaneous free surface (i.e., $z_{p\mp} \le z_{\mp}$). When $z_{p\mp} > z_{\mp}$, the pressure is p_0. The closeness of the points $(\mp\frac{1}{2}l, z_{p\mp})$ to the free surface means that $z_{p\mp} \approx z_{\mp}$, namely, $z_{p\mp}/l \le O(\varepsilon^{1/3})$. The latter fact must be accounted for in the spatial expressions of formula (8.70) when looking for the pressure at points $(\mp\frac{1}{2}l, z_{p\mp})$ under the condition $z_{p\mp} \le z_{\mp}$. We must substitute $y = \mp\frac{1}{2}l$ and $z = z_{p\mp}$ into expression (8.70), use a Taylor-series expansion as previously described, and neglect the $o(\varepsilon)$-terms. Furthermore, we must remember that, for the tank motion, $\eta_i = O(\varepsilon)$. The result is

$$
\begin{aligned}
p\left(\mp\tfrac{l}{2}, z_{p\mp}\right)
&= p_0 - \rho g l \left(\tilde{z}_{p\mp} \mp \tfrac{1}{2}\eta_4\right) \\
&\quad - \rho l^2 \left(\mp\tfrac{1}{2}\ddot{\eta}_2/l + \ddot{\eta}_4 l^{-2}\Omega_{01}(\mp\tfrac{1}{2}l, 0) + \sum_{i=1}^{\infty}(\pm 1)^i \dot{R}_i\right) \\
&\quad - \rho l^2 \left[\pm\ddot{\beta}_1\tilde{z}_{p\mp} \left(1 + \frac{\pi}{\tanh(\pi\bar{h})}\left(\tfrac{1}{2}\tilde{z}_{p\mp} \mp \beta_1\right)\right) \right. \\
&\quad \left. + \ddot{\beta}_2\tilde{z}_{p\mp} + \dot{\beta}_1^2\left(\tfrac{1}{2} - \frac{\pi(\tilde{z}_{p\mp} \pm \beta_1)}{\tanh(\pi\bar{h})}\right) \pm \dot{\beta}_1\dot{\beta}_2 \right],
\end{aligned}
\tag{8.72}
$$

where $\tilde{z}_{p\mp} = z_{p\mp}/l$.

8.3.2.2 Hydrodynamic force

The hydrodynamic force follows from Lukovsky formula (7.22) and the fact that the instantaneous mass center (x_{lC}, y_{lC}, z_{lC}) can be expressed explicitly in terms of modal functions β_i. From eq. (7.20) it follows that

$$
x_{lC}(t) = 0; \quad y_{lC}(t) = -\frac{l^2}{\pi^2 h}\sum_{i=1}^{\infty}\beta_i(t)\frac{1+(-1)^{i+1}}{i^2},
$$

$$
z_{lC}(t) = -\tfrac{1}{2}h + \frac{l^2}{h}\sum_{i=1}^{\infty}\beta_i^2(t), \tag{8.73}
$$

where the z-coordinate can be simplified by neglecting higher-order terms; that is,

$$
z_{lC}(t) = -\tfrac{1}{2}h + l^2 h^{-1}\left[\beta_1^2 + o(\varepsilon)\right]. \tag{8.74}
$$

Neglecting the $o(\varepsilon)$-terms gives the following components of the hydrodynamic force:

$$
\begin{aligned}
F_1(t) &= 0, \\
F_2(t) &= m_l \left(-g\eta_4(t) - \tfrac{1}{2}h\ddot{\eta}_4(t) - \ddot{\eta}_2(t) \right. \\
&\qquad \left. + \frac{l^2}{\pi^2 h}\sum_{i=1}^{\infty}\ddot{\beta}_i(t)\frac{1+(-1)^{i+1}}{i^2} \right), \\
F_3(t) &= -m_l\left(\ddot{\eta}_3(t) + \frac{2l^2}{h}\left[\ddot{\beta}_1(t)\beta_1 + \dot{\beta}_1^2(t)\right]\right).
\end{aligned}
\tag{8.75}
$$

Because the horizontal component of the mass center, $y_{lC}(t)$, is a linear combination of the modal functions, the expression for the horizontal force $F_2(t)$ is the same as in the linear modal theory. The vertical force is composed of a "frozen" liquid effect $-m_l\ddot{\eta}_3(t)$ and a nonlinear sloshing effect. If steady-state oscillations are assumed and $\beta_1 \propto \cos(\sigma t + \alpha)$, the time dependence of the sloshing contribution to the vertical force is $\cos[2(\sigma t + \alpha)]$ (i.e., the time-averaged vertical force over one oscillation period is zero).

8.3.2.3 Hydrodynamic moment relative to origin 0

The modal expression for the hydrodynamic moment based on formula (7.32) is more complicated than the Lukovsky formula for the force. The moment $F_4(t)$ about the Ox-axis is the only nonzero moment component. We start by neglecting terms of $o(\varepsilon)$ in the Lukovsky formula and note that eq. (7.32) includes the squares of the forcing terms $(\boldsymbol{\omega} \times \boldsymbol{v}_O)$, $(\boldsymbol{\omega} \times (\boldsymbol{J}^1 \cdot \boldsymbol{\omega}))$, which are of $O(\varepsilon^2)$. Furthermore, the time derivatives of $\boldsymbol{J}^1$, $\boldsymbol{l}_\omega$ and $\boldsymbol{l}_{\omega t}$ are proportional to β_j and their derivatives and, therefore, are of $O(\varepsilon^{1/3})$. Therefore, $(d^*\boldsymbol{J}^1/dt) \cdot \boldsymbol{\omega}$ and $\boldsymbol{\omega} \times (d^*\boldsymbol{l}_\omega/dt - \boldsymbol{l}_{\omega t})$ are of $O(\varepsilon^{4/3})$ and can be neglected. The remaining expression for the F_4-component is

$$F_4(t) = m_l\left[y_{lC}(g_3 - \ddot{\eta}_3) - z_{lC}(g_2 - \ddot{\eta}_2)\right] - J^1_{011}\ddot{\eta}_4 - \frac{\partial}{\partial t}\left\{\dot{l}_{1\omega} - l_{1\omega t}\right\},$$

where the gravity acceleration components are defined by eq. (5.13). Furthermore, we may exclude the higher-order terms from the square-bracket terms. Because $g_2 = -g\eta_4$, the $O(1)$-term of z_{lC}, $z_{lC_0} = -\frac{1}{2}h$, matters only asymptotically. The y_{lC}-coordinate of eq. (8.73) is proportional to $O(\varepsilon^{1/3})$ and, therefore, we must neglect $y_{lC}\eta_3$. The final result is

$$F_4(t) = \underline{-m_l\tfrac{1}{2}h\left[g\eta_4 + \ddot{\eta}_2\right] - m_l g y_{lC} - J^1_{011}\ddot{\eta}_4} - \frac{\partial}{\partial t}\left\{\dot{l}_{1\omega} - l_{1\omega t}\right\}, \tag{8.76}$$

where the underlined part also appears in the linear expression. The term $m_l g y_{lC}$ is the same as $g\sum\lambda_{2j}\beta_j$ in eq. (5.89) and the linear component of $\partial\{\dot{l}_{1\omega} - l_{1\omega t}\}/\partial t$ is $\sum\lambda_{01j}\ddot{\beta}_j$. However, we require nonlinear terms of $\partial\{\dot{l}_{1\omega} - l_{1\omega t}\}/\partial t$ up to cubic order in β_j. The details are given by Faltinsen and Timokha (2001). The final result is

$$\begin{aligned} F_4(t) &= -m_l\tfrac{1}{2}h\left[g\eta_4 + \ddot{\eta}_2\right] - m_l g y_{lC} - J^1_{011}\ddot{\eta}_4 \\ &\quad + \tfrac{1}{2}\rho l^4\left[\sum_{i=1}^{\infty}\ddot{\beta}_i L_i^{(0)} + L_{2,1}^{(1)}\ddot{\beta}_1\beta_2 + L_{1,2}^{(1)}\ddot{\beta}_2\beta_1 \right.\\ &\quad \left. + \left(L_{2,1}^{(1)} + L_{1,2}^{(1)}\right)\dot{\beta}_1\dot{\beta}_2 + L_{1,1,1}^{(2)}\left(\ddot{\beta}_1\beta_1^2 + 2\dot{\beta}_1^2\beta_1\right)\right], \end{aligned} \tag{8.77}$$

where $L_i^{(0)} = -4[(-1)^i - 1]\tanh(\frac{1}{2}\pi\bar{h}i)/(\pi i)^3$ corresponds to the linear term $-\sum\lambda_{01j}\ddot{\beta}_j$ in eq. (5.89). Furthermore,

$$\begin{aligned} L_{1,2}^{(1)} &= \frac{4}{\pi^2}\left[\frac{5}{9} - \frac{\tanh\left(\frac{1}{2}\pi\bar{h}\right) + \frac{1}{9}\tanh\left(\frac{3}{2}\pi\bar{h}\right)}{\tanh(2\pi\bar{h})}\right], \\ L_{2,1}^{(1)} &= \frac{4}{\pi^2}\left[\frac{5}{9} + \frac{\tanh\left(\frac{1}{2}\pi\bar{h}\right) - \frac{1}{9}\tanh\left(\frac{3}{2}\pi\bar{h}\right)}{\tanh\left(\pi\bar{h}\right)}\right], \\ L_{1,1,1}^{(2)} &= -\frac{4}{3\pi\tanh(\pi\bar{h})} \\ &\quad + \frac{1}{\pi}\left[\tanh\left(\tfrac{1}{2}\pi\bar{h}\right) + \tfrac{1}{9}\tanh\left(\tfrac{3}{2}\pi\bar{h}\right)\right] \\ &\quad \times\left\{-1 + \frac{4}{\tanh(2\pi\bar{h})\tanh(\pi\bar{h})}\right\}. \end{aligned} \tag{8.78}$$

8.3.2.4 Nonimpulsive hydrodynamic loads on internal structures

Hydrodynamic loads on internal structures such as pump towers can be calculated by means of a generalized Morison equation where the modal solution is the ambient flow. We need to express the ambient cross-sectional components of velocity and acceleration at the cross-sectional centers along the internal structure. The ambient pressure is also needed at junctions between structural members. When the internal structure is free-surface piercing, the ambient flow velocities must be expanded via a Taylor series about the mean free surface.

A generalized Morison equation in a tank-fixed coordinate system was presented in Section 6.4. Let us keep in mind the analysis in Section 6.4.1 for a circular cylinder that stands on the tank bottom and penetrates the free surface.

The cylinder is vertical when the tank is at rest in an upright position. The ambient flow is two-dimensional and the tank has rigid-body sway and roll motions. The expression for lateral force on a strip of the cylinder involves the relative lateral liquid velocity component between the ambient flow and the tank-fixed cylinder. We can express this component as $v_r = \partial\Phi/\partial y - \dot{\eta}_2 + z\dot{\eta}_4$, where Φ is the total velocity potential given by eq. (8.69). When deriving the Morison equation, ambient pressure was assumed to vary slowly across the cylinder. Because the ambient wave field is composed of many natural modes, we must require that the lowest modes, with the largest wavelengths, have a dominant effect and we limit the number of modes by requiring that the wavelength of an included mode is, for example, larger than five times the cylinder diameter. Such a criterion can be derived by considering a vertical cylinder in linear regular waves in a liquid of infinite horizontal extent and infinite depth. The linear horizontal force within potential flow theory can be exactly expressed by the MacCamy–Fuchs (1954) theory. The criterion follows by comparing with the mass term in Morison's equation. Special care should be shown in evaluating v_r and the relative acceleration, $a_r = \partial v_r/\partial t$, correctly to $O(\varepsilon)$ in the free-surface zone (i.e., for the points (y_m, z), $|z| = O(\varepsilon^{1/3})$, where y_m is the y-coordinate of the cylinder axis). The result for the velocity is

$$\begin{aligned} v_r = \dot{\eta}_4(t)\frac{\partial\Omega_{01}(y,0)}{\partial y} - l\pi\Bigg\{ & R_1(t)\sin\left(\frac{\pi}{l}\left(y_a + \tfrac{1}{2}l\right)\right) \\ & \times\left[1 + \frac{\pi}{l}\tanh(\pi\bar{h})\cdot z + \frac{\pi^2}{2l^2}z^2\right] \\ & + 2R_2(t)\sin\left(\frac{2\pi}{l}\left(y_a + \tfrac{1}{2}l\right)\right) \\ & \times\left[1 + \frac{2\pi}{l}\tanh(2\pi\bar{h})\cdot z\right] \\ & + \sum_{n=3}^{\infty} nR_n(t)\sin\left(\frac{\pi n}{l}\left(y_a + \tfrac{1}{2}l\right)\right)\Bigg\}. \end{aligned} \quad (8.79)$$

Furthermore, we can write $a_r = \dot{v}_r$ and use eq. (8.79) to get the acceleration correctly to $O(\varepsilon)$.

The approximation is based on multimodal theory with one dominant mode. This theory is applicable for infinite and finite liquid depths, except in the vicinity of the critical depth in a frequency range close to the lowest natural frequency. Later in this chapter we introduce other versions of the multimodal method with more than one dominant mode that are applicable to other depth conditions. Because the ambient free-surface slope is assumed small, special analysis is needed when a steep wave front impacts the cylinder. This scenario is, for instance, relevant for shallow-liquid conditions with hydraulic jumps and is addressed in Section 8.9.

8.3.3 Coupled ship motion and sloshing

We discuss how the previously described nonlinear sloshing theory can be used in determining motion of a ship with partially filled tanks in a seaway. To simplify the presentation, we consider the case presented in Section 3.6.5; that is, a two-dimensional ship model in beam-sea regular waves. The ship model is restrained from oscillating in modes of motions other than sway. The sway amplitude results presented in Figure 3.23 with nonlinear sloshing are a consequence of the presented analysis. The unknowns in the problem are the sway motion $\eta_2(t)$ and the generalized coordinates $\beta_i(t)$ $(i = 1, \ldots, N)$ for the free-surface elevation in the tank, where N is the number of sloshing modes used in the analysis. Equations (8.43) and (8.44) provide N differential equations for $\beta_i(t)$. Rognebakke and Faltinsen (2003) included viscous damping terms $2\xi_i\sigma_i\dot{\beta}_i$ on the left-hand sides of the equations, where $\xi_i = 1$ represents critical damping for a linear mass–spring system. The damping term was based on laminar boundary-layer flow analysis with free decay as described in Section 6.3. The damping ratio between the damping and critical damping, ξ_i, is in reality less than 0.004 for the dominant mode in the cases examined by Rognebakke and Faltinsen (2003). Its important role is to damp the effect of initial conditions in long-term simulations. Because the damping ratio is so small, it does not influence the magnitude of the response at resonance in steady-state conditions. Nonlinear potential flow effects are the cause of finite amplitude at resonance.

The differential equations for β_i can now be expressed as

$$\begin{aligned} &(\ddot{\beta}_1 + 2\xi_1\sigma_1\dot{\beta}_1 + \sigma_1^2\beta_1) + d_1(\ddot{\beta}_1\beta_2 + \dot{\beta}_1\dot{\beta}_2) \\ &\quad + d_2(\ddot{\beta}_1\beta_1^2 + \dot{\beta}_1^2\beta_1) + d_3\ddot{\beta}_2\beta_1 = \tilde{K}_1(t), \end{aligned}$$

$$
\begin{aligned}
&(\ddot{\beta}_2 + 2\xi_2\sigma_2\dot{\beta}_2 + \sigma_2^2\beta_2) + d_4\ddot{\beta}_1\beta_1 + d_5\dot{\beta}_1^2 = 0, \\
&(\ddot{\beta}_3 + 2\xi_3\sigma_3\dot{\beta}_3 + \sigma_3^2\beta_3) + q_1\ddot{\beta}_1\beta_2 + q_2\ddot{\beta}_1\beta_1^2 \\
&\quad + q_3\ddot{\beta}_2\beta_1 + q_4\dot{\beta}_1^2\beta_1 + q_5\dot{\beta}_1\dot{\beta}_2 = \tilde{K}_3(t); \\
&\ddot{\beta}_i + 2\xi_i\sigma_i\dot{\beta}_i + \sigma_i^2\beta_i = \tilde{K}_i(t), \quad i = 4, \ldots, N.
\end{aligned}
\tag{8.80}
$$

In addition, we have an equation that follows from applying Newton's second law to the ship, which, based on eq. (3.68), can be written as

$$
\begin{aligned}
&[M + A_{22}(\infty)]\,\ddot{\eta}_2 + B_{22}^{\text{visc}}\dot{\eta}_2|\dot{\eta}_2| + C_{22}\eta_2 \\
&\quad + \int_0^t h_{22}(\tau)\dot{\eta}_2(t-\tau)\mathrm{d}\tau \\
&= F_2^{\text{exc}} + F_2^{\text{slosh}} + F_2^{\text{bearing}}\dot{\eta}_2/|\dot{\eta}_2|.
\end{aligned}
\tag{8.81}
$$

From eq. (8.75) it follows that the lateral sloshing force F_2^{slosh} in eq. (8.81) can be expressed as

$$
\begin{aligned}
&F_2^{\text{slosh}}(t) \\
&= m_l L_t\left(-\ddot{\eta}_2(t) + \frac{l^2}{\pi^2 h}\sum_{i=1}^{N}\ddot{\beta}_i(t)\frac{1+(-1)^{i+1}}{i^2}\right),
\end{aligned}
\tag{8.82}
$$

where L_t is the tank length in the x-direction. We refer to the detailed explanation of the other terms in eq. (8.81) in the text associated with eq. (3.68). An interesting difference is present in the expressions of the external and internal hydrodynamic loads. The external hydrodynamic loads due to ship motion have a memory effect expressed in terms of the convolution integral in eq. (8.81), whereas the hydrodynamic loads due to sloshing given by eq. (8.82) depend on the instantaneous response of the ship and the sloshing. The reason for the memory effect in the external load expression has to do with wave-radiation damping, which is not present for sloshing.

We may generalize the procedure to analyze a ship in waves with six degrees of rigid-body motion by considering the six components of hydrodynamic forces and moments caused by the tank and including them in the six equations of rigid-body ship motion. If the ship has several partially filled tanks, we just include a set of generalized coordinates for the free surface in each tank.

The described analysis is not limited to regular incident waves. We can study the effect of an irregular sea by representing the incident waves as a sum of regular waves of different amplitudes, phases, and directions, where the amplitudes and directions can be described in terms of a directional wave spectrum in a short-term sea state defined by a significant wave height, mean wave period, and mean wave direction. The phases of the individual waves are stochastically determined as described in Section 3.3 for long-crested waves. However, none of the forcing frequencies may be in the vicinity of higher natural sloshing frequencies, which in practice may indicate a truncated wave spectrum. Stochastic distributions of response variables follow from the time simulations. Very long time series are needed to establish satisfactory approximations of different sloshing response variables. The described procedure requires very small CPU time relative to normal CFD methods and the numerical damping is negligible.

8.3.4 Applicability: effect of higher modes and secondary resonance

Modal system (8.43) and (8.44) assumed that the lowest natural mode has the lowest asymptotic order, and single dominant ordering (8.33) is satisfied. This assumption fails due to (1) critical depth, (2) large-amplitude response, and (3) secondary resonance. Because the Moiseev type of ordering causes an infinite steady-state response at the lowest natural frequency at the critical depth, something is obviously wrong. It does not help to assume that the lowest-order mode is $O(\varepsilon^{1/5})$ instead of $O(\varepsilon^{1/3})$, as Waterhouse (1994) did. The energy must be redistributed differently between the modes. The only possibility to achieve this outcome within an asymptotic theory is to assume that more than one dominant mode exists.

Large-amplitude response occurs also for conditions other than those at the critical depth when the forcing is large in resonant conditions. This scenario also necessitates the reordering of the modes. Secondary resonance implies resonant amplification of β_2 or of β_2, β_3, and so on due to nonlinear effects that cause energy transfer from the primary excited first mode to the second, third, and possibly higher natural modes. The amplification can occur for both steady-state and transient conditions. How this transfer happens for transient sloshing was demonstrated by the example illustrated in Figure 8.16. A

Table 8.1. *Values of σ/σ_1 associated with the secondary resonance for four modes ($k = 2, 3, 4, 5$) and depth ratios $\bar{h} = h/l$ corresponding to the experimental conditions of the examples in this chapter*

$\bar{h}$	σ/σ_1 (second mode)	σ/σ_1 (third mode)	σ/σ_1 (fourth mode)	σ/σ_1 (fifth mode)
0.5	0.74	0.60	0.52	0.47
0.4	0.76	0.63	0.54	0.49
0.35	0.78	0.64	0.56	0.5
0.29	0.81	0.68	0.59	0.53
0.17	0.9	0.79	0.71	0.64
0.12	0.94	0.87	0.79	0.73
0.08	0.97	0.93	0.88	0.83
0.05	0.99	0.97	0.95	0.92

necessary but insufficient condition for secondary resonance can mathematically be expressed as $n\sigma = \sigma_n$ ($n > 1$). As we showed in Section 4.3.1.1, this condition leads to

$$\frac{\sigma}{\sigma_1} = i_n^{-1}(\bar{h}) = j_n(\bar{h}) = \sqrt{\frac{\tanh(n\pi\bar{h})}{n\tanh(\pi\bar{h})}}. \qquad (8.83)$$

Equation (8.83) gives a relationship between $\bar{h} = h/l$ and σ/σ_1 for a given value of n as presented in Figure 8.18. Table 8.1 gives the forcing frequencies that satisfy this relationship for selected values of $\bar{h}$. When using Figure 8.18 to assess the possibility of secondary resonance, σ/σ_1 (corresponding to secondary resonance) must not be in a range where linear theory adequately describes the response. The range where nonlinearities matter in the vicinity of $\sigma/\sigma_1 = 1$ is a function of the excitation amplitude characterized by ε. The larger the value of ε, the broader the frequency range where nonlinearities matter. Some guidance can be provided by using Moiseev-like ordering for the forcing frequency, eq. (8.24), rewritten as

$$K = \left(1 - \sigma^2/\sigma_1^2\right)/\varepsilon^{2/3} = O(1). \qquad (8.84)$$

Therefore, K must be a finite number (depending on the forcing frequency) for small fixed ε (forcing amplitude). Obviously, K becomes asymptotically large away from the primary resonance and small ε. When this happens, nonlinearities caused by the primary resonance do not matter. Using eq. (8.84) gives condition $|1 - \sigma^2/\sigma_1^2| \lesssim \varepsilon^{2/3}$,

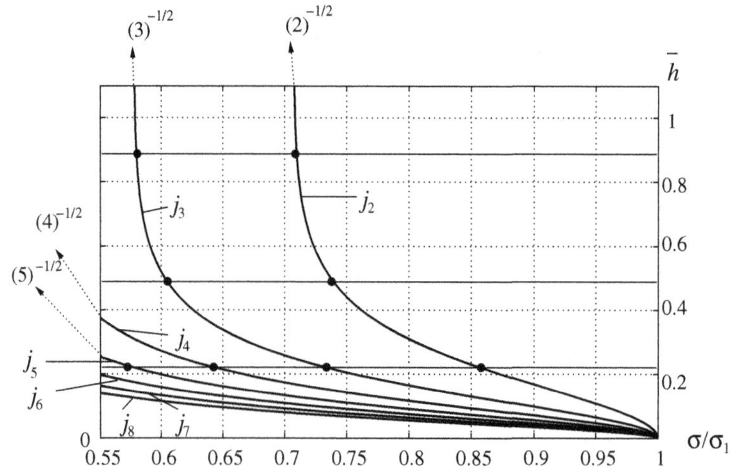

Figure 8.18. Intersections of $j_k(\bar{h})$ ($k = 2, \ldots, 8$) with horizontal lines identifying the values of σ/σ_1 at which there may be amplification of the corresponding natural mode k due to the secondary resonance for given values of $\bar{h} = h/l$.

which determines the effective frequency domain of the primary resonance. The larger ε, the wider the effective frequency domain.

Figure 8.18 and Table 8.1 show that, when $\overline{h}$ is finite, values of j_n $(n > 1)$ are far from the primary resonance at $\sigma/\sigma_1 = 1$. For instance, if $\overline{h} = 0.5$, the condition $\sigma/\sigma_1 \approx j_2 = 0.74$ is a necessary condition for secondary resonance of the second mode. We can use eq. (8.84) to assess if the secondary resonance point $\sigma/\sigma_1 = 0.74$ belong to the effective frequency domain of the primary resonance. Using eq. (8.84) with $K = 1$ (as a representative value of $O(1)$) gives $\varepsilon = 0.3$ at $\sigma/\sigma_1 = j_2$. This ε-value is not consistent with ε being a small value, implying that nonlinearities are not important in this frequency range.

When the ratio of liquid depth to tank breadth, $\overline{h}$, becomes small, Figure 8.18 and Table 8.1 show that the values of σ/σ_1 that express the necessary conditions for secondary resonance become close to 1 for many modes. This conditions implies that many modes should be assumed dominant in shallow-liquid theory. Furthermore, both ε and $\overline{h}$ must be introduced as small parameters in deriving the asymptotic theory for small depths, which we discuss in Section 8.6.

The fact that many modes must be considered in a shallow-liquid condition can be understood by considering a hydraulic jump traveling back and forth in the tank. A Fourier-series representation of the hydraulic jump for a given time instant would need many terms to adequately describe the free-surface shape. Because the free-surface representation in a modal approach is a Fourier series in which each mode contributes with its modal function β_i, it follows that many modes must have equal order.

To describe sloshing associated with the critical depth, large-amplitude response, and secondary resonance, we should reject the Moiseev-like ordering for modal functions (8.33). More complicated asymptotic intermodal relationships are needed that reflect the actual ordering dependent on the frequency and amplitude domains. The number of dominant modes can change from case to case and depends on, for instance, what kind of secondary resonance occurs. Bearing in mind a modal system, which may include all the terms due to different possible asymptotics, Faltinsen and Timokha (2001, 2002) proposed two general infinite-dimensional modal systems of polynomial structure in β_i, R_n and their time derivatives in the modal differential equations (see Sections 8.4 and 8.6). The first system is derived for *finite liquid depths* with $\overline{h} \gtrsim 0.2$. Its finite-dimensional forms can describe sloshing at the critical depth (see Section 8.5). The second system is applicable to resonant sloshing with $0.05 \lesssim \overline{h} \lesssim 0.2$. Details are given in Sections 8.7 and 8.8.

8.4 Adaptive asymptotic modal system for finite liquid depth

8.4.1 Infinite-dimensional modal system

The derivation of a polynomial-type system from modal system (7.14) and (7.15) needs the Taylor expansions of all the l- and ρ-scaled integrals D_n and D_{nk} in terms of β_i. The forcing term is assumed to be of the highest considered order and, therefore, is associated with $\tilde{K}_m(t)$ in the modal equations.

The Taylor-series expansion of D_n can be expressed as

$$D_n = \int_{Q_0} \overline{\varphi}_n \mathrm{d}Q + \int_{-\frac{1}{2}}^{\frac{1}{2}} \overline{\varphi}_n \zeta \,\mathrm{d}x + \frac{1}{2}\int_{-\frac{1}{2}}^{\frac{1}{2}} \left.\frac{\partial \overline{\varphi}_n}{\partial z}\right|_{z=0} \zeta^2 \mathrm{d}x + \frac{1}{6}\int_{-\frac{1}{2}}^{\frac{1}{2}} \left.\frac{\partial^2 \overline{\varphi}_n}{\partial z^2}\right|_{z=0} \zeta^3 \mathrm{d}x + o(\zeta^3). \tag{8.85}$$

By inserting the expression for the natural modes (8.34) and modal representation of the free surface (7.1) into eq. (8.85) we get, correctly to third order in the modal functions, that

$$D_n = \tfrac{1}{2}\left\{\beta_n + \tfrac{1}{2}\tilde{E}_n \Lambda^{(1)}_{nij}\beta^i\beta^j + \tfrac{1}{3}C_{nn}\Lambda^{(2)}_{nijk}\beta^i\beta^j\beta^k\right\}, \tag{8.86}$$

where

$$\tilde{E}_n = \tfrac{1}{2}\pi n \tanh(\pi n \overline{h}); \quad C_{nk} = \tfrac{1}{8}\pi^2 nk, \tag{8.87}$$

where the set of tensors $\Lambda^{(k)}$ is defined by the following recurrence formulas:

$$\Lambda^{(0)}_{ij} = \left|\begin{cases} 2, & i = j = 0, \\ 1, & i = j \neq 0, \\ 0, & \text{otherwise}, \end{cases}\right| ,$$

$$\Lambda^{(1)}_{nkj} = \Lambda^{(0)}_{|n-k|j} + \Lambda^{(0)}_{|n+k|j},$$

$$\Lambda^{(2)}_{nkjp} = \Lambda^{(1)}_{|n-k|jp} + \Lambda^{(1)}_{|n+k|jp}.$$

Note that, in eq. (8.86) and henceforth in the derivations, we use *tensor expressions* with standard notations for the formulas. The *repeated upper and lower indices* in eq. (8.86) mean *summation*. The superscripts that are used for β must not be misunderstood as powers of β. Furthermore, the tensor summation in eq. (8.86) and in the majority of the forthcoming expressions is finite, because the auxiliary tensors in these sums are zero when one of the summation indices exceeds the sum of the other indices (for example, $\Lambda^{(2)}_{nkjp} = 0$ when $n > k + j + p$). The explicit expressions for $\dot{D}_n$ and $\partial D_n/\partial\beta_\mu$ are

$$\dot{D}_n = \tfrac{1}{2}\left\{\dot{\beta}_n + \tilde{E}_n\Lambda^{(1)}_{nij}\dot{\beta}^i\beta^j + C_{nn}\Lambda^{(2)}_{nijk}\dot{\beta}^i\beta^j\beta^k\right\}, \tag{8.88}$$

$$\frac{\partial D_n}{\partial\beta_\mu} = \tfrac{1}{2}\left\{\delta_{n\mu} + \tilde{E}_n\Lambda^{(1)}_{ni\mu}\beta^i + C_{nn}\Lambda^{(2)}_{nij\mu}\beta^i\beta^j\right\}. \tag{8.89}$$

The expansion of D_{nk} must be made up to second order. The Taylor series gives

$$\begin{aligned} D_{nk} = {} & \int_{Q_0} \nabla\overline{\varphi}_n \cdot \nabla\overline{\varphi}_k \mathrm{d}Q \\ & + \int_{-\frac{1}{2}}^{\frac{1}{2}} (\nabla\overline{\varphi}_n \cdot \nabla\overline{\varphi}_k)\Big|_{z=0} \zeta\,\mathrm{d}x \\ & + \frac{1}{2}\int_{-\frac{1}{2}}^{\frac{1}{2}} \frac{\partial(\nabla\overline{\varphi}_n \cdot \nabla\overline{\varphi}_k)}{\partial z}\Big|_{z=0} \zeta^2\,\mathrm{d}x + o(f^2) \end{aligned}$$

and, therefore,

$$D_{nk} = \tfrac{1}{2}\left(\delta_{nk}2\tilde{E}^k + \Pi^{(1)}_{nk,i}\beta^i + \Pi^{(2)}_{nk,ij}\beta^i\beta^j\right), \tag{8.90}$$

where the comma separates symmetric sets of indices (e.g., $\Pi^{(2)}_{nk,ij} = \Pi^{(2)}_{kn,ji}$). The tensors Π are

$$\begin{aligned} \Pi^{(1)}_{nk,i} &= 4C_{nk}\Lambda^{(-1)}_{nk,i} + 2\tilde{E}_n\tilde{E}_k\Lambda^{(1)}_{nki}, \\ \Pi^{(2)}_{nk,ij} &= 2C_{nk}(\tilde{E}_n + \tilde{E}_k)\Lambda^{(-2)}_{nk,ij} \\ &\quad + 2(C_{nn}\tilde{E}_k + C_{kk}\tilde{E}_n)\Lambda^{(2)}_{nkij} \end{aligned}$$

with

$$\begin{aligned} \Lambda^{(-1)}_{nk,i} &= \Lambda^{(0)}_{|n-k|i} - \Lambda^{(0)}_{|n+k|i}, \\ \Lambda^{(-2)}_{nk,ij} &= \Lambda^{(1)}_{|n-k|ij} - \Lambda^{(1)}_{|n+k|ij}. \end{aligned}$$

The partial derivatives of D_{nk} with respect to β_μ have the following form:

$$\frac{\partial D_{nk}}{\partial\beta_\mu} = \tfrac{1}{2}\left[\Pi^{(1)}_{nk,\mu} + 2\Pi^{(2)}_{nk,i\mu}\beta^i\right]. \tag{8.91}$$

As noted at the end of Section 7.2.3, system (7.14) may be considered a linear system of algebraic equations with respect to the functions R^k; that is,

$$D_{nk}R^k = \dot{D}_n, \tag{8.92}$$

where the "matrix" $\{D_{nk}\}$ is a function of $\{\beta_i\}$. In the asymptotic limit $\beta_i \to 0$, the diagonal elements of the matrix are $O(1)$ for finite liquid depths:

$$D_{nn} = \tilde{E}_n = \tfrac{1}{2}\pi n \tanh(\pi n\overline{h}), \tag{8.93}$$

but the nondiagonal ones tend to zero (see eqs. (8.87) and (8.90)). This fact makes possible an asymptotic solution of linear "algebraic" equations (8.92) with respect to R^k; R^k is then expressed as the following third-order polynomials in β^i and $\dot{\beta}^i$:

$$R^k = \dot{\beta}^k/(2\tilde{E}_k) + V^{2,k}_{i,j}\dot{\beta}^i\beta^j + V^{3,k}_{i,j,p}\dot{\beta}^i\beta^j\beta^p, \tag{8.94}$$

where the tensors V are found by substituting eq. (8.94) into (8.92) and collecting similar terms in β_i (i.e., β^i according to the introduced notation). The introduced tensors V have no symmetry between index i and other indices (i corresponds to the summation of $\dot{\beta}^i$). The commas between i, j, and p are used in eq. (8.94) because there is no guarantee of symmetry between indices i, j, and p. The calculations give

$$\begin{aligned} V^{2,n}_{a,b} &= \tfrac{1}{2}\Lambda^{(1)}_{nab} - (4\tilde{E}_n\tilde{E}_a)^{-1}\Pi^{(1)}_{na,b}, \\ V^{3,n}_{a,b,c} &= (2\tilde{E}_n)^{-1}C_{nn}\Lambda^{(2)}_{nabc} - (4\tilde{E}_n\tilde{E}_a)^{-1}\Pi^{(2)}_{na,bc} \\ &\quad - (2\tilde{E}_n)^{-1}V^{2,k}_{a,b}\Pi^{(1)}_{nk,c}. \end{aligned}$$

The time derivative of R^k is

$$\begin{aligned} \dot{R}^k = {} & (2\tilde{E}_k)^{-1}\ddot{\beta}_k + \ddot{\beta}^i(V^{2,k}_{i,j}\beta^j + V^{3,k}_{i,j,p}\beta^j\beta^p) \\ & + \dot{\beta}^i\dot{\beta}^j(V^{2,k}_{i,j} + 2\overline{V}^{3,k}_{i,jp}\beta^p), \end{aligned}$$

where $\overline{V}^{3,k}_{i,jp} = \tfrac{1}{2}(V^{3,k}_{i,j,p} + V^{3,k}_{i,p,j})$.

Finally, substitution of the preceding expressions into modal equations (8.42) and remembering tensor summation rule and that forcing is assumed to be of the highest order correctly gives to cubic terms in β_i that

$$\begin{aligned} & \ddot{\beta}^a(\delta_{a\mu} + d^{1,\mu}_{a,b}\beta^b + d^{2,\mu}_{a,b,c}\beta^b\beta^c) + \dot{\beta}^a\dot{\beta}^b(t^{0,\mu}_{a,b} + t^{1,\mu}_{a,b,c}\beta^c) \\ & + \sigma^2_\mu\beta_\mu = \tilde{K}_\mu(t), \quad \mu \geq 1, \end{aligned} \tag{8.95}$$

where

$$d_{a,b}^{1,\mu} = 2\tilde{E}_\mu \left(\tfrac{1}{2}\Lambda_{ab\mu}^{(1)} + V_{a,b}^{2,\mu} \right),$$

$$d_{a,b,c}^{2,\mu} = 2\tilde{E}_\mu \Big((2\tilde{E}_a)^{-1} C_{aa} \Lambda_{abc\mu}^{(2)} + \tilde{E}_n \Lambda_{nc\mu}^{(1)} V_{a,b}^{2,n} + V_{a,b,c}^{3,\mu} \Big), \quad (8.96)$$

$$t_{a,b}^{0,\mu} = 2\tilde{E}_\mu \big(V_{a,b}^{2,\mu} + (8\tilde{E}_a\tilde{E}_b)^{-1}\Pi_{ab,\mu}^{(1)}\big),$$

$$t_{a,b,c}^{1,\mu} = 2\tilde{E}_\mu \big(2\overline{V}_{a,b,c}^{3,\mu} + V_{a,b}^{2,n}\tilde{E}_n\Lambda_{nc\mu}^{(1)} + (4\tilde{E}_a\tilde{E}_b)^{-1}\Pi_{ab,\mu c}^{(2)} + (2\tilde{E}_a)^{-1}V_{b,c}^{2,n}\Pi_{an,\mu}^{(1)}\big). \quad (8.97)$$

System (8.95) can be rewritten in the form

$$\left.\begin{aligned}
&\sum_{a=1}^{N}\ddot{\beta}_a\left(\delta_{am} + \sum_{b=1}^{N}\beta_b D1^m(a,b) + \sum_{b=1}^{N}\sum_{c=1}^{b}\beta_b\beta_c D2^m(a,b,c)\right)\\
&+ \sum_{a=1}^{N}\sum_{b=1}^{a}\dot{\beta}_a\dot{\beta}_b T0^m(a,b)\\
&+ \sum_{a=1}^{N}\sum_{b=1}^{a}\sum_{c=1}^{N}\dot{\beta}_a\dot{\beta}_b\beta_c T1^m(a,b,c)\\
&+ \sigma_m^2\beta_m = \tilde{K}_m(t), \quad m = 1,2,\ldots; \quad N \to \infty,
\end{aligned}\right\} \quad (8.98)$$

where

$$D1^m(a,b) = d_{a,b}^{1,m},$$

$$D2^m(a,b,c) = \begin{cases} d_{a,b,b}^{2,m}, & b = c,\\ d_{a,b,c}^{2,m} + d_{a,c,b}^{2,m}, & b \neq c,\end{cases}$$

$$T0^m(a,b) = \begin{cases} t_{a,a}^{0,m}, & a = b,\\ t_{a,b}^{0,m} + t_{b,a}^{0,m}, & a \neq b,\end{cases}$$

$$T1^m(a,b,c) = \begin{cases} t_{a,a,c}^{1,m}, & a = b,\\ t_{a,b,c}^{1,m} + t_{b,a,c}^{1,m}, & a \neq b.\end{cases}$$

8.4.2 Hydrodynamic force and moment

Modal system (8.95) accounts formally for an infinite set of modal functions $\beta_j(t)$, but requires a generalization of the formulas from Sections 8.3.2.2 and 8.3.2.3. Faltinsen and Timokha (2001) presented this generalization and gave the required expressions for hydrodynamic force and moment (consistent with ordering (8.95)) in a tensor form. The formulas use eq. (8.73) for the liquid mass center and the Lukovsky formulas from Chapter 7. The expression for hydrodynamic force $(0, F_2(t), F_3(t))$ in the two-dimensional case can be written as

$$F_2(t) = m_l g_2 - m_l(\ddot{\eta}_2 - \ddot{\eta}_4 z_{lC_0} + \ddot{y}_{lC}),$$
$$F_3(t) = m_l g_3 - m_l(\ddot{\eta}_3 + \ddot{\eta}_4 y_{lC_0} + \ddot{z}_{lC}), \quad (8.99)$$

where m_l is the liquid mass and g_2 and g_3 are linear components of the gravitational acceleration in the $Oxyz$-system defined by eq. (5.13). The nondimensional static center of gravity $(0, y_{lC_0}, z_{lC_0})$ is $(0, 0, -\frac{1}{2}\overline{h})$ and the mobile liquid mass center $(0, y_{lC}(t), z_{lC}(t))$ is given by modal formulas (8.73).

The expression for the hydrodynamic moment relative to the Ox-axis is far more complicated. As long as this expression follows from Lukovsky formula (7.32), we need a way to compute the quantity $\dot{l}_{1\omega} - l_{1\omega t}$. Faltinsen and Timokha (2001) gave the third-order approximation of this quantity in the tensor form $\dot{l}_{1\omega} - l_{1\omega t} = -\frac{1}{2}\rho l(\dot{\beta}^m L_m^{(0)} + \dot{\beta}^m\beta^p L_{p,m}^{(1)} + \dot{\beta}^m\beta^n\beta^p L_{k,p,m}^{(2)})$, where $L^{(0)}$, $L^{(1)}$, $L^{(2)}$ are presented in the original publication as functions of $\overline{h}$ (which are very lengthy and we refer the readers to the literature). The roll moment about O in two-dimensional flow, $M_O = (F_4(t), 0, 0)$, can be written as

$$F_4(t) = m(y_{lC}(g_3 - \ddot{\eta}_3) + z_{lC}(-g_2 + \ddot{\eta}_2)) - \rho l\ddot{\eta}_4 J^{(0)} + \tfrac{1}{2}l\rho\Big(\ddot{\beta}^m L_m^{(0)} + \ddot{\beta}^m\beta^p L_{p,m}^{(1)} + \ddot{\beta}^m\beta^k\beta^p L_{k,p,m}^{(2)} + L_{p,m}^{(1)}\dot{\beta}^m\dot{\beta}^p + 2\overline{L}_{kp,m}^{(2)}\dot{\beta}^m\dot{\beta}^k\beta^p\Big),$$

where $J^{(0)}$ is given by eq. (5.83).

8.4.3 Particular finite-dimensional modal systems

As shown in Section 8.3.4, amplification of higher modes is expected to occur due to secondary resonance. The relations $\beta_1 \sim \cdots \sim \beta_n$ for an integer $n \geq 2$ may be required for this amplification. Reordering of the modes is also needed at the critical depth and for large-amplitude forcing at the primary resonance frequency.

The frequency range where secondary resonance has to be considered depends on the forcing amplitude. Moreover, ranges of the effective frequency domain of the secondary resonance, generally speaking, are not clearly identified so we do not know exactly where and how a small variation of σ/σ_1 changes the asymptotics from that for the secondary resonance to the third-order Moiseev asymptotics and vice versa. To solve this problem, Faltinsen and Timokha (2001) proposed an adaptive modal approach. The approach handles both possible asymptotics by the *same* asymptotic modal system. The requirement is

$$\beta_1 \sim \cdots \sim \beta_n = O(\varepsilon^{1/3}), \quad \beta_i = O(\varepsilon), \quad i \geq n+1. \tag{8.100}$$

Using condition (8.100) in modal equations (8.98) and keeping the $O(\varepsilon)$-terms retains the nonlinear quantities from modal system (8.43) and those caused by the secondary resonance. The corresponding finite-dimensional nonlinear modal systems are called "Model n," where n denotes number of dominant modes.

Let us exemplify the adaptive modal approach for the case

$$\beta_1 \sim \beta_2 = O(\varepsilon^{1/3}), \quad \beta_i = O(\varepsilon), \quad i \geq 3 \tag{8.101}$$

and derive the finite-dimensional "Model 2." Considering this ordering in the basic polynomial system (8.98) causes six modal equations to describe nonlinear intermodal interaction. The first two nonlinear equations couple β_1 and β_2. When $j = 3, 4, 5, 6$, the equations are linear in β_j, but contain nonlinear terms in β_1 and β_2. The nonlinear equations for β_1 and β_2 are

$$\begin{aligned}(\ddot\beta_1 + \sigma_1^2\beta_1) &+ d_1(\ddot\beta_1\beta_2 + \dot\beta_1\dot\beta_2) + d_2(\ddot\beta_1\beta_1^2 + \dot\beta_1^2\beta_1)\\ &+ d_3\ddot\beta_2\beta_1 + \tilde d_1\ddot\beta_1\beta_2^2 + \tilde d_2\ddot\beta_2\beta_2\beta_1 + \tilde d_3\dot\beta_2^2\beta_1\\ &+ \tilde d_4\dot\beta_1\dot\beta_2\beta_2 = \tilde K_1(t),\end{aligned} \tag{8.102}$$

$$\begin{aligned}(\ddot\beta_2 + \sigma_2^2\beta_2) &+ d_4\ddot\beta_1\beta_1 + d_5\dot\beta_1^2 + \tilde d_5\ddot\beta_1\beta_1\beta_2 + \tilde d_6\ddot\beta_2\beta_1^2\\ &+ \tilde d_7(\ddot\beta_2\beta_2^2 + \dot\beta_2^2\beta_2) + \tilde d_8\dot\beta_1^2\beta_2 + \tilde d_9\dot\beta_1\dot\beta_2\beta_1 = 0,\end{aligned} \tag{8.103}$$

where the new coefficients

$$\tilde d_1 = D2^1(1,2,2) = -4C_{11} + 4\frac{C_{11}^2}{\tilde E_1^2} + 36\frac{C_{11}^2}{\tilde E_1\tilde E_3},$$

$$\tilde d_2 = D2^1(2,2,1) = -8\frac{C_{11}^2}{\tilde E_1\tilde E_2} + 72\frac{C_{11}^2}{\tilde E_2\tilde E_3},$$

$$\tilde d_3 = T1^1(2,2,1) = -16\frac{C_{11}^2}{\tilde E_2^2} - 144\frac{C_{11}^2\tilde E_1}{\tilde E_2^2\tilde E_3} + 16\frac{C_{11}\tilde E_1}{\tilde E_2} - 8\frac{C_{11}^2}{\tilde E_1\tilde E_2} + 72\frac{C_{11}^2}{\tilde E_2\tilde E_3},$$

$$\tilde d_4 = T1^1(1,2,2) = -8C_{11} + 8\frac{C_{11}^2}{\tilde E_1^2} + 72\frac{C_{11}^2}{\tilde E_1\tilde E_3},$$

$$\tilde d_5 = D2^2(1,1,2) = -8\frac{C_{11}^2}{\tilde E_1^2} + 72\frac{C_{11}^2}{\tilde E_1\tilde E_3},$$

$$\tilde d_6 = D2^2(2,1,1) = -16C_{11} + 16\frac{C_{11}^2}{\tilde E_1\tilde E_2} + 144\frac{C_{11}^2}{\tilde E_2\tilde E_3},$$

$$\tilde d_7 = D2^2(2,2,2) = -8C_{11} + 256\frac{C_{11}^2}{\tilde E_2\tilde E_4},$$

$$\tilde d_8 = T1^2(1,1,2) = -4\frac{C_{11}^2\tilde E_2}{\tilde E_1^3} - 36\frac{C_{11}^2\tilde E_2}{\tilde E_1^2\tilde E_3} + 4\frac{C_{11}\tilde E_2}{\tilde E_1} - 8\frac{C_{11}^2}{\tilde E_1^2} + 72\frac{C_{11}^2}{\tilde E_1\tilde E_3},$$

$$\tilde d_9 = T1^2(1,2,1) = -32C_{11} + 32\frac{C_{11}^2}{\tilde E_1\tilde E_2} + 288\frac{C_{11}^2}{\tilde E_2\tilde E_3}$$

have been introduced and where C_{11} and $\tilde E_n$ are given by eq. (8.87). The forcing term $\tilde K_1(t)$ and the coefficients d_n are defined by eq. (8.45) and eq. (8.47), respectively.

A derivation of these coefficients from the tensor formulas by Faltinsen and Timokha (2001) is tedious work, for which we used Maple™. Subsystem (8.102) and (8.103) contains cubic terms in β_1, β_2 and their derivatives. It includes all the nonlinear quantities from single-dominant modal equations (8.43). In addition, the third-order terms similar to β_2^3, $\beta_2^2\beta_1$, and so forth are included to describe a switch from Moiseev-type asymptotics to Model 2 during transients when $\beta_1 \sim \beta_2 = O(\varepsilon^{1/3})$. The third, fourth, fifth, and sixth modes are not included in these equations. These modes are excited by β_1, $\dot\beta_1$, $\ddot\beta_1$, β_2, $\dot\beta_2$, $\ddot\beta_2$ and the tank motions; that is,

$$\ddot\beta_i + \sigma_i^2\beta_i + \mathsf{F}(\ddot\beta_1, \ddot\beta_2; \dot\beta_1, \dot\beta_2; \beta_1, \beta_2) = \tilde K_i(t), \quad 3 \leq i \leq 6, \tag{8.104}$$

where F is a polynomial function of the arguments (up to the third order). The higher modes ($i \geq 7$) are handled by the linear modal equations.

In the forthcoming sections, we show that *large-amplitude forcing* (with finite liquid depth) needs only Model 2 for an adequate description of the nonlinear steady-state sloshing with σ in a small vicinity of σ_1. However, in a critical depth case for σ slightly lower than the

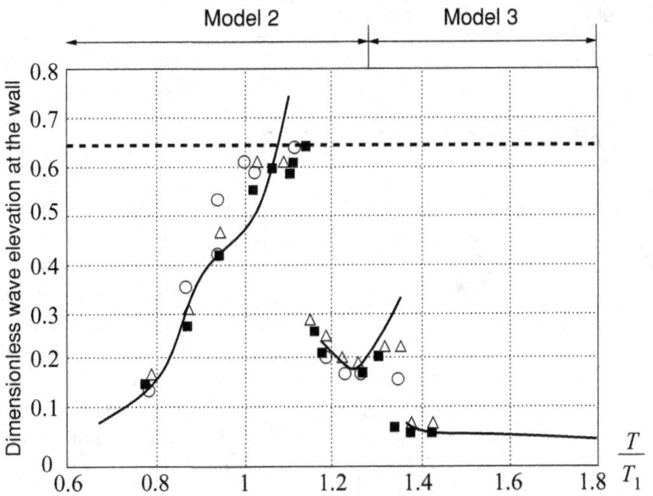

Figure 8.19. Theoretical and experimental predictions of maximum steady-state wave elevation at a tank wall divided by tank breadth l during harmonic sway excitation with period T in a rectangular tank with ratio of water depth to tank breadth of 0.35 and two-dimensional flow conditions. The forcing amplitude is $\eta_{2a}/l = 0.05$. Experimental measurements of the elevations were done by Olsen and Johnsen (1975) (Δ) and Colagrossi *et al.* (2003) (○). The solid line shows results based on the multidominant multimodal theory (Model 2 and Model 3) in the corresponding frequency domains as discussed in Section 8.4.3 (based on computations by Faltinsen & Timokha, 2001). The horizontal dashed line shows the roof level in the experimental tank. The square (■) indicates numerical values from the SPH method.

primary resonant zone and subharmonic waves, the modeling requires "Model 3," which assumes $\beta_1 \sim \beta_2 \sim \beta_3 = O(\varepsilon^{1/3})$, $\beta_k \sim O(\varepsilon)$ $(k \geq 4)$. This model suggests amplification of the third mode. Other, more complicated models are not needed, because these higher modes are highly damped in practice and, therefore, their nonlinear contribution to resonant sloshing can be neglected.

8.5 Critical depth

The single-dominant modal theory fails to predict steady-state resonant sloshing in the vicinity of the critical depth $h/l = 0.3368\ldots$; the adaptive modal modeling from previous sections is used instead. We describe typical experimental wave phenomena at the critical-depth ratio and compare them with theoretical predictions. Two sets of independent experiments are presented in Figures 8.19 and 8.20. The first comparison is presented in Figure 8.19, where we see the maximum wave elevation at the tank wall during two-dimensional sloshing in a rectangular tank with a ratio of water depth to tank breadth of 0.35. The ratio between the forced sway amplitude and the tank breadth is $\eta_{2a}/l = 0.05$. The numerical results in Figure 8.19 are obtained by using the SPH method (see Section 10.6) and the multidominant multimodal method with Model 2 and Model 3. The SPH method is based on solving the Euler equations and does not consider turbulence (i.e., it cannot describe turbulent energy dissipation). A plunging, breaking wave occurring in the experiments can be described. The domains of applicability of Models 2 and 3 are also shown in Figure 8.19. In practice, the domains overlap. The domain of applicability cannot be determined a priori; rather it is a consequence of doing simulations and realizing that a mode blows up, which necessitates that an additional mode has to be assumed dominant. Jumps occur in the solutions for $T/T_1 \approx 1.12$ and $T/T_1 \approx 1.33$. The jump at $T/T_1 \approx 1.33$ is associated with secondary resonance (i.e., $2\sigma = \sigma_2$, which means $\sigma/\sigma_1 = 0.78$ (see Table 8.1) or $T/T_1 = 1.28$). Because the secondary resonance by the third mode $(3\sigma = \sigma_1)$ is expected at $T/T_1 = 1.56$, the calculations for $T/T_1 > 1.33$ need to switch to Model 3. Neglecting the dominant behavior of the third mode (using Model 2 at $T/T_1 > 1.33$) makes the modal

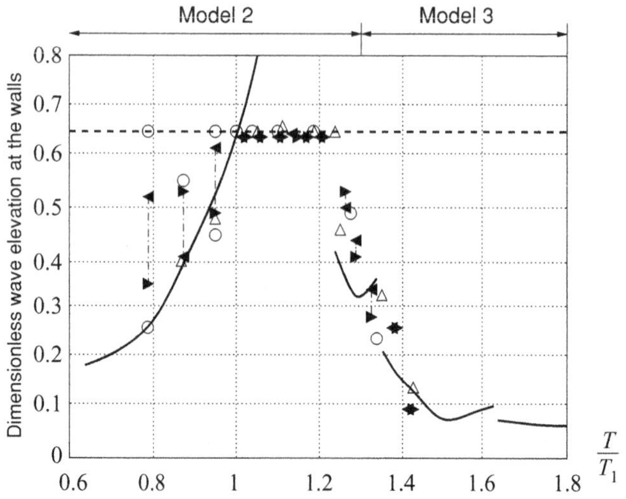

Figure 8.20. The same as in Figure 8.19, but for $\eta_{2a}/l = 0.1$. Whereas Olsen and Johnsen (1975) and the multimodal method show T-periodic steady-state results in the whole frequency domain, the experimental measurements by Colagrossi *et al.* (2003) (○) showed subharmonic steady-state oscillations for $T/T_1 < 1$. In this range, maximum elevations near one wall are clearly different for the opposite wall. The same solution behavior was confirmed by SPH simulations: ◀, results for the "left" wall: ▶, values for the "right" wall.

calculations unstable. A jump is expected at $T/T_1 = 1.56$ due to secondary resonance by the third mode. However, no jump appears. The energy is concentrated in the first two modes. We see in the case of Figure 8.20 that the jump at $T/T_1 = 1.56$ may only occur with increasing the forcing amplitude.

Good general agreement exists among modal predictions, SPH simulations, and experimental results. The single-dominant theory for critical depth has a behavior at the resonance frequency similar to that of a linear theory (i.e., the response becomes infinite at $\sigma = \sigma_1$ with no damping). However, the adaptive modal theory and the experiments show soft-spring behavior (see Figure 8.4). The response curve based on the adaptive modal method changes the sign of the curvature twice in the vicinity of $T/T_1 = 0.9$. A single-dominant theory does not show this change in curvature.

Figure 8.20 presents similar results when the ratio between the sway amplitude and the tank breadth has increased to 0.1. The agreement is satisfactory for $T/T_1 > 1$ and the third mode is clearly sensitive to the secondary resonance at $T/T_1 \approx 1.56$. However, differences and inconsistencies are present in the results when $0.75 \lesssim T/T_1 \lesssim 1.05$. Both the experiments and the SPH predictions by Colagrossi *et al.* (2003) agree that the maximum wave elevations are not the same at the two opposite tank walls. We have no information from the report by Olsen and Johnsen (1975) on whether this was true for their experiments. Furthermore, the multimodal method from Faltinsen and Timokha (2002) did not recognize this fact. The differences in wave amplitudes on the two tank walls are illustrated in Figure 8.20 by triangles with different orientation and dashed-and-dotted lines between. The different maximum elevations at opposite walls in Figure 8.20 cannot be explained by a steady-state theory with T-periodic solutions.

Subharmonic oscillations occur; "subharmonic solutions" means that the steady-state response contains oscillation frequencies that are smaller than the forcing frequency. A detailed investigation of the experimental sloshing behavior in this frequency domain was presented by Colagrossi *et al.* (2006). Investigated cases are presented in Table 8.2. All cases marked "II" showed steady-state oscillations with subharmonic behavior with a period three times the forcing period. We first concentrate on a case with small excitation amplitude and show that a subharmonic solution was clearly evident during steady-state conditions. The results are repeated with a period three times the forcing period. Furthermore, the wave elevation had different behaviors at the two tank walls (e.g., the maximum wave elevations were not the same). Case 484, with $T/T_1 = 0.87$, $\eta_{2a}/l = 0.03$, is illustrated in Figure 8.21.

In all type II cases except for case 484, the steady-state behavior was interrupted by transient phases because energetic breaking waves caused a lowering of the wave elevation. Wave breaking alternated on the two sides of the tank.

Table 8.2. *Experimental test conditions: Forced harmonic sway oscillations with amplitude η_{2a} and period T of a rectangular tank with ratio of water depth to tank breadth of 0.35 and two-dimensional flow conditions*

		η_{2a}/l			
T(s)	T/T_1	0.03	0.05	0.07	0.1
1.0	0.79	183(I)	174(I)	206(II)	414(III)
1.1	0.87	484(II)	175(II)	207(IV)	415(IV)
1.2	0.95	186(IV)	176(II)	208(II)	416(IV)
1.3	1.03	189(II)	178(II)	210(IV)	418(IV)

Note: The tank breadth l is 1 m (Colagrossi *et al.*, 2006).

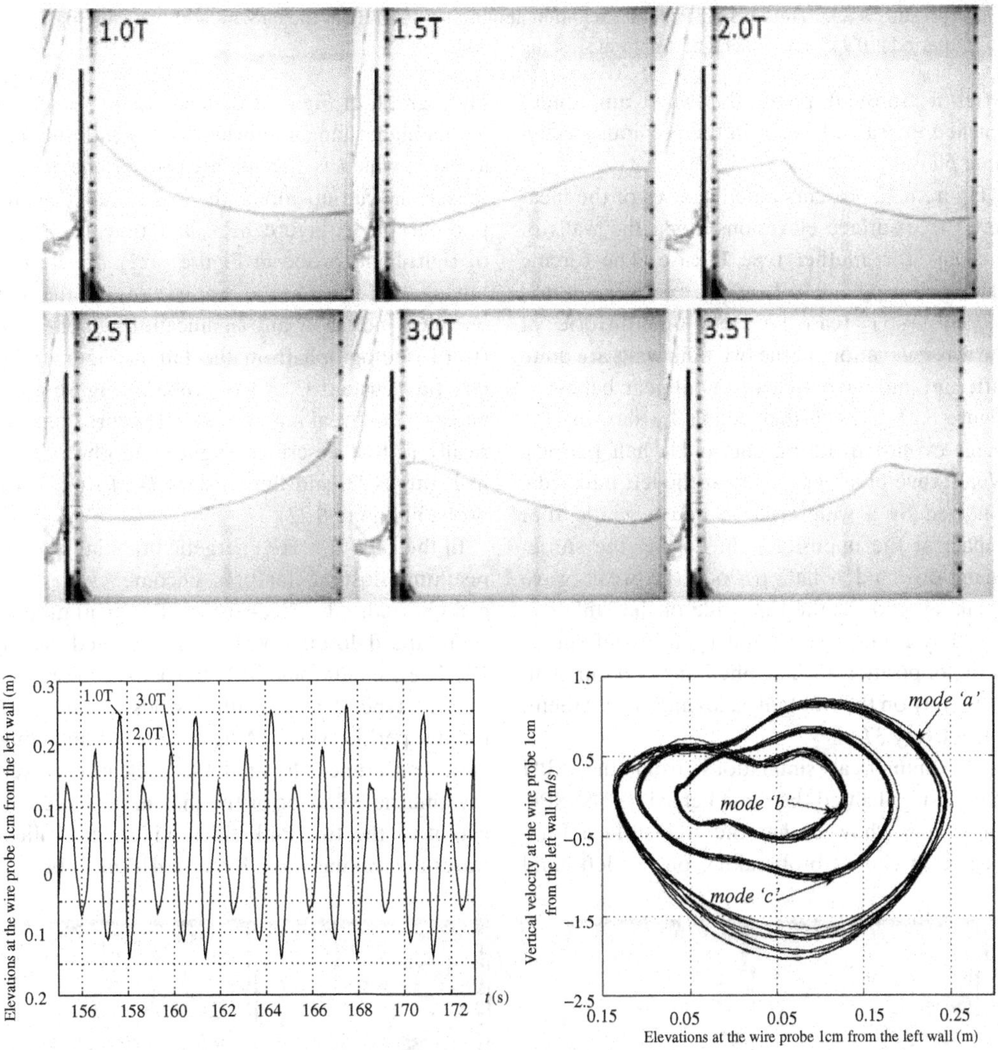

Figure 8.21. Experimental prediction of steady-state free-surface elevation during harmonic sway excitation with period T in a rectangular tank with ratio of water depth to tank breadth of 0.35 and two-dimensional flow conditions. Case 484 of Table 8.2 ($T/T_1 = 0.87$, $\eta_{2a}/l = 0.03$; Lugni *et al.*, 2006a). Six snapshots from time $1.0T$ to $3.5T$ (top), the corresponding experimental measurements of the free-surface elevation with a wire probe 1 cm from the left vertical wall and the phase plane of the measurements during the end of the experiments (Lugni *et al.*, 2006a). Loop mode b denotes the T-periodic cycle with lowest amplitude (see $2.0T$ point for the measured elevations), mode c implies the next T-periodic cycle (see $3.0T$ point for the measured elevations), and mode a is the the T-periodic cycle denoted by the point $1.0T$ for the measured elevations.

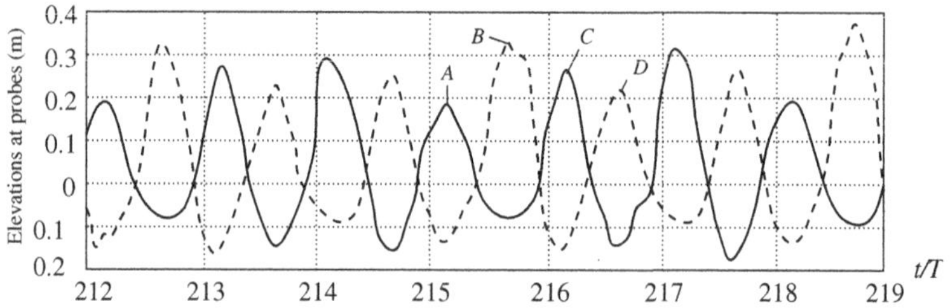

Figure 8.22. Free-surface elevation from the experiments by Colagrossi *et al.* (2004) at the two side-walls of a rectangular tank for case 176 of Table 8.2. The solid line denotes elevation at the wire probe at the left wall, and the dashed line means elevation at the right wall. The ratio of water depth to tank breadth is 0.35. The harmonic lateral excitation amplitude is 0.05 times the tank breadth. The forcing period is 0.945 times the highest natural sloshing period.

After a transient phase the wave amplitudes reached the same level as in the previous steady-state phase.

Figure 8.22 presents time histories of the measured free-surface elevations along the walls of the tank for another type II case. The forcing amplitude is $\eta_{2a} = 0.05\, l$ and the excitation period is $T = 0.945T_1$ (case 176). The time histories of the wave elevations at the two tank walls are quite different and have a clearly nonlinear behavior. Figure 8.23 gives further details by showing the water evolution during one and a half periods. Weak wave breaking occurs at the left tank side, followed by a water run-up and a gentle roof impact at the opposite wall. During the subsequent one and a half periods the weak wave breaking starts at the right side of the tank, followed by a water run-up and a gentle roof impact at the opposite wall. The phenomena occur symmetrically on the two tank sides and repeat themselves every $3T$.

The numerical simulations from the SPH method highlighted the same behavior. The SPH simulations showed that the free-surface fragmentation caused by breaking on the left-hand photograph of Figure 8.23 leads to a shear layer with a higher tangential velocity beneath the free surface, which is the main cause of the higher maximum run-up along the right wall (second photograph in Figure 8.23, and time instant *B* of right-hand probe in Figure 8.22). In the following half-period, the water rise on the left wall does not show any significant fragmentation (third photograph from the left in Figure 8.23, and time instant *C* of left probe in Figure 8.22) and another breaking wave event occurs symmetrically to that described (right-hand photograph in Figure 8.23, and time instant *D* of right-hand probe in Figure 8.22).

In the case of more energetic breaking events, resulting sloshing features become even more complicated. The free-surface fragmentation is pronounced during sloshing characterized as type III. The subharmonic behavior is more complex than in type II cases with periods *N* times the forcing period, where *N* is a large number that varies with time. The model tests clearly showed asymmetric behavior of the maximum wave elevation at the two tank sides. Figure 8.24 illustrates a breaking wave for case 414 in Table 8.2

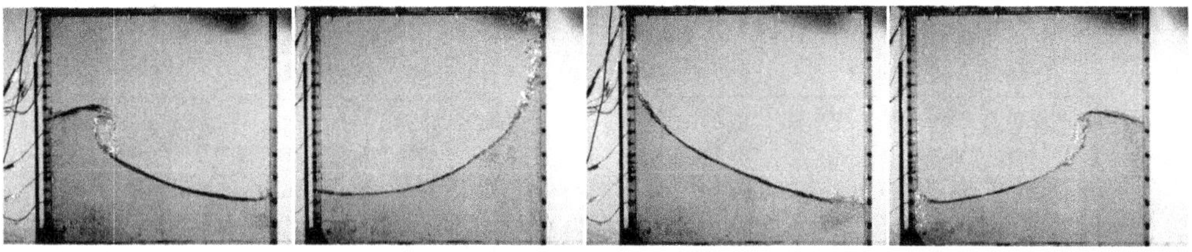

Figure 8.23. Visualization of the water behavior in two-dimensional sloshing experiments with a rectangular tank. The ratio of water depth to tank breadth is 0.35. The harmonic lateral excitation amplitude is 0.05 times the tank length. The forcing period is 0.945 times the highest natural sloshing period. Time increases from left to right with time instants $t/T = 215.1$, $t/T = 215.1 + 1/2$, $t/T = 215.1 + 1$, $t/T = 215.1 + 3/2$, which correspond to labels A, B, C, and D in Figure 8.22, respectively. Case 176 of Table 8.2 (Colagrossi *et al.*, 2004).

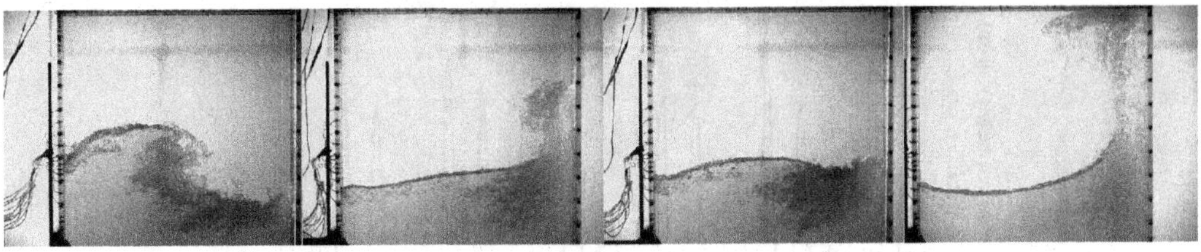

Figure 8.24. The same as Figure 8.23, but the harmonic lateral excitation amplitude is 0.1 times the tank length and the forcing period is 0.787 times the highest natural sloshing period. Case 414 (III) of Table 8.2 by Colagrossi *et al.* (2004).

corresponding to $T/T_1 = 0.79$ and $\eta_{2a} = 0.1\ l$. For about twenty periods, the right-hand probe experiences a larger amplitude of oscillation, after which the situation is reversed and the left sidewall shows a higher wave elevation than the one to the right. The physical mechanisms behind this behavior are related to rather violent wave-breaking events that repeat themselves only on one tank side. In Figure 8.24 the breaking toward the right wall of the tank starts from the left. Any of these breaking events causes free-surface fragmentation and the subsequent water flow is not able to organize a new plunging phenomenon on the other side. The water interaction with the opposite wall results in a large water run-up with a subsequent tank roof impact (see right-hand photograph of Figure 8.24), which sets up the required flow conditions for the occurrence of the next breaking event on the other side. The numerical simulations of the SPH method highlighted sensitivity to initial conditions.

The type IV scenario is characterized by the formation of a local splashing jet at a distance from the wall, leading to a continuous fragmentation of the free surface (see Figure 8.25). No well-defined subharmonics exist for this case.

Can the multimodal method predict the previously mentioned subharmonic behavior? All previous cases with steady-state solutions based on the multimodal method have been periodic with the forcing period and have included only superharmonic components in addition to the fundamental harmonic component. If we solve the problem in the time domain and breaking waves and free-surface fragmentation are not dominant, we should expect that a case with steady-state subharmonic behavior (e.g., case 484), could be handled. The nature of the steady-state solution predicted by the multimodal method depends on how many modes are activated and on the transient damping behavior. To get the T-periodic solution, Faltinsen and Timokha (2001) used Model 2. Furthermore, they modeled the transient phase by using different damping terms: the initial series used theoretical viscous boundary-layer damping coefficients from Section 6.3.1 multiplied by a factor varying from 5 to 10 and, after achieving the approximate T-periodic solution, these coefficients were consequently decreased down to 10^{-4} times the theoretical values. No physical basis exists for this choice of damping coefficients; it was a necessity to damp out high-amplitude free-surface motion during the transient phase. This approach is not suitable to predict a subharmonic solution and must be modified.

The revision of the multimodal method should involve the following two effects. First, the instant wave profiles in Figure 8.21 show the importance of the third mode. This fact follows by comparing the last photograph with the profile of the third mode in Figure 4.3. Model 3, therefore, is needed. Second, the initial damping coefficients that are required to reach a periodic solution

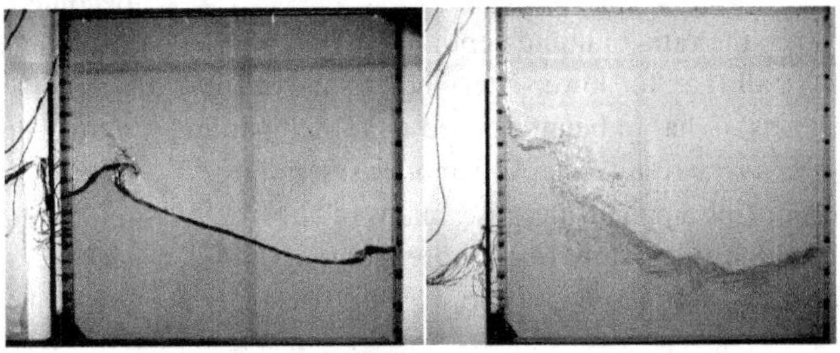

Figure 8.25. Examples of case IV from Table 8.2: (left) case 207 ($T/T_1 = 0.87$, $\eta_{2a}/l = 0.07$); (right) case 416 ($T/T_1 = 0.95$, $\eta_{2a}/l = 0.1$), with clear fragmentation of the free surface.

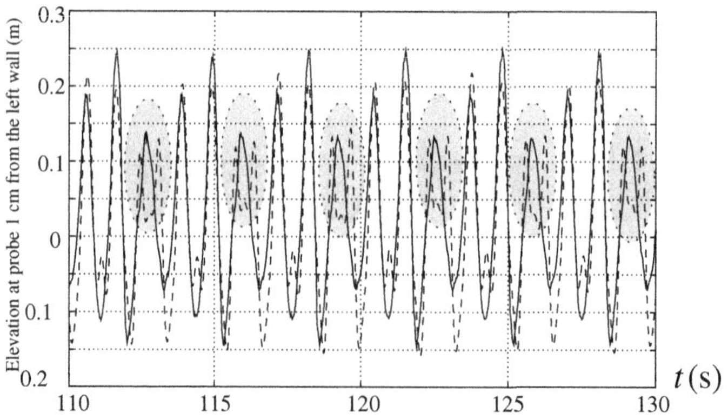

Figure 8.26. Experimental (solid line) and multidominant multimodal (dashed line) predictions by Model 3 of steady-state free surface elevation with a wire probe 1 cm from the left vertical wall during harmonic sway excitation with period $T = 1.147$s in a rectangular tank with breadth 1 m and ratio of water depth to tank breadth 0.35 and two-dimensional flow conditions. Case 484 of Table 8.2 ($T/T_1 = 0.87$, $\eta_{2a}/l = 0.03$) documented in Figure 8.21. The shaded zone marks the difference.

should be smaller (i.e., comparable with the viscous boundary-layer damping). The subharmonic wave motions are shown in Figure 8.26. A total of 1,000 forcing periods were simulated. The transient phase was about 30 forcing periods. The agreement with SPH simulations and the experiments by Colagrossi *et al.* (2003) is good except for the domains marked by ellipses. Generally, increasing the number of dominant modes (consider Models 4, 5, etc.) shows that the wave behavior inside these domains is very sensitive to the number of dominant modes.

The comparisons show that the SPH method can describe well the cases when steady-state oscillations are interrupted with wave breaking. To the authors' knowledge, no rational ways exist to introduce new "initial" conditions in the multimodal method (as well as other analytically oriented methods) due to plunging–breaking wave interruption of the flow.

8.6 Asymptotic modal theory of Boussinesq-type for lower-intermediate and shallow-liquid depths

8.6.1 Intermodal ordering

When the ratio of liquid depth to tank breadth, $\bar{h}$, is small (i.e., for lower-intermediate and shallow depths), $\bar{h}$ has to be introduced as a small parameter. As a result, adaptive modal system (8.95) cannot be applied, because many of the coefficients of the system (e.g., those from eq. (8.47)) tend to infinity as $\bar{h} \to 0$. So we must use general modal system (7.14) and (7.15) and introduce the smallness of $\bar{h}$ in the multimodal method. A variant for intermediate and shallow depths was proposed by Faltinsen and Timokha (2002).

The analysis is based on the modal solution for the velocity potential, eq. (7.2), where φ_i are natural sloshing modes. For finite liquid depths, the analysis involves the Taylor expansions of the integrals D_n and D_{nk} (defined by eqs. (7.12)) in terms of the free-surface elevation. Therefore, the lowest-order approximations of these integrals are

$$(D_n)_0 = \rho \int_{-\frac{1}{2}l}^{\frac{1}{2}l} \left\{ \int_{-h}^{0} \varphi_n \mathrm{d}z \right\} \mathrm{d}x = O(1),$$

$$(D_{nk})_0 = \int_{-\frac{1}{2}l}^{\frac{1}{2}l} \left\{ \int_{-h}^{0} \nabla\varphi_n \cdot \nabla\varphi_k \mathrm{d}z \right\} \mathrm{d}x = O(1).$$

The forcing is assumed to have the highest asymptotic order ε, but the free-surface elevations are of lower order than $O(\varepsilon)$. Furthermore, $\bar{h}$ is introduced as a small parameter by performing Taylor expansions of the integrals D_n and D_{nk} relative to the tank bottom instead of relative to the mean free surface (i.e., $(D_n)_0$ and $(D_{nk})_0$ become Taylor series in $\bar{h}$).

A key problem is to find the asymptotic relationships between $\bar{h}$, β_i, and R_n and their asymptotical link with nondimensional forcing amplitude ε. To get the relationships for intermediate and shallow liquid depth, Faltinsen and Timokha (2002) required that the resulting modal equations should match the multidominant modal approximation by Faltinsen and Timokha (2001) for finite depth as well as with the shallow-liquid theory by Ockendon and Ockendon (1973) and Ockendon *et al.* (1986). The ordering of generalized coordinates in the finite-depth multidominant multimodal method is $\beta_i \sim R_i \sim \varepsilon^{1/3}$. The mentioned shallow-liquid theory assumes that $\overline{h} \sim \zeta(x,t) \sim \varepsilon^{1/4}$ and that the horizontal liquid velocity $u(x,t) \sim \varepsilon^{1/2}$; that is, the modal functions and liquid depth should have the following orders of magnitude:

$$R_i = O(\varepsilon^{1/2}); \quad O(\overline{h}) = O(\beta_i) = O(\varepsilon^{1/4}), \quad i \geq 1. \tag{8.105}$$

The matching of these asymptotics in the multimodal method means that the desired modal system has to include all quantities following from these finite and shallow-depth approximations in the modal functions. Analyzing the sum $\sum D_{nk} R_n R_k$ in eq. (7.15) for the finite-depth asymptotics and eq. (8.105), we see that eq. (8.105) does not provide the products $R_i R_n \beta_k$. However, this product is present for finite-depth asymptotics. Faltinsen and Timokha (2002) remarked the need to impose

$$R_i \sim \beta_i = O(\overline{h}) = O(\varepsilon^{1/4}), \quad i \geq 1 \tag{8.106}$$

to consistently keep the products of modal functions following from the matched asymptotics between finite and small depths. This asymptotic modal system accounts for the smallness of $\overline{h}$, keeps the quantities of adaptive modal system (8.95), and, with decreasing $\overline{h}$, includes all necessary quantities following from eq. (8.105).

Madsen *et al.* (2003) classified and listed the different Boussinesq methods occurring in coastal engineering in terms of nondimensional parameters $\delta = O(a_0/h)$, a dimensionless amplitude, and $\mu = O(2\pi h/\lambda)$, a dispersion parameter, where a_0, h, and λ mean characteristic wave amplitude, depth, and wavelength, respectively. Originally it was assumed that δ and μ were small parameters linked by the relationship $\delta = O(\mu^2)$. However, since Wei *et al.* (1995), all newer Boussinesq models have been "fully nonlinear"; that is, they are derived with $\delta = O(1)$, so that only the power series in μ is involved in the derivation of the problem. The old assumption that $\delta = O(\mu^2)$ is therefore old-fashioned. If we set $O(\lambda) = O(l)$ in our sloshing problem, the multimodal theory for intermediate and shallow depths corresponds to $\delta = O(1)$ and $\mu = O(\varepsilon^{1/4})$, which means that it has similarities with the actual concept of Boussinesq-type methods and so is referred to as Boussinesq-type multimodal method.

8.6.2 Boussinesq-type multimodal system for intermediate and shallow depths

Based on relation (8.106), Faltinsen and Timokha (2002) derived an infinite-dimensional asymptotic modal system. The starting point is system (8.37)–(8.42) in which the matrices $\partial D_n/\partial \beta_i$ and $\partial D_{nk}/\partial \beta_i$ may be explicitly given as polynomial expressions in β_i (see Faltinsen & Timokha, 2001, 2002) starting from their general definitions in eqs. (7.14) and (7.15). When $\overline{h}$ is small, the scaled natural modes are approximated as

$$\overline{f}_i(y) = \cos\left(\pi i(\overline{y} + \tfrac{1}{2})\right),$$
$$\overline{\varphi}_i(\overline{y}, \overline{z}) = \cos\left(\pi i(\overline{y} + \tfrac{1}{2})\right)\cosh(\pi i(\overline{z} + \overline{h})), \tag{8.107}$$

that is, they assume that $\cosh(\pi i \overline{h}) \approx 1$ but do not approximate $\cosh(\pi i(\overline{z} + \overline{h}))$ in the finite-depth expressions of the natural modes.

Within the nondimensional framework, the scaling with respect to l and ρ retains the structure of eq. (8.37) with the corresponding expressions for D_n and D_{nk}, but eq. (8.42) is rewritten in the form

$$\dot{R}^n \frac{\partial D_n}{\partial \beta_i} + \tfrac{1}{2}\frac{\partial D_{nk}}{\partial \beta_i} R_n R_k + \frac{(-1)^i - 1}{(i\pi)^2} \times \left(\frac{2}{(i\pi)}\tanh\left(\tfrac{1}{2} i\pi \overline{h}\right)\ddot{\eta}_4 + \frac{\ddot{\eta}_2}{l} + \frac{g}{l}\eta_4\right) + \frac{g}{2l}\beta_i = 0, \quad i \geq 1. \tag{8.108}$$

In contrast to the finite-depth theory, the Boussinesq-type approximation considers $\partial D_n/\partial \beta_i$ and $\partial D_{nk}/\partial \beta_i$ polynomials in β_i and $\overline{h}$ (up to fourth order, due to ordering (8.106)). Faltinsen and Timokha (2002) Taylor-expanded $\partial D_n/\partial \beta_i$ and $\partial D_{nk}/\partial \beta_i$ with respect to β_i and $\overline{h}$ and presented the final results in tensor form.

The multimodal method for intermediate and shallow depths differs from that for a shallow-liquid theory, which can be seen by studying the natural frequencies, σ_n. The linear shallow-liquid approximation is

$$\sigma_n^2 = g l^{-1} (\pi n)^2 \overline{h} \tag{8.109}$$

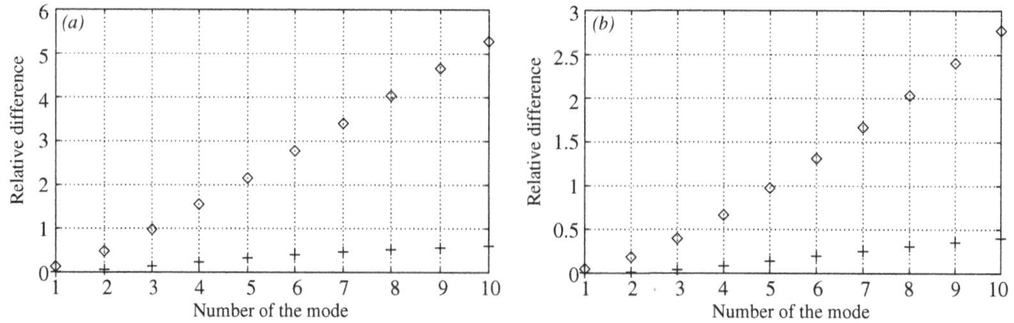

Figure 8.27. The relative difference $|\sigma^2_{n(\text{approximate})} - \sigma_n^2|/\sigma_n^2$ between the exact values of the natural frequencies in a rectangular tank (eq. (4.11)) and the approximations given by eq. (8.112) for intermediate and shallow depths (denoted +) and linear shallow-liquid approximation (8.109) (◇): (a) $h/l = 0.2$; (b) $h/l = 0.12$.

(see also Section 8.8.2). The corresponding expression by the multimodal method is found by extracting the linear terms in β_i from the modal system without forcing. The result is

$$\left(\frac{\partial D_n}{\partial \beta_k}\right)_0 \dot{\beta}^k = (A_{nj})_0 \dot{R}^j, \quad n \geq 1,$$
$$\dot{R}^n \left(\frac{\partial D_n}{\partial \beta_i}\right)_0 + \frac{g}{2l}\beta_i = 0, \quad i \geq 1, \tag{8.110}$$

where matrices $(\partial D_n/\partial \beta_i)_0$ and $(D_{nj})_0$ are expressed by evaluating the integrals based on eq. (8.107) and keeping the polynomial quantities in $\overline{h}$ up to the corresponding order. Only their diagonal terms are nonzero and may be expressed as

$$\begin{aligned}(D_{nn})_0 &= \int_{Q_0} (\nabla \overline{\varphi}_n)^2 \, dQ = \int_{-\frac{1}{2}}^{\frac{1}{2}} \overline{\varphi}_n \frac{\partial \overline{\varphi}_n}{\partial z} dy \\ &= \tfrac{1}{2}\pi n \sinh(\pi n \overline{h}) \cosh(\pi n \overline{h}) \\ &= \tfrac{1}{2}(\pi n)^2 \overline{h}(1 + \tfrac{2}{3}(\pi \overline{h} n)^2) + O(\overline{h}^4), \\ \left(\frac{\partial D_n}{\partial \beta_n}\right)_0 &= \frac{\partial}{\partial \beta_n}\left(\int_{-\frac{1}{2}}^{\frac{1}{2}} \int_{-\overline{h}}^{\sum_{i=1}^{\infty} \beta_i \overline{f}_i(y)} \overline{\varphi}_n dz \, dy\right) \\ &= \int_{-\frac{1}{2}}^{\frac{1}{2}} \overline{f}_n^2(y) \cosh(\pi n \overline{h}) dy \\ &= \tfrac{1}{2} + \tfrac{1}{4}(\pi n \overline{h})^2 + O(\overline{h}^4).\end{aligned} \tag{8.111}$$

In the derivation of $(D_{nn})_0$ we have used the fact that $(\nabla \overline{\varphi}_n)^2 = \nabla \cdot (\overline{\varphi}_n \nabla \overline{\varphi}_n)$, applied the divergence theorem (see eq. (A.1)), and used that $\overline{\varphi}_n$ satisfies the zero-Neumann boundary condition on the tank surface below the mean free surface.

Using eqs. (8.111) in eq. (8.110) leads to the following expression for the natural frequencies:

$$\sigma_n^2 = \frac{g}{l}(\pi n)^2 \overline{h} \frac{1 + \frac{2}{3}(\pi n \overline{h})^2}{\left(1 + \frac{1}{2}(\pi n \overline{h})^2\right)^2}. \tag{8.112}$$

Furthermore, expression (8.111) shows that $D_{nn} = O(\overline{h})$ and, therefore, the determinant of matrix D_{nk} tends to zero as $\overline{h} \to 0$. Therefore, we cannot organize an asymptotic solution on R_i from *kinematic subsystem* (8.37) as was possible for finite depth. The latter procedure is related to an inversion of matrix D_{nk}.

The result of eq. (8.112) should be compared with the exact linear theory of eq. (4.11) and the linear shallow-liquid limit given by eq. (8.109). We see that eq. (8.112) is the ratio of two polynomials (i.e., by definition, a rational function of $\overline{h}$). In contrast, expression (8.109) only accounts for the first term in a Taylor-series expansion with $\overline{h}$ as a small parameter. Equation (8.112) can be considered a "compromise" between shallow- and finite-liquid depth approximations in the intermediate depth range. This case is demonstrated in Figure 8.27, where the relative difference in approximating the exact natural frequencies by eqs. (8.112) and (8.109) is illustrated for $\overline{h} = 0.2$ and $\overline{h} = 0.12$. For these liquid depths, the difference is large for shallow-liquid approximation (8.109) except for the lowest frequency. In contrast, the Boussinesq-type approximation, eq. (8.112), leads to quite good agreement with the exact solution for several of the lowest natural frequencies.

8.6.3 Damping

Small damping is typical for a rectangular tank with $\overline{h} = h/l \gtrsim 0.4$ and infinite tank roof height. Generally speaking, damping increases as the ratio of liquid depth to tank breadth, $\overline{h}$, decreases. Damping sources are run up along the walls with subsequent overturning of the free surface (see Figure 8.1), breaking waves away from the walls and viscous boundary-layer flows. The damping due to breaking waves is mainly caused by turbulent energy dissipation (see Section 8.1).

The importance of viscous boundary-layer damping increases with decreasing $\overline{h}$. This fact is associated with damping of higher dominant modes that are needed in a shallow-liquid analysis. Figure 8.18 shows, by using potential flow theory without damping, that the higher modes are progressively activated in the limit $\overline{h} \to 0$ (intersection points of j_k and the horizontal lines move toward the origin). Because of higher damping with increasing mode number, the number of modes is limited and the infinite-dimensional system of eqs. (8.37) and (8.108) can be truncated with dimension N in both subsystems. Therefore, a practical use of the modal theory requires damping rates for the derived modal systems (as discussed in Chapter 6).

The dissipative effect on sloshing may be introduced in several ways. A strategy is to estimate damping rates for natural linear sloshing described by modal system (8.110). The damping term, $\xi_i'\dot{\beta}_i$, should then be substituted in the second of eqs. (8.110). We can relate the damping coefficients ξ_i' to the damping ratios ξ_i of the natural modes by rewriting system (8.110) in terms of the modal equation $\ddot{\beta}_i + 2\xi_i\sigma_i\dot{\beta} + \sigma_i^2\beta_i = 0$ as in Section 6.3.2, which leads to $\xi_i' = \xi_i g/l\sigma_i^2$ and, therefore, eq. (8.108) takes the following form:

$$\ddot{R}^n \frac{\partial D_n}{\partial \beta_i} + \xi_i \frac{g}{l\sigma_i^2}\dot{\beta}_i + \tfrac{1}{2}\frac{\partial D_{nk}}{\partial \beta_i}\dot{R}_n\dot{R}_k + \frac{(-1)^i - 1}{(i\pi)^2} \times \left(\frac{2}{(i\pi)}\tanh(\tfrac{1}{2}i\pi\overline{h})\ddot{\eta}_4 + \frac{\ddot{\eta}_2}{l} + \frac{g}{l}\eta_4\right) + \frac{g}{2l}\beta_i = 0, \quad i \geq 1. \tag{8.113}$$

The damping ratios, ξ_i, can be split into terms associated with different physical phenomena; that is, $\xi_i = \xi_i^{\text{tank surface}} + \xi_i^{\text{bulk}} + \xi_i^{\text{others}}$, where $\xi_i^{\text{tank surface}}$ is associated with viscous energy dissipation in the boundary layer and ξ_i^{bulk} is the viscous energy dissipation outside the boundary layer in the case of nonbreaking waves. Finally, the damping component ξ_i^{others} accounts, for instance, for turbulent energy dissipation in breaking waves. However, we have no theoretical means to estimate ξ_i^{others} and our numerical studies consider only $\xi_i^{\text{tank surface}}$ and ξ_i^{bulk}; $\xi_i^{\text{tank surface}}$ gives a clearly larger contribution than ξ_i^{bulk}.

Although the dissipation function gives the correct expression in the linearized case, it causes nonlinearities in eq. (8.113). These nonlinearities are associated with the inversion of the tensor $\partial D_n/\partial \beta_i$ for each time step. In addition, when substituting $\dot{\beta}_i$ we find additional nonlinearities in terms of R_n and β_i. This drawback is typical of incorporating linear damping terms in nonlinear conservative mechanical systems. Each formal structure of the dissipation function leads to different nonlinearities, but all alternative equations agree with each other in their linearized form. Even though linear viscous damping contributes only partially to the dissipation, the presence of damping terms in eq. (8.113) with $\xi_i = \xi_i^{\text{tank surface}} + \xi_i^{\text{bulk}}$ is relevant to obtain physical results (i.e., satisfactory agreement with experiments). This relevance is shown in the forthcoming sections.

8.7 Intermediate liquid depth

Faltinsen and Timokha (2002) validated experimentally the multimodal system for intermediate and shallow depths. The tank in Figure 8.9 was used. The experimental setup typically needed up to 5 to 10 s to reach the maximum amplitude of excitation. During this time the free-surface elevation generally remained small. Although nonzero initial conditions may affect different transient flows, their influence is small relative to other physical effects. Therefore, we use zero initial conditions.

Some appropriate cases can be found in the experiments by Faltinsen and Timokha (2002) for $h/l = 0.173$ (highest natural period is T_1 2.115 s). This series contains the measurements of wave elevations at FS2 and FS3 (see Figure 8.9) for the excitation periods $T = 1.1$, 1.17, 1.8, 1.9, 2.0, 2.1, 2.2, 2.3, and 2.4 s. Because the records for $T =$ 1.9, 2.0, and 2.1 s showed steady-state periodic solutions after two forcing periods, the damping must be very high. These cases are discussed later.

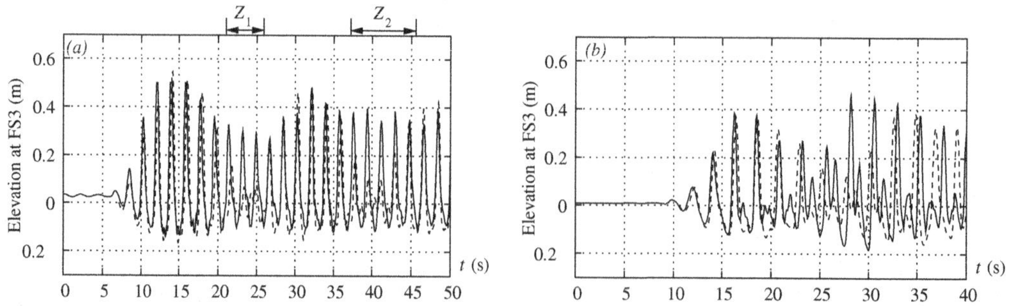

Figure 8.28. The measured and calculated wave elevations near the wall at wave probe FS3 (see Figure 8.9) for horizontal forcing ($h/l = 0.173$, $\eta_{2a}/l = 0.028$). The solid and dashed lines correspond to experiments (Faltinsen & Timokha, 2002) and the Boussinesq-type multimodal method, respectively. The highest natural period is $T_1 = 2.115$ s: (a) $T = 1.8$ s and (b) $T = 2.3$ s.

Typical transients are demonstrated in Figure 8.28. These cases were extensively discussed by Faltinsen *et al.* (2000) and later by Faltinsen and Timokha (2001) as examples of single-dominant modal system failure (see calculations in Figure 8.16). The present calculations are in agreement with the experimental data. The discrepancy observed during the initial forcing periods (e.g., at the time intervals Z_1 and Z_2 shown in Figure 8.28(a)) and as time increases is believed to be connected with run-up at the tank walls.

Video recordings have been used to clarify the types of physical phenomena that have been ignored by multimodal theory for intermediate and shallow depths. Corresponding photos are shown in Figure 8.29. These instantaneous wave profiles are typical for experiments at intermediate depths. There are a series of local nonlinear phenomena with breaking waves (panel c), overturning (panels b and d), tank roof impact (panels a and b), or thin jet (run-up) along the vertical wall. The modal representation of the

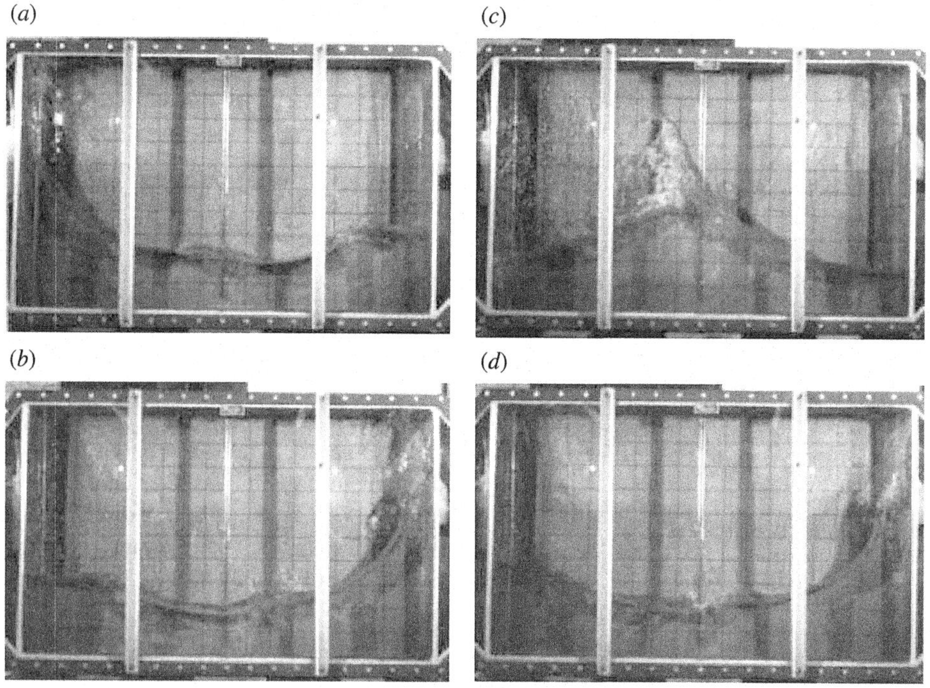

Figure 8.29. Experimental instantaneous water configurations. The excitation period is $T = 2.3$ s (highest natural period is $T_1 = 2.115$ s), the excitation amplitude is $\eta_{2a}/l = 0.028$, and the nondimensional depth is $h/l = 0.173$. Measured and calculated wave elevations at FS3 (see Figure 8.9) are given in Figure 8.28(b).

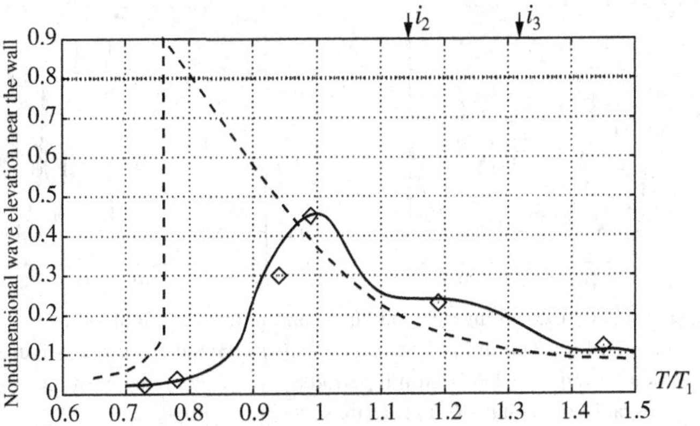

Figure 8.30. Wave elevation amplitude near the wall divided by the tank breadth versus "period-first natural period ratio" for $\overline{h} = 0.2$. The angular forcing has amplitude $\eta_{4a} = 0.1$ rad, $T/T_1 = i_2 = 1.1446$ for secondary resonance of the second mode, and $T/T_1 = i_3 = 1.3226$ for secondary resonance of the third mode. The solid line represents results by the Boussinesq-type multimodal theory and the dashed line gives predictions by Faltinsen's (1974) Moiseev-based theory; the diamond symbols (◇) are experiments by Olsen and Johnsen (1975). The dotted line at 0.8 marks the tank roof.

free surface guarantees convergence only in a mean-squared sense. This means that, if $\zeta_{\text{exact}}(x, t)$ is the exact solution of the free-surface elevation and $\zeta_{\text{modal}}(x, t) = \sum_{i=1}^{N} \beta_i(t) f_i(x)$ is its modal approximation, $\int_{-\frac{1}{2}l}^{\frac{1}{2}l} (\zeta_{\text{exact}} - \zeta_{\text{modal}})^2 \, dx \to 0$ as $N \to \infty$. The mean-squared convergence does not guarantee that $|\zeta_{\text{exact}} - \zeta_{\text{modal}}| \to 0$ for a fixed point x in the interval $[-\frac{1}{2}l, \frac{1}{2}l]$ with increasing N. In particular, it cannot provide convergence at the walls $x \to \pm\frac{1}{2}l$ when wave profiles are not perpendicular to the vertical walls. Uniform convergence is also impossible for breakers spilling away from the tank walls even though the free surface is a single-valued function of the lateral tank coordinate. The contribution of the higher modes during breaking is large and they must be accurately accounted for by the modal approximation. Furthermore, the modal system does not account for the turbulent energy dissipation associated with breaking waves.

Faltinsen and Timokha (2002) also reported on the importance of damping of the higher modes to reach a steady-state solution. They used the experimental data by Olsen and Johnsen (1975) for angularly forced steady-state waves with $\overline{h} = 0.2$. The sloshing in the rectangular tank was harmonically excited with roll amplitude 0.1 rad. The roll axis goes through the center of the mean liquid volume. The experimental measurements and numerical results are compared in Figure 8.30. The two values of T/T_1 indicated in the figure, i_2 and i_3, correspond to secondary resonance. The figure shows that theory and experiments agree well. Damping terms were important to reach periodic solutions. Some additional damping had to be added to the theoretical damping ratios $\xi_i = \xi_i^{\text{tank surface}} + \xi_i^{\text{bulk}}$; one reason is that energy dissipation during transient conditions is believed to be larger than for steady-state motions. The calculations started with higher damping coefficients. The theoretical damping ratios, ξ_i, were multiplied by the factor $1 + \delta(\pi i)^2$. The total simulation time was divided into a series of time intervals: $\delta = 1$ was used in the initial time interval; then the parameter δ was divided by a factor of 5; at the final time interval $\delta = 10^{-4}$. Therefore, steady-state values are consistent with predictions of linear viscous damping.

8.8 Shallow liquid depth

8.8.1 Use of the Boussinesq-type multimodal method for intermediate and shallow depths

8.8.1.1 Transients

Comparisons of the multimodal method for intermediate and shallow depths with experimental data show the importance of run-ups along the tank walls with thin jets as well as breaking waves. The jets and associated recordings of the wave elevation are illustrated in Figures 8.31 and 8.32. Figure 8.31 shows experimental and theoretical time histories of the wave elevation at wave

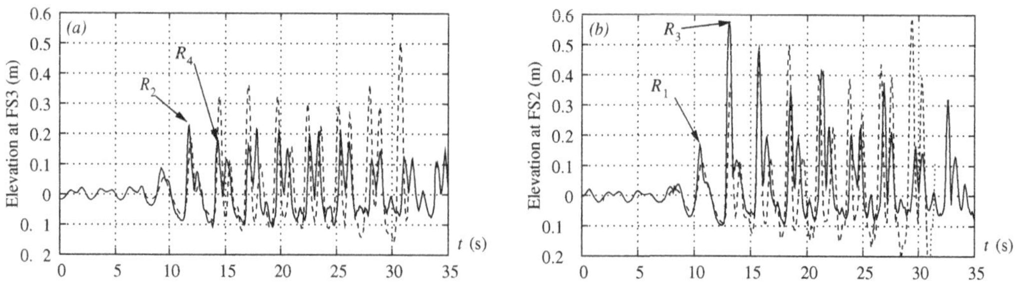

Figure 8.31. Experimental measurements (solid line) and numerical simulations by the Boussinesq-type multimodal method (dashed line) at wave probes (a) FS3 and (b) FS2 in Figure 8.9 for $h/l =$ 0.116 with excitation period $T = 2.7$ s (natural period is $T_1 = 2.25$ s) and excitation amplitude η_{2a}/l = 0.028. The photographs of run-ups R_1 and R_2 are shown in Figure 8.32.

probes FS2 and FS3 (see Figure 8.9) from the test series with $\overline{h} = h/l = 0.116$ and excitation period $T = 2.7$ s. Satisfactory agreement exists for about 20 s of real time. However, the calculations caused a "dry" bottom after 30 s of real time and therefore the calculations had to be stopped. The experiments did not result in a dry bottom. Furthermore, a clear difference can be seen between maximum elevations at FS3 and those at FS2. FS2 is at the wall and FS3 is only 5 cm away from the wall in a tank 1.73 m wide. The difference between the measurements at FS2 and FS3 indicates the presence of run-ups of thin liquid layers. Run-ups occur at the time instants R_1, $R_2, \ldots$ indicated in Figure 8.31. The corresponding upper–lower arrows in Figure 8.32 denote the instantaneous height of the jets at the walls. The numerical results show a difference at FS2 and FS3 but not as large as in the experiments. The reason is that the Fourier modal presentation requires the free surface to be perpendicular to the vertical walls, which contradicts the surface shapes in Figure 8.32.

8.8.1.2 Steady-state regimes

The multimodal method is efficient and accurate in predicting the steady-state response as long as h/l is not too small and the experiments do not detect strongly breaking waves, run-ups, or free-surface fragmentation. The validation of the dissipative modal theory requires long time simulations to reach a periodic solution. The calculation procedure is based on two steps. In the first step we start with zero initial conditions until the periodic solution is reached. Then the excitation period is changed, starting from the steady-state numerical solution obtained from the first step, which makes it possible to detect multiple solutions occurring at the same forcing frequency and corresponding jumps in the response curves. The first example is chosen from the experimental shallow-water results of Chester and Bones (1968). The tank breadth, height, and length are 0.6096, 0.155, and 0.1524 m, respectively.

Because the forcing amplitudes were very small and Chester and Bones (1968) did not

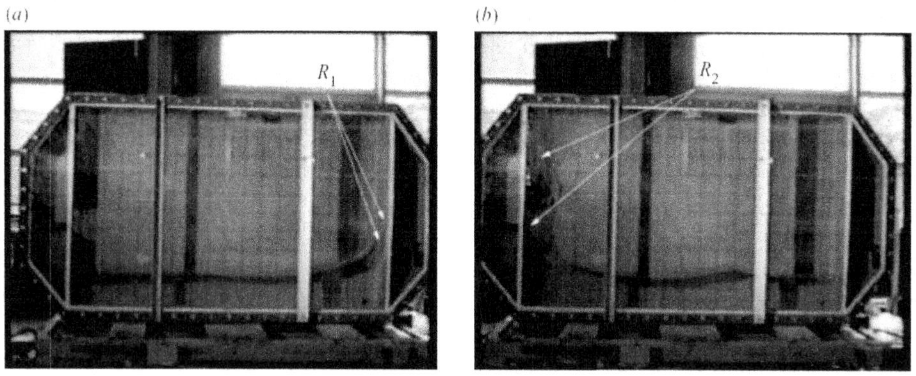

Figure 8.32. Experimental photos for the series of the first run-ups occurring at the right- and left-hand walls. The case is discussed in Figure 8.31.

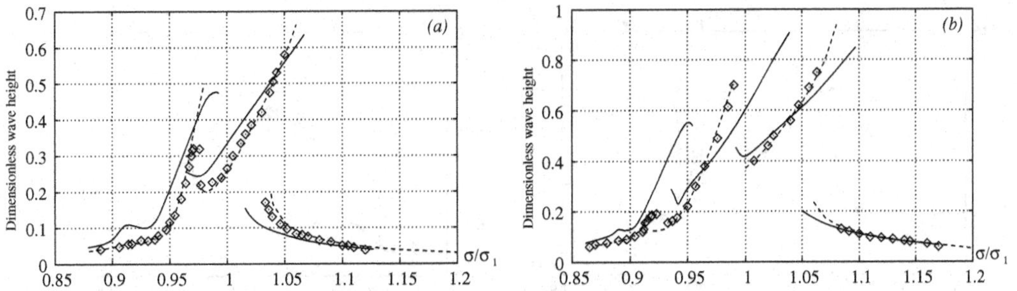

Figure 8.33. Dimensionless steady-state wave height (scaled by h) versus the excitation frequency. Rectangular tank with ratio of water depth to tank breadth $h/l = 0.08333$. Horizontal harmonic excitations are present. The calculated data are for freshwater with $\nu = 1.1 \cdot 10^{-6}$ m^2 s^{-1}. ◇, experiments by Chester and Bones (1968); dashed line, Boussinesq-type multimodal theory; and solid line, theory by Chester (1968). (a) $\eta_{2a}/l = 0.001254$ and (b) $\eta_{2a}/l = 0.002583$.

mention anything about observed run-up and wave breaking, one can expect that only viscous boundary-layer and bulk damping, as discussed in Section 6.3, matter. Due to small damping it took up to 100 forced periods to reach a steady-state solution. Chester and Bones (1968) presented experimental and theoretical results of wave amplitude response near the wall for sway-excited sloshing with $h/l = 0.083333$ and 0.041667. The nondimensional sway amplitudes were $\eta_{2a}/l = 0.001254$ and 0.002583. Results for $h/l = 0.083333$ and $\eta_{2a}/l = 0.001254$ are presented in Figure 8.33(a). The Boussinesq-type multimodal method agrees well with experiments and clearly gives better theoretical results than those presented by Chester (1968). Figure 8.33(b) shows good agreement between the multimodal theory and experiments with a forcing amplitude twice larger. The theoretical and numerical results in Figures 8.33(a) and 8.33(b) demonstrate only three jumps, which are associated with the primary resonance and secondary resonances of the second and third modes. The secondary resonance of the fourth mode is predicted near $\sigma/\sigma_1 = 0.91$, but we see a smooth curve instead of jumps. The reason is damping, which is larger for higher modes. The numerical results are based on using only eight modes. However, increasing the number of modes led to a relative error of less than 10^{-4} implying good convergence.

When comparing with the experimental results by Chester and Bones (1968) conducted at a smaller depth of $\overline{h} = 0.0416667$, we got convergence problems. Reasonable agreement exists with the experiments with eight modes. If number of modes is further increased, convergence with relative error 10^{-4} is not reached. Because of the exponentially increasing calculation time and stiffness with increasing modal dimension of the system, the modal method is not able to deal with very large dimensions.

Figure 8.2(a) presents additional experimental results that are compared with the Boussinesq-type multimodal theory. The forced sway amplitude η_{2a} is small (i.e., $\eta_{2a}/l = 0.01$). The effect of liquid viscosity is clearly present. The first comparison is for "freshwater" ($\nu = 1.0 \cdot 10^{-6}$ m^2 s^{-1}). The effect of liquid viscosity is accounted for in the bulk damping ratio ξ_i^{bulk} and in the viscous boundary-layer damping ratio $\xi_i^{\text{tank surface}}$. The measured and calculated lateral forces presented in Figure 8.34(a) show generally good agreement. The steady-state wave amplitude response has several branches and the possibility of multiple solutions. The branches theoretically imply four superharmonic resonances (jumps) in the response, which are very small and not as clearly seen as for shallower depths. We denoted them as j_1, j_2, and j_3. The jump j_1 is associated with the primary resonance. The jumps j_2 and j_3 are caused by secondary resonance of the second and third modes, respectively, and are estimated as being near $T = 2.37$ s (i_2) and $T = 2.61$ s (i_3), within the framework of our approximation of the natural frequencies. The predicted jump j_0 was originally considered an error but a similar jump was also detected in the experiments. We do not have a reasonable physical explanation of the jump. The average time to reach periodic solutions with viscous damping coefficients was between 400 and 600 forcing periods. This corresponds to approximately 20 to 25 minutes of real

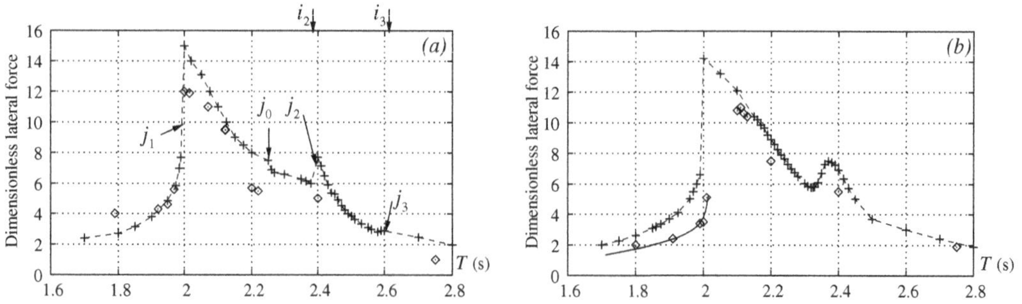

Figure 8.34. Dimensionless lateral force $1000F_y/(\rho g l^2 B)$ versus excitation period for the prismatic tank in Figure 4.22(a) with $h/l = 0.12$. Horizontal excitation with $\eta_{2a}/l = 0.01$. The measured and calculated data are for (a) "freshwater" with $\nu = 1.1 \cdot 10^{-6}$ m^2 s^{-1} and (b) "glycerol/water 85%" with $\nu = 1.1 \cdot 10^{-4}$ m^2 s^{-1}). Dashed line with +, Boussinesq-type multimodal theory for intermediate and shallow depths; ◇, experiments. The solid line in part (b) accounts for the shear force at the bottom.

sloshing time in model scale. However, we have no experimental results to support this long transient phase.

A discrepancy exists for 2.0 s $\le T \le$ 2.1 s, which has no clear explanation. Conversely, we can clarify some disagreement in the response and in the value of corresponding jumps between j_0 and j_2. The disagreement may be influenced by the lower chamfers. Because the chamfers increase the natural periods, i_2 and i_3 would move to higher periods. This drift may also explain the measured value for the highest excitation period as shown in Figure 8.34(a).

The second comparison is for glycerol–water 85% ($\nu = 1.1 \cdot 10^{-4}$ m^2 s^{-1}). The measured and calculated values are presented in Figure 8.34(b). This comparison also confirms the effect of secondary resonance of the second mode. The experimental results show a clear influence of viscosity for $T <$ 2.0 s. This influence can be partly explained by adding viscous shear stresses due to boundary-layer flow at the bottom and the side walls parallel to our two-dimensional potential flow. Because the mode amplitudes were small for $T <$ 2.0 s, linear theory can be used for the flow outside the boundary layer in combination with Stokes linear laminar boundary-layer flow (see Section 6.3.1). The solid line in Figure 8.34(b) shows that the disagreement between theory and experiment disappears by accounting for shear forces on the tank's bottom and the tank's wall. The effect of lateral shear force may also be important for other excitation periods. However, these ranges are associated with strong nonlinearities and a nonlinear boundary-layer model may be needed.

The effect of the boundary-layer inflow/outflow on the pressure loads was not accounted for in the calculations in Figure 8.34(b). The contributions to the total force from the shear stresses and latter viscous pressure loads are equal for attached laminar flow past a circular cylinder (see Section 6.2.2). However, the viscous shear stresses are expected to dominate over viscous pressure loads in our case. One reason is that the area of the wetted tank surface where viscous shear stresses contribute to the horizontal force is far larger than the area of the end walls, where viscous pressure loads contribute to the horizontal force. Furthermore, the boundary-layer inflow/outflow at the end walls causing the viscous pressure loads is a consequence of the vertical gradient $\partial w/\partial z$ of the vertical potential flow velocity at the end walls. However, $\partial w/\partial z$ is relatively small in shallow-liquid conditions.

Finally, Figure 8.2(b) presents the lateral hydrodynamic forces for $\eta_{2a}/l = 0.1$. Abramson *et al.* (1974) photographed breaking waves, hydraulic jumps, and heavy liquid impacts on the walls for this case. We believe that the damping due to breaking waves is important. Because the experimental results do not depend strongly on the viscosity, the damping due to breaking waves must have a different reason, such as turbulence. Therefore, our previous predictions of the damping ratio ξ_i, mainly based on laminar boundary-layer flow, are only lower bounds. Although our Fourier series of the free-surface elevation is not

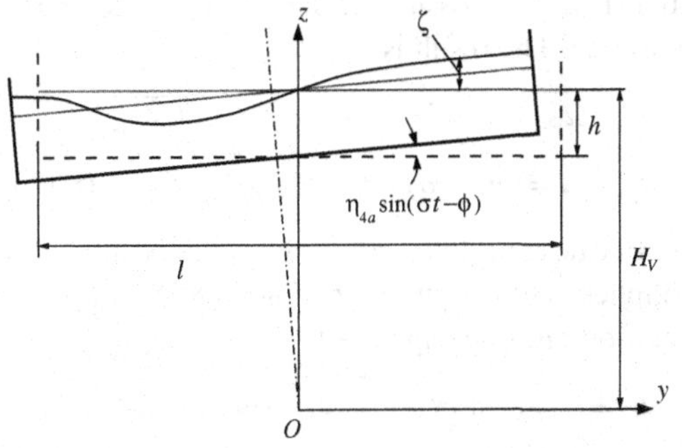

Figure 8.35. Liquid oscillating in a container (Verhagen & van Wijngaarden, 1965).

able to describe the breaking waves and bores, we tried to use it to calculate lateral hydrodynamic force. The reason is that the formula for hydrodynamic force is based on an integral over the Fourier series representing the free surface, and, even though the Fourier series has only weak (mean square) convergence, the integral can imply uniform convergence. Therefore, even a few of the lowest terms in the modal approximation may give adequate estimations of the force. Seven modes are used in the numerical modal calculations presented in Figure 8.2(b). The first three modes were damped by using the damping ratio $\xi_i = \xi_i^{\text{tank surface}} + \xi_i^{\text{bulk}}$, but the higher modes were damped overcritically (i.e., $\xi_i \geq 1$). The calculations are generally in satisfactory agreement with experiments. The discrepancy is largest in the vicinity of $T = 2.0$ s, where the steady-state solution reaches the maximum amplitude response and severe breaking waves are expected.

8.8.2 Steady-state hydraulic jumps

Hydraulic jumps may be formed around resonance in the shallow-liquid case. They can result in very high impact pressures at the end walls. Verhagen and van Wijngaarden (1965) derived a two-dimensional shallow-liquid theory for a rectangular tank of breadth l and liquid depth h. An analytical steady-state (periodic) solution for forced harmonic roll oscillations of small amplitude η_{4a} and with frequency σ near the lowest natural frequency σ_1 was derived. Because of the shallow-liquid approximation $h/l \ll 1$, the first natural frequency (see Section 4.3.1.1) is $\sigma_1 = \pi\sqrt{gh}/l$. The forced roll angle of the tank is expressed as

$$\eta_4 = \eta_{4a}\sin(\sigma t - \phi), \quad 0 < \eta_{4a} \ll 1. \qquad (8.114)$$

We introduce a Cartesian coordinate system Oyz with origin in the roll axis (see Figure 8.35). The z-axis is positive upward and is in the tank's centerplane when the tank is in an upright position. The z-coordinate of the mean free surface with no tank oscillations is denoted H_V. The case of forced harmonic sway oscillations can be obtained by letting H_V go to infinity and η_{4a} go to zero, so that η_{4a} multiplied by H_V approaches the forced sway amplitude η_{2a}.

The steady-state theory by Verhagen and van Wijngaarden (1965) is based on the nonlinear shallow-liquid potential flow theory described by Stoker (1992). An important assumption is that the hydrodynamic pressure is hydrostatic below the instantaneous free surface. The body boundary condition at $y = \pm\frac{1}{2}l$ is $v = -\eta_{4a}H_V\sigma\cos(\sigma t - \phi)$, where v is the lateral liquid velocity. The surface level relative to the tank bottom is expressed as $\lambda(y,t) = h + \zeta(y,t) - y\eta_{4a}\sin(\sigma t - \phi)$, where $z = \zeta(y,t)$ denotes the wave elevation relative to the mean free surface. Verhagen and van Wijngaarden wrote their solution as a power series in the small parameter

$$\delta = \sqrt{l\eta_{4a}/\pi h}, \qquad (8.115)$$

which expresses the relation between small values of η_{4a} and h/l. The forcing frequency is close to the lowest natural frequency (i.e., $(\sigma - \sigma_1)/\sigma_1 \ll 1$). Terms of $O(\delta)$ were included.

When a steady-state solution exists, a hydraulic jump travels back and forth between the tank walls. The path of the hydraulic jumps in the y–t-plane is illustrated as solid zigzag line in Figure 8.36. The hydraulic jump is at the left wall ($y = -\frac{1}{2}l$) at the time instant $t = 0$ corresponding to the point P in Figure 8.36. When $t = \pi/\sigma$, the hydraulic jump is at the right wall ($y = \frac{1}{2}l$, point Q). The hydraulic jump path separates different regions I_j ($j = 0, 1, \ldots$) and II_j ($j = -1, 0, \ldots$) in the y–t-plane, as indicated in the figure. Conservation of mass and momentum across the hydraulic jump is used to relate the energies in regions I_j and II_j. Because we look for periodic solutions, the solution in I_j and II_j can be

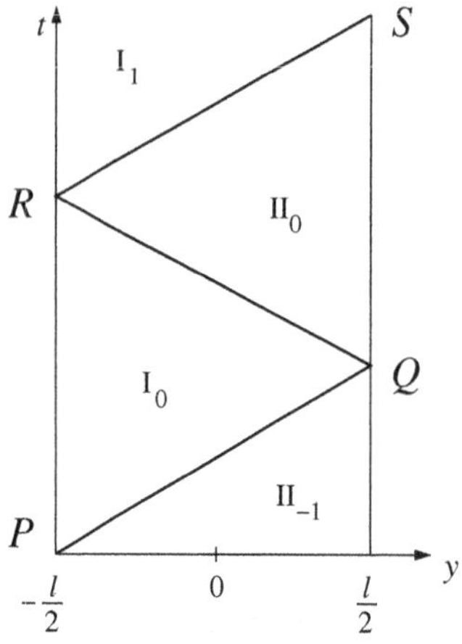

Figure 8.36. Paths of hydraulic jump in the y–t-plane.

obtained from the solutions in I_0 and II_0, respectively, by introducing the time shift $t = t + 2\pi j/\sigma$, where j is an integer.

The phase ϕ of the roll angle in eq. (8.114) relative to the hydraulic jump position at $t = 0$ is the root of the equation

$$\sin\left(\tfrac{1}{2}\phi + \tfrac{1}{4}\pi\right) = \frac{\pi}{6A\delta}\left(1 - \frac{\sigma}{\sigma_1}\right) \quad \text{with}$$

$$A = \sqrt{\frac{2}{3\pi}\left[1 + \pi^2\frac{hH_V}{l^2}\right]}. \qquad (8.116)$$

If a solution of eq. (8.116) does not exist, no hydraulic jump is possible. The solvability condition can be expressed as

$$|\Omega| \le 1, \quad \Omega = \pi\,(1 - \sigma/\sigma_1)\,/(6A\delta). \qquad (8.117)$$

From eqs. (8.115) and (8.116) it follows that Ω can be expressed as

$$\Omega = \frac{(\sigma_1 - \sigma)}{\sigma_1}\pi^2\sqrt{\frac{h}{24\eta_{4a}l\,(1 + \pi^2 hH_V/l^2)}}$$

$$= (\sigma_1 - \sigma)\sqrt{\frac{\pi^2 l}{24\eta_{4a}g\,(1 + \pi^2 hH_V/l^2)}}. \qquad (8.118)$$

When eq. (8.117) is satisfied, the hydraulic jump solution exists and the phase is defined by the formula

$$\phi = -\tfrac{1}{2}\pi + 2\arcsin\Omega, \quad -\tfrac{3}{2}\pi \le \phi < \tfrac{1}{2}\pi. \qquad (8.119)$$

The value of Ω for sway excitation with amplitude η_{2a} is obtained by letting $\eta_{4a}H_V \to \eta_{2a}$ as $H_V/l \to \infty$ and $\eta_{4a} \to 0$ in eq. (8.118). We stress that it is not possible to let $H_V/l \to -\infty$ in the formula. The result is

$$\Omega = \Omega_{\text{sway}} = (1 - \sigma/\sigma_1)\,\pi\sqrt{l/(24\eta_{2a})}$$

$$= (\sigma_1 - \sigma)\sqrt{l^3/(24\eta_{2a}gh)}. \qquad (8.120)$$

Because positive roll angle and positive H_V implies negative sway (see Figure 8.35), the sway motion must be expressed as

$$\eta_2 = -\eta_{2a}\sin(\sigma t - \phi) = \eta_{2a}\sin(\sigma t - \phi + \pi), \qquad (8.121)$$

where the phase is defined by eq. (8.119) with $\Omega = \Omega_{\text{sway}}$ given by eq. (8.120).

A hydraulic jump solution is, according to eq. (8.117), only possible if

$$|1 - \sigma/\sigma_1| \le \pi^{-2}\sqrt{24}\sqrt{l\eta_{4a}/h + \pi^2 H_V\eta_{4a}/l} \qquad (8.122)$$

for roll excitation and if

$$|1 - \sigma/\sigma_1| \le \pi^{-1}\sqrt{24\eta_{2a}/l} \qquad (8.123)$$

for sway excitation; that is, the hydraulic jump condition is independent of the depth for forced sway motion, whereas this is not true for forced roll.

Lugni (personal communication, 2008) experimentally investigated hydraulic jump condition (8.123) for a ratio of water depth to tank breadth of $\bar{h} = 0.125$ by choosing η_{2a}/l as 0.03, 0.04, 0.05, and 0.1. The wave systems were categorized as proposed by Olsen and Johnsen (1975):

- *Wave system I.* Standing waves (see Figure 8.37(a)).
- *Wave system II.* Progressive nonbreaking waves (see Figure 8.37(b)) that are traveling back and forth in the tank.
- *Wave system III.* Progressive waves (see Figure 8.37(c)) traveling back and forth in the tank and eventually breaking near the vertical walls.
- *Wave system IV.* Hydraulic jumps (see Figure 8.37(d)) traveling back and forth in the tank.

The different wave systems are shown in Figure 8.38 together with Verhagen and van Wijngaarden's theoretical borderline for hydraulic jumps to occur. When $T/T_1 = 1$, the theory predicts hydraulic jumps for any forcing amplitude. However, no hydraulic jumps are experimentally observed for the lowest experimental forcing

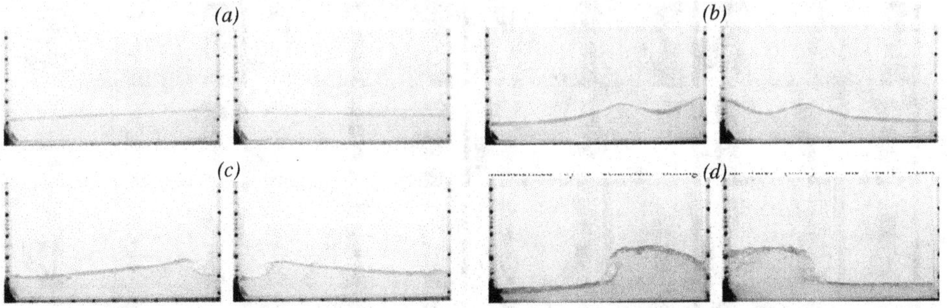

Figure 8.37. Wave systems I, II, III, and IV for $h/l = 0.125$: (a) wave system I, $\eta_{2a}/l = 0.03$, $T/T_1 = 1.55$; (b) wave system II, $\eta_{2a}/l = 0.03$, $T/T_1 = 1.25$; (c) wave system III, $\eta_{2a}/l = 0.03$, $T/T_1 = 1.0$; (d) wave system IV, $\eta_{2a}/l = 0.1$, $T/T_1 = 1.0$ (Lugni, personal communication, 2008).

amplitude in Figure 8.37(c). For $\eta_{2a}/l \geq 0.04$, the hydraulic jumps occur in a period range that is not as broad as theoretically predicted. One possible reason is that the ratio of water depth to tank breadth may not be sufficiently small for the shallow-depth assumption to be fully valid.

Waves do not necessarily break at the tank wall. A different scenario where they start to break in the middle part of the tank is illustrated in Figure 8.39. The condition is $\overline{h} = 0.06$, $\eta_{2a}/l = 0.07$, $T/T_1 = 0.725$. A splash-up occurs after the overturning wave impacts the free surface. The splash-up splits up into two parts. The first is reversed relative to the progressive wave and the second causes a new splash-up with a lower intensity than the primary generated splash-up. The latter scenario is not associated with large impact loads on the tank walls. However, it could be of concern if an interior structure such as a pump tower is sufficiently close to the breaking position.

Large slamming loads are caused on the tank wall instead if the sway amplitude is $\eta_{2a}/l = 0.05$, as illustrated in Figure 8.40. The other conditions, $\overline{h} = 0.06$ and $T/T_1 = 0.725$, are the same as in Figure 8.39. The wave approaches the tank wall but does not impact against it. A flip-through of the free surface happens instead at the tank wall (see Sections 11.1, 11.3.3, and 11.4 for further discussion of flip-through). The flow is locally strongly accelerated with accompanying large pressures. A further consequence is a very thin vertical jet flow along the tank wall.

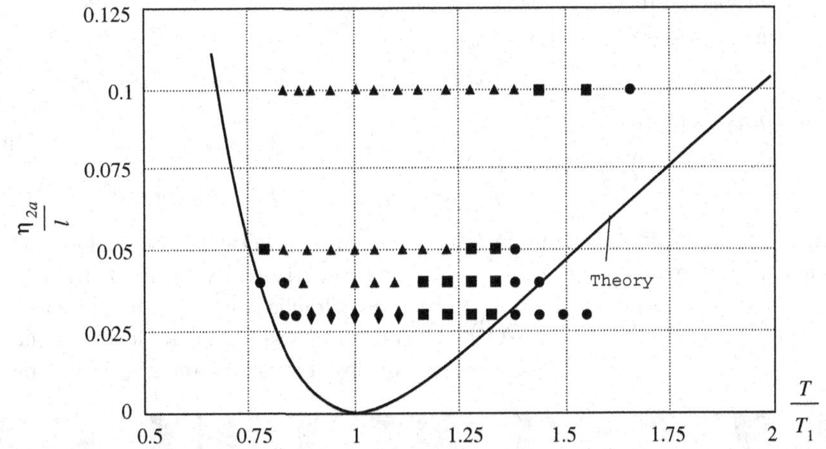

Figure 8.38. Occurrence of experimentally observed wave systems I, II, III, and IV for forced sway oscillations of two-dimensional flow in a rectangular tank with $h/l = 0.125$ shown together with theoretical line of Verhagen and van Wijngaarden (1965) for the occurrence of hydraulic jumps. Nondimensional sway amplitude, η_{2a}/l, is presented as a function of the ratio between the forcing period T and the highest natural sloshing period T_1 (Lugni, personal communication, 2008) (•, system I; ■, system II; ♦, system III; ▲, system IV).

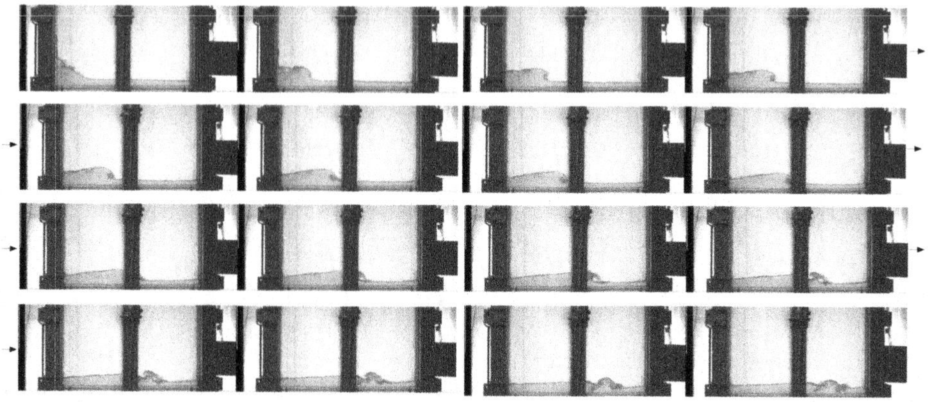

Figure 8.39. Overturning waves with splash-up occurring in the middle part of the tank. The conditions are $h/l = 0.06$, $\eta_{2a}/l = 0.07$, and $T/T_1 = 0.725$ (Lugni, personal communication, 2008).

Verhagen and van Wijngaarden (1965) found the steady-state solution in domains I_0 and II_0 in Figure 8.36. The surface elevations correct to $O(\delta)$ are expressed as

$$\begin{aligned}\frac{\zeta(y,t)}{h} &= \tfrac{4}{3}\frac{\sigma-\sigma_1}{\sigma_1} + 4A\delta\cos\left(\tfrac{1}{2}\sigma_1 t - \tfrac{1}{2}\phi - \tfrac{1}{4}\pi\right)\\ &\quad\times \cos\left[\tfrac{1}{2}\pi\left(y+\tfrac{1}{2}l\right)/l\right], \quad (t,y)\in I_0,\\ \frac{\zeta(x,t)}{h} &= \tfrac{4}{3}\frac{\sigma-\sigma_1}{\sigma_1} + 4A\delta\sin\left(\tfrac{1}{2}\sigma_1 t - \tfrac{1}{2}\phi - \tfrac{1}{4}\pi\right)\\ &\quad\times \sin\left[\tfrac{1}{2}\pi\left(y+\tfrac{1}{2}l\right)/l\right], \quad (t,y)\in II_0.\end{aligned} \tag{8.124}$$

The lateral liquid particle velocities in regions I_0 and II_0 are

$$\begin{aligned}v(y,t) &= 4\sqrt{gh}A\delta\sin\left(\tfrac{1}{2}\sigma_1 t - \tfrac{1}{2}\phi - \tfrac{1}{4}\pi\right)\\ &\quad\times \sin\left[\tfrac{1}{2}\pi\left(y+\tfrac{1}{2}l\right)/l\right], \quad (t,y)\in I_0,\end{aligned} \tag{8.125}$$

$$\begin{aligned}v(y,t) &= 4\sqrt{gh}A\delta\cos\left(\tfrac{1}{2}\sigma_1 t - \tfrac{1}{2}\phi - \tfrac{1}{4}\pi\right)\\ &\quad\times \cos\left[\tfrac{1}{2}\pi\left(y+\tfrac{1}{2}l\right)/l\right], \quad (t,y)\in II_0.\end{aligned} \tag{8.126}$$

The height of the hydraulic jump is time-independent and given by

$$h_{\text{jump}} = 4h\sqrt{A^2\delta^2 - \tfrac{1}{36}\pi^2\left(\sigma/\sigma_1 - 1\right)^2} + O(\delta^2). \tag{8.127}$$

The position of the jump traveling from left to right is

$$\begin{aligned}y_{\text{jump}} &= -\tfrac{1}{2}l + V_{g0}t + \delta\left[\frac{l}{\pi}\frac{\sigma-\sigma_1}{\delta}t\right.\\ &\quad + \frac{2A}{\pi}\left(V_{g0}t - l\right)\sin\left(\tfrac{1}{2}\phi+\tfrac{1}{4}\pi\right)\\ &\quad \left. - 2\frac{Al}{\pi}\sin\left(\tfrac{1}{2}\sigma_1 t - \tfrac{1}{2}\phi - \tfrac{1}{2}\pi\right)\cos\left(\tfrac{1}{2}\sigma_1 t\right)\right]\\ &\quad + O(\delta^2) \quad \text{along } PQ,\end{aligned} \tag{8.128}$$

where $V_{g0} = \sqrt{gh}$ is both the phase and the group velocity according to linear shallow-liquid theory. Furthermore, A is defined by eq. (8.116). The same jump moves from right to left according to the expression

$$\begin{aligned}y_{\text{jump}} &= \tfrac{3}{2}l - V_{g0}t - \delta\left[\frac{l}{\pi}\frac{\sigma-\sigma_1}{\delta}t\right.\\ &\quad + \frac{2A}{\pi}\left(V_{g0}t - 2l\right)\sin\left(\tfrac{1}{2}\phi+\tfrac{1}{4}\pi\right)\\ &\quad \left. + 2\frac{Al}{\pi}\cos\left(\tfrac{1}{2}\sigma_1 t - \tfrac{1}{2}\phi - \tfrac{1}{2}\pi\right)\sin\left(\tfrac{1}{2}\sigma_1 t\right)\right]\\ &\quad + O(\delta^2) \quad \text{along } QR.\end{aligned} \tag{8.129}$$

The paths PQ and QR are defined in Figure 8.36.

The horizontal velocity of the free surface of the hydraulic jump, V_{jump}, is an important parameter in assessing impact loads on the tank wall or on internal structures such as a pump tower.

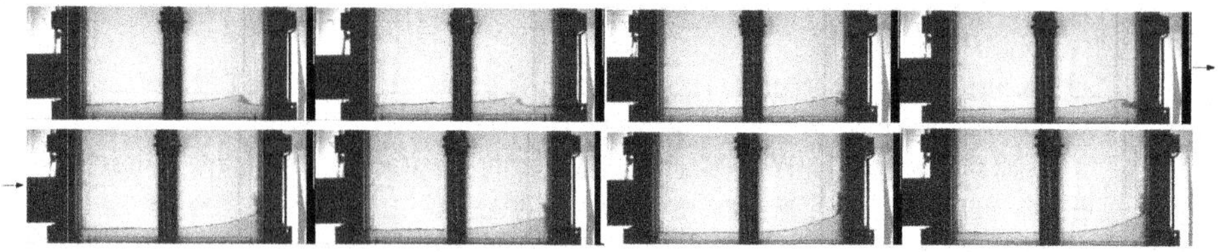

Figure 8.40. Steep waves approaching vertical tank wall with subsequent 'flip-through'. The conditions are $h/l = 0.06$, $\eta_{2a}/l = 0.05$, and $T/T_1 = 0.725$ (Lugni, personal communication, 2008).

By time-differentiating eq. (8.128) it follows that

$$V_{\text{jump}} = V_{g0} + \delta\left[\frac{l}{\pi}\frac{\sigma-\sigma_1}{\delta} + \frac{2AV_{g0}}{\pi}\sin\left(\tfrac{1}{2}\phi + \tfrac{1}{4}\pi\right) - \frac{Al\sigma_1}{\pi}\cos\left(\sigma_1 t - \tfrac{1}{2}\phi - \tfrac{1}{2}\pi\right)\right] + O(\delta^2). \tag{8.130}$$

When the hydraulic jump is at the tank wall at $y = \frac{1}{2}l$ (i.e., $t = \pi/\sigma$), V_{jump} can be expressed as

$$V_{\text{jump}} = \sqrt{gh}\left[1 + \sqrt{\frac{\eta_{4a}}{3\pi^2}}\sqrt{\left(\frac{l}{h} + \pi^2\frac{H_V}{l}\right)}\right] \quad \text{and}$$

$$V_{\text{jump}} = \sqrt{gh}\left[1 + \sqrt{\frac{\eta_{2a}}{3l}}\right]$$

for roll and sway, respectively.

Verhagen and van Wijngaarden (1965) compared their theory with experimental data of instantaneous wave profiles for forced harmonic roll motion. The roll axis was at the intersection of the bottom and the centerplane of the tank. The tank breadth and the ratio of water depth to tank breadth were $l = 1.20$ m and $\bar{h} = 0.075$, respectively. The forced roll amplitudes in radians were $\eta_{4a} = \pi/180$, $\pi/90$, $\pi/60$, and $\pi/45$ and the ratios σ/σ_1 between the forcing frequency and the lowest natural sloshing frequency were 0.89, 1.0, and 1.14. The agreement between theoretical predictions and experimental values was partial. However, the theoretical prediction of wave profiles close to the middle of the tank for $\sigma/\sigma_1 = 1$ and roll amplitude $\pi/90$ was unsatisfactory; the maximum experimental wave elevation was about 2.5 times the theoretical value. Higher harmonic components, not predicted theoretically, were evident in the experiments.

We also compare Verhagen and van Wijngaarden's theory with the experimental results presented in Figure 8.37(d), where $\bar{h} = 0.125$, $\eta_{2a}/l = 0.1$, and $T/T_1 = 1.0$. The theoretical values correspond to the fact that the hydraulic jump moves along the line QR in Figure 8.36. The free-surface elevation can be expressed as $\zeta = H(y^{QR}_{\text{jump}} - y)\zeta_{I_0} + H(y - y^{QR}_{\text{jump}})\zeta_{II_0}$, where $H(x)$ is the Heaviside function, which has values of 1 and 0 for $x > 0$ and $x < 0$, respectively. Furthermore, y^{QR}_{jump} is the expression for y_{jump} given by eq. (8.129). The free-surface elevations ζ_{I_0} and ζ_{II_0} are the values given for regions I_0 and II_0 by eq. (8.124). A three-dimensional drawing of the free-surface elevation as a function of $y \in [-\frac{1}{2}l, \frac{1}{2}l]$ and $t \in [\pi/\sigma_1, 2\pi/\sigma_1]$ is presented in Figure 8.41. A hydraulic jump starts at $y = \frac{1}{2}l$ at $t = \pi/\sigma_1$ and ends up at $y = -\frac{1}{2}l$ at $t = 2\pi/\sigma_1$. The height of the hydraulic jump is $h_{\text{jump}} = 0.046l$ according to eq. (8.127). An estimate based on Figure 8.37(d) gives $h_{\text{jump}} = 0.13l$ (i.e., about 2.8 times higher than theoretical values). For this case the value of h/l may be too high for the shallow-liquid theory to be consistent.

The phase angle ϕ is $-\frac{1}{2}\pi$ according to eqs. (8.119) and (8.120), which means the sway motion and velocity are $-\eta_{2a}\cos\sigma t$ and $\sigma\eta_{2a}\sin\sigma t$, respectively, according to eq. (8.121). When, for instance, $t = \frac{3}{2}\pi/\sigma_1$ and the hydraulic jump is close to the middle of the tank, the sway motion is zero, the sway velocity is at a minimum, and the horizontal force is at a maximum.

Because the hydrodynamic pressure is hydrostatic below the instantaneous free surface, the hydrodynamic force and moment can be expressed in terms of the free-surface elevations given by eq. (8.124). The wave motions are steady state, so we need to present the shallow-water approximation of the lateral force only in the time range $0 < t < 2\pi/\sigma$. The natural frequency σ_1 is replaced by the forcing frequency σ in eq. (8.124). The error in doing so is $O(\delta^2)$ and can be neglected. The resulting force is

$$\begin{aligned} F_2(t) &= \frac{\rho g h^2}{2}\left[\left(\frac{\zeta_{y=l/2}}{h} + 1\right)^2 - \left(\frac{\zeta_{y=-l/2}}{h} + 1\right)^2\right] \\ &= \frac{\rho g h^2}{2}\left[\left(\begin{Bmatrix} \left[\zeta^{II_{-1}}_{y=l/2}/h\right], & 0 < t < \pi/\sigma \\ \left[\zeta^{II_0}_{y=l/2}/h\right], & \pi/\sigma < t < 2\pi/\sigma \end{Bmatrix} + 1\right)^2 - \left(\frac{\zeta^{I_0}_{y=-l/2}}{h} + 1\right)^2\right] \\ &= -\rho g l^2 \left(\frac{h}{l}\right)^2 \underbrace{F_m}_{4\sqrt{2}A\delta} \begin{cases} \sin\left(\tfrac{1}{2}\sigma t - \tfrac{1}{2}\phi\right) + O(\delta), & 0 < t < \pi/\sigma, \\ \cos\left(\tfrac{1}{2}\sigma t - \tfrac{1}{2}\phi\right) + O(\delta), & \pi/\sigma < t < 2\pi/\sigma, \end{cases} \end{aligned} \tag{8.131}$$

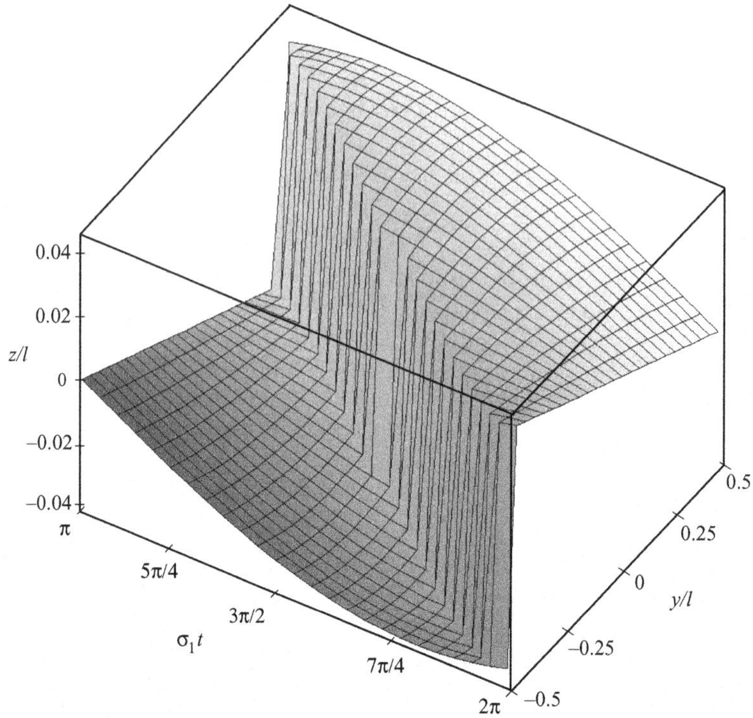

Figure 8.41. Free-surface elevation predicted by Verhagen and van Wijngaarden's two-dimensional theory for a rectangular tank harmonically forced in sway. The conditions are similar to those for the case presented in Figure 8.37(d): $h/l = 0.125$, $\eta_{2a}/l = 0.1$, and $T/T_1 = 1.0$. The free-surface elevation is presented as a function of $y \in [-\frac{1}{2}l, \frac{1}{2}l]$ and $t \in [\pi/\sigma_1, 2\pi/\sigma_1]$.

where the amplitude coefficient is

$$F_m = \begin{cases} \dfrac{8}{\pi}\sqrt{\dfrac{l\eta_{4a}}{3h}}\left(1 + \dfrac{\pi^2 h H_V}{l^2}\right) & \text{for roll,} \\ 8\sqrt{\dfrac{\eta_{2a}}{3l}} & \text{for sway} \end{cases} \qquad (8.132)$$

and phase angle ϕ is defined in connection with eqs. (8.119) and (8.121). The force is not a smooth function of time and a jump occurs at each one-half of the forcing period. Equation (8.132) shows that the amplitude coefficient does not depend on the forcing frequency and increases with the square root of the roll amplitude and of the sway amplitude for forced roll and sway, respectively.

Formula (8.131) gives the lowest-order approximation of the hydrodynamic force. The $O(\delta^2)$-terms are neglected, which is consistent with Verhagen and van Wijngaarden's results for the free-surface elevation eq. (8.124) as well as with corresponding asymptotic solutions for other sloshing characteristics obtained within the framework of this theory. Formula (8.131) can be used to estimate the nondimensional maximum lateral force present in Figure 8.2. The values are lower than the experimental ones and, moreover, are constant in a wide frequency range around the primary resonance. The formula gives nondimensional lateral force of about 6.65 in the range 1.9 s $< T <$ 2.4 s and nondimensional lateral force of about 21.0 in the range 1.6 s $< T <$ 3.4 s for $\eta_{2a} = 0.01\, l$ and $\eta_{2a} = 0.1\, l$, respectively. The definition of the nondimensional force is given in the caption of Figure 8.2. Abramson *et al.* (1974) did not report a hydraulic jump for $\eta_{2a} = 0.01\, l$. When $\eta_{2a} = 0.1\, l$, a hydraulic jump occurs for 1.6 s $< T <$ 2.95 s, in reasonable agreement with the theory.

In a similar way, we can find the moment $F_4(t)$ relative to an axis at the *middle of the tank bottom*. The contribution to the roll moment by the pressure on the tank bottom is

$$F_4^{\text{bottom}}(t) = -\rho g \int_{-\frac{1}{2}l}^{\frac{1}{2}l} \zeta y dy$$

$$= \rho g l^3 \underbrace{M_m}_{16\pi^{-2}\delta Ah/l} \begin{cases} \{[-1+\cos(\frac{1}{2}\sigma t)+\sin(\frac{1}{2}\sigma t)]\cos(\frac{1}{2}\phi+\frac{1}{4}\pi)+[-\frac{1}{2}\sigma t+\frac{1}{4}\pi-\cos(\frac{1}{2}\sigma t) \\ \quad +\sin(\frac{1}{2}\sigma t)]\sin(\frac{1}{2}\phi+\frac{1}{4}\pi)\}+O(\delta), \quad 0<t<\pi/\sigma; \\ \{[1+\cos(\frac{1}{2}\sigma t)-\sin(\frac{1}{2}\sigma t)]\cos(\frac{1}{2}\phi+\frac{1}{4}\pi)+[\frac{1}{2}\sigma t-\frac{3}{4}\pi+\cos(\frac{1}{2}\sigma t) \\ \quad +\sin(\frac{1}{2}\sigma t)]\sin(\frac{1}{2}\phi+\frac{1}{4}\pi)\}+O(\delta), \quad \pi/\sigma<t<2\pi/\sigma, \end{cases} \tag{8.133}$$

where the amplitude parameter M_m is defined by the formula

$$M_m = \begin{cases} \dfrac{16}{\pi^3}\left(\dfrac{h}{l}\right)\sqrt{\dfrac{2\eta_{4a}l}{3h}\left(1+\dfrac{\pi^2 hH_V}{l^2}\right)} & \text{for roll,} \\ \dfrac{16}{\pi^2}\left(\dfrac{h}{l}\right)\sqrt{\dfrac{2\eta_{2a}}{3l}} & \text{for sway.} \end{cases} \tag{8.134}$$

Expression (8.133) only accounts for the lowest-order terms. The amplitude parameter M_m is independent of the forcing frequency, but, because the maximum of the discontinuous function $F_4(t)$ depends on the phase $\phi = \phi(\Omega)$, the maximum hydrodynamic moment varies with σ. Equation (8.134) shows that the amplitude parameter increases with the square root of the roll amplitude and the square root of the sway amplitude for forced roll and sway, respectively. Although $\zeta(y,t)$ is not a smooth function, the roll moment due to the pressure on the bottom is an integral over ζ and, as one can check from the direct expressions, $F_4^{\text{bottom}}(t)$ is a periodic smooth function of time.

The contribution from the pressure along the tank walls to the roll moment relative to an axis at the middle of the tank bottom can be expressed as

$$F_4^{\text{walls}}(t)$$

$$= \rho g \left[\int_0^{\lambda(\frac{1}{2}l,t)} z\left(z-\lambda\left(\tfrac{1}{2}l,t\right)\right)dz - \int_0^{\lambda(-\frac{1}{2}l,t)} z\left(z-\lambda\left(-\tfrac{1}{2}l,t\right)\right)dz\right]$$

$$= -\frac{\rho g h^3}{6}\left[\left(\frac{\zeta_{y=l/2}}{h}+1\right)^3 - \left(\frac{\zeta_{y=-l/2}}{h}+1\right)^3\right]$$

$$= \tfrac{1}{2}\rho g l^2 h \left[-\underbrace{\frac{F_2(t)}{\rho g l^2}}_{(\frac{h}{l})^2 O(\delta)} + \left(\frac{h}{l}\right)^2 O(\delta^2)\right],$$

where the lowest-order term is the same as $-F_2^{\text{walls}}(t)$ times the arm $\frac{1}{2}h$ to the static center of the liquid mass.

The roll moment relative to the roll axis is the sum of the two components of the roll moment relative to the bottom center plus the product of the force $F_2^{\text{walls}}(t)$ and the arm $(H_V - h)$; that is,

$$F_4^{\text{center}}(t) = F_4^{\text{bottom}}(t) + F_4^{\text{walls}}(t) - (H_V - h)F_2(t)$$
$$= F_4^{\text{bottom}}(t) - (H_V - \tfrac{1}{2}h)F_2(t), \tag{8.135}$$

where $(H_V - \frac{1}{2}h)$ is the arm between the roll axis and the liquid mass center in the static state.

Formulas (8.131) and (8.135) can be used to evaluate the force and moment. The hydrodynamic tank force and moment may also be included in a linear ship motion analysis in the frequency domain by an equivalent linearization procedure. This procedure involves extraction of the lowest Fourier component $\sin(\sigma t - \phi - \psi)$ from the right-hand sides of eqs. (8.131) and (8.133) so that they are approximated as $F_2(t) = F_{2a}\sin(\sigma t - \phi - \psi_{F_2})$ and $F_4^{\text{bottom}}(t) = F_{4a}\sin(\sigma t - \phi - \psi_{F_4})$. The phase shifts ψ_{F_2} and ψ_{F_4} and amplitudes F_{2a} and F_{4a} are functions of the forcing frequency and amplitude. Direct derivations lead to

$$\psi_{F_2} = -\phi + \arctan\left(\frac{\Omega}{2\sqrt{1-\Omega^2}}\right),$$

$$F_{2a} = -\frac{4\sqrt{2}\rho g h^2}{3\pi} F_m \sqrt{4 - 3\Omega^2}, \tag{8.136}$$

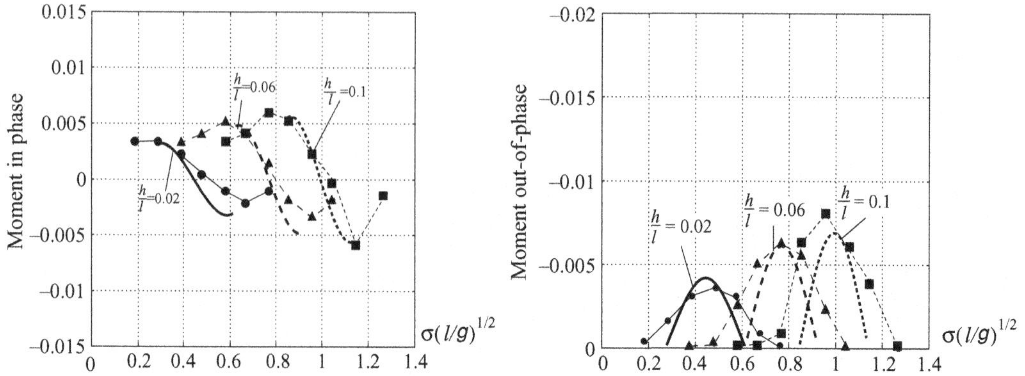

Figure 8.42. Nondimensional moment (scaled by $\rho g l^3$) in phase and out of phase for $s/l = -0.2$ ($s = H_V - h$), $\eta_{4a} = 0.03333$ rad, and different liquid depths. The symbols • ($h/l = 0.02$), ▲ ($h/l = 0.06$), and ■ ($h/l = 0.1$) connected by lines represent experimental data by van den Bosch and Vugts (1966). The corresponding lines without symbols represent the theoretical results by Verhagen and van Wijngaarden's (1965) theory.

where Ω is given by eqs. (8.117) and (8.120) for roll and sway, respectively. Furthermore,

$$\psi_{F_4} = -\phi + \arcsin\left(\frac{\Omega}{\sqrt{4-3\Omega^2}}\right),$$
$$F_{4a} = \frac{4\rho g l^3}{3\pi} M_m \sqrt{\left(1-\tfrac{3}{4}\Omega^2\right)}, \qquad (8.137)$$

where M_m and Ω are given by the corresponding expressions from eqs. (8.134), (8.118), and (8.120). The roll moment parameters ψ_{F_4} and F_{4a} were obtained in the original paper by Verhagen and van Wijngaarden (1965) and, later on, with some corrections by Journee (2000).

In the more general case with a finite distance H_V between the roll axis and the mean free surface, by using formula (8.135) for the roll moment relative to the roll axis in Figure 8.35 we get

$$F_4^{\text{center}}(t) = \rho g l^3 \left[\underbrace{\frac{F_{4a}}{\rho g l^3}}_{\frac{h}{l}O(\delta)} \sin(\sigma t - \phi - \psi_{F_4}) - \underbrace{\frac{F_{2a}}{\rho g l^2}}_{\left(\frac{h}{l}\right)^2 O(\delta)} \left(\frac{H_V}{l} - \frac{1}{2}\frac{h}{l}\right) \sin(\sigma t - \phi - \psi_{F_2}) \right]$$
$$= F_{4a}^{\text{center}} \sin\left(\sigma t - \phi - \psi_{F_4}^{\text{center}}\right). \qquad (8.138)$$

The lowest-order approximations of F_{4a}^{center} and $\psi_{F_4}^{\text{center}}$ are generally given by F_{4a} and ψ_{F_4}. However, the second term in the brackets contributes when hH_V/l^2 is of $O(1)$. Both terms in the brackets are needed in the following study of the effect of the roll-axis position.

Verhagen and van Wijngaarden (1965) demonstrated good agreement with experimental data on the roll moment when the roll axis is in the middle of the tank bottom and the second term in the brackets in eq. (8.138) is neglected. Van den Bosch and Vugts (1966) presented systematic model test results for roll moment amplitudes and phases of a rectangular tank with two-dimensional shallow-water flow. Three roll amplitudes ($\eta_{4a} = 0.033$, 0.067, and 0.100 rad), four centers of roll axis ($-0.40 \le s/l \le 0.20$), and five water depths ($0.02 \le \overline{h} \le 0.10$) were used in their tests, where $s = H_V - h$. Figures 8.42, 8.43, and 8.44 show examples of comparisons between the shallow-liquid theory and the experiments. The first harmonic of the roll moment based on Verhagen and van Wijngaarden's (1965) theory is used. The in-phase and out-of-phase moment components relative to the roll motion are presented as a function of the forcing frequency for different ratios of water depth to tank breadth, $\overline{h} = h/l$, in each figure. The roll amplitude is $\eta_{4a} = 0.03333$ rad for all cases. Each figure corresponds to different values of s/l. The absolute value of the out-of-phase component of the moment can be related to the roll damping caused by the sloshing in a ship motion analysis in a seaway. Taking into account that the theory handles only the lowest-order asymptotic term, the figures

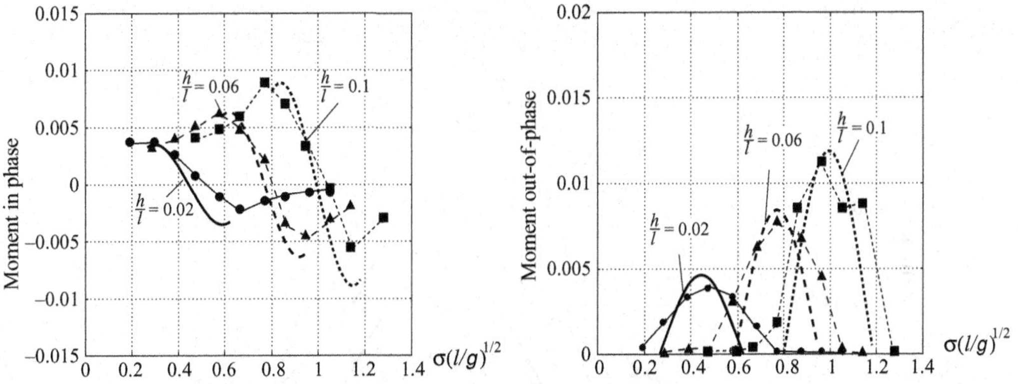

Figure 8.43. Nondimensional moment (scaled by $\rho g l^3$) in phase and out of phase for $s/l = 0.0$ ($s = H_V - h$), $\eta_{4a} = 0.03333$ rad, and different liquid depths. The symbols • ($h/l = 0.02$), ▲ ($h/l = 0.06$), and ■ ($h/l = 0.1$) connected by lines represent experimental data by van den Bosch and Vugts (1966). The corresponding lines without symbols represent the theoretical results by Verhagen and van Wijngaarden's (1965) theory.

establish quite good agreement between theory and experiments.

Other comparisons between Verhagen and van Wijngaarden's theory and van den Bosch and Vugts' experimental data are presented in Figure 8.45. The roll amplitude is $\eta_{4a} = 0.1$ rad and the ratio of water depth to tank breadth is $\bar{h} = 0.06667$. The out-of-phase moment as a function of the forcing frequency is presented for different positions of the roll axis. The theoretical maximum absolute value of the out-of-phase moment occurs at $\sigma = \sigma_1$, while the experimental data clearly show a "hard-spring" behavior for the moment component (i.e., the maximum occurs for larger values of the forcing frequency than of the linear resonance frequency). Theoretical and experimental values for the maximum amplitude of the roll moment are presented in Figure 8.46 as a function of s/l for the same case as in Figure 8.45. The roll damping and the roll moment amplitude increase with increasing s/l according to Figures 8.45 and 8.46. If an uncoupled roll analysis is made of a ship with an antirolling tank in waves, the smallest resonant roll motion consequently occurs when the tank is situated as high as possible above the roll axis.

Finally, van den Bosch and Vugts stressed that the experimental phase is always lower than

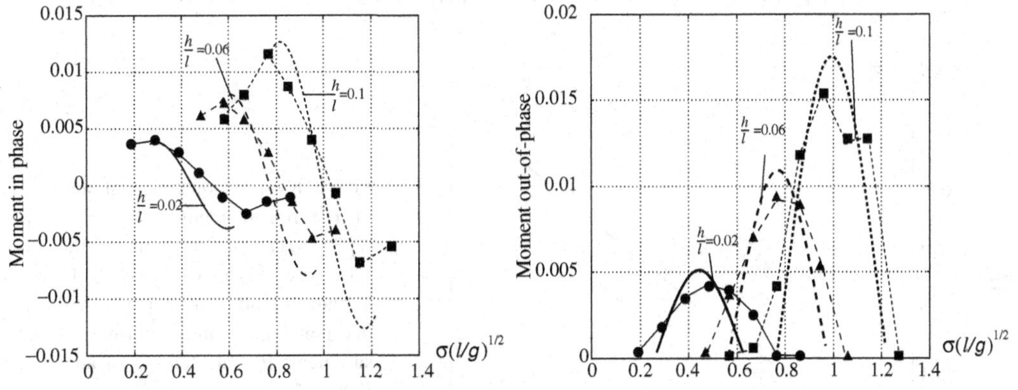

Figure 8.44. Nondimensional moment (scaled by $\rho g l^3$) in phase and out of phase for $s/l = 0.2$ ($s = H_V - h$), $\eta_{4a} = 0.03333$ rad, and different liquid depths. The symbols • ($h/l = 0.02$), ▲ ($h/l = 0.06$), and ■ ($h/l = 0.1$) connected by lines represent experimental data by van den Bosch and Vugts (1966). The corresponding lines without symbols represent the theoretical results by Verhagen and van Wijngaarden's (1965) theory.

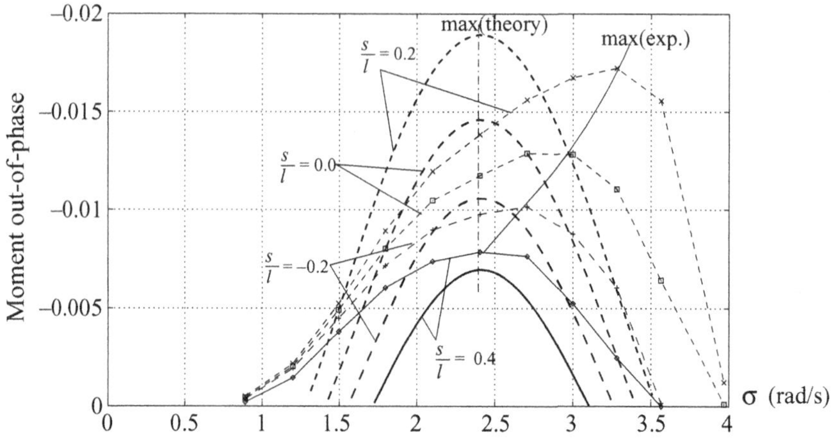

Figure 8.45. Nondimensional out-of-phase moment (scaled by $\rho g l^3$) for different position of the roll center, $s = H_V - h$, for $\eta_{4a} = 0.1$ rad, $h/l = 0.06667$ ($l = 0.741$ m). Comparison between theory (solid and dashed lines without symbols) and experiments (the corresponding lines with symbols).

$\frac{1}{2}\pi$ for $\sigma = \sigma_1$ and it decreases with increasing s/l. However, Verhagen and van Wijngaarden's theory predicts that the phase is $\frac{1}{2}\pi$ when $\sigma = \sigma_1$.

We may obviously also use CFD to describe the hydrodynamic effect of an antirolling tank. Van Daalen *et al.* (2000) showed good agreement between a volume of fluid-based Navier–Stokes solver and the experiments by van den Bosch and Vugts (1966).

When the tank force and moments are introduced in the frequency-domain equations of ship motion, we must divide the loads into two parts, where one part is proportional to the forced tank velocity with time dependence, $\cos(\sigma t - \psi)$, and the other is proportional to the forced tank acceleration with time dependence, $\sin(\sigma t - \psi)$. This condition implies

$$\begin{aligned} F_2(t) &= F_{2a}\sin(\sigma t - \phi)\cos\psi_{F_2} \\ &\quad - F_{2a}\cos(\sigma t - \phi)\sin\psi_{F_2}, \\ F_4(t) &= F_{4a}\sin(\sigma t - \phi)\cos\psi_{F_4} \\ &\quad - F_{4a}\cos(\sigma t - \phi)\sin\psi_{F_4}. \end{aligned} \quad (8.139)$$

We limit our detailed discussion to sway and roll excitation. However, we should also account for yaw. We may formally express the tank loads in terms of two-dimensional added mass (a^t_{ij}) and damping (b^t_{ij}) coefficients:

$$\begin{aligned} F_2(t) &= -a^t_{22}\ddot{\eta}_2 - b^t_{22}\dot{\eta}_2 - a^t_{24}\ddot{\eta}_4 - b^t_{24}\dot{\eta}_4, \\ F_4(t) &= -a^t_{42}\ddot{\eta}_2 - b^t_{42}\dot{\eta}_2 - a^t_{44}\ddot{\eta}_4 - b^t_{44}\dot{\eta}_4. \end{aligned} \quad (8.140)$$

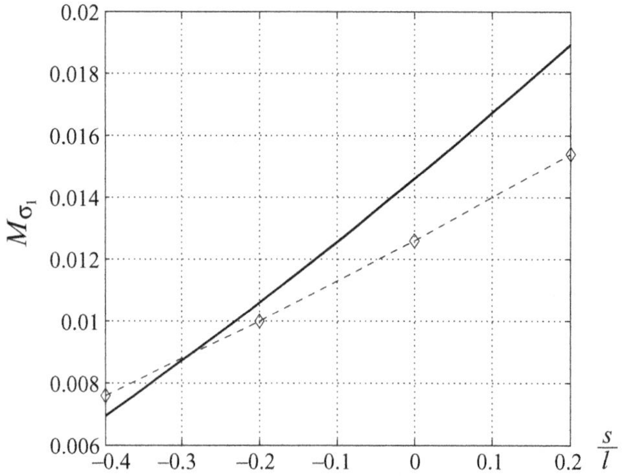

Figure 8.46. Nondimensional amplitude of the roll moment at $\sigma = \sigma_1$, $M_{\sigma_1} = F^{\text{center}}_{4a}/(\rho g l^3)$ as function of the ratio $s/l = (H_V - h)/l$. Comparison between Verhagen and van Wijngaarden's theory (solid line) and experimental measurements by van den Bosch and Vugts (1966), ◊.

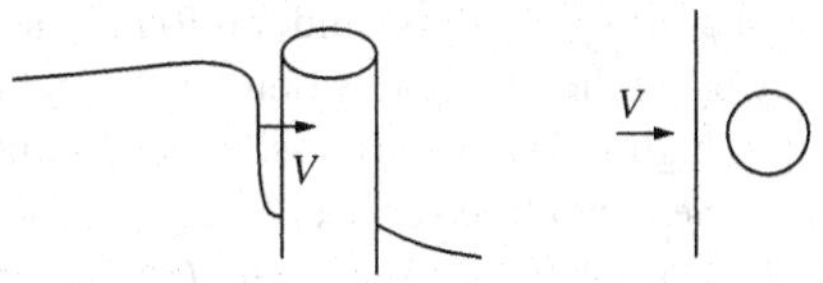

Figure 8.47. Impact loads against a vertical cylinder from waves with a vertical front.

From eq. (8.139) it follows that

$$a_{ij}^t = \frac{F_{ia}^{(j)} \cos \psi_{F_i}^{(j)}}{\sigma^2 \eta_{ja}}, \quad b_{ij}^t = \frac{F_{ia}^{(j)} \sin \psi_{F_i}^{(j)}}{\sigma \eta_{ja}}, \quad (8.141)$$

where we have used the superscript (j) on ψ_{F_i} and F_{ia} to indicate if the forcing is sway ($j = 2$) or roll ($j = 4$). The three-dimensional added mass and damping coefficients are simply obtained by multiplying the two-dimensional coefficients with the tank length. When the three-dimensional generalizations of eq. (8.140) are included in the equations of ship motion (i.e., eq. (3.26)), we have to solve the equations of motion by an iterative procedure due to the nonlinear dependence of the tank loads on sway and roll.

The effect of yaw, η_6, can be included by expressing the local sway motion of the tank as $\eta_2 + x_t\eta_6$, where η_2 is the sway at the origin of the global coordinate system used in the ship motion calculations. Furthermore, x_t is the global x-coordinate of the center of the tank liquid. We must also incorporate the hydrodynamic yaw moment due to sloshing in the equations of motion. The yaw moment is simply x_t times the transverse hydrodynamic sloshing force when the tank length is short relative to the tank breadth.

8.9 Wave loads on interior structures in shallow liquid depth

In Section 6.4 we described how to generalize the empirically based Morison equation to evaluate hydrodynamic loads on interior vertical tank structures. An assumption was that the incident wave slope is small. How should we account for a nearly vertical wall of liquid impacting the internal structure? The problem is illustrated in the case of a vertical circular cylinder in Figure 8.47. Because the additional flow caused by the initial part of the impact is not affected by gravity, the problem is initially similar to that of water impact of a horizontal cylinder on the free surface. Later on gravity becomes important and nonviscous flow separation may occur, as illustrated in Figure 8.48. A similar phenomenon with nonviscous flow separation happens when a vertical wave front impacts a vertical internal structure (see Figure 8.49). So how should we handle this case in a pragmatic way for a vertical internal circular cylinder? We could define a phase from when the impact starts until the ambient free surface has reached the maximum width of the

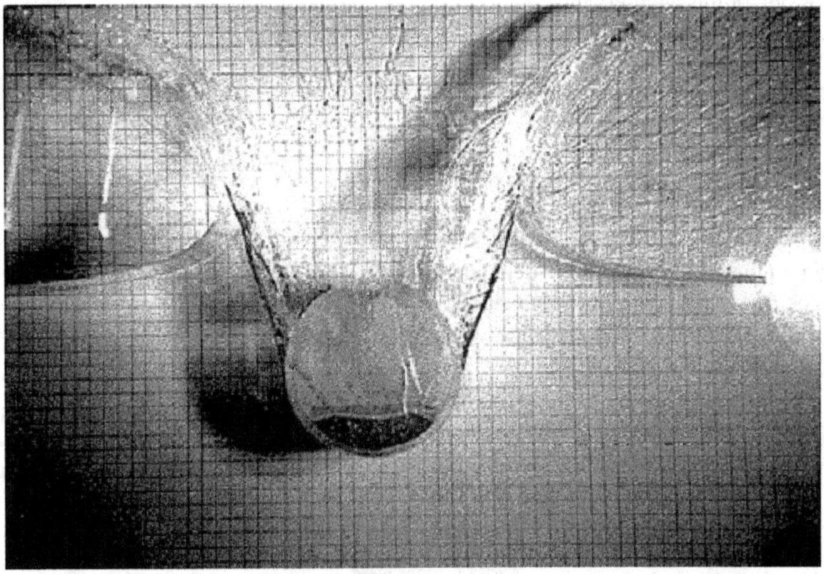

Figure 8.48. Water entry of a horizontal cylinder (Greenhow & Lin, 1983).

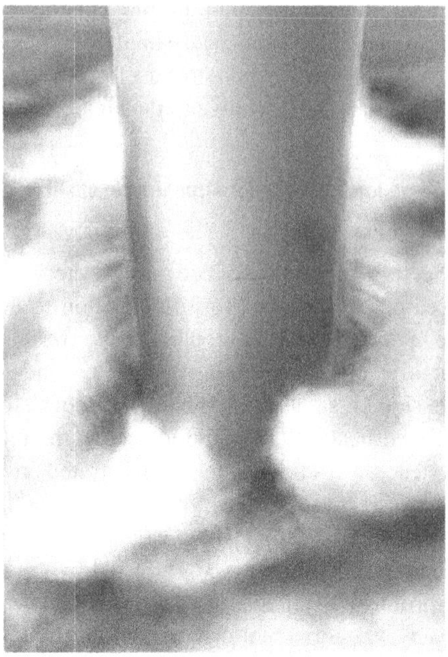

Figure 8.49. Artist's impression of a steep wave impacting a column and subsequent nonviscous flow separation (Artist: Bjarne Stenberg).

cylinder. During this phase the lateral force per unit vertical length acting on the part of the cylinder facing the steep wave can be divided into two parts:

- the lateral Froude–Kriloff force, which requires that we know the ambient pressure distribution at the position of the cylinder; and
- a lateral slamming force.

The lateral slamming force per unit vertical length, F_s^{2D}, can be expressed as

$$F_s^{2D} = \rho R C_s V^2 \qquad (8.142)$$

for constant water entry velocity V, where R is the cylinder radius and C_s is a time-dependent slamming coefficient. The experimental values by Campbell and Weynberg (1980) for vertical water entry of a horizontal circular cylinder are presented in Figure 8.50 together with theoretical values based on a von Karman (1929) method. A von Karman method assumes that the wetted surface is unaffected by the flow caused by the cylinder (i.e., we find the wetted surface as the geometrical intersection between the body surface and the incident free surface). The velocity potential φ caused by the cylinder is found numerically by using the exact body boundary condition and the high-frequency free-surface condition $\varphi = 0$ on the mean free surface. The experimental values by Campbell and Weynberg (1980) can be expressed as

$$C_s = 5.15/(1 + 9.5Vt/R) + 0.275Vt/R, \qquad (8.143)$$

where $t = 0$ corresponds to the initial impact time. Buoyancy effects have not been subtracted from this formula. Campbell and Weynberg estimated the error due to buoyancy effects to be from 0.05 to 0.54. As already stated we should question the formula after nonviscous flow separation occurs.

A formula such as eq. (8.142) is used in coastal engineering to predict impact forces due to breaking waves on a vertical cylinder. Goda *et al.*'s (1966) formula $C_s = \pi(1 - Vt/R)$, $0 \le Vt/R \le 1$, is commonly adopted. The initial impact coefficient π is consistent with von Karman's method. However, from Figure 8.50 we see that the von Karman method differs at later time instants. Wienke and Oumeraci (2005) showed that an initial slamming coefficient $C_s = 2\pi$ based on Wagner's (1932) method agrees better with model tests of breaking waves incident on a vertical cylinder. The Wagner method accounts for the influence of the body on the free-surface shape. Campbell and Weynberg's empirical formula gives that C_s is initially 5.15.

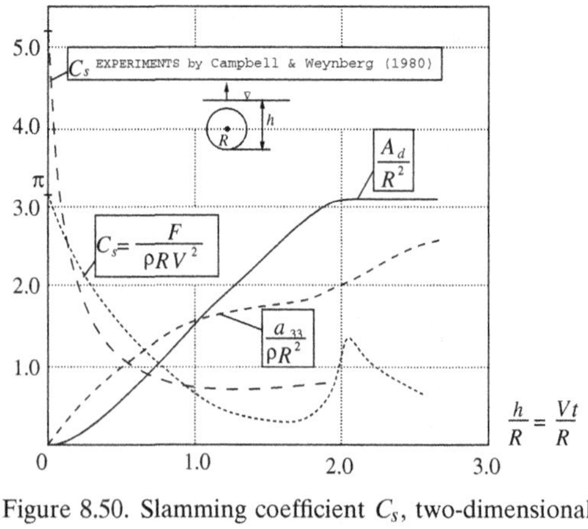

Figure 8.50. Slamming coefficient C_s, two-dimensional heave infinite-frequency added mass a_{33}, and displaced area A_d relative to the mean free surface of a circular cylinder as a function of submergence h; F is the vertical force, V is the constant downward velocity of the cylinder, t is the time variable with $t = 0$ corresponding to initial time of impact.

One possible slamming model for the case when a hydraulic jump impacts a vertical cylinder is to neglect the Froude–Kriloff force and express the slamming force as

$$F_s = \rho R h_{\text{jump}} C_s V_{\text{jump}}^2, \tag{8.144}$$

where h_{jump}, V_{jump}, and C_s are given by eqs. (8.127), (8.130), and (8.143), respectively. We define V_{jump} to correspond to the initial impact time. The time interval for which eq. (8.144) is a good approximation depends on the Froude number, $V_{\text{jump}}/\sqrt{gR}$. To be on the safe side we suggest that the formula is restricted to the time interval $0 \le t \le R/V_{\text{jump}}$.

The impacting breaking wave does not need to be a hydraulic jump. Different impact load cases for coastal engineering applications have been experimentally investigated by Wienke and Oumeraci (2005).

Because the slamming loads vary rapidly, the internal structural response may have to be calculated by a dynamic hydroelastic analysis. An important parameter for dynamic effects is the ratio T_d/T_n between the duration of the slamming loads, T_d, and the highest natural structural period, T_n. The previous discussion suggests that $T_d = R/V_{\text{jump}}$. If T_d/T_n is larger than ~5, the structural response is, in general, quasi-steady.

After the cross-section of the internal structure is completely wetted we may adopt our formulation of the generalized Morison equation from Section 6.4 with proper ambient flow velocities and accelerations. However, we should realize that viscous flow separation takes time to develop for cross-flow past structures without a sharp corner (Schlichting, 1979). This fact influences, for instance, the drag coefficient and is Reynolds-number–dependent. The time-dependent experimental drag coefficient for startup flow with almost impulse initial condition past a circular cylinder in infinite fluid is presented by Sarpkaya (1966) in the subcritical Reynolds-number range. Further studies are needed to estimate proper mass and drag coefficients and to experimentally document the described procedure.

8.10 Mathieu instability for vertical tank excitation

In Section 5.2 we showed that vertical tank motion cannot excite sloshing by a linear theory. However, nonlinearities may cause hydrodynamic instability of the free surface and result in significant resonant sloshing at some forcing frequencies. Because Faraday (1831) was the first to observe the instabilities, the resulting waves are referred to as Faraday waves. Lord Rayleigh (Rayleigh, 1883) made a further series of experiments and supported Faraday's results. The problem has been investigated by many other researchers. Reviews are presented, for example, by Ibrahim (2005, Chap. 6), Nevolin (1984), and Perlin and Schwarz (2000).

The instabilities are explained by using the modal approach. Nonlinear modal system (8.95) for a rectangular tank with two-dimensional flow takes the following form for vertical excitations with $\eta_i = 0$, $i \neq 3$, and $\eta_3 \neq 0$:

$$\begin{aligned}&\ddot{\beta}^a\left(\delta_{a\mu} + d^{1,\mu}_{a,b}\beta^b + d^{2,\mu}_{a,b,c}\beta^b\beta^c + \cdots\right)\\ &\quad + \dot{\beta}^a\dot{\beta}^b\left(t^{0,\mu}_{a,b} + t^{1,\mu}_{a,b,c}\beta^c + \cdots\right)\\ &\quad + \left[\sigma_\mu^2 + \ddot{\eta}_3 l^{-1}\pi\mu\tanh(\pi\mu\bar{h})\right]\beta_\mu = 0,\\ &\qquad\qquad \mu = 1, 2, \ldots. \end{aligned}\tag{8.145}$$

The nonlinear differential equations do not have forcing terms on the right-hand sides of the equations. Possible solutions are $\beta_\mu = 0$, which corresponds to a hydrostatic liquid shape. A linear stability analysis of this trivial solution follows by introducing small perturbations of the generalized free-surface coordinates β_μ and linearizing eq. (8.145). When we do that, we should consider $[\sigma_\mu^2 + \ddot{\eta}_3 l^{-1}\pi\mu\tanh(\pi\mu\bar{h})]$ as a time-dependent coefficient and not involve $\ddot{\eta}_3$ in the linearization process. This consideration leads to a set of uncoupled modal equations:

$$\ddot{\beta}_\mu + \underbrace{\left[\sigma_\mu^2 + \ddot{\eta}_3 l^{-1}\pi\mu\tanh(\pi\mu\bar{h})\right]}_{\sigma_\mu^2[1+\ddot{\eta}_3/g]}\beta_\mu = 0, \quad \mu = 1, 2, \ldots. \tag{8.146}$$

The time-dependent coefficient associated with the β_μ-term in the preceding equations is expressed as $\sigma_\mu^2\,[1 + \ddot{\eta}_3/g]$ to reflect the fact that the heave acceleration $\ddot{\eta}_3$ acts physically as a modification of the acceleration of gravity in the system. By noting this fact we can also generalize the result to other tank shapes with other natural frequencies σ_μ. One case of practical importance is the vertical excitation of a propellant tank in space applications (Abramson, 1966). Because forcing is included in the parameter $\sigma_\mu^2\,[1 + \ddot{\eta}_3/g]$,

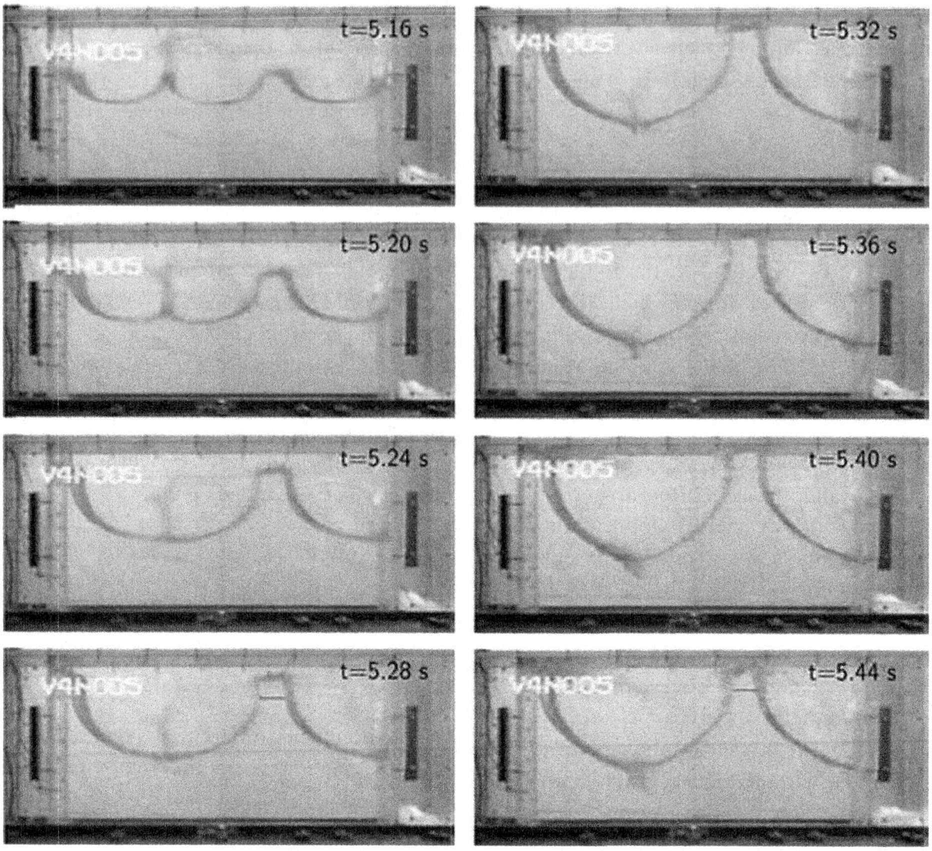

Figure 8.51. Time sequence of free-surface shapes due to vertical harmonic excitation of a rectangular tank with water depth $h = 0.4$ m and tank breadth $l = 1.48$ m. The forcing period T is 0.4 s and the heave amplitude η_{3a} is 0.015 m (Bredmose *et al.*, 2003).

one talks about *parametric excitation*. If $\eta_3 = \eta_{3a}\cos(\sigma t)$, eq. (8.146) is a particular case of Mathieu (1868) equation (3.48) with zero damping ξ and $\delta = \sigma^2\eta_{3a}/g$.

Figure 3.18 presents the stability diagram of the Mathieu equation. In our problem we have to study the stability in the $((\sigma_m/\sigma)^2, \eta_{3a}\sigma_m^2/g)$-plane. The instabilities occur in the vicinities of $(\sigma_m/\sigma)^2 = \frac{1}{4}, 1, \frac{9}{4}, \ldots$. The widest and most dangerous instability region appears for $(\sigma_m/\sigma)^2 = \frac{1}{4}$; that is, $T = \frac{1}{2}T_m$ $(m = 1, \ldots)$, where T is the forcing period and T_m is the natural period of eigenmode m. In the literature the instabilities are referred to as *parametric resonance*, where the word "resonance" does not have the same meaning as for a mass–spring system, which is forced with a frequency equal to a resonance frequency.

Frandsen (2004) analyzed vertical forcing of a rectangular tank and checked the stability results by using two-dimensional CFD simulations based on a fully nonlinear formulation. Furthermore, she constructed third-order asymptotic solutions starting from approximation (8.146). She showed that the prediction of the stable regions is in good agreement when the forcing parameter is small. If the excitation amplitude grows, nonlinearities due to intermodal interaction have to be considered.

Bredmose *et al.* (2003) described experimentally steep standing waves generated by vertical excitations of a rectangular tank with liquid depth $h = 0.4$ m and tank breadth $l = 1.48$ m. Figure 8.51 shows a sequence of video snapshots with flat-topped waves at an incipient breaking stage. The input signal conditions for vertical forcing are chosen to excite the third natural mode, which implies that the forcing period is $T = 0.4$ s $\approx \frac{1}{2}T_3$. From Figure 3.18 we see that the linear instability depends on the damping ratio, ξ. An estimate can be made by assuming laminar flow conditions. The damping ratio $\xi_3 = \xi_3^{bulk} + \xi_3^{tank\ surface}$ of the third mode can be estimated by the expressions for ξ_3^{bulk} and $\xi_3^{tank\ surface}$ given in Section 6.3. For the input parameters ($B/l = \overline{B} = 0.27$), the damping ratio is $\xi_3 = 0.002$. The amplitude of the forced tank motion was small (i.e., $\eta_{3a}/l = 0.01$). However, the corresponding forced acceleration amplitude with the forcing period $T = \frac{1}{2}T_3$ is large:

$$\delta = \ddot{\eta}_{3a}/g = \eta_{3a}\,(2\sigma_3)^2\,/g$$
$$= (\eta_{3a}/l)\,12\pi \tanh(3\pi\overline{h}) = 0.372.$$

With this excitation, the damping ratio is too small to prevent the Mathieu instability (parametric resonance).

Our next step is to consider if a value of $\ddot{\eta}_{3a}/g = 0.372$ with the given forced oscillation period is realistic when scaled to a full-scale ship. We show that it may be relevant in connection with springing of a ship. Springing is steady-state resonant elastic vibration of a ship due to continuous wave loading. If external slamming on the ship hull occurs frequently, springing can be difficult to distinguish from whipping (i.e., a transient response). The reason is that the small damping causes slow decay of the whipping-induced response. Both springing and whipping in head sea are mainly related to the two-node vertical bending of the hull. Both linear and nonlinear external wave loads have to be considered in analyzing springing. Storhaug (2007) documented that springing may contribute to approximately 50% of the accumulated fatigue damage based on full-scale measurements of a 300-m-long bulk carrier.

We choose as an example a 300-m-long vessel with a natural period of two-node vertical bending equal to 2 s. The model test results are scaled to full scale by Froude scaling; that is, the model test results of $\ddot{\eta}_{3a}/g = 0.372$ also apply to full scale. This acceleration level is possible in connection with springing. Froude scaling of the period gives a full-scale period $\sqrt{L_p/L_m}$ times the model scale period, where L_p and L_m are the ship length in full scale (prototype) and model scale, respectively. We assume that the ratio between the tank length, L_t, and the ship length, L, is 0.123 to obtain a full-scale forced oscillation period of 2 s. The chosen value of L_t/L is realistic according to Table 1.1. Even though it is possible that sloshing with parametric resonance may be excited due to springing, we still have to consider the contribution from sloshing to fatigue, which needs further investigation.

Another concern is if the tank has a sufficiently high filling ratio for the flat-topped waves shown in Figure 8.51 to impact the tank roof. An impact with a small angle between the impacting free surface and the tank surface can cause important slamming load effects (see Chapter 11).

8.11 Summary

8.11.1 Nonlinear multimodal method

A nonlinear multimodal method has been applied to two-dimensional irrotational flow of an incompressible liquid in a rigid rectangular tank with infinite tank roof height. An essential assumption is that the free-surface elevation is a single-valued function of the tank-fixed lateral coordinate. Finite, intermediate, and shallow depths were considered. The applicability and limitations of the multimodal method were extensively documented by comparisons with model tests.

The free-surface elevation was expressed in a Fourier series as a function of the tank-fixed lateral coordinate. The Fourier series representation implies that the free surface is perpendicular to the tank walls. The latter fact prevents adequate prediction of run-up in terms of a thin liquid layer along the walls.

The velocity potential was represented as a sum of terms proportional to the linear natural sloshing modes. The unknowns are generalized time-dependent coordinates β_j and R_k for the free-surface elevation and the velocity potential, respectively. Forced sway and roll with frequency close to the lowest natural frequency are examined. The studies examined one single forcing frequency, but forcing with general time dependence can be handled as long as primary steady-state resonance of higher natural sloshing frequencies does not occur. It is shown that the nonlinear transfer of energy between natural modes plays an important role in limiting the response at resonance.

When a time-domain solution is considered, it is necessary to introduce damping to reach steady-state conditions. The viscous damping associated with a laminar boundary-layer flow can be theoretically estimated. However, we have no theoretical means to estimate damping due to breaking waves.

The ordering of the natural modes in terms of the small parameter ε characterizing the roll amplitude and the ratio between the forced sway amplitude and the tank breadth l is essential. When the liquid depth is either intermediate or shallow, the ratio $\overline{h}$ between the liquid depth and the tank breadth is also introduced as a small parameter in deriving ordinary differential equations for the time-dependent generalized coordinates of the free surface and the velocity potential. All the coefficients in the differential equations are analytically expressed. The equations can be used as a basis for analytical studies of steady-state response and physical stability. Numerical solution of the time-domain equations requires very little CPU time relative to CFD methods.

When the liquid depth is finite and larger than $\sim$0.4 l, only the lowest mode is dominant and of $O(\varepsilon^{1/3})$. Because this is of lower order than the excitation, resonant amplification occurs. The amplitude A of the lowest-order solution is determined by the equations that are of the same ε order as the forcing. If a Moiseev-type method is used, the equation for A follows as a consequence of a secularity (solvability) condition requiring finite response, leading to a cubic equation for A, which may have either one or three real solutions for a given forcing frequency. When three real solutions exist, only two of the solutions are hydrodynamically stable. The characteristics of the solutions resemble the solutions of the Duffing equation describing a mass–spring system with a linear and cubic spring term. The hydrodynamic response has either a "hard-spring" or a "soft-spring" response, similar to the Duffing equation. The response changes from hard-spring to soft-spring behavior at the so-called critical depth, $h = 0.3368\ldots l$. The perturbation scheme, following a Moiseev scheme, fails at the critical depth and predicts an infinite response when the forcing frequency is equal to lowest natural sloshing period. Except for the critical depth, the maximum response occurs at a frequency different from the lowest natural frequency. However, we must account for damping to predict the frequency of maximum steady-state response. The behavior of steady-state response shows that jumps between solution branches and hysteresis effects are possible. A time-domain solution is needed to properly predict hysteresis and jumps between solution branches.

A time-domain solution based on the nonlinear multimodal method with one dominant mode is first presented. The derivation is not limited to the fact that the forcing frequency should be asymptotically close to the lowest natural frequency. Furthermore, a secularity condition, as needed in the Moiseev approach, is not necessary. The derivation of the multimodal method is based on the Bateman–Luke variational formulation, which leads to an infinite system of nonlinear ordinary differential equations with the generalized coordinates β_j and R_k as unknowns. The equation system is made finite by introducing the small parameter ε to characterize the orders of magnitude of β_j and R_k. How to do this ordering is based on physical considerations. Furthermore, the ratio of tank liquid depth to tank breadth, $\overline{h}$, is introduced as a small parameter for lower-intermediate and shallow depth conditions.

The assumption that only the lowest order is dominant fails due to (1) critical depth, (2) large-amplitude response, and (3) secondary resonance. A necessary but insufficient condition for secondary resonance can mathematically be expressed as $n\sigma = \sigma_n$ $(n > 1)$, where $n\sigma$ expresses superharmonic terms generated due to hydrodynamic nonlinearities. Furthermore, σ_n is the natural frequency for the nth eigenmode. It is important that σ/σ_1 satisfying $n\sigma = \sigma_n$ $(n > 1)$ is close to 1. The range where nonlinearities matter in the vicinity of $\sigma/\sigma_1 = 1$ is a function of the excitation amplitude characterized by ε. The larger the value of ε, the broader the frequency range becomes where nonlinearities matter. When the ratio of liquid depth to tank breadth is small, the values of σ/σ_1 expressing necessary conditions for secondary resonance become close to 1 for many modes. We see this fact by noting that $\sigma_n \to n\sigma_1$ when $\overline{h} = h/l \to 0$.

An adaptive modal approach for finite and nonsmall intermediate depths is derived based on the ordering $\beta_1 \sim \cdots \sim \beta_n = O(\varepsilon^{1/3})$, $\beta_i = O(\varepsilon)$, $i \geq n+1$. The corresponding finite-dimensional nonlinear modal system is called Model n, where n is chosen by considering the possibility of secondary resonance. The generalized coordinates for the velocity potential, R_k, can be explicitly expressed in terms of β_j.

When the liquid depth is shallow or in the lower-intermediate range, it is assumed that $R_i \sim \beta_i = O(\bar{h}) = O(\varepsilon^{1/4})$ $(i \geq 1)$. We cannot separate the equations for R_k and β_j in this case.

Once the time-dependent generalized coordinates β_j for the free-surface elevation are determined, integrated hydrodynamic force and moment on the tank can be expressed in terms of the Lukovsky formulas. This fact facilitates coupling of sloshing with wave-induced ship motion in a seaway. The coupled sloshing–ship motion analysis is not limited to regular incident waves on a ship; for instance, the effect of a sea state described by a wave spectrum can be analyzed. However, we must require that no forcing frequency exists in the vicinity of higher natural sloshing frequencies.

The evaluation of nonimpulsive pressure in the free-surface zone where the wetted tank surface changes with time requires consistent Taylor-series expansions of the velocity potential about the mean free surface. Contributions from higher natural modes give unrealistically high exponential contributions when the velocity potential is evaluated above the mean free surface.

8.11.2 Subharmonics

Experiments and numerical calculations show that subharmonic steady-state behavior is possible. The studied case corresponds to a ratio of liquid depth to tank breadth of 0.35. Both the forcing amplitude and the period influence subharmonic behavior. As an example the results may repeat themselves with a period three times the forcing period; however, other types of subharmonic behavior are possible.

8.11.3 Damping

When the liquid depth is finite and larger than $\sim 0.4\, l$, the damping is very small for a clean tank and mainly due to viscous energy dissipation in the boundary layer along the tank surface. The consequence is that a large number of oscillation periods is needed to allow transients to die out. The importance of spilling and plunging breaking waves and associated damping due to turbulent energy dissipation increases with decreasing liquid depth and increasing ε.

8.11.4 Hydraulic jumps

Hydraulic jumps may be created in shallow-liquid conditions when the excitation amplitude is higher than a threshold value and the forcing frequency is close to the lowest natural frequency. Verhagen and van Wijngaarden's steady-state theory for hydraulic jumps provides guidance on when hydraulic jumps are formed as a function of the excitation amplitude and the forcing frequency in the vicinity of the lowest natural sloshing frequency. However, comparisons with model tests show that the amplitude and frequency range of hydraulic jumps are narrower than theoretically predicted.

8.11.5 Hydrodynamic loads on interior structures

The hydrodynamic loads on interior structures were discussed, and a method was presented for estimating impact loads in shallow-liquid conditions with hydraulic jumps traveling back and forth in the tank. Empirical time-dependent slamming coefficients for a single vertical circular cylinder were provided. A generalized Morison equation was recommended for nonimpulsive loads on interior structures. A method was discussed for correct prediction of the loads on surface-piercing interior structures in the free-surface zone by properly expanding the velocities and accelerations predicted by the multimodal method via Taylor series. A key point was also how to obtain empirical mass and drag coefficients that account, for instance, for the Reynolds-number dependence and the hydrodynamic interaction between structural members.

8.12 Exercises

8.12.1 Moiseev's asymptotic solution for a rectangular tank with infinite depth

Consider two-dimensional flow in a rectangular tank with breadth l and infinite liquid depth. A Cartesian coordinate system Oyz with origin at the intersection between the tank's centerplane and the mean free surface is introduced with the z-axis vertical and positive upward. The tank is forced to oscillate horizontally with velocity $\dot{\eta}_2 = -\sigma\eta_{2a}\sin\sigma t$ and with a circular frequency σ close to the lowest natural sloshing frequency, σ_1. We introduce the small parameter $\varepsilon = \eta_{2a}/l$ and expand the velocity potential Φ and the free-surface elevation ζ as

$$\Phi = \phi_1\varepsilon^{1/3} + \phi_2\varepsilon^{2/3} + \phi_3\varepsilon - \sigma\eta_{2a}y\sin\sigma t,$$
$$\zeta = \zeta_1\varepsilon^{1/3} + \zeta_2\varepsilon^{2/3} + \zeta_3\varepsilon.$$

The solution is presented in Section 8.2.2.

(a) Express the roll moment and the horizontal and vertical hydrodynamic force by means of the Lukovsky formula by first expressing the Fourier coefficients of the free-surface elevation that are needed in the formula.

(b) Describe how to evaluate the pressure distribution in the free-surface zone correctly to $O(\varepsilon)$.

8.12.2 Mean steady-state hydrodynamic loads

Assume that the single-dominant nonlinear modal theory can be used to describe the sloshing in a two-dimensional tank. Consider a steady-state condition where the flow has the same period as the tank excitation in sway and roll. Use the Lukovsky formulas to express the mean hydrodynamic force components and roll moment acting on the tank in terms of the generalized coordinates of the free-surface elevation, β_j.

8.12.3 Simulation by multimodal method

(a) Use modal system (8.43) to compare the transient wave simulations with those in Figures 8.10, 8.11, 8.12, and 8.15. (*Hint*: Rewrite the corresponding modal system in the standard form $M(t, \boldsymbol{x})\dot{\boldsymbol{x}} = F(\boldsymbol{x}, t)$ $(x_1 = \beta_1,\ x_2 = \dot{\beta}_1, \dots,\ x_{2n-1} = \beta_n,\ x_{2n} = \dot{\beta}_n, \dots)$ and use a fourth-order Runge–Kutta solver.) When comparing the results, note that the graphs may not agree perfectly because simulations of these strongly resonant waves are partly sensitive to the numerical procedure, its precision, and to the significant digits of the forcing frequency, whose values in the captions are restricted to either three or four digits.

(b) Hermann and Timokha (2008) used a numerical method to study periodic solutions of Model 2 by eqs. (8.102) and (8.103) with liquid depths that are in a neighborhood of the critical value $h = 0.3368\dots l$ of the single-dominant multimodal method. They showed that Model 2 predicts a decreasing theoretical critical-depth value with increasing forcing amplitude.

By using directly the long time-series simulation with small damping as described in Sections 8.5 and 8.8.1.2, confirm that the steady-state solutions of eqs. (8.102) and (8.103) impose soft-spring behavior at the theoretical critical depth $h = 0.3368\dots l$. (*Hint*: For steady-state solutions, draw the points in the plane $(\sigma/\sigma_1, ||\beta_1||)$, where $||\beta_1||$ is the amplitude of the first mode, and show that the maximum occurs to the left of the linear resonance $\sigma/\sigma_1 = 1$. Use the forcing amplitude $\eta_{2a}/l = 0.002$, which is close to the mean experimental value of Fultz (1962).)

8.12.4 Force on a vertical circular cylinder for shallow depth

Consider a rectangular tank with breadth $l = 40$ m and ratio of liquid depth to tank breadth of 0.1. A vertical circular cylinder 3 m in diameter stands on the tank bottom with its axis in the tank's centerplane. Assume two-dimensional flow conditions in the tank when the cylinder is not present. The tank is harmonically forced in the horizontal direction with sway amplitude $\eta_{2a} = 0.1\,l$ and a frequency equal to the lowest natural sloshing frequency without the presence of the cylinder. Assume steady-state conditions and apply the hydraulic jump theory by Verhagen and van Wijngaarden (1965) to evaluate the inflow to the cylinder.

(a) Use the empirical formula by Campbell and Weynberg (1980) to evaluate the slamming coefficient. Present the slamming force divided by the mass density of the liquid, ρ, as a function of time.

(b) When the slamming phase is over, use the generalized Morison equation to calculate the hydrodynamic force on the cylinder divided by ρ. You should justify your choice of mass and drag coefficients and discuss when it is appropriate to say that the slamming phase is over.

8.12.5 Mathieu-type instability

Estimate the instability region in Figure 3.18 for $(\sigma_m/\sigma)^2 = 1/4$ and zero damping.

(*Hint*: Assume that you look for the solution at the borderline and, therefore, present the $2T$-periodic function as $\beta_m(t) = \sum_{k=1}^{\infty} \{a_k \sin(\frac{1}{2}k\sigma t) + b_k \cos(\frac{1}{2}k\sigma t)\}$, where coefficients a_k and b_k are unknown. Substitute the solution into eq. (8.146) and derive an infinitely coupled system of linear algebraic equations. The result is

$$\left(1+\frac{\sigma^2\eta_{3a}}{2g}-\frac{\sigma^2}{4\sigma_m^2}\right)a_1-\frac{\sigma^2\eta_{3a}}{2g}a_3=0,$$

$$\left(1-\frac{\sigma^2}{\sigma_m^2}\right)a_2-\frac{\sigma^2\eta_{3a}}{2g}a_4=0,\ \dots,$$

$$\left(1-\frac{k^2\sigma^2}{4\sigma_m^2}\right)a_k-\frac{\sigma^2\eta_{3a}}{2g}(a_{k-2}+a_{k+2})=0,\quad k\ge 3;$$

$$\left(1-\frac{\sigma^2\eta_{3a}}{2g}-\frac{\sigma^2}{4\sigma_m^2}\right)b_1-\frac{\sigma^2\eta_{3a}}{2g}b_3=0,$$

$$\left(1-\frac{\sigma^2}{\sigma_m^2}\right)b_2-\frac{\sigma^2\eta_{3a}}{2g}b_4=0,\ \dots,$$

$$\left(1-\frac{k^2\sigma^2}{4\sigma_m^2}\right)b_k-\frac{\sigma^2\eta_{3a}}{2g}(b_{k-2}+b_{k+2})=0,\quad k\ge 3.$$

Restricting the analysis to $k = 1$ with the requirement to obtain a nontrivial subharmonic solution ($a_1^2 + b_1^2 \neq 0$) leads to the two-dimensional homogeneous linear algebraic system. Its solvability condition leads to the required function coupling η_{3a} and σ.)

Answer:

$$\left(1+\frac{\sigma^2\eta_{3a}}{2g}-\frac{\sigma^2}{4\sigma_m^2}\right)\left(1-\frac{\sigma^2\eta_{3a}}{2g}-\frac{\sigma^2}{4\sigma_m^2}\right)=0$$

$$\Rightarrow\left(1-\frac{\sigma^2}{4\sigma_m^2}\right)^2=\left(\frac{\sigma^2\eta_{3a}}{2g}\right)^2.$$

9 Nonlinear Asymptotic Theories and Experiments for Three-Dimensional Sloshing

9.1 Introduction

Our focus in this chapter is on analytically based nonlinear hydrodynamic studies and experiments of laterally forced three-dimensional rectangular, upright circular cylindrical and spherical tanks with forcing frequency close to the lowest natural sloshing frequency.

Two-dimensional flow conditions are commonly assumed in experimental and numerical studies of partially filled prismatic ship tanks. However, nonlinear three-dimensional flow may occur when the tank length is close to the tank breadth and when the forcing frequency is in the vicinity of the lowest natural sloshing frequency. When the tank excitation is parallel to a tank wall and the flow is two-dimensional, three-dimensional behavior is triggered by instabilities. Generally speaking, these instabilities lead to nonlinear interaction between three-dimensional liquid motion modes and play an important role for all types of tank motion excitations. Possible stable steady-state waves for longitudinal excitation along a tank wall of square-base and nearly square-base tanks are planar, swirling, and nearly diagonal waves. Swirling and purely diagonal waves can occur for diagonal excitation of a square-base tank. Stable steady-state sloshing is not always possible, unlike for two-dimensional flow. The fact that a steady-state condition is not reached during the evolution is referred to as *chaos* or *irregular waves* in the following discussion.

In the case of upright circular cylindrical tanks we present theoretical and experimental studies on the wave regimes due to lateral excitations at the lowest natural frequencies. One can distinguish two different stable steady-state wave regimes (i.e., planar waves and swirling), as well as chaos. In the case of spherical tanks, we have to rely on model tests to identify planar and swirling wave domains and to determine hydrodynamic forces because we are not aware of any available nonlinear analytically based methods. Experimental results for wave loads on a tower inside a spherical tank are also discussed.

A spherical pendulum has been used as an equivalent mechanical system to describe resonant three-dimensional sloshing in a vertical circular cylindrical tank and a spherical tank (Ibrahim, 2005; Miles, 1962). Two-dimensional sloshing has also been described by a mathematical pendulum (Ibrahim, 2005, and references therein). It is difficult to justify pendulum models from a hydrodynamic point of view. Exercise 9.6.2 provides some insights into pendulum models. If the pendulum models should be correct, their mathematical statement should be similar to the multimodal system.

In this chapter the analytically based multimodal method is formulated for finite liquid depths. Lower-intermediate and shallow liquid depths are not examined. Section 8.8.2 documented a steady-state solution for hydraulic jumps based on shallow liquid theory; a similar investigation can be performed in three dimensions. The three-dimensional shallow-water equations in the body reference frame are discussed in Exercise 9.6.4 and must be solved numerically. An example of their application for sloshing in ship tanks can be found in Dillingham and Falzarano (1986). This approach is also common when simulating green water on a ship deck instead of using more complex computational fluid dynamics methods.

Classifications of three-dimensional sloshing are performed in terms of the forcing amplitude and frequency domains in which the steady-state wave regimes are stable or not. Zones exist where all steady-state regimes are unstable, where irregular (chaotic) motions occur. This classification is a primary focus of the forthcoming sections.

9.1.1 Steady-state resonant wave regimes and hydrodynamic instability

9.1.1.1 Theoretical treatment by the two lowest natural modes

In Section 4.3.2, we discussed two- and three-dimensional waves by operating with the two lowest conjugate natural modes for nearly square-base and circular upright tanks. For square-base tanks, Section 4.3.2.1 distinguishes two-dimensional, *planar* motions associated with Stokes freestanding waves ($ij = 0$ in eq. (4.31)), *swirling* (rotary) waves (eq. (4.32)), and *nearly diagonal* waves associated with the patterns given by eq. (4.35). A Cartesian coordinate system $Oxyz$ was defined with origin at the center of the mean free surface. The Oz-axis is positive upward and the Ox- and Oy-axes are parallel to the tank walls. For a case where liquid depth is finite, the forcing frequency is close to the lowest natural frequency, and the roll, pitch, and ratio of the lateral forcing amplitude to the tank breadth are small, Faltinsen *et al.* (2003) showed that the lowest mode in the x- and y-directions dominates for steady-state wave motions. Planar, purely diagonal, nearly diagonal, and swirling steady-state wave motions are possible. If the forcing is along the Ox-axis of a nearly square base tank, the following approximations of the steady-state free-surface elevation $\zeta(x, y, t)$ are possible for planar waves,

$$\zeta(x, y, t) = Af_1^{(1)}(x)\cos\sigma t + o(A), \quad A \neq 0, \tag{9.1}$$

for swirling,

$$\zeta(x, y, t) = Af_1^{(1)}(x)\cos\sigma t \pm Bf_1^{(2)}(y)\sin\sigma t + o(A, B), \quad AB \neq 0, \tag{9.2}$$

and for nearly diagonal waves,

$$\zeta(x, y, t) = Af_1^{(1)}(x)\cos\sigma t \pm \overline{B}f_1^{(2)}(y)\cos\sigma t + o(A, \overline{B}), \quad A\overline{B} \neq 0, \tag{9.3}$$

where $f_1^{(1)}(x) = \cos[\pi(x + \frac{1}{2}L_1)/L_1]$ and $f_1^{(2)}(y) = \cos[\pi(y + \frac{1}{2}L_2)/L_2]$, where L_1 and L_2 are the tank lengths in the x- and y-directions, respectively. The $\pm$ sign in eq. (9.2) means that, for each amplitude parameter A, two amplitude parameters $\pm B$ exist of the same modulus but with different signs. Physically, the $\pm$ sign expresses that the steady-state rotation direction of swirling can be either clockwise or counterclockwise depending on initial and transient conditions. The $\pm$ sign in the expression for nearly diagonal waves reflects the involvement of one of the two diagonal directions.

Solution (9.1), with dominant amplitude parameter A, is the same as that adopted in Chapter 8 for asymptotic sloshing analysis in a two-dimensional rectangular tank. For longitudinal forcing along the Ox-axis, the amplitude parameter A remains nonzero in eqs. (9.2) and (9.3), in which we see two additional nonzero amplitude parameters, B and $\overline{B}$, of the lowest cross mode.

Similar to square-base tanks, a circular upright shape and other shapes of revolution (sphere, cone, etc.) are also characterized by the two lowest conjugate modes having the same natural frequency. Section 4.3.2.2 gives more details on these modes for a vertical circular cylinder and discusses *planar* motions and *swirling* that may occur for this geometry. Miles (1984a, 1984b) and Lukovsky (1990) proved that only these wave regimes are possible for resonant longitudinal excitation of the lowest mode with small forcing amplitude (relative to the radius) and $\sigma \approx \sigma_{1,1}$. In terms of definitions from Section 4.3.2.2, these wave regimes are

$$\zeta(r, \theta, t) = A\frac{J_1(\iota_{1,1}r/R_0)}{J_1(\iota_{1,1})}\cos\theta\cos\sigma t + o(A), \quad A \neq 0 \tag{9.4}$$

(for planar waves, in the Ox-plane) and

$$\zeta(r, \theta, t) = A\frac{J_1(\iota_{1,1}r/R_0)}{J_1(\iota_{1,1})}\cos\theta\cos\sigma t \pm B\frac{J_1(\iota_{1,1}r/R_0)}{J_1(\iota_{1,1})}\sin\theta\sin\sigma t + o(A, B), \quad AB \neq 0 \tag{9.5}$$

(for swirling), where (r, θ, z) are polar coordinates with origin in the center of the mean free surface and with z positive upward; $\iota_{1,1} = 1.841\ldots$; R_0 is the cylinder radius, and J_1 is a Bessel function of the first kind. Resonant forcing of the two lowest conjugate modes for conical tanks also leads to planar and swirling regimes (see the analyses by Gavrilyuk *et al.*, 2005; Lukovsky & Timokha, 2002).

When the corresponding steady-state regimes are not stable, irregular "chaotic" wave motions occur. Experiments by Abramson (1966) and Royon-Lebeaud *et al.* (2007) support this theoretical conclusion for circular upright cylindrical tanks, and experimental data from Faltinsen *et al.* (2003, 2005a, 2005b) confirm this conclusion for a square-base tank. Because the case of a square-base tank is characterized by a larger number of steady-state waves, we focus on that fact in detail.

9.1.1.2 Experimental observations and measurements for a nearly square-base tank

Two different experimental series for lateral harmonic excitations with frequency close to the first natural frequency were reported by Faltinsen *et al.* (2003, 2005a). An objective was to detect and classify the steady-state resonant regimes. In the main experimental series, a cubic tank was used with length, breadth, and height of 0.6 m.

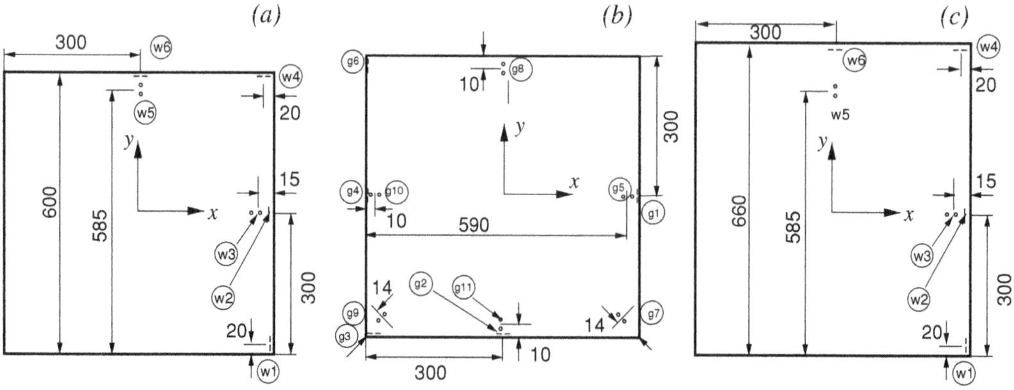

Figure 9.1. Top view of the square-base tanks with wave probes for (a) the first and (b) the second model test series. Part (c) shows configurations and measuring sensors for experiments with a nearly square-base tank (Faltinsen *et al.*, 2003, 2005a, 2005b, 2006b).

It was partially filled with freshwater at room temperature. Different finite depth filling levels were examined corresponding to the water depth-to-tank length ratios equal to 0.508, 0.5, 0.34 and 0.27. The tank walls are made of 20-mm-thick acrylic with a steel frame added for support. The weight of the empty tank and frame is 124 kg.

The instrumentation consisted of wave probes, force gauges, accelerometers, a steering system with velocity feedback monitoring, and a digital video camera. The location of wave-measuring sensors and the typical dimensions of the tanks are shown in Figure 9.1 for square-base (panels a and b) and nearly square base (panel c) tanks. In the figures, w1, w2, w4, w6, g1, g2, g3, g4, and g6 denote parallel copper tapes fixed to the wall. Sensors w3, w5, g5, g7, g8, g9, g10, and g11 consist of two lengths of parallel wire stretched vertically between the tank bottom and the roof. The wires have a diameter of 0.6 mm and the center distance is around 10 and 14 mm. Both types of wave probes, copper tape and wires, are capacitance probes. Because of surface tension, the wire probes have accuracy on less than the wire diameter, in this case < 0.6 mm, when the instantaneous free surface, except the meniscus, is perpendicular to the wires. If the free surface is steep or curved, the accuracy may be lower.

Four force gauges, referred to as guide towers, support the frame that encloses the tank model. Each guide tower includes three strain gauges and facilitates force measurements along all three axes. The accelerometers are aligned with the x- and y-axes and are situated on the top of the tank model. They are used to measure the tank motion and are a supplement to the measured servomotor velocity from the feedback steering system. The data analysis was performed in Matlab. A filtering frequency of 50 Hz was applied for both forces and accelerations when accelerometer readings were used to calculate the horizontal forces excluding the inertia force due to the tank mass. A zero-crossing analysis of the tank motion provided the exact forcing frequency. Fourier analysis of the forced motions was conducted, showing that the mechanical system, including the servomotor actuators, provides an accurate sinusoidal acceleration. The lowest frequency to disturb the prescribed regular motion was around 9 Hz, which is considered to be sufficiently far away from the input frequency, about 1 Hz. A special damping device was used to reduce the time between each test. A horizontal mesh of metal was lowered onto and through the free surface to suppress free-surface motion.

Each model test lasted about 3–5 minutes or approximately 120–200 forcing periods. The experimental observations established a relatively long transient phase, but after 100 forcing periods, the model tests demonstrated either "almost periodic" waves or completely irregular wave motions.

As long as the forcing frequency was equal or very close to the lowest natural frequency, video recordings typically showed three-dimensional surface waves for both transient and steady-state phases for the considered water fillings of the square and nearly-square base tanks in Figure 9.1. Two-dimensional or diagonal sloshing was observed only for longitudinal or diagonal excitation when the forcing frequency was not in the vicinity of the lowest natural frequency. Because the main interest of the experiment was classification of the almost periodic, steady-state sloshing, Faltinsen *et al.* (2005b) conducted experiments of up to 80 forcing periods. However, even after that, the experiments demonstrated beating which indicates that the effect of initial conditions has not died out. Beating was smaller for the cases with strongly breaking waves. The reason is that damping is very small when no wave breaking occurs and is mainly associated with the boundary-layer flow, which is why the three-dimensional wave phenomena were characterized as nearly steady state. The contribution from higher modes to the instantaneous wave profiles matters, as we see later in the text. Section 9.2.6 studies this contribution in detail for nearly steady-state swirling.

A classification of the steady-state regimes may be done by analyzing the behavior of the lowest primary-excited modes, as discussed in Section 9.1.1.1. Because the video recordings demonstrated a significant run-up in terms of a thin film of water at the wall, only wave probes placed at a small distance from the wall should be involved in the analysis. Typical treatment of the signals is presented in Figures 9.2 and 9.3. When we have longitudinal excitations (forcing along two parallel walls) with a frequency σ away from the primary resonance, the measured elevations are as in Figure 9.2(a) with $\sigma/\sigma_{1,0} = \sigma/\sigma_1 = 0.92$, which indicates planar waves (eq. (9.1)) in the plane of horizontal excitation. Due to the lowest-mode prediction (9.1), the measured elevations at w5 should be zero. The recorded values are much smaller than those at w3. The nonzero values at w5 are caused by nonlinearities that amplify higher modes. Furthermore, we illustrate the drawing in the (w3,w5)-plane based on the measurements after 100 s. At this stage the wave motions are close to steady-state conditions. Based on theoretical treatment (9.1), we expect that the response curve in the (w3,w5)-plane is a finite straight line, reported in the right upper corner of Figure 9.2(a). The experiments show slightly more complex behavior because the free-surface measurements at w5 were not exactly zero.

Figure 9.2(b) illustrates the transition to swirling. The elevation measured at w5 is small during the first beating periods (i.e., the wave motions are close to being two-dimensional). However, the instability of the cross-waves activates the cross-modes perpendicular to the forcing and the measurements at w5 become comparable with those at w3 after about 25 s. The modal analysis of eq. (9.2) predicts for swirling an ellipse as the response curve in the (w3,w5)-plane, which was discussed in Section 4.3.2.1. The ellipse is illustrated in the upper corner of Figure 9.2(b). The experimental measurements show a nearly closed trajectory but deviate from the ellipse due to nonlinearities that are of primary importance for three-dimensional waves. The quantitative description of swirling also needs to account for contributions from several higher modes as we show in Section 9.2.6. Figure 9.2(c) gives the measured elevations at the same two probes for the case when a steady-state regime is not achieved and the wave motions are of irregular ("chaotic") nature for the entirety of the studied experimental time.

Finally, for diagonal excitations and, in some cases, for the longitudinal excitations, the experimental records in the nearly steady-state time-domain were as in Figure 9.3 (for longitudinal forcing). The experimental behavior is consistent with theoretical prediction (9.3) for nearly diagonal waves.

The types of resonant wave regimes did not change for the nearly square-base tank shown in Figure 9.1(c) when the excitation direction was along the x-axis (shorter walls). The 10% perturbations of the aspect ratio of the tank bottom caused an influence on the effective frequency domains. In particular, the experiments established a smaller range with irregular motions and an increased swirling range. Faltinsen *et al.* (2006a) studied this fact theoretically. We report such studies in Section 9.2.5.

Experimental data on swirling and nearly diagonal and irregular waves exhibited steep wave patterns, local breaking at the walls, and irregular fluctuations of the recorded wave amplitudes. Recurrent wave breaking is initiated as a thin vertical jet, which overturns and forms droplets falling under gravity onto the underlying free surface. Experimental observations of breaking waves were reported by Royon-Lebeaud *et al.* (2007) for a vertical circular cylindrical tank. The local run-up and water films covering the vertical walls are not believed to significantly affect the

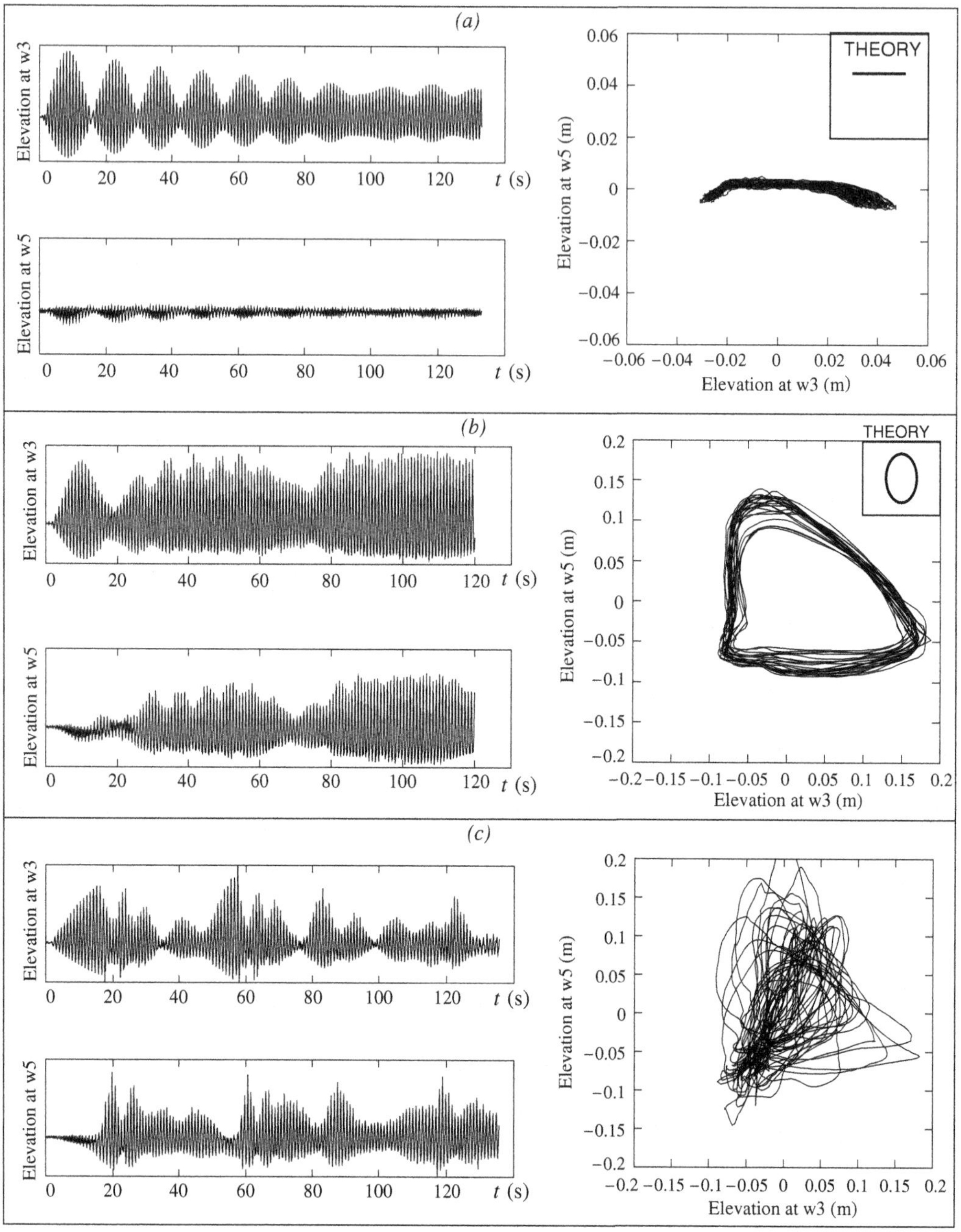

Figure 9.2. Example of measurements at wave probes w3 and w5 and corresponding parametric curves. The square-base tank was excited longitudinally with $\overline{h} = 0.508$ and $\eta_{1a}/L_1 = 0.0078$. Figure 9.1 defines positions of w3 and w5; (a) planar wave with $\sigma/\sigma_1 = 0.92$, (b) swirling with $\sigma/\sigma_1 = 1.011$, and (c) irregular waves with $\sigma/\sigma_1 = 0.945$; σ_1 is the lowest natural frequency equal to $\sigma_{1,0} = \sigma_{0,1}$ for the square base tank.

hydrodynamic forces. The measured wave elevations at sensors w3, w5, g5, g7, g8, g9, g10, and g11 slightly away from the wall were not significantly affected by local wall effects. During transients and swirling, the wave profiles were particularly steep near the corners and, although the forcing amplitude was small, all three-dimensional waves demonstrated significant local near-wall phenomena in the form of run-up at the walls accompanied by splashing/overturning with possible drop

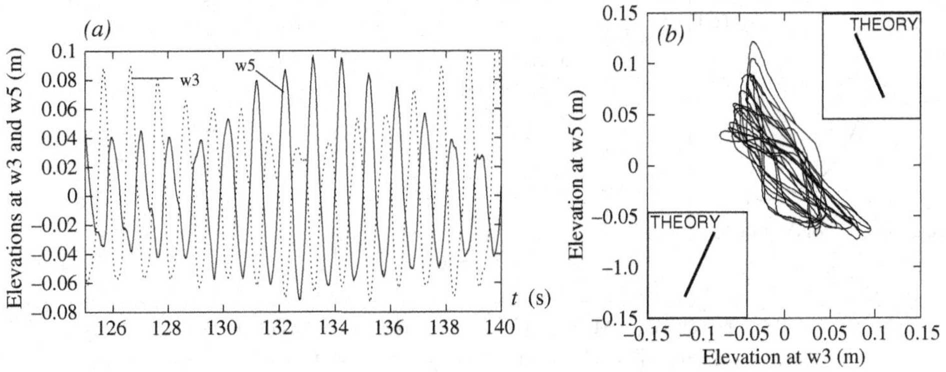

Figure 9.3. Example of measurements at wave probes w3 and w5 and corresponding parametric curve for "nearly diagonal" waves. The square-base tank was excited longitudinally with $\overline{h} = 0.34$, $\eta_{1a}/L_1 = 0.0078$, and $\sigma/\sigma_1 = 0.98$; σ_1 is the lowest natural frequency equal to $\sigma_{1,0} = \sigma_{0,1}$ for the square base tank.

formation. Typical instantaneous wave shapes near the walls are shown in Figure 9.4. The situation is very similar to the description by Abramson *et al.* (1974) for sloshing in a vertical circular cylindrical tank due to horizontal excitation.

Steep wave profiles indicate implicitly a considerable amplification of higher modes and strong nonlinearities. The irregular amplitude fluctuations in almost periodic swirling can partly be explained by three-dimensional wave breaking, which is mostly localized at the tank corners. In contrast to initial transients and irregular motions, local breaking involves insignificant water mass. However, the falling droplets may cause nonnegligible irregular perturbations of higher modes and randomize the damping, thereby extending the transient phase.

9.1.2 Bifurcation and stability

The nonlinear analysis of resonant two-dimensional sloshing in finite depth discussed in Chapter 8 shows frequency domains with either one or three steady-state solutions. One or two of these solutions are stable; which of the stable steady-state solutions occurs in reality depends on transient scenarios. This chapter extends the asymptotic methods of Chapter 8 to three-dimensional sloshing for finite liquid depths and establishes more types of steady-state solutions.

As earlier, we consider a finite liquid depth and a small ratio of lateral forcing amplitude to tank breadth, ε. The two-dimensional analysis in Section 8.3.1 showed that the dominant amplitude parameter A (scaled by the tank

Figure 9.4. Local phenomena in the corner during three-dimensional wave motions with swirling.

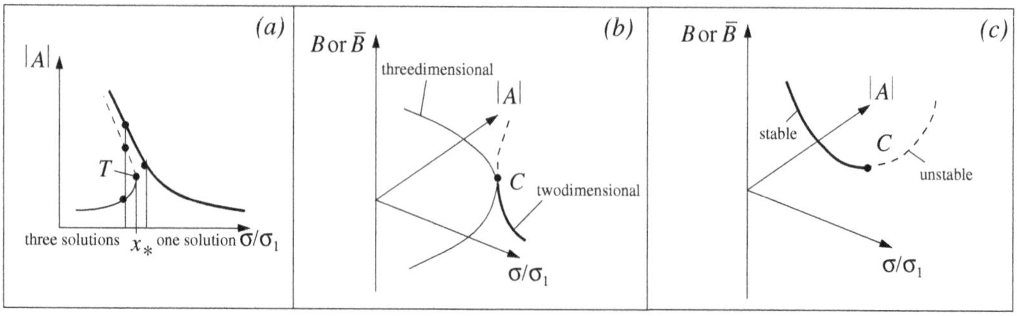

Figure 9.5. (a) Response curve for Moiseev-type solution of two-dimensional sloshing; (b, c) types of bifurcation points C appearing in the forthcoming analysis of sloshing in a three-dimensional rectangular tank.

breadth) is of lower order than ε (i.e., $A \gg \varepsilon$) and is determined by a cubic algebraic equation in A (see eq. (8.54)). Furthermore, the σ/σ_1-value (i.e., the ratio between the forcing frequency σ and the lowest natural frequency σ_1) was interpreted as an input parameter, which is henceforth called a *bifurcation parameter.* Changing this input parameter in a neighborhood of $\sigma/\sigma_1 = 1$, we treated solutions of the algebraic equation for A as *planar* response curves in the $(\sigma/\sigma_1, |A|)$-plane. In Section 8.3.1.1 we discussed the existence of a value $x_* = \sigma/\sigma_1$ of the bifurcation parameter at which "a small change of σ/σ_1 causes 'sudden' *qualitative* (or topological) changes in its long-term dynamic behavior (here steady-state) of the system." This definition of bifurcation is the most general from *bifurcation theory* (Seydel, 1994). The x_*-value is the abscissa of the so-called turning point T in Figure 8.4. Both Figure 8.4(a) and Figure 9.5(a) show a unique solution for $\sigma/\sigma_1 > x_*$. When σ/σ_1 decreases and becomes equal to x_*, we get two solutions, whereas $\sigma/\sigma_1 < x_*$ leads to three steady-state solutions. Therefore, the steady-state solution splits, or *bifurcates,* into many solutions at $\sigma/\sigma_1 = x_*$. The turning point T in each figure is an example of a *local bifurcation point.* By definition (Seydel, 1994), "a local bifurcation occurs when the bifurcation parameter change causes the stability of an 'equilibrium' (steady-state solution) to change." Indeed, the point T divides the corresponding response curve into subcurves that imply stable and unstable solutions. Types of local bifurcations are saddle-node bifurcation (the case in Figure 9.5(a) belongs to this type), transcritical bifurcation, pitchfork bifurcation, period-doubling (flip) bifurcation, Hopf bifurcation, and Neimark bifurcation.

For brevity, we do *not* relate the local bifurcations to special mathematical terminology. Interested readers may find more details in the books by Seydel (1994), Hassard *et al.* (1981), and Chow and Hale (1982). In contrast to Chapter 8, we must now consider *three* amplitude parameters: A, B, and $\overline{B}$. If the forcing is longitudinally along a tank wall, A is the wave amplitude parameter in the excitation plane in phase with forcing, but $\overline{B}$ and B correspond to in- and out-of-phase components for the cross-waves. The bifurcation parameter is the same as in Section 8.3.1.1 if we denote $\sigma_1 = \sigma_{1,0} = \sigma_{0,1}$ for a square-base tank and $\sigma_1 = \sigma_{1,1}$ for axisymmetric tanks. We later see that three algebraic equations couple the behavior of A, B, and $\overline{B}$. The corresponding response curves must then be considered in the $(\sigma/\sigma_1, |A|, |B|, |\overline{B}|)$-space. Only for the planar regime (9.1) with $B = \overline{B} = 0$ do the curves belong to the $(\sigma/\sigma_1, |A|)$ -plane. Swirling ($AB \neq 0$, $\overline{B} = 0$) corresponds to curves in the $(\sigma/\sigma_1, |A|, |B|)$-subspace and the nearly diagonal wave regime ($A\overline{B} \neq 0$, $B = 0$) is described by three-dimensional curves in the $(\sigma/\sigma_1, |A|, |\overline{B}|)$-subspace. The joint points of the last two subspaces belong to the $(\sigma/\sigma_1, |A|)$-plane, in which we expect planar regimes. Therefore, swirling and nearly diagonal wave regimes may only bifurcate (split) at a point C from the planar regime as illustrated by an example in Figure 9.5(b). We have four real different solutions emerging from C. Two of them are in the planar regime (i.e., in the $(\sigma/\sigma_1, |A|)$-plane). The others correspond to a swirling or diagonal regime with the

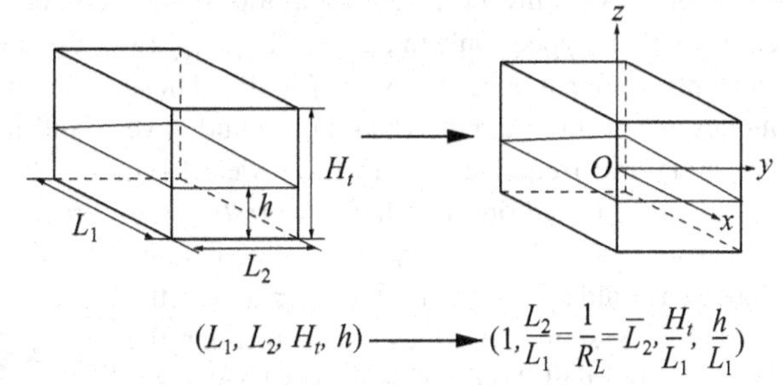

Figure 9.6. Geometrical scaling of a rectangular-base tank.

same amplitude parameter of opposite signs ($\pm B$ or $\pm\overline{B}$). Starting with the planar wave regime, we can say that a "new" three-dimensional wave regime emerges from C.

Another local bifurcation point appearing in the forthcoming analysis is shown by the three-dimensional curve in Figure 9.5(c). The solid (stable) and dashed (unstable) branches imply a steady-state solution, which changes its stability property at point C. No other steady-state solutions (other curves) are associated with *real numbers* A and $B(\overline{B})$ that bifurcate from point C. Bifurcation implies a change in the stability property of the solution. In our case this change occurs through point C, because C gives rise to two solutions with complex conjugate amplitude parameters A and $B(\overline{B})$. These amplitude parameters are physically irrelevant but exist mathematically. The appearance of the complex conjugate solutions causes a stability change when going through C along the response curve. In the mathematical literature, this case is often called a Hamiltonian–Hopf bifurcation, and we refer interested readers to Hassard *et al.* (1981).

9.2 Rectangular-base tank with a finite liquid depth

9.2.1 Statement and generalization of adaptive modal system (8.95)

Let us consider a rigid open parallelepipedal tank with length L_1 and breadth L_2. The filling depth is h. Henceforth, we assume potential flow of an incompressible liquid and make sizes nondimensional by scaling with L_1. Therefore, the forthcoming analysis considers a tank with length 1 and breadth $1/R_L = L_2/L_1 = \overline{L}_2$. A particular consequence of the scaling is also that the generalized coordinates β_i are assumed to be nondimensional. The transformation is demonstrated in Figure 9.6. Because the analysis assumes no roof impact, H_t is not a parameter in our studies.

Similar to the analysis in Section 8.3.1, our analysis starts with modal system (7.14) and (7.15). The corresponding natural modes in dimensional form are given in Section 4.3.2.1; that is, we can write in L_1-scaled form that

$$\overline{\varphi}_{i,j}(x, y, z) = \overline{f}_i^{(1)}(x)\overline{f}_j^{(2)}(y)\frac{\cosh[\lambda_{i,j}(z+\overline{h})]}{\cosh(\lambda_{i,j}\overline{h})};$$

$$\lambda_{i,j} = \pi\sqrt{i^2 + R_L^2 j^2},$$

$$\sigma_{i,j}^2 = \frac{g}{L_1}\lambda_{i,j}\tanh(\lambda_{i,j}\overline{h}), \quad i, j \geq 0, \quad i+j \neq 0,$$

$$R_L = L_1/L_2 \qquad (9.6)$$

where $\sigma_{i,j}$ are the natural frequencies, $\overline{f}_i^{(1)}(x) = \cos[i\pi(x+\frac{1}{2})]$, and $\overline{f}_j^{(2)}(y) = \cos[j\pi(R_L y + \frac{1}{2})]$. Projections of $\overline{\varphi}_{i,j}$ on the mean free surface $z = 0$ introduce the shapes of standing waves $\overline{f}_{i,j}(x, y) = \overline{\varphi}_{i,j}|_{z=0}$ as

$$\overline{f}_{i,j}(x, y) = \overline{\varphi}_{i,j}(x, y, 0) = \overline{f}_i^{(1)}(x)\overline{f}_j^{(2)}(y). \qquad (9.7)$$

The shapes may be classified in terms of subclasses as discussed in Section 4.3.2.1.

Furthermore, we assume small-amplitude excitations of the tank with frequencies close to the lowest natural frequency. The lowest natural frequency is then associated with either $\sigma_{1,0}$ or $\sigma_{0,1}$, which are equal for a square-base tank. When the z-axis of the coordinate system $Oxyz$ goes through the center of the tank liquid, linear modal equations (5.127), (5.129), and (5.131) show that the *lowest* mode can only be excited by surge, sway, roll, and pitch. The heave motions may lead to a parametric resonance (as discussed in Section 8.10), but this is not expected to be important when the forcing frequency is in a vicinity of the lowest frequency. Therefore, we can disregard the effect of heave. Furthermore, yaw motions associated with η_6 appear only in eq. (5.131) (i.e., yaw can only excite modes $\overline{\varphi}_{i,j}$ with $ij \neq 0$). These modes are of three-dimensional nature and have larger natural frequencies than the lowest from $\sigma_{1,0}$ and $\sigma_{0,1}$. Based on these facts, we *restrict our analysis to the case* $\eta_3 = \eta_6 = 0$. When doing that, we should remember that, if the z-axis of the coordinate system $Oxyz$ does not go through the center of the tank liquid, yaw causes local surge and sway of the tank so that it excites the lowest modes.

An infinite-dimensional modal system for adaptive modeling of two-dimensional sloshing in a rectangular tank was derived in Section 8.4.1. Faltinsen *et al.* (2003) have formally used the same derivation scheme for the three-dimensional case. The only difference is that the derivation replaces each integer index by the pair (i, j) and the corresponding expressions for the hydrodynamic coefficients of the modal system are more lengthy. The result can be written in the following form:

$$\sum_{\substack{a,b=0\\a+b\neq 0}}^{\infty} \ddot{\beta}_{a,b}\left[\delta_{ia}\delta_{jb} + \sum_{\substack{c,d=0\\c+d\neq 0}}^{\infty} d^{1,(i,j)}_{(a,b),(c,d)}\beta_{c,d} + \sum_{\substack{c,d=0\\c+d\neq 0}}^{\infty}\sum_{\substack{e,f=0\\e+f\neq 0}}^{\infty} d^{2,(i,j)}_{(a,b),(c,d),(e,f)}\beta_{c,d}\beta_{e,f}\right] + \sigma^2_{i,j}\beta_{i,j}$$
$$+ \sum_{\substack{a,b=0\\a+b\neq 0}}^{\infty}\sum_{\substack{c,d=0\\c+d\neq 0}}^{\infty} \dot{\beta}_{a,b}\dot{\beta}_{c,d} \times \left[t^{0,(i,j)}_{(a,b),(c,d)} + \sum_{\substack{e,f=0\\e+f\neq 0}}^{\infty} t^{1,(i,j)}_{(a,b),(c,d),(e,f)}\beta_{e,f}\right]$$
$$+ \delta_{i0}P_{0,j}\left[\frac{\ddot{\eta}_2}{L_1} + \overline{S}_{0,j}\ddot{\eta}_4 + \frac{g\eta_4}{L_1}\right]$$
$$+ \delta_{j0}P_{i,0}\left[\frac{\ddot{\eta}_1}{L_1} - \overline{S}_{i,0}\ddot{\eta}_5 - \frac{g\eta_5}{L_1}\right] = 0, \quad i+j \geq 1, \tag{9.8}$$

where $P_{i,j}$ are defined by eqs. (5.128) and (5.130), δ_{ij} is the Kronecker delta, and $\overline{S}_{0,j}$ and $\overline{S}_{i,0}$ are the L_1-scaled coefficients of eqs. (5.128) and (5.130). Furthermore, the tensors d and t are explicitly given in the original paper by Faltinsen *et al.* (2003) as functions of $\overline{h}$ and R_L. The derivation of eq. (9.8) is particularly tedious and analytically difficult relative to the corresponding analysis for two-dimensional sloshing. Hydrodynamic forces and moments can also be found in terms of $\beta_{i,j}$ via the Lukovsky formulas. We refer interested readers to the original publications by Faltinsen *et al.* (2003, 2005a, 2005b).

9.2.2 Moiseev-based modal system for a nearly square-base tank

We concentrate in this section on *finite* liquid depths and resonant excitation of the two lowest modes that have almost equal natural frequencies. These modes are associated with two-dimensional standing waves occurring along the Ox- and Oy-axes parallel to the tank walls.

We consider a transition to square geometry, implying $L_1 \approx L_2$ ($R_L \to 1$); therefore, the pair of primary natural modes $f_1^{(1)}(x)$ and $f_1^{(2)}(y)$ degenerates (i.e., they have equal natural frequencies). In the asymptotic limit $\sigma \to \sigma_{0,1}$ they are directly excited and of leading magnitude. We now assume that the generalized coordinates $\beta_{i,j}$ for the free-surface elevation in eq. (9.8) have the following orders of magnitude:

$$\begin{aligned}
&\beta_{1,0} \sim \beta_{0,1} = O(\varepsilon^{1/3}),\\
&\beta_{2,0} \sim \beta_{1,1} \sim \beta_{0,2} = O(\varepsilon^{2/3}),\\
&\beta_{3,0} \sim \beta_{2,1} \sim \beta_{1,2} \sim \beta_{0,3} = O(\varepsilon),\\
&\beta_{i,j} \leq O(\varepsilon), \quad i+j \geq 4,
\end{aligned} \tag{9.9}$$

where the small parameter ε is associated with the L_1-scaled amplitude of surge and sway and with the roll and pitch amplitudes.

Relationships (9.9) are formally consistent with the Moiseev-type third-order assumptions leading to single-dominant modal theory (8.43) for two-dimensional sloshing. Faltinsen *et al.* (2003) showed that asymptotic ordering (9.9), together with neglect of the $o(\varepsilon)$-terms, reduces modal system (9.8) to a finite-dimensional system of nonlinear ordinary differential equations coupling $\beta_{i,j}$ ($i+j \leq 3$). Other modes ($i+j \geq 4$) are governed by linear sloshing theory. For brevity we denote

$$\begin{aligned}
&\beta_{1,0} = a_1, \quad \beta_{2,0} = a_2, \quad \beta_{0,1} = b_1,\\
&\beta_{0,2} = b_2, \quad \beta_{1,1} = c_1, \quad \beta_{3,0} = a_3,\\
&\beta_{2,1} = c_{21}, \quad \beta_{1,2} = c_{12}, \quad \beta_{0,3} = b_3,
\end{aligned} \tag{9.10}$$

so that modal system (9.8) can be rewritten in the following form:

$$\left[\ddot{a}_1 + \sigma_{1,0}^2 a_1 + d_1(\ddot{a}_1 a_2 + \dot{a}_1\dot{a}_2) + d_2(\ddot{a}_1 a_1^2 + \dot{a}_1^2 a_1) + d_3\ddot{a}_2 a_1 + P_{1,0}\left(\frac{\ddot{\eta}_1}{L_1} - \overline{S}_{1,0}\ddot{\eta}_5 - \frac{g\eta_5}{L_1}\right)\right] + d_6\ddot{a}_1 b_1^2 + \ddot{b}_1(d_7 c_1 + d_8 a_1 b_1) + d_9\ddot{c}_1 b_1 + d_{10}\dot{b}_1^2 a_1 + d_{11}\dot{a}_1\dot{b}_1 b_1 + d_{12}\dot{b}_1\dot{c}_1 = 0, \quad (9.11)$$

$$\left[\ddot{b}_1 + \sigma_{0,1}^2 b_1 + \overline{d}_1(\ddot{b}_1 b_2 + \dot{b}_1\dot{b}_2) + \overline{d}_2(\ddot{b}_1 b_1^2 + \dot{b}_1^2 b_1) + \overline{d}_3\ddot{b}_2 b_1 + P_{0,1}\left(\frac{\ddot{\eta}_2}{L_1} + \overline{S}_{0,1}\ddot{\eta}_4 + \frac{g\eta_4}{L_1}\right)\right] + \overline{d}_6\ddot{b}_1 a_1^2 + \ddot{a}_1(\overline{d}_7 c_1 + \overline{d}_8 a_1 b_1) + \overline{d}_9\ddot{c}_1 a_1 + \overline{d}_{10}\dot{a}_1^2 b_1 + \overline{d}_{11}\dot{a}_1\dot{b}_1 a_1 + \overline{d}_{12}\dot{a}_1\dot{c}_1 = 0, \quad (9.12)$$

$$\left[\ddot{a}_2 + \sigma_{2,0}^2 a_2 + d_4\ddot{a}_1 a_1 + d_5\dot{a}_1^2\right] = 0, \quad (9.13)$$

$$\left[\ddot{b}_2 + \sigma_{0,2}^2 b_2 + \overline{d}_4\ddot{b}_1 b_1 + \overline{d}_5\dot{b}_1^2\right] = 0, \quad (9.14)$$

$$\ddot{c}_1 + \hat{d}_1\ddot{a}_1 b_1 + \hat{d}_2\ddot{b}_1 a_1 + \hat{d}_3\dot{a}_1\dot{b}_1 + \sigma_{1,1}^2 c_1 = 0, \quad (9.15)$$

$$\left[\ddot{a}_3 + \sigma_{3,0}^2 a_3 + \ddot{a}_1(q_1 a_2 + q_2 a_1^2) + q_3\ddot{a}_2 a_1 + q_4\dot{a}_1^2 a_1 + q_5\dot{a}_1\dot{a}_2 + P_{3,0}\left(\frac{\ddot{\eta}_1}{L_1} - \overline{S}_{3,0}\ddot{\eta}_5 - \frac{g\eta_5}{L_1}\right)\right] = 0, \quad (9.16)$$

$$\ddot{c}_{21} + \sigma_{2,1}^2 c_{21} + \ddot{a}_1(q_6 c_1 + q_7 a_1 b_1) + \ddot{b}_1(q_8 a_2 + q_9 a_1^2) + q_{10}\ddot{a}_2 b_1 + q_{11}\ddot{c}_1 a_1 + q_{12}\dot{a}_1^2 b_1 + q_{13}\dot{a}_1\dot{b}_1 a_1 + q_{14}\dot{a}_1\dot{c}_1 + q_{15}\dot{a}_2\dot{b}_1 = 0, \quad (9.17)$$

$$\ddot{c}_{12} + \sigma_{1,2}^2 c_{12} + \ddot{b}_1(\overline{q}_6 c_1 + \overline{q}_7 a_1 b_1) + \ddot{a}_1(\overline{q}_8 b_2 + \overline{q}_9 b_1^2) + \overline{q}_{10}\ddot{b}_2 a_1 + \overline{q}_{11}\ddot{c}_1 b_1 + \overline{q}_{12}\dot{b}_1^2 a_1 + \overline{q}_{13}\dot{a}_1\dot{b}_1 b_1 + \overline{q}_{14}\dot{b}_1\dot{c}_1 + \overline{q}_{15}\dot{a}_1\dot{b}_2 = 0, \quad (9.18)$$

$$\left[\ddot{b}_3 + \sigma_{0,3}^2 b_3 + \ddot{b}_1(\overline{q}_1 b_2 + \overline{q}_2 b_1^2) + \overline{q}_3\ddot{b}_2 b_1 + \overline{q}_4\dot{b}_1^2 b_1 + \overline{q}_5\dot{b}_1\dot{b}_2 + P_{0,3}\left(\frac{\ddot{\eta}_2}{L_1} + \overline{S}_{0,3}\ddot{\eta}_4 + \frac{g\eta_4}{L_1}\right)\right] = 0. \quad (9.19)$$

The higher modes are governed by the following linear equations (see Section 5.4.3.1):

$$\ddot{\beta}_{i,j} + \sigma_{i,j}^2\beta_{i,j} + \delta_{0i}P_{0,j}\left[\frac{\ddot{\eta}_2}{L_1} + \overline{S}_{0,j}\ddot{\eta}_4 + \frac{g\eta_4}{L_1}\right] + \delta_{0j}P_{i,0}\left[\frac{\ddot{\eta}_1}{L_1} - \overline{S}_{i,0}\ddot{\eta}_5 - \frac{g\eta_5}{L_1}\right] = 0, \quad i+j \geq 4. \quad (9.20)$$

The coefficients of the nonlinear quantities are functions of $\overline{h}$ and R_L. They can be computed by using the formulas for the tensors d and t from the paper by Faltinsen *et al.* (2003). For *square cross-section*, $d_i = \overline{d}_i$ and $q_i = \overline{q}_i$; their values are presented in Tables 9.1 and 9.2, respectively. Note that the square geometry leads to $d_{12} = d_7$ and $\hat{d}_2 = \hat{d}_3$ in the modal equations.

The terms in the square brackets of eqs. (9.11)–(9.20) are associated with two-dimensional flows in either the Oxz- or the Oyz-plane. For these terms, the coefficients of the nonlinear terms are exactly the same as those in eq. (8.43) derived in Section 8.3.1 for the Moiseev-based theory in a two-dimensional rectangular tank, but the characteristic size is L_1 instead of l. Other terms and additional equations for c_1, c_{21}, and c_{12} appear due to three-dimensional intermodal interaction. Subsystem (9.11)–(9.15) couples a_1, b_1, a_2, b_2, and c_1 but does not depend on a_3, c_{21}, c_{12}, and b_3 calculated from eqs. (9.16)–(9.19). Subsystem (9.16)–(9.19) is linear in a_3, c_{21}, c_{12}, and b_3 and depends nonlinearly on a_1, b_1, a_2, b_2, and c_1. Expressions for hydrodynamic forces and moments can be obtained by using the Lukovsky formulas derived in Section 7.3.

In the following paragraphs, asymptotic modal system (9.11)–(9.20) is our main mathematical tool. Once we know the forcing terms and initial (or periodic) conditions explicitly, we can integrate its linear (infinite) subsystem (9.20). The solution of the modal system may be studied either analytically or numerically. Validation of modal system (9.11)–(9.20) is typically done for resonant harmonic excitation of the lowest modes.

Faltinsen *et al.* (2003, 2005a) showed that this finite-dimensional modal theory for finite depth liquid has limitations in quantitative predictions similar to its two-dimensional analogy, eqs. (8.43). The reason is that both theories use Moiseev intermodal ordering, which assumes only the

Table 9.1. *Coefficients d_i and $\hat{d}_i$ versus depth-to-breadth ratio $\overline{h}$ for a square-base tank ($\bar{d}_i = d_i$)*

$\overline{h}$	d_1	d_2	d_3	d_4	d_5	d_6	d_7	d_8	d_9	d_{10}	d_{11}	$\hat{d}_1$	$\hat{d}_3$
0.3	3.290	4.551	−0.488	−1.266	−5.533	0.512	1.157	5.447	−0.120	4.935	1.025	−0.401	−4.668
0.4	3.183	3.414	−0.256	−0.595	−4.290	−0.589	1.335	4.346	0.159	4.935	−1.177	0.500	−3.195
0.5	3.153	2.933	−0.136	−0.295	−3.721	−1.040	1.441	3.895	0.303	4.935	−2.079	0.914	−2.511
0.6	3.145	2.706	−0.072	−0.152	−3.441	−1.245	1.500	3.690	0.378	4.935	−2.490	1.110	−2.180
0.7	3.143	2.592	−0.039	−0.079	−3.299	−1.344	1.533	3.591	0.417	4.935	−2.688	1.205	−2.014
0.8	3.142	2.533	−0.021	−0.042	−3.225	−1.393	1.550	3.541	0.438	4.935	−2.787	1.253	−1.931
0.9	3.142	2.502	−0.011	−0.022	−3.186	−1.418	1.560	3.516	0.448	4.935	−2.837	1.276	−1.887
1.0	3.142	2.486	−0.006	−0.012	−3.165	−1.431	1.565	3.503	0.454	4.935	−2.863	1.288	−1.865
1.1	3.142	2.477	−0.003	−0.006	−3.154	−1.438	1.568	3.497	0.457	4.935	−2.876	1.295	−1.853
1.2	3.142	2.473	−0.002	−0.003	−3.148	−1.441	1.569	3.493	0.458	4.935	−2.883	1.298	−1.847
1.3	3.142	2.470	−0.001	−0.002	−3.145	−1.443	1.570	3.491	0.459	4.935	−2.887	1.299	−1.844
1.4	3.142	2.469	0.000	−0.001	−3.143	−1.444	1.570	3.491	0.460	4.935	−2.889	1.300	−1.842
1.5	3.142	2.468	0.000	−0.001	−3.143	−1.445	1.571	3.490	0.460	4.935	−2.890	1.301	−1.841
1.6	3.142	2.468	0.000	0.000	−3.142	−1.445	1.571	3.490	0.460	4.935	−2.890	1.301	−1.841
1.7	3.142	2.468	0.000	0.000	−3.142	−1.445	1.571	3.490	0.460	4.935	−2.890	1.301	−1.841
1.8	3.142	2.468	0.000	0.000	−3.142	−1.445	1.571	3.490	0.460	4.935	−2.891	1.301	−1.840
1.9	3.142	2.467	0.000	0.000	−3.142	−1.445	1.571	3.489	0.460	4.935	−2.891	1.301	−1.840
2.0	3.142	2.467	0.000	0.000	−3.142	−1.445	1.571	3.489	0.460	4.935	−2.891	1.301	−1.840

Note: The coefficients needed in eqs. (9.11)–(9.15) for this case are $d_{12} = d_7$ and $\hat{d}_2 = \hat{d}_3$.

Table 9.2. *Coefficients q_i versus depth-to-breadth ratio $\overline{h}$ for a square-base tank ($\bar{q}_i = q_i$)*

$\overline{h}$	q_1	q_2	q_3	q_4	q_5	q_6	q_7	q_8	q_9	q_{10}	q_{11}	q_{12}	q_{13}	q_{14}	q_{15}
0.3	−1.720	2.379	−0.255	23.168	−13.31	−0.856	5.446	2.554	6.600	0.241	−0.420	6.956	34.705	−8.451	−4.026
0.4	−0.836	0.896	−0.067	13.827	−11.22	−0.208	0.923	3.279	4.128	0.608	−0.041	−0.614	20.529	−6.808	−3.088
0.5	−0.426	0.397	−0.018	10.387	−10.313	0.081	−0.601	3.587	3.213	0.706	0.095	−3.131	15.305	−6.098	−2.720
0.6	−0.223	0.191	−0.005	8.883	−9.880	0.221	−1.208	3.732	2.826	0.732	0.146	−4.123	13.080	−5.769	−2.558
0.7	−0.117	0.097	−0.001	8.163	−9.662	0.292	−1.477	3.804	2.650	0.739	0.167	−4.557	12.051	−5.609	−2.481
0.8	−0.062	0.050	0.000	7.800	−9.550	0.329	−1.604	3.842	2.567	0.741	0.175	−4.761	11.552	−5.529	−2.442
0.9	−0.033	0.026	0.000	7.612	−9.491	0.349	−1.666	3.861	2.526	0.741	0.178	−4.860	11.303	−5.488	−2.422
1.0	−0.018	0.014	0.000	7.514	−9.460	0.359	−1.698	3.871	2.505	0.742	0.179	−4.910	11.176	−5.468	−2.412
1.1	−0.008	0.007	0.000	7.461	−9.444	0.365	−1.714	3.877	2.495	0.742	0.180	−4.936	11.110	−5.457	−2.406
1.2	−0.005	0.004	0.000	7.434	−9.435	0.367	−1.722	3.880	2.489	0.742	0.180	−4.949	11.076	−5.451	−2.403
1.3	−0.003	0.002	0.000	7.419	−9.430	0.369	−1.726	3.881	2.487	0.742	0.180	−4.956	11.058	−5.448	−2.402
1.4	−0.001	0.001	0.000	7.411	−9.428	0.370	−1.729	3.882	2.485	0.742	0.180	−4.959	11.049	−5.447	−2.401
1.5	−0.001	0.001	0.000	7.407	−9.426	0.370	−1.730	3.883	2.484	0.742	0.180	−4.961	11.044	−5.446	−2.400
1.6	0.000	0.000	0.000	7.405	−9.426	0.371	−1.730	3.883	2.484	0.742	0.180	−4.962	11.042	−5.445	−2.400
1.7	0.000	0.000	0.000	7.404	−9.425	0.371	−1.731	3.883	2.484	0.742	0.180	−4.963	11.040	−5.445	−2.400
1.8	0.000	0.000	0.000	7.403	−9.425	0.371	−1.731	3.883	2.484	0.742	0.180	−4.963	11.039	−5.445	−2.400
1.9	0.000	0.000	0.000	7.403	−9.425	0.371	−1.731	3.883	2.484	0.742	0.180	−4.963	11.039	−5.445	−2.400
2.0	0.000	0.000	0.000	7.402	−9.425	0.371	−1.731	3.883	2.484	0.742	0.180	−4.963	11.039	−5.445	−2.400

Note: The coefficients are needed in eqs. (9.16)–(9.19).

lowest mode to be dominant in two dimensions and the lowest mode in the x- and y-directions to be dominant in three dimensions. The Moiseev-based theory is limited in describing nonlinear sloshing when

- the liquid depth is close to some critical value, causing infinite resonant response for a steady-state regime (see discussion on the critical depth for two-dimensional resonant sloshing in a rectangular tank in Section 8.5);
- resonant excitation of the lowest modes leads to amplification of higher natural modes due to secondary resonance, where nonlinearity yields higher harmonics that are close to the natural frequencies of the higher modes;
- values of the forcing parameter, ε, are increased;
- the liquid depth is shallow or in the lower range of intermediate depths so that the ratio between the liquid depth and the tank breadth (length) must be introduced as a small parameter in the development of the multimodal method; and
- transient sloshing is excited in such a way that steep wave patterns occur.

For two-dimensional sloshing we showed that, when the finite depth assumption fails due to the aforementioned reasons, one must assume *additional dominant* natural modes (i.e., similar to the primary excited one). Faltinsen *et al.* (2005a, 2006b) demonstrated that the same is true for the three-dimensional modal method. Moreover, by considering experimental cases with a duration of up to 300 forcing periods, they pointed out that clearly periodic swirling in a square-base tank with realistic forcing amplitudes is never achieved. Their measurements show fluctuations of the response amplitudes and amplification of higher modes during the whole time evolution. As a consequence, an accurate description of the nearly steady-state swirling needs multidimensional nonlinear modal systems with many dominant modes. Section 9.2.6 focuses on how to model this system.

9.2.3 Steady-state resonance solutions for a nearly square-base tank

We consider steady-state three-dimensional resonant sloshing and limit ourselves to horizontal translatory forcing either in the diagonal direction of a square-base tank or parallel to a tank wall for a square-base and nearly square-base tank. The forced surge and sway motions are expressed as $\eta_1(t) = \eta_{1a}\cos(\sigma t)$ and $\eta_2(t) = \eta_{2a}\cos(\sigma t)$. Diagonal translatory forcing of a square-base tank corresponds to $\sqrt{\eta_{1a}^2 + \eta_{2a}^2}\cos(\sigma t)$ with $\eta_{1a} = \eta_{2a}$.

The modal system with eqs. (9.11)–(9.19) can be asymptotically integrated to find the $2\pi/\sigma$-periodic solution that satisfies condition (9.9) with eqs. (9.10). The solution method is the same as in Section 8.3.1.1, but with much more tedious derivations. The procedure is explained in this section. The representation

$$\begin{aligned} a_1(t) &= A\cos\sigma t + \overline{A}\sin\sigma t + o(A, \overline{A}),\\ b_1(t) &= \overline{B}\cos\sigma t + B\sin\sigma t + o(B, \overline{B}),\\ &\qquad A, \overline{A}, B, \overline{B} \sim \varepsilon^{1/3} \end{aligned} \tag{9.21}$$

is adopted and introduced into eqs. (9.13)–(9.15) to find the $O(\varepsilon^{2/3})$-modal functions associated with the constant and $\cos(2\sigma t)$ Fourier terms. The next step is to use the second-order functions a_2, b_2, and c_1 in eqs. (9.11), (9.12), and (9.16)–(9.19) to detect the $O(\varepsilon)$ quantities associated with the Fourier terms $\cos\sigma t$, $\sin\sigma t$, $\cos(3\sigma t)$, and $\sin(3\sigma t)$.

The coupled algebraic equations that determine the dominant amplitude parameters A, $\overline{A}$, B, and $\overline{B}$ follow by gathering all the quantities at $\cos\sigma t$ and $\sin\sigma t$ in eqs. (9.11) and (9.12). Because each harmonic term must be zero, requiring that the terms oscillating with $\cos\sigma t$ and $\sin\sigma t$ are zero gives

$$\left.\begin{aligned} &A\left((\overline{\sigma}_{1,0}^2 - 1) + m_1(A^2 + \overline{A}^2) + m_2\overline{B}^2 + m_3B^2\right)\\ &\quad + (m_2 - m_3)\overline{A}\,\overline{B}B - P_{1,0}\eta_{1a}/L_1 = 0,\\ &\overline{A}\left((\overline{\sigma}_{1,0}^2 - 1) + m_1(A^2 + \overline{A}^2) + m_2B^2 + m_3\overline{B}^2\right)\\ &\quad + (m_2 - m_3)AB\overline{B} = 0,\\ &\overline{B}\left((\overline{\sigma}_{0,1}^2 - 1) + \overline{m}_1(B^2 + \overline{B}^2) + \overline{m}_2A^2 + \overline{m}_3\overline{A}^2\right)\\ &\quad + (\overline{m}_2 - \overline{m}_3)\overline{A}AB - P_{0,1}\eta_{2a}/L_1 = 0,\\ &B\left((\overline{\sigma}_{0,1}^2 - 1) + \overline{m}_1(B^2 + \overline{B}^2) + \overline{m}_2\overline{A}^2 + \overline{m}_3A^2\right)\\ &\quad + (\overline{m}_2 - \overline{m}_3)\overline{A}A\overline{B} = 0, \end{aligned}\right\} \tag{9.22}$$

where the coefficients m_i and $\overline{m}_i$ depend on the hydrodynamic coefficients in modal equations (9.11)–(9.15) and the σ-scaled natural sloshing frequencies, $\overline{\sigma}_{i,j} = \sigma_{i,j}/\sigma$ (see Faltinsen *et al.*, 2003, for detailed expressions of m_i and $\overline{m}_i$). Because both the hydrodynamic coefficients and

$\sigma_{i,j}$ are functions of $\overline{h}$ and R_L, the coefficients m_i and $\overline{m}_i$ depend on $\overline{h}$, R_L, and σ. When introducing

$$\overline{\sigma}_1 = \sigma_{1,0}/\sigma, \tag{9.23}$$

we get

$$m_i = m_i(\overline{h}, R_L, \overline{\sigma}_1), \quad \overline{m}_i = \overline{m}_i(\overline{h}, R_L, \overline{\sigma}_1). \tag{9.24}$$

In the particular case of a square-base tank with $R_L = 1$, eq. (9.24) gives

$$R_L = 1 \Rightarrow \sigma_{i,j} = \sigma_{j,i} \text{ and } m_i = \overline{m}_i = m_i(\overline{h}, \overline{\sigma}_1). \tag{9.25}$$

The system of nonlinear algebraic equations (9.22) generalizes the single algebraic equation in A for planar sloshing in Section 8.3.1. The coefficient m_1 is the same for both cases.

The asymptotic steady-state solution correct to $O(\varepsilon)$ is presented by Faltinsen *et al.* (2003). To study the stability of the periodic asymptotic solution we can follow the procedure described for two-dimensional sloshing in Section 8.3.1.1. The scheme introduces the slowly varying time parameter $\tau = \frac{1}{2}\varepsilon^{2/3}\sigma t$ and expresses the infinitesimally perturbed dominant solutions as

$$a_1 = (A + \alpha(\tau))\cos\sigma t + (\overline{A} + \overline{\alpha}(\tau))\sin\sigma t + o(\varepsilon^{1/3}),$$
$$b_1 = (\overline{B} + \overline{\beta}(\tau))\cos\sigma t + (B + \beta(\tau))\sin\sigma t + o(\varepsilon^{1/3}), \tag{9.26}$$

where A, $\overline{A}$, B, and $\overline{B}$ are the solutions of eqs. (9.22) and $\alpha(\tau)$, $\overline{\alpha}(\tau)$, $\beta(\tau)$, and $\overline{\beta}(\tau)$ are the perturbations of the steady-state solutions. Inserting eqs. (9.26) into the original modal system (9.11)–(9.15), gathering the terms of lowest asymptotic order, and keeping linear terms in α, $\overline{\alpha}$, β, and $\overline{\beta}$ leads to the following linear system of ordinary differential equations:

$$\frac{d\mathbf{c}}{d\tau} + \mathbf{C}\mathbf{c} = 0, \tag{9.27}$$

where the matrix $\mathbf{C}$ has dimension 4×4 (in the two-dimensional case the dimension is 2×2). The solution of eq. (9.27) has the time dependence $\exp(\eta\tau)$, where η represents solutions of the eigenvalue problem $\det[\eta\mathbf{E} + \mathbf{C}] = 0$ and $\mathbf{E}$ is the identity matrix. The derivations by Faltinsen *et al.* (2003) gave the following characteristic polynomial:

$$\eta^4 + C_1\eta^2 + C_0 = 0, \tag{9.28}$$

where C_0 is the determinant of $\mathbf{C}$ and C_1 is a complicated function of the elements of $\mathbf{C}$.

As in Sections 8.2.1.2 and 8.3.1.1 we say that a steady-state solution is stable when small perturbations of the steady-state solution remain bounded at any time. Instability means that at least one eigenvalue η has a positive real part. Solutions of eq. (9.28) are expressed as $\pm\sqrt{x_{1,2}}$, where $x_{1,2}$ are the two roots of the equation $x^2 + C_1x + C_0 = 0$. The presence of both positive and negative signs means that instability occurs when $\sqrt{x_{1,2}}$ has a nonzero real part. Stability requires $\sqrt{x_{1,2}}$ to be purely imaginary (i.e., x_1 and x_2 are negative real numbers). To find the conditions for C_0 and C_1, providing $x_{1,2} = \frac{1}{2}(-C_1 \pm \sqrt{C_1^2 - 4C_0}) < 0$, we note that $x_1x_2 = C_0$, which means that C_0 must be positive. Furthermore, it is necessary that the discriminant $C_1^2 - 4C_0$ is positive to avoid complex roots. Then if we also require C_1 to be positive, we see that x_1 and x_2 are negative real numbers. Therefore, the conditions for stable steady-state solutions are

$$C_0 > 0, \quad C_1 > 0, \quad C_1^2 - 4C_0 > 0. \tag{9.29}$$

The change in the signs of C_0, C_1, or $C_1^2 - 4C_0$ indicates a possible alteration of the stability property of the considered solution. This alteration indicated a local bifurcation in Section 9.1.2. Therefore, we should study the sign changes of C_0, C_1, or $C_1^2 - 4C_0$ to detect possible bifurcation points.

9.2.4 Classification of steady-state regimes for a square-base tank with longitudinal and diagonal excitations

The algebraic equation system (9.22) may have no solution or multiple solutions, depending on the values of m_i and $\overline{m}_i$. In this section, we focus on a square-base tank ($R_L = 1$) whose conditions satisfy eq. (9.25). The excitation is either *longitudinal* along the x-axis ($P_1^\varepsilon = P_{1,0}\eta_{1a}/L_1 \neq 0$, $P_2^\varepsilon = P_{0,1}\eta_{2a}/L_1 = 0$) or *diagonal* ($P_1^\varepsilon = P_2^\varepsilon \neq 0$). The analysis is based on Faltinsen *et al.* (2003, 2005a). When $\sigma \approx \sigma_1 = \sigma_{0,1} = \sigma_{1,0}$ (i.e., $\overline{\sigma}_1 \approx 1$), $m_i = \overline{m}_i = m_i(\overline{h}, \overline{\sigma}_1)$, with $i = 1, 2, 3$, can be approximated as $m_i^{(0)}(\overline{h}) = m_i(\overline{h}, 1)$. Figure 9.7 shows $m_i^{(0)}$ and some of their linear combinations versus $\overline{h}$. Critical depths are denoted as h_i^* in the figure, and points H_i in the graphs. We note that values of $m_i^{(0)}$ are approximately constant for $1 < \overline{h}$ and vary slowly with $\overline{h}$ in the range $1 > \overline{h} > h_1^*$, where $h_1^* = 0.3368\ldots$ is associated with the change from

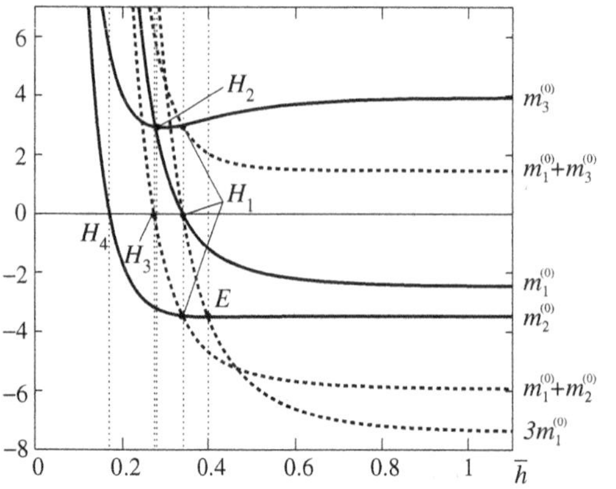

Figure 9.7. Graphs representing $m_i(\overline{h}, 1) = m_i^{(0)}(\overline{h})$, $i = 1, 2, 3$, and some of their combinations versus the ratio of liquid depth to breadth, $\overline{h}$, for a square-base tank. The point $H_1(m_1^{(0)} = 0, h_1{}^* = 0.3368\ldots)$ denotes the change from soft- to hard-spring behavior for "planar" steady-state waves. The point H_2 ($m_1^{(0)} = m_3^{(0)}$) defines another critical depth $h_2{}^* = 0.274\ldots$, where the solvability condition for the swirling mode is not satisfied. The point H_3 denotes the critical depth $h_3^* = 0.27\ldots$ with the soft/hard spring changes in response of a nearly diagonal mode, and the point H_4 denotes the critical depth $h_4^* = 0.17\ldots$, where $m_2^{(0)} = 0$. In addition, the auxiliary point E corresponds to the critical depth $h_5^* = 0.4\ldots$, obtained from the condition $m_2^{(0)} = 3m_1^{(0)}$.

a soft- to a hard-spring type of "planar" wave response. This critical depth and the meaning of soft and hard springs are extensively discussed in Sections 8.2.2, 8.3.1.1, and 8.5. The lower critical depths $h_2^* = 0.274\ldots$(point H_2, $m_1^{(0)} = m_3^{(0)}$), $h_3^* = 0.27\ldots$(point H_3, $m_1^{(0)} + m_2^{(0)} = 0$) and $h_4^* = 0.17\ldots$(point H_4, $m_2^{(0)} = 0$) are discussed later in the text.

9.2.4.1 Longitudinal excitation

Let us first show that longitudinal excitation with $P_1^\varepsilon = P_{1,0}\eta_{1a}/L_1 \neq 0$ and $P_2^\varepsilon = P_{0,1}\eta_{2a}/L_1 = 0$ implies $\overline{A} = 0$ and $A \neq 0$. For this purpose, we multiply the first and second of eqs. (9.22) by $\overline{A}$ and A, respectively, and the third and fourth equations by B and $\overline{B}$, respectively. By combining the difference between the first and second result with the difference between the third and fourth result, it follows that

$$P_1^\varepsilon \overline{A} = (m_2 - m_3)\left[A\overline{A}(\overline{B}^2 - B^2) + B\overline{B}(\overline{A}^2 - A^2)\right] = 0. \tag{9.30}$$

As long as $m_2 - m_3 = m_2^{(0)} - m_3^{(0)} + O(\overline{\sigma}_1^2 - 1) \neq 0$ (the graphs in Figure 9.7 confirm that $m_2^{(0)} - m_3^{(0)} \neq 0$), eq. (9.30) with $\sigma \approx \sigma_1$ has meaning only when $\overline{A} = 0$. Substituting $\overline{A} = 0$ into eq. (9.30) and the first of eqs. (9.22) gives the following solvability condition of system (9.22):

$$\overline{A} = 0, \quad A \neq 0, \quad B\overline{B} = 0. \tag{9.31}$$

This solvability condition leads to three possible cases when the system has a real solution. These yield the corresponding three types of the steady-state waves, eqs. (9.1)–(9.3). Swirling and nearly diagonal motions imply three-dimensional wave motions. The ± sign of the amplitude parameter B in the swirling expression refers to clockwise or counterclockwise "rotation." The same sign at $\overline{B}$ implies the possibility that the waves can occur approximately along either of the two diagonals of the tank. The initial conditions and the transient phase determine if plus or minus is realized for these regimes.

Let us first consider the *planar* two-dimensional wave regime and study its stability ranges for different depths. For this case, system (9.22) transforms to the following single equation:

$$A\left((\overline{\sigma}_1^2 - 1) + m_1 A^2\right) = P_1^\varepsilon. \tag{9.32}$$

It is the same algebraic equation as eq. (8.57) for sloshing in a two-dimensional rectangular tank. Equation (9.32) defines two branches in the $(\sigma/\sigma_1, |A|)$-plane, which have "soft-" and "hard"-spring behavior for $\overline{h} > h_1^*$ and $\overline{h} < h_1^*$, respectively. Stable and unstable solution branches when the horizontal cross-section is far from square-shaped (i.e., R_L is large or small and the flow is two-dimensional) are illustrated in Figure 8.7. Unless $\overline{h}$ is close to h_1^*, the planar waves are stable in the vicinity of $\sigma/\sigma_1 = 1$. Let us see what happens to the stability of this regime for a square-base tank.

We start with soft-spring response curves (two branches, P_1P and P_2P, meeting each other at infinity) as demonstrated in Figure 9.8(a) for $\overline{h} = 0.508$. When the excitation is along the Ox-axis, the unstable planar waves are associated with points on the two branches that are located between the two auxiliary curves γ_{m_1} : $\overline{\sigma}_1^2 - 1 + m_1 A^2 = 0$ (backbone) and γ_{3m_1} : $\overline{\sigma}_1^2 - 1 + 3m_1 A^2 = 0$. These curves coincide with those given by eq. (8.65). The curve γ_{3m_1} intersects the

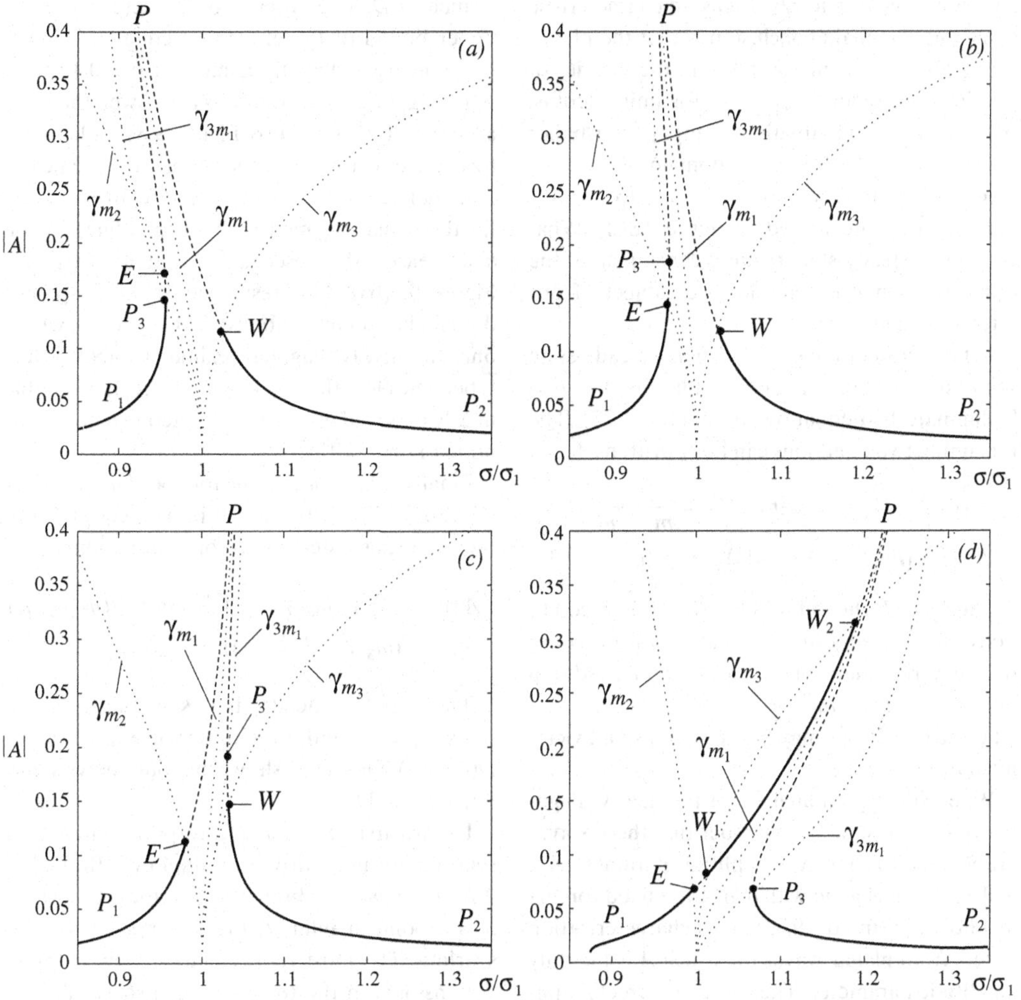

Figure 9.8. Steady-state "planar" wave amplitude response (σ/σ_1, $|A|$) for different ratios of liquid depth to breadth, $\bar{h}$, of a square-base tank. Horizontal excitation along the Ox-axis with $\eta_{1a}/L_1 = \varepsilon = 0.0078$. Bold solid lines denote stable steady-state planar sloshing: (a) $\bar{h} = 0.508$, (b) $\bar{h} = 0.39$, (c) $h_2^* < \bar{h} < h_1^*$ (calculations were made for $\bar{h} = 0.3$), and (d) $h_4^* < \bar{h} < h_2^*$ (calculations were made for $\bar{h} = 0.2$).

branch P_1P at the point P_3 (the so-called turning point T in Figure 8.7). Now, we operate with stability condition (9.29), which accounts for three-dimensional perturbations.

Then we find analytically an additional zone of instability for the points on branches P_1P and P_2P that are situated between the two other auxiliary curves $\gamma_{m_2} : \bar{\sigma}_1^2 - 1 + m_2A^2 = 0$ and $\gamma_{m_3} : \bar{\sigma}_1^2 - 1 + m_3A^2 = 0$. Figure 9.7 shows that $m_3^{(0)} > 0$ for all depths, whereas $m_2^{(0)}$ changes sign at $h_4^* = 0.17\ldots$. Because $m_i \approx m_i^{(0)}$ as $\sigma \approx \sigma_1$, the curve γ_{m_3} always demonstrates hard-spring behavior and intersects the branch P_2P at point W for the case in Figure 9.8(a)–(c). For $\bar{h} > h_4^*$ the curve γ_{m_2} has a soft-spring character. In addition, it is situated between the curves γ_{m_1} and γ_{3m_1} for $\bar{h} > h_5^*$ and lies under γ_{3m_1} for $\bar{h} < h_5^*$, where h_5^* is associated with the condition $3m_1^{(0)} = m_2^{(0)}$ (point E in Figure 9.7). Therefore, for $\bar{h} > h_5^*$ the stable planar steady regimes are associated with the subbranches P_1P_3 and WP_2 in Figure 9.8(a). However, when $h_4^* < \bar{h} < h_5^*$, E is situated to the left of P_3 and the left stable subbranch is then P_1E instead of P_1P_3. This condition is demonstrated in Figure 9.8(b) and (c). Figure 9.8(b)–(d) presents the "planar" response for different liquid depths and shows the previously discussed characteristic behavior. The main conclusion is that the planar waves always become unstable in the vicinity of the linear resonance $\sigma/\sigma_1 = 1$ for finite depths, (i.e., $\overline{h} > h_2^*$). The situation changes for smaller depths, where the frequency domain of unstable planar waves is very narrow relative to that for finite depth as illustrated in Figure 9.8(d). What new stable steady-state regimes can appear at the primary resonance zone are the subject of our forthcoming analysis.

Let us consider the *nearly diagonal* steady-state solution (9.3). For this case, conditions $\overline{A} = B = 0$ eliminate two equations of system (9.22). The remaining two equations can be rewritten as

$$A\left[(\overline{\sigma}_1{}^2 - 1) + (m_1 + m_2)A^2\right] = \frac{m_1}{m_1 - m_2} P_1{}^{\varepsilon},$$
$$\overline{B}^2 = -\left(m_2 A^2 + (\overline{\sigma}_1{}^2 - 1)\right)/m_1 \geq 0. \qquad (9.33)$$

They need the additional solvability conditions $m_1 \neq m_2$ and $m_1 \neq 0$. Taking into account the numerical data in Figure 9.7 and condition $m_2 - m_3 = m_2^{(0)} - m_3^{(0)} + O(\overline{\sigma}_1^2 - 1)$, we see that all these conditions may fail only in a small vicinity of h_1^*.

When starting an analysis of the nearly diagonal waves, we should remember that these waves are characterized by two amplitude parameters, A and $\overline{B}$, which change with σ/σ_1 for a fixed forcing amplitude P_1^{ε} in eq. (9.33). This characterization differs from planar waves for which A is the only amplitude parameter. Therefore, a correct graphical representation of the response curves should be in the $(\sigma/\sigma_1, |A|, |\overline{B}|)$-space. Figure 9.9 illustrates typical response curves for nearly diagonal regimes in the $(\sigma/\sigma_1, |A|, |\overline{B}|)$-space as well as projections of these curves on the $(\sigma/\sigma_1, |A|)$-plane. The bold solid line denotes a stable nearly diagonal regime. Projection of this line on the σ/σ_1-axis detects the frequency domain in which the steady-state regime can be realized. Figure 9.9 shows that nearly diagonal waves correspond to points on the two branches D_1D and DE (where D is a point at infinity). Point E is the bifurcation point, an intersection between γ_{m_2} and the curve P_1P responsible for the planar waves. We described this class of bifurcation points in Section 9.1.2. Figure 9.9(a) presents the position of these two branches for a fairly deep liquid. Branch DE implies unstable nearly diagonal waves. A "stable" subbranch U_1D belongs to the branch D_1D. The abscissa of U_1 determines the upper bound of the effective frequency domain of the nearly diagonal regime. For $\overline{h} > 0.4$ (point E in Figure 9.7) this abscissa is lower than the abscissa of P_3 (see also Figure 9.8) and, therefore, the effective frequency domain of the nearly diagonal regime is inside the effective domain for the planar regime (nonbold solid line). When $\overline{h}$ decreases, the abscissa of U_1 tends to 1 (see Figure 9.9(b)). This result yields the frequency domain between the abscissas of E and U_1, where only the nearly diagonal regime is stable. When $\overline{h}$ becomes less than h_3^* (Figure 9.9(c)), the stable nearly diagonal waves are detected in a neighborhood of $\sigma/\sigma_1 = 1$.

Finally, the conditions for *swirling* to exist ($\overline{A} = \overline{B} = 0$, $A \neq 0$, $B \neq 0$) reduce eq. (9.22) to the following system of algebraic equations:

$$A\left[(\overline{\sigma}_1^2 - 1) + (m_1 + m_3)A^2\right] = m_1 P_1^{\varepsilon}/(m_1 - m_3),$$
$$B^2 = -\left(m_3 A^2 + (\overline{\sigma}_1^2 - 1)\right)/m_1 \geq 0. \qquad (9.34)$$

This system needs the solvability conditions $m_1 \neq m_3$ and $m_1 \neq 0$. Because $m_i = m_i^{(0)} + O(\overline{\sigma}_1^2 - 1)$, Figure 9.7 shows that our analysis may fail at h_1^* and h_2^*.

Let us first see what the response curves for swirling are in a fairly deep liquid depth. Figure 9.10(a) shows two branches S_1S and SW (where S is a point at infinity) that are responsible for swirling. The most representative view is three-dimensional in the $(\sigma/\sigma_1, |A|, |B|)$-space. Projection on the $(\sigma/\sigma_1, |A|)$-plane is used to identify the forcing frequencies for which a stable swirling regime exists. Appearance of the branch SW is associated with the bifurcation point W formed by intersection of γ_{m_3} and the branch PP_2 corresponding to the planar regime (see Figure 9.8). Branch SW is unstable. The effective frequency domain of swirling is determined by the abscissa of the point V_1. The abscissa is less than 1 for $\overline{h} > h_1^*$, which means we have a frequency domain between this abscissa and the abscissa of the point W where swirling is the only stable steady-state regime. Figure 9.10(b)–(d) illustrates how swirling-responsible curves change with decreasing $\overline{h}$. In particular we see that the abscissa of V_1 increases with decreasing nondimensional liquid depth. Therefore, the effective frequency domain for swirling moves upward away from the main resonance.

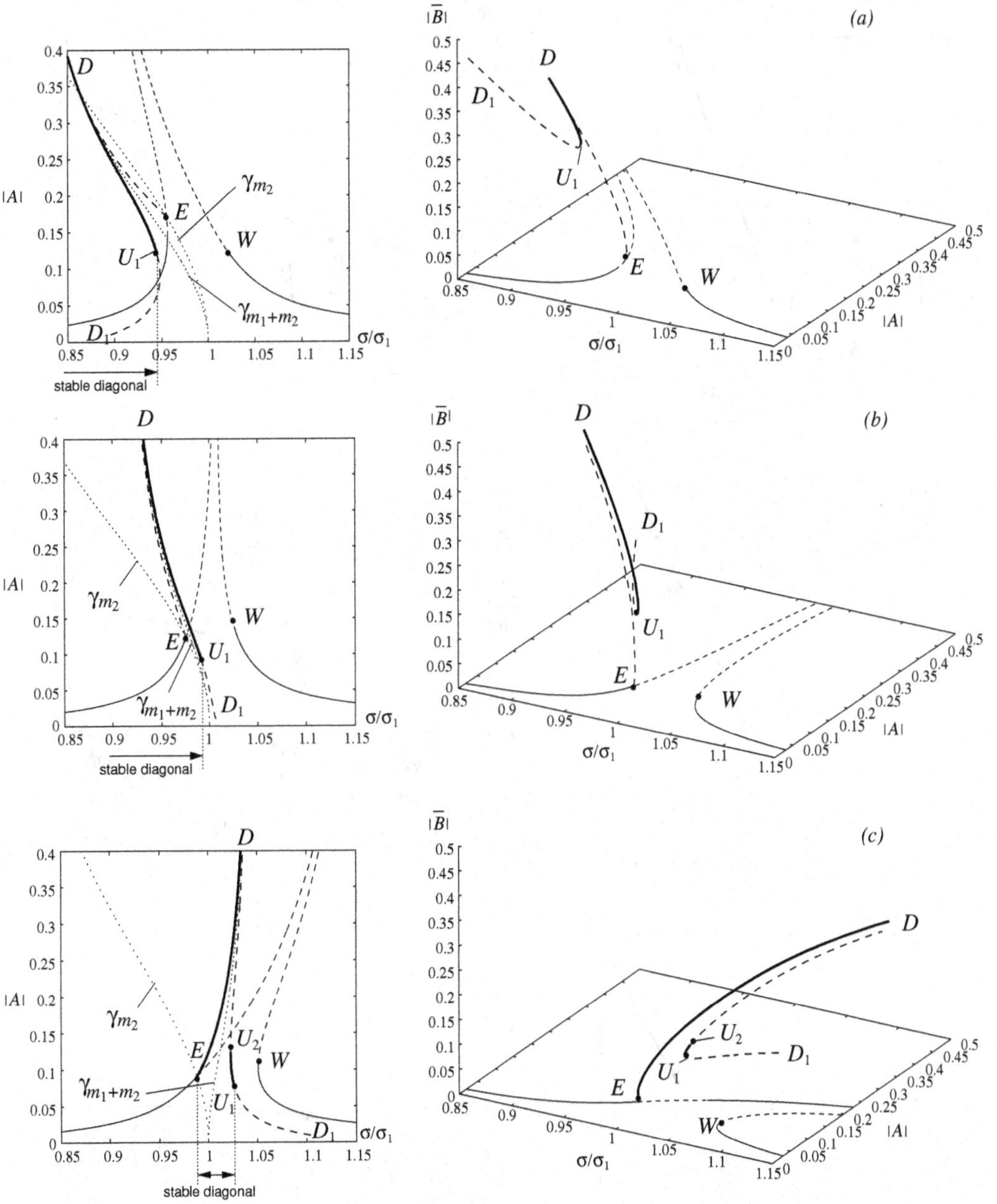

Figure 9.9. Steady-state wave amplitude components $|A|$ and $|\overline{B}|$ of nearly diagonal regimes versus σ/σ_1 for different ratios of liquid depth to breadth, $\overline{h}$, of a square-base tank. Horizontal excitation is along the Ox-axis with $\eta_{1a}/L_1 = \varepsilon = 0.0078$. Solid thin lines denote stable planar waves, whereas bold solid lines imply stable nearly diagonal wave regimes. Dashed lines correspond to unstable regimes. (a) $\overline{h} = 0.508$, (b) $\overline{h} = 0.33$, (c) and $\overline{h} = 0.25$.

The change with $\overline{h}$ of the frequency domains for the stable steady-state solutions is summarized in Figure 9.11 for four different forcing amplitudes. The arrows indicate the region of stable steady-state motion (the corresponding steady regimes disappear or become unstable in the direction of the arrows).

The figure shows that the region of stable planar waves is always away from the primary resonance $\sigma/\sigma_1 = 1$. The region where planar

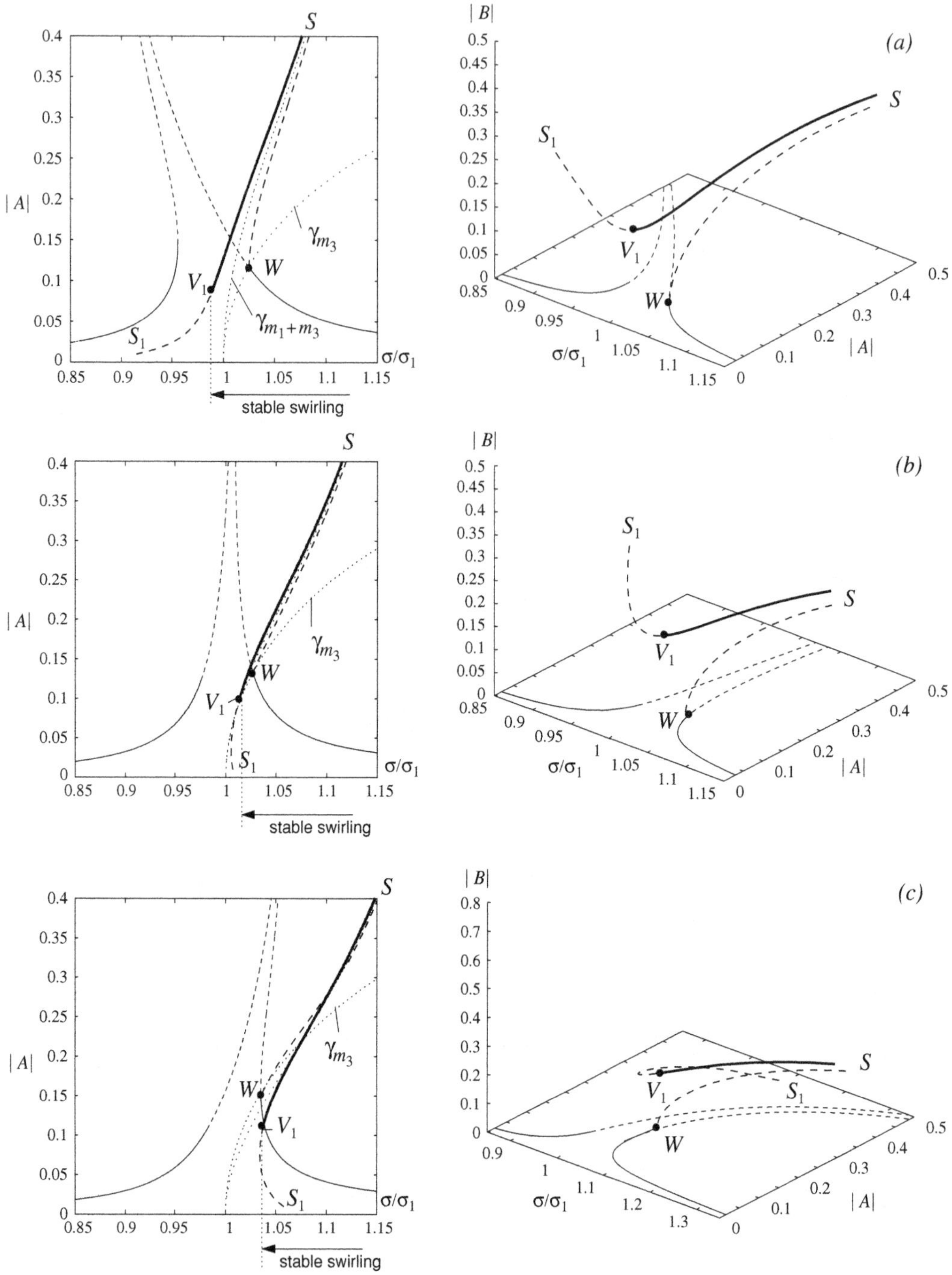

Figure 9.10. Steady-state wave amplitude components $|A|$ and $|B|$ of swirling versus σ/σ_1 for different ratios of liquid depth to breadth, $\bar{h}$, of a square-base tank. Horizontal excitation is along the Ox-axis with $\eta_{1a}/L_1 = \varepsilon = 0.0078$. Solid thin lines denote stable planar waves, whereas bold solid lines imply stable swirling regimes. Dashed lines correspond to unstable regimes. (a) $\bar{h} = 0.508$, (b) $\bar{h} = 0.33$, (c) and $\bar{h} = 0.29$.

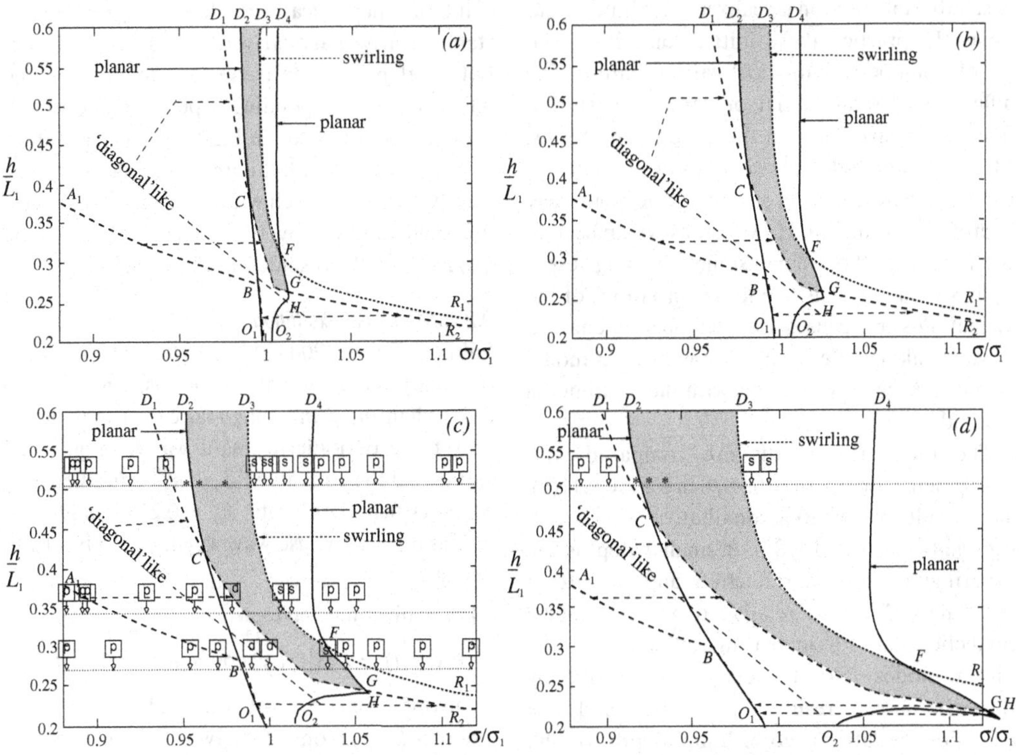

Figure 9.11. Theoretical frequency domains of stable resonant steady-state motion as a function of the ratio of liquid depth to breadth, $\overline{h}$, versus σ/σ_1 for different forcing amplitudes of a square-base tank. Longitudinal excitations with amplitudes (a) $\eta_{1a}/L_1 = 0.001$, (b) $\eta_{1a}/L_1 = 0.0025$, (c) $\eta_{1a}/L_1 = 0.0078$, and (d) $\eta_{1a}/L_1 = 0.025$. No stable steady-state waves exist and irregular chaotic motions occur in the shaded area. Comparisons with experimental observations are made in (c) for the liquid depths $\overline{h} = 0.508$, 0.34, and 0.27 and in (d) for $\overline{h} = 0.508$. Experimental steady-state regimes are denoted "p" for planar waves, "s" for swirling, "d" for nearly diagonal, and * for chaos.

waves are unstable is denoted $D_2D_4FGHO_2O_1$. This region becomes wider with increasing forcing amplitude. A two-dimensional analysis does not give this instability region, which confirms it is caused by three-dimensional wave perturbations. Geometrically, $D_2D_4FGHO_2O_1$ falls into three subregions. The first, O_1CGHO_2, corresponds to an unstable planar regime, but nearly diagonal waves are stable. This region appears only for smaller depths and is absent for fairly deep fillings. The second region, FD_4D_3, corresponds to the case when stable swirling exists, but planar waves are unstable (the region disappears at $\sigma/\sigma_1 = 1$ for small depths). No stable steady-state solutions exist and chaotic motion is possible for the region D_2D_3FHC, which disappears for small depths.

Away from the region $D_2D_4FGHO_2O_1$, the planar waves are stable and may coexist with stable swirling (region R_1FD_4) or nearly diagonal waves (region A_1BCD_1). No regions exist where stable swirling coexists with stable nearly diagonal waves.

Swirling is stable right up to the border D_3FR_1. The region of stable swirling is away from $\sigma/\sigma_1 = 1$ for smaller $\overline{h}$, while the effective domain of stable nearly diagonal waves, $A_1BO_1O_2HCD_1$, drifts left of the main resonance with increasing $\overline{h}$. The nearly diagonal waves coexist with planar waves when $\overline{h}$ is larger than the ordinate of C. Even though the quantitative applicability of the theory is disputable for small liquid depths, Figure 9.11 indicates that chaos and swirling do not occur at the primary resonance. However, the primary resonance conditions for shallow liquid depths cause planar or nearly-diagonal waves as in simulations by Wu *et al.* (1998). Theoretically, the initial conditions determine what kind of steady-state motion is realized after initial transients. In practice these initial conditions consist of small perturbations of various nature. The influence of small initial or random perturbations was unavoidable in the experimental tests. Ideally, the time-domain simulations with damping coefficients and zero initial conditions may address this problem. However, as we analyze in Section 9.2.6, three-dimensional sloshing, especially swirling, is sensitive to perturbations of the system. Possible causes are local phenomena and wave breaking. We do not know how to model effects of such perturbations with the multimodal method.

The theoretical results are compared with experiments for $\eta_{1a}/L_1 = 0.0078$ in Figure 9.11(c). The results show good qualitative agreement, especially for $\overline{h} = 0.508$. For smaller depths the experiments show a reduction and a shift of the "chaotic" region relative to the theoretical prediction due to increasing influence of the higher modes. This tendency is illustrated by the experimental results for $\overline{h} = 0.34$ and $\overline{h} = 0.27$. For the latter ratio of liquid depth to tank length, the experiments observe swirling waves in region D_2CHGFD_3. No chaotic waves are detected, as predicted by the theory. Analyzing the larger discrepancy for smaller depths, we should remember that the damping and amplification of higher modes are important in this case, as discussed in Sections 8.3.4 and 8.6.3 for two-dimensional sloshing. Furthermore, the qualitative agreement with experiments on the classification of the steady-state regimes in Figure 9.11 does not guarantee quantitative agreement by using the Moiseev-based theory for smaller depth and larger forcing amplitudes. Improvements of the asymptotic modal theory to obtain quantitative agreement could be achieved through modifications of the asymptotic ordering to account for nonlinear interaction of some of the higher modes. This approach is also explicitly confirmed both by our direct numerical simulations and by experiments (see Section 9.2.6), which documented significant contribution from nonprimary modes.

Because the effective frequency domain of nearly diagonal solutions always overlaps that for planar waves for $\overline{h} = 0.508$, Faltinsen *et al.* (2005a) were strongly motivated to find solutions in observations for lower depths in experimental cases with $\overline{h} = 0.34$ and 0.27. For these solutions, theoretically, the frequency domain for stable nearly diagonal waves is small and does not overlap with the frequency domain for planar waves. Corresponding experimental observations are denoted "d" in Figure 9.11(c) and show good agreement with theoretical predictions. The theory also agrees well with experimental data for $\overline{h} = 0.508$ and relatively large forcing amplitude $\eta_{1a}/L_1 = 0.025$ presented in Figure 9.11(d).

9.2.4.2 Diagonal excitation

Faltinsen *et al.* (2003) have extended the previous analysis to diagonal excitations. They showed that, when $m_1 \neq m_3$, diagonal excitations define only three periodic resonant waves. (Figure 9.7 shows that the equality $m_1^{(0)} = m_3^{(0)}$ is fulfilled only at the critical depth ratio $h_2^* = 0.274\ldots$ and $m_i \approx m_i^{(0)}$ as $\sigma \approx \sigma_1$.) These wave regimes can be classified as

(i) *pure* diagonal waves,

$$\xi(x, y, t) = B_1 S_2^1(x, y) \cos \sigma t + o(\varepsilon^{1/3}), \quad (9.35)$$

occurring for $B = -\overline{A} = 0$ and $A = \overline{B} = B_1$, where $S_2^1(x, y)$ is one of the two "diagonal" modes defined by eq. (4.34) involving the pairs of the lowest-order ($i = 1$) Stokes waves (and the other diagonal mode associated with $i = 1$ is denoted $S_2^1(x, y)$);

(ii) *nearly* diagonal waves,

$$f = \left[\pm A_1 S_1^1(x, y) + B_1 S_2^1(x, y)\right] \cos \sigma t + o(\varepsilon^{1/3}), \quad (9.36)$$

which describe an in-phase amplification of the pair S_1^1 and S_2^1 (where $B_1 = \frac{1}{2}(A + \overline{B}), A_1 = \frac{1}{2}(A - \overline{B})$) occurring for $B = -\overline{A} = 0$ and $A \neq \overline{B}$; and, finally,

(iii) swirling,

$$f = B_1 S_2^1(x, y) \cos \sigma t \mp A_1 S_1^1(x, y) \sin \sigma t + o(\varepsilon^{1/3}), \quad (9.37)$$

which occurs for $A_1 = \overline{A} = -B \neq 0$ and $B_1 = A = \overline{B}$.

Because the analysis is mathematically equivalent to the previous asymptotic scheme, the detailed derivation is omitted. Response curves in the $(\sigma/\sigma_1, |B_1|)$-plane are presented in Figure 9.12 for two different liquid depths. A main conclusion is that the nearly diagonal regimes are always unstable for finite depths. This conclusion follows by observing the branches L_1LW_1. Furthermore, on the left and right of the primary resonance at $\sigma/\sigma_1 \approx 1$, two zones exist in

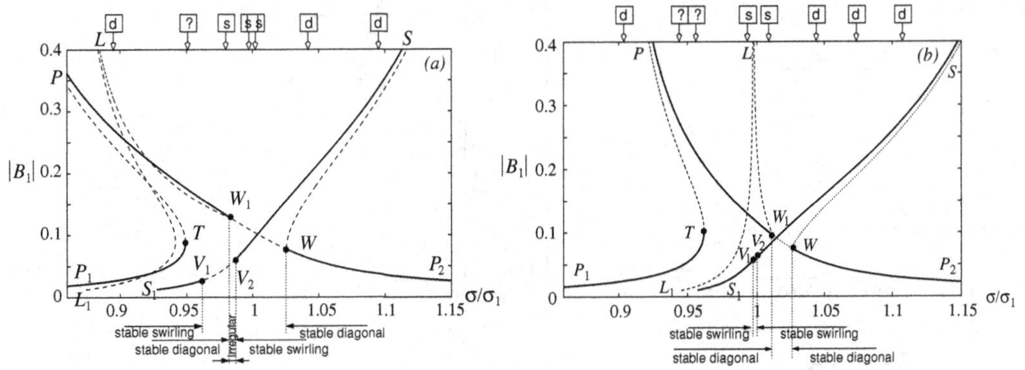

Figure 9.12. Steady-state wave amplitude response B_1 versus σ/σ_1 for diagonal forcing with $\sqrt{\eta_{1a}^2 + \eta_{2a}^2}/L_1 = \varepsilon = 0.0078$ of a square-base tank: (a) $\bar{h} = 0.508$ and (b) $\bar{h} = 0.34$; P_1PP_2 indicates pure-diagonal waves, L_1LW_1 indicates nearly diagonal waves, and S_1SW indicates swirling. Bold solid lines denote stable wave motions. Theoretical results and experimental classification of the types of wave motions are presented on the bottom and the top, respectively; d, stable diagonal waves; s, stable swirling waves; ?, not clear whether steady-state waves or irregular waves are present (Faltinsen *et al.*, 2003).

which stable swirling coexists with stable diagonal waves. No theoretical chaos (irregular waves) is predicted for the considered case in Figure 9.12(b) with $\bar{h} = 0.34$. However, as we see in panel (a), increasing $\bar{h}$ to 0.508 leads to the appearance of a narrow frequency domain (between abscissas W_1 and V_2), in which irregular wave motions (chaos) are theoretically possible. The probability of chaos increases by further increasing $\bar{h}$ and the forcing amplitude.

Comparisons with experiments by Faltinsen *et al.* (2003) show generally good agreement. However, experimental results were found with clearly non-steady-state waves, denoted by a question mark in Figure 9.12. This finding contradicts the results from Moiseev-type theory, which predicts for the tested frequencies, either diagonal or swirling stable waves, depending on the transient scenario. The results are explained by considering the influence of higher modes in Section 9.2.6.3.

The theoretical instability domains are summarized in Figure 9.13. The instability of diagonal waves and swirling is between the solid and dashed lines, respectively. The instability zones become narrower with lower forcing amplitudes (note the abscissas in parts (a)–(d)). Both regimes are simultaneously unstable only when the zones overlap each other, which is only possible for smaller depths (and we excluded that from the analysis in Figure 9.13, because the small depths need another theory) and depends on the forcing amplitudes for $\bar{h} \gtrsim 0.5$. Another interesting point is that two different stable steady regimes (diagonal and swirling) coexist for $\bar{h} > 0.27$ in the vicinity of the main resonance $\sigma/\sigma_1 = 1$. Depending on the transient and initial perturbation scenarios, either diagonal or swirling waves are excited.

9.2.5 Longitudinal excitation of a nearly square-base tank

Based on the asymptotic technique from the previous sections, Faltinsen *et al.* (2006a) studied steady-state wave regimes in a nearly square-base tank due to *longitudinal* forcing along the Ox-axis. They found the same three types of waves, (i.e., eqs. (9.1)–(9.3)), but the stability ranges of these wave regimes are functions of $R_L = L_1/L_2$, where L_1 and L_2 are the tank length in the x-direction and the tank breadth in the y-direction, respectively. The small parameter μ is introduced. The definition and ordering of μ is

$$R_L = 1 + \mu;\quad 1/R_L = 1 - \mu + o(\mu),\quad \mu = O(\varepsilon^{2/3}). \tag{9.38}$$

By introducing this parameter, the analysis needs Moiseev ordering for the frequencies:

$$\bar{\sigma}_{1,0}^2 - 1 \sim \bar{\sigma}_{0,1}^2 - 1 = O(\varepsilon^{2/3}) \quad \text{or} \\ \sigma = \sigma_{1,0} + O(\varepsilon^{2/3}) = \sigma_{0,1} + O(\varepsilon^{2/3}). \tag{9.39}$$

As a consequence,

$$\bar{\sigma}_{1,0}^2 = \bar{\sigma}_{0,1}^2 - \mu\sigma_0 + o(\mu);\quad \sigma_0 = 1 + \frac{2\pi\bar{h}}{\sinh(2\pi\bar{h})}. \tag{9.40}$$

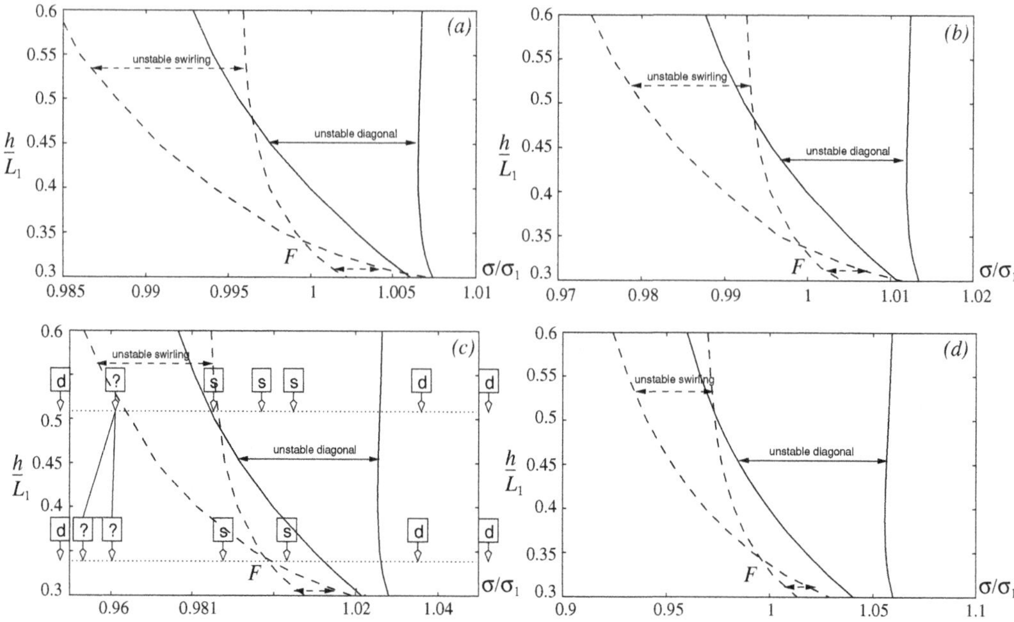

Figure 9.13. Theoretical frequency domains of stable resonant steady-state motions caused by resonant diagonal excitations and presented in the $(\sigma/\sigma_1, \overline{h})$-plane for different forcing amplitudes of a square-base tank: (a) $\sqrt{\eta_{1a}^2+\eta_{2a}^2}/L_1=\varepsilon=0.001$, (b) $\sqrt{\eta_{1a}^2+\eta_{2a}^2}/L_1=\varepsilon=0.0025$, (c) $\sqrt{\eta_{1a}^2+\eta_{2a}^2}/L_1=\varepsilon=0.0078$, and (d) $\sqrt{\eta_{1a}^2+\eta_{2a}^2}/L_1=\varepsilon=0.025$. The instability of diagonal waves and swirling is expected between solid and dashed lines, respectively. Comparisons with experimental data from Figure 9.12 are made in part (c) for the depths $\overline{h}=0.508$ and $\overline{h}=0.34$; d, stable diagonal waves; s, stable swirling waves; ?, not clear whether steady-state waves or irregular waves are present (Faltinsen *et al.*, 2005a).

The coefficients m_i and $\overline{m}_i$ $(i=1,2,3)$ in eqs. (9.22) depend on the aspect ratio R_L but the ordering (9.38) of R_L means

$$m_i(\overline{h},\sigma/\sigma_{1,0}) = m_i^0(\overline{h}) + O(\varepsilon^{2/3}) \quad \text{and}$$
$$\overline{m}_i(\overline{h},\sigma/\sigma_{0,1}) = m_i^0(\overline{h}) + O(\varepsilon^{2/3}). \qquad (9.41)$$

When introducing eqs. (9.40) and (9.41) in eqs. (9.22) and neglecting the $o(\varepsilon)$-terms, we obtain

$$\left.\begin{aligned}
&A\big((\overline{\sigma}_{1,0}^2-1)+m_1^{(0)}(A^2+\overline{A}^2)+m_2^{(0)}\overline{B}^2\\
&\quad+m_3^{(0)}B^2\big)+\big(m_2^{(0)}-m_3^{(0)}\big)\overline{A}\overline{B}B-P_1^{\varepsilon}=0,\\
&\overline{A}\big((\overline{\sigma}_{1,0}^2-1)+m_1^{(0)}(A^2+\overline{A}^2)+m_2^{(0)}B^2\\
&\quad+m_3^{(0)}\overline{B}^2\big)+\big(m_2^{(0)}-m_3^{(0)}\big)AB\overline{B}=0,\\
&\overline{B}\big((\overline{\sigma}_{1,0}^2-1)+\mu\sigma_0+m_1^{(0)}(B^2+\overline{B}^2)\\
&\quad+m_2^{(0)}A^2+m_3^{(0)}\overline{A}^2\big)+\big(m_2^{(0)}-m_3^{(0)}\big)\overline{A}AB=0,\\
&B\big((\overline{\sigma}_{1,0}^2-1)+\mu\sigma_0+m_1^{(0)}(B^2+\overline{B}^2)\\
&\quad+m_2^{(0)}\overline{A}^2+m_3^{(0)}A^2\big)+\big(m_2^{(0)}-m_3^{(0)}\big)\overline{A}A\overline{B}=0
\end{aligned}\right\} \qquad (9.42)$$

for longitudinal excitations. The appearance of the new parameter R_L (or μ) in eqs. (9.42) considerably changes the response curves. Faltinsen *et al.* (2006a) tested numerous combinations of $\overline{h}$, ε, and R_L and, after analyzing them, found them to be qualitatively the same for each of the ranges $\overline{h}>h_1^*$, $h_2^*<\overline{h}<h_1^*$, $h_3^*<\overline{h}<h_2^*$, $h_4^*<\overline{h}<h_3^*$, and $\overline{h}<h_4^*$.

When $R_L\le 1$, the response curves representing the nearly diagonal wave motions versus $\sigma/\sigma_{1,0}$ are qualitatively similar to curves found for a square-base tank. New types of branching are established only for $R_L>1$. To illustrate this point we give four examples in Figure 9.14, ordered with increasing R_L. Because the computations of resonant longitudinal waves involve nonzero A, the three-dimensional curves in the $(\sigma/\sigma_{1,0}, |A|, |\overline{B}|)$-frame are accompanied by their projections in the $(\sigma/\sigma_{1,0}, |A|)$-plane. All definitions related to points E, W, and so on are presented in the caption of Figure 9.14. The stability analysis makes it possible to distinguish stable

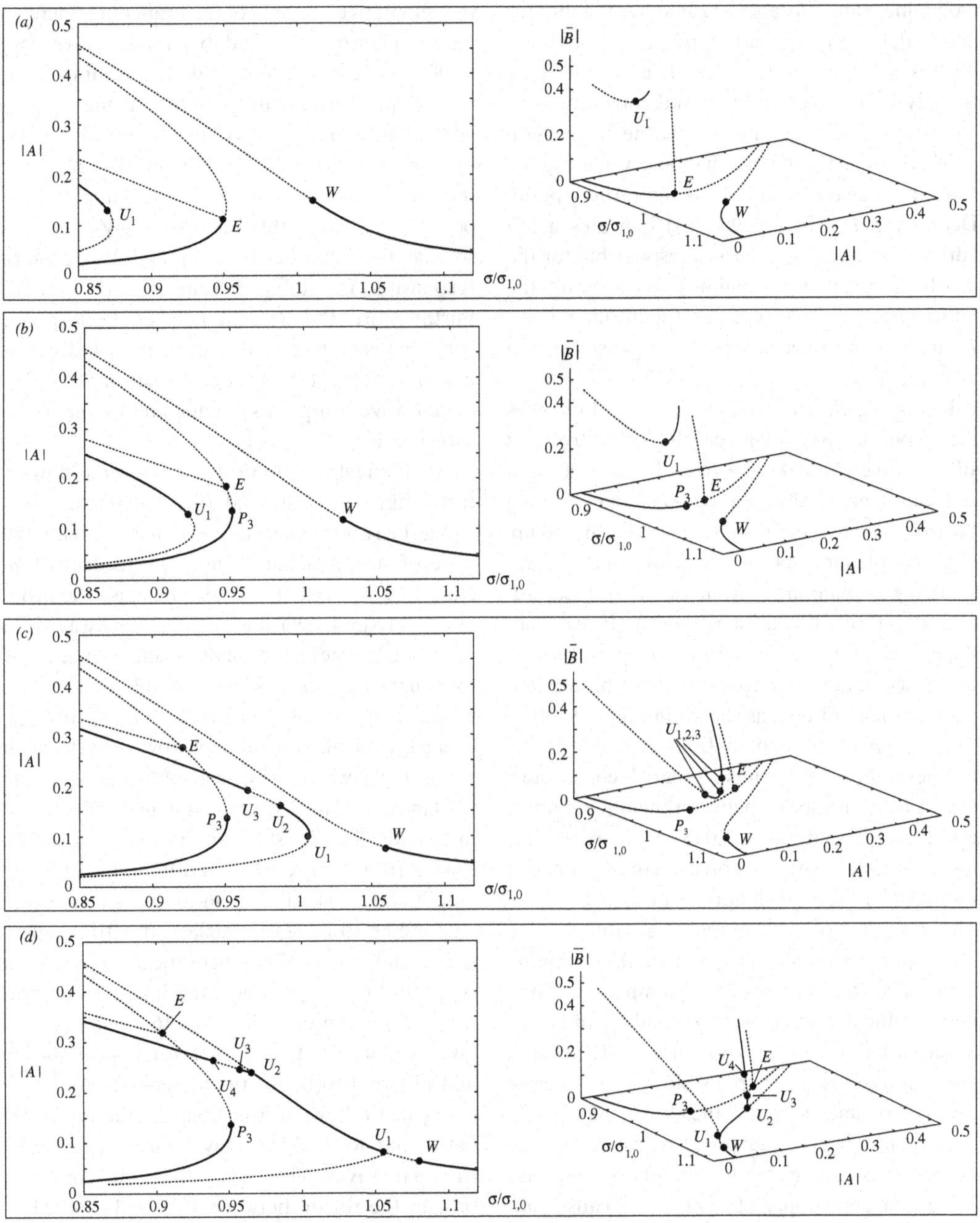

Figure 9.14. Response curves responsible for planar and nearly diagonal waves for different $R_L = L_1/L_2$: (a) $R_L = 0.95$, (b) $R_L = 1$ (square-base tank), (c) $R_L = 1.065$, and (d) $R_L = 1.1$. The tank oscillates harmonically in the longitudinal direction along the tank wall with length L_1. The calculations were made with $\bar{h} = 0.5$ and $\eta_{1a}/L_1 = 0.008$; the branching is typical for $\bar{h} > h_1^*$. The solid lines imply stable solutions and the dashed lines indicate instability of the corresponding solutions. The point W is related to bifurcations leading to swirling, but other points are caused by the nearly diagonal wave regimes: E implies the bifurcation point yielding three-dimensional nearly diagonal waves from the planar regime, and P_3 corresponds to the turning point on the branch responsible for the planar regime. The points U_1, U_2, U_3, and U_4 evolve from a single turning point U_1 when R_L is increased.

(solid lines) and unstable (dashed lines) solutions and to detect the bifurcation points on the three-dimensional response curves. The starting case is analyzed in Figure 9.14(b) and corresponds to R_L =1 (square-base tank); E is the bifurcation point yielding the nearly diagonal wave regime from the planar one and U_1 is the turning point. Decreasing R_L (Figure 9.14(a)) indicates a left "drift" of the response curves responsible for the nearly diagonal wave regimes. As a result the bifurcation point E replaces the turning point P_3 and becomes a separator between stable and unstable solutions.

Increasing R_L from 1 (with $L_1 > L_2$) changes the response curves significantly. A positive, but still relatively small, $\mu = R_L - 1$ yields a right shift of the nearly diagonal response curves along the $(\sigma/\sigma_{1,0})$-axis. As a result the point E moves up, along the planar response curve, so that the corresponding $\overline{B}$-component at the point U_1 decreases (see the three-dimensional views). If we continue increasing μ, the branch responsible for the nearly diagonal waves has three bifurcation points instead of one, as shown in Figure 9.14(c). These points are denoted U_i (i = 1, 2, 3). The two new bifurcation points U_2 and U_3 constitute a new zone of instability between them. One interesting fact is that R_L = 1.065 in Figure 9.14(c) implies stable nearly diagonal regimes (between U_2 and U_1) for a small zone around $\sigma/\sigma_{1,0} = 1$. The existence of such a zone was not possible for a square-base tank with finite $\overline{h}$. A continuing increase of R_L decreases the $\overline{B}$-component of the point on the branch between U_1 and U_2 and leads to the situation depicted in Figure 9.14(d), where the response curve responsible for the nearly diagonal regime "touches" the $(\sigma/\sigma_{1,0}, |A|)$-plane. As a consequence, the point between U_1 and U_2 becomes responsible for the planar response curve. The appearance of U_i ($i = 1, 2$) causes stable planar waves in the vicinity of $\sigma/\sigma_{1,0} = 1$. The latter is impossible for a square-base tank.

Three-dimensional response curves (in the $(\sigma/\sigma_{1,0}, |A|, |B|)$-coordinate system) and their projections (in the $(\sigma/\sigma_{1,0}, |A|)$-plane) responsible for planar waves and swirling are shown in Figure 9.15. To understand the changes due to base-ratio perturbations, we must consider the case $R_L = 1$ presented in Figure 9.15(c). In contrast to the nearly diagonal wave regimes, both increasing and decreasing R_L around 1 can change the qualitative features of the branching. Increasing R_L (Figure 9.15(a) and (b)) moves the abscissa of V_1 to the right and yields the turning point W_1 on the swirling curve (note that the appearance of U_2 and U_3 is associated with nearly diagonal waves). When $R_L < 1$, W moves left along the corresponding planar response curve, but the point V_1 falls into three points, V_1, V_2, and V_3, so that the edge between V_1 and V_2 becomes responsible for stable swirling (Figure 9.15(d)). Furthermore, the B-component of the response curve between V_1 and V_2 decreases with decreasing R_L and this stability zone transforms to stable planar waves for points bounded by V_1 and V_2, as shown in Figure 9.15(e).

An interesting *physical conclusion* following from Figure 9.15(d) and (e) is that excitation along the shorter walls ($R_L < 1$) may increase the zone of stable planar regime due to drift of W close to $\sigma/\sigma_{1,0}$ =1. It may also increase the probability of stable swirling regimes far away from $\sigma/\sigma_{1,0} = 1$, where the analysis and experiments for square-base tanks show no stable steady-state waves. For the cases in Figure 9.15(d) and (e), chaos is expected only in a narrow zone between V_2 and V_3, which will probably disappear due to damping. The experimental tests (described in Section 9.1.1.2) support the conclusions following from Figures 9.14 and 9.15. Two forcing amplitudes, 0.048 and 0.096 m, were tested corresponding to $\eta_{1a}/L_1 = 0.008$ and 0.016, respectively, and R_L = 60/66. Theoretical wave-motion types and corresponding experimental observations are presented in Figure 9.16. Figure 9.16(a) gives results for the smaller forcing amplitude and Figure 9.16(b) for the larger one. An interesting fact following from consideration of panel (a) is that R_L = 60/66 may reduce the zone of irregular waves. The theory finds this condition only in the region between V_2 and V_3, which is too narrow to be captured in experiments. One must remember that, for an analogous forcing amplitude of a 60 × 60 cm square-base tank (Figure 9.11(c)), we detected a significant domain of irregular waves. The larger excitation amplitude increases the zone of chaos (which is now situated between e and w).

Another *fascinating fact* is the experimental confirmation of the theoretical predictions that for $R_L < 1$ the swirling regime can be expected to be located far away from the primary resonance.

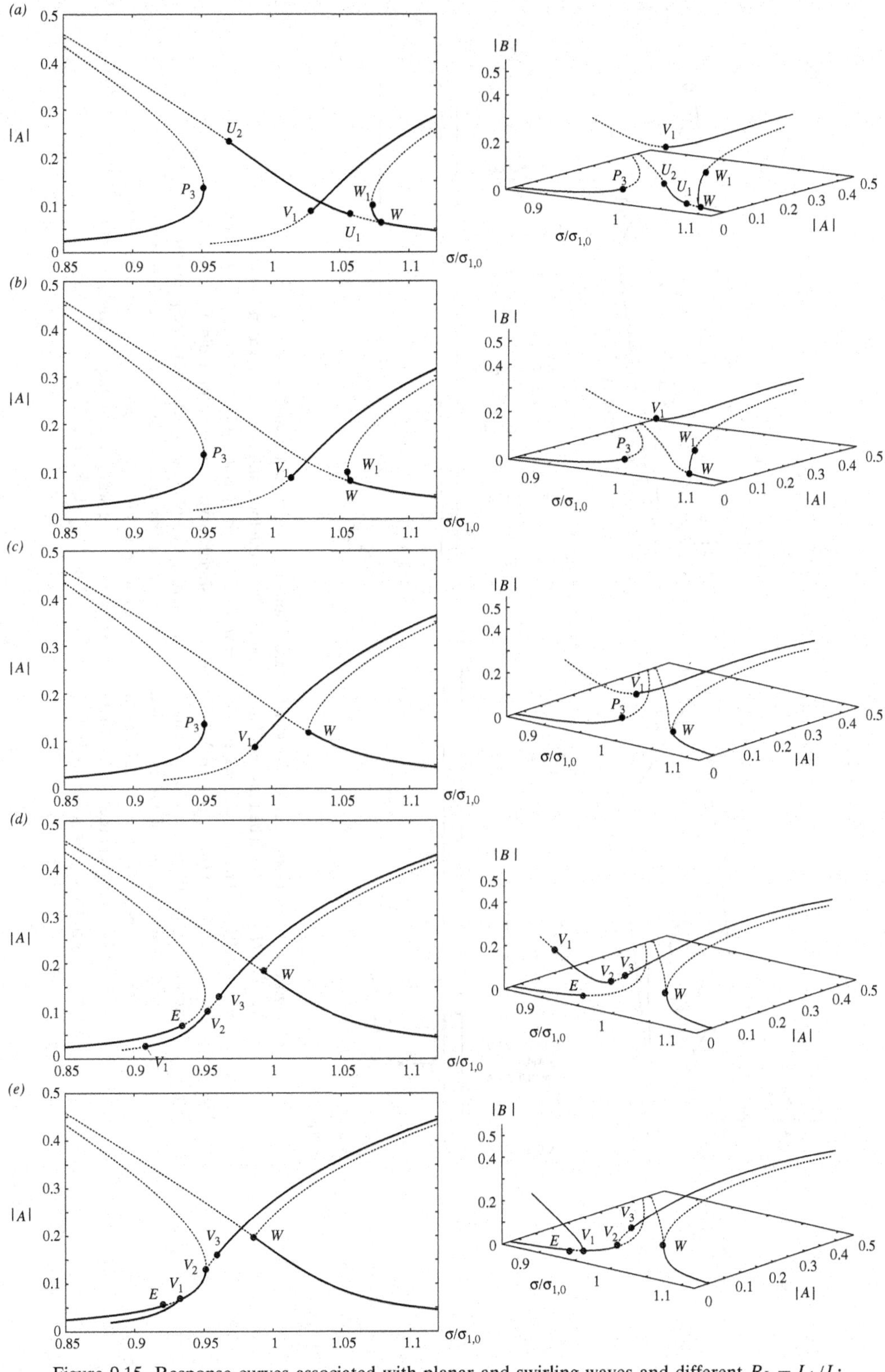

Figure 9.15. Response curves associated with planar and swirling waves and different $R_L = L_1/L$: (a) $R_L = 1.1$, (b) $R_L = 1.065$, (c) $R_L = 1$ (square-base tank), (d) $R_L = 0.9$, and (e) $R_L = 0.87$. The tank oscillates harmonically in the longitudinal direction along the tank wall with length L_1. The calculations are made for $\overline{h} = 0.5$ and $\eta_{1a}/L_1 = 0.008$; the branching is typical for finite depths ($\overline{h} > h_1^*$). Solid lines imply stable solutions and the dashed lines indicate the instability of the corresponding branches. Point E is related to bifurcations leading to the nearly diagonal wave regime (see details in Figure 9.14) and P_3 is the turning point on the planar wave curve response.

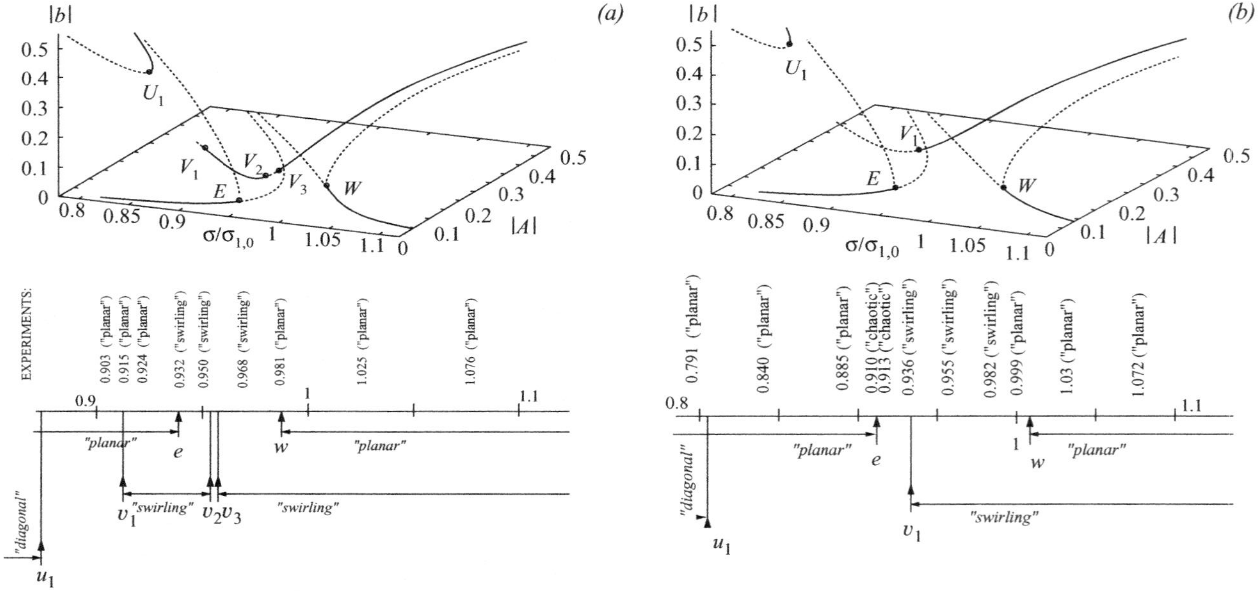

Figure 9.16. Response curves for steady-state resonant sloshing in a nearly square-base tank with $R_L = L_1/L_2 = 60/66$. The tank oscillates harmonically in the longitudinal direction along a tank wall with length L_1. The calculations are made for $\overline{h} = 0.5$. The graphs are given in the $(\sigma/\sigma_{1,0}, |A|, b)$-coordinate system ($b = \sqrt{B^2 + \overline{B}^2}$). The three-dimensional curves represent three-dimensional wave motion, but for two-dimensional branches in the $(\sigma/\sigma_{1,0}, |A|)$-plane that determine the planar wave regime. Points on the solid lines correspond to stable solutions, and the dashed curves denote instability. The meaning of the bifurcation points is explained in Figures 9.14 and 9.15. The abscissa of these points intersects the ranges where different steady-state waves are stable. Resulting stability frequency domains $\sigma/\sigma_{1,0}$ are drawn for each case: (a) $\eta_{1a}/L_1 = 0.008$ and (b) $\eta_{1a}/L_1 = 0.016$.

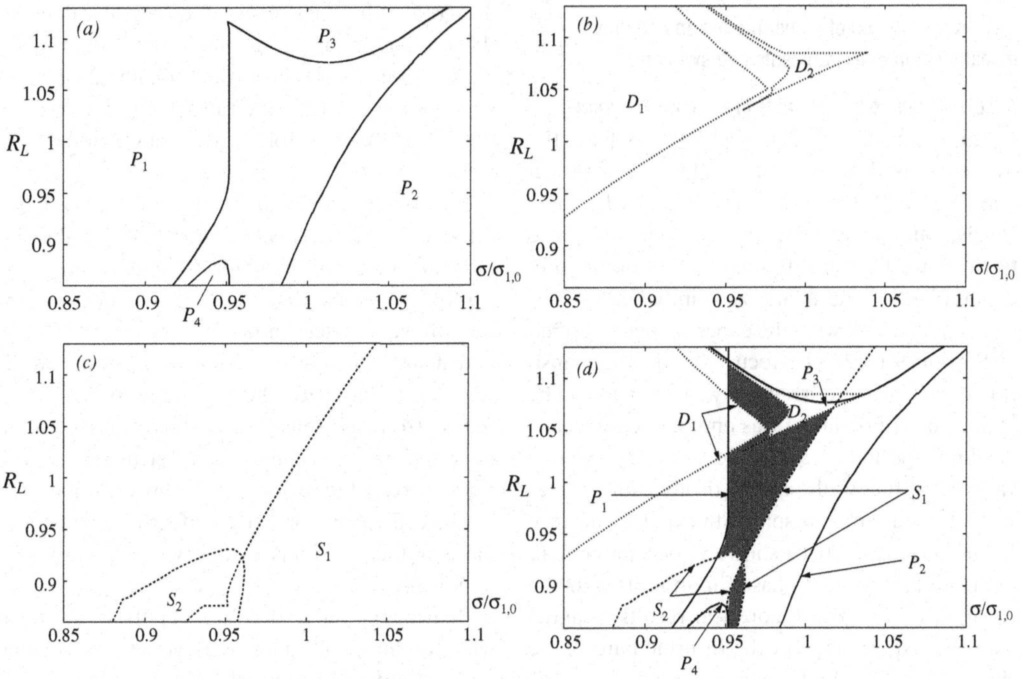

Figure 9.17. The domains of stability and instability of the steady-state motions in the $(\sigma/\sigma_{1,0}, R_L)$-plane for $\overline{h} = 0.5$, $\eta_{1a}/L_1 = \varepsilon = 0.008$: (a) stability areas of "planar" waves ($P_i, i = 1, 4$), (b) stability areas for "diagonal" waves (D_1 and D_2), and (c) stability domain S of "swirling" waves. The domains are superposed in panel (d), where chaotic waves correspond to the shaded area.

The experimental observations of swirling at $\sigma/\sigma_{1,0} = 0.932$ (Figure 9.16(a)) and $\sigma/\sigma_{1,0} = 0.936$ (Figure 9.16(b)) correspond to the frequency domain where the excitations of square-base tanks resulted in planar waves.

In general, a small disturbance $\mu = O(\varepsilon^{2/3})$ of the base ratio $R_L = \mu + 1$ has a considerable effect on the effective frequency domains of stable steady-state waves and on the existence regions of irregular (chaotic) waves. Variation of the effective frequency domains versus R_L (in the $(\sigma/\sigma_{1,0}, R_L)$-plane) is exemplified in Figure 9.17. Figure 9.17(a) shows four effective domains P_i $(i = 1, 2, 3, 4)$ and an internal area corresponding to unstable planar waves. From this area, planar motions remain unstable at $\sigma/\sigma_{1,0} = 1$ for R_L very close to 1 (for $0.92 < R_L < 1.09$ in our case), but become stable for $R_L < 0.92$ and $R_L > 1.09$.

Figure 9.17(b) shows two effective areas of nearly diagonal waves: D_1 and D_2. The numerical analysis shows that "diagonal" resonant waves at $\sigma/\sigma_{1,0} = 1$ are only possible for a small "island" D_2. Moreover, if $R_L < 1$, the effective frequency domain (associated with D_1) shifts to the left of the primary resonance and the resonant nearly diagonal regime is not realized at all. In contrast, Figure 9.17(c) predicts the dominant character of swirling at $\sigma/\sigma_{1,0} = 1$ and $R_L \leq 1$ (in the area S_1) and a new zone of stable swirling associated with S_2.

To estimate the occurrence of chaotic waves when varying R_L and $\sigma/\sigma_{1,0}$, we overlapped the areas of instability from Figure 9.17(a)–(c) and marked the result by the shadow area in Figure 9.17(d). This subfigure shows that chaotic waves appear for small variations of R_L around 1. However, because of the dominant character of planar (for increasing R_L) and swirling (for decreasing R_L) regimes, the frequency domain of irregular motion gets narrow for $R_L > 1.05$ and $R_L < 0.92$. Therefore, changing to a nonsquare base can lead to the disappearance of irregular motions.

The classification of frequency domains becomes more complicated for $\overline{h} < h_1^*$. Also in this case the region of chaotic waves enlarges by increasing the forcing amplitude. However, as mentioned in Section 9.2.6, both smaller depths and larger forcing amplitudes require modification of Moiseev-type asymptotic ordering.

9.2.6 Amplification of higher modes and adaptive modal modeling for transients and swirling

9.2.6.1 Adaptive modal modeling and its accuracy

Faltinsen *et al.* (2005b, 2006b) applied the Moiseev-based modal system (9.11)–(9.19) for a *quantitative* description of transient and almost steady-state waves with forcing frequency close to the lowest natural frequency. Different forcing amplitudes and nearly zero initial conditions were applied following the experiments described in Section 9.1.1.2. The focus was on the transition to swirling and irregular waves (chaos). The importance of damping was emphasized to reach an almost-periodic wave response. The wave elevations at the walls and hydrodynamic forces were studied. Comparisons with experiments (see Faltinsen *et al.*, 2005b) showed good agreement in the initial transient phase for about two to five forcing periods. The theoretical results disagree with the experiments in the intermediate phase (between 5 and 15 forcing periods) for *nearly planar* and *diagonal* waves but are quite accurate close to nearly steady-state conditions. When *irregular* waves (chaos) are expected, simulations made with the Moiseev-based modal theory disagreed significantly with the experiments after a short initial transient phase (one to three forcing periods). The simulation may even break down due to unrealistic amplification of some modes that are assumed nondominant.

Special efforts were made by Faltinsen *et al.* (2005b) to simulate *swirling* with modal system (9.11)–(9.19). The numerical results were consistent with the measurements during the initial two to five forcing periods, but unrealistic amplification of higher modes occurred at a later stage. Furthermore, a steady-state solution correct to $O(\varepsilon)$ showed clearly lower values than the experimental wave elevations and hydrodynamic forces during steady-state swirling. The relative errors of the asymptotic model were between 30 and 100%. When the time-domain model used the steady-state solution correct to $O(\varepsilon)$ as an initial condition and introduced small disturbances during the evolution, a strong excitation of the second-order modes occurred, followed by a numerical breakdown. This fact can be partly explained by the asymptotic character of the modal system. The amplitude-to-breadth ratio has to be less than 0.001 to avoid numerical breakdown. When results from modal system (9.11)–(9.19) disagree with experiments, the critical value for the forcing amplitude is much lower than analogous values for planar and diagonal steady-state motions (where the ratio is estimated to be around 0.01). Therefore, even though modal theory (9.11)–(9.19) gives good qualitative prediction of all steady-state regimes, it is quantitatively reliable for effective amplitudes below a threshold value that is *different* for *different regimes*.

Other factors that undermine the use of the Moiseev-based modal theory for transient and nearly steady-state three-dimensional wave regimes (relative to the two-dimensional case) follow from the sensitivity of the modal system to initial perturbations. Small perturbations are always present in experiments. One reason is that the forced velocity is not a perfectly periodic signal. Another reason is local wave breaking at the tank walls.

Bearing in mind that ship applications deal with forcing amplitudes as large as 10% of the tank breadth and remembering the earlier two-dimensional results for the adaptive modal systems, we further follow Faltinsen *et al.* (2005a, 2006b) and modify the modal approach by introducing a larger set of dominant modes. The altered system also accounts for damping to reach steady-state conditions. Because both the experiments and the simulations based on eqs. (9.11)–(9.19) confirm a significant contribution of higher modes, many dominant modes are assumed (as in Section 8.4 for two-dimensional sloshing). This result implies that Moiseev-type ordering (9.9) must be replaced. Solutions based on ordering (9.9) should coexist in the same frequency domain with solutions based on the multidominant technique. To match both types of asymptotic solutions and transients between them by using a single nonlinear modal system, Faltinsen *et al.* (2005b) proposed an adaptive ordering, which is quite different from the adaptive ordering adopted for the two-dimensional case. It requires

$$\begin{aligned} \beta_{i,j} &= O(\varepsilon^{1/3}), \quad i+j \le N, \\ \beta_{i,j} &= O(\varepsilon^{2/3}), \quad N+1 \le i+j \le 2N, \\ \beta_{i,j} &\le O(\varepsilon), \quad 2N+1 \le i+j. \end{aligned} \tag{9.43}$$

Ordering (9.43) transforms the infinite-dimensional modal system (9.8) into a finite-dimensional structure by neglecting the terms of $o(\varepsilon)$ for a fixed finite number N related to the number of dominant modes. By introducing ordering (9.43) in eq. (9.8) and increasing N, we get a set of "embedded" modal systems, where the case $N = 1$ corresponds to Moiseev-based modal theory. There are two dominant modes for $N = 1$, five dominant modes for $N = 2$, nine dominant modes for $N = 3$, and so on.

Faltinsen *et al.* (2005b) also discussed the difficulties and differences of adaptive modal modeling based on ordering (9.43) and the adaptive modal approach for two-dimensional sloshing in a rectangular tank from Section 8.4. The main difficulty in the three-dimensional case is associated with types of wave motions (e.g., planar waves and swirling) that coexist in the same frequency domain. The different types of wave motion require different numbers of dominant modes to be accurately handled. In the two-dimensional case it is possible to predict a priori how many dominant modes are needed for fixed values of forcing frequency and amplitude, which is not generally true for three-dimensional sloshing in a square-base tank. The numbers may be different for swirling, nearly diagonal, and planar wave motions as we discussed earlier when examining Moiseev-based theory. Faltinsen *et al.* (2005b) showed that the experimental cases from Section 9.1.1.2 need at least nine dominant modes for swirling, but planar motions can be simulated with two dominant modes (i.e., by eqs. (9.11)–(9.19)). For corresponding two-dimensional sloshing with the same ratio of liquid depth to tank breadth, $\overline{h} = 0.5$, the number of dominant modes required to get quantitatively good results for nearly steady-state waves is up to 3, even for a forcing amplitude that is 10% of the breadth. A way to find the number N required for reliable results consists of testing results obtained with N and $N + 1$. If N corresponds to a consistent number of dominant modes, results obtained with N and $N + 1$ should differ only by $O(\varepsilon)$. Due to this asymptotic treatment of the convergence, the modal functions of $O(\varepsilon)$ are unnecessary and therefore the adaptive modal systems can be restricted to the modal functions $\beta_{i,j}$ $(i + j \leq 2N)$.

The handling of the damping represents a problem, at least for the lower modes. In the simulations of Section 8.4, for two-dimensional sloshing, we needed a linear damping to get a steady-state regime after the transient. However, the final simulations of the steady-state solution were performed with very small damping, because only the lowest three modes were assumed dominant. For the three-dimensional case, the number of dominant modes increases and we cannot reduce the damping ratios to zero. Damping in the theoretical three-dimensional model is introduced by the linear terms $2\xi_{i,j}\sigma_{i,j}\dot{\beta}_{i,j}$ in general modal system (9.8). We can only estimate the damping in a rational way if the damping ratio is small and due to bulk flow ($\xi_{i,j}^{\text{bulk}}$) and boundary-layer flow ($\xi_{i,j}^{\text{tank surface}}$; see Exercise 6.11.1). The latter fact is true for the lower modes when breaking waves do not matter. Some guidance about the damping of higher modes can be obtained from the theoretical studies by Krein (1964). He found that a finite number of natural modes exist with nonzero natural frequencies for linear natural sloshing of a viscous liquid. Furthermore, an infinite set of modes exists with pure decaying behavior. The latter fact can be approximated by assuming that the very high modes are critically damped. Therefore, Faltinsen *et al.* (2005b, 2006b) used the viscous damping ratios $\xi_{i,j}^{\text{bulk}}$ and $\xi_{i,j}^{\text{tank surface}}$ in the modal equations associated with the dominant modes $\beta_{i,j}$ with $1 \leq i + j \leq N$, whereas the second-order modes $\beta_{i,j}$, with $N + 1 \leq i + j \leq 2N$, are critically damped (i.e., $\xi_{i,j} = 1$). However, it is easy to criticize this damping model.

Because the relative error of the adaptive method is $O(\varepsilon)$, each point on a response curve (e.g., elevation, force, etc.) is associated with a circle of radius ε to which the actual theoretical values can belong. Experimental steady-state amplitudes should target a domain formed by the set of these circles moving along the numerical response curves as demonstrated in Figure 9.18. This finding is true when the nondimensional units on the axes are the same. The circles are replaced by ellipses when the units of the horizontal and vertical axes are unequal, as shown in the second part of Figure 9.18.

A limitation of the adaptive modal systems is the inadequate description of local phenomena, which are particularly important for complex transient and irregular flows in the frequency domains with no stable wave solutions.

9.2.6.2 Transient amplitudes

Even though the objective in the experiments by Faltinsen *et al.* (2003) was to start with a liquid at rest, in practice small motions initially

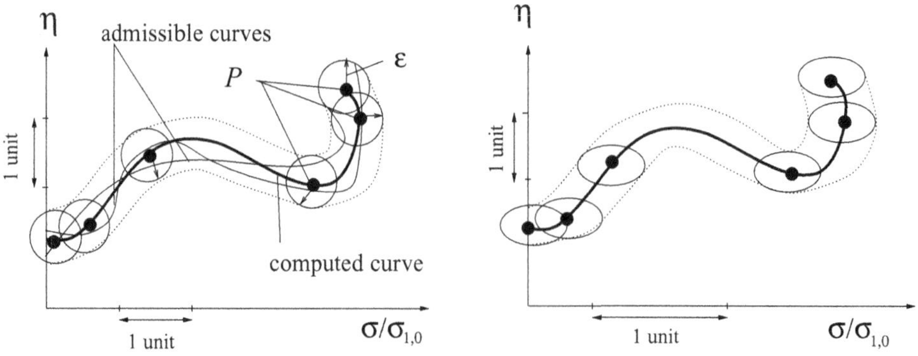

Figure 9.18. Computed steady-state response curves (bold solid lines) of the adaptive modal technique in the $(\sigma/\sigma_{1,0}, \eta)$-plane. P is an arbitrary point on the curve and η is either the wave amplitude or the hydrodynamic force amplitude. The relative error of the adaptive method is $O(\varepsilon)$ and, therefore, the actual responses belong to an $O(\varepsilon)$ neighborhood of the computed data visualized as circles or ellipses, depending on the ratio between the units of the axes.

exist. The latter fact influences the occurrence of three-dimensional waves. If longitudinal excitation is theoretically studied, transverse perturbations are needed to obtain three-dimensional sloshing (e.g., swirling). Otherwise, modal theory gives pure two-dimensional sloshing. The magnitude of the initial conditions should be consistent with linear sloshing theory (i.e., equal to the order of the forcing). Long simulations with these small initial perturbations should either lead to steady-state solutions or exhibit irregular (chaotic) motions. The calculations by Faltinsen *et al.* (2006b) showed that modeling the transients by the adaptive model technique goes through a maximum of five relatively long time intervals (enumerated as phases I–V), where different wave behaviors are displayed. Phases III–V are shown in Figure 9.19. Phase I usually lasts about 30–40 forcing periods. Faltinsen *et al.* (2005a) studied this phase and established good agreement with experiments. If there are no stable steady-state solutions, the experiments demonstrate considerable breaking waves, which limit the agreement to 10–15 forcing periods.

Furthermore, both the simulations and the experiments show that phase I is followed by the strongly irregular waves of phase II, which last up to 10 forcing periods before the transition to planar and diagonal regimes and up to 30–40 forcing periods for swirling to be established. Faltinsen *et al.* (2005a, 2006b) showed that the transients of phase II cannot be precisely approximated by the adaptive method. A reason is that the liquid flow in phase II is very sensitive to small disturbances. Increasing dimension of adaptive systems, small variations of damping, and even $O(\varepsilon)$ variations of initial conditions influence the duration and sloshing behavior in phase II. All these disturbances, and physical factors involving breaking phenomena and near-wall jets, remind us that "gentle touches to a stone rolling down a mountain can be decisive in determining its future trajectory."

If the modal system predicts transition to swirling, phase II is followed by phase III of considerable duration. However, phases III and IV are absent for planar and diagonal cases. Beginning from phase III, numerical swirling does not change rotation direction and is characterized by a modulation with a fixed beating period, which is exemplified in Figure 9.19. The duration of phase III, as well as the maximum (A_M) and minimum (A_m) amplitude values, may be influenced by the time history in phase II, which is determined by the initial conditions. Another decisive factor is the dimension of the adaptive modal system (i.e., the feedback of higher modes and the total damping). The arithmetic mean values of A_m and A_M remain stable but prolonged calculations with $N \geq 5$ become numerically stiff. A *stiff equation* is a differential equation for which certain numerical methods for solving the equation are numerically unstable, unless the step size taken is extremely small (see mathematical details in, for instance, the book by Hairer and Wanner, 1996). The mean value of A_m and A_M corresponding to the upper dotted line in Figure 9.19 is clearly larger than the

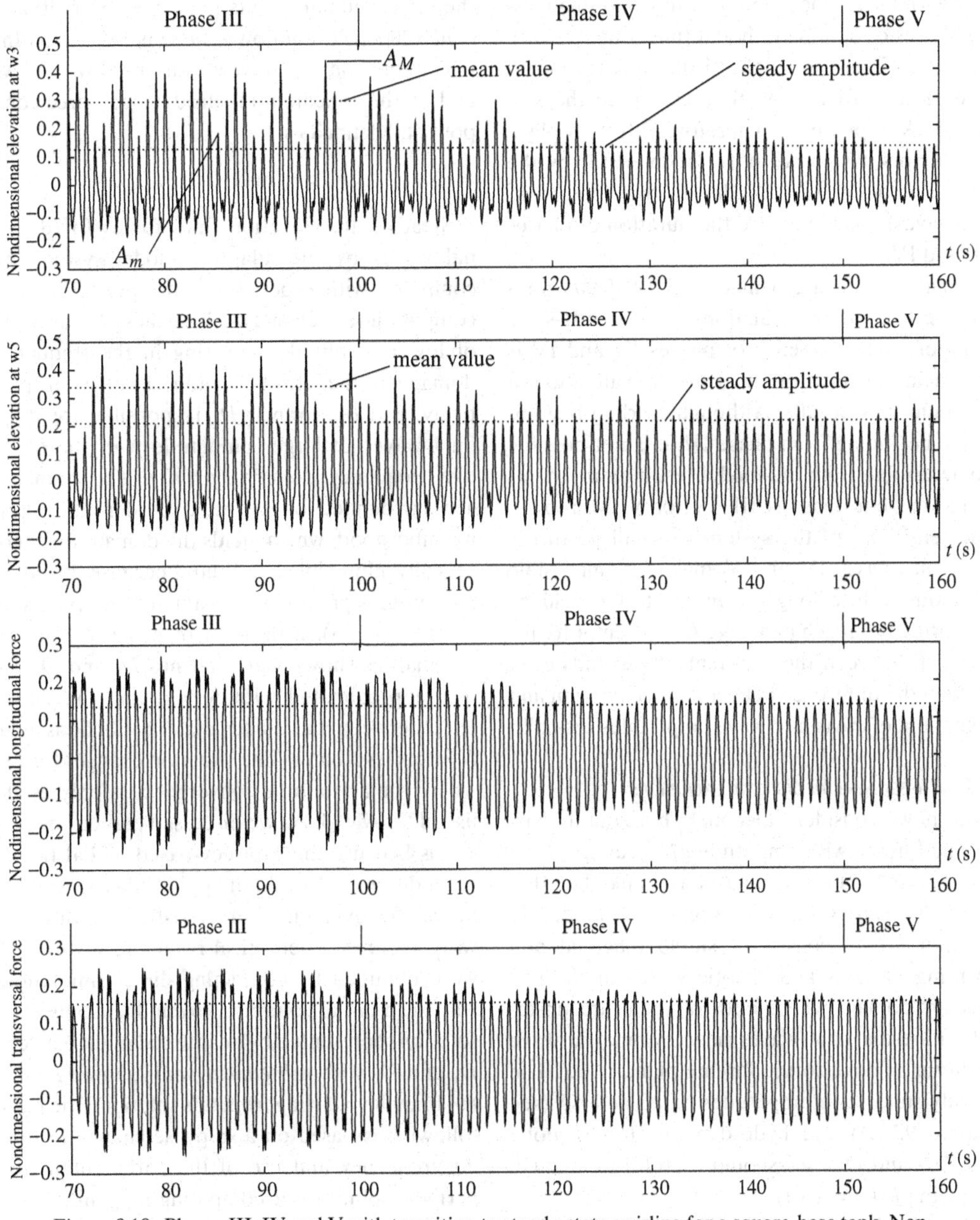

Figure 9.19. Phases III, IV and V with transition to steady-state swirling for a square-base tank. Nondimensional wave amplitudes at w3 and w5 (see Figure 9.1) and hydrodynamic forces (normalized by $M_l g$) are included. Numerical data are included for longitudinal forcing with $\bar{h} = 0.5$, $\eta_{1a}/L_1 = 0.00817$, $\eta_{2a} = 0$, and $\sigma/\sigma_1 = 0.99$. The forcing period is 0.9247 s. The lower dotted line represents the steady-state amplitude, and A_M and A_m indicate maximum and minimum amplitude, respectively, in phase III, whose mean value is marked by the upper horizontal line (Faltinsen *et al.*, 2006b).

steady-state amplitude represented by the lower dotted line.

The next typical time period, phase IV, is characterized by increased duration of the beating period of modulated waves as well as a monotonic decrease of the mean amplitude. The length of the beating period stabilizes only in phase V, where the mean amplitude becomes close to the steady-state value. The duration of phase IV is very sensitive to small variations of damping, feedback of higher modes to the dominant ones, and small disturbances of dominant modal functions $\beta_{i,j}$, with $i+j \leq N$. Even though the disturbances are small and of the order $O(\varepsilon)$, they can force the system back to phase III. Therefore, different physical factors, which are not accounted for by the modal modeling (e.g., feedback from the breaking waves), may influence the duration of phases III and IV.

Finally, after approximately 3,000–5,000 forcing periods, the calculations give steady-state solutions. The presence of phases III and IV is the main difficulty for experimental validation of the numerical results. Although swirling is stabilized, beginning from phase III, it is not possible to determine from the model test measurements if the current phase is III, IV, or V. Moreover, the sensitivity of the system to small perturbations in phases IV and V makes it impossible to estimate how long the model test should be performed to reach phase V. Any attempt to distinguish between the transient phases fails even though the difference between the maximum and minimum of measured wave amplitudes is small.

9.2.6.3 Response for diagonal excitations

We now consider diagonal horizontal harmonic forcing with amplitude εL_1 (i.e., $\eta_{1a}/L_1 = \eta_{2a}/L_1 = \varepsilon/\sqrt{2}$) for a square-base tank. The theoretical steady-state wave regimes described in Section 9.2.4.2 consist of stable diagonal and swirling motions and chaotic waves in the frequency domain where both of them are unstable. Pure-diagonal waves have symmetric wave patterns relative to the forcing plane $y = x$. The steady-state wave amplitudes at w3 and w5 (see Figure 9.1(a)) and hydrodynamic forces along the *Ox*- and *Oy*-axes should therefore be equal, implying $F_1(t) = F_2(t)$.

For *swirling* excited by the diagonal forcing, the maximum wave elevations at w3 and w5 and the maximum steady-state horizontal forces along the *Ox*- and *Oy*-axes, $\max|F_1|$ and $\max|F_2|$, *are not equal.* This fact was extensively discussed by Faltinsen *et al.* (2006b) from both theoretical and experimental points of view. The direction that experiences the largest elevations (at w3 or w5) and forces during steady-state swirling depends on the rotation direction. The swirling direction depends on the initial conditions and transient perturbations. Because the rotation direction in the experimental results presented by Faltinsen *et al.* (2006b) is unknown, we concentrate on the maximum from the wave elevations at w3 and w5 and on the "maximum hydrodynamic force component," defined as

$$F_{\max} = \max\left(|F_1|, |F_2|\right). \tag{9.44}$$

Steady-state elevations and forces for diagonal waves by the adaptive modal system are compared with experiments in Figure 9.20. The computations detect two branches, D_1 and D_2, of lower amplitude occurring in the frequency domains to the left of p_1 and right of p_2, respectively, and the branch D_3 responsible for diagonal waves of larger amplitude with the effective frequency domain (p_4, p_5). The branches are surrounded by a dotted border, the $O(\varepsilon)$-neighborhood, which yields the domain for theoretically admissible amplitudes because the adaptive systems provide approximations of points on D_i ($i = 1, 2, 3$), with an error of $O(\varepsilon)$. Numerical analysis shows that branches D_1 and D_2 are well approximated with $N = 1$ and the difference between the steady-state predictions with $N = 2$ is less than ε. The situation changes for D_3, where the asymptotic convergence is only established for $N = 3$. As we discussed earlier, the reason is because the Moiseev-based modal theory has a different domain of applicability to approximate the dynamic characteristics for different wave regimes. Theoretical results in Figure 9.20 were obtained by combining direct simulations with zero initial conditions to find a steady-state solution represented by an internal point between D_1 and D_2 with *subsequent path-following* along the branches starting from this solution. The path-following is based on a stepwise change in forcing frequency and use of the earlier numerical periodic solution for computing new initial conditions at t_0. This procedure corresponds to an artificial model test where the forcing frequency changes very slowly to avoid considerable transients just after reaching a steady-state motion. The subbranch D_3 was in general obtained by the path-following, which started with a single numerical series for $\sigma/\sigma_1 = \sigma/\sigma_{1,0}$ in a middle point of (p_1, p_5). The series used zero initial conditions. Similar series for $\sigma/\sigma_1 < p_1$ always lead to solutions on D_1. The endpoints p_4 and p_5 of D_3 were also detected by path-following. The Moiseev-based theory predicts the corresponding

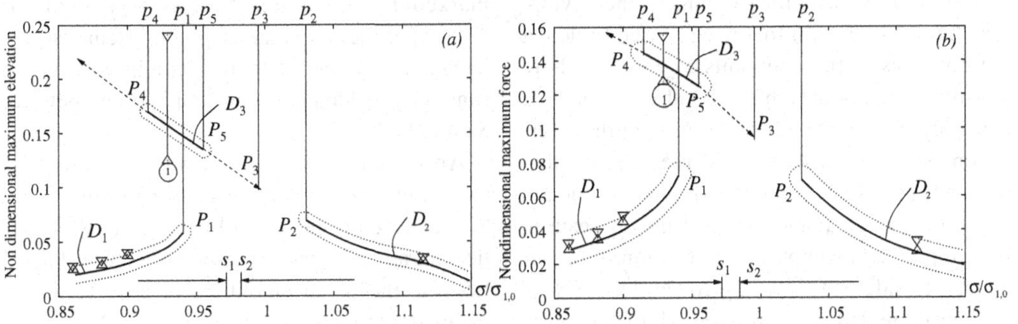

Figure 9.20. Diagonal resonant excitations with $\overline{h} = 0.5$ and $\varepsilon = \sqrt{\eta_{1a}^2 + \eta_{2a}^2}/L_1 = 0.00817$ for a square-base tank. Response curves are for stable diagonal waves. Solid lines (curves D_1, D_2, and D_3) imply theoretical steady-state wave elevations (maxima from w3 and w5 probes) scaled by L_1 and the maximum hydrodynamic force component $F_{\max}$ by eq. (9.44) (normalized by $M_l g$) versus $\sigma/\sigma_{1,0}$. Measured amplitudes (A_M and A_m as explained in Figure 9.19) are represented by minima (Δ) and maxima (∇) during the last 20 s of the model tests. The figure includes dotted lines that encompass the ε-domain for the expected measured steady-state amplitudes. Points P_i and their abscissas p_i ($i \neq 3$) denote the endpoints of the theoretical response curves, where the diagonal regime changes stability properties. The dashed line shows extra subbranches, which exist in the asymptotic limit $\varepsilon \to 0$ but disappear in the adaptive modal method. Analogously, the range (s_1, s_2) implies a theoretical estimate of irregular (chaotic) motions as $\varepsilon \to 0$.

branch for $\sigma/\sigma_1 \le p_3$ but not for $p_4 \le \sigma/\sigma_1 \le p_5$. The new left bound p_4 is most probably associated with damping, but the right bound p_5 is due to amplification of higher modes (convergence requires $N \ge 3$). Abscissas of p_1 and p_2, also determined by path-following, are generally consistent with Moiseev-based results.

Numerical results are compared with measured maximum hydrodynamic force and wave elevations during the last 20 s of a total model test duration of 360 s. The A_M- and A_m-values of the measured elevations are denoted by the pairs of inverted triangles in Figure 9.20. Solutions on D_1 and D_2 show very good agreement between theory and experiments. The point 1 ($\sigma/\sigma_1 = 0.929$) from the model test involves considerable beating, but the mean measured amplitude is close to the value on D_3, especially for the hydrodynamic force.

Figure 9.21 completes Figure 9.20 by giving results on swirling occurring inside the range (p_1, p_2). The theoretical maximum steady-state elevations of stable swirling based on the adaptive modal system are represented by the solid line S, which determines the effective frequency domain (s_2, s_3). The dashed lines give the shapes

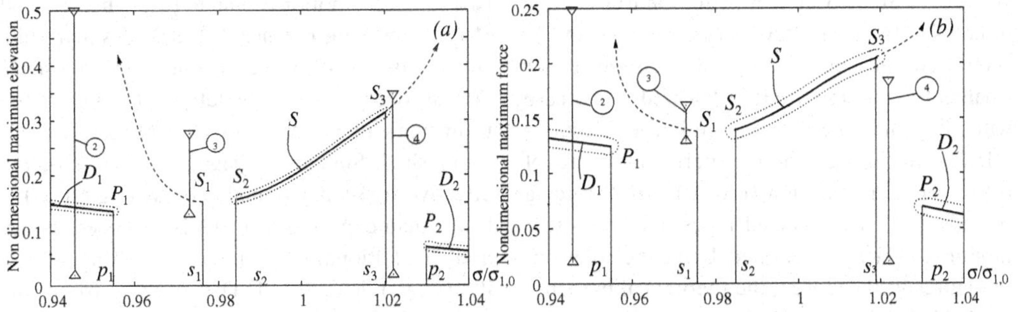

Figure 9.21. The same as in Figure 9.20, but for swirling. Solid lines (curves D_2 and D_1) are parts of the branches in Figure 9.20. The curve S with endpoints S_2 and S_3 corresponds to stable swirling, so that (s_2, s_3) is its effective frequency domain. Dashed lines correspond to additional frequency domains of stable swirling established by single-dominant theory. However, adaptive modal theory predicts irregular motions in the ranges (p_1, s_2) and (s_3, p_2).

of response curves with swirling in the asymptotic limit $\varepsilon \to 0$. Path-following is used to detect the positions of the endpoints S_2 and S_3, where swirling becomes unstable. The abscissa of S_2 is generally consistent with the abscissa of the bifurcation point established by Moiseev-based theory in Section 9.2.2, but the frequency domain (s_3, p_2) of irregular motions is a novelty caused by damping and complex amplification of higher modes. Besides, the swirling to the left of s_1 is also unstable within the framework of the adaptive modal modeling.

In the case of Figure 9.21, only three test runs were performed in the model tests around the interesting frequency domain (points 2, 3, and 4). All these runs were unfortunately performed away from (s_2, s_3) and therefore do not demonstrate swirling but rather irregular (chaotic) motions. Point 4 is very useful because it confirms the new zone of hydrodynamic instability between s_3 and p_2. The irregular motion at 2, where a diagonal wave is expected instead, can be easily explained by the initial scenario of the experiments. In this case, path-following to the right along D_1 indeed identifies a stable diagonal regime, but the calculations with small initial conditions lead to chaotic motions appearing as a perpetual switching of the swirling direction. Point 3 has been extensively discussed by Faltinsen *et al.* (2003, 2006b). In this case, the model test demonstrates swirling with a relatively small fluctuation of the amplitude between 90 and 200 s, but after 200 s the fluctuation increases (see the recordings in Figure 9.22), which suggests hydrodynamic instability.

To validate the effective frequency domain of swirling related to branch S, to confirm chaotic motions in the frequency domain (p_1, s_2), and to show that the zone (s_3, p_2) may disappear for smaller forcing amplitudes, the model test case with diagonal forcing was used from Sections 9.1.1.2 and 9.2.4.2. These experiments were of relatively short duration (up to 120–160 forcing periods), but the expected types of steady-state motions are clearly identified. Using the notations in Figures 9.20 and 9.21, the theoretical branches D_i $(i = 1, 2, 3)$ and S are compared with measured values of $F_{\max}$ during the last 20 s in Figure 9.23. Three model test results (points 6, 7, and 8), which imply stable swirling, support the existence of the branch S if the admissible $O(\varepsilon)$ neighborhood, marked by the dotted line, is accounted for. The experimental value of point 5 demonstrates irregular motions and thereby confirms instability ranges embedded between S and D_1 and between S and D_3.

An analysis of measured steady-state values for swirling concludes that the measurements at point 8 are most probably made in phase V, identified numerically as discussed in Section 9.2.6.2, because the mean amplitude is close to S. The mean amplitudes at points 6 and 7 are larger than the corresponding steady-state predictions, which indicates that such points belong to phase III or IV. This hypothesis has been confirmed by computing the mean amplitude in phase III for point 7. Its dimensionless value is equal to 0.225 and clearly larger than the mean value measured at the last 20 s.

9.2.6.4 Response for longitudinal excitations

Longitudinal excitation of the tanks in Figure 9.1(a) and (b) has been performed. The sensors w and g of the two tanks are situated at distances from the wall that differ by only about 3 mm (0.5% of the tank breadth). If the forcing is the same for the two tanks, the measured wave elevations at w3 and g10 (and g5), and at w5 and g8, must be approximately the same.

Comparison of the theoretical results and experimental data are presented in Figure 9.24. Solid lines represent theoretical values of stable steady-state regimes, but inverted triangles give upper (A_M) and lower (A_m) bounds of the amplitudes in the model test data from the last 20 s of measurements (see explanation in Figure 9.19). The adaptive modal technique detects only two types of motions: planar (branches P_1 and P_2) and swirling (branch S_1). Stable square-like (nearly diagonal) wave regimes, which should appear over and to the left of P_1 within the framework of the Moiseev asymptotics, are not established. Similar to diagonal excitations, the effective frequency domain of swirling (s_1, s_2) is lower than expected from Moiseev-based theory and an additional zone of hydrodynamic instability, (s_2, p_2), is detected. The effective frequency domain for planar motions generally agrees with earlier predictions.

The agreement between theory and experiments for the effective frequency domains is very good. Measured amplitudes are also well

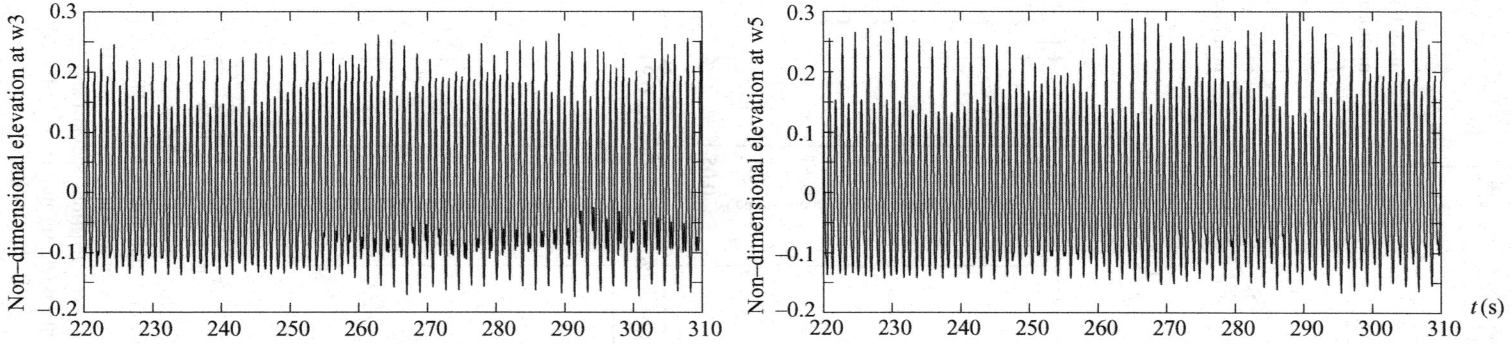

Figure 9.22. Later stages of model test recordings at w3 and w5 (see Figure 9.1) for a square-base tank. Diagonal forcing with $\overline{h} = 0.5$, $\sqrt{\eta_{1a}^2 + \eta_{2a}^2}/L_1 = 0.00817$, and $\sigma/\sigma_{1,0} = 0.973$.

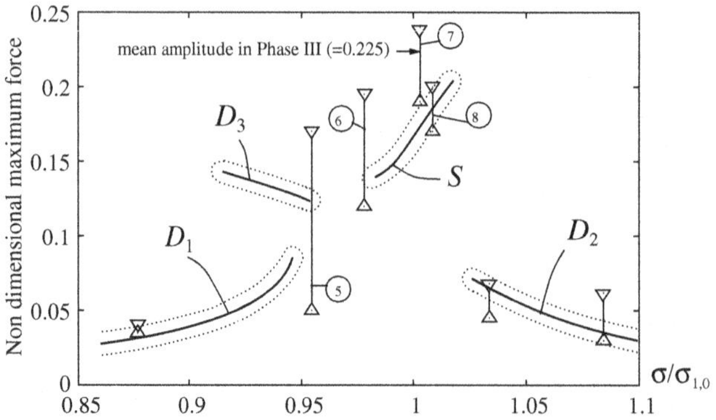

Figure 9.23. Theoretical results and model test data for stable steady-state motions. The maximum force component F_{max} defined by eq. (9.44) is normalized by $M_l g$. Diagonal forcing with $\overline{h} = 0.508$ and horizontal forcing amplitude-to-breadth ratio equal to 0.0078. The notation is as in Figures 9.20 and 9.21.

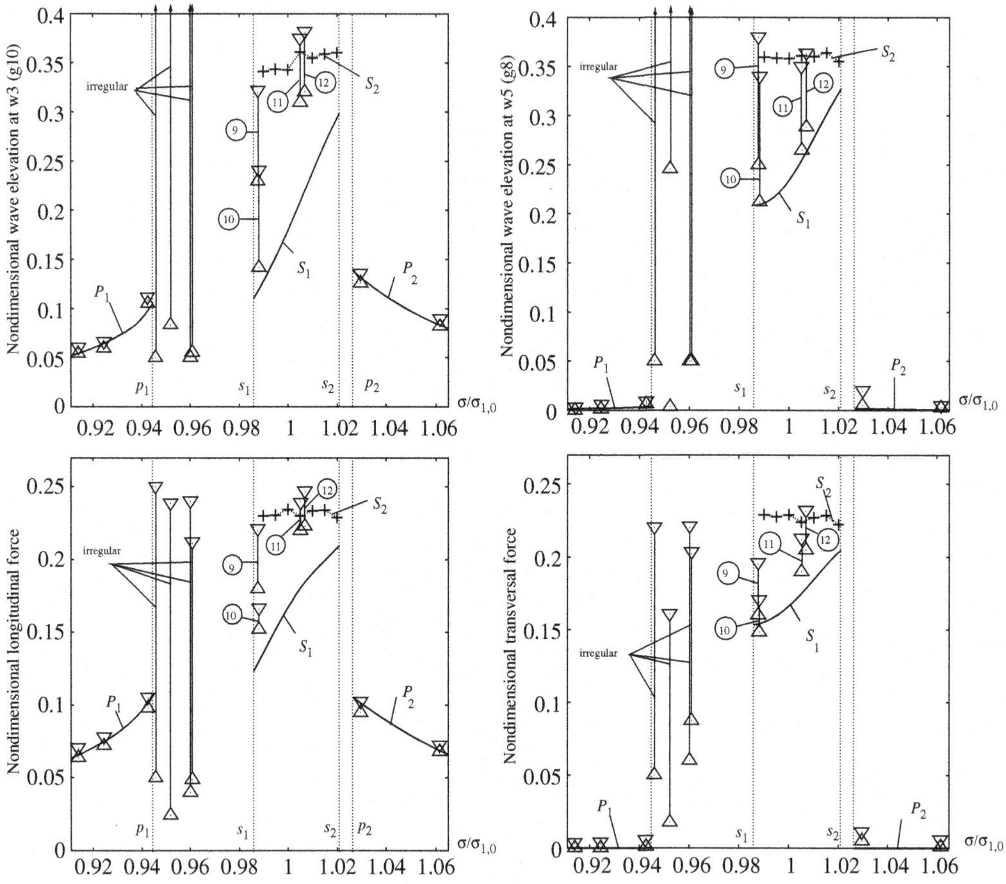

Figure 9.24. Longitudinal resonant excitation with $\overline{h} = 0.5$ and $\varepsilon = \eta_{1a}/L_1 = 0.00817$ for a square-base tank. Response curves and model test data are included. The solid lines (curves P_1, P_2, and S_1) are theoretical steady-state values for maximum elevations at w3/g10 and w5/g8 in Figure 9.1 scaled by L_1 and maximum hydrodynamic forces $|F_1|$ (longitudinal) and $|F_2|$ (transversal) normalized by $M_l g$ versus $\sigma/\sigma_{1,0}$. The frequency domains (p_1, s_1) and (s_2, p_2) correspond to irregular (chaotic) wave motions. The mean amplitudes in phase III are computed and drawn as the branch S_2 (cross-and-dotted line). Measured amplitudes are represented by minima (Δ) and maxima (∇) during the last 20 s of the model tests.

approximated for the planar regimes. The model tests that exhibit swirling are characterized by significant amplitude fluctuations and give much larger results than the theoretical steady-state predictions. This fact, and differences in experimental data obtained approximately for the same forcing parameters (see points 9 and 10), implies that the measurements probably correspond to phases III and IV; therefore, the mean model test amplitudes should be between the mean amplitudes at phase III and the steady amplitude. The mean amplitudes were computed and drawn as the branch S_2 (crest-and-dotted line). Figure 9.24 confirms that the mean measured amplitudes are located between S_1 and S_2.

Doubling the forcing amplitude relative to the previously described experiments (i.e., $\varepsilon = \eta_{1a}/L_1 = 0.01634$) leads to considerable local breaking with steep waves hitting the wave probes as documented by the video recording published by Faltinsen *et al.* (2006b). This event may result in unpredictable recordings for the wave elevation at the walls and makes it difficult to obtain good agreement between theoretical and experimental results. However, because the local phenomena give little influence on integral hydrodynamic characteristics (i.e., forces and moments on the tank), the adaptive modal technique should model such variables adequately. This modeling is confirmed by the results in Figure 9.25 (notations are the same as in Figure 9.24), where a disagreement is only established for model test case 14. The model test shows clear irregular motions for a duration of at least 5 minutes while the steady-state theory predicts planar waves for $\sigma/\sigma_1 < p_1$ and $\sigma/\sigma_1 > p_2$. As we remarked for a similar disagreement for case 2 in Figure 9.21, the experimental behavior can be handled by performing simulations with initial conditions that are not exactly zero. Such simulations lead to irregular waves, whereas steady-state solutions are only obtained by a path-following procedure along branches P_1 and P_2. The value of p_1 computed by means of path-following along branch P_1 does not coincide with the actual right bound $\sigma/\sigma_1 < p_1^*$ of the planar regime estimated by direct simulations with small initial conditions. Another test case, case 13, is very close to the limit p_1^* and, therefore, experiments demonstrate very long three-dimensional transient lasting for about 3 minutes before reaching planar motions.

The main consequence of increasing the excitation amplitudes consists of changing the effective frequency domain for stable swirling. It includes two intervals, (s_1, s_2) and (s_3, s_4), corresponding to the branches S_1 and S_3, respectively. The zone (s_2, s_3) corresponds to an extended domain (s_2, p_2) in Figure 9.24 and implies chaotic waves. Model test 17 confirms this fact. The effective domain of swirling, (s_1, s_2), is validated by experiments 15 and 16, but no appropriate model test is available to validate domain (s_3, s_4). Finally, branch S_2, corresponding to the mean amplitude at phase III, justifies the conclusion that the mean amplitudes of swirling must lie between S_1 and S_2.

In summary, the following conclusions can be drawn for the adaptive modal approach and nonlinear three-dimensional wave phenomena for square-base and nearly square-base tanks:

1. The number of dominant modes may be different for the same forcing amplitude and frequency. The number depends on the type of steady-state regime and the response branch to which the wave regime belongs.
2. Three-dimensional resonant steady-state sloshing, especially swirling, is very sensitive to transient conditions that may be influenced by different kinds of perturbations and by breaking waves. Swirling has several transient phases, which demonstrate beating waves with different beating periods and maximum and minimum amplitudes. The minima for the transient phases may be larger than the steady-state amplitude. This case is the opposite of two-dimensional sloshing, where the mean value of the maximum and minimum of the beating wave is close to the theoretical steady-state amplitude. Swirling may change rotation direction during the transient phase.

9.3 Vertical circular cylinder

Large-amplitude liquid motions in a circular-base tank was a main focus for spacecraft applications in the 1950s and 1960s. This field motivated detailed experimental and theoretical studies on resonantly forced sloshing with frequency in the vicinity of the lowest natural sloshing

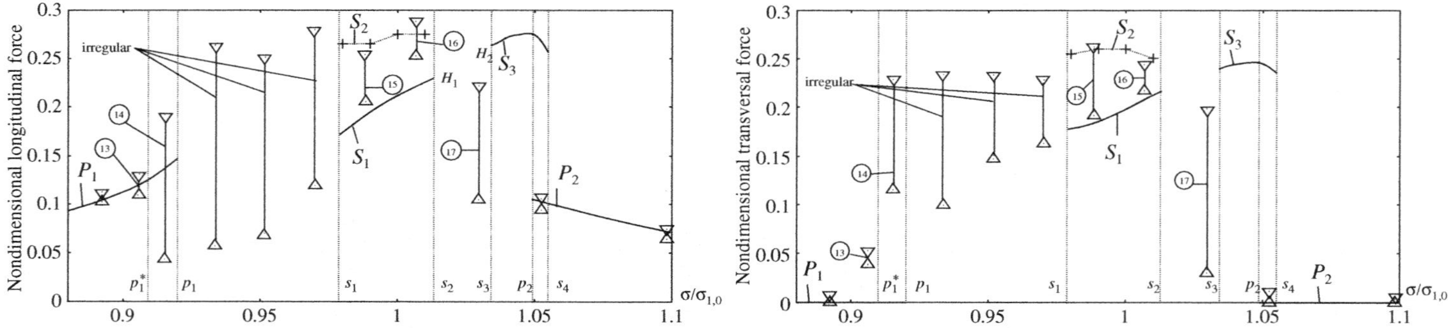

Figure 9.25. The same as in Figure 9.24, but for the nondimensional forcing amplitude $\eta_{1a}/L_1 = 0.01634$. The frequency domains (p_1, s_1) and (s_2, s_3) imply theoretical estimates of irregular (chaotic) wave motions. The estimate of p_1 changes to p_1^* if calculations are made with small initial conditions.

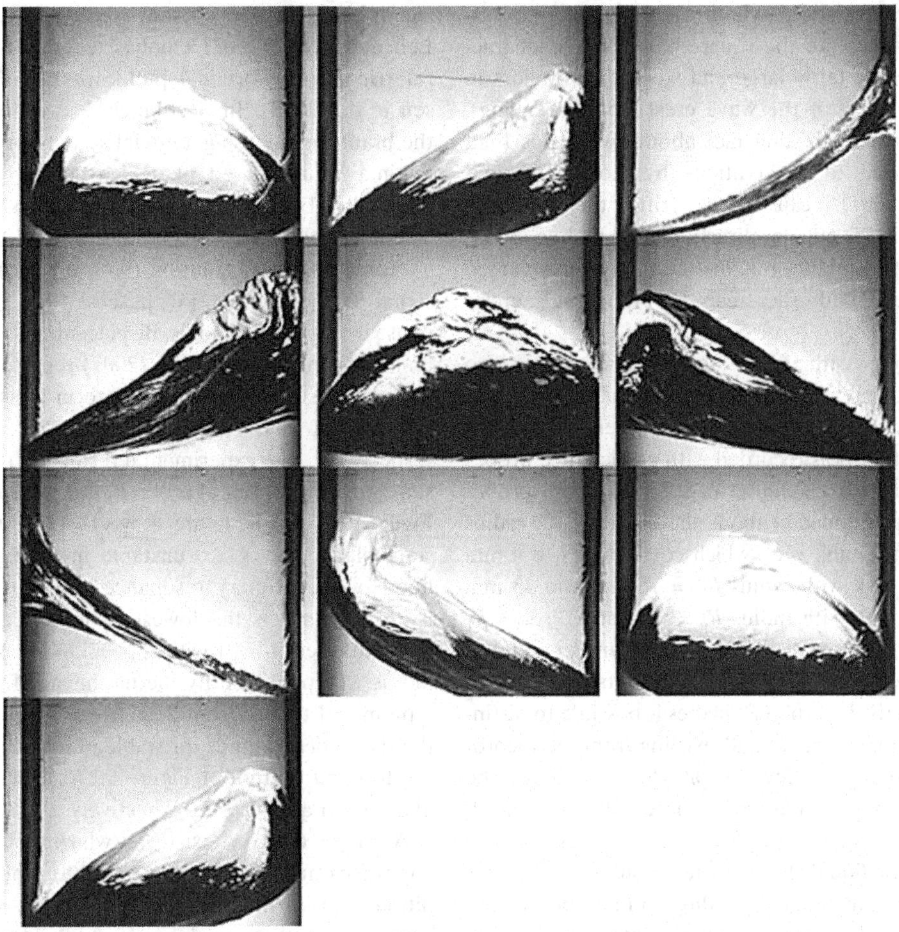

Figure 9.26. Images of a swirling wave in a vertical circular cylinder of radius 150 mm, partially filled with water. Views are in the direction normal to the tank motion ($h/R_0 = 1.2$, $\sigma/\sigma_1 = 1.02$, $\eta_{2a}/R_0 = 0.023$). The forcing period is 570 ms and the time between two images is 67 ms (Royon-Lebeaud *et al.*, 2007).

frequency. An important investigation was made by Abramson (1966), who is well known for outlining spacecraft-motivated experimental studies. His report also gives some details of the extended modal theory by Hutton (1963; see also Abramson *et al.*, 1966) that was further developed by Miles (1984a, 1984b). Abramson et al. (1966) experimentally investigated the effect of a vertical splitter plate in the forcing direction to suppress swirling. A review of other results on nonlinear sloshing in a circular-base tank can be found in the books by Lukovsky (1990) and Ibrahim (2005).

This section concentrates on the experimental results related to steady-state resonant waves (with emphasis on the investigations by Royon-Lebeaud *et al.*, 2007) and on the multimodal method for an upright circular cylinder. The use of the multimodal technique is, in our opinion, the best way to classify steady-state motions.

9.3.1 Experiments

Based on the experimental data by Abramson (1966) and Royon-Lebeaud *et al.* (2007), one can establish two types of stable steady-state wave regimes due to resonant lateral excitations with frequency in the vicinity of the lowest natural sloshing frequency. As mentioned in Section 9.1.1, these regimes are planar and swirling. Royon-Lebeaud *et al.* (2007) presented photos of the swirling (Figure 9.26) similar to those in Figure 1.4 for swirling in a square-base tank. Royon-Lebeaud *et al.* (2007) wrote that swirling in upright circular cylindrical tanks is very robust and, similar to the square-base tank, is characterized by fairly large and steep disturbances in the vicinity of the wave crest (local breaking). Another interesting fact about swirling is that, according to observations by Royon-Lebeaud *et al.* (2007), it causes mean drift of the liquid particles in the angular direction. This finding is consistent with Hutton (1964) and may be explained by the second-order analysis outlined in Exercise 9.6.3.

Along with these observations and video recordings, Royon-Lebeaud *et al.* (2007) presented wave amplitude measurements. The wave amplitude was recorded with capacitance probes that have a resolution of 0.2 mm. These probes were positioned at about one-eighth of the radius from the tank wall, which corresponds to 4 mm for a tank with radius $R_0 = 78$ mm and 13 mm for a tank with radius $R_0 = 150$ mm. Probes P_1 and P_2 were placed along a line parallel and perpendicular to the tank's horizontal movement, respectively. Probe P_2 makes it possible to distinguish planar waves and swirling from the recordings (i.e., the elevation at P_2 is small for the planar regime but comparable with those at P_1 for swirling and irregular (chaotic) motion). The ratio of liquid depth to tank radius, h/R_0, was larger than 1, corresponding to finite depth conditions. However, the experimental cylinder radii are relatively small and surface tension matters. The reason is that the Bond number for the experiments (water/air) with $R_0 = 78$ mm was equal to 820, but $R_0 = 150$ mm causes $Bo = 3{,}031$, while, according to the standard classification, the surface tension (and associated meniscus effect) may be neglected for Bond numbers greater than about 10,000 (Myskis et al., 1987). However, good agreement is later shown between the experiments and the multimodal method in classification of the wave system, which indicates that surface tension does not have a dominant effect.

When the forcing frequency is lower than the linear resonance frequency, the experiments demonstrate stable planar waves. This condition is the same as for longitudinal forcing in a square-base tank. Increasing the frequency gives a range of irregular waves and, at $\sigma/\sigma_1 = 1$, swirling occurs. Qualitatively, the branching is similar to that for the square-base tank in Figure 9.10(a), but a nearly diagonal steady-state regime is not realized. Royon-Lebeaud *et al.* (2007) emphasized this fact by referring to the original publication by Faltinsen *et al.* (2003). In Section 9.3.3, we describe the branching by using a modal system for sloshing in a circular-base tank, which was derived by Lukovsky (1990) and modified by Lukovsky and Timokha (1995) and Gavrilyuk *et al.* (2000).

From the planar solution (9.4), the measured signal at P_1 is either in phase or 180° out of phase with the harmonic displacements of the tank. Royon-Lebeaud *et al.* (2007) reported that the planar steady-state waves are in phase and 180° out of phase for $\sigma/\sigma_1 < 1$ and $\sigma/\sigma_1 > 1$, respectively. The experimental results for steady-state wave elevations at probe P_1 are presented in Figure 9.27(a). The figure shows that the steady-state planar waves are unstable in a neighborhood of the primary resonance $\sigma = \sigma_1$ (henceforth, $\sigma_1 = \sigma_{1,1}$ is the lowest natural frequency defined by eq. (4.39) with $\iota_{1,1} = 1.841\ldots$). This fact is well predicted by the nonlinear Moiseev-type modal theory from Section 9.3.2. We show the theoretical bounds for stable planar waves at the top and bottom of Figure 9.27(a) for larger and smaller amplitudes, respectively.

A range of σ/σ_1 exists in which both planar waves and swirling are unstable. This condition leads to irregular waves (chaos), whose occurrence and properties were studied by Miles (1984a, 1984b). This range is also well predicted by the modal theory that we present in the forthcoming sections. Royon-Lebeaud *et al.* (2007) detected swirling at $\sigma/\sigma_1 = 1$, which is the same as in the experiments by Abramson (1966). In accordance with the modal theory, which generally predicts well where stable steady-state swirling occurs, a narrow frequency domain exists where swirling is the only stable wave regime. In Figure 9.27 we see that, except at a small vicinity of the primary resonance, swirling coexists with planar waves. To study the steady-state amplitude of swirling when it coexists with planar waves, Royon-Lebeaud *et al.* (2007) proposed an experimental "path-following" procedure, starting with a forcing frequency close to $\sigma/\sigma_1 = 1$. After a steady state was reached, they increased the forcing frequency by a small increment and got a new steady-state swirling with a larger amplitude response. This experimental path-following corresponds to the

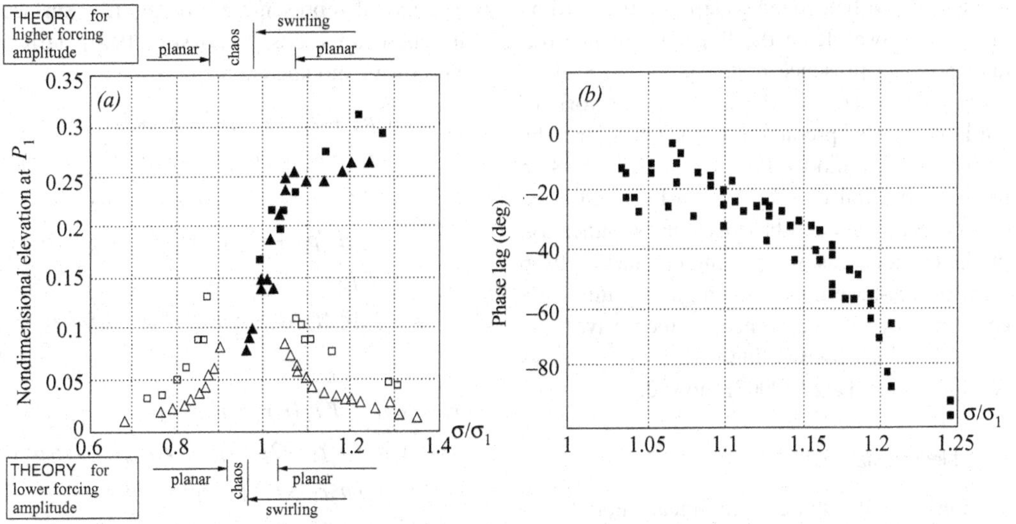

Figure 9.27. Experimental results by Royon-Lebeaud *et al.* (2007) for (a) the steady-state amplitudes of wave elevations at probe P_1 for planar waves (Δ and □) and swirling (▲ and ■) and (b) the phase between harmonic displacements of the tank and wave elevations at P_1 for swirling. The tank radius is $R_0 = 0.15$ m and the water depth is $h = 1.5R_0$. The symbols Δ and ▲ correspond to the measured data for the lower excitation amplitude $\eta_{2a}/R_0 = 0.023$, but □ and ■ denote the measured data for the larger amplitude, $\eta_{2a}/R_0 = 0.045$. The measured elevations were scaled by Royon-Lebeaud *et al.* (2007) by the wavelength of the lowest modes, $\lambda = 2\pi/\kappa_{1,1} = 2\pi R_0/\iota_{1,1} = 3.411 \cdot R_0$. The modal theory from Sections 9.3.2 and 9.3.3 gives good predictions for the frequency domains where stable planar waves and swirling are present. The theoretical wave domains are indicated on the bottom for $\eta_{2a}/R_0 = 0.023$ and on the top for $\eta_{2a}/R_0 = 0.045$ (see panel a). The theoretical estimates of "chaos" agree with the experiments. For planar waves, the phase between the input displacements and steady-state elevations at P_1 is zero when $\sigma/\sigma_1 < 1$ and 180° when $\sigma/\sigma_1 > 1$. Panel (b) presents the phase for swirling, which changes from 0 to −180°.

numerical path-following procedure elaborated for a square-base tank in Section 9.2.6. The results for the amplitude response at P_1 are presented in Figure 9.27(a) for two forcing amplitudes. Royon-Lebeaud *et al.* (2007) experimentally detected the aforementioned overlapping between the frequency domains of swirling and planar waves. The largest forcing frequency at which steady-state swirling remained stable was estimated to be approximately $\sigma/\sigma_1 = 1.3$. When the forcing frequency is further increased by a small amount, the swirling suddenly collapses and the wave motion switches to a small-amplitude out-of-phase planar wave (Figure 9.27(a)). This behavior is the same as that for swirling predictions in a square-base tank by the Moiseev-type theory in Section 9.2.4.1. However, Section 9.2.6.4 shows that for a square-base tank these forcing amplitudes lead to significant amplification of the higher modes, which affects the effective frequency domains of swirling and planar waves.

Figure 9.27(a) indicates two distinct dependencies of the swirling wave amplitude on the forcing frequency for a fixed forcing amplitude and slowly changing forcing frequency ("path-following"). In the range $1 < \sigma/\sigma_1 < 1.08$, the wave amplitude increases nearly linearly with the forcing frequency, as predicted by the multimodal theory for a square-base tank. In this case, breaking waves are of a very local nature and may be neglected. However, for larger forcing frequencies, and up to the collapse at $\sigma/\sigma_1 \approx 1.3$, the experiments reported local wave breaking near the wave crest. Furthermore, the output signal at P_1 is initially nearly in phase with the forcing as $1 < \sigma/\sigma_1 < 1.08$. This finding is consistent with the lowest-order prediction (9.5), because $\theta = 0$ for probe P_1. However, when the forcing frequency is increased, the experimental phase at P_1 decreases from zero to approximately $-\frac{1}{2}\pi$ (Figure 9.27(b)), which is consistent with reports by Abramson (1966) concerning conditions just before the collapse of the swirling mode (i.e., away from the linear resonance frequency). The presence of the phase $-\frac{1}{2}\pi$ at P_1 cannot be described within the framework of the lowest-order prediction (9.5). Therefore, the Moiseev-based model fails for $\sigma/\sigma_1 > 1.08$ in the experimental case by Royon-Lebeaud *et al.* (2007). This result may be caused by amplification of higher modes, as we explained in Section 9.2.6 for a square-base tank. This finding is implicitly confirmed by the occurrence of steep waves and strong breaking for swirling with $\sigma/\sigma_1 > 1.08$, as Royon-Lebeaud *et al.* (2007) showed.

9.3.2 Modal equations

Section 5.4.4 presented the linear modal system for sloshing in an upright circular cylindrical tank, which can be extended to describe nonlinear resonant motions (Lukovsky, 1990). The modal theory is also based on the Moiseev ordering between two primary excited modes and secondary modes. The latter modes are for a circular geometry associated with $m = 0$ and $m = 2$ in eq. (4.38). The third-order modes are related to modes with $m = 3$. The Moiseev ordering in terms of the modal function is

$$R_{1,1,i} \sim \beta_{1,1,i} \sim \varepsilon^{1/3},$$
$$R_{0,j} \sim \beta_{0,j} \sim R_{2,j,i} \sim \beta_{2,j,i} \sim \varepsilon^{2/3},$$
$$R_{3,j,i} \sim \beta_{3,j,i} \sim \varepsilon, \quad i = 1, 2; \quad j = 1, 2, \ldots. \tag{9.45}$$

The higher modes are either $O(\varepsilon)$ or $o(\varepsilon)$.

In contrast to the corresponding ordering for a rectangular tank (9.9), the Moiseev-type ordering operates with an infinite set of second-order modes, meaning that the corresponding modal system formally has an infinite number of nonlinear equations. However, Lukovsky (1990) and Miles (1984a, 1984b) showed that the second-order contribution for a circular base is associated basically with three modes: (0,1), (2,1,1), and (2,1,2). Based on this finding, Lukovsky (1990) derived a five-dimensional nonlinear modal system that couples the first- and second-order modes, as shown in the following text.

The analysis assumes that all geometric dimensions are scaled by the radius R_0; we redefine

$$p_1 = \beta_{1,1,1}, \quad r_1 = \beta_{1,1,2}, \quad p_0 = \beta_{0,1},$$
$$p_2 = \beta_{2,1,1}, \quad r_2 = \beta_{2,1,2}, \tag{9.46}$$

in the modal representation of the free surface. With these notations, Lukovsky (1990) derived the following modal system:

$$\begin{aligned}\ddot{p}_1 &+ \sigma_1^2 p_1 + d_1^* p_1\left(p_1\ddot{p}_1 + \dot{p}_1^2 + r_1\ddot{r}_1 + \dot{r}_1^2\right)\\ &+ d_2^*\left(r_1^2\ddot{p}_1 + 2r_1\dot{r}_1\dot{p}_1 - r_1 p_1\ddot{r}_1 - 2p_1\dot{r}_1^2\right)\\ &+ d_3^*\left(p_2\ddot{p}_1 + r_2\ddot{r}_1 + \dot{r}_1\dot{r}_2 + \dot{p}_1\dot{p}_2\right)\\ &- d_4^*\left(p_1\ddot{p}_2 + r_1\ddot{r}_2\right) + d_5^*\left(p_0\ddot{p}_1 + \dot{p}_1\dot{p}_0\right)\\ &+ d_6^* p_1\ddot{p}_0\\ &= -P_1[\ddot{\eta}_1(t)/R_0 - g\eta_5(t)/R_0 - \overline{S}_1\ddot{\eta}_5(t)],\end{aligned} \tag{9.47}$$

$$\begin{aligned}\ddot{r}_1 &+ \sigma_1^2 r_1 + d_1^* r_1\left(r_1\ddot{r}_1 + \dot{r}_1^2 + p_1\ddot{p}_1 + \dot{p}_1^2\right)\\ &+ d_2^*\left(p_1^2\ddot{r}_1 + 2p_1\dot{r}_1\dot{p}_1 - r_1 p_1\ddot{p}_1 - 2r_1\dot{p}_1^2\right)\\ &- d_3^*\left(p_2\ddot{r}_1 - r_2\ddot{p}_1 + \dot{r}_1\dot{p}_2 - \dot{p}_1\dot{r}_2\right)\\ &+ d_4^*\left(r_1\ddot{p}_2 - p_1\ddot{r}_2\right) + d_5^*\left(p_0\ddot{r}_1 + \dot{r}_1\dot{p}_0\right)\\ &+ d_6^* r_1\ddot{p}_0\\ &= -P_1[\ddot{\eta}_2(t)/R_0 + g\eta_4(t)/R_0 + \overline{S}_1\ddot{\eta}_4(t)],\end{aligned} \tag{9.48}$$

$$\ddot{p}_0 + \sigma_0^2 p_0 + d_{10}^*\left(r_1\ddot{r}_1 + p_1\ddot{p}_1\right) + d_8^*\left(\dot{r}_1^2 + \dot{p}_1^2\right) = 0, \tag{9.49}$$

$$\ddot{p}_2 + \sigma_2^2 p_2 + d_9^*\left(r_1\ddot{r}_1 - p_1\ddot{p}_1\right) + d_7^*\left(\dot{r}_1^2 - \dot{p}_1^2\right) = 0, \tag{9.50}$$

$$\ddot{r}_2 + \sigma_2^2 r_2 - d_9^*\left(r_1\ddot{p}_1 + p_1\ddot{r}_1\right) - 2d_7^*\dot{r}_1\dot{p}_1 = 0, \tag{9.51}$$

where the coefficients P_j and $\overline{S}_1 = S_1/R_0$ are defined by eq. (5.156) and the nondimensional values of the hydrodynamic coefficients d_j^* are presented in Table 9.3, which collects numerical results by Lukovsky (1990) and Gavrilyuk *et al.* (2000). The coefficients become almost the same for $h/R_0 > 2.4$ (i.e., close to a deep-liquid approximation for nonlinear sloshing).

Based on general expressions from Section 7.3, Lukovsky (1990) presented the asymptotic formulas for hydrodynamic force and moment. Whereas the formula for the moment is very complicated due to the last quantities in eq. (7.32) associated with $\boldsymbol{l}_\omega$ and $\boldsymbol{l}_{\omega t}$ requiring an asymptotic solution (in terms of the generalized coordinates) of the problem on Stokes–Joukovski potential (7.3) – Lukovsky adopted the corresponding solution from the book by Narimanov *et al.* (1977) and did not validate it – the formula for the hydrodynamic force is of very simple form. The reason is that, according to eq. (7.22), the hydrodynamic force contains only the terms $d^*\boldsymbol{v}_O/dt$,

Table 9.3. *Coefficients in eqs. (9.47)–(9.51) for an upright circular cylindrical tank calculated by Lukovsky (1990) and Gavrilyuk* et al. *(2000)*

h/R_0	$d_1^* R_0^2$	$d_2^* R_0^2$	$d_3^* R_0$	$d_4^* R_0$	$d_5^* R_0$	$d_6^* R_0$	$d_7^* R_0$	$d_8^* R_0$	$d_9^* R_0$	$d_{10}^* R_0$
0.2	5.4405	2.9244	1.0668	0.8799	2.7616	−0.952	4.4734	−3.1305	2.7851	−1.2775
0.4	1.6064	0.2886	0.9222	0.2708	1.8764	−0.389	2.0079	−1.4145	0.7429	−0.4146
0.6	0.9226	−0.1782	0.9556	0.0532	1.7018	−0.1959	1.2877	−0.9382	0.1290	−0.1756
0.8	0.7053	−0.3249	0.9909	−0.0437	1.6567	−0.1094	1.0122	−0.7609	−0.0979	−0.0888
1.0	0.6198	−0.3820	1.0123	−0.0885	1.6427	−0.0685	0.8951	−0.6858	−0.1898	−0.0528
1.2	0.5827	−0.4063	1.0237	−0.1096	1.6377	−0.0487	0.8425	−0.6517	−0.2294	−0.0366
1.4	0.5658	−0.4171	1.0293	−0.1195	1.6360	−0.0393	0.8181	−0.6362	−0.2474	−0.0291
1.6	0.558	−0.4223	1.0321	−0.1240	1.6353	−0.0347	0.8070	−0.6287	−0.2553	−0.0256
1.8	0.5543	−0.4246	1.0333	−0.1262	1.6349	−0.0327	0.8016	−0.6253	−0.2590	−0.0240
2.0	0.5526	−0.4258	1.0340	−0.1274	1.6348	−0.0316	0.7987	−0.6236	−0.2611	−0.0232
2.2	0.5518	−0.4264	1.0344	−0.1280	1.6348	−0.031	0.7972	−0.6226	−0.2621	−0.0227
2.4	0.5514	−0.4265	1.0346	−0.1282	1.6347	−0.0208	0.7968	−0.6223	−0.2624	−0.0226
2.6	0.5514	−0.4265	1.0346	−0.1281	1.6346	−0.0308	0.7969	−0.6223	−0.2624	−0.0226
2.8	0.5513	−0.4265	1.0345	−0.1281	1.6346	−0.0308	0.7970	−0.6223	−0.2624	−0.0226
3.0	0.5511	−0.4268	1.0345	−0.1281	1.6346	−0.0308	0.7971	−0.6222	−0.2624	−0.0226

$\dot{\omega} \times r_{lC} = \dot{\omega} \times r_{lC_0} + o(\varepsilon)$, and $d^{*2} r_{lC}/dt^2$, representing the forcing and sloshing correctly to $O(\varepsilon)$. Including only nonlinearly coupled generalized coordinates from eqs. (9.47)–(9.51), we get the approximate result in dimensional form as follows:

$$F_1 = M_l\left(g\eta_5 - \ddot{\eta}_1 + \tfrac{1}{2}h\ddot{\eta}_5\right) - C_1 R_0 \ddot{p}_1,$$
$$F_2 = M_l\left(-g\eta_4 - \ddot{\eta}_2 - \tfrac{1}{2}h\ddot{\eta}_4\right) - C_1 R_0 \ddot{r}_1,$$
$$F_3 = -M_l g - \tfrac{1}{2}C\left(r_1\ddot{r}_1 + p_1\ddot{p}_1 + \dot{r}_1^2 + \dot{p}_1^2\right), \quad (9.52)$$

where $C_1 = \lambda_{1(1,1,1)} = \lambda_{2(1,1,2)}$ (see eq. (5.153)) and

$$C = \rho\pi R_0^2 \int_0^{R_0} r R_{1,1}^2(r)\,\mathrm{d}r = \tfrac{1}{2}\rho\pi R_0^4\left(\iota_{1,1}^2 - 1\right)/\iota_{1,1}^2.$$

Lukovsky (1990) reported that formulas (9.52) are supported by experiments.

9.3.3 Steady-state solutions

Following the assumptions from Section 9.2.3, we consider the case when the tank oscillates harmonically with a frequency σ and no phase shift occurs between the tank's motion modes $\eta_j(t)$, namely, $\eta_j(t) = \eta_{ja}\cos(\sigma t)$ (j = 1, 2, 4, 5). This type of harmonic excitation implies forcing for which trajectories of any fixed point of the solid body belongs to a vertical plane that includes the axis of the upright circular cylinder. When we study forced sway and roll motion in the Oyz-plane, the only homogeneous term appears in eq. (9.48). The right-hand side of eq. (9.48) then takes the form

$$\sigma^2 P_1^\varepsilon \cos\sigma t$$
$$= -\left[\sigma^2(P_1(-\eta_{2a}) + \eta_{4a}(-S_1 + g/\sigma^2))/R_0\right]\cos\sigma t. \quad (9.53)$$

Based on eq. (9.53), modal system (9.47)–(9.51) admits an asymptotic steady-state analysis following the scheme developed in Section 9.2.3. This analysis assumes the two dominant modes to have generalized coordinates

$$r_1(t) = A\cos(\sigma t) + \overline{A}\sin(\sigma t) + o(\varepsilon^{1/3}),$$
$$p_1(t) = \overline{B}\cos(\sigma t) + B\sin(\sigma t) + o(\varepsilon^{1/3}), \quad (9.54)$$

and, after substitution into eqs. (9.49)–(9.51), the remaining modal functions are computed as

$$p_0(t) = c_0 + c_1\cos(2\sigma t) + c_2\sin(2\sigma t) + o(\varepsilon^{2/3}),$$
$$p_2(t) = s_0 + s_1\cos(2\sigma t) + s_2\sin(2\sigma t) + o(\varepsilon^{2/3}),$$
$$r_2(t) = e_0 + e_1\cos(2\sigma t) + e_2\sin(2\sigma t) + o(\varepsilon^{2/3}), \quad (9.55)$$

where

$$c_0 = l_0(A^2 + \overline{A}^2 + B^2 + \overline{B}^2),$$
$$c_1 = h_0(A^2 - \overline{A}^2 - B^2 + \overline{B}^2),$$
$$c_2 = 2h_0(A\overline{A} + B\overline{B}),$$
$$s_0 = l_2(A^2 + \overline{A}^2 - B^2 - \overline{B}^2),$$
$$s_1 = h_2(A^2 - \overline{A}^2 + B^2 - \overline{B}^2),$$
$$s_2 = 2h_2(A\overline{A} - B\overline{B}),$$
$$e_0 = -2l_2(A\overline{B} + B\overline{A}),$$
$$e_1 = 2h_2(\overline{A}B - A\overline{B}),$$
$$e_2 = -2h_2(AB + \overline{A}\overline{B}),$$

$$h_0 = \frac{d_{10}^* + d_8^*}{2\left(\overline{\sigma}_0^2 - 4\right)}, \quad h_2 = \frac{d_9^* + d_7^*}{2\left(\overline{\sigma}_2^2 - 4\right)},$$
$$l_0 = \frac{d_{10}^* - d_8^*}{2\overline{\sigma}_0^2}, \quad l_2 = \frac{d_9^* - d_7^*}{2\overline{\sigma}_2^2}.$$

Substitution of eqs. (9.55) into eqs. (9.47) and (9.48) gives a solvability condition similar to eq. (9.22), with formal equivalence $m_1 = m_2$, as follows:

$$\left.\begin{aligned}
&A((\overline{\sigma}_1^2 - 1) + m_1(A^2 + \overline{A}^2 + \overline{B}^2) + m_3 B^2) \\
&\quad + (m_1 - m_3)\overline{A}B\overline{B} - P_1^\varepsilon = 0, \\
&\overline{A}((\overline{\sigma}_1^2 - 1) + m_1(A^2 + \overline{A}^2 + B^2) + m_3\overline{B}^2) \\
&\quad + (m_1 - m_3)AB\overline{B} = 0, \\
&\overline{B}((\overline{\sigma}_1^2 - 1) + m_1(B^2 + \overline{B}^2 + A^2) + m_3\overline{A}^2) \\
&\quad + (m_1 - m_3)\overline{A}AB = 0, \\
&B((\overline{\sigma}_1^2 - 1) + m_1(B^2 + \overline{B}^2 + \overline{A}^2) + m_3 A^2) \\
&\quad + (m_1 - m_3)\overline{A}A\overline{B} = 0,
\end{aligned}\right\} \quad (9.56)$$

where $\overline{\sigma}_1 = \sigma_{1,1}/\sigma$ and

$$\begin{aligned}
m_1 = m_2 &= d_5^*\left(\tfrac{1}{2}h_0 - l_0\right) - d_3^*\left(\tfrac{1}{2}h_2 - l_2\right) \\
&\quad - 2d_6^* h_0 - 2d_4^* h_2 - \tfrac{1}{2}d_1^*, \\
m_3 &= -d_3^*\left(l_2 + \tfrac{3}{2}h_2\right) - d_5^*\left(l_0 + \tfrac{1}{2}h_0\right) \\
&\quad + 2d_6^* h_0 - 6d_4^* h_2 + \tfrac{1}{2}d_1^* - 2d_2^*.
\end{aligned} \quad (9.57)$$

System (9.56) has the same structure as eq. (9.22) for a square-base tank in which we require condition (9.25), longitudinal forcing, and $m_1 = m_2$. When repeating the analysis from Section 9.2.4.1, we can conclude the existence of the two types of motion: planar waves ($\overline{A} = B = \overline{B} = 0$) and swirling ($\overline{A} = \overline{B} = 0$). Nearly diagonal wave motions do not exist, because these motions require the condition $m_1 \neq m_2$, which is not fulfilled.

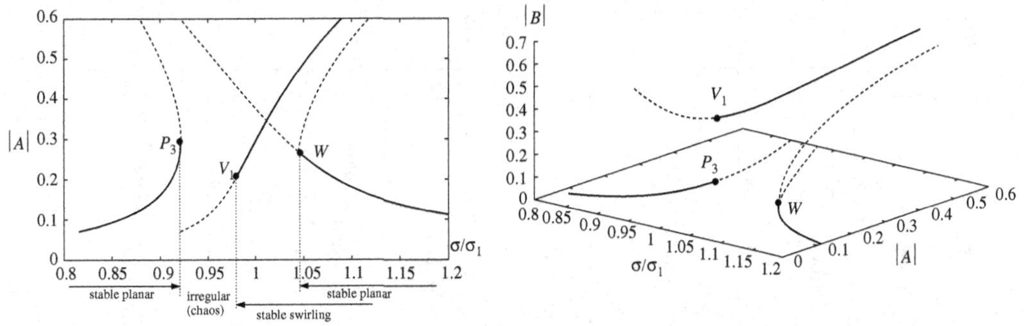

Figure 9.28. Response curves for nonlinear sloshing in an upright circular cylindrical tank excited by horizontal harmonic forcing with $\eta_{2a}/R_0 = 0.023$ and $h/R_0 = 1.5$. Notations are explained in Section 9.2.4.1 for the case of sloshing in a square-base tank due to longitudinal excitations. Solid lines correspond to stable steady-state waves. Dashed lines imply hydrodynamic instability. The case $B = 0$ (planar curves from three-dimensional view) means a planar regime, but three-dimensional curves correspond to swirling. A range exists between the abscissas of V_1 and P_3, where the theory predicts irregular waves (chaos). The theoretical predictions for stable steady-state regimes are compared with experimental data in Figure 9.27.

Based on the two-time-scale technique that was applied for a square-base tank in Section 9.2.3, one can study the stability of planar and swirling wave regimes for an upright circular cylindrical tank. Typical branching for the experimental case by Royon-Lebeaud *et al.* (2007) with $\eta_{2a}/R_0 = 0.023$ and $h/R_0 = 1.5$ is presented in Figure 9.28. If we exclude the branches responsible for the nearly diagonal regime, the branching is qualitatively similar to that for a longitudinally forced square-base tank with either finite or infinite liquid depth. For the studied case, the planar regime is characterized by a soft-spring response. Swirling demonstrates a hard-spring behavior. It is of interest to study the critical depths. The corresponding analysis, based on modal theory (9.47)–(9.51), establishes a zero of $m_1^{(0)} = m_1|_{\sigma=\sigma_1}$ when $h/R_0 = 0.5059$. At this depth, the planar waves change from soft- to hard-spring behavior. The result is consistent with that of Miles (1984a), who gave the same value. Furthermore, the sign of $m_1 + m_3$ is always positive, which means that swirling keeps a hard-spring character for $h/R_0 \geq 0.2$, as was shown in Section 9.2.4.1.

The theoretical response curves and the stability of steady-state waves are of primary interest in the context of the experimental data discussed in connection with Figure 9.27. One can see a range with only stable planar waves to the left of the abscissa of P_3. Furthermore, swirling is the only possibility in the domain between the abscissas of V_1 and W. We can expect either planar waves or swirling for values of σ/σ_1 to the right of the abscissa of W. Finally, chaos is expected in a range between P_3 and V_1. These conclusions, as well as the theoretical bounds of the frequency domains, are generally in agreement with the experimental data. The corresponding comparisons are shown in Figure 9.27.

The change of the effective frequency domains associated with stable steady-state regimes and chaos versus forcing amplitude and liquid depth

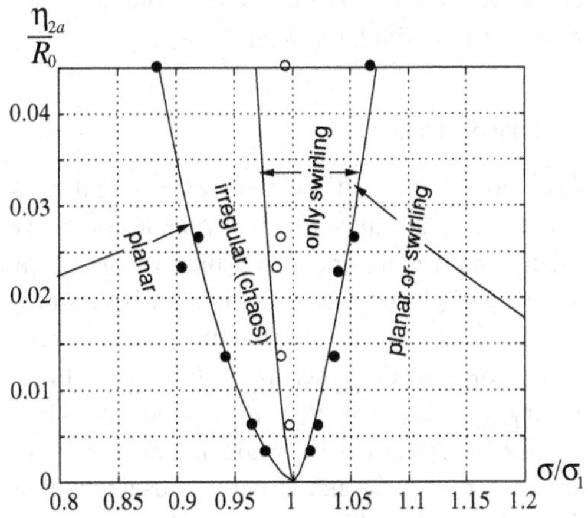

Figure 9.29. Effective frequency domains for stable steady-state waves and chaos in the $(\sigma/\sigma_1, \eta_{2a}/R_0)$-plane for an upright circular cylindrical tank with $h/R_0 = 1.5$ and horizontal excitation. The bounds between the domains are obtained by using modal theory (9.47)–(9.51) (solid lines). Experimentally predicted bounds are taken from Royon-Lebeaud *et al.* (2007); •, experimental bounds for planar waves; ∘, experimental bound for swirling.

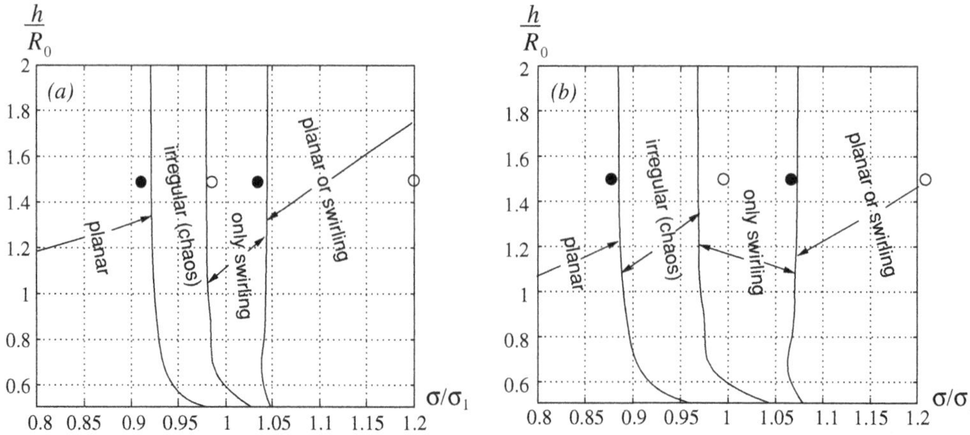

Figure 9.30. Effective frequency domains for stable steady-state waves and chaos in the $(\sigma/\sigma_1, h/R_0)$-plane for an upright circular cylindrical tank with (a) $\eta_{2a}/R_0 = 0.023$ and (b) $\eta_{2a}/R_0 = 0.045$. The bounds between the domains are obtained by using modal theory (9.47)–(9.51) (solid lines). Experimentally predicted bounds are taken from Figure 9.27 with the same notations.

are illustrated in Figures 9.29 and 9.30. The influence of the forcing amplitude is generally supported by Royon-Lebeaud *et al.*'s (2007) experimental results. One important fact is that the range of hydrodynamic instability (chaos) increases with increasing η_{2a}/R_0. From Figure 9.30 we note that the frequency domain with chaos decreases with decreasing h/R_0. (Figure 9.30 presents the results for $h/R_0 \geq 0.5059$.) Furthermore, the frequency domains for different wave motions become almost the same for $h/R_0 \geq 1$ (i.e., similar as for infinite depth).

9.4 Spherical tank

The most important load on a spherical tank is the sloshing-induced hydrodynamic force, the predominant component of which is the lateral force. Swirling occurs readily so there is generally a force component perpendicular to the forced oscillation direction. According to potential flow theory, angular tank velocities about axes in a coordinate system with origin in the geometrical center of the tank do not cause any flow in a spherical tank without interior structures. Because viscous effects are secondary, translatory tank velocities are most important, particularly lateral tank velocities. An important part of the vertical dynamic force due to vertical excitation is the inertial force of the frozen liquid. Our focus in the following text is on experimental steady-state hydrodynamic forces due to harmonic excitations. Because of the nonlinear nature of sloshing, we cannot generalize the results to a tank onboard a ship in a realistic seaway. However, the results can give an idea about the magnitude of the hydrodynamic forces by considering regular design wave conditions. The strength criteria for different parts of the shell of a spherical tank are summarized in Figure 9.31. Ideally we should know the hydrodynamic pressure distribution, instead of integrated forces, when assessing the strength.

To our knowledge, no analytical studies of nonlinear liquid sloshing in a spherical tank have been performed like those for rectangular- and circular-base cylindrical tanks. However, the literature contains experimental investigations. For instance, Abramson *et al.* (1966) did experiments for a half-filled spherical tank with a vertical splitter plate installed in the tank parallel to the excitation direction. The splitter plate suppresses swirling. The experiments show the jump phenomena between steady-state solution branches discussed in Chapter 8 for two-dimensional rectangular tanks. Furthermore, the larger the excitation amplitude, the longer the period of maximum response.

Olsen and Hysing (1974) presented extensive experimental results for forced harmonic sway excitation of a spherical tank with a tower. Some of these results have also been presented by Hysing (1976). The instrumentation is shown in Figure 9.32. The ratio of tower diameter to tank diameter was $D/D_0 = 1/12$. By tank diameter we mean the diameter relative to the inner tank

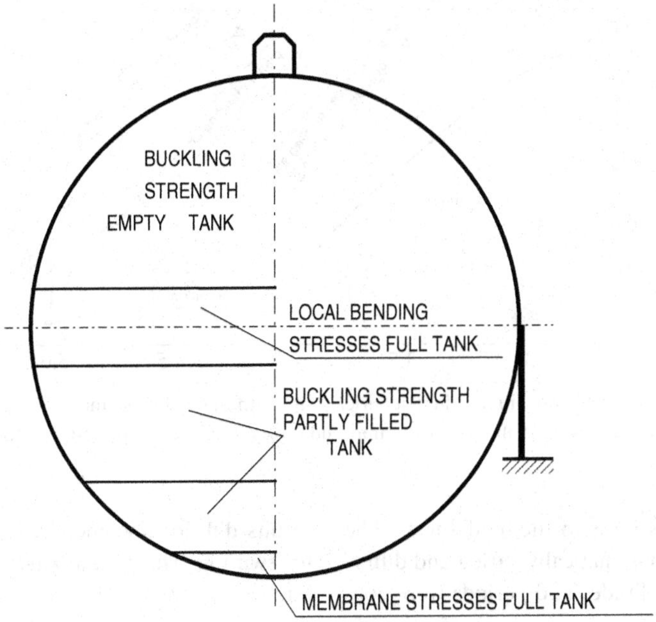

Figure 9.31. Strength calculations of spherical tanks (Hysing, 1976).

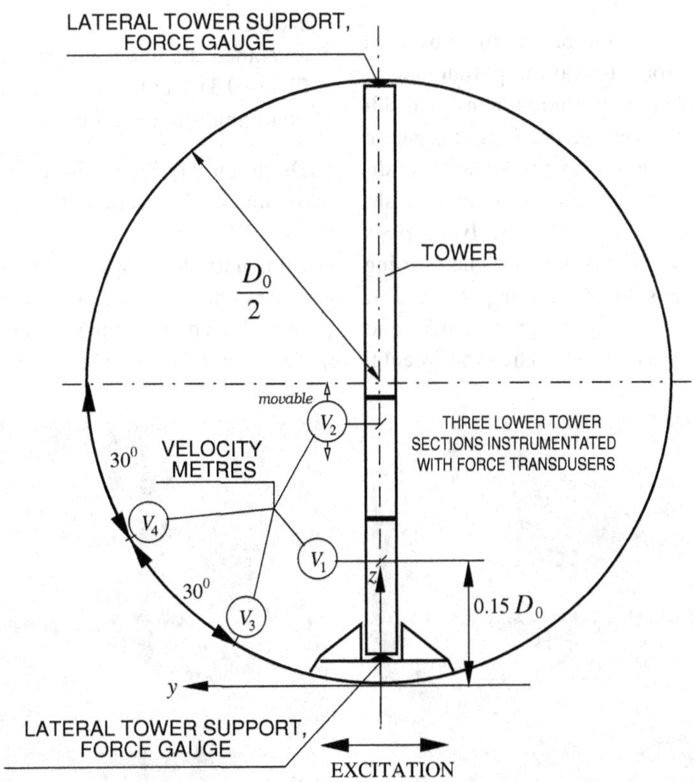

Figure 9.32. Instrumentation in the model tests of the spherical tank with tower by Olsen and Hysing (1974).

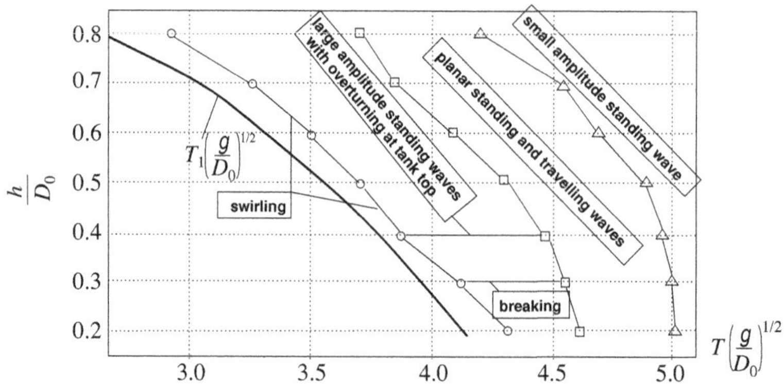

Figure 9.33. Observed wave modes in a spherical tank with diameter D_0 and different filling levels h. Forced harmonic sway oscillations with amplitude $\eta_{2a} = 0.05D_0$ and period T (Olsen & Hysing, 1974).

surface, which was 1.8 m in the model tests. The filling depth was systematically varied and different excitation amplitudes and periods were considered.

9.4.1 Wave regimes

Figure 9.33 gives an example of the observed wave modes for forced sway amplitude $\eta_{2a} = 0.05D_0$ as a function of the nondimensional filling level h/D_0 and forcing period $T\sqrt{g/D_0}$, where h is the maximum liquid depth in static conditions. The highest nondimensional natural sloshing period $T_1\sqrt{g/D_0}$, based on McIver (1989; see also Table 4.4), is also plotted. The swirling domain corresponds to excitation periods $T < T_{\text{swlim}}$, where T_{swlim} is slightly higher than T_1 and decreases with increasing h/D_0. The experimental results did not document a lower period bound for swirling. The planar wave modes are divided into

- large-amplitude standing waves with overturning at the tank top occurring for $h/D_0 \geq 0.4$,
- breaking waves occurring for smaller liquid levels $h/D_0 \leq 0.3$,
- combined standing and traveling waves (see Figure 9.34), and
- small-amplitude standing waves.

Olsen and Hysing (1974) also reported on the wave mode observations for $\eta_{2a} = 0.01D_0$ and $0.08D_0$. When $\eta_{2a} = 0.01D_0$, the border between swirling and planar waves corresponded to periods very close to the highest natural sloshing period, T_1. The examined filling depth ratios for $\eta_{2a} = 0.08D_0$ were $h/D_0 = 0.3$, 0.4, and 0.7.

Figure 9.34. Combined planar standing and traveling waves in a spherical tank with filling level $h/D_0 = 0.5$. Forced harmonic sway oscillations with amplitude $\eta_{2a} = 0.05\,D_0$ and period $T\sqrt{g/D_0} = 4.79$ (Olsen & Hysing, 1974).

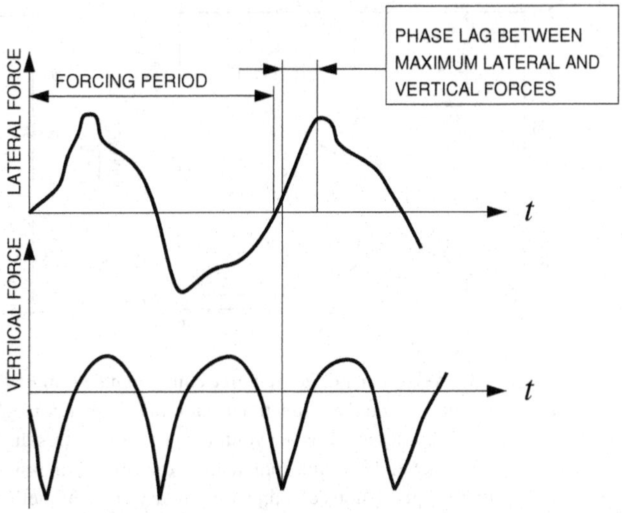

Figure 9.35. Characteristic time history of vertical and lateral sloshing forces on a spherical tank during steady-state periodic conditions (experimental results by Hysing, 1976).

When $h/D_0 = 0.3$ and 0.4, the border between swirling and planar waves was similar to that for $\eta_{2a} = 0.05D_0$. When $h/D_0 = 0.7$, swirling occurred for $T\sqrt{g/D_0} \leq 2.95$ (i.e., for periods lower than T_1).

Hysing (1976) examined the dependence of experimental vertical and transverse hydrodynamic forces on the filling height and the forced oscillation amplitude and period. An example of the time history of steady-state horizontal and vertical hydrodynamic forces is shown in Figure 9.35. The periodicity of the vertical force is half the forcing period, which is consistent with eq. (5.36) showing that the linear β_i-components of the vertical force component are zero and the lowest-order term of the vertical force is related to $d^{*2}\boldsymbol{r}_{C_l}/dt^2$ (due to eq. (7.22)). The latter is quadratic in the modal functions.

Maximum steady-state transverse force as a function of the oscillation period is presented in Figure 9.36(a) for the ratios of filling depth to tank diameter $h/D_0 = 0.3$, 0.5, and 0.7. The sway forcing amplitude η_{2a} is $0.05D_0$. Because of swirling, the *transverse force component* perpendicular to the forcing direction in the horizontal plane is not zero. We can use Figure 9.33 as a guide when swirling occurs. The maximum of the frozen liquid force $\rho V_l \ddot{\eta}_2$ is also included in

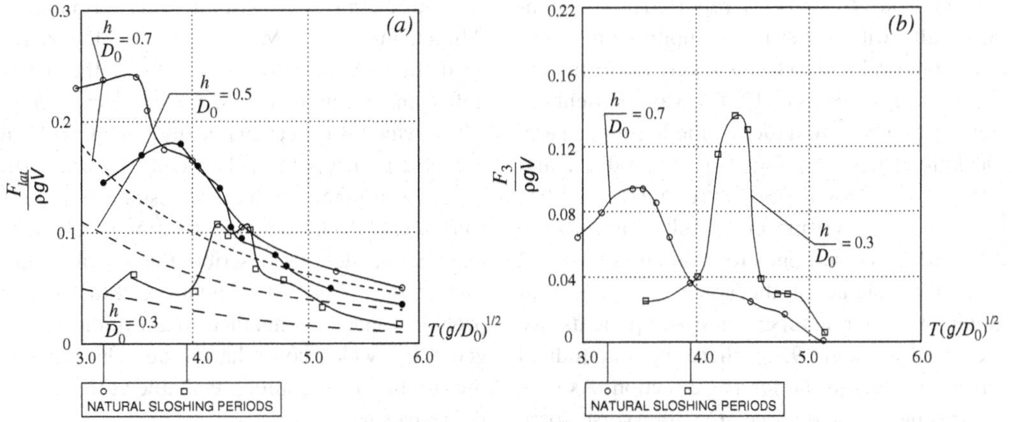

Figure 9.36. (a) Maximum lateral force F_{lat} and (b) maximum vertical dynamic force components versus nondimensional forced oscillation period $T\sqrt{g/D_0}$ for different ratios of filling depth to tank diameter, h/D_0, of a spherical tank. Sway amplitude $\eta_{2a} = 0.05\,D_0$, V is the volume of sphere. Note that the lateral force direction does not generally coincide with the forcing direction. The natural sloshing periods are specified on the bottom of the figures: $h/D_0 = 0.7(\circ)$ and 0.3 ($\square$). Dashed lines represent the "frozen" liquid force (Hysing, 1976).

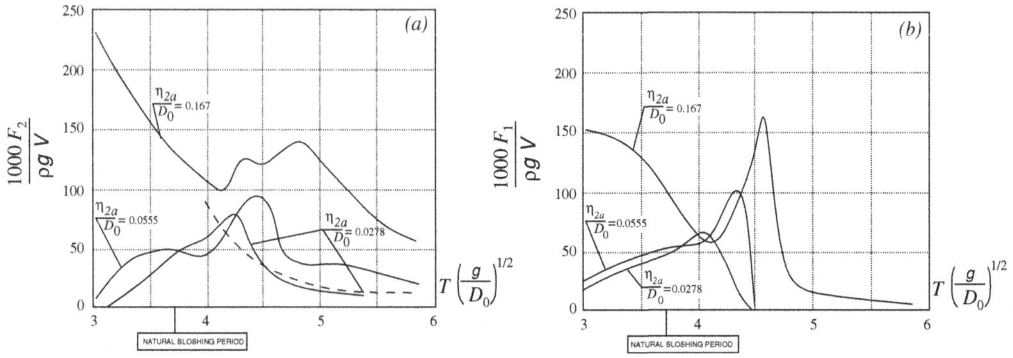

Figure 9.37. Maximum horizontal force components F_2 and F_1 (a) aligned with and (b) perpendicular to the forcing direction as a function of nondimensional forced oscillation period $T\sqrt{g/D_0}$ for different sway amplitudes η_{2a}; V is the volume of the sphere (results by Abramson *et al.*, 1974; obtained for $h/D_0 = 0.289$). Solid lines interpolate the experimental measurements, but the dashed line represents the theoretical predictions of a linear boundary element method (Abramson *et al.*, 1974).

the figure and illustrates the fact that sloshing is important. When the forcing period is clearly higher than the highest natural sloshing period, the frozen-liquid force asymptotically approaches the hydrodynamic force. Even though the vertical hydrodynamic force is, in general, smaller than the transverse hydrodynamic force, it is not negligible (see Figure 9.36(b)). Figure 9.36(a) shows that the period of maximum lateral force decreases with increasing filling depth for the given sway motion. However, one should realize that a shorter period may imply a shorter design value for sway. The largest difference between the hydrodynamic force and the frozen liquid force occurs at $h/D_0 = 0.5$.

Abramson *et al.* (1974) reported experimental results with no swirling suppression for a clean spherical tank. The tests were performed at Det norske Veritas in 1972 for Moss-Rosenberg Verft A/S. The amplitudes of the harmonic sway oscillations were $\eta_{2a}/D_0 = 0.0278$, 0.0555, and 0.167. The filling depths were $h/D_0 = 0.289$, 0.5, and 0.65. We present results for $h/D_0 = 0.289$, which corresponds to a liquid volume 0.2 times the volume of the sphere, V. The longitudinal and transverse force components are presented in Figure 9.37, where the longitudinal direction refers to the forcing direction. As long as swirling is not realized, the transverse force component is theoretically zero. Figure 9.37(b) shows that this is not the case in a certain frequency domain. The swirling hinders clear observation of the resonance phenomena, especially for the largest excitation amplitude. The maximum response occurs for much smaller periods than those in the range where maximum response is expected when swirling is suppressed. If $\eta_{2a}/D_0 = 0.167$, the largest horizontal force component in the forced oscillation direction, F_2, occurs at the nondimensional period $T\sqrt{g/D_0} = 3.0$ (see Figure 9.37(a)) with magnitude 1.2 times the weight of the liquid.

A linear boundary element method was used by Abramson *et al.* (1974) to predict tank forces in the oscillation direction. Because swirling is a nonlinear phenomenon, it could not be numerically predicted. For excitation periods sufficiently larger than the natural period, where not much swirling occurs, the linear results agree well with experiments. The calculated linear natural period is plotted and is consistent with formulas by Abramson (1966). McIver's (1989) theoretical prediction of the natural period for the lowest antisymmetric mode is $T_1\sqrt{g/D_0} = 3.95$ for $h/D_0 = 0.3$, which is in reasonable agreement with the numerical calculations. The convergence could be an issue for the numerical results by Abramson *et al.* (1974) meaning that more elements may be needed to describe the mean wetted surface in their BEM calculations. Interior flow problems need a detailed description of the geometry, which could hardly be achieved with the computer capabilities available at the time of their investigation.

9.4.2 Tower forces

The hydrodynamic force distribution along a pumping tower in a spherical tank was also experimentally investigated by Olsen and

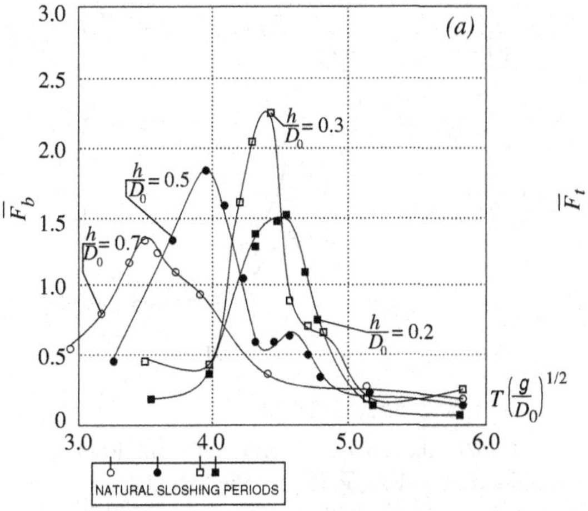

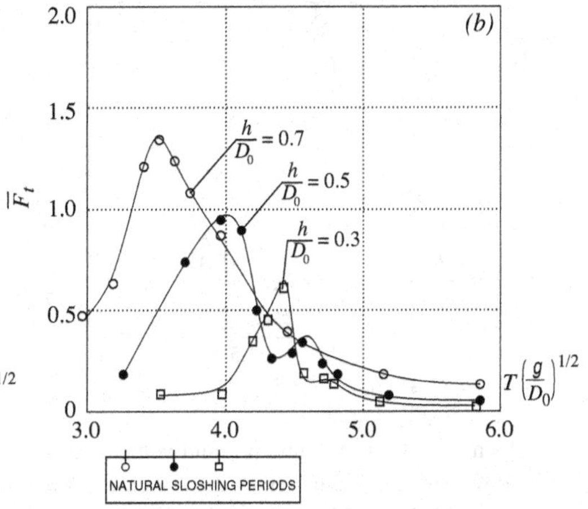

Figure 9.38. (a) Nondimensional bottom tower support force amplitude $\overline{F}_b = 5F_b/(\frac{1}{4}\rho g\pi D^2 D_0)$ and (b) nondimensional top tower support force amplitude $\overline{F}_t = 5F_t/(\frac{1}{4}\rho g\pi D^2 D_0)$ for several filling levels h of a spherical tank. The ratio of tower diameter to tank diameter is $D/D_0 = 1/12$. Forced harmonic sway oscillations with amplitude $\eta_{2a} = 0.05D_0$ and nondimensional periods $T\sqrt{g/D_0}$. The symbols for the highest natural sloshing periods are the same as for the response curves (Olsen & Hysing, 1974).

Hysing (1974). Figure 9.38 presents the nondimensional tower support forces on the top and bottom, respectively (see Figure 9.32 for definitions). An estimate of the maximum hydrodynamic force on the tower is obtained by adding the results in Figure 9.38. The maximum hydrodynamic force amplitude in the considered period interval, $F_{\text{hyd}}^{\text{tower}}$, is nearly constant for $0.3 \le h/D_0 \le 0.7$ and corresponds to $F_{\text{hyd}}^{\text{tower}} \approx 0.15\rho g\pi D^2 D_0$. If we consider liquefied natural gas with density $\rho = 474$ kg m^{-3} and a spherical tank with diameter $D_0 = 36$ m and a tower diameter $D = 3$ m, this gives $F_{\text{hyd}}^{\text{tower}} \approx 710$ kN. The inertial force due to the weight of the tower is on the order of 200 kN (Hysing, 1976). The previous application of the expression for $F_{\text{hyd}}^{\text{tower}}$ assumes that Froude scaling is valid (i.e., that the scale effect due to Reynolds number was neglected).

Forces on the tower sections were also measured, as illustrated in Figure 9.32. The sectional forces can be estimated by the generalized Morison equation given by eq. (6.66). For this case we need to know the transverse relative velocity and acceleration components at the tower axis when the tower is not there. Olsen and Hysing (1974) presented results for the maximum relative inflow velocity at $0.15D_0$ above the base and $0.05D_0$ below the static liquid surface (see Figure 9.32 for a definition of measurement points). Results are presented in Figure 9.39 for $\eta_{2a} = 0.05D_0$. By relating the results to Figure 9.33 we note that nonzero velocities occur in the period range categorized as swirling in Figure 9.33. Because pure swirling cannot cause nonzero ambient flow velocities at the tower position, it means that a combination of swirling and planar waves must have been present. The maximum nondimensional relative inflow velocities $U/\sqrt{gD_0}$ at $0.05D_0$ below the static liquid surface are as high as 0.62 for $\eta_{2a} = 0.05D_0$, which corresponds to 11.7 m s^{-1} for a full-scale tank with $D_0 = 36$ m. A qualitative estimate of the maximum relative inflow acceleration can be obtained by multiplying the inflow velocity by $2\pi/T$. However, generally speaking, nonlinearities imply superharmonic oscillations. When applying Morison's equation we also need to know the ambient free-surface position at the tower as well as how the mass and drag coefficients depend on the Keulegan–Carpenter number ($KC = UT/D$), the Reynolds number ($Rn = UD/\nu$), and the surface roughness (see discussion in Section 6.4.3). We choose the values $U/\sqrt{gD_0} \approx 0.6$ at $T\sqrt{g/D_0} = 4.5$ from Figure 9.39 to estimate roughly the Keulegan–Carpenter number. This estimate gives $KC = UT/D = 32.4$ with $D/D_0 = 1/12$. The Reynolds number Rn, with $U/\sqrt{gD_0} = 0.6$, $D_0 = 1.8$ m, $D = 0.15$ m, and $\nu = 10^{-6}$ m^2 s^{-1} for water, is $3.8 \cdot 10^5$ in the model test conditions. This

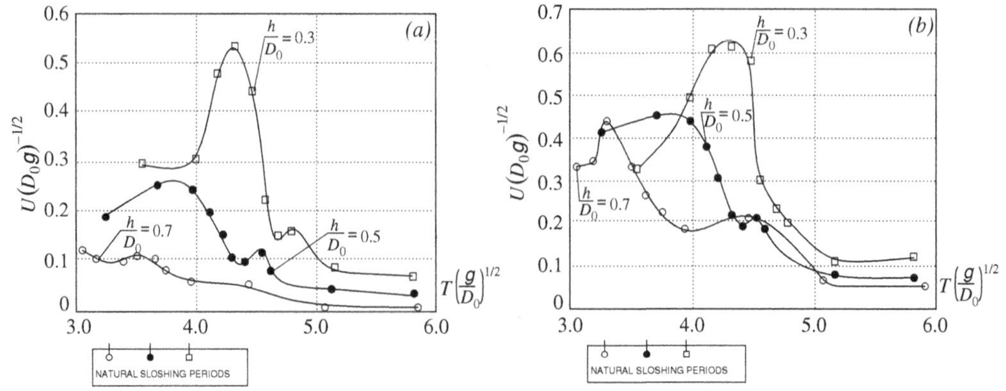

Figure 9.39. (a) Maximum liquid velocity U at tower $0.15D_0$ above base for different filling levels h of a spherical tank versus the nondimensional excitation period $T\sqrt{g/D_0}$; (b) the same at tower $0.05D_0$ below static liquid surface. Forced harmonic sway oscillations with amplitude $\eta_{2a} = 0.05D_0$. The symbols for the natural sloshing periods are the same as for the response curves (Olsen & Hysing, 1974).

value is in the so-called critical Reynolds-number range for a smooth cylinder surface and the drag coefficient is relatively low for steady ambient flow (see Figure 6.11). Olsen and Hysing (1974) do not give information about the surface roughness, so we assume a smooth cylinder surface in the following discussion. Froude scaling to $D = 36$ m gives a Reynolds number of $3.4 \cdot 10^7$. Figure 6.11 indicates that the drag coefficient is higher in full-scale than in model-scale conditions. Because the hydrodynamic force is drag-dominant at high KC (e.g., $KC = 30$), and the drag coefficient is higher in full scale than in model scale, a Froude scaling of the hydrodynamic force to full scale is not conservative. The experimental results in Figure 6.10 suggest that it is advantageous to use surface roughness to minimize the difference between the model-scale and full-scale drag coefficients. The drawback is that the roughness influences the drag coefficient.

A natural structural period T_n^{tower} of the tower that is close to the vortex shedding period T_v is an indication of vortex-induced vibration (VIV). Because KC of about 30 is relatively high, we can apply a quasi-steady approach by assuming a steady inflow velocity as a first approximation. For steady ambient flow, T_v can be expressed in terms of the Strouhal number, St, as $T_v = D/(U \cdot St)$; see Section 6.9. A more correct approach is to follow the procedure by Verley (1982) and Bearman *et al.* (1984) for ambient oscillatory flow (see also Faltinsen, 1990, pp. 248–9, 254). The Strouhal number is Reynolds-number–dependent. We simply set $St = 0.25$ (see Figure 6.28) for full-scale conditions and use $D = 3$ m and $D_0 = 36$ m with $U = 11.7$ m s^{-1}. This gives $T_v = 1.03$ s. Because a natural structural period of the tower may be 1 s in full-scale conditions according to Olsen and Hysing (1974), VIV in the cross-flow direction should be considered.

Olsen and Hysing (1974) also reported on measurements of velocities at the points of the spherical tank surface indicated in Figure 9.32, for nondimensional filling depths $h/D_0 = 0.2$ and 0.3 for sway amplitude $\eta_{2a} = 0.05D_0$. The maximum velocities were twice as high as those measured near the tower.

9.5 Summary

Our focus has been on forced lateral harmonic oscillations with frequency σ close to the lowest natural frequency σ_1 of three-dimensional rectangular, spherical, and upright circular cylindrical tanks. The multimodal method was used assuming that the excitation amplitude is small relative to the horizontal dimension of the tank. Even though the tested excitation amplitudes are generally lower than realistic magnitudes of the ship tank motions, we are able to list important aspects for the different tank types as follows.

9.5.1 Square-base tank

Possible stable steady-state waves for *longitudinal excitation* along a tank wall are planar waves, swirling, and nearly diagonal waves. Stable steady-state sloshing is not always possible as it is for two-dimensional flow. The fact that the flow does not reach a steady-state condition is referred to as *chaos*. Chaos and types of steady-state waves depend on

- forcing frequency relative to the lowest natural frequency,
- forcing amplitude,
- forcing direction, and
- ratio of liquid depth to tank breadth.

The wave regimes for different stable steady-state waves have overlaps; that is, more than one type of stable steady-state wave is possible for a given condition. The wave type that occurs depends on transient conditions.

Different stable wave regimes and chaos were examined with a Moiseev-type multimodal method, varying the longitudinal excitation amplitude η_{1a} and the ratio of liquid depth to tank length, h/L_1. The chaotic wave motion occurs in a frequency domain near $\sigma/\sigma_1 = 1$. The frequency range of chaotic motion increases with the forcing amplitude. Generally good agreement was shown with experimental results for $h/L_1 = 0.508$, 0.5 and 0.34; however, there was only qualitative agreement for $h/L_1 = 0.27$. Because a Moiseev-type method assumes that only the two lowest conjugate modes are dominant, we should expect decreasing accuracy with decreasing h/L_1.

Only pure-diagonal and swirling waves can be stable for *diagonal excitation*. However, instabilities occur for σ/σ_1 close to 1. Generally good agreement was found between the experiments and the theory.

The Moiseev-type multimodal method has to be modified by accounting for a large number of dominant modes to satisfactorily predict the wave amplitude and the hydrodynamic force with decreasing h/L_1. The number of dominant modes depends on the type of wave motion. Generally speaking, more dominant modes are needed for swirling than for planar and diagonal waves. Furthermore, the number of dominant modes increases with the forcing amplitude and decreasing liquid depth. The case of lower-intermediate and shallow-liquid depths was not examined. In this case, h/L_1 has to be introduced as a small parameter in a similar way as shown in Chapter 8 for two-dimensional sloshing.

Time-domain simulations by the multimodal method show that many transient phases occur before nearly steady-state conditions are achieved. The obtained results can be very sensitive to perturbations, both in the transient and in the nearly steady-state regimes. The latter fact has consequences for applications to sloshing in a ship in a stochastic sea state. A very large number of realizations of a given sea state are needed to obtain reliable probability density functions of response variables.

9.5.2 Nearly square-base tanks

Longitudinal forcing of nearly square-base tanks was discussed, varying $R_L = L_1/L_2$, where L_1 is the tank length in the forcing direction and L_2 is the tank breadth. The theoretical domains of stable and unstable steady-state wave motions were consistent with model tests. Chaos and planar waves, swirling, and nearly diagonal steady-state waves are possible. The main conclusions were similar to those for longitudinal forcing of a square-base tank. However, the effective frequency domains for chaos and stable wave modes are clearly a function of R_L. In some cases we may note a particularly strong sensitivity to R_L.

When R_L is either much smaller than 0.9 or much larger than 1.1 we should expect that the two-dimensional nonlinear multimodal methods described in Chapter 8 should be applicable.

9.5.3 Circular base

Only swirling and planar waves are possible stable steady-state wave modes for an upright circular cylindrical tank; chaos is also possible. The dependence of chaos and stable wave modes on excitation amplitude and ratio of liquid depth to cylinder radius was discussed. Results based on a Moiseev-type multimodal theory and experiments for the finite liquid depths were presented and showed good agreement.

9.5.4 Spherical tank

No analytically based method exists to describe nonlinear sloshing in a spherical tank. Experimental results are used to show how steady-state swirling and planar waves depend on the ratio of liquid depth to tank diameter and the lateral excitation amplitude. No chaos was reported. However, we cannot conclude that chaos is not possible. A theory would be very useful in this context.

Experimental results for hydrodynamic forces on the tank and a tower inside the tank were presented. Viscous flow separation plays an important role for the tower forces and can lead to VIV. Scale effects due to the different Reynolds numbers in model and full scale make it difficult to scale the tower forces. It was shown that we are not guaranteed to obtain conservative tower force results based on Froude scaling of model tests.

9.6 Exercises

9.6.1 Multimodal methods for square- and circular-base tanks

Although an accurate description of long-time transient waves generally needs the adaptive multimodal scheme from Section 9.2.6, the multimodal method could be applied, too, by using eqs. (9.11)–(9.20) and Tables 9.1 and 9.2 (square-base tank with ratio of liquid depth to breadth $>\sim 0.4$) as well as eqs. (9.47)–(9.51) and Table 9.3 (an upright circular cylindrical tank with ratio of liquid depth to tank radius >1.5). The long-term simulations are expected to be robust for excitation amplitudes less than or equal to $0.001L_1$ for square-base tanks and less than or equal to $0.005R_0$ for circular-base tanks. Small damping coefficients have to be added to provide passage to the steady-state regime.

Confirm the results for steady-state regimes in Figures 9.11, 9.13, and 9.29 by performing direct simulations of the transients to the steady-state regimes by the modal systems. (*Hint*: Rewrite the corresponding modal system in the standard form $M(t,\boldsymbol{x})\dot{\boldsymbol{x}} = F(\boldsymbol{x},t)$ ($x_1 = \beta_1, x_2 = \dot{\beta}_1, \ldots, x_{2n-1} = \beta_n, x_{2n} = \dot{\beta}_n, \ldots$) and use a fourth-order Runge–Kutta solver. Choose any small initial perturbations (compared with the nondimensional forcing amplitude) for the two lowest natural modes for three-dimensional wave motions to occur. Note that numerical problems are possible in simulations of long-term transients with relatively large forcing amplitude and initial conditions due to amplification of higher-order modes. Therefore, the studied case requires adaptive multimodal modeling from Section 9.2.6. Use a method as described in Section 9.1.1.2 for a square-base tank to identify what type of wave motion occurs.)

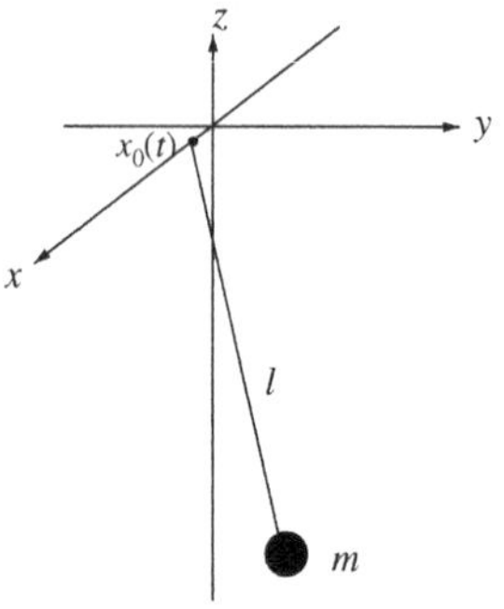

Figure 9.40. Spherical pendulum.

9.6.2 Spherical pendulum, planar, and rotary motions

A conical (spherical) pendulum has several instabilities that are similar to sloshing in a vertical circular cylinder. The mechanical system consists of a point mass m on the end of a light inextensible string of length l suspended at a point that has translatory motion $x_0(t)$ along the Ox-axis (see Figure 9.40). The mass can move in any direction (as long as the string remains taut) so the pendulum is free to swing in any plane. When the spherical pendulum performs small-amplitude motions relative to its static state, the motions of mass m can be studied in a linear approximation. It is well known that the natural frequency of a linear pendulum is $\sigma_0 = \sqrt{g/l}$. When the suspension point performs small-amplitude (relative to l) harmonic motions with frequency $\sigma \to \sigma_0$, the linear analysis may fail. However, one can use a third-order asymptotic technique to describe what kind of steady-state regimes are typical for the spherical pendulum.

Consider the pendulum in Figure 9.40 with a suspension point that oscillates along the Ox-axis with the translatory motion $x_0(t) = \varepsilon l \cos(\sigma t)$. Under this condition, the coordinates $(x(t), y(t), z(t))$ of the point mass are restricted to

$$(x - x_0)^2 + y^2 + z^2 = l^2. \qquad (9.58)$$

(a) Derive the equations of motion for $x(t)$ and $y(t)$ using a Lagrangian formulation and restriction (9.58). Assume $x \sim y \sim l\varepsilon^{1/3}$ and save terms up to $O(\varepsilon)$. (*Hint*: The Lagrangian is

$$L = \tfrac{1}{2}m(\dot{x}^2 + \dot{y}^2 + \dot{z}^2) - gmz.$$

Substitute $z(t)$ from eq. (9.58) and save terms up to $O(\varepsilon^{4/3})$.)

Answer:

$$\ddot{x} + \sigma_0^2 x + l^{-2}x(x\ddot{x} + y\ddot{y} + \dot{x}^2 + \dot{y}^2) + \tfrac{1}{2}l^{-2}x\sigma_0^2(x^2+y^2) = \sigma_0^2 l\varepsilon\cos(\sigma t),$$
$$\ddot{y} + \sigma_0^2 y + l^{-2}y(x\ddot{x} + y\ddot{y} + \dot{x}^2 + \dot{y}^2) + \tfrac{1}{2}l^{-2}y\sigma_0^2(x^2+y^2) = 0. \quad (9.59)$$

(b) Express the linearized equations and discuss the solutions.

(c) Assume the Moiseev-type asymptotic solution that $\overline{\sigma}_0^2 - 1 = O(\varepsilon^{2/3})$, $\overline{\sigma} = \sigma_0/\sigma$, and steady-state solutions

$$x(t)/l = A\cos\sigma t + \overline{A}\sin\sigma t + o(A, \overline{A}),$$
$$y(t)/l = \overline{B}\cos\sigma t + B\sin\sigma t + o(B, \overline{B}), \quad (9.60)$$

where $A \sim \overline{A} \sim B \sim \overline{B} = O(\varepsilon^{1/3})$. Use the Moiseev technique to find the secularity equations.

Show that the secularity conditions take the form of eqs. (9.56). Find m_1 and m_3 in eq. (9.56).

Answer: $m_1 = -\frac{1}{8}$, $m_3 = \frac{5}{8}$.

(d) Why are only planar and swirling motions possible? Argue this by using the analysis from Section 9.2.4.1. Is the planar response soft- or hard-spring? Why?

(e) Draw the response curves in the $(\sigma/\sigma_0, |A|)$-plane. Using eqs. (9.27)–(9.29), show that the planar regime is unstable at $\sigma = \sigma_0$.

9.6.3 Angular Stokes drift for swirling

When it comes to resonant swirling represented by the lowest-order modal term (9.5), the amplitude components A and B are generally not equal. Based on this fact, solution (9.5) can be presented as

$$\zeta(r,\theta,t) = \underbrace{(A \pm B)\frac{J_1(\iota_{1,1}r/R_0)}{J_1(\iota_{1,1})}\cos\theta\cos(\sigma t)}_{\text{planar}} \underbrace{\mp B\frac{J_1(\iota_{1,1}r/R_0)}{J_1(\iota_{1,1})}\cos(\theta \pm \sigma t)}_{\text{pure swirling}} + o(A,B), \quad (9.61)$$

where a pure-swirling (rotary) wave is described in detail in Section 4.3.2.2.

"Stokes drift" is a second-order effect associated with regular propagating long-crested waves in water with infinite horizontal extent and either infinite or constant finite depth. Studying the drift, we should start with the first-order free-surface elevation given by $\cos(kx \pm \sigma t + \theta_s)$. This expression is similar to the angular component of pure-swirling component in representation (9.61), in which the angular coordinate replaces k times the x-coordinate of the Stokes analysis.

We now study a similar effect associated with swirling in a vertical circular cylinder of radius R_0 and liquid depth h. The cylindrical coordinate system $Or\theta z$, as shown in Figure 4.9, is introduced. A Moiseev-type analysis and associated perturbation technique can be pursued where the forcing is harmonic with frequency σ and of $O(\varepsilon)$. Therefore, we adopt eq. (9.5) or (9.61) and consider the lowest-order term as the first-order approximation of the free-surface elevation, $\zeta_1(r,\theta,t)$, of order $O(\varepsilon^{1/3})$, namely, with $A \sim B \sim \varepsilon^{1/3}$. The velocity potential φ_1 is then

$$\varphi_1(r,\theta,z,t) = C\sigma\frac{J_1(\iota_{1,1}r/R_0)}{J_1(\iota_{1,1})}\frac{\cosh\left[\iota_{1,1}(z+h)/R_0\right]}{\cosh(\iota_{1,1}h/R_0)} \times (A\cos\theta\sin(\sigma t) - B\sin\theta\cos(\sigma t)),$$

where C is a constant.

(a) Find the constant C by using the linear dynamic free-surface condition $(g\zeta_1 + (\sigma_{1,1}^2/\sigma^2)\,\partial\varphi_1/\partial t|_{z=0} = 0,\ \ \sigma_{1,1}^2 = (g/R_0)\iota_{1,1} \times \tanh(\iota_{1,1}h/R_0))$ following from the Moiseev asymptotic scheme. *Answer*: $C = -g/\sigma_{1,1}^2$.

(b) Study the mean mass flux to $O(\varepsilon^{2/3})$ through the meridional plane with a constant θ-value: for the meridional plane, r varies from 0 to R_0 and z from $-h$ to the instantaneous free surface with $z = 0$ as the mean free surface. Perform the analysis to second-order approximation of the velocity potential (i.e., up to $O(\varepsilon^{2/3})$).

Show that the first- and second-order approximations caused by the velocity potential give zero contribution. (*Hint*: Use the general representation of the absolute velocity potential (7.2) with the natural modes (4.37).)

Evaluate the liquid mass flux through the meridional plane by introducing the velocity along the θ-coordinate, $v_\theta = r^{-1}\partial\varphi_1/\partial\theta$. Explain why the mean mass flux can be expressed as

$$\rho \int_0^{R_0} \overline{\zeta_1 \frac{\partial \varphi_1/\partial \theta|_{z=0}}{r}}\, dr,$$

where the bar indicates the time average.

Find the mass flux and explain why it is independent of θ. *Answer*:

$$-\tfrac{1}{2} ABC\sigma\rho \int_0^{R_0} r^{-1} \left(\frac{J_1(\iota_{1,1} r/R_0)}{J_1(\iota_{1,1})} \right)^2 dr.$$

(c) The result in (b) shows that a liquid particle will rotate around the cylinder axis. Confirm this fact numerically by choosing a liquid particle with an initial position at the tank surface and other parameters according to your own choice and by only using φ_1.

(d) Does the obtained result contradict the fact that the flow is irrotational?

9.6.4 Three-dimensional shallow-liquid equations in a body-fixed accelerated coordinate system

The nonlinear shallow-water equations are used to analyze the effect on ship motion of green water on deck (Dillingham & Falzarano, 1986). They can similarly be applied in connection with water ingress/egress through a hole of a damaged ship onto a deck. These applications and sloshing in ship tanks provide background scenarios for the following exercise.

We assume the liquid depth is small so that we can make a shallow-liquid approximation. The resulting equations are nonlinear and must be solved numerically for a three-dimensional problem. We define a body-fixed coordinate system $Oxyz$ relative to the tank (or deck) with $z = 0$ as the mean free surface and $z = \eta$ as the free-surface elevation. Furthermore, the liquid velocity components relative to the tank are denoted u, v, w along the x-, y-, and z-axes, respectively.

(a) Show that

$$\frac{\partial}{\partial x}\int_{-h}^{\eta} u\, dz + \frac{\partial}{\partial y}\int_{-h}^{\eta} v\, dz = -\frac{\partial \eta}{\partial t}$$

by using the continuity equation $\partial u/\partial x + \partial v/\partial y + \partial w/\partial z = 0$ and the kinematic free-surface condition

$$\left(\frac{\partial \eta}{\partial t} + u\frac{\partial \eta}{\partial x} + v\frac{\partial \eta}{\partial y} - w \right)\bigg|_{z=\eta} = 0.$$

(*Hint*: Integrate the continuity equation from $z = -h$ to $z = \eta$. Show that

$$\frac{\partial}{\partial x}\left[u(\eta + h)\right] + \frac{\partial}{\partial y}\left[v(\eta + h)\right] = -\frac{\partial \eta}{\partial t}$$

by using the shallow-liquid approximation that u and v are independent of z.)

(b) Consider the Euler equations in a body-fixed coordinate system following from eq. (2.56) by setting the viscosity term equal to zero. We only include linear terms involving the rigid-body motion modes η_i ($i = 1, \ldots, 6$) and define the translatory modes relative to the origin of the coordinate system. The rigid-body motions are defined relative to an inertial system in a similar way as described in Chapter 2.

We start with the z-component of the Euler equations. Neglect w and show, by using the dynamic free-surface condition, that the pressure can be expressed as $p = -\rho(g + \ddot{\eta}_3 + y\ddot{\eta}_4 - x\ddot{\eta}_5 + 2v\dot{\eta}_4 - 2u\dot{\eta}_5)(z - \eta)$.

Show, by keeping linear terms in η_i and integrating the x- and y-components of the Euler equations from $z = -h$ to $z = \eta$, that

$$\begin{aligned}
&\frac{\partial u}{\partial t} + u\frac{\partial u}{\partial x} + v\frac{\partial u}{\partial y} \\
&\quad = -\left(g + \ddot{\eta}_3 + y\ddot{\eta}_4 - x\ddot{\eta}_5 + 2v\dot{\eta}_4 - 2u\dot{\eta}_5\right)\frac{\partial \eta}{\partial x} \\
&\qquad + g\eta_5 - \ddot{\eta}_1 + h\ddot{\eta}_5 + y\ddot{\eta}_6 + 2v\dot{\eta}_6 \\
&\frac{\partial v}{\partial t} + u\frac{\partial v}{\partial x} + v\frac{\partial v}{\partial y} \\
&\quad = -\left(g + \ddot{\eta}_3 + y\ddot{\eta}_4 - x\ddot{\eta}_5 + 2v\dot{\eta}_4 - 2u\dot{\eta}_5\right)\frac{\partial \eta}{\partial y} \\
&\qquad - g\eta_4 - \ddot{\eta}_2 - h\ddot{\eta}_4 - x\ddot{\eta}_6 - 2u\dot{\eta}_6.
\end{aligned}$$

(c) What are the boundary conditions when an intact ship tank is considered? Use pressure integration to express the hydrodynamic force and moment acting on the ship due to sloshing in the tank.

(d) We consider a submerged "small" hole with cross-sectional area $\mathcal{A}$ in the ship side, which may be caused by ship collision. The flow rate Q_r through the hole is based on a quasi-steady analysis; that is, we write $Q_r = C\mathcal{A}\sqrt{|\Delta p|/\rho}$, where $|\Delta p|$ is the absolute value of the ambient pressure jump at the hole (i.e., with accounting for the flow through the hole). C is a coefficient that depends on the size and shape of the hole as well as on $|\Delta p|$ (Le Conte, 1926). Furthermore, the sign of C decides if ingress or egress occurs. For instance, C may be 0.6 for ingress.

Show how to use the formula for our problem by assuming the pressure is hydrostatic relative to the instantaneous free surface when we neglect the flow through the hole. You can assume linear theory for the exterior problem.

How does the expression for Q_r enter as a boundary condition in the nonlinear shallow-liquid equations?

(e) Assume finite-amplitude ship motions and define rigid-body velocity and acceleration components in the body-fixed accelerated coordinate system $Oxyz$. We must now leave the formulation with η_i, where corresponding velocities and accelerations are respectively first- and second-order time derivatives of η_i (Etkin, 1959; Faltinsen, 2005). Thereby we introduce the Euler angles Φ, Θ, and Ψ, representing respectively the roll, pitch, and yaw angle of the tank (see Figure 2.2). We use the notation $\boldsymbol{v}_O = (v_{O1}, v_{O2}, v_{O3})$ for the translatory velocity vector of the origin O of the tank-fixed coordinate system and introduce the angular velocity $\boldsymbol{\omega} = (\omega_1, \omega_2, \omega_3)$ of the tank-fixed coordinate system $Oxyz$ relative to the inertial system $O'x'y'z'$ defined in connection with Figure 2.2. In addition, we have the acceleration term $\boldsymbol{a_O} = (a_{O1}, a_{O2}, a_{O3})$ in the Euler equations following from eq. (2.56). The acceleration-of-gravity vector $\boldsymbol{g} = (g_1, g_2, g_3)$ is defined in terms of the Euler angles.

You can now follow a similar procedure as in the linear rigid-body motion case in part (b). We concentrate on the centrifugal acceleration term $-\boldsymbol{\omega} \times (\boldsymbol{\omega} \times \boldsymbol{r})$. Show that this term causes the pressure contribution

$$p_{\text{centri}} = -\rho\,(\omega_1\omega_3 x + \omega_2\omega_3 y)\,(z - \eta) + \tfrac{1}{2}\rho\left(\omega_1^2 + \omega_2^2\right)\left(z^2 - \eta^2\right).$$

Show that the centrifugal acceleration causes the following contribution to the right-hand side of the x-component of the shallow-liquid equation:

$$f_x^{\text{centri}} = -\left(\omega_1\omega_3 x + \omega_2\omega_3 y\right)\frac{\partial \eta}{\partial x} + \left(\omega_1^2 + \omega_2^2\right)\eta\frac{\partial \eta}{\partial x} + \left(\omega_2^2 + \omega_3^2\right)x - \omega_1\omega_2 y - \omega_1\omega_3\eta.$$

Then show that the contribution to the right-hand side of the y-component of the shallow-liquid equation is

$$f_y^{\text{centri}} = -\left(\omega_1\omega_3 x + \omega_2\omega_3 y\right)\frac{\partial \eta}{\partial y} + \left(\omega_1^2 + \omega_2^2\right)\eta\frac{\partial \eta}{\partial y} + \left(\omega_1^2 + \omega_3^2\right)y - \omega_1\omega_2 x - \omega_2\omega_3\eta.$$

Show that the final result for the pressure is

$$p = -\rho(-g_3 + a_{O3} + \omega_1 v_{O2} - \omega_2 v_{O1} + y\dot{\omega}_1 - x\dot{\omega}_2 + 2v\omega_1 - 2u\omega_2)\,(z - \eta) + p_{\text{centri}}.$$

The shallow-liquid approximations of the Euler equations are

$$\begin{aligned}\frac{\partial u}{\partial t} + u\frac{\partial u}{\partial x} + v\frac{\partial u}{\partial y} &= -(-g_3 + a_{O3} + \omega_1 v_{O2} - \omega_2 v_{O1} + y\dot{\omega}_1 - x\dot{\omega}_2 + 2v\omega_1 - 2u\omega_2)\frac{\partial \eta}{\partial x}\\ &\quad + g_1 - a_{O1} - \omega_2 v_{O3} + \omega_3 v_{O2} + h\dot{\omega}_2 + y\dot{\omega}_3 + 2v\omega_3 + f_x^{\text{centri}},\\ \frac{\partial v}{\partial t} + u\frac{\partial v}{\partial x} + v\frac{\partial v}{\partial y} &= -(-g_3 + a_{O3} + \omega_1 v_{O2} - \omega_2 v_{O1} + y\dot{\omega}_1 - x\dot{\omega}_2 + 2v\omega_1 - 2u\omega_2)\frac{\partial \eta}{\partial y}\\ &\quad + g_2 - a_{O2} - \omega_3 v_{O1} + \omega_1 v_{O3} - h\dot{\omega}_1 - x\dot{\omega}_3 - 2u\omega_3 + f_y^{\text{centri}}.\end{aligned}$$

9.6.5 Wave loads on a spherical tank with a tower

(a) We presented in Section 5.4.7.2 a formula for linear hydrodynamic force on a spherical tank for different ratio of liquid depth to tank diameter based on using a simplified dipole solution for the flow. Compare with the experimental results in Section 9.4.

(b) Section 9.4.2 presents experimental results for the hydrodynamic force on a tower in a spherical tank. Estimate the force by using the Morison equation and your own estimates of mass and drag coefficients where you account for the Reynolds and the Keulegan–Carpenter numbers. Assume a smooth cylinder surface. Before using the Morison equation, you need to assess the inflow velocities and accelerations along the tower based on the measurements of velocities presented in Figure 9.39. To what extent can you use the dipole solution presented in eq. (4.73) with singularity positions given in Table 4.4 to predict the ratio of the inflow velocities at the two measurement points?

(c) Make estimates of the hydrodynamic force on the tower for $h/D = 0.3$ at the period that gives maximum force for $\eta_{2a} = 0.05D$. Discuss the relative importance of the different terms in Morison's equation.

(d) Discuss scale effects due to Reynolds number between model and full-scale conditions.

10 Computational Fluid Dynamics

10.1 Introduction

It has become popular to use computational fluid dynamics (CFD) to solve fluid flow problems. This methodology is also used to solve the violent sloshing phenomena considered in this book. The construction of CFD software comprises many different aspects that consist of the following main tasks:

- Choose mathematical formulation of the problem (i.e., the governing equations and boundary conditions).
- Develop a numerical formulation of the equations and the boundary conditions.
- Solve the equations in spatial coordinates as the fluid flow evolves in time. The evolution is computed by using some time-stepping procedure.

The types of governing equations that we want to solve depend on the assumptions and simplifications made (i.e., compressible or incompressible fluid, viscous or inviscid fluid, laminar or turbulent viscous flow conditions, rotational or irrotational flow). A broad variety of numerical methods exists. An overview is given in Figure 10.1, where different abbreviations are used:

- boundary element method (BEM),
- constrained interpolation profile (CIP),
- finite difference method (FDM),
- finite element method (FEM),
- finite volume method (FVM),
- level set (LS),
- marker and cell (MAC),
- moving particle semi-implicit (MPS),
- smoothed particle hydrodynamics (SPH), and
- volume of fluid (VOF).

We can divide types of methods into two classes. One class of methods comprises *potential flow methods* and the other class comprises *Navier–Stokes methods*. *Hybrid* methods also exist, which combine potential flow and Navier–Stokes methods. One example is the slamming analysis discussed in Chapter 11, where the details of the flow caused by the impact are described by potential flow while the inflow to the impact may be described by a Navier–Stokes method. The potential flow methods assume irrotational flow, but that does not mean that the flow in all parts of the flow domain has to be irrotational. One can account for the dynamics of thin free shear layers that separate from, for instance, baffles. The flow inside the free shear layer is rotational and the potential flow method describes the flow outside the free shear layer by accounting for the presence of the free shear layer. The procedure is described for hydrofoils in Faltinsen (2005) and exemplified for moonpool resonance in connection with Figures 3.32 and 3.33. The flow is also rotational in a boundary layer and the effect of the boundary layer on the potential flow domain can be accounted for by an inflow/outflow from the boundary layer. If the liquid is treated as incompressible, one often uses a BEM to solve the liquid flow. Rankine sources on their own or with normal dipoles are usually distributed over the instantaneous wetted body surface and the free surface in a fully nonlinear formulation. The dynamic and kinematic conditions on the free surface can be used to track its velocity potential and position by following individual particles. This approach is referred to as a mixed Eulerian–Lagrangian method in the literature (see, e.g., Faltinsen, 1977; Longuett-Higgins & Cokelet, 1976). It is possible to include the effect of gas pockets. The effect of flow separation in terms of thin free shear layers can be accounted for by using a distribution of normal dipoles along a free shear layer. It is also possible to use a distribution of potential flow vortices. We give further details about the BEM in Section 10.2. A potential flow problem could also be solved by other numerical schemes and we exemplify in Sections 10.4 and 10.5 how the FVM and FEM can be used to solve potential flow problems.

Because the Euler equations ignore viscosity in the Navier–Stokes equations, their solution can be considered a special case of the Navier–Stokes methods. Vorticity is generated by the Euler equations as well as by the Navier–Stokes equations. An issue to be considered is if the

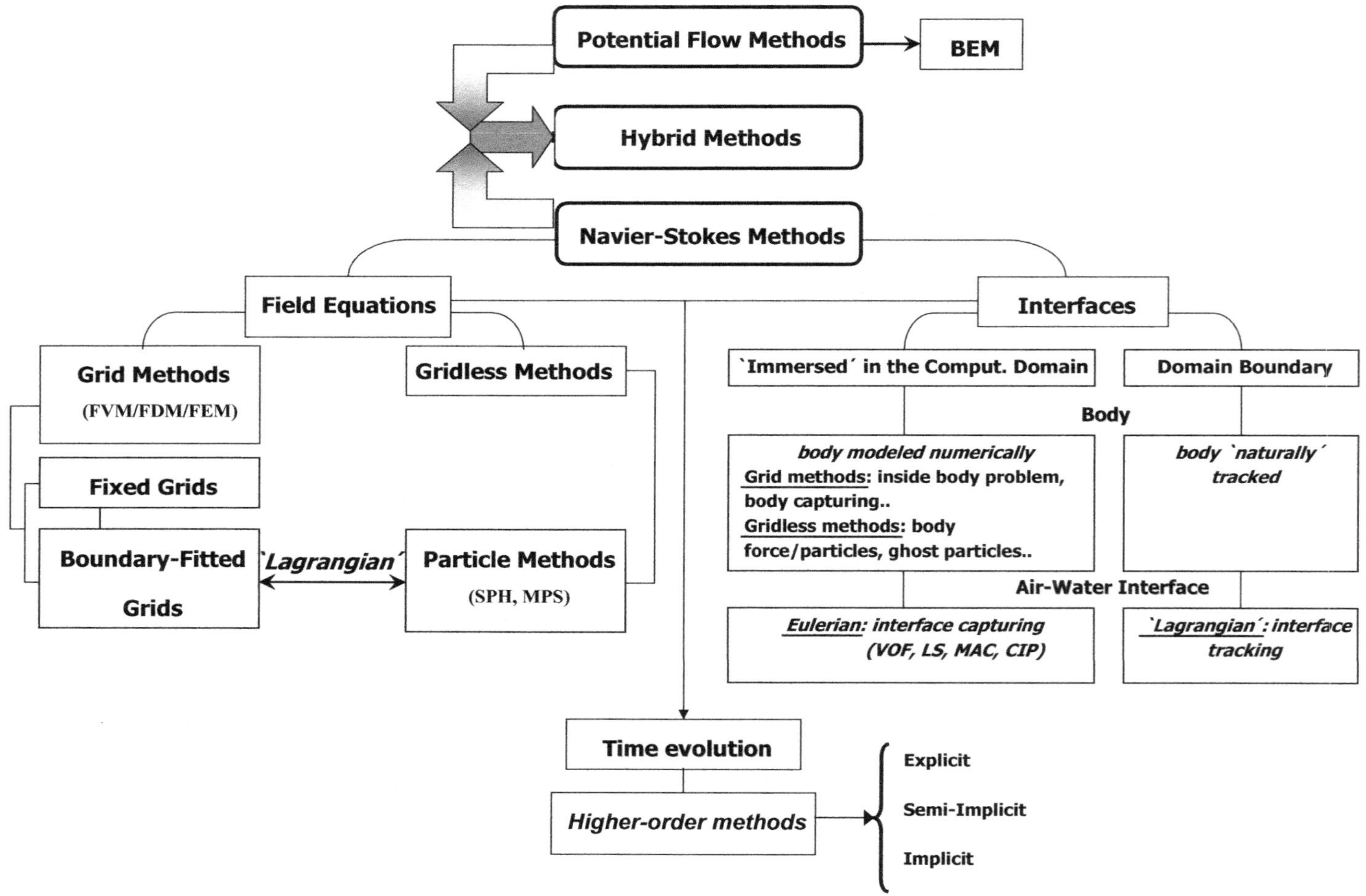

Figure 10.1. Overview of numerical methods in fluid dynamics with emphasis on possible solution strategies within Navier–Stokes solvers. For instance, the field equations in Navier–Stokes methods may be solved by either grid methods or gridless methods. The figure also shows how interfaces between air (gas) and water (liquid) are numerically handled and different ways of accounting for the presence of a body (Greco, personal communication, 2008).

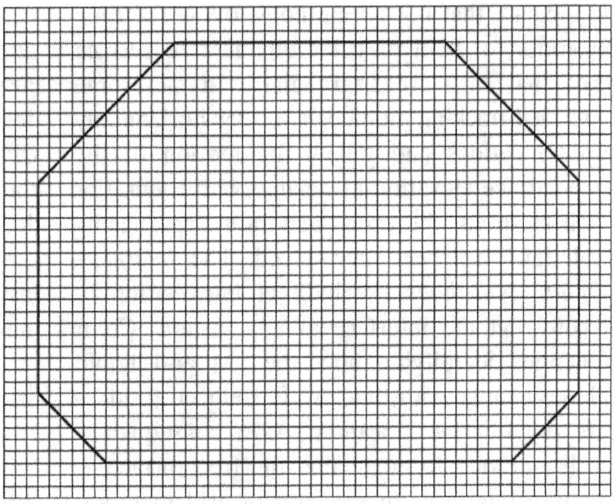

Figure 10.2. Example of a grid cell system used in combination with FDM to describe two-dimensional flow in the prismatic tank illustrated in Figures 1.17 and 1.19. The mesh is uniform. The tank surface and the free surface are embedded in the grid system.

viscous flow is laminar or turbulent; it will certainly be turbulent at full scale. However, many computer codes are based on laminar flow. The effect of viscosity is secondary, for instance, if the boundary-layer flow does not separate during sloshing in the tank. The effect of viscosity and turbulence is not secondary for separated flow past interior structures without sharp corners, such as a circular cylinder. The flow separation points (lines) depend on laminar or turbulent boundary-layer flow upstream of separation. Another issue to be considered is the effect of compressibility. However, it is common to assume incompressible gas and liquid in numerical simulations. If the changes in the gas density affect the two-phase flow evolution, the gas has to be considered compressible. For instance, a gas cavity has a natural frequency due to the gas compressibility. Depending on the cavity size, the compressibility may have an important effect on the liquid behavior.

The Navier–Stokes equations are numerically solved by either *grid* methods or *gridless* methods. The FDM, FVM, and FEM require a grid system that may be boundary-fitted or -fixed. Figure 10.2 illustrates a fixed-grid system referred to in Section 10.3 about the FDM. The tank boundaries and the free surface (not shown) are immersed in the computational domain. A boundary-fitted grid might need regridding according to the motions involved. A fixed grid, as in Figure 10.2, is generated only once. Those facts make a fixed grid computationally more efficient than a boundary-fitted grid with moving boundaries. However, a boundary-fitted grid may achieve better accuracy.

The MPS and SPH methods are examples of grid-less methods where individual fluid particles are followed in time. Koshizuka *et al.* (1998) used the MPS method to study breaking waves. In later sections we give more details about FDM, FVM, FEM, and SPH. Our objective here is to give an introduction to CFD methods. We are *mostly dealing with two-dimensional flow* even though three-dimensional flow aspects may be important, as demonstrated in Chapter 9. Several textbooks deal extensively with CFD. A main difference between FDM and FVM is in terms of the way governing equations are discretized. FDM starts with the Navier–Stokes equations and continuity equation as they are formulated and then makes a finite difference approximation to the derivatives. FVM uses the equations following from conservation of fluid momentum and mass of a cell (see, e.g., Section 2.2.4) as a basis and then makes a finite difference approximation.

There are two classes of methods used to track the shape of the free surface, both for grid and grid-less methods. One class follows the free surface in time and as consequence treats the free surface as a sharp interface (*interface-tracking methods*). This method is, for instance, used in combination with BEM. If a grid method is used, it requires a boundary-fitted grid. The SPH method is implicitly an interface tracking method by the fact that all fluid particles are followed in time. In Figure 10.1 we have written "Lagrangian" in connection with interface tracking. The use of the quotation marks is meant to indicate that not all methods follow the time evolution of fluid particles on the free surface (i.e., that they are Lagrangian). In fact some interface-tracking methods consider horizontally fixed locations and follow the motion of the free-surface elevation associated with them. For such methods, the term "arbitrary Lagrangian–Eulerian" has been introduced. The *interface-capturing methods* use indirect information to reconstruct the interface at any instant. Color functions identifying how the gas and liquid phases change with time are introduced to identify the interface location. As a consequence, such methods do not define the free surface as a sharp interface. In the case of grid methods, interface-capturing methods are combined with fixed grids and require the extension of the latter beyond the free surface for a proper interface reconstruction. Dynamic conditions requiring continuity in the stresses are also required at the free surface. However, if the liquid and gas are handled as a single fluid with a continuous change of the fluid properties between the liquid and the gas, dynamic conditions are implicitly satisfied.

We may ask: what are the advantages and disadvantages of using CFD relative to analytically based methods such as the multimodal method? As an advantage, the flow features associated with generally shaped sloshing tanks, any filling depth, and generic excitation, may, in principle, be handled. In Sections 8.1 and 8.5 we provided examples where the multimodal method fails, whereas a CFD solver furnishes reliable results in the case of sloshing involving breaking-wave phenomena. Furthermore, CFD methods may provide good flow visualization, including details such as the vorticity distribution. Flow separation around internal structures can be simulated. It seems generally accepted that CFD codes have difficulties in predicting impact loads when the angle between the impacting free surface and the tank surface is small. Conversely, we should not overemphasize the difficulties of CFD methods in predicting high impact pressures that are highly concentrated in time and space. More suitable parameters for assessing their capabilities and accuracy in such scenarios are the slamming force impulse and the induced structural strains, whose proper estimate requires the coupling of the hydrodynamic analysis with a structural analysis. In fact, usually in the computations the structure is assumed rigid, but in the circumstances discussed hydroelasticity plays an important role, as described in Chapter 11. A disadvantage is that CFD methods are time consuming, which makes statistical estimates of tank response variables difficult.

Some methods may not be robust enough. For instance, BEM breaks down when an overturning wave hits the underlying free surface. Numerical problems may also arise with a BEM at the intersection between the free surface and the tank boundary. Landrini *et al.* (1999) discussed numerical problems associated with BEM and sloshing. When a BEM works, it is generally a fast and accurate method.

CFD methods based on the Navier–Stokes and Euler equations are often very robust. However, care must be shown that the solutions are true physical solutions. If sufficient care is not shown, some of the methods may, for instance, numerically lose or generate liquid mass on a long time scale. Conservation of liquid mass is of particular concern for sloshing. Because the highest natural period of the liquid motion is strongly dependent on the liquid mass, an unphysical numerical simulation can result. Verification and validation are mandatory. By *verification* is meant that the solutions are consistent with the governing equations and initial and boundary conditions that have been used. *Validation* indicates comparisons with model tests and full scale trials. Roache (1997) has stated this succinctly as follows:

- Are we solving the equations right? (*verification*)
- Are we solving the right equations? (*validation*)

Model tests and full-scale trials are, of course, not free of errors. Therefore, it is important that the physical experiments are accompanied by an error analysis. Verification can be done by temporal and spatial convergence tests and by checking whether conservation of global mass, momentum, and energy are satisfied (see Section 2.2.4). Procedures to establish convergence rates are discussed by Ferziger and Peric (2002). Comparisons should be made with benchmark numerical results and

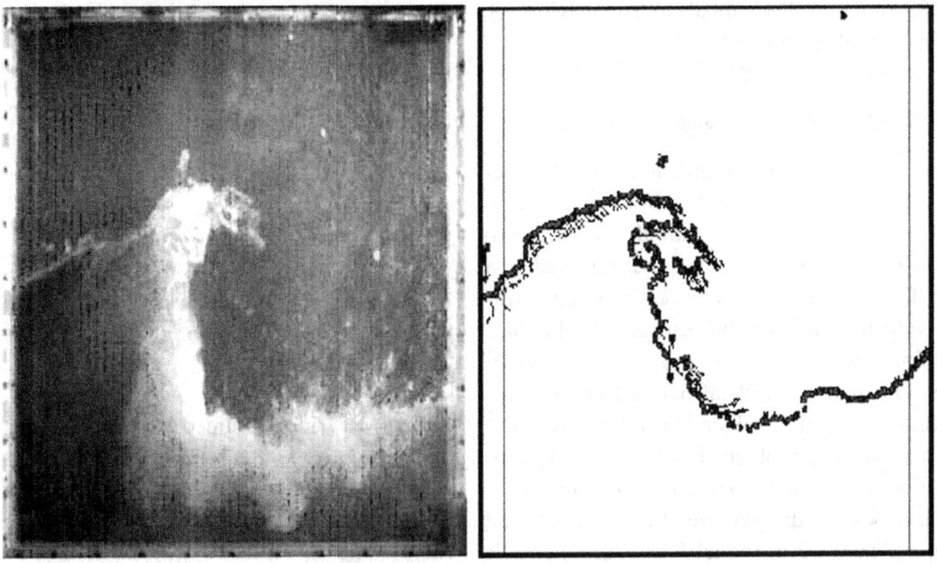

Figure 10.3. Experiment (left) and SPH simulation (right) of sloshing in a tank (Landrini *et al.*, 2003).

analytical methods such as the linear potential flow theory and the multimodal method.

Because commercial CFD codes are generic in nature and special physical features are specific for the different application fields, one cannot necessarily trust the documentation of verification and validation in applications beyond those tested by the developers. Solaas (1995) illustrated the grid dependence and large sensitivity to parameters used in numerical differentiation and iteration procedures in a commercial code (see discussion related to Figure 8.3(a)). A primary difficulty was to satisfy mass conservation at resonant condition. Setting the numerical parameters equal to the values suggested by the code developers did not give the best results.

CFD methods may provide important insights into flow particulars. We illustrate this by an example, where the SPH has been used in two-dimensional flow conditions. This example illustrates the importance of mutual interaction among analysis, CFD, and experiments. However, let us first give some background for the story. One of the authors of this book was consulted in the development of an SPH code at INSEAN. He then wanted the code developer to check his results against the multimodal method and existing experiments. For example, he wanted to see how well jump phenomena in the steady-state solution were predicted. The SPH code did well. However, some results for excitation periods from about 0.8 T_1 to 0.95 T_1 did not agree well with either the multimodal method or the experiments, where T_1 is the highest natural period. The excitation was horizontal. After a lengthy check of the code without finding errors, it was decided to do new experiments at INSEAN and those experiments did agree with the SPH simulations. Some of these results are reported in Section 8.5. The objective was to study when the initial transient phase is complete and the flow in the tank achieves features that globally appear periodic (i.e., nearly steady-state conditions). The ratio of liquid depth to tank length, h/l, was 0.35, which is close to the *critical depth*, $h/l = 0.3368\ldots$, where a change in flow conditions occurs due to nonlinear effects (Section 8.5).

Figure 10.3 shows both experimental and SPH results illustrating very steep overturning waves. The model tests also show strong mixing between air and liquid. Because the SPH assumed one-phase flow, the mixing of air and liquid was not predicted. The numerical simulations demonstrated a subsequent decrease in wave elevation with a later increase to what is shown in Figure 10.3. This change happens over a longer time than the forced oscillation period. The consequence was the presence of subharmonic terms in the wave elevation during what we previously referred to as nearly steady-state conditions (see Section 8.5). The subharmonic behavior was experimentally confirmed.

10.2 Boundary element methods

When the liquid is incompressible and the flow is irrotational, BEMs (also called panel methods) may be used to solve liquid flow phenomena with nonlinear free-surface behavior inside a tank. The flow variables can be expressed in terms of a velocity potential that satisfies the Laplace equation and appropriate tank and free-surface boundary conditions. We also need to specify initial conditions if the problem is transient. Because a potential flow problem inside a tank has no damping, transient behavior due to initial conditions does not die out over time. An artificial damping may be introduced to reach periodic conditions. We describe that approach in detail later. A more physical approach is to couple the potential flow with a boundary-layer flow in a similar way to that described in Section 6.3.2. The boundary-layer flow is a function of the potential flow and causes an inflow/outflow to the potential flow domain.

The basis of the BEMs to be described in the following text is Green's second identity, given by eq. (A.4). An alternative BEM may, for instance, be the vortex-lattice method, which follows from Biot–Savart's law (Faltinsen, 2005).

We first consider a two-dimensional variant and two functions, Φ and ψ. The first function satisfies $\nabla^2\Phi = 0$ everywhere in the fluid domain and

$$\psi\,(x, y; x_1, y_1) = \ln r,\qquad r = [(x - x_1)^2 + (y - y_1)^2]^{1/2} \quad (10.1)$$

with the point (x_1, y_1) inside the fluid volume (see Figure 10.4). We have to be careful how we handle the singular point in eq. (10.1). We note that ψ is the velocity potential for a source (Rankine type) of strength 2π located at (x_1, y_1). Outside the singular point, ψ satisfies the Laplace equation. Equation (A.4) therefore applies if we integrate over $S_Q + S_1$; that is,

$$\int_{S_Q+S_1}\left(\psi\frac{\partial\Phi}{\partial n} - \Phi\frac{\partial\psi}{\partial n}\right)\mathrm{d}S = 0.$$

Note that eq. (A.4) assumes the surface normal direction to be out of the liquid domain, but the

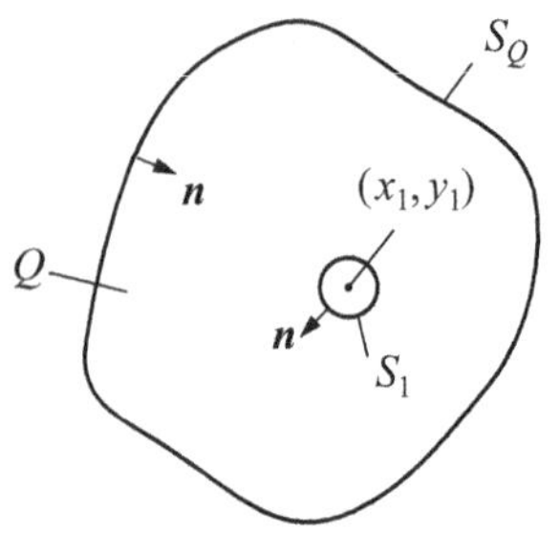

Figure 10.4. Integration surfaces used in applying Green's second identity: (x_1, y_1) is the singular point of eq. (10.1).

normal $\boldsymbol{n}$ is here assumed to be in the *positive direction into the liquid.* This information is not relevant at this stage but is required by the further analysis. The surface of the circle, S_1, has a small radius r and the center at (x_1, y_1). Along S_1 we can write $\frac{\partial\psi}{\partial n} = \frac{\partial\psi}{\partial r} = 1/r$. The limit $r \to 0$ implies

$$\int_{S_1}\Phi\frac{\partial\psi}{\partial n}\mathrm{d}S = \Phi(x_1, y_1)\int_0^{2\pi}\frac{1}{r}r\,\mathrm{d}\theta = 2\pi\Phi(x_1, y_1)\ \text{ and }\ \int_{S_1}\frac{\partial\Phi}{\partial n}\psi\,\mathrm{d}S = 0. \quad (10.2)$$

As a result, we can write the velocity potential for two-dimensional flows as

$$2\pi\Phi\,(x_1, y_1) = \int_{S_Q}\left(\frac{\partial\Phi(s)}{\partial n}\ln r - \Phi(s)\frac{\partial}{\partial n}\ln r\right)\mathrm{d}S(x, y) \quad (10.3)$$

with geometric notations from Figure 10.4.

The expression $\partial(\ln r)/\partial n$ in the preceding equation is in fact a dipole in the normal direction of S_Q. Equation (10.3) therefore states that we can represent the velocity potential in a fluid domain by means of a combined source and dipole distribution along the surface enclosing the liquid domain, S_Q. If we study the liquid flow in the tank, this means that S_Q consists of the wetted tank surface $S(t)$ and the free surface $\Sigma(t)$. However, we need to know the values of $\partial\Phi/\partial n$ and Φ along S and Σ and the instantaneous position of Σ to determine the velocity potential Φ at (x_1, y_1) in the fluid domain. What is known on S is $\partial\Phi/\partial n$ from the tank boundary condition that requires no flow through the tank surface. This condition is expressed by eq. (2.63) in the case of a rigid tank structure. On the free surface we have the dynamic and kinematic boundary conditions, which lead to a numerical time integration process that provides both the instantaneous position of the free surface and the instantaneous values of Φ on the free surface. We describe these equations in detail later in the text. The unknowns are then Φ on S and $\partial\Phi/\partial n$ on Σ, which are obtained by formulating an integral equation by letting (x_1, y_1) approach S_Q. This integral equation has to be numerically solved (see Section 10.2.3).

It is possible to show that the representation in terms of sources and dipoles is not unique. However, the solution is unique. An alternative is to use only a distribution of sources along S_Q. Let us show this solution. We define an artificial exterior problem with a velocity potential φ_e outside the liquid. The exterior domain has infinite extent. Then eq. (A.4) is applied to φ_e with the source expression ψ given by eq. (10.1). Because the singular point of ψ is outside the exterior domain and if we require that φ_e goes to zero at infinity sufficiently fast, we are left with the following condition:

$$\int_{S_Q} \left(\psi \frac{\partial \varphi_e}{\partial n} - \varphi_e \frac{\partial \psi}{\partial n} \right) \mathrm{d}S = 0.$$

Subtraction of this condition from eq. (10.3) gives

$$2\pi\Phi\,(x_1, y_1) = \int_{S_Q} \left(\frac{\partial\,[\Phi(s) - \varphi_e(s)]}{\partial n} \ln r - [\Phi(s) - \varphi_e(s)] \frac{\partial}{\partial n} \ln r \right) \mathrm{d}S(x, y). \tag{10.4}$$

Because φ_e represents an artificial problem, we have the freedom to select its boundary conditions on S_Q. Let us set $\varphi_e(s) = \Phi(s)$; then eq. (10.4) leads to source distribution only. Even though the exterior problem is artificial, we must of course require that a nontrivial solution exists. This is a matter of concern if our physical problem is an exterior problem with a free-surface-piercing body. An infinite number of discrete frequencies (*irregular frequencies*) exist that cause the method to break down (John, 1950).

A numerical method based on representing the fluid flow through a distribution of singularities (source, dipoles, vortices) along boundary surfaces is called a *boundary element method* (BEM).

The theoretical and numerical procedure just outlined can be generalized to three-dimensional flows. We start with eq. (A.4) and choose

$$\psi = 1/R,$$
$$R = [(x - x_1)^2 + (y - y_1)^2 + (z - z_1)^2]^{1/2}, \tag{10.5}$$

where (x_1, y_1, z_1) is inside the fluid volume. Therefore, a three-dimensional Rankine source in infinite fluid is used instead of a two-dimensional source. Because eq. (10.5) represents a sink, our use of the word *source* is not precise. However, it is common practice in source distribution methods to not distinguish between the words source and sink. Generalizing the derivation for two-dimensional flow to three dimensions gives

$$\Phi(x_1, y_1, z_1) = \frac{1}{4\pi} \int_{S+\Sigma} \left(\Phi \frac{\partial}{\partial n}\left(\frac{1}{R}\right) - \frac{1}{R}\frac{\partial \Phi}{\partial n} \right) \mathrm{d}S. \tag{10.6}$$

10.2.1 Free-surface conditions

The free surface is the interface between the gas and the liquid. We neglect surface tension. On the free surface we must then require the liquid and gas pressures to be the same. This condition is the *dynamic free-surface condition*, which can be expressed by eq. (2.64). In addition, the *kinematic free-surface condition* states that particles on the free surface will remain there at any time; that is, within a fully nonlinear unsteady Lagrangian BEM, the velocities of liquid particles on the free surface are calculated and their position tracked in time to follow the free-surface evolution. This formulation is different than the one used by the multimodal method and associated with eq. (2.68). It is convenient in connection with solving a nonlinear free-surface problem with the BEM to reexpress the dynamic free-surface condition in terms of $D\Phi/Dt$ (i.e., the time rate of change of Φ as we follow a liquid particle on the free surface). We can do that by starting out with the Bernoulli equation in an *inertial system* (2.23) and noting that as a matter of definition of the material derivative $D\Phi/Dt = \partial\Phi/\partial t + \nabla\Phi \cdot \nabla\Phi$, where $\partial\Phi/\partial t$ means the time derivative of Φ for a fixed point in the inertial system $Ox'y'z'$. A key point in deriving eq. (2.64) was that we had to express the time derivative of Φ for a fixed point in the tank-fixed coordinate system $Oxyz$. This step is avoided in the following derivation. The time-dependent constant $C(t)$ in eq. (2.23) can be determined in the same way that we derived the Bernoulli equation for the mobile tank-fixed coordinate system $Oxyz$ defined in Figure 2.2. We let the inertial system $Ox'y'z'$ and the tank-fixed coordinate systems coincide at the initial time $t = 0$, when no sloshing occurs and the tank is an upright position with $z = 0$ corresponding to the free surface. Formally we can express $\nabla\Phi$ as the absolute liquid velocity field in both the $Ox'y'z'$- and $Oxyz$-coordinate systems. The final result is that the pressure in the liquid, p, can be expressed as

$$p = p_0 - \rho\left[\frac{D\Phi}{Dt} - \tfrac{1}{2}\nabla\Phi\cdot\nabla\Phi - \boldsymbol{g}\cdot\boldsymbol{r}\right], \quad (10.7)$$

where p_0 is the ambient gas pressure, $\boldsymbol{g}\cdot\boldsymbol{r} = g_1 x + g_2 y + g_3 z$, and $\boldsymbol{g}$ is the gravity acceleration vector with components g_i $(i = 1, 3)$ in the $Oxyz$-coordinate system.

By requiring $p = p_0$ on the free surface, the case of spatially constant gas pressure gives the dynamic free-surface condition as

$$\frac{D\Phi}{Dt} = \tfrac{1}{2}\nabla\Phi\cdot\nabla\Phi + \boldsymbol{g}\cdot\boldsymbol{r} \quad \text{on } \Sigma(t). \quad (10.8)$$

A consequence of using potential flow theory inside a tank is that transients due to initial conditions do not die out. In reality an energy dissipation exists due to viscous effects that will damp out transients due to initial conditions. Energy dissipation due to viscous effects in the boundary layers and due to separated flow past internal structures was discussed in Chapter 6. If no internal structures exist inside the tank, the viscous damping is small. However, even though it is small, it must be considered for the sloshing problem for a longer period of time. The way that we propose to account for this fact is similar to that presented by Faltinsen (1978). A fictitious small term, $-\mu\nabla\Phi$, is added to the right-hand side of the Euler equations. We may treat μ as a kind of viscosity coefficient. If the velocity terms in the modified Euler equations are expressed in terms of the velocity potential and the equation is integrated in space, we get a modified Bernoulli equation. The consequence is that an additional term $\mu\Phi$ appears within the brackets on the right-hand side of eq. (10.7). Equation (10.8) can, therefore, be modified to

$$\frac{D\Phi}{Dt} = \tfrac{1}{2}\nabla\Phi\cdot\nabla\Phi + \boldsymbol{g}\cdot\boldsymbol{r} - \mu\Phi. \quad (10.9)$$

The kinematic free-surface condition is satisfied by calculating the liquid velocity for the fluid particles on the free surface and thereby tracking their position as a function of time. If a gas cushion occurs, for instance in connection with water impact, we must modify eq. (10.9) on the gas cavity surface to be consistent with the gas pressure. The gas flow in the cushion and the surface tension are neglected. The cushion pressure $p_c(t)$ is time dependent as a consequence of variation in the cushion volume, $\Omega(t)$. We start with eq. (10.7) and set the pressure p equal to $p_c(t)$ in the cushion interface. Hence, eq. (10.9) is modified to

$$\frac{D\Phi}{Dt} = \tfrac{1}{2}\nabla\Phi\cdot\nabla\Phi + \boldsymbol{g}\cdot\boldsymbol{r} - \mu\Phi + p_0 - p_c(t) \quad (10.10)$$

on the cushion interface. We now need additional equations to determine $p_c(t)$ to account for the fact that the gas is compressible. By assuming no leakage or flow into the gas cushion, we can write the continuity of mass for the gas cushion as

$$\frac{d\left[\rho_c(t)\Omega(t)\right]}{dt} = 0, \quad (10.11)$$

where $\rho_c(t)$ is the mass density of the gas in the cushion. It follows by assuming an adiabatic pressure–density relationship that

$$\frac{p_c(t)}{p_g} = \left[\frac{\rho_c(t)}{\rho_g}\right]^{\kappa}, \quad (10.12)$$

where $\kappa = 1.4$ for air and ρ_g and p_g are the mass density and pressure of the gas at closure of the cushion, respectively.

Because eq. (10.11) means that $\rho_c(t)\Omega(t)$ is a constant, we can rewrite eq. (10.12) as

$$\frac{p_c(t)}{p_g} = \left[\frac{\Omega_0}{\Omega(t)}\right]^{\kappa},$$

where Ω_0 is the initial volume of the cushion at the closure. To find $\Omega(t)$ we use the relationship $\dot{\Omega} = \oiint_{S_c} U_n \, dS$, where S_c is the surface enclosing the cushion volume and $\boldsymbol{n}$ represents the normal direction to this surface pointing out of the cushion; U_n is the normal velocity at S_c, which is equal to $\partial\Phi/\partial n$ at the interface and is $\boldsymbol{n}\cdot(\boldsymbol{v}_O + \boldsymbol{\omega}\times\boldsymbol{r})$ if parts of S_c contain the tank surface.

This procedure was followed by Greco *et al.* (2003), where bottom slamming with local hydroelastic effects for a very long barge with shallow draft was studied.

10.2.2 Generation of vorticity

When a plunging breaker has hit the underlying liquid and a contact zone is established between the breaker and the underlying liquid, vorticity is present at the contact zone. The presence of vorticity is a consequence of the fact that there is a jump in velocity potential between the impacting breaker and the underlying touching zone on the liquid. The reason is simply because the underlying liquid cannot instantaneously adjust its potential to that on the tip of the breaker. The consequence of the jump of the velocity potential is that a jump of the tangential liquid velocity occurs at the contact zone. For this reason, one can say a vortex sheet exists in the liquid along the interface. The situation is similar to the interface of two-layer liquids. This fact has implications to the numerical method. If one uses BEM, a singularity distribution with normal dipoles is needed on the vortex sheet. This procedure is similar to what is done, for instance, in describing the effect of separating free shear layers from a hydrofoil (Faltinsen, 2005) or to what was done in connection with Figure 3.33.

A trick used in exterior flow problems to avoid the impact between a plunging breaker and the underlying liquid is to cut away liquid mass from the tip of the breaker so that it does not hit the underlying liquid. However, conservation of mass is critical for proper flow simulation in an interior flow problem. Therefore, we cannot recommend such a procedure.

10.2.3 Example: numerical discretization

We are going to solve the two-dimensional sloshing problem for forced harmonic sway motion, $\eta_2 = \eta_{2a} \sin \sigma t$. We select a tank-fixed coordinate system (x, y) with $y = 0$ in the mean free surface and y positive upward in this and the following sections. The tank is rectangular with length $l = 2a$. The mean liquid depth is h. The vertical sides of the tank are at $x = \pm a$. There is no tank roof and the possibility of liquid impact on the tank walls is not discussed. The boundary condition on the wetted tank surface is

$$\frac{\partial \Phi}{\partial n} = \mp \eta_{2a} \sigma \cos \sigma t, \quad x = \pm a;$$
$$\frac{\partial \Phi}{\partial n} = 0, \quad y = -h. \tag{10.13}$$

The dynamic free-surface condition is given by eq. (10.9). In addition the kinematic free-surface condition also applies. Initial conditions also need to be specified; they could, for instance, be that the free-surface elevation and the velocity potential on the free surface are initially zero.

The velocity potential in the liquid can be written as a distribution of sources over the wetted tank surface S and the instantaneous free surface Σ as

$$\Phi(x_1, y_1) = \int_{S+\Sigma} q(x, y) \ln \sqrt{(x - x_1)^2 + (y - y_1)^2} \, ds(x, y), \tag{10.14}$$

where (x_1, y_1) is a point in the liquid. A low-order panel method is used in the example, meaning that the tank surface and the free surface are approximated by straight line segments. The source density and the velocity potential are assumed constant over each segment and the boundary conditions are satisfied at the midpoints $(\bar{x}_j, \bar{y}_j)$ of each line segment.

If the subdivision of the surface $S + \Sigma$ in Figure 10.5 is used as an example, we can set up the following linear algebraic equations to determine $q(x, y)$:

$$\sum_{j=1}^{18} \mu_{ij} q(\bar{x}_j, \bar{y}_j) = B_i, \quad i = 1, 18, \tag{10.15}$$

where

$$\mu_{ij}\,(i = 1, 5; j = 1, 18) = \int_{S_j} \ln \sqrt{(x - \bar{x}_i)^2 + (y - \bar{y}_i)^2} \, ds(x, y),$$

$$\mu_{ij}\,(i = 6, 18; j = 1, 18) = \lim_{(x_1, y_1) \to (\bar{x}_i, \bar{y}_i)} \frac{\partial}{\partial n(x_1, y_1)} \times \int_{S_j} \ln \sqrt{(x - \bar{x}_i)^2 + (y - \bar{y}_i)^2} \, ds(x, y),$$

$$B_i\,(i = 1, 5) = \varphi(\bar{x}_i, \bar{y}_i),$$
$$B_i\,(i = 6, 10) = -\eta_{2a} \sigma \cos \sigma t,$$
$$B_i\,(i = 11, 15) = 0,$$
$$B_i\,(i = 16, 18) = \eta_{2a} \sigma \cos \sigma t. \tag{10.16}$$

We would use many more elements in practical calculations. A general rule is that neighboring elements should not have a large difference in size. Generally speaking, one needs a higher density of elements for an interior problem than for an exterior problem. One should also note that

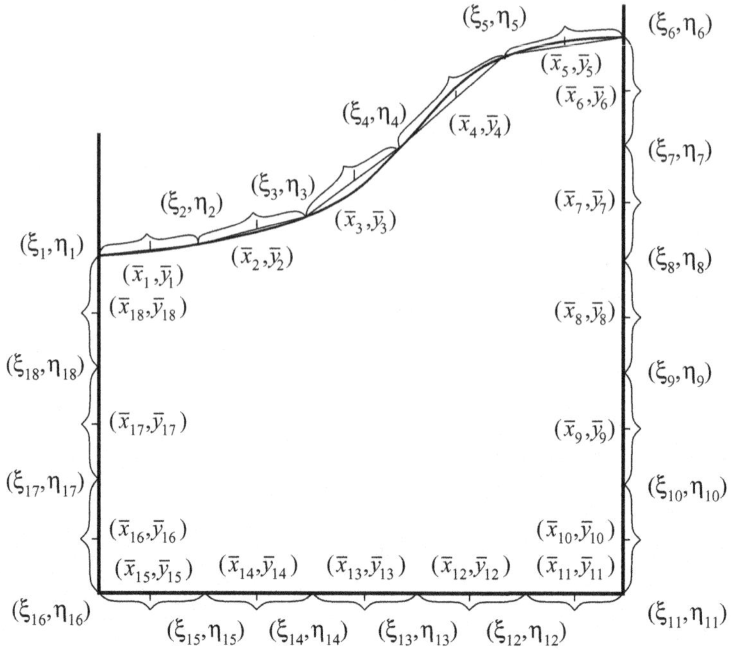

Figure 10.5. Approximation of body surface and free surface by line segments.

redistribution of elements may be necessary at each time instant or after every several time steps.

We now derive expressions for

$$\int_{s_1}^{s_2} \ln\sqrt{(x-x_1)^2+(y-y_1)^2}\,\mathrm{d}s(x,y), \tag{10.17}$$

where s_1 and s_2 are the endpoints of the line segment (see Figure 10.6). We write $r=\sqrt{(x-x_1)^2+(y-y_1)^2}=\sqrt{n^2+s^2}$. Thereby eq. (10.17) can be rewritten as

$$\begin{aligned}\int_{s_1}^{s_2} \ln\sqrt{n^2+s^2}\,\mathrm{d}s \\ = \tfrac{1}{2}\{s_2\ln(n^2+s_2^2)-2s_2+2n\ \mathrm{arctg}(s_2/n) \\ -s_1\ln(n^2+s_1^2)+2s_1-2n\ \mathrm{arctg}(s_1/n)\}.\end{aligned} \tag{10.18}$$

To evaluate the foregoing expressions, we need expressions for n, s_1, and s_2. We can write

$$n=\frac{(x_1-\xi_1)(\eta_2-\eta_1)-(y_1-\eta_1)(\xi_2-\xi_1)}{\sqrt{(\eta_2-\eta_1)^2+(\xi_2-\xi_1)^2}},$$

$$s_1=-\left[\frac{(x_1-\xi_1)(\xi_2-\xi_1)+(y_1-\eta_1)(\eta_2-\eta_1)}{\sqrt{(\eta_2-\eta_1)^2+(\xi_2-\xi_1)^2}}\right],$$

$$s_2=-\left[\frac{(x_1-\xi_2)(\xi_2-\xi_1)+(y_1-\eta_2)(\eta_2-\eta_1)}{\sqrt{(\eta_2-\eta_1)^2+(\xi_2-\xi_1)^2}}\right].$$

To evaluate the velocity in the liquid, we need expressions for $\partial\phi_p/\partial x_1$ and $\partial\phi_p/\partial y_1$, where

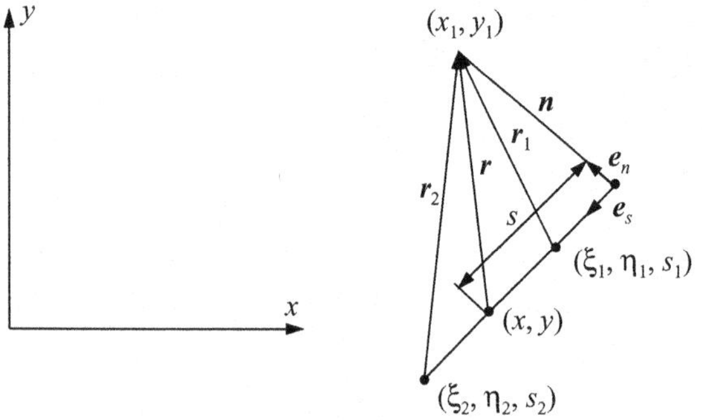

Figure 10.6. Local coordinate system for a line segment.

$\phi_p = \int_{s_1}^{s_2} \ln\sqrt{n^2+s^2}\, ds$. Using eq. (10.18) we can write

$$\frac{\partial \phi_p}{\partial x_1} = \frac{1}{2}\left\{\frac{\partial s_2}{\partial x_1}\ln\left(n^2+s_2^2\right) + 2\frac{\partial n}{\partial x_1}\left[\operatorname{arctg}\left(\frac{s_2}{n}\right) - \operatorname{arctg}\left(\frac{s_1}{n}\right)\right] - \frac{\partial s_1}{\partial x_1}\ln\left(n^2+s_1^2\right)\right\},$$

$$\frac{\partial \phi_p}{\partial y_1} = \frac{1}{2}\left\{\frac{\partial s_2}{\partial y_1}\ln\left(n^2+s_2^2\right) + 2\frac{\partial n}{\partial y_1}\left[\operatorname{arctg}\left(\frac{s_2}{n}\right) - \operatorname{arctg}\left(\frac{s_1}{n}\right)\right] - \frac{\partial s_1}{\partial y_1}\ln\left(n^2+s_1^2\right)\right\},$$

where

$$\frac{\partial s_2}{\partial x_1} = \frac{\partial s_1}{\partial x_1} = \frac{-(\xi_2-\xi_1)}{\sqrt{(\eta_2-\eta_1)^2+(\xi_2-\xi_1)^2}},$$

$$\frac{\partial n}{\partial x_1} = \frac{(\eta_2-\eta_1)}{\sqrt{(\eta_2-\eta_1)^2+(\xi_2-\xi_1)^2}},$$

$$\frac{\partial s_2}{\partial y_1} = \frac{\partial s_1}{\partial y_1} = \frac{-(\eta_2-\eta_1)}{\sqrt{(\eta_2-\eta_1)^2+(\xi_2-\xi_1)^2}},$$

$$\frac{\partial n}{\partial y_1} = \frac{-(\xi_2-\xi_1)}{\sqrt{(\eta_2-\eta_1)^2+(\xi_2-\xi_1)^2}}.$$

Let us consider a special case of a source distribution along the x-axis between 0 and 1, as illustrated in Figure 10.7. The source density q, as defined in eq. (10.14), is set equal to 1. The vertical velocity $\partial\phi/\partial y$ is plotted in Figure 10.7 along the line $x = 0.5$. Far away from the segment we see that the behavior is similar to a source singularity at (0.5,0). However, close to the segment the behavior differs, and we see that $\partial\phi/\partial y \rightarrow \pi$ when y is approaching a point on the segment from positive values. The consequence is that the value of μ_{ij} given by eq. (10.16) is π when $i = j$. It follows from inspecting the mathematical expressions that $\partial\phi/\partial y \rightarrow \infty$ at the endpoints (0,0) and (0,1) of the segment. This is not a physical phenomenon; it is partly a consequence of assuming that the source density is constant over the element and is discontinuous from one element to the next. However, using continuous (e.g., piecewise-linear) distribution of sources does not completely remove the singularity near the edges of the segments, if the discretized contour remains not smooth. What the example with straight line segments and piecewise constant source density tells us is that we must not evaluate the velocity potential and the velocity at any other points on the boundary than where the boundary conditions are satisfied (i.e., at the midpoint of a segment). We must also be careful in evaluating flow variables in the vicinity of the boundary.

Similar three-dimensional expressions for plane quadrilateral elements can, for instance, be found in Hess and Smith (1962) and Newman (1985).

10.2.4 Linear frequency-domain solutions

We start by illustrating how BEM can be used to find natural frequencies and modes for sloshing inside any tank shape. Our detailed description assumes two-dimensional flow. The velocity potential is represented in terms of sources and normal dipoles (see eq. (10.3)). A periodic solution with frequency σ is considered. The linear free-surface condition $-\sigma^2\Phi + g\partial\Phi/\partial y = 0$ is applied on the mean free surface. Natural frequencies and modes correspond to nontrivial flow

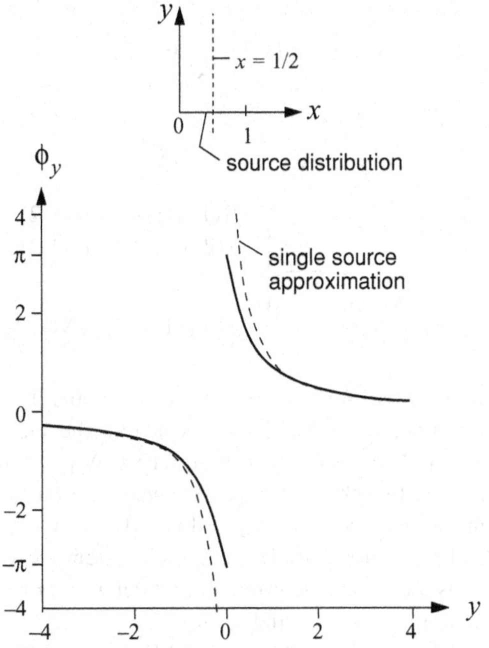

Figure 10.7. The y-variation of the vertical velocity $\partial\phi/\partial y = \phi_y$ induced at $x = 0.5$ due to a line distribution of two-dimensional sources of constant strength 2π distributed between $x = 0$ and $x = 1$ on the x-axis.

solutions with no tank excitation. Therefore, no forcing terms are present from the body boundary conditions. The mean free surface and the mean wetted body surface are divided into a total of N_{EL} straight line segments. We define

$$\mu(j,i) = \int_{S_i} \frac{\partial \psi(j,i)}{\partial n(x,y)} \, ds(x,y),$$

$$B(j,i) = \int_{S_i} \psi(j,i) \, ds(x,y), \qquad (10.19)$$

where $\psi(j,i) = \ln[(x-x_j)^2 + (y-y_j)^2]^{1/2}$ and (x_j, y_j) is the midpoint of segment number j; $\mu(i,j)$ and $B(j,i)$ can be expressed analytically in a way similar to that described in Section 10.2.3. One should note that in eq. (10.19) we differentiate ψ with respect to the x- and y-coordinates for the body surface (i.e., not with respect to the field point coordinates as we did in expressing the tank boundary conditions with a source distribution in the previous section).

We denote the natural frequencies and modes as σ_n and φ_n, respectively. We now start with eq. (10.3) and replace $\partial\varphi_n/\partial n$ with $(\sigma_n^2/g)\varphi_n$ on the mean free surface and zero on the mean wetted tank surface. Using a low-order panel gives the following equation system:

$$\begin{bmatrix} 2\pi + \mu(1,1) & \mu(1,2) & \cdots & \mu(1,N_{EL}) \\ \mu(2,1) & 2\pi + \mu(2,2) & \cdots & \mu(2,N_{EL}) \\ \vdots & \vdots & \vdots & \vdots \\ \mu(N_{EL},1) & \mu(N_{EL},2) & \cdots & 2\pi + \mu(N_{EL},N_{EL}) \end{bmatrix} \begin{bmatrix} \varphi_n(1) \\ \varphi_n(2) \\ \vdots \\ \varphi_n(N_{EL}) \end{bmatrix}$$

$$= -\frac{\sigma_n^2}{g} \begin{bmatrix} B(1,1)\gamma_1 & B(1,2)\gamma_2 & \cdots & B(1,N_{EL})\gamma_{N_{EL}} \\ B(2,1)\gamma_1 & B(2,2)\gamma_2 & \cdots & B(2,N_{EL})\gamma_{N_{EL}} \\ \vdots & \vdots & \vdots & \vdots \\ B(N_{EL},1)\gamma_1 & B(N_{EL},2)\gamma_2 & \cdots & B(N_{EL},N_{EL})\gamma_{N_{EL}} \end{bmatrix} \begin{bmatrix} \varphi_n(1) \\ \varphi_n(2) \\ \vdots \\ \varphi_n(N_{EL}) \end{bmatrix},$$

where γ_j is 1 for a free-surface element and 0 for the other elements. So let us denote the number of elements on the free surface as N_{FREE} and assume that the first N_{FREE} elements are on the mean free surface; then $\gamma_j = 1$ $(j = 1, \ldots, N_{\text{FREE}})$ and zero otherwise. This equation system represents an eigenvalue problem that determines the natural frequencies and modes.

Solaas (1995) and Solaas and Faltinsen (1997) used a similar BEM formulation combined with the Moiseev approach to solve nonlinear free-surface problems for general two-dimensional tank shapes. The described eigenvalue problem is then part of the solution. Section 8.2.2 also introduces a series of inhomogeneous free-surface problems that has to be solved. Many elements were needed for the higher-order solutions of $O(\varepsilon^{2/3})$ and $O(\varepsilon)$ in the Moiseev approach, where ε is a parameter expressing the smallness of the tank motions relative to the tank breadth. The numerical accuracy of the method at the intersection between the mean free surface and the tank walls was particularly challenging.

Commercial BEM codes developed for linear frequency-domain solutions of wave-induced loads and motions on stationary platforms and ships can incorporate in their analysis the linear flow in a tank (see Section 3.6.6). It is common in exterior flow problems to use a source expression that satisfies the linear free-surface and radiation conditions. A consequence is that it is sufficient to have a distribution of singularities over the mean wetted tank surface. The radiation condition means that outgoing waves are guaranteed in the exterior problem. Even though this has no meaning for an interior problem, it does no harm for the interior flow formulation to include a source with a radiation condition formulation.

10.3 Finite difference method

This section describes how an FDM can be used to solve the gas and liquid behavior in a tank. The SOLA scheme by Hirt *et al.* (1975) is an extensively used finite difference scheme for Navier–Stokes equations for an incompressible confined fluid. Additional techniques have to be considered to handle the presence of a free surface. The effect of the free surface has, for instance, been studied using the MAC method (Welch *et al.*, 1966), the SURF method (Hirt *et al.*, 1975), and the VOF method (Hirt & Nichols, 1981) method. In the SURF approach, the free-surface elevation is followed, implying that the free surface must be a single-valued function of the horizontal tank coordinates, so that overturning waves cannot be described. The MAC method and the VOF method are described in some detail in Section 10.3.3.

Arai *et al.* (1992a, 1992b) and Kim (2001, 2007) used the SOLA scheme in simulating three-dimensional sloshing. The free surface is tracked by the SURF scheme. The dynamic free-surface condition of hydrodynamic pressure being equal to the gas pressure at the free surface is used. Special considerations were made by Arai *et al.* (1992a, 1992b) and Kim (2001, 2007) for when the liquid is expected to impact on the tank ceiling. A buffer zone with a mixed free surface and tank boundary condition is introduced within a layer near the tank ceiling.

We want to emphasize that different FDM schemes exist and an essential part is how boundary conditions on the tank boundary and the interface between the gas and the liquid are handled. We describe one of several methods published in the literature. The following text is partly based on descriptions by Greco (personal communication, 2008) and Berthelsen (see Berthelsen & Faltinsen, 2008). Our described procedure differs, for instance, from the work by Arai *et al.* (1992a, 1992b) and Kim (2001, 2007) in how the free-surface conditions are incorporated. We limit ourselves to two dimensions when the details of the numerical scheme are described in this chapter. However, the procedure can be generalized to three dimensions. When the FDM is described, we introduce procedures, which are not limited to the FDM, on how to numerically solve the Navier–Stokes equations and how to apply interface capturing. We give some general preliminaries about the FDM before introducing the governing equations.

10.3.1 Preliminaries

FDM is based on approximations of the derivatives by using differences of the variable at some finite distance Δx (or time Δt). We obtain the usual definition of differentiation if we make this distance infinitely small, but in numerical work this difference is finite, giving rise to the name *finite difference method*. We may deduce finite difference approximations by starting with the well-known Taylor expansion around the point x, where we want to express the derivatives:

$$u(x+h) = u(x) + h\frac{du}{dx} + \tfrac{1}{2}h^2\frac{d^2u}{dx^2} + \tfrac{1}{6}h^3\frac{d^3u}{dx^3} + O(h^4), \quad (10.20)$$

$$u(x-h) = u(x) - h\frac{du}{dx} + \tfrac{1}{2}h^2\frac{d^2u}{dx^2} - \tfrac{1}{6}h^3\frac{d^3u}{dx^3} + O(h^4). \quad (10.21)$$

A Taylor expansion requires that the considered function is smooth at the point x (i.e., that no singularity exists at that point). By summation of the preceding equations, we get

$$u(x+h) + u(x-h) = 2u(x) + h^2\frac{d^2u}{dx^2} + O(h^4),$$

which gives an expression for the second-order derivative

$$\frac{d^2u}{dx^2} = \frac{u(x+h) - 2u(x) + u(x-h)}{h^2} + O(h^2). \quad (10.22)$$

If we subtract eq. (10.21) from eq. (10.20), we get

$$u(x+h) - u(x-h) = 2h\frac{du}{dx} + O(h^3),$$

which leads to

$$\frac{du}{dx} = \frac{u(x+h) - u(x-h)}{2h} + O(h^2).$$

The last equation is the approximation of the first-order derivative by the *central difference*. If we use eq. (10.20) alone, we may approximate the first derivative by the *forward difference*:

$$\frac{du}{dx} = \frac{u(x+h) - u(x)}{h} + O(h). \quad (10.23)$$

By using eq. (10.21) alone, we may also approximate the first-order derivative by the *backward difference*:

$$\frac{du}{dx} = \frac{u(x) - u(x-h)}{h} + O(h).$$

10.3.2 Governing equations

A single-fluid formulation is used to describe the whole tank, where the density and viscosity are varying across the interface between gas and liquid. Therefore, the interface is inside the fluid domain and the continuity of the velocity and of the tangential and normal stresses is automatically fulfilled. The thickness of the artificial layer where the density and viscosity vary must be thin but finite. The position of the interface can be found by an interface-capturing technique. At this stage we assume that the location of the interface is known. Further details are given in Section 10.3.3.

Both the gas and the liquid are assumed incompressible. This is, in general, a good assumption for the liquid. However, if gas pockets occur, one must account for the compressibility of the gas. A practical reason for neglecting the compressibility is the required CPU time for properly resolving the acoustic waves. The time step must then be related to the speed of sound. If a fluid is incompressible, the sound speed is infinite and no longer a parameter of the problem.

Furthermore, the flow is assumed to be laminar. A tank-fixed Cartesian coordinate system (x, y, z) is used, as shown in Figure 2.2. The unit vectors along the x-, y-, and z-axes are denoted $\boldsymbol{e}_1$, $\boldsymbol{e}_2$, $\boldsymbol{e}_3$, respectively. The rigid-body translational velocity and acceleration of the origin of the $Oxyz$-system relative to a nonaccelerated (inertial) coordinate system are denoted $\boldsymbol{v}_O$ and $\boldsymbol{a}_O$, respectively. Furthermore, the instantaneous angular velocity of the $Oxyz$ -system is $\boldsymbol{\omega}(t)$.

It follows by generalizing eq. (2.56) that the Navier–Stokes equations for a single fluid in the tank-fixed Cartesian coordinate system (x, y, z) can be expressed as

$$\frac{\partial^* \boldsymbol{v}_r}{\partial t} + \boldsymbol{v}_r \cdot \nabla \boldsymbol{v}_r = -\frac{\nabla p}{\rho} + \frac{\nabla \cdot (\tau_{ij} \boldsymbol{e}_i \boldsymbol{e}_j)}{\rho} + \boldsymbol{f}, \tag{10.24}$$

where $\boldsymbol{v}_r$ is the fluid velocity relative to the $Oxyz$-system. Furthermore, $\partial^* \boldsymbol{v}_r / \partial t$ means that we time-differentiate $\boldsymbol{v}_r$ for a fixed point in the tank-fixed coordinate system and that we then must not differentiate the unit vectors with respect to time. Other terms in eq. (10.24) are defined as follows:

$$\begin{aligned}\tau_{ij} &= \mu \left(\frac{\partial u_i^r}{\partial x_j} + \frac{\partial u_j^r}{\partial x_i} \right),\\ \boldsymbol{f} &= \boldsymbol{g} - \boldsymbol{a}_O - (\boldsymbol{\omega} \times \boldsymbol{v}_O) - \left(\frac{d\boldsymbol{\omega}}{dt} \times \boldsymbol{r} \right)\\ &\quad - 2(\boldsymbol{\omega} \times \boldsymbol{v}_r) - \boldsymbol{\omega} \times (\boldsymbol{\omega} \times \boldsymbol{r}),\end{aligned} \tag{10.25}$$

where u_i^r are components of $\boldsymbol{v}_r$. When the tank has angular velocity, the components of acceleration of vector $\boldsymbol{g}$ due to gravity are time-dependent relative to the tank-fixed coordinate system.

The fact that the fluid is incompressible can be expressed as

$$\nabla \cdot \boldsymbol{v}_r = 0. \tag{10.26}$$

To solve the problem we need to specify tank boundary conditions. The requirement of no slip at the tank boundary gives $\boldsymbol{v}_r = 0$. By using the no-slip tank boundary condition it follows from eq. (10.24) that the tank boundary condition for the pressure is

$$\begin{aligned}\nabla p &= \nabla \cdot (\tau_{ij} \boldsymbol{e}_i \boldsymbol{e}_j) + \rho \boldsymbol{f};\\ \boldsymbol{f} &= \boldsymbol{g} - \boldsymbol{a}_O - (\boldsymbol{\omega} \times \boldsymbol{v}_O) - \left(\frac{d\boldsymbol{\omega}}{dt} \times \boldsymbol{r} \right)\\ &\quad - \boldsymbol{\omega} \times (\boldsymbol{\omega} \times \boldsymbol{r}), \quad \boldsymbol{x} \in S_B,\end{aligned} \tag{10.27}$$

where S_B is the tank boundary. In addition to the boundary conditions, initial conditions are also needed.

10.3.3 Interface capturing

Interface-capturing methods do not normally define the free surface as a sharp interface and require a fixed grid to extend beyond the free surface. The so-called color functions, which identify how the gas and liquid phases change with time, are introduced. These functions enable us to reconstruct the time evolution of the interface. Different methods exist to capture the interface between liquid and gas (Scardovelli & Zaleski, 1999).

The VOF method solves a transport equation as eq. (10.28) for the volume fraction α of the occupied liquid in each grid cell (Hirt & Nichols, 1981). The values of the fractional volumes in the filled, partially filled, and empty cells are unity, between zero and unity, and zero, respectively. In the VOF method, only one scalar value of the fractional volume is required for each cell, and the fractional volume at the current time step in each cell is calculated using the velocity field and fractional volume at the previous time step. The VOF method is a very time-efficient method for analyzing transient fluid flow with a free surface. However, it has drawbacks in that the position of the free surface is predicted only by the scalar fractional volume value and the filling state of its neighboring cells, defined as cells sharing a common side.

The MAC scheme (Harlow & Welch, 1965) spreads particles both in the liquid and in the gas across the interface. The particles have no mass and are marked to recognize if they are liquid or gas particles. At any time step one lets them move in a Lagrangian fashion and then one can find out how many liquid or gas particles are inside each cell. In this way one can reconstruct the liquid/gas fraction for each cell. The latter information is used to reconstruct the interface in a way similar to that of the VOF method. The MAC method requires sufficient memory for the markers to be distributed in the fluid region. A drawback of MAC is the distortion of the free-surface in flows with large shear in the free-surface area (El Moctar, personal communication, 2008).

Another method is to define density (color) functions $\phi_m (m = 1, 2, 3)$ to recognize different phases. In this case $m = 1, 2, 3$ denotes liquid, gas, and solid phase, respectively. The density functions have values between 0 and 1. If $\phi_1 = 1$ for a cell, the cell is completely occupied by liquid. The density functions for each computational cell have the relationship $\sum_{m=1}^{3} \phi_m = 1$ and satisfy the following transport equation:

$$\frac{\partial \phi_m}{\partial t} + u_i^r \frac{\partial \phi_m}{\partial x_i} = 0. \qquad (10.28)$$

These equations express the fact that ϕ_m does not change with time when we follow a fluid particle. When the tank position is known, it is not necessary to use eq. (10.28) for ϕ_3. Because $\sum_{m=1}^{3} \phi_m = 1$, eq. (10.28) is only necessary to have $m = 1$. The VOF method is a special version of this approach, where $\phi_1 = \alpha$.

10.3.3.1 Level-set technique

Another interface-capturing method is the LS technique. This method was used for the first time for incompressible flows by Sussman *et al.* (1994). A complete description can be found in Sethian (1999). An LS function ϕ is introduced with the following properties:

$\phi(P) < 0$ *if* $\boldsymbol{P}$ *is in the liquid (phase 1).*

$\phi(P) > 0$ *if* $\boldsymbol{P}$ *is in the gas (phase 2).*

$\phi(P) = 0$ *if* $\boldsymbol{P}$ *is on the interface.*

The function ϕ represents the normal positive or negative distance of any fluid point P from the interface. An illustration is given in Figure 10.8.

The approach is named to reflect the fact that the interface is the set of points where the function ϕ has the level zero. An important advantage of using the distance function is that it directly defines the interface position, through $\phi = 0$, and allows us to evaluate easily its normal vector, through $\boldsymbol{n} = \nabla\phi / |\nabla\phi|$, and its curvature through $\kappa = \nabla \cdot (\nabla\phi / |\nabla\phi|)$. The latter is needed when surface tension matters. The method assumes that the mass density, ρ, and the dynamic viscosity coefficient are only functions of ϕ; that is, $\rho = \rho(\phi)$, $\mu = \mu(\phi)$. No rule exists for what these functional relationships should be and what one should choose. A trigonometric behavior is commonly selected. However, an exponential function has, for instance, also been used (Colicchio, 2004).

When a fluid particle is followed, ϕ does not change with time; therefore,

$$\frac{D\phi}{Dt} \equiv \frac{\partial \phi}{\partial t} + \boldsymbol{v}_r \cdot \nabla\phi = 0, \qquad (10.29)$$

where D/Dt is the material derivative. It is necessary in practical calculations to reinitialize ϕ at period intervals for ϕ to preserve its character as a distance function (Sethian, 1999). Due consideration must then be given to mass conservation (Colicchio, 2004).

When ϕ has been determined at any time instant t, it is possible to reconstruct the position of the interface and its geometrical characteristics. The LS function needs only to be exactly defined in a narrow band 2α (exact-ϕ region) centered at the interface. Outside this region it is kept constant and equal to $-\alpha$ and α in phases 1 and 2, respectively (see top sketch of Figure 10.8). To avoid numerical difficulties this cutoff is performed in a smooth way. The band $[-\alpha, \alpha]$ has to contain the region across the interface, say $[-\delta, \delta]$, where the smoothing of the fluid properties is performed (see bottom sketch in Figure 10.8). The latter should be very small to recover the sharp behavior of the gas–liquid interface. However, it cannot be smaller than a threshold value; otherwise, numerical instabilities may occur. Special attention has to be given to proper smoothing of the variables inside the transition area, because they are linked to each other nonlinearly through the governing equations, and inconsistent smoothing can result in unphysical solutions outside the transition region. The resulting error

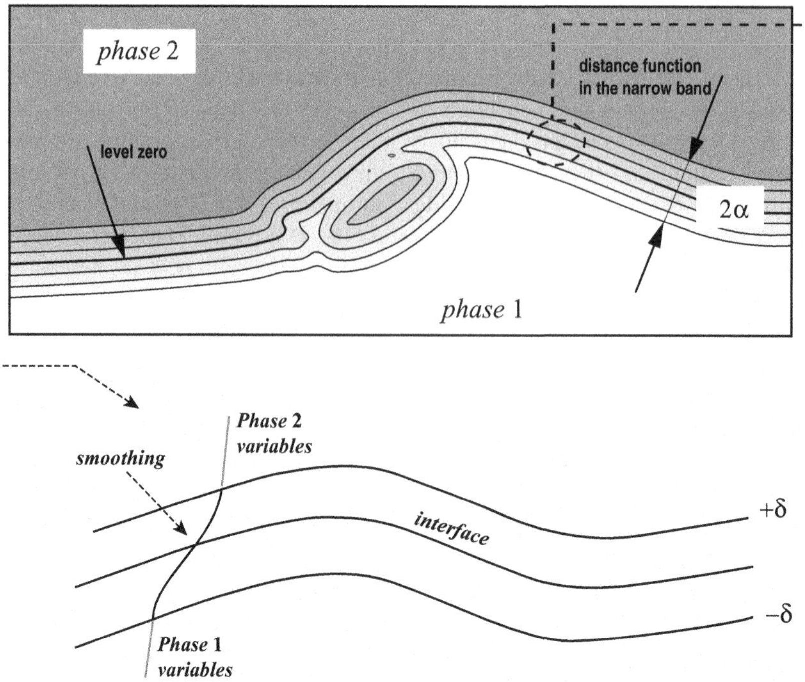

Figure 10.8. Level-set technique. At the top, the definition of a distance function ϕ in a narrow band around the interface represented by the zero level. At the bottom is a sketch of the isocontours delimiting the region across the air–water interface where the smoothing is applied (Colicchio, 2004).

is associated with δ, which is a function of the grid size. When the grid size goes to zero, a sharp interface is recovered in the limit.

If water impact occurs, a layer consistent with the LS technique and the body boundary conditions remains. What we implicitly say is that the pressure does not vary much across the LS layer. This can be argued in the same way as for thin boundary-layer and thin free-shear layer theory. The LS layer is an artificial thin free shear layer. If the angle between the impacting free surface and the body surface is very small, the impact pressure will vary strongly in time and space when no air cushions are formed. To properly solve this problem, stringent requirements on having sufficiently small time steps and grid sizes as well as the time integration procedure are necessary.

10.3.4 Introduction to numerical solution procedures

Higher-order time integration methods are needed for long time simulations to avoid excessive numerical diffusion. Some details on a numerical method for solving incompressible two-phase flow in tank geometries using the LS method are presented next. This scheme is based on the projection method of Chorin (1968). The two steps in Chorin's procedure for each time step are as follows:

1. An intermediate artificial velocity field is first obtained by neglecting the pressure term in momentum eq. (10.24).
2. Then an elliptic equation (Poisson equation) is solved for the pressure, which follows from enforcing eq. (10.26) (i.e., the fact that the fluid is incompressible).

The time-stepping of the governing equations in the following text is done explicitly by using a second-order predictor–corrector method. The spatial derivatives are approximated by finite difference discretization on a *staggered* grid. By "staggered grid," we mean that not all flow variables are defined on the same set of nodes. A higher-order *upwind* scheme is used to treat the convective terms in eqs. (10.24) and (10.29), whereas central schemes are used for all other terms. By "upwind" we mean that due consideration is given to the direction that the flow comes from (convected or advected from) relative to a grid cell.

10.3.5 Time-stepping procedures

Common time-stepping procedures assume that the integrated variable is a smooth function of time. If an impact between the liquid and the body surface occurs with zero relative angle between the free surface and the body surface, Wagner's (1932) theory shows that the solution is then singular at the initial impact time, which is a consequence of the physical approximations made in Wagner's theory. A singularity would not exist in physical reality due to the liquid–gas interaction and the fluid compressibility. However, a solution closer to the physics than Wagner's theory shows a tendency toward a singular behavior at the initial impact time and, therefore, numerical time-integration problems can occur with standard integration methods (Zhu, 2006).

The following predictor–corrector procedure is a Runge–Kutta method based on the trapezoidal rule, also known as Heun's method. The two steps of the projection method are done twice at each time step to ensure that only a divergence-free velocity field is used in the different terms of eqs. (10.24) and (10.29). At the predictor stage, an Euler step is taken to advance the solution with time step Δt. An Euler step is consistent with the forward difference scheme given by eq. (10.23). Let

$$F(\mathbf{v}_r, \phi, t) = -\mathbf{v}_r \cdot \nabla \mathbf{v}_r + \nabla \cdot (\tau_{ij}\mathbf{e}_i\mathbf{e}_j)/\rho + \mathbf{f}(t) \tag{10.30}$$

and $G(\mathbf{v}_r, \phi, t) = -\mathbf{v}_r \cdot \nabla\phi$; then, using a time-discrete form, we may write

$$\mathbf{v}_r^* = \mathbf{v}_r^n + \Delta t F(\mathbf{v}_r^n, \phi^n, t^n) \quad \text{and}$$
$$\overline{\phi}^{n+1} = \phi^n + \Delta t G(\mathbf{v}_r^n, \phi^n, t^n), \tag{10.31}$$

where the superscript n refers to time step n, $\overline{\phi}^{n+1}$ is the predicted LS function at time step $n+1$, and $\mathbf{v}_r^*$ is an interim approximation of the relative fluid velocity. Because the pressure gradient is neglected, $\mathbf{v}_r^*$ cannot, in general, be a true approximation of the fluid velocity; $\mathbf{v}_r^*$ just represents an intermediate step in the time integration of the relative fluid velocity, so, in general, $\nabla \cdot \mathbf{v}_r^* \neq 0$. We now return to the complete Navier–Stokes equations and approximate $\partial^* \mathbf{v}_r/\partial t$ as $(\overline{\mathbf{v}}_r^{n+1} - \mathbf{v}_r^n)/\Delta t = [(\overline{\mathbf{v}}_r^{n+1} - \mathbf{v}_r^*) + (\mathbf{v}_r^* - \mathbf{v}_r^n)]/\Delta t$, where $(\mathbf{v}_r^* - \mathbf{v}_r^n)/\Delta t$ is in balance with the convective acceleration, viscous term, and acceleration $\mathbf{f}(t)$ as given by eqs. (10.30) and (10.31). A first approximation of the relative fluid velocity at time instant $n+1$ is therefore

$$\overline{\mathbf{v}}_r^{n+1} = \mathbf{v}_r^* - \nabla\overline{p}\Delta t/\overline{\rho}. \tag{10.32}$$

By imposing the continuity equation on $\overline{\mathbf{v}}_r^{n+1}$, $\overline{p}$ can be obtained from solving

$$\nabla \cdot \left(\frac{1}{\overline{\rho}}\nabla\overline{p}\right) = \frac{1}{\Delta t}\nabla \cdot \mathbf{v}_r^*, \tag{10.33}$$

where $\overline{\rho} = \rho\left(\overline{\phi}^{n+1}\right)$.

The predicted velocities, $\overline{\mathbf{v}}_r^{n+1}$, and LS function, $\overline{\phi}^{n+1}$, are then used in the corrector step to obtain the solution at time t^{n+1}:

$$\mathbf{v}_r^{**} = \mathbf{v}_r^n + \tfrac{1}{2}\Delta t\big[F(\mathbf{v}_r^n, \phi^n, t^n) + F(\overline{\mathbf{v}}_r^{n+1}, \overline{\phi}^{n+1}, t^{n+1})\big], \tag{10.34}$$

$$\phi^{n+1} = \phi^n + \tfrac{1}{2}\Delta t\big[G(\mathbf{v}_r^n, \phi^n, t^n) + G(\overline{\mathbf{v}}_r^{n+1}, \overline{\phi}^{n+1}, t^{n+1})\big], \tag{10.35}$$

and

$$\mathbf{v}_r^{n+1} = \mathbf{v}_r^{**} - \frac{\Delta t}{\rho^{n+\frac{1}{2}}}\nabla p^{n+\frac{1}{2}}, \tag{10.36}$$

where $p^{n+1/2}$ is obtained from solving

$$\nabla \cdot \left(\frac{1}{\rho^{n+\frac{1}{2}}}\nabla p^{n+\frac{1}{2}}\right) = \frac{1}{\Delta t}\nabla \cdot \mathbf{v}_r^{**} \tag{10.37}$$

and $\rho^{n+\frac{1}{2}} = \rho(\phi^{n+\frac{1}{2}})$ for $\phi^{n+\frac{1}{2}} = \frac{1}{2}(\phi^n + \phi^{n+1})$. The argument is as before: we take the divergence of eq. (10.36) and use the fact that $\nabla \cdot \mathbf{v}_r^{n+1} = 0$. Note that the pressure lags in time.

Appropriate numerical boundary conditions for Poisson equation (10.37) can be obtained by rearranging eq. (10.36) as follows:

$$\frac{1}{\rho^{n+\frac{1}{2}}}\nabla p^{n+\frac{1}{2}} = \frac{\left(\mathbf{v}_{r\Gamma}^{**} - \mathbf{v}_{r\Gamma}^{n+1}\right)}{\Delta t} \quad \text{at } \mathbf{x} \in S_B, \tag{10.38}$$

where S_B as well as the subscript Γ denote the tank boundary. By setting $\mathbf{v}_{r\Gamma}^{**} = \mathbf{v}_{r\Gamma}^{n+1}$, boundary conditions (10.27) are implicitly satisfied, leading to a simpler homogeneous boundary condition for the pressure when solving eq. (10.37):

$$\nabla p^{n+\frac{1}{2}} = 0 \quad \text{at } \mathbf{x} \in S_B. \tag{10.39}$$

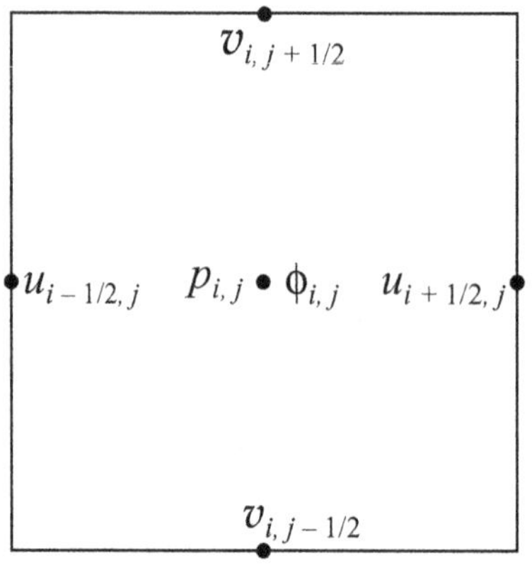

Figure 10.9. Grid cell (i, j) showing where velocity components u, v, pressure p, and LS function ϕ are defined. The total grid system is illustrated in Figure 10.2.

10.3.6 Spatial discretizations

We describe the spatial discretization by assuming two-dimensional flow. The handling of baffles is computationally demanding due to the fact that the baffles have small cross-sectional dimension. A complication is that the flow tends to be rapidly varying at the sharp corners of the baffles; however, the flow is actually singular there if potential flow is assumed without flow separation. A detailed procedure to handle thin plates in combination with local grid refinements is described by Berthelsen and Faltinsen (2008) and is not discussed here.

A grid cell system is assumed, as illustrated in Figure 10.2. The tank boundary is embedded in the grid system. Each grid cell is equally sized. In reality, we need sufficient grid resolution to describe the boundary layer along the tank surface, which can be accomplished by a local grid refinement approach (Berthelsen & Faltinsen, 2008). The x- and y-components of $\boldsymbol{v}_r$ are denoted u and v, respectively. The governing equations are discretized using a staggered grid.

The velocity components are defined at the appropriate cell faces, $u_{i+\frac{1}{2},j}$ and $v_{i,j+\frac{1}{2}}$, whereas the pressure, $p_{i,j}$, and the LS function, $\phi_{i,j}$, are defined at the cell center (see Figure 10.9). The mass density and the dynamic viscosity, denoted ρ and μ, respectively, are determined by the LS function. In this case, the subscript i, j indicates the grid cell (i, j) in the index system or $\mathbf{x}_{i,j} = (x_i, y_j)$ in the physical space, the subscript $i+\frac{1}{2}, j$ indicates the cell face separating cell (i, j) and $(i+1, j)$ at $\mathbf{x}_{i+\frac{1}{2},j}$, and $i, j+\frac{1}{2}$ indicates the cell face separating cell (i, j) and $(i, j+1)$ at $\mathbf{x}_{i,j+\frac{1}{2}}$. The uniform grid cell spacing is denoted Δx and Δy in the x- and y-direction, respectively. Where it is appropriate, we have dropped the superscript n for the ease of notation in the remaining sections.

10.3.7 Discretization of the convective and viscous terms

For now, let us assume that the grid nodes involved in the following discretization are located far away from the tank boundary so that the finite difference schemes are all well-defined and valid.

We define a set of cell-center velocities by simple averaging:

$$u_{i,j} = \tfrac{1}{2}\left(u_{i-\frac{1}{2},j} + u_{i+\frac{1}{2},j}\right) \quad \text{and}$$
$$v_{i,j} = \tfrac{1}{2}\left(v_{i,j-\frac{1}{2}} + v_{i,j+\frac{1}{2}}\right). \tag{10.40}$$

Then the convective term $\boldsymbol{v}_r \cdot \nabla\phi$ in eq. (10.29) can be discretized using a higher-order upwind scheme. For instance, the ENO2 scheme (Shu & Osher, 1988, 1989) gives the following discretization for the derivative in the x-direction at $\mathbf{x}_{i,j}$ (omitting the j index):

$$\phi_x^- = \frac{\phi_i - \phi_{i-1}}{\Delta x} + \tfrac{1}{2}\Delta x \times \mathrm{m}\left(\frac{\phi_{i+1} - 2\phi_i + \phi_{i-1}}{\Delta x^2}, \frac{\phi_i - 2\phi_{i-1} + \phi_{i-2}}{\Delta x^2}\right) \tag{10.41}$$

$$\phi_x^+ = \frac{\phi_{i+1} - \phi_i}{\Delta x} - \tfrac{1}{2}\Delta x \times \mathrm{m}\left(\frac{\phi_{i+1} - 2\phi_i + \phi_{i-1}}{\Delta x^2}, \frac{\phi_{i+2} - 2\phi_{i+1} + \phi_i}{\Delta x^2}\right), \tag{10.42}$$

where the minmod limiter m (Harten & Osher, 1987; http://en.wikipedia.org/wiki/Flux_limiter) is given by

$$\mathrm{m}(a, b) = \begin{cases} a & \text{if} \quad |a| \le |b|, \quad ab > 0, \\ b & \text{if} \quad |a| > |b|, \quad ab > 0 \\ 0 & \text{if} \quad ab < 0. \end{cases} \tag{10.43}$$

The objective of the minmod function is to minimize the local gradient to stabilize the solution. The following upwind condition determines which of eqs. (10.41) and (10.42) are used:

$$\phi_x = \begin{cases} \phi_x^- & \text{if} \quad u_{i,j} > 0, \\ \phi_x^+ & \text{if} \quad u_{i,j} < 0, \\ 0 & \text{otherwise.} \end{cases}$$

The ENO2 scheme is second-order accurate. It is essential to employ a higher-order scheme to avoid excessive numerical diffusion. For improved accuracy we may use the third-order ENO3 scheme or the fifth-order WENO scheme instead (Jiang & Peng, 2000; Jiang & Shu, 1996). We refer to these publications for further details.

To update u and v on the appropriate cell faces, we also need to define u at the cell face $(i,\ j+\frac{1}{2})$ and v at $(i+\frac{1}{2},\ j)$. Simple averaging gives

$$u_{i,j+\frac{1}{2}} = \tfrac{1}{2}(u_{i,j} + u_{i,j+1}) \quad \text{and}$$

$$v_{i+\frac{1}{2},j} = \tfrac{1}{2}(v_{i,j} + v_{i+1,j}). \tag{10.44}$$

The convective terms $\boldsymbol{v}_r \cdot \nabla u$ and $\boldsymbol{v}_r \cdot \nabla v$ can then be discretized in a similar fashion as described earlier for $\boldsymbol{v}_r \cdot \nabla\phi$.

For the viscous term we consider the x-component of the momentum equation only, meaning that we consider $(2\mu u_x)_x$ and $(\mu(u_y + v_x))_y$. The first derivatives of the velocity components are discretized by standard central differencing; that is,

$$(u_x)_{i,j} = \frac{u_{i+\frac{1}{2},j} - u_{i-\frac{1}{2},j}}{\Delta x},$$

$$(u_y)_{i+\frac{1}{2},j+\frac{1}{2}} = \frac{u_{i+\frac{1}{2},j+1} - u_{i+\frac{1}{2},j}}{\Delta y},$$

$$(v_x)_{i+\frac{1}{2},j+\frac{1}{2}} = \frac{v_{i+1,j+\frac{1}{2}} - v_{i,j+\frac{1}{2}}}{\Delta x}.$$

Then, using central differencing, we may write the second derivative as

$$[(2\mu u_x)_x]_{i+\frac{1}{2},j} = 2\frac{[\mu u_x]_{i+1,j} - [\mu u_x]_{i,j}}{\Delta x}$$

and

$$[(\mu(u_y + v_x))_y]_{i+\frac{1}{2},j} = \frac{[\mu(u_y + v_x)]_{i+\frac{1}{2},j+\frac{1}{2}} - [\mu(u_y + v_x)]_{i+\frac{1}{2},j-\frac{1}{2}}}{\Delta y}$$

for $\mu_{i,j} = \mu(\phi_{i,j})$ and $\mu_{i+\frac{1}{2},j+\frac{1}{2}} = \mu(\phi_{i+\frac{1}{2},j+\frac{1}{2}})$, where

$$\phi_{i+\frac{1}{2},j+\frac{1}{2}} = \tfrac{1}{4}\left(\phi_{i,j} + \phi_{i+1,j} + \phi_{i,j+1} + \phi_{i+1,j+1}\right).$$

10.3.8 Discretization of the Poisson equation for pressure

To advance the solution to a divergence-free velocity field at time t^{n+1}, we need to solve a Poisson equation for the pressure. For a staggered grid arrangement we use a so-called exact projection method: the discrete divergence constraint is exactly enforced. As already shown, the projection step is performed twice, one for each step of the time-integration procedure. For brevity, because both projection steps are identical, we only focus on the pressure Poisson equation (10.37) in the corrector step.

We try to find a pressure, $p_{i,j}$, such that the following constraint is satisfied:

$$\nabla \cdot \boldsymbol{u}_{i,j}^{n+1} = 0, \tag{10.45}$$

where the divergence operator is defined by the central scheme (see Section 10.3.1)

$$\nabla \cdot \boldsymbol{u}_{i,j}^{n+1} = \frac{u_{i+\frac{1}{2},j}^{n+1} - u_{i-\frac{1}{2},j}^{n+1}}{\Delta x} + \frac{v_{i,j+\frac{1}{2}}^{n+1} - v_{i,j-\frac{1}{2}}^{n+1}}{\Delta y}.$$

Furthermore, using eq. (10.36) we may write

$$u_{i+\frac{1}{2},j}^{n+1} = u_{i+\frac{1}{2},j}^{**} - \frac{\Delta t}{\rho_{i+\frac{1}{2},j}}\left(\frac{p_{i+1,j} - p_{i,j}}{\Delta x}\right),$$

$$u_{i-\frac{1}{2},j}^{n+1} = u_{i-\frac{1}{2},j}^{**} - \frac{\Delta t}{\rho_{i-\frac{1}{2},j}}\left(\frac{p_{i,j} - p_{i-1,j}}{\Delta x}\right),$$

$$v_{i,j+\frac{1}{2}}^{n+1} = v_{i,j+\frac{1}{2}}^{**} - \frac{\Delta t}{\rho_{i,j+\frac{1}{2}}}\left(\frac{p_{i,j+1} - p_{i,j}}{\Delta y}\right),$$

$$v_{i,j-\frac{1}{2}}^{n+1} = v_{i,j-\frac{1}{2}}^{**} - \frac{\Delta t}{\rho_{i,j-\frac{1}{2}}}\left(\frac{p_{i,j} - p_{i,j-1}}{\Delta y}\right),$$

where ∇p is approximated by central differences and $\rho_{i+\frac{1}{2},j} = \rho(\phi_{i+\frac{1}{2},j})$ for $\phi_{i+\frac{1}{2},j} = \frac{1}{2}(\phi_{i+1,j} + \phi_{i,j})$. Substitution into eq. (10.45) gives standard discrete Poisson equation (10.37) for the pressure:

$$\frac{\beta_{i+\frac{1}{2},j}(p_{i+1,j} - p_{i,j}) - \beta_{i-\frac{1}{2},j}(p_{i,j} - p_{i-1,j})}{\Delta x^2} + \frac{\beta_{i,j+\frac{1}{2}}(p_{i,j+1} - p_{i,j}) - \beta_{i,j-\frac{1}{2}}(p_{i,j} - p_{i,j-1})}{\Delta y^2}$$
$$= \frac{1}{\Delta t}\left(\frac{u_{i+\frac{1}{2},j}^{**} - u_{i-\frac{1}{2},j}^{**}}{\Delta x} + \frac{v_{i,j+\frac{1}{2}}^{**} - v_{i,j-\frac{1}{2}}^{**}}{\Delta y}\right),$$

where $\beta = 1/\rho$. If ρ is a constant, we note that the formulation of the left-hand side is consistent with the second-order central difference scheme given by eq. (10.22).

The resulting system of linear equations can be solved using most types of iterative methods, such as the BiCGSTAB or GMRES method with ILU(k) preconditioning (Saad, 2003).

10.3.9 Treatment of immersed boundaries

The immersed boundary method, sometimes referred to as a Cartesian grid method, represents a simple way of treating complex geometries without having the complexity of generating a high-quality grid. In this approach the boundary simply intersects with the underlying Cartesian grid. Most standard numerical schemes can be used, although some modifications to the discretization are required in the vicinity of the immersed boundary. Several different immersed boundary methods can be found in the literature. A comprehensive review covering the most important features of the immersed boundary methods is given by Mittal and Iaccarino (2005) and is not repeated here. However, we briefly outline the concept of ghost cells.

Ghost cells represent an artificial continuation of the solution outside the fluid domain and introduce an alternative way of imposing the body boundary conditions. They are updated by extrapolating values from the flow field and the boundary. In this way the numerical operators do not need to be reformulated near the boundary. Instead the boundary conditions are implicitly incorporated through the ghost cells. Values can be extrapolated into the ghost cells in numerous ways. A local flow variable is commonly represented in terms of a (higher-order) polynomial that is used to evaluate the ghost point. The accuracy of the ghost cell depends on the order of the interpolation scheme used to obtain this polynomial (Tseng & Ferziger, 2003). Higher-order polynomials are more accurate but also known to be more sensitive to numerical instabilities.

To remedy potential instabilities, it is common to introduce an image point I, which is inside the fluid domain along the normal to the boundary and passing through the ghost node G as in Figure 10.10. The image point can be constructed by using two-dimensional interpolation involving fluid nodes in the vicinity of the boundary and points on the boundary. The values at this image point are then extrapolated using the boundary condition to update the ghost cell value. If we,

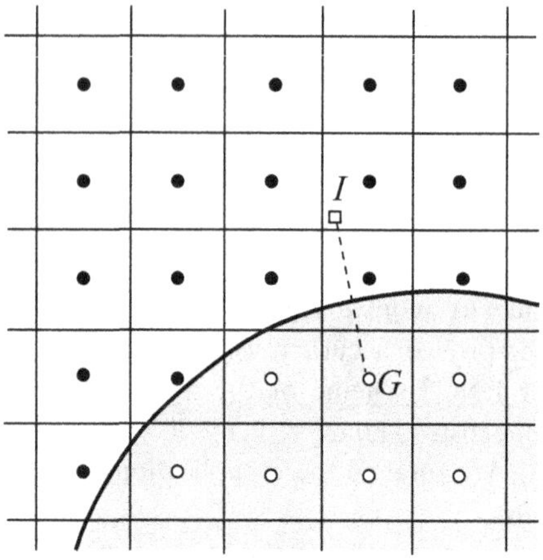

Figure 10.10. Ghost cells. Point I is in the fluid domain and G is a ghost node outside the fluid.

for instance, consider a plane surface, then we can, as the simplest approach, linearly extrapolate velocities from a node of a fluid cell next to the boundary to a node at the ghost cell in such a way that the relative velocity is zero at the boundary. Those two nodes are on a line perpendicular to the boundary. The pressure value at the ghost cell center and the fluid cell center next to the boundary would be the same with the boundary condition given by eq. (10.39). The ghost cell values are needed in the finite difference scheme for the cells that contain fluid. The concept of using an image point in the wall-normal direction has been widely used in the literature (e.g., Ghias *et al.*, 2004; Majumdar *et al.*, 2001; Tseng & Ferziger, 2003; among others), and it is also suited for general tank geometry without baffles.

Obtaining a wall-normal direction is not always straightforward; instead a weighted combination of one-dimensional extrapolations can be used to update the ghost cells. Tremblay and Friedrich (2000) used such a multidirectional approach. Their weighting coefficients depend only on the distance of the cells from the wall, where the direction closest to the boundary is given the largest weighting. An alternative to the wall-normal or multidirectional approach is the local directional ghost cell method presented by Berthelsen and Faltinsen (2008). The ghost cell value is obtained by one-dimensional extrapolation along the same direction as the discretization for which it will be used. This approach is specially constructed for highly irregular boundaries with sharp corners.

In a ghost cell approach the grid cells outside the fluid domain (solid cells) are marked as inactive, which means no governing equations are solved for these grid cells. However, not all inactive grid cells are ghost cells. Typically, only a band of cells near the solid boundary are ghost cells. The boundary itself can be represented by the zero-level set of a signed distance function, similar to the LS method for the interface. Then the normal direction can be easily obtained directly from the distance function. Conversely, the LS formulation fails to describe sharp corners and infinitely thin plates (e.g., baffles). In those cases it is necessary to resolve to elements to describe the solid boundary. More details on the numerical formulation and implementation of the ghost cell method can be found in the aforementioned publications and the references therein.

10.3.10 Constrained interpolation profile method

The CIP method was pioneered by Yabe and Wang (1991). It is basically an FDM. The essential feature is in the time-stepping, where a pure advection (convection) step is made and handled in a special way. As a base for our discussion let us use the same governing equations as in Section 10.3.2, but note that the method is not limited to incompressible fluid. Chorin's (1968) projection method is used. This projection method involves two steps, as described in Section 10.3.5. The first step neglects the pressure terms in the Navier–Stokes equations and an interim artificial velocity is calculated. The second step includes the pressure terms and a fluid velocity is obtained after solving the Poisson equation for the pressure. The convective (advective) acceleration terms in the first step needs special numerical techniques to account for upwind conditions. The essential feature of the CIP method is that the first step in Chorin's projection method is divided into two steps. The first of these two steps is called the advection step and considers only the advection terms. The advection step means for eq. (10.26) that the solution of

$$\frac{\partial^* \boldsymbol{v}_r}{\partial t} + \boldsymbol{v}_r \cdot \nabla \boldsymbol{v}_r = 0 \tag{10.46}$$

is studied numerically. If the LS technique is combined with the CIP method, then governing equation (10.29) for the LS function ϕ is in the form of an advection equation. The following describes how this advection step is handled numerically in the CIP method.

We consider a variable χ, which may be one of the relative velocity components in eq. (10.46) or the LS function ϕ if the LS method is used for interface tracking. The advection step means that we want to solve numerically the following equations for χ and the spatial derivatives of χ:

$$\frac{\partial \chi}{\partial t} + u_i \frac{\partial \chi}{\partial x_i} = 0, \quad i = 1, 2, 3, \tag{10.47}$$

$$\frac{\partial(\partial_\xi \chi)}{\partial t} + u_i \frac{\partial(\partial_\xi \chi)}{\partial x_i} = 0, \quad \xi = x_1, x_2, x_3, \tag{10.48}$$

where u_i are velocity components and $\partial_\xi \chi = \partial \chi / \partial x_i$. The spatial derivatives are introduced as variables to ensure better accuracy. Let us limit ourselves to the two-dimensional case and consider a grid point (i, j). (Note that "grid point" now means one of the corners of a grid cell.) We can find an upwind cell with the four grid points (i, j), (iw, j), (i, jw), and (iw, jw), where $iw = i - \text{sign}(u_{i,j})$, $jw = j - \text{sign}(v_{i,j})$, and u and v are used as notation for velocity components instead of u_1 and u_2, respectively. Then a cubic polynomial that approximates the spatial distribution of χ in the upwind cell can be written as

$$\begin{aligned} F^n(\xi, \eta) = {} & C_{30}\xi^3 + C_{21}\xi^2\eta + C_{12}\xi\eta^2 + C_{03}\eta^3 \\ & + C_{20}\xi^2 + C_{11}\xi\eta + C_{02}\eta^2 + C_{10}\xi \\ & + C_{01}\eta + C_{00}, \end{aligned} \tag{10.49}$$

where $\xi = x - x_i$, $\eta = y - y_j$. There are 10 unknown coefficients C_{mn} in eq. (10.49), which can be determined by the known values of χ^n, χ^n_x, and χ^n_y at grid points (i, j), (iw, j), and (i, jw), and the value of χ^n at the grid point (iw, jw). The superscript n denotes the current time instant. Once the interpolation function is determined, it follows that the solution at time step Δt after time instant n is

$$\begin{aligned} \chi^*(\boldsymbol{x}) &= \chi^n(\boldsymbol{x} - \boldsymbol{u}^n \Delta t), \\ (\partial_i \chi)^*(\boldsymbol{x}) &= \frac{\partial \chi^n}{\partial x_i}(\boldsymbol{x} - \boldsymbol{u}^n \Delta t), \end{aligned} \tag{10.50}$$

where the superscript $*$ denotes the intermediate time level after the advection step. The procedure is illustrated in Figure 10.11.

Then we must continue with intermediate steps consistently with Chorin's method to obtain an

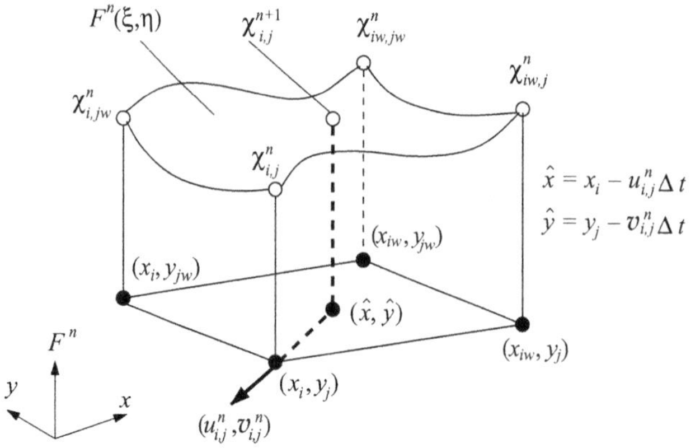

Figure 10.11. Illustration of the advection step in the CIP method.

approximation of the true values at the next time step. Hu *et al.* (2006) applied the CIP method to problems in marine hydrodynamics, including sloshing.

10.4 Finite volume method

10.4.1 Introduction

The FVM assumes the spatial computational domain is divided into a finite number of contiguous control volumes (CVs) of polyhedral shape. The CVs are typically smaller in regions of rapid flow variation, such as in the boundary layer.

Two sets of governing equations express the time rate of change of the mass and momentum of the control volume. The basis for the formulation is continuity of fluid mass and the Navier–Stokes equations. We show how these equations are derived by making the following assumptions:

- The fluid is compressible.
- The viscous flow is laminar.
- The reference coordinate system is an inertial system.

However, we may generalize the formulation by introducing a body-fixed accelerated coordinate system. We assume in the derivation that the surface of the control volume has a normal velocity U_{sn}, which is positive out of the control volume, and U_{sn} has nothing a priori to do with the fluid velocity. We denote ΔQ as the volume of the CV and $S_{\Delta Q}$ as the surface that encloses ΔQ.

We start by expressing the time rate of change of the mass of the CV. By noting that the volume ΔQ changes with time, we can apply eq. (A.2) with ΔQ, $S_{\Delta Q}$, and integrand ρ. The next step is to express $\partial\rho/\partial t$ by the continuity equation given by eq. (2.27) and to rewrite the volume integral by the generalized Gauss theorem given by eq. (A.1), which gives

$$\frac{d}{dt}\int_{\Delta Q} \rho \, \mathrm{d}Q + \int_{S_{\Delta Q}} \rho\,(u_n - U_{sn})\,\mathrm{d}S = 0,$$

where u_n is the normal component of the fluid velocity at the surface $S_{\Delta Q}$.

We now follow a similar procedure for the time rate of change of momentum of the CV, giving

$$\frac{d}{dt}\int_{\Delta Q} \rho \boldsymbol{v} \, \mathrm{d}Q = \int_{\Delta Q} \frac{\partial\,(\rho\boldsymbol{v})}{\partial t}\mathrm{d}Q + \int_{S_{\Delta Q}} \rho\boldsymbol{v} U_{sn}\,\mathrm{d}S.$$

Our next step is to write $\partial\,(\rho\boldsymbol{v})\,/\partial t = \boldsymbol{v}\partial\rho/\partial t + \rho\partial\boldsymbol{v}/\partial t$ and use continuity equation (2.27) to find $\boldsymbol{v}\partial\rho/\partial t$ and the Navier–Stokes equations given by eqs. (2.18) and (2.28) to find $\rho\partial\boldsymbol{v}/\partial t$. Finally, the generalized Gauss theorem gives

$$\frac{d}{dt}\int_{\Delta Q} \rho\boldsymbol{v}\,\mathrm{d}Q + \int_{S_{\Delta Q}} \rho\boldsymbol{v}(u_n - U_{sn})\mathrm{d}S$$
$$= -\int_{S_{\Delta Q}} p\,\boldsymbol{n}\,\mathrm{d}S + \int_{S_{\Delta Q}} \boldsymbol{n}\cdot\tau_{ij}\boldsymbol{e}_i\boldsymbol{e}_j\,\mathrm{d}S + \int_{\Delta Q} \rho\boldsymbol{g}\,\mathrm{d}Q.$$

For a compressible fluid we also need to relate the pressure p to the mass density ρ. This relationship is discussed for an adiabatic process in

Section 2.2.2.2. When the grid is moving, one must also take into account the space conservation law:

$$\frac{d}{dt}\int_{\Delta Q} dQ - \int_{S_{\Delta Q}} U_{sn}\, dS = 0.$$

One way to account for the free surface is to use the VOF method. The following equation for the time rate of change of the volume fraction c of the gas fraction in a CV is used by Peric *et al.* (2007) for incompressible fluid:

$$\frac{d}{dt}\int_{\Delta Q} c\, dQ + \int_{S_{\Delta Q}} c(u_n - U_{sn})\, dS = 0.$$

Liquid and gas can be considered two immiscible components of a single fluid. Within each CV we can write that the mass density ρ and the dynamic viscosity coefficient μ can be expressed as $\rho = \rho_g c + \rho_l(1-c)$, $\mu = \mu_g c + \mu_l(1-c)$. The subscripts g and l denote gas and liquid, respectively. Therefore, a one-fluid description that incorporates both the gas and the liquid is used. The consequence is that the dynamic free-surface condition requiring continuity of stresses at the interface is implicitly included in the governing equations. The kinematic boundary condition of the interface is accounted for by the VOF method.

The body boundary condition requiring no flow through the body can be directly enforced for the CV adjacent to the body by requiring $u_n = U_{sn}$ on the surface of the CV that is common with the body boundary. A no-slip boundary condition must also be imposed. The initial conditions are also needed.

Numerical details related to FVM can be found in Muzaferija and Peric (1999) and Ferziger and Peric (2002). Applications of FVM to sloshing in liquefied natural gas tanks are presented by Peric *et al.* (2007). Their basic equations differ from the previous equations by the fact that Reynolds-averaged Navier–Stokes equations are solved. Furthermore, they include coupling with ship motions. A summary of the numerical solution techniques is provided and we extract the following description of relevance for our governing equations. All integrals are approximated by the midpoint rule; that is, the value to be integrated is first evaluated at the center of the integration domain (CV face centers for surface integrals, CV center for volume integrals, and time level for time integrals) and then multiplied by the integration range (face area, cell volume, or time step). The approximations have second-order accuracy. Because variable values are computed at CV centers, interpolation is used to compute values at face centers. Most often this is achieved by linear interpolation. However, first-order upwind interpolation has to be used sometimes for numerical stability reasons. To compute the fluxes associated with the viscous stress term τ_{ij}, gradients of the fluid velocity are also needed at the cell faces.

Convective fluxes require special treatment in the equation for the volume fraction of the gas phase to achieve a sharp interface between the liquid and the gas within one cell (see Muzaferija & Peric, 1999). The scheme ensures that the volume fraction is always bounded by 0 and 1 to avoid nonphysical solutions.

An iterative method is used to solve the Navier–Stokes equations. The linearized momentum component equations are solved first, using prevailing pressure and mass fluxes through cell faces (inner iterations), followed by solving the pressure-correction equation derived from the continuity equation (SIMPLE algorithm; see Ferziger & Peric, 2002, for more details). Thereafter the equation for volume fraction is solved. The sequence is repeated until the equations are satisfied within a prescribed tolerance, after which the process is repeated for the next time step.

Either a tank-fixed grid or a grid moving relative to an inertial system may be used. A tank-fixed grid requires that the governing equations are formulated in an accelerated coordinate system; that is, we must use the Navier–Stokes equations as formulated in Section 2.4.1 when deriving expressions for the time rate of change of fluid momentum within a CV. Additional volume integrals then appear and are associated with, for instance, Coriolis and centrifugal accelerations. Because Peric *et al.* (2007) studied coupled sloshing and ship motion, they preferred to use a moving-coordinate system (i.e., the grid moves). Their argument was that the tank motion is not known a priori and has to be computed as part of the global solution that incorporates both internal and external hydrodynamic loads in the equations of rigid-body ship motion. The grid must then be updated after every time step.

10.4.2 FVM applied to linear sloshing with potential flow

We show how to apply the FVM to two-dimensional linear steady-state potential flow sloshing of an incompressible liquid in a rectangular tank that is forced to oscillate harmonically and horizontally with motion $\eta_1 = \eta_{1a} \exp(i\sigma t)$ (Guo, personal communication, 2008). The absolute velocity potential is denoted Φ. A normalized velocity potential ϕ defined by $\Phi = \phi\dot{\eta}_1$ is introduced. We select a Cartesian coordinate system Oxy, where $y = 0$ corresponds to the mean free surface and the y-axis is positive upward. The liquid depth is denoted h and the tank walls have coordinates $x = \pm\frac{1}{2}l$. The governing equation and boundary conditions can be expressed as

$$\nabla^2\phi = 0, \quad \left(-\sigma^2\phi + g\frac{\partial\phi}{\partial y} = 0\right)_{y=0},$$

$$\left.\frac{\partial\phi}{\partial x}\right|_{x=\pm\frac{1}{2}l} = 1, \quad \left.\frac{\partial\phi}{\partial y}\right|_{y=-h} = 0.$$

The Laplace equation must be reexpressed to apply the FVM. We consider a CV, integrate the Laplace equation over the CV, and apply the divergence theorem, which gives

$$\int_{S_{\Delta Q}} \frac{\partial\phi}{\partial n} dS = 0, \tag{10.51}$$

where $S_{\Delta Q}$ is the surface of the CV and n is in the direction of the normal vector of $S_{\Delta Q}$. Equation (10.51) expresses the continuity of liquid mass for the control volume.

We can now start the numerical approximation and divide the mean liquid domain into N_{CV} *equal* rectangular CVs. The length and height of each CV are denoted δx and δy, respectively. The equations are expressed in terms of values of ϕ at the geometrical center of each CV. The boundary conditions are imposed on the CVs adjacent to the boundaries.

We start by focusing on the immediate neighborhood of one CV, as illustrated in Figure 10.12. A low-order numerical method is used in the integration and differentiation; we approximate the velocity as a constant on each side of the CV. Equation (10.51) can then be approximated as

$$\delta y\left(\frac{\partial\phi}{\partial x}\right)_e - \delta y\left(\frac{\partial\phi}{\partial x}\right)_w + \delta x\left(\frac{\partial\phi}{\partial y}\right)_n - \delta x\left(\frac{\partial\phi}{\partial y}\right)_s = 0. \tag{10.52}$$

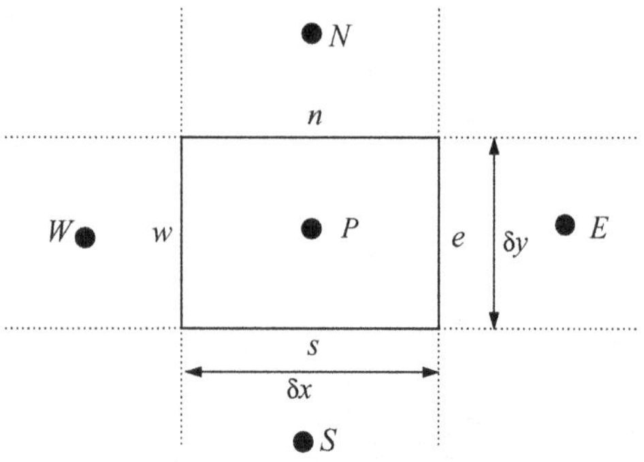

Figure 10.12. A rectangular-shaped CV with geometrical center at P. The sides of the CV are denoted as west (w), north (n), east (e), and south (s). The centers W, N, E, and S of the adjacent CVs are also indicated.

We have used the subscripts e, w, n, and s to indicate that we are on the east, west, north, and south side of the CV with a midpoint P as indicated in Figure 10.12. How to further approximate eq. (10.52) depends on whether the CV is adjacent to the tank boundary and/or the mean free surface. We illustrate the procedure by classifying the different CVs by numbers as indicated in Figure 10.13.

We start with an interior CV, which is denoted by 1. The velocities at the sides of the CV are expressed in terms of the velocity potential at the geometrical centers of the CVs adjacent to the CV with midpoint P (see Figure 10.12). A central difference scheme then gives

$$\left.\frac{\partial\phi}{\partial x}\right|_e = \frac{\phi_E - \phi_P}{\delta x}, \quad \left.\frac{\partial\phi}{\partial x}\right|_w = \frac{\phi_P - \phi_W}{\delta x},$$
$$\left.\frac{\partial\phi}{\partial y}\right|_n = \frac{\phi_N - \phi_P}{\delta y}, \quad \left.\frac{\partial\phi}{\partial y}\right|_s = \frac{\phi_P - \phi_S}{\delta y}. \tag{10.53}$$

We have used capital letters E, W, N, and S to indicate that we express the velocity potential ϕ at the centers of the CV that are respectively to the east, west, north, and south of the CV with a center at P (see Figure 10.12). Equation (10.51) can then be expressed as

$$\frac{\delta y}{\delta x}\phi_E + \frac{\delta y}{\delta x}\phi_W + \frac{\delta x}{\delta y}\phi_N + \frac{\delta x}{\delta y}\phi_S = 2\left(\frac{\delta y}{\delta x} + \frac{\delta x}{\delta y}\right)\phi_P.$$

For a CV denoted by 2 in Figure 10.12 we use the tank boundary condition (i.e., $(\partial\phi/\partial x)_w = 1$). Using eq. (10.53) for the rest of the sides of the

6	5	5	5	5	5	7
2	1	1	1	1	1	3
2	1	1	1	1	1	3
2	1	1	1	1	1	3
8	4	4	4	4	4	9

Figure 10.13. Classification of CVs according to whether the CV is adjacent to boundaries.

CV gives eq. (10.51) approximated as

$$\frac{\delta y}{\delta x}\phi_E + \frac{\delta x}{\delta y}\phi_N + \frac{\delta x}{\delta y}\phi_S = \left(\frac{\delta y}{\delta x} + 2\frac{\delta x}{\delta y}\right)\phi_P + \delta y.$$

It follows by using the tank boundary condition, that is, $(\partial\phi/\partial x)_e = 1$ for a CV denoted by 3 in Figure 10.13, that eq. (10.51) can be approximated as

$$\frac{\delta y}{\delta x}\phi_W + \frac{\delta x}{\delta y}\phi_N + \frac{\delta x}{\delta y}\phi_S = \left(\frac{\delta y}{\delta x} + 2\frac{\delta x}{\delta y}\right)\phi_P - \delta y.$$

Similarly it follows by using the tank boundary condition, $(\partial\phi/\partial y)_s = 0$, for the CVs denoted by 4 in Figure 10.13 that

$$\frac{\delta y}{\delta x}\phi_E + \frac{\delta y}{\delta x}\phi_W + \frac{\delta x}{\delta y}\phi_N = \left(2\frac{\delta y}{\delta x} + \frac{\delta x}{\delta y}\right)\phi_P.$$

The free-surface condition is imposed for a CV denoted by 5 as

$$\left.\frac{\partial\phi}{\partial y}\right|_n = \frac{\sigma^2}{g}\phi_n = \frac{\sigma^2}{g}\left(\tfrac{3}{2}\phi_P - \tfrac{1}{2}\phi_S\right). \qquad (10.54)$$

The expression for ϕ_n has been obtained by using the facts that $\phi_P = \frac{1}{2}(\phi_n + \phi_s)$ and $\phi_s = \frac{1}{2}(\phi_P + \phi_S)$. Equation (10.51) can then be approximated as

$$\frac{\delta y}{\delta x}\phi_E + \frac{\delta y}{\delta x}\phi_W + \left(\frac{\delta x}{\delta y} - \tfrac{1}{2}\delta x\frac{\sigma^2}{g}\right)\phi_S$$
$$= \left(2\frac{\delta y}{\delta x} + \frac{\delta x}{\delta y} - \tfrac{3}{2}\delta x\frac{\sigma^2}{g}\right)\phi_P.$$

Both the free-surface condition and the tank boundary condition are imposed for the CV denoted by 6, which leads to

$$\frac{\delta y}{\delta x}\phi_E + \left(\frac{\delta x}{\delta y} - \tfrac{1}{2}\delta x\frac{\sigma^2}{g}\right)\phi_S$$
$$= \left(\frac{\delta y}{\delta x} + \frac{\delta x}{\delta y} - \tfrac{3}{2}\delta x\frac{\sigma^2}{g}\right)\phi_P + \delta y.$$

Similarly for the CV denoted by 7, we have

$$\frac{\delta y}{\delta x}\phi_W + \left(\frac{\delta x}{\delta y} - \tfrac{1}{2}\delta x\frac{\sigma^2}{g}\right)\phi_S$$
$$= \left(\frac{\delta y}{\delta x} + \frac{\delta x}{\delta y} - \tfrac{3}{2}\delta x\frac{\sigma^2}{g}\right)\phi_P - \delta y.$$

The body boundary condition on the tank wall and the tank bottom is imposed for the CV denoted by 8, which leads to

$$\frac{\delta y}{\delta x}\phi_E + \frac{\delta x}{\delta y}\phi_N = \left(\frac{\delta y}{\delta x} + \frac{\delta x}{\delta y}\right)\phi_P + \delta y.$$

Similarly for the CV denoted by 9, we have

$$\frac{\delta y}{\delta x}\phi_W + \frac{\delta x}{\delta y}\phi_N = \left(\frac{\delta y}{\delta x} + \frac{\delta x}{\delta y}\right)\phi_P - \delta y.$$

We have now shown how to set up the linear algebraic equation system that determines the normalized velocity potential at the geometrical centers of the CVs. The wave elevation follows by using the dynamic free-surface condition $\zeta = -(\partial\Phi|_{y=0}/\partial t)/g = \sigma^2\phi|_{y=0}\eta_1/g$. The finite volume approximation of $\phi|_{y=0}$ is obtained in a way similar to that for the combined dynamic and kinematic free-surface condition and is approximated by eq. (10.54).

The horizontal linear hydrodynamic force F_1 acting on the tank follows by properly integrating the hydrodynamic pressure $-\rho\partial\Phi/\partial t$ over the mean wetted tank walls. The result is $F_1 = \rho\sigma^2\eta_1\int_{-h}^{0}(\phi_R - \phi_L)\,dy$. The subscripts R and L indicate the right and left tank wall, respectively. The finite volume approximations of ϕ_R and ϕ_L can be obtained in a way similar to the evaluation of ϕ on the mean free surface in connection with eq. (10.54).

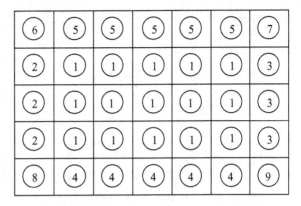

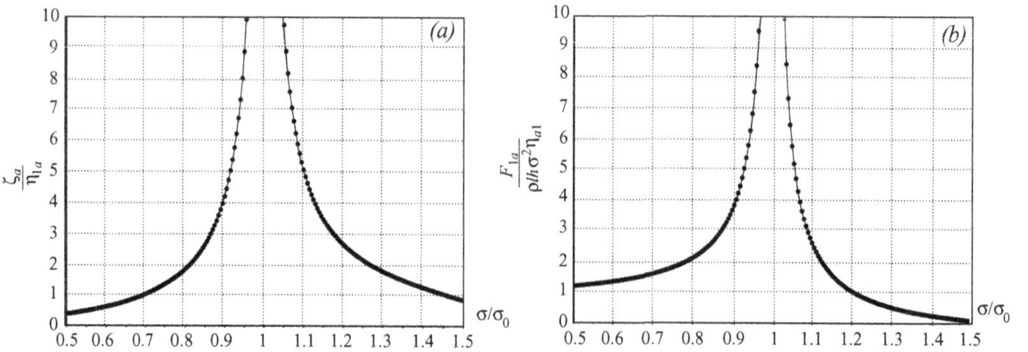

Figure 10.14. (a) Steady-state nondimensional wave amplitude ζ_a/η_{1a} at the vertical walls and (b) steady-state nondimensional horizontal force amplitude $F_{1a}/(\rho l h \sigma^2 \eta_{1a})$ versus nondimensional forcing frequency σ/σ_0 for sloshing in a two-dimensional rectangular tank with ratio of liquid depth to tank breadth of 0.3. The tank with breadth l and filling depth h is excited horizontally with forcing amplitude η_{1a}. The number of control volumes (grid cells) is 2,500. The solid line represents the analytical solution; the circles indicate results by FVM (Guo, personal communication, 2008).

10.4.2.1 Example

We study forced harmonic horizontal motion $\eta_{1a} \exp(i\sigma t)$ under steady-state conditions of a two-dimensional rectangular tank that has a ratio of liquid depth to tank length of $h/l = 0.3$ in the vicinity of the lowest natural sloshing frequency σ_0. Values of δx and δy do not vary and are chosen so that $\delta x/\delta y = 10/3$. However, the construction of the node system generally depends on both σ and h/l, bearing in mind at what depths nonnegligible flow exists. Furthermore, when we choose the node density we must also consider how the flow changes along the free surface. Similar results for the FEM are presented in Section 10.5.4.2.

Our numerical results are compared with the analytical results for wave elevation and horizontal force per unit length following from Section 5.2 by linear modal theory. The damping is set equal to zero in the analytical results. The numerical and analytical results for the nondimensional wave amplitude ζ_a/η_{1a} at the vertical walls are presented in Figure 10.14(a) as a function of σ/σ_0 ($\sigma_0 = \sigma_1$ is the lowest natural frequency). The agreement is very good. The same can be said about the results for the two-dimensional nondimensional horizontal force amplitude $F_{1a}/(\rho l h \sigma^2 \eta_{1a})$ presented in Figure 10.14(b). The figure illustrates that sloshing near resonant conditions clearly causes larger force than the inertial force $\rho l h \sigma^2 \eta_{1a}$ of the material mass of the liquid. When $\sigma/\sigma_0 \to 0$, we note that $F_{1a} \to \rho l h \sigma^2 \eta_{1a}$.

Figure 10.15 presents a convergence study of the predicted horizontal force amplitude as a

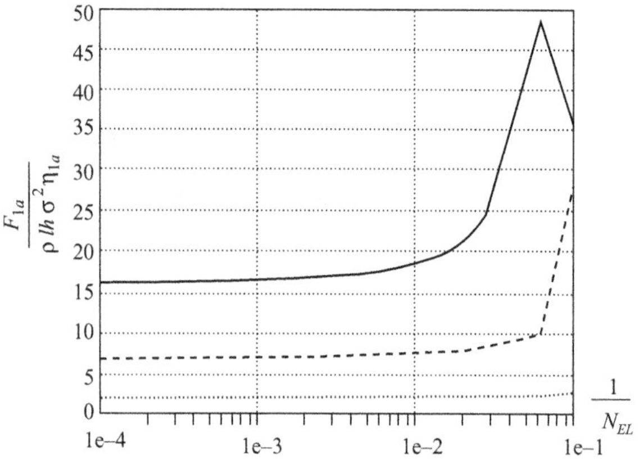

Figure 10.15. Convergence test as a function of total number of control volumes (grid cells), N_{EL}, for steady-state nondimensional horizontal force amplitude $F_{1a}/(\rho l h \sigma^2 \eta_{1a})$ at selected nondimensional forcing frequency σ/σ_0 (dotted line, $\sigma/\sigma_0 = 0.8$; dashed line, $\sigma/\sigma_0 = 0.95$; and solid line, $\sigma/\sigma_0 = 0.98$). (Guo, personal communication, 2008). Similar results from FEM are presented in Figure 10.23.

function of the total number of elements, N_{EL}, for selected frequencies. The convergence rate increases with increasing difference between the forcing frequency and the resonance frequency.

10.5 Finite element method

10.5.1 Introduction

The FEM is popular in many types of engineering applications. Several sloshing studies with FEM are reported by Solaas (1995). Wu *et al.* (1998) used FEM to solve nonlinear sloshing problems in a three-dimensional rectangular tank by assuming potential flow. The popularity of using FEM in marine hydrodynamics is increasing, particularly when solving potential flow problems. Ma *et al.* (2001a, 2001b), Wu and Hu (2004), Wang and Wu (2007), and Wang *et al.* (2007) have, for instance, used FEM for three-dimensional nonlinear external-body–free-surface interaction problems within potential flow theory. FEM is more complex to use in solving Navier–Stokes and Euler equations than FDM or FVM. Herfjord (1996) and Tønnessen (1999), for instance, used FEM with good results to study unsteady viscous flow around two-dimensional blunt bodies in infinite fluid.

FEM is well described in, for instance, the textbooks by Johnson (1995), Huebner *et al.* (1995), and Zienkiewics and Taylor (1991). The following presentation is only of an introductory nature. FEM is rather different from FDM and FVM. Our approach is based on the *Galerkin* method. The liquid domain is divided into elements, and shape functions are used over the elements. The Galerkin method implies that integrals over each element of the discretization have to be evaluated.

Let us introduce the *weighted residual method*, which is the basis for the Galerkin method. We start with a differential equation written in general form:

$$L(u) + f = 0 \qquad (10.55)$$

defined on a domain Ω bounded by Γ, where boundary conditions are imposed. The unknown function is denoted u while the function f is known. The following presentation concentrates on two-dimensional problems. However, the procedure may be extended to three dimensions. A Cartesian coordinate system Oxy is used.

The first step in the *weighted residual method* is to approximate u as

$$u \approx \hat{u} = \sum_{i=1}^{m} C_i(t) N_i(x, y), \qquad (10.56)$$

where $C_i(t)$ are unknowns that may or may not depend on time t. The m known functions N_i may satisfy the global boundary conditions; however, this is not normally the case.

When the approximate (*trial*) solution $\hat{u}$ is inserted into eq. (10.55), the equation is normally not exactly satisfied. We get

$$R_\Omega = L\hat{u} + f, \qquad (10.57)$$

where R_Ω is called the residual or the error. The next step is to multiply eq. (10.57) with the *weighting functions* W_i, integrate the expression over Ω, and require that

$$\int_\Omega W_i(x, y) R_\Omega \, d\Omega = 0, \quad i = 1, 2, \ldots, m. \qquad (10.58)$$

This procedure gives a set of m equations for the m unknowns $C_i(t)$. The weighting functions W_i are chosen to be equal to N_i in the Galerkin method.

10.5.2 A model problem

The following problem is given:

$$\frac{d^2u}{dx^2} + f = 0 \quad \text{on } \Omega, \quad u(1) = g, \quad -\left.\frac{du}{dx}\right|_{x=0} = h. \qquad (10.59)$$

This problem is one-dimensional, and we define the domain Ω to be a part of the x-axis ($\Omega = [0, 1]$); g and h are given constants. The weighted residual method may now be used on this problem. In doing this we insert a trial solution $\hat{u}$ into the differential equation, multiply it with a weighting function, integrate the result over the domain, and subsequently set it equal to zero:

$$\int_0^1 w \frac{d^2\hat{u}}{dx^2} dx + \int_0^1 w f \, dx = 0. \qquad (10.60)$$

Integration by parts of the first term in eq. (10.60) gives

$$\int_0^1 w \frac{d^2\hat{u}}{dx^2} dx = -\int_0^1 \frac{dw}{dx}\frac{d\hat{u}}{dx} dx + w \left.\frac{d\hat{u}}{dx}\right|_0^1. \qquad (10.61)$$

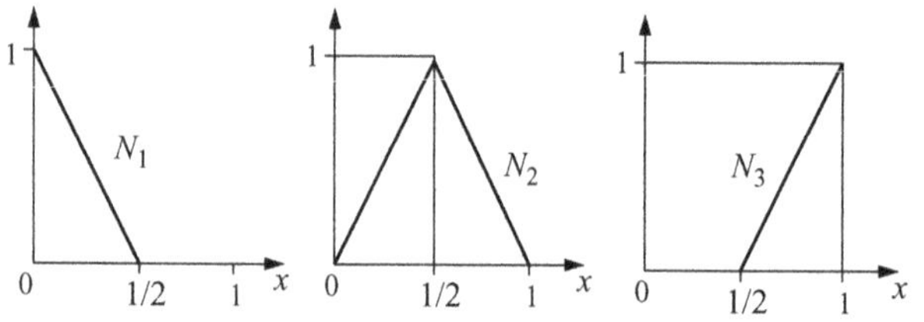

Figure 10.16. Basis functions for a model problem with two subdomains.

A reason behind the integration by parts is that if we choose piecewise linear trial functions, they vanish when differentiated twice. By performing the integration by parts we may choose functions of low polynomial order.

We can now write eq. (10.60) as

$$w(1)\frac{d\hat{u}}{dx}\bigg|_1 + w(0)\,h - \int_0^1 \frac{dw}{dx}\frac{d\hat{u}}{dx}\,\mathrm{d}x + \int_0^1 wf\,\mathrm{d}x = 0. \qquad (10.62)$$

The discretization of the problem means that we divide the domain (0,1) into subdomains or elements.

10.5.2.1 Numerical example

We illustrate the Galerkin method via a simple example where $\Omega = [0, 1]$ is only divided into two equal subdomains, elements 1 and 2. This enables us to carry out the analysis by hand. We define

$$N_1(x) = \begin{cases} 1 - 2x, & 0 \le x \le 0.5, \\ 0, & 0.5 \le x \le 1, \end{cases}$$

$$N_2(x) = \begin{cases} 2x, & 0 \le x \le 0.5, \\ 2(1-x), & 0.5 \le x \le 1, \end{cases}$$

$$N_3(x) = \begin{cases} 0, & 0 \le x \le 0.5, \\ 2x - 1, & 0.5 \le x \le 1. \end{cases} \qquad (10.63)$$

The functions are shown in Figure 10.16.

We write $\hat{u} = C_1N_1 + C_2N_2 + C_3N_3$, where we note that for $C_3 = g$ the boundary condition at $x = 1$ is satisfied due to the fact that $N_1(1) = 0$, $N_2(1) = 0$, $N_3(1) = 1$. We assume that f is equal to a constant p. We now apply eq. (10.62) with $w = N_1$, which gives $h - \int_0^{\frac{1}{2}} (-2)[C_1 \cdot (-2) + C_2 \cdot 2]\,\mathrm{d}x + p\int_0^{\frac{1}{2}} (1-2x)\,\mathrm{d}x = 0$; that is,

$$-2C_1 + 2C_2 = -\tfrac{1}{4}p - h. \qquad (10.64)$$

Then we repeat the procedure with $w = N_2$, which gives

$$-\int_0^{\frac{1}{2}} 2\,[C_1 \cdot (-2) + C_2 \cdot 2]\,\mathrm{d}x - \int_{\frac{1}{2}}^1 (-2)\,[C_2 \cdot (-2) + g \cdot 2]\,\mathrm{d}x + \tfrac{1}{2}p = 0,$$

that is,

$$2C_1 - 4C_2 = -\tfrac{1}{2}p - 2g. \qquad (10.65)$$

The solution of eqs. (10.64) and (10.65) is

$$C_1 = \tfrac{1}{2}p + h + g, \quad C_2 = \tfrac{3}{8}p + \tfrac{1}{2}h + g. \quad (10.66)$$

The exact solution of eq. (10.59) is

$$u(x) = g + (1-x)\,h + \tfrac{1}{2}p(1-x^2). \qquad (10.67)$$

We see that $\hat{u}$ takes the exact value at all three nodes, that the approximate FEM solution is continuous at the intersection between the elements, and that the first derivative is discontinuous.

Formally we should have set up another equation by using $w = N_3$. However, we used the boundary condition at $x = 1$ directly to find $C_3 = g$ and thereby reduced the problem to two unknowns, C_1 and C_2.

10.5.3 One-dimensional acoustic resonance

Acoustic waves in a closed container are a well-known phenomenon in pipe flow (i.e., the water hammer effect). Furthermore, sound waves are generated inside a tank during sloshing. We normally neglect this effect in a sloshing analysis. It may matter for mixture between gas and liquid. We use resonant acoustic waves inside a tank as our basis for another application of FEM. One-dimensional potential flow with velocity potential Φ is assumed. The problem is linearized and the dynamic pressure is expressed as $p = -\rho\partial\Phi/\partial t$, where ρ means the mean mass density. The governing equation is the wave equation:

$$\frac{\partial^2 \Phi}{\partial x^2} - \frac{1}{c_0^2}\frac{\partial^2 \Phi}{\partial t^2} = 0. \qquad (10.68)$$

(see eq. (2.32)). The speed of sound is denoted as c_0. We set the boundary conditions $\partial\Phi/\partial x = u_0$ at $x = 0$ and L. Initial conditions also have to be specified. A relevant practical problem is one-dimensional resonant acoustic waves, either in gas or liquid, between two vertical walls separated by a distance L. The tank is forced to oscillate horizontally with a prescribed velocity u_0. However, because gravity does not enter the problem, the words "vertical walls" and "horizontal excitation" are used only for exemplifying the problem. No energy dissipation occurs in this system, which means that transients due to initial conditions do not die out over time.

We start by showing how this problem can be solved via FEM. The number of nodes and unknowns is M and the distance between two neighboring nodes is $a = L/(M-1)$. The trial solution of Φ is expressed as

$$\hat{\Phi} = \sum_{j=1}^{M} \Phi_j(t) N_j(x). \qquad (10.69)$$

We introduce $X = x - (j-1)a$ $(j = 1, \ldots, M)$ and define

$$N_j(x) = \begin{cases} 1 + X/a, & -a \le X \le 0, \\ 1 - X/a, & 0 \le X \le a, \\ 0, & \text{otherwise.} \end{cases} \qquad (10.70)$$

Therefore, Φ_j in eq. (10.69) is the value of Φ at $x = (j-1)a$. The Galerkin method requires

$$\int_0^L N_i(x)\left[\frac{\partial^2}{\partial x^2}\sum_{j=1}^{M}\Phi_j(t)N_j(x) - \frac{1}{c_0^2}\frac{\partial^2}{\partial t^2}\sum_{j=1}^{M}\Phi_j(t)N_j(x)\right]\mathrm{d}x = 0. \qquad (10.71)$$

Integration of eq. (10.71) by parts gives

$$[N_i(L) - N_i(0)]\,u_0 - \int_0^L \frac{dN_i(x)}{dx}\sum_{j=1}^{M}\Phi_j(t)\frac{dN_j(x)}{dx}\mathrm{d}x - \frac{1}{c_0^2}\int_0^L N_i(x)\sum_{j=1}^{M}\frac{\partial^2\Phi_j(t)}{\partial t^2}N_j(x)\mathrm{d}x = 0. \qquad (10.72)$$

We now need to evaluate the integrals in eq. (10.72). For a given j, only a nonzero contribution exists for $i = j-1, j$, and $j+1$. We study first the cases when i differs from 1 and M.

For case A ($i = j - 1$):

$$\int_0^L N_iN_j\,\mathrm{d}x = \int_{-a}^0\left(1 - \frac{X+a}{a}\right)\left(1 + \frac{X}{a}\right)\mathrm{d}X = \frac{a}{6},$$

$$\int_0^L \frac{dN_i}{dx}\frac{dN_j}{dx}\mathrm{d}x = \int_{-a}^0\left(-\frac{1}{a}\right)\left(\frac{1}{a}\right)\mathrm{d}X = -\frac{1}{a}.$$

For case B ($i = j$):

$$\int_0^L N_iN_j\,\mathrm{d}x = \int_{-a}^0\left(1 + \frac{X}{a}\right)^2\mathrm{d}X + \int_0^a\left(1 - \frac{X}{a}\right)^2\mathrm{d}X = \frac{2a}{3},$$

$$\int_0^L \frac{dN_i}{dx}\frac{dN_j}{dx}\mathrm{d}x = \int_{-a}^a\left(\frac{1}{a}\right)^2\mathrm{d}X = \frac{2}{a}.$$

For case C ($i = j + 1$):

$$\int_0^L N_iN_j\,\mathrm{d}x = \int_0^a\left(1 + \frac{X-a}{a}\right)\left(1 - \frac{X}{a}\right)\mathrm{d}X = \frac{a}{6},$$

$$\int_0^L \frac{dN_i}{dx}\frac{dN_j}{dx}\mathrm{d}x = \int_0^a\left(\frac{1}{a}\right)\left(-\frac{1}{a}\right)\mathrm{d}X = -\frac{1}{a}.$$

When $i = j = 1$ and $i = j = M$, we get $\int_0^L N_iN_i\mathrm{d}x = \frac{1}{3}a$ and $\int_0^L \frac{dN_i}{dx}\frac{dN_i}{dx}\mathrm{d}x = 1/a$. Furthermore, case B does not apply for $i = 1$ and $i = M$. We are now ready to express eq. (10.72) as a system of linear equations, which we write in matrix form as

$$\mathbf{M}\begin{bmatrix}\ddot{\Phi}_1 \\ \ddot{\Phi}_2 \\ \vdots \\ \ddot{\Phi}_{M-1} \\ \ddot{\Phi}_M\end{bmatrix} + \mathbf{C}\begin{bmatrix}\Phi_1 \\ \Phi_2 \\ \vdots \\ \Phi_{M-1} \\ \Phi_M\end{bmatrix} = \begin{bmatrix}-u_0 \\ 0 \\ \vdots \\ 0 \\ u_0\end{bmatrix},$$

where

$$\mathbf{M} = \frac{a}{6c_0^2}\begin{bmatrix} 2 & 1 & 0 & 0 & \cdots & 0 \\ 1 & 4 & 1 & 0 & \cdots & \vdots \\ 0 & \ddots & \ddots & \ddots & \ddots & \vdots \\ \vdots & \ddots & \ddots & \ddots & \ddots & 0 \\ \vdots & \ddots & \ddots & 1 & 4 & 1 \\ 0 & 0 & 0 & 0 & 1 & 2 \end{bmatrix};$$

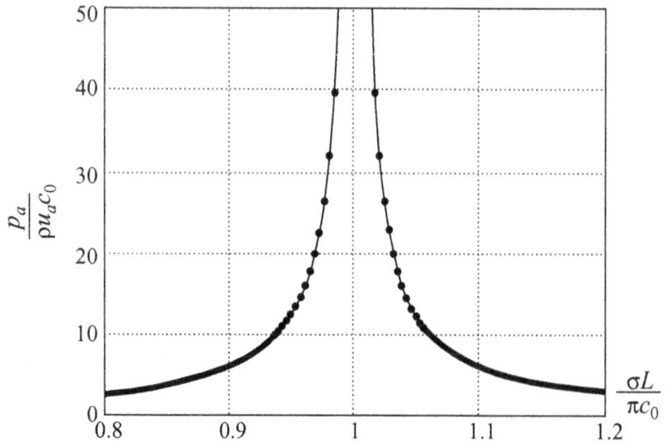

Figure 10.17. Steady-state pressure amplitudes p_a at $x = 0$ and L for one-dimensional acoustic resonance in a chamber of length L. The forcing velocity amplitude is u_a. The forcing frequency is σ. The speed of sound and the mean mass density are c_0 and ρ, respectively. The solid line represents the analytical solution, but the circles indicate results by FEM (Zhu, personal communication, 2007).

$$\mathbf{C} = \frac{1}{a}\begin{bmatrix} 1 & -1 & 0 & 0 & \cdots & 0 \\ -1 & 2 & -1 & 0 & \cdots & \vdots \\ 0 & \ddots & \ddots & \ddots & \ddots & \vdots \\ \vdots & \ddots & \ddots & \ddots & \ddots & 0 \\ \vdots & \ddots & \ddots & -1 & 2 & -1 \\ 0 & 0 & 0 & 0 & -1 & 1 \end{bmatrix},$$

This equation system can be numerically integrated in time if initial conditions are given. An alternative is to study periodic solutions, which means that the forcing velocity u_0 is harmonically oscillating with circular frequency σ and $\ddot{\Phi}_j = -\sigma^2 \Phi_j$.

Nondimensional pressure amplitude $p_a/(\rho u_a c_0)$ at $x = 0$ and L is presented in Figure 10.17 as a function of nondimensional frequency $\sigma L/(\pi c_0)$ in the vicinity of the lowest natural frequency, where u_a is the amplitude of the forcing velocity u_0. Good agreement is shown relative to the analytical solution considered in Exercise 10.8.1. The convergence rate of the FEM solution as a function of number of elements, N_{EL}, is demonstrated in Figure 10.18 by comparing with the analytical solution for selected frequencies. The convergence rate increases with increasing difference between the forcing frequency and the resonance frequency.

10.5.4 FEM applied to linear sloshing with potential flow

The governing equation is the two-dimensional Laplace equation for the velocity potential Φ. That means $\partial^2\Phi/\partial x^2 + \partial^2\Phi/\partial y^2 = 0$ in the liquid domain. We assume a steady-state condition where the flow is oscillating with the forced frequency σ. The linear free-surface condition on the mean free surface $y = 0$ is $-\sigma^2\Phi + g\partial\Phi/\partial y = 0$, where the y-coordinate is positive upward. The boundary condition on the mean wetted tank surface is expressed as $\partial\Phi/\partial n = U_n$ and the normal

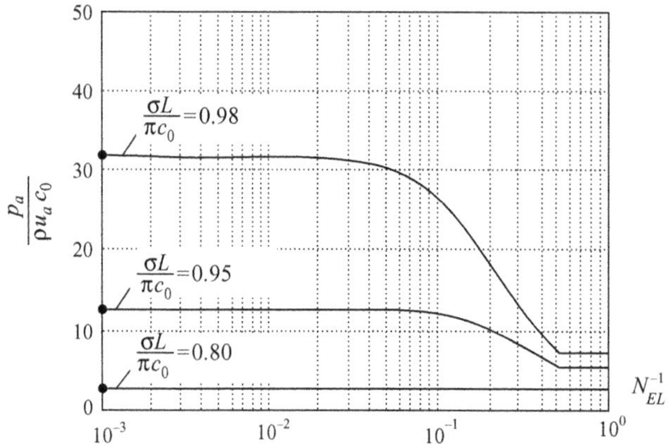

Figure 10.18. Convergence as a function of number of elements, N_{EL}, of steady-state pressure amplitudes p_a at $x = 0$ and L for one-dimensional acoustic resonance in a chamber of length L. The forcing velocity amplitude is u_a. The forcing frequency is σ. The speed of sound and the mean mass density are c_0 and ρ, respectively (Zhu, personal communication, 2007).

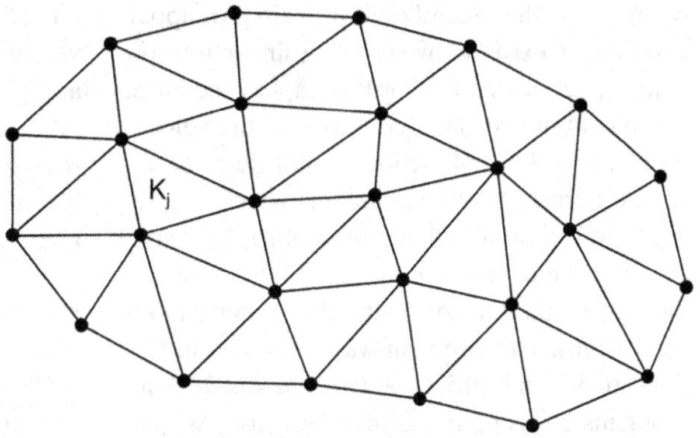

Figure 10.19. Two-dimensional liquid domain divided into triangular elements (Johnson, 1995).

vector to the tank surface, $\boldsymbol{n}$, is positive into the liquid domain.

We now introduce the mesh system that is used in our solution by means of FEM. It is a triangular mesh system as illustrated in Figure 10.19. It covers the whole liquid domain. The triangles K_j are nonoverlapping; no vertex of one triangle lies on the edge of another triangle. The corners of the triangles are called nodes. Furthermore, we use shape functions $N_j\,(x, y)$ that are linearly varying over each element, as shown in Figure 10.20. The function $N_j(x, y)$ is equal to 1 at node number j, and it is 0 at the neighboring nodes of the node number j. Furthermore, it is 0 outside the elements that have node number j as a vertex. If we concentrate on one element, then there exist three nonzero basis functions for that element.

Zienkiewicz and Taylor (1991) showed that the integration of any multiplication of the linear interpolation functions $N_j(x, y)$ ($j = 1, 2, 3$, i.e., one for each of the three linear functions in the triangular element) to any power over the element domain can be expressed as

$$\int_{\Omega_e} N_1^a N_2^b N_3^c \mathrm{d}\Omega = \frac{a!b!c!}{(a+b+c+2)!} 2\Omega_e,$$

where Ω_e denotes the area of the element and an exclamation point indicates factorial.

Other element types are used. For instance, quadrilateral elements with bilinear interpolation functions were used by Tønnessen (1999) in his studies of unsteady viscous flow past two-dimensional blunt bodies in an infinite fluid.

A trial solution of Φ is expressed as $\hat{\Phi}(x, y; t) = \sum_{j=1}^{M} \Phi_j(t) N_j(x, y)$, where M is number of nodes. Application of the Galerkin method gives

$$\int_{Q_0} N_i\,(x, y) \left[\nabla^2 \sum_{j=1}^{M} \Phi_j N_j\,(x, y) \right] dx\, dy = 0, \tag{10.73}$$

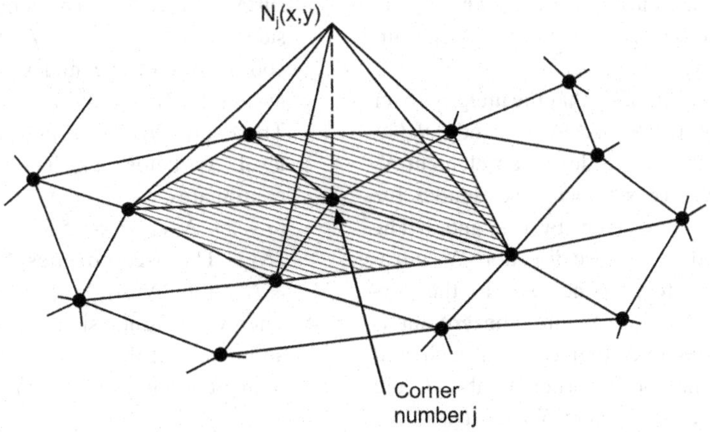

Figure 10.20. Example of basis function $N_j(x, y)$ for a triangular mesh. The function is 1 at corner number j and is zero at neighboring corners. It varies linearly within an element and is zero outside the shaded area shown in the figure (Johnson, 1995).

where Q_0 is the mean liquid domain. An apparent conflict exists between using linear interpolation functions and using the Laplace equation which contains second derivatives of the velocity potential. A linear function cannot describe the second derivative of a variable. However, we see in the following text how this conflict is avoided by using the divergence theorem, which reduces the differentiation from order 2 to order 1 in the expressions. This problem was addressed in Sections 10.5.2 and 10.5.3 for two one-dimensional problems by using integration by parts. We can write

$$\nabla \cdot \left[N_i(x,y)\nabla\hat{\Phi}\right] = \nabla N_i(x,y)\cdot\nabla\hat{\Phi} + N_i(x,y)\nabla^2\hat{\Phi}.$$

Furthermore, the divergence theorem (A.1) gives

$$\int_{Q_0} \nabla\cdot\left[N_i(x,y)\nabla\hat{\Phi}\right]\mathrm{d}x\,\mathrm{d}y = -\int_{\Gamma_0} N_i(x,y)\frac{\partial\hat{\Phi}}{\partial n}\mathrm{d}s,$$

where Γ_0 is the surface surrounding Q_0. This means that eq. (10.73) can be rewritten as

$$\int_{Q_0} \nabla N_i(x,y)\cdot\sum_{j=1}^{M}\Phi_j\nabla N_j(x,y)\,\mathrm{d}x\,\mathrm{d}y + \int_{S_B} N_i(x,y)\,U_n\mathrm{d}s - \frac{\sigma^2}{g}\int_{\Sigma_0} N_i(x,y)\,\hat{\Phi}\,\mathrm{d}s = 0, \tag{10.74}$$

where we have used the tank boundary and free-surface conditions. Furthermore, S_B and Σ_0 are the mean wetted tank surface and mean free surface, respectively.

We note from eq. (10.74) that the integrals over Q_0 involve $\partial N_i/\partial x$, $\partial N_j/\partial x$, $\partial N_i/\partial y$, and $\partial N_j/\partial y$. We express them by considering an element K_j in Figure 10.19. For simplicity we denote the three corners of the element by 1, 2, and 3. The numbering is in the clockwise direction. We introduce a z-coordinate perpendicular to the x–y-plane, and N_1, N_2, and N_3 are nonzero interpolation functions over that element such that N_k has z-coordinate 1 at corner number k and zero at the other two corners. We start with N_1, which has the coordinates $(x'_1, y'_1, 1)$, $(x'_2, y'_2, 0)$, $(x'_3, y'_3, 0)$ at the three corners of the element in the coordinate system (x, y, z). We then define an infinite plane that contains N_1. The normal vector of this plane is $\mathbf{n} = \mathbf{a}\times\mathbf{b}/|\mathbf{a}\times\mathbf{b}|$, where $\mathbf{a} = (x'_2 - x'_3)\,\mathbf{i} + (y'_2 - y'_3)\,\mathbf{j}$ and $\mathbf{b} = (x'_1 - x'_3)\,\mathbf{i} + (y'_1 - y'_3)\,\mathbf{j} + \mathbf{k}$ and $\mathbf{i}, \mathbf{j}$, and $\mathbf{k}$ are unit vectors along the x-, y-, and z-axes, respectively. The equation for the infinite plane is $(\mathbf{r} - \mathbf{r}_2)\cdot\mathbf{n} = 0$, where $\mathbf{r}_2 = x'_2\mathbf{i} + y'_2\mathbf{j}$ and $\mathbf{r} = x\mathbf{i} + y\mathbf{j} + z\mathbf{k}$. Therefore,

$$z = \frac{x'_2 - x'_3}{(x'_2 - x'_3)(y'_1 - y'_3) - (x'_1 - x'_3)(y'_2 - y'_3)}(y - y'_2) - \frac{y'_2 - y'_3}{(x'_2 - x'_3)(y'_1 - y'_3) - (x'_1 - x'_3)(y'_2 - y'_3)}(x - x'_2),$$

which implies that

$$\frac{\partial N_1}{\partial x} = -\frac{y'_2 - y'_3}{(x'_2 - x'_3)(y'_1 - y'_3) - (x'_1 - x'_3)(y'_2 - y'_3)},$$
$$\frac{\partial N_1}{\partial y} = \frac{x'_2 - x'_3}{(x'_2 - x'_3)(y'_1 - y'_3) - (x'_1 - x'_3)(y'_2 - y'_3)}, \tag{10.75}$$

and $\partial N_2/\partial x$ and $\partial N_2/\partial y$ are obtained by letting subscripts 1, 2, and 3 be replaced by 2, 3, and 1. Similarly, $\partial N_3/\partial x$ and $\partial N_3/\partial y$ follow by replacing the subscripts 1, 2, and 3 with 3, 1, and 2.

Let us elaborate further by considering the example of a rectangular tank of length L and mean liquid depth h. We assume forced surge velocity u_0 and divide the liquid volume Q_0 into a set of triangles (see Figure 10.21) as follows:

1. We define a number of depth levels, N_h. This means that the heights of the triangles are $\Delta y = h/(N_h - 1)$.
2. The number of nodes on the mean free surface Σ_0 is N_{FS}. The lengths of triangle sides on Σ_0 are $\Delta x = L/(N_{FS} - 1)$. The x-coordinates of the nodes on Σ_0 are $x_j = (j-1)\,\Delta x$, $(j = 1, N_{FS})$.
3. There are $N_{FS} + 1$ nodes at the depth level Δy below Σ_0. The nodes have x-coordinates $x_{N_{FS}+1} = 0$, $x_{2N_{FS}+1} = L$, $x_j = (j - 2 - N_{FS})\,\Delta x + 0.5\Delta x$, $\quad j = N_{FS}+2, 2N_{FS}$. The y-coordinates are $y_j = -\Delta y, j = N_{FS}+1, 2N_{FS}+1$.
4. Then we continue similarly at lower depths and note that the number of nodes on subsequent depth levels are N_{FS} and $N_{FS}+1$.

10.5.4.1 Matrix system

Formally we write the equation system following from eq. (10.74) as $(\mathbf{C} - \sigma^2 g^{-1}\mathbf{M})$

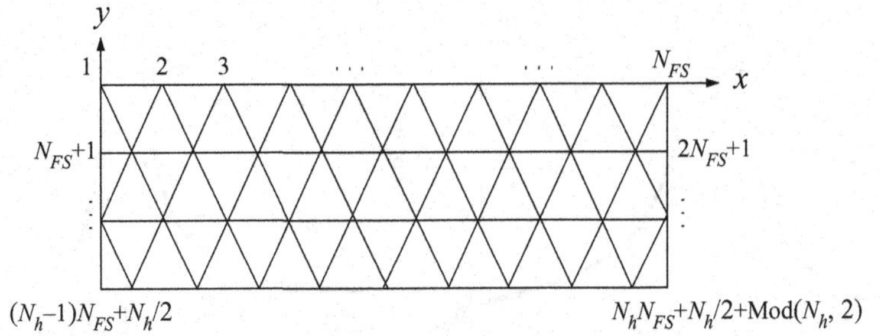

Figure 10.21. Division of the mean fluid volume of a rectangular tank. The value $\frac{1}{2}N_h$ accounts only for the integer part of the division; Mod$(N_h, 2)$ is 1 or 0 when N_h is an odd or even number, respectively.

$[\Phi_1, \Phi_2 \dots \Phi_M]^T = \mathbf{F}^T$. The matrices $\mathbf{C}$ and $\mathbf{M}$ are based on the Q_0 integral and the Σ_0 integral in eq. (10.74), respectively. The matrices have dimensions $M \times M$. The superscript T means transposed. The $\mathbf{F}$ vector has dimension M and follows from the S_B integral in eq. (10.74).

We show how the matrix $\mathbf{C}$ can be evaluated and start with the first row in the equation system, $i = 1$ in eq. (10.74). We can divide C_{11} into two terms. The first term is

$$C_{11}^1 = \tfrac{1}{4}\Delta y \Delta x\left(\frac{\partial N_1}{\partial x}\frac{\partial N_1}{\partial x} + \frac{\partial N_1}{\partial y}\frac{\partial N_1}{\partial y}\right),$$

where we evaluate $\partial N_1/\partial x$ and $\partial N_1/\partial y$ by using eq. (10.75) and noting that $(x_1', y_1') = (0, 0)$, $(x_2', y_2') = (x_{N_{FS}+2}, -\Delta y)$, $(x_3', y_3') = (x_{N_{FS}+1}, -\Delta y)$. The second term is

$$C_{11}^2 = \tfrac{1}{2}\Delta y \Delta x\left(\frac{\partial N_1}{\partial x}\frac{\partial N_1}{\partial x} + \frac{\partial N_1}{\partial y}\frac{\partial N_1}{\partial y}\right),$$

where we evaluate the derivative terms by using $(x_1', y_1') = (0, 0)$, $(x_2', y_2') = (x_2, 0)$, $(x_3', y_3') = (x_{N_{FS}+2}, -\Delta y)$ in eq. (10.75).

Similarly

$$C_{12} = \tfrac{1}{2}\Delta y \Delta x\left(\frac{\partial N_1}{\partial x}\frac{\partial N_2}{\partial x} + \frac{\partial N_1}{\partial y}\frac{\partial N_2}{\partial y}\right).$$

The derivative terms are evaluated by using $(x_1', y_1') = (0, 0)$, $(x_2', y_2') = (x_2, 0)$, $(x_3', y_3') = (x_{N_{FS}+2}, -\Delta y)$. The other nonzero elements in the first row of the matrix $\mathbf{C}$ are $C_{1,N_{FS}+1}$ and $C_{1,N_{FS}+2}$. Then we have to continue with the other rows.

Let us describe a procedure on how to construct the elements in the matrix $\mathbf{C}$ by choosing an arbitrary row number of the matrix, i. This means that we study node number i. We must first identify how many elements are associated with node number i. There were two elements when we considered C_{11}. The maximum possible number with our choice of elements is 6. We then consider one element and introduce a local numbering system from 1 to 3 for the element, which is similar to the procedure used for eq. (10.75). Without losing generality we can assume that node i has local number 1. Therefore, we study the integral part $C_{11}^{(e)}\Phi_1^{(e)} + C_{12}^{(e)}\Phi_2^{(e)} + C_{13}^{(e)}\Phi_3^{(e)}$ of the Q_0 integral in eq. (10.74). The superscript (e) indicates that we operate with local numbering. Furthermore,

$$C_{11}^{(e)} = \Omega^{(e)}\left(\frac{\partial N_1^{(e)}}{\partial x}\frac{\partial N_1^{(e)}}{\partial x} + \frac{\partial N_1^{(e)}}{\partial y}\frac{\partial N_1^{(e)}}{\partial y}\right),$$

$$C_{12}^{(e)} = \Omega^{(e)}\left(\frac{\partial N_1^{(e)}}{\partial x}\frac{\partial N_2^{(e)}}{\partial x} + \frac{\partial N_1^{(e)}}{\partial y}\frac{\partial N_2^{(e)}}{\partial y}\right),$$

$$C_{13}^{(e)} = \Omega^{(e)}\left(\frac{\partial N_1^{(e)}}{\partial x}\frac{\partial N_3^{(e)}}{\partial x} + \frac{\partial N_1^{(e)}}{\partial y}\frac{\partial N_3^{(e)}}{\partial y}\right),$$

where $\Omega^{(e)}$ means the area of the element. The final step is to transfer $C_{ij}^{(e)}$ to the global index system to obtain C_{ij}.

The matrix $\mathbf{M}$ is expressed in eq. (10.76). The upper-left submatrix has the dimension $N_{FS} \times N_{FS}$. The derivation is similar to that shown for cases A, B, and C in Section 10.5.3:

$$\mathbf{M} = \frac{\Delta x}{6}\left|\begin{array}{ccccc|c} 2 & 1 & 0 & \cdots & 0 & \\ 1 & 4 & 1 & \ddots & \vdots & \\ 0 & \ddots & \ddots & \ddots & 0 & 0 \\ \vdots & \ddots & 1 & 4 & 1 & \\ 0 & \cdots & 0 & 1 & 2 & \\ \hline & & 0 & & & 0 \end{array}\right|. \tag{10.76}$$

The vector $\mathbf{F}$ contains only nonzero elements when the row number corresponds to a node at

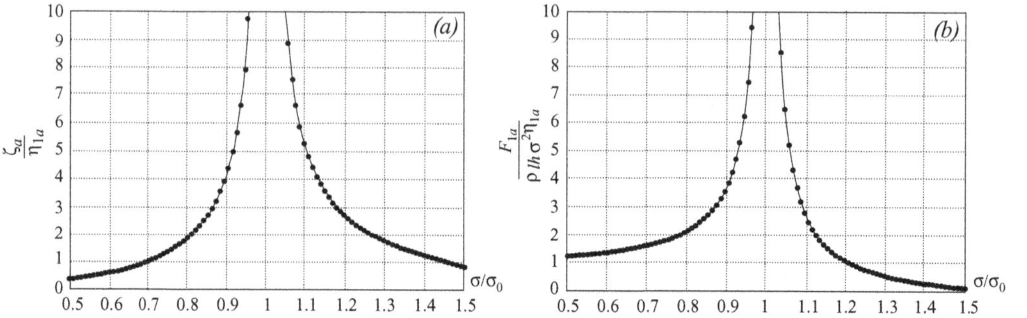

Figure 10.22. (a) Steady-state nondimensional wave amplitude ζ_a/η_{1a} at the vertical walls and (b) steady-state nondimensional horizontal force amplitude $F_{1a}/(\rho l h \sigma^2 \eta_{1a})$ versus nondimensional forcing frequency σ/σ_0 for sloshing in a two-dimensional rectangular tank with ratio of liquid depth to tank length of 0.3. The tank is excited horizontally with forcing amplitude η_{1a}. The number of elements is 1,962. Similar results from FVM are presented in Figure 10.14. The solid line represents the analytical solution, but the circles indicate results from FEM (Zhu, personal communication, 2007).

the vertical tank walls, which gives $F_1 = -\frac{1}{2}u_0\Delta y$, $F_{N_{FS}} = \frac{1}{2}u_0\Delta y$, $F_{N_{FS}+1} = -u_0\Delta y$, $F_{2N_{FS}+1} = u_0\Delta y$, $F_{2N_{FS}+2} = -u_0\Delta y$, and so on.

10.5.4.2 Example

We study forced harmonic horizontal motion $\eta_{1a}\exp(i\sigma t)$ under steady-state conditions of a two-dimensional rectangular tank with ratio of liquid depth to tank length of $h/l = 0.3$ in the vicinity of the lowest natural sloshing frequency, σ_0; Δx and Δy are chosen to be approximately equal. However, the construction of the node system generally depends on both σ and h/l, bearing in mind at what depths nonnegligible flow occurs. Furthermore, we must also consider how the flow changes along the free surface when we choose the node density. Similar results are presented in Section 10.4.2.1 for the FVM.

Our numerical results are compared with the analytical results for wave elevation and horizontal force per unit length following from Section 5.2. The damping is set equal to zero in the analytical results. The numerical and analytical results for nondimensional wave amplitude ζ_a/η_{1a} at the vertical walls are presented in Figure 10.22(a) as a function of σ/σ_0. The agreement is very good. The same can be said about the results for the nondimensional horizontal force amplitude $F_{1a}/(\rho l h \sigma^2 \eta_{1a})$ presented in Figure 10.22(b).

Figure 10.23 presents a convergence study of the predicted horizontal force amplitude as a function of the total number of elements, N_{EL}, for selected frequencies. The convergence rate increases with increasing difference between the forcing frequency and the resonance frequency.

10.6 Smoothed particle hydrodynamics method

In the SPH method the fluid is modeled as a finite number of particles, each with local mass and other properties that remain with them during their evolution. A rigorous derivation of the SPH equation in fluid dynamics was given by Bicknell (1991). Monaghan (1992) gives a general presentation of SPH and discusses various applications.

A compressible liquid in a tank with two-dimensional flow conditions is assumed in the following presentation. The gas flow is neglected. Basic equations are the continuity equation, the momentum equation, and an equation $p = f(\rho)$ to relate pressure p and the mass density ρ. The pressure–density relationship assumes adiabatic conditions (i.e., no heat exchange). The Euler equations are used to express the momentum equation, while SPH can be applied to the viscous flow. Because liquid particles are followed in time and space, the continuity and Euler equations are written in terms of material derivatives D/Dt. Furthermore, Monaghan's (1992) golden rule, "Formulae should be rewritten with the density placed inside operators," is also respected. A tank-fixed coordinate system is used and $\boldsymbol{v}_r$ is the fluid velocity relative to the tank. It follows from eq. (2.27) that the continuity equation can

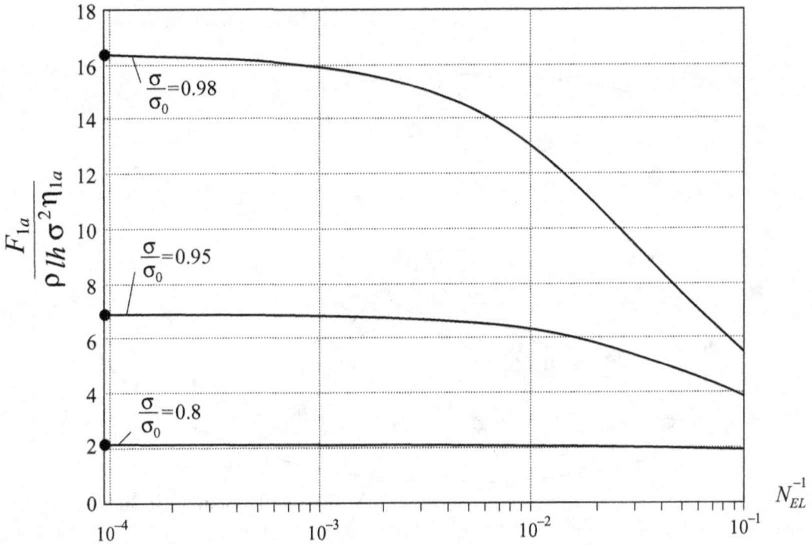

Figure 10.23. Convergence test as a function of total number of elements, N_{EL}, for steady-state nondimensional horizontal force amplitude $F_{1a}/\left(\rho l h \sigma^2 \eta_{1a}\right)$ at selected nondimensional forcing frequency σ/σ_0 for sloshing in a two-dimensional rectangular tank with ratio of liquid depth to tank length of 0.3. The tank is excited horizontally with forcing amplitude η_{1a}. Similar results from FVM are presented in Figure 10.15 (Zhu, personal communication, 2007).

be expressed as

$$\frac{D\rho}{Dt} = -\rho \nabla \cdot \boldsymbol{v}_r \equiv -\nabla \cdot (\rho \boldsymbol{v}_r) + \boldsymbol{v}_r \cdot \nabla \rho. \quad (10.77)$$

Neglecting viscosity and using eq. (2.56) gives

$$\frac{D^* \boldsymbol{v}_r}{Dt} = -\frac{\nabla p}{\rho} + \boldsymbol{f} \equiv -\nabla\left(\frac{p}{\rho}\right) - \frac{p}{\rho^2}\nabla \rho + \boldsymbol{f}. \quad (10.78)$$

The material time derivative in eq. (10.78) does not involve differentiation of the unit vectors of the tank-fixed coordinate system, which is indicated by a superscript asterisk. We have preferred such a formulation to the one using the operator D/Dt. As a result the expression of the term $\boldsymbol{f}$ is

$$\boldsymbol{f} = \boldsymbol{g} - \boldsymbol{a}_O - (\boldsymbol{\omega} \times \boldsymbol{v}_O) - \left(\frac{d\boldsymbol{\omega}}{dt} \times \boldsymbol{r}\right) - 2(\boldsymbol{\omega} \times \boldsymbol{v}_r) - \boldsymbol{\omega} \times (\boldsymbol{\omega} \times \boldsymbol{r}). \quad (10.79)$$

The kinematic equation for a liquid particle relative to the tank is

$$\frac{D^* \boldsymbol{x}}{Dt} = \boldsymbol{v}_r, \quad (10.80)$$

where $\boldsymbol{x}$ is the same as $\boldsymbol{r}$ in eq. (10.79). According to the Tait equation (see, e.g., Cole, 1948) the pressure–density relationship can be expressed as

$$p(\boldsymbol{r}) = \frac{\rho_0 c_0^2}{\kappa}\left\{\left[\frac{\rho(\boldsymbol{r})}{\rho_0}\right]^{\kappa} - 1\right\} \quad (10.81)$$

with local speed of sound defined as

$$c(\boldsymbol{r}) = \sqrt{\frac{\partial p}{\partial \rho}} = c_0\left[\frac{\rho(\boldsymbol{r})}{\rho_0}\right]^{\frac{\kappa-1}{2}}, \quad (10.82)$$

where ρ_0 is a constant, c_0 is the speed of sound, and κ is the heat ratio, which is about 7 for water. A similar relationship was used by Colagrossi and Landrini (2003).

A good compromise between efficiency and accuracy can be obtained by assuming a sound speed smaller than reality but larger than ten times the maximum local fluid speed. In this way we can allow larger time steps in the computations and still ensure approximate incompressible behavior for the liquid.

As a first step, one way to find initial values for the mass density and the pressure is to consider static conditions with constant mass density ρ_0 and let the pressure be equal to zero on the free surface. Actually the pressure is equal to the gas pressure at the free surface; however, adding a constant pressure in the formulation does not

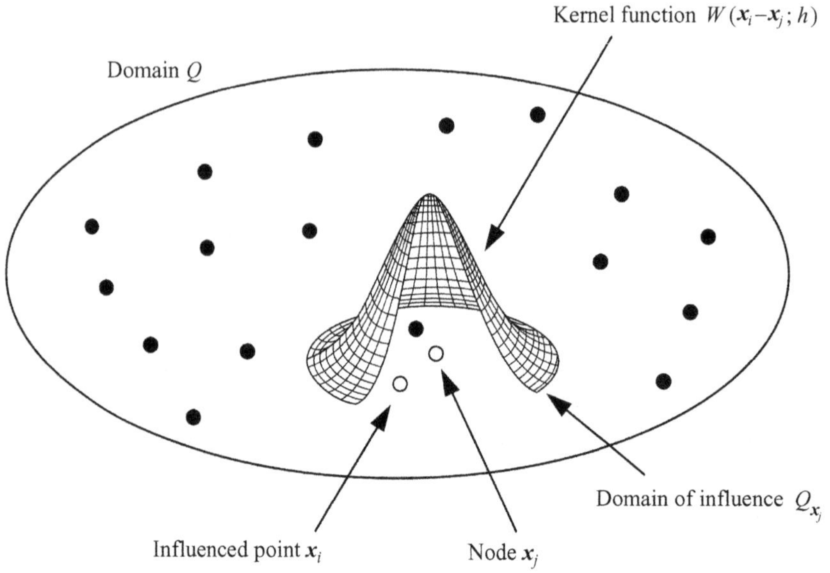

Figure 10.24. Sketch of the kernel function (Colagrossi, 2005).

change the flow. The pressure is, therefore, initially hydrostatic relative to the free surface with $p = 0$ on the free surface. The corresponding initial values for the mass density can be found by solving eq. (10.81) with given hydrostatic pressure distribution. Because eq. (10.81) gives $p = 0$ when $\rho = \rho_0$ and $p = 0$ is the dynamic free-surface condition, a free-surface particle keeps the mass density ρ_0.

The SPH formulation of the continuity, momentum, and kinematic equations for particle i is

$$\frac{D\rho_i}{Dt} = -\rho_i \sum_j M_{ij}, \tag{10.83}$$

$$\frac{D^*\boldsymbol{v}_i}{Dt} = -\frac{1}{\rho_i}\sum_j \boldsymbol{F}_{ij} + \boldsymbol{f}_i, \tag{10.84}$$

$$\frac{D^*\boldsymbol{x}_i}{Dt} = \boldsymbol{v}_i. \tag{10.85}$$

The subscript r for the flow velocities has been omitted from eqs. (10.84) and (10.85). The interaction terms M_{ij} and F_{ij}, and the body force per unit mass, $\boldsymbol{f}_i$, are derived next. But before doing that we need to discuss the interpolation integral:

$$\langle \boldsymbol{v}(\boldsymbol{x})\rangle = \int_Q \boldsymbol{v}(\boldsymbol{y}) W(\boldsymbol{x}-\boldsymbol{y};h)\,\mathrm{d}Q_y. \tag{10.86}$$

This integral gives an interpolation of the solution everywhere in the fluid domain Q from particles where we know the solution. The function $W(\boldsymbol{x}-\boldsymbol{y};h)$ is called a *smoothing function* or *kernel* and has the following properties:

- $W(\boldsymbol{x}-\boldsymbol{y};h) \geq 0$ for $\boldsymbol{x} \in Q_y \subset Q$ and zero otherwise, where Q_y is the domain of influence of the point y, as illustrated in Figure 10.24.
- $\int_Q W(\boldsymbol{x}-\boldsymbol{y};h)\mathrm{d}Q_y = 1$.
- $W(\boldsymbol{x}-\boldsymbol{y};h)$ decreases monotonously as $|\boldsymbol{x}-\boldsymbol{y}|$ increases.

When the parameter $h \to 0$, W becomes the Dirac delta function so that

$$\lim_{h\to 0}\int_Q \boldsymbol{v}(\boldsymbol{y})\, W(\boldsymbol{x}-\boldsymbol{y};h)\mathrm{d}Q_y \equiv \boldsymbol{v}(\boldsymbol{x}). \tag{10.87}$$

One example on a smoothing function is the Gaussian kernel

$$W(s,h) = \frac{1}{\pi h^2}\exp[-(s/h)^2], \quad s = |\boldsymbol{x}-\boldsymbol{y}|. \tag{10.88}$$

Colagrossi and Landrini (2003) introduced a cutoff limit δ of eq. (10.88) and normalized W so that the integral of W from $s = 0$ to δ is approximately 1, which gives

$$W(s,h,\delta) = \frac{\exp[-(s/h)^2] - \exp[-(\delta/h)^2]}{2\pi\int_0^\delta s\{\exp[-(s/h)^2] - \exp[-(\delta/h)^2]\}\,\mathrm{d}s}, \quad s = |\boldsymbol{x}-\boldsymbol{y}|. \tag{10.89}$$

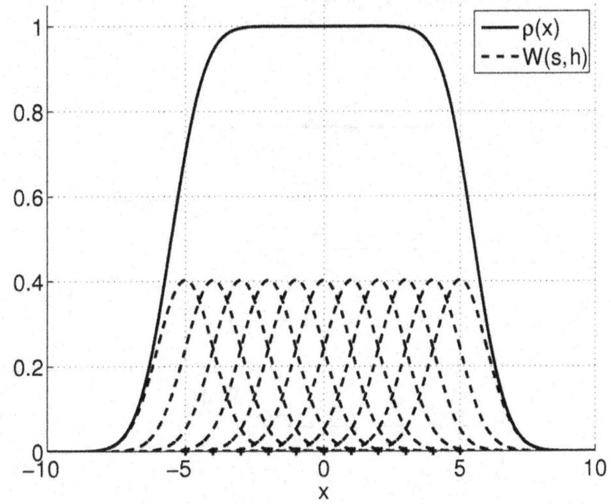

Figure 10.25. Example of prediction of density in one dimension by using 11 particles.

A typical cutoff distance is $\delta = 3h$. Colagrossi and Landrini (2003) proposed that $h/\Delta x = 1.33$, where Δx is the original distance between particles. If a tank of length l is considered, then $l/\Delta x$ may, for instance, be 250. If the ratio of liquid depth to tank length is 0.5, we get 31,250 particles. Equation (10.86) is numerically approximated as

$$\langle \boldsymbol{v}_i \rangle = \sum_j \boldsymbol{v}_j W(\boldsymbol{x}_i - \boldsymbol{x}_j; h) \Delta Q_j. \qquad (10.90)$$

A sketch of a kernel function is shown in Figure 10.24. The figure illustrates how other particles are influenced by any one particle. Let us illustrate what the number of particles might be.

Each particle carries a constant mass m_j during the SPH simulation, which implies that ΔQ_j in eq. (10.90) is replaced by $\Delta Q_j = m_j/\rho_j$. Furthermore, it is assumed that $\langle \boldsymbol{v}_i \rangle \simeq \boldsymbol{v}_i$. When the gradient of $\boldsymbol{v}_i$ is evaluated, no need exists to use grids and finite differences. We just differentiate the kernel function: $\nabla \boldsymbol{v}_i = \sum_j \boldsymbol{v}_j \left(m_j/\rho_j\right) \nabla W(\boldsymbol{x}_i - \boldsymbol{x}_j; h)$. We then introduce the SPH formulation into eq. (10.77) and express the density ρ_i for particle number i as

$$\rho_i = \sum_j m_j W(\boldsymbol{x}_i - \boldsymbol{x}_j; h). \qquad (10.91)$$

Equation (10.91) is exemplified in one dimension in Figure 10.25 by using the kernel function $W = \exp(-|\boldsymbol{x}_i - \boldsymbol{x}_j|^2/h^2)/(h\sqrt{\pi})$. The objective is to create a constant value of $\rho = 1$ in the interval $-5 \leq x \leq 5$, which is accomplished in this case by using 11 particles. The choice of m_j depends on the number of particles.

Furthermore, we can write

$$\rho_i \boldsymbol{v}_i = \sum_j \boldsymbol{v}_j m_j W(\boldsymbol{x}_i - \boldsymbol{x}_j; h), \qquad (10.92)$$

which means

$$\begin{aligned} \rho_i \nabla \cdot \boldsymbol{v}_i &\equiv \nabla \cdot (\rho_i \boldsymbol{v}_i) - \boldsymbol{v}_i \cdot \nabla \rho_i \\ &= \sum_j (\boldsymbol{v}_j - \boldsymbol{v}_i) m_j \cdot \nabla_i W(\boldsymbol{x}_i - \boldsymbol{x}_j; h). \end{aligned} \qquad (10.93)$$

The subscript i on the gradient operator means differentiation with respect to $\boldsymbol{x}_i$. The interaction term in eq. (10.83) can now be defined as

$$M_{ij} = (\boldsymbol{v}_j - \boldsymbol{v}_i) \cdot \nabla_i W(\boldsymbol{x}_i - \boldsymbol{x}_j; h) \frac{m_j}{\rho_i}. \qquad (10.94)$$

Following a similar procedure for Euler equations (10.78) gives the interaction term

$$\boldsymbol{F}_{ij} = \rho_i \left(\frac{p_j}{\rho_j^2} + \frac{p_i}{\rho_i^2} \right) \nabla_i W(\boldsymbol{x}_i - \boldsymbol{x}_j; h) m_j \qquad (10.95)$$

in eq. (10.84). Furthermore, the body force per unit mass of particle number i is

$$\begin{aligned} \boldsymbol{f}_i = \boldsymbol{g} - \boldsymbol{a}_O - (\boldsymbol{\omega} \times \boldsymbol{v}_O) - \left(\frac{d\boldsymbol{\omega}}{dt} \times \boldsymbol{r}_i \right) \\ - 2(\boldsymbol{\omega} \times \boldsymbol{v}_i) - \boldsymbol{\omega} \times (\boldsymbol{\omega} \times \boldsymbol{r}_i). \end{aligned} \qquad (10.96)$$

In practice, the Euler equations are modified by including an artificial viscosity term. This modification is needed to ensure that transients due to initial conditions are damped out, but it also helps to improve the stability of the

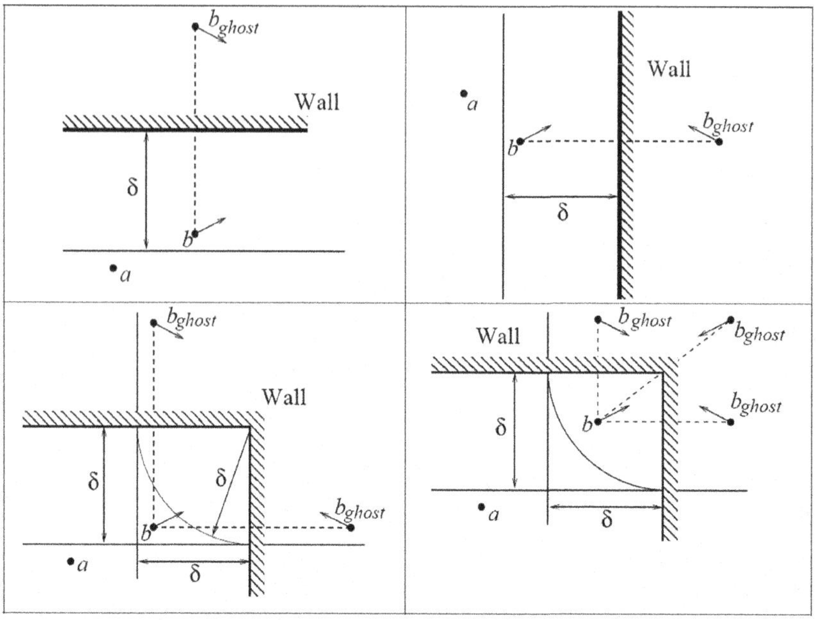

Figure 10.26. Introduction of ghost particles to satisfy the body boundary conditions. Ghost particles are chosen relative to liquid particles within a distance δ as illustrated in the figure (Pakozdi, personal communication, 2008).

numerical algorithm. Monaghan (1992) reformulated eq. (10.95) as $\boldsymbol{F}_{ij} = \rho_i(p_j/\rho_j^2 + p_i/\rho_i^2 + \Pi_{ij})$ $\nabla_i W(\boldsymbol{x}_i - \boldsymbol{x}_j; h)m_j$ with

$$\Pi_{ij} = \begin{cases} -\alpha\mu_{ij}\dfrac{(c_{s,i}+c_{s,j})}{(\rho_i+\rho_j)} & \text{if } (\boldsymbol{v}_i - \boldsymbol{v}_j)\cdot(\boldsymbol{x}_i - \boldsymbol{x}_j) < 0, \\ 0 & \text{otherwise,} \end{cases}$$

where the speed of sound, c_s, follows from the state equation, and the equivalent kinematic viscosity coefficient has the form $\alpha c_s h$ (Monaghan & Gingold, 1983). Furthermore,

$$\mu_{ij} = h\frac{(\boldsymbol{v}_i - \boldsymbol{v}_j)\cdot(\boldsymbol{x}_i - \boldsymbol{x}_j)}{(|\boldsymbol{x}_i - \boldsymbol{x}_j|^2 + \varepsilon h^2)},$$

where $\varepsilon \sim 0.001$. The term εh^2 is introduced to avoid a singularity in the expression. Colagrossi and Landrini (2003) used a modified version of the expression for μ_{ij}.

Even though a viscous term is introduced, a slip condition is used at the body boundary. One way to enforce the body boundary conditions is to use ghost particles outside the fluid domain next to the body boundary and at positions that are symmetric to fluid particles at the boundary. This procedure is followed for all fluid particles within a distance of δ of $O(h)$ from the boundary and is illustrated in Figure 10.26. For instance, δ is chosen as $3h$ with the kernel function given by eq. (10.89). Because we operate with relative velocities in our formulation, each ghost particle is given a velocity component normal to the tank boundary, which is opposite the normal velocity component of its mirrored fluid particle. Special care must be shown at the corners, where the normal direction is not uniquely defined. Figure 10.26 shows a practical way to deal with this problem. The ghost particles must also be assigned pressures and densities. The normal pressure gradient at the wall must be consistent with eq. (10.78). We start by expressing the substantive derivative of the relative velocity $\boldsymbol{v}_r$ as in eq. (10.24) and recall that $\boldsymbol{v}_r \cdot \boldsymbol{n} = 0$ on the tank boundary, where $\boldsymbol{n}$ is the normal vector to the tank surface. A consequence is that both $\partial(\boldsymbol{v}_r \cdot \boldsymbol{n})/\partial t$ and the variation of $\boldsymbol{v}_r \cdot \boldsymbol{n}$ along the tank boundary are zero. It follows that $D(\boldsymbol{v}_r \cdot \boldsymbol{n})/Dt = 0$ and therefore $\partial p/\partial n = \boldsymbol{n} \cdot \boldsymbol{f}$ at the tank boundary. The described procedure can also be applied for curved bodies. This condition is associated with the fact that a ghost particle is placed close to the tank surface so that the surface appears locally flat as a first approximation from the liquid particle and its ghost particle's point of view. More than one ghost particle is considered for each physical particle near a tank corner in the strategy shown in Figure 10.26.

Both kinematic and dynamic free-surface conditions are automatically satisfied by the SPH method. A free-surface particle stays at the free surface and its pressure is consistent with the gas pressure. However, the free-surface conditions are only approximately satisfied due to the finite number of particles and the inherent numerical approximations.

Colagrossi and Landrini (2003) used a fourth-order Runga–Kutta method to integrate the equations in time. The time step δt was chosen according to a Courant–Friedrich–Levy-type condition (Heath, 2002).

No numerical methods are without problems. The body boundary condition requiring no flow through the boundary is only approximate. If special care is not taken, particles may cross the tank boundaries.

Numerical difficulties occur for the SPH method when particles cluster too close together. The reason for the attraction between the particles is negative pressures and, consequently, instabilities can arise. Monaghan (2000) suggested using a small repulsive corrective pressure term when particles are very close. However, this does not mean that negative pressures are unphysical. For instance, cavitation may occur, which means that the total pressure is equal to the vapor pressure. The total pressure is the sum of the gas (ullage) pressure above the liquid in calm conditions and the pressure that we predict with the described numerical scheme. The vapor pressure may be close to zero, which means that our predicted pressure should be close to minus the gas pressure for inception of cavitation. Because the described method is not a multiphase model, we cannot predict the behavior of any cavitating flows.

Particles can get stuck to the tank roof due to negative pressures. One pragmatic way to overcome this numerical problem is to set the force term due to the ghost particle associated with the particle to zero. In this way the particle is able to leave the tank roof and as a consequence fall under gravity in a natural way.

Another problem is unphysical pressure oscillations with a much higher frequency than the forcing frequency of the tank. One reason is numerical: a higher number of particles reduces the high-frequency oscillation amplitudes. Another reason is false acoustic wave effects due to weak compressibility. The unphysical pressure oscillations are often numerically filtered out as part of a simulation. Standing resonant acoustic waves are actually possible within a tank.

10.7 Summary

The types of governing equations that we want to solve with CFD depend on the assumptions and simplifications made (i.e., compressible or incompressible fluid, viscous or inviscid fluid, laminar or turbulent viscous flow, rotational or irrotational flow). If the changes in the gas affect the two-phase flow evolution, it is essential to consider the gas compressible. However, the liquid may, for most purposes, be considered incompressible. It is difficult from a computational time point of view to account for a compressible gas flow. It is of less significance to account for the fact that the flow may be turbulent. However, generation of vorticity may matter, for instance, in describing the effect of breaking waves. The fact that violent sloshing is highly nonlinear makes the computational task very difficult.

A general difficulty with a CFD method is to accurately describe slamming loads when the relative angle between the impacting free surface and the tank surface is small. The hydroelasticity may then play an important role as described in Chapter 11. For such scenarios the slamming force impulse becomes a more important parameter than the maximum pressure.

A disadvantage is that CFD methods are time-consuming, which makes statistical estimates of tank response variables difficult.

It is important to ensure that continuity of the liquid mass is satisfied; otherwise, unphysical solutions may occur.

A general overview of CFD methods has been given (see Figure 10.1). Particular attention was paid to BEM, FDM, FVM, FEM, and SPH methods. A short description of special features, advantages, and disadvantages of these methods are given as follows.

Our presentation of the BEM assumes an incompressible liquid and an irrotational flow so that a velocity potential satisfying Laplace equation can be used to describe the liquid flow. Rankine sources and normal dipoles, or only Rankine sources, are typically distributed over the instantaneous wetted body surface and the free surface in a completely nonlinear formulation. The dynamic and kinematic free-surface conditions can be used to track the position of the free surface by following individual particles on the free surface. It is possible to include the effect of compressible gas pockets.

The BEM breaks down when an overturning wave hits the underlying free surface. Numerical problems may also arise with a BEM at the intersection between the free surface and the tank boundary. When a BEM works, it is generally a fast and accurate method.

The FDM is based on approximations of the derivatives by using differences of the flow variables at some finite distance Δx. We want to emphasize that different FDM schemes exist for solving Navier–Stokes equations, and an essential part is how boundary conditions on the tank boundary and the interface between the gas and the liquid are handled. The method requires a grid system that may be boundary-fitted or fixed. A fixed grid is computationally more efficient than a boundary-fitted grid. However, a boundary-fitted grid may achieve better accuracy. Our detailed description for fixed grids included the use of ghost cells to satisfy the tank boundary conditions. A one-fluid formulation including the gas and the liquid phases was introduced, which necessitated a finite but thin layer between the gas and the liquid, where there is a continuous change of fluid properties. The artificial layer between the gas and the liquid represents an error source. Numerical diffusion may occur over long time simulations. A main reason is the need for interpolation between grid points, which necessitates a high-order time integration method. The advantage of using the artificial layer is that the dynamic free-surface condition is automatically satisfied. An interface technique has to be used to find the position of the free surface. Some capturing techniques, such as the LS method, were described in detail. The description of the grid system and the interface-capturing methods also applies to other grid methods such as the FVM and FEM.

The FVM divides the fluid volume into control volumes (CVs). The governing equations are expressed for each CV. The consequence is that lower-order derivatives are involved relative to the governing equations for the FDM. The use of CVs makes it easy to incorporate the body boundary conditions.

The FEM is rather different from the FDM. Our approach is based on the Galerkin method. The liquid domain is divided into elements and interpolating functions are used over the elements. The Galerkin method implies that integrals over each element have to be evaluated. The FEM has been applied to two linear potential flow problems. One is one-dimensional acoustic resonance and the other is two-dimensional sloshing in a rectangular tank of an incompressible liquid in the vicinity of the highest natural period. Good convergence to analytical solutions was demonstrated. The same sloshing problem, with similar good results, was also studied by the FVM.

The SPH method is an example of a grid-less method. It can give a very good description of violent sloshing (see the Section 10.1). Individual mass particles are followed in time. The Euler equations of a compressible liquid are the basis of our formulation; however, Navier–Stokes equations may also be used. It is possible, but numerically demanding from a computational time point of view, to include the gas flow. In practice, the Euler equation is modified by including an artificial viscosity term, which is needed for very long time simulations to ensure that transients due to initial conditions are damped out and also for numerical stability issues. The use of ghost particles is one way to satisfy the tank boundary conditions while the dynamic and kinematic free-surface conditions are automatically satisfied.

Particles can get stuck to the tank roof because of negative pressures. One pragmatic way to deal with this is to get rid of the ghost particle after a liquid particle gets stuck.

A drawback of the method is the artificial relation between the pressure and the liquid density, which causes unphysical high-frequency pressure oscillations that are often filtered out.

Numerical difficulties occur for the SPH when particles cluster too close together. A small repulsive corrective pressure term is needed when particles are very close.

10.8 Exercises

10.8.1 One-dimensional acoustic resonance

We continue to study the problem addressed in Section 10.5.3, both by an analytical method and by means of FDM and FVM. One-dimensional potential flow with velocity potential Φ is assumed. The governing equation for the analytical method, FDM, and FVM is wave equation (10.68). The boundary conditions are $\partial\Phi/\partial x = u_0$

at $x = 0$ and L. Initial conditions have to be additionally specified.

(a) We derive an analytical solution based on an expansion of the solution in terms of normal modes. We therefore start by finding the normal modes, which means nontrivial solution when the excitation is zero. Show that the normal modes of the velocity potential, φ_n, and associated natural frequencies σ_n can be expressed as

$$\varphi_n = \cos(\sigma_n x/c_0); \quad \sigma_n = n\pi c_0/L, \quad (10.97)$$

where c_0 is the speed of sound.

Then express the velocity potential with forced excitation as

$$\Phi = u_0\left(x - \tfrac{1}{2}L\right) + \sum_{n=1}^{\infty} a_n(t)\,\varphi_n. \quad (10.98)$$

Use the Galerkin method to determine the differential equations for $a_n(t)$. The answer for the steady-state solution with $u_0 = u_a \exp(i\sigma t)$ is that the pressure can be expressed as

$$p = -\rho\left[i\sigma\left(x - \tfrac{1}{2}L\right) u_a \exp(i\sigma t) + \sum_{n=1}^{\infty} \frac{i\sigma K_{na}\exp(i\sigma t)}{-\sigma^2 + \sigma_n^2}\varphi_n(x)\right],$$

$$K_{na} = \frac{2\sigma^2 L u_a}{(n\pi)^2}\left[(-1)^n - 1\right],$$

where ρ is the mean mass density.

(b) Assume steady-state (periodic) solutions. Consider the nondimensional pressure amplitude at $x = 0$ and L as presented in Figure 10.17 and use the FDM, FVM, and the analytical method derived in part (a) to solve the problem. Discuss the convergence of the FDM and the FVM solutions by comparing with the analytical solution as was done for the FEM solution in Figure 10.18.

10.8.2 BEM applied to steady flow past a cylinder in infinite fluid

We study a fixed circular cylinder with radius R in infinite fluid and constant ambient flow velocity. We define polar coordinates (r, θ) where $r = 0$ is the cylinder axis and $\theta = 0$ corresponds to the ambient flow direction. Furthermore, a Cartesian coordinate system (x, y) is introduced so that $x = r\cos\theta$, $y = r\sin\theta$. Potential flow of an incompressible fluid is assumed. The classical result for the velocity potential is the sum of the ambient flow and a dipole in the center of the cylinder with dipole direction following the ambient flow. Therefore, the velocity potential with ambient flow along the x-axis can be expressed as

$$\varphi = Ux + \frac{UR^2}{r}\cos\theta, \quad (10.99)$$

where U is the constant ambient flow velocity.

In parts (a), (b), and (c) you will apply the basic formulation of the BEM without doing numerical calculations. It requires analytical expressions of integrals.

(a) Start with an expression similar to eq. (10.3), where the velocity potential due to the cylinder is represented as a source and dipole distribution over the cylinder surface. Show that the solution is consistent with eq. (10.99).

(*Hints:*
1. Represent the velocity potential on the cylinder surface caused by the presence of the cylinder as a Fourier series in θ.
2. You may find it helpful that, for $r \geq R$,

$$\int_\alpha^{\alpha+2\pi} \cos\theta \ln[R^2 + r^2 - 2Rr\cos(\theta - \alpha)]d\theta = -\frac{2\pi R}{r}\cos\alpha,$$

$$\int_\alpha^{\alpha+2\pi} \cos\theta \frac{R - r\cos(\theta-\alpha)}{[R^2 + r^2 - 2Rr\cos(\theta - \alpha)]}d\theta = -\frac{\pi\cos\alpha}{r}.)$$

(b) Start with an expression similar to eq. (10.14), where the velocity potential due to the cylinder is represented as a source distribution over the cylinder surface. Show that the solution is consistent with eq. (10.99).

(c) The source distribution can be related to a fictitious interior problem. Solve the interior problem and show that the source density is consistent with the exterior and fictitious interior solutions.

(d) You should now try to solve the problem numerically by a source distribution method, where the sources are inside the cylinder but the collocation points where you satisfy the body boundary conditions are on the body surface. What are the numerical conditions in terms of number of sources and distance from the cylinder surface so that you can use discrete sources instead of sources integrated over segments?

10.8.3 BEM applied to linear sloshing with potential flow and viscous damping

Sections 6.2.2 and 6.3.2 discussed how to calculate the effect of a laminar steady-state boundary-layer flow on the potential flow solution. The effect of the boundary layer was in terms of an inflow/outflow velocity distribution that depends on the tangential velocity at the body surface. Follow this procedure in combination with a BEM for linear steady-state sloshing inside a tank and do calculations for a rectangular two-dimensional tank. Compare the predicted damping with the damping rate analyzed in Section 6.3.1 for selected cases.

10.8.4 Application of FEM to the Navier–Stokes equations

Assume two-dimensional laminar flow past a body in infinite fluid and adopt a time-stepping procedure based on Chorin's projection method. One example of a time-integration procedure is the predictor–corrector method presented in Section 10.3. You should follow that procedure with constant mass density and viscosity coefficient.

(a) Use the x-component of eq. (10.31) and apply the weighted residual method to the equation. Apply the divergence theorem and rewrite the integral involving second-order derivatives in terms of first-order derivatives.
(b) Do similar as in (a) with the Poisson equation for the pressure.

10.8.5 SPH method

(a) The kernel function in the SPH method can be approximated as in eq. (10.89) for two-dimensional flow. Use the same cutoff limit δ and show what the exact expression should be.
(b) Study eq. (10.91) in one dimension by using the kernel function $W = \exp(-|\boldsymbol{x}_i - \boldsymbol{x}_j|^2/h^2)/(h\sqrt{\pi})$. Select a given domain and study the behavior in the limit of increasing number of particles and decreasing value of h. Show that the properties of a Dirac delta function are satisfied (i.e., similar to eq. (10.87)).
(c) A fictitious Rayleigh viscosity term was introduced in the section on BEM. Do the same with the SPH method. We rewrite eq. (10.78) as

$$\frac{D^* \boldsymbol{v}_r}{Dt} = -\frac{\nabla p}{\rho} + \boldsymbol{f} + \frac{\mu}{\rho}\boldsymbol{v}_r.$$

Show how to modify eq. (10.84) with the fictitious Rayleigh viscosity term.

11 Slamming

11.1 Introduction

Slamming denotes the impact between a liquid surface and a solid boundary (e.g., a tank bulkhead). Different physical effects occur during slamming. A gas cushion may be formed between the liquid and tank surface if the impact angle between the incident free surface and the body boundary is small at the impact point (line). Gas compressibility influences the cushion dynamics, which in turn influences the liquid flow. The gas cushion generates gas bubbles when it collapses. The *ullage* pressure influences the presence and behavior of gas bubbles. The word "ullage" refers to the space above a liquid in a tank. The ullage gas density does also matter. Local hydrodynamic effects can cause vibrations of the tank structure that trigger ventilation and cavitation. When analyzing slamming, one must always have the structural response in mind in terms of maximum stresses. An important consideration is the time scale of any particular hydrodynamic effect, such as liquid compressibility, relative to wet natural periods for structural modes that may contribute significantly to large structural stresses. If the time scale of a particular hydrodynamic effect is very small relative to the important structural natural periods, the details of the particular hydrodynamic effect can be neglected. When hydrodynamic loading occurs over a time scale of important structural periods, hydroelasticity must be considered.

A design procedure for the slamming load effect must account for the probability of different sea states in realistic ship routes (long-term approach). The coupled effect of sloshing and ship motion has to be considered. The filling height is an important parameter. The fine details of the slamming loads are secondary in the coupled analysis of sloshing and ship motions. Because slamming is a strongly nonlinear process associated with violent sloshing, many realizations of the same storm are needed relative to a linear process to obtain reliable predictions of the probability density function for the slamming loads and response. The methodology for sloshing assessment of membrane LNG vessels by Gaztransport & Technigaz (GTT) is described by Gervaise *et al.* (2009). A basis is a long-term approach.

The Det Norske Veritas (DNV) Rules for Classification of Ships (2007), in covering hull structural design of ships with length 100 meters or more, consider impact pressures in the upper parts of tanks and in the lower parts of smooth tanks. A *smooth tank* means the tank has no interior load-carrying structures such as girders or baffles obstructing the flow. The fact that the tank has a cooling tower or a corrugated tank surface does not make the tank nonsmooth. The rule states that "tanks with sloshing length $0.13L < l_s < 0.16L$ or with free sloshing breadths $b_s > 0.56B$ will generate an impact pressure on horizontal and inclined surfaces adjacent to vertical surfaces in the upper part of the tank due to high liquid velocities meeting these surfaces." The sloshing length and breadth are defined in the rules (DNV Rules for Classification of Ships, 2007, Newbuildings, hull and equipment main class, part 3, chapter 1, Hull structural design with length 100 metres and above, page 34) and are effective lengths to account for the tank geometries and contribution from internal structures. Furthermore, L is the "length of the ship in m defined as the distance on the summer load waterline from the fore side of the stem to the axis of the rudder stock. L shall not be taken less than 96% and need not to be taken greater than 97% of the extreme length on the summer load waterline. For ships with unusual stern and bow arrangements, the length L will be especially considered." B is the "greatest moulded breadth in m, measured at the summer waterline." Examples of tank dimensions are given in Table 1.1. The fact that the impact causes a flow implies that the pressure on the tank surface in the vicinity of the impact area must also be considered. We consider, for instance, an example in Section 11.9.4 with tank roof impact of an oil/bulk/ore (OBO) carrier. The tank roof is very stiff so the impact loads are of minor concern for the roof structure. An area vulnerable to the effect of tank roof impact is that of the vertical stiffeners

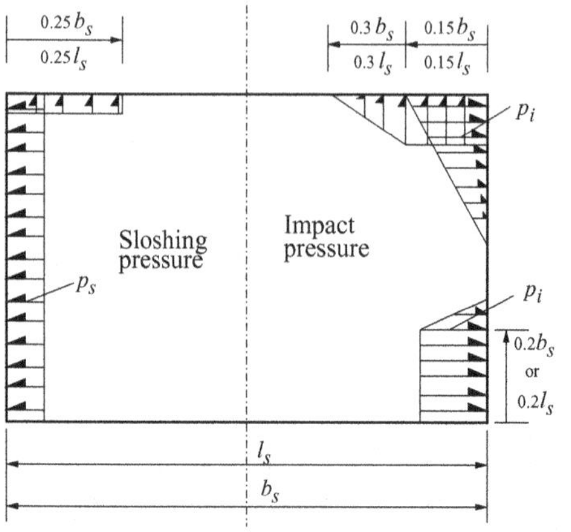

Figure 11.1. Pressure distribution in a tank according to DNV's *Rules for Classification of Ships* (2007) (Det Norske Veritas AS).

at the tank wall adjacent to the impact. The largest impact pressures are said in the rules to occur either within $0.15l_s$ from transverse wash or end bulkheads or within $0.15b_s$ from longitudinal wash bulkheads and tank sides. Outside $0.15l_s$ and $0.15b_s$ the impact pressure may be reduced to zero at $0.3l_s$ and $0.3b_s$, respectively (see Figure 11.1). The distributions of the impact pressures in the lower part of the tank, and the sloshing pressures, are also shown in Figure 11.1. By sloshing pressure, we mean the nonimpact pressure as discussed, for instance, in Section 5.2. We refer to the rules of classification societies for details on the parameter dependence of the pressures and on how to express pressures in tank corners at transverse bulkheads, for instance. Our approach is to follow a physically based discussion of impact pressures and resulting structural stresses.

Pastoor *et al.* (2005) analyzed the sloshing impact loads in tank No. 2 counted from the bow of a 138,000 m^3 liquefied natural gas (LNG) carrier based on model tests. The tank has a prismatic form. The worst filling level as a percentage of the tank height was found to be around 95% in head-sea conditions. However, they referred also to large pressures for a filling level of 30% in beam and head sea. The tank roof is most exposed for higher fillings in head sea. The upper chamfer corners are most exposed for 50–70% fillings in beam seas. The lower corner of the upper chamfer and the tank sides are most exposed for filling levels less than 50%. Pastoor *et al.*'s conclusions are based on Froude scaling of probability distributions of maximum pressure in realistic sea states.

The inflow condition for slamming depends on the filling ratio. Three different scenarios from experimental studies with tank roof impact are illustrated in Figure 11.2. All of them refer to cases where the flow in the tank is two-dimensional. Figure 11.2(a) illustrates a sudden flip-through of the free surface at the tank wall where a jet with high velocity impacts the tank roof. The filling is high and waves propagate toward the wall before the flip-through. The word "flip-through" refers to a condition where a steep wave approaches a wall without impacting. A jet flow is instead generated at the wall. If the wave front is close to vertical, very high pressure gradients occur as a consequence at the jet root. A flip-through is in the latter case characterized by large time rates of change of the free-surface slope at the tank wall. Section 11.5 presents experimental results for the case in Figure 11.2(a). Cases in Figure 11.2(b) and (c) with flat impact and impact with oscillating gas cavity are associated with finite liquid depth conditions (i.e., the filling does not need to be high).

When tank flow resonance occurs in shallow- and lower-intermediate liquid conditions, steep waves (e.g., hydraulic jumps) may occur as discussed in Section 8.8.2. The details of the impact may involve flip-through and gas cavities. Three-dimensional flow aspects may be important for slamming; this is true, for instance, when swirling occurs in a square-base or nearly square-base rectangular tank in finite depth. Because the

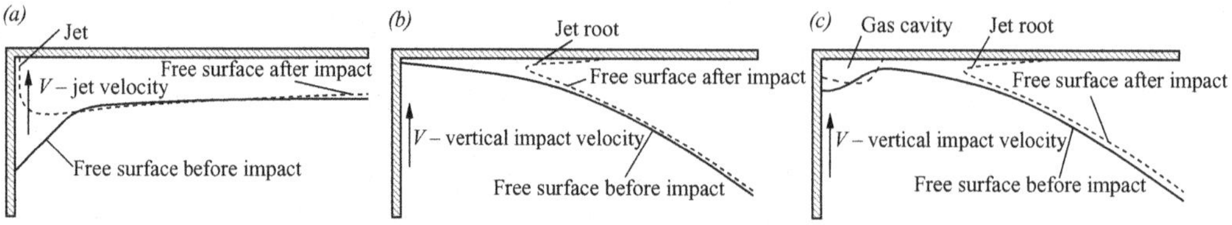

Figure 11.2. Three different scenarios of impact against the tank roof considered by Rognebakke and Faltinsen, 2005: (a) high-curvature free-surface impact with a high velocity jet, (b) flat impact, and (c) impact with oscillating gas cavity.

free-surface elevation without impact is largest at the corners (see Figure 1.4), that is where the probability of impact is highest.

Very high slamming pressures may occur when the angle between the impacting free surface and the tank surface is small. The pressure is sensitive to the details of the impacting free surface. Abramson *et al.* (1974) showed experimentally that slamming pressures can have a stochastic nature even though the tank is excited harmonically. It is always important to find the structural reaction in terms of maximum stresses. Even though the pressures may be very high and have a stochastic nature, the stress may not have a similar behavior (Faltinsen, 1997, 2005). Slamming pressures are a bad measure when they are very high, because such loads are normally connected with the fact that the high pressures are strongly limited in space and time. The slamming force impulse is then a more reliable parameter even though it is difficult to precisely quantify.

Model tests are commonly used to assess slamming loads for the design of tank structures. Many flow parameters ideally should be the same in full and model scale based on the similarity requirement. Obviously the inner tank dimensions and liquid depth must be geometrically similar in the two scales. Because sloshing is associated with gravity waves, we require that the Froude number is the same in model and full scales. Other flow parameters, such as the ratio between the ullage gas density and liquid density, the Reynolds, Bond, Euler, cavitation, and Cauchy numbers, and the scaling of hydroelasticity are discussed in Section 11.2 and in later sections in this chapter. The Reynolds-number effect is also illustrated later in this section.

It is also common to do computational fluid dynamics (CFD) calculations to assess slamming loads. However, CFD calculations may be difficult at the initial impact stage if the relative angle between the impacting free surface and the tank surface is small. When the relative angle is zero and the impacting free surface has large radius of curvature at initial impact, a Wagner type of solution (Wagner, 1932) gives an infinite rate of change of the wetted surface. This solution is not compatible with the numerical time-integration procedures that are normally used in CFD calculations. However, the stretched coordinate system method of Wu (2007) can resolve this incompatibility. The force impulse during a small initial time may be satisfactorily predicted even though the initial pressures are poorly predicted (Zhu, 2006). A difficulty is also due to the fact that some CFD methods assume an artificial gradual change in the density between the liquid and the gas. Furthermore, present-day computer capabilities prohibit the use of CFD for the long time simulations that are needed to obtain reliable information about the probability density function of slamming loads. Conversely, we should not overemphasize the difficulties of CFD methods in predicting high impact pressures that are highly concentrated in time and space. More suitable parameters for assessing their capabilities and accuracy in such scenarios are the slamming force impulse and the induced structural strains, whose proper estimate requires coupling of the hydrodynamic analysis with a structural analysis. In fact, it is common in the computations that the structure is assumed rigid, but under the discussed circumstances, hydroelasticity plays an important role.

The nonlinear multimodal method described in Chapters 8 and 9 does not predict slamming pressures. If the filling ratio (i.e., the ratio between the liquid volume and the tank volume) is not too close to 1, let us say smaller than 0.7, the method may be combined with a separate slamming calculation based on, for instance, a Wagner-type approach to predict slamming loads on the tank roof. This analytically based method provides a robust procedure for impact loads but must be validated for cases where impact pressures are relevant for the local structural response. Abramson *et al.* (1974) presented experimental slamming predictions for the LNG tank shown in Figure 11.3, including statistical distributions. The objective was to create two-dimensional flow conditions. The pressure transducer location is indicated as P_2 in Figure 11.3. Viscosity seemed to be important when the forced sway amplitude η_{2a} was 0.01 times the tank breadth l but not for $\eta_{2a} = 0.1l$. The kinematic viscosity coefficients ν of the considered liquids (see Table 11.1) range from 10^{-6} m^2 s^{-1} for water to $435 \cdot 10^{-6}$ m^2 s^{-1} for reginol oil. Figure 11.3 shows computed and experimental pressures for $\eta_{2a} = 0.1l$. The computations are for nearly steady-state liquid motions. The experimental values represent 10% exceedance limits for the pressure. The computed values represent the most frequent occurrence in a long time

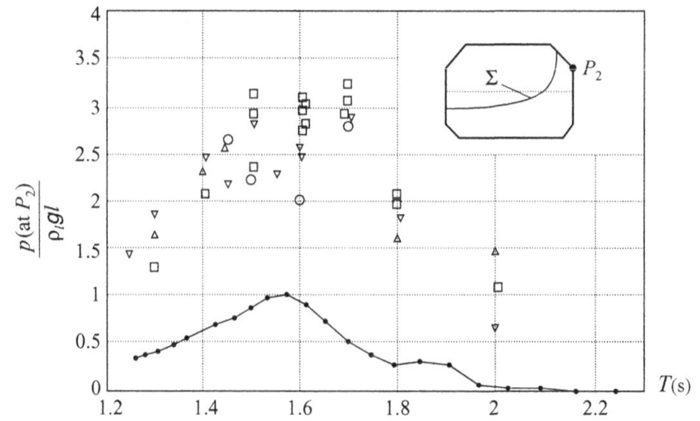

Figure 11.3. Measured (Abramson *et al.*, 1974) impact pressures p (10% exceedance value) at the location P_2 in the prismatic LNG tank presented as a function of forced oscillation period T, where ρ_l is the liquid density. The tank geometric proportions are shown in Figure 4.22(a). The forcing amplitude is $\eta_{2a} = 0.1l$ and the liquid depth is $h = 0.4l$. Experimental measurements were performed for (□) freshwater, (○) 63% glycerol/water, (Δ) 85% glycerol/water, and (∇) reginol oil (where liquid properties are defined in Table 11.1). Computational steady-state (time-periodic) impact pressures (—•—) are predicted with the multimodal method in combination with Wagner theory (see also Figure 8.3) (Faltinsen & Rognebakke, 2000).

simulation; the results depend on the time series. We can certainly obtain the level of pressure shown in Figure 11.3 in the transient phase of our computations. However, we would need the complete time series in the experiments to make a quantitative estimate of the 10% exceedance limit. The way that the data of Abramson *et al.* (1974) were presented suggests that they thought the process was stochastic. However, in our opinion, this particular type of impact on a chamfered tank roof is deterministic during harmonic excitation of the tank. The impact on a horizontal tank roof is another matter. In that case the impact pressure may very well have a stochastic behavior, but, as previously discussed, the maximum pressure may be an irrelevant parameter for local structural response.

Different inflow scenarios for slamming and physical effects are systematically studied in this chapter. Theoretical analysis is partly used; however, the state of the art does not permit a purely theoretical analysis. Model test results are usually the basis for structural design. We start, therefore, by discussing the scaling laws for model testing. After that we systematically examine different physical effects. We start out by assuming potential flow theory of an incompressible liquid. The tank structure is considered rigid. First the gas is assumed not to influence the liquid flow, then we incorporate the effect of the compressibility of gas cavities and acoustic effects in the liquid are considered. The final sections are about hydroelastic slamming. The presented results assume two-dimensional flow.

The two-dimensional analysis of slamming makes partial use of a complex-variable formulation. We therefore consider these two-dimensional problems in the *Oxy*-coordinate system with the notation φ for the velocity potential. The adopted notation for the velocity potential in three dimensions is Φ.

11.2 Scaling laws for model testing

Scaling of the model test results is an important issue. In the following discussion we assume a *rigid* tank and that the slamming pressure can be applied in a quasi-steady structural analysis. However, *hydroelastic* effects may play an important realistic role. The scaling laws are derived via the Pi theorem. The Pi theorem is due to Buckingham (1915) and elaborated in detail by Rouse (1961).

The Pi theorem states:

Let a physical law be expressed in terms of n physical quantities, and let k be the number of fundamental units needed to measure all quantities. Then the law can be re-expressed *as a relation among $(n-k)$ dimensionless quantities.*

It is necessary that we make correct assumptions about which physical quantities matter in describing the slamming pressure. The Pi theorem does not help us do this; the Pi theorem tells us only how we can construct nondimensional relationships.

The following derivation assumes that the tank is forced harmonically with a period T and a transverse motion amplitude η_a. We assume that the maximum slamming pressure p at a certain point is a function of the following parameters:

L: length, breadth, diameter, or another characteristic dimension of the tank;
U: characteristic velocity of the tank (e.g., L/T);
h: liquid depth;
S_i: other pertinent tank dimensions ($i = 1, 2, \ldots, N$);
ν: kinematic viscosity of the liquid;
ρ_l: liquid density;
ρ_0: ullage gas density;
T_s: surface tension;
E_l: bulk modulus of the liquid;
p_0: ullage pressure; and
p_v: liquid vapor pressure.

Our problem has three fundamental units: mass, length, and time. It follows from the Pi theorem that

$$\frac{p}{\rho_l U^2} = f\left[\frac{U}{\sqrt{gL}}, \frac{UL}{\nu}, \frac{p_0}{\rho_l U^2}, \frac{\rho_l U^2}{E_l}, \frac{\rho_0}{\rho_l}, \frac{p_0 - p_v}{\frac{1}{2}\rho_l U^2}, \frac{\rho_l U^2 L}{T_s}, \frac{\eta_a}{L}, \frac{h}{L}, \frac{S_i}{L}, \ldots\right]. \quad (11.1)$$

where $U/\sqrt{gL}$ is the Froude number, UL/ν is the Reynolds number, $p_0/\rho_l U^2$ is the Euler number, $\rho_l U^2/E_l$ is the Cauchy number, $(p_0 - p_v)/(\frac{1}{2}\rho_l U^2)$ is the cavitation number, and $\rho_l U^2 L/T_s$ is the Weber number. We can combine the aforementioned cavitation number and the Froude number into the modified cavitation number, $C_v = (p_0 - p_v)/(\rho_l g L)$, which was used by Olsen and Hysing (1974). Because the Froude number is kept the same in model and full scale, C_v expresses the same physical effect as $(p_0 - p_v)/(\frac{1}{2}\rho_l U^2)$. Similarly, we can introduce the modified Euler number $C_E = p_0/(\rho_l g L)$. The procedure can be generalized to any forced tank motion by introducing nondimensional translatory tank motion with respect to L and by Froude-scaling the time scale. Angular tank motions remain the same in model and full scales. If only Froude scaling and similarity in forcing and geometry are assumed, it follows from eq. (11.1) that $p/(\rho_l g L)$ is the same in model and full scales.

Properties of different liquids used in model experiments and of different cargo liquids are presented in Tables 11.1 and 11.2, respectively. Table 11.1 gives weight percentages between glycerol and water, which refer to the ratio between the weight of glycerol and the weight of water in the mixture. The values in Tables 11.1 and 11.2 are not exactly the same as those of Olsen and Hysing (1974). Because the density is used in normalizing Olsen and Hysing's measured slamming pressures, attention should be paid to the fact that the density values in Table 11.1 are the same as in Olsen and Hysing (1974) except for glycerol/water 85%. We operate with $\rho_l = 1.23 \cdot 10^3$ kg m^{-3}, whereas Olsen and Hysing (1974) specify $\rho_l = 1.22 \cdot 10^3$ kg m^{-3}.

Only Froude and geometric scaling have traditionally been considered important in model testing. Abramson *et al.* (1974) presented an example illustrating the importance of Froude scaling for slamming pressures. A number of pressure gauges were installed at different locations in an OBO tank carrying water ballast, and simultaneous recordings of pressures and the roll motion of the ship were taken with different filling heights during a voyage from Japan to the Persian Gulf. As part of a subsequent model-test program, recorded roll motions of the ship were imposed on a tank model (scale 1:30), and the pressures at corresponding locations were measured. Extremely high impact pressures were recorded at the underside of the top wing tank. Generally it was found that fewer impacts occurred in the model than in the prototype. However, the magnitudes of the pressure peaks were quite similarly distributed. An example that represents the condition that could be best simulated in the model is given in Table 11.3. The pressures obtained at model scale have been scaled according to Froude's law of similarity (i.e., in proportion to the linear scale ratio). The percentage of all peaks in a sample that lie within different pressure intervals is tabulated.

As long as the numbers of actual impacts (i.e., the probability of occurrence) in model and prototype are of the same order of magnitude,

Table 11.1. *Properties of different liquids used in the model experiments by Olsen and Hysing (1974)*

Liquid	Temperature (°C)	Density, $\rho_l \cdot 10^{-3}$ (kg m^{-3})	Kinematic viscosity $\nu \cdot 10^6$ (m^2 s^{-1})	Sound speed c_0 (m s^{-1})	Vapor pressure p_v (kPa)	Cavitation number,[a] $\frac{p_0 - p_v}{\rho_l g L}$	Euler number,[a] $\frac{p_0}{\rho_l g L}$	Reynolds number,[a] $L^{3/2} g^{1/2} \nu^{-1}$
Reginol oil	20	0.89	435.0				8.41	$0.012 \cdot 10^6$
Glycerol/ water 85%	20	1.23	110.0	1741.5[b]	≈ 0	6.13	6.13	$0.046 \cdot 10^6$
Glycerol/ water 73%	20	1.19	27.5				6.29	$0.18 \cdot 10^6$
Glycerol/ water 63%	20	1.16	11.5				6.45	$0.44 \cdot 10^6$
Water	20	0.998	1.00	1481.3	2.3	7.39	7.50	$5.1 \cdot 10^6$
Water	95	0.962	0.31	1546.2	84.5	1.29	7.78	$16.9 \cdot 10^6$
Water	100	0.958	0.29	1542.7	101.3	0	7.81	$17.5 \cdot 10^6$

Note: The ullage pressure p_0 is atmospheric pressure (i.e., 101.3 kPa). The data except for water are taken from Olsen and Hysing (1974).
[a] $L = 1.38\,\text{m}$.
[b] The bulk modulus used in the calculation of the speed of sound is based on 100% glycerol values.

the prediction of extreme values is not seriously affected and model test results can be taken as a basis for such predictions. It should be noted that both model and prototype liquids were water and, therefore, the question of modeling LNG with water is not resolved. It should be noticed that the highest measured full-scale pressure exceeded the corresponding highest pressures measured for the model. However, this discrepancy may be a question of data sample size.

We could have combined the assumed variables affecting the slamming pressures into other nondimensional parameters. For instance, if the gas–liquid volume fraction is zero, the speed of sound in the liquid is $c_0 = \sqrt{E_l/\rho_l}$ and we can rewrite the Cauchy number in terms of the Mach number: $Ma = U/c_0$. The effect of the surface tension is sometimes characterized by the Bond number instead of the Weber number. The Bond number, $Bo = \rho_l g L^2/T_s$, is the Weber number

Table 11.2. *Properties of different cargo liquids*

Liquid	Temperature (°C)	Density, $\rho_l \cdot 10^{-3}$ (kg m^{-3})	Kinematic viscosity, $\nu \cdot 10^6$ (m^2 s^{-1})	Sound speed, c_0 (m s^{-1})	Ullage pressure, p_0 (kPa)	Euler number,[a] $\frac{p_0}{\rho_l g L}$	Reynolds number,[a] $L^{3/2} g^{1/2}/\nu$
LNG	−162	0.474	0.29	1709.9	101.3	0.54	$2.7 \cdot 10^9$
Methane (CH_4)	−162	0.423	0.28	1340.4	101.3	0.61	$2.8 \cdot 10^9$
Ethylene (C_2H_4)	−104	0.568	0.31	1306.8	101.3	0.45	$2.6 \cdot 10^9$
Ethylene (C_2H_4)	−92	0.551	0.27	1219.9	202.7	0.94	$2.9 \cdot 10^9$
Ethane (C_2H_6)	−90	0.546	0.31	1326.0	101.3	0.47	$2.6 \cdot 10^9$
Ethane (C_2H_6)	−75	0.526	0.27	1209.9	202.7	0.98	$2.9 \cdot 10^9$
Propane (C_3H_8)	−43	0.582	0.34	1165.1	101.3	0.44	$2.3 \cdot 10^9$
Propane (C_3H_8)	−14	0.547	0.27	975.0	304.0	1.42	$2.9 \cdot 10^9$
Butane (C_4H_10)	−3	0.604	0.35	1053.2	101.3	0.43	$2.3 \cdot 10^9$
Ammonia (NH_3)	−34	0.683	0.38	1774.2	101.3	0.38	$2.1 \cdot 10^9$

Note: The vapor pressure p_v is approximately equal to the ullage pressure p_0 (i.e., the cavitation number is zero; NIST Chemistry Webbook at http://webbook.nist.gov/chemistry). The data for LNG are from Abramson *et al.* (1974).
[a] $L = 40\,\text{m}$.

Table 11.3. *Comparison of impact pressures in model and prototype of an OBO tank*

Pressure (atm)	Percentage of peaks in pressure range: Model	Prototype	Test condition
0–6	83.3	86.0	$h/l = 0.215$
6–12	13.6	10.0	Random rolling
12–18	3.1	2.5	(Maximum roll angle 7.4°
18–24		1.5	RMS of roll angle 2.9°)

divided by the square of the Froude number. The Bond number represents the ratio of inertia to surface tension forces and, if greater than $\sim 10^4$, surface tension and the associated meniscus effect can be neglected when modeling a free surface (Myskis *et al.*, 1987). If we consider, for instance, freshwater in a tank with $L = 1$ m and surface tension $T_s = 0.074$ N m^{-1}, the Bond number is $1.32 \cdot 10^5$. If a tank of length $L = 30$ m is filled with LNG and $T_s = 0.504$ N m^{-1}, the Bond number is $8.30 \cdot 10^6$. Because the surface tension value for LNG is unknown, the maximum value for liquids has been used. The magnitude of these Bond numbers indicates that surface tension effects can be neglected once mixing processes are not considered. For instance, the surface tension of LNG may be important. It governs the size of gas bubbles mixed with LNG. Eigenfrequencies of gas bubbles depend on their radius. These frequencies contribute to time scales of hydrodynamic loads.

The effect of the liquid viscosity is represented by the Reynolds number. We can combine the Reynolds and Froude numbers into the following modified Reynolds number: $\overline{Rn} = L^{3/2}g^{1/2}/\nu$. Examples of its values are given in Tables 11.1 and 11.2. We note that the values of $\overline{Rn}$ are very different in model and full scales. For both Reynolds and Froude number scaling to be satisfied, the ratio between the length in model and full scale is given by

$$L_m/L_p = (\nu_m/\nu_p)^{2/3} \quad \text{for} \quad g_m = g_p, \qquad (11.2)$$

where the subscripts m and p indicate model and full scale (prototype), respectively. With realistic values of L_m/L_p it is impossible to find a model liquid that satisfies eq. (11.2). Therefore, we cannot properly model viscous effects in model tests. However, viscosity, generally speaking, does not have a dominant influence on slamming pressures and integrated hydrodynamic loads for a smooth tank. The influence of viscosity on slamming loads is discussed in connection with Figures 11.3 and 11.13. The influence of viscosity on the lateral hydrodynamic force is shown in Figures 8.2 and 8.3. The liquids listed in Table 11.1 are used in the experimental studies.

Maillard and Brosset (2009) performed model tests with forced harmonic surge motions of a rigid rectangular tank filled either 81% or 90% with water and showed that the density ratio between the ullage gas and the liquid matters for the statistical values of the slamming pressure. Most presented statistical pressure values were based on using a return period of 1/10th of the total test duration. The statistical sample considers the maximum pressure irrespective of the sensor for a given impact. Two-dimensional flow conditions were attempted. The cavitation number was kept equal to zero in one of the test series by varying the ullage pressure and temperature and by using water vapor as the ullage gas. The density ratio between gas and liquid varied from 0.00005 to 0.0058. The density ratio between natural gas and LNG within an LNG tank is around 0.004 depending on the quality of LNG and the temperature of the gas, while it is around 0.0012 during model tests with air and water at room temperature. Zero cavitation number is relevant for LNG and the other cargo liquids listed in Table 11.2. The vapor can condense at zero cavitation number due to overpressure during impact. The condensation in LNG tanks may not be as quick as the condensation of water vapor (Maillard & Brosset, 2009). Another test series was done at atmospheric pressure with helium to get a density ratio of 0.0005 and with different mixtures of sulphur hexafluoride and nitrogen to get density ratios of 0.0036 and 0.0046. A third test series was done at atmospheric pressure at different temperatures with air and density ratios from 0.0008 to 0.0012. The general trend in the test series is that the slamming pressure decreases with increasing density ratio. For instance, the decrease of the statistical pressure between a density ratio of 0.0012 and 0.0036 is around 50% for the tests with a non-condensable gas, while this decrease is around 65% for the vapor tests. Their argument is that a larger share of the energy is transferred from the liquid to the gas for a heavier gas than for a lighter gas and thereby reduces the impact velocity of the liquid. Their results at a density ratio around 0.004 show that it is conservative to assume atmospheric pressure conditions. A main contributing factor to lower pressures with water vapor as a gas at high density ratios, that is, 0.004, is believed to be condensation. Maillard and Brosset (2009) stated, based on their own experiments and numerical investigations by Braeunig *et al.* (2009), that the ullage pressure (Euler number) is not an important parameter. However, the experimental variation of the ullage pressure at constant density ratio was small. A broader range of ullage pressure was obtained in the numerical studies. However, the impact flow was idealized by considering a liquid with initially rectangular shape and horizontal bottom falling downward and impacting on a rigid horizontal surface. Their scenario has similarities with Figure 11.25. It is shown in Section 11.6 that the Euler number is an important parameter when the geometry of the ambient impacting free surface generates a gas cavity at non-small cavitation number. Maillard and Brosset (2009) studied number of gas pocket and aerated impact events when no condensation occurs. An aerated impact involves bubbles of small gas pockets. Number of gas pocket and aerated impact events increases with the density ratio between gas and liquid, that is, the number of events relative to the total number of events was around 30% with a density ratio of 0.0036. Maillard and Brosset (2009) state that even though not easily detectable, there is no reason to believe that this trend is different with water vapor tests. An important aspect is how much gas pocket and aerated impact events contribute to the statistical pressure relative to jet and Wagner-type impact events in the particular cases presented by Maillard and Brosset (2009). Only one harmonic forcing condition was studied by Maillard and Brosset (2009). Number and severity of gas pocket and aerated impact events depend, for instance, on the time history of the tank motion and the filling ratio.

In general, we cannot rule out the influence of other flow parameters than introduced in eq. (11.1). For instance, the bulk modulus of the gas and the temperature may matter.

As mentioned earlier, an important consideration is the time scale of a physical effect relative to wet structural natural periods associated with large structural stresses. Maillard and Brosset (2009) presented experimental probability density functions of the rise time of the slamming pressures in their model tests with a rigid model. The rise time is defined as twice the time from the half peak value to the peak value. There is a general slowdown of sloshing impacts when increasing the density ratio. They related the rise time with dynamic amplification factors (DAFs) determined through Finite Element Analysis for each Limit State of the Cargo Containment System (see also discussion of DAF in connection with Figure 11.58 for Wagner-type of tank roof impact with no gas pockets in a non-membrane tank). The DAF curve was Froude scaled back to model scale. Their results show that dynamic amplification matters. However, this procedure is approximate and not truly hydroelastic because the model tests are done with a rigid model. The scaling of the time from model to full scale may not always Froude scale. For instance, it is shown in Section 11.6.5 by a theoretical model of a gas cushion how the time duration of slamming pressures scales with the Froude number and the Euler number for non-small cavitation number. We need also to know the spatial distribution of impact pressure as a function of time. Maillard and Brosset (2009) describe a strategy on how to experimentally obtain slamming loads on a specific tank surface area.

The different flow parameters are discussed in later sections. Scaling of *hydroelasticity* for steel structures is considered and the complexity of correctly modeling the hydroelastic effects of membrane tanks is also examined.

11.3 Incompressible liquid impact on rigid tank roof without gas cavities

This section considers different theoretical potential flow models for incompressible tank roof impact without gas influence. The tank is assumed rigid. The first scenario assumes a nearly horizontal impacting free surface with a high radius of curvature (see Figure 11.2(b)). Wagner's (1932) theory is used. The focus is first on the hydrodynamic loads, after which we show how to apply the predicted mass flux and kinetic energy flux in the jet caused by the impact to estimate damping of sloshing. The following subsection presents numerical results for the impact of a liquid wedge that are relevant for Figure 11.2(a).

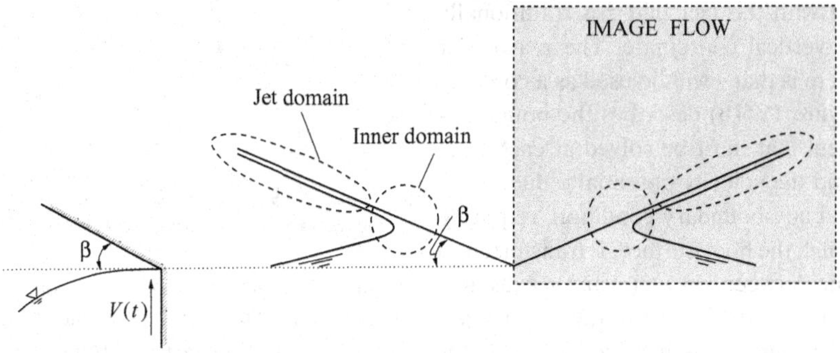

Figure 11.4. Impact of a free surface against a chamfered tank roof and its image flow about the tank wall.

Two-dimensional flow is assumed. A formula on how to estimate the integrated hydrodynamic loads due to three-dimensional flow is finally presented.

11.3.1 Wagner model

The case in Figure 11.2(b), where a nearly horizontal free surface impacts the tank roof, can be studied theoretically by Wagner's (1932) method when no hydroelastic effect occurs. Potential flow of an incompressible liquid with no gas influence is assumed. The procedure requires that the profile and the velocity of the impacting free surface are known at the impact time instant, for instance from particle image velocimetry measurements. We generalize the procedure by accounting for the fact that the tank roof may have a chamfer (see Figure 11.4). Even though Wagner's slamming model assumes local small deadrise (chamfer) angle β, the method is useful because it provides simple analytical results. These results can be used to assess how slamming pressures depend on structural form and time-dependent liquid entry velocity.

The influence by the other wall and the bottom is small (Faltinsen & Rognebakke, 1999). Our next step is to introduce an image flow about the tank wall as illustrated in Figure 11.4, where we have introduced the concepts of jet domains and inner domains at the spray roots.

Wagner's detailed description of the flow at the intersections between the free-surface and the body surface are discussed in Section 11.3.1.2. This local flow describes a jet flow which in practice ends up as spray. We focus now on what is called the outer flow domain, which is located below (outside) the inner and jet domains shown in Figure 11.4. We have no details on the spatially rapidly varying flow at the spray roots (inner domain). The predicted intersections between the free-surface and the body surface in the outer flow domain model are very close to the spray roots.

Figure 11.5(a) presents the impacting symmetric body and the free surface in the outer flow domain (see Faltinsen, 1990, for details). The free surface moves up with a uniform speed $V(t)$ (see Figure 11.4). The latter fact represents an approximation of the incident flow field. The reason why we could assume a uniform incident speed V is the fact that we consider a local scenario. The vertical velocity of the free surface is approximately constant on the length scale of the wetted body length due to the impact. We should note that we have used y as a vertical coordinate, which

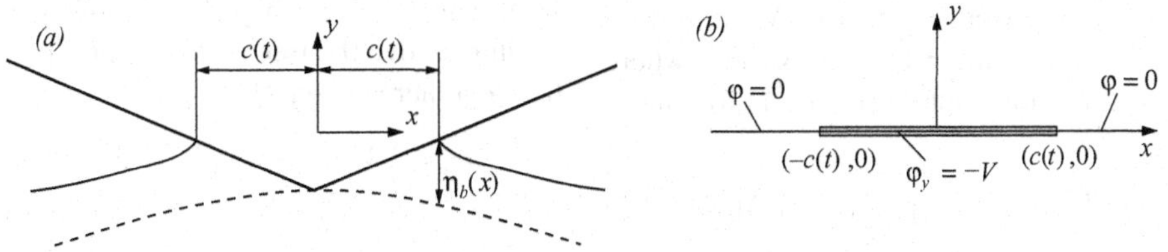

Figure 11.5. Definition of parameters in the analysis of impact forces and pressures on a two-dimensional body by means of (a) Wagner's outer flow domain solution and (b) the corresponding boundary-value problem in Wagner's outer domain analysis ($\varphi_y = \partial\varphi/\partial y$).

is in conflict with the fact that z is traditionally used as the vertical coordinate. The reason for doing this here is that z will be used as a complex variable. Figure 11.5(b) describes the boundary-value problem that must be solved at each time instant to find the velocity potential φ due to the impact. The body-boundary condition, requiring no flow through the body surface, is transferred to a straight line between $x = -c(t)$ and $c(t)$ using a Taylor expansion with respect to the penetration depth and ignoring the higher-order terms. This procedure can be done when the local relative angle between the body surface and the impacting free surface is small. The endpoints $x = \pm c$ correspond to the instantaneous intersections between the outer flow free surface and the body surface (see Figure 11.5(a)). We note that the free-surface condition $\varphi = 0$ on $y = 0$ has been used in Figure 11.5(b). This use is a consequence of the fact that liquid accelerations in the vicinity of the body dominate over gravitational acceleration during impact of a blunt body and the nonlinear term in the dynamic boundary condition is of higher order (see Faltinsen, 2005). Furthermore, the free-surface condition has, by ignoring the higher-order terms, been transferred to the straight line $y = 0$.

The solution to the boundary-value problem shown in Figure 11.5(b) can be found in many textbooks (e.g., Kochin *et al.*, 1964). The complex variable $z = x + \mathrm{i}y$, where $\mathrm{i} = \sqrt{-1}$ is the complex unit, is introduced. The *complex velocity potential* W can be expressed as

$$W = \varphi + \mathrm{i}\psi = \mathrm{i}Vz - \mathrm{i}V(z^2 - c^2)^{1/2}, \qquad (11.3)$$

where φ is the *velocity potential* and ψ is the *stream function*. The complex velocity is

$$\frac{dW}{dz} = u - \mathrm{i}v = \mathrm{i}V - \mathrm{i}V\frac{z}{(z^2 - c^2)^{1/2}}. \qquad (11.4)$$

Let us check that the boundary conditions are satisfied. Care must be shown in evaluating the complex function $(z^2 - c^2)^{1/2}$, which has a branch cut along the line from $z = -c$ to c. We introduce $z - c = r_1 \exp(\mathrm{i}\theta_1)$ and $z + c = r_2 \exp(\mathrm{i}\theta_2)$, where θ_1 and θ_2 vary from $-\pi$ to π (Figure 11.6). Therefore,

$$(z^2 - c^2)^{1/2} = \sqrt{r_1 r_2}\exp\left[\mathrm{i}\tfrac{1}{2}(\theta_1 + \theta_2)\right]. \qquad (11.5)$$

We can write $\theta_1 = -\pi$ and $\theta_2 = 0$ when $|x| < c$ and $y = 0^-$, which gives $(z^2 - c^2)^{1/2} = -\mathrm{i}(c^2 - x^2)^{1/2}$ for $|x| < c$, $y = 0^-$, and $y = 0^-$ corresponds to the underside of the body. When $x > c$ and $y = 0$, both θ_1 and θ_2 are zero; that is,

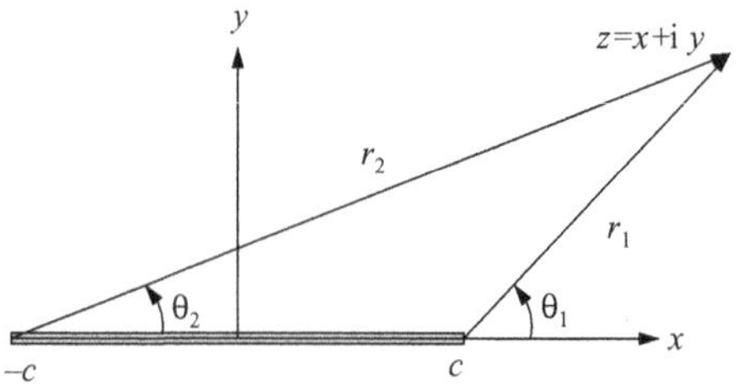

Figure 11.6. Definition of polar coordinates (r_1, θ_1) and (r_2, θ_2) used in evaluating the complex function $(z^2 - c^2)^{1/2}$. The angles θ_i vary from $-\pi$ to π.

$$(z^2 - c^2)^{1/2} = (x^2 - c^2)^{1/2} \quad \text{for} \quad x > c, \quad y = 0. \qquad (11.6)$$

Furthermore, $x < -c$ and $y = 0^{\pm}$ means that $\theta_1 = \theta_2 = \pm\pi$; that is,

$$(z^2 - c^2)^{1/2} = -(x^2 - c^2)^{1/2} \quad \text{for} \quad x < -c, \quad y = 0. \qquad (11.7)$$

Equation (11.3) then gives $\varphi = 0$ for $|x| > c$ on $y = 0$. Furthermore, eq. (11.4) gives

$$\frac{dW}{dz} = u - \mathrm{i}v = \mathrm{i}V + Vx(c^2 - x^2)^{-1/2} \quad \text{for} \quad |x| < c \quad \text{on} \quad y = 0^-. \qquad (11.8)$$

Because $v = -\mathrm{Im}(dW/dz) = \partial\varphi/\partial y$, we see from eq. (11.8) that the body boundary condition, $\varphi_y = -V$, is satisfied. Furthermore, eq. (11.4) gives that the fluid velocity goes asymptotically to zero when $|z| \to \infty$.

Equation (11.3) gives $W = \varphi + \mathrm{i}\psi = \mathrm{i}Vx - V(c^2 - x^2)^{1/2}$ for $|x| < c(t)$, $y = 0^-$. We can then write the velocity potential on the body as

$$\varphi = -V(c^2 - x^2)^{1/2}, \quad |x| < c(t). \qquad (11.9)$$

The hydrodynamic pressure can be approximated as $p - p_0 = -\rho_l \partial\varphi/\partial t$ (see Faltinsen, 2005, pp. 305, 306), where p_0 is the ullage pressure, which is the ambient pressure in the gas medium above the free surface. This approximation together with eq. (11.9) gives

$$p - p_0 = \rho_l Vc(c^2 - x^2)^{-1/2}\dot{c} + \rho_l \dot{V}(c^2 - x^2)^{1/2}. \qquad (11.10)$$

The first term in expression (11.10) is denoted the slamming pressure. It is associated with the rate of change of the wetted surface, $2\dot{c}$. The second term is called the added mass pressure. The reason for this is more evident when one considers the resulting hydrodynamic force. We note that the slamming pressure is infinite at $x = \pm c$; however, this condition is not physically feasible. A detailed analysis near the spray roots (inner domain solution) is needed to find the correct pressure near $x = \pm c$ (see Section 11.3.1.2). If V is constant, we get a maximum pressure of $p - p_0 = \frac{1}{2}\rho_l \dot{c}^2$ at $x = \pm c$. Because eq. (11.10) is integrable with respect to x and we are interested in pressure integration over an area in the structural analysis, the singularity appearing in the outer domain solution is not serious. We should note that $c(t)$ is still unknown.

The loads on the walls due to tank roof impact also matter. We start by expressing eq. (11.5) along the wall ($y < 0$), where $r_1 = r_2 = \sqrt{c^2 + y^2}$, $\theta_1 + \theta_2 = -\pi$, which gives $(z^2 - c^2)^{1/2} = -\mathrm{i}\sqrt{c^2 + y^2}$. From eq. (11.3) it follows then that the velocity potential on the tank wall adjacent to the tank roof impact can be expressed as $\varphi_{\text{wall}} = -V[(c^2 + y^2)^{1/2} + y]$, where $y < 0$ and $\varphi_{\text{wall}} \to 0$ as $y \to -\infty$. The corresponding pressure loads are given by $p - p_0 = -\rho_l \partial\varphi_{\text{wall}}/\partial t$. This result is needed in the hydroelastic analysis in Section 11.9.

11.3.1.1 Prediction of wetted surface

We now show how to find the half-wetted length $c(t)$ based on the outer domain solution. To do this, we need to follow liquid particles on the free surface and see when they intersect with the body surface. Since $\varphi = 0$ on the free surface, the horizontal velocity $\partial\varphi/\partial x$ is zero on the free surface. Equations (11.4), (11.6), and (11.7) can be used to express the vertical velocity $v = \partial\varphi/\partial y$ on the free surface:

$$\frac{\partial\varphi}{\partial y} = \frac{V\,|x|}{\sqrt{x^2 - c^2(t)}} - V \quad \text{on} \quad y = 0, \quad |x| > c(t). \tag{11.11}$$

It should be stressed that this expression does not apply on $y = 0$ and $|x| < c(t)$; there $\partial\varphi/\partial y = -V$. Equation (11.11) is the vertical velocity caused by the impact. The vertical velocity V of the incident free surface has to be added to find the global vertical velocity of a particle on the free surface. We now focus on one fluid particle with a given x that at time t intersects the body surface, which is $c(t) = |x|$. From the impact instant to t, this particle has moved a vertical distance $\eta_b(x)$ (Figure 11.5(a)). Therefore,

$$\eta_b(x) = \int_0^t \frac{V\,|x|}{\sqrt{x^2 - c^2(t)}}\,\mathrm{d}t, \tag{11.12}$$

where $\eta_b(x)$ is the distance between the incident free surface and the chamfered tank roof at the impact instant $t = 0$. This function is a known function of x. Equation. (11.12) is an integral equation for the unknown $c(t)$. We now derive the details. We consider positive x and rewrite eq. (11.12) as

$$\eta_b(x) = \int_0^x \frac{x\mu(c)\mathrm{d}c}{\sqrt{x^2 - c^2}}, \quad \text{where} \quad \mu(c)\mathrm{d}c = V\mathrm{d}t. \tag{11.13}$$

We do not know $\mu(c)$. Equation (11.13) is therefore an integral equation that determines $\mu(c)$. When $\mu(c)$ is found and $V(t)$ is known, we can find c as a function of time. We try to find an approximate solution of eq. (11.13) by guessing that $\mu(c) \approx A_0 + A_1 c$, where A_0 and A_1 are unknown constants. By integrating the right-hand side of eq. (11.13), it follows that

$$\eta_b(x) = \tfrac{1}{2}A_0\pi x + A_1 x^2. \tag{11.14}$$

If $\eta_b(x)$ is given as a second-order polynomial, we can determine A_0 and A_1 from eq. (11.14). In our case we approximate the impacting free surface as a parabola with radius of curvature R. Therefore,

$$\eta_b(x) = x\tan\beta + \tfrac{1}{2}x^2/R. \tag{11.15}$$

Comparing eqs. (11.14) and (11.15), we conclude that $A_0 = 2\tan\beta/\pi$ and $A_1 = \frac{1}{2}R^{-1}$ in eq. (11.14). Having determined $\mu(c)$ we can now find c as a function of time. Taking into account that $c(0) = 0$, we find the following quadratic equation:

$$A_0 c + \tfrac{1}{2}A_1 c^2 = \int_0^t V\mathrm{d}t. \tag{11.16}$$

Assuming a linearly changing impact velocity, $V(t) = V_0 + V_1 t$, the solution for c becomes

$$c(t) = \left[-A_0 + \sqrt{A_0^2 + 2A_1\left(V_0 t + \tfrac{1}{2}V_1 t^2\right)}\right] \Big/ A_1. \tag{11.17}$$

A more general solution of integral equation (11.13) is derived next. This derivation was also considered by Howison *et al.* (1991) by following the derivation by Tollmien (1934). We use the fact that the solution of

$$\int_a^x \frac{\mu(c)\mathrm{d}c}{\sqrt{g(x)-g(c)}} = f(x), \qquad \frac{dg(x)}{dx} > 0$$

is

$$\mu(x) = \frac{1}{\pi}\frac{d}{dx}\int_a^x \frac{f(c)}{\sqrt{g(x)-g(c)}}\frac{dg(c)}{dc}\mathrm{d}c$$

according to Polyanin and Manzhirov (1998). Therefore, for our case, $f(x) = \eta_b(x)/x$, $g(x) = x^2$, $a = 0$, and

$$\mu(x) = \frac{2}{\pi}\frac{d}{dx}\int_0^x \frac{\eta_b(c)}{\sqrt{x^2-c^2}}\mathrm{d}c.$$

We substitute $c = x\sin\theta$ and get

$$\mu(x) = \frac{2}{\pi}\frac{d}{dx}\int_0^{\frac{1}{2}\pi} \eta_b(x\sin\theta)\,\mathrm{d}\theta.$$

Integration of eq. (11.13) gives

$$\frac{2}{\pi}\int_0^{\frac{1}{2}\pi} \eta_b(c\cdot\sin\theta)\,\mathrm{d}\theta = \int_0^t V\mathrm{d}t. \qquad (11.18)$$

By substituting eq. (11.14) we can confirm that eq. (11.18) is consistent with eq. (11.16). The previous case with $\eta_b(x)$ as a second-order polynomial enabled us to express $c(t)$ as an explicit function of time by means of eq. (11.17). In general, this cannot be done by means of eq. (11.18). However, if we assume a linearly changing impact velocity, $V(t) = V_0 + V_1 t$, and consider $t(c)$, eq. (11.18) gives a quadratic algebraic equation in time that enables us to find the time as a function of the half-wetted length c.

Wagner's method does not work for water exit (i.e., a diminishing wetted surface), which is a consequence of the free-surface condition $\varphi = 0$, which is not valid in this case because, for instance, the liquid accelerations are no longer dominant relative to gravitational acceleration. In principle one should use the exact free-surface conditions given in Chapter 2. This procedure requires a numerical method, which is by no means trivial to apply. However, a different simplified method is to use the von Karman (1929) method, which assumes geometrical intersection between the undisturbed free surface and the body surface. Von Karman's method provides a solution during water exit, but the correctness of the solution is not clear. Our experience shows that the applicability of von Karman's method depends on the duration T_d of the sum of the water entry and exit phases relative to a characteristic time scale. If we consider wave impact in a tank, this characteristic time scale is the sloshing period T_e, which in practice is the highest natural period of sloshing (i.e., $T_e = T_1$). The ratio T_d/T_e should be small. We can exemplify what small means by referring to two examples. Ge (2002) showed good agreement between a von Karman model and experimental results of vertical forces during wetdeck slamming on a catamaran at forward speed. The wetdeck is the lowest part of the cross-structure connecting two adjacent side hulls of a multihull vessel. In Ge's case, T_e is the wave encounter period. The ratio T_d/T_e was less than 0.2. Baarholm (2001) also examined wetdeck slamming loads. However, the time duration of the water exit phase was not satisfactorily predicted by a von Karman model. The ratio T_d/T_e was about 0.65 in this case.

11.3.1.2 Spray root solution

Armand and Cointe (1986), Howison *et al.* (1991), and Zhao and Faltinsen (1993) showed how to match Wagner's (1932) inner and outer domain solutions. A composite expression for the pressure that is valid in both domains can then be constructed. Cointe (1991) also studied the details of the solution in the jet domains defined in Figure 11.4. We focus on the inner domain (spray root) illustrated in Figure 11.7. First of all, we should recall that Wagner's theory is an asymptotic theory for the deadrise angle $\beta \to 0$. It implies that the outer part of the jet can be approximated as being parallel with the free surface below. The spray root area moves with velocity $u_c = \dot{c}$. A jet thickness δ is defined. This jet thickness is assumed constant in the total jet part of Wagner's inner domain solution. The velocity $u_a = 2\dot{c}$ in the jet does not vary across the jet and can be derived as follows. We introduce a reference frame OXY that moves with the velocity $u_c = \dot{c}$. The incident velocity at infinity outside the jet area is consequently $-u_c$ along the X-axis in the relative reference frame. The problem is steady in this reference frame. Because we neglect gravity and assume potential flow, the Bernoulli equation can be expressed as $p - p_0 = -\frac{1}{2}\rho_l(u^2+v^2) + \frac{1}{2}\rho_l u_c^2$. We can confirm this fact by noting that the pressure is equal to p_0 at infinity outside the jet

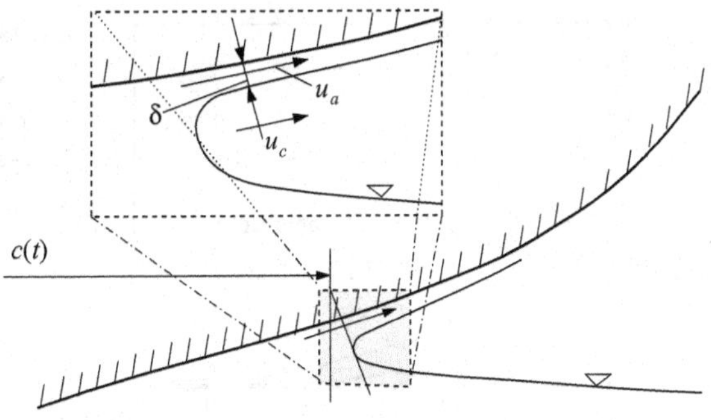

Figure 11.7. Details of the inner domain (spray root) shown in Figure 11.4.

area. Because the dynamic free-surface condition states that $p = p_0$ on the free surface, it follows that $u^2 + v^2 = u_c^2$ on the free surface. Since the jet is thin, no pressure gradient is present across the jet and, hence, v is zero in the jet in the relative reference frame. Furthermore, because the jet boundary is parallel to the body boundary outside the spray root, u is constant and equal to $u_c = \dot{c}$ across the jet in the relative reference frame. Since the jet flow is in the positive X-direction, it follows that the velocity u_a is equal to $2\dot{c}$ in the jet in the Earth-fixed coordinate system Oxy.

We now show how to perform the matching and start by presenting the inner domain solution derived by Wagner (1932). He follows the steady reference frame OXY that moves with velocity $u_c = \dot{c}$. The solution is found via a Schwarz–Christoffel transformation. The details of the derivation are not needed in the matching and are omitted in this presentation. The velocity potential on the body surface can be expressed as

$$\varphi = -U\delta\pi^{-1}(1 + \ln|\tau| - |\tau|). \quad (11.19)$$

The parameter τ determines the coordinate along the body surface by the relationship

$$X \equiv x - c = \delta\pi^{-1}\left(-\ln|\tau| - 4\sqrt{|\tau|} - |\tau| + 5\right). \quad (11.20)$$

If we let $|\tau| \to \infty$, eqs. (11.19) and (11.20) give

$$X \sim -\delta\pi^{-1}|\tau|, \quad \varphi \sim U\delta\pi^{-1}|\tau|. \quad (11.21)$$

Equation (11.21) shows that $\varphi \sim -UX$ when $X \to -\infty$, which means that a flow with velocity U exists along the negative X-axis. This flow confirms that we are following a coordinate system that moves with the velocity $U = dc/dt$. In the Earth-fixed coordinate system the velocity potential can be expressed as $\phi_1 = \varphi + UX = -U\delta\pi^{-1}(1 + \ln|\tau| - |\tau|) + UX$.

We let $|\tau| \to \infty$ and keep one more term in the expansion of X and the velocity potential. We can write $X \sim \delta\pi^{-1}(-4\sqrt{|\tau|} - |\tau|)$. The velocity potential can be approximated as

$$\begin{aligned}\phi_1 &\sim U\delta\pi^{-1}|\tau| + UX \\ &= U\delta\pi^{-1}|\tau| + U\delta\pi^{-1}\left(-4\sqrt{|\tau|} - |\tau|\right) \\ &= -4U\delta\pi^{-1}\sqrt{|\tau|} \sim -4U\delta\pi^{-1}\sqrt{|X|\pi\delta^{-1}},\end{aligned} \quad (11.22)$$

which is called the outer expansion of the inner domain solution. This approximation should match with the inner expansion of the outer domain solution. The outer domain solution of the velocity potential on the body surface is given in the Oxy-system by eq. (11.9); that is, $\varphi = -V\sqrt{c^2 - x^2} = -V\sqrt{(c+x)(c-x)}$. The inner expansion near $x = c$ is

$$\varphi \sim -V\sqrt{2c(c-x)}, \quad x < c. \quad (11.23)$$

Noting that $U = \dot{c}$ and $|X| = c - x$ and requiring eqs. (11.22) and (11.23) to be equal gives that $4\sqrt{\delta/\pi} \cdot \dot{c} = V\sqrt{2c}$. Thereby the thickness of the jet is

$$\delta = \tfrac{1}{2}\pi c(V/\dot{c})^2. \quad (11.24)$$

Strictly speaking, we must show that the matching works elsewhere than on the body surface, but we omit those details here. Wagner's inner-domain hydrodynamic pressure on the body surface is given as

$$p - p_0 = 2\rho_l\dot{c}^2\frac{\sqrt{|\tau|}}{(1+\sqrt{|\tau|})^2}, \quad (11.25)$$

where the coordinate along the body surface can be expressed by eq. (11.20). The maximum value of eq. (11.25) is $p_{\max} - p_0 = \frac{1}{2}\rho_l\dot{c}^2$, which occurs when $|\tau| = 1$ or $x = c$. It is possible to construct a composite pressure distribution that is valid in both the outer and the inner domains. Constant V

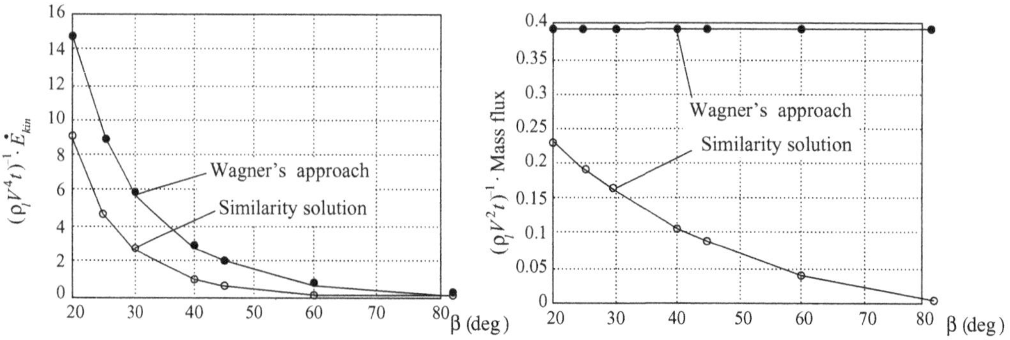

Figure 11.8. Kinetic energy and mass flux for a wedge with deadrise angle β and constant entry velocity V. The similarity solution results are based on calculations by Zhao and Faltinsen (1993) using the theory by Dobrovol'skaia (1969).

is implicitly assumed. We start by taking an outer expansion of eq. (11.25) when $X \to -\infty$. By using eqs. (11.21) and (11.24), we can write

$$p - p_0 \sim 2\rho_l \dot{c}^2 \sqrt{|\tau|^{-1}} = 2\rho_l \dot{c}^2 \sqrt{\delta |X|^{-1} \pi^{-1}} = \rho_l \dot{c} V \frac{c}{\sqrt{2c(c-x)}}. \tag{11.26}$$

This solution is exactly the same as we obtain by taking the inner expansion of the outer solution given by eq. (11.10). The composite solution for the hydrodynamic pressure on the body surface is then $p - p_0 = p_{\text{outer}} + p_{\text{inner}} - p_{\text{common}}$, where p_{outer}, p_{inner}, and p_{common} are given by eqs. (11.10), (11.25), and (11.26), respectively.

The mass flux M_{flux} and kinetic energy flux $\dot{E}_{\text{kin}}$ in the jet are

$$\begin{aligned} M_{\text{flux}} &= \rho_l \delta (u_a - u_c) = \rho_l \delta \dot{c}, \\ \dot{E}_{\text{kin}} &= \tfrac{1}{2} \rho_l u_a^2 \delta (u_a - u_c) = \tfrac{1}{2} \rho_l u_a^2 \delta \dot{c}. \end{aligned} \tag{11.27}$$

Let us assume as an example that the impact velocity V of the free surface is time-independent and the radius of curvature, R, is infinite. It follows then from eqs. (11.16) and (11.24) that $c = \frac{1}{2}\pi Vt/\tan(\beta)$, $\delta = \frac{1}{4}\tan(\beta)Vt$. Equation (11.27) gives the following mass flux and kinetic energy flux: $M_{\text{flux}} = \frac{1}{8}\rho_l \pi V^2 t$, $\dot{E}_{\text{kin}} = \frac{1}{16}\rho_l V^4 \pi^3 \tan^{-2}(\beta) t$. The similarity solution by Dobrovol'skaia (1969) gives more accurate results for nonsmall β-values. Figure 11.8 shows the difference in kinetic energy and mass flux for Wagner's approach and the similarity solution by Dobrovol'skaia (1969). The numerical results are from Rognebakke and Faltinsen (2000) and are based on analyzing the numerical results by Zhao and Faltinsen (1993). The similarity solution assumes a constant entry velocity V of a wedge. By making the coordinates nondimensional with respect to Vt, and by making the flow velocity divided by V a function of these nondimensional coordinates, it is possible to show that the problem involving nondimensional variables does not depend explicitly on time. This is what is implied by a similarity solution. When $\beta \to 0$, the numerical results of the similarity solution agree with Wagner's results.

11.3.2 Damping of sloshing due to tank roof impact

The jet flow caused by a liquid impact on the tank roof results in damping of sloshing. The hypothesis is that when the liquid in the jet flow later impacts the underlying liquid, the kinetic and potential energy in the jet flow are dissipated. Figure 11.9 shows the evolution of an impact in the upper-left corner of an LNG ship tank. The formation and overturning of the jet is evident. This energy loss is related to the total energy in the system, E. When studying one oscillation period, the previous loss of kinetic and potential energy is subtracted from E. Only the liquid entry phase where the wetted tank roof surface increases is considered in this procedure.

Faltinsen and Rognebakke (1999) combined this idea with the Moiseev-type multimodal method described in Section 8.3.1. The linear damping term $2\xi\sigma_i\dot{\beta}_i$, as described in Section 6.3.2, is included in each of the equations (i.e., the damping ratio ξ is assumed the same for each mode). The damping is found as an equivalent damping so that the energy ΔE removed from the system during one full cycle is equal to the kinetic and potential energy lost in the impact; that is, $\xi = \Delta E/(4\pi E)$. The total energy of the system,

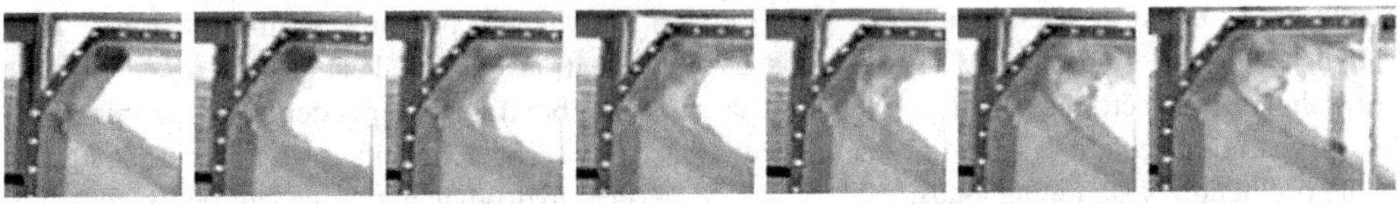

Figure 11.9. Photographs of the upper-left corner of an LNG tank during impact.

E, is found from $\dot{E} = F_2(t)\dot{\eta}_2$ for forced sway motion η_2 (see Section 2.2.4.2), where $F_2(t)$ is the horizontal hydrodynamic force acting on the tank.

An iterative procedure is followed. A simulation over one period is started with no damping. A first estimate of ξ is found. The simulation is repeated, resulting in a new ΔE and thereafter a new ξ. This process is done for iteration $i > 1$ as $\frac{1}{2}(\Delta E_i + \Delta E_{i-1})/E = 4\pi\xi$. Typically, five iterations are sufficient for convergence with a relative error of 0.01–0.02. The tank is assumed rigid, so possible hydroelastic effects are ignored. The uniform impact velocity V is set equal to the value of V at the tank wall adjacent to the impact area. Because we are studying the response on a small time scale relative to the sloshing period, the impact velocity is approximated by a linear function, $V(t) = V_0 + V_1 t$ ($t = 0$ is the time of impact). The impacting surface is approximated by a parabola with radius of curvature R. The wetted length $c(t)$ follows from Wagner's integral equation (see eq. (11.17)). This solution can be corrected by accounting for the tank walls and bottom. Details can be found in Faltinsen and Rognebakke (1999).

The kinetic energy flux through the jet in Wagner's theory is based on eq. (11.27). The potential energy flux is calculated as the mass flux M_{flux} through the jet times the product of acceleration of gravity and the tank roof height relative to the mean free surface. Therefore, the potential energy flux through the jet is expressed as $\dot{E}_{\text{pot}} = g(H_t - h)M_{\text{flux}}$, where H_t and h are the tank height and the mean liquid depth, respectively, and M_{flux} is given by eq. (11.27) in the case of Wagner's theory. The similarity solution results presented by Zhao and Faltinsen (1993), which are valid for large angles, are applied to correct the energy estimates (see Figure 11.8). The energy estimates obtained from Wagner's analysis are multiplied by reduction factors that are obtained using the results in Figure 11.8 even though those results assume a constant "entry" velocity of the horizontal free surface. In Figure 11.10 the free-surface elevation is compared with experimental results for a heavy-impact case. The details of the interaction with the tank roof are not shown; only the tank roof damping is incorporated. The tank is rectangular with $l = 1.73$ m, a filling height $h = 0.5$ m, and a total height of $H_t = 1.02$ m. The tank roof is horizontal. The period and

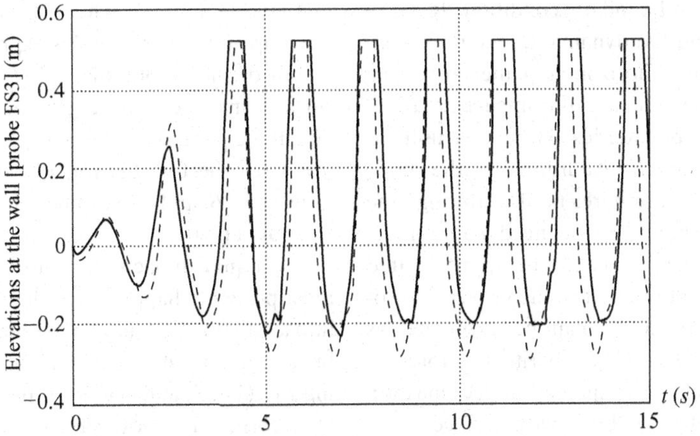

Figure 11.10. Free-surface elevation for a case of heavy roof impact on a horizontal tank roof in a rectangular tank. The details of the interaction with the tank roof are not shown. (Solid line denotes experiments, dashed line denotes theory; wave probe FS3 is shown in Figure 8.9.)

amplitude of the sway excitation are $T = 1.71$ s and $\eta_{2a} = 0.05$ m, respectively. The figure suggests that a satisfactory estimate of the impact velocity can be calculated. This value is important in the prediction of slamming loads.

The effect of tank roof damping on the horizontal hydrodynamic force amplitude F_{2a} on the LNG tank with chamfer angle 45° presented in Figure 4.22(a) is shown in Figure 8.3. Both smaller and larger horizontal excitation amplitudes are considered. The experimental results can be found in Abramson *et al.* (1974), but the theoretical values are obtained by using the multimodal theory with and without tank roof damping. The force calculations neglect the horizontal force caused directly by the impact. The additional force has a magnitude of approximately 10% of the total force for a heavy impact in the LNG tank. The impulse of the contribution from the impact is close to zero. The results show that tank roof damping clearly improves the calculations.

11.3.3 Three-dimensional liquid impact

Three-dimensional flow aspects may be important for slamming. We present a formula for the impact force on a horizontal tank roof. A uniform inflow velocity $V(t)$ is assumed. We define outer, inner, and jet domains as in Figure 11.4 for the two-dimensional case. We concentrate on the outer-domain solution and denote the corresponding time-dependent wetted area of the tank roof as $S_{wr}(t)$. As explained earlier, the impact causes a flow that can be described by a three-dimensional velocity potential Φ that satisfies the body boundary condition $\partial\Phi/\partial z = -V$ on $S_{wr}(t)$ and the dynamic free-surface condition $\Phi = 0$ on a horizontal plane coinciding with the tank roof outside the impact area and within the tank boundaries. Here z is used as a vertical coordinate. Additionally, the body boundary condition requires no flow through the other wet tank boundaries. The dynamic pressure caused by the impact can be approximated as $p - p_0 = -\rho_l \partial\Phi/\partial t$; that is, the vertical hydrodynamic force on $S_{wr}(t)$ can be expressed as $F_3 = -\rho_l \int_{S_{wr}} \partial\Phi/\partial t dS$. We rewrite the force in terms of the infinite-frequency added mass in heave, $A_{33}(t)$, of a flat body with surface $S_{wr}(t)$ and with a similar position inside the tank as the impact area (i.e., $A_{33}(t)$ is influenced by all wet tank boundaries). We start by relating $A_{33}(t)$ to the velocity potential Φ and introduce $\overline{\Phi}$ by the relationship $\Phi = V\overline{\Phi}$. For the time being, we let S_{wr} be time independent and imagine that S_{wr} has vertical velocity $-V$. The corresponding vertical hydrodynamic force can be expressed as $-\rho_l \dot{V} \int_{S_{wr}} \overline{\Phi} dS$. Because no damping exists for the considered problem, we can as a matter of definition (see Section 3.5.2) express the hydrodynamic force as $A_{33}\dot{V}$; that is, $A_{33} = -\rho_l \int_{S_{wr}} \overline{\Phi} dS$. We now let S_{wr} be time dependent and use that $\frac{d}{dt}\int_{S_{wr}} \Phi dS = \int_{S_{wr}} \frac{\partial\Phi}{\partial t} dS + \int_C \Phi U_n ds$ (see eq. (A.2)), where C is the boundary curve of S_{wr} and U_n is the normal velocity of C with outward positive direction. Because $\overline{\Phi} = 0$ on C, it follows that $d(\int_{S_{wr}} \Phi dS)/dt = \int_{S_{wr}} \partial\Phi/\partial t dS$; that is,

$$F_3 = -\rho_l \frac{d}{dt}\int_{S_{wr}} \overline{\Phi} V dS = \frac{d}{dt}[A_{33}(t)V(t)]. \quad (11.28)$$

If the instantaneous contact line between the body and the free surface is elliptic and we neglect the influence of other wet tank boundaries with respect to the impact area, we can express the three-dimensional heave added mass as

$$A_{33}(t) = \tfrac{2}{3}\pi\rho_l a^2 b/E(e) \quad (11.29)$$

(see Scolan & Korobkin, 2001), where $a(t)$ and $b(t)$ are the shortest and longest semi-axes of the ellipse, respectively. Furthermore, $e = \sqrt{1-(a/b)^2}$ is the ellipse eccentricity and E is a complete elliptic integral of the second kind (Abramowitz & Stegun, 1964). In the particular case of a circular disk (i.e., $a = b$), we get that $A_{33} = \frac{4}{3}\rho a^3$. When $b \to \infty$, $A_{33}/b \to \frac{2}{3}\rho_l a^3$. This result can also be obtained by strip theory (i.e., by expressing the two-dimensional added mass in heave as $\rho_l \pi c^2/2$, where $2c$ is the breadth of the strip and integrating this expression along the length of the elliptic disk). If the incident free-surface shape can be approximated as part of an elliptic paraboloid $z = -x^2/A^2 - y^2/B^2 + h(t)$, $\dot{h} = V(t)$, the formula for $a(t)$ and $b(t)$ as functions of A, B, and $h(t)$ can be found in Scolan and Korobkin (2001).

The liquid impacts on a horizontal tank roof most probably happen near the side walls. If the impact does not occur at a corner, we can mirror-image the flow about the tank wall before applying eqs. (11.28) and (11.29). If the impact happens at a corner, the flow should be mirror-imaged about the two adjacent tank walls. Using eqs. (11.28) and (11.29) in combination with a von Karman method is straightforward and provides a

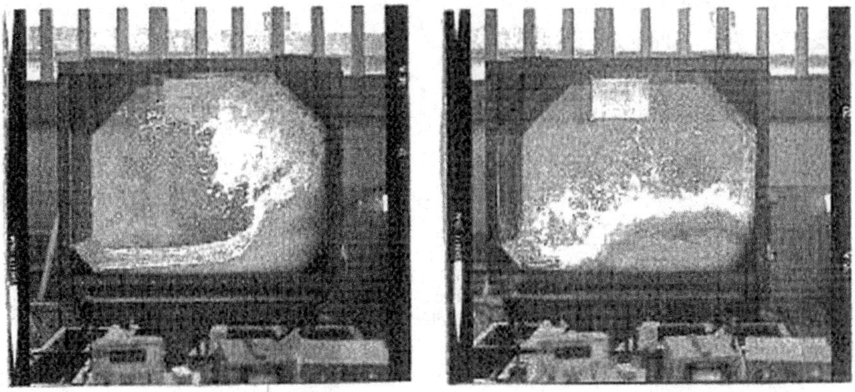

Figure 11.11. Hydraulic jump profile for $h/l = 0.12$, $\eta_{2a} = 0.1l$, $T\sqrt{g/l} = 5.33$ (Abramson *et al.*, 1974).

simple way to assess qualitatively the importance of three-dimensional flow effects during impact.

11.4 Impact of steep waves against a vertical wall

The impact of steep waves against vertical tank walls can occur in shallow-liquid conditions associated, for instance, with hydraulic jumps that are formed when the forcing frequency σ is in the vicinity of the lowest natural frequency, σ_1 and the horizontal forcing amplitude is beyond a threshold value (see Section 8.8.2). Very high impact pressures can occur. The shape of the impacting wave has a significant effect on the impact pressure induced on the wall.

Abramson *et al.* (1974) presented experimental slamming pressures for the prismatic LNG tank shown in Figure 4.22(a) in the case of forced harmonic sway motions and $h = 0.12l$. The experimental results for hydrodynamic lateral force for the tank are presented in Figures 8.2 and 8.34 with results from the multimodal method. Our focus is on forced sway amplitude $\eta_{2a} = 0.1l$, which corresponds to realistic extreme sway motion of a ship. When the excitation period is in the range $1.65\text{ s} < T < 2.95\text{ s}$ (i.e., $2.1\text{ rad/s} < \sigma < 3.8\text{ rad/s}$), the experiments show a hydraulic jump traveling back and forth in the tank (see Figure 11.11). Equation (8.123) states that hydraulic jumps should occur when $1.5\text{ s} < T < 3.7\text{ s}$, which is in reasonable agreement with the experiments. Excitation periods in the range $2.95\text{ s} < T < 3.25\text{ s}$ ($1.93\text{ rad/s} < \sigma < 2.1\text{ rad/s}$), produced a single traveling wave in the model tests.

The impact pressures occurring when the wave front hits the vertical bulkhead are of particular interest. Experiments show that, even under harmonic oscillations, the pressure variation is neither harmonic nor periodic because the magnitude of the pressure peaks vary from cycle to cycle. A typical histogram for the distribution of peaks is shown in Figure 11.12. The most frequently occurring pressure peaks reach 0.4 times the pressure level exceeded by 10% of all peaks. The 1% exceedence limit is two to three times the 10% exceedence level. However, the prediction of extreme values is not reliable.

Experimental values for the 10% exceedence level of impact pressures as a function of excitation period are shown in Figure 11.13. The pressures are made nondimensional by $\rho_l g l$. Four liquids (freshwater, 63% glycerol/water, 85% glycerol/water, and reginol oil) were used in the experiments. All the liquids seem to give approximately the same nondimensional maximum value for the 10% exceedence level. If we define a pressure coefficient as $C_p = p/(\frac{1}{2}\rho_l V^2)$, approximate the impact velocity as $V = \sqrt{gh}$, and set the maximum pressure based on Figure 11.13 as $p/(\rho_l g l) = 2$, we obtain $C_p = 33.3$. For a full-scale ship with tank breadth $l = 40$ m and $\rho_l = 10^3\text{ kg m}^{-3}$, this maximum value would correspond to approximately $p = 8$ bar. If we use the density values in Table 11.2 for typical cargo liquids, the maximum slamming pressure based on Figure 11.13 is in the range from 42% to 68% of 8 bar. Two experimental resonance periods appear in Figure 11.13. The lowest resonance period is lower than the highest sloshing period, $T_1 = 2.216$ s, based on linear theory. This value is as expected according to the nonlinear potential flow theory described in Chapter 8. We note that the lower resonance period in Figure 11.13 is influenced by viscosity. Reginol oil, which

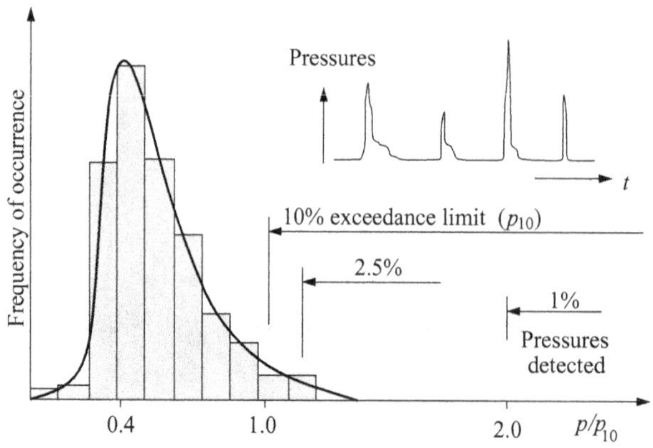

Figure 11.12. Probability density function of impact pressure peaks (Abramson *et al.*, 1974).

has the highest kinematic viscosity coefficient of the considered liquids, has the highest value of the lowest resonance period. Because the highest resonant period may correspond to the secondary resonance by the lowest natural mode, we have plotted the periods corresponding to secondary resonance in the figure. However, we cannot conclude that secondary resonance is the cause.

A comprehensive description of the research concerned with wave impacts on a wall is given in the review by Peregrine (2003). Hattori *et al.* (1994) experimentally studied the impact pressures on vertical walls due to breaking waves and identified the following four scenarios:

1. "Flip-through" condition with no air bubbles.
2. Collision of a vertically flat wave front with entrapment of small air bubbles. The impact pressure had a single peak.
3. Collision of plunging breaker with a thin air pocket.
4. Collision of a fully developed plunging breaker with a thick air pocket.

The flip-through condition is characterized by the incident wave front being inclined from the wall and that the wave front does not impact with the wall; a jet flow is instead generated at the wall. Air cavities may occur later. Figure 11.2(a) illustrates that flip-through can also occur for high filling ratios. Numerical computations by Cooker and Peregrine (1990, 1991) and experimental studies by Chan and Melville (1988) showed that

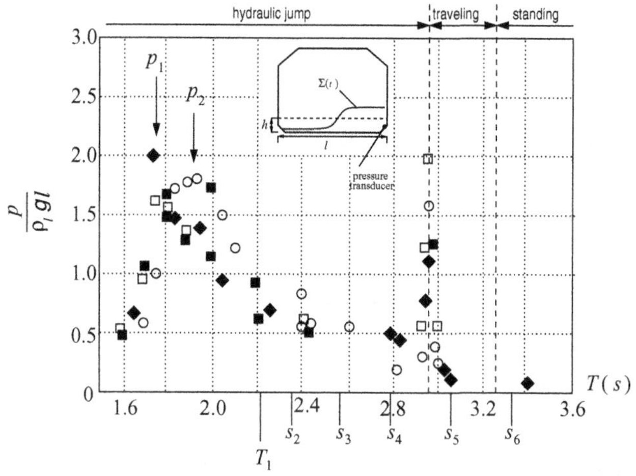

Figure 11.13. Impact pressure (10% exceedence limit) with shallow depth and large excitation amplitude (Abramson *et al.*, 1974). Liquid properties are defined in Table 11.1. The prismatic tank model is illustrated in Figure 4.22(a). The experimental data of Abramson *et al.* (1974) show the influence of viscosity: ■, freshwater; ○, reginol oil; □, glycerol/water 63%; ♦, glycerol/water 85%; $h/l = 0.12$ and $\eta_{2a}/l = 0.1$. The horizontal hydrodynamic force for the same input parameters are presented in Figure 8.2(b). The linear resonance is expected at $T_1 = 2.216$ s, and the s_k-values represent the points when we expect the secondary resonance by the kth mode.

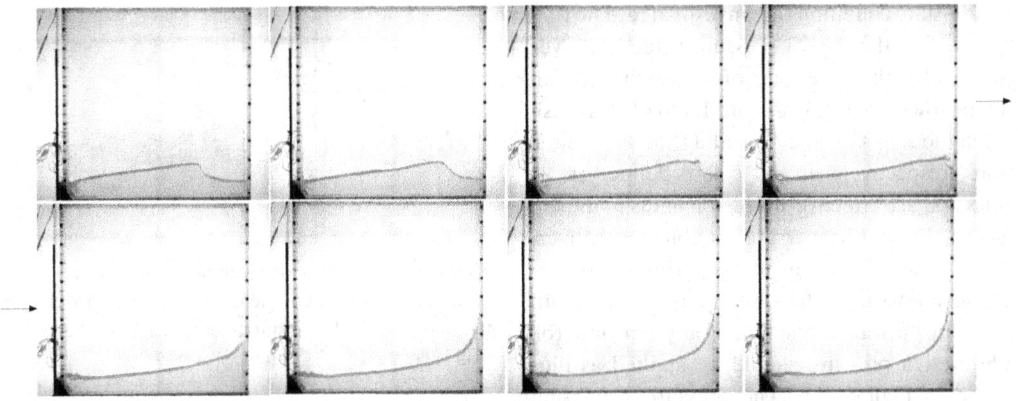

Figure 11.14. Sequence of images illustrating the global evolution of the free surface during a flip-through, from left to right and from top to bottom; $\Delta t = 1/23$ s (Lugni *et al.*, 2006b).

the largest pressures occur for the flip-through condition. Systematic studies by Hull and Mueller (2002) of breaker shapes and wave-impact pressures on a wall suggested that the maximum pressures occur at the still water level and are induced by a plunging breaker engulfing a large air pocket (i.e., the flip-through condition does not cause the largest pressures). Damped impact-pressure oscillations occur in conditions 3 and 4 with a similar behavior for tank roof impact with a gas cavity as described in Section 11.6. Bagnold (1939) proposed a gas pocket model when the wave front is inclined toward the wall. It is common to refer to Bagnold-type impact as a tribute to his pioneering work. Topliss *et al.* (1992) derived the theoretical formula

$$f_{ap} = \frac{1}{2\pi}\sqrt{\frac{2\kappa p_a\left(1+0.5\lambda_d^2 r^2\right)}{\rho_l r^2\left[\log(0.5\lambda_d r\tan(\lambda_d d_b)) + 0.25\lambda_d^2 r^2\right]}}$$

for the natural frequency f_{ap} in Hertz of a single two-dimensional gas pocket with semicircular cross-section of radius r, where p_a is the atmospheric pressure or the ullage pressure in our applications; ρ_l the liquid density; κ is the ratio of the specific heat; d_b and d_a are the distances of the gas-pocket center below the free surface and above the bottom, respectively; and $\lambda_d = \frac{1}{2}\pi/(d_a + d_b)$.

When the flip-through phenomenon occurs, the concave face of the wave approaches the wall with the crest moving forward and the trough rapidly rising at the wall (Oumeraci *et al.*, 1993). As highlighted by Cooker and Peregrine (1992), the presence of the wall delays wave breaking and causes the rise of the leading wave trough. The latter focuses with the wave front, giving intense acceleration to the flow and turning it in the focusing area to form a vertical jet.

Lugni *et al.* (2006b) experimentally studied the occurrence of flip-through during sloshing in a two-dimensional tank with mean ratio of water depth to tank breadth of 0.125. The tank breadth is $l = 1$ m. Photographs from their tests are presented in Figure 11.14 and show the global behavior of the wave propagating back and forth during a run with flip-through. We cannot see the details of the flip-through in the figure. Three different flip-through modes were investigated. No air entrainment characterizes a "mode a" flip-through; engulfment of a single, well-formed air bubble is typical of a "mode b" event; and the generation of fine-scale air–water mixing occurs for a "mode c" event. Upward accelerations of the flip-through jet exceeding 1,500 g were measured. These details are challenging to be predicted numerically. However, such fine details may not matter in finding the maximum structural stresses, which, conversely, necessitate a hydroelastic analysis.

Detailed experimental pressure results were presented by Lugni *et al.* (2006b). Eight pressure probes were placed along the wall at heights from 0.05 to 0.21 m above the tank bottom. The experiments were repeated seven times. Let us try to relate the experimental pressures for a "mode a" flip-through event to the design pressures provided by the DNV Rules for Classification of Ships (2007) for ships with length 100 m or more. The forcing period and sway amplitude are 1.6 s and 0.03 m, respectively. Global nearly steady-state conditions are investigated. The rules specify that the impact pressure values on a vertical wall in the lower part of a smooth tank cannot be lower than $1.5\rho_l g l_s$ and $1.42\rho_l g b_s$ for sloshing in the ship's longitudinal and transverse directions, respectively; l_s and b_s are defined in the rules and are effective sloshing lengths in the longitudinal and transverse direction, respectively. The sloshing length in the experiments is 1 m. These minimum values are in the rules multiplied by factors accounting for pitch and roll. The design pressure should be applied as illustrated in Figure 11.1. The pressure is constant along the wall from the bottom up to a height of either $0.2l_s$ or $0.2b_s$. A pressure of 1.5 $\rho_l g l_s$ corresponds in our case to $12\rho_l gh$ – lower than the maximum value of $30\rho_l gh$ presented by Lugni *et al.* (2006b) for the wave probe positioned at a height of $0.19l_s$ above the tank bottom. Because the ratio of forcing amplitude to tank length was $\varepsilon = 0.03$ and a more realistic design value could be $\varepsilon = 0.1$, higher pressures are likely to occur in reality for a ship in a seaway. The measured high impact pressures are highly localized in space relative to the rules. Furthermore, the rules specify a steady value while the measured high impact pressures last a short time, on the order of 0.01 s. If we Froude-scale the duration to a full-scale tank of length 36 m, the duration is six times the duration in model scale. The only way we can relate the slamming experiments to the rules is by focusing on resulting maximum stresses in the structure, which is done by a static structural analysis by following the rules. However, the rapid time variation of the impact suggests a hydroelastic (dynamic) analysis. To assess if a hydroelastic analysis is needed, we need to know the wet natural periods of the structural modes that contribute to maximum structural stresses (see similar discussion in Section 11.9.6). Hydroelasticity does not matter if the duration of large pressures is high relative to important structural natural periods. However, generally speaking, the obtained experimental results suggest that a hydroelastic analysis is needed, which poses another complication. Hydroelasticity means that the structural vibrations affect the liquid flow and that the experiments should be performed with a properly scaled structural model instead of using a rigid tank. However, an engineering approach would be to apply the experimental pressure records as excitation in a three-dimensional FEM structural model. The effect of the structural vibrations on the liquid flow in terms of generalized added mass coefficients should be part of the numerical analysis (see a similar procedure in Section 11.9.2).

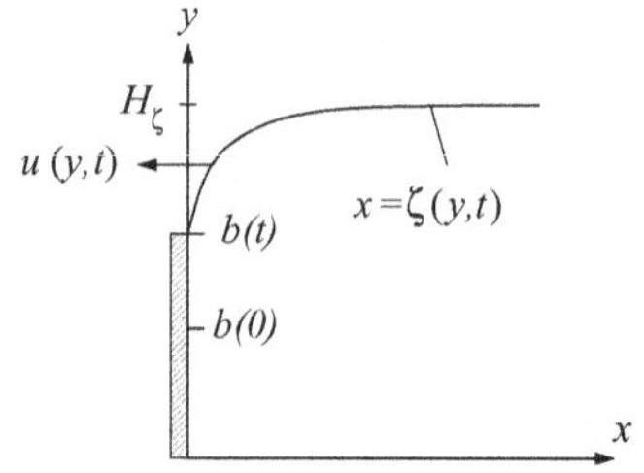

Figure 11.15. Impact of steep wave against a vertical tank wall. The height of the wetted wall is $b(0)$ at the initial impact time. The instantaneous height of the wetted wall is $b(t)$.

We should also be concerned about the lack of repeatability in the pressure measurements. However, if we look upon the structural response in terms of maximum stresses, the maximum pressures may be a bad indicator for the studied cases (see Section 11.9). As previously mentioned it may be more relevant to study the pressure impulse. The pressure impulse has much better repeatability than the pressure. Lugni *et al.* (2006b) were able to show good correlation between the experiments and the analytical pressure impulse model by Cooker and Peregrine (1995).

11.4.1 Wagner-type model

Korobkin (2008) generalized the Wagner theory to impact against a vertical wall. The incident steep wave against a vertical tank wall is inclined from the wall. A sketch of the scenario is shown in Figure 11.15. The outer-domain solution, with no details of the jet flow at the intersection between the free surface and the wall, was considered. Part of the tank wall with height $b(t)$ is wetted near the tank bottom. The initial wetted height of the tank wall is $b(0)$. The horizontal distance between the approaching steep wave and the vertical tank wall is denoted $\zeta(y, t)$. The horizontal inflow velocity of the steep wave is denoted $u(y, t)$. The *liquid depth outside the steep wave* part accounts for the wave elevation and is set equal to a constant, H_ζ.

The total velocity potential is expressed as $\phi(x, y, t) = \phi_0(x, y) + \varphi(x, y, t)$, where φ accounts for the impact. Subscript 0 is used to indicate the

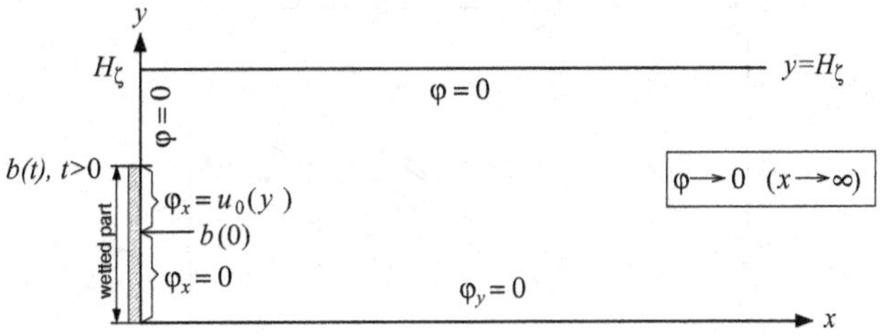

Figure 11.16. Boundary conditions for the velocity potential φ caused by impact on the vertical tank wall at $x=0$; $\varphi_x = \partial\varphi/\partial x$, $\varphi_y = \partial\varphi/\partial y$.

values of $\zeta(y,t)$ and $u(y,t)$ without the influence of the impact. The horizontal distance between the tank wall and the steep wave part is assumed to be a monotonic function of y and, therefore, can be approximated as $\zeta(y,t) = \zeta_0(y) - u_0(y)t + \xi_1(y,t)$, where $\zeta_0(y)$ is the value of $\zeta(y,t)$ at the initial time and $u_0(y) = -\partial\phi_0/\partial x|_{(0,y)}$ is an approximation of the horizontal incident velocity of the steep part of the free surface toward the wall. Furthermore, $\xi_1(y,t)$ is the effect of the impact on the steep free-surface shape. We now transfer the boundary conditions for the velocity potential φ to the boundaries of the domain $x \geq 0$, $0 \leq y \leq H_\zeta$. The governing equation is the two-dimensional Laplace equation and the boundary conditions are illustrated in Figure 11.16. The dynamic free-surface condition associated with φ is expressed as $\varphi = 0$ on $b(t) \leq y \leq H_\zeta$, $x = 0$ and at $y = H_\zeta$. This boundary condition is similar to the one used by Wagner (1932) for water entry and is discussed in Section 11.3.1. The other boundary conditions at $x = 0$ are $\partial\varphi/\partial x = u_0(y)$, $b(0) \leq y \leq b(t)$, and $\partial\varphi/\partial x = 0, 0 \leq y \leq b(0)$. Furthermore, we have $\partial\varphi/\partial y = 0$ on the bottom $y = 0$ and $\varphi \to 0$ $(x \to \infty)$. The instantaneous wetted height $b(t)$ of the tank wall is unknown and has to be found as part of the solution. The kinematic free-surface condition is then needed.

Korobkin (personal communication, 2008) solved the boundary-value problem by means of the displacement potential $\psi(x,y,t) = \int_0^t \varphi(x,y,\tau)d\tau$. Corresponding boundary conditions for the horizontal and vertical displacements $X(x,y,t) = \partial\psi/\partial x$ and $Y(x,y,t) = \partial\psi/\partial y$ were formulated. We leave out the details of the solution technique. Korobkin showed that the wetted tank wall height is determined by the equations

$$\bar{t}_c \equiv \frac{Ut_c}{H_\zeta} = \frac{1}{M(\bar{b})}\int_{\bar{b}(0)}^{\bar{b}} \frac{\bar{\zeta}_0(\bar{y})d\bar{y}}{\sqrt{\cos\pi\bar{y} - \cos\pi\bar{b}}},$$

$$M(\bar{b}) = \int_{\bar{b}(0)}^{\bar{b}} \frac{\bar{u}_0(\bar{y})d\bar{y}}{\sqrt{\cos\pi\bar{y} - \cos\pi\bar{b}}}, \qquad (11.30)$$

where t_c is the time when the wetted tank height is b. Furthermore, U is a characteristic velocity used in nondimensionalizing the incident flow velocity u_0. An overbar indicates a nondimensional value (i.e., $\bar{b} = b/H_\zeta$, $\bar{y} = y/H_\zeta$, $\bar{\zeta}_0 = \zeta_0/H_\zeta$, $\bar{u}_0 = u_0/U$). Figure 11.17 considers two wave impact conditions. The corresponding solutions of b/H_ζ by means of eq. (11.30) are shown in Figure 11.18(a) as a function of Ut_c/H_ζ. Case 2 corresponds to impact of the liquid wedge illustrated in Figure 11.23. If a Wagner type of analysis is performed, it leads to $b = \frac{1}{2}\pi Ut/\tan(\frac{1}{2}\pi - \beta)$, where β is defined in Figure 11.23 and $U = u_0$ corresponds to the constant incident flow velocity. The studied liquid wedge corresponds to $\tan(\frac{1}{2}\pi - \beta) = 1$; that is, $b/H_\zeta = \frac{1}{2}\pi(Ut_c/H_\zeta)$. Case 1 is more relevant than case 2 for our impact problem. Case 2 is only used for comparisons. The values of b/H_ζ for cases 2 and 1 show a clear difference for $Ut_c/H_\zeta \gtrsim 0.4$ (i.e., when $b/H_\zeta \gtrsim 0.6$). Strictly speaking, a β-value of 45° is too low for a Wagner-type analysis to be accurate. However, the results for other β-values wqith a wave front that is a straight line can be obtained by replacing Ut_c/H_ζ on the horizontal axis in Figure 11.18(a) with $Ut_c \tan\beta/H_\zeta$.

The nondimensional pressure distribution on the wetted tank wall is given by

$$\frac{p(0,y,t)}{\rho_l U^2} = \frac{M(\bar{b})\sin\pi\bar{b}}{\sqrt{\cos\pi\bar{y} - \cos\pi\bar{b}}}\frac{d\bar{b}}{d\bar{t}}.$$

The simplest way to obtain $d\bar{b}/d\bar{t}$ is to differentiate numerically $\bar{b}$ with respect to $\bar{t} \equiv Ut/H_\zeta$.

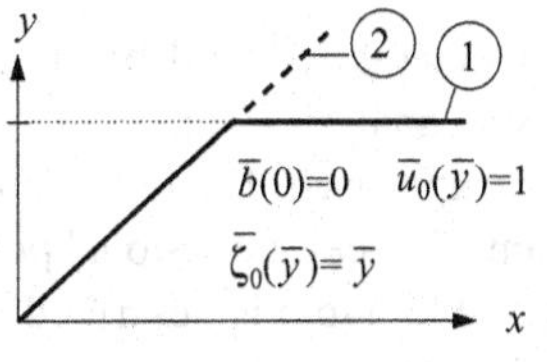

Figure 11.17. Two incident wave conditions (1 and 2) for impact against a vertical wall.

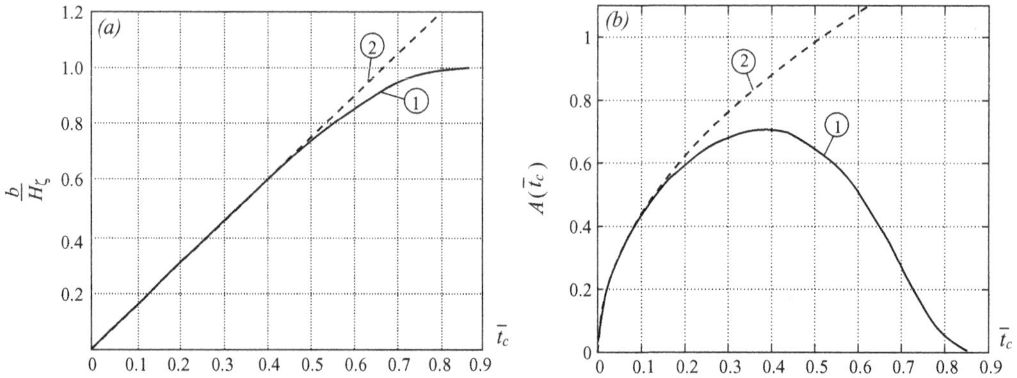

Figure 11.18. (a) Wetted wall height b as a function of nondimensional time $\bar{t}_c = Ut_c/H_\zeta$ and (b) the A-value of the pressure distribution at the spray root defined by eq. (11.31) for the two wave impact conditions defined in Figure 11.17. The results are presented as a function of nondimensional time $\bar{t}_c$.

We note that the pressure tends to zero as $\bar{b} \to 1$. The pressure has a square-root singularity at $y = b$, which is expressed as

$$\frac{p(0, y, t)}{\rho_l U^2} \approx \frac{A(\bar{b})}{\sqrt{\bar{b} - \bar{y}}},$$

$$A(\bar{b}) = \pi^{-\frac{1}{2}} \sin^{\frac{1}{2}}(\pi \bar{b}) M(\bar{b}) \frac{d\bar{b}}{d\bar{t}}. \qquad (11.31)$$

The nondimensional values of $A(\bar{t}_c)$ given by eq. (11.31) for the two different inflow conditions defined in Figure 11.17 are presented in Figure 11.18(b). The analysis for case 2 gives, via Wagner's theory,

$$p = \rho_l \dot{b} U \frac{b}{\sqrt{2b(b-y)}} = \tfrac{1}{4}\rho_l U^2 \pi^{3/2} \sqrt{\frac{Ut_c}{b-y}}.$$

The latter expression implies that $A = \frac{1}{4}\pi^{3/2}\sqrt{Ut_c/H_\zeta}$. The values of A for cases 2 and 1 show a clear difference when $Ut_c/H_\zeta \gtrsim 0.2$.

We concentrate now on the fine details of the flow in the vicinity of $x = 0$, $y = b$. Because the outer-domain solution has a square-root singularity at $y = b$, we can match the outer-domain solution with Wagner's inner-domain (spray-root) solution by following a procedure similar to that described in Section 11.3.1.2. Therefore, for instance, the maximum pressure relative to the ullage pressure is $p = \frac{1}{2}\rho_l \dot{b}^2$.

Korobkin's method should be combined with a hydroelastic analysis. Ten *et al.* (2008) followed such an approach. The inflow condition is a hydraulic jump. Three-dimensional potential flow of a compressible liquid is described analytically and coupled with a three-dimensional structural elastic analysis by means of a commercial FEM code. We come back to the flow description in Section 11.8.3.

11.4.2 Pressure-impulse theory

Cooker and Peregrine (1992, 1995) presented a pressure-impulse theory for shallow-liquid impact of a very steep wave front against a vertical wall. We choose a scenario as described in Section 11.4.1 and a Cartesian coordinate system Oxy as shown in Figure 11.19. The *pressure impulse* is defined as

$$P(x, y) = \int_{t_b}^{t_a} p(x, y, t)\,dt, \qquad (11.32)$$

where t_b and t_a are the times immediately before and after the impact, respectively, and p is the pressure due to the impact. In addition there is an incident flow and a corresponding horizontal impact velocity U_0 of the steep wave front from $y = -\mu H_\zeta$ to $y = 0$, where H_ζ is as in the previous section the *liquid depth relative to the incident wave top*.

The boundary-value problem for P is shown in Figure 11.19 and can be derived by introducing the velocity potential φ associated with the slamming pressure p and by approximating the slamming pressure as $p = -\rho\partial\varphi/\partial t$. Because φ satisfies the Laplace equation, P satisfies the Laplace equation. Furthermore, the free-surface condition $P = 0$ on $y = 0$ is a consequence of $p = 0$. The body conditions $\partial P/\partial x = 0$ on $x = 0$, $-H_\zeta \le y \le -\mu H_\zeta$ and $\partial P/\partial y = 0$ on $x \ge 0$, $y = -H_\zeta$ is a consequence of zero normal liquid velocity at

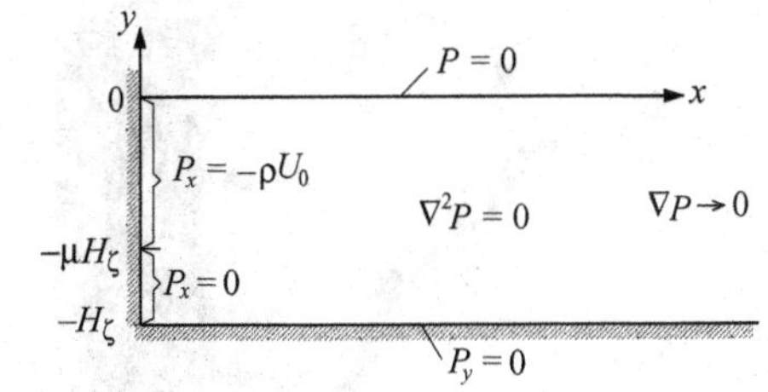

Figure 11.19. Pressure-impulse theory: a two-dimensional boundary-value problem for the pressure impulse $P\,(\partial P/\partial x = P_x)$.

the latter boundaries. The boundary condition on $x = 0, -\mu H_\zeta \le y \le 0$ is $\partial\varphi/\partial x = U_0$ on the instantaneous wetted tank wall. It follows by using eq. (11.32) that $\partial P/\partial x = -\rho U_0$ on $x = 0, -\mu H_\zeta \le y \le 0$. Finally, we must require that the disturbance due to the impact dies out as $x \to \infty$ (i.e., $\nabla P \to 0$ when $x \to \infty$). We can solve the boundary-value problem by using separation of variables. The result is

$$P(x, y) = 2\rho U_0 H_\zeta \sum_{n=0}^{\infty} \frac{[\cos(\mu\lambda_n) - 1]}{\lambda_n^2} \times \sin(\lambda_n y/H_\zeta)\exp(-\lambda_n x/H_\zeta), \quad (11.33)$$

where $\lambda_n = (n + \frac{1}{2})\pi$. Figure 11.20 presents the nondimensional pressure impulse $P/(2\rho U_0 H_\zeta)$ for $\mu = 1$ along the vertical wall $x = 0$ as a function of y/H_ζ. Here $\mu = 1$ means that the impacting wave is over the whole height from the tank bottom to the top of the wave. The pressure impulse does not vary strongly for $-1 \le y/H_\zeta <\sim -0.6$. The other graph in the figure presents $P/(2\rho U_0 H_\zeta)$ as a function of x/H_ζ on the bottom. We note that the influence of the impact on the vertical wall becomes small for $x/H_\zeta >\sim 2$.

A difficulty is to precisely define t_a. If the slamming pressure increases linearly to a maximum value $p_{\max}$ and then decreases linearly to zero at $t = t_a$, we can express the maximum slamming pressure as

$$p_{\max} = 2P/\Delta t, \quad (11.34)$$

where $\Delta t = t_a - t_b$.

Because gravity does not matter in the slamming analysis, the incident flow is not restricted to be horizontal. We will actually apply the pressure impulse theory to vertical jet impact in the next section.

11.5 Tank roof impact at high filling ratios

High filling ratios can lead to high impact pressure, as already mentioned in the introduction. Allers (2004) included high filling ratios in his sloshing tests. The tank was rectangular with breadth B_t and height H_t equal to 0.60 m. The width of the tank was 0.1 m (see Figure 11.21). The case that we examine had a filling height $h = 0.586$ m. The tank was harmonically forced with a sway amplitude $\eta_{2a} = 0.04$ m in the breadth direction and a period of 1.33 s. Figure 11.21 shows photographs from the experiments. The experiments were intended to be two-dimensional. However, we notice small amplitude three-dimensional waves, which we believe are caused by surface tension. Meniscus effects on the wall are generated in a similar way as illustrated in Figure 8.8. The time dependence of the meniscus generates three-dimensional capillary waves. We disregard the latter effect in our further discussion. We see from Figure 11.21 that a steep wave front is approaching

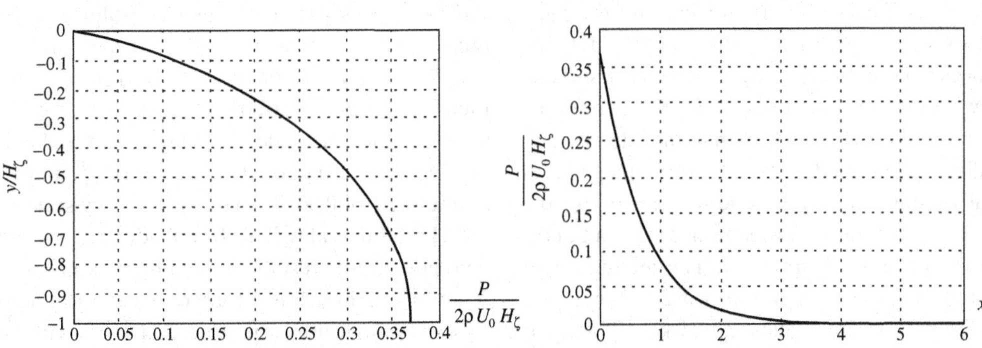

Figure 11.20. Nondimensional pressure impulse $P/(2\rho U_0 H_\zeta)$ following from eq. (11.33) for $\mu = 1$. The left-hand graph is for $x = 0$ and the right-hand graph is for $y = -H_\zeta$.

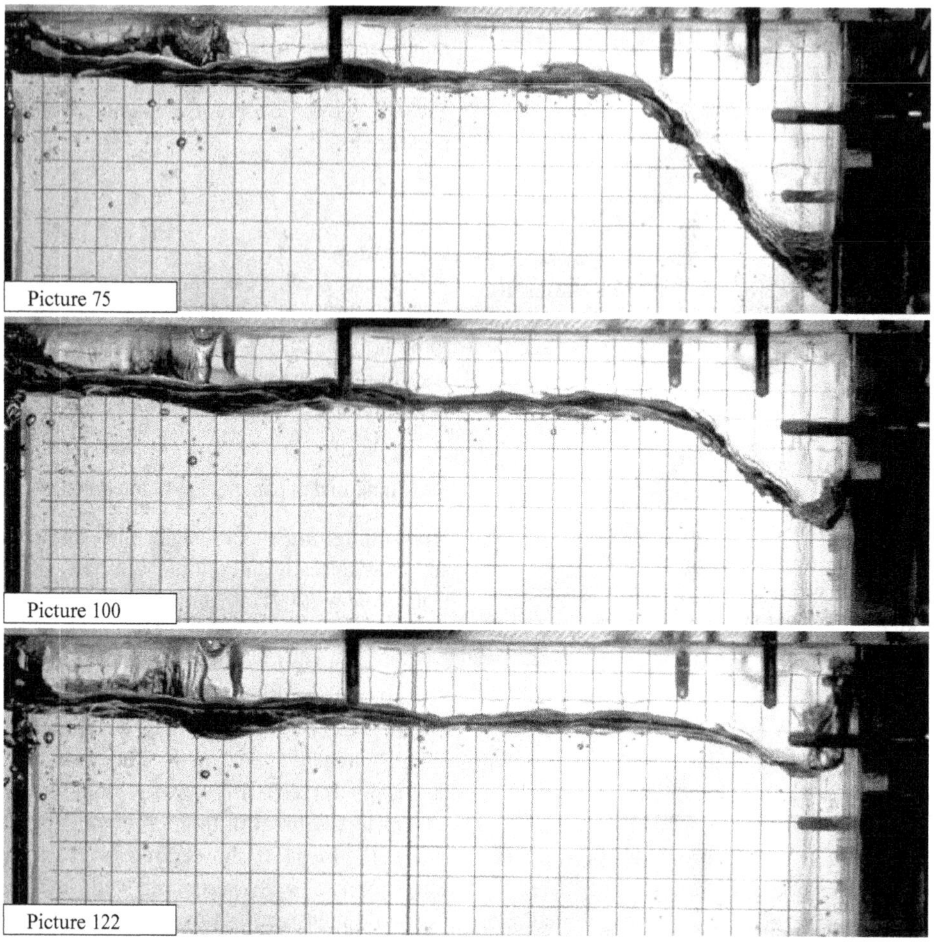

Figure 11.21. Photos of the flow in a tank with filling ratio 0.586/0.600 before tank roof impact occurs. The tank has breadth and height of 0.600 m. The grids shown in the photos are 10 × 10 mm. Pictures 75, 100, and 122 are shown. The time difference between each frame is 1/1250 s (Allers, 2004).

the wall with subsequent flip-through of the free surface at the wall. The resulting jet flow at the wall is first wedge-shaped. The top of the jet becomes nearly horizontal before impact at the tank roof. Theoretical predictions of the time history of the impact pressure require detailed information about the impacting liquid surface. If we rotate the photograph by 90°, it presents a similar impact scenario that may happen in shallow-liquid conditions with very steep waves approaching a tank wall. The time history of the resulting roof impact pressure near the tank corner is shown in Figure 11.22. The measurement area of the pressure transducer is 2 × 2 mm. The maximum pressure is close to 25 kPa. If we, for instance, consider a tank length of 45 m in full scale and use Froude scaling, with a ratio between the density of LNG and water equal to 0.474, it corresponds to a maximum pressure of 889 kPa or close to 9 bar. However, as we argue several times in this chapter, it is important to consider slamming pressures in terms of resulting maximum structural stresses. Both the time duration and the spatial extent of the high pressure are then important. If, based on Figure 11.22, we use 0.001 s as the time scale of the high pressure loading, Froude scaling to a tank length of 45 m gives a time scale of 0.0087 s (i.e., a frequency scale of 115 Hz in full scale). This frequency scale should be related to the results for membrane tanks presented in Section 11.9.6 for wet natural frequencies associated with large structural stresses in the structure. This comparison shows that the time scale of the loading may matter for structural

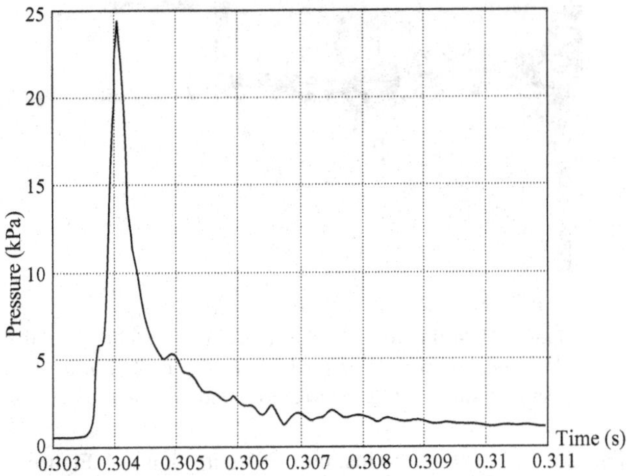

Figure 11.22. Slamming pressure measurements in the tank roof next to the corner due to the jet flow impact illustrated in Figure 11.21. Time 0.304 s is estimated to correspond to picture 130 (Allers, 2004).

stresses. Another factor is the spatial extent of the slamming pressure. Let us say that the thickness of the jet is about 4 mm and use that as a rough estimate of the spatial extent of the slamming pressure. Froude scaling to a tank length of 45 m gives a spatial extent of the slamming pressure equal to 0.3 m. This length scale must be considered relative to the structural dimensions. For instance, Section 11.9.6 refers to slamming studies with a panel that has lateral dimensions 3,300 × 840 mm. Obviously, a more thorough study is needed to evaluate the structural importance of the measured pressure. Other forcing histories of the tank must also be considered.

The jet velocity at impact is estimated to be $V = 2.3 \text{ m s}^{-1}$; that is, the slamming pressure coefficient $p_{\max}/(\rho_l V^2)$ associated with a maximum pressure $p_{\max}$ of 25 kPa and liquid density ρ_l of 10^3 kg m^{-3} is 4.7. If we had used theoretical values for impact of liquid wedges with small internal angles presented in Figure 11.23, we could see that the slamming coefficient would be far smaller.

Another possibility is to use the pressure-impulse theory described in Section 11.4.2. Because we do not know the details of the impact-ing top, we assume a horizontal top (i.e., $\mu = 1$ in eq. (11.33)). We should ideally have integrated the pressure impulse across the measurement area of the pressure transducer but select to evaluate the pressure impulse P at $x = 0$, $y = -H_\zeta$. Using $H_\zeta = 0.004$ m, $U_0 = 2.3 \text{ m s}^{-1}$, $\rho = 10^3 \text{ kg m}^{-3}$ gives $P = 0.00683$ kPa s. We now need to estimate when the slamming has ended to use eq. (11.34) to estimate $p_{\max}$. We note from Figure 11.22 that the pressure decays rapidly initially after reaching its maximum value. However, following the rapid initial decay is a phase with a relatively small decay of the pressure. This phase is associated with a jet flow in a corner as illustrated in

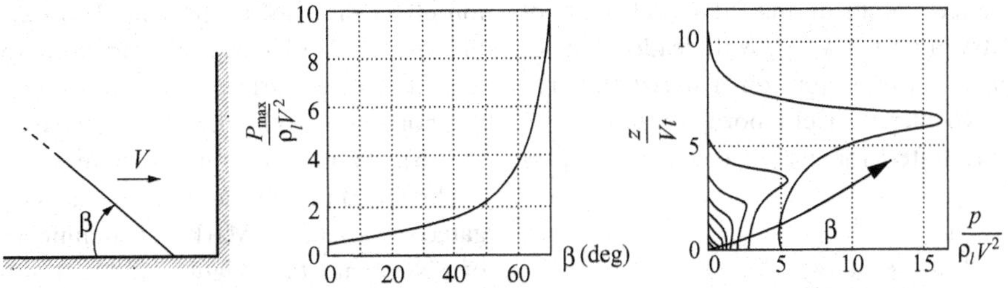

Figure 11.23. Sketch of the equivalent problem of a liquid (half) wedge impacting a flat wall at 90° (left), maximum pressure on a wall due to the liquid impact (center), and pressure distribution along the vertical wall for $5^\circ \leq \beta \leq 75^\circ$ with increment $\Delta\beta = 10^\circ$ (right). The results are numerically obtained by neglecting gravity and using the similarity solution by Zhang *et al.* (1996; see also Greco, 2001).

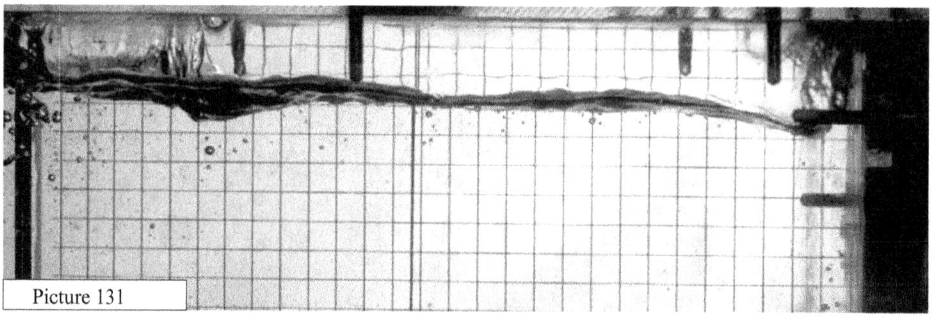

Figure 11.24. Photo of the flow in a tank with filling ratio 0.586/0.600 after tank roof impact occurs. The tank has breadth and height of 0.600 m. The grids shown in the photos are 10 × 10 mm. The picture frame is estimated to correspond to time 0.305 s in Figure 11.21 (Allers, 2004).

Figure 11.24 at the time instant 0.305 s in Figure 11.22 and is not directly related to the impact. Therefore, the end of the impact should be selected before this time instant. Of course, an ambiguity exists for how to do that. If we choose $\Delta t = 0.0005$ s in eq. (11.34), it gives $p_{max} =$ 27.3 kPa. The fact that we should have integrated the theoretical pressure across the measurement area of the pressure transducer means that the estimated pressure should be lower. Figure 11.20 gives an indication of this effect by examining the variation of $P/(2\rho U_0 H_\zeta)$ for $-1 \leq y/H_\zeta \leq -\frac{1}{2}$ on $x = 0$. The results show that the pressure-impulse method describes the correct physics. The fact that high slamming pressures occur for high fillings may hardly be noticed on the horizontal force (see results in Figure 5.10 for the moment).

11.6 Slamming with gas pocket

In Section 11.4 we discussed the formation of a gas pocket when a plunging breaker hits a vertical wall. Other cases are now presented. Between the interface of liquid and body surface, air (gas) cushion effects occur during the initial phase of water entry of a rigid wedge with deadrise angles less than 2°–3° impacting on a horizontal free surface (Koehler & Kettleborugh, 1977). This case is illustrated for a zero deadrise angle in

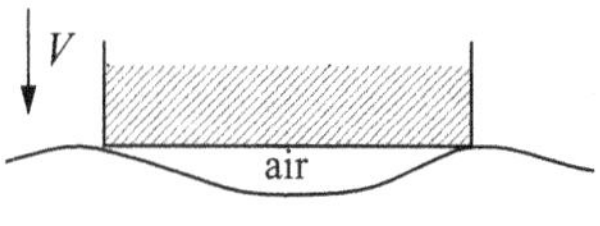

Figure 11.25. Deformation of the free surface and formation of an air pocket during entry of a rigid body with a horizontal flat bottom. The thickness of the pocket is exaggerated.

Figure 11.25. A rigid body is assumed. However, hydroelasticity ought to be considered in a scenario such as this. The air flow causes the water to rise at the corners of the flat bottom and enclose an air pocket. Miyamoto and Tanizawa (1985) filmed the details of the process. An air–water mixture occurs approximately $\frac{1}{3}L$ from the edges where L is one-half of the bottom width. The mixture touches the bottom very early and later air escapes from the edge. Chuang (1966) showed experimentally that the cushioning effect of the air pocket reduces the pressure during water entry of wedges with small deadrise angles. Verhagen (1967) analyzed the two-dimensional flow during impact between a flat plate and a water surface both before and after closure of the air cushion. The pressure was assumed spatially constant in the closed air cushion. This case is similar to the assumption of the theoretical model discussed later in this section. The same behavior occurs with any liquid and gas. Even though the local deadrise angle is very small initially for the case of water entry of a circular cross-section through a horizontal free surface, the air cushion effect may not be present. The water entry of a circular cross-section has similarities with the impact of a free surface with finite curvature on a flat horizontal tank roof as illustrated in Figure 11.2(b). Another example that may create gas pockets, and thereby cushion effects, is the corrugated surface of the Mark III containment system of LNG tanks (see Figure 11.60). However, the effect on the maximum structural stresses is more important than the effect on the maximum pressure. This result is a function of the ratio between the duration of the cushioning and the relevant natural periods for the structural response. If this ratio is very small, the consequence is minor. The

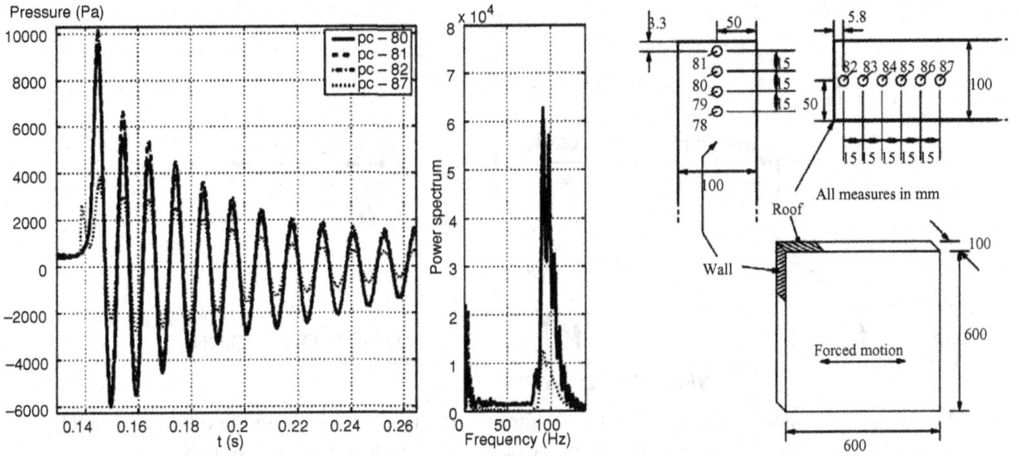

Figure 11.26. Slamming pressure measurement when an air pocket is generated. The figure also shows the power spectrum of the pressure recording and the pressure-gauge locations.

lack of correlation between maximum pressure and maximum structural stresses for large impact pressures that are limited in time and space was demonstrated in the hydroelastic studies by Faltinsen (1997) with drop tests of horizontal steel and aluminum plates (see also Section 11.9). Chung *et al.* (2006) presented experimental results from drop tests of specimens with length 1.655 m, breadth 1.655 m, and height 0.5 m. A rigid steel model, Mark III with flat membrane, and Mark III with corrugated membrane were used. The inclination angles of the specimen were varied from 0° to 15°. The magnitude of the pressure is reduced due to the corrugation. The structural elasticity clearly reduces the pressure for inclination angles 0° and 2°. Because only one specimen was tested, it is difficult to apply these results to slamming inside a tank. For instance, the flow due to the square-based test specimens should be highly three-dimensional and does not account for the effect of neighboring tank structural parts. Shin *et al.* (2003) concluded similarly to Chung *et al.* (2006) on the cushioning effect due to corrugation of the tank surface. Not all containment systems for LNG have a corrugated tank surface. Another important issue is the scaling of model test results with a gas cushion to full scale, which is addressed in Section 11.6.5.

The studies that follow in this chapter consider two-dimensional flow and a flat rigid surface. The entrapped gas occurs mainly due to the geometry of the impacting free surface and not due to the effect of the gas flow on the liquid flow. Theoretical analysis is used to a large extent. However, it is conceivable that a numerical method, for instance, the boundary element method, may be used, which Greco *et al.* (2003) did in analyzing local hydroelastic bottom impact with an air cavity for a very large floating structure with shallow draft.

The following discussion concentrates on the scenario in Figure 11.2(c), where an air pocket is trapped at the tank corner as a consequence of the impacting free-surface geometry. An assumption is that the ullage pressure is not close to the vapor pressure. The compressibility of the gas causes a natural frequency for the gas pocket. Resonant oscillations of the gas cushion are excited during the impact. The consequence of these oscillations is that the pressure oscillates with the same frequency, which is illustrated in Figure 11.26 in terms of slamming pressure measurements at the tank roof with information about tank dimensions and positions of the pressure transducers. Several pressure measurements are shown, indicating that the pressure is nearly uniform within the air cushion. However, a quantitative discussion of this fact requires the position of the air cavity. The precision error in experiments such as this is low. The natural frequency of the gas pocket oscillations changes with time; we discuss this in Sections 11.6.1 and 11.6.3.2. The power spectrum of the pressure recording shows that the spectral energy of the pressure signal is concentrated around a narrow frequency range that is related to the natural frequency. A particular concern is that the time-dependent behavior of the gas cushion can cause hydroelastic effects. If the ullage

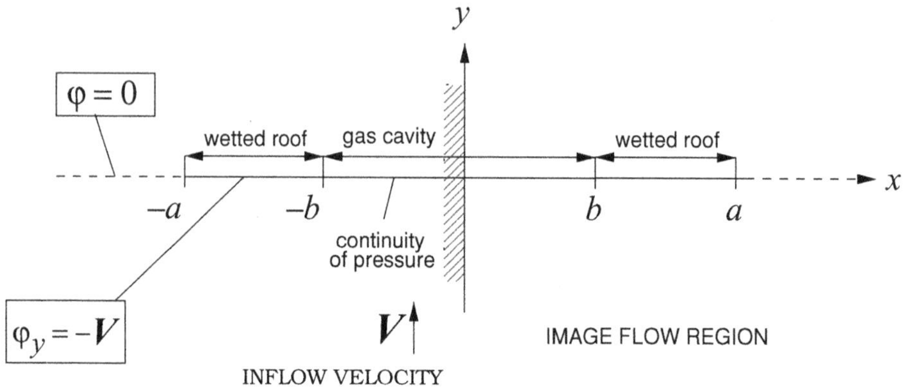

Figure 11.27. Boundary-value problem for the impact with gas cavity, where φ is the velocity potential caused by the impact, $\varphi_y = \partial\varphi/\partial y$. The inflow velocity V and its direction are also shown.

pressure is close to the vapor pressure, the oscillatory behavior shown in Figure 11.26 is highly damped (Maillard & Brosset, 2009; Lugni, personal communication, 2009). Since Maillard and Brosset (2009) used water vapor as the ullage gas, the damping was explained to be due to condensation. Because Lugni used water and air in his experiments, condensation is not believed to be a factor.

Let us explain in more detail why the natural frequency oscillations for the gas pocket are excited. The tank surface is assumed rigid. The gas must be considered compressible. We assume spatially constant pressure inside the gas cushion and that the gas cushion is thin so that we can linearize the problem. Furthermore, we assume no heat exchange, which means an adiabatic process for the gas. The dynamic gas density and the dynamic gas pressure are unknown variables. The changing volume of the gas cushion influences the liquid flow, which can be described by the potential flow theory of an incompressible fluid. On the interface of the gas and the liquid we require that the pressure is continuous. Thus, we have the following three unknowns in the problem:

1. the dynamic gas density,
2. the dynamic gas pressure, and
3. the velocity potential of the liquid flow.

When we have three unknowns, we need three equations:

1. the mass continuity equation for the gas cushion,
2. a relationship between the gas pressure and the gas density, and
3. the Laplace equation with boundary conditions for the velocity potential in the liquid.

First we consider the flow in the liquid caused by the impact. In addition, we have the incident flow, which is assumed known. It is assumed that the tank has only forced horizontal velocities; there is no upper chamfer. However, extending the analysis to a situation where the tank has a more general excitation velocity does not cause any difficulty. The flow relative to a tank-fixed coordinate system is studied. A local coordinate system Oxy is introduced with origin at the corner between the tank wall and the tank roof in the impact region, where y is positive upward. The wetted tank roof corresponds to $-a \le x \le -b$. The incident impacting free surface is assumed to have a time-dependent vertical velocity $V(t)$ that is uniform in space in the analysis. The corresponding velocity potential is Vy. Some simplifications are made in the analysis of the flow caused by the impact: The tank bottom and the tank wall opposite the impact area are assumed far away relative to the length of the impact area, so the tank bottom and the opposite tank wall are assumed to be at infinity in the analysis. The boundary condition at the tank wall adjacent to the impact area is accounted for by mirroring the flow about the tank wall. The boundary-value problem for the velocity potential φ in the liquid caused by the impact is illustrated in Figure 11.27. The boundary conditions on the cavity surface and the free surface outside the impact are transferred to the x-axis. The dynamic free-surface condition is $\varphi = 0$ on $y = 0$ and $x < -a$, which is a common free-surface condition in impact problems (see discussion in Section 11.3).

The idea is to model the liquid flow by a distribution of potential-flow vortices on $y = 0$ between $x = -a$ and $x = a$ (see a similar solution technique for flow around a hydrofoil in Newman (1977)). The vortex distribution causes only a vertical velocity on the x-axis outside x-values between $-a$ and a (i.e., for $y = 0$ and $|x| > a$), which is consistent with the dynamic free-surface condition that states that the velocity potential is constant on $y = 0$. On the wetted tank roof and its image about the tank wall we have $\partial\varphi/\partial y = -V$. Because the total flow relative to the tank in the impact area is described by the velocity potential $Vy + \varphi$, this boundary condition ensures that there is no flow through the wetted tank roof. The vertical velocity acting on the gas cavity is unknown a priori and is found by solving the described problem.

The notation $u_\pm$ and $v_\pm$ are used for the values of the horizontal and vertical velocities u and v for $-a \le x \le a$ at $y = 0^+$ and $y = 0^-$, respectively. We can now express the velocities $v_\pm$ and $u_\pm$ along the air cushion and tank roof as

$$v_\pm(x) = -\frac{1}{2\pi} PV \int_{-a}^{a} \frac{\gamma(\xi)}{\xi - x} d\xi, \quad u_\pm(x) = \mp\tfrac{1}{2}\gamma(x), \tag{11.35}$$

where γ is the vortex density and PV denotes a principal value integral. The consequence of eq. (11.35) is that $v_+ = v_-$; therefore, we drop the subscripts $\pm$ for v. The solution of eq. (11.35) can be found, for instance, in Newman (1977). Because eq. (11.35) has a nontrivial solution $\gamma(x) = C_1/\sqrt{a^2 - x^2}$ when the left-hand side is equal to zero, we need to specify an additional condition. When the flow around a hydrofoil is considered and the linear flow due to the hydrofoil is expressed as a vortex distribution, then the Kutta condition is needed as the additional condition. The Kutta condition reflects the fact that the flow separates smoothly at the trailing edge of a hydrofoil and has no relevance in our problem. However, we need an additional condition to make the solution unique. In our case we must require that $u_\pm(x) = -u_\pm(-x)$; that is, the horizontal velocity is antisymmetric with respect to $x = 0$. The consequence is that $C_1 = 0$. We can now express the solution of eq. (11.35) as (Newman, 1977)

$$\gamma(x) = \frac{2}{\pi} \frac{1}{\sqrt{a^2 - x^2}} PV \int_{-a}^{a} \frac{v(\xi)\sqrt{a^2 - \xi^2}}{\xi - x} d\xi. \tag{11.36}$$

11.6.1 Natural frequency for a gas cavity

We want to study the possible eigenvalues for the pressure and density oscillations in the gas cushion and the associated flow in the liquid. This implies a boundary-value problem with no excitation (i.e., $v_+ = 0$ for $-a \le x \le -b$, $b \le x \le a$). Furthermore, the harmonically oscillating pressure is assumed uniform in the gas cavity. The dynamic pressure in the cavity is equal to the hydrodynamic pressure at the boundary of the cavity. If linear theory is used, we can express the hydrodynamic pressure as $-\rho_l \partial\varphi/\partial t$, where φ is the velocity potential for the liquid flow and ρ_l is the liquid density. Because the pressure does not vary along the cavity, the velocity potential does not change along the cavity. The horizontal liquid velocity is therefore zero along the cavity. From eq. (11.35), we have $\gamma(x) = 0$ for $-b \le x \le b$. From eq. (11.36) and the fact that $v_+ = 0$ for $-a \le x \le -b$, $b \le x \le a$ it follows then that

$$0 = \frac{2}{\pi} \frac{1}{\sqrt{a^2 - x^2}} PV \int_{-b}^{b} \frac{v(\xi)\sqrt{a^2 - \xi^2}}{\xi - x} d\xi \quad \text{for} \quad -b \le x \le b. \tag{11.37}$$

Because eq. (11.35) with $v_\pm(x) = 0$ has the nontrivial solution $\gamma(x) = C_1/\sqrt{a^2 - x^2}$, the nontrivial solution of eq. (11.37) is

$$v(x) = \frac{C(t)}{\sqrt{b^2 - x^2}\sqrt{a^2 - x^2}} \quad \text{for} \quad -b \le x \le b. \tag{11.38}$$

We note that eq. (11.38) is consistent with the symmetry property $v(x) = v(-x)$.

We now want to determine $\gamma(x)$ for $-a \le x \le -b$, $b \le x \le a$. Equation (11.36) gives

$$\gamma(x) = \frac{2}{\pi} \frac{C(t)}{\sqrt{a^2 - x^2}} \int_{-b}^{b} \frac{1}{\sqrt{b^2 - \xi^2}(\xi - x)} d\xi.$$

This integral can be integrated analytically by first substituting $\xi = b\cos\theta$ and using θ as an integration variable. From expression 3.613 in Gradshtein and Ryzhik (1965) it follows that $\gamma(x) = -\frac{2C(t)\mathrm{sign}(x)}{\sqrt{a^2 - x^2}\sqrt{x^2 - b^2}}$. Function $\gamma(x)$ is linked to $\partial\varphi/\partial x$ through

$$\left.\frac{\partial \varphi}{\partial x}\right|_{y=0^-} = \tfrac{1}{2}\gamma(x). \tag{11.39}$$

By integrating eq. (11.39) and using the free-surface condition $\varphi = 0$, it follows that the velocity potential on the cavity surface is

$$\varphi_{\text{cav}} = C(t)\int_{-a}^{-b} \frac{\mathrm{d}x}{\sqrt{a^2 - x^2}\sqrt{x^2 - b^2}} = \frac{C(t)}{a} K\left[\sqrt{1-(b/a)^2}\right], \tag{11.40}$$

where $K(k) = \int_0^1 ((1-x^2)(1-k^2x^2))^{-1/2}\,\mathrm{d}x$ is the complete elliptic integral of the first kind as defined by Gradshtein and Ryzhik (1965). Both a and b may be functions of time.

The dynamic pressure in the cavity can be expressed as

$$p_D = -\rho_l \frac{\partial \varphi_{\text{cav}}}{\partial t} = -\rho_l \frac{d}{dt}\left[C(t)K\left[\sqrt{1-(b/a)^2}\right]\Big/ a\right]. \tag{11.41}$$

We have until now used only one of the equations for solving our problem (i.e., the Laplace equation for the liquid flow). The next equation to be considered is the relationship between the gas pressure and the gas density in the gas cushion. We express the gas pressure as $p_{\text{gas}} = p_0 + p_D$ and the mass density in the gas as $\rho_{\text{gas}} = \rho_0 + \rho_D$, where p_0 and ρ_0 are values of p and ρ at the initial time of creation of the gas pocket. We neglect the fact that there is a gas flow before the closure of the gas cavity. Therefore, p_0 and ρ_0 are what we have referred to as the ullage pressure and gas density, respectively. However, the gas flow before closure may matter. The relationship between the gas pressure and the gas density can for adiabatic processes be expressed as

$$p_{\text{gas}}/p_0 = (\rho_{\text{gas}}/\rho_0)^{\kappa}, \tag{11.42}$$

where κ is the ratio of specific heats, which for diatomic gases such as air is $\kappa = 1.4$. In the following it will be assumed that $p_D/p_0 \ll 1$, $\rho_D/\rho_0 \ll 1$ so that the problem can be linearized. By using a Taylor-series expansion of eq. (11.42), it follows that

$$\frac{\rho_{\text{gas}}}{\rho_0} \approx 1 + \frac{1}{\kappa}\frac{p_D}{p_0}. \tag{11.43}$$

Because we assume no leakage and inflow to the gas pocket, the continuity equation for the gas cushion can be written as

$$\rho_{\text{gas}}\dot{\Omega} + \frac{d\rho_{\text{gas}}}{dt}\Omega = 0, \tag{11.44}$$

where Ω is the gas-cushion volume. The time rate of change of Ω is

$$\dot{\Omega} = -\int_{-b}^{0} v(x)\mathrm{d}x = -C(t)\int_{-b}^{0}\frac{\sqrt{a^2-x^2}}{\sqrt{b^2-x^2}}\mathrm{d}x = -C(t)\frac{K(b/a)}{a}. \tag{11.45}$$

We now linearize eq. (11.44) and use eq. (11.43) to express $d\rho_{\text{gas}}/dt$. Equation (11.44) becomes

$$\rho_0\dot{\Omega} + \frac{\rho_0}{\kappa p_0}\frac{dp_D}{dt}\Omega_0 = 0, \tag{11.46}$$

where Ω_0 is the initial gas-cushion volume. If we introduce a new variable

$$C_1(t) = C(t)K\left[\sqrt{1-(b/a)^2}\right]\Big/ a \tag{11.47}$$

from eq. (11.41) it follows that the cavity pressure is

$$p_D = -\rho_l \dot{C}_1(t). \tag{11.48}$$

From eqs. (11.41), (11.45), (11.46), and (11.47) it follows that

$$\left(\frac{\Omega_0\rho_l}{\kappa p_o}\right)\ddot{C}_1(t) + \frac{K(b/a)}{K[\sqrt{1-(b/a)^2}]}C_1(t) = 0. \tag{11.49}$$

For the time being we assume that a and b are time-independent. Then eq. (11.49) is a classical mass–spring system without damping and forcing. The undamped natural frequency σ_n in radians per second is found by assuming that $C_1(t)$ has the harmonic time dependence $\exp(i\sigma_n t)$, which gives

$$\sigma_n^2\left(\frac{\Omega_0\rho_l}{\kappa p_0}\right) = \frac{K(b/a)}{K[\sqrt{1-(b/a)^2}]}. \tag{11.50}$$

Figure 11.28 presents the nondimensional natural frequency $\sigma_n\sqrt{\Omega_0\rho_l/(\kappa p_0)}$ as a function of b/a. Due to eq. (11.48) and the "linearized hydrodynamic model," the results assume thin pockets.

The behavior of the natural frequency when $b/a \to 1$ and $b/a \to 0$ can be found from the asymptotic behavior of the elliptical integral. From expressions 17.3.11 and 17.3.26 in Abramowitz and Stegun (1964), it follows that

$$\lim_{b/a\to 0} K\left(\frac{b}{a}\right) = \tfrac{1}{2}\pi,$$

$$\lim_{b/a\to 1} K\left(\frac{b}{a}\right) = \tfrac{1}{2}\ln\left[\frac{16}{1-(b/a)^2}\right]. \tag{11.51}$$

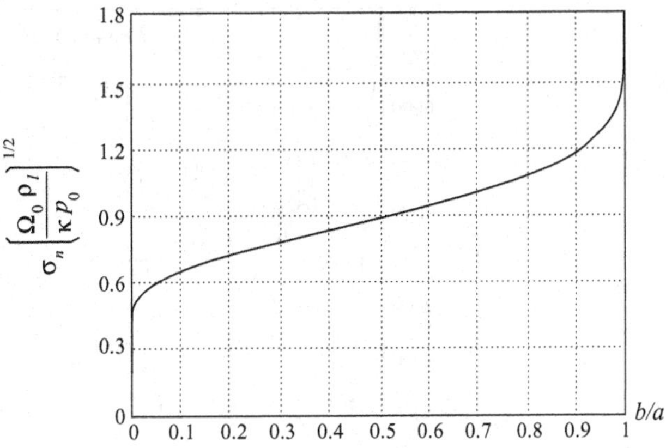

Figure 11.28. Nondimensional natural frequency $\sigma_n\sqrt{\Omega_0\rho_l/(\kappa p_0)}$ for the gas cavity in Figure 11.27 as a function of the length ratio b/a defined in Figure 11.27.

The consequence is that σ_n^2 goes logarithmically to infinity when $b/a \to 1$. Physically, ventilation may occur when $b/a \to 1$. Another consequence of eq. (11.51) is that $\sigma_n \to 0$ when $b/a \to 0$.

11.6.1.1 Simplified analysis

We assume now that $v(x) = -V_n$ is a constant over the gas cushion (i.e., the gas cushion acts as a piston). This approximation simplifies the analysis. We may then write $\dot{\Omega} \sim V_n b$. Equation (11.46) gives

$$\rho_0 V_n b + \frac{\rho_0}{\kappa p_0}\dot{p}_D\Omega_0 = 0, \tag{11.52}$$

where V_n and p_D are unknown and can be related additionally as follows. We consider a boundary-value problem for flow in the liquid where the cushion is forced to oscillate with V_n. We could say this problem is the same as the problem for a heaving flat plate between $x = -b$ and $x = b$ in combination with a free surface from $x = -\infty$ to $x = -a$ and from $x = a$ to $x = \infty$ and fixed flat plates from $x = -a$ to $x = -b$ and from $x = b$ to $x = a$. The free-surface condition is as in Figure 11.27. We can solve this problem for unit V_n by using a vortex distribution between $x = -a$ and $x = a$. We do not show the details on how to solve such a problem, but we recall instead from Chapter 3 how added mass is defined. Forced oscillation of the plate with velocity V_n causes a vertical force on the plate that can be expressed as $a_{33}\dot{V}_n$, where a_{33} is the two-dimensional added mass in heave (note that the sign of the force is consistent with the fact that positive V_n is in the negative y-direction). This force comes from integrating a dynamic pressure. This pressure is the same as p_D, which is approximated as a uniform pressure from the force; that is,

$$p_D = \tfrac{1}{2}a_{33}b^{-1}\dot{V}_n. \tag{11.53}$$

We can express a_{33} in a nondimensional way as follows:

$$a_{33}^* = \frac{a_{33}}{\rho_l\frac{1}{2}\pi b^2}, \tag{11.54}$$

where ρ_l is the liquid density.

We substitute eq. (11.53) into eq. (11.52) and assume harmonic oscillations, which gives the eigenfrequency

$$\sigma_n = \sqrt{4\kappa p_0/\left(\pi a_{33}^*\rho_l\Omega_0\right)}, \tag{11.55}$$

where a_{33}^* is defined by eq. (11.54). Because a_{33}^* is a function of a/b, σ_n in eq. (11.55) is consistent with the results in Figure 11.28, that is, that $\sigma_n\sqrt{\Omega_0\rho_l/(\kappa p_0)}$ is a function of a/b. However, we should not expect the same values because of the simplifications of assuming constant value V_n along the gas cushion, which requires that a is not close to b. Equation (11.55) also gives the time scaling; that is, we should present results in nondimensional form as a function of nondimensional time $t^* = t\sqrt{p_0/(\rho_l L^2)}$, where L is a length scale of the tank.

11.6.2 Damping of gas cavity oscillations

Thermal conductivity, shear viscosity, and acoustic wave radiation are examples of damping sources. Furthermore, the leakage and inflow to a cavity are a well-known damping mechanism for resonant fluid motion in the air cushion of a surface effect ship (SES); see Faltinsen (2005).

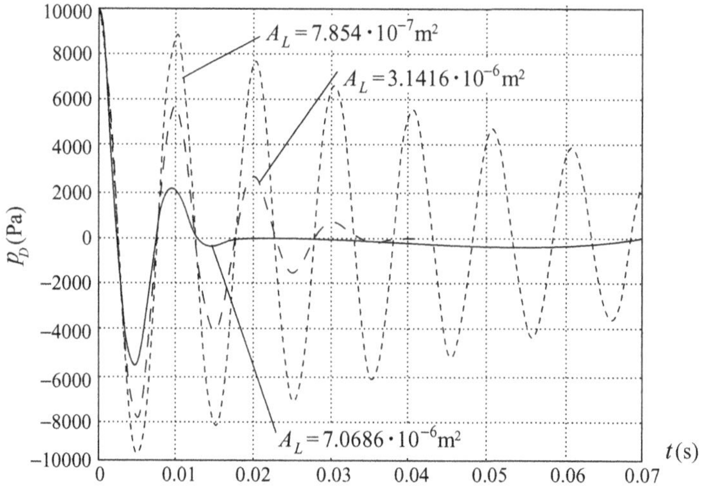

Figure 11.29. Free vibration of dynamic pressure p_D in an air cavity with $\Omega_0 = 4.32 \cdot 10^{-4}\ \text{m}^2$, $a = 0.085$ m, and $b = 0.07$ m. The tank width is 0.1 m. The effect of the leakage/inflow through an area A_L is investigated (Abrahamsen, personal communication, 2008).

Leighton (1994) and Medwin and Clay (1998) discussed how thermal conductivity, shear viscosity, and acoustic wave radiation affect damping of resonant gas cavity oscillations of a radially oscillating spherical bubble in infinite fluid as a function of bubble diameter and resonance frequency. They presented their results in terms of the damping constant $\delta = 2\xi$. According to their results, thermal conductivity and shear viscosity should have a minor influence in the case of bubble (cavity) dimensions in our application. The role of acoustic wave radiation is difficult to assess. The damping constant ξ due to acoustic wave radiation for water and a gas bubble diameter with a size of interest is on the order of 0.01. Our environment for the acoustic wave radiation differs in many aspects from Leighton's or Medwin and Clay's considerations. For instance, the spherical bubble has source behavior, whereas our gas cavity has dipole behavior. This fact should reduce the wave radiation for our gas cavity. Conversely, our problem is two-dimensional, which should increase the acoustic wave radiation for an infinite fluid problem. An important effect is believed to be the fact that we consider a tank. If the acoustic waves cannot escape from the tank, the damping according to potential flow theory must be zero unless the wave energy is transmitted through the tank walls and the free surface.

We show how to incorporate the leakage and inflow effect in our mathematical model, starting with modifying continuity equation (11.44) for the gas cavity to account for leakage and inflow. The modified continuity equation becomes

$$\rho_{\text{gas}}\dot{\Omega} + \dot{\rho}_{\text{gas}}\Omega = \rho_0 Q_{\text{in}} - \rho_0 Q_{\text{out}}, \quad (11.56)$$

where $\rho_0 Q_{\text{in}}$ and $\rho_0 Q_{\text{out}}$ are the gas mass flow per unit time due to inflow and leakage, respectively; Q_{out} through a small leakage/inflow area A_L is estimated for an SES by a quasi-steady formulation. We may write

$$Q_{\text{in}} - Q_{\text{out}} = -0.61 A_L \text{sign}(p_D)\sqrt{2|p_D|/\rho_0}, \quad (11.57)$$

where a contraction coefficient of 0.61 for the escaping jet flow is used as is common to do for an SES. In reality, this use depends on the local details of the structure at the leakage area. A critical issue is what A_L represents in our problem; A_L may be time-dependent and associated with three-dimensional flow effects.

We have applied the presented leakage damping model to the experimental results in Figure 11.26. The two-dimensional air-cushion volume $\Omega_0 = 4.32 \cdot 10^{-4}\ \text{m}^2$ and $a = 0.085$ m and $b = 0.07$ m (see Figure 11.27). The width of the tank is 0.1 m and has been accounted for in the calculations. No excitation is applied (i.e., we study free vibrations). The initial value of the dynamic pressure p_D in the cushion is 10 kPa. The results with different leakage/inflow area are presented in Figure 11.29. We note that, even with the small leakage/inflow area $A_L = 7.854 \cdot 10^{-7}\ \text{m}^2$, damping is noticeable. No attempt was made to identify if the damping was due to leakage/inflow in the experimental results presented in Figure 11.26.

Abrahamsen (personal communication, 2008) recently performed experiments with tank roof impact that showed a decaying pressure in the air cushion similar to that in Figure 11.26. However, he could not observe any leakage. Furthermore, he investigated the effect of leakage by drilling holes into the air cushion. The leakage/inflow model could then explain the additional decay of the pressure oscillations. The presented model in previous paragraphs cannot fully explain the behavior of the air cushion. In Section 11.6.3.2 we explain that some of the initial decay of the pressure amplitude in the air cushion is due to nondissipative effects. Error sources in the analysis are the assumed horizontal free surface and the fact that the air flow before closure affects the initial conditions for the oscillations of the air cavity. Neglecting the gas(air) flow implies that the density ratio between the ullage gas and the liquid is not properly accounted for. The density ratio was shown in Section 11.2 to matter.

11.6.3 Forced oscillations of a gas cavity

The previous analysis can be generalized to a liquid impact with a gas cushion. The air flow between the tank roof and the free surface before the closure of the gas cavity is neglected. A one-dimensional nonlinear compressible flow model can approximate the gas flow before closure (Verhagen, 1967). We denote the vertical velocity of the impacting free surface as $V(t)$ (i.e., the velocity is constant in space). The velocity $V(t)$ has to be determined either by a CFD method or by experiments. The use of the linear multimodal method described in Chapter 5 is an efficient way to provide guidance on which tank motions create an air pocket. The impact causes an additional flow with velocity components u and v along the x- and y-axes, respectively; see Figure 11.27, which also defines the wetted part and the gas cushion part of the tank roof. An image flow about the tank wall adjacent to the impact is introduced to account for the boundary condition on this tank wall. We can express the solution of the additional flow problem by a vortex distribution with density $\gamma(x)$ from $x=-a$ to $x=a$. We explained in Section 11.6.1 that $\gamma(x)=0$ from $x=-b$ to $x=b$ (i.e., in the range of the gas cushion and its image). From eq. (11.35) and the boundary condition on the wetted part of the tank roof it follows that

$$V=\frac{1}{2\pi}PV\int_{-a}^{a}\frac{\gamma(\xi)}{\xi-x}d\xi,\quad -a<x<-b,\ b<x<a.$$

We can solve this integral equation for γ as described in connection with eq. (11.36), which should present no difficulties. However, it turns out that the solution procedure becomes cumbersome with complicated final expressions and therefore we follow a different procedure. The construction of the solution of the boundary-value problem can be found in Gakhov (1966) and Sedov (1965). We do not show the details here but rather present the solution and confirm that it satisfies the necessary boundary conditions. The flow is represented by complex function theory (Newman, 1977). The complex velocity potential is denoted $W=\phi+i\psi$, where ϕ is the sum of the velocity potential Vy due to the inflow velocity $V(t)$ and the velocity potential φ caused by the impact. The stream function is denoted ψ. The complex velocity can be expressed as

$$\frac{dW}{dz}\equiv u-iv=\frac{-i[V(t)z^2-C(t)]}{\sqrt{(z^2-a^2)(z^2-b^2)}} \tag{11.58}$$

(see Korobkin, 1996a), where $i=\sqrt{-1}$ is the complex unit, $z=x+iy$, and u and v are the flow velocity components along the x- and y-axes, respectively. We note that for $z\to\infty$, $dW/dz\to -iV(t)$, which means that we have an inflow velocity $V(t)$ along the y-axis.

To properly describe the behavior of $I_1=\sqrt{(z^2-a^2)(z^2-b^2)}$, we introduce $z+a=r_1\exp(i\theta_1)$, $z+b=r_2\exp(i\theta_2)$, $z-b=r_3\exp(i\theta_3)$, $z-a=r_4\exp(i\theta_4)$ (see Figure 11.30) so that $I_1=\sqrt{r_1r_2r_3r_4}\exp[i\frac{1}{2}(\theta_1+\theta_2+\theta_3+\theta_4)]$, where $-\pi<\theta_j<\pi$. It can then be shown that I_1 has branch cuts along the x-axis from $-a$ to $-b$ and from b to a. The consequence is that I_1 and hence dW/dz is discontinuous across these branch cuts.

Let us elaborate more on expression (11.58) for dW/dz and confirm that boundary conditions are satisfied. We start with $y=0^-$, $-a<x<-b$, $b<x<a$, and find that $v=0$ and

$$u=\frac{\mathrm{sgn}(x)[V(t)x^2-C(t)]}{\sqrt{(a^2-x^2)(x^2-b^2)}}. \tag{11.59}$$

This finding confirms that the body boundary condition is satisfied on the wetted part of the tank roof and its image about the adjacent vertical tank wall.

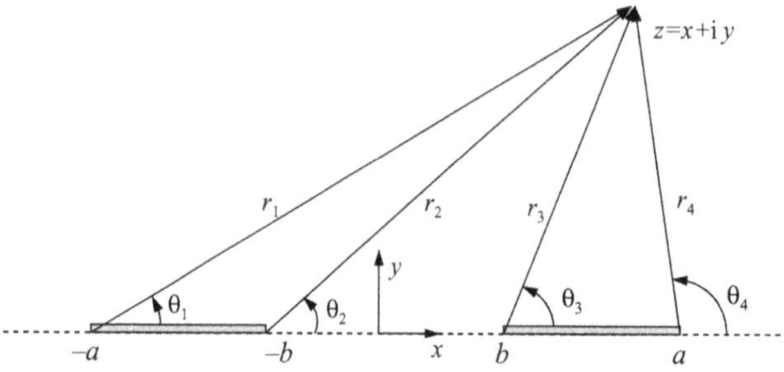

Figure 11.30. Definitions of polar coordinate systems (r_j, θ_j) in the complex plane $z = x + iy$.

Furthermore, we find that for $y = 0$, $-b < x < b$ that $u = 0$ and

$$v = -\frac{V(t)x^2 - C(t)}{\sqrt{(a^2 - x^2)(b^2 - x^2)}}. \qquad (11.60)$$

This expression for v is needed in the evaluation of the volume change of the gas cavity.

Equation (11.58) gives

$$u = 0, \quad v = \frac{V(t)y^2 + C(t)}{\sqrt{(a^2 + y^2)(b^2 + y^2)}} \qquad (11.61)$$

along the adjacent vertical tank wall, where we have used that $\theta_1 + \theta_4 = -\pi$, $\theta_2 + \theta_3 = -\pi$ so that $I_1 = -\sqrt{r_1 r_2 r_3 r_4}$. Equation (11.61) is consistent with the fact that there is no flow through the tank wall.

When $y = 0$, $|x| > a$, we find that

$$u = 0, \quad v = \frac{V(t)x^2 - C(t)}{\sqrt{(x^2 - a^2)(x^2 - b^2)}},$$

which can be used to find how a changes with time.

We can now use eq. (11.59) and the fact that $u = \partial\varphi/\partial x$ and $\varphi = 0$ for $x \le -a$ to express the velocity potential at the gas cavity as

$$\varphi_{\text{cav}} = \frac{C(t)}{a} K\left[\sqrt{1 - (b/a)^2}\right]$$
$$-aV(t)E\left[\sqrt{1 - (b/a)^2}\right], \qquad (11.62)$$

where $E(m) = \int_0^1 \frac{\sqrt{1-m^2t^2}}{\sqrt{1-t^2}}dt$ is the complete elliptical integral of the second kind while $K(k) = \int_0^1 ((1 - x^2)(1 - k^2x^2))^{-1/2}dx$ is the complete elliptical integral of the first kind as defined by Gradshtein and Ryzhik (1965). We proceed by generalizing the procedure in Section 11.6.1. A generalization of eq. (11.45) gives the time rate of change of the cushion volume Ω:

$$\dot{\Omega} = -\int_{-b}^{0} v(x)dx, \qquad (11.63)$$

where $v(x)$ is the sum of the incident flow velocity and the vertical liquid flow velocity at the gas cushion caused by the impact. This result can be expressed by eq. (11.60). The result is that

$$\dot{\Omega} = V(t) \cdot a[K(b/a) - E(b/a)] - a^{-1}C(t)K(b/a). \qquad (11.64)$$

We now insert the expressions into eq. (11.46) and introduce $C_1(t)$ by eq. (11.47), which gives the following differential equation:

$$\left(\frac{\Omega_0 \rho_l}{\kappa p_0}\right)\ddot{C}_1(t) + \frac{K(b/a)}{K[\sqrt{1 - (b/a)^2}]}C_1(t) = F(t),$$

where

$$F(t) = \dot{\Omega}_{exc} + \frac{\Omega_0}{\kappa p_0}\dot{p}_{1exc},$$

$$\dot{\Omega}_{exc} = V(t) \cdot a\left[K\left(\frac{b}{a}\right) - E\left(\frac{b}{a}\right)\right],$$

$$\dot{p}_{1exc} = \rho_l \frac{d^2}{dt^2}\left\{aV(t)E\left[\sqrt{1 - \left(\frac{b}{a}\right)^2}\right]\right\}.$$

The initial conditions for $C_1(t)$ and $\dot{C}_1(t)$ are derived by requiring that the initial velocity potential φ_{cav} and dynamic pressure p_D in the cavity are zero. The pressure in the gas cavity is $p_D = -\rho_l \partial\varphi_{\text{cav}}/\partial t$, where φ_{cav} is given by eq. (11.62), which gives

$$p_D = -\rho_l \frac{d}{dt}\left\{C_1(t) - aV(t)E\left[\sqrt{1 - (b/a)^2}\right]\right\},$$
$$-b \le x \le 0. \qquad (11.65)$$

The initial conditions are then

$$C_1(0) = a(0)V(0)E\left[\sqrt{1 - [b(0)/a(0)]^2}\right],$$

$$\dot{C}_1(0) = \frac{d}{dt}\left\{aV(t)E\left[\sqrt{1 - (b/a)^2}\right]\right\}\Big|_{t=0}. \qquad (11.66)$$

The pressure on the wetted tank roof follows from using the fact that $u = \partial\varphi/\partial x$, the expression for u given by eq. (11.59), and that the hydrodynamic pressure can be expressed as $p_D = -\rho_l \partial\varphi/\partial t$:

$$p_D = -\rho_l \frac{\partial}{\partial t}\left[\int_{-a}^{x} \frac{-V(t)x^2 + C(t)}{\sqrt{(a^2 - x^2)(x^2 - b^2)}}\mathrm{d}x\right],$$
$$-a \le x \le -b,$$

where $C(t)$ is given in terms of $C_1(t)$ by means of eq. (11.47). The velocity potential at the vertical tank wall adjacent to the impact follows by using the equality $v = \partial\varphi/\partial y$ and the expression for v given by eq. (11.61), which gives

$$\varphi = \varphi_{cav} - \int_{y}^{0} \frac{V(t)y^2 + C(t)}{\sqrt{(a^2 + y^2)(b^2 + y^2)}}\mathrm{d}y.$$

The pressure is given by

$$\begin{aligned} p_D &= -\rho_l \frac{\partial\varphi}{\partial t} \\ &= -\rho_l \frac{\partial}{\partial t}\left[\varphi_{cav} - \int_{y}^{0} \frac{V(t)y^2 + C(t)}{\sqrt{(a^2 + y^2)(b^2 + y^2)}}\mathrm{d}y\right]. \end{aligned} \quad (11.67)$$

Damping could be incorporated in the analysis in the same way as it is described in Section 11.6.2.

11.6.3.1 Prediction of the wetted surface

We also need to know how a and b depend on time. One way of doing this is to follow a von Karman procedure, which means simply to find the geometrical intersection between the incident free surface and the tank roof. This procedure likely underestimates the loading because in reality the impact causes a flow that affects the free-surface elevation.

We may generalize the procedure in Section 11.3.1.1 for how to find the wetted surface of the tank roof. Because linear theory is assumed when considering the rate of change of the gas cavity volume, we may assume b to be time-independent. We consider the time of the initial impact and define $\eta_0(x)$ as the vertical distance between the tank roof and the free surface outside the impact point. Equation (11.12) can then be generalized as

$$\eta_0(x) = \int_{0}^{t} \frac{V(t)x^2 - C(t)}{\sqrt{(x^2 - a^2)(x^2 - b^2)}}\mathrm{d}t, \quad (11.68)$$

which gives us an integral equation with the unknown time-dependent variable $a(t)$. A different approach to determine a and b by accounting for the effect of the impact is described by Korobkin (1996b). In the following example we use experimentally determined values of a and b. It turns out that the predicted pressures are not sensitive to a and b.

11.6.3.2 Case study

We use the experimental results presented in Figure 11.26 as a basis for comparisons between experiments and theory. The model tests have been conducted with forced horizontal regular oscillatory motions of a narrow tank measuring $0.6 \times 0.6 \times 0.1$ m. Uncertainties exist related to the measurements of inflow conditions. The vertical inflow velocity is expressed as $V(t) = V_0 + V_1 t$, where $t = 0$ is the time of initial impact. The average upward velocity of the free surface along the length of the air cushion is estimated to be 0.6 m s^{-1} at the time of impact and the acceleration V_1 is -6.51 m s^{-2}. The two-dimensional air cushion volume Ω_0 starts out at $4.32 \cdot 10^{-4}$ m^2. In addition, we set $b = 0.0695$ m, $\kappa = 1.4$, $p_0 = 10^5$ Pa, $\rho_l = 10^3$ kg m^{-3}, and $a = 0.07 + 0.781t$(m). The numerical results are presented in Figure 11.31 by both the linear method described in this section and the nonlinear analysis described in Section 11.6.4. The effect of nonlinearities is not significant. We note an overprediction of the maximum pressure relative to the experiments, which could be caused by inaccuracies in the estimated inflow conditions or due to the fact that we have assumed that the inflow velocity is not spatially varying. Another error source is the assumed, nearly horizontal, free surface. A parameter variation has shown that the maximum pressure is nearly linearly dependent on the initial impact velocity and is associated with the change of the air cushion volume. The rate of change of the wetted surface plays a minor role.

By comparing the present calculations with experimental results, we observe that the number of oscillations during the considered time interval agrees with the experimental results. Furthermore, the oscillation period changes with time as in Figure 11.26. The reason for the change in the oscillation period is associated with the change of b/a. This change is illustrated in Figure 11.32, where the natural period based on eq. (11.50) is

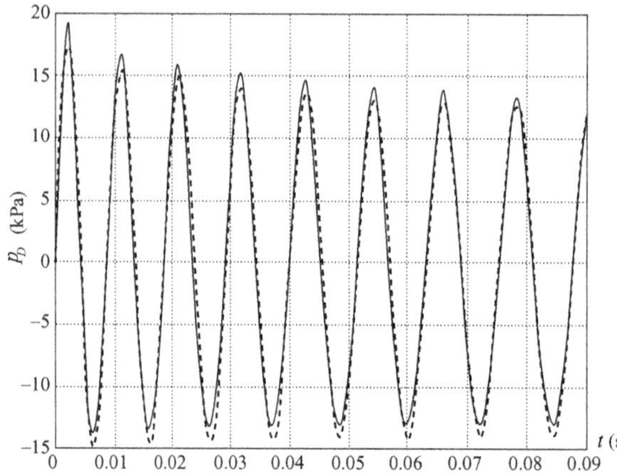

Figure 11.31. Numerical time-domain simulations of the dynamic pressure oscillations in the air cavity for the same conditions as in Figure 11.29. The damping coefficient is zero. Solid and dashed lines correspond to the nonlinear theory and the linear theory, respectively (Opedal, personal communication, 2007).

plotted for the same time interval as the results in Figure 11.26. A quasi-steady assumption of the effect of b/a has been made. We note a reasonable agreement between the experimental and theoretical values of the natural period as a function of time.

From Figure 11.26 we note a strong time decay of the pressure amplitudes in the experiments. We are only able to predict an initial decay over the duration of the time period of the changing natural period.

11.6.4 Nonlinear gas cavity analysis

Our linear analysis assumed $p_D/p_0 \ll 1$, $\rho_D/\rho_0 \ll 1$. This condition may not be fulfilled in full-scale conditions. Therefore, we do not linearize the continuity equation for the mass in the gas pocket. However, it is assumed that the quadratic velocity term can be neglected in Bernoulli's equation for the liquid flow. Gas leakage and gas inflow to the cavity are neglected. We start with eq. (11.44), which can be expressed as $\rho_{gas}\Omega = \rho_0\Omega_0$. By using the adiabatic pressure–density relationship given by eq. (11.42) in combination with the continuity equation we find that

$$p_{gas} = p_0(\Omega_0/\Omega)^{\kappa}. \qquad (11.69)$$

From eq. (11.65) it follows that

$$p_{gas} = p_0 - \rho_l \frac{d}{dt}\left\{C_1(t) - aV(t)E\left[\sqrt{1-(b/a)^2}\right]\right\}. \qquad (11.70)$$

Equations (11.69) and (11.70) can be combined as

$$\dot{C}_1(t) = \frac{p_0}{\rho_l}\left[1 - (\Omega_0/\Omega)^{\kappa}\right] + \frac{d}{dt}\left\{aV(t)E\left[\sqrt{1-(b/a)^2}\right]\right\}. \qquad (11.71)$$

Furthermore, from eqs. (11.64) and (11.47) it follows that

$$\dot{\Omega}(t) = -K(b/a)C_1(t)/K\left(\sqrt{1-(b/a)^2}\right) + V(t)\cdot a[K(b/a) - E(b/a)]. \qquad (11.72)$$

Equations (11.71) and (11.72) are two coupled nonlinear differential equations for $\Omega(t)$ and $C_1(t)$ that can be solved numerically. The initial conditions are $\Omega(0) = \Omega_0$ and $C_1(0)$ given by eq. (11.66).

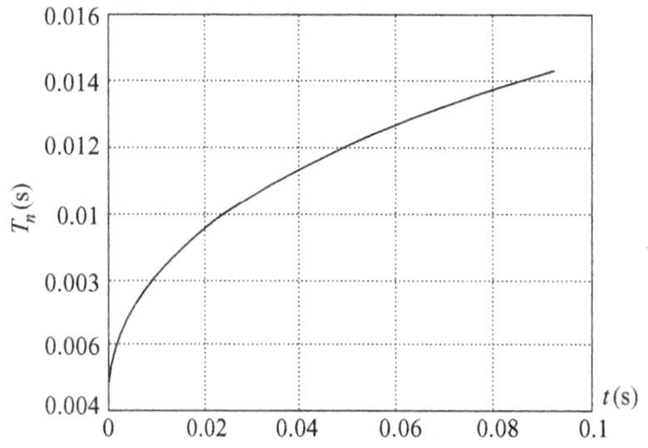

Figure 11.32. Theoretical natural period of the air cavity as a function of time spanning the duration of the experiments presented in Figure 11.26.

11.6.5 Scaling

We show that both the Froude number, $Fn = U/\sqrt{Lg}$, and the Euler number, $Eu = p_0/(\rho_l U^2)$, matter for slamming with gas cavity oscillations. In this case U is a characteristic velocity of the tank excitation (e.g., $U = L/T$, where T is the forcing period); L is a characteristic length such as the tank breadth and g is the acceleration of gravity. Furthermore, p_0 is the ullage pressure (i.e., the pressure of the gas in the tank outside any gas cavity). We start with the fact that Section 11.6.1 showed that $\sigma_n\sqrt{\Omega_0\rho_l/p_0}$ is a nondimensional natural frequency of the gas pocket oscillations. This quantity must be the same in model and full scales. We assume that the impact is associated with Froude-scale effects (i.e., we neglect, for instance, acoustic flow effects). The consequence is that σ_n must also be Froude scaled. Therefore, $\sigma_n\sqrt{L/g}$ must be the same in model and full scales. We then rewrite $\sigma_n\sqrt{\Omega_0\rho_l/p_0}$ as

$$\sigma_n\sqrt{\Omega_0\rho_l/p_0} = (\sigma_n\sqrt{L/g})\sqrt{(\Omega_0/L^2)(\rho_l U^2/p_0)(Lg/U^2)},$$

where Ω_0/L^2 is the same in model and full scales due to geometrical similarity. We then see that $\sigma_n\sqrt{\Omega_0\rho_l/p_0}$ is the same in model and full scales if the Froude number and the Euler number are the same in model and full scales.

Let us study the scaling of the dynamic pressure p_D based on the simplified analysis in Section 11.6.1, where we assumed a constant vertical velocity U_n along the gas cavity. The positive direction of U_n is downward. An eigenvalue problem was considered in Section 11.6.1.1. We now consider the case of an impact and express $U_n = U_{na}\cos\sigma_n t$, where U_{na} is the initial normal velocity at $t = 0$. This velocity is Froude scaled. By means of eqs. (11.53), (11.54), and (11.55), we can write that

$$\frac{p_D}{\sqrt{\rho_l g L p_0}} = -\sqrt{\frac{\pi}{4}a_{33}^*\kappa\left(\frac{b}{L}\right)^2\frac{L^2}{\Omega_0}}\frac{U_{na}}{\sqrt{gL}}\sin\sigma_n t. \tag{11.73}$$

Let us consider a model test based on Froude scaling of the impact velocity. Calculation of $U_{na}/\sqrt{gL}$, b/L, and Ω_0/L^2 would be the same in model and full scales. Equation (11.73) then states that $p_D/\sqrt{\rho_l g L p_0}$ is the same in model and full scales. If we call L_m the model-scale length and L_p the prototype (full-scale) length and assume ρ_l and p_0 are the same in model and full scales, the pressure in full scale is $(L_p/L_m)^{\frac{1}{2}}$ times the pressure in model scale. If Froude scaling of pressure had been used, the pressure at full scale would be L_p/L_m times the pressure at model scale (see discussion following eq. (11.1)). Therefore, Froude scaling is clearly conservative for linear gas cavity oscillations when slamming pressures associated with gas cavities are scaled. This fact was also documented numerically by Greco *et al.* (2003). Their studied case also showed that the linear behavior is appropriate in model tests, but the oscillations of the air cushion had a strong nonlinear behavior at full scale when only Froude scaling was assumed.

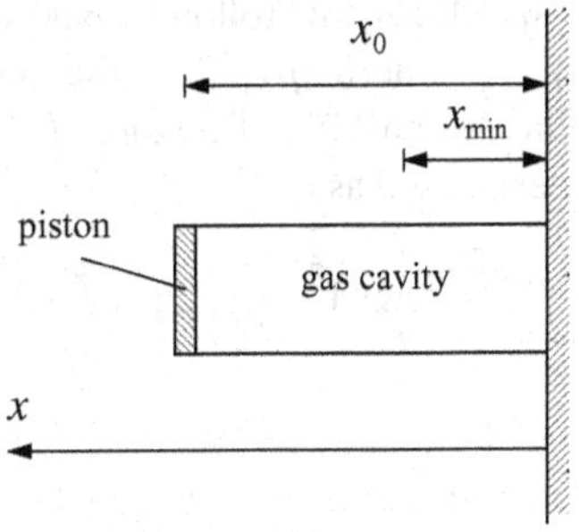

Figure 11.33. Definitions in a Bagnold-type gas cushion model.

We now generalize the scaling procedure by accounting for nonlinear cushion effects. Our focus is on the maximum pressure. The method is not limited to tank roof impact; we can also analyze the effect of a closed cavity due to impact of a steep wave against a tank wall. The procedure has similarities with the analyses of Mitsuyasu (1969), Lundgren (1969), and Bredmose and Bullock (2008), who studied wave impact on a vertical wall. The behavior of the gas cavity is approximated by a Bagnold-type model. The compression of the gas cavity is approximated by a piston that moves toward the wall (see Figure 11.33); that is, we do not account for the fact that the normal velocity of the cavity surface is nonconstant in space. We denote the distance of the piston from the tank boundary by the coordinate x, which is positive in the direction of the piston. The initial and final x-coordinates of the piston are x_0 and $x_{\min}$, respectively. The cross-sectional area of the cushion, A_c, is assumed constant. From eq. (11.69) it follows that the gas cavity pressure is

$$p_{\text{gas}} = p_0(x_0/x)^\kappa. \tag{11.74}$$

The work W done in moving the piston from x_0 to the position of maximum compression, $x_{\min}$, is

$$W = A_c\int_{x_0}^{x_{\min}}\left[p_0\left(\frac{x_0}{x}\right)^\kappa - p_0\right]d(-x) = \frac{p_0 A_c x_0}{\kappa-1}\left[\left(\frac{x_{\min}}{x_0}\right)^{1-\kappa} + (\kappa-1)\frac{x_{\min}}{x_0} - \kappa\right]. \tag{11.75}$$

From eq. (11.74) it follows that $x_{\min}/x_0 = (p_{\max}/p_0)^{-1/\kappa}$, where $p_{\max}$ is the maximum pressure in the cushion. Equation (11.75) can then be reexpressed as

$$W = \frac{p_0 A_c x_0}{\kappa - 1}\left[\left(\frac{p_{\max}}{p_0}\right)^{\frac{\kappa-1}{\kappa}} + (\kappa-1)\left(\frac{p_{\max}}{p_0}\right)^{-\frac{1}{\kappa}} - \kappa\right]. \quad (11.76)$$

We postulate that this work is equal to the initial kinetic energy associated with the impact. The assumption can be defended for linear gas cushion oscillations. If the gas cushion oscillations are nonlinear, the assumption is appropriate when the inflow is one-dimensional or radial in two- and three-dimensional flow. However, the assumption needs to be justified for a general inflow (Korobkin, personal communication, 2008). We can formally set up

$$W = \tfrac{1}{2}\rho_l \int_{\Omega_I} u^2 \mathrm{d}\Omega, \quad (11.77)$$

where Ω_I symbolizes a volume and u is the liquid velocity at the initial time of compression of the gas cavity. We do not know, without detailed studies, what the kinetic energy of the right-hand side of eq. (11.77) is. What is important for our further studies is that the kinetic energy is associated with the initial time of the compression and, therefore, Froude-scales as long as we consider a potential flow model of an incompressible liquid with gravity. We now nondimensionalize eq. (11.77) and show that both the Euler number and the Froude number determine the maximum pressure. A characteristic length L and a characteristic velocity U (see Section 11.2) are introduced. Equation (11.77) can then be rewritten as

$$\left(\frac{A_c}{L^\alpha}\right)\left(\frac{x_0}{L}\right)\frac{Eu}{\kappa-1}\left[\left(\frac{p_{\max}}{\rho_l U^2}Eu^{-1}\right)^{\frac{\kappa-1}{\kappa}} + (\kappa-1)\left(\frac{p_{\max}}{\rho_l U^2}Eu^{-1}\right)^{-\frac{1}{\kappa}} - \kappa\right] = \frac{1}{2L^{\alpha+1}}\int_{\Omega_I}\left(\frac{u}{U}\right)^2 d\Omega, \quad (11.78)$$

where $Eu = p_0/(\rho_l U^2)$ is the Euler number. Furthermore, α is 1 for a two-dimensional problem and 2 for a three-dimensional problem. Because the right-hand side of eq. (11.78) is a function of the Froude number, $Fn = U/\sqrt{gL}$, we have shown that $p_{\max}/\rho_l U^2$ is a function of Fn and Eu. The derivation implies implicitly that we consider geometrically similar tanks and a given nondimensional tank excitation.

The ratio between eq. (11.77) in full and model scales is

$$\frac{p_{0_p}\left[(p_{\max_p}/p_{0_p})^{\frac{\kappa-1}{\kappa}} + (\kappa-1)(p_{\max_p}/p_{0_p})^{-\frac{1}{\kappa}} - \kappa\right]}{p_{0_m}\left[(p_{\max_m}/p_{0_m})^{\frac{\kappa-1}{\kappa}} + (\kappa-1)(p_{\max_m}/p_{0_m})^{-\frac{1}{\kappa}} - \kappa\right]} = \frac{\rho_{l_p} L_p}{\rho_{l_m} L_m}. \quad (11.79)$$

We have assumed that κ is the same for the model- and full-scale gas which is generally not true. The subscripts p and m in eq. (11.79) denote full (prototype) and model scales, respectively. If we know $p_{\max}$ from model tests, eq. (11.79) determines $p_{\max}$ in full scale.

If $p_{\max} - p_0$ is small relative to p_0, eq. (11.79) can be simplified. This fact follows from Taylor-series expansion of the expression for W given by eq. (11.76) in terms of $\Delta = (p_{\max} - p_0)/p_0$, which gives

$$W = \tfrac{1}{2} p_0 x_0 A_c \kappa^{-1}\Delta^2 + O(\Delta^3). \quad (11.80)$$

Using eq. (11.80) in eq. (11.79) and neglecting terms of $O(\Delta^3)$ gives

$$p_{\max_p} - p_{0_p} = (p_{\max_m} - p_{0_m})\sqrt{(p_{0_p}/p_{0_m})(\rho_{l_p}/\rho_{l_m})(L_p/L_m)}$$

If only Froude scaling had been applied, $p_{\max_p} - p_{0_p}$ would be $\rho_{0_p} L_p/\rho_{0_m} L_m$ times $p_{\max_m} - p_{0_m}$. Therefore, it is conservative to use Froude scaling when $p_{0_p} = p_{0_m}$; however, we should realize that $(p_{\max_p} - p_{0_p})/p_{0_p}$ may not be small even though $(p_{\max_m} - p_{0_m})/p_{0_m}$ is small.

Another way to show how to scale model test results with gas cavity oscillations is to introduce nondimensional variables in the nonlinear governing equations and the initial conditions defined in Section 11.6.4. We define the following nondimensional variables:

$$\Omega^* = \frac{\Omega}{L^2},\ a^* = \frac{a}{L},\ b^* = \frac{b}{L}\ t^* = t\sqrt{\frac{g}{L}},$$
$$V^* = \frac{V}{\sqrt{gL}},\ C_1^* = \frac{C_1}{L\sqrt{gL}},\ p_0^* = \frac{p_0}{\rho_l g L}. \quad (11.81)$$

We divide eq. (11.71) by gL, which gives

$$\frac{dC_1^*}{dt^*} = p_0^*[1 - (\Omega_0^*/\Omega^*)^\kappa] + \frac{d}{dt^*}\left\{a^* V^* E\left[\sqrt{1-(b^*/a^*)^2}\right]\right\}.$$

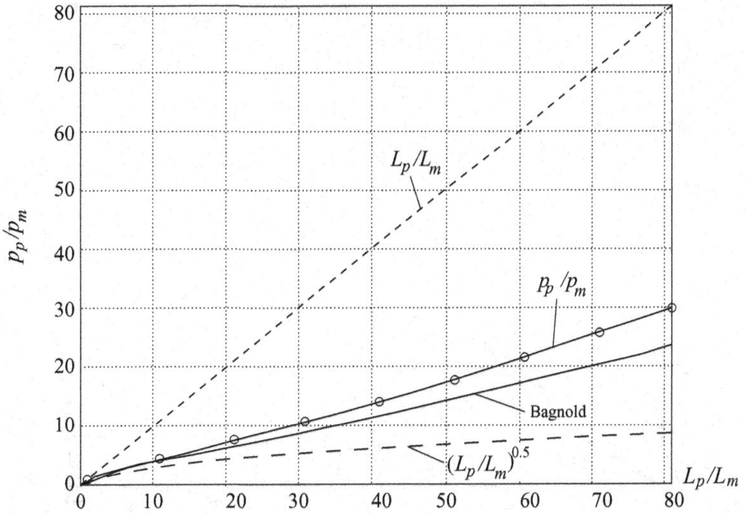

Figure 11.34. Calculated ratio p_p/p_m between maximum *dynamic* gas cavity pressure in full scale and model scale presented as a function of the scaling factor L_p/L_m between full- and model-scale lengths. The model scale is based on results for the same conditions presented in Figure 11.31. The calculations are based on a nonlinear analysis. Only Froude scaling has been used. The ullage pressure does not vary with L_p/L_m. Note that the scaling shown is not general and depends on the model-scale conditions (Abrahamsen, personal communication, 2008).

Dividing eq. (11.72) by $L\sqrt{gL}$ gives

$$\frac{d\Omega^*}{dt^*} = -K(b^*/a^*)\,C_1^*/K\left(\sqrt{1-(b^*/a^*)^2}\right) + V^*a^*\left[K(b^*/a^*) - E(b^*/a^*)\right].$$

The initial conditions are $\Omega^*(0) = \Omega_0^*$ and $C_1^*(0) = a^*(0)V^*(0)E\{\sqrt{1-[b^*(0)/a^*(0)]^2}\}$ in accordance with eq. (11.66). According to eq. (11.70), the nondimensional gas pressure can be expressed as

$$\frac{p_{gas}}{\rho_l g L} = p_0^* - \frac{d}{dt^*}\left\{C_1^* - a^*V^*E\left[\sqrt{1-(b^*/a^*)^2}\right]\right\}. \quad (11.82)$$

These equations show that if V is Froude-scalable, Ω_0 scales with L^2, a and b scale with L, and p_0^* is constant, then $p_{gas}/(\rho_l g L)$ does not vary with changing L. The fact that p_0^* is a constant means that p_0 must vary linearly with L. Therefore, the ullage pressure in model scale must be lower than in full scale. Because of the fact that the pressure is normally atmospheric outside the tank, the static pressure difference between the exterior and interior of the tank in model scale would be large. This condition requires a consideration of sufficient stiffening of the tank in model scale, which is challenging when the effect of hydroelasticity must also be properly modeled.

We should note that we can write $p_0^* = Eu \cdot Fn^2$ and that eqs. (11.81) and (11.82) are consistent when both the Euler number and the Froude number are the same in model and full scales, which is due to the fact that p_0 must vary linearly with L. The consequence is that the left-hand side of eq. (11.73) expresses the fact that the pressure is varying linearly with L.

We elaborate more on the fact that it is normal in tank testing to only preserve Froude scaling. We assume in the following that p_0 is the same in model and full scales. We illustrate the consequence of this fact in terms of scaling the pressure by using the nonlinear governing equations.

Figure 11.34 shows how the ratio p_p/p_m between maximum *dynamic* air cavity pressure in full scale and model scale behaves as a function of the ratio $\Lambda_L = L_p/L_m$ between full-scale and model-scale length. The previous model test conditions are used. Froude scaling of the inflow conditions has been made, which means that velocities in full scale have been obtained by multiplying model-scale velocities by $\sqrt{\Lambda_L}$. Length scales at full scale have been obtained by multiplying model-scale length by Λ_L. Accelerations are the same in model and full scales. If the dynamic pressure should be in Froude scale, then the dynamic pressure in full scale should be

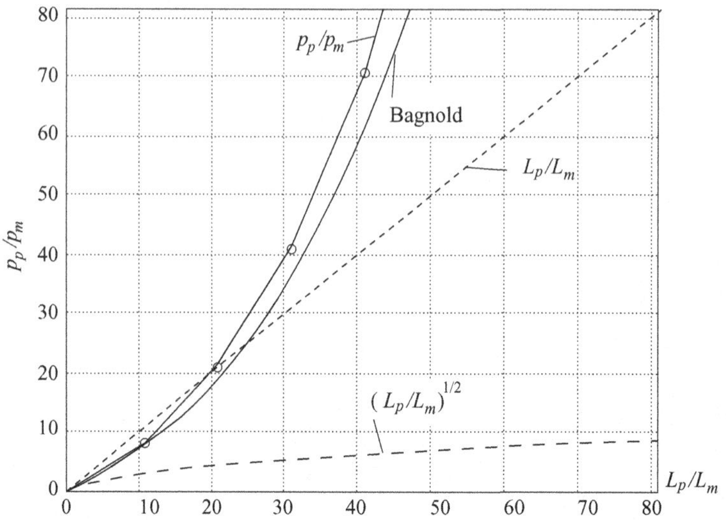

Figure 11.35. Calculated ratio p_p/p_m between maximum *dynamic* gas cavity pressure in full and model scales presented as a function of the scaling factor L_p/L_m between full- and model-scale lengths. The model scale is based on results for the same conditions presented in Figure 11.31 except that the air cushion volume is $0.1\Omega_0$. The calculations are based on a nonlinear analysis. Only Froude scaling has been used. The ullage pressure does not vary with L_p/L_m. (Abrahamsen, personal communication, 2008).

obtained by multiplying the model-scale pressure by Λ_L. Figure 11.34 shows that Froude scaling provides a conservative estimate for this particular case. Furthermore, our simplified linear analysis showed that the dynamic pressure should scale with $\sqrt{\Lambda_L}$, which gives nonconservative results. The results of using a Bagnold-type analysis based on eq. (11.79) are also presented in Figure 11.34. The results are in reasonable agreement. However, we should not expect the same results with a Bagnold-type model and a model that accounts for the fact that the normal velocity of the cavity surface is nonconstant in space.

The scaling shown in Figure 11.34 is not general (e.g., Froude scaling does not need to be conservative). To illustrate this fact we have chosen the initial gas cavity volume equal to 10% of the volume presented in Figure 11.31. The consequence is a higher maximum pressure. The results are presented in Figure 11.35. The fact that Froude scaling does not need to be conservative was recognized by Bredmose and Bullock (2008).

Nonlinearities play a more significant role in the full scale than in the model scale. We illustrate this fact with an example by choosing Λ_L equal to 81. The previous model test conditions are mainly used. However, the deceleration term V_1 in the inflow velocity $V(t) = V_0 + V_1 t$, where the initial impact time is $t = 0$, has been changed to -1.35 m s^{-2} to avoid $V(t)$ becoming negative during the selected full-scale simulation time. Froude scaling of the inflow conditions has been done. The results for the total pressure from the linear and nonlinear analyses are presented in Figure 11.36(a). The nonlinear and linear models give very different time histories for full-scale conditions. We note that the linear model predicts nonphysical negative total pressures during certain intervals, whereas the nonlinear model gives significant reduction in the pressure amplitude with time even though damping, as discussed in Section 11.6.2, has been excluded. The maximum predicted total pressure from the nonlinear model is about 8 bar in full-scale conditions. However, we should recall that the scaled results are based on using water with a density $\rho = 10^3$ kg m^{-3} and that the liquid density influences the magnitude of the pressure. We note that the pressure changes rapidly with time in the initial phase.

The corresponding air cavity volume predicted from the nonlinear model is presented in Figure 11.36(b). The presence of nonlinearities is less evident in the air cushion volume than in the pressure.

How the initial gas cushion volume Ω_0 influences the results is shown in Figure 11.37. All

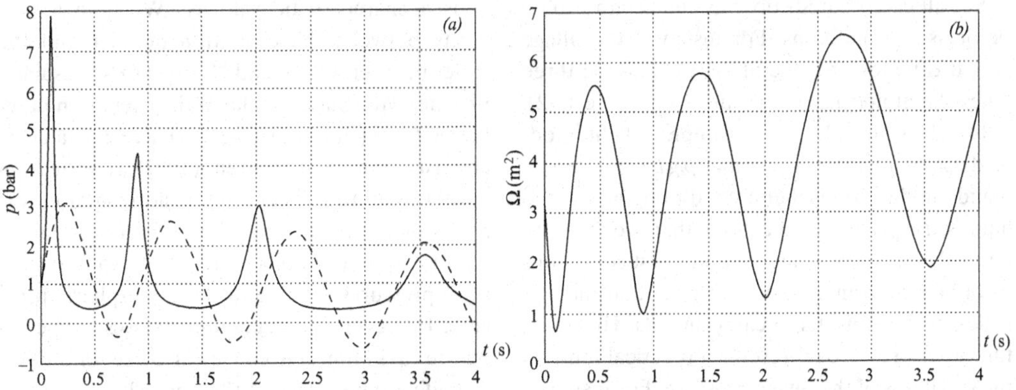

Figure 11.36. (a) *Total* air cavity pressure and (b) air cavity volume in full-scale conditions based on scaling the model scale conditions presented in Figure 11.31 with length ratio L_p/L_m equal to 81; $b = 5.6295$ m, $a = 5.67 + 7.029t$ (m), $\Omega_0 = 2.8344$ m^2, $V_0 = 5.4$ m s^{-1}, $V_1 = -1.35$ m s^{-2}. The solid line denotes the results of nonlinear theory. The dashed line in part (a) corresponds to linear theory (Opedal, personal communication, 2007).

other parameters remain unchanged. The smaller the value of Ω_0, the larger the maximum pressure. Furthermore, the oscillation period is, as expected, influenced by Ω_0.

The time history of the pressure determines the influence of hydroelasticity on structural stresses. From the analysis of a simple mass–spring system it is well known that the ratio between the duration of the loading and the natural period plays an important role in determining the dynamic amplification factor. There are many natural structural periods to be considered for a membrane structure. The relatively rapid variation of the pressure during the initial phase should be of particular concern for dynamic amplification.

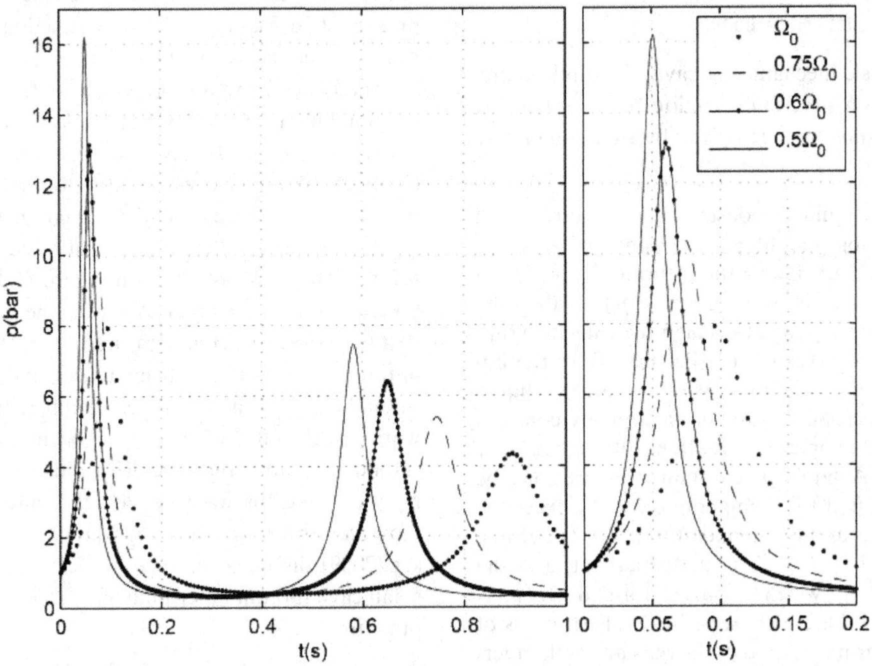

Figure 11.37. The effect of the initial air cushion volume on the *total* air cavity pressure in full-scale conditions based on scaling the model-scale conditions presented in Figure 11.31 with length ratio L_p/L_m equal to 81; $b = 5.6295$ m, $a = 5.67 + 7.029t$ (m), $\Omega_0 = 2.8344$ m^2, $V_0 = 5.4$ m s^{-1}, $V_1 = -1.35$ m s^{-2} (Abrahamsen, personal communication, 2008).

Not all cargo liquids operate under atmospheric pressure conditions. For instance, the ullage pressure p_0 for propane at $-10°C$ may be three times the atmospheric pressure (see Table 11.2). Tables 11.1 and 11.2 give examples of the modified Euler number, $C_E = p_0/(\rho_l g L)$, for model liquids under atmospheric conditions and cargo liquids, respectively. We note that we have to lower the ullage pressure in model test conditions to obtain the same C_E-value in model and full scales of the considered cargo liquids. However, the vapor pressure represents a practical limit on the smallness of the ullage pressure. For instance, the vapor pressure for water is 1.704, 2.337, and 3.167 kPa at 15°, 20°, and 25°C, respectively.

Can we apply the Bagnold-type model in scaling model test results for the case of small cavitation numbers when the gas cavity oscillations are heavily damped? The answer is yes if it is appropriate to assume an adiabatic pressure-density relationship during the compression of the gas cavity. Condensation is an example of when this relationship is questionable. Maillard and Brosset (2009) report that during water vapor tests, the condensation occurs quickly due to the overpressure.

11.7 Cavitation and boiling

The basic mechanics of cavitation and boiling is similar. A precise distinction between cavitation and boiling has been given by Brennen (1995) as follows:

1. A liquid at constant temperature could be subjected to a decreasing pressure, p, which falls below the saturated vapor pressure, p_v. The value of $(p_v - p)$ is called the tension, Δp, and the magnitude at which rupture occurs is the tensile strength of the liquid, Δp_C. The process of rupturing a liquid by decrease in pressure at roughly constant liquid temperature is often called cavitation.
2. A liquid at constant pressure may be subjected to a temperature, T, in excess of the saturation temperature, T_S. The value of $\Delta T = T - T_S$ is the superheat, and the point at which vapor is formed, ΔT_C, is called the critical superheat. The process of rupturing a liquid by increasing the temperature at roughly constant pressure is often called boiling.

The cavitation number is defined as $(p_0 - p_v)/(\frac{1}{2}\rho_l U^2)$, where p_0 is the ullage pressure and U is a characteristic velocity. We note a similarity between the cavitation number and the Euler number. Olsen and Hysing (1974) assumed Froude scaling and redefined the cavitation number as $C_v = (p_0 - p_v)/(\rho_l g L)$, where L is a characteristic length. Faltinsen (1997) reported on cavitation during drop tests of horizontal elastic plates on a free surface. The cavitation occurred because the vibrations of the plate caused negative pressures relative to the atmospheric pressure. However, the maximum stresses happened before cavitation started (see also Section 11.9).

Boiling takes place in the upper layer of LNG and the other cargo liquids listed in Table 11.2. Because $p_0 = p_v$, the cavitation number is zero. All the cargo liquids listed in Table 11.2 operate at zero cavitation number, whereas it is only boiling water of the listed liquids in Table 11.1 that can achieve zero cavitation number in model experiments under atmospheric conditions. Olsen and Hysing (1974) presented experimental slamming pressure results with water at 20, 95, and 100°C. If we use the tank breadth l as the characteristic length, the cavitation number C_v is 7.39, 1.29, and 0, respectively. The slamming pressure coefficient $p/(\rho_l g l)$ for the lower corner of the upper chamfer with a 10% exceedance level is presented in Figure 11.38 as a function of the nondimensional forcing period $T\sqrt{g/l}$. The ratio of liquid depth to tank breadth is $h/l = 0.4$ and the nondimensional sway amplitude η_{2a}/l is 0.10. Correponding results for the slamming coefficients at the upper corner of the lower chamfer for $h/l = 0.12$ showed a similar variation with the cavitation number. Because the different results for water at 20°C presented in Figure 11.38 show a variation that is comparable with the variation as a function of the temperature, that is, the cavitation number, it is difficult to conclude about the influence of the cavitation number. In other words, such influence seems to be limited. We should also note that the Reynolds number is not the same for water at 20°, 95°, and 100°C. However, the experiments by Maillard and Brasset (2009) discussed in Section 11.2 show that a small cavitation number influences the slamming pressure.

11.8 Acoustic liquid effects

The *Cauchy number*, $Ca = \rho_l U^2/E_l$, characterizes the effect of compressibility on the flow with no

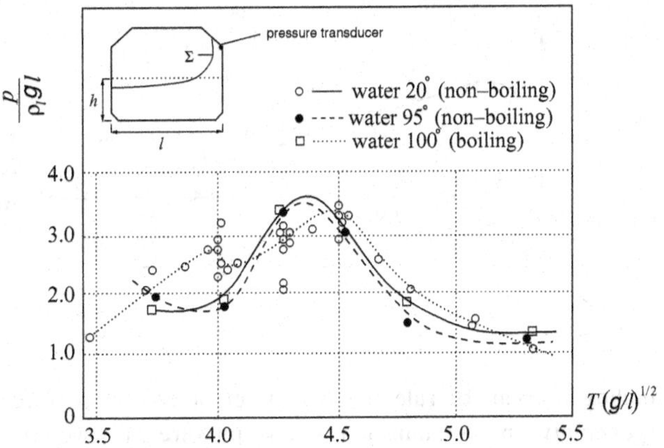

Figure 11.38. Influence of cavitation: 10% exceedance values of the slamming pressure coefficient $p/(\rho_l g l)$ versus nondimensional forcing period $T\sqrt{g/l}$. The ratio of liquid depth to tank breadth is $h/l = 0.4$. The nondimensional sway amplitude is $\eta_{2a}/l = 0.1$ (Olsen & Hysing, 1974) (Det Norske Veritas AS).

mixture between gas and liquid. In this case E_l is the bulk modulus for elasticity of the liquid. The effect of liquid compressibility on slamming is, in general, unimportant in a liquid with no gas bubbles. We say "in general, unimportant," referring to the discussion in Section 11.9.6, where it is shown that a natural period of a structural mode causing large structural stresses during impact in a membrane tank can be of the order of the time scale of acoustic effects in a liquid with no gas bubbles. Liquid compressibility always has an effect on the slamming pressure in an initial time. Because the speed of sound is high in a liquid (from 975 to 1,774 m s^{-1} for the cargo liquids listed in Table 11.2), this initial time is very small and sufficiently small, in general, for the tank structure to not react in terms of large stresses.

Excessive boiling leads to a change of the compressibility properties of the liquid. The speed of sound in a liquid–gas mixture can be substantially lower than that in a separate gas or liquid. Let us illustrate this by using Wood's (1930) formula. The sound speed can be expressed as

$$c_0 = \sqrt{\frac{E_l E_g}{[(1-\chi)\rho_l + \chi\rho_g][(1-\chi)E_g + \chi E_l]}}, \tag{11.83}$$

where χ is the void fraction equal to 0 in the case of no gas bubbles suspended in the liquid and equal to 1 in the case there is only gas and no liquid; E_l and E_g mean the bulk modulus of the liquid and gas, respectively. Furthermore, ρ_l and ρ_g are the mass density of the liquid and gas, respectively. We exemplify the formula for water and air at the temperature 20°C and choose $E_l = 2.2 \cdot 10^9$ Pa, $E_g = 1.42 \cdot 10^5$ Pa, $\rho_l = 999$ kg m^{-3}, and $\rho_g = 1.293$ kg m^{-3}. When there is no mixture between air and water, this gives a sound speed of $\sqrt{E_l/\rho_l} = 1,484$ m s^{-1} and $\sqrt{E_g/\rho_g} = 332$ m s^{-1} for water and air, respectively. Figure 11.39 illustrates the effect of mixture between air and water and shows how the speed of sound depends on the void fraction χ based on using eq. (11.83) and the aforementioned values of density and bulk modulus. When $0.0025 < \chi < 1$, the sound speed is lower than the sound speed of air. The minimum value of the speed of sound is only 23.828 m s^{-1}, which occurs at $\chi = 0.5$. We would certainly have to take the effect of compressibility into consideration for slamming when the speed of sound is this low.

Because a mixture of gas and liquid can happen during violent sloshing and because LNG is

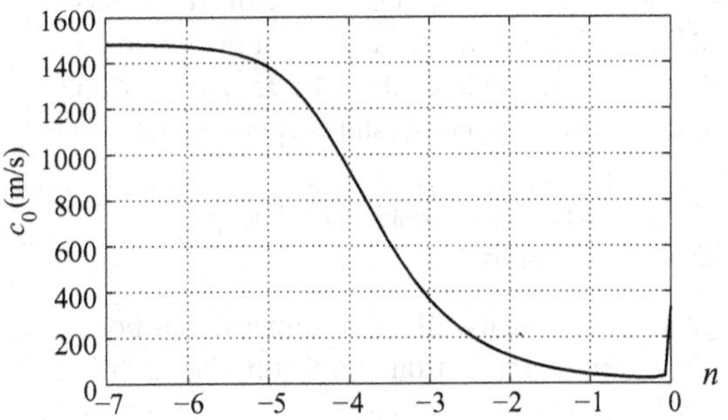

Figure 11.39. The speed of sound, c_0, in meters per second for a mixture between water and air as a function of the void fraction $\chi = 10^n$.

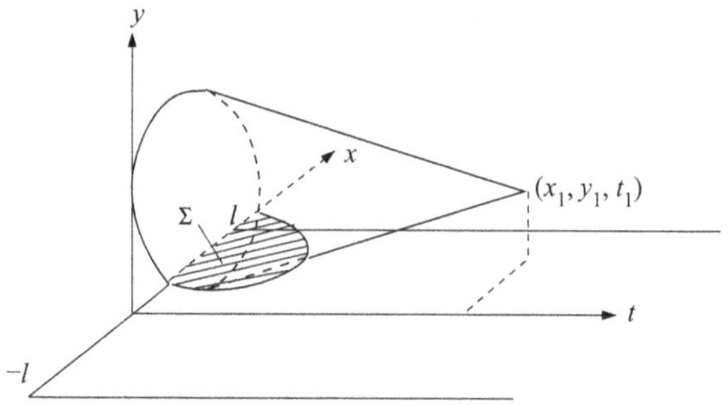

Figure 11.40. Representation in the (x, y, t)-space of an impacting two-dimensional body with horizontal bottom (Ogilvie, 1963).

boiling, we cannot rule out the effect of compressibility on slamming loads. The mixture of LNG and gas is not uniform (i.e., the same at any point). It is only a layer next to the free surface of LNG that is boiling. Because sloshing and impacts affect the pressure distribution in the fluid and condensation can occur due to overpressure, the void fraction may vary in time and space. In evaluating the slamming loads we must also account for the fact that the density of the mixture is

$$\rho_{\text{mix}} = (1 - \chi)\rho_l + \chi\rho_g. \tag{11.84}$$

The solution of the sloshing/slamming problem can be approximated by assuming irrotational flow (i.e., there exists a velocity potential φ). Furthermore, we can linearize the problem (see, e.g., Ogilvie, 1963), which leads to the wave equation $\nabla^2\varphi - c_0^{-2}\varphi_{tt} = 0$, where $\varphi_{tt} = \partial^2\varphi/\partial t^2$ (see Section 2.2.2.2). The dynamic free-surface condition is $\varphi = 0$, that is, similar to what we specified in solving the Wagner problem in Section 11.3. The body boundary condition follows from requiring no flow through the body; that is, the normal component $\partial\varphi/\partial n$ of the liquid velocity at the body surface is equal to the normal component of the body velocity. A review of acoustic effects during liquid-entry of a body is presented by Korobkin (1996c). Our following discussion of acoustic effects starts out by assuming constant sound speed and density. Sections 11.8.1 and 11.8.2 consider liquid entry of bodies with acoustic effects and have relevance for the impact of the liquid on the tank surface in a similar way as discussed in Section 11.3 for incompressible liquid impact.

11.8.1 Two-dimensional liquid entry of body with horizontal bottom

Ogilvie (1963) analyzed a two-dimensional body with a horizontal bottom impacting on a free surface with a constant water-entry velocity V. If we introduce a horizontal coordinate x, then the flat bottom corresponds to $-l \le x \le l$. The liquid domain is $y \ge 0$ and the body boundary condition is $\partial\varphi/\partial y = V$ on $-l \le x \le l$, $y = 0^+$. The procedure neglects the fact that a gas cushion is initially created between the bottom of the body and the free surface (see Section 11.6 and Figure 11.25). Furthermore, the rigid-body assumption may be questionable in practice.

Because the considered differential equation is hyperbolic (Stoker, 1992), the velocity potential at an arbitrary point (x_1, y_1, t_1) is associated with a "domain of dependence," which is the conical region $c_0 t < c_0 t_1 - \sqrt{(x - x_1)^2 + (y - y_1)^2}$ (see Figure 11.40). The flow at (x_1, y_1, t_1) depends only on events occurring at times and places in the interior of the cone. Ward (1955) shows that the solution can be expressed as

$$\varphi(x_1, y_1, t_1) = -\frac{c_0}{\pi}\iint_\Sigma \left(\frac{\partial\varphi}{\partial y}\right)_{y=0} \frac{\mathrm{d}t\,\mathrm{d}x}{\sqrt{c_0^2(t - t_1)^2 - (x - x_1)^2 - y_1^2}},$$

where Σ lies in the y-plane and is bounded by the line $t = 0$ and the hyperbola which results from the intersection of the y-plane with the cone. When Σ includes only a region where $-l < x < l$, disturbances outside $-l < x < l$ are not felt at (x_1, y_1, t_1). The local behavior must be the same as if the whole x-axis was subjected to the same boundary condition, $\varphi_y = V$. The consequence is that the solution does not depend on x and the wave equation becomes $\varphi_{yy} - c_0^{-2}\varphi_{tt} = 0$. A solution satisfying the wave equation and the body boundary condition is $\varphi = (y - c_0 t)V$. When $y - c_0 t$ is equal to a constant, the value of φ does not vary (i.e., constant values

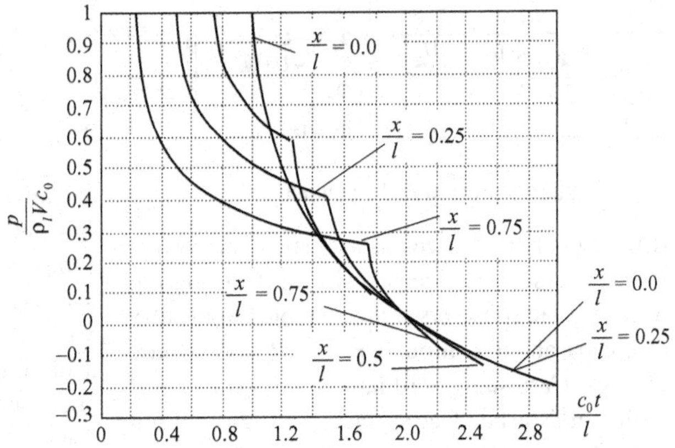

Figure 11.41. Pressure p on the body surface during liquid entry of a two-dimensional flat-bottomed body with constant velocity V. The flat bottom has x-values between $-l$ and l. The effect of liquid compressibility is considered; c_0 is the speed of sound (Ogilvie, 1963).

of φ propagate with velocity c_0 in the positive y-direction). The solution $\varphi = (y + c_0 t)V$ also satisfies the wave equation and the body boundary condition. However, it corresponds to a propagation of constant values of the velocity potential in the negative y-direction, which is not possible. Therefore, our solution is $\varphi = (y - c_0 t)V$. The pressure is $p = -\rho_l \partial\varphi/\partial t = \rho_l V c_0$ according to the linearized Bernoulli equation without gravity.

Figure 11.41 presents the theoretical pressure distribution on the flat bottom as a function of time. We note that the pressure at the center of the bottom, $x/l = 0$, remains equal to $\rho_l V c_0$ (i.e., the so-called acoustic pressure) until $t = l/c_0$ (where $t = 0$ is the initial time of the impact). At this time instant relief waves starting from the edges of the bottom have reached the middle of the bottom. They are called "relief waves" because they reduce the pressure. The time history of the pressure for $x/l = 0.25, 0.5$, and 0.75 is also given in the figure. The time when the pressure becomes smaller than the acoustic pressure corresponds to $t = (l - x)/c_0$ (i.e., the time it takes for acoustic waves starting from the closest edge of the bottom to propagate up to the given x-value). It can be shown that the pressure becomes smaller than the ambient pressure at $c_0 t/l = 2$ (see Figure 11.41). At this time instant the separation of the liquid from the plate may start. This effect was studied by Korobkin (1993).

The previous results show that the considered acoustic problem has a time scale l/c_0. To find out if acoustic effects are important, we should relate this time scale to the natural periods of the tank's structural vibrations causing large structural stresses. If the acoustic time scale l/c_0 is very small relative to important structural natural periods, we can neglect the effect of liquid compressibility; that is, we can use the Laplace equation instead of the wave equation to solve our problem. However, the compressible liquid effects are not negligible at the time l/c_0. In a later example with liquid entry of a parabolic contour, we quantify the time required for negligible acoustic effects for that particular problem.

The acoustic pressure is very high for a liquid without any entrained gas bubbles. For instance, if we consider the case with water and air presented in Figure 11.39, the speed of sound for $\chi = 0$ is $1,484\ \mathrm{m\,s^{-1}}$. When $V = 10\ \mathrm{m\,s^{-1}}$, the acoustic pressure is 146 times the atmospheric pressure. If we choose as an example $l = 1\,\mathrm{m}$, we see that the time scale of the acoustic effects is $7 \cdot 10^{-4}$ s. The presence of entrained gas bubbles can have a significant effect on both the acoustic pressure and the time scale.

If we had solved the previous problem by assuming an incompressible liquid, the slamming force by a Wagner method without flow separation at the edges would be infinite at the impact instant and zero just after. The reason is that the slamming loads of such a method are associated with the time rate of change of the wetted surface, which changes initially from zero to $2l$ instantly and does not vary afterward. Another difference between an incompressible and a compressible liquid is that the impact is felt immediately in the whole liquid for an incompressible liquid (i.e., the speed of sound is infinite), whereas the disturbances propagate with a finite speed of sound in a compressible liquid.

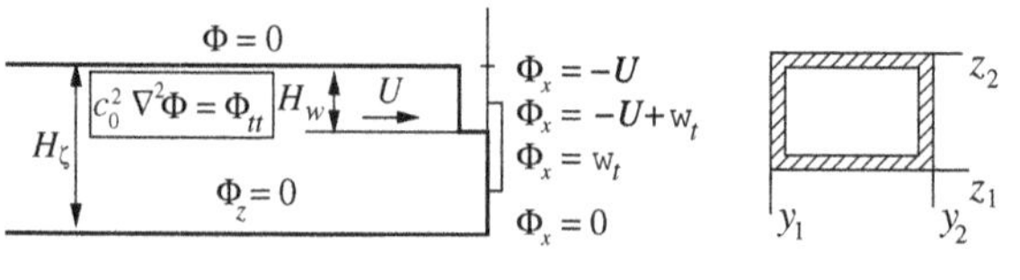

Figure 11.42. Formulation of the acoustic problem with a hydraulic jump impacting on a partly elastic wall (Ten *et al.*, 2008); $\Phi_{tt} = \partial^2\Phi/\partial t^2$.

11.8.2 Liquid entry of parabolic contour

Korobkin (1996c) discusses acoustic effects when the wetted surface changes with time during liquid entry. He considers as an example liquid entry of the parabolic contour $y = \frac{1}{2}x^2/R - Vt$, where Oxy is a Cartesian coordinate system with $y = 0$ as the undisturbed free surface. The constant liquid entry velocity, V, is along the negative y-axis and $t = 0$ corresponds to the initial impact time. The intersections between the contour and $y = 0$ are $x_\pm = \pm\sqrt{2RVt}$ and move with speed $\dot{x}_+ = \sqrt{RV/2t}$. When $dx_+/dt > c_0$, no disturbances can reach the free surface. This condition is referred to as the supersonic stage. The duration

$$T_d = \frac{1}{2}RVc_0^{-2} \tag{11.85}$$

of the supersonic stage follows by setting $\dot{x}_+ = c_0$. The dimension of the wetted part of the body after T_d is $2x_+ = 2RV/c_0$. When there is initially zero angle between the impacting free surface and the body surface (e.g., liquid entry of a parabolic contour), the initial pressure on the body for an incompressible liquid is according to Wagner's (1932) theory infinite while it is equal to the product of entry speed, density, and speed of sound (acoustic pressure) for a compressible liquid. Both the theoretical pressures in an incompressible and compressible liquid will decay with time and come closer to each other. Korobkin (1996c) illustrates the time it takes for the acoustic solution for a compressible liquid to reach the Wagner solution for an incompressible liquid by considering the pressure on the lowest point of the body. The pressures are quite close after about $6T_d$. The studied case also has relevance for roof impact of a wave with a parabolic free-surface shape.

11.8.3 Hydraulic jump impact

Ten *et al.* (2008) analyzed the acoustic effect of a hydraulic jump with a vertical front impacting a vertical wall. The scenario is illustrated in Figure 11.42. Three-dimensional flow effects and hydroelasticity are considered. A Cartesian coordinate system $Oxyz$ is introduced with origin at the intersection between the wall and the flat rigid bottom. Positive z is upward and the hydraulic jump propagates with velocity U in the positive x-direction. The liquid has a height H_ζ and a width L and semi-infinite extent in the negative x-direction. Before the impact, $t < 0$, a portion of the liquid boundary $x = 0, 0 \le z \le H_\zeta - H_w$, is in contact with the vertical wall. Ten *et al.* (2008) considered that a region of the wall, $S \equiv [y_1, y_2] \times [z_1, z_2]$, is elastic. The deflection of the wall in the x-direction is denoted w. The boundary-value problem for the velocity potential Φ caused by the impact is shown in Figure 11.42. The pressure is given by $p_D = -\rho_l \partial\Phi/\partial t$. When solving the problem, nondimensional variables are introduced. We now switch notation and use a prime to indicate dimensional variables. The relationships between dimensional variables and nondimensional variables are as follow:

$$x' = H_\zeta x, \quad y' = H_\zeta y, \quad z' = H_\zeta z, \quad \Phi' = UH_\zeta\Phi,$$
$$H_w = H_\zeta h_w, \quad t' = \frac{H_\zeta}{c_0}t, \quad p' = \rho_l U c_0 p,$$
$$w' = \frac{H_\zeta U}{c_0}w, \quad L = H_\zeta l.$$

The wave equation in nondimensional variables becomes $\nabla^2\Phi = \partial^2\Phi/\partial t^2$. The boundary conditions are $\partial\Phi/\partial y = \Phi_y = 0$ at $y = 0$ and $y = l$. The dynamic free-surface condition and the bottom condition are $\Phi = 0$ at $z = 1$ and $\partial\Phi/\partial z = \Phi_z = 0$ at $z = 0$, respectively. The boundary condition at $x = 0$ can be formulated as $\Phi_x = -\chi_1 + w_t\chi_2$, where

$$\chi_1(y, z) = \begin{cases} 1, & 0 < y < l, \quad 1 - h_w = h_1 < z < 1, \\ 0, & 0 < y < l, \quad 0 < z < h_1 = 1 - h_w, \end{cases}$$

where $\chi_2(y, z)$ is 1 when $(y, z) \in S$ and zero otherwise. The initial conditions are zero Φ and Φ_t at $t = 0$. The solution to the problem is found by expressing the velocity potential as

$$\Phi(x, y, z, t) = \sum_{\substack{i=0 \\ j=1}}^{\infty} X_{ij}(x, t)V_{ij}(y, z), \tag{11.86}$$

where

$$V_{ij} = N_{ij}\cos[\lambda_i y]\cos[\mu_j z],$$
$$\lambda_i = \frac{\pi i}{l} \quad (i = 0, 1, 2, \ldots),$$

$$\mu_j = \tfrac{1}{2}\pi(2j-1)\ (j=1,2,\ldots),$$
$$N_{ij} = \begin{cases} 2/\sqrt{l} & \text{for } i \geq 1, \\ \sqrt{2/l} & \text{for } i = 0, \end{cases}$$

and V_{ij} satisfies the dynamic free-surface condition at $z = 1$, the bottom condition at $z = 0$, and the wall conditions at $y = 0$ and $y = l$. Substituting eq. (11.86) into the wave equation gives partial differential equations for $X_{ij}(x, t)$ that must satisfy the boundary conditions at $x = 0$ and the initial conditions, which in our case take the following form:

$$\frac{\partial \Phi}{\partial x}(0, y, z, t) = -\chi_1(y, z) + w_t(y, z, t)\chi_2(y, z);$$
$$\Phi(x, y, z, 0) = \frac{\partial \Phi}{\partial t}(x, y, z, 0) = 0. \qquad (11.87)$$

Separation of the spatial variables in the wave equation gives

$$X''_{ij} - \alpha_{ij}^2 X_{ij} - \ddot{X}_{ij} = 0, \quad \alpha_{ij}^2 = \lambda_i^2 + \mu_j^2, \quad (11.88)$$

where the prime is used for differentiation by x and a dot denotes the time derivative. Furthermore, it follows from the orthogonality of V_{ij} that the boundary condition at $x = 0$ can be expressed as

$$X'_{ij}(0, t) = F_{ij}(t)$$
$$= \int_0^1 \int_0^l [-\chi_1(y, z) + w_t(y, z, t)\chi_2(y, z)] \times V_{ij}(y, z)\, dy\, dz. \qquad (11.89)$$

The zero initial conditions (11.87) mean

$$X_{ij}(x, 0) = \dot{X}_{ij}(x, 0) = 0. \qquad (11.90)$$

Problem (11.88) and (11.89) can be solved by Laplace transform, which images any function $f(t)$ (*original*) to $\tilde{f}(p) = \int_0^\infty f(t)\exp(-pt)dt$ (*image*), where p is the complex variable with $\mathrm{Re}(p) \geq 0$ (Bateman, 1954). Accounting for eq. (11.90) in computing the images of the time derivatives implies that the Laplace transform of $\ddot{X}_{ij}$ is $p^2\tilde{X}_{ij}$. Taking the Laplace transform of eqs. (11.88) and (11.89) gives the following problem:

$$\tilde{X}''_{ij} - r_{ij}^2 \tilde{X}_{ij} = 0 \quad \left(r_{ij}^2 = \alpha_{ij}^2 + p^2\right), \quad x < 0;$$
$$\tilde{X}'_{ij}(0, p) = \tilde{F}_{ij}(p). \qquad (11.91)$$

Because X_{ij} has to go to zero when $x \to -\infty$, the solution of differential problem (11.91) must also vanish when $x \to -\infty$. The solution is

$$\tilde{X}_{ij}(x, p) = \tilde{F}_{ij}(p)\left[\frac{\exp(r_{ij}x)}{r_{ij}}\right]. \qquad (11.92)$$

We now apply the convolution theorem to express X_{ij} by first finding the originals of $\tilde{F}_{ij}$ and the term in the square bracket of eq. (11.92). Obviously, the original of the function $\tilde{F}_{ij}$ is F_{ij}, but the square-bracket expression of eq. (11.92) has the original

$$\begin{cases} 0 & \text{for } 0 < t < -x, \\ J_0\left(\alpha_{ij}\sqrt{t^2 - x^2}\right) & \text{for } t > -x, \end{cases}$$

according to Bateman (1954, Table 5.6), where J_0 is the Bessel function of the first kind and of zero order.

It follows from the convolution theorem (Bateman, 1954, Table 4.1) that the original of eq. (11.92) for $t > -x$ is

$$X_{ij}(x, t) = \begin{cases} \displaystyle\int_0^{t+x} F_{ij}(\tau) J_0\left(\alpha_{ij}\sqrt{(t-\tau)^2 - x^2}\right) d\tau & \text{for } t + x > 0, \\ 0 & \text{otherwise.} \end{cases}$$

When $t < -x$, the solution is zero; that is, there exists for a given x a disturbance due to the impact only when $t > -x$ in nondimensional variables or $c_0 t > -x$ in dimensional variables. The pressure distribution on the wall at $x = 0$ is

$$p(y, z, t) = -\sum_{\substack{i=0 \\ j=1}}^{\infty} \left\{ \frac{d}{dt} \int_0^t F_{ij}(\tau) J_0\left[\alpha_{ij}(t - \tau)\right] d\tau \right\} V_{ij}(y, z). \qquad (11.93)$$

To account for hydroelasticity (i.e., nonzero deflections w), we must also simultaneously solve the equations describing the structural elastic vibrations. The latter is considered in detail in Section 11.9 by considering an incompressible liquid and assuming that the structural behavior can be described by the beam equation.

11.8.4 Thin-layer approximation of liquid–gas mixture

Korobkin (2006) and Iafrati and Korobkin (2006) analyzed two-dimensional steep-wave impact on a vertical elastic wall with an aerated layer in the impact region. The wave front of the incident wave is vertical and is wetting the vertical wall over the complete liquid depth at the initial time. The liquid is incompressible outside the aerated layer. A thin-layer approximation is used to account for the mixture between liquid and gas. Let us use Figure 11.15 as a reference for the coordinate system. The ratio D/H_ζ between the thickness of the aerated layer and the liquid depth H_ζ is assumed small. We consider the effect of the impact on the flow. The velocity potential φ_m in the aerated layer from $x = 0$ to $x = D$ is described by the wave equation $\nabla^2\varphi_m - c_0^{-2}\partial^2\varphi_m/\partial t^2 = 0$, where the speed of sound, c_0, is given by eq. (11.83) and depends on the void fraction χ and the density and bulk modulus of the liquid and the gas. The velocity potential in the incompressible liquid domain is denoted φ. An essential part in the derivation of the solution is to express the boundary conditions for φ at $x = D$ in terms of the boundary condition for $\partial\varphi_m/\partial x$ at $x = 0$. The matching conditions $\partial\varphi_m/\partial x = \partial\varphi/\partial x$ and $\rho_{\text{mix}}\varphi_m = \rho_l\varphi$ at $x = D$ are used, where $\rho_{\text{mix}} = (1-\chi)\rho_l + \chi\rho_g$ (see eq. (11.84)). The condition $\rho_{\text{mix}}\varphi_m = \rho_l\varphi$ is a consequence of continuity of the pressure at $x = D$; that is, $-\rho_{\text{mix}}\partial\varphi_m/\partial t = -\rho_l\partial\varphi/\partial t$ and the fact that the initial values of φ_m and φ are zero. The matching condition $\rho_{\text{mix}}\varphi_m = \rho_l\varphi$ follows by time-integrating the pressure continuity condition.

A first step is to integrate the wave equation with respect to x from $x = 0$ to $x = D$. Because D/H_ζ is small, the velocity potential φ_m is weakly dependent on x. The result is

$$\frac{\partial^2\varphi_m}{\partial t^2}(D, y, t)D = c_0^2\left[\frac{\partial\varphi_m}{\partial x}(D, y, t) - \frac{\partial\varphi_m}{\partial x}(0, y, t) + \frac{\partial^2\varphi_m}{\partial y^2}(D, y, t)D\right].$$

It follows from the matching conditions at $x = D$ that

$$\frac{\partial\varphi}{\partial x}(D, y, t) = \frac{\partial\varphi_m}{\partial x}(0, y, t) + \frac{\rho_l D}{\rho_{\text{mix}}c_0^2}\left(\frac{\partial^2\varphi}{\partial t^2}(D, y, t) - c_0^2\frac{\partial^2\varphi}{\partial y^2}(D, y, t)\right), \tag{11.94}$$

where $\partial\varphi_m(0, y, t)/\partial x$ is the given boundary condition at $x = 0$ in terms of the velocity of the incident waves and the deflection velocity of the wall. Because D/H_ζ *is* small, we can apply condition (11.94) as a first approximation at $x = 0$ instead of at $x = D$. What we have achieved is that it is unnecessary to solve the wave equation for φ_m. We only need to solve the Laplace equation for φ with the derived boundary condition at $x = 0$.

Other boundary conditions have to be specified as described in previous examples. The solution for the velocity potential φ in $0 \le x \le \infty$, $0 \le y \le H_\zeta$ can be expressed as

$$\varphi = \sum_{k=1}^{\infty} b_k(t)\exp(-\mu_k x/H_\zeta)\cos(\mu_k y/H_\zeta),$$
$$\mu_k = \tfrac{1}{2}\pi(2k-1). \tag{11.95}$$

We note that the two-dimensional Laplace equation and the boundary conditions $\partial\varphi/\partial y = 0$ on $y = 0$, $\varphi = 0$ on $y = H_\zeta$, and $\varphi \to 0$ *as* $x \to \infty$ are satisfied. The coefficients $b_k(t)$ in eq. (11.95) can be obtained by applying eq. (11.94) at $x = 0$.

The effect of the elasticity properties of the vertical wall is accounted for by the beam equation. The pressure loading is given by $-\rho_l\partial\varphi(0, y, t)/\partial t$. A case study of the maximum structural strain was presented for $D/H_\zeta = 0.01, 0.1, 0.25$ as a function of the void fraction χ between 0 and 30%. If we disregard small-amplitude oscillatory behavior, the result with zero aeration is conservative and can be considered a first approximation. However, the aerated layer matters, particularly in terms of the hydrodynamic loads on the wall.

11.9 Hydroelastic slamming

Hydroelasticity may matter during slamming. Hydroelasticity implies that the analysis of the hydrodynamic flow and structural reaction in terms of deflections and stresses cannot be separated. A mutual interaction exists whereby the structural vibrations cause hydrodynamic loads and vice versa.

Theoretical and experimental studies of wave impact on horizontal elastic plates of steel and aluminum were presented by Faltinsen (1997, 2005). The plates in the experiments were dropped vertically on calm water as well as on waves. The rigid-body downward impact velocity is denoted V. Theoretical studies were made assuming two-dimensional beam theory for strips of the plates. The test sections were divided into three parts: one measuring section with a dummy section on each side. The measured nondimensional maximum strains $\varepsilon_m\sqrt{\mathtt{EI}/(\rho L)}/(z_a V)$ in the middle of the plate are presented in Figure 11.43 as a function of nondimensional impact velocity $V\sqrt{\rho L^3/\mathtt{EI}}$, where $\mathtt{E}$ is the Young modulus of elasticity of the plate and $\mathtt{EI}$ is the bending

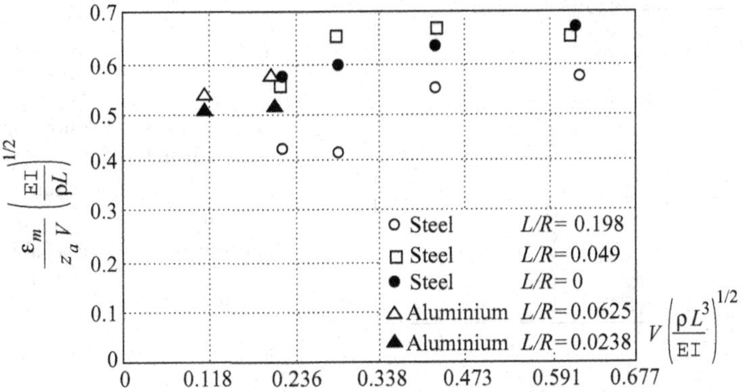

Figure 11.43. Measured maximum nondimensional strain amplitude ε_m in the centers of the horizontal steel and aluminum plates as function of nondimensional water entry velocity. V is water entry velocity, ρ is the mass density of water, and R is the radius of curvature of the waves at the impact position (Faltinsen, 2005).

stiffness of the beam model. Therefore, $\mathtt{I}$ is the cross-sectional-area moment of inertia about the neutral axis divided by the cross-sectional plate breadth B. Furthermore, ρ is the water density, L is the length of the plate, and z_a is the distance from the neutral axis of the beam to the point of maximum stress. The scaling assumes that maximum strain is proportional to V and that the highest wet natural period T_{n1} for the elastic plate is the time scale. One should note that it is only the water density and not also the mass density of the material that is involved in the nondimensional expression.

All experimental data in Figure 11.43 refer to cases with the following impact scenario. Initially a wave crest with radius of curvature R (or flat water with $R = \infty$) hits the plate between its edges. The impact position and R change for the different cases. The measurements show that a smaller nondimensional impact velocity leads to a slightly lower level of the maximum nondimensional strain. If the largest value of L/R is disregarded, the effect of L/R on the nondimensional maximum strain appears rather limited. The results from the asymptotic theory for vertical impact of a horizontal elastic beam with springs at the beam ends by Faltinsen (1997) agree well with the experiments. The theory gives nondimensional maximum strain for a given structural mass and connecting spring parameter that is independent of impact speed and wave characteristics. The results are not shown in the figure.

The physics associated with Faltinsen's (1997) asymptotic analysis can be explained as follows. Because it takes time to build up deformations of the plate, w, the pressure loads from either the water or an air cushion initially balance the structural inertial force of the plate, which is why it is called the *structural inertia phase.* During this phase we can formally express the vertical velocity of the plate as $\partial \mathrm{w}/\partial t - V$, where V is the downward plate velocity at the impact instant before the interaction between the plate and the liquid starts. The plate experiences a large force impulse during a small time relative to the highest natural period for the plate vibrations in the structural inertia phase. This force impulse causes elastic vibration velocity $\partial \mathrm{w}/\partial t$ to be equal to the water entry velocity, V, at the end of the initial phase. Although V is uniform in space, w depends on the position along the plate; as a result the initial condition for $\partial \mathrm{w}/\partial t$ for the new phase has to be expressed in a space-averaged way. If w is represented in terms of a finite number N of dry normal modes ψ_i (i.e., $\mathrm{w} = \sum_{i=1}^{N} a_i \psi_i$), a Galerkin method can be used to express the initial conditions for the generalized coordinates of the normal modes. This implies that the initial condition $\partial \mathrm{w}/\partial t = V$ is multiplied by the N different values of ψ_k and integrated over the plate length. The whole plate is wetted at the start of the *free-vibration phase.* The plate then starts to vibrate in a manner similar to the free vibration of a wet beam with an initial space-averaged vibration velocity and zero initial deflection. Maximum strains occur during the free-vibration phase.

The details of the pressure distribution during the structural inertia phase are not important, but the impulse of the impact is important. Very large

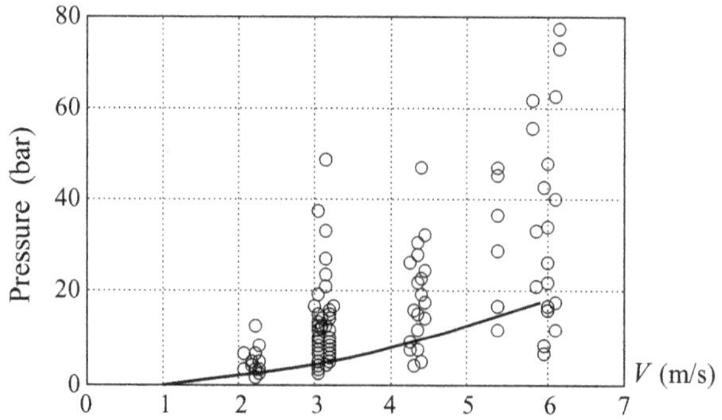

Figure 11.44. Measured maximum pressure (○) from different drop tests of horizontal elastic plates as a function of the water entry velocity, V. $C_p = (p - p_a)/(\frac{1}{2}\rho V^2)$ is the pressure coefficient (Faltinsen, 2005). Solid line represents $C_p = 100$.

pressures that are sensitive to small changes in the physical conditions may occur in this initial phase, which can be seen from the collection of measured maximum pressures during the tests. They appear to be stochastic in nature (as documented by Figure 11.44).

The measured maximum strains showed very small scatter for given impact velocity of the plate even though the maximum pressure varied strongly. Because a pressure gauge measures the average pressure over an area, the diameter of the pressure cell matters when the pressures are very concentrated in space. The latter is expected to be true for the high pressures reported in Figure 11.44. The largest measured pressure was approximately 80 bar for V, equal to 6 m s^{-1}. This value is close to the acoustic pressure $\rho c_0 V \approx 1000 \cdot 1500 \cdot 6 \approx 90$ bar. Because the maximum slamming pressure cannot be higher than the acoustic pressure, much larger pressures are not expected using smaller pressure gauges. The diameter of each pressure cell was 4 mm. A sampling frequency up to 500 kHz was used. These results document that it can be misleading from a structural point of view to measure the peak pressures for the effect of hydrodynamic impact on aluminum and steel structures when hydroelasticity matters. We may say that the structural elastic response acts as a filter on many hydrodynamic flow parameters.

The hydrodynamic pressure in the free-vibration phase is caused by the plate vibrations. The lowest structural mode is dominant. If we limit the description to the lowest structural mode with a wet natural period T_{n1}, the hydrodynamic pressure is positive for $0 < t < \frac{1}{2}T_{n1}$, where $t = 0$ is the initial time of the free-vibration phase. When the hydrodynamic pressure becomes negative, we must recognize the facts that the total pressure is the sum of the hydrodynamic pressure and the ambient pressure and that the total pressure cannot be negative and must be related to the vapor pressure. Because the submergence of the plate is very small, the ambient pressure on the plate can be set equal to the atmospheric pressure. When a one-mode structural model is used, the magnitude of the negative hydrodynamic pressure is largest in the middle of the plate. If the total pressure is equal to the vapor pressure, cavitation can be triggered. This fact was evident in the experimental results presented by Faltinsen (1997) and started in the middle of the plate. Because the submergence of the plate is very small, ventilation follows and causes the plate to oscillate as if it was in air. The experiments confirmed this fact. Since the largest structural stresses occurred at approximately $\frac{1}{4}T_{n1}$, cavitation and ventilation are not factors to be considered for ultimate limit state assessments.

Faltinsen (1999) studied the relative importance of hydroelasticity for an elastic hull with wedge-shaped cross-sections penetrating an initially calm water surface (see Figure 11.45). Wagner's theory was generalized to include elastic vibrations. A stiffened plating between two rigid transverse frames was examined (see Figure 11.46). A hydrodynamic strip theory in combination with orthotropic plate theory was used. The water-entry velocity was assumed constant.

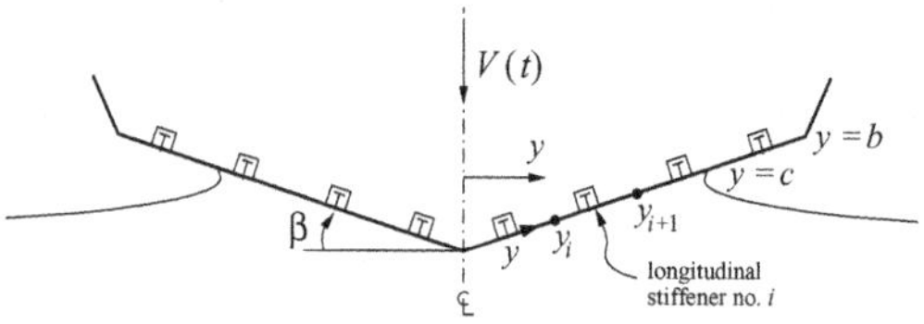

Figure 11.45. Water entry of a wedge-shaped elastic cross-section.

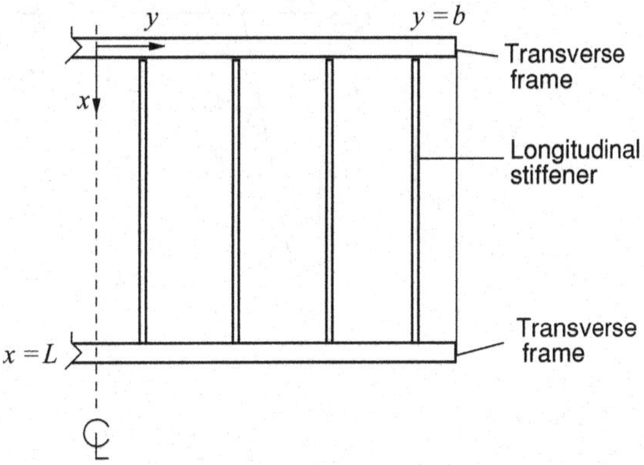

Figure 11.46. Stiffened plating consisting of plate and longitudinal stiffeners. ℄ denotes keel line.

The importance of hydroelasticity for the local slamming-induced maximum stresses increased with decreasing deadrise angle β and increasing impact velocity V. The nondimensional parameter $\xi = \tan\beta/[V(\rho_l L^3/\mathtt{EI})^{1/2}]$ was introduced, where L is the length of the analyzed longitudinal stiffener between two transverse frames, and $\mathtt{EI}$ is the bending stiffness per width of the longitudinal stiffener including the effective plate flange. Furthermore, ρ_l is the liquid density. The parameter ξ is proportional to the ratio between the wetting time of the rigid wedge and the highest natural period of the longitudinal stiffener. We can see this situation as follows. If Wagner's theory is used, the wetting time of a rigid wedge with beam B and constant V is $B\tan\beta/(\pi V)$. Because $\sigma_w\sqrt{\rho_l L^5/\mathtt{EI}}$ is a constant and $T_w = 2\pi/\sigma_w$ is the wet stiffener natural period, T_w is proportional to $\sqrt{\rho_l L^5/\mathtt{EI}}$. Assuming B/L is a constant gives the desired result. We can associate the wetting time of the wedge with the duration of the loading. If we make an analogy to a simple mechanical system consisting of a mass and a spring, then we know that the duration of the loading relative to the natural period characterizes the dynamic effects of a transient system (Clough and Penzien, 1993). Nondimensional results are presented in Figure 11.47. Results are also shown based on quasi-steady analysis and asymptotic hydroelastic analysis for small deadrise angles β (see Figure 11.4). The quasi-steady analysis assumes that the structure is rigid in the hydrodynamic calculations. The pressure and, therefore, the maximum strain is then proportional to V^2. The analysis of the structural deformations due to the water impact shows that $\varepsilon_m \mathtt{EI}\tan\beta/(z_a V^2\rho_l L^2)$ is independent of the abscissa $\tan\beta/(V\sqrt{\rho_l L^3/\mathtt{EI}})$ in Figure 11.47. The asymptotic hydroelastic analysis is based on writing the ordinate in Figure 11.47 as a function of the abscissa; that is,

$$\varepsilon_m \frac{\mathtt{EI}\cdot\tan\beta}{z_a V^2 \rho_l L^2} = E_{HE}\frac{\tan\beta}{V\sqrt{\rho_l L^3/\mathtt{EI}}},$$

where $E_{HE} = \varepsilon_m/(z_a V)\sqrt{\mathtt{EI}/(\rho_l L^3)}$ is estimated by Faltinsen's (1997) hydroelastic analysis for $\beta = 0$, and z_a is the distance from the neutral axis to the stress point and ε_m is the maximum strain. A representative value of E_{HE} equal to 0.7 was used in presenting the results in Figure 11.47. This is based on setting $M_B/(\rho_l L) = 0.015$ and $k_\theta L/(2\mathtt{EI}) = 3.0$, where M_B is an average mass of the plate per unit area and k_θ is a spring stiffness that is related to a restoring moment $-k_\theta\theta_{be}$ at $x = 0$ and L (see Figure 11.46); θ_{be} is the rotation angle at the ends. Because the nondimensional strain is proportional to $\tan\beta/(V\sqrt{\rho_l L^3/\mathtt{EI}})$, the results from the asymptotic hydroelastic analysis appear as a straight line in Figure 11.47.

This particular way of nondimensioning the results gives small explicit dependence on the dimensionless impact velocity, $V_{ND} = V\sqrt{\rho_l L^3/\mathtt{EI}}$.

Figure 11.47 illustrates that hydroelastic effects are present when $\tan\beta <\sim 1.5V(\rho_l L^3/\mathtt{EI})^{1/2}$ for the studied stiffened plating. The stress from the hydroelastic case may also exceed the stress from the quasi-steady case. A large influence of hydroelasticity occurs when $\tan\beta <\sim 0.25V(\rho_l L^3/\mathtt{EI})^{1/2}$. By varying independently the parameters in $\xi = \tan\beta/(V\sqrt{\rho_l L^3/\mathtt{EI}})$, we see that small ξ-values are obtained when

- the deadrise angle β is small;
- the liquid-entry velocity V is large;

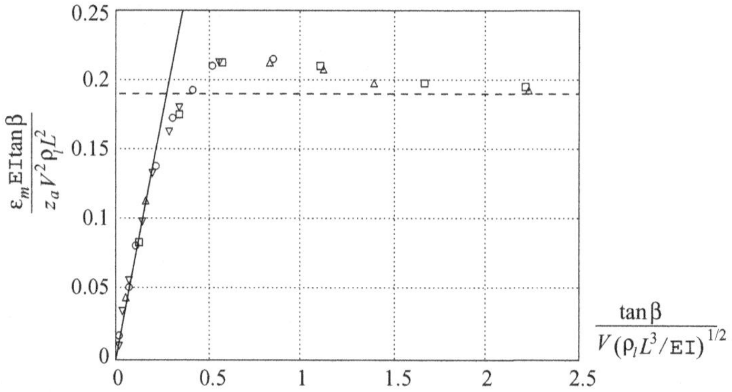

Figure 11.47. Maximum strain ε_m in the middle of the longitudinal stiffener number second from the keel. The strain is presented as a function of a parameter that is proportional to the ratio between wetting time of a rigid wedge and natural period of the longitudinal stiffener based on beam theory. Different nondimensional constant impact velocities, $V_{ND} = V\sqrt{\rho_l L^3/\mathrm{EI}}$, are given; β is the deadrise angle. Calculations by hydroelastic orthotropic plate theory (Faltinsen, 1999) are shown. The solid line corresponds to the hydroelastic beam theory ($\beta \approx 0$) and the dashed line represents the quasi-steady orthotropic theory; □, $V_{ND} = 0.089$; Δ, $V_{ND} = 0.178$; ○, $V_{ND} = 0.467$; ∇, $V_{ND} = 0.715$.

- $\sqrt{\rho_l L^3/\mathrm{EI}}$ is large. Since $\sigma_w L\sqrt{\rho_l L^3/\mathrm{EI}}$ is a constant, large $\sqrt{\rho_l L^3/\mathrm{EI}}$ corresponds to small values of $\sigma_w L$, where σ_w is the lowest wet natural frequency. If L is a constant, large $\sqrt{\rho_l L^3/\mathrm{EI}}$ corresponds to a high wet natural period $2\pi/\sigma_w$.

In the following we demonstrate the effect of hydroelasticity during slamming in tanks by experimental and theoretical studies. First, hydroelastic sloshing experiments are described, including some results. No air cushions are present. The second part considers an elastic beam model that is used to calculate the tank wall response based on the pressure load from the Wagner theory. The slamming pressures acting over the whole beam are important. An added mass approach is applied. A simplification is made by the fact that no nonlinear interaction occurs between the flow caused by the beam vibrations and the slamming loads. A reason why we can do this is that the wetted area of the beam does not change with time. Comparisons are then made with the experiments. Then we use the theoretical model for parametric studies. The procedure for scaling model tests with hydroelastic effects of a steel structure is also discussed. Finally, we highlight the complexities of hydroelastic slamming in a membrane tank.

11.9.1 Experimental study

Figure 11.48 shows the steel tank positioned in the Marintek sloshing test rig. An aluminum plate along the vertical wall at the upper right corner is used in the hydroelastic studies. The tank model is relatively narrow to achieve two-dimensional flow. Stiffened steel plates ensure the rigidity of the tank outside the aluminum plate. A high-speed camera is used to capture the free-surface motion. The camera is seen in Figure 11.48 pointing at the plexiglass observation window. A total of nine pressure cells are positioned in the tank roof close to the right corner, and five strain gauges are mounted on the aluminum plate (see Figure 11.49). The tank is forced to oscillate harmonically in the horizontal plane with a period close to the highest natural period for the water motion. The aluminum plate is clamped in the upper and lower part and is free to move along the vertical sides. Flexible and thin rubber membranes keep the 0.2-mm slits watertight. The plate thickness is chosen so that the scaled lowest natural frequency is comparable to a similar structural element in a tank onboard a typical OBO carrier. The sampling frequency is 19.2 kHz, and an analog filter with a cutoff frequency of about 5 kHz is used on all channels.

The clamped-end condition of the aluminum plate is achieved by pressing the ends flat between

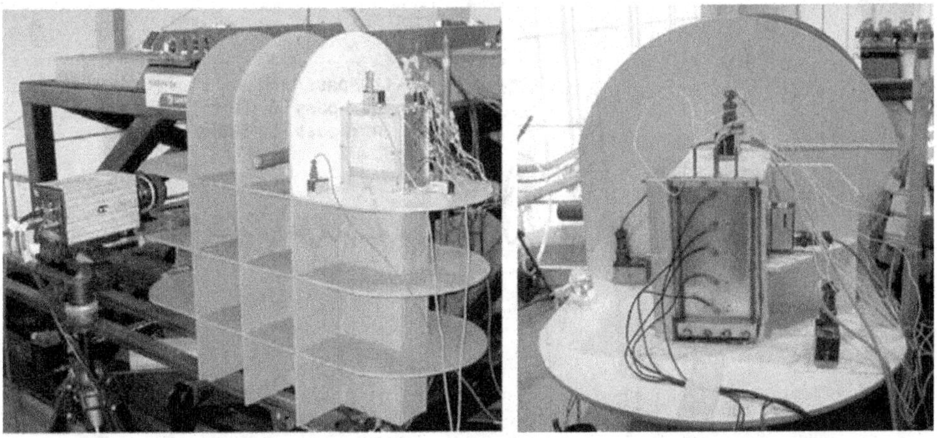

Figure 11.48. Steel tank and high-speed camera. Detailed view of the aluminum plate.

an aluminum rod and the steel tank by bolts. A good check is a comparison of the lowest calculated eigenfrequency in air and the measured value (Figure 11.50). For the 2-mm plate, the measured dry eigenfrequency is about 330 Hz, whereas the calculated value is 362 Hz. The lowest measured wet eigenfrequency (5 mm from free surface to tank roof) is approximately 105 Hz, and calculated as 99.3 Hz. The strain gauges are calibrated by fixing the flexible plates at both short ends in sequence and applying weights on the resulting cantilever beam. Figure 11.50 shows free oscillation of the wet plate and can be used to estimate the damping of the system. Band-pass filtering is used to investigate the strain response from the two lowest modes. The damping of the lowest mode is found to be approximately 7% of the critical damping. The level of damping for the next mode is more difficult to estimate, but it seems to be of the same order and therefore the same factor is applied.

We compare the experiments with theory after the theory has been presented in the next section.

11.9.2 Theoretical hydroelastic beam model

We assume tank roof impact with no gas influence. The structural response of a clean tank roof can be analyzed as described by Kvålsvold and Faltinsen (1995) and Faltinsen (1997).

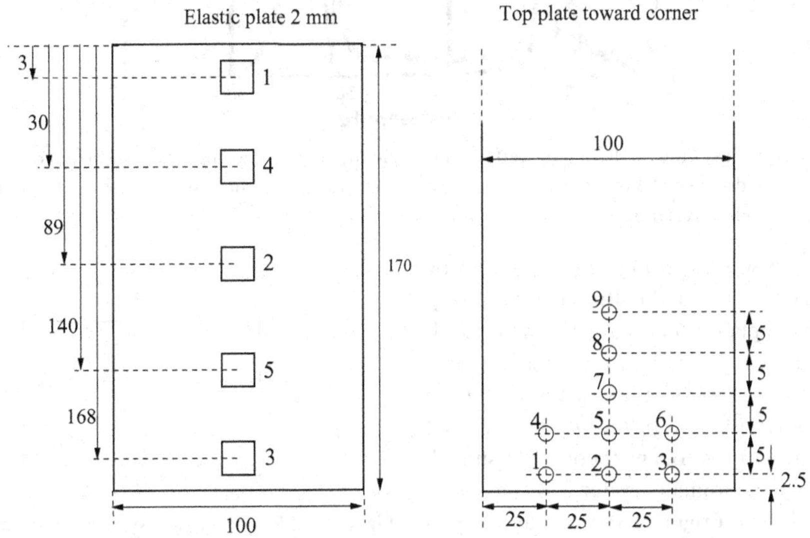

Figure 11.49. Strain-gauge positions for the elastic panel and pressure-sensor locations on the tank top plate. Dimensions are in millimeters.

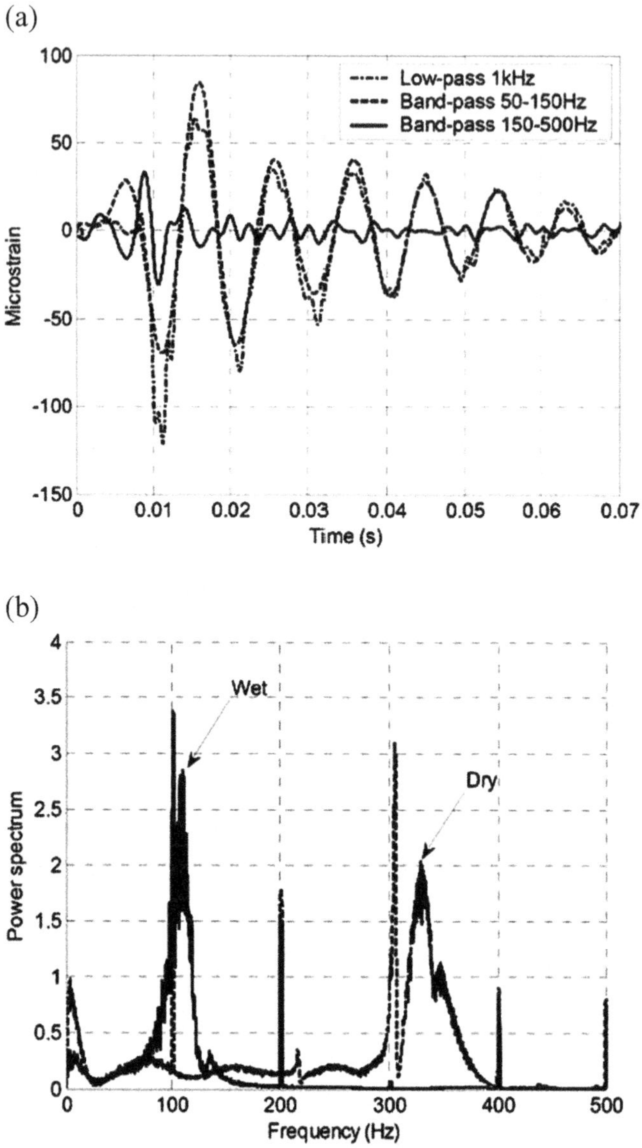

Figure 11.50. (a) Free oscillation of wetted 2-mm aluminum plate and (b) power spectrum of strains at a vertical distance of 3 mm from the top end of the elastic panel. The results for dry and wet plates are used to identify natural periods.

This approach was experimentally validated by Kvålsvold *et al.* (1995) and Faltinsen *et al.* (1997). A clean tank implies no internal structures including, for instance, corrugations on the membrane. In this section we focus on the structural response of the wall adjacent to the portion of the tank roof where impact occurs. Beam theory is assumed. We choose a coordinate system as shown in Figure 11.51. The origin of the coordinate system is in the tank roof. The length and the height of the tank are l and H_t, respectively. Tank roof

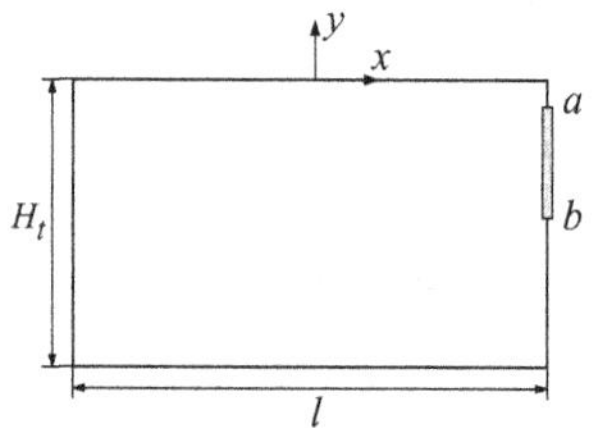

Figure 11.51. Coordinate system used in analyzing beam response in the vertical tank wall adjacent to tank roof impact position.

impact is supposed to happen at the right corner. The response of a structural beam element with end coordinates $(\frac{1}{2}l, b)$ and $(\frac{1}{2}l, a)$ in the tank wall is analyzed (see Figure 11.51). A similar procedure can be applied for impact close to $x = -\frac{1}{2}l$. We write the beam equation as

$$M_B \frac{\partial^2 \mathtt{w}}{\partial t^2} + \mathtt{EI}\frac{\partial^4 \mathtt{w}}{\partial y^4} = p_{\text{exc}} + p_{\text{adm}}, \qquad (11.96)$$

where M_B is the structural mass per unit length and breadth of the beam, $\mathtt{w}(y, t)$ is the elastic beam deflection, t is the time variable, y is an axial coordinate along the length $L = a - b$ of the beam, $\mathtt{EI}$ is the bending stiffness, so that $\mathtt{E}$ is Young's modulus and $\mathtt{I}$ is the area moment of inertia of the beam cross-section divided by the breadth of the beam. Furthermore, p_{exc} is the excitation pressure caused by tank roof impact and p_{adm} is the "added mass" pressure due to the beam vibration. We write the deflections as $\mathtt{w} = \Sigma a_k(t)\psi_k(y)$, where the eigenfunctions $\psi_k(y)$ satisfy $\partial^4\psi_k/\partial y^4 = p_k^4\psi_k$ together with boundary conditions at the beam ends. The eigenfunctions $\psi_k(y)$ can be written as

$$\psi_k = A_k \sin p_k y + B_k \cos p_k y + C_k \sinh p_k y + D_k \cosh p_k y; \quad p_k^4 = M_B\sigma_k^2/\mathtt{EI}, \qquad (11.97)$$

where σ_k is the dry natural frequency for mode k; A_k, B_k, C_k and D_k are constants that are determined by the beam-end conditions. The solution of the eigenfunctions is not unique. We can multiply the solution with any constant.

We define the velocity potential associated with vibrations in mode k as $\phi_k\dot{a}_k(t)$, where $\dot{a}_k(t)\psi_k$ is the deformation velocity due to mode k. We can formally express p_{adm} as $-\rho_l\partial[\sum\phi_k\dot{a}_k(t)]/\partial t$ by the linearized Bernoulli equation, where ρ_l is the liquid density.

We multiply eq. (11.96) by ψ_m, integrate from b to a, and define

$$\int_b^a p_{\text{adm}}\psi_m dy = -A_{mm}\ddot{a}_m - \sum_{k\neq m} A_{km}\ddot{a}_k,$$

where $A_{km} = \rho_l \int_b^a \phi_k\psi_m dy$ is the generalized added mass. We later show how to evaluate the added mass. By using orthogonality properties, we get

$$(M_{mm} + A_{mm})\ddot{a}_m + \sum_{k\neq m} A_{km}\ddot{a}_k + \sigma_m^2 M_{mm}a_m = \int_b^a p_{\text{exc}}\psi_m\, dy, \qquad (11.98)$$

where $M_{mm} = M_B\int_b^a \psi_m^2(y)\, dy$ can be expressed analytically. The right-hand side of eq. (11.98) is the generalized excitation force. The theoretical approach in Section 11.3 is used to express p_{exc}. By neglecting the influence of the other tank wall and the tank bottom, it follows that

$$p_{\text{exc}} = \rho_l\frac{\partial}{\partial t}\left(V(c^2 + y^2)^{1/2} + Vy\right), \qquad (11.99)$$

where $V(t)$ is the impact velocity of the liquid on the tank roof and $c(t)$ is the wetted surface of the tank roof. In the following text we assume that the impacting free surface can be approximated as a parabola with radius of curvature R; we use eq. (11.15) with $\beta = 0$ to express $\eta_b(x)$ defined in Figure 11.5(a). Furthermore, $V(t)$ is assumed linearly varying so that $c(t)$ is given by eq. (11.17). The procedure can be generalized to the case of tank roof impact with a gas cavity by using eq. (11.67) to calculate the excitation pressure.

Equation (11.99) can also be expressed as

$$p_{\text{exc}} = \rho_l\dot{V}\left[\sqrt{c^2 + y^2} + y\right] + \rho_l V^2\frac{2R}{\sqrt{c^2 + y^2}}. \qquad (11.100)$$

This approach assumes that the tank roof is rigid. The generalized excitation force is evaluated numerically using Simpson's integration.

The added mass coefficients can be determined analytically, which is shown in the following text. First we determine normalized velocity potentials ϕ_k associated with mode k satisfying a two-dimensional Laplace equation in the domain $-\frac{1}{2}l \le x \le \frac{1}{2}l$, $-H_t \le y \le 0$. The boundary conditions on $y = 0$, $-\frac{1}{2}l \le x \le \frac{1}{2}l$ are approximated by neglecting the fact that part of the tank roof is wetted; that is, we apply the high-frequency free-surface condition $\phi_k = 0$ on $y = 0$, $-\frac{1}{2}l \le x \le \frac{1}{2}l$. The argument for doing this is pragmatic; it simplifies the calculations. A requirement for the approximation to be valid is that the length of the wetted tank roof is small relative to the beam length. Additional boundary conditions are $\partial\phi_k/\partial n = 0$ on the vertical wall $x = -\frac{1}{2}l$, the bottom $y = -H_t$, and the part of the vertical wall at $x = \frac{1}{2}l$ outside the structural beam. Here $\partial/\partial n$ is the normal derivative to the surface. The positive normal direction is taken to be out of the liquid. The boundary condition on the beam is $\partial\phi_k/\partial x = \psi_k$. The velocity potential

$$\phi_k = \sum_{n=0}^{\infty} a_{nk} \sin((2n+1)\pi y/(2H_t)) \times \cosh((2n+1)\pi\left(x+\tfrac{1}{2}l\right)/(2H_t))$$

satisfies the Laplace equation, and the boundary conditions on the free surface, the bottom, and the wall opposite the impact are satisfied. The coefficients a_{nk} are determined from the body boundary condition at $x = \frac{1}{2}l$, by

$$\frac{\partial\phi_k}{\partial x} = \sum_{n=0}^{\infty} a_{nk}\left(\frac{2n+1}{2H_t}\pi\right)\sinh\left(\frac{2n+1}{2H_t}\pi l\right) \times \sin\left(\frac{2n+1}{2H_t}\pi y\right) = \begin{cases}\psi_k & \text{for } b<y<a,\\ 0 & \text{otherwise,}\end{cases} \tag{11.101}$$

which implies that the normal velocity of the wall $x = \frac{1}{2}l$ given by the right-hand side of eq. (11.101) must be expressed as a Fourier series in y, as for the left-hand side. It follows then that

$$a_{nk} = \frac{4[A_k I_{1nk} + B_k I_{2nk} + C_k I_{3nk} + D_k I_{4nk}]}{(2n+1)\pi\sinh((2n+1)\pi l/(2H_t))}, \tag{11.102}$$

where

$$I_{1nk} = \frac{\sin(c_{nk}a)}{2c_{nk}} - \frac{\sin(c_{nk}b)}{2c_{nk}} - \frac{\sin(d_{nk}a)}{2d_{nk}} + \frac{\sin(d_{nk}b)}{2d_{nk}}, \tag{11.103}$$

$$I_{2nk} = -\frac{\cos(c_{nk}a)}{2c_{nk}} + \frac{\cos(c_{nk}b)}{2c_{nk}} - \frac{\cos(d_{nk}a)}{2d_{nk}} + \frac{\cos(d_{nk}b)}{2d_{nk}}, \tag{11.104}$$

$$\begin{aligned} I_{3nk} &= e_{nk}\cosh(p_k a)\cdot\sin(g_n a) - e_{nk}\cosh(p_k b)\sin(g_n b) \\ &\quad - f_{nk}\sinh(p_k a)\cos(g_n a) + f_{nk}\sinh(p_k b)\cos(g_n b), \\ I_{4nk} &= e_{nk}\sinh(p_k a)\sin(g_n a) - e_{nk}\sinh(p_k b)\sin(g_n b) \\ &\quad - f_{nk}\cosh(p_k a)\cos(g_n a) + f_{nk}\cosh(p_k b)\cos(g_n b) \end{aligned}$$

and

$$c_{nk} = g_n - p_k;\quad d_{nk} = g_n + p_k;\quad e_{nk} = \frac{p_k}{p_k^2 + g_n^2},$$

$$f_{nk} = \frac{g_n}{p_k^2+g_n^2};\quad g_n = \frac{2n+1}{2H_t}\pi.$$

When $c_{nk} = 0$, eqs. (11.103) and (11.104) are found by a limiting process. The added mass coefficients A_{km} (see definition ahead of eq. (11.98)) are given by

$$\frac{A_{km}}{\rho_l} = \int_b^a \phi_k\psi_m\,dy = 4\sum_{n=0}^{\infty}\frac{(A_kI_{1nk}+B_kI_{2nk}+C_kI_{3nk}+D_kI_{4nk})(A_mI_{1nm}+B_mI_{2nm}+C_mI_{3nm}+D_mI_{4nm})}{(2n+1)\pi\tanh[(2n+1)\pi l/(2H_t)]}. \tag{11.105}$$

Equation (11.105) is based on directly integrating pressure due to forced oscillation in mode k. Equation (11.105) can be derived alternatively by using Green's first identity (see eq. (A.5)). Green's first identity gives

$$\iint_Q \nabla\phi_k\cdot\nabla\phi_m \mathrm{d}Q = \oint_{S+\Sigma}\phi_k\frac{\partial\phi_m}{\partial n}\mathrm{d}S,$$

where Q is the liquid volume, Σ is the free surface, $y = 0$, and S is the total wetted body surface. By using the boundary conditions it follows that

$$\int_b^a \phi_k\psi_m \mathrm{d}s = \int_{-\frac{1}{2}\ell}^{\frac{1}{2}\ell}\int_{-H}^{0}\left[\frac{\partial\phi_k}{\partial x}\frac{\partial\phi_m}{\partial x} + \frac{\partial\phi_k}{\partial y}\frac{\partial\phi_m}{\partial y}\right]\mathrm{d}y\,\mathrm{d}x. \tag{11.106}$$

By substituting the series expansion for ϕ_k and ϕ_m ahead of eq. (11.101) together with expression (11.102) it follows that the right-hand side of eq. (11.106) gives the same answer as eq. (11.105). Equation (11.105) shows that $A_{km} = A_{mk}$ for $k \neq m$, which can also be obtained by using Green's second identity (see eq. (A.4)) with ϕ_k and ϕ_m as functions.

When L/l and L/H_t are small, a simplified way to estimate A_{11} is as follows. We assume a constant modal value of 1 over the beam, neglect the influence of tank bottom and the other tank wall, and assume that the beam is part of a semi-infinite vertical wall. Furthermore, we assume $b = -L$,

$a = 0$. We can write the velocity potential on the wall for $-L \le y \le 0$ as

$$\varphi = -\frac{1}{\pi}\int_{-L}^{0}[\ln|y-\eta| - \ln|y+\eta|]\,d\eta$$
$$= -\frac{1}{\pi}[(y+L)\ln(y+L) - 2y\ln|y| - (L-y)\ln(L-y)],$$

which gives the following added mass:

$$a_{11} = \rho_l \int_{-L}^{0}\varphi dy = \frac{2}{\pi}\ln 2 \cdot \rho_l L^2. \qquad (11.107)$$

This result can be roughly related to A_{11} by imagining that the beam oscillates with an average value $\overline{\psi_1^2}$ of ψ_1^2 over the beam, which gives $a_{11}\overline{\psi_1^2} = A_{11}$.

We have now shown how all terms of eq. (11.98) are found. The equations may be solved, for instance, numerically by a fourth-order Runge–Kutta method. When a_n are found, the strains are calculated as

$$\varepsilon = -z_a\frac{\partial^2 w}{\partial y^2}$$
$$= z_a\sum_{k=1}^{\infty} a_k p_k^2(A_k \sin p_k y + B_k \cos p_k y - C_k \sinh p_k y - D_k \cosh p_k y), \qquad (11.108)$$

where z_a is the distance from the neutral axis. A quasi-steady analysis means that

$$a_k = \int_b^a p_{\text{exc}}\psi_k dy / (\sigma_k^2 M_{kk}).$$

11.9.3 Comparisons between theory and experiments

The calculations are made with the same tank geometry and elastic plate properties as used in the experiments described in Section 11.9.1. A rectangular tank with length $l = 0.8$ m and height $H_t = 0.8$ m is used. The structural response of a beam with end coordinates $y = b = -0.17$ m and $y = 0$ in the vertical wall at $x = \frac{1}{2}l$ is studied. Clamped-end conditions are assumed. The eigenfunctions are

$$\psi_k(y) = \cosh u - \cos u - \frac{\cosh\lambda_k - \cos\lambda_k}{\sinh\lambda_k - \sin\lambda_k}(\sinh u - \sin u), \qquad (11.109)$$

where $u = p_k(y-b)$, $\lambda_k = p_k L$, and p_k are determined by the roots of the equation $\cos\lambda_k \cosh\lambda_k = 1$; and L is the beam length. Aluminum is used as a basic material so $E = 70 \cdot 10^9$ Pa and the density is 2700 kg m^{-3}.

The analysis is done with the two lowest modes. Experience from hydroelastic impact of a horizontal beam on a flat or nearly flat surface has shown this to be sufficient (Faltinsen, 1997; Faltinsen *et al.*, 1997; Kvålsvold & Faltinsen, 1995; Kvålsvold *et al.*, 1995). Important reasons are that the highest natural periods are not close and the fact that the higher modes are highly structurally damped. The maximum strains occur on the time scale of the highest wet eigenperiod. In reality, structural damping is present also for the lower modes. Equation (11.98) is modified to include this damping. The level of damping is found from decay tests performed as part of the experiments (see Figure 11.5). Equal damping is applied for both modes.

Figure 11.52 shows the two lowest mode shapes. In the added mass analysis the deformation of the whole wall is approximated as a Fourier series (see the wall condition given by eq. (11.101)). The eigenfunctions ψ_k in the representation of the wall deflection w are expressed as $\psi_k = \sum c_n \sin((2n+1)\pi y/(2H_t))$ for $-H_t \le y \le 0$. Figure 11.52 also illustrates the small difference between the actual mode shape and the Fourier-series approximation of the whole tank wall and illustrates that the Fourier series closely approximates zero tank wall motion outside the structural beam. Figure 11.53 shows the strain response in the middle and the two ends of the beam. The highest absolute value occurs at the beam ends. It is approximately 200 microstrain for the damped case. The calculated natural frequencies for the first and second modes are 99.3 and 413 Hz, respectively.

Figure 11.54 shows calculated and measured strains for the first impact of one the test cases. The agreement is good, especially if the static strain component due to the hydrostatic added pressure on the panel is subtracted. Zero strain is measured when the panel is dry. The velocity of the impacting surface can be found relatively accurately from the measurements, but the radius of curvature at the intersection with the vertical wall is more difficult to estimate.

In the calculations we have neglected acoustic liquid effects. The relative importance of liquid compressibility can be assessed by using eq. (11.85) to estimate the duration of the supersonic

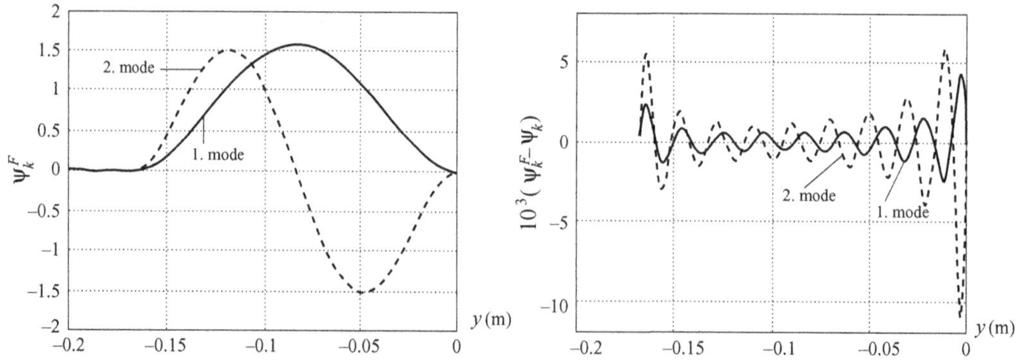

Figure 11.52. Mode shapes with clamped-end conditions and difference between mode shape ψ_k^F based on Fourier-series representation and actual mode shape ψ_k. Fifty Fourier-series components are used.

stage and multiplying eq. (11.85) by 6 to find an approximation of the time it takes for the acoustic solution to reach the Wagner solution for an incompressible liquid. Using $c_0 = 1481.3$ m s^{-1}, $R = 1$ m, and $V = 0.2$ m s^{-1} gives $3 \cdot 10^{-5}$ s; that is, the time scale of acoustic liquid effects is very small relative to the time scale of the structural response shown in Figure 11.54. The consequence is that the effect of liquid compressibility can be neglected. An implicit assumption is no mixture between gas and liquid. Figure 11.39 illustrates that the speed of sound may be considerably lower in a mixture of gas and liquid, with the consequence that the time scale of liquid compressibility is strongly increased.

11.9.4 Parameter study for full-scale tank

The hydroelastic tank roof impact model described in Sections 11.9.2 and 11.9.3 is applied in this section to a full-scale tank. The effects on the maximum structural stresses of the radius of curvature of the incident wave as well as the inflow velocity and acceleration are systematically investigated. It is also shown how the Young modulus of elasticity, E, influences the results. The E-values include data that are relevant for steel and aluminum as well as lower values. Furthermore, it is demonstrated how a chamfered tank roof reduces the structural response. Dynamic amplification factors of the impact are presented.

A rectangular tank with length $l = 51.2$ m and height $H_t = 29$ m without chamfered tank roof is first used. We have in mind the upper part of the transverse bulkhead in a large OBO carrier as illustrated in Figure 11.55. Our tank model is not identical because we do not consider the effect of interior baffles. The tank roof is in reality very stiff, so impact loads on the tank roof are of minor

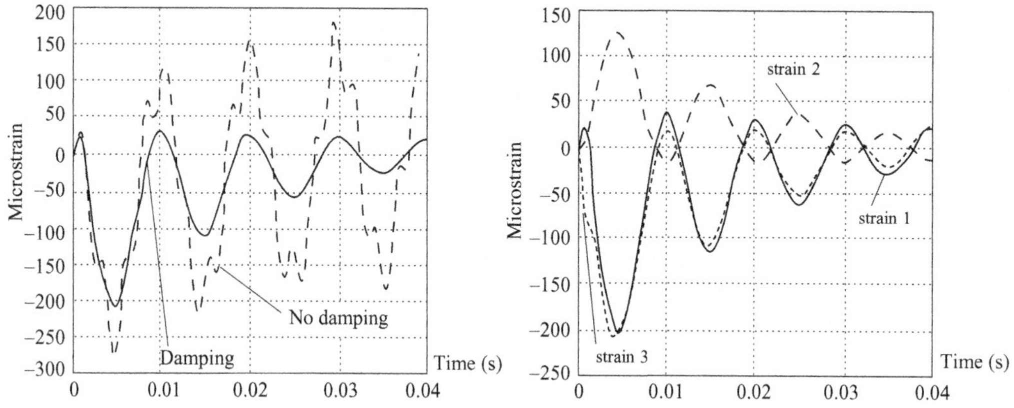

Figure 11.53. Calculated strains for the aluminum plate when the impact is modeled by $V(t) = \frac{1}{2} - 5t$ m s^{-1} and $R = 0.5$ m. The strain positions refer to Figure 11.49.

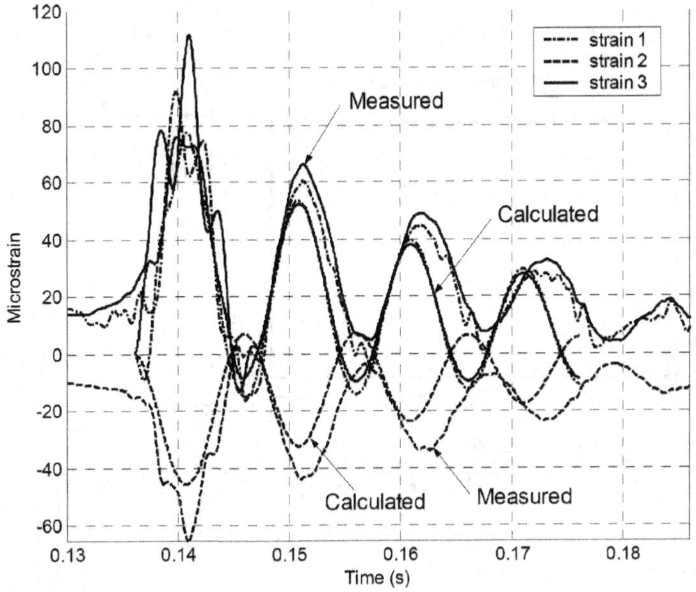

Figure 11.54. Comparison between measured and calculated strains for $V(t) = 0.2 - 1.05t$ m s^{-1} and $R = 1.0$ m. The strain positions refer to Figure 11.49.

concern for the roof structure. An area vulnerable to the effect of tank roof impact is that of the vertical stiffeners at the tank wall adjacent to the impact.

The structural response of a beam of uniform thickness with end coordinates $y = b = -9.5$ m and $y = a = 0$ in the vertical wall at $x = \frac{1}{2}l$ is studied. Clamped-end conditions are assumed. The eigenfunctions are given by eq. (11.109). Steel is used as a basic material so $\mathtt{E} = 210 \cdot 10^9$Pa. The other beam parameters are found by setting them equal to the corresponding values for the vertical stiffeners and flanges in Figure 11.55. The effective flange was chosen as the distance between the vertical stiffeners (i.e., 0.91 m). From Figure 11.55 we find that the cross-sectional area of a stiffener with plate flange is 0.024315 m^2. Therefore, the mass per length of the cross-section of the vertical stiffener and the flange divided by the width of the flange is $M_B/\rho_l = 0.2084$ m. The center of gravity (neutral axis) of the flange and the vertical stiffener is 0.2488 m from the interior tank side of the flange. The maximum distance $z_a = 0.5102$ m from the neutral axis to the top of the stiffener flange is used in the calculation of bending stresses. The cross-sectional moment of inertia about the neutral axis divided by the length of the flange is $\mathtt{I} = 0.002307$ m^3. Furthermore, we use $\rho_l = 10^3$ kg m^{-3}.

As before, the analysis is done with the two lowest modes. An important reason is that the highest natural periods are not close. The maximum strains occur on the time scale of the highest eigenperiod. Modes higher than the second were difficult to observe in the experimental studies on beam impact. Structural damping, which is not included in the present study, should, in reality, damp higher modes significantly during a cycle of the lowest mode.

Case studies of the structural response are presented in the following. Necessary information on the impact velocity $V(t)$ and the radius of curvature, R, of the impacting free surface are obtained by the method described by Faltinsen and Rognebakke (1999); $V(t)$ is approximated by $V(t) = V_0 + V_1 t$ during the impact. Faltinsen and Rognebakke (1999) validated via model tests that the free surface elevation was well predicted.

We use information based on Froude scaling of model test data. The mean water depth of the tank is 17.6 m in the first case. The tank is forced to oscillate harmonically in the longitudinal direction with amplitude 0.746 m and period 9.09 s. The highest natural period for the liquid motion is 9.0929 s. Many impacts occurred. We chose one of the most severe impacts, where $V_0 = 9.081$ m s^{-1}, $V_1 = -7.668$ m s^{-2}, and $R = 7.647$ m.

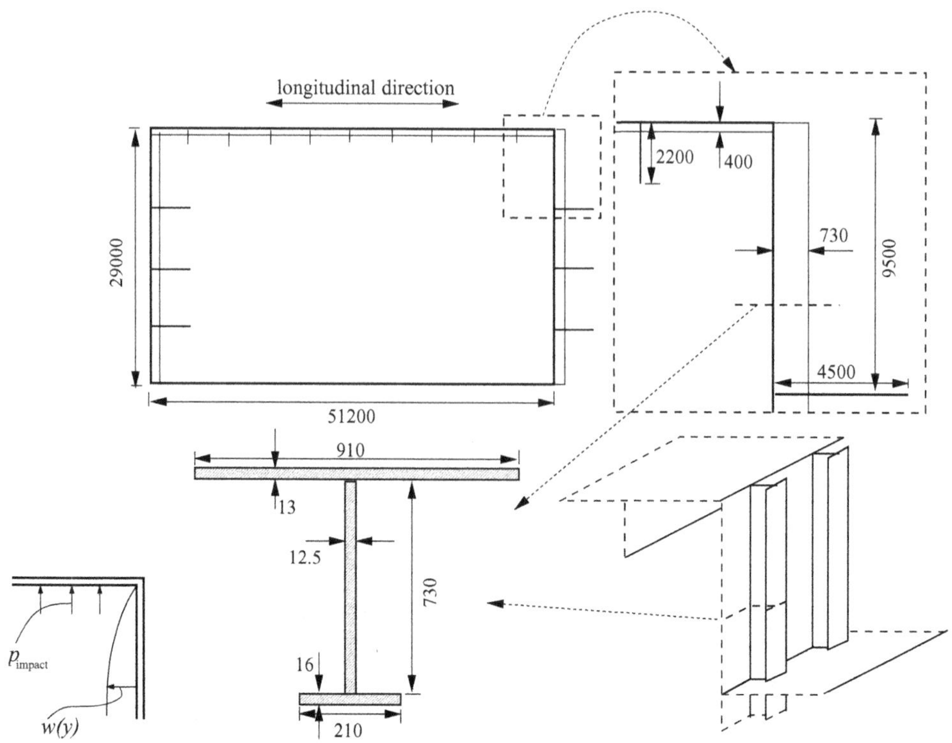

Figure 11.55 Details of structural arrangement in the tank of an OBO carrier. Dimensions are in millimeters. The top of tank is assumed rigid. Structural details are only shown for the vertical stiffeners (clamped ends). The liquid impacts a rigid tank roof in the examples. The resulting elastic deformations $w(y)$ of the adjacent vertical tank wall are considered (see left picture).

Figure 11.56 shows strain response at $y = -9.5$ m, –4.75, and 0 m (i.e., in the middle and at the two ends of the beam). Both hydroelastic and quasi-steady responses are plotted. Hydroelasticity is important and causes higher strains than a quasi-steady analysis. The highest absolute value, which is approximately 4,000 μs, occurs at the beam ends. This corresponds to a bending stress of 840 MPa, which is clearly beyond the yield strength of realistic steel materials (i.e., 235–360 MPa) and happens for the relatively low longitudinal tank motion amplitude of 0.746 m. The analysis assumes no internal structures. Internal structures cause flow-separation damping that lowers the impact velocity on the tank roof and resultant stresses in the tank wall. Flow-separation damping can be included empirically in terms of drag coefficients in the analytical model, but this requires further systematic studies.

In the calculations we have neglected the influence of acoustic liquid effects. The relative importance of the liquid compressibility can similarly as in Section 11.9.3 be assessed by using eq. (11.85) to estimate the duration of the supersonic stage and to multiply eq. (11.85) by 6 to find an approximation of the time it takes for the acoustic solution to reach the Wagner solution for an incompressible liquid. We assume no mixture between gas and liquid and use the speed of sound of freshwater at 20°C. Using $c_0 = 1481.3 \text{ m s}^{-1}$, $R = 7.647$ m, and $V = 9.081 \text{ m s}^{-1}$ gives 10^{-4} s; the time scale is very small relative to the time scale of the structural response shown in Figure 11.56. The consequence is that the effect of liquid compressibility can be neglected.

The hydroelastic response oscillates with the natural periods of the beam vibration. The natural periods of the two lowest modes are 0.0749 and 0.0182 s. The uncoupled natural period in air of the lowest mode is 0.0166 s. It illustrates that added mass has a dominating influence. We estimated

$$\frac{A_{11}}{0.25\rho_l \pi L^2} = 0.538, \qquad \frac{A_{22}}{0.25\rho_l \pi L^2} = 0.236,$$

$$\frac{A_{12}}{0.25\rho_l \pi L^2} = 0.072.$$

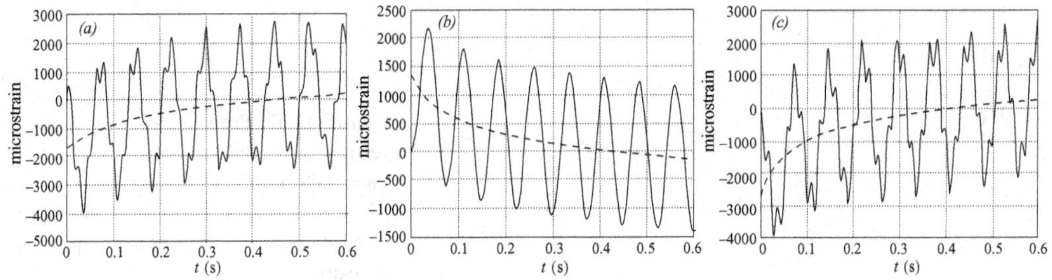

Figure 11.56. Elastic strain response of an adjacent vertical wall due to tank roof impact with $V_0 = 9.081\,\mathrm{m\,s^{-1}}$, $V_1 = -7.668\,\mathrm{m\,s^{-2}}$, and $R = 7.647\,\mathrm{m}$. The solid line denotes hydroelastic solution. The dashed line corresponds to quasi-static solutions: (a) $y = -9.5\mathrm{m}$, (b) $y = -4.75$ m, and (c) $y = 0$.

The added mass values are dependent on the amplitude of the mode shapes (see eqs. (11.101) and (11.102)). The mode shapes are defined by eq. (11.109) and normalized so that

$$\overline{\psi_m^2} = \int_b^a \psi_m^2 \mathrm{d}y/L = 1, \quad m = 1, 2.$$

The simplified calculation of A_{11} described in Section 11.9.2 (see the text in connection with eq. (11.107)) gives $A_{11} = 0.56 \cdot (\frac{1}{4}\rho_l \pi L^2)$, which agrees well with the exact aforementioned value. The latter analysis neglects the influence of the tank bottom and the other tank wall. Table 11.4 illustrates that this is a good approximation by varying systematically the length l and the height H of the tank.

Figure 11.57 shows both a space-averaged ($\overline{p}$) and an equivalent excitation pressure (p_{eq}) over the beam as a function of time; p_{eq} is found by considering which averaged loading over the beam causes the same strain in the middle of the beam as the strain (ε_{qsm}) calculated by quasi-steady theory:

$$p_{eq} = 24\mathrm{EI} \cdot \varepsilon_{qsm}/(z_a L^2)\,, \qquad (11.110)$$

which is not the same as the space-averaged excitation pressure $\overline{p}$ over the beam. $\overline{p}$ is logarithmically singular when $t = 0$, which can be seen from the last term of eq. (11.100) by noting that $c \to 0$ when $t \to 0$. However, the generalized force $\int_b^a p_{exc}\psi_1 \mathrm{d}y$ causing the strain in the middle of the beam is finite when $t = 0$. The reason is that ψ_1 is proportional to y for small y, which means p_{eq} is nonsingular when $t = 0$. Figure 11.57 shows that p_{eq} and $\overline{p}$ are quite close except for a very short initial time. The pressure is influenced by the beam vibrations because the pressure term causes added mass contributions in the equations of motion and is not included in p_{eq} or $\overline{p}$.

It is common in design to do a quasi-steady structural analysis. The maximum p_{eq} shown in Figure 11.57 is then not the design pressure for the particular examined condition. Dynamic effects must be accounted for so that the maximum strain obtained by the quasi-steady analysis is the same as the maximum strain obtained by the dynamic analysis. For the particular examined condition, therefore, the maximum p_{eq} in Figure 11.57 should be multiplied by 1.52.

Table 11.4. *Influence of tank bottom and the other wall on the generalized added mass coefficients A_{11} and A_{22} of vibrating beam in a tank wall*

l(m)	H_t(m)	$\frac{l}{L}$	$\frac{H_t}{L}$	$\frac{A_{11}}{0.25\rho_l\pi L^2}$	$\frac{A_{22}}{0.25\rho_l\pi L^2}$
51.2	29.0	5.39	3.05	0.538	0.236
20.0	29.0	2.11	3.05	0.557	0.240
10.0	29.0	1.05	3.05	0.626	0.253
5.0	29.0	0.53	3.05	0.848	0.289
2.0	29.0	0.21	3.05	1.728	0.446
1.0	29.0	0.11	3.05	3.327	0.775
0.5	29.0	0.05	3.05	6.589	1.487
0.1	29.0	0.01	3.05	32.837	7.327
100.0	29.0	10.50	3.05	0.538	0.236
51.2	20.0	5.39	2.11	0.545	0.238
51.2	15.0	5.39	1.58	0.557	0.241
51.2	12.0	5.39	1.26	0.573	0.246
51.2	11.0	5.39	1.16	0.583	0.249
51.2	10.0	5.39	1.05	0.596	0.255
51.2	9.6	5.39	1.01	0.604	0.258
51.2	9.5	5.39	1.00	0.606	0.259
51.2	50.0	5.39	5.26	0.536	0.236

Note: l is the tank length, H_t is the tank height, and L is the beam length.

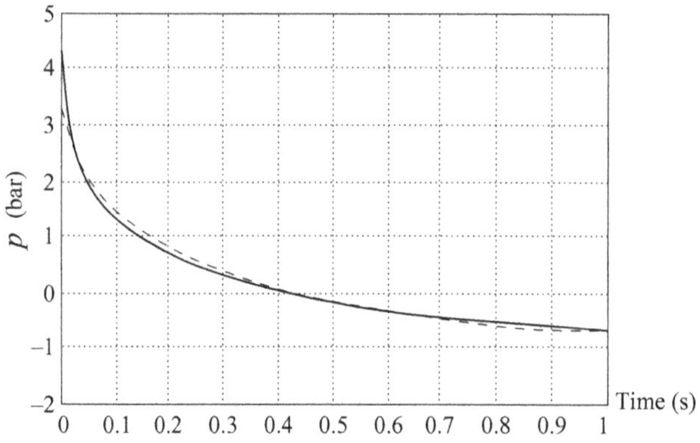

Figure 11.57. Pressure on an adjacent vertical wall due to tank roof impact: $V_0 = 9.081\,\mathrm{m\,s^{-1}}$, $V_1 = -7.668\,\mathrm{m\,s^{-1}}$, $R = 7.647$. Solid line represents average pressure, dashed line represents equivalent pressure.

This discussion shows that great care has to be taken in sloshing measurements with hydroelastic effects. If pressures are measured, flexibility of the structure matters. Very high localized pressures can be measured initially. In both cases there are difficulties in interpreting how the measured pressure should be used in a structural response analysis. Using force panels instead of small pressure gauges is better and filters out a highly concentrated pressure in space. However, because the structural response is dynamic, it is best to measure strains directly, which requires special scaling in model tests and is discussed in Section 11.9.5.

Introducing nondimensional values is a compact way to present the parametric dependence of the strain response. Equation (11.100) suggests that $\rho V_0^2 R/L$ can be used to nondimensionalize excitation pressure. A combination of this with eqs. (11.96) and (11.108) suggests that a good way is to use the following nondimensional expression for quasi-steady strains:

$$\bar{\varepsilon} = \varepsilon \mathrm{EI}/\left(z_a V_0^2 R \rho_l L\right), \qquad (11.111)$$

which is also used to make nondimensional hydroelastic strain response. The ratio T_R between the impact duration $-V_0/V_1$ and the highest natural period T_{n1} of the beam vibrations is then introduced as a parameter: $T_R = -V_0/(V_1 T_{n1})$. Table 11.5 presents results for maximum nondimensional absolute values of strain in the middle of the beam. The different values of V_0, V_1, and R are calculated based on Faltinsen and Rognebakke (1999). The main tank dimensions were previously presented in this section and in Figure 11.55. Case 1 corresponds to the same forced longitudinal tank motion amplitude A, forced period T, and mean water depth h as used in Figure 11.56. For case 2, $A = 0.947$ m, $T = 8.16$ s, and $h = 17.4$ m; for case 3, $A = 1.48$ m, $T = 9.3$ s, and $h = 14.5$ m; for case 4, $A = 2.5$ m, $T = 7.62$ s, and $h = 14.5$ m. The highest natural periods for the liquid motion are 9.12 and 9.60 s for cases 2 and 4, respectively. The E-value has been varied. Table 11.5 illustrates that eq. (11.111) is a very good way to make a nondimensional quasi-steady response. A reason is that maximum loading and response occur initially (i.e., when the impact velocity is V_0).

Figure 11.58 presents the dynamic amplification factor $D = \varepsilon_{mD}/\varepsilon_{mS}$ as a function of the relative time parameter T_R, where ε_{mD} and ε_{mS} are maximum absolute values of dynamic strain and quasi-steady strain, respectively. The same results are given in Table 11.5. When T_R is small, ε_{mD} occurs during the free vibrations. When T_R is large, ε_{mD} occurs initially. This result is expected from vibration theory. When T_R is small, the force impulse matters. By integrating eq. (11.99) in time and noting that $V\sqrt{c^2 + y^2} + Vy$ is zero both initially (when $c = 0$) and at the end of the impact (assumed when V becomes 0), it follows that the impulse is zero. Therefore, the hydrodynamic pressure is positive initially but negative by the end of the impact stage. The fact that the total pressure obtained by adding the ambient pressure and the hydrodynamic pressure should be related to the vapor pressure has not been accounted for in the analysis. The loading continues after $t = -V_0/V_1$ (i.e., during the water-exit phase when the wetted surface on the tank roof diminishes). It is questionable to use the same expression for pressure for the water-exit phase as for during the impact phase; it causes an error in our predictions when T_R is very small. It will not have any

Table 11.5. *Maximum nondimensional absolute values of calculated strain in the middle of the beam*

Case	$\mathrm{E}\cdot 10^{-9}$ (Pa)	$V_0(\mathrm{m\,s^{-1}})$	$V_1(\mathrm{m\,s^{-2}})$	R (m)	$\dfrac{\varepsilon_{mS}\mathrm{EI}}{z_a V_0^2 R\rho_l L}$	$\dfrac{\varepsilon_{mD}}{\varepsilon_{mS}}$	$-\dfrac{V_0}{V_1 T_{n1}}$
1	210	9.081	−7.668	7.647	0.224	1.52	15.8
1	210	4.072	−8.965	7.801	0.224	1.54	6.1
1	210	7.0	−8.202	7.831	0.224	1.54	11.4
1	70	7.0	−8.202	7.831	0.224	1.38	6.6
2	210	3.464	−9.804	9.946	0.224	1.48	4.7
2	210	5.023	−9.537	9.576	0.224	1.50	7.0
2	210	2.520	−9.956	9.959	0.224	1.44	3.4
2	210	1.906	−10.033	10.196	0.224	1.37	2.5
2	210	0.945	−10.071	10.230	0.224	1.42	1.25
2	70	0.945	−10.071	10.230	0.224	1.42	0.72
3	210	13.253	−11.139	4.792	0.224	1.53	15.9
3	210	9.631	−11.940	5.146	0.224	1.55	10.8
3	70	9.631	−11.940	5.146	0.224	1.39	6.2
4	210	6.023	−14.42	7.114	0.224	1.49	5.6
4	210	8.148	−13.199	7.73	0.224	1.48	8.3
4	210	2.858	−15.144	8.635	0.224	1.35	2.5
4	210	3.07	−14.801	9.21	0.224	1.37	2.8
4	70	3.07	−14.801	9.21	0.224	1.17	1.6
4	50	3.07	−14.801	9.21	0.224	1.25	1.35
4	30	3.07	−14.801	9.21	0.224	1.30	1.05
4	20	3.07	−14.801	9.21	0.224	1.34	0.86
4	10	3.07	−14.801	9.21	0.224	1.03	0.6
4	8	3.07	−14.801	9.21	0.224	0.89	0.54
4	5	3.07	−14.801	9.21	0.224	0.63	0.43
4	3	3.07	−14.801	9.21	0.224	0.40	0.33

Note: ε_{mS} is the strain based on quasi-steady analysis; ε_{mD} is the strain based on hydroelastic analysis.

effect when T_R is large and the largest loading and response of the beam occurs. We can indirectly see this by comparing Figure 11.57 with Figure 11.56. The time-dependent space-averaged excitation pressure is a measure of the loading. Initially the loading is large (see Figure 11.57). It is then dominated by the last term in eq. (11.100). Later the largest strains occur. As time increases, V decreases and the first term proportional to the acceleration $\dot{V} = V_1$ increases in importance. Because V_1 is negative, this term is negative. This term finally causes negative hydrodynamic loading, which we can see in Figure 11.57. We have not shown the loading during the whole time $-V_0/V_1$ in Figure 11.57. If we had, it would have been evident that the time integral of the

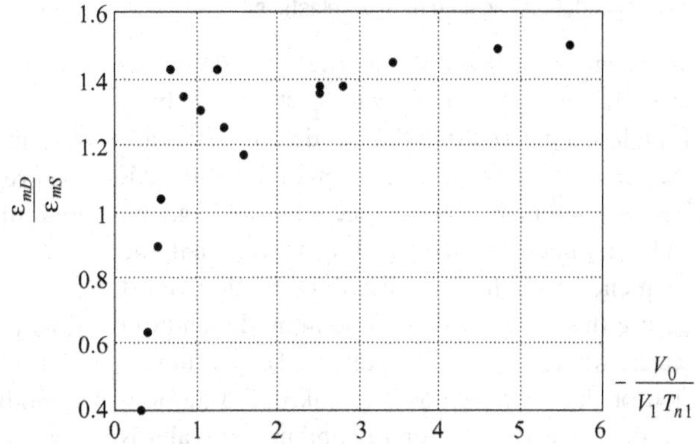

Figure 11.58. Dynamic amplification factor $D = \varepsilon_{mD}/\varepsilon_{mS}$ due to tank roof impact presented as a function of relative time parameter $T_R = -V_0/(V_1 T_{n1})$.

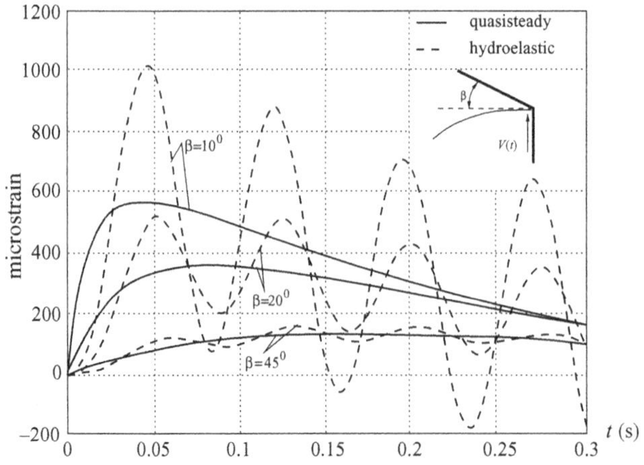

Figure 11.59. Elastic strain response of adjacent vertical wall due to tank roof impact as a function of chamfer angle β, $V_0 = 9.081$ m s^{-1}, $V_1 = -7.668$ m s^{-1}, $R = 7.647$, and $y = -4.75$ m.

negative hydrodynamic loading cancels the time integral of the positive hydrodynamic loading (i.e., the impulse is zero).

Figure 11.58 illustrates that, for intermediate values of T_R between ~0.7 and 1.3, T_R is not the only important parameter to characterize the dynamic response. Other relevant nondimensional parameters are $V_0 T_{n1}/L$, $V_1 T_{n1}^2/L$, and R/L.

A chamfered tank roof reduces the structural response. We illustrate this fact by considering the same example as in Figures 11.57 and 11.58, but let the chamfer angle β be different from zero (see Figure 11.59). The same values of V_0, V_1, and R are used for different β-values; that is, the effect of β on the global flow has been neglected. A higher β-value for a given excitation means in reality a lower tank roof damping and higher global sloshing motion. Theoretically β should be small, but β up to 45° has been investigated to show a trend.

11.9.5 Model test scaling of hydroelasticity

When model tests with hydroelastic slamming and sloshing are done, we must check both Froude scaling and that the elastic properties are properly scaled. In the discussion of possible scaling parameters, we have neglected some (e.g., the Euler number). Froude scaling means that the frequency σ of the forced tank oscillations must ensure that $\sigma\sqrt{L/g}$ is the same in model and full scales, where L is a characteristic length dimension of the tank such as the tank breadth. Geometric similarity between full and model scales is assumed. Let us use the subscript m and p to indicate model and full scale (prototype). Thereby Froude scaling gives

$$\sigma_m = \sigma_p (\Lambda_L)^{1/2}, \tag{11.112}$$

where $\Lambda_L = L_p/L_m$ is the geometrical scale factor. To find out how to scale elastic properties associated with bending stiffness $\mathtt{EI}$, we can use eq. (11.98) together with eq. (11.97) as a basis and set the excitation equal to zero. Furthermore, we neglect coupling effects. The natural frequency of mode n can be expressed as

$$\sigma_n = \sqrt{\frac{\mathtt{EI} \cdot p_n^4 \cdot M_{nn}}{M_B \cdot (M_{nn} + A_{nn})}}. \tag{11.113}$$

We note that $M_B/(\rho_l L)$, $M_{nn}/(\rho_l L^2)$, $A_{nn}/(\rho_l L^2)$, and $p_n L$ are nondimensional. Then from eq. (11.113) it follows that $\sigma_n\sqrt{\rho_l L^5/\mathtt{EI}}$ is a nondimensional frequency associated with elastic vibrations due to bending stiffness. Therefore,

$$\sigma_m = \sigma_p \sqrt{\frac{\rho_p (\mathtt{EI})_m}{\rho_m (\mathtt{EI})_p} \Lambda_L^5}. \tag{11.114}$$

To satisfy eqs. (11.112) and (11.114), we must require that $(\mathtt{EI})_m \rho_p = (\mathtt{EI})_p \rho_m / \Lambda_L^4$. We must also require that $M_B/(\rho_l L)$ is the same in model and full scales.

If $\mathtt{I}$ is geometrically scaled and $\rho_p = \rho_m$, then $\mathtt{E}_m/\mathtt{E}_p = L_m/L_p$. However, scaling $\mathtt{I}$ geometrically can lead to unrealistically small cross-sectional dimensions of the beam. In any case the $\mathtt{E}$-module has to be smaller in model than in full scale. It is also pertinent that resulting stresses in

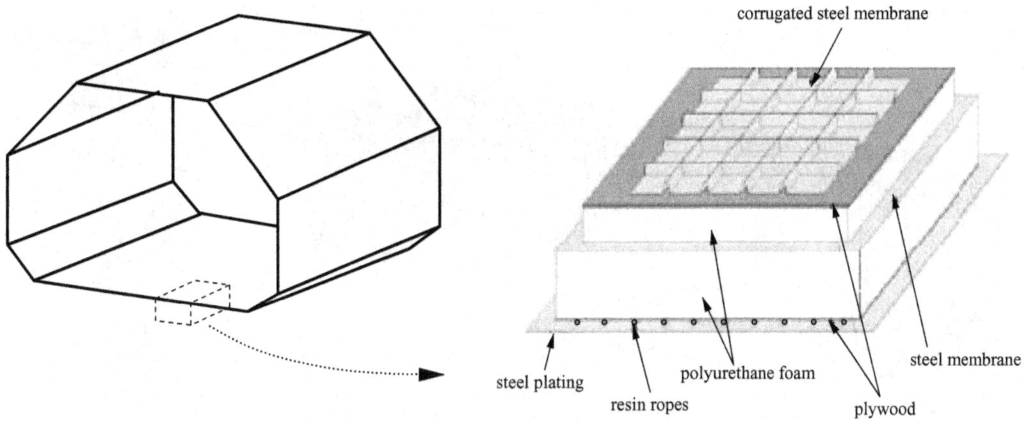

Figure 11.60. Schematic illustration of the Mark III containment system (Graczyk, 2008).

the model beam should be lower than the yield stress.

11.9.6 Slamming in membrane tanks

From a structural point of view a membrane tank is a more complicated structure to analyze than a steel structure. A schematic illustration of the Mark III containment system is shown in Figure 11.60 with a detailed view of the containment system. A corrugated stainless steel membrane (primary barrier) faces the tank interior. The support of the containment system consists of stiffened steel plating with crossing stiffeners and girders with the steel plating adjacent to the containment system. Unidirectional resin (mastic) ropes lay between this steel plating and a plywood plate.

A main insulating material of the Mark III system is reinforced polyurethane foam. The first layer is a 1.2-mm-thick corrugated stainless steel membrane, attached by tack welding to the weld protection strips embodied in the plywood underneath. The plywood sheet is integrated with a pad of cryogenic foam. It is cut perpendicularly to the surface to accommodate the contractive forces associated with the low temperatures. A secondary membrane of fiberglass fabric and aluminum foil laminate called Triplex is placed beneath the pad. Underneath is a prefabricated foam panel and a plywood sheet. A standoff space between panel and hull is held by mastic ropes. The layers are mechanically and adhesively joined to each other and to the hull. The insulation space below the containment system is kept inert with nitrogen and enables monitoring to detect natural gas or water. It also accommodates a tolerance for hull deformations during installation of the panels.

A major concern is the failure of the plywood plate adjacent to the resin ropes, particularly its local bending between the ropes and the shear near the ropes. Rupture of the foam next to the latter plywood plate must be avoided. It is common in a structural analysis of the containment system to assume stiff girders and stiffeners. The steel plating in the support must be assumed elastic.

Graczyk (2008) numerically studied slamming load effects on a part of the Mark III containment system. Typical main dimensions of the tank could be a length of 43 m, a breadth of 37 m, and a height of 27 m. Because the corners complicate the analysis and structural details were not available, the studied segment of the containment system was not adjacent to corners. The lateral dimensions of the panel were 3300 × 840 mm, which corresponds to an assumed span of girders and stiffeners in two perpendicular directions. One-quarter of the panel is illustrated in Figure 11.61. The thickness of the segment was approximately 300 mm. The resin ropes, the steel plate, and two layers of plywood with a layer of foam in between were included. These components are the most important in a dynamic analysis.

Figure 11.62 shows the numerically calculated eigenfrequencies with different assumed wetted areas. Simplistic calculations of the added mass were done by neglecting the influence of other parts of the tank. We note a significant effect of the wetted area: the estimated added mass. The

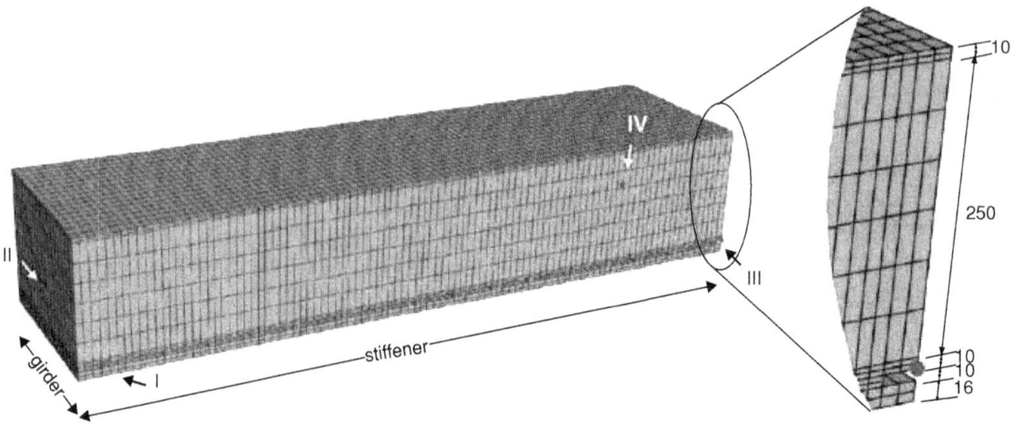

Figure 11.61. Investigated model of the Mark III containment system: one-quarter of a 3300 × 840 mm panel and the element sizes are shown. The Roman numbers indicate locations that are referred to in Figure 11.63. Note that all the points I, II, III, and IV are in the plywood next to the resin ropes (Graczyk, 2008).

lowest modes are governed by the steel-plating response. The bending response of the plywood next to the resin ropes matters from about 650 Hz for the dry model and from about 300 Hz for the model with the largest wetted area. Because the steel bending causes tension or compression of the plywood, an important plywood response is associated with the lowest mode.

A slamming case is analyzed numerically in terms of normalized response spectra in Figure 11.63. An average slamming pressure of 10 bar acting on the considered segment was assumed. The time duration of the loading was 3 ms. The effect of added mass was included by assuming a wetted area of 2.772 m^2 (see Figure 11.62). The maximum response values were of significance for the evaluation of the structural strength. Four different locations indicated on Figure 11.61 were studied. We note that it is not only the lowest modes, governed by the steel response, that

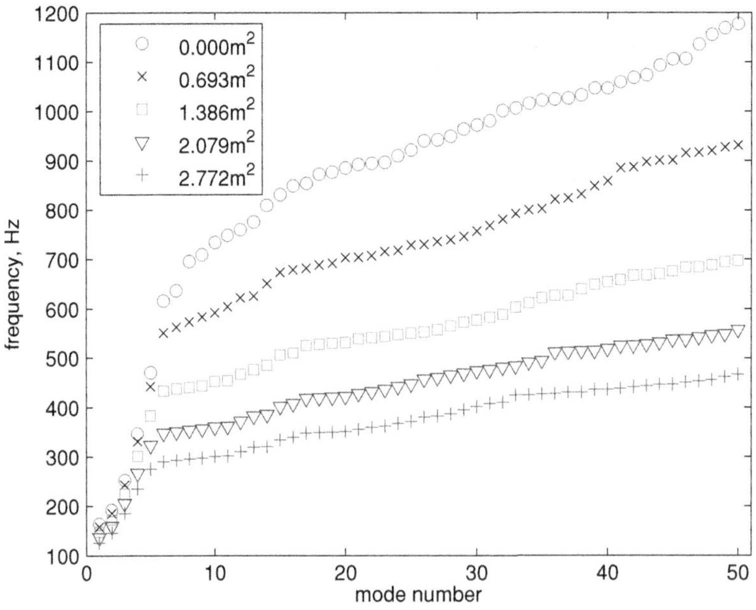

Figure 11.62. The first fifty eigenfrequencies for the flexible model shown in Figure 11.61. The numbers represent different values of the wetted area that are used as a basis in the estimation of added mass (Graczyk, 2008).

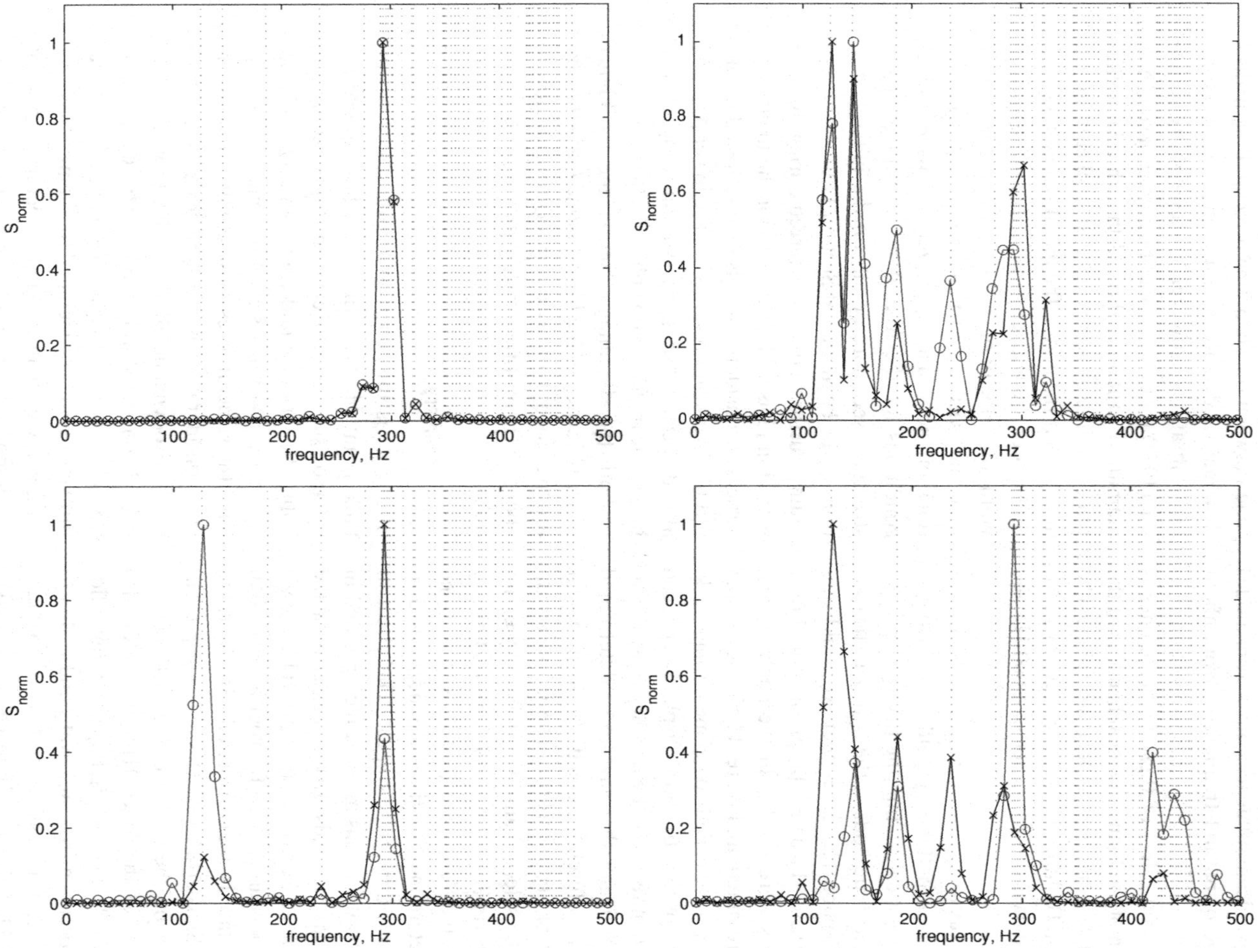

Figure 11.63. Normalized spectra of the bending stresses in the lowest (○) and the highest (×) fiber of the plywood next to the resin ropes. The effect of added mass with a wetted area of 2.772 m^2 is included. The natural frequencies are marked as vertical lines: left upper panel = location I, right upper panel = location II, left lower panel = location III, and right lower panel = location IV (see Figure 11.61) (Graczyk, 2008).

matter. A significant influence comes from modes with a range of natural frequencies from about 100 to 500 Hz.

An important effect of these higher modes is compression of the foam and local bending of the plywood plate adjacent to the resin ropes. The effect of liquid compressibility is believed to matter for frequencies of the order of 1,000 Hz and higher.

A hydroelastic analysis that accounts for the liquid flow in a better way could, for instance, be done by first calculating the dry modes using a finite element method. In the hydroelastic case studied in Section 11.9.2 for steel, we could simplify the structural side of the problem and only a few modes were needed in the analysis. The example presented in Figure 11.63 shows that more modes are needed for a membrane tank. The solution of the structural displacements can be expanded in terms of the dry modes with generalized coordinates that have to be determined by a set of differential equations in a similar way to that of Section 11.9.2. The inflow conditions must be selected and determined by either CFD or experimental results. A more analytically oriented method can then be followed in some cases to determine the time-dependent hydrodynamic loads during the impact. The effect of the structural vibrations must be incorporated in the hydrodynamic loads. If the impact is of the Wagner type as described in Figure 11.2(b), we can follow such an analysis for roof impact when the tank surface is not corrugated. However, the studies by Shin *et al.* (2003) show a gas cushioning effect on the pressure due to the corrugated surface of the Mark III system (see also Section 11.6). The slamming impact at lower liquid fillings occurs at the tank walls and is due to impacting steep waves as illustrated, for instance, in Figure 11.11. This type of impact is described in Section 11.4. The impact analysis should ideally incorporate the effect of gas density, boiling, and the compressibility of the liquid and gas (see Sections 11.2, 11.7 and 11.8).

Important parameters in the slamming analysis are the ratios between the impact duration and important natural periods. If this ratio is small, the fine details of the hydrodynamics are not needed in describing the maximum structural stresses. Actually, it is the slamming force impulse that matters. This force impulse is difficult to define accurately because of no clear border in time between the slamming force and the hydrodynamic (added mass) force due to the structural vibrations. However, the analysis by Faltinsen (1997) for the hydroelastic impact of steel and aluminum structures showed that the slamming-force impulse was not needed explicitly in the analysis. At the end of the initial slamming phase the sum of the impacting velocity and the elastic vibration velocity of the structure was zero. This outcome, with the fact that the initial deformations of the structure are zero, provided initial conditions for the free elastic vibrations of the structure (see more details on the initial conditions in the beginning of Section 11.9). The maximum structural stresses occurred during this free vibration phase. The situation for the membrane structure considered in this section would be different. Significant response of the lower plywood already occurs during the slamming impact (i.e., before the free-vibration phase). This result is due to both the slamming duration and the higher eigenfrequency of the lowest mode (125–165 Hz) relative to a typical lowest eigenfrequency of about 50 Hz of stiffened steel plating.

11.10 Summary

Slamming is important in the design of a ship tank. Many physical effects may have to be considered, such as gas cushion, liquid compressibility, boiling of liquid cargo, and hydroelasticity. When analyzing slamming, one must always have the structural response in mind. An important consideration is the time scale of a particular hydrodynamic effect relative to wet natural periods for structural modes contributing significantly to large structural stresses. More structural modes may be included for membrane structural analyses than for steel structures. Some of the important structural modes for membrane structures may have relatively lower natural periods than for steel structures. If the time scale of a hydrodynamic effect (for instance, acoustic effects) is *very* small relative to important structural natural periods, the details of the particular hydrodynamic effect can be neglected. When the hydrodynamic loads occur on the time scale of important structural modes, hydroelasticity must be considered. This condition implies that the fluid (liquid, gas) flow must be solved simultaneously with the dynamic elastic structural reaction.

Slamming in a ship tank is associated with violent liquid motion and many possible impact scenarios have to be considered. For instance, large filling ratios can cause important slamming loads. Examples include the tank-roof impact of a nearly horizontal free surface and impact with an oscillating gas cavity. Another possible scenario is a sudden flip-through of the free surface at the tank wall. A jet flow will, as a consequence, impact the tank roof. A chamfered tank roof reduces the severity of slamming for large filling ratios.

One must recognize that slamming does not cause important loads only in the impact area. This fact is demonstrated with a case where the liquid impacts a rigid tank roof. Because the adjacent tank wall was more flexible than the tank roof, significant stresses occurred as a consequence of the tank roof impact.

Steep waves impacting a vertical tank wall represent an important concern for shallow and lower-intermediate liquid depths. An example is a hydraulic jump that can be formed at resonant conditions, resulting in high-pressure loading on the tank wall. The pressure is sensitive to small changes of an impacting steep free surface. However, when high pressures of small duration and limited spatial extent occur, it is important to not overemphasize the pressure level. As earlier stated it is the structural response that matters. The maximum structural stresses may not be sensitive to small changes in the impacting conditions.

It is common to analyze slamming by assuming a two-dimensional flow. However, swirling may cause an important impact against the tank roof corners for nonsmall liquid depths in a nearly square based tank (see Chapter 9 for more details on swirling).

CFD methods incur general difficulties in accurately predicting slamming effects for severe impact situations. This chapter instead focused on analytically based two-dimensional methods which, for given inflow conditions to the impact, can provide valuable information. This situation was studied in detail for tank roof slamming by starting with the impact of a nearly horizontal free surface with no gas influence and use of the Wagner method. The effects of a gas cushion, liquid compressibility, and hydroelasticity for steel tanks were also analyzed. Furthermore, shallow liquid impact with a steep wave was considered theoretically with and without the effect of liquid compressibility.

It is common in tank design to do model experiments for sloshing-induced slamming effects by means of forced oscillation tests. However, the scaling of the model test results represents a challenge due to the many physical effects that may matter, in particular, in a membrane tank with liquefied gas.

Because sloshing is associated with gravity waves, we must require that the Froude number is the same in model and full scales. Furthermore, the wave-induced ship motion that excites sloshing is also Froude scaled. If harmonically forced oscillation of the tank with frequency σ is considered, Froude scaling implies that $\sigma\sqrt{L/g}$ must be the same in model and full scales. The parameter L is a characteristic tank dimension such as the tank breadth. A conventional model test approach only considers the effect of Froude and geometrical scaling. However, other scaling parameters of possible importance are summarized.

If *hydroelasticity* matters during impact, we must ensure that the relevant natural frequencies for the elastic structural vibrations are Froude scaled. For instance, let us consider a steel tank. The bending stiffness matters for the natural elastic frequencies of importance. Furthermore, the length of the elastic plate must be geometrically similar in model and full scales. The scaling of the natural frequency may be achieved by having different bending stiffness in model and full scales. The bending stiffness may be properly scaled by considering a different material or properly changing the thickness of the material. Because the main interest is to find slamming-induced structural stresses, care must also be taken in scaling structural stresses (Faltinsen, 2005).

If slamming is associated with the formation of gas pockets, the *Euler number* must be the same in model and full scales. The Euler number is defined as $Eu = p_0/(\rho_l U^2)$, where p_0 is the ullage pressure (i.e., the pressure in the ambient gas inside the tank). It is generally necessary to lower the ullage pressure in model tests to achieve the same Euler number in model and full scales. However, a limitation exists due to cavitation on the lowest value of the ullage pressure.

A gas pocket has a natural frequency associated with the compressibility of the gas and a generalized added mass due to the liquid oscillations caused by gas cavity oscillations. Even though the gas cavity oscillations may be linear at model scale they may be strongly nonlinear at full scale. If the time scale associated with this natural frequency is very small relative to important wet natural structural periods, the effect of a gas pocket can be neglected. If only Froude scaling is used, it is shown in an example that Froude scaling is conservative. However, another example shows that we cannot guarantee that Froude scaling is conservative. A Bagnold-type model provides a simple mean to scale model test results of maximum pressure due to a gas pocket. Because an adiabatic pressure-density relationship is assumed, the method is questionable when condensation occurs at zero cavitation number.

The *cavitation number* is defined as $(p_0 - p_v)/(\frac{1}{2}\rho_l U^2)$ where p_v is the vapor pressure and p_0 is the ullage pressure. We note a similarity between the cavitation number and the Euler number. The basic mechanics of boiling and cavitation is similar. Boiling takes place in a layer next to the free surface of LNG and the other cargo liquids listed in Table 11.2. Boiling corresponds to $p_0 = p_v$ (i.e., zero cavitation number). The cavitation number must be considered when scaling model test results to full-scale values for boiling liquid cargoes.

It is discussed by model test results in Section 11.2 that the density ratio between the ullage gas and the liquid matters. The heavier the ullage gas is, the more the liquid impact is slowed down.

When there is no mixture between liquid and gas, the *Cauchy number*, $Ca = \rho_l U^2/E_l$, characterizes the effect of compressibility on the liquid flow, where E_l is the bulk modulus for elasticity. The effect of liquid compressibility is less important in the case of a liquid with no gas bubbles and may be disregarded for a steel tank. This result depends on the important wet structural natural periods. However, the speed of sound can be substantially lower in the case of a gas–liquid mixture. A consequence is an increased characteristic time scale for the effect of liquid compressibility. Because a mixture of gas and liquid can occur during violent sloshing and because LNG is boiling, we cannot rule out the effects of compressibility on slamming loads. The mixture of LNG and gas is not homogeneous. Because sloshing and impact affect the pressure distribution in the fluid and the pressure determines if cavitation occurs, the amount of bubbles is time dependent. Further condensation due to overpressure is a factor. The void fraction is therefore a function of time and space.

Surface tension and *viscosity* are believed to be less important effects during slamming. The model test results presented by Abramson *et al.* (1974) using liquid of different viscosity showed that viscosity did not have a dominant effect for large excitations that are realistic as design conditions.

Figure 11.64 gives a final summary flowchart of sloshing and slamming. The first step is to identify which natural structural periods are associated with large structural stresses. Then we must identify different inflow conditions for slamming. The different physical effects to consider are then listed. Their importance depends on both the inflow conditions and relevant natural structural periods. Finally we must assess if hydroelasticity matters.

11.11 Exercises

11.11.1 Impact force on a wedge

Consider a wedge with deadrise angle β that enters with constant velocity V a calm liquid with mass density ρ_l. Use Wagner's method to analyze the problem.

(a) Show that the two-dimensional vertical hydrodynamic force F_3 based on integrating the pressure in the outer domain solution can be expressed as $F_3 = \rho_l \pi V c\dot{c}$. Express c as a function of time.

(b) Derive the force expression given in part (a) by using conservation of energy. You must then consider the kinetic energy going into the jets as well as the kinetic energy in the outer flow domain. (*Hint*: Use the divergence theorem to rewrite the expression for the kinetic energy in the outer flow domain.)

11.11.2 Prediction of the wetted surface by Wagner's method

An essential part in finding the wetted surface in Wagner's method was a solution of the integral equation given by eq. (11.13). A general solution of this integral equation is given in Section 11.3.1.1. Consider the impact of a sinusoidal standing wave against a horizontal tank roof. Thereby $\eta_b(x)$ defined by Figure 11.5(a) is

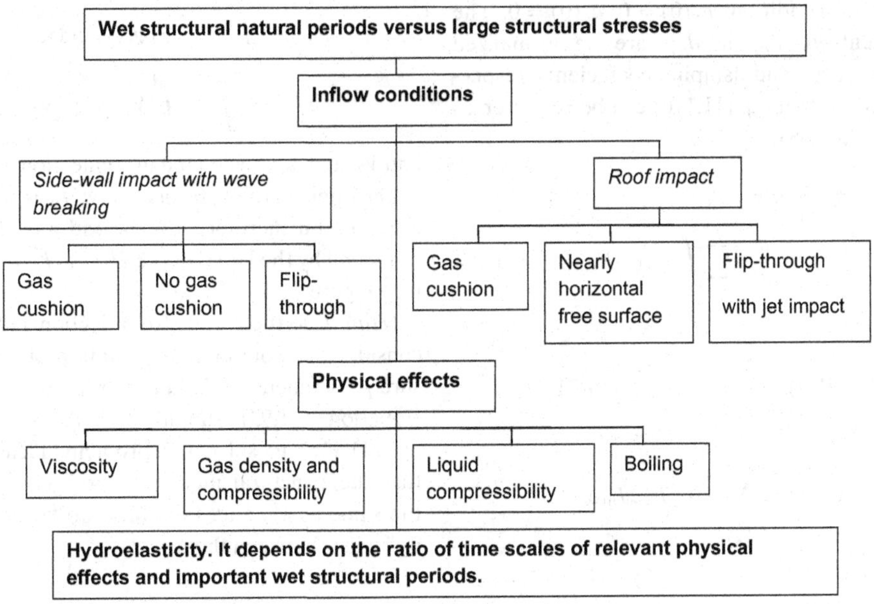

Figure 11.64. Summary flowchart of slamming and sloshing.

expressed as

$$\eta_b(x) = A[1 - \cos(kx)]. \qquad (11.115)$$

Assume constant water-entry velocity V and compare the solutions of $c(t)$ and $\dot{c}(t)$ obtained by using eq. (11.115) with the solutions obtained by approximating $\eta_b(x)$ as a second-order polynomial (see eq. (11.15)). Make your own choice of parameters.

11.11.3 Integrated slamming loads on part of the tank roof

Consider a two-dimensional rigid rectangular tank as illustrated in Figure 11.3. Assume that the impacting free surface can be approximated as a parabola with radius of curvature R and approximate the impacting velocity as $V(t) = V_0 + V_1 t$, where $t = 0$ is time of impact. Consider a segment of the tank roof that is in the impact region. The ends of the segment have x-coordinates x_i and x_{i+1}. Use Wagner's outer domain solution to set up expressions for the force on the panel as a function of time. When does the maximum force occur and what is its expression?

11.11.4 Impact of a liquid wedge

Figure 11.23 presents results for pressure loads due to impact of a liquid wedge with interior angle β. Use the composite solution based on the inner and outer domain solutions by Wagner to derive the results when β is close to 90°. Compare the results with Figure 11.23 in terms of both maximum pressure and pressure distribution.

11.11.5 Acoustic impact of a hydraulic jump against a vertical wall

(a) We refer to definitions and formulations in Figure 11.42. Assume the vertical wall is rigid and use eq. (11.93) for the pressure distribution on the wall. Present similar results as in Figure 11.41 for the case that $H_w = H_\varsigma$.

(b) Express the mass flux through the plane $x = -1$ and discuss how it varies with time. Assume a rigid wall as in part (a).

(c) Assume that part of the wall is elastic with a surface S (see Figure 11.42). The initial conditions for the wall deflections are $w(y, z, 0) = w_t(y, z, 0) = 0$. Express the wall deflections as $w = \sum_{n=1}^{\infty} a_n(t)\Psi_n(y, z)$, where Ψ_n is associated with the structural eigenmodes. The orthogonality of the structural eigenmodes implies the following equation for the modal amplitudes $a_n(t)$:

$$m_n\ddot{a}_n + d_n(a_n + \gamma_n\dot{a}_n) = \int_S p(y, z, t)\Psi_n(y, z)\mathrm{d}S \qquad (11.116)$$

with initial conditions $a_n(0) = 0$, $\dot{a}_n(0) = 0$. The coefficients m_n, d_n, and $d_n\gamma_n$ are the normalized mass, stiffness, and damping coefficients, respectively. Show that eq. (11.116) can be rewritten as (Ten *et al.*, 2008)

$$m_n\ddot{a}_n + d_n(a_n + \gamma_n\dot{a}_n)$$
$$= P_n(t) - \frac{d^2}{dt^2}\sum_{m=1}^{\infty}\int_0^t a_m(\tau)K_{nm}(t-\tau)\mathrm{d}\tau,$$

where

$$P_n(t) = -\sum_{\substack{i=0\\j=1}}^{\infty} v_{ij}T_{ij,n}J_0(\alpha_{ij}t),$$

$$K_{nm}(t) = \sum_{\substack{i=0\\j=1}}^{\infty} T_{ij,m}T_{ij,n}J_0(\alpha_{ij}t),$$

$$T_{ij,n} = \int_S V_{ij}(y,z)\Psi_n(y,z)\mathrm{d}S,$$

$$v_{ij} = -\int_{1-h_w}^{1}\int_0^l V_{ij}(y,z)\mathrm{d}y\,\mathrm{d}z$$

and V_{ij} and α_{ij} are defined in connection with eqs. (11.86) and (11.88), respectively; J_0 is a Bessel function. Furthermore, l is the width of the tank L divided by the liquid depth H_ζ, and $h_w = H_w/H_\zeta$ (see Figure 11.42).

(*Hint*: Use the results from Section 11.8.3 and consider the Fourier representation of the pressure p by functions V_{ij} in the form of eq. (11.93). Equation (11.93) should be substituted into eq. (11.116) to solve the problem. Please note that the structural modes Ψ_n are generally not the same as V_{ij} and, therefore, no "orthogonality" exists between them.)

APPENDIX

Integral Theorems

The derivations in the book need several integral formulas. First of all, we extensively use a generalized Gauss theorem, which we base on Brand (1964). He states: "If $\boldsymbol{X}$ is a continuously differentiable tensor point function over the region Q bounded by a closed surface S_Q, whose external normal $\boldsymbol{n}$ is sectionally continuous, then the integral of $\nabla\boldsymbol{X}$ over the volume Q is equal to the integral of $\boldsymbol{nX}$ over the surface S." Here, by definition, $\nabla\boldsymbol{X} = \mathbf{i}\frac{\partial \boldsymbol{X}}{\partial x} + \mathbf{j}\frac{\partial \boldsymbol{X}}{\partial y} + \mathbf{k}\frac{\partial \boldsymbol{X}}{\partial z}$. Furthermore, Brand (1964) remarks that if $\boldsymbol{X}$ is a vector or tensor of higher valence, one can place a dot (scalar product) or cross (vector product) between terms of each integrand. This gives the mathematical expression of a generalized Gauss theorem:

$$\int_Q \nabla \circ \boldsymbol{X} \mathrm{d}Q = \int_{S_Q} \boldsymbol{n} \circ \boldsymbol{X} \,\mathrm{d}S, \tag{A.1}$$

where $\circ$ denotes a dot or cross or an ordinary multiplication. If $\boldsymbol{X}$ is a vector and $\circ$ denotes a dot, the generalized Gauss theorem is referred to as the *divergence theorem*. If $\circ$ is an ordinary multiplication and $\boldsymbol{X}$ is a scalar, then the theorem is known as a *Gauss theorem*. We apply the divergence and Gauss theorems in the main text. Furthermore, we use the generalized Gauss theorem in the case that $\boldsymbol{X}$ is a dyad (pair of vectors) and $\circ$ means dot multiplication.

Furthermore, the Reynolds transport theorem is often applied to a scalar or a vector $\boldsymbol{X} = \boldsymbol{X}(x, y, z, t)$ in the form presented, for example, by Wesseling (2001). The theorem should consider the control volume $Q(t)$ and its change with time as shown in Figure A.1. Then

$$\frac{d}{dt}\int_{Q(t)} \boldsymbol{X}\,\mathrm{d}Q = \int_{Q(t)} \frac{\partial \boldsymbol{X}}{\partial t}\mathrm{d}Q + \int_{S_Q(t)} \boldsymbol{X} U_{sn}\,\mathrm{d}S, \tag{A.2}$$

where U_{sn} is the normal component of the velocity of the surface S_Q. The positive normal direction is *out* of the volume. The derivation follows by starting with the definition of a derivative; that is, $df(t)/dt = \lim\{[f(t+\Delta t) - f(t)]/\Delta t\}$. The last integral is the outcome of integrating over the shaded area in Figure A.1 and letting Δt be small (i.e., go to zero).

We also use Green's first and second identities, which are powerful tools to rewrite expressions involving velocity potentials and are used several times in the text (e.g., in Section 4.7.1 when deriving formulas for the effect of interior structures on the natural sloshing frequencies, in Section 6.2.3 when considering the pressure force on a cylinder due to inflow/outflow of a boundary layer, in Section 10.2 when presenting the boundary element method, and in Section 11.9.2 in showing symmetry properties of generalized added mass coefficients used in hydroelastic slamming analysis). Green's second identity states

$$\iiint_Q (\varphi\nabla^2\psi - \psi\nabla^2\varphi)\,\mathrm{d}Q = \iint_{S_Q}\left(\varphi\frac{\partial\psi}{\partial n} - \psi\frac{\partial\varphi}{\partial n}\right)\mathrm{d}S, \tag{A.3}$$

where S_Q is the surface enclosing the fluid volume Q. It is necessary that the scalar functions φ and ψ have continuous first and second derivatives in Q. The normal direction n is out of the fluid region. If $\nabla^2\varphi = 0$ and $\nabla^2\psi = 0$ everywhere in the fluid domain, it follows from eq. (A.3) that

$$\iint_S \left(\psi\frac{\partial\varphi}{\partial n} - \varphi\frac{\partial\psi}{\partial n}\right)\mathrm{d}S = 0. \tag{A.4}$$

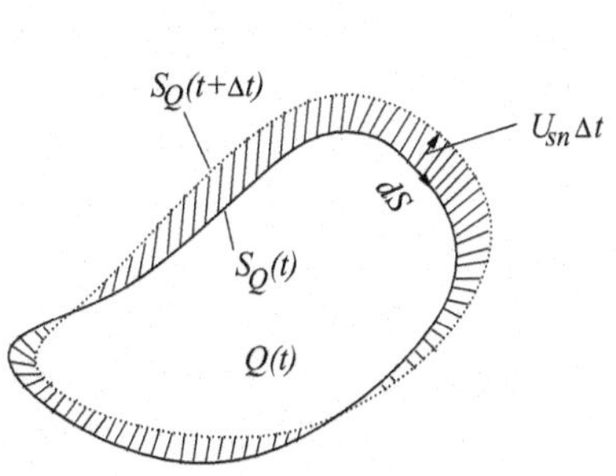

Figure A.1. Illustration of how the control volume $Q(t)$ changes in a time increment Δt; U_{Sn} is the normal component of the velocity of the surface $S_Q(t)$.

Green's first identity is

$$\iiint_Q (\varphi \nabla^2 \psi + \nabla\varphi \cdot \nabla\psi)\, \mathrm{d}Q = \iint_{S_Q} \varphi \frac{\partial \psi}{\partial n} \mathrm{d}S. \tag{A.5}$$

Finally, in connection with the derivation of the Lukovsky formula for the hydrodynamic tank moment, we use

$$\int_{S_Q} \boldsymbol{n} \times (f\boldsymbol{r})\, \mathrm{d}S = \int_Q \nabla \times (f\boldsymbol{r})\, \mathrm{d}Q = \int_Q (\nabla f) \times \boldsymbol{r}\, \mathrm{d}Q, \tag{A.6}$$

where f is a continuously differentiable function and $\boldsymbol{r}$ is the position vector. Equation (A.6) follows from using eq. (A.1) and vector algebra.

Bibliography

AASHTO Green Book (1990) *A Policy on Geometric Design of Highways and Streets-1990.* American Association of State Highway and Transportation Officials. Washington, D.C.

Abramowitz, M., and Stegun, I.A. (1964) *Handbook of Mathematical Functions with Formulas, Graphs and Mathematical Tables.* New York: Dover.

Abramson, H.N. (1966) *The Dynamic Behavior of Liquids in a Moving Container.* Technical Report SP-106, NASA, Washington, D.C.

Abramson, H.N., and Garza, L.R. (1965) Some measurements of liquid frequencies and damping in compartmented cylindrical tanks. *Journal of Spacecraft and Rockets* 2, 453–455.

Abramson, H.N., Chu, W.H., and Garza, L.R. (1963) Liquid sloshing in spherical tanks. *AIAA Journal* 1, no. 2, 384–389.

Abramson, H.N., Chu, W.H., and Kana, D.D. (1966) Some studies of nonlinear lateral sloshing in rigid containers. *Journal of Applied Mechanics* 33, no. 4, 66–74.

Abramson, H.N., Bass, R.L., Faltinsen, O.M., and Olsen, H.A. (1974) Liquid slosh in LNG Carriers. In *Tenth Symposium on Naval Hydrodynamics*, June 24–28, 1974. Cambridge, Massachusetts: ACR-204, 371–388. Published also as Faltinsen, O.M., Olsen, H., Abramson, H.N., and Bass, R.L. (1974) *Liquid Slosh in LNG Carriers.* Technical Report 85, Det Norske Veritas.

Achenbach, E. (1971) Influence of surface roughness on the cross-flow around a circular cylinder. *Journal of Fluid Mechanics* 46, pt. 2, 321–335.

Achenbach, E., and Heinecke, E. (1981) On vortex shedding from smooth and rough cylinders in the range of Reynolds numbers 6 × 103–5 × 106. *Journal of Fluid Mechanics* 109, 239–251.

Adegeest, L.J.M. (1995) *Nonlinear Hull Girder Loads.* Ph.D. Thesis, Delft University of Technology, Faculty Mechanical Engineering and Marine Technology, Delft.

Allers, J.M. (2004) *Experimental investigation of high filling sloshing induced impacts for two-dimensional flow conditions.* M.Sc. thesis, Department of Marine Technology, Norwegian University of Science and Technology.

Arai, M., Cheng, L.-Y., and Inoue, Y. (1992a) 3D numerical simulation of impact loads due to liquid cargo sloshing. *Journal of Society of Naval Architects of Japan* 171, 177–184.

Arai, M., Cheng, L.-Y., Inoue, Y., Sasaki, H., and Yamagishi, N. (1992b) Numerical analysis of liquid sloshing in tanks of FPSO. In *Proceedings of the Second International Offshore and Polar Engineering Conference* (ISOPE'92). San Francisco, 1992. Vol. 3, pp. 383–390.

Armand, J.L., and Cointe, R. (1986) Hydrodynamic impact analysis of a cylinder. In *Proceedings of the Fifth International Offshore Mechanics and Arctic Engineering Symposium*, Vol. 1, pp. 609–634. ASME.

Aubin, J.-P. (1972) *Approximation of Elliptic Boundary-Value Problems.* New York-London-Sydney: Wiley-Interscience.

Baarholm, R.J. (2001) *Theoretical and Experimental Studies of Wave Impact Underneath Decks of Offshore Platforms.* Dr.Ing. thesis, Department of Marine Hydrodynamics, NTNU, Trondheim.

Bagnold, R.A. (1939) Interim report on wave pressure research. *Journal of the Institite of Civil Engineers, London* 12, 202–225.

Barkowiak, K., Gampert, B., and Siekmann, J. (1985) On liquid motion in a circular cylinder with horizontal axis. *Acta Mechanica* 54, 207–220.

Batchelor, G.K. (1970) *An Introduction to Fluid Dynamics.* Cambridge: Cambridge University Press.

Bateman, H. (1954) *Tables of Integral Transforms.* Vol. 1. New York, Toronto, London: McGraw-Hill.

Bauer, H.F. (1982) Sloshing in conical tanks. *Acta Mechanica* 43, no. 3–4, 185–200.

Bearman, P.W. (1985) Vortex trajectories in oscillatory flow. In *Proc. Separated Flow around Marine Structures*, pp. 133–154, Trondheim: Norwegian Institute of Technology.

Bearman, P.W. (1988) Wave loading experiments on circular cylinders at large scale. In *Proceedings of the International Conference on Behaviour of Offshore Structures* (BOSS '88), ed. T. Moan, N. Janbu, and O. Faltinsen, Vol. 2, pp. 471–487. Trondheim: Tapir Publishers.

Bearman, P.W., Graham, J.M.R., and Sing, S. (1979) Forces on cylinders in harmonically oscillating flow. In *Proc. Mechanics of Wave-Induced Forces on Cylinders*, ed. T.L. Shaw, pp. 437–449. London: Pitman.

Bearman, P.W., Graham, J.M.R., and Obasaju, E.D. (1984) A model equation for the transverse force on cylinders in oscillatory flows. *Applied Ocean Research* 6, no. 3, 166–172.

Bearman, P.W., Downie, M.J., Graham, J.M.R., and Obasaju, E.D. (1985) Forces on cylinders in viscous oscillatory flow at low Keulegan-Carpenter numbers. *Journal of Fluid Mechanics* 154, 337–356.

Berdichevsky, V.L. (1983) *Variational Principles of Continuum Mechanics*. Moscow: Nauka (in Russian).

Berstad, A.J., Faltinsen, O.M., and Larsen, C.M. (1997) Fatigue crack growth in side longitudinals. In *Proc. NAV&HSMV*, pp. 5.3–15, Naples: Dipartimento Ingeneria-Università di Napoli "Federico II".

Berthelsen, P.A., and Faltinsen, O.M. (2008) A local directional ghost cell approach for incompressible viscous flow problems with irregular boundaries. *Journal of Computational Physics* 227, 4354–4397.

Bicknell, G.V. (1991) The equation of motion of particles in a smoothed particle hydrodynamics. *SIAM Journal of Scientific and Statistical Computing* 12, 1198–1206.

Birkhoff, G. (1960) *Hydrodynamics*. Princeton, New Jersey: Princeton University Press.

Blevins, R.D. (1990) *Flow Induced Vibrations*. Malabar, FL: Krieger.

Blevins, R.D. (1992) *Applied Fluid Dynamics Handbook*. Malabar, FL: Krieger.

Bogoljubov, N.N., and Mitropolski, Yu.A. (1961) *Asymptotic Methods in the Theory of Nonlinear Oscillations*. New York: Gordon and Breach.

Bogomaz, G.I. (2004) *Dynamics of the Railway Cysterns*. Kiev: Naukova Dumka (in Russian).

Bogomaz, G.I., and Sirota, S.A. (2002) *Oscillations of a Liquid in Containers: Methods and Results on Experimental Studies*. Dnepropetrovsk: National Space Agency of Ukraine (in Russian).

Bogomaz, G.I., Markova, O.M., and Chernomashentseva, Y.G. (1998) Mathematical modelling of vibrations and loading of railway tanks taking into account the liquid cargo mobility. *Vehicle System Dynamics: International Journal of Vehicle Mechanics and Mobility* 30, no. 3, 285–294.

Braeunig, J.-P., and Brosset, L. (2009) Phenomenological study of liquid impacts through 2D compressible two-fluid numerical simulations. In *Proceedings of the Nineteenth International Offshore and Polar Engineering Conference* (ISOPE 2009). Osaka, Japan, June 21-26, 2009, Paper No. 2009-TPC-653.

Brand, L. (1964) *Vector and Tensor Analysis*. New York: John Wiley & Sons.

Bredmose, H., and Bullock, G.N. (2008) Scaling of wave-impact pressures in trapped air pockets, In *Proc. 23rd IWWWFB*, Jeju, Korea, 13–16 April 2008.

Bredmose, H., Brocchini, M., Peregrine, F.H., and Thais, L. (2003) Experimental investigation and numerical modelling of steep forced water waves. *Journal of Fluid Mechanics* 490, 217–249.

Brennen, C.E. (1995) *Cavitation and Bubble Dynamics*. Oxford Engineering Science Series 44, New York: Oxford University Press.

Bryant, J.P. (1983) Waves and wave groups in deep water. In *Nonlinear Waves*, ed. L. Debnath, Ch. 6, pp. 100–115. Cambridge: Cambridge University Press.

Bryant, P.J. (1989) Nonlinear progressive waves in a circular basin. *Journal of Fluid Mechanics* 205, 453–467.

Buckingham, E. (1915) Model experiments and the forms of empirical equations. *Transactions of ASME* 37, 263–296.

Budiansky, B. (1960) Sloshing of liquids in circular canals and spherical tanks. *Journal of Aerospace Sciences* 27, no. 3, 161–172.

Campbell, I.M.C., and Weynberg, P.A. (1980) *Measurents of parameters affecting slamming*. Final Report, Rep. No. 440, Technology Reports Centre No. OT-R-8042. Southampton University: Wolfson Unit for Marine Technology.

Caughey, T.K. (1960) Classical normal modes in damped linear systems. *Journal of Applied Mechanics* 27, 269–271.

Chan, E.S., and Melville, W.K. (1988) Deep-water plunging wave pressures on a vertical plane wall. *Proceedings of the Royal Society of London. Series A, Mathematical and Physical Sciences* 417, 95–131.

Chaplin, J.R. (1984) Nonlinear forces on a horizontal cylinder beneath waves. *Journal of Fluid Mechanics* 147, 449–464.

Chaplin, J.R. (1988) Nonlinear forces on horizontal cylinders in the Inertia Regime in Waves at High Reynolds numbers. In *Proceedings of the International Conference on Behaviour of Offshore Structures, (BOSS '88)*, ed. T. Moan, N. Janbu, and O. Faltinsen, Vol. 2, pp. 505–518. Trondheim: Tapir Publishers.

Chester, W. (1968) Resonant oscillation of water waves. I. Theory. *Proceedings of the Royal Society of London. Series A* 306, 5–22.

Chester, W., and Bones, J.A. (1968) Resonant oscillation of water waves. II. Experiment. *Proceedings of the Royal Society of London. Series A* 306, 23–30.

Chhabra, R.P. (2006) *Bubbles, Drops, and Particles in Non-Newtonian Fluids. 2nd Edition*. Boca Raton, FL: CRC Press, Taylor & Francis Group.

Chhabra, R.P., and Richardson, J.F. (1999) *Non-Newtonian Flow: Fundamentals and Engineering Applications*. Oxford: Butterworth-Heinemann, Elsevier.

Chorin, A.J. (1968) Numerical solution of the Navier-Stokes equations. *Mathematics of Computation* 22, 742–762.

Chow, S.-N., and Hale, J.K. (1982) *Methods of Bifurcation Theory*. New York: Springer.

Christoffersen, J.B., and Jonsson, I.G. (1985) Bed friction and dissipation in combined current and wave motion. *Ocean Engineering* 12, 387–423.

Chu, W.H. (1964) Fuel sloshing in a spherical tank filled to an arbitrary depth. *AIAA Journal* 2, 1972–1979.

Chuang, S.L. (1966) Experiments on flat-bottom slamming. *Journal of Ship Research* 10, no. 1, 10–17.

Chung, J.Y., Lee, J.H., Kwon, S.H., Ha, M.K., Bang, C.S., Lee, J.N., and Kim, J.J. (2006) Wet drop tests for LNG insulation system. In *Proceedings of the 16th International Offshore and Polar Engineering Conference*, San Francisco, CA.

Clough, R.W., and Penzien, J. (1993) *Dynamics of Structures*, 2nd edition. New York: McGraw-Hill.

Cointe, R. (1991) Free surface flows close to a surface-piercing body. In *Mathematical Approaches in Hydrodynamics*, ed. T. Miloh, pp. 319–334. Philadelphia, PA: Society for Industrial and Applied Mathematics.

Colagrossi, A. (2005) *A meshless Lagrangian method for free-surface and interface flow*. Ph.D. thesis. University of Rome, La Sapienza.

Colagrossi, A., and Landrini, M. (2003) Numerical solutions of interfacial flows by smoothed particle hydrodynamics. *Journal of Computational Physics* 191, 448–475.

Colagrossi, A., Lugni, C., Dousset, V., Bertram, V., and Faltinsen, O.M. (2003) Numerical and experimental study of sloshing in partially filled rectangular tanks. In *Proceedings of the 6th Numerical Towing Tank Symposium*, Rome.

Colagrossi, A., Lugni, C., Greco, M., and Faltinsen, O.M. (2004) Experimental and numerical investigation of 2D sloshing with slamming. In *Proceedings of the 19th International Workshop on Water Waves and Floating Bodies*, Cortona, Italy.

Colagrossi, A., Palladino, F., Greco, M., Lugni, C., and Faltinsen, O.M. (2006) Experimental an numerical investigation of 2D sloshing: Scenarios near the critical filling depth. In *Proceedings of the 21st International Workshop on Water Waves and Floating Bodies*, Loughborough University, England.

Cole, R.H. (1948) *Underwater Explosions*. Princeton, New Jersey: Princeton University Press.

Colicchio, G. (2004) *Violent disturbance and fragmentation of free surfaces*. Ph.D. thesis, School of Civil Engineering and the environment, University of Southampton.

Cooker, M.J., and Peregrine, D.H. (1990) Violent water motion at breaking wave impact. In *Proceedings of the 22nd International Conference on Coastal Engineering*, Delft, ASCE, New York, pp. 164–176.

Cooker, M.J., and Peregrine, D.H. (1991) Wave breaking and wave impact pressures. In *Developments in Coastal Engineering*, Univ. of Bristol, pp. 47–64.

Cooker, M.J., and Peregrine, D.H. (1992) Wave impact pressure and its effect upon bodies lying on the sea bed. *Coastal Engineering* 18, 205–229.

Cooker, M.J., and Peregrine, D.H. (1995) Pressure-impulse theory for liquid impact problems. *Journal of Fluid Mechanics* 297, 193–214.

Courant, R., and Hilbert, D. (1953) *Methods of Mathematical Physics, Vol I*. New York: Interscience Press.

Cummins, W.E. (1962) The impulse response function and ship motions. *Schiffstechnik* 9, no. 47, 101–109.

den Hartog, J.P. (1984) *Mechanical Vibrations*. New York: Dover.

Dillingham, J., and Falzarano, J.M. (1986) A numerical method for three-dimensional sloshing. In *Proceedings of the SNAME Spring Meeting/Technical and Research Symposium (STAR)*, Portland, OR, pp. 75–85.

DNV (Det Norske Veritas) (2007) *Recommended Practice DNV-RP-C205; Environmental Conditions and Environmental Loads*. Høvik, Norway: DNV.

DNV Classification (2006) *Sloshing Analysis of LNG Membrane Tanks. DNV Classification Notes*. No. 30.9, June 2006. Høvic: DNV.

DNV Guidelines (1998) *Guidelines No. 14 on free spanning pipelines*. June 1998, Høvik, Norway: DNV.

DNV Rules for Classification of Ships (2007) *Newbuildings, Hull and equipment main class, Part 3, chapter 1*, Høvik, Norway: DNV.

Dobrovol'skaya, Z.N. (1969) On some problems of fluid with a free surface. *Journal of Fluid Mechanics* 36, no. 4, 805–829.

Dodge, F.T. (2000) *The New "Dynamic Behavior of Liquids in Moving Containers."* San Antonio, TX: Southwest Research Institute.

Dodge, F.T., Kana, D.D., and Abramson, H.N. (1965) Liquid surface oscillations in longitudinally excited rigid cylindrical containers. *AIAA Journal* 3, 685–695.

Dokuchaev, L.V. (1976) On questions about derivation of nonlinear equations for motions of a body with a cavity partially filled by a liquid. In *Oscillations of Elastic Constructions with a Liquid*, Moscow, Volna, pp. 149–156 (in Russian).

Drake, K.R. (1999) The effect of internal pipes on the fundamental frequency of liquid sloshing in a circular tank. *Applied Ocean Research* 21, 133–143.

Drimer, N., Glozman, M., Stiassnie, M., and Zilman, G. (2000) Forecasting the motion of berthed ships in harbors. In *Proceedings of the 15th International Workshop on Water Waves and Floating Bodies*, Faculty of Engineering, Tel-Aviv University, Israel.

Eastham, M. (1962) An eigenvalue problem with parameter in the boundary condition. *Quarterly Journal of Mathematics* 13, 304–320.

Eatock Taylor, R., Sun, L., and Taylor, P.H. (2008) Gap resonances in focused wave groups. In *Proceedings of the 23rd IWWWFB*, Jeju, Korea.

Emi, H., Fujitani, T., and Abe, A. (2004) Liquefied gas carriers. In *Ship Design and Construction*, ed. T. Lamb, chap. 32. Society of Naval Architects and Engineers (SNAME).

Ersdal, S., and Faltinsen, O.M. (2006) Normal forces on cylinders in near-axial flow. *Journal of Fluids & Structures* 22, 1057–1077.

Etkin, B. (1959) *Dynamics of Flight: Stability and Control*. New York: John Wiley & Sons.

Evans, D.V., and McIver, P. (1987) Resonant frequencies in a container with vertical baffles. *Journal of Fluid Mechanics* 175, 295–305.

Fage, A., and Warsap, J.H. (1929) *The Effects of Turbulence and Surface Roughness on the Drag of Circular Cylinders*. Aero Res. Comm., Report and Memo No. 1283.

Faltinsen, O.M. (1974) A nonlinear theory of sloshing in rectangular tanks. *Journal of Ship Research* 18, 224–241.

Faltinsen, O.M. (1977) Numerical solutions of transient nonlinear free-surface motion outside or inside moving bodies. In *Proceedings of the 2nd International Conference Numerical Ship Hydrodynamics*. University of California, Berkley, 19–21 September, 1977, pp. 347–357.

Faltinsen, O.M. (1978) A numerical method of sloshing in tanks with two-dimensional flow. *Journal of Ship Research* 22, no. 3, 193–202.

Faltinsen, O.M. (1990) *Sea Loads on Ships and Offshore Structures*. New York: Cambridge University Press.

Faltinsen, O.M. (1997) The effect of hydroelasticity on ship slamming. *Philosophical Transactions of Royal Society of London, Series A* 355, 575–591.

Faltinsen, O.M. (1999) Water entry of a wedge by hydroelastic orthotropic plate theory. *Journal of Ship Research* 43, no. 3, 180–193.

Faltinsen, O.M. (2005) *Hydrodynamics of High-Speed Marine Vehicles*. New York: Cambridge University Press.

Faltinsen, O.M., and Rogenbakke, O.F. (1999) Sloshing and slamming in tanks. In *Proc. HYDRONAV'99-MANOEUVRING'99*, Gdansk-Ostrada, Poland.

Faltinsen, O.M., and Rognebakke, O.F. (2000) *Sloshing. In NAV 2000. Proceeding of the International Conference on Ship and Shipping Research*. Venice, 19–22 September 2000, Italy.

Faltinsen, O.M., and Timokha, A.N. (2001) Adaptive multimodal approach to nonlinear sloshing in a rectangular tank. *Journal of Fluid Mechanics* 432, 167–200.

Faltinsen, O.M., and Timokha, A.N. (2002) Asymptotic modal approximation of nonlinear resonant sloshing in a rectangular tank with small fluid depth. *Journal of Fluid Mechanics* 470, 319–357.

Faltinsen, O.M., Newman, J.N., and Vinje, T. (1995) Nonlinear wave loads on a slender vertical cylinder. *Journal of Fluid Mechanics* 289, 179–198.

Faltinsen, O.M., Kvålsvold, J., and Aarsnes, J.V. (1997) Wave impact on a horizontal elastic plate. *Journal of Marine Science & Technology* 2, no. 2, 87–100.

Faltinsen, O.M., Rognebakke, O.F., Lukovsky, I.A., and Timokha, A.N. (2000) Multidimensional modal analysis of nonlinear sloshing in a rectangular tank with finite water depth. *Journal of Fluid Mechanics* 407, 201–234.

Faltinsen, O.M., Rognebakke, O.F., and Timokha, A.N. (2003) Resonant three-dimensional nonlinear sloshing in a square base basin. *Journal of Fluid Mechanics* 487, 1–42.

Faltinsen, O.M., Rognebakke, O.F., and Timokha, A.N. (2005a) Classification of three-dimensional nonlinear sloshing in a square base tank with finite depth. *Journal of Fluids and Structures* 20, no. 1, 81–103.

Faltinsen, O.M., Rognebakke, O.F., and Timokha, A.N. (2005b) Resonant three-dimensional sloshing in a square base basin. Part 2. Effect of higher modes. *Journal of Fluid Mechanics* 523, 199–218.

Faltinsen, O.M., Rognebakke, O.F., and Timokha, A.N. (2006a) Resonant three-dimensional nonlinear sloshing in a square-base basin. Part 3. Base ratio perturbations. *Journal of Fluid Mechanics* 551, 93–116.

Faltinsen, O.M., Rognebakke, O.F., and Timokha, A.N. (2006b) Transient and steady-state amplitudes of resonant three-dimensional sloshing in a square tank with a finite fluid depth. *Physics of Fluids* 18, Art. no. 012103.

Faltinsen, O.M., Rognebakke, O., and Timokha, A.N. (2007) Two-dimensional resonant piston-like sloshing in a moonpool. *Journal of Fluid Mechanics* 575, 359–397.

Faraday, M. (1831) On a peculiar class of acoustical figures. On certain forms assumed by groups of particles upon vibrating elastic surfaces. *Philosophical Transactions of the Royal Society of London* 121, 299–340.

Ferziger, J.H., and Peric, M. (2002) *Computational Methods for Fluid Dynamics*, 3rd rev. ed. Berlin: Springer.

Feschenko, S.F., Lukovsky, I.A., Rabinovich, B.I., and Dokuchaev, L.V. (1969) *Methods for Determining Added Fluid Mass in Mobile Cavities*. Kiev: Naukova Dumka (in Russian).

Fischer, F.D., and Rammerstorfer, F.G. (1999) A refined analysis of sloshing effects in seismically excited tanks. *International Journal of Pressure Vessels and Piping* 76, 693–709.

Fox, D.W., and Kuttler, J.R. (1983) Sloshing frequencies. *ZAMP* 34, 668–696.

Frandsen, J.B. (2004) Sloshing motions in excited tanks. *Journal of Computational Physics* 196, 53–87.

Fredsøe, J., and Deigaard, R. (1993) *Mechanics of Coastal Sediment Transport*. Singapore: World Scientific.

Fridman, A.L. (1998) *Calculations for Fishing Gear Designs*. Farnham, UK: Fishing News Books.

Fultz, D. (1962) An experimental note on finite-amplitude standing gravity waves. *Journal of Fluid Mechanics* 13, no. 2, 193–212.

Gakhov, F.D. (1966) *Boundary Value Problems*. Pergamon Press.

Ganiev, R.F. (1977) Nonlinear resonant oscillations of bodies with a liquid. *Soviet Applied Mechanics* 13, no. 10, 978–984.

Garza, L.R. (1964) *Measurements of liquid natural frequencies and damping in compartmented cylindrical tanks*. Technical Report No. 8. Southwest Research Institute, San Antonio, TX.

Gavrilyuk, I.P., Lukovsky, I.A., and Timokha, A.N. (2000) A multimodal approach to nonlinear sloshing in a circular cylindrical tank. *Hybrid Methods in Engineering* 2, no. 4, 463–483.

Gavrilyuk, I.P., Kulik, A.V., and Makarov, V.L. (2001) Integral equations of the linear sloshing in an infinite chute and their discretisation. *Computational Methods in Applied Mathematics* 1, 39–61.

Gavrilyuk, I.P., Lukovsky, I.A., and Timokha, A.N. (2005) Linear and nonlinear sloshing in a conical tank. *Fluid Dynamics Research* 37, 399–429.

Gavrilyuk, I., Lukovsky, I., Trotsenko, Yu., and Timokha, A. (2006) Sloshing in a vertical circular cylindrical tank with an annular baffle. Part 1. Linear fundamental solutions. *Journal of Engineering Mathematics* 54, 71–88.

Ge, C. (2002) *Global hydroelastic response of catamarans due to wetdeck slamming*. Dr.Ing. thesis, Department of Marine Technology, NTNU, Trondheim.

Gervaise, E., de Sèze, P.E., and Maillard, S. (2009) Reliability-based methodology for sloshing assessment of membrane LNG vessels. In *Proceedings of the Nineteenth International Offshore and Polar Engineering Conference* (ISOPE 2009). Osaka, Japan, June 21–26, 2009, Paper No. 2009-FD-17.

Ghias, R.R., Mittal, R., and Lund, T.S. (2004) A non-body conformal grid method for simulation of compressible flows with complex immersed boundaries. In *Proceedings of the 42nd AIAA Aerospace Sciences Meeting and Exhibit*, Reno, NV, 5–8 January.

Gjelsvik, A. (1983) *On the Damping of Floating Vessel Motions by Means of Positioning Thrusters*. Gjettum, Norway: MAROTEC.

Goda, Y., Haranaka, S., and Kitahata, M. (1966) Study on impulsive breaking wave forces on piles. *Report Port and Harbour Technical Research Institute* 5, 1–30 (in Japanese).

Graczyk, M. (2008) *Experimental investigation of sloshing loading and load effects in membrane LNG tanks subjected to random excitation*. Ph.D. thesis, Department of Marine Technology, NTNU, Trondheim.

Gradshtein, I., and Ryzhik, I. (1965) *Tables of Integrals, Series and Products, Fourth ed.* London and New York: Academic Press.

Graham, J.M.R. (1980) The forces on sharp-edged cylinder in oscillatory flow at low Keulegan–Carpenter numbers. *Journal of Fluid Mechanics* 97, 331–346.

Graham, E.W., and Rodriguez, A.M. (1952) The characteristics of fuel motion which affect airplane dynamics. *ASME Journal of Applied Mechanics* 19 no. 3, 381–388.

Greco, M. (2001) *A two-dimensional study of green water loading*. Dr.Ing. thesis, Department of Marine Hydrodynamics, NTNU, Trondheim.

Greco, M., Landrini, M., and Faltinsen, O.M. (2003) Local hydroelastic analysis on a VLFS with shallow draft, In *Proc. Hydroelasticity in Marine Technology*, ed. R. Eatock Taylor, pp. 201–214. Oxford: University of Oxford.

Greenhill, A.G. (1887) Wave motion in hydrodynamics. *American Journal of Mathematics* 9, 62–112.

Greenhow, M. (1986) High- and low-frequency asymptotic consequences of the Kramers-Kronig relations. *Journal of Engineering Mathematics* 20, 293–306.

Greenhow, M., and Li, Y. (1987) Added masses for circular cylinders near or penetrating fluid boundaries – review, extension and application to water-entry, -exit and slamming. *Ocean Engineering* 14, no. 4, 325–348.

Greenhow, M., and Lin, W.-M. (1983) *Non-linear free surface effects: experiments and theory*. Report No. 83-19, Department of Ocean Engineering, Cambridge, Mass. Inst. Tech.

Greenhow, M., and White, S.P. (1997) Optimal heave motion of some axisymmetric wave energy devices in sinusoidal waves. *Applied Ocean Researches* 19, 141–159.

Haberman, W.L., Jaski, E.J., and John, J.E. (1974) A note on the sloshing motion in a triangular tank. *ZAMP* 25, 292–293.

Hairer, E., and Wanner, G. (1996) *Solving Ordinary Differential Equations II: Stiff and Differential-Algebraic Problems, 2nd ed.* Berlin: Springer-Verlag.

Hansen, H.R. (1976) Damage experience, potential damages, current problems involving slosh considerations. In *Seminar on Liquid Sloshing*. Høvik: Det Norske Veritas.

Harlow, F.H., and Welch, J.E. (1965) Numerical calculation of time dependent viscous incompressible flow with free surface. *Physics of Fluids* 8, 2182–2189.

Harten, A., and Osher, S. (1987). Uniformly high-order accurate nonoscillatory schemes: I. *SIAM Journal of Numerical Analysis* 24, 279–309.

Haslum, H.A. (2000) *Simplified methods applied to non-linear motion of Spar platforms.* Dr.Ing. thesis, Department of Marine Hydrodynamics, Norwegian University of Science and Technology, Trondheim.

Hassard, B.D., Kazarinoff, N.D., and Wan, Y.-H. (1981) *Theory and Applications of Hopf Bifurcation.* New York: Cambridge University Press.

Hatayama, K. (2008) Lessons from the 2003 Tokachi-oki, Japan earthquake for prediction of long-period strong ground motions and sloshing damage to oil storage tanks. *Journal of Seismology* 12, no. 2, 255–263.

Hattori, M., Arami, A., and Yui, T. (1994) Wave impact pressure on vertical walls under breaking waves of various types. *Coastal Engineering* 22, 79–114.

Hawthorne, W.R. (1961) The early development of the Dracone flexible barge. *Proceedings of the Institution of Mechanical Engineers* 175, 52–83.

Heath, M.T. (2002) *Scientific Computing. An Introductory Survey. Second edition.* New York: McGraw-Hill.

Henrici, P., Troesch, B.A., and Wuytack, L. (1970) Sloshing frequencies for a half-space with circular or stripe-like apertude. *ZAMP* 21, no. 3, 285–315.

Herfjord, K. (1996) *A study of two-dimensional separated flow by a combination of the Finite Element Method and Navier-Stokes equations.* Dr.Ing. thesis, Department of Marine Hydrodynamics, Norwegian University of Science and Technology, Trondheim.

Hermann, M., and Timokha, A. (2008) Modal modelling of the nonlinear resonant fluid sloshing in a rectangular tank II. Secondary resonance. *Mathematical Models and Methods in Applied Sciences* 18, no. 11, 1845–1867.

Hess, J.L., and Smith, A.M.O. (1962) *Calculation of non-lifting potential flow about arbitrary three-dimensional bodies.* Report No. E. S. 40622, Douglas Aircraft Division, Long Beach, CA. (See also *Journal of Ship Research* (1964), 8, 2, 22–44.)

Hirt, C.W., and Nichols, B.D. (1981) Volume of fluid (VOF) method for dynamics of free boundaries. *Journal of Computational Physics* 39, 201–221.

Hirt, C.W., Nichols, B.D., and Romero, N.C. (1975) *SOLA – A numerical solution algorithm for transient fluid flows.* Los Alamos Scientific Laboratory, Report LA-5852.

Hogben, N., and Lumb, F.E. (1967) *Ocean Wave Statistics.* National Physical Laboratory, U.K.

Housner, G.W. (1957) Dynamic pressures on accelerated fluid containers. *Bulletin of Seismic Society of America* 47, 15–35.

Housner, G.W. (1963) The dynamic behavior of water tanks. *Bulletin of Seismic Society of America* 53, 381–387.

Howison, S.D., Ockendon, J.R., and Wilson, S.K. (1991) Incompressible water-entry problems at small deadrise angles. *Journal of Fluid Mechanics* 222, 215–230.

Hu C., Kishev Z., Kashiwagi M., Sueyoshi, M., and Faltinsen, O. (2006) Application of CIP method for strongly nonlinear marine hydrodynamics. *Ship Technology Research* 53, no. 2, 74–87.

Huebner, K.H., Thornton, E.A., and Byrom, T.G. (1995) *The Finite Element Method for Engineers.* New York: John Wiley & Sons.

Hull, P., and Mueller, G. (2002) An investigation of breaker heights, shapes and pressures. *Ocean Engineering* 29, no. 1, 59–79.

Hutton, R.E. (1963) *An investigation of resonant, nonlinear, non-planar, free surface oscillations of a fluid.* NASA Technical Note D-1870 (Washington, DC).

Hysing, T. (1976) Loads related to sloshing in spherical tanks. In *Seminar on Liquid Sloshing,* Det Norske Veritas, Høvik, Norway.

Iafrati, A., and Korobkin, A.A. (2006) Breaking wave impact onto vertical wall. In *Proceedings of 4th International Conference on Hydroelasticity in Marine Technology,* 10–14 September, 2006, Wuxi, China, pp. 139–148.

Ibrahim, R.A. (2005) *Liquid Sloshing Dynamics.* Cambridge University Press.

Ikeda, Y., Komatsu, K., Himeno, Y., and Tanaka, N. (1977) On roll damping force of ship: Effects of hull surface pressure created by bilge keels. *Journal of Kansai Society of Naval Architect, Japan* 165, 31–40.

Isaacson, M., and Premasiri, S. (2001) Hydrodynamic damping due to baffles in a rectangular tank. *Canadian Journal of Civilian Engineering* 28, 608–616.

Janssens, P. (2006) The development of the first energy bridge regasification vessel. In *Proceedings of the 2006 Offshore Technology Conference (OTC),* Houston, TX, Paper OTC 18398.

Jensen, B.L., Sumer, B.M., and Fredsøe, J. (1989) Turbulent oscillatory boundary layers at high Reynolds numbers. *Journal of Fluid Mechanics* 206, 265–297.

Jeyakumaran, R., and McIver, P. (1995) Approximations to sloshing frequencies for rectangular tanks with internal structures. *Journal of Engineering Mathematics* 29, 537–556.

Jiang, G.-S., and Peng, D. (2000) Weighted ENO schemes for Hamilton-Jacobi equations, *SIAM Journal of Scientific Computing* 21 no. 6, 2126–2143.

Jiang, G.-S., and Shu, C.-W. (1996) Efficient implementation of weighted ENO schemes. *Journal of Computational Physics* 126, 202–228.

John, F. (1950) On the motion of floating bodies II. *Communications on Pure and Applied Mathematics* 3, 45–101.

Johnson, C. (1995) *Numerical Solution of Partial Differential Equations by the Finite Element Method.* Cambridge University Press.

Jonsson, I.G. (1978) *A new approach to oscillatory rough turbulent boundary layers*. Series Paper 17, Institute of Hydrodynamic and Hydraulic Engineering, Technical University of Denmark, Lyngby.

Joukowski, N. (1885) On motions of a rigid body with cavity filled by homogeneous liquid. *Journal of Russian Physical-Mathematical Society* XVI, 30–85 (in Russian).

Journee, J.M.J. (1983) *Roll moments due to free surface cargo tanks*. Delft University of Technology, Report 583-O.

Journee, J.M.J. (2000) Fluid tanks and ship motions. In *Report 1237*, Delft University of Technology, Ship Hydromechanics Laboratory, Delft, The Netherlands, 27 October 2000.

Kareem, A., Kijewski, T., and Tamura, Y. (1999) Mitigation of motions of tall buildings with specific examples of recent applications. *Wind and Structures* 2, no. 3, 201–251.

Kato, H. (1966) Effect of bilge keels on the rolling of ships. *Mem. of the Defence Academy, Japan* IV, no. 3, 369–384.

Kelland, H. (1840) On the theory of waves. Part 1. *Transactions of the Royal Society, Edinburg* XIV, 497–545.

Kelland, H. (1844) On the theory of waves. Part 2. *Transactions of the Royal Society, Edinburg* XV, 101–144.

Keulegan, G.H. (1959) Energy dissipation, in standing waves in rectangular basins. *Journal of Fluid Mechanics* 6, 33–50.

Kim, Y. (2001) Numerical simulation of sloshing flows with impact loads. *Applied Ocean Research* 23, 53–62.

Kim, Y. (2002) A numerical study on sloshing flows coupled with ship motion. The anti-rolling tank problem. *Journal of Ship Research* 46, no. 1, 52–62.

Kim, Y. (2007) Experimental and numerical analysis of sloshing flows. *Journal of Engineering Mathematics* 58, 191–210.

Kinsman, B. (1965) *Wind Waves*. Englewood Cliffs, NJ: Prentice Hall.

Kirchhoff, G. (1879) Ueber stehende Schwingungen einer schweren Fluessigkeit. In *Berliener Monatsbericht*, May 1879 [Gesammelte Abhandlungen, Barth., Leipzig, 1882, p. 428].

Klotter, K. (1978) *Technische Schwingungslehre*. Erster Band: Einfache Schwinger. Teil A: Lineare Schwingungen. Berlin, Heidelberg, and New York: Springer-Verlag.

Kochin, N.E., Kibel, I.A., and Roze, N.V. (1964) *Theoretical Hydromechanics*. New York, London, and Sydney: Interscience–John Wiley & Sons.

Koehler, B.R., and Kettleborough, C.F. (1977) Hydrodynamic impact of a falling body upon a viscous incompressible fluid. *Journal of Ship Research* 21, no. 3, 165–181.

Kolmogorov, A.N., and Fomin, S.V. (1961) *Measure, Lebesgue Integrals and Hilbert Space*. New York and London: Academic Press.

Komarenko, A.N. (1989) Equivalence of the Hamilton-Ostrogrdasky principle and a boundary problem of the liquid dynamics in a container. In *Stability of Motions of Solid Bodies and Deformable Systems*, ed. I. Lukovsky. Kiev: In-t of Mathematics of AN UkrSSR, pp. 52–60 (in Russian).

Komarenko, A.N., and Lukovsky, I.A. (1974) Nonlinear oscillations of a liquid in a cylindrical vessel under low gravity. In *Dynamics and Stability of Multidimensional Systems*, Inst. Mathem. Akad. Nauk. Ukr. SSR, pp. 86–97 (in Russian).

Komarenko, A.N., Lukovsky, I.A., and Feschenko, S.F. (1965) On an eigenvalue problem with spectral parameter in boundary conditions. *Ukrainian Mathemaical Journal* 17, no. 6, 22–30 (in Russian).

Kopachevsky, N.D., and Krein, S.G. (2001) *Operator Approach to Linear Problems of Hydrodynamics. Volume 1: Fluids Self-adjoint Problems for an Ideal Fluid. Operator Theory: Advances and Applications.* Birkhauser Verlag, Basel.

Kopachevsky, N.D., and Krein, S.G. (2003) *Operator Approach to Linear Problems of Hydrodynamics. Volume 2: Nonself-adjoint Problems for Viscous Fluids. Operator Theory: Advances and Applications.* Birkhauser Verlag, Basel.

Korobkin, A.A. (1993) Low-pressure zones under a liquid-solid impact. In *Bubble Dynamics and Interface Phenomena: Proceedings of the IUTAM Symposium*. Birmingham, UK, 6–9 September 1993 (eds. J.R. Blake, J.M. Boulton-Stone, and N.H. Thomas). Dorndrecht: Kluwer Academic, pp. 375–381.

Korobkin, A.A. (1996a) Entry problem for body with attached cavity. In *Proceedings of the 11th Workshop on Water Waves and Floating Bodies*, Hamburg.

Korobkin, A.A. (1996b) *Elastic effects on slamming*. Report NAOE-96-39, University of Glasgow, 134 pp.

Korobkin, A.A. (1996c) Water impact problems in ship hydrodynamic. Chapter 7 in *Advances in Marine Hydrodynamics* (ed. M. Okhusu). Southampton, UK; Boston: Computational Mechanics Publications.

Korobkin, A.A. (2006) Two-dimensional problem of the impact of a vertical wall on a layer of a partially aerated liquid. *Journal of Applied Mechanics and Technical Physics* 47, no. 5, 643–653.

Korobkin, A.A. (2008) Wagner theory of steep wave impact. In *Proceedings of the 23rd International Workshop on Water Waves and Floating Bodies*, Jeju, Korea.

Koshizuka, S., Nobe, A., and Oka, Y. (1998) Numerical analysis of breaking waves using the moving particle semi-implicit method. *International Journal of Numerical Methods in Fluids* 26, 751–769.

Kotik, J., and Mangulis, V. (1962) On the Kramers-Kronig relations for ship motions. *International Shipbuilding Progress* 9, no. 97, 361–368.

Kraus, H. (1967) *Thin Elastic Shells*. New York: John Wiley.

Krein, S.G. (1964) On oscillations of viscous liquid inside vessel. *Doklady Akademii Nauk SSSR* 159, Issue 2, 262–265.

Krein, S.G., and Langer, H. (1978a) On some mathematical principles in the linear theory of damped oscillations of continua. *Integral Equations Operator Theory* 3, 364–399.

Krein, S.G., and Langer, H. (1978b) On some mathematical principles in the linear theory of damped oscillations of continua. *II. Integral Equations Operator Theory* 3, 539–566.

Kristiansen, T., and Faltinsen, O.M. (2008) Application of a vortex tracking method to the piston-like behaviour in a semi-entrained vertical gap. *Applied Ocean Research* 30, no. 1, 1–16.

Kuttler, J.R., and Sigillito, V.G. (1984) Sloshing of liquids in cylindrical tanks. *AIAA Journal* 22, no. 2, 309–311.

Kvålsvold, J. (1994) *Hydroelastic Modelling of Wetdeck Slamming on Multihull Vessels*. Dr.Ing. thesis, Department of Marine Hydrodynamics, Norway Institute of Technology, Trondheim.

Kvålsvold, J., and Faltinsen, O.M. (1995) Hydroelastic modelling of wet deck slamming on multihull vessels. *Journal of Ship Research* 39, 225–229.

Kvålsvold, J., Faltinsen, O.M., and Aarsnes, J.V. (1995) Effect of structural elasticity on slamming against wetdecks of multihull vessels. In *Proceedings of the 6th International Symposium on the Practical Design of Ships and Offshore Mobile Units, (PRADS)*, Seoul, South Korea, pp. 1684–1699.

La Rocca, M., Sciortino, G., and Boniforti, M.A. (2000) A fully nonlinear model for sloshing in a rotating container. *Fluid Dynamics Research* 27, 23–52.

La Rocca, M., Sciortino, G., and Boniforti, M. (2002) Interfacial gravity waves in a two-fluid system. *Fluid Dynamics Research* 30, 31–66.

Lamb, H. (1945) *Hydrodynamics*. Cambridge: Cambridge University Press.

Landau, L.D., and Lifschitz, E.M (1959) *Fluid Mechanics*. London: Pergamon.

Landrini, M., Colagrossi, A., and Faltinsen, O.M. (2003) Sloshing and slamming in 2-D flows by the SPH Method. In *Proceedings of the 8th International Conference on Numerical Ship Hydrodynamics*, Busan, Korea.

Landrini, M., Grytøyr, G., and Faltinsen, O.M. (1999) A B-spline based BEM for unsteady free-surface flows. *Journal of Ship Research* 13, no. 1, 13–24.

Langbein, D. (2002) *Capillary Surfaces*. New York: Springer.

Laws, E.M., and Livesey, J.L. (1978) Flow through screens. *Annual Review of Fluid Mechanics* 10, 247–66.

Le Conte, J.N. (1926) *Hydraulics*. New York: McGraw-Hill.

Lee, W.T., and Bales, S.L. (1985) Environmental data for design of marine vehicles, In *Ship Structure Symposium '84*, pp. 197–209, New York: Society of Naval Architects and Marine Engineers.

Leighton, T.G. (1994) *The Acoustic Bubble*. London: Academic Press.

Leonard, H.W., and Walton, W.C. (1961) *An investigation of the natural frequencies and mode shapes of liquids in oblate spheroidal tanks*. NASA TN-D-904.

Lide, D.R. (2008) *Handbook of Chemistry and Physics, 88th edition*. Boca Raton, FL: CRC Press.

Limarchenko, O.S. (1978a) Variational formulation of the problem on the motion of a tank with a fluid. *Dopovidi Akademii Nauk Ukrains'koi RSR, Series A* no. 10, 903–907 (in Ukrainian).

Limarchenko, O.S. (1978b) Direct method for solution of nonlinear dynamic problem on the motion of a tank with fluid. *Dopovidi Akademii Nauk Ukrains'koi RSR, Series A* no. 11, 999–1002 (in Ukrainian).

Limarchenko, O.S. (1983) Application of a variational method to the solution of nonlinear problems of the dynamics of combined motions of a tank with a fluid. *Soviet Applied Mechanics* 19, no. 11, 1021–1025.

Limarchenko, O.S., and Yasinskii, V.V. (1996) *Nonlinear Dynamics of Constructions with a Fluid*. Kiev Polytechnical University (in Russian).

Lin, W.M., and Yue, D.K.P. (1990) Numerical solution for large-amplitude ship motions in time domain. In *Proceedings of the 17th Symposium on Naval Hydrodynamics*, Ann Arbor, MI.

Lloyd, A.R.J.M. (1989) *Seakeeping: Ship Behaviour in Rough Weather*. Chichester, UK: Ellis Horwood Limited.

Løland, G., and Aarsnes, J.V. (1994) Fabric as construction material for marine applications. In *Proceedings of the International Conference on Hydroelasticity in Marine Technology*, Trondheim, Norway.

Longuet-Higgins, M.S. (1953) Can sea cause microseism? In *Proc. Symp. on Microseism*, publ. no. 306, pp. 74–93, New York: U.S. National Academy of Sciences, National Research Council.

Longuett-Higgins, M.S., and Cokelet, E.D. (1976) The deformation of steep surface waves on water. I. A numerical method of computation. *Proceedings of the Royal Society of London, Series A* 350, 1–26.

Lugni, C., Colicchio, G., and Colagrossi, A. (2006a) Investigation of sloshing phenomena near the critical filling depth through the Hilbert-Huang Transformation. In *Proceedings of the 10th Numerical Towing Tank Symposium (NuTTS '06)*, Le Croisic, France, Ecole Centrale de Nantes.

Lugni, C., Brocchini, M., and Faltinsen, O.M. (2006b) Wave impact loads: The role of the flip-through. *Physics of Fluids* 18, Art. no. 122101.

Lukovsky, I.A. (1975) *Nonlinear Sloshing in Tanks of Complex Geometrical Shape*. Kiev: Naukova Dumka (in Russian).

Lukovsky, I.A. (1976) Variational method in the nonlinear problems of the dynamics of a limited liquid volume with free surface. In book *Oscillations of Elastic Constructions with Liquid*. Moscow: Volna, pp. 260–264 (in Russian).

Lukovsky, I.A. (1990) *Introduction to Nonlinear Dynamics of a Solid Body with a Cavity including a Liquid*. Kiev: Naukova dumka (in Russian).

Lukovsky, I.A., and Timokha, A.N. (1995) *Variational Methods in Nonlinear Dynamics of a Limited Liquid Volume*. Kiev: Institute of Mathematics (in Russian).

Lukovsky, I.A., and Timokha, A.N. (2002) Modal modelling of nonlinear sloshing in tanks with non-vertical walls: Non-conformal mapping technique. *International Journal of Fluid Mechanics Research* 29, no. 2, 216–242.

Lukovsky, I.A., Barnyak, M.Ya., and Komarenko, A.N. (1984) *Approximate Methods of Solving the Problems of the Dynamics of a Limited Liquid Volume*. Kiev, Naukova Dumka (in Russian).

Lundgren, H. (1969) Wave shock forces: An analysis of deformations and forces in the wave and in the foundation. In *Proc. Symp. Res. Wave Action*, Vol. II, Paper 4, Delft, The Netherlands.

Ma, Q.W., Wu, G. X., and Eatock Taylor, R. (2001a) Finite element simulation of fully non-linear interaction between vertical cylinders and steep waves. Part 1: Methodology and numerical procedure. *International Journal of Numerical Methods in Fluids* 36, 265–285.

Ma, Q.W., Wu, G.X., and Eatock Taylor, R. (2001b) Finite element simulation of fully non-linear interaction between vertical cylinders and steep waves. Part 2: Numerical results and validation. *International Journal of Numerical Methods in Fluids* 36, 287–308.

MacCamy, R.C., and Fuchs, R.A. (1954) *Wave forces on piles: A diffraction theory*. U.S. Army Corps of Engineers, Beach Erosion Board, Technical Memo no. 69, Washington, DC.

Macdonald, H.M. (1894) Waves in canals. *Proceedings of London Mathematical Society* 25, 101–111.

Macdonald, H.M. (1896) Waves in canals and a sloping bank. *Proceedings of London Mathematical Society* 27, 622–632.

Madsen, P.A., Bingham, H.B., and Schaeffer, H.A. (2003) Boussinesq-type formulations for fully nonlinear and extremely dispersive water waves-Derivation and analysis. *Proceedings of the Royal Society of London, Series A* 459, 1075–1104.

Maillard, S., and Brosset, L. (2009) Influence of density ratio between liquid and gas on sloshing model test results. In *Proceedings of the Nineteenth International Offshore and Polar Engineering Conference* (ISOPE 2009). Osaka, Japan, June 21–26, 2009, Paper No. 2009-FD-14.

Majumdar, S., Iaccarino, G., and Durbin, G.P. (2001) RANS solvers with adaptive structured boundary non-conforming grids. *Annual Research Briefs*. Center for Turbulence Research, pp. 353–366.

Maruo, H. (1960) The drift of a body floating in waves. *Journal of Ship Research* 4, no. 3, 1–10.

Mathieu, É. (1868) Mémoire sur le mouvement vibratoire d'une membrane de forme elliptique. *Journal des Mathématiques Pures et Appliquées* 13, 137–203.

McCarty, J.L., and Stephens, D.G. (1960) *Investigation of the natural frequencies of fluid in spherical and cylindrical tanks*. NASA TN D-252.

McIver, P. (1989) Sloshing frequencies for cylindrical and spherical containers filled to an arbitrary depth. *Journal of Fluid Mechanics* 201, 243–257.

McIver, P., and McIver, M. (1993) Sloshing frequencies of longitudinal modes for a liquid contained in a trough. *Journal of Fluid Mechanics* 252, 525–541.

Medwin, H., and Clay, C.S. (1998) *Fundamentals of Acoustical Oceanography*. Boston: Academic Press.

Mikelis, N.E., Miller, J.K., and Taylor, K.V. (1984) Sloshing in partially filled liquid tanks and its effect on ship motions: numerical simulations and experimental verification. In *The Royal Institution of Naval Architects*. Spring Meeting 1984. Paper no. 7. 11 pp.

Mikishev, G.N., and Churilov, G.A. (1977) Some results on experimental hydrodynamic coefficients for a cylinder with ribs. In *Oscillations of Elastic Constructions with Liquid*. Tomsk: Tomsk University, pp. 31–37 (in Russian).

Mikishev, G.N., and Dorozhkin, N. (1961) Experimental investigation of free oscillations of a liquid in tanks. *Izvestiya Akademii Nauk SSSR. Otdelenie Tekhnicheskikh Nauk i Mashinostroenie* 4, 46–53 (translated into English by D.D. Kana, SwRI, June 30, 1963).

Miles, J.W. (1962) Stability of forced oscillations of a spherical pendulum. *Quarterly Applied Mathematics* 20, 21–32.

Miles, J.W. (1972) On the eigenvalue problem for fluid sloshing in a half-space. *ZAMP* 23, 861–869.

Miles, J.W. (1976) Nonlinear surface waves in closed basins. *Journal of Fluid Mechanics* 75, 419–448.

Miles, J.W. (1984a) Internally resonant surface waves in a circular cylinder. *Journal of Fluid Mechanics* 149, 1–14.

Miles, J.W. (1984b) Resonantly forced surface waves in a circular cylinder. *Journal of Fluid Mechanics* 149, 15–31.

Miles, J.W. (1994) Faraday waves: Rolls versus squares. *Journal of Fluid Mechanics* 269, 353–371.

Miles, J.W., and Henderson, D.M. (1998) A note on interior vs. boundary-layer damping of surface waves in a circular cylinder. *Journal of Fluid Mechanics* 364, 319–323.

Mitsuyasu, H. (1969) Shock pressure of breaking wave. In *Proceedings of the 10th International Conference on Coastal Engineering*, Tokyo, Vol. 1, ASCE, pp. 268–283.

Mittal, R., and Iaccarino, G. (2005) Immersed boundary methods. *Annual Review of Fluid Mechanics* 37, 239–261.

Miyamoto, T., and Tanizawa, K. (1985) A study of the impact on ship bow (2nd Report). *Journal of Society of Naval Architect of Japan* 158, 270–279.

Moderassi-Tehrani, K., Rakheja, S., and Sedaghati, R. (2006) Analysis of the overturning moment caused by transient liquid slosh inside a partly filled moving tank. *Proc. IMechE, Part D: Journal of Automobile Engineering* 220, 289–301.

Modi, V.J., and Akinturk, A. (2002) An efficient liquid sloshing damper for control of wind-induced instabilities. *Journal of Wind Engineering & Industrial Aerodynamics* 90, 197–198.

Moiseev, N.N. (1958) To the theory of nonlinear oscillations of a limited liquid volume of a liquid. *Applied Mathematics and Mechanics (PMM)* 22, 612–621 (in Russian).

Moiseev, N.N., and Petrov, A.A. (1965) *Numerical Methods of Calculation of the Natural Frequencies of Vibrations of a Bounded Volume of Fluid.* Moscow: Academy of Sciences of USSR, Press (in Russian).

Moisev, N.N., and Rumyantsev, V.V. (1968) *Dynamic Stability of Bodies Containing Fluid.* Berlin: Springer.

Molin, B. (2001) On the piston and sloshing modes in moonpools. *Journal of Fluid Mechanics* 430, 27–50.

Monaghan, J.J. (1992) Smoothed particle hydrodynamics. *Annual Review of Astronomy and Astrophysics* 30, 543–574.

Monaghan, J.J. (2000) SPH without a tensile instability. *Journal of Computational Physics* 52, 374–389.

Monaghan, J.J., and Gingold, R.A. (1983) Shock simulation by the particle method SPH. *Journal of Computational Physics* 52, 374–389.

Morand, H.J.-P., and Ohayon, R. (1995) *Fluid Structure Interaction. Applied Numerical Methods.* Chichester, New York, Brisbane, Toronto, Singapore: John Wiley & Sons.

Morison, J.R., O'Brien, M.P., Johnson, J.W., and Schaaf, S.A. (1950) The force exerted by surface waves on piles. *Petroleum Transactions, AIME* 189, 149–154.

Muzaferija, S., and Peric, M. (1999) Computation of free surface flows using interface-tracking and interface-capturing methods. Chapter 2 in *O. Nonlinear Water Wave Interactions* (ed. Mahrenholtz and M. Markiewicz.), pp. 59–100, WIT Press, Southampton.

Myskis, A.D., Babskii, V.G., Kopachevskii, N.D., Slobozhanin, L.A., and Tiuptsov, A.D. (1987) *Low-Gravity Fluid Mechanics: Mathematical Theory of Capillary Phenomena.* Berlin and New York: Springer-Verlag.

Nam, B.W., Kim, Y., and Kim, D.W. (2006) Nonlinear effects of sloshing flows on ship motion. In *Proceedings of the 21st International Workshop on Water Waves and Floating Bodies*, Loughborough University, UK.

Narimanov, G.S. (1957) Movement of a tank partly filled by a fluid: The taking into account of non-smallness of amplitude. *Applied Mathematics and Mechanics (PMM)* 21, 513–524 (in Russian).

Narimanov, G.S., Dokuchaev, L.V., and Lukovsky, I.A. (1977) *Nonlinear Dynamics of Flying Apparatus with a Liquid.* Moscow: Mashinostroenie (in Russian).

Nevolin, V.G. (1984) Parametric excitation of surface waves. *Journal of Engineering Physics and Thermophysics* 47, no. 6, 1482–1494.

Newland, D.E. (1984) *An Introduction to Random Vibration and Spectral Analysis, Second Edition.* Longman Group Ltd.

Newman, J.N. (1962) The exciting forces on fixed bodies in waves. *Journal of Ship Research* 11, no. 1, 51–60.

Newman, J.N. (1977) *Marine Hydrodynamics.* Cambridge: MIT Press.

Newman, J. N. (1985) Distributions of sources and normal dipoles over a quadrilateral panel. *Journal of Engineering Mathematics* 20, no. 2, 113–126.

Newman, J.N. (2005) Wave effects on vessels with internal tanks. In *Proceedings of the 20th International Workshop on Water Waves and Floating Bodies*, Lonyearbyen, Svalbard.

Newman, J.N., Sortland, B., and Vinje, T. (1984) Added mass and damping of rectangular bodies close to the free surface. *Journal of Ship Research* 28, no. 4, 219–225.

Nordenstrøm, N. (1973) *A method to predict long-term distributions of waves and wave-induced motions and loads on ships and other floating structures.* DNV Publications No. 81, Det Norske Veritas, Høvik, Norway.

Ochi, M.K. (1982) Stochastic analysis and probability distribution in random seas. *Advances in Hydroscience* 13, 217–375.

Ogilvie, T.F. (1963) Compressibility effects in ship slamming. *Schiffstechnik* 53, 147–154.

Ogilvie, T.F. (1964) Recent progress towards the understanding and prediction of ship motions. In *Proceedings of the Fifth Symposium on Naval Hydrodynamics*, pp. 3–128. Washington, DC: Office of Naval Research, Department of the Navy.

Ockendon, H., and Ockendon, J. (2001) Nonlinearity in fluid resonances. *Meccanica* 36, 297–321.

Ockendon, H., Ockendon, J.R., and Johnson, A.D. (1986) Resonant sloshing in shallow water. *Journal of Fluid Mechanics* 167, 465–479.

Ockendon, H., Ockendon, J.R., and Waterhouse, D.D. (1996) Multi-mode resonance in fluids. *Journal of Fluid Mechanics* 315, 317–344.

Ockendon, J.R., and Ockendon, H. (1973) Resonant surface waves. *Journal of Fluid Mechanics* 59, 397–413.

Olsen, H. (1976) What is sloshing? In *Seminar on Liquid Sloshing*, Høvik: Det Norske Veritas.

Olsen, H., and Hysing, T. (1974) *A study of dynamic loads caused by liquid sloshing in LNG tanks. DNV Report No.* 74–276-C.

Olsen, H., and Johnsen, K.R. (1975) *Nonlinear sloshing in rectangular tanks. A pilot study on the applicability of analytical models.* Report no. 74–72-S, Vol. 2, 24 July 1975. Det Norske Veritas, Hovik, Norway.

Oumeraci, H., Klammer, P., and Partenscky, H.W. (1993) Classification of breaking wave loads on vertical structures. *Journal of Waterway, Port, Coastal, Ocean Engineering* 119, 381.

Overvik, T. (1982) *Hydroelastic motion of multiple risers in a steady current.* Dr.Ing. thesis, Division of Port and Ocean Engineering. Norwegian Institute of Technology, Trondheim.

Ovsyannikov, L.V., Makarenko, N.I., and Nalimov, V.I. (1985) *Nonlinear Problems of the Theory of Surface and Interior Waves*. Novosibirsk: Nauka (in Russian).

Packham, B.A. (1980) Small-amplitude waves in a straight channel of uniform triangular cross-section. *Quarterly Journal of Mechanics and Applied Mathematics* 33, 179–187.

Pastoor, W., Østvold, T.K., Byklum, E., and Valsgård, S. (2005) *Sloshing loads and response in LNG carriers for new designs and new trades.* Gastech 2005, March 14–17, 2005, Bilbao, Spain.

Peregrine, D.H. (2003) Water wave impact on walls. *Annual Review of Fluid Mechanics* 35, 23–43.

Perez, T. (2002) *Patrol Boat Stabilizers.* Technical Report, 11 Oct 2002, ADI-Limited, Carrington NSW 2294, Australia.

Peric, M., Zorn, T., el Moctar, O., Schellin, T.E., and Kim, Y.-S. (2007) Simulation of sloshing in LNG-tanks. In *Proceedings of the 26th International Conference on Offshore Mechanics and Arctic Engineering (OMAE2007)*, San Diego, CA.

Perko, L.M. (1969) Large-amplitude motions of liquid-vapour interface in an accelerating container. *Journal of Fluid Mechanics* 35, 77–96.

Perlin, M., and Schwarz, W.W. (2000) Capillary effects on surface waves. *Annual Review of Fluid Mechanics* 32, 241–274.

Petrov, A.A. (1964) Variational formulation of the problem on motions of liquid in a bounded reservoir. *Applied Mathematics and Mechanics (PMM)* 28, no. 4, 754–758 (in Russian).

Platzman, G.W. (1972) Two-dimensional free oscillations in natural basins. *Journal of Physical Oceanography* 2, 117–138.

Polyanin, A.D., and Manzhirov, A.V. (1998) *Handbook of Integral Equations.* Boca Raton, FL: CRC Press.

Price, W.G., and Bishop, R.E.D. (1974) *Probabilistic Theory of Ship Dynamics.* London: Chapman and Hall.

Quarteroni, A., and Valli, A. (1999) *Domain Decomposition Methods for Partial Differential Equations.* Oxford: Oxford University Press.

Rao, D.B. (1966) Free gravitational oscillations in rotating rectangular basins. *Journal of Fluid Mechanics* 25, 523–555.

Rayleigh, Lord (1883) On the crispations of fluid resting on a vibrating support. *Philosophical Magazine* 16 (5&58) (*Scientific Papers*, vol. 2, pp. 212–219, Cambridge University Press, 1900).

Rayleigh, J.W.S., and Lindsay, R.B. (1945) *The Theory of Sound. 1st American ed.* New York: Dover.

Reeder, J., and Shinbrot, M. (1976) The initial value problem for surface waves under gravity. II. The simplest 3-dimensional case. *Indiana University Mathematical Journal* 25, no. 11, 1049–1071.

Reeder, J., and Shinbrot, M. (1979) The initial value problem for surface waves under gravity. III. Uniformly analytic initial domains. *Journal of Mathematical Analysis and Applications* 67, 340–391.

Ridley, P.A. (2004) Floating production, storage and offloading (FPSO) vessels. In *Ship Design and Construction*, Vol. II, Chapter 29 (ed. T. Lamb.). Jersey City, NJ: SNAME.

Roach, P.E. (1987) The generation of nearly isotropic turbulence by means of grids. *Heat and Fluid Flow* 8, no. 2, 82–92.

Roache, P.J. (1997) Quantification of uncertainty in computational fluid dynamics. *Annual Review of Fluid Mechanics* 29, 123–160.

Rognebakke, O.F. (2002) *Sloshing in rectangular tanks and interaction with ship motions.* Dr.Ing. thesis, Rep. MTA2002–151, Department of Marine Technology, NTNU, Trondheim.

Rognebakke, O.F., and Faltinsen, O.M. (2000) Damping of sloshing due to tank roof impact. In *Proceedings of the 15th International Workshop on Water Waves and Floating Bodies*, Caesarea, Israel.

Rognebakke, O.F., and Faltinsen, O.M. (2003) Coupling of sloshing and ship motions. *Journal of Ship Research* 47, no. 3, 208–221.

Rognebakke, O.F., and Faltinsen, O.M. (2005) Sloshing induced impact with air cavity in rectangular tank with a high filling ratio. In *Proceedings of the 20th International Workshop on Water Waves and Floating Bodies*, Svalbard, Norway.

Romero J. A., Hildebrand, R., Martinez, M., Ramirez, O., and Fortanell, J.A. (2005) Natural sloshing frequencies of liquid cargo in road tankers. *International Journal of Heavy Vehicle Systems* 2, no. 2, 121–138.

Ronæss, M. (2002) *Wave induced motions of two ships advancing on parallel course.* Dr.Ing. thesis, Department of Marine Hydrodynamics, NTNU, Trondheim. MTA report 2002:155.

Rouse, H. (1961) *Fluid Mechanics for Hydraulic Engineers.* New York: Dover.

Royon-Lebeaud, A., Hopfinger, E.J., and Cartellier, A. (2007) Liquid sloshing and wave breaking in circular and square-base cylindrical containers. *Journal of Fluid Mechanics* 577, 467–494.

Saad, Y. (2003) *Iterative Methods for Sparse Linear Systems*, 2nd edition. Philadelphia, PA: Society for Industrial and Applied Mathematics.

Salvesen, N., Tuck, E.O., and Faltinsen, O.M. (1970) Ship motions and sea loads. *Transactions of the SNAME* 78, 250–287.

Sarpkaya, T. (1966) Separated flow about lifting bodies and impulsive flow about cylinders. *AIAA Journal* 44, 414–20.

Sarpkaya, T. (1978) Fluid forces on oscillating cylinders. *Journal of Waterway, Port, Coastal and Ocean Division of ASCE* 104, 275–290.

Sarpkaya, T. (1985) Past progress and outstanding problems in time-dependent flows about ocean structures. In *Proc. Conf. Separated Flow around Marine Structures*, pp. 1–36, Trondheim: Norwegian Institute of Technology.

Sarpkaya, T. (1986) Force on a circular cylinder in viscous oscillatory flow at low Keulegan–Carpenter numbers. *Journal of Fluid Mechanics* 165, 61–71.

Sarpkaya, T., and Isaacson, M. (1981) *Mechanics of Wave Forces on Offshore Structures.* New York: Van Nostrand Reinhold.

Sarpkaya, T., and O'Keefe, J.L. (1996) Oscillating flow about two and three-dimensional bilge keels. *Journal of Offshore Mechanics & Arctic Engineering* 118, 1–6.

Scardovelli, R., and Zaleski, S. (1999) Direct numerical simulation of free surface and interfacial flow. *Annual Review of Fluid Mechanics* 31, 567–603.

Schlichting, H. (1979) *Boundary-Layer Theory.* New York: McGraw-Hill.

Schwartz, L.W. (1974) Computer extension and analytic continuation of Stokes' expansion for gravity waves. *Journal of Fluid Mechanics* 62, 553–578.

Sclavounos, P. (1996) Computation of wave ship interactions. In *Advances in Marine Hydrodynamics*, ed. M. Ohkusu, chap. 4, pp. 177–231. Southampton: Computational Mechanics Publications.

Sclavounos, P.D., and Borgen, H. (2004) Seakeeping analysis of a high-speed monohull with a motion control bow hydrofoil. *Journal of Ship Research* 28, no. 2, 77–117.

Scolan, Y.-M., and Korobkin, A.A. (2001) Three-dimensional theory of water impact, part 1, inverse Wagner problem. *Journal of Fluid Mechanics* 440, 293–326.

Sedov, L.I. (1965) *Two-Dimensional Problems in Hydrodynamics and Aerodynamics.* New York: Interscience.

Sen, B.M. (1927) Waves in canals and basins. *Proceedings of London Mathematical Society* 26, no. 2, 363–376.

Senjanovic, I., Rudan, S., and Ljustina, A.M. (2004) Remedy for misalignment of bilobe cargo tanks in liquefied petroleum gas carriers. *Journal of Ship Production* 20, no. 3, 133–146.

Sethian, J.A. (1999) *Level Set Methods and fast Marching Methods: Evolving Interfaces in Computational Geometry, Fluid Mechanics, Computer Vision and Materials Science.* New York: Cambridge University Press.

Seydel, R. (1994) *Practical Bifurcation and Stability Analysis. From Equilibrium to Chaos. Second Edition.* Springer Interdisciplinary Applied Mathematics, Vol. 5. New York: Springer.

Shankar, P.N., and Kidambi, R. (2002) A modal method for finite amplitude, nonlinear sloshing. *Pramana – Journal of Physics* 59, no. 4, 631–651.

Shin, Y., Kim, J.W., Lee, H., and Hwang, C. (2003) Sloshing impact of LNG cargoes in membrane containment systems in the partially filled condition. In *Proceedings of the 13th ISOPE Conference*, May, 25–30, 2003, Honolulu, Hawaii.

Shinbrot, M. (1976) The initial value problem for surface waves under gravity. I. The simplest case. *Indiana University Mathematical Journal* 25, no. 3, 281–300.

Shu, C.-W., and Osher, S. (1988) Efficient implementation of essentially non-oscillatory shock capturing schemes. *Journal of Computational Physics* 77, 439–471.

Shu, C.-W., and Osher, S. (1989) Efficient implementation of essentially non-oscillatory shock capturing schemes II. *Journal of Computational Physics* 83, 32–78.

Skop, R.A., Griffin, O.M., and Ramberg, S.E. (1977) Streaming predictions for the Seacon II Experimental Mooring., In *Proceedings of the Offshore Technology Conference (OTC)*, vol. 3, pp. 61–66. Houston: Offshore Technology Conference.

Smith, J.D. (1977) Modelling of sediment transport on continental shelves. In *The Sea*, vol. 6. New York: Wiley Interscience.

Smith, P. (1998) *Explaining Chaos*. Cambridge University Press.

Solaas, F. (1995) *Analytical and numerical studies of sloshing in tanks*. Dr.Ing. thesis, Trondheim: Norwegian Institute of Technology, Faculty of Marine Technology.

Solaas, F., and Faltinsen, O.M. (1997) Combined numerical and analytical solution for sloshing in two-dimensional tanks of general shape. *Journal of Ship Research* 41, no. 2, 118–129.

Sorensen, R.M. (1993) *Basic Wave Mechanics for Coastal and Ocean Engineers*. New York: John Wiley & Sons.

Sortland, B. (1986) *Force measurements on oscillating ship sections and circular sections in a U-tube water tank*. Dr.Ing. thesis, Department of Marine Technology, NTNU, Trondheim.

Stigter, C. (1966) The performance of U-tanks as passive anti-rolling device. *International Shipbuilding Progress* 13, 144–150.

Stofan, A.J., and Armstead, A.L. (1962) *Analytical and experimental investigation of forces and frequencies resulting from liquid sloshing in a spherical tank*. NASA TN D-1281.

Stoker, J.J. (1992) *Water Waves The Mathematical Theory with Applications*. New York: Wiley Classic Library.

Stokes, G.G. (1851) On the effect of the internal friction of fluids on the motion of pendulums. *Transactions of Cambridge Philosophical Society* 92, 8–106 (also *Mathematical and Physical Papers* 3, 1–141, Johnson Reprint Corporation, 1966).

Stolbetsov, V.I. (1967a) On oscillations of a fluid in the tank having the shape of rectangular parallelepiped. *Mechanika Zhidkosyi i Gaza (Fluid Dynamics)*, no. 1, 67–76 (in Russian).

Stolbetsov, V.I. (1967b) On non-small oscillations of a fluid in an upright circular cylinder. *Mechanika Zhidosti i Gaza (Fluid Dynamics)*, no. 2, 59–66 (in Russian).

Storchi, E. (1949) Legame fra la forma del pelo libero a quella recipiente nell oscillazioni di un liquiido. *Istituto Lombardo di Scienze e Lettere, Classe di Scienze Matematiche e Naturali* 13, no. 82, 95–112.

Storchi, E. (1952) Piccole oscillazioni dell'acqua contenuta da pareti piane. *Atti della Accademia Nazionale dei Lincei. Classe di Scienze Fisiche, Matematiche e Naturali* 8, no. 12, 544–552.

Storhaug, G. (2007) *Experimental investigation of wave induced vibrations and their effect on the fatigue loadings of ships*. Dr.Ing. thesis, Department of Marine Technology, Norwegian University of Science and Technology, Trondheim.

Strandberg, L. (1978) *Lateral Stability of Road Tankers. Vol. 1-Main Report*. Statens Vaeg-och Trafikinstitut, Linkoeping, Sweden. Statens Trafiksaekerhetsverk, Solna, Sweden; Transportforskningsdelegationen, Stockholm, Sweden. Report No. 138A.

Sumer, B.M., and Fredsøe, J. (1997) *Hydrodynamics around Cylindrical Structures*. Singapore: World Scientific.

Sussman, M., Almgren, A.S., Bell, J.B., Colella, P., Howell, H., and Welcome, M. (1994) An adaptive Level Set approach for incompressible two-phase flows. *Journal of Computational Physics* 148, 81–124.

Svendsen, I.A., and Jonsson, I.G. (1976) *Hydrodynamics of Coastal Regions*. Technical University of Denmark.

Tait, M.J., El Damatty, A.A., and Isyumov, N. (2004a) Testing of tuned liquid damper with screens and development of equivalent TMD model. *Wind and Structures* 7, no. 4, 215–234.

Tait, M.J., El Damatty, A.A., Isyumov, N., and Siddique, M.R. (2004b) Numerical flow models to simulate tuned liquid dampers (TLD) with slat screens. *Journal of Fluids and Structures* 20, 1007–1023.

Tait, M. J., El Damatty, A.A., and Isyumov, N. (2005) An investigation of tuned liquid dampers equipped with damping screens under 2D excitation. *Earthquake Engineering and Structural Dynamics* 34, 719–735.

Tanaka, N. (1961) A study on the bilge keels. (Part 4. On the eddy-making resistance to the rolling of a ship hull.) *Journal of the Society of Naval Architects of Japan* 109, 205–212.

Taylor, G.I. (1944) *Air resistance of a flat plate of very porous material*. Aeronautical Research Council Reports and Memoranda no. 2236.

Ten, L., Korobkin, A.A., Malenica, S., De Lauzon, J., and Mravak, Z. (2008) Steep wave impact onto a complex 3D structure. In *Proceedings of the 23rd IWWWFB*, Jeju, Korea.

Timokha, A.N. (2002) Note on ad hoc computing based on a modal basis. *Proceedings of the Institute of Mathematics of NASU* 44, 269–274.

Tollmien, W. (1934) Zum Landestoss von Seeflugzeugen. *ZAMM* 14, 251.

Topliss, M., Cooker, M.J., and Peregrine, D.H. (1992) Pressure oscillations during wave impact on vertical walls. In *Proceedings of the 23rd International Conference on Coastal Engineering*, Venice. New York: ASCE, pp. 1639–1650.

Tønnessen, R. (1999) *A finite element method applied to unsteady viscous flow around 2D blunt bodies with sharp corners*. Dr.Ing. thesis, Norwegian University of Science and Technology, Department of Marine Hydrodynamics, Trondheim, Norway.

Tremblay, F., and Friedrich, R. (2000) An algorithm to treat flows bounded by arbitrarily shaped surfaces with Cartesian meshes. In *Proceedings of the AGSTAB Conference*, University of Stuttgart, Germany, 15–17 November.

Tritton, D.J. (1959) Experiments on the flow past a circular cylinder at low Reynolds number. *Journal of Fluid Mechanics* 6, no. 4, 547–567.

Troesch, B.A. (1960) Free oscillations of a fluid in a container. In *Boundary Problems in Differential Equations*, pp. 279–299. Madison, WI: University of Wisconsin.

Troesch, B.A., and Weidman, P.D. (1972) Containers with isochronous free oscillations. *SIAM Journal of Applied Mathematics* 23, 447–489.

Tseng, Y.-H., and Ferziger, J.H. (2003) A ghost-cell immersed boundary method for flow in complex geometry. *Journal of Computational Physics* 192, 593–623.

Urm, H.S., and Shin, J.G. (2004) Bulk carriers. In *Ship Design and Construction*, Vol. II, Chapter 33 (ed. T. Lamb.). Jersey City, NJ: SNAME.

van Daalen, E.F.G., Gerrits, J., Loots, G.E., and Veldman, A.E.P. (2000) Free surface anti-roll tank simulations with a volume of fluid based Navier-Stokes solver. In *Proceedings of the 15th IWWWFB*, Caesarea, Israel.

Van Den Bosch, J.J., and Vugts, J.H. (1966) *Roll Damping by Free Surface Tanks*. Report No. 83S, Netherlands Ship Research Centre, Delft.

Venkatesan, S.K. (1985) Added mass of two cylinders. *Journal of Ship Research* 29, no. 4, 234–240.

Vera, C., Paulin, J., Suarez, B., and Gutierrez, M. (2005) Simulations of freight trains equipped with partially filled tank containers and related resonance phenomenon. *Proceedings of the IMechE, Part F: Journal of Rail and Rapid Transit* 219, 245–259.

Verhagen, J.H.G. (1967) The impact of a flat plate on a water surface. *Journal of Ship Research* 11, no. 4, 211–223.

Verhagen, J.H.G., and van Wijngaarden, L. (1965) Nonlinear oscillations of fluid in a container. *Journal of Fluid Mechanics* 22, no. 4, 737–751.

Verley, R.L.P. (1982) A simple model of vortex-induced forces in waves and oscillating currents. *Applied Ocean Research 4*, no. 2, 117–120.

Vinberg, E.B. (2003). *A Course in Algebra*. Providence, RI: American Mathematical Society.

Vinje, T. (1972) *On vibration of spherical shells interacting with fluid*. Report No. SK/M 23, Division of Ship Structures, Norwegian Technical University, Trondheim.

Vinje, T. (1973) On calculation of the characteristic frequencies of fluid-filled spherical shells. *Norwegian Maritime Research* 1, no. 1, 15–26.

von Karman, T. (1929) *The impact of seaplane floats during landing*. NACA TN 321, Washington, DC.

Vugts, J.H. (1968) *A comparative study on four different passive roll damping tanks, Part 1*. Report no. 109 S, Netherlands Ship Research Centre TNO.

Wagner, H. (1932) Über Stoss- und Gleitvorgänge and der Oberflache von Flussigkeiten. *ZAMM* 12, 193–235.

Walton, P. (1986) *The hydrodynamics of floating bodies*. Ph.D. thesis, University of Manchester.

Wang, C.-Y. (1968) On high-frequency oscillating viscous flows. *Journal of Fluid Mechanics* 32, 55–68.

Wang, C.Z., and Wu, G.X. (2007) Time domain analysis of second-order wave diffraction by an array of vertical cylinders. *Journal of Fluids and Structures* 23, 605–631.

Wang, C.Z., Wu, G.X., and Drake, K.R. (2007) Interactions between nonlinear water waves and non-wall-sided 3D structures. *Ocean Engineering* 34, 1182–1196.

Ward, G.N. (1955) *Linearized Theory of Steady High-Speed Flow*. Cambridge University Press.

Warnitchai, P., and Pinkaew, T. (1998) Modelling of liquid sloshing in rectangular tanks with flow-damping devices. *Engineering Structures* 20, 593–600.

Waterhouse, D.D. (1994) Resonant sloshing near a critical depth. *Journal of Fluid Mechanics* 281, 313–318.

Watson, G.N. (1944) *A Treatise on the Theory of Bessel Functions*, 2nd edition. Cambridge University Press.

Wehausen, J.V., and Laitone, E.V. (1960) *Handbuch der Physik IX*. Berlin: Springer.

Wei, G., Kirby, J.T., Grilli, S.T., and Subramanya, R. (1995) A fully nonlinear Boussinesq model for surface waves. Part 1. Highly nonlinear unsteady waves. *Journal of Fluid Mechanics* 294, 71–92.

Welch, J.E., Harlow, F.H., Shannon, J.P., and Daly, B.J. (1966) *The MAC method: A computing technique for viscous, incompressible, transient fluid-flow problems involving free surfaces*. Los Alamos Scientific Laboratory, Report LA-3425.

Werner, M. (2004) Chemical tankers. In *Ship Design and Construction*, Vol. II, Chapter 31 (ed. T. Lamb). Jersey City, NJ: SNAME.

White F.M. (1974) *Viscous Fluid Flow*. New York: McGraw-Hill.

Wienke, J., and Oumeraci, H. (2005) Breaking wave impact force on a vertical and inclined slender pile-theoretical and large-scale model investigations. *Coastal Engineering* 5, 435–462.

Williamson, C.H.K. (1996) Three-dimensional wake transition. *Journal of Fluid Mechanics* 328, 345–407.

Winkler, C. (2000) Rollover of heavy commercial vehicles. *UMTRI Research Review* 31, no. 4, 1–20.

Wood, A.B. (1930) *A Textbook of Sound*, 1st ed. New York: MacMillan.

Wu, G.X. (2007) Fluid impact on a solid boundary. *Journal of Fluids and Structures* 23, 755–765.

Wu, G.X., and Hu, Z.Z. (2004) Simulation of nonlinear interactions between waves and floating bodies through a finite-element-based numerical tank. *Proceedings of the Royal Society of London. Series A, Mathematical and Physical Sciences* 460, 2797–2817.

Wu, G.X., Ma, Q.W., and Eatock Taylor, R. (1998) Numerical simulation of sloshing waves in a 3D tank based on the Finite Element Method. *Applied Ocean Research* 20, no. 6, 337–355.

Yabe, T., and Wang, P.Y. (1991) Unified numerical procedure for compressible and incompressible fluid. *Journal of Physical Society of Japan* 60, 2105–2108.

Zalar, M., Cambos, P., Besse, P., Le Gallo, B., and Mravak, Z. (2005) Partial filling of membrane type LNG carriers. In *Proceedings of the 21st GASTECH*, Bilbao, Spain.

Zdravkovich, M.M. (1997) *Flow Around Circular Cylinders. Vol. 1. Fundamentals*. Oxford: Oxford University Press.

Zhang, A., and Suzuki, K. (2007) A comparative study of numerical simulations for fluid-structure interaction of liquid-filled tank during ship collision. *Ocean Engineering* 34, 645–652.

Zhang, S., Yue, D.K.P., and Tanizawa, K. (1996) Simulation of plunging wave impact on a vertical wall. *Journal of Fluid Mechanics* 327, 221–254.

Zhao, R. (1995) A complete linear theory for a two-dimensional floating and liquid-filled membrane structure in waves. *Journal of Fluids and Structures* 9, 937–956.

Zhao, R., and Faltinsen, O.M. (1993) Water entry of two-dimensional bodies. *Journal of Fluid Mechanics* 246, 593–612.

Zhao, R., and Triantafyllou, M. (1994) Hydroelastic analysis of a long flexible tube in waves. In *Proceedings of International Conference on Hydroelasticity in Marine Technology*, pp. 287–300. Trondheim, Norway.

Zhu, X. (2006) *Application of the CIP method to strongly nonlinear wave-body interaction problems*. Ph.D. thesis, Department of Marine Technology, Norwegian University of Science and Technology, Trondheim, Norway.

Ziegler, F. (1995) *Mechanics of Solids and Fluids*, 2nd edition. Berlin: Springer.

Zienkiewicz, O.C., and Taylor, R.L. (1991) *The Finite Element Method*. New York: McGraw-Hill.

Index

Printed in the United States
By Bookmasters